ALLE ZEIT WACH
1842

K. Kitazawa   T. Ishiguro (Eds.)

Advances in Superconductivity

Proceedings of the 1st International Symposium on Superconductivity (ISS '88), August 28-31, 1988, Nagoya

With 938 Figures

Springer-Verlag
Tokyo Berlin Heidelberg New York London Paris

Prof. Dr. Koichi Kitazawa
Department of Industrial Chemistry
The University of Tokyo
Tokyo, 113 Japan

Prof. Dr. Takehiko Ishiguro
Department of Physics
Kyoto University
Kyoto, 606 Japan

ISBN 4-431-70039-0 Springer-Verlag Tokyo Berlin Heidelberg New York
ISBN 3-540-70039-0 Springer-Verlag Berlin Heidelberg New York Tokyo
ISBN 0-387-70039-0 Springer-Verlag New York Berlin Heidelberg Tokyo

Printed in Japan

Printing/Binding: Kowa Art Printing, Tokyo

Preface

These proceedings are a referred account of the First International Symposium on Superconductivity (ISS '88) held on August 28 to 31 in Nagoya, Japan under the sponsorship of the International Superconductivity Technology Center (ISTEC). Approximately 43 invited papers and 111 contributed papers were presented at the meeting. Most of the invited papers include reviews, ranging from mechanisms of superconductivity, its history, developments of new materials, to prospects of its application. Up-to-date developments are mostly included in the poster presentation. However, time and space limitations prevented the Program Committee from having more invited and poster presentations. This volume contains 112 invited and contributed papers from Japanese industries and hence a good overview of their current activities.

The practical application of superconducting materials progressed forward at a rather slow pace until two years ago. The revolutionary breakthrough in the critical temperature of superconductivity brought about by Bednorz and Müller then not only stirred the studies on high temperature superconductors but also prompted practical usage of conventional metallic superconductors. Therefore, this symposium covered the possible fields in which the application of the metallic superconductors is expected as well as the developments in the basic studies on high temperature superconductors.

Standing at what may prove to be a turning point of scientific history, we are overwhelmed by the enthusiasm shown by the people who are pushing this history ahead and sparing some time to get together to discuss their latest findings.

It is impossible to summarize the current status of such fascinating fields in short phrases, but certain progress has been achieved, especially in the various processing methods of the materials and devices and in the resulting superconducting properties. In the meantime, the problems associated with the oxide superconductors (which are thought to be inherent in these materials) have been clarified to some degree. We hope that steady effort will someday solve these problems and open the bright future of superconducitivity.

Finally, we are grateful to Aichi Prefecture and Nagoya City for the financial support and to the symposium committees and the ISTEC office for their great devotion during the past several months.

Koichi Kitazawa
Takehiko Ishiguro

Organization of ISS '88

Sponsored by:
International Superconductivity Technology Center (ISTEC)

Co-Sponsored by:
Aichi Prefectural Government
City of Nagoya
Chubu Economic Federation
Nagoya Chamber of Commerce and Industry
The Chubu Industrial Advancement Center

Supported by:
Ministry of International Trade and Industry
Nagoya Bureau of International Trade and Industry
Ceramic Society of Japan
Japan Chemical Industry Association
Japan Electronic Industry Development Association
Japan Mining Industry Association
Japan Fine Ceramics Association
New Energy Development Organization
Physical Society of Japan
Research and Development Association for Future Electron Devices
The Chemical Society of Japan
The Cryogenic Society of Japan
The Federation of Electric Power Companies
The Institute of Electrical Engineers of Japan
The Institute of Electronics Information and Communication Engineers, Japan
The Iron and Steel Institute of Japan
The Japan Electrical Manufacturers' Association
The Japanese Electric Wire & Cable Makers' Association
The Japan Institute of Metals
The Japan Iron and Steel Federation
The Japan Key Technology Center
The Japan Society of Applied Physics

Organization Committee

Chairman:
GAISHI HIRAIWA, President, ISTEC (Vice Chairman, Keidanren)

Members:
KAZUO FUEKI, Professor, Science University of Tokyo
KOZO IIZUKA, Director General, AIST, MITI
TETSURO KAWAKAMI, Deputy President, ISTEC
(President, Sumitomo Electric Industries, Ltd.)
KATSUSHIGE MITA, Deputy President, ISTEC (President, Hitachi, Ltd.)
NAOKI KURODA, Director General, Nagoya Bureau of International Trade and Industry, MITI
YOSHIO MUTO, Professor, Tohoku University
SADAO NAKAJIMA, Professor, Tokai University
TAKEYOSHI NISHIO, Mayor, Nagoya City
SHINROKU SAITO, Professor Emeritus, Tokyo Institute of Technology
REIJI SUZUKI, Governor, Aichi Prefecture
KOTARO TAKEDA, President, Nagoya Chamber of Commerce and Industry
SEIICHI TANAKA, Chairman, Chubu Economic Federation
(Chairman, The Chubu Industrial Advancement Center)
SHOJI TANAKA, Vice President, ISTEC (Professor, Tokai University)
SAKAE YAMAMURA, Professor Emeritus, The University of Tokyo

International Advisory Committee

B. BATLOGG, AT & T Bell Laboratories
M.R. BEASLEY, Stanford University
H.K. BOWEN, MIT
P. CHAUDHURI, IBM Watson Research Center
C.W. CHU, Houston University
V.J. EMERY, Brookhaven National Laboratory
J.E. EVETTS, University of Cambridge
Φ. FISHER, University of Geneva
D.K. FINNEMORE, AMES Laboratory
F.Y. FRADIN, Argonne National Laboratory
T.H. GEBALLE, Stanford University
D. JÉROME, Universite de Paris-Sud
M.B. MAPLE, University of California
K.A. MÜLLER, IBM Zürich
H.R. OTT, ETH Zürich
B. RAVEAU, Universite de Caen
A.W. SLEIGHT, Du Pont
M. SUENAGA, Brookhaven National Laboratory

Steering Committee

Chairman:
Shoji Tanaka, ISTEC

Members:
Kazuo Fueki, Science University of Tokyo
Hidetoshi Fukuyama, The University of Tokyo
Hisao Hayakawa, Nagoya University
Takashi Hirayama, Tokyo Electric Power Company
Tohru Inoue, Nippon Steel Corporation
Takehiko Ishiguro, Kyoto University
Shoei Kataoka, Sharp Corporation
Koichi Kitazawa, The University of Tokyo
Shigeru Maekawa, Electrotechnical Laboratories
Yoshio Muto, Tohoku University
Hiroyasu Ogiwara, Toshiba Corporation
Daizaburo Shinoda, NEC Corporation
Yasutsugu Takeda, Hitachi, Ltd.
Tetsuya Uchino, Asahi Glass Co., Ltd.

Program and Editorial Committee

Chairman:
Koichi Kitazawa, The University of Tokyo

Subchairman:
Takehiko Ishiguro, Kyoto University

Members:
Junji Fujie, Railway Technical Research Institute
Hidetoshi Fukuyama, The University of Tokyo
Yasuo Hashimoto, Mitsubishi Electric Corporation
Shinya Hasuo, Fujitsu Laboratories, Ltd.
Hisao Hayakawa, Nagoya University
Osamu Horigami, Toshiba Corporation
Keizo Ishikawa, Mitsubishi Heavy Industries, Ltd.
Yuzo Katayama, NTT Corporation
Ushio Kawabe, Hitachi, Ltd.
Yoichi Kimura, Electrotechnical Laboratory
Takeshi Kobayashi, Osaka University
Osamu Kohno, Fujikura, Ltd.
Sadamichi Maekawa, Nagoya University
Eisuke Masada, The University of Tokyo
Shinpei Matsuda, Hitachi, Ltd.
Masato Murakami, Nippon Steel Corporation
Yoshio Muto, Tohoku University
Koshichi Noto, Tohoku University
Setsuo Takezawa, Japan Foundation for Shipbuilding Advancement
Yasuzo Tanaka, The Furukawa Electric Co., Ltd.
Jaw-Shen Tsai, NEC Corporation
Shinichi Uchida, The University of Tokyo
Tokuo Wakiyama, Tohoku University
Kiyotaka Wasa, Matsushita Electric Industrial Co., Ltd.
Atsushi Yamaguchi, The Kansai Electric Power Co., Ltd.
Shuji Yazu, Sumitomo Electric Industries, Ltd.

General Affairs Committee

Chairman:
HISAO HAYAKAWA, Nagoya University

Members:
YOSHIO FURUTO, The Furukawa Electric Co., Ltd.
YASUO HASHIMOTO, Mitsubishi Electric Corporation
KAORU INAGAKI, Nagoya City
HIROO KINOSHITA, Toyota Motor Corporation
KANJI OHYA, NGK Spark Plug Co., Ltd.
HIDEAKI OKINO, Nagoya Bureau of International Trade and Industry, MITI
SAKAE TANEMURA, Government Industrial Research Institute, Nagoya
KAZUHIKO YAMADA, Aichi Prefecture
HIROSHI YOSHIDA, Chubu Electric Power Co., Ltd.
SHIZUKA YOSHII, Ube Industries, Ltd.

Contents

1 History of Superconductivity

2 Prospects for Application of Superconductivity

3 Organic Superconductors

4 Oxide Superconductors

4.1 Mechanisms of Superconductivity

4.2 Crystal Chemistry and Electronic Structure

4.3 Phase Diagram and Crystal Growth

4.4 Processing and Microstructure

4.5 Tapes and Thick Films

4.6 Wires and Coils

4.7 Coherence Length, Magnetic Properties, and Critical Current

4.8 Irradiation Effect

4.9 Thin Film Processing and Properties: Part 1

4.10 Thin Film Processing and Properties: Part 2

4.11 Chemical Reactions and Superconductor/Substrate Interaction

4.12 Devices and Applications

4.13 Tunneling and Tunneling Junction

4.14 Bi- and Tl-Based Cuprate Superconductors

4.15 110K Phase of Bi-Sr-Ca-Cu-O Fabrication and Microstructure

5 Research Policy and Technology Trends

Closing Remarks

* invited paper

1
History of Superconductivity

A History of Superconductivity

M.R. BEASLEY

Department of Applied Physics, Stanford University, Stanford, CA 94305, USA

This first International Symposium on Superconductivity, ISS-88, is an historic event. It brings together for the first time a truly international collection of scientists and engineers to discuss the full range of issues—from basic science to practical technology—raised by the discovery of the new high-temperature superconductors. It celebrates the founding of ISTEC, the first major research organization in the world to be formed for the expressed purpose of advancing the science and technology of these new materials. The international emphasis of both ISTEC and this Symposium is a clear reflection of the times in which we live—times in which the international character of science and technology is fully being appreciated, not only by scientists and engineers, but the general populace as well.

It seems fitting, therefore, to begin this Symposium by recounting the history of our field—to see how we got here. Professor Schrieffer in the previous paper has given a personal account of the days of the BCS theory. My task is to describe the larger history, including the days before and after BCS. The story begins with the discovery of superconductivity by Onnes in 1911 and ends with the discovery of high-temperature superconductivity by Bednorz and Müller in 1986. The history spans much of the 20th century. Clearly superconductivity is one of the major scientific stories of this century. But history is best in the telling, so let us begin.

Like the development of the science of superconductivity itself, this paper is woven from three historical threads—each with its own logic and imperatives, but interacting in fascinating and historically crucial ways. The first thread deals with the phenomena of superconductivity and the early phenomenological theories that provided the first level of understanding of the subject. It begins with the discovery of superconductivity and ends with the Ginzburg-Landau theory, which to this day provides us with some of our deepest insights into the nature of superconductivity as a macroscopic quantum phenomenon. The second thread traces the history of superconducting materials. Here we see how an insistence on understanding empirically the occurrence of superconductivity and its range of properties in real materials has profoundly impacted the field. Finally, we trace the development of the microscopic theory and our understanding of superconductivity in terms of the electron-phonon mechanism.

The Beginning – Zero Resistance

As we all know, the field of superconductivity began with the discovery of zero resistance by Kammerlingh Onnes in 1911. The famous figure from his original paper is reproduced in Fig. 1. But let us not forget that the discovery of superconductivity was built on events that trace back into the late 19th century. During those times one of the great challenges of physics was to liquify all of the rare gases and thereby to achieve lower and lower temperatures. Onnes won this race by first liquifying helium in 1908. His reward was the discovery of superconductivity three years later. It is perhaps ironic that much of the current excitement over the new high-temperature superconductors revolves around the easier refrigeration requirements they permit. It seems that superconductivity and issues of cryogenic refrigeration have always been and likely always will be inexorably linked.

So that's the beginning. Zero resistance, one of the key properties of superconductors, was discovered. Also, the first superconducting material, mercury, was known. Two of our threads begin with this seminal event.

The Meissner Effect

Although the fact of zero resistance is spectacular, it took a new experiment and new concepts derived from that experiment to provide the next advance in our understanding of superconductivity. In 1933 Meissner discovered that not only did superconductors exhibit zero resistance, but they also spontaneously expel all magnetic flux when cooled through the superconducting transition—that is, they are also perfect diamagnets. We call this the Meissner effect. As all students of superconductivity learn, the Meissner effect is not a consequence of zero resistance and Lenz' law. The flux is expelled as the superconductor is cooled in constant magnetic field. There is no time rate of change of the magnetic induction. Lenz' law does not apply.

Perfect diamagnetism is an independent property of superconductors and shows that superconductivity involves a change of thermodynamic state, not just a spectacular change in electrical resistance. In retrospect, we now

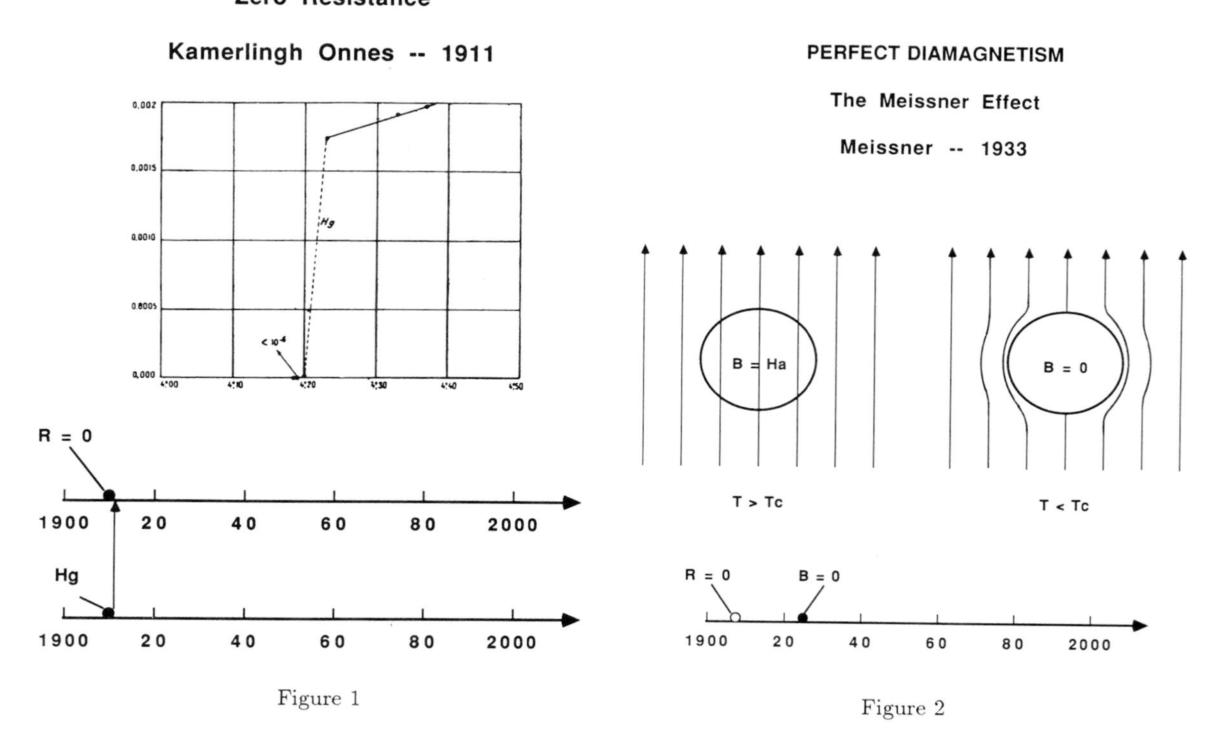

Figure 1

Figure 2

appreciate that it is one of the conceptually deepest properties of superconductivity. Figure 2 shows the textbook illustration of the Meissner effect that is one of the popular symbols of superconductivity. Also the time line shown on the figure indicates how our first thread in the development of superconductivity is beginning to evolve.

The London Theory

Given a knowledge of the Meissner effect, F. and H. London introduced the first phenomenological theory of superconductivity in 1934. Appropriately it is known as the London theory. The equations of this theory are shown in Fig. 3, along with an indication of their physical content. In brief, the Londons conceived of a set of equations that accounted for the observed properties of zero resistance and perfect diamagnetism. The physics embodied in these equations was that of a classical charged fluid flow in which there was no friction and in which the flow was basically irrotational, i.e., with no vorticity. The solutions of the equations yielded the Meissner effect and gave an explicit expression for the magnetic penetration depth of the superconductor (see the figure).

While the Londons' equations accounted for zero resistance and perfect diamagnetism, they were not explicitly based on fundamental concepts. Phenomenological theories of this sort play a tremendously important role in the advancement of science. They contain the distillate or essence of a collection of experimental facts and reduce them to a deeper set of questions. The work of the Londons was the first time this happened in the conceptual development of superconductivity, and F. London is clearly the first theoretical hero in the history of superconductivity.

Eventually the London theory was extended to include the two-fluid concept—the idea that a superconductor consists of two interpenetrating fluids, a normal fluid that exhibits resistance and a superconducting fluid that obeys the London equations. The concept is illustrated in the lower half of Fig. 3. To think for a moment in electrical engineering terms, the two-fluid concept can be represented by a parallel circuit with the normal electron behaving as a normal resistor and the superconducting electron behaving like a perfect inductor. This inductance is associated with the kinetic energy of the superconducting electrons and is known as the kinetic inductance of the superconductor. Even today this approach to superconductivity is useful in the engineering applications of superconductivity, for example in rf applications. But our concern here is with the science not applications, so let us return to our story.

THE LONDON THEORY

F. and H. London -- 1934

$$J_s = n_s e v_s$$

$$\frac{dJ_s}{dt} = \frac{n_s e^2}{m} E \leftrightarrow \frac{dv_s}{dt} = \frac{eE}{m} \leftrightarrow \text{ no scattering}$$

$$\nabla \times J_s = \frac{-n_s e^2}{m} B \leftrightarrow \nabla \times v_s = \frac{-eB}{m} \leftrightarrow \text{ irrotational flow } (B = 0)$$

$$\Rightarrow R = 0 \qquad B = 0 \qquad \lambda = \frac{m}{\mu_o n_s e^2}$$

Two-Fluid Concept

$$J = J_s + J_n$$

J_n

J_s

$$\sigma = \sigma_s + \sigma_n = \frac{n_s e^2}{j\omega m} + \sigma_n$$

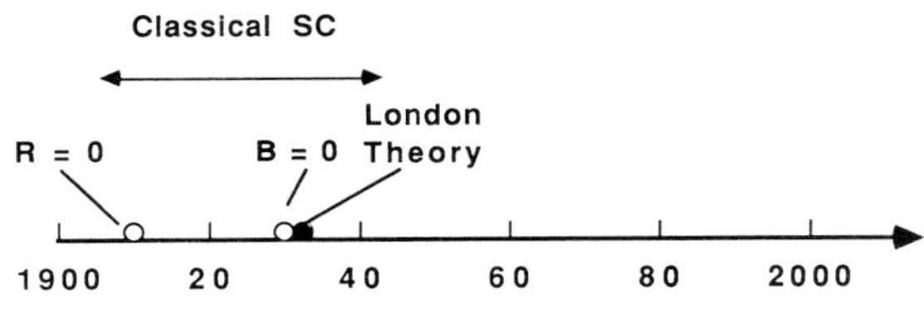

Figure 3

THE GINZBURG-LANDAU THEORY

A Macroscopic Quantum Theory of SC

Ginzburg and Landau -- 1950

$$\Psi = |\Psi| e^{i\phi} \leftrightarrow \text{ macroscopic quantum wavefunction}$$

$$n_s^* = |\Psi|^2 \qquad v_s = \frac{1}{m^*}(\hbar \nabla \phi - e^* A)$$

$$F_{GL} = \int dv \left\{ \alpha |\Psi|^2 + \frac{\beta}{2} |\Psi|^4 + \frac{1}{2m^*} \left| \left(\frac{\hbar \nabla}{i} - e^* A \right) \Psi \right|^2 + \frac{B^2}{2\mu_o} \right\}$$

$$\alpha \Psi + \beta |\Psi|^2 \Psi + \frac{1}{2m^*} \left(\frac{\hbar \nabla}{i} - e^* A \right)^2 \Psi = 0$$

$$J_s = \frac{e^* \hbar}{2m^* i} (\Psi^* \nabla \Psi - \Psi \nabla \Psi^*) - \frac{e^{*2}}{m^*} |\Psi|^2 A$$

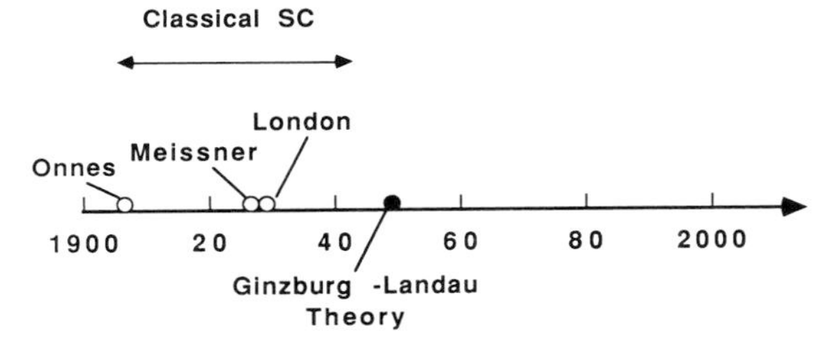

Figure 4

The Ginzburg – Landau Theory

In the years following the introduction of the London theory, F. London articulated the deeper idea that the theory followed somehow from the fact that the superconducting electrons were in a macroscopic quantum state. However, it was not until Ginzburg and Landau introduced their famous phenomenological theory of superconductivity that this idea was firmly embodied in a formal theory. Ginzburg and Landau postulated the existence of a macroscopic wave function (or order parameter) to describe the behavior of the superconducting electrons. The London theory follows directly from this postulate. Thus, the Ginzburg-Landau theory accounts for zero resistance and the Meissner effect, but now in much more fundamental terms. The theory also accounted for the second order (or continuous) nature of the superconducting phase transition in zero magnetic field.

The Ginzburg-Landau free energy, and the equations that follow from a variational analysis of the free energy, are shown in Fig. 4. In the figure we show the correspondence between the Ginzburg-Landau wave function and the superconducting electron density and velocity that appear in the London theory. But the Ginzburg-Landau theory did more than just give a deeper meaning to the London theory. It provided a set of equations with which the superconducting macroscopic wave function could be calculated under many circumstances, in particular in the presence of applied currents and magnetic fields. The predictions of these equations so correctly describe the qualitative behavior of superconductors that they are the subject of continuing study to this very day.

The fact that superconductivity was a quantum phenomenon described by a macroscopic quantum wave function implied that there was a phase involved. This means that there is the possibility of quantization effects (e.g., flux quantization) and quantum interference, although these possibilities were not fully appreciated at the time. In any event, with the introduction of the Ginzburg-Landau theory, the classical period of superconductivity with Onnes, Meissner, and the Londons came to a close, and the era of superconductivity as a macroscopic quantum phenomenon began. Still, like the London theory, the Ginzburg-Landau theory was a phenomenological theory, and its deeper origins in a microscopic theory were yet to be established.

TYPE II SUPERCONDUCTIVITY

Abrikosov - 1957

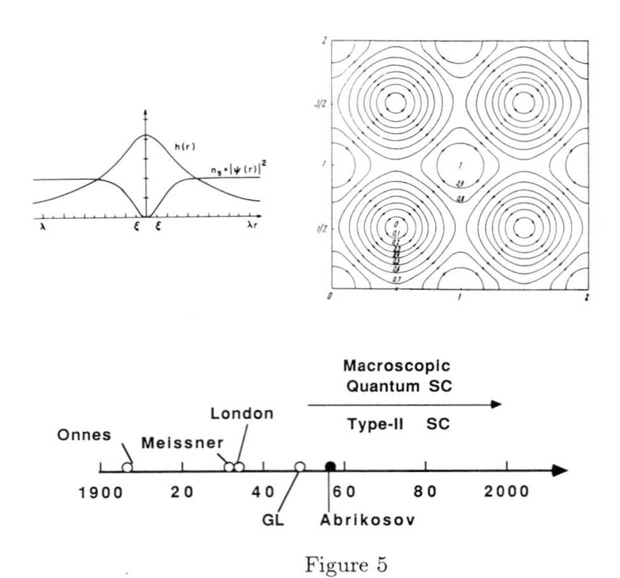

Figure 5

Type II Superconductivity

One of the most dramatic consequences of the Ginzburg-Landau theory was published by Abrikosov in 1957. Abrikosov, who was working in the Landau group, recognized that for a particular range of material parameters the Ginzburg-Landau equations contained a peculiar solution, one in which the magnetic induction inside the superconductor was not zero. Rather, in the presence of a sufficiently large magnetic field there were vortices inside the superconductor—regions of circulating currents around a singularity in the amplitude of the macroscopic quantum wave function. Associated with this current was a peak in the magnetic field that decayed from the center over a length scale given by the magnetic penetration depth. These vortices were found to form a regular lattice. Each contained exactly one quantum of flux.

Figure 5 illustrates the nature of this new solution. The contour diagram shows contours of the amplitude of the wave function and is from Abrikosov's original paper. Of course, this predicted behavior was very peculiar. No one had seen, or at least recognized the fact that they had seen, a superconductor behave this way. We now call such superconductors type II superconductors. As I understand the story, even Landau himself was reluctant to accept Abrikosov's work. Moreover, initially Abrikosov's work had no impact on our understanding of superconductivity, at least outside the Soviet Union. This was perhaps one of the many unfortunate aspects of the Cold War, which was intense at the time. It took some later events to propel Abrikosov's theory into the mainstream of understanding of superconductivity.

But even setting aside the work of Abrikosov, the late 1950's was not an ordinary time in the history of superconductivity. Many other threads in the flow of the history of our subject were coming together and ushering in what young people have come to refer to as the Golden Age of superconductivity. We need to go back now and describe some of the other ideas and discoveries that were leading up to this marvelous period. So let us deviate from our story of the phenomena and phenomenological theories of superconductivity and ask what was going on in the area of superconducting materials.

Superconducting Materials

We began our history with Onnes' discovery of superconductivity in mercury, the first superconducting material. Fortunately, some scientists like to study materials for their own sake. They like to study phenomena as they occur in a wide range of real materials, not just model systems. In particular, there has been a long and distinguished history of people seeking new superconductors and trying to establish the rules governing their occurrence. At the same time, they established the actual range of the parameters that govern the superconducting behavior of all these materials. Not surprisingly, particular fascination revolves around the transition temperature of a superconductor and how one might increase it.

SYSTEMATIC SEARCH FOR SUPERCONDUCTIVITY AMONG THE ELEMENTS AND BINARY ALLOYS AND COMPOUNDS

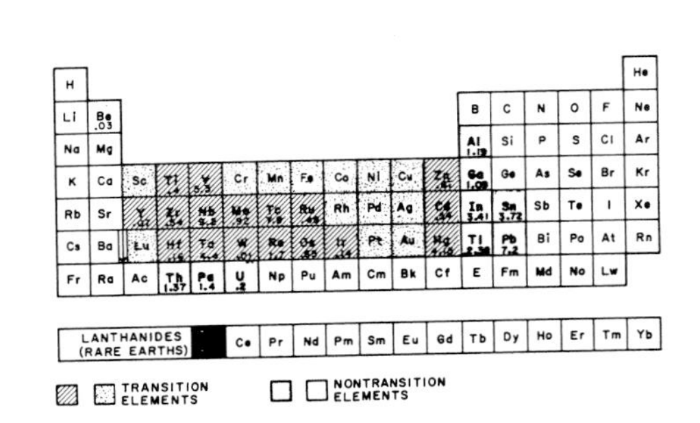

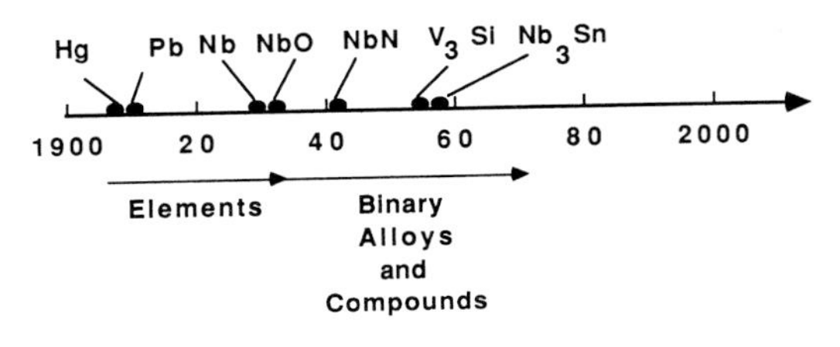

Figure 6

THE STEADY RISE IN Tc

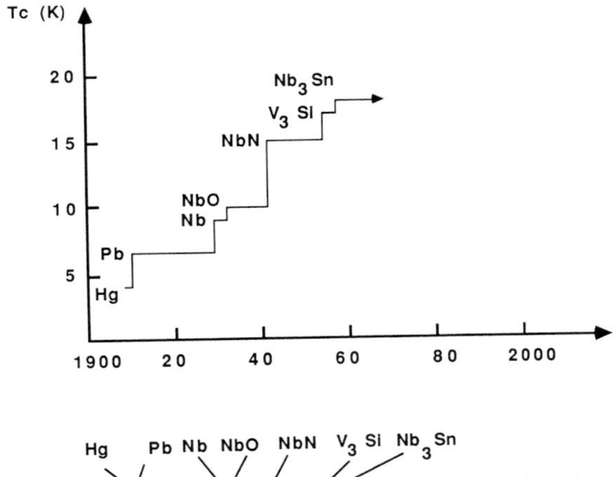

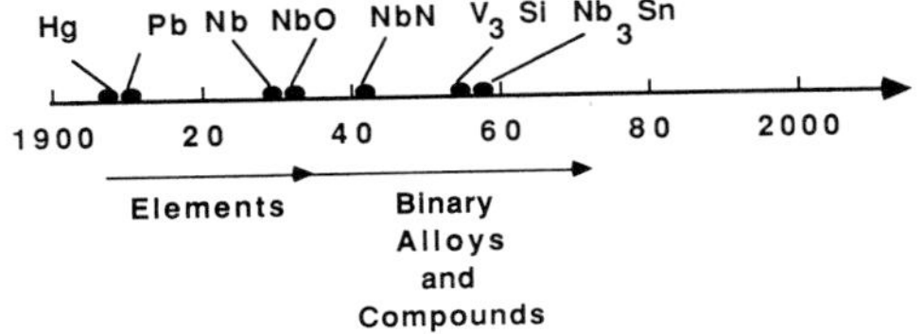

Figure 7

In the early part of the century, superconductivity was explored systematically throughout the elements of the periodic table. Somewhat later, people began to look at binary systems including both alloys and compounds. Some of the more famous discoveries are shown in Fig. 6, which shows the time line associated with the second main thread of our history of the science of superconductivity. Time does not permit us to trace these developments in detail, except to say that most of the interesting superconductors of the time involved the transition metals and their alloys. Matthias and his colleagues Hulm and Geballe were key figures in this work, along with many others. Matthias, in particular, challenged us to understand superconductivity in terms of the periodic table. Also shown in Fig. 6 is a copy of the periodic table indicating the occurrence of superconductivity among the elements. This representation of the periodic table, like the picture of the Meissner effect in Fig. 2, is one of the popular symbols of superconductivity. There is no doubt that Matthias' challenge has profoundly affected the field of superconductivity and has led us to many wonderful and practical superconducting materials. It also helped establish a tradition of the empirical search for new superconducting materials that continues up to the present day.

Associated with this systematic search for new superconducting materials, slowly but surely the superconducting transition temperature was increased. This evolution is illustrated in Fig. 7. From the point of view of our history, during the 1950's the search for higher transition temperature reached an important branch point with the discovery of superconductivity in V_3Si by Hulm and Hardy. This discovery opened up the whole era of the so-called A15 superconductors. By the late 1950's the transition temperature had been increased to 18K with the discovery of superconductivity in Nb_3Sn.

Despite all this success, these developments had limited impact on the basic understanding of superconductivity and virtually no impact on the limited superconductive technologies of the time. It was perhaps too early. There was being developed, however, a wonderful data base and collection of superconducting materials. There was simply no theoretical understanding of it all. On the other hand, studies of the detailed properties of some of the simpler, elemental superconductors were beginning to shed light on the microscopic origins of superconductivity.

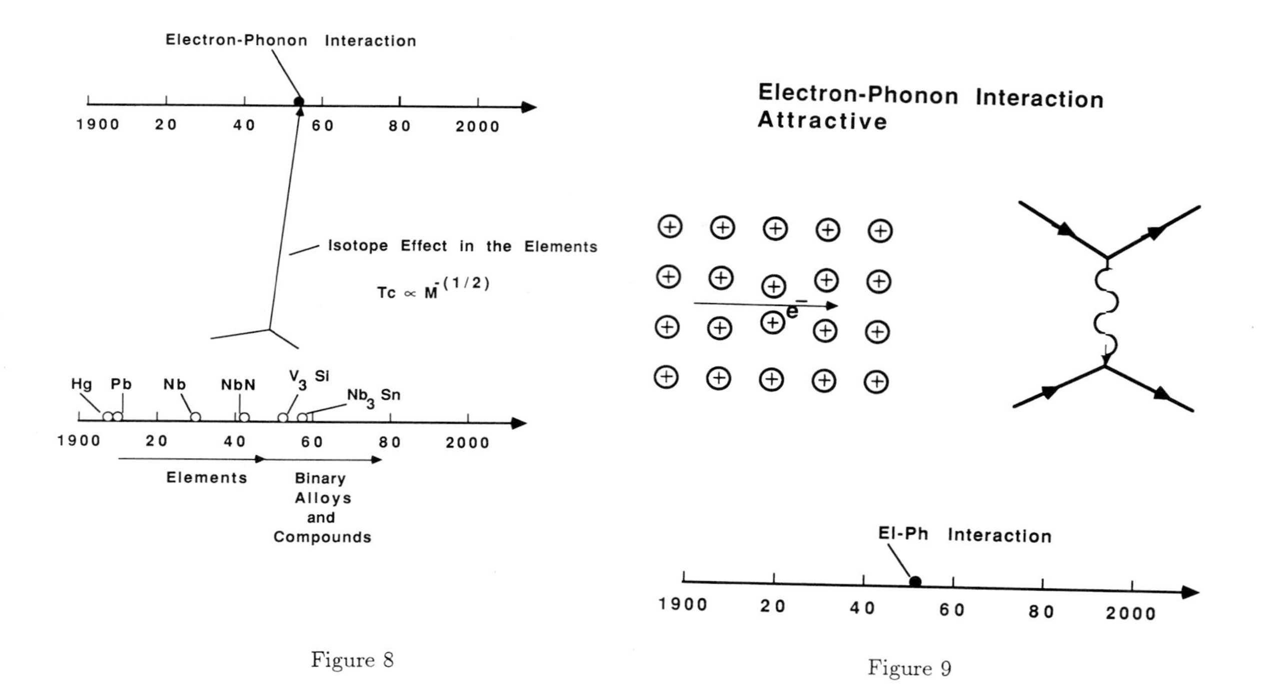

Figure 8

Figure 9

The Mechanism of Superconductivity

In the 1950's experiments were going on in which people examined the effect of changing the isotopic mass of a superconductor on its transition temperature. As illustrated in Fig. 8, one of the first important interactions among the various threads in the development of superconductivity was taking place. In these famous experiments, the transition temperature was found to go up as the inverse square root of the isotopic mass of the element. This suggested that vibrations of the ionic lattice (or phonons) were playing a role in the superconductivity. The implication was that the electron-phonon interaction must be causing the superconductivity.

More generally during this period, it was becoming understood that when an electron moves through a crystal it polarizes the lattice in a dynamic way, leaving a positive (or attractive) potential for a second electron coming along later. This is the attractive electron-phonon interaction between two electrons. As Professor Schrieffer pointed out in his talk, Frölich and Bardeen were beginning to appreciate the possible connection of this interaction and superconductivity. Apparently Frölich had come to this view even without knowledge of the isotope effect. This realization, depicted in Fig. 9, both in literal terms and in the form of a Feynman diagram showing the virtual exchange of a phonon between two electrons, represented the beginnings of the third thread in our history of superconductivity. In any event, for the sake of the intellectual history of our field, it is interesting to note that the experimental community and the theoretical community began to converge on the idea that the electron-phonon interaction may be responsible for superconductivity.

So we have now about 1957 quite a situation. We find ourselves at a unique time in our history. To a substantial degree independently, the phenomena of superconductivity were being better understood on the basis of the phenomenological theories, the occurrence of superconductivity in materials was being better explored, and the importance of the electron-phonon interaction was being recognized. The three threads in the history of superconductivity—phenomena, materials and mechanisms—were poised to come together to create an era in which the conceptual and practical understandinng of superconductivity were about to advance explosively. The late 1950's and early 1960's were that era. Everything was primed, although I doubt that anybody really knew it.

From a personal point of view, I was an undergraduate at Cornell University at the time, and was just being introduced to superconductivity. I would ultimately go on in graduate school to do my thesis in the field. Given the excitement of the time, you can understand my continuing fascination with superconductivity.

The BCS Theory

As Professor Schrieffer has described, out of all these events came the BCS theory. First Cooper showed that, in the presence of an attractive interaction, no matter how small, two independent electrons above a filled Fermi sphere are unstable toward the formation of a so-called bound Cooper pair. Then Bardeen, Cooper and Schrieffer, in an astonishing achievement in many-body physics, solved the problem of attracting electrons not for just two electrons but for all the conduction electrons in a solid. A schematic representation of these two problems is shown in Fig. 10, along with two of the famous equations that resulted from their solution.

A physical interpretation of the BCS theory and a listing of its many achievements have been presented by Professor Schrieffer. There is no need to repeat all that here. There are, however, a few aspects of this theory that need to be emphasized from the point of view of our larger history of the science of superconductivity. In Fig. 10 we show the famous BCS wave function. As is well known, it is in the form of a product of creation operators for Cooper pairs. The important point for us is that the coefficients that enter this wave function, the u's and v's, are complex numbers. They have a phase. This is explicitly shown in the equation by the inclusion of a factor $\exp(i\phi_k)$ with each v_k. The point is that in principle each Cooper pair can have a different phase.

But in superconductivity that is not what happens. What happens is that all those phases lock to the same value, i.e, $\phi_k = \phi$ for each k. There is a phase ordering of all the Cooper pairs. As we now know, the resulting overall phase of the BCS wave functions is just the phase of the Ginzburg-Landau wave function. This connection was made explicit through the work of Gor'kov, who showed that one could derive the Ginzburg-Landau theory from BCS. Hence BCS contained the macroscopic quantum nature of superconductivity, although this was not so clear at the time. Of course, none of these issues play a role in identifying the mechanism of superconductivity. But they are fundamental to our ultimate understanding of the subject. The work of Gor'kov also made the first explicit connection between the phenomenological thread of the history of superconductivity and that spun from microscopic theory. The cloth was beginning to become whole.

One final consequence of BCS theory is essential for our story. The BCS theory predicts that there is an energy gap in the single particle density of states of a superconductor. As shown in Fig. 10, this density of states has a characteristic dependence on energy, with a sharp peak just above the gap followed by a smooth decrease to the

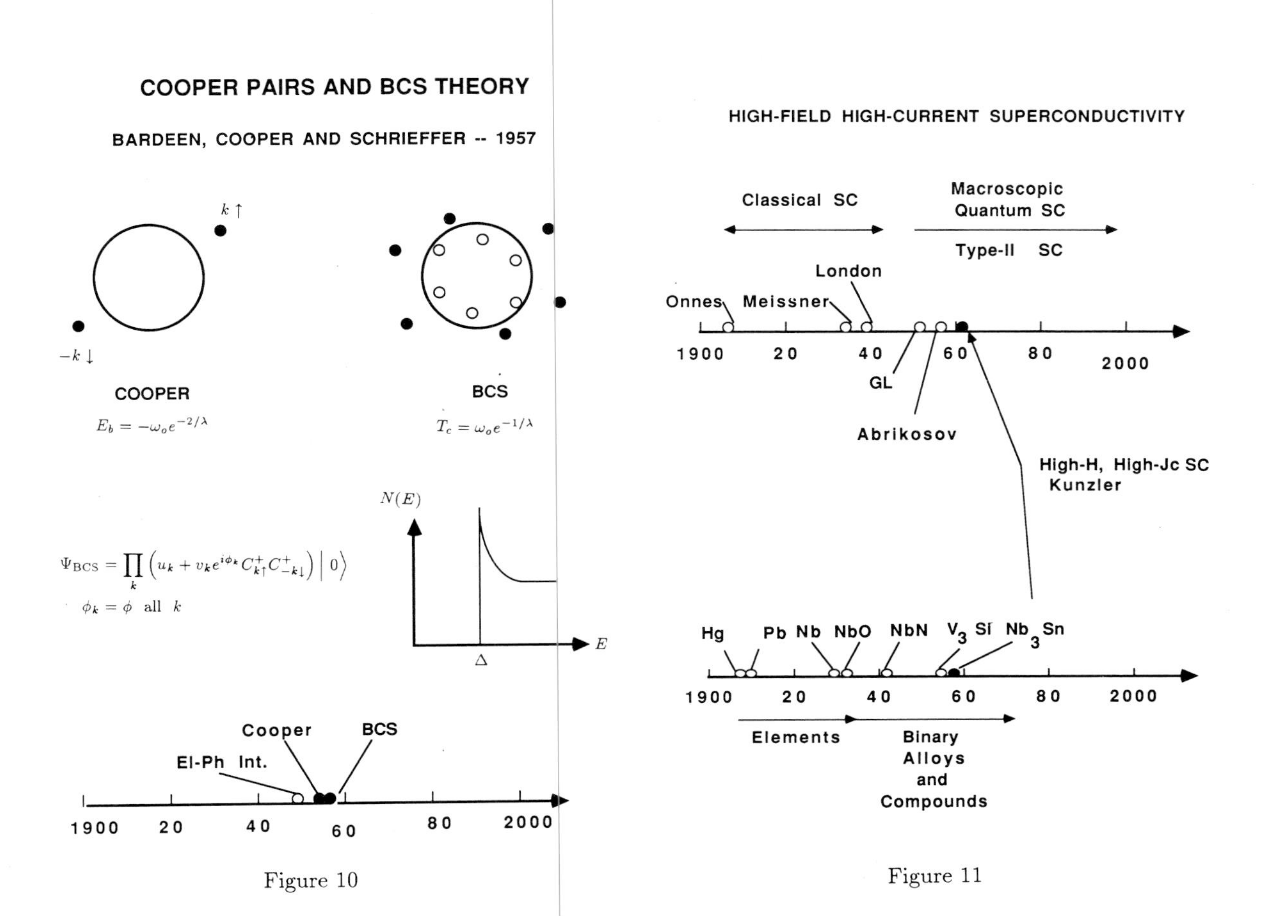

Figure 10

Figure 11

DISCOVERY OF HIGH-FIELD HIGH-CURRENT SUPERCONDUCTIVITY IN Nb_3Sn

Kunzler -- 1961

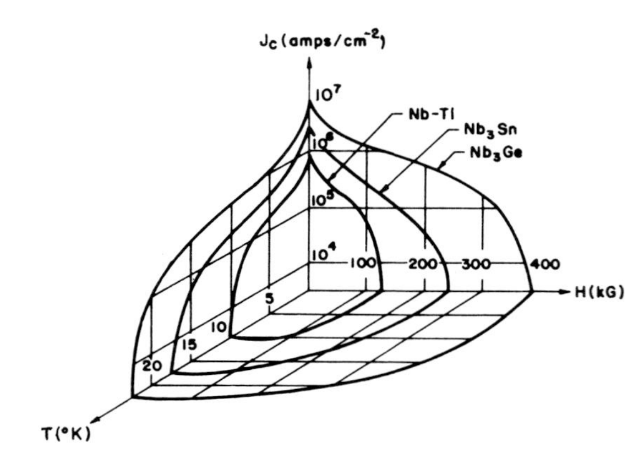

Figure 12

TYPE II SUPERCONDUCTIVITY APPRECIATED

MAGNETIC FIELD
Hc2(T)
VORTEX STATE B≠0, R≠0
MEISSNER STATE B=0, R=0
Hc1(T)
Tc
TEMPERATURE

Classical SC
Macroscopic Quantum SC
Type-II SC
Onnes
Meissner
London
GL
Abrikosov
Kunzler
1900 20 40 60 80 2000

Figure 13

normal state value. The details of the energy dependence of this density of states played a very important role in unraveling the microscopic origin of superconductivity, as we shall shortly see. But there is more going on in this Golden Age of superconductivity than BCS and the revelation of the mechanism of superconductivity.

High – Field, High – Current Superconductivity

Quite independently of BCS and all the experimental work that was going on to test its validity, another event of considerable importance in the history of superconductivity took place. Recall that we stopped our story about superconducting materials with the discovery of Nb_3Sn. While studying the properties of Nb_3Sn, Kunzler and his colleagues discovered that Nb_3Sn not only had a high transition temperature, but that it also sustained superconductivity to very high current densities and very high magnetic fields. The practical field of superconducting magnets and all the other large-scale applications of superconductivity grew out of this discovery. More importantly, from the point of view of the history of the science of superconductivity, Kunzler's discovery led ultimately to the realization that yes, Abrikosov was right: there are two types of superconductors. Superconductors do exist in which flux penetrates in the form of quantized flux lines, just as he predicted. Moreover, it is only because flux does penetrate that high-field superconductivity is possible at all. As an historical footnote, we should add that Shubnikov in the USSR had found earlier that superconducting alloys had peculiar magnetic properties that we now associate with type II superconductors. Nonetheless, as the history played out, it was Kunzler's work that first caught the attention of the world.

This chain of events is illustrated in Figs. 11, 12 and 13. The time lines in Fig. 11 show how once again the threads of our story intertwine. Fig. 12 shows the current status of the revolution that the discovery of type II superconductivity has meant in terms of practical superconducting materials. The figure is yet another one of the many popular symbols of superconductivity—the famous current-field-temperature diagram of high-field, high-current superconductivity. And Fig. 13 shows the well known phase diagram of a type II superconductor in which the Meissner state gives way to Abrikosov's vortex state as the field is increased. Also shown in Fig. 13 is an image of the vortex lattice, obtained by decorating the superconductor with small magnetic particles that are subsequently imaged with an electron microscope. There can be no doubt about the reality of the vortex lattice.

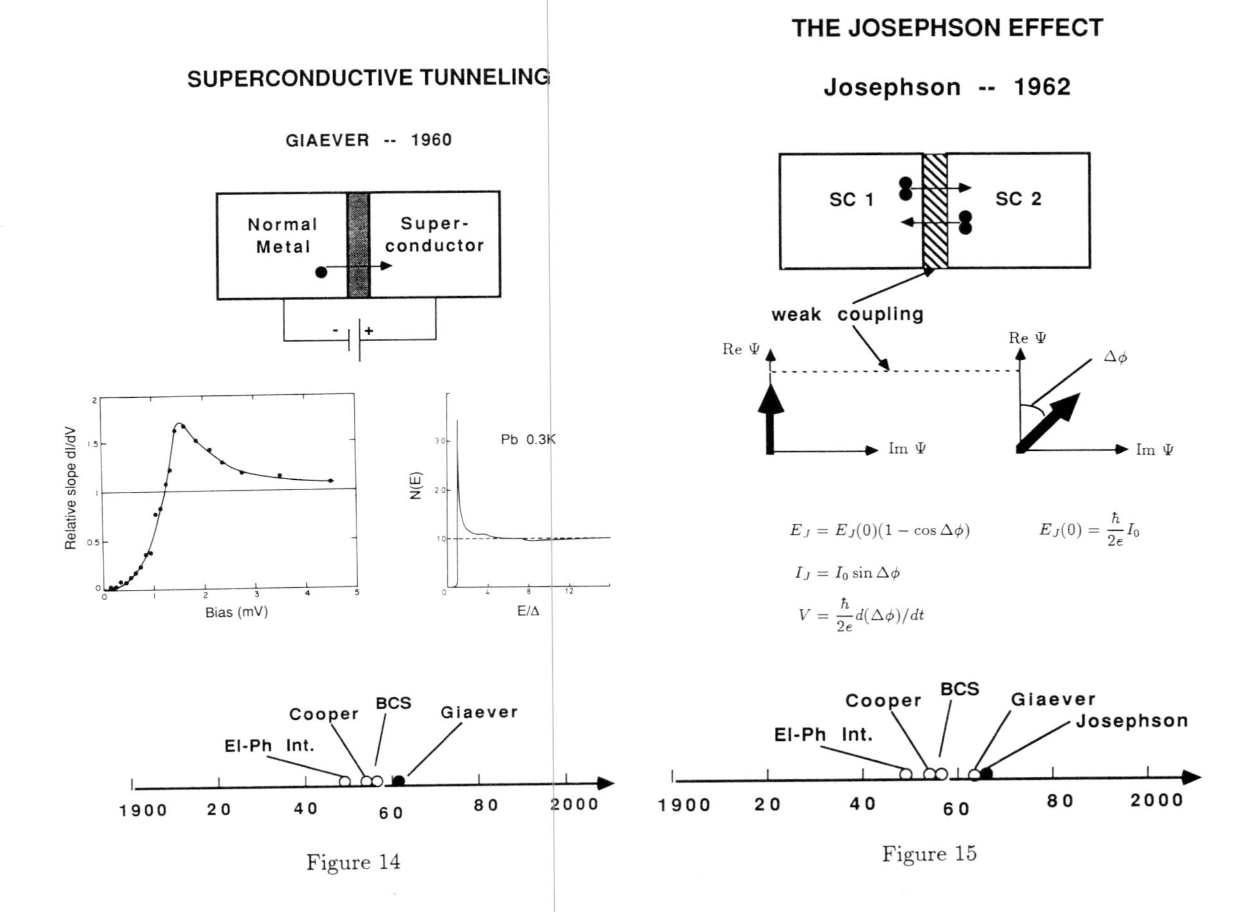

Figure 14

Figure 15

Superconductive Tunneling

For all its success, BCS theory by itself does not directly illuminate the mechanism of superconductivity. As Professor Schrieffer pointed out, BCS is a generic theory. It may have been motivated by the electron-phonon interaction, but it does not explicitly depend on it. Any attractive interaction will do. How did we begin to know that it was really the electron-phonon interaction that produced superconductivity? The first critical step was the discovery of superconducting tunneling by Giaever. Giaever found that it was possible to tunnel from a normal metal through an insulating barrier into a superconductor. More importantly, as illustrated in Fig. 14, he found that the differential conductance of the junction, dI/dV, as a function of bias voltage had a particular shape, a shape we now know to be the thermally-smeared density of states in the superconductor. Giaever's work confirmed the existence of an energy gap and provided a practical, quantitive way to measure the density of states of a superconductor.

This was not all, however. Upon careful examination of the tunneling density of states, Giaever noted that there were distinct "wiggles" on the curve that came at energies characteristic of the phonons in the superconductor. (See the right-hand side of Fig. 14.) These results gave pretty intimate evidence that superconductivity was in fact reflecting the electron-phonon interaction. The machinery necessary to extract this information quantitatively, and to connect it to microscopic theory, was not yet at hand. That was yet to come. In addition, other aspects of superconductive tunneling were rapidly developing.

The Josephson Effect

In the early 1960's Josephson was examining the theory of superconducting tunneling using BCS theory. He found that not only could single electrons tunnel, but that pairs could tunnel as well. This work led to the discovery of the Josephson effect and ultimately to all the Josephson effect devices of superconductive electronics. The fundamental equations of the Josephson effect are shown in Fig. 15, along with an illustration of the Josephson effect as a phase coupling of two weakly-coupled superconductors. This is not the usual textbook representation of the Josephson effect, but it is the most fundamental. This fact was not lost on Josephson, who in his early review papers describes

CONFIRMATION OF THE ELECTRON-PHONON MECHANISM

THE ELECTRON-PHONON INTERACTION SPECTRAL FUNCTION

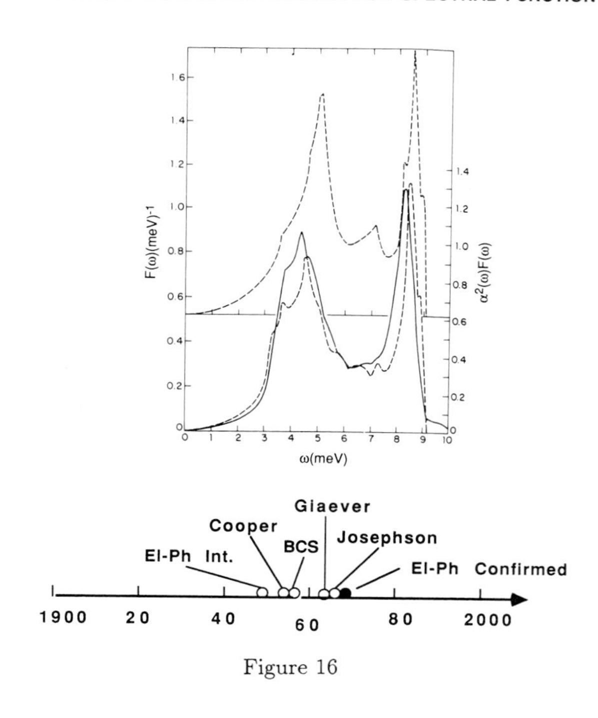

Figure 16

NOVEL MECHANISMS AND EXOTIC SC's

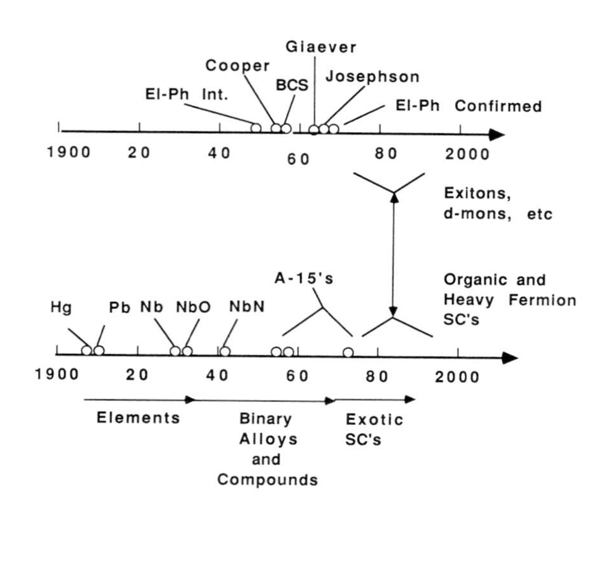

Figure 17

his effect clearly in these terms. The view of the Josephson effect as a phase coupling of two weakly-coupled superconductors make explicit the connection between the Josephson effect and the macroscopic quantum nature of superconductivity.

Hence Josephson, like Gor'kov, was able to make an explicit connection between BCS and the macroscopic quantum nature of superconductivity. So again we see the melding, the joining, the intellectual unification of the microscopic and the phenomenological macroscopic quantum understandings of superconductivity. Clearly it was a Golden Age.

Confirmation of the Electron – Phonon Mechanism

Let us now return to the "wiggles" first seen by Giaever, and the mechanism of superconductivity. The final chapter of the story is illustrated in Fig. 16. As part of the development of the theory of superconductivity that followed BCS, Eliashberg developed a formalism that permitted calculation of the properties of a superconductor in terms of the basic spectrum of the electron-phonon interaction. McMillan and Rowell combined that formalism and the experimental side of superconducting tunneling (i.e., the study of the "wiggles") into a machinery with which one could extract the electron-phonon interaction spectral function $\alpha^2F(\omega)$ from tunneling data. This function describes the strength of the electron-phonon interaction in the most basic terms. The success of this approach is illustrated in Fig. 16 for the superconductor Pb. The solid curve is the experimental determination of $\alpha^2F(\omega)$ from tunneling measurements. The lower dashed curve is a neutron scattering measurement of the phonon density of states of Pb. The upper dashed curve is a theoretical calculation of the phonon density of states $F(\omega)$ from first principles. These curves are all remarkably similar. One can argue about the differences, but such arguments are about our ability to calculate or measure precisely, not about the origins of superconductivity. That the superconductivity of Pb arises from the electron-phonon interaction is not to be denied. In fact, for all cases where such analysis has been definitively carried out, including most of the materials shown in Fig. 7, the electron-phonon mechanism of superconductivity has been confirmed.

So here we have it. The mechanism of superconductivity was established. The Golden Age was coming to a close. The large-scale electrical and the small-scale electronic applications that this age made possible were being pursued aggressively, hoping for their own Golden Age. The cloth of the science of superconductivity was clearly whole,

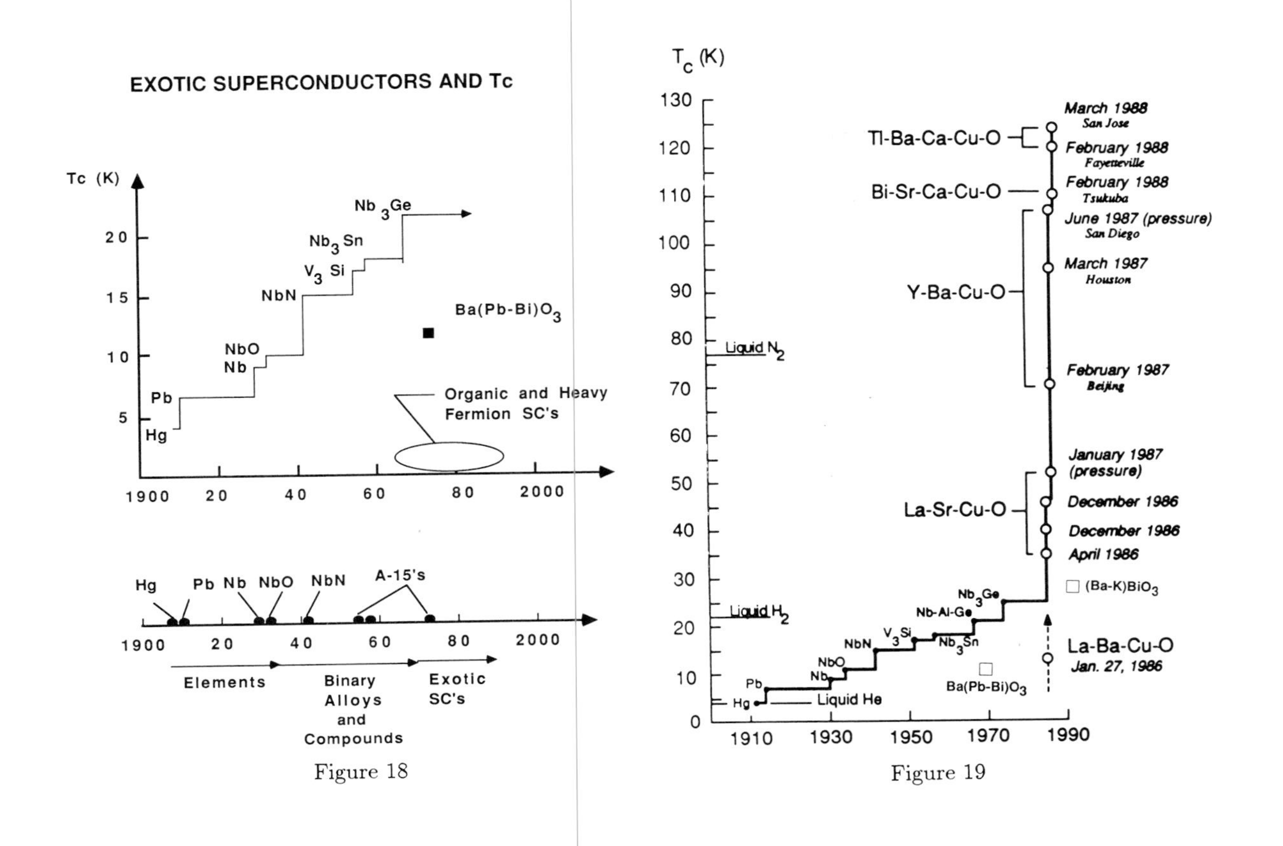

Figure 18

Figure 19

with all its principal threads firmly included. But, of course, there are always people who persist and say that there have to be new ideas, that there are always new threads, that there has to be a future. So what happened in this period after the Golden Age, the period that brings us up to today?

Novel Mechanisms and Exotic Superconductors

After the confirmation of the electron-phonon mechanism in conventional superconductors, some bold souls turned their attention to novel, non electron-phonon mechanisms, and to exotic superconductors where there was reason to believe the underlying physics might be different. As illustrated by the time lines in Fig. 17, challenges were being exchanged between theorists and experimentalists in a lively manner. Theoretically-minded people conceived of new ways of getting superconductivity, and invented new pairing interactions, excitons, plasmons, d-mons—all kinds of things. The superconducting materials community sought higher transition temperatures and superconductivity in unusual materials. Out of this came organic superconductors, heavy fermion superconductors, and new low-electron-density superconductors such as $Ba(Pb\text{-}Bi)O_3$, all of which remain exotic to this day.

But what was happening to the transition temperature during this period? As shown in Fig. 18, the T_c of conventional superconductors had been pushed up to 23K in Nb_3Ge. Niobium remained the favored base material. The exotic superconductors, for all their interesting properties, were not doing as well, which is not to say the physics was not interesting; it may well have been prophetic.

High – Temperature Superconductivity

Out of the search for novel mechanisms, out of the search for exotic superconductors, out of the search for higher transition temperatures came Bednorz and Müller. They were brave people out there where most people fear to tread, looking for superconductivity in very strange places based primarily on intuition and courage. And they had their competitors/allies, such as Chu and his colleagues. Everyone at this Symposium knows what happened. The result is shown in Fig. 19, which is surely the current symbol of superconductivity, and hopefully represents the dawn of the second Golden Age. But my charter was a history of the science of superconductivity, not prognostication. So it is time to sum up, to end our story.

WHERE DO WE STAND NOW ?

SUPERCONDUCTIVITY AND MATERIALS

THE VALUE OF COMPLEX MATERIALS

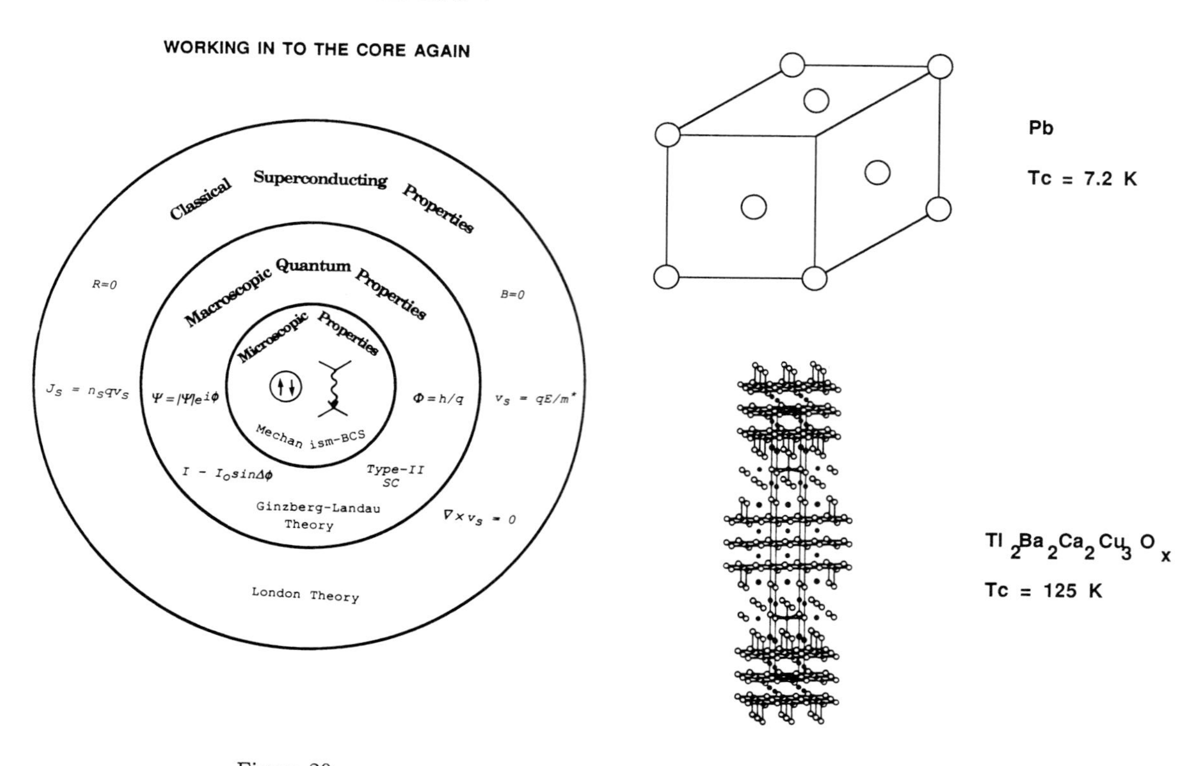

Figure 20

Figure 21

Where Do We Stand

So where have we come, and how does it relate to today's developments? Fig. 20 is an attempt to put our entire story into perspective. It shows a schematic representation of the conceptual understanding of superconductivity that has developed over the past 75 years or so. It shows that understanding organized in three layers going down to a central core. The outermost layer contains the classical superconducting properties and the London theory. Inside that is the macroscopic properties of superconductivity and the Ginzburg-Landau theory. At the core is the microscopic theory and the mechanism of superconductivity. Each layer provides a powerful level of understanding. Each layer derives its more fundamental justification from those below it.

Where do we stand with the new high-temperature superconductors as regards these levels of understanding? In a real sense we are working our way once again down to the core. There is no doubt that these new superconductors have zero resistance and exhibit a Meissner effect. Almost certainly they can be described by an appropriately generalized form of the London theory into which anisotropy has been incorporated. Also, it seems clear that they can be described by a macroscopic quantum wave function with charge $2e$. They can be formed into superconducting quantum interference devices. But there is the question of the extremely short superconducting coherence lengths in these materials, and the associated possiblity of critical fluctuations. Critical fluctuations are generally negligible in conventional superconductors. And there may be some radically new twists. For example, the Ginzburg-Landau order parameter may not be simply a complex number. It may have lower symmetry, such as in the case of ^{3}He, or, as is strongly expected, in some of the heavy fermion superconductors. Such lower symmetries would reflect a non-BCS-type pairing or something else exotic. Thus there is much interesting physics to be gleaned at this level, provided the right questions are asked.

In the core, it is anybody's guess. Clearly we just don't know. It is not appropriate here to go into the battle of phonons versus spin bags versus RVB versus excitons versus whatever. And in any event, it is experiment that will ultimately instruct us. But clearly it is here at the core where the greatest challenges and the greatest interest lie from the basic point of view. We shall see.

Let us end our history with some questions. In Fig. 21 we show the world's best understood superconductor along with perhaps the world's most interesting superconductor, certainly the world's best superconductor as gauged by its transition temperature. The first is the superconductor Pb, which has a transition temperature of 7.2 K. It contains one element and has a simple face centered cubic structure. The other is $Tl_2Ba_2Ca_2Cu_3O_x$, which has a transition temperature of 125 K. It contains five elements and has a very complicated structure. Are there some deeper meanings to be drawn from all of this? Are the structural and chemical complexities of these new superconductors essential? No one really knows. But it does seem clear that complex materials are not bad, nor merely complicated. They provide some of the simplest physical systems from a conceptual point of view, e.g, quasi-two-dimensionality. They provide the ability to tune physical properties over dramatic ranges by artful chemical substitution, e.g., from insulation to conduction, from magnetism to superconductivity. Manifestly they permit spectacular extremes of physical behavior. They clearly raise new questions of physics, chemistry, and materials science. They may be the basis of new technologies. What will we make of all this beyond the wonders of superconductivity itself? What will we make of it for interdisciplinary science and education? For science as a whole? These are questions for the future, and for real historians.

Acknowledgement

I would like to thank Donn Forbes of *Supercurrents* for providing an edited transcript of my talk from which this paper was written. I am deeply grateful to him for enormously simplifying the task of putting it into a coherent whole, and for a critical reading of the final manuscript. I would also like to emphasize that this is at best an amateur history. There has been no attempt to research detail. Errors of fact may be present. Errors of omission certainly exist. A personal choice of emphasis is freely admitted. The goal was to provide some feel for how we arrived at where we are today and to capture something of the way in which the various communities of superconductivity, along with their intellectual traditions, interacted along the way to bring us to our current understanding, and to the current era of high-temperature superconductivity. I hope that I have succeeded, if only a little.

2
Prospects for Application of Superconductivity

Application to Electric Power System

TSUNEO MITSUI

The Tokyo Electric Power Co., Inc., 1-1-3, Uchisaiwai-cho, Chiyoda-ku, Tokyo, 100 Japan

Abstract

The commercialization of high-temperature superconductivity will have a great impact on electric power systems. This paper begins by taking a look at an electric power system, and goes on to describe the characteristics of generators, underground cables and energy storage of the electric power system when superconductivity technology is applied.

Introduction

In the field of electric power, there has for some time been great hope that superconductivity at room temperature can be achieved. Since the year before last when high-temperature superconductivity was first achieved, research by scientists has brought its practical realization closer, and our hopes are heightening.

One of the problems we have to overcome in operating electric power systems is the treatment of heat. We have developed elaborate new cooling systems for generators and underground cables, and have attempted to reduce their size and increase their transmission capacities. If superconductivity technology is established, heat will no longer be a problem.

Configuration of electric power system

A schematic diagram of an electric power system is shown in Fig. 1.

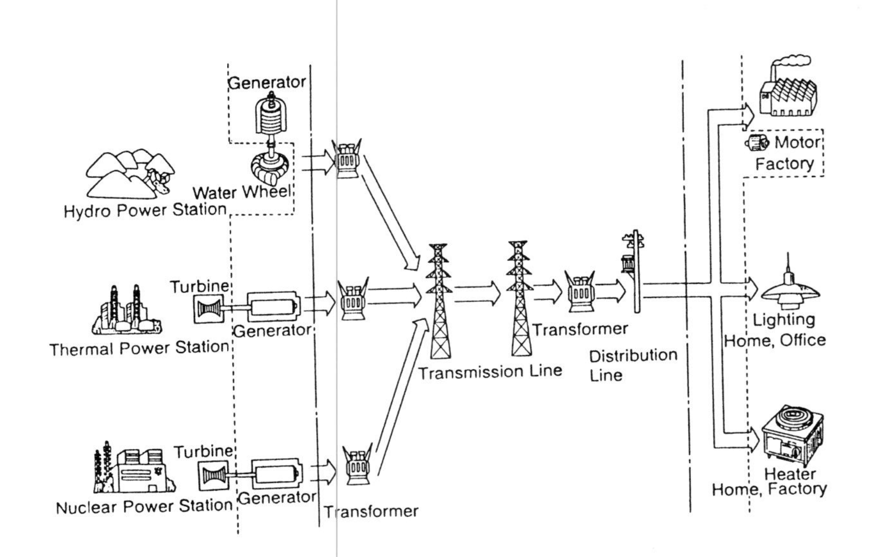

FIG.1 DIAGRAM OF POWER SYSTEM

Power loss which occurs in transmission and distribution equipment of an electric power system is 5.5% at present (Fig. 2). Let us hypothesize a system which begins with mechanical energy at a power station and ends with mechanical energy of a motor used by a customer. Fig. 3 shows the power loss which occurs in the system. If loss in a generator and at the customer's end are combined with loss during transmission and distribution, total loss reaches 19.4%. Fig. 4 shows the points at which loss occurs. Loss in the generator, and in transmission and distribution equipment accounts for one-third of total loss. If resistance loss indicated by oblique-line shading in Fig. 3 is totalled, it amounts to 7.8%. If, however, the conductors were to be replaced by superconductors, power loss in these sections could be greatly reduced.

Application to electric power systems

If superconductivity technology is applied to an electric power system, it will reduce power loss and raise efficiency, thus enabling equipment to be made more compact and with improved functions. However, there are design problems inherent in the use of superconductivity.

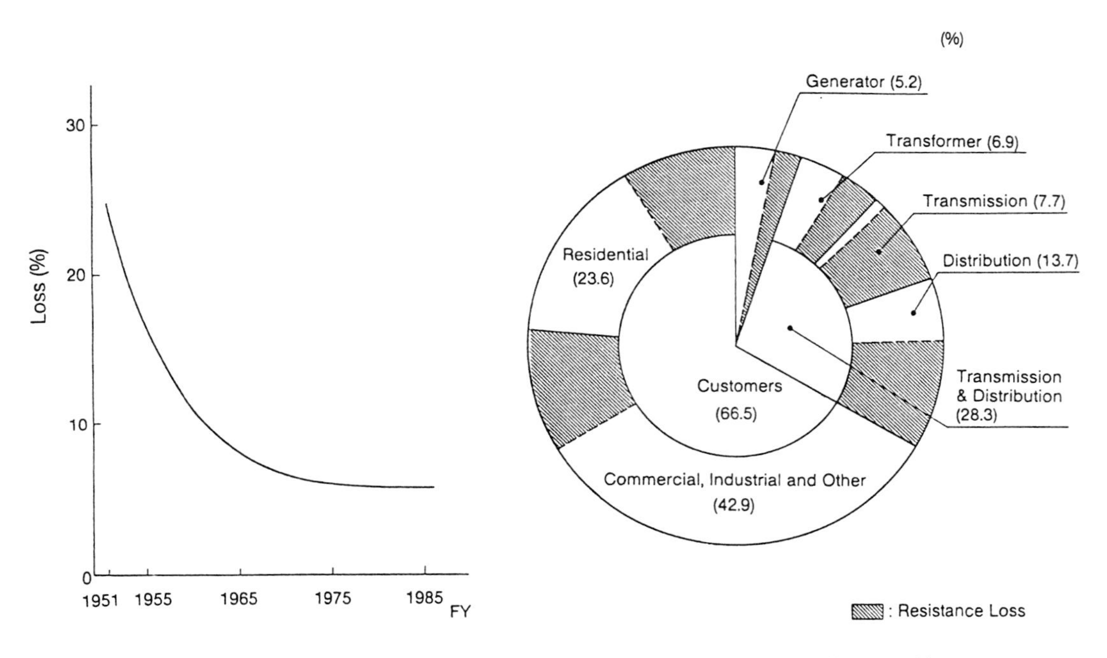

FIG.2 TRANSMISSION & DISTRIBUTION LOSS RATE (TEPCO)

FIG.4 SHARE OF TOTAL LOSS

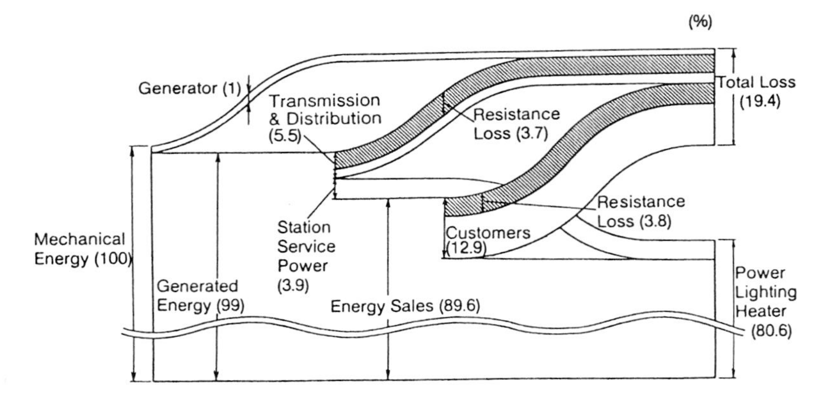

FIG.3 DIAGRAM OF LOSS FLOW

(1) Generators

Fig. 5 compares the structures of a superconducting generator and a conventional generator. In a superconducting generator, superconducting wire is used in place of copper wire as field winding.

(a) A superconducting generator can be made about 50% smaller in size and weight than a conventional generator (Fig. 6).

(b) Since power loss in the field winding is less, efficiency can be improved by about 0.5%.

(c) As reactance is lower, power system stability improves.

(d) The application of a ultra high-response excitation system helps stability power limit rise to 1.7 times that of a conventional generator.

(e) As field winding must be cooled, a vacuum structure and a cooling system to supply coolant are needed.

If such a superconducting generator were to be installed in a thermal or nuclear power plant, the generator would occupy only 5% of the turbine floorage (Fig. 7).

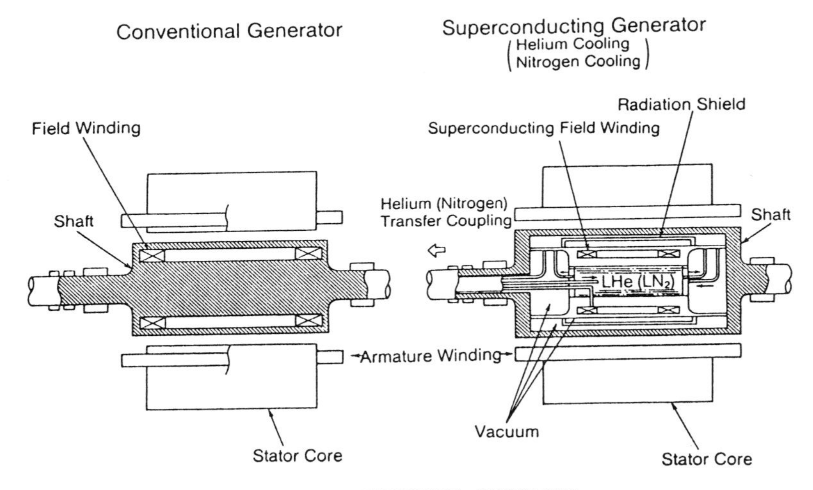

FIG.5 SUPERCONDUCTING GENERATOR

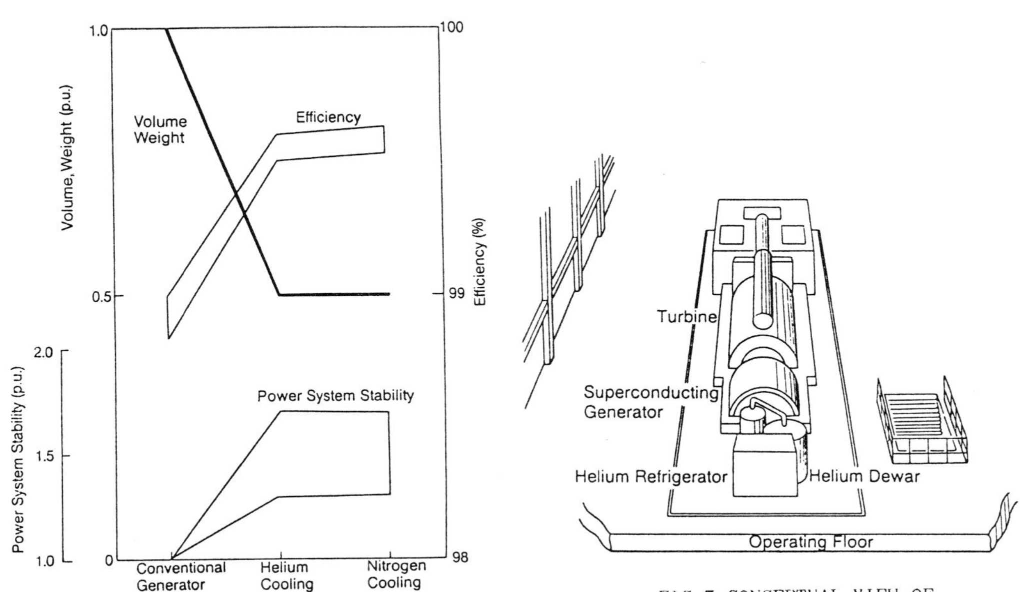

FIG.6 CHARACTERISTICS OF SUPERCONDUCTING GENERATOR

FIG.7 CONCEPTUAL VIEW OF SUPERCONDUCTING GENERATOR INSTALLATION

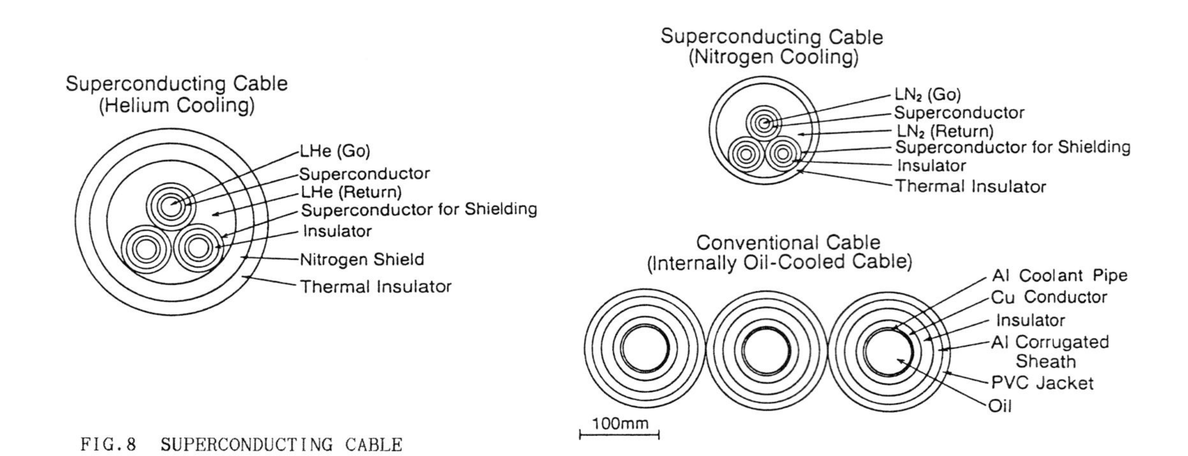

FIG.8 SUPERCONDUCTING CABLE

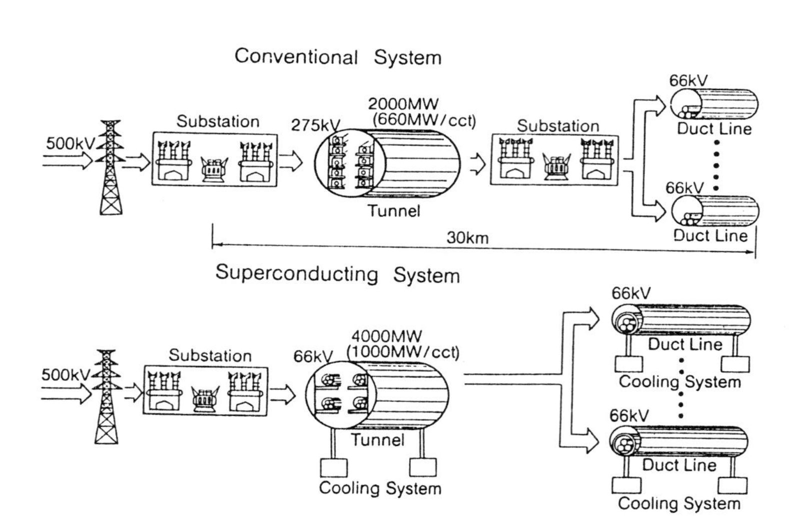

FIG.9 METROPOLITAN TRANSMISSION SYSTEM

On balance, the ability of a superconducting generator to improve functions such as increase of stability power limit due to its ultra high-response excitation would seem to be a greater advantage than that of size.

(2) Underground cables

Shown in Fig. 8 is a conceptual drawing of a superconducting underground cable.

(a)In a conventional underground cable, the increase in temperature limits its capacity. With a superconducting cable, transmission capacity can be increased.

(b)For this reason, transmission of electric power even as high as 1,000 MW would not require extra high voltage; 66 kV would suffice.

(c)A liquid nitrogen-cooled cable can be made more compact than a helium-cooled cable, which requires a cooling medium, or an internal oil-cooled cable, thus enabling size to be reduced. In an underground cable, tunnel and duct construction accounts for a large share of construction cost. Compactization would significantly reduce this cost.

Using a hypothetical power system, we considered what would happen if such a superconducting cable were to be applied. We supposed the development of a liquid nitrogen-cooled AC superconducting cable, with a current density of 10^5 A/cm^2.

As shown in the upper part of Fig. 9, 2,000 MW is transmitted to central Tokyo over a 275 kV underground cable from a 500 kV substation in the outskirts of the city. We then suppose that power consumption increases, necessitating the transmission of 4,000 MW, or twice the normal

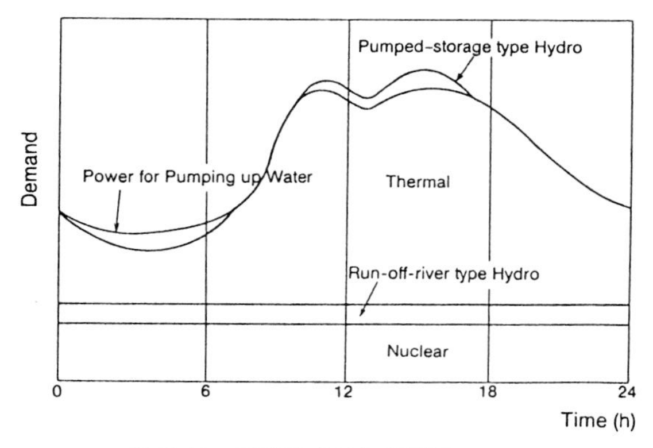

FIG.10 DAILY LOAD CURVE

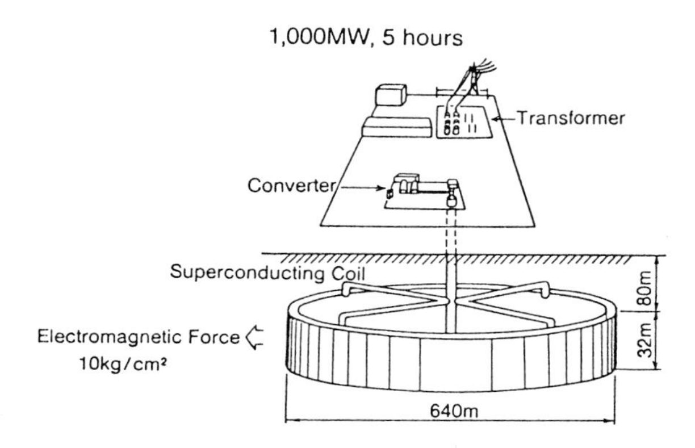

FIG.11 SUPERCONDUCTING MAGNETIC ENERGY STORAGE SYSTEM (SMES)

level. A normal internally oil-cooled cable requires the addition of a tunnel. If it is a superconducting cable, however, a 66 kV cable can transmit 4,000 MW as far as the city center and can be housed in the existing tunnel or duct. The need for an additional tunnel or duct no longer exists, and in-between substations can be omitted. It is difficult at this point to estimate with accuracy the economies to be made. However, total cost has already been demonstrated to be in the region of 60 ~ 70% that incurred by using ordinary cables.

There are underground cables of more than 66 kV extending over about 5,000 km in the Tokyo area. If these cables were to be replaced with superconducting cables, power transmission capacity would rise dramatically. Total power transmission capacity of 275 kV cables from the outskirts to central Tokyo is about 8,000 MW at present. If these were replaced with superconducting cables, that figure would rise to 25,000 ~ 30,000 MW.

(3) Storage of energy

Fig. 10 shows a load curve in a power system. As power consumption drops around midnight, a power plant's output is lowered, and energy is stored by pumped hydraulic power for use during the following day's peak period.

If a more economical means of energy storage than this pumped hydraulic power system were to be developed, the amount of generation equipment required to handle peak consumption periods could be reduced.

A superconducting magnetic energy storage (SMES) stores electricity as magnetic energy in a superconducting coil for transmission when necessary. Fig. 11 shows a coil with a storage capacity of 5,000 MWh, which is capable of storing 1,000 MW for five hours. The diameter of this coil is 640m.

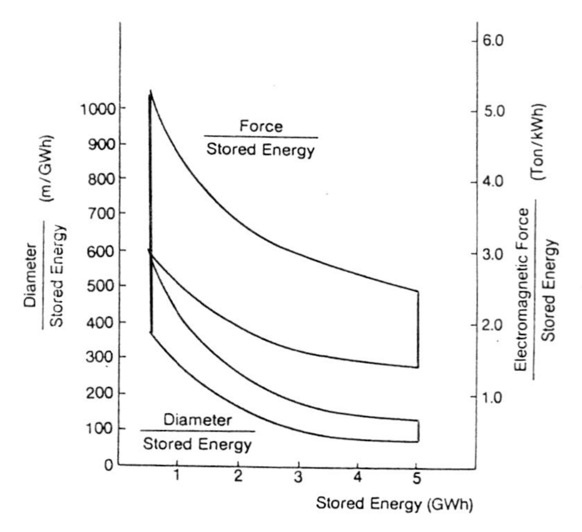

FIG.12 CHARACTERISTICS OF SMES

General characteristics of a SMES are as follows:

(a)The smaller the storage capacity, the larger the diameter of a coil per unit energy. As the storage capacity becomes smaller, the volume of a conductor per unit energy tends to be larger. Thus, a SMES of large storage capacity is more advantageous.

(b)The current of a SMES produces electromagnetic force radially and axially. Radial electromagnetic force of a 5,000 MWh SMES is about 7,000,000 tons. Electromagnetic force per unit energy also rises when storage capacity is reduced (Fig.12). As this electromagnetic force becomes repetitive with the inputting and outputting of electrical energy, a highly durable structure is needed. Usually an underground structure supported by firm base rock as shown in Fig. 11 is considered.

(c)As current of a SMES is direct current, an AC/DC converter is installed to control inputting and outputting of electric energy.

As for energy storage systems other than SMES, advanced types of storage battery which would store energy in the form of chemical energy are under development. A comparison of these two systems with regard to energy density per unit volume, reveals that a superconducting coil has a density far lower than that of a battery.

The problem which will face SMES in the future will be how to achieve greater economy of design in areas such as the underground structure and support.

SMES can be used to cut power that passes through transmission lines and raise the stability power limit of transmission lines. The research and development of these new opportunities are expected.

Conclusion

To commercialize superconducting technology in power systems, the first important steps will be to improve critical current density and critical magnetic field, and to develop wire-processing technology.

When high-temperature superconductivity is used, cooling equipment is required, which should be both economical and reliable. Furthermore, means should be taken to protect against quench of superconductivity.

Should superconductivity be realized at room temperature, of course, cooling equipment would not be necessary, and applications would increase, including overhead transmission lines. Great hopes are placed on basic research now under way elsewhere which aims to realize higher temperature superconductivity.

In addition to this form of electric power system, other new equipment and systems incorporating superconducting technology will emerge to enhance the effectiveness of electric power usage. Already we can point to the linear motor car MAGLEV and the linear motor catapult for launching vehicles into space. New technologies usually go far beyond our earliest anticipations.

Applicability of Superconductivity in the Electric Power System: A Canadian Perspective

FRANK Y. CHU[1] and RAYMOND ROBERGE[2]

[1] Ontario Hydro Research Division, Toronto, Ontario, M8Z 5S4, Canada
[2] Hydro Quebec, Vice Presidence Recherche, Varennes, Quebec, J0L 2P0, Canada

ABSTRACT

The recent discovery of high temperature superconducting (HTSC) materials and the development of low loss (in alternating current application), low temperature multifilamentary superconductors have the potential to benefit the electric power companies. The potential benefits of these materials, and the public interest in HTSC, have motivated electrical utilities to examine the technologies closely. This paper presents a Canadian view on prospective applications of superconductivity in a power system.

INTRODUCTION

The applications of conventional superconducting materials such as NbTi and Nb_3Sn in the electric power system were examined in the sixties and seventies. However, most studies concluded that large scale exploitation of the materials was not economically feasible [1]. The recent discovery of high temperature superconducting (HTSC) materials [2,3] could change the economic picture in that refrigeration would be much simplified and less costly. With the intense public interest and media focus on HTSC applications in the power system, utilities must respond to many questions regarding this emerging technology. Similarly, development of low loss materials [4] has opened the door to applications not technically feasible before. What are the promises and limitations of these materials? What is the impact on electric utilities? How can utilities exploit the opportunities offered?

Both Ontario Hydro and Hydro Quebec, the latter under contract to the Canadian Electrical Association, have been reexamining previous conclusions in the light of the recent developments. The objective of this paper is to present a preliminary perspective on possible applications and future developments of these materials. It should be mentioned that new advances are occurring regularly which could easily change the present assessment of applicability.

SYSTEM CONSIDERATIONS

Table I presents some relevant data on Canada's power systems. In assessing the applicability of superconductivity to Canada's power systems, two basic questions should be addressed:

1) How might these new technologies affect the generation, transmission and distribution of electric power?

2) What impact might the application of these materials, devices and systems have on the demand for electricity?

The impact of superconducting materials on the supply side of the power system was assessed in many studies in the sixties and seventies when NbTi and Nb_3Sn superconductors were developed. Most studies concluded that although the application of superconducting materials in a power system did indeed in some cases lead to improved efficiencies and equipment size reduction, unfortunately economic competitiveness with conventional technologies was only possible at very large unit sizes. The major impact of the recently discovered HTSC materials is that the refrigeration energy cost can be significantly lowered if the material could be operated at liquid nitrogen temperature instead of liquid helium temperature. The key question is whether the savings in refrigeration energy cost is enough to make HTSC applications in the system competitive with other options. Low loss superconductors could extend the range of possible applications.

Generation

Electric generators are very efficient machines. By rewinding the rotor of the generator with superconducting wire, the size and weight of the generator can be reduced by up to 50% and efficiencies can be improved by 0.5 to 1.0%. It is expected that further small improvements in efficiency can be realized with the HTSC materials because of the less stringent cooling requirements. A 1% increase in efficiency of a 500-MW generator can translate into $25 million saving over the 30-year life of the generator. Still further increases in efficiency could be obtained by replacement of the armature windings with the new low loss superconductors.

Despite the many potential advantages of superconducting generators, their introduction into a power system will depend on the successful development and testing of prototype machines. The reduced demand for new power generation equipment in the past decade has significantly slowed the development of superconducting rotating machines. The new HTSC technology may lower cooling requirements, reduce the need for extensive thermal insulation in the rotor and simplify the rotor construction. We foresee that the development of superconducting machines will grow at a modest pace based on NbTi with a gradual transition to the HTSC materials, if they prove to be reliable. We also foresee that the time frame for prototype application is in the 10-15 years range.

TABLE I

Canadian Utilities Energy Analysis (1986)

	Ontario Hydro	Hydro Quebec	Canada
Installed Capacity (MW)	27 000	28 000	98 500
Energy Generated & Received (GWh)	126 619	148 260	421 856
Transmission Loss (GWh)	5 100		
Distribution Loss (GWh)	4 000		

Transmission

Power transmission is usually carried out at high voltages in overhead transmission lines. Losses are about 4-6% for a typical utility. At Ontario Hydro, the transmission loss is about 4% of the total energy generated on an annual basis. A superconducting cable would not be more energy efficient. Its use would be as a replacement for overhead transmission lines in highly congested urban areas where right-of-way is expensive or unavailable. Table II compares the costs of three types of transmission system, viz, overhead, underground superconducting, and pipe type underground cable.

Even for underground transmission, superconducting cables would have to be more economical than other cables such as oil filled pipe type cables or SF_6 cables. Previous analyses have suggested that conventional superconducting cables can only compete with oil filled or SF_6 cables if their capacity exceeds 5000 MVA. At this capacity level, the loss of the superconducting cable is about 10% of the loss in oil filled cables. Without assumptions on material and fabrication costs, it is difficult at this stage to compare the cost of a low temperature superconducting cable with a HTSC cable. The question of whether HTSC cable may be able to compete with oil-filled or SF_6 cable at a lower MVA rating in the range of 500 to 3000 MVA has not been answered. However, one thing is clear, the critical current density of the material will affect the final cost of the cable. A preliminary analysis based on the material cost of the yttrium based compounds has shown that the cost of a HTSC cable could vary from $800 to $13 000/MVA/km for critical current densities of 10^5 and 10^3 A/cm^2 respectively. Clearly a high critical current conductor must be developed.

Contrary to popular belief, there are some losses in ac superconducting cables, partly hysteresis in the superconductor, and partly heat leak and dielectric losses. Due to the nature of the existing power system which is predominantly ac, most future HTSC cables will be ac for system compatibility. The conversion cost from ac to dc and back to ac will affect the cost effectiveness of the dc superconducting cables. For bulk power long distance transmission, a dc overground pipeline type cable may be considered in the future. We foresee that if HTSC materials can satisfy all the technical requirements (see Table 3), a prototype underground cable in the 1000-MVA range cooled by liquid nitrogen, may be developed in about 10 years. Technology already existed for cryoresistive cables, which would also be cooled by liquid nitrogen, can be extensively used in such cable development.

TABLE II

Preliminary Cost Comparison
100-km Transmission System 10 000 MVA*

	US$(M)	Ratio
500 kV Overhead	700	1.0
230 kV SC Underground	1600	2.3
500 kV Underground	3000	4.3

* Brookhaven National Laboratory data [5].

Superconducting Magnetic Energy Storage (SMES)

Another application receiving considerable attention from utilities is the use of superconducting coils to store energy. Such a coil would be charged during off peak hours by using power from baseload generators, then discharged during peaks, thus reducing the need for expensive cycling power plants. A study carried out by Bechtel International Inc and GA Technologies [6] showed that the capital cost for a 5000 MWh SMES plant operating at liquid helium temperatures may be about $1.0 billion. For load levelling, this cost is still competitive with other storage schemes such as pumped hydro storage. The applications of HTSC materials in a

SMES system may lead to a reduction of 5 to 10% capital cost as compared to low temperature superconducting SMES, thus lowering the capital cost to about $950 M.

Despite SMES's many attractive features, any such system would pose unique engineering challenges. A 1000 MW, 5000 MWh design would require an underground storage space of about 1 km diameter. The outer boundary for such a facility would have to be more than 5 km in diameter because of the high magnetic field generated. To support the large magnetic forces, the SMES coil must be installed in a trench of solid bedrock for stability. The high magnetic field would also pose some environmental problems. Its size would be comparable to a conventional pumped storage hydro system.

A much smaller scale SMES may be used in a power system as a reactor or system stabilizer for short term energy storage purposes. These devices would have an energy storage capacity in the range of 30 MJ to 1000 MJ (0.3 MWh) as compared to the load leveling scheme of 18 TJ (5000 MWh). When used for dynamic or transient stabilization in the power system, these devices can temporarily store large blocks of energy, equivalent to 10-500 MW for a few seconds, for smoothing any disturbance or transients in the system. A HTSC SMES may lower the capital cost, and permit better utilization of the transmission system. Several Canadian utilities are watching the development of SMES technology closely for possible future implementation.

DEMAND SIDE IMPACT

The impact of HTSC applications on the demand of electrical energy is more difficult to assess. Magnetic levitated trains, magnetic separators, electric motors and magnetic resonance imaging machines could benefit from the discovery of HTSC materials. In general, the efficiencies of any machine or device which uses electromagnets can be improved by employing superconducting windings. The introduction of a superconducting magnetic levitated train is unlikely ,to be affected by the discovery of HTSC since the major cost is for track installation. It has been pointed out that the superconducting magnet accounts for less than 10% of the capital cost of the total system. The impact on the demand for electricity as a result of the introduction of magnetic levitation trains is probably very slight. Magnetic resonance imaging machines are becoming more and more widespread. Total numbers are , however, rather modest. In Canada, there are at present about 10 units in operation and hence the energy requirement is negligible.

The introduction of HTSC materials in motor design may have a significant impact on the utilization of electrical energy. A preliminary assessment indicates that a HTSC superconducting motor in the size of 1 MW may prove to be economically feasible. However, the actual improvement in efficiency is difficult to assess. Both the low temperature superconductor and HTSC could be more energy efficient. In Quebec, it is estimated that 10% of the total electrical energy consumption is in large motors. Savings of a few percent may thus be feasible by employing superconductors, traditional or HTSC materials in all large motors.

Many scientists believe that one of the early applications of HTSC will be in electronics and computers. Although these two areas may not have a direct impact on the demand or supply of electrical energy, they may have an indirect impact. Superconducting switches using the principles of Josephson junctions, can process information about 10 times faster than conventional silicon circuits. Adoption of Josephson junction technology to make faster computers could, therefore, affect utility operation because the electric power industry is the third largest user of computers. Supercomputers may help system operators to acquire system parameters, optimize and control the power system on a real time basis to achieve maximum operating efficiency for improved energy management.

TECHNICAL REQUIREMENTS AND RESEARCH NEEDS

The prospects for various applications in a power system depend, of course, on the material's properties. Many properties are important, including the transition temperature, critical current density, critical field, mechanical properties, interfacial interactions and contacts, and ac losses. Furthermore, material aging characteristics and environmental compatibility are also important factors affecting the applicability of the material in a utility environment. Table III lists some of the HTSC material requirements for superconducting applications. Applications in areas such as generation and transmission may require different properties because of the different operating conditions.

TABLE III

Technical Requirements

* Tc	>100 K, LN_2 based
* Hc	=<10 T
* Jc	10^4 - 10^5 A/cm^{2}φ
* A.C. Loss	<200 mW/m at 60 Hz, 4000A & 77K
* Mechanical	0.2% strain tolerance
* Stability	High

While the base of experimental knowledge on the 90-K yttrium based material is growing rapidly, there is as yet no generally accepted theoretical explanation of its behaviour. Much research is needed to optimize the 90-K material's properties so that it can be used under practical utility conditions. The material's low critical current density may be a major hurdle in its acceptance. The fabrication and processing challenges presented by the HTSC material suggest that the period of development will probably be long-term and may extend for a decade or more. Table IV lists some of the material's engineering concerns which need to be resolved before large scale applications can be considered. Despite the less than optimum material properties for power system application, material scientists and engineers may eventually be able to optimize and fabricate the material according to the set requirements.

Due to the complexity of the power system, application of materials with unique characteristics will lead to changes in system design and operating practices. Reliable cryogenic systems are needed to cool superconducting cables to below the critical temperature. In the SMES area, significant research effort will be needed to protect personnel from the magnetic field generated by the magnet. Much geotechnical engineering research will be required to design the underground support structure to a high degree of mechanical stability. Table V lists some examples of peripheral technologies which are needed for successful application of HTSC material and devices in the future. The continued development of these technologies to mesh with the HTSC material development for practical applications will challenge the utilities and the research community in the coming years.

TABLE IV

Major Material & Engineering Problems to be Solved

* Improve critical current densities to greater than $10^5 A/cm^2$
* Measure ac losses and develop suitable conductors
* Improve mechanical properties such as fracture toughness
* Improve material stabilities
* Investigate and resolve aging and compatibility problem
* Develop more efficient and cost effective fabrication techniques
* Investigate environmental issues, eg magnetic field effects
* Solve problem of electrical joints and contacts

TABLE V

Peripheral Technologies

(In Addition to Material and Conductor Technologies)

* Cryogenic Technology - Cooling System
  - Control and Instrumentation
* Thermal Insulation Technology
* Low Temperature Electrical Insulation and Dielectrics
* Magnetic Field - Environmental
  - Measurement
* Geotechnical - UG Cable
  - UG Storage Ring
* Mechanical - Structural Support (Short Circuit)
* Diagnostics and Instrumentation

CANADIAN UTILITIES RESPONSE

The reaction of Ontario Hydro and Hydro Quebec to the discovery of HTSC material is that of excitement and cautious optimism. We recognize the potential advantages offered by the HTSC materials over low temperature superconductors; however, we also feel that the transfer of the material from the laboratory to practical application will not be easy and may take many years. While the applications discussed above may have an impact to Canadian utilities in the future, the immediate impact derives from the immense public interest.

In order to address the question of material application in a power system and to respond to external interests, several Canadian utilities, including Ontario Hydro and Hydro Quebec, have established work programs and joined consortia with universities and research institutions to investigate material properties for conductor development and to assess the potential applications in the power system. The Canadian Electrical Association, an organization consisting of major Canadian utilities, has contracted Hydro Quebec to study the possible opportunities offered by the HTSC materials. Ontario Hydro publishes a newsletter periodically to inform its employees and the public about superconductor technology developments which may have significant relevance to the power industry.

CONCLUSIONS

The impact of HTSC materials, devices and components on a power system has been discussed. While the benefits to various aspects of power system operations can be assessed, the cost for implementation is not clear. This is partly due to the lack of adequate information about material properties and fabrication cost. Serious consideration for major component development and implementation will wait for further improvements of material properties that are compatible with the utility operating environment. Despite the long term effort required to bring the technology from the laboratory to application, the potential payoff can be substantial. Utilities in Canada are prepared to work with the research community to meet the challenges for exploiting the potential advantages of applying HTSC materials in the electric power system.

ACKNOWLEDGEMENT

The authors would like to thank members of the Ontario Hydro HTSC Task Force and the Hydro Quebec Research Division for their helpful discussions. Hydro Quebec acknowledges the financial contribution of the Canadian Electrical Association.

REFERENCES

[1] G. Bogner, "Large Scale Applications of Superconductivity", ed. B.B. Schwartz and S. Foner (Plenum Press, New York 1976).

[2] J.D. Bednorz and K.A. Muller, "Possible High Tc Superconductivity in the Ba-La-Cu-O System", Z. Phys. B 64, 189 (1986).

[3] M.K. Wu, J.R. Ashburn, C.J. Torng, P.H. Hor, R.L. Meng, L.Gao, Z.J. Huang, Y.Q. Wang and C.W. Chu, "Superconductivity at 93 K in a New Mixed Phase Y-Ba-Cu-O Compound at Ambient Pressure", Phys. Rev. Lett. 58, 908 (1987).

[4] J.L. Sabrie, "La Cryoelectricite", Revue Alsthom, No.5, 31 (1986).

[5] R.A. Thomas and E.B. Forsyth, "Preliminary Economic Analysis of an HTSC Power Transmission System",Argonne National Lab Report, ANL/CNSV 64, January 1988.

[6] R.J. Loyd, G.F. Moyer, J.R. Purcell, J. Alcorn, "Conceptual Design and Cost of a Superconducting Magnetic Energy Storage Plant", EPRI Report, EM-3457, April 1984.

Superconductive Magnetic Energy Storage (SMES) for Electric Utilities

J.M. PFOTENHAUER and R.W. BOOM

Applied Superconductivity Center, University of Wisconsin, Madison, WI 53706, USA

ABSTRACT

SMES system designs, component developments, and utility usage studies at Wisconsin will be reviewed. The SMES design criteria for superfluid helium cooled NbTi solenoids is outlined and compared to the probable design criteria for the future potential use of high T_c ceramic conductor solenoids cooled by liquid nitrogen.

INTRODUCTION

Superconducting magnetic energy storage (SMES) provides a very real potential for the largest market of superconductivity in the immediate future. Of the various proposed uses for superconductivity, to date the most commercially successful product has been the MRI units. A recent publication [1] noted that in 1987 there were ~ 800 MRI units in operation, which at a conservative average price of \$0.25 M each, represented a total revenue flow of \$200 M into applied superconductivity. A conservative estimate of \$800 M has been made for the price of a <u>single</u> 5000 MWH SMES system [2]. The conductor cost alone totals \$145 M.

In spite of this large capital investment, SMES systems remain attractive to the electrical utilities primarily due to their anticipated usefulness in diurnal load leveling. The high 95% storage efficiency of SMES makes possible effective large scale load leveling and avoidance of more peak generation units. Overall cost comparisons [2,3,4] of SMES with other power generation and storage methods reveal obvious benefits of SMES systems with storage capacities of about 5000 MWH. In addition [5], smaller units may also prove useful for regulation, ramping and energy sales and purchases.

SMES SYSTEM DESCRIPTION

Research and development on SMES began at the University of Wisconsin with the introduction by Peterson and Boom [6] of a storage system consisting of a superconducting solenoid charged and discharged by highly efficient Graetz bridge ac/dc converters. Since then, work has been ongoing in the form of analyses and experiments in magnet design, energy conversion, structural support, thermal insulation, superconducting materials, aluminum properties, refrigeration, cryogenics, and protection. This work has produced the following preferred system design options:

- Efficient Graetz bridge energy conversion between ac (transmission line) and dc (superconductive solenoid)
- Single and double layer low aspect ratio solenoids mounted in surface trenches in bedrock
- Rippled or non-rippled solenoids designed to accommodate cooldown and warmup stresses
- Radial forces transmitted from the 1.8 K solenoid to the outer trench wall by reinforced epoxy insulating struts
- Axial forces balanced internally by aluminum alloy structure in the superfluid helium bath
- Protection discharge system to expel the helium and dissipate the storage energy as heat in axial structure
- Refrigeration system including liquid nitrogen and liquid neon cooled shrouds and an annular bath of superfluid helium at 1.8 K and 0.1 MPa
- Composite conductor of NbTi in high purity Al; cooled in the superfluid helium bath at 1.8 K
- Identification of a long term "cryogenic stability" in the SMES superfluid helium open bath

An artist's conception of a design incorporating the above features is shown in Figure 1. Extensive descriptions of each of these design options are given elsewhere [4,7].

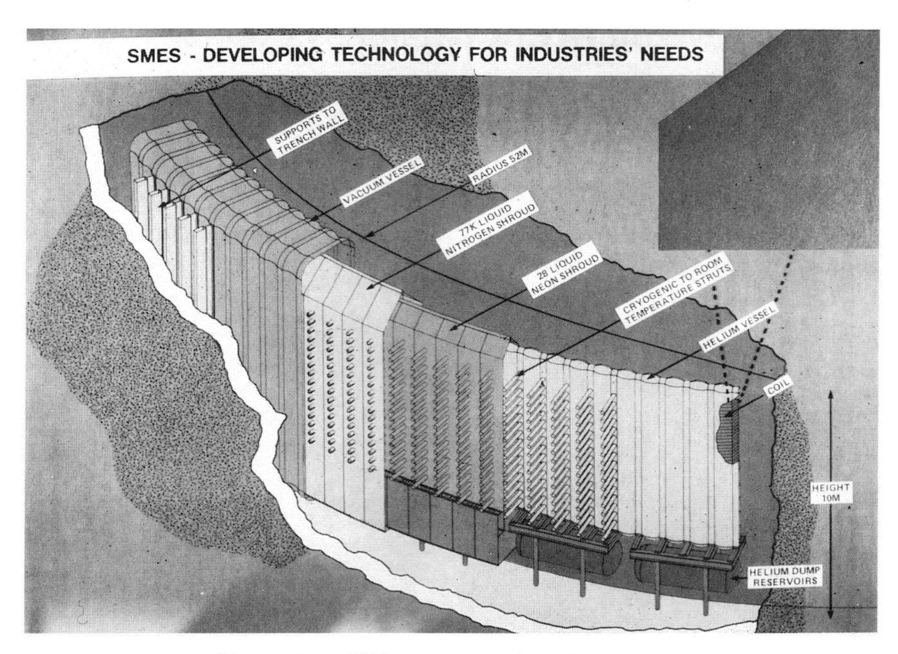

Figure 1: SMES system design features

STABILITY

A priority in the Wisconsin SMES designs has always been to provide a system with a high degree of reliability. This feature is essential for success in the market place. In addition to the fact that reliability is critical to the first few units, one finds that it is also a necessity for all subsequent units because of the large investment and potential involved. Reliability in the Wisconsin SMES design is provided to a large extent by the use of He II for thermal stability of the superconducting coil.

Very early in the design process, the decision was made to operate the magnet at 1.8 K and 0.1 MPa (1atm), thus using subcooled He II to provide the necessary cryogenic stability. Other possible methods for providing stability include pool boiling with saturated, 4.2 K helium, and forced flow cooling using supercritical helium. However, as Van Sciver [8] has pointed out, the use of these methods in a SMES system result in either unacceptably low current densities or excessively large pumping requirements. The use of He II and the associated low temperatures provide the benefit of attaining larger current densities in the superconductor; consequently less superconductor is required to obtain a given current. This cost savings is slightly offset by the higher refrigeration power required to produce 1.8 K as compared to that required to produce 4.2 K. However, whereas the superconductor cost can be reduced by about 33%, this far outweighs the increase in refrigeration cost. In spite of the significantly lower temperature required, the increased refrigeration demand remains low due to the fact that the heat load is also taken up by additional 77 K and 28 K refrigeration systems.

Beyond the superconductor cost savings, the use of subcooled He II is advantageous for the thermal stability it provides to the superconductor. It is necessary to use the liquid helium for absorbing any thermal disturbance since the heat capacities of metals at liquid helium temperatures are small. For example, ρC_p for aluminum is approximately 1 mJ/cm^3 K at 4K while typical values of ρC_p for liquid helium are closer to 400 mJ/cm^{3}K. Thus it is prudent to use the liquid helium both as a means for obtaining the cold temperatures and for absorbing any thermal disturbances. In this respect He II also has definite advantages over He I since the large effective thermal conductivity of the He II rapidly diffuses the heat over a large volume. A closer examination of the issues related to stability reveals the advantages of the He II design in the SMES system.

Thermal stability of a superconductor in fact depends on four factors: 1) the nature of the thermal disturbance (E) including amplitude, spatial extent, and duration; 2) the time dependent generation of heat in the conductor; 3) thermal conduction along the conductor; and 4) the time dependent cooling provided by the helium. Although it is difficult to control the first of these factors, the second two can be greatly determined by the design of the composite conductor. The last factor is determined by the geometry of the liquid helium volume. Stability can obviously be enhanced by maximizing the cooling provided by the liquid helium.

In some magnet designs, a channel of helium running adjacent to the conductor allows for one dimensional heat flow in the diection normal to the conductor surface and across the width of the channel. Analytical solutions for one dimensional, linear, transient heat transfer in He II have been obtained [9,10] and show that $q_H \propto t^{-1/4}$ when the linear dimension is infinite. With the limited amount of helium available in a channel of finite width, q_H falls more rapidly. Such rapid drop in q_H results in very short recovery times t (10's of milliseconds) and small allowed initial heat pulses E. Longer recovery times t and larger energy pulses E can be obtained uniquely for the two-layer designs in which conductors share the same large volume of helium available to each conductor. Here the "open bath" design makes the best use of the rapid "diffusion" of heat in He II and it provides a large supply of enthalpy reserve in direct contact with any conductor hot spot.

The exact form of the time dependent heat transfer into He II is very geometry dependent and analytical solutions are not available when the heat flow has dimensionality greater than one. In the Wisconsin SMES design heat flow from any warm region is essentially two-dimensional (see Figure 2) since the width (w) of the annular layer of He II is small (from 3 to 6 cm). In this case

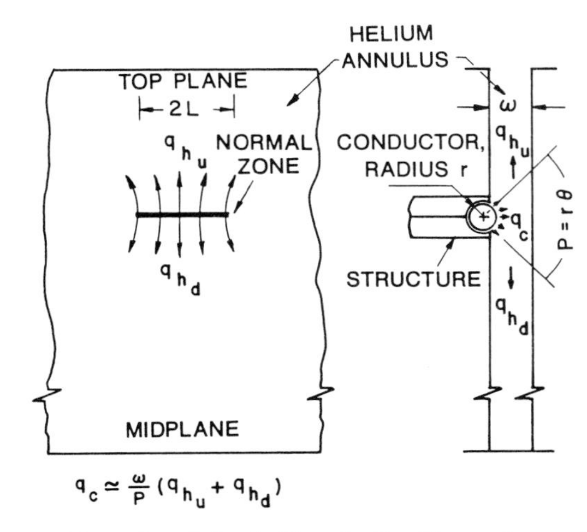

Figure 2.. Two-dimensional heat flux in He II annulus. A normal zone of length 2L generates heat flux q_C off conductor surface. Resultant heat flux in the He II annulus of width w is q_{hu} in the upw rd direction and q_{hd} in the downward direction.

solutions for q_H, the heat flow in the He II, are found by a finite difference technique which is outlined in a previous paper [11] and has recently been verified by experimental measurements [12]. The computational approach begins with the assumptions that the temperature gradients ΔT along the streamlines are described by the Gorter-Mellink relationship:

$$\vec{q} = [f^{-1}(T)\ ]^{1/3} \frac{\vec{\nabla}T}{|\nabla T|^{2/3}} \tag{1}$$

where $f^{-1}(T)$ is an effective thermal conductivity function (see Van Sciver [13]), and that q is parallel to $\vec{\nabla}T$. These assumptions, coupled with the boundary conditions given by the normal zone length and adiabatic or isothermal boundary conditions at the vertical end of the annulus, determine the exact form of q_H in both the up and down directions from the heated normal zone. As outlined in Figure 2, the resultant heat flux removed from the conductor by the helium is given by:

$$q_C = \frac{w}{p}\ (q_{HU} + q_{HD}) \tag{2}$$

where p is the wetted perimeter of the conductor. Results from this finite difference technique are shown in Figure 3 for the case of normal zones at the midheight of a 20 m tall He II annulus. Here the average heat flux into the He II annulus in one direction is plotted against time. The allowable surface heat generation off of the conductor as determined by equation (2) would then be $q_C = (2\ w/p)q_H$. An important feature gained by the open bath geometry which is evident in

Figure 3 is the very slow decay of q_H with time. This feature can afford long time stability by designing the steady state heat generation off of the conductor appropriately. For example if w = p, then by limiting the maximum current in the conductor so that $q_c \leq 2.4$ w/cm^2, a normal zone of 10 m or less can exist indefinitely as long as the refrigeration system can maintain a bath temperature of 1.8 K at the top of the annulus. In a worst case scenario of no refrigeration (i.e. $q_{top} = 0$) the large open bath still affords over 100 seconds of time before generation would exceed cooling.

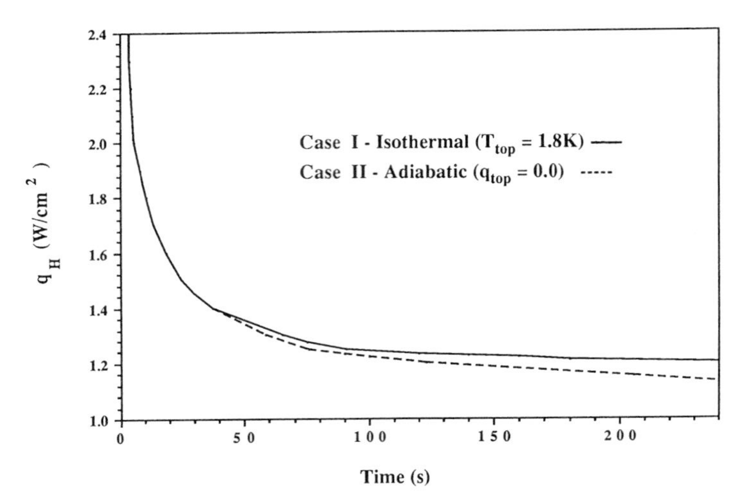

Figure 3. Results of finite difference computations describing transient heat transfer in He II annulus. Normal zone length 2L = 10 meters at the midplane of 20 m tall annulus. Heat flux values represent average value in one vertical direction.

DESIGN CRITERIA FOR SMES WITH HIGH T_C MATERIALS

The discovery of high temperature superconductors has prompted recent assessments of their use for magnets in general [14] and for SMES systems in particular [15,16]. In all cases these have begun with the assumption that the new ceramics will achieve in field current densities equivalent to those of conventional alloy superconductors.

The issue of conductor stability reveals that whereas with the conventional superconductors, large amounts of liquid helium are necessary to absorb thermal disturbances, the necessary inventory of liquid nitrogen can be much smaller since the specific heat of conductor stabilizer material (such as copper or aluminum) is a factor of 2000 larger at 77 K than at 4 K. In fact the quantity of heat absorbing metal included a high T_c magnet design is crucial but it should be determined either on the dubious premise of eliminating the possibility of a normal zone altogether, or for the more realistic purpose of thermal mass in a protection scheme. Notions of cryogenic stability will be prohibitive due to the fact that although the cooling provided by liquid nitrogen is equivalent to that provided by He II (~ 3 W/cm^2), heat generation will be a factor of 25 times larger (for the same amount of aluminum) than in present SMES designs owing to the relatively large values of electrical resistivity (2 nΩm at 77 K compared with 0.08 nΩm at 1.8 K). Thus a cryogenically stable conductor would require 25 times more aluminum stabilizer than in present designs. In contrast, the quantity of aluminum required for protection would need to be increased by only 5% since the enthalpy rise between 1.8 K and 77 K is only 8.4 J/g, while that between 1.8 K and 300 K is 170 J/g.

Other SMES components which would be affected by the high T_c materials include the struts, current leads, and the refrigeration system. The findings of our recent publication [15] are summarized below.

The refrigeration power requirements for SMES magnets operating at liquid nitrogen temperature are so low that there is no need for heat intercepts on the mechanical supports. Losses at low temperature, such as eddy current, coupling losses, radiation loads and lead losses are insignificant (see Fig. 4) for 77 K coils with either 50 kA or 230 kA power leads. At 77 K, units as small as 10 MWh require only 4% refrigeration energy daily compared to the stored energy. At 1.8 K, SMES units must be larger than 200 MWh to be comparably efficient [15]. Thus, the high T_c materials make smaller SMES systems more attractive.

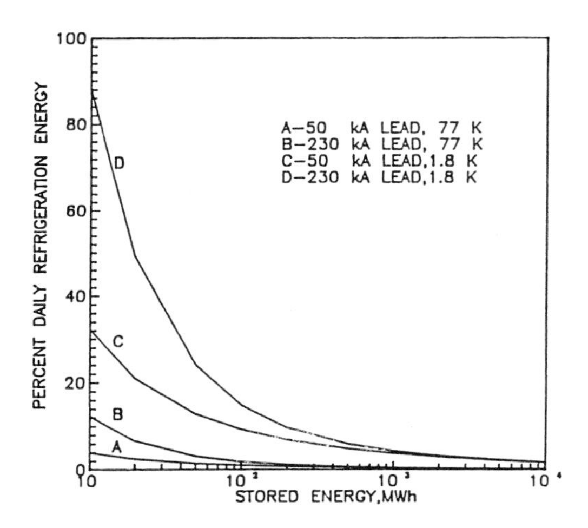

Figure 4. Daily total refrigeration of SMES system

The design, construction and operation of SMES with HTSC should be less costly and less complex than for helium refrigeration. This simplicity will improve maintainability and constructability since superinsulation and thermal shield needs are substantially reduced and the need for a tight helium system disappears. The structural forces, cool down contraction, cyclic operation, and warmup during an energy dump are essentially the same at 77 K and 1.8 K and are primarily a function of SMES geometry and energy stored. It is expected that the overall structural concept of SMES units will be nearly identical for either HTSC or NbTi designs, namely: the conductors are supported by a cold structure which accommodates thermal, magnetic and structural loads; the coil and structure are cooled in a liquid cryogen; expanding radial magnetic forces are carried to the trench wall via thermal insulating load bearing struts; and the magnet is mounted on surface trench walls.

RECENT DEVELOPMENTS - THE ETM

It is recognized [2,5] that the next necessary step toward commercialization of SMES technology is to build a relatively small scale unit which will nevertheless demonstrate the technology of a full scale system. Such a unit is appropriately named the Engineering Test Model (ETM) since many of the component designs, which have to date been tested on a small laboratory scale, will be incorporated and tested in the ETM. In addition, the ETM is to test/demonstrate its functional ability for a variety of power and energy storage services.

Presently, a project for a 400 MW, 20 MWH ETM is being pursued in the U.S. with SDIO government funds. Although a SMES unit of this size will not provide the large diurnal load leveling capability expected of a 5000 MWH system, a Wisconsin utility - MG&E - has suggested several ways in which they could benefit from the 20 MWH unit [5]. These include regulation (maintaining generation to within ± 5 MW of demand), ramping (smoothing of the power step sizes inherent in initiating generation units) and the sales and purchases of energy (storage). It is hoped that construction of an ETM will begin within the next few years.

Summary

Superconducting magnetic energy storage shows very real promise of providing a large, near future market for superconductivity. The electric utilities will benefit from the use of full scale SMES systems for diurnal load leveling. Smaller systems can be used for regulation, ramping, and sales and purchase of power. The technology of SMES is backed by nearly twenty years of research and development in areas including electrical, structural, superconducting materials, and cryogenics. With respect to the last of these, the annular layer of HeII provides long time stability for the necessary high reliability of the SMES system. High T_c superconductors hold potential for simplifying SMES systems and for decreasing costs - noticeably improving the attractiveness of smaller units. The present state of SMES commercialization hopefully awaits construction in the near future of a 20 MWH Engineering Test Model funded by SDIO-DNA and EPRI.

REFERENCES

1. Schwall, R.E., "MRI-Superconductivity in the Marketplace," IEEE Trans. Mag. MAG-23 No. 2, pp. 1287-1993, (1987).

2. Loyd, R.J. et. al., "SMES Engineering Test Model: Preliminary Definition, Test Plan and Development Program," Final Report to EPRI, April (1987).

3. Rodriguez, L.A., et. al., "Comparison of Energy Storage Plants in Generation System Expansion," Int. Sym. on Dynamic Benefits of Energy Storage Plant Operation, DOE and EPRI, p. 218 (1984).

4. Boom, R.W., et. al., "Superconductive Magnetic Energy Storage (SMES) System Studies for Electrical Utility Usage at Wisconsin," presented at the High Temperature Superconductivity Workshop - Tokyo, Janpan, October 17, 1987.

5. "Superconducting Magnetic Energy Storage (SMES): The Next Step (an Engineering Test Model)", private communication from D. J. Helfrecht, Pres., Madison Gas and Electric Co.

6. Boom, R.W. and Peterson, H.A., "Superconductive Energy Storage for Power Systems," IEEE Trans. Mag. Mag-8 No. 3, pp. 701-703 (1972).

7. Boom, R. W. et al. Vols. I, II, III and IV (1974-1981) of the Wisconsin Superconductive Energy Storage Project, UW-Madison; and Boom, R. W., et al., "Cryogenic Aspects of Inductor-Converter Superconductive Magnetic Energy Storage," Proc. Ninth Int. Cryo. Eng. Conf., Kobe (1982), pp. 731-744.

8. Van Sciver, S.W., "HeII Cooling for Superconductive Magnetic Energy Storage," in Superconductive Energy Storage, Proceedings of the United States-Japan Workshop on Superconductive Magnetic Energy Storage, University of Wisconsin: Madison, Wisconsin (1981).

9. Dresner, L., "Transient Heat Transfer in Superfluid Helium - Part II," Adv. Cryo. Eng., Vol. 29, Plenum Press, New York (1980).

10. Dresner, L., "A Rapid Semi-Emperical Method of Calculating the Stability Margins of Superconductors Cooled with Subcooled He II," IEEE Trans. Mag. MAG 23, No. 2, pp. 918-921 (1987).

11. Eyssa, Y.M., et al., "Heat Transfer in Helium II for Two-Layer Energy Storage Magnets," IEEE Trans. Mag. MAG 23, No. 2, pp. 561-564 (1987).

12. Pfotenhauer, J.M. and Huang, X., "Two-Dimensional Transient Heat Transfer in HeII," presented at the Applied Superconductivity Conference, San Francisco, Aug. (1988).

13. Van Sciver, S.W., Helium Cryogenics, Plenum Press: New York, Chapter 5, p. 144 (1986).

14. Iwasa, Y., "Design and Operational Issues for 77 K Superconducting Magnets," IEEE Trans. Mag, Vol 24, No. 2, pp. 1211-1214 (1988).

15. Eyssa, Y., "The Potential Impact of Developing High T_c Superconductors on Superconductive Magnetic Energy Storage (SMES)," Adv. Cryo. Eng., Vol. 33, Plenum Press: New York (1988).

16. Yoshihara, T., "Design Study of SMES System Using High Temperature Superconductors," IEEE Trans. Mag., Vol. 24, No. 2, pp. 891-894 (1988).

Current Situation of R&D on Superconductive Power Application Technologies Carried out by Engineering Research Association for Superconductive Generation Equipment and Materials

K. UYEDA

Engineering Research Association for Superconductive Generation Equipment and Materials (Super-GM), 5-14-10, Nishitenma, Kita-ku, Osaka, 530 Japan

ABSTRACT

The feasibility studies on superconducting generator and superconductors were carried out by Super-GM in 1987. Conceptual designs of two types of superconducting generators, two types of refrigeration systems, and superconductor for each type of generator were conducted. Study on the measuring methods for characteristics of oxide superconducting materials was also conducted.

INTRODUCTION

The feasibility studies on the equipment and materials for superconducting generator were conducted in 1985 and 1986 within the framework of the moonlight project sponsored by the Agency of Industrial Science and Technology, Ministry of International Trade and Industry.The Engineering Research Association for Superconductive Generation Equipment and Materials was established on October 1,1987. In its first year, 1987, the Association conducted researches shown below to identify the research and development tasks over the 8-year national project to be started in 1988 and to establish the research and development plan.

1)Study on the conceptual design of superconducting generator and the feasibility of advanced type
2)Conceptual design of refrigeration systems and study on reliability evaluation procedures
3)Conceptual design and characteristics testing of superconducting wires materials for generator
4)Study on the measuring methods for characteristics of oxide superconducting materials
5)Study on optimum constants and testing methods of reliability assurance of superconducting generator

Major items in these studies are described below.

SUPERCONDUCTING GENERATOR

In the feasibility study, conceptual design of pilot generator was further detailed based on conceptual designs established in the studies in 1985 and 1986. Two types of generators for thermal power plant (2 poles) were studied including: low response excitation type superconducting generator (referred to as low response machine) and quick response excitation type superconducting generator (referred to as quick response machine).

The low response machine is applicable to small and medium power plants constructed near load center areas. It can contribute to voltage stability of power system in addition to general advantages of superconducting generators, e.g., improved efficiency. Possible application of the quick response machine is the large scale power generation with long distance transmission line. This generator will be effective for improvement of power system stability.

Since the low response machine has slow transient response control, no strict requirement will be posed upon the amplitude and rate of changes in the current and magnetic field of superconducting field windings. On the contrary, the damper of this machine should have a high shielding effect to maintain levels of external disturbance (armature reaction) low but still compatible with slow forcing. For this purpose, double dampers were equipped for warm and cold temperatures. Windings of the high stability and high current density are possible under these conditions because the effective magnetic flux density of the field can be set to high levels and the AC loss of conductors is minimized. Therefore, synchronous reactance can also be reduced.

For the effective exciting control of the quick response machine, amplitude and changing rate of the current and field must be high. As the damping characteristics of damper must be lowered to meet these requirements, only warm damper was employed and the cold damper was excluded. Conductors with favorable AC characteristics are required as the field variation rate and superimposed AC magnetic field strength are high.

Specifications and a number of basic data items have been established by these conceptual designs of pilot generator, which were conducted at similar levels with those of previous years and partially more detailed levels.

REFRIGERATION SYSTEM

Refrigeration loads were established for various operating environments of superconducting generator in the 1987 study, based on results obtained from studies in the previous years. Researches including conceptual design and reliability evaluation were conducted for full-fledged research and development in 1988 and subsequent years based on conventional systems with improved reliability for the model generators and oil-free low temperature compression systems without generation of impurities for the pilot generators.

Whereas the superconducting generators require high reliability at the same level as the conventional generators, similar reliability is required for refrigeration systems to maintain very low temperature environment for a long period of time.

Improvement of reliability of major components (e.g., main compressor and expansion turbine) is one of the most important measures for improved reliability of conventional refrigeration systems. Thorough removal of impurities generated in the warm part of the system is also seriously required to assure normal operation of systems without any failure due to solidified impurities on the cold part.

These measures for high reliability of conventional refrigeration systems, however , are not permanent solutions. As far as the oil lubrication is employed, all measures are palliative. The advanced refrigeration system is intended to develop and employ components without oil lubrication.

High reliability oil-free compressors to be used with advanced refrigeration system include the screw type (volume type) and turbo type (centrifugal type). Helium gas will be maintained at low temperature to facilitate compression in oil-free compressor. Wet turbine will be developed instead of the conventional JT valve to improve the liquidizing efficiency.

METALLIC SUPERCONDUCTING MATERIALS

In 1987, conceptual design of superconducting field windings for pilot generator was made based on the revised design standard of wires design flow and stability standard. Study of application of the Nb_3Sn wires to generator was also performed.

The low response machine has low variation of field current even if transmission line faults occur in the power system because of low ceiling voltage and slow response to field circuit control system. Consequently, the rated current density can be set to high levels relatively because the small difference of rate between the maximum current and the rated current can be set. The relatively tolerant wires design can also be made because the AC loss due to the change of field current is smaller than the quick response machine. However, if transmission line faults occur, a magnetic field rises with the current and conductor temperature rises with AC loss. In this case, the tolerance between the maximum current and the critical current in the magnetic field is replaced with temperature margin 1K.

The quick response machine has the large variation of field current and flux because exciting control is performed for rapid load change if an accident occurs in the power system. The ratio of the maximum current to rated current is inevitably large. The small loss type of wires with the primary strand and strand insulation must therefore be used because AC loss increases with the current change rate. In this case, the maximum current at the occurrence of faults must be set to 80% or less of the critical current on the load line as a stability criterion and 1K or more is allocated as a temperature margin.

Alternatively, for the case that the Nb_3Sn wires can be applied to the field windings, the resulting performance and safety improvements of the field coils (e.g., reducing generator size and weight due to higher current density of the conductor and extending possible exciting control range due to increased temperature margin) are considered. However, the Nb_3Sn wires have a disadvantage of great deterioration of the characteristics due to strain. To achieve the application to the generators, mechanical characteristics improvements must be carried out to endure various kinds of stress during coil manufacturing and operation. Therefore, study of application to the field windings of the Nb_3Sn wires was performed to achieve;

- improvement of stress-strain characteristics
- lower heat treatment temperature
- connection technology development
- evaluation of strain induced by diffusion reaction.

OXIDE SUPERCONDUCTING MATERIAL

The key to practical use of the oxide superconducting material, that has much higher critical temperature than metallic superconducting materials, depends on whether it can acquire the sufficient current density under the condition of liquid netrogen temperature or higher.

This material has many unknown characteristics unlike the existing superconducting materials. In the material research, high performance monocrystal and polycrystal are synthesized to obtain more precise data and macroscopic and microscopic material evaluations (crystal structure analysis, compositional analysis, electronic structure analysis, and others) are performed.

In addition, the study for practical use of the oxide superconducting material is performed. The purpose of this study is to develop the reliable and operable materials. As a measure of it, the intergrated evaluation criterion must be established based on the machine design.

As the material development and application comes to this level, it becomes important to set up the common standard for critical temperature and critical current density measurement. To do it, a round robin test was conducted using the common samples so that discussion can be made on the common basis. Furthermore, the oxide superconducting materials were produced subject to the purposes and conditions of individual research and development institutes.

In the research and development using the common samples, measurement method was specified, but problems of sample treatment, maintenance, terminal installation, and others were left out. The common samples have the nearly similar characteristics, but are not microscopically equalized. It, therefore, is difficult to identify the cause of the characteristics difference. Further research and development must be continued for voltage and current terminal installation, sample installation, and sample storing. The electrode connection technology must also be studied as a common technology because it is required when other materials are mixed.

In the material measurement using the individual samples, study of electrode installation, sample installation, measurement conditions, material processing, and others have made progress. Data acquisition was made in the different lines of the research using the common samples and the results that are helpful for further research and development were obtained.

TOTAL SYSTEM

Study on appropriate constant of pilot generator was performed after the power system stability improvement and voltage increase suppression effects during load shedding were analyzed through computer simulation.

The annual cost was investigated on appropriate constant of pilot generator, considering;

- the power system stability improvement effect
- the operation cost reduction currently obtained by efficiency improvement
- manufacturing cost increase accompanied with lower synchronous reactance.

The research and study of the superconducting generator, superconducting magnetic energy storage, superconducting cables, superconducting transformer, and other electric power systems were performed from the viewpoints of their technology and development. In addition, investigation and discussion of other technologies were made from the viewpoints of efficiency, energy saving, characteristics, and others for final evaluation. The study of introduction effect of superconducting magnetic energy storage to power system and fault analysis and protection was performed by computer simulation to confirm power system stability improvement effect and seek out problems on stability and reliability of superconducting coils.

ACKNOWLEDGEMENT

The full-scale research and development that has started from 1988 is performed for "superconducting application technology" for practical use of various kinds of superconducting application equipment including sperconducting generator.

The evaluation and study of previous research and development results, addition of new research and development items, and research modification must be checked and reviewed. In intermediate evaluation, the state of element technology research and development on the generators will be checked and addition of research and development of element technology for pilot generator will be welcomed if necessary.

The author is much indebted to Moonlight Project Promotion Office, Agency of Industrial Science and Technology and the following members of Engineering Research Association for Superconductive Generation Equipment and Materials (Super-GM).

Members of Super-GM

- Tokyo Electric Power Company
- The Chubu Electric Power Co.,Inc.
- The Kansai Electric Power Co.,Inc.
- Central Research Institute of Electric Power Industry
- Hitachi,Ltd.
- Toshiba Corporation
- Mitsubishi Electric Corporation
- Sumitomo Electric Industries,Ltd.
- The Furukawa Electric Co.,Ltd.
- Fujikura Ltd.
- Ishikawajima-Harima Heavy Industries Co.,Ltd.
- Mayekawa MFG.Co.,Ltd.
- Japan Fine Ceramics Center
- Hitachi Cable,Ltd.

Total 14 members

Development of Superconducting Magnet for Fusion Power

S. SHIMAMOTO

Japan Atomic Energy Research Institute, Naka-machi, Naka-gun, Ibaraki, 311-01 Japan

ABSTRACT

This paper describes firstly the present status of fusion plasma and secondary necessity of superconducting magnet for fusion reactor. As further explanation, basic technologies, which are indispensable for superconducting magnet are listed up and the present status of each technology is described. A superconducting magnet system for fusion experimental reactor is large and heavy compared with those for other application. Therefore, logical step of scaling-up in the technology is required. The project work for tokamak machine is mainly divided into three major components : toroidal coil program, poloidal coil program and refrigerator program. The present development status of these components are explained. On the other hand, medium size superconducting tokamaks for plasma experiment already constructed or under construction are described. Finally, future perspective on magnet development is written.

INTRODUCTION

The thermonuclear fusion with D-T or D-D by magnetic confinement is one of energy extraction means which should support variety of resources for stable energy supply. The development of fusion reactor requires a number of high technologies which make many spin-offs to other industrial applications : high electric power control technology by thrystor, high vacuum & high speed evacuation technology, ion beam technology, material technology against high temperature, cryogenic & superconducting magnet technology and so on. From view point of reactor system technology, superconducting magnet technology is one of key technologies because the weight of the magnet system is about half of the reactor, for example, more than 10,000 tons, and the cost of the magnet system takes major part of the total.

On the contrary, if we generally overview all the aspect of cryogenic or superconducting technology, application of the magnet system to fusion reactor is one of the most important since it requires large volume, high field, etc. which do not correspond to any other application.

REQUIREMENT OF SUPERCONDUCTING MAGNET FOR FUSION

Among several types of fusion machines, tokamak is the most advanced one from view point of plasma confinement which satisfies critical plasma condition. The next step of this type of machine could be experimental reactor which demonstrates ignition or energy extraction. Therefore, this paper is focused on superconducting magnet only for tokamak machine. The bird's-eye view of tokamak Fusion Experimental Reactor (FER), which was designed by Japan Atomic Energy Research Institute [1] is shown in Fig. 1. The outer diameter is around 23 m, the height is 20 m and the weight is 20,000 tons. In a tokamak machine, magnet system can be divided into two: toroidal and poloidal. If we use a conventional magnet system for both, total electric power due to joule heating would be more than 2,000 MW which may be about five times of thermal power by fusion reaction. By using a superconducting magnet system, electric power required for helium refrigerator would be around 30 MW. In order to get net electric power, thus superconducting magnet system is absolutely indispensable.

Table 1 Parameters required for the Fusion Experimental Reactor

	Toroidal Coil	Poloidal Coil
Inner Bore	10 m	2 - 20 m
Peak Field	12 T	12 T
Operating Current	30 kA	30 - 100 kA
Operating Mode	DC	Pulsed (~10T/s)
Insulation Voltage	20 kV	20 kV
Stored Energy	30 GJ	12 GJ

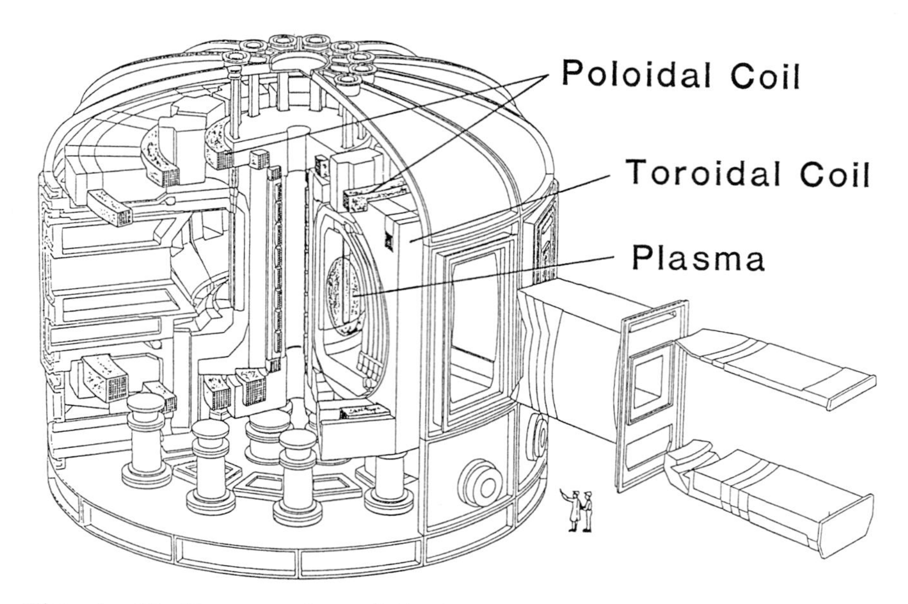

Fig. 1 Bird's-eye view of Fusion Experimental Reactor

The magnet parameters required for the FER are shown in Table 1. Such a large, high field superconducting magnet cannot be built without intensive development step by step since the technique is different from that for other purposes such as high field energy physics, MHD (Magneto Hydro Dynamics) power generation.

BASIC TECHNOLOGIES

The technical issues of superconducting magnet system is not at all superconducting material technology itself. For practical magnetic field over 10 T, industrial production technologies are almost completely established on Nb_3Sn compound or Nb-Ti alloy. For a large superconducting magnet system such as for fusion application, key issues are complicated superposition of advanced techniques and realization of large & heavy machine. The advanced techniques which are required for the magnet system are listed in Fig. 2. All these techniques are superimposed simultaneously. For example, stress value in the toroidal coil of FER becomes easily around 80 kg/mm^2 even in normal operation (not fault condition) at liquid helium temperature and high vacuum. In such a condition, if there is a necessity of energy dumping, high electric voltage will appear in all the electric system in the same circumstances. Thus, system development step by step is required for a large magnet.

DEVELOPMENT STEP AND TARGET

TOROIDAL COIL DEVELOPMENT : This coil system is the most important one in tokamak since it is indispensable for plasma confinement and much bigger than poloidal coil system. Therefore, The superconducting magnet development for fusion was started from this coil. Fig. 3 shows the present status (black colored coils) compared with FER coil. In this figure, T-15(still under construction), Tore Supra and Triam are practical superconducting tokamaks which include plasma with other auxiliary equipments. The details on these machines are described later. On the other hand,

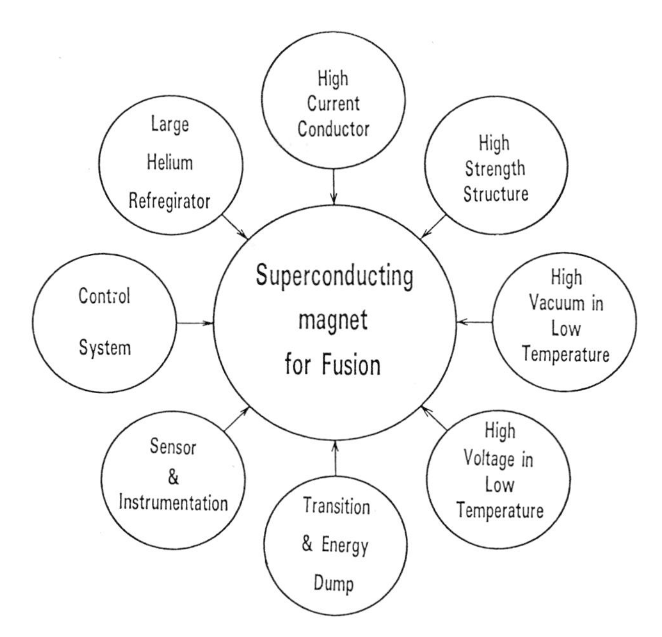

Fig. 2
Basic technologies which support superconducting magnet for fusion

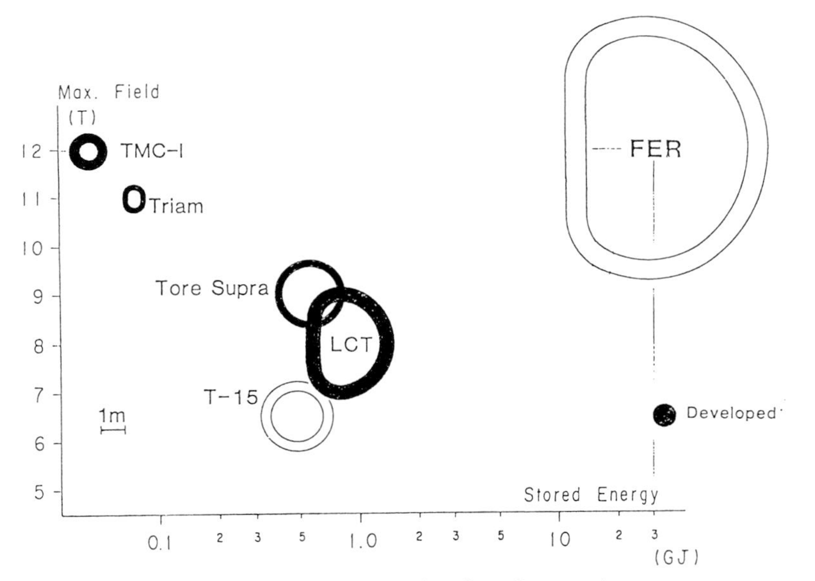

Fig. 3 Toroidal coil development

TMC-1 and LCT were really undertook as the technological demonstration for FER, high field and large scale respectively.

POLOIDAL COIL DEVELOPMENT : The development of this coil system is behind that of toroidal coil. One reason is that while toroidal coil is operated with DC field, poloidal coil is operated with pulsed field which provides electric loss both in conductor and structure. At early stage of superconducting magnet development, it was not at all easy to realize low pulsed loss conductor. Another reason is that a huge pulsed power supply is in order to operate large pulsed coil. It is not easy to get several hundreds MVA pulsed power either directly from line or with generator. Fig. 4 shows similar chart on poloidal coil to Fig. 3. Any superconducting coil have not been integrated in a practical tokamak. Thus, compared with toroidal coil development, much further way can be seen to FER in poloidal coil development.

HELIUM REFRIGERATOR DEVELOPMENT : An intensive helium refrigerator development is indispensable since large refrigerator, which requests completely different

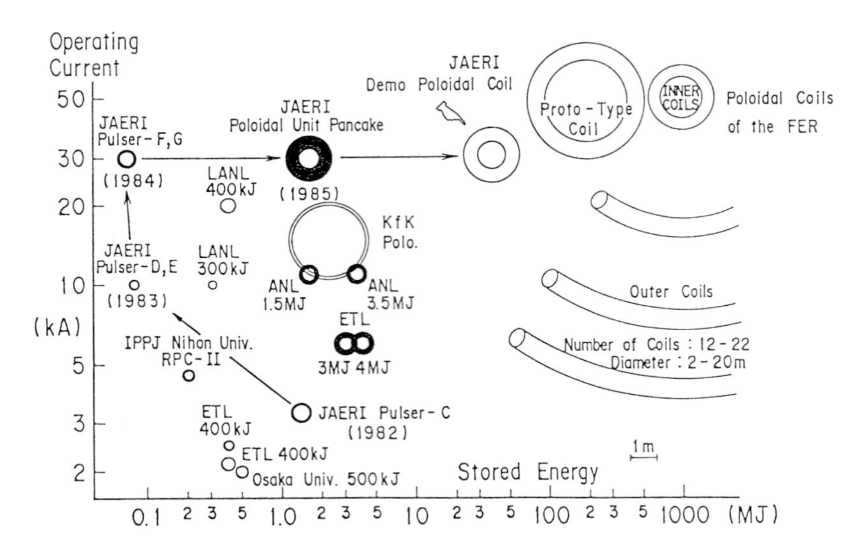

Fig. 4 Poloidal coil development

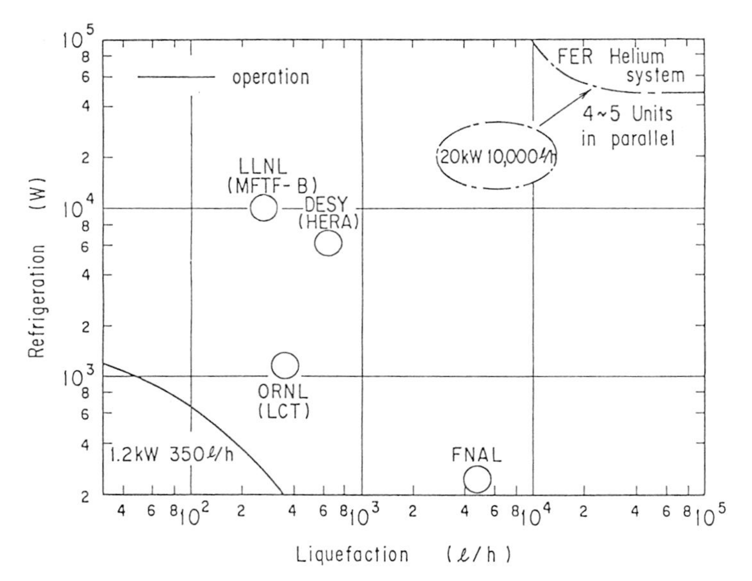

Fig. 5 Helium refrigerator development

components from small one, is required for corresponding large magnet system. This requirement is not only for steady state heat load but also for cool-down of heavy weight in a reasonable time. Usually the latter condition dominates to specify the capacity. As shown in Fig. 5, FER requests about 100 kW at 4 k refrigerator. However, practical realization up to now is 6 kW as unit for particle accelerator. Especially, development of large helium compressor and large turbine expander is needed.

DEVELOPMENT STATUS IN THE WORLD

The toroidal coil development has been highly advanced since 10 years ago. One of the objectives is to demonstrate 12 T field generation with filamentary Nb_3Sn conductor. This was carried out by one of JAERI's projects as shown in Fig. 3, TMC-1. At the design and verification stage, the critical current of this conductor was confirmed as 10 kA at 10 T in short sample. This coil has 1 m average diameter and react & wind method was applied for the purpose of further coil scaling-up. As the result of extended test, 12 T with 6.5 kA was obtained without any appearance of normalcy. This experience was transferred to Triam tokamak , as shown in Fig. 3

Fig. 6 Six LCT coils arranged in the vacuum tank at ORNL in USA

in Kyushu University in Japan. The conductor of similar type was exported and utilized in MFTF-B, other type of fusion machine in USA, to generate 12 T.

Another objective is to demonstrate operation of full toroidal coil system. This project have been carried out as an international collaboration under auspices of International Energy Agency (IEA). USA have been the operating agent, fabricating 3 coils and EURATOM, Switzerland, Japan have supplied other one coil by each. One coil has 5.5 m height, 3.5 m width and 0.8 m thickness. Three coil used pool-boiling cooling ant the others forced-flow cooling. Fig. 6 shows six coils arranged in torus configuration in the 11 m diameter vacuum tank at ORNL in October 1985. 320 tons weight, of which 6 coils weight 240 tons, was cooled down in one month and kept at 4 K for 19 months. Although the rated field was 8 T, in the course of extended condition test, 9 T was obtained in September 1987 in full symmetric toroidal field. At that moment, a quenching began in one of the six coils with initial stored energy of 944 MJ. By means of energy dumping, 110 MJ was consumed in the vacuum tank and 834 MJ extracted outside. Since the design of the six coil are different, a lot of engineering data have been provided for toroidal coil system of experimental reactor from this project[2, 3].

Regarding poloidal coil development, as shown in Fig. 4, KfK is now carrying out Polo program and JAERI Demo Poloidal Coil program (DPC). The former is for development of equilibrium coil (large ring type) which has a stored energy of 1.5 MJ. The latter [7] is for development of ohmic heating coil (transformer type) which has a stored energy of 40 MJ with pulsed field rate of 7 T/s. Both programs are now in the final stage of coil fabrication and in 1989 testings will be started almost simultaneously.

The development of large helium refrigerator and auxiliary equipments is usually carried out as facility of large project. For example, cold compressors have been developed for Tore Supra and JAERI's DPC between 15 g/s and 60 g/s around 4 K. A supercritical helium circulation pump has been realized for DPC with flow rate of 500 g/s and pressure drop of 1 atm. For development of large compressor and expansion turbine, an investment is highly requested.

Fig. 7 Superconducting tokamak:Tore Supra at Cadarache in France

SUPERCONDUCTING TOKAMAK

Tokamak machines with superconducting toroidal coil system are now under operation. These are Tore Supra in France and Triam in Japan. Tore Supra [4] has stored energy of 600 MJ with maximum field of 9 T. In order to get 9 T with Nb-Ti, this machine uses superfluid helium of 1.8 K at 1 atm. The current capacity of the conductor is 1,400 A with pool cooled type. The poloidal coil system are water cooled. This tokamak has been in operation with plasma current since this spring. Fig. 7 shows Tore Supra at the moment of construction completion.

Triam [6] has stored energy of 76 MJ with maximum field of 11 T. The react and wind Nb_3Sn was used in this machine with current capacity of 6,000 A in bath of 4.4 K saturated boiling helium. The poloidal coil in this machine uses also copper conductor cooled with water. Triam started the operation two years ago with plasma current of 0.5 MA.

The third tokamak with superconducting toroidal coil is T-15 in USSR [5] which is under construction. The size of this machine is similar to Tore Supra with Nb_3Sn cooled by two phase forced flow helium at 4.5 K. The current capacity is 5.6 kA.The operation is expected to be in one year.

CONCLUSION

The superconducting magnet technology for fusion has been highly advanced since ten years ago. For example, a heavy and complicated international collaboration, the Large Coil Task, was undertook and successfully completed with a number of hardware data on toroidal coil system for reactor. If we look at superconducting tokamak, there exist two or three medium size tokamaks only with superconducting toroidal coil.

Regarding poloidal coil system, development works are still under way and no practical application exists in tokamak system. However, with data which will be obtained in a few years, the poloidal coil system with copper conductor in existing tokamaks may be replaced with superconductor.

In parallel with coil development and practical system operation, large scale helium refrigerator and the auxiliary equipments, such as helium circulation pump and cold compressor, will be developed.

The design work of International Thermonuclear Experimental Reactor (ITER), of which mission is to demonstrate ignition, is now being carried out as international collaboration by four major countries under auspices of IAEA. A part of R & D program is included in this collaboration. The output will be obtained in a few years.

REFERENCES

1. FER Design Team : Conceptual Design Study of Fusion Experimental Reactor(FY87FER)-Summary Report- (publication in 1988 in Japanese), JAERI-M 88-090
2. S. Shimamoto et al :Experiments on Large Superconducting Toroidal Coil in LCT, Journal of the Atomic Society of Japan, Vol. 30 (1988) p.488 (in Japanese)
3. LCT community : The IEA Large Coil Task - Development of Superconducting Toroidal Field Magnets for Fusion Power, Fusion Engineering and Design (Special Issue) to be published in September, 1988
4. R. Aymar et al : Conceptual Design of a Superconducting Tokamak : Tore II Supra, IEEE Trans. MAG - 15(1979) p.542
5. B. Kadomtsev et al : T-15 Installation, the Main Chracteristics, Electromagnetic System, Proc. of 12th SOFT(1983) p.208
6. S. Itho et al : High -field Tokamak TRIAM-1M with Superconducting Toroidal Magnets, Proc. of 11th European Conf. on Controlled Fusion and Plasma Physics (1983) A04
7. S. Shimamoto et al : Development of a Large Forced-flow-cooled Pulsed Coil for the Next Tokamak, Proc. of 11th Symp. on Fusion Eng. (1986) p.1017

R&D Superconductive Electro-Magnetic Propulsion Ship

SETSUO TAKEZAWA

Japan Foundation for Shipbuilding Advancement, 1-15-16, Toranomon, Minato-ku, Tokyo, 105 Japan

ABSTRACT

Superconductive electro-magnetic propulsion ship is drawing the world's attention as an imaginative application of superconductive technology. The Japan Foundation for Shipbuilding Advancement (JAFSA) inaugurated an R&D Committee in June 1985 in order to expedite research into an innovative ship that can be propelled by use of superconductive electro-magnetic thrusters. Our project team is now engaged in construction of a 150-ton testcraft, which will be launched in 1990.

INTRODUCTION

The discovery and subsequent development of high-temperature superconductive materials has given good hope to those engaged in the research of superconductive technology. Although it is expected to take a considerable length of time before the new high-temperature superconductive materials being developed at present will become available as practical conductor filaments for superconductive magnets, their significance suggesting the feasibility of superconductive technology without relying on liquid helium is great indeed.

However, even in the event that practical use of high-temperature superconductive materials is realized, it will probably take still more time before we benefit from the superconductive technology which relies on high temperature superconductive materials. Superconductivity, which is recognized as the last technological innovation in the 20th century, cannot be achieved if the application techniques promoted within the cryogenic temperature zone at present have not satisfactorily been established. Still, many technical problems stand in the way of practical use of this frontier technology.

PRINCIPLE OF ELECTRO-MAGNETIC PROPULSION SHIP

The concept of electro-magnetic propulsion was proposed by W. A. Rice of the United States in 1961 as an inverse idea of electro-magnetic pump of liquid metal. The principle is based on Fleming's left-hand rule which is one of the basic rules of electro-magnetics. When a magnetic field is formed in sea water by electro-magnets provided on board a ship and an electoric current is applied in the direction at right angles with magnetic field, the sea water is pushed away by electro-magnetic force. The principle is to propel the ship forward by utilizing the reaction force exerted by the sea water.

For utilizing electro-magnetic force to propel a large bulky object like a ship, three essential conditions must simultaneously be satisfied: a strong magnetic field; current; and crossing of the magnetic field and electric current at right angles.

To generate a strong magnetic field by electro-magnets, it is necessary to increase the amperage of current flowing through coils and the number of coil turns. However, due to the electric resistance in copper and aluminum used for the electro-magnets, no electric current of high magnetic field density can be applied to the coils. Moreover, even if the ampereturns are increased, certain limits are reached because of poor efficiency in generating a magnetic field. For these reasons, the use of superconductive magnets is prerequisite. Since superconductive magnets are capable of continuously generating a current of high magnetic field density without attenuation, they can produce an extremely strong permanent magnetic field at a high efficiency.

The most frequently used superconductive materials at present are NbTi and NbSn3. Each must be cooled down to a cryogenic temperature of -270°C. Liquid helium is used for cooling purposes.

TYPES OF ELECTRO-MAGNETIC PROPULSION SYSTEMS

There are two different systems in electro-magnetic propulsion: one is the external magnetic field system in which electro-magnetic force is generated in the sea water at the peripheral space of ship's bottom, and the other is the internal magnetic field system where electro-magnetic force is generated in the sea water within the longitudinal ducts.

In the case of the external magnetic field system, a magnetic field is generated around the ship's bottom and an electric current flowing in the direction at right angles with magnetic field must be applied. Accordingly, its environmental impacts must be taken into account. By contrast, the magnetic field and current generated by the internal magnetic field system is confined within the duct space and such environmental impacts can be controlled to an almost negligible extent. In terms of higher propulsive efficiency and speed, the internal magnetic field system is considered superior the other system.

RESEARCH AND DEVELOPMENT OF SUPERCONDUCTIVE ELECTRO-MAGNETIC PROPULSION SHIPS

The Japan Foundation for Shipbuilding Advancement, in 1985, organized a Research and Development Committee for Superconductive Electro-Magnetic Propulsion Ships (Chairman: Yohei Sasakawa) with a subsidy given by The Japan Shipbuilding Industry Foundation (Chairman: Ryoichi Sasakawa) and began research and development. An R&D project is now underway at its Tsukuba Institute in Tsukuba Science City, in cooperation with related industrial concerns.

Emphases in the project are laid on developing the optimum hull for an electro-magnetic propulation ship, measures for improving performance and weight reduction of superconductive magnet coils. Substantial achievements have been made one after another through both theoretical and experimental approaches. As an electro-magnetic propulsion system, the internal magnetic field system has been employed and the ship type with two propelling units will be adopted considering desired weight reduction, in other words, horsepower and better arrangement of equipment. At an early stage of the R&D project, trial weight calculation of superconductive magnet coils was made where we obtained a calculated weight more than twice the scheduled displacement of the ship. As this weight was large enough to cause an immediate sinking of the ship, attempts were then made to shift the priority of studies to weight reduction of the hull whereby total weight of the superconductive magnet coils including the cooling system was successfully reduced to half the displacement tonnage of the prototype ship.

According to our schedule, in fiscal 1988 the manufacture of propelling system and related equipment will be commenced. In fiscal 1989 the hull will be outfitted after installing the propelling system and equipment. Test navigation to prove seaworthiness and practical ability of the completed testcraft will be carried out in 1990. The planned principal dimensions and features of the testcraft are: displacement = about 150t; length overall = about 22m; breadth mould = about 10m; hull type = monohull with an estimated thrust of 8,000 Newton. (See Table I, II and Fig. 1, 2)

Table I

Principal items of the propelling system of the experimental ship

Type: twin propelling units, each consists of 6 dipole coils.	
Inside diameter of the duct	0.260m
Magnetic flux density at center of ducts	4T
Max. electric current density inside ducts	4,000A/m2
Length of electrodes	2.50m
Lorentz force:	about 8,000N x 2
Thrust:	about 4,000N x 2
Capacity of main generator	3,800kw
Total weight of the system	about 100ton

Table II

Principal items of the experimental ship.

Displacement:	about 150ton
Length:	about 22m
Breadth:	about 10m
Thrust:	about 8,000N
Speed:	about 8kts

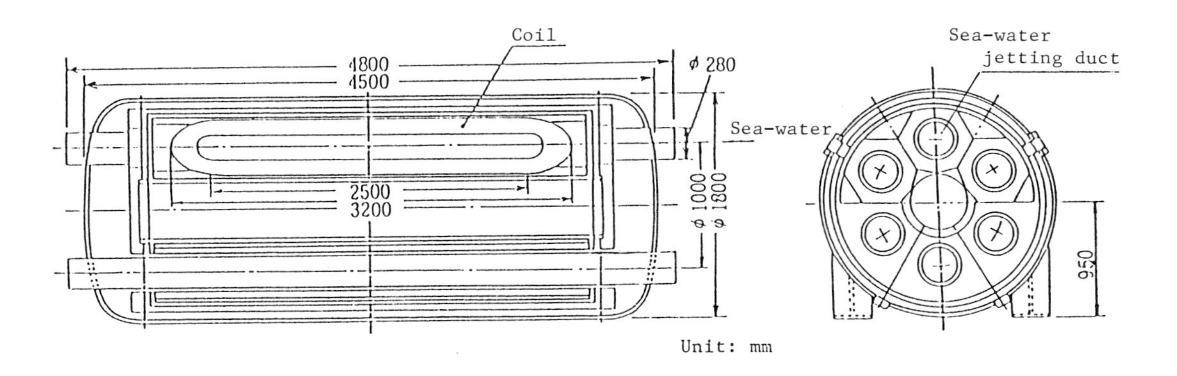

Fig. 1 Thruster of experiment ship

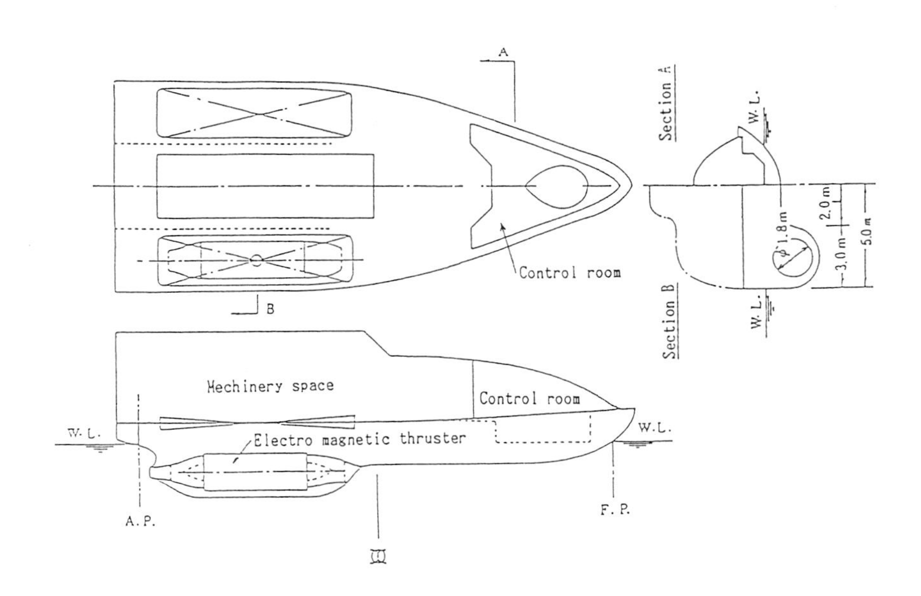

Fig. 2 Rough arrangement of the experiment ship

What is described above is the first phase target of the R&D project. The second phase target is to construct a ship utilizing lightweight superconductive magnets capable of generating a strong magnetic field workable at the temperature of liquid nitrogen. The target period of the R&D project is solely dependent on when high temperature superconductive materials will be available that are under development throughout the world. It will be a couple of years later than the success in the manufacture of practical conductors of high temperature superconductive materials.

APPLICATION OF SUPERCONDUCTIVE TECHNOLOGY

The superconductive ship propulsion techniques described here may well be applied to off-shore structures and oceanographic survey ships without any modifications.

However, the application of superconductive technology to ships is subject to stringent conditions compared with other applications. Superconductive electro-magnetic propulsion systems must be light in weight and serviceable in severe marine environments such as briny wind. Therefore, our project will have to be promoted as an integrated project embracing such fields as materials, manufacture, shipbuilding and oceanography. Once a project of such an integrity initiates treading its step, however, days will come much earlier than expected, when applications of superconductive technology are realized and ships propelled by superconductive electro-magnetic propulsion systems proceed through the oceans of the world.

Superconducting Magnets for Magnetic Resonance Imaging

T. YAMADA[1], S. YAMAMOTO[1], T. MATSUDA[1], M. MORITA[1], M. TAKECHI[2], and Y. SHIMADA[2]

[1] Mitsubishi Electric Corporation, 8-1-1, Tsukaguchi-honmachi, Amagasaki, 661 Japan
[2] Mitsubishi Electric Corporation, 651, Tenwa, Ako, 678-02 Japan

ABSTRACT

A High homogeneous and stable field, the reduction of a fringe field, the minimization of cryogen consumption are necessary for MRI magnets. This paper presents key technologies to satisfy these requirements. Joints of conductors, effects of superconducting shim coils on error fields caused by irons, iron yoke shield, and a low heat load cryostat equipped with a shield refrigerator are mentioned.

INTRODUCTION

Technical requirements for a magnetic resonance imaging (MRI) magnet performance are a high homogeneouos field, a stable field, the reduction of a fringe field, the minimization of the support necessary for cryogens.

MRI magnets are operated in persistent current mode to attain very stable field. Joints of conductors with low resistance are necessary in persistent current circuits. Especially in superconducting shim coils, joints with ultra low resistance are required because inductances of shim coils are very small. Superconducting shim coils are very powerful to compensate large error fields. During the operation of the magnet,they decrease the change of the field homogeneity caused by irons near the magnet automatically.

Iron yoke shield provides efficient shielding for the fringe field from MRI magnet and reduce the interaction of the magnet with other medical equipments.

Low heat load cryostats and shield refrigerators are most important technologies to apply superconducting magnets to the magnetic resonance imaging in hospitals.

This paper describes these key technologies; joints of conductors, the suppression of the change of the field homogeneity due to irons near a MRI magnet, a iron yoke shield, the low heat load cryostat equipped with a shield refrigerator.

JOINTS OF CONDUCTORS

The field stability of better than 0.1 ppm/h is required in a MRI magnet. Since the inductance of the smallest superconducting shim coil is about 5 mH, the joint resistance must be less than 1.4×10^{-13} Ω. The spot weld joint of bared superconducting filaments were developed to attain this low resistance(Fig. 1).[1] Quench currents Iq's of spot weld joints are plotted in Fig. 2. Quench current is larger than 30% of critical current of the conductor. As joints can be set in the area of low field lower than 1 T, quench current is much larger than the operating current of a magnet. The persistent current in the 1 turn loop containing this joint was monitored by a flux meter. The joint resistance was estimated from the time constant of the current decay. Figure 3 shows results. It is concluded that the joint resistance is much smaller than 1×10^{-13} Ω.

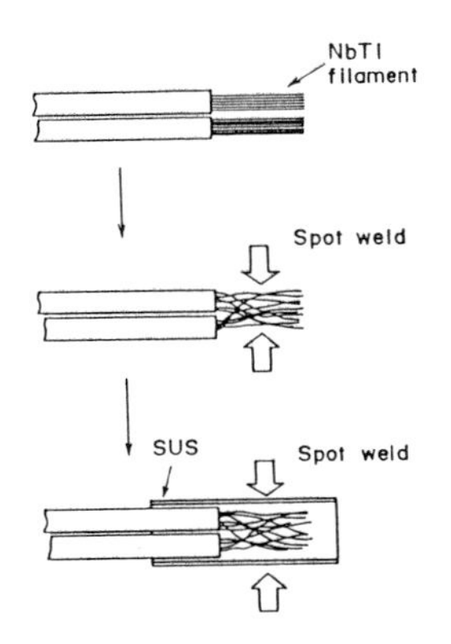

Fig. 1 Joint of conductors.

The different type of joints were made and tested. Bared superconducting filaments were crimped together in an ofhc copper sleeve. The joint resistance was also much smaller than $1X10^{-13}\Omega$.

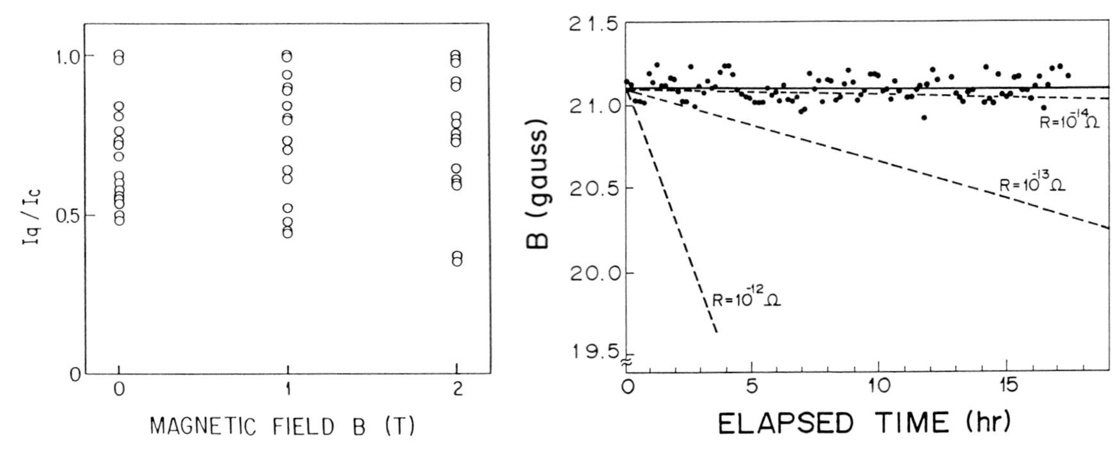

Fig. 2 Quench currents of joints.

Fig. 3 Decay of the flux generated by the 1-turn loop which contains a joint.

CHANGES OF FIELD HOMOGENEITY CAUSED BY IRONS

The effects of irons near a MRI magnet on the field homogeneity were investigated. The position of a iron and co-ordinates are shown in Fig. 4. A columnar iron or a spherical iron is set up on the z-axis or y-axis, and the change of field homogeneity is measured in a imaging volume. Results are plotted in Fig.5. The change of homogeneity is not very large. If the weight of a iron is less than 10 kg and the position of a iron and the center of a magnet is apart longer than 2 m, then the change of homogeneity is small enough (< 1 ppm).

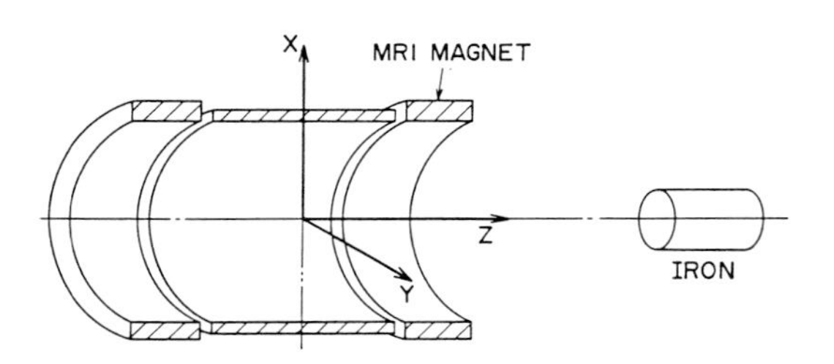

Fig. 4 MRI magnet and an iron. Length/diameter ratio of the iron is 2.

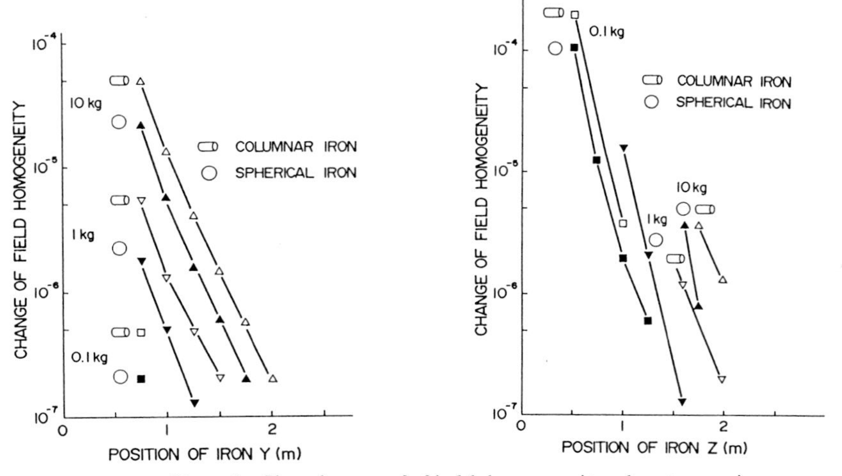

Fig. 5 The change of field homogeneity due to an iron.

In passing, the iron on the z-axis causes larger change of field homogeneity than one on the y-axis. Superconducting shim coils suppress the change of field homogeneity automatically. When superconducting shim coils was not operated in persistent current mode, that is, they were connected to current sources, the change of homogeneity increased more than three times.

IRON YOKE SHIELD

The MRI magnet with iron yoke shield was developed (Fig. 6). Magnetic performances are shown in Table 1. The error field due to iron yoke shield was compensated by superconducting shim coils. Figure 7 shows the distribution of fringe field. The volume within 5 gauss envelope is reduced to about 7 % by the adoption of the iron yoke shield. It is possible to install a MRI magnet in a small space avoiding the interaction of other medical equipments.

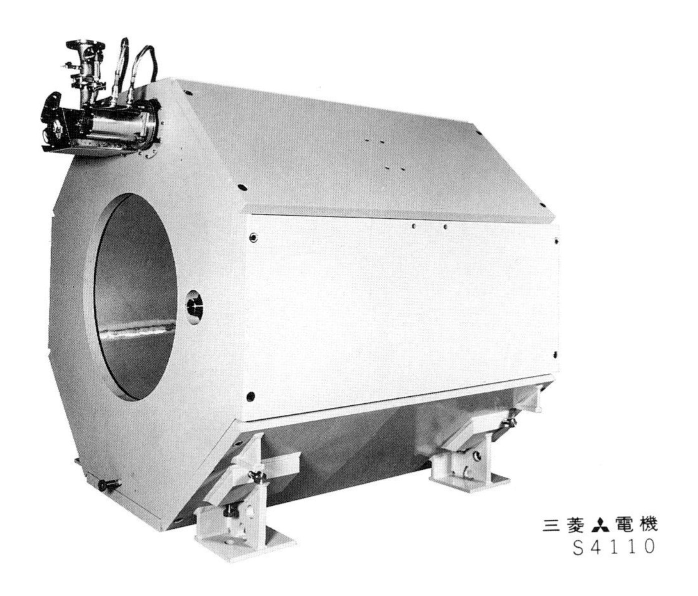

Fig. 6 A MRI magnet with iron yoke shield (0.6 T).

Table 1 Magnetic features of a MRI magnet

field strength	0.6 T
homogeneity	< 10 ppm/35cm DSV
stability	< 0.1ppm/h
bore diameter	1 m
shim coils	superconductive

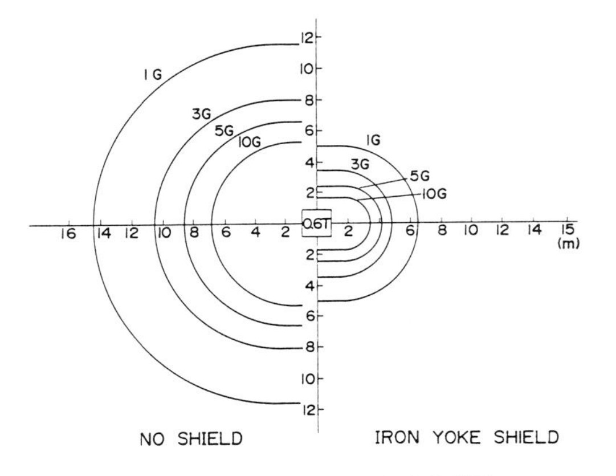

Fig. 7 Fringe field of a 0.6 T MRI magnet.

CRYOSTAT

The structure of a cryostat is shown in Fig. 8. A service port and a shield refrigerator are attached on the end of the cryostat. This constitution reduces the total height of a MRI magnet. The shield refrigerator keeps the temperatures of two thermal shields at about 20 K and about 77 K respectively. Cold structures are supported by glass-fiber-reinforced epoxy rods. Cold surfaces are covered with material has high electrical conductivity to reduce thermal emissivity.[2] Liquid helium boil-off is less than 0.1ℓ/h. Liquid nitrogen is not used. Cryogenic features are summarized in Table 2.

Table 2 Cryogenic features of a MRI magnet

liquid helium boil-off	$< 0.1 \ell/h$
liquid helium holding time	> 4 months
a shield refrigerator is used.	
liquid nitrogen is not used.	

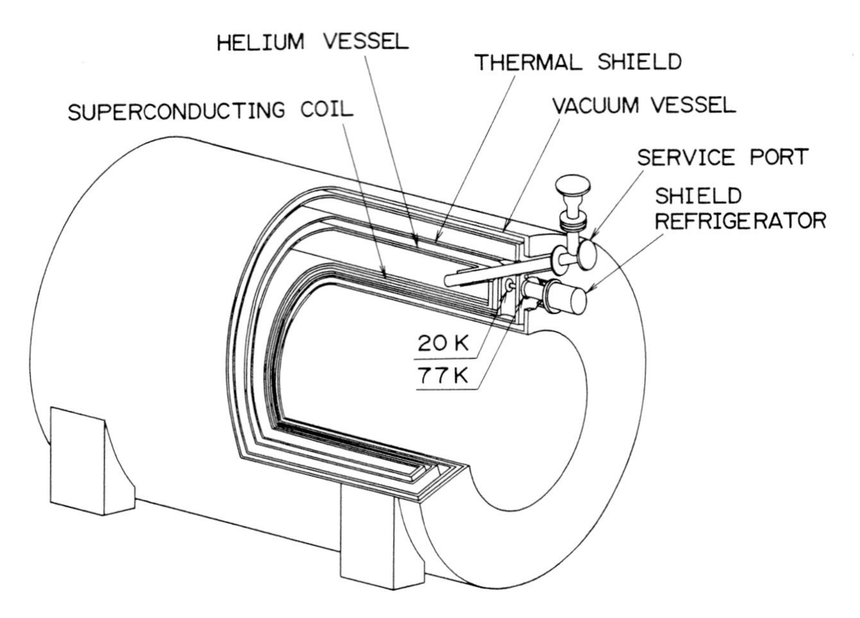

Fig. 8 The structure of the cryostat.

CONCLUSIONS

The spot weld joint of bared superconducting filaments was developed. The joint resistance was less than $1X10^{-13}\Omega$. This joint technology enabled us the usage of superconducting shim coils in MRI magnets, and high homogeneous and stable magnetic field was obtained. It was found that superconducting shim coils compensated error field caused by irons.

The volume within 5 gauss envelope is reduced to about 7 % by the adoption of iron yoke shield. It is possible to install a MRI magnet in a small space avoiding the interaction of other medical equipments.

The low heat load cryostat equipped with a shield refrigerator was developed. The consumption of liquid helium is less than 0.1 ℓ/h and liquid nitrogen is not used.

REFERENCES

[1] M.Morita, et al., IEE of Japan Static Apparatus Research Meeting, Nov. 1983, AS-83-66, pp 29-33.
[2] T.Amano,et al.,Proceedings of 25th National Heat Transfer Symposium of Japan,Jun 1988,pp358-360.

Application of Superconductivity to Transportation: Magnetically Levitated Train

HISASHI TANAKA

Railway Technical Research Institute, 2-8-38, Hikari-cho, Kokubunji, Tokyo, 185 Japan

ABSTRACT

Taking over from the defunct JNR, the Railway Technical Research Institute continues to be engaged in the maglev development project. Now, the safety, reliability and practicality of the system are being confirmed using an MLU 002 test vehicle on the Miyazaki test track. The vehicle has superconducting coils composed of niobium-titanium alloy series wire of the copper ratio 1, impregnated with epoxy resin and having a magnetomotive force of 700 kA. This paper also refers to the conceptual design of the commercial vehicle, inauguration time of commercial operation, and so on.

INTRODUCTION

Taking over from the defunct JNR as of April 1987, the Railway Technical Research Institute (reorganized as research foundation) continues to be engaged in the maglev development project. At present a test run of the MLU 002 test vehicle is under way on the Miyazaki test track of 7 km in length. Meanwhile the Ministry of Transport has started a preliminary investigation about further extension of the test track, moving a step forward in the direction of realizing a commercial maglev line. This paper purports to introduce the present status and future prospects of maglev development.

WHY SUPERCONDUCTIVITY IS NEEDED

The maglev project originated from the idea that in future a superspeed mode of land transport in the order of 500 km/h will be necessitated. The conventional form of railway will hit the speed limit in terms of (1) adhesion drive, (2) wheel/rail support and guidance and (3) current collection by overhead wire/pantograph system; further speedup must rely on entirely different system. Higher speed will not be the only requirement. There are other important conditions to be fulfilled such as no pollution with gas emissions or noise, and maintenance work. From this standpoint, the desirable solution will be an electric system featuring an absolute non-contact between vehicle and guideway. Thus in terms of (1) the linear motor is recommendable; in terms of (2) the magnetic levitation is recommendable; and in terms of (3) the active track system is recommendable. When additionally the installation of the guideway and the possible seismic displacements are considered, the vehicle-guideway gap is preferably set at about 100 mm, which will call for presence of powerful superconducting magnets. Superconducting magnets which function in the persistent current mode need no energizing source mounted on the vehicle and, being devoid of the iron core, they contribute to reducing the vehicle weight.

UTILIZATION OF SUPERCONDUCTIVITY

Levitation is effected by repulsion between the same poles of the magnets. When the guideway too is constituted of superconducting magnets, however, it is uneconomical. This is avoided by arranging normal-conducting loop coils, instead of superconducting magnets, on the guideway. As the vehicle moves, the magnetic flux of the superconducting coils cuts across the ground coil and thereby a counteracting current flows in the ground coil by induction, causing the two coils to repel each other (Fig. 1). Use of inductive current makes it necessary that the vehicle, which will cease to be levitated while standing still, be equipped with auxiliary support wheels; but the vehicle will never drop to the ground, even if a power failure happens while it is running levitated. Meanwhile any control will not be needed to keep the vehicle afloat with natural stability.

Propulsion is effected by a linear synchronous motor which represents a linearly-unfolded rotary synchronous motor. As illustrated in Fig. 2, under the arrangement of the superconducting magnets on the vehicle and the three-phase propulsion coils (armature coils) on the ground, a magnetic field is created by passage of the current through the propulsion coils and thereby the force acting between these coils and the superconducting magnets aboard the vehicle propels the vehicle. In order that the vehicle can be propelled with a constant force, the magnetic field of the propulsion coil must be moved following the running vehicle. For this purpose the phase (frequency) of the three-phase a.c. flowing in the propulsion coil must be changed. On the other hand for the purpose

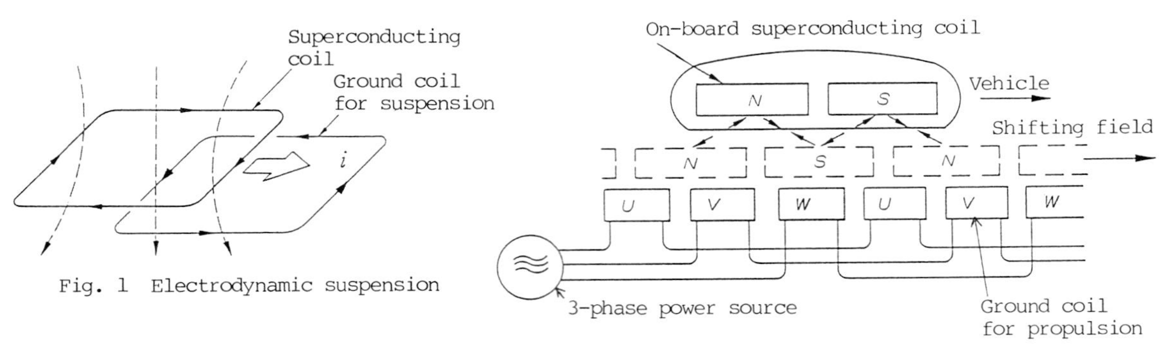

Fig. 1 Electrodynamic suspension

Fig. 2 Linear synchronous motor

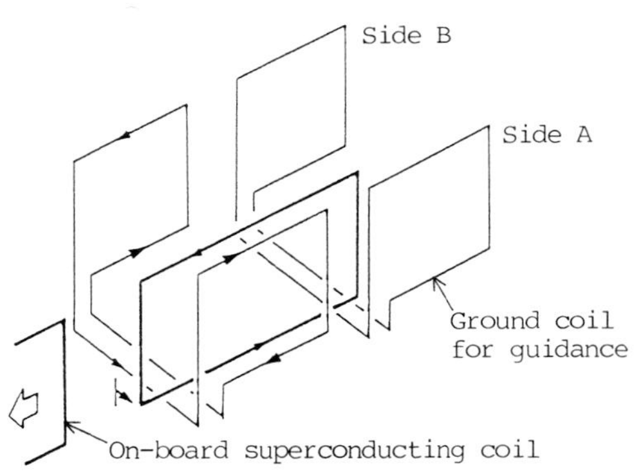

Fig. 3 Null-flux magnetic guidance

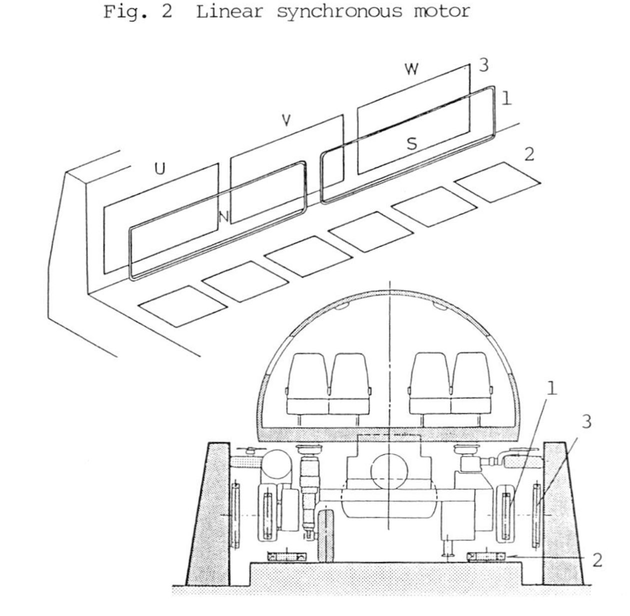

1: Superconducting coil
2: Ground coil for suspension
3: Ground coil for propulsion and guidance

Fig. 4 Coil arrangement

of speed control, the propulsion force must be freely variable. This is done by changing the intensity of the magnetic field through variation of the magnitude (amplitude) of the flowing current. For this purpose, a variable-voltage variable-frequency power source must be provided, which in the case of the Miyazaki test track is a cyclo-converter. The linear synchronous motor utilizing the superconducting magnet is characterized by excellence of electric performance even with a large gap between vehicle and guideway.

Guidance is done on the same principle as levitation. However, a so-called "null-flux" coil arrangement is adopted to avoid a running resistance caused as a result of energy loss suffered on account of the current induced in the ground coil. As seen in Fig. 3, when the vehicle runs along the center of the guiding coils on the ground, no current is induced in the coils on the ground and there occurs no loss. When the vehicle happens to deviate from the center, a current is induced matching the amount of the deviation and then a vehicle-righting force proportional to the amount of the deviation is generated.

The present Miyazaki test track is arranged such that a single superconducting coil can serve concurrently for levitation, for propulsion and for guidance, while the ground coil can serve concurrently for both propulsion and guidance (Fig. 4).

The brake adopted is a regenerative one in which the braking force is obtained regeneratively by reversing the phase of the current flowing in the propulsion/guidance coil 180 degree to the phase in time of powering; but a dynamic brake which connects the propulsion/guidance coil to the resistor is also provided as a backup.

The energy required for illumination, air-conditioning and helium refrigerator of the vehicle is obtained as a current induced in the current-collection coils installed on the vehicle by a harmonic magnetic field which is created in the ground coils for levitation.

PROCESS OF DEVELOPMENT

The maglev development started in 1970 when the generation of a levitating force by an electrodynamic suspension system was confirmed by means of a rotary type simulation device newly designed for the testing. Since then the work proceeded to a running experiment, which began using the most simplified version of a two-pole superconducting coil and continued on a steadily enlarged scale

Table 1 Main features of test vehicles

Name	System			Length	Mass	Number of coils	Speed	Year
LSM 200	EDS	LSM	MG	4.0 m	2.0 t	320 kA x 2	50 km/h	1972
ML 100	EDS	LIM	MG	7.0 m	3.5 t	250 kA x 4	60 km/h	1972
ML 100A	EDS	LSM	NFG	5.0 m	3.6 t	160 kA x 4, 450 kA x 4	60 km/h	1975
ML 500	EDS	LSM	NFM	13.5 m	10.0 t	250 kA x 8, 450 kA x 8	517 km/h	1977
ML 500R	EDS	LSM	NFM	12.6 m	12.7 t	250 kA x 8, 450 kA x 8	204 km/h	1979
MLU 001	EDS	LSM	NFM	10.0 m (8.2 m)	10.0 t	700 kA x 8	405 km/h	1980-82
MLU 002	EDS	LSM	NFM	22.0 m	17.0 t	700 kA x12	420 km/h*	1987

EDS Electrodynamic suspension
LSM Linear synchronous motor
LIM Linear induction motor
MG Mechanical guidance
NFM Null-flux magnetic guidance
( ) In case of the mid vehicle
* Planned

with sophisticated content. Table 1 gives the details of the experimental vehicles so far manufactured and tested.

In LSM 200 a superconducting coil was employed for both levitation and propulsion, but a roller was substituted for guidance. ML 100 was designed on such a scale that it could seat 4 passengers and in this type a linear induction motor was employed for propulsion, but a slide shoe was substituted for guidance. ML 100A is a predecessor to ML 500 and it was the first one to implement the absolute non-contact run with "null-flux" magnetic guidance. In ML 100A, the guideway was designed with an inverted-T section and the propulsion/guidance coils were installed on the projecting edge of the guideway. The superconducting coils were provided in two types, i.e., for levitation and for guidance, which were housed in a cryostat with L-section. All the experimental vehicles mentioned above have been submitted to testing on the premises of the Railway Technical Research Institute.

The Miyazaki test track was laid with a guideway of an inverted-T section. ML 500 attained in December 1979 a maximum speed of 517 km/h, demonstrating the feasibility of a high-speed run at over 500 km/h by maglev. Prior to this, a test run was done using ML 500R which was ML 500 provisionally mounted with helium refrigerator, to verify the possibility of carring a refrigerator on the vehicle.

Subsequently the guideway has been modified to a more practical U-shape section and thereby with a successful development of a superconducting coil with a magnetomotive force of 700 kA it was become possible to make a single coil serve the three purpose of levitation, guidance and propulsion (see Fig. 4). Three units of the MLU 001 experimental vehicle have been manufactured for U-shape guideway use and employing them it has been confirmed that there is no problem with multi-unit operation of the maglev vehicles and that these vehicles are comfortable to ride. In February 1987, the vehicles successfully developed a maximum speed of 400.8 km/h carrying three passengers. To collect data on limitations in track installation, a test run was performed on the guideway deliberately designed with an angular bend of 30 mm and a joint with 20 mm stagger. In this test MLU 001 vehicle cleared these obstacles which would be insurmountable by the conventional railway vehicle, with an ample gap left between the vehicle and guideway, proving the superiority of maglev system with a large levitation. Meanwhile a mechanical braking test with an emergency landing device (skid shoe) was executed supposing a contingency of the electric brake turning ineffective as a result of a sudden "quenching" of all the superconducting coils and thereby it has been confirmed that the vehicle can be safely braked to a halt from the speed of 300 km/h; from the speed beyond this value, the air drag will be utilized.

Under the circumstances calling for simultaneous development of the magnetic levitation system and the superconducting magnets required for it, the initial test vehicle was equipped with an open type superconducting magnet, but in time for MLU 001 the magnet combined with an on-board refrigerator has been perfected.

MLU 002 TEST VEHICLE

MLU 002 test vehicle has a body length of 22 m, that is, close to the dimension for commercial service and it features the superconducting magnets mounted in fewer number and located only at the trucks, reflecting the improved performance achieved. The magnetomotive force of the superconducting coil is 700 kA, the same as in MLU 001. To make available the ground coils as they are on the Miyazaki test track, the superconducting coils are installed in three poles on one side of each truck (Table 2, Figs. 5,6).

Only one unit of MLU 002 has been built, it is streamlined at both end and the seating capacity is limited to 44 to permit instrumentation on the floor. Learning from the experience with MLU001, the optimum body form has been selected through wind tunnel testing, the window recess has been eliminated and the side panels have been welded instead of riveted to smoothen the body surface so that the air resistance of the body, the aerodynamic instability and the aerodynamic noise generated can be minimized; furthermore, the auxiliary support wheels have been designed retreatable completely within the housing under levitation, and they are provided with a cover, and the vehicle body is designed airtight. The mid-portion of the body is fabricated of aluminum but the curved 5 m

Table 2 Main features of MLU 002

Vehicle dimensions	
Length x Width x Height	22.0 m x 3.0 m x 3.7 m
Mass	17 t
Seating capacity	44
Superconducting coil	
Number of coils	6 poles x 2 rows
Magnetomotive force	700 kA
Pole pitch	2.1 m
Suspension	
Lift	196 kN
Effective gap	110 mm
Guidance	
Guidance force	83.3 kN at 50 mm shift
Effective gap	more than 150 mm
Propulsion	
Thrust	0-79.4 kN
Phase	3
Frequency	0-28 Hz
Voltage	5800 V
Current	900 A
Maximum speed	420 km/h

Fig. 5 MLU 002

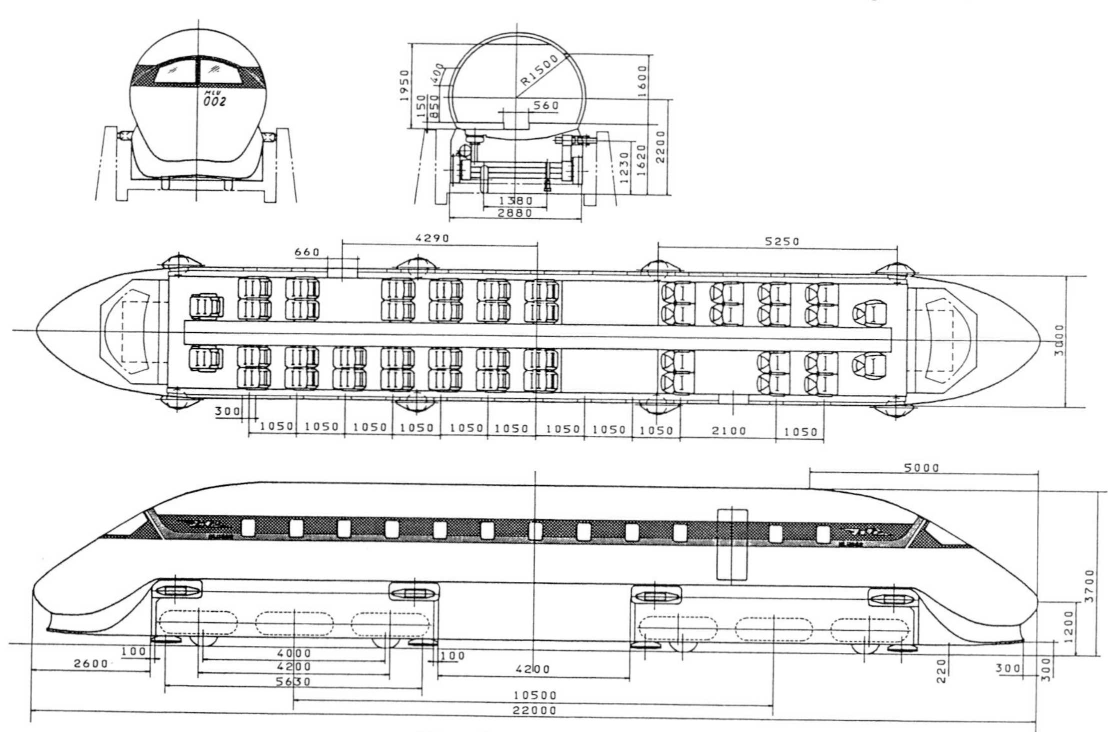

Fig. 6 Plan of MLU 002

portions at both ends are made of carbon fiber reinforced plastics.

At present the safety,reliability and practicality of the system are being confirmed using MLU 002 on the Miyazaki test track. As of August 1988, the maximum speed attained is 375 km/h and a further speedup is contemplated.

Composition of the superconducting magnet employed in MLU 002 is illustrated in Fig. 7 and the details of the superconducting coil are given in Table 3. The wire employed are niobium-titanium alloy series as before but with the copper ratio reduced to 1 from 2 in MLU 001, the weight has been substantially reduced. The coil is reinforced by epoxy resin impregnation just as that of MLU 001.

The inner vessel of the cryostat is made of stainless steel to improve the low-temperature strength, while the outer vessel of it is made of aluminum alloy to decrease the weight. The load-bearing members are designed as multi-cylinders of FRP and CFRP to minimize the heat penetration.

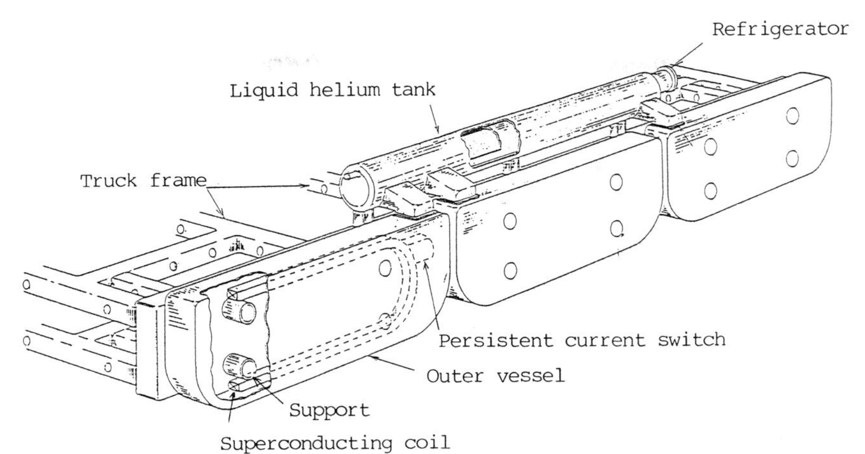

Fig. 7 Superconducting magnet for MLU 002

Table 3 Main features of superconducting coil

Coil	Dimension(Length x Height)	1.7 m x 0.5 m
	Mass	77 kg
	Cross section(Thickness x Width)	45 mm x 71 mm
	Magnetomotive force	700 kA
	Number of turns	1167
	Self inductance	3.04 H
	Current density	219 A/mm^2
	Stored energy	550 kJ
	Maximum field	5.1 T
Wire	Copper ratio	1.06
	Cross section(Thickness x Width)	1.05 mm x 2.12 mm
	Number of filaments	2382
	Diameter of filaments	23 μm
	Twist pitch	49 mm

The persistent current switch is a thermal type of alternate lap-windings of superconducting wires and heater wires. Improvements of the superconducting coil and the persistent current switch have enabled a high speed excitation of 10 A/s. Meanwhile a great stride has been taken toward realization of a superconducting magnet independent of the ground base, which is the target for commercial operation, by reduction of shunting to the persistent current switch and suppression of gaseous helium generation. Between the inner and outer vessels is provided a thermal shield wound with liquid nitrogen piping. The coil portion and the liquid helium tank portion are structually designed separable for convinience of access for repair. In spite of such a construction, the heat leak into the inner vessel is limited to about 3 W.

As the on-board refrigerator system, the arrangement proved to be superior with MLU 001 in which compact refrigerators are distributed to each magnet is adopted. For the purpose of comparison, Claude cycle refrigerators and reversed Stirling cycle ones are selectively mounted on the trucks.

FUTURE PROSPECT

Commercial vehicles will be designed in an articulated type with the superconducting magnets located at the couplings only and between the magnets the floor will be lowered to provide the passenger rooms (Fig. 8). By this designing the vehicle height and the cross-section will be reduced, resulting in a decreased air resistance. Since the passenger room is isolated from the magnets, the magnetic flux density of the passenger room will be made low enough to protect the watches or pacemakers worn by the passengers from exposure to magnetism. The magnetic shield has only to be provided around the magnets. The standard train makeup will be 14 units with a total passenger capacity of 950 (Table 4). Thus more than 10,000 passengers per hour one way can be transported. The radius of curvature will be set at 6000 m for a designed curve passing speed of 500 km/h and the steepest gradient will be set at 10 ‰, while the continuous gradient will be at about 6 ‰.

Already the superconducting magnet to be employed in future operation has been tentatively built and submitted to the stationary test. This magnet, shown in Fig. 9, using niobium-titanium alloy series wires of copper ratio 1 generates a magnetomotive force of 700 kA, the same as the one for MLU 002 does, but it has the pole pitch increased from 2.1 m to 2.7 m.

Running tests of the substations cross-over control and high-speed turnouts presently under research and development are scheduled for 1989 on the Miyazaki test track with them assembled therein.

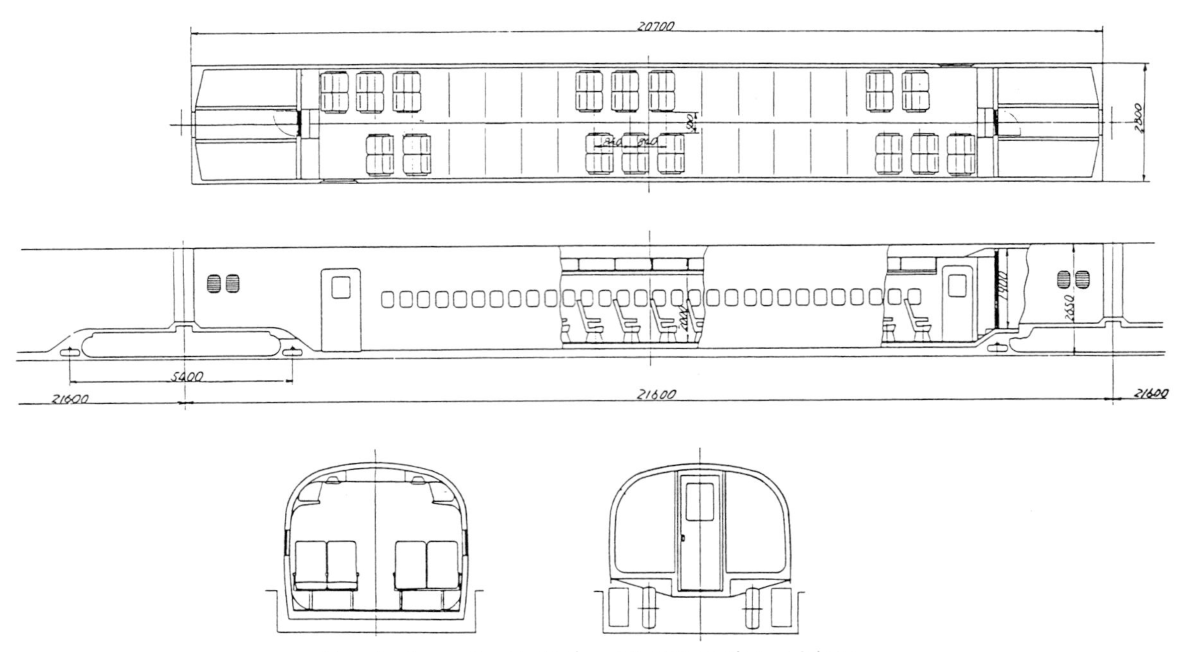

Fig. 8 Conceptual design of commercial vehicle

Table 4 Main features of commercial system

Length x Width x Height	28.0 m x 2.8 m x 2.65 m (E) 21.6 m x 2.8 m x 2.65 m (M)
Mass	27 t (E), 18 t (M)
Seating capacity	950 (14 vehicles)
Curve radius	6000 m at 500 km/h
Gradient	Maximum 10 %, Continuous 6 %
Levitation height	100 mm
Maximum speed	500 km/h

Fig. 9 New superconducting magnet

A short-distance type consisting of a single track about 50 km now under consideration has already reached a technological level enabling its commercial operation 5-6 years hence. Meanwhile for realization of the long distance version, it would be necessary to perform running tests of trains passing each other or in tunnels on the demonstration track of several tens kilometers, targeted at the inauguration of commercial operation in the year 2000.

CONCLUDING REMARKS

It seems that the practical application of liquid helium refrigeration to superconducting magnets is feasible enough. Moreover, it goes without saying that, if high temperature superconductors become available, the structures of magnets or refrigerator systems will be drastically simplified. Research for high temperature superconductor is yet technologically far from being immediately available as the coil material; and for the time being, the applicability for magnetic shielding is being investigated. When the high temperature superconductor becomes practically available, there will be no problem at all in replacing the liquid helium-cooled magnet with this material.

Potential Benefits of Superconductivity to Transportation in the United States*

DONALD M. ROTE and LARRY R. JOHNSON

Energy and Environmental Systems Division, Argonne National Laboratory, 9700 S. Cass Ave., Argonne, IL 60439, USA

ABSTRACT

Research in U.S. transportation applications of superconductors is strongly motivated by a number of potential national benefits. These include the reduction of dependence on petroleum-based fuels, energy savings, substantially reduced air and noise pollution, increased customer convenience, and reduced maintenance costs. Current transportation technology offers little flexibility to switch to alternative fuels, and efforts to achieve the other benefits are confounded by growing congestion at airports and on urban roadways. A program has been undertaken to identify possible applications of the emerging superconducting applications to transportation and to evaluate potential national benefits. The current phase of the program will select the most promising applications for a more detailed subsequent study. Transportation modes being examined include highway and industrial vehicles, as well as rail, sea, and air transport and pipelines. Three strategies are being considered: (1) replacing present components with those employing superconductors, (2) substituting new combinations of components or systems for present systems, and (3) developing completely new technologies. Distinctions are made between low-, medium-, and near-room-temperature superconductors. The most promising applications include magnetically levitated passenger and freight vehicles; replacement of drive systems in locomotives, self-propelled rail cars, and ships; and electric vehicles inductively coupled to electrified roadways.

INTRODUCTION

Research in applications of superconductors to transportation in the United States is strongly motivated by the potential for major national benefits, including reduced dependence on petroleum-based fuels, reduced energy needs, substantially reduced air and noise pollution, reduced costs associated with airport and highway congestion, increased customer convenience, and reduced maintenance costs. Transportation currently accounts for 27% of total U.S. energy consumption and 62% of U.S. petroleum use [1]. The present transportation system and technology offers little flexibility for switching to alternative fuels. Further efforts to reduce energy consumption and noise and pollution emissions from conventional modes of transport are likely to incur diminishing returns and are being counteracted simply by the growth in transportation activity.

In light of these concerns, and to aid in deciding on research priorities, Argonne National Laboratory undertook a program to investigate ways that developments in high-temperature superconductor technology might benefit transportation. The program has several objectives:

- Identify large-scale applications of the emerging high-temperature superconductivity technology to various transportation modes.
- Evaluate the requirements for, and the potential benefits of, such applications and estimate how those benefits might change with critical temperature (T_c) and with other material properties.
- Select the most worthwhile concepts for more detailed study and possible implementation in the future.

It is important to point out that the properties of high-temperature superconducting materials are still not well understood -- indeed even the potential limits of the critical parameters are not known -- and experience with their use in practical devices is essentially nonexistent. This analysis should therefore be regarded as a preliminary assessment of the benefits of high-temperature superconductivity in transportation. Nevertheless, the general direction, if not the

*Work supported in part by the U.S. Department of Energy, Assistant Secretary for Conservation and Renewable Energy, under Contract W-31-109-Eng-38, and in part by the Electric Power Research Institute.

quantification, should be sufficiently valid to determine the most promising applications, given the current state of knowledge.

Despite the difficulties, some relatively obvious considerations may help to judge the suitability of superconductor applications to transportation. First, superconducting devices generally require some auxiliary equipment (extra insulation, refrigeration, power electronics, quench protection, magnetic shielding, etc.). This equipment will have adverse impacts such as added size, weight, energy consumption, complexity, maintenance costs, etc. Second, for an application to be both technically feasible and economically attractive, the device must offer at least enough benefits to more than compensate for the adverse impacts of the auxiliary equipment. This suggests that the larger the scale of the application, the more likely it is to be economically attractive; it also suggests that the more superconducting components that can be incorporated into a system and therefore share the auxiliary components, the greater the system's efficiency. Finally, because the adverse impacts of the auxiliary equipment are likely to diminish with increasing T_c, it suggests that the viability of an application should increase with T_c. Hence, it is expected at the onset that drive motors for large ships will be more cost-effective than for automobiles and that there is a greater potential in transportation for high-T_c applications than for low-T_c. However, at exactly what scale and at what T_c value a particular application becomes more cost-effective than a conventional device remains very uncertain.

All of the traditional transportation modes, including highway vehicles, railroads, ships, aircraft, and pipelines were considered, along with some new technologies such as electric vehicles, inductive-power-coupled electric vehicles, magnetically levitated (maglev) vehicles, and magnetohydrodynamically propelled ships. A summary of several of the more interesting and promising applications follows.

MAGLEV VEHICLES

The term *vehicles* is used here deliberately to emphasize the inappropriateness of referring to such vehicles as "trains." Although it has become commonplace to consider magnetic levitation as another railroad technology, maglev vehicles have relatively little in common with conventional steel-wheel-on-steel-rail trains. In fact, maglev vehicles more closely resemble aircraft fuselages without wings than they do railroad locomotives or cars. The issue is much deeper than this, however; viewed as another railroad technology, maglev has two formidable competitors, the traditional railroad industry and the commercial airline and aerospace industries. Viewed as an aerospace technology that can be incorporated into the existing airline "hub-and-spoke" system, maglev vehicles create a niche in the U.S. transportation system that complements rather than competes with both air and rail transport (see Ref. 2 for a more detailed analysis). By taking advantage of maglev technology's high speed and other attractive features, maglev routes could readily replace short-haul (160-1000 km) commercial airline flights that connect hub and regional airports. This would substantially reduce airport congestion and open airport slots for the longer-range flights that are more efficient and profitable for the airlines. Reduced airport congestion alone could save 1.5-5.5% of aircraft fuel presently consumed. Such maglev routes would also transfer some ridership from intercity highways, thus reducing road congestion as well. Figure 1 illustrates some of the potential U.S. markets that could be served by maglev routes. Table I summarizes various aspects and potential benefits of maglev vehicles.

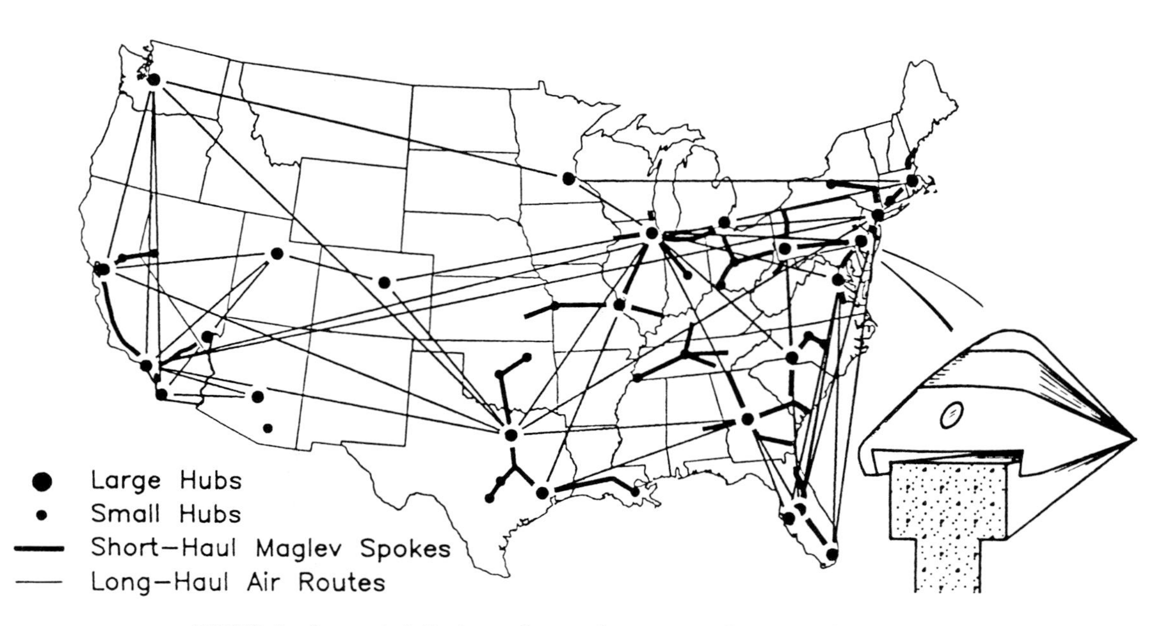

FIGURE 1 Potential Markets for Maglev Routes in the United States

TABLE I Maglev Vehicles Are the Most Promising Applications for High-T_c Superconductivity

Superconducting Components
- On-board levitation, guidance, and propulsion coils
- Regenerative braking, plus wayside or on-board superconducting magnetic energy storage (SMES)
- Magnetic shielding and magnetocaloric refrigeration

Benefits of Low-T_c Superconductivity
- High field strength
- Large energy gap
- High power factor

Benefits of High-T_c Superconductivity
- Increased reliability
- 10-15% reduction in vehicle weight
- 5-10% reduction in on-board energy needs
- 5-15% reduction in propulsion energy needs
- Potential reduction in substation capital and/or operating costs

Requirements
- T_c = 77K
- $J \geq$ 10k A/cm^2 for levitation coils; 10-100k A/cm^2 for SMES
- B ≈ 5 T for levitation coils; 15-20 T for SMES

Petroleum Savings
- Diesel and aircraft fuel replaced by utility energy sources: 194,000-233,000 bbl/day, which includes:
  - Replacement of some short-range flights: 127,000 bbl/day
  - Replacement of some intercity auto travel: 52,000 bbl/day
  - Reduction of airport congestion: 15,000-54,000 bbl/day

Technical Barriers
- High-T_c materials needed to meet requirements listed above
- Lightweight magnetic shielding needed that does not adversely affect magnet performance

Market Size
- 10-20 networks in the U.S., connecting hub airports with "spoke" cities

SUPERCONDUCTING SHIP PROPULSION

Because of their scale, ships could maximize the technical benefits of superconducting applications. As illustrated in Figure 2, there are two general approaches. First, the more conventional approach is replacement of certain components, e.g., drive motors, generators, power electronics, and refrigeration systems with superconducting components. Given the potential for significantly reduced space and weight requirements for the motor, and to a lesser extent the generator and magnetocaloric refrigeration systems [3], there is considerable added flexibility -- especially when compared with mechanical drive systems -- for optimal component placement and hull design. For example, in certain applications it may be desirable to avoid penetrating the hull with a large rotating shaft. This can be achieved by placing the motors and propellers in pods outside the hull.

The second and more radical approach to incorporating superconductor technology into ship propulsion utilizes a magnetohydrodynamic (MHD) system. This system uses long superconducting coils to produce very large magnetic fields (B). Electrically energized plates set up a current J in the sea water so that the Lorentz force ($J \times B$) provides the thrust against the seawater to propel the ship. This scheme, however, must cope with the very high resistivity of seawater (≈25 Ω/cm). According to an analysis currently in progress by Holtz and Krazinski at Argonne National Laboratory [4], efficiencies in the range of 20-65% are achievable with field strengths in the range of 10-20 T for small submersibles. Efficiencies as high as 70% or more can be achieved with large vessels, but require a field strength of ≈20 T. In view of the high field and very large capacity power supply requirements needed to achieve efficiencies approaching those of conventional pumps, the MHD approach may not be viable except under special circumstances. The potential benefits of superconductors applied to ship propulsion are summarized in Table II.

SUPERCONDUCTOR APPLICATIONS IN LOCOMOTIVES

Almost all U.S. intercity passenger and freight cars are pulled by diesel-electric locomotives; in the current population of more than 21,000 locomotives, only about 60 are all-electric. Hence, only diesel-electric locomotives represent a large enough U.S. market for superconducting components. There are several very good reasons for growth in the all-electric locomotive population; however, until the necessary investments are made in route electrification, little change can be expected. If a significant changeover in locomotive stock does occur, development of superconducting components for all-electric locomotives would also be justified.

The use of superconducting components in a diesel-electric locomotive is conceptually illustrated in Figure 3. The benefits to diesel-electric and all-electric locomotives, as well as to self-propelled railcars, are summarized in Table III. Perhaps the most noteworthy benefit to diesel-electric locomotives would be the substantial increase in tractive power, due to the potentially greater power density of superconducting traction motors. Conventional motors are constrained in size (and thus power) because of space restrictions.

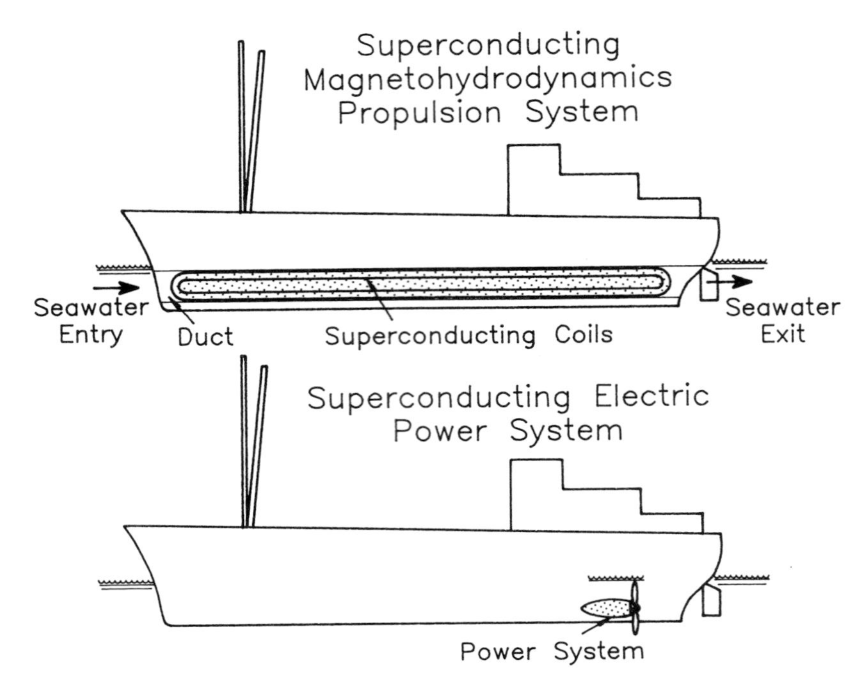

FIGURE 2 Two Approaches to Use of Superconductivity in Ships

TABLE II Ship Propulsion Systems Are Potentially the Largest and Most Cost-Effective Users of Superconductivity

Superconducting Components
- Generators
- Magnetocaloric refrigeration
- Drive motor units or magnetohydrodynamic (MHD) propulsion systems

Benefits of Low-T_c Superconductivity
- Weight and volume reduced by factor of 2-5
- 50% reduction in waste heat
- 4-6% reduction in energy needs
- Flexibility in component placement (e.g., no need for hull-penetrating propeller shaft)

Benefits of High-T_c Superconductivity
- Reduced need for cooling energy
- Higher reliability
- Plentiful supply of coolant
- Cost-effective at lower scale
- Higher field strengths make MHD propulsion more feasible

Petroleum Savings
- 5-10% reduction in petroleum use: 34,000-69,000 bbl/day

Requirements
- T_c = 4K or 77K
- $J > 10k\ A/cm^2$
- $B \approx 5$ T for superconducting electric drive; ≈ 20 T for MHD

Technical Barriers
- High-T_c materials needed for motors
- High-current brushes needed for homopolar motors
- High-kVA power supplies needed for MHD
- Suitable designs needed for synchronous motors

SUPERCONDUCTOR APPLICATIONS IN ELECTRIC VEHICLES

Two different electric vehicle technologies were considered as candidates for superconducting components. The first is the conventional battery-powered electric vehicle in which the batteries are charged in the stationary mode. The second is the inductive-power-coupled electric vehicle that derives propulsion power from an inductive power coupling linked to active elements embedded in an electrified roadway and that is driven by batteries while off the electrified roadway [5,6].

Inductive power coupling offers considerable advantages in vehicle weight and energy savings and in extension of vehicle range over battery-only electric vehicles. Because of the small scale of this application, however, it is unclear whether superconductivity offers any net benefits after the impacts of the required auxiliary equipment are considered.

Based on present limitations of low-T_c superconductors with DC power, the only applications of higher-T_c superconductors are likely to be in traction motors and possibly regenerative braking with on-board superconducting magnetic energy storage (SMES). According to an ongoing analysis at

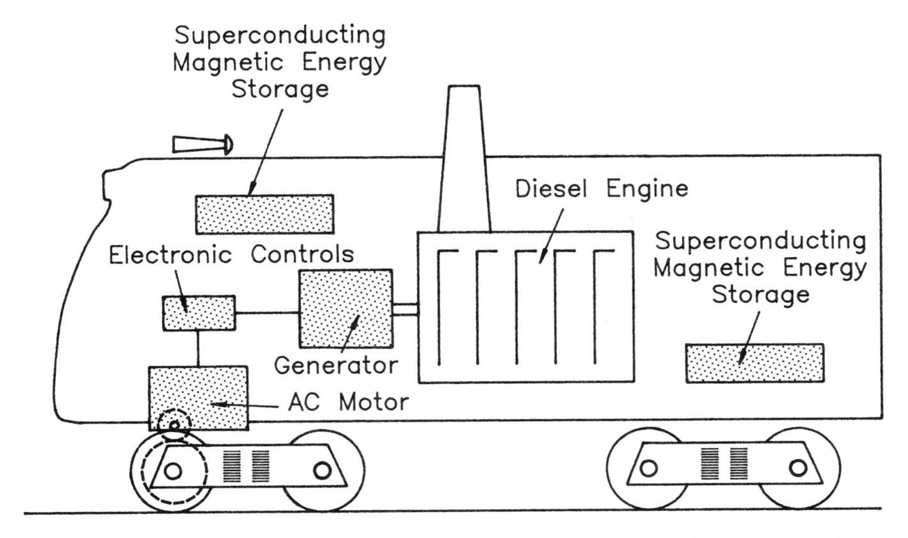

FIGURE 3 Potential Use of Superconducting Components (shaded areas) in a Diesel-Electric Locomotive

TABLE III Superconducting Applications in Diesel and All-Electric Locomotives and Self-Propelled Railcars Appear Very Promising

Superconducting Components
- Traction motors (and possibly generators)
- Regenerative braking and SMES for self-propelled railcars
- Magnetic energy storage
- Magnetocaloric refrigeration

Benefits of Low-T_c Superconductivity
(probably not applicable)

Benefits of High-T_c Superconductivity
- Enables application to locomotives
- Possible 25-50% increase in diesel-electric tractive power
- 5-10% reduction in energy needs

Petroleum Savings
- Diesel-electric locomotives: 5-10% (36,000-71,000 bbl/day)

Requirements
- T_c = 77K
- $J \simeq$ 10k A/cm^2 for motors; 10-100k A/cm^2 for railcar SMES
- $B \simeq$ 3-5 T for motors; 15-20 T for railcar SMES

Technical Barriers
- High-T_c materials needed
- Compact motor (and possibly generator) needed

Market
- 21,000 diesel-electric locomotives
- 61 all-electric locomotives
- If extensive electrification occurs, perhaps 50% of diesels could be replaced by all-electrics in 15 years

Idaho National Engineering Laboratory [7], SMES has potential applications in a wide variety of vehicle sizes. It offers several important advantages over storage batteries that are particularly attractive for regenerative braking. These include rapid charge and discharge capability, higher efficiency, and longer lifetimes under repeated rapid charging cycles. Specific energy densities are comparable to or slightly lower than those of batteries. However, a complete analysis of a superconducting system using both traction motors, regenerative braking, and SMES has not yet been carried out. A conceptual illustration of a superconducting electric vehicle is shown in Figure 4; benefits are summarized in Table IV.

CONCLUSIONS

If high-T_c superconductors become a practical reality, they could have a major impact on U.S. transportation. Not only could a significant amount of petroleum be saved by displacing some of the petroleum-based vehicle drive systems with utility-powered drive systems, but -- due to efficiency improvements -- energy could also be saved. Furthermore, high-T_c systems could have a significant effect on designs of power transmission systems for ships and locomotives. Finally, maglev vehicles show the greatest potential for use of high-T_c systems and also offer the greatest potential benefits to transportation problems in the United States.

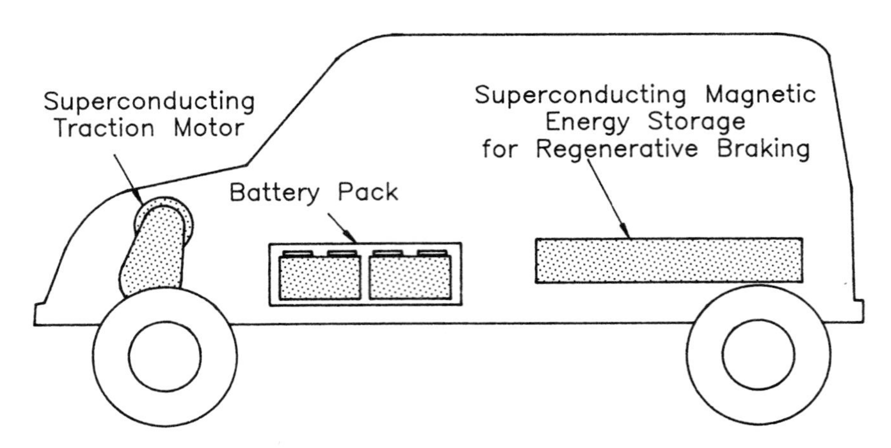

FIGURE 4 Schematic of a Possible Superconducting Electric Vehicle

TABLE IV Traction Motor Replacement in Electric Vehicles Is Advantageous if Specific Mass of the Superconducting Motor, Controller, and Cooling System is Less than 1.6 kg/kW

Superconducting Components
- Traction motor and controller
- Regenerative braking system
- Magnetic energy storage

Benefits of High-T_c Superconductivity
- Application possible in highway vehicles
- 5% reduction in energy consumption
- 14-17% reduction in annual user cost
- 20-26% increase in vehicle range

Requirements
- T_c = 77K
- $J_c \approx$ 10k A/cm^2 for motors; 10-100k A/cm^2 for SMES
- B ≈ 3-5 T for motors; 15 T for SMES

Petroleum Savings
- At 2% market penetration: ≈ 94,000 bbl/day

Technical Barriers
- Compact, lightweight, high-T_c motor and cooling system needed (overall specific mass < 1.6 kg/kW)

Market Size
- ≈1% of personal highway vehicles
- ≈10% of commercial fleet vehicles
- If inductive-power-coupled technology is used, the market for personal highway vehicles could be much larger

REFERENCES

1. Holcomb, M.C., S.D. Floyd, and S.L. Cagle, *Transportation Energy Data Book: Edition 9*, Oak Ridge National Laboratory Report ORNL-6325 (April 1987).

2. Johnson, L.R., Written testimony for U.S. Senate Committee on Environment and Public Works, Subcommittee on Water Resources, Transportation, and Infrastructure, Honorable Daniel Patrick Moynihan, Chairman, Washington, D.C. (Feb. 26, 1988).

3. Green, G., et al., *Magnetocaloric Refrigeration*, David W. Taylor Naval Ship Research and Development Center Report DTNSRDC/87/032 (March 1987).

4. Holtz, R.E., and J.L. Krazinski, Argonne National Laboratory, Argonne, Illinois 60439, personal communication (1988).

5. Bolger, J.G., et al., *Test of the Performance and Characteristics of a Prototype Inductive Power Coupling for Electric Highway Systems*, Lawrence Berkeley Laboratory Report LBL-7522 (July 1978).

6. Bolger, J.G., *Power and Control from the Roadway: Status, Potential, and Constraints*, Inductran Corp., 1325 Ninth St., Berkeley, Calif. 94710 (Oct. 1986).

7. Herring, S., Idaho National Engineering Laboratory, Idaho Falls, Idaho 83415, personal communication (1988).

Superconducting Magnet for Magnetically Levitated Vehicle

M. YAMAJI[1] and H. NAKASHIMA[2]

[1] Toshiba Corporation, 1, Toshiba-cho, Fuchu, Tokyo, 183 Japan
[2] Railway Technical Research Institute, 1-8-38, Hikari-cho, Kokubunji, Tokyo, 185 Japan

SUMMARY

In March, 1987, the prototype MLU001 magnetically levitated (maglev) vehicle using superconducting magnets made its debut on Miyazaki Test Track of the Railway Technical Research Institute. Since then repeated tests have confirmed the MLU002's high speed running capability over 300 km/h. The most important component of maglev transportation is the superconducting magnet on which development was started in 1970, and whose performance has since been continually improved in parallel with the progress of superconducting technology.

In particular, a number of developments here have contributed to continuing this progress:
(1) Development of intrinsically stabilized superconducting wire, improvement of epoxy impregnated coils, coil mounting and inner vessel construction have led to superconducting coils of high current density and high magnetomotive force.
(2) Along with the development of the superconducting coils, improvements of construction methods, heat insulation techniques, supports and current leads have led to reduced weight and size with low heat leakage.
(3) Development of a small on-board refrigerator system which pre-cools the superconducting magnet, and maintains the liquid helium reserves, liquefying the vaporized helium.
(4) Development of a high resistance persistent current switch has enabled high speed energizing and de-energizing of the superconducting magnet to improve the magnet operation.

As a result of these development steps, the stage for manufacturing the superconducting magnet for future operation line use is near.

DEVELOPMENT HISTORY OF MAGNETICALLY LEVITATED VEHICLES

Investigation of a next-generation super-express vehicle was started by the Japanese National Railways early in the 1960s and various systems were examined and compared. It was then believed that a speed of 500 km/h could be safely achieved with a large non-contact gap using the strong magnetic field generated by superconducting magnets. This concept would also reduce track maintainance and non-contact levitation would greatly reduce noise and vibration.

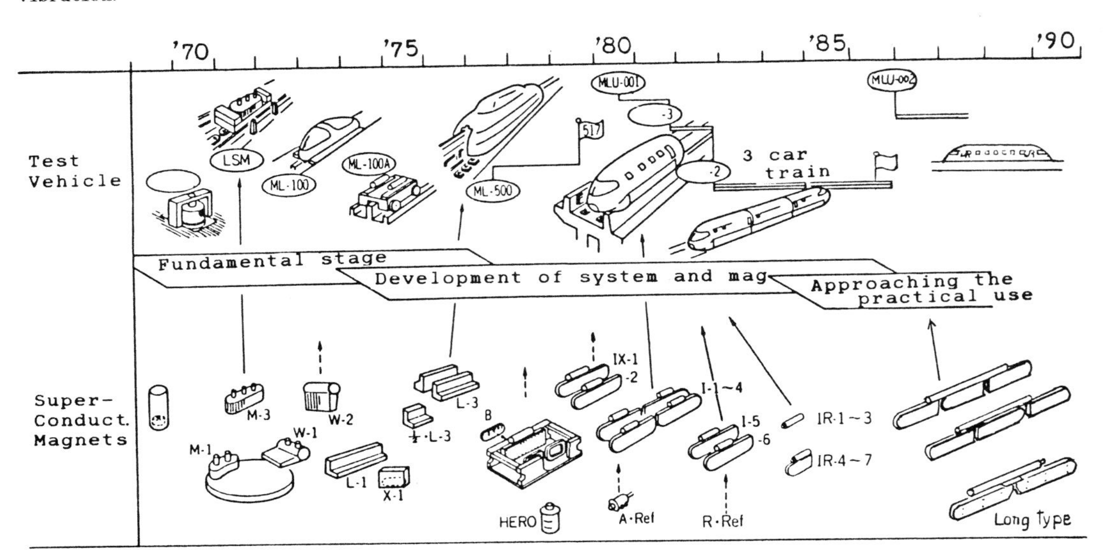

Fig 1 History of Development of Maglev Vehicles and Superconducting Magnets

The first facility for basic testing of superconductive magnetic levitation was installed in the JNR technical research center in 1970. The ML vehicle operation was demonstrated by the small-sized test vehicle, LSM-200, in 1972. In the autumn of that year the second small-sized test vehicle, ML-100, was exhibited.

Construction of a full-scale test track in Hyuga, Miyazaki, was started in 1976, and in 1977, operating tests were started using the ML-500 super high-speed test vehicle. The test track had an inverted T-shaped cross-section, and the vehicle sat astride the track with symmetrically mounted L-type superconducting magnets. The 13.5 m long 10 ton weight ML-500 recorded a speed of 517 km/h on the 7 km test track in December, 1979.

The track was remodeled to the U-shaped cross-section and the MLU001 manned test vehicle was delivered in 1980. I-type superconducting magnets were mounted on both sides of the bogie and a cabin was installed on the bogie. In 1981, the MLU001 was coupled to consist of two cars train, and next year three cars train arrangement.

After many confirmation tests were performed, including the composite running characteristics test, and ride quality improvement test, the manned MLU001 vehicle composed of two cars achieved levitated running at 400 km/h in February, 1987. In total the MLU001 test vehicle ran 40,000 km.

During these tests, development of the superconducting magnet also continued, and as a result of weight and size reductions, levitation force per superconducting coil could be increased, the number of coils reduced. Then, the prototype vehicle, MLU002, was completed in March, 1987.

REQUIREMENTS FOR ON-BOARD SUPERCONDUCTING MAGNETS

The following characteristics are required for on-board superconducting magnets:

(1) Light weight. Light weight of a ML vehicle reduces required for levitation force and magnetomotive force of a superconducting coil, so that the superconducting magnet size can be reduced. Light weight is required as much or more than an air craft.

Table 1 Features of Superconducting Coils and Wires

Test Vehicle Manufactured Year	ML-500 1977	MLU001 1980	MLU002 1987
Coil Dimension Racetrack-shape	1.65mL, 0.5mW	1.7mL, 0.5mW	1.7mL, 0.5mW
Weight of Wire	116 kg	100 kg	77 kg
Cross-section of Coil Winding	71mmH, 71mmW	49mmH, 78mmW	46mmH, 72mmW
Magnetomotive Force	450 kA	700 kA	700 kA
Number of Coiling Turns	560 Turns	1000 Turns	1167 Turns
Operating Current	804 A	700 A	600 A
Coil Inductance	0.61 H	2.61 H	3.0 H
Electric Current Density	89 A/mm2	183 A/mm2	210 A/mm2
Stored Energy	196 KJ	522 kJ	550 kJ
Maximum Magnetic Field	2.9 Tesla	4.7 Tesla	5.1 Tesla
Type of Winding	pancake type with cooling channel	solenoid type and impregnated	solenoid type and impregnated
SC wire Material	Cu;Nb-Ti	Cu;Nb-Ti	Cu;Nb-Ti
Copper Ratio	7.0	2.0	1.0
Cross-section	2.0mmH, 3.0mmW	1.2mmH, 2.6mmW	1.05mmH, 2.12mmW
Number of Core Wires	253	2257	2689
Dimension h	55 mm	42 mm	30 mm
Cross-section of Coil and Inner Vessel	h	h	h

(2)Compact structure. When a magnetic field generated by the superconducting magnet on a high speed running vehicle moves over normal conducting coils on the ground, electromagnetically induced current flows in the coils, so that an induced magnetic force is generated to levitate the vehicle. The electromagnetic force is greater the smaller the gap between the superconducting coil on the vehicle and the normal conducting coil on the ground. To reduce the distance between the coils with space required for safe running, it is most important to reduce the distance between the center of the supperconducting magnetic coils, acting against the ground coil, and the external surface of the outer vessel.

(3) Durability against vibration and shock. Although vibration and shock to the superconducting magnet are greatly reduced because of non-contact, the magnet always receives electromagnetic force from the ground coil, the size, direction, and field frequency of which varies over a wide range. It is necessary that no failure results from the influence of these disturbances. If a failure should occur, e.g., one superconducting magnet should fail, it is important to safely stop the vehicle without causing a chain reaction leading to a catastrophic accident.

(4) Reliable operation. It is necessary that the superconducting magnet must be cooled by liquid helium to maintain the superconductivity state, and for it to be energized to give the persistent current state. These requirments must be efficiently achieved for reliable operation of the levitated vehicle and system.

SUPERCONDUCTING COIL DEVELOPMENT

Improved characteristics of the superconducting coil

In the L-type superconducting magnet for the ML-500, the superconducting coil had a winding cross section structure which formed a cooling channel around the superconducting wire by inserting spacers between layers. When the superconducting coil is electrically charged, a strong magnetic field generated in the coil causes a very strong electromagnetic force to work the superconducting wire in the central direction of the winding cross section. The superconducting wire needed enough bending strength and thickness to endure this electromagnetic force. This was one of the factors which prevented the superconducting coil from having high current density.

When the superconducting coil for the MLU001 was developed, the wire performance improved as the 'intrinsic stabilized wire' with a low copper ratio and fine multi-filament structure was gradually developed. A new structure has integrated the intrinsic stabilized wire and the coil by impregnating the wire wound structure of the coil with epoxy resin. In this structure, since the electromagnetic force between superconducting wires in the coil winding cross section causes compressive power between cross sections, the superconducting wire is prevented from deforming and the coil characteristics become more stable. Consequently, high current density has been achieved and a superconducting coil with smaller cross section and high magnetomotive force has been developed.

When a superconducting coil of racetrack shape is energized, the electromagnetic force generated forms a stable circle. This electromagnetic force reaches 20 to 30 tons per meter between two parallel sides in a straight section of the racetrack shape. The problems caused by the large magnetomotive force in the newly developed superconducting coils were solved by

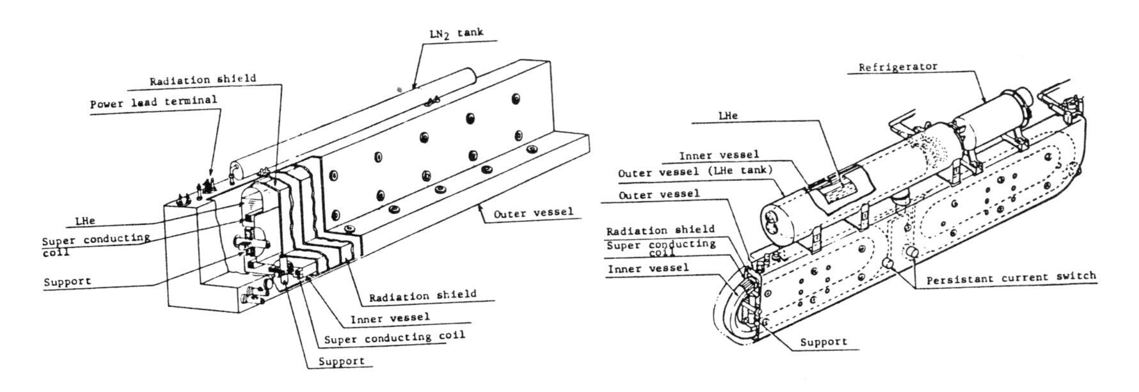

Fig.2 The L-type Magnet for ML-500

Fig.3 The I-type Magnet for MLU001

Table 2 Advanced Features of Superconducting Magnet

Item	ML-500	MLU001	MLU002
Cross-sectional shape	L shaped	I shaped	I shaped
Superconducting coil	Each 2 coils	2 coils	3 coils
Magnetomotive force	(G&P) 450 kA (L) 250 KA	(G&P&L) 700 kA	(G&P&L) 700 kA
Cupper ratio of S.C. wire	5.0 or 6.0	2.0	1.0
Support	Steel column	FRP column	FRP column
Radiation shield	Duble shield	Single	Single
Liq. herium tank	Built-in type	Separated	Separated
Heat leakage to inn. vessel	10 Watts	2.5 Watts	3.0 Wats
Weight of magnet	1,400 kg	650 kg	950 kg
Maglev force / One Magnet	2,500 kg	2,500 kg	4,250 kg
Maglev force / Mag. weight	1.8	3.9	4.5
Heat leak. / Maglev force	4.9 W/ton	1.0 W/ton	0.7 W/ton

developing a new structure including (A) an electromagnetic force transfer structure on the surface of the integral impregnated coil, (B) a tight fitting configuration for dispersing the electromagnetic force and transferring it to the inner vessel, and (C) a thin-tube-shaped inner vessel for supporting the electromagnetic force by integrating it with the superconducting coil.

The development of this integral impregnated coil structure using intrinsic stabilized wire with a low copper ratio has enabled high current density and high magnetomotive force to be generated, and one type of the superconducting coil can now give the three directional forces for vehicle levitation, guidance, and propulsion. Consequently, the weight, size, and stability characteristics were remarkably improved. The superconducting magnet for the MLU002 is being improved by reducing the copper ratio and size using the same development procedures. Table 1 shows the features of these superconducting coils and wires.

Superconducting magnet structure developments

Fig.2 shows the L-type superconducting magnet for the ML-500.
(1) Cross section has an L-shaped configuration that combines the superconducting coils for guiding and propulsion, and the superconducting coils for levitation against the gruond coils on the inverted T-shaped track.
(2) Superconducting coils are fixed in a box-type inner vessel having the liquid helium tank in the upper section.
(3) A supporting column of stainless steel with a multi-tube structure firmly connected the inner vessel to the outer vessel, and the electromagnetic force is transferred to the car body via the outer vessel's mounting eye for the bogie frame.
(4) A two-layer thermal shield plate for intermediate thermal insulation with evaporated helium and liquid nitrogen used for cooling.

The structure of the I-type superconducting magnet for the MLU001 has made further progress, so that lightweight, small-size, and low heat leakage have been achieved. (fig.3)
(1) High magnetomotive force has enabled the superconducting coils to be configured in a single shape, so that the coil configuration was simplified as an I-shaped cross section having a separate helium tank in the upper section, and the entire cross-sectional area was reduced.
(2) Radiation heat transmitted between the inner vessel surface and the thermal shield plate was reduced, and intermediate thermal insulating performed by the one-layer thermal shield plate's liquid nitrogen cooling.
(3) A multi-layered supporting column was developed, using fiber reiforced plastic (FRP) for low thermal conduction and high strength, resulting in a simplified support structure with greatly reduced heat leakage to the inner vessel.
(5) Phosphorus deoxidized copper with comparatively low thermal conductivity in low temperature was used as the current lead conductor material, and a good heat energy transfer balance was achieved.

On-board refrigration system developments

The basic principle of the on-board refrigeration system is 'cooling the superconducting magnet without discharging helium to the outside'. This is for reducing the refrigerating and liquefication facility at the ground site, and for improving vehicle system reliability at low temperatures. This concept is important for the vehicle system efficiency and for the configuration and scale of the ground facility.

Now, the functions were extended to the 'direct refrigeration system', operated as follow: (1) After being cooled through the liquid nitrogen heat exchanger in the on-board refrigerator, gas helium is transferred to the superconducting coil section and pre-cools the superconducting coil. (2)Then the on-board refrigerator starts and cools the superconducting coil at a much lower temperature. (3) Helium is liquefied and stored in the liquid helium tank. (4) After that, the normal operation maintains a constant amount of liquid helium in the tank.

Energizing and de-energizing system

The magnetically levitated vehicle uses a superconducting coil in the persistent current state. In this persistent current mode, the vehicle runs separately from the ground energizing power supply. As a result, the vehicle requires no continuing power supply for levitation, and is levitated and guided by electromagnetically induced forces once it starts running. This is the advantage of the magnetic levitation using superconductivity.

The conventional persistent current switch system is a thermal switch with a structure which winds superconducting wire and heater wire together, and is operated by turning the heater on and off. When the wire is heated, superconductivity changes to normal conductivity and electric resistance is generated. Generating resistance in a superconductive circuit without any electric resistance is equal to opening the circuit. This is the switch off state. When the heater is turned off, the coil is re-cooled and superconductivity is recovered, i.e., the switch on state.

When the superconducting coil is energized at high speed, the energizing voltage is increased, and current is shunted to the persistent current switch, in the switch off state, is also increased. Exothermic transfer increases the evaporation of liquid helium. The most effective way to reduce the consumption of the liquid helium is to increase the electrical resistance when the persistent current switch is turned off. The development was the result of improving the critical current characteristics using the new manufacturing technology for the superconducting wire containing the new copper-nickel alloy as the base material, and improving the structure of the persistent current switch winding.

SUPERCONDUCTING MAGNET CHARACTERISTICS FOR THE MLU002

Summarizing the characteristics of the superconducting magnet for the MLU002 :

(1) Divided configuration of the magnet. In the connecting box, three superconducting coils each having an individual outer vessel are connected to the liquid helium tank with a built-in on-board refrigerator. This configuration is used because: (A) if a vacuum accident should occur in one coil the influence is prevented from spreading over the remaining superconducting coils, and reducing the scale of the failure: (B) when a failure occurs, this configuration is advantageous because only the given superconducting coil section needs to be replaced: (C) a standard small-sized superconducting coil can readily be manufactured.

(2) Light weight and compact coil. The cross section of the superconducting coil was reduced by reducing the copper ratio from 2.0 to 1.0 for the MLU002. A square inner vessel in the shape of the impregnated coil was adoped. As a result, the thickness from the coil center of the superconducting magnet to the outer surface was reduced, magnetic levitation force was increased,the number of superconducting coils reduced and the coils configured in a concentrated arrangement.

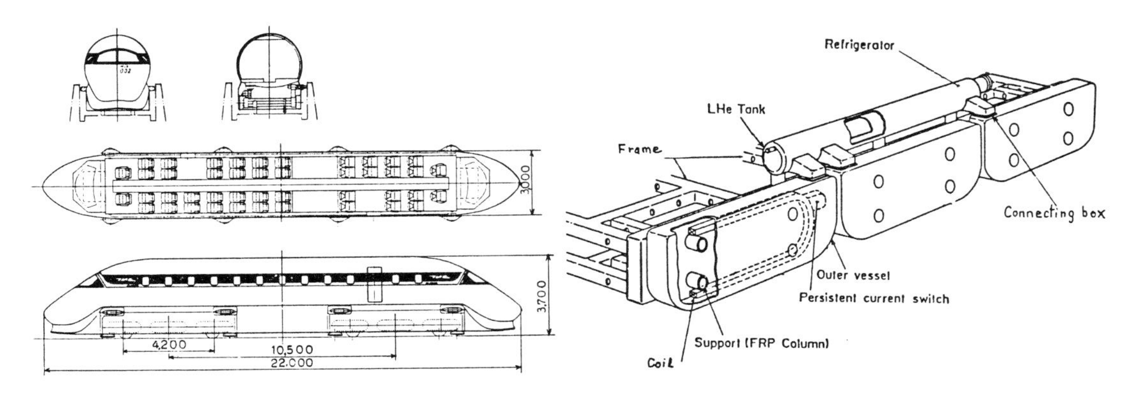

Fig.4 Prototype Maglev Vehicle, MLU002

Fig.5 SuperConducting Magnet for MLU002

(3) High speed energizing system. As a result of supplying high resistance (50 ohms) to the persistent current switch, a 700 kA magnetomotive and 600 A current superconducting coil could be energized at a speed of 10 A/S for one minute. The energizing and de-energizing operation was remarkably improved as well as a switching time reduction.

(4) Direct cooling on-board refrigeration system. The on-board refrigerator was equipped for pre-cooling the superconducting coil and for direct liquefying of helium to extend the system functions. As a result, a new system configuration has become applicable to practical future vehicle operation.

CONCEPTUAL OPERATIONAL VEHICLE AND SUPERCONDUCTING MAGNET

The concept of the operational vehicle is largely determined by the following factors:

(1) A low floor vehicle whose cross sectional area can be minimized. The use of this structure enables the running resistance, including air resistance, to be reduced, and it is expected to lower the construction costs by reducing the tunnel cross sectional area.
(2) A bogie is placed in the joint section between cars. The superconducting magnets are concentrated in this bogie. As a result, the cabin is separated from the superconducting magnet and the magnetic field in the cabin can be reduced.
(3) Two poles/ two lines superconducting coils are placed in each bogie, reducing the number of the on-board superconducting coils to effectively lighten the vehicle. Related ground facilities can also be reduced.
(4) The pole pitch of the superconducting coil is extended from 2.1 m to 2.7 m. Although the superconducting coil on-board becomes longer, the number of ground coils is reduced and the output frequency of the linear synchronous propulsion power source is lowered.

This concept for an operational line vehicle considers prospects for further development of superconducting magnets. To immediately develop and verify the superconducting magnet for an operational line vehicle the following three items should be developed.
(A) Confirming the characteristics of long-span superconducting coils and establishing mass-production technology for superconducting magnets.
(B) Verifying superconducting magnet stability under the increased electromagnetic force from the concentrated magnet arrangement on the bogie.
(C) Confirming the reliability of the on-board refrigerating system in long-term operation.

CONCLUSION

As a result of this 20-years development of the superconducting magnets for magnetic levitation, the superconducting magnet for an operational line vehicle can soon be manufactured.
At the present, expansion of magnetic levitation development is considered based on the discovery of high-temperature superconductivity. Although it is difficult to immediately apply the new high-temperature superconductivity to the large-scale superconducting magnets, the discovery encourages the further development of magnetic levitation.
We will continue developments for manufacturing superconducting magnets for an operational line vehicle and an early realization of magnetic levitation, with due attention to the development of the high-temperature superconductivity.

REFERENCES

(1) Y.Nakayama, M.Yamaji, et al., "Superconducting Magnets for Magnetically Levitated Train", TOSHIBA REVIEW, pp641-647, Vol.36, No.7, 1981.

(2) M.Yamaji, "Maglev Superconducting Coil", The 2nd International Seminar on Superconductive Magnetic Levitation Train, pp49-56, 1982.

(3) Y.Furuta, "Cryostat for Magnetic Levitated Train", The 2nd International Seminar on Superconductive Magnetic Levitation Train, pp57-64, 1982.

(4) Y.Jizo, H.Nakashima, et al., "SUPERCONDUCTING MAGNET FOR MAGLEV TRAIN", International Conference on Maglev Transport '85, pp185-192, 1985.

(5) T.Iwahana, K.Nemoto, "Development of Superconducting Coils for Maglev", The Journal of the Society of Mechanical Engineers, pp19-25, Vol.89, No.817, 1986

(6) H.Tanaka, "Maglev Approaches toward Practical Use", Japanese Railway Engineering, pp2-6, No.102, June 1987.

Prospects for Superconducting Electronics

YASUTSUGU TAKEDA

Central Research Laboratory, Hitachi, Ltd., 1-280, Higashi-koigakubo, Kokubunji, Tokyo, 185 Japan

ABSTRACT

Information systems in the 21st century are spotlighted and superconducting electronics are stressed as one of the key technologies. Present status of superconducting electronics is reviewed with some historical backgrounds. "Quantum flux parametron" and three terminal devices are given as examples of 2nd generation technology in superconducting electronics. A tentative goal and the state of the art of high-Tc superconductors are summarized for superconducting LSIs at Liq. N_2 temperature.

INTRODUCTION

The third industrial revolution as we move into the 21st century, which we are experiencing, differs fundamentally from the previous two in that it is oriented on humanity, culture, and the globe and space. Let us first ask "What kind of information system technologies are essential to the mind of the revolution?". An answer to the question is described in Fig.1. The information system technologies of the 21st century will consist of three technologies which are closely connected each other. They are the broadband ISDN network, as a global infrastructure for world communication, an information base storing knowledge and culture of the mankind, and a "Considerator" instead of a computer [1]. A considerator is defined as a machine executing not only computation but also consideration; such as translation, inference, forecast, planning and decision making.

The relationship among the three technologies can be understood by its similarity to the information system mechanism of a human being. Roughly speaking, the eyes, ears and mouth of a human being perform B-ISDN terminal functions and the brain, to be focused on in the following part of the paper, performs the functions of the "Considerator" and the information base. The scale of the brain's function can be estimated from the number of neurons and synapse per neuron. They are known to be 10^{10} and 10^3, respectively. When a brain is to be simulated by a machine, the total number of synapses of the brain, 10^{13}, can corresponds to the same number of logic gates of the "Considerator" or the 1TB memories of the information base. Figure 2 explains the basic reasons superconducting LSI technology is promising for logic gates or memories of the "Considerator" or the information base. The figure shows the relation between switching time and power dissipation per logic gate of various technologies. It is clear that a total power dissipation of 10^{13} gates reaches 10^{10}W by conventional transistor technology. This is clearly unfeasible since 10^{10}W (10^4MW) is almost equal to the amount of power generated by a typical large power station. To reduce power dissipation by the order of 10^5 - 10^6, superconducting technology seems to be the only choice for the breakthrough required. The minimal switching time of the superconducting devices definitizes the choice.

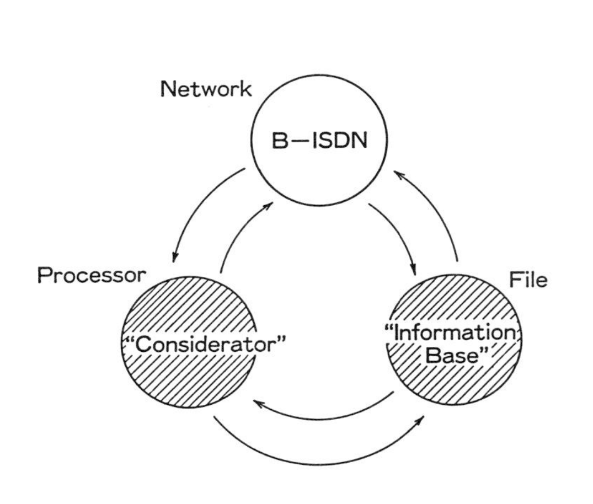

Fig.1 Information System Technology in the 21st Century

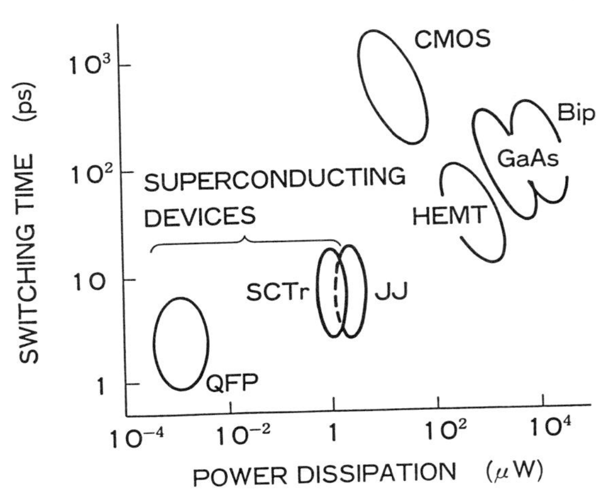

Fig.2 Relationship between Switching Time and Power Dissipation per Logic Gate

FIRST GENERATION TECHNOLOGY OF SUPERCONDUCTING ELECTRONICS

Several superconducting device technologies are undergoing research. The Josephson junction device and its circuit technology have been developed over these 25 years. DC and AC effects of superconducting tunneling current between two superconductors were predicted by Josephson in 1962 [2], and experimentally verified by a group at BTL [3]. The Josephson effect was first applied to a switching element by a group at IBM [4]. The device has merits of high-speed switching in the pico-second range and very small power dissipation at the μW level. Several logic or memory devices were recently announced by Hitachi [5], ETL [6], Fujitsu [7] and NEC [8] as accomplishments of a supercomputer project supported by MITI. In Fig.3, a Josephson logic gate array with 3264 gates, fabricated at CRL, Hitachi, is shown with the switching waveform. A power dissipation of 14mW/chip (=4μW/gate) and a switching time of 10ps/gate were observed. Josephson junction circuits are essentially two terminal devices and the operating power should be alternating. The complexity of the driving system must limit future application to some extent, nevertheless practical application of superconducting LSI technology will begin with the Josephson junction circuits. Therefore, I refer to the JJ circuits as first generation technology in the micro-electronic application of superconducting technologies.

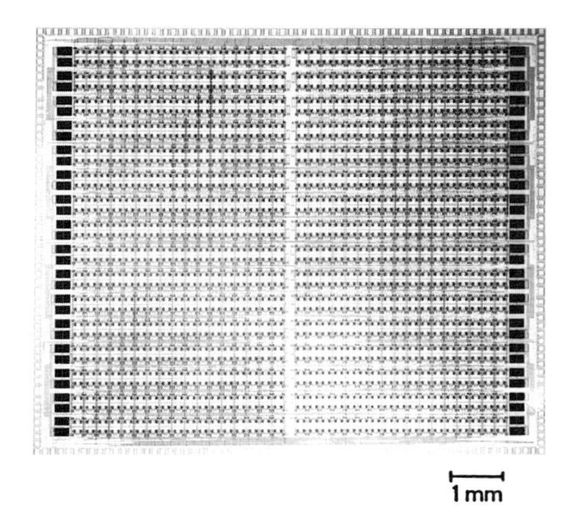

(a) Gate Array Chip

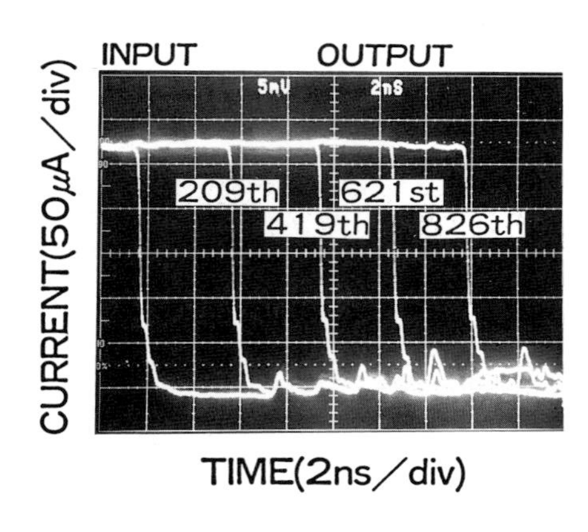

(b) Switching Waveform of an 826 OR-Gate Chain

Fig.3 Josephson Logic Gate Array[5]

SECOND GENERATION TECHNOLOGY OF SUPERCONDUCTING ELECTRONICS

Far advancing technologies beyond the JJ circuits have been proposed and investigated these several years mainly in Japan. The first example is the quantum flux parametron invented by Professor Goto, the University of Tokyo and the Institute of Physical and Chemical Research, in 1984 [9,10]. The quantum flux parametron is a circuit composed of twin JJs, coupling and loading inductors, and an excitation line as shown in Fig.4a. The output signal comes on the loading inductor when the input signal breaks the circulating current of a loop through off-switching action of one of the twin JJs caused by superposition of exciting and signal current on the device. Features of the operation of the circuit are explained by the inventor as follows [9]:

1) The circuit consists of elements of minimal (almost zero) power dissipation with three-dimensional connection possible with coupling inductors. This realizes an ultra fast processor in a compact package.
2) The circuit can amplify signals with latching action. This circuit is suitable for pipe-line and parallel architecture of an ultra fast processor system.

Basic operation of up to 10 GHz was verified through computer simulation. Simple QFP integrated circuits were fabricated with the normal NbN/Pb alloy process. Basic operation with a current gain of more than 3.6 was observed in 1985 [11]. The data is shown in Fig.4b. These works were joint efforts between IPCR and CRL, Hitachi. They have been promoted by the ERATO project supported by JRDC.

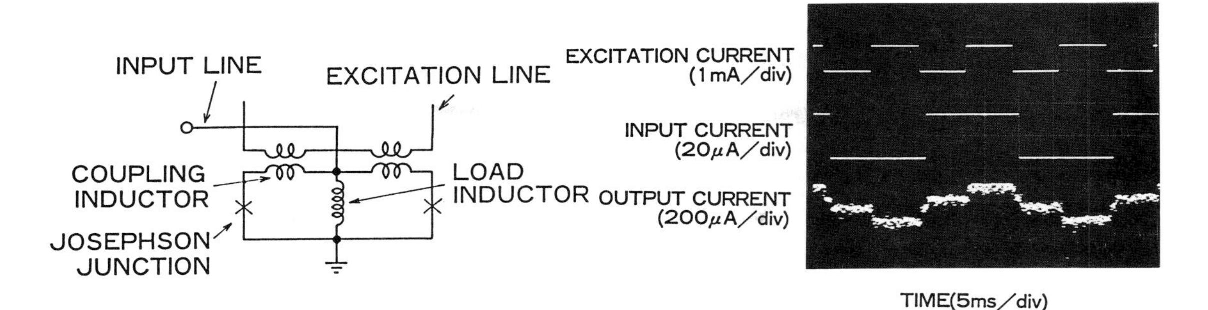

(a) Principle of QFP

(b) Basic Operation of QFP

Fig.4 Quantum Flux Parametron (QFP)[10]

The second example is the superconducting transistor. The principle of the superconducting transistor was proposed by T. D. Clark, Sussex Univ. in 1980 [12]. The successful experiments were reported by a group at CRL, Hitachi, and a group at NTT independently in 1985 [13,14]. Quite recently, the former group succeeded in operating the superconducting transistor fabricated on a silicon crystal with 0.1μm gate length. Fundamental characteristics for drain current vs. drain voltage are shown with the structure in Fig.5 [15]. Operational features of the superconducting transistor can be explained as follows:

1) The circuit has almost the same triode structure as the Si MOS transistor. This makes it possible for system designers to use or extend conventional Si LSI design methodology.

2) The circuit can be operated by direct current so there is no need to install a complex power supply.

The above mentioned technologies are far more advanced than the JJ circuits but they are still in the fundamental research stage. I refer to these as second generation technologies in "superconducting electronics". Their industrial application will start around the beginning of the next century.

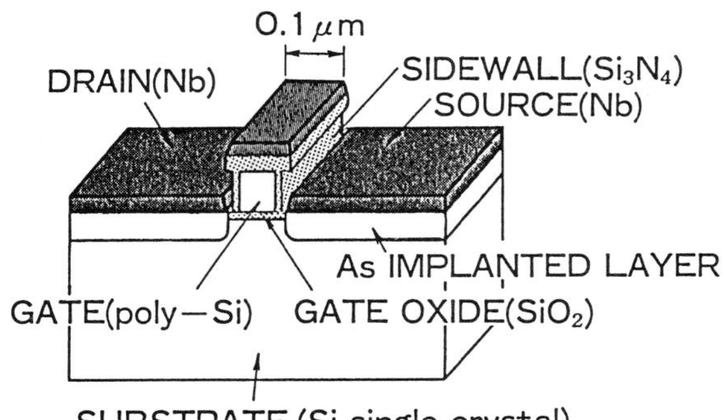

(a) Structure

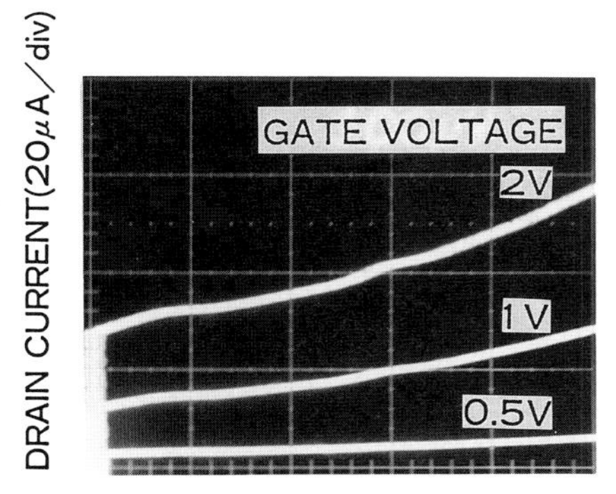

(b) Current-Voltage Characteristics

Fig.5 Superconducting Transistor[15]

HIGH-Tc SUPERCONDUCTORS AND THEIR GOAL FOR SUPERCONDUCTING ELECTRONICS

Without doubt, high-Tc superconductors are required to industrialize superconducting LSIs. There has been lots of research on this topic in the past year and a half. Bismuth oxide compound was reported to show a critical temperature around 110K by National Research Institute for Metals [16]. Thallium oxide compounds showing 125K critical temperature were reported first by the Univ. of Arkansas [17] and then by IBM [18]. Formation of a single crystal film of yttrium oxide compound was reported by Kyoto Univ. to show a critical temperature of 90K and a critical current density of $4X10^6$ A/cm^2 [19]. These data are exciting indeed, however, more intensive research work is necessary to satisfy our tentative goal of a superconducting LSI at Liq. N_2 temperature. Even for this device target, superconductors should have a critical temperature around 150K and a critical current density above 10^7 A/cm^2 with high stability and reliability. Moreover, the thin film crystal should be formed in a temperature range below 400°C to avoid unfavorable reaction between the film and the base material during processing. When the above mentioned devices and the circuit technology of second generation superconducting electronics combine with high-Tc superconductors, they can provide great impact upon various information systems.

Characteristics of such present high-Tc superconductors seem to be more promising in application to simple devices as sensors or connectors. It is widely known that high-Tc superconductors are favorable for almost all sorts of connecting wires among devices and chips of electronic systems. Fundamental experiments on SQUID as a flux sensor expected for medical application were reported by several institutes in 1987 [20,21]. One experiment done at CRL, Hitachi, is shown in Fig.6 [21]. Voltage vs. magnetic-flux characteristics of a DC SQUID using an yttrium oxide compound film are described in the figure. The DC SQUID was operated up to 72K. The SQUID sensitivity characteristics have to be improved by the order of 10^3 for measurement of magnetic waves from the human brain. There are notable reports of fundamental experiments on microwave or light wave sensors fabricated with high-Tc superconductors [22,23]. These are promising for communication system application in the B-ISDN age.

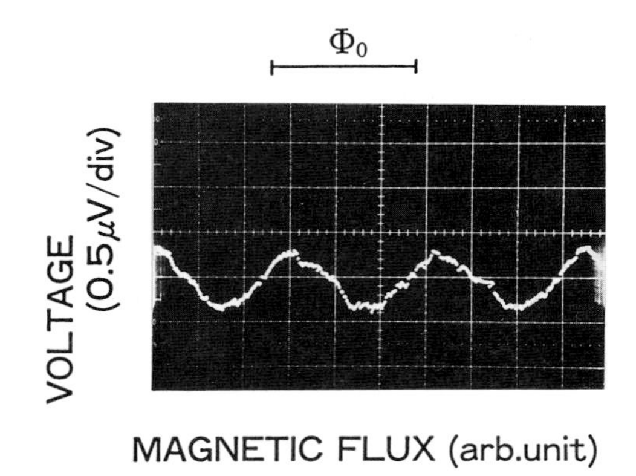

Fig.6 Voltage-Magnetic Flux Characteristics of a DC SQUID[21]

CONCLUSIONS

1) Superconducting electronics is a key technology for future information systems.
2) Josephson junction devices and the ICs are being developed for 1st generation superconducting electronics.
3) Quantum flux parametrons and superconducting transistors are under research for 2nd generation superconducting electronics.
4) Prototypes of high-Tc superconducting devices have been fabricated. Further progress of material and process technology is necessary for IC or LSI applications.
5) When 2nd generation superconducting electronics combines with high-Tc superconductors, there will be a great impact on various information systems.

ACKNOWLEDGEMENT

The present research effort concerning the Josephson logic gate array is part of the National Research and Development Program on the "Scientific Computing System", conducted under a program set by the Agency of Industrial Science and Technology, Ministry of International Trade and Industry, Japan.

REFERENCES

[1] Y. Takeda, Abstracts of Optoelectronics: 1990 and Beyond, Killarney 25 (1988) (Ireland).
[2] B. D. Josephson, Phys. Lett., 1, 251 (1962).
[3] P. W. Anderson and J. M. Rowell, Phys. Rev. Lett., 10, 230 (1963).
[4] J. Matisoo, Proc. IEEE., 55, 172 (1967).
[5] S. Yano et al., IEEE Trans. Magn., MAG-23, 1472 (1987).
[6] H. Nakagawa et al., Extended Abstracts of the 18th Conference on Solid State Devices and Materials, Tokyo, 751 (1986).
[7] S. Kotani et al., ISSCC 88 Digest of Technical Papers, 150 (1988).
[8] Y. Wada et al., ISSCC 88 Digest of Technical Papers, 84 (1988).
[9] E. Goto, 1st Riken Symp. of Josephson Electronics, 48, Mar. (1984) (in Japanese).
[10] K. Loe and E. Goto, IEEE Trans. Magn., MAG-21, 884 (1985).
[11] Y. Harada et al., IEEE Trans. Magn., MAG-23, 8301, (1987).
[12] T. D. Clark et al., J. Appl. Phys., 51, 2736 (1980).
[13] H. Takayanagi et al., Phys. Rev. Lett., 54, 2449 (1985).
[14] T. Nishino et al., IEEE Electron Dev. Lett., EDL-6, 297 (1985).
[15] T. Nishino et al., submitted to IEEE Electron Dev. Lett.
[16] H. Maeda et al., Jpn. J. Appl. Phys., 27, L209 (1988).
[17] Z. Z. Sheng et al., Appl. Phys. Lett., 52, 1738 (1988).
[18] S. S. P. Parkin et al., Phys. Rev. Lett., 60, 2539 (1988).
[19] T. Terashima et al., Jpn. J. Appl. Phys., 27, L91 (1988).
[20] R. H. Koch et al., Appl. Phys. Lett., 51, 200 (1987).
[21] H. Nakane et al., Jpn. J. Appl. Phys., 26, L1925 (1987).
[22] J. S. Tsai et al., Jpn. J. Appl. Phys., 26, L701 (1987).
[23] T. Nishino et al., Jpn. J. Appl. Phys., 26, L1320 (1987).

Josephson Digital and Analog Devices with Niobium Junctions

S. HASUO

Fujitsu Laboratories Ltd., Atsugi, Fujitsu Limited, 10-1, Morinosato-Wakamiya, Atsugi, 243-01 Japan

ABSTRACT

This paper describes recent progress on digital and analog circuits using all niobium (Nb/AlOx/Nb) Josephson junctions. A niobium junction has excellent characteristics with a low leakage current, and is stable with respect to thermal cycling and long term storage. We used the niobium junctions in a variety of digital and analog circuits. This paper firstly describes the logic gate family, named the MVTL (Modified Variable Threshold Logic) gate family. The lowest experimentally obtained MVTL-OR gate delay was only 2.5 ps with a power consumption of 17 μW/gate. The gate family was applied to various circuits, such as 8-bit shift registers, 16-bit ALUs (Arithmetic Logic Unit), and 4-bit microprocessors. We confirmed the high speed operation of less than 10 ps per gate on the average for these circuits.

We also developed a new high-sensitivity magnetic sensor using the SQUID (Superconducting QUantum Interference Device). We call it a single-chip SQUID magnetometer, because the feedback circuit, which is operated at room temperature in a conventional SQUID system, has been integrated on the same chip as the SQUID sensor itself.

INTRODUCTION

Performance of Josephson integrated circuits has been dramatically improved since niobium junctions were introduced[1]-[3]. Before niobium junctions, lead-alloy junctions were mainly used for these integrated circuits. Large scale integrated (LSI) circuits, however, seldom worked well. Almost all the reasons for these difficulties were originated from the unstable characteristics of lead-alloy junctions.

At the end of 1983, niobium junctions became available for use in integrated circuits and lead-alloy junctions were abandoned. After niobium junctions were introduced, various kinds of circuits operating much higher speeds than those using lead-alloy junctions were made. The higher speeds were due to the small scattering of the junction characteristics, and the inherent high performances of Josephson junctions were realized.

At present, we make LSI-level circuits which include several thousands of junctions. Small scale Josephson computer operating at speeds of more than one order of magnitude faster than semiconductor computers has become a distinct possibility. In this paper, we describe our recent progress in Josephson IC technologies developed at our laboratory.

FABRICATION PROCESS

This section describes the fabrication technology of circuits constructed with Nb/AlOx/Nb Josephson junctions.

Recently the characteristics of niobium junctions have been greatly improved. Nb/AlOx/Nb junctions, whose excellent characteristics were demonstrated by Gurvitch et al.[4], and then further improved by Morohashi et al.[5],[6], are expected to be used in high-speed digital circuits and analog applications. The controllability, stability, uniformity, and reproducibility gained through the use of Nb/AlOx/Nb are much better than that from the use of lead-alloys. We thus used the Nb/AlOx/Nb junction to fabricate various high-speed digital and analog circuits.

The materials and gases for the reactive ion etching (RIE) technique are listed in Table 1. Metals were deposited by dc magnetron sputtering in an Ar gas atmosphere at deposition rates of 200 nm/min for Nb, 130 nm/min for Mo, and 7 nm/min for Al. SiO_2 was deposited by rf sputtering in an Ar atmosphere at a deposition rate of 8 nm/min. The Al film surface was oxidized in an Ar + 10 % O_2 gas ambient at room temperature. A typical oxidation time was 60 minutes. We controled the critical current density from 500 to 5000 A/cm^2 by changing the gas pressure from 100 to 40 Pa. The thickness of the Mo resistor was typically 100 nm and its sheet resistance was 1.5 $\Omega/\square$.

MVTL GATE FAMILY

Josephson logic circuits are usually composed of OR and AND gates and are operated in a dual-rail manner since it is difficult to construct INVERTER without a timing signal. An AND gate is usually driven by the output signals of OR gates because an AND gate by itself cannot isolate the output signal from the input signal. The MVTL gate family we designed consists of OR, AND, and 2/3 MAJORITY gates. The TIMED INVERTER is sometimes combined with an OR gate and another single junction.

The MVTL OR gate has an asymmetric interferometer and a magnetically coupled control line. The control current is injected into the interferometer after magnetic coupling. Figure 1 shows the equivalent circuit of the MVTL OR gate. Using the single junction J_3 and a resistor R_i, the output current is isolated from the injected control current. We fabricated 4 types of the MVTL OR gates. The fastest gate speed obtained was 2.5 ps for the gate with a minimum junction diameter of 1.5 µm. The power consumption was 17 µW/gate[7]. This gate is the fastest of all logic (including semiconductor) gates. The relation between the fastest gate delay and the minimum junction diameter is shown in Fig. 2. It must be noted that Josephson logic gates can attain the gate delay of less than 10 ps/gate without using submicron process technology. Figure 2 suggests that sub-ps/gate delay may be achieved if we fabricate 0.6 µm minimum diameter junctions.

The AND gate is constructed with a single junction, and is always driven by OR gate output signals. This is because the AND gate cannot isolate the output signal from the input signal. Unit cells are combined using two OR gates and an AND gate. The gate delay of the unit cell was 16 ps for the minimum junction diameter of 4 µm and 11.5 ps for a diameter of 2.5 µm. We also desinged and tested a 2/3 MAJORITY gate and a TIMED INVERTER (TI).

DIGITAL AND ANALOG CIRCUITS

Using the MVTL gate family, we fabricated various logic circuits to test the high-speed operation of these gates. They are a 16-bit ALU[1], an 8-bit shift register[2], and a 4-bit microprocessor[3]. The performance of these circuits is described here. We also fabricated a single-chip SQUID magnetometer[8]. The basic idea of the SQUID is described here.

Table 1 Materials for Josephson integrated circuits

Layers	Materials	Gases for RIE
Josephson junction	Nb/AlOx/Nb	$CF_4 + O_2$
Wiring	Nb	$CF_4 + O_2$
Insulation	SiO_2	$CHF_3 + O_2$
Resistor	Mo	$CF_4 + O_2$

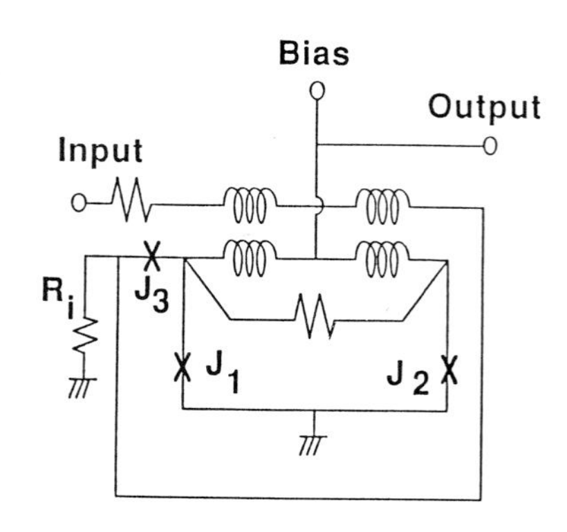

Fig.1 An equivalent circuit on an MVTL OR gate

A. 16-bit ALU

We fabricated a 16-bit ALU, which performed eight arithmetic and four logic functions. Figure 3 shows the block diagram of our 1-bit ALU. This device integrates 18 unit cells. The unit cell used in this circuit is the same as that described in the previous section. A multiple-bit ALU can be achieved by serially connecting multiple blocks of the 1-bit ALU. For circuit layout simplicity, no special high-speed operation algorithm such as carry-look-ahead has been used in the ALU. Therefore, the delay time on the critical path of the ALU is the sum of the delays in carry signal propagation in the adder mode. The 16-bit ALU chip is shown in Fig.4. The dimensions of the circuit size are 0.85 mm x 8.2 mm. There are 900 gates in the ALU, including the 36 gates needed to measure the critical path delay.

The critical path delay was measured to be 0.86 ns. The signal path during this operation covered 83 stages of the MVTL OR and AND gates. Since the propagation delay in the interconnecting lines on the signal path was calculated to be about 95 ps, the average gate delay was estimated to be 9.2 ps/gate. The total power consumption of the chip was 10.1 mW or an average of 11.3 µW/gate.

We also fabricated a 16-bit multiplier critical path model[9]. The model includes 828 MVTL gates, which are extracted along the critical path from the multiplier in order to estimate a multiplication time. The observed multiplication time was 1.1 ns.

B. 8-bit shift register

We designed an 8-bit shift register using MVTL gates. It is capable of SHIFT, LOAD, HOLD, and CLEAR functions. The circuit for a 1-bit shift register is shown in Fig. 5. Here S, L, and H represent the control signals for SHIFT, LOAD, and HOLD. DS and DL represent the data for SHIFT and LOAD.

Five unit cells, which have the same size as those in the ALU and the mulitiplier model, and one TI in each 1-bit shift register were used. They were supplied with three-phase power ϕ_1, ϕ_2, and ϕ_3. Their waveforms are sinusoidal with dc offsets and the phases are 120°

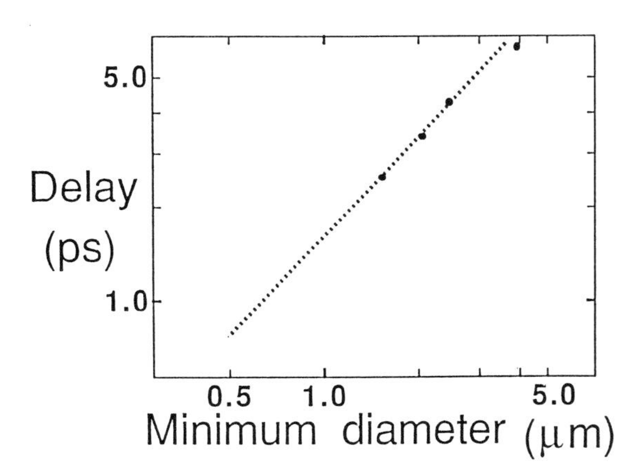

Fig.2 Relation between the fastest gate delay and the minimum junction diameter for the MVTL OR gate

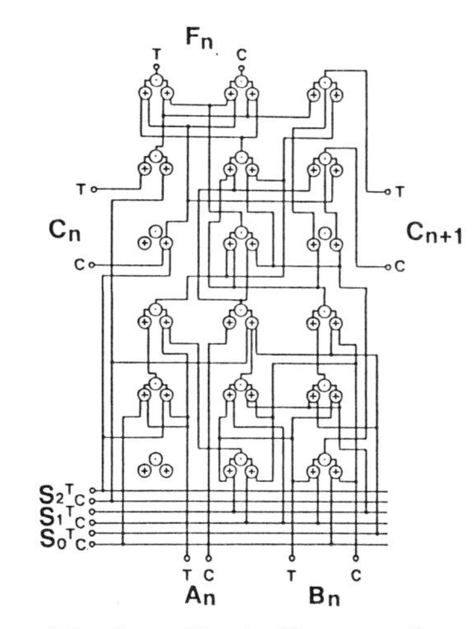

Fig.3 Block diagram of a 1-bit ALU

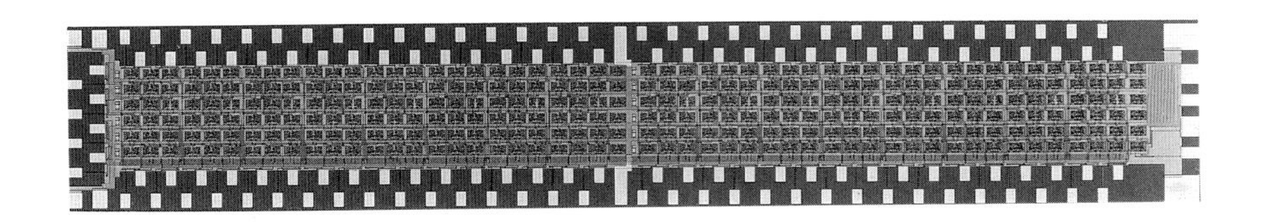

Fig.4 Photomicrograph of 16-bit ALU

apart as shown in Fig. 5. These sinusoidal waveforms can be replaced by trapezoidal ones. The operating margin is slightly larger for the trapezoidal waveforms. However, it is easier to supply sinusoidal waves at high frequencies.

The fabricated chip contains 112 gates. The circuit dimensions are 1.1 mm x 2.1 mm. The critical current density was 1700 A/cm^2, while the design value was 2100 A/cm^2. We confirmed that the 8-bit shift register operated correctly for all stages of all the control signals at an 80 μs clock. High-speed operation was tested. The SHIFT function was correctly operated up to a 2.3 GHz clock. The total power consumption was 1.8 mW.

We also developed a pseudorandom bit sequence generator[10]. The circuit is constructed with 9 stages of the one-bit shift register described above, and its output signal is fed back to the 5th stage through an exclusive-OR gate. Thus it can generate a pseudorandom number with a 511-bit sequence. We confirmed its correct operation up to 2.2 GHz.

C. 4-bit microprocessor

We have fabiricated a 4-bit microprocessor[3]. This is the first instance to our knowledge, of application of Josephson devices to a microprocessor, so we wanted to verify the feasibility of the chip in comparison with a typical microprocessor constructed with semiconductor devices. We selected chip functions that were similar to those of the Am 2901 microprocessor made by Advanced Micro Devices Inc. This microprocessor has come to be regarded as the standard four-bit microprocessor slice.

Figure 6 is a block diagram of our microprocessor. It has a dual memory set which is used as a 16-word by 4-bit two-port RAM with a RAM shifter, an eight-function ALU, a Q register with a Q shifter, and several controllers. This circuit is driven by three-phase power, ϕ_1, ϕ_2, and ϕ_3. Dual-rail logic was adopted in the ALU and the controllers of the microprocessor, and complement signals are made from the input signals by TIs powered by ϕ_1. Decoding operations are run in gates powered by ϕ_1, reading memory data by ϕ_2, and modifying and writing data by ϕ_3.

Both the minimum junction diameter and line width are 2.5 μm. The interconnecting lines are 4 μm wide. Figure 7 is a photograph of the chip. The basic gate is MVTL, as mentioned above, and the total number of gates is 1841. All functions and source combinations were confirmed at a clock frequency up to 100 MHz, the limit of the maximum clock of the word pattern generator. The operation along the critical path of the chip was tested using the high-speed pulse generator, and confirmed to operate correctly up to a clock frequency of 770 MHz. The gate power dissipation was 3.6 μW/gate, and the total power of the chip was 5 mW.

We verified that the Josephson microprocessor operated with a clock that was one order of magnitude faster, and consumed three orders of magnitude less power than semiconductor microprocessor. Performances of the AM2901 type microprocessors for three different materials are compared in Table 2.

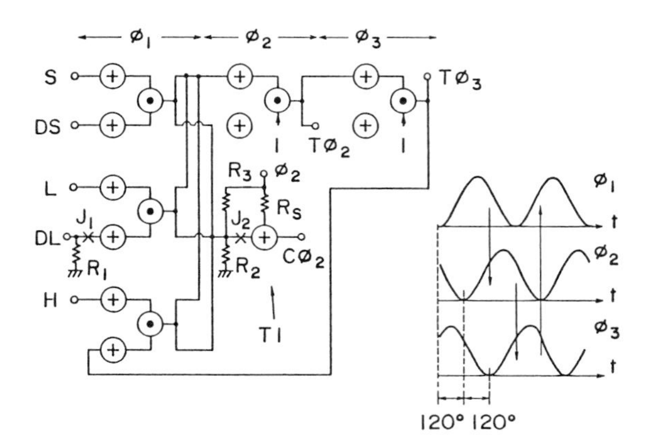

Fig.5 Circuit diagram of 1-bit sift register

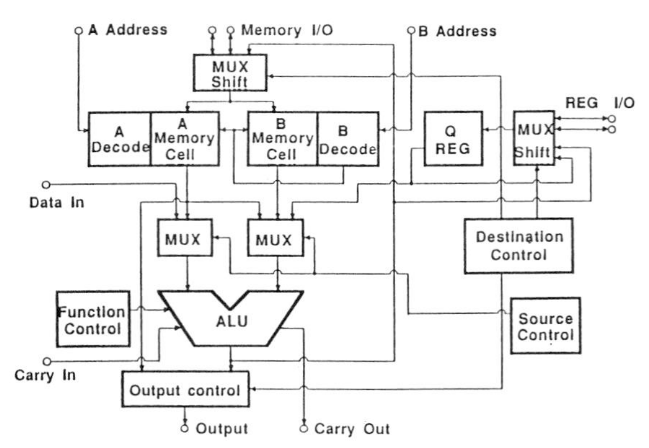

Fig.6 Block diagram of a 4-bit microprocessor

D. A single-chip SQUID magnetometer

The SQUID magnetometer is a very high-sensitive magnetic sensor, that is expected to use as an image sensor for medical and other applications. We fabricated a single-chip SQUID magnetometer[8], which included entire circuits such as the pickup coil, SQUID sensor, and feedback circuit. Conventional rf and dc SQUID magnetometers use analog feedback circuits, such as lock-in-amplifiers that make integration difficult. We introduced a digital feedback circuit and a superconducting storage loop. This made it possible to integrate the SQUID magnetometer into a single chip.

The single-chip SQUID magnetometer requires only an AC bias and produces a digital output, with no peripherals, at room temperature. The output pulse can be processed by a digital processor or applied to a display instrument through a counter to directly monitor input magnetic field waveforms.

Figure 8 diagrams the circuit. The pickup coil transmits the magnetic flux to be measured to the SQUID sensor through coupling coils. The digital feedback circuit is fabricated using a superconducting storage loop and an interferometer as a write gate. The write gate receives a pulse sequence and writes a positive or negative flux quantum to the storage loop when a pulse arrives.

We fabricated the single-chip SQUID magnetometer and tested it. The magnetic flux coupled to the SQUID sensor was measured as low as 7×10^{-5} $\Phi_0/\sqrt{Hz}$, Φ_0 being the flux quantum (2.07×10^{-15} Wb). This corresponds to a magnetic field of 4.7×10^{-12} T/$\sqrt{Hz}$, and the magnetic field gradient of 4.5×10^{-9}T/m$\sqrt{Hz}$ at the pickup coil. In our experiment, the sensitivity was believed to have been limited by environmental noise, not by device noise. Therefore we believe that the sensitivity can be further improved. In any case, this device is more sensitive than magnetocardiograms, which can only measure fields on the order of 10^{-11} T.

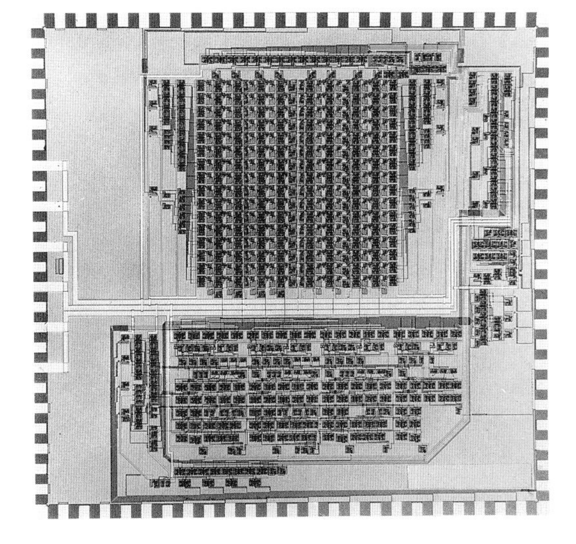

Fig.7 Photomicrograph of a 4-bit microprocessor

Table 2 Performance of 4-bit microprocessor

Device	Si 1)	GaAs 2)	Josephson
Maximum clock (MHz)	30	72	770
Power (W)	1.4	2.2	0.005

1) AMD, 1985 data book
2) Vitesse, 1987 GaAs IC Symposium

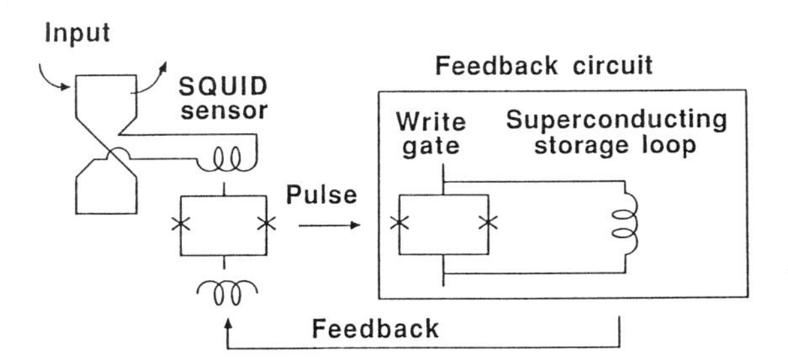

Fig.8 A circuit of the single-chip SQUID

CONCLUSION

We described our recent progress on digital and analog circuits with Josephson IC technologies. Progress has been rapid since we changed the junction material from the lead-alloy to niobium. We can operate LSI circuits with a few thousands gates. The operating speed is more than one order of magnitude faster and the power consumption is more than two orders of magnitude smaller as compared with semiconductor circuits. Although not described here, but Josephson memory circuit is also feasible up to 4 K bit with half a nanosecond access time has also been developed[11]. As a result of our research, we feel that Josephson LSI with tens of thousands of junctions on a chip are feasible. We also developed a single-chip SQUID magnetometer. This will be widely used for medical and other applications. Various kinds of digital and analog superconducting circuits will exploit new fields of superconducting electronics.

ACKNOWLEDGEMENT

The progress on Josephson digital analog devices described here is based on the work done by my colleagues, T. Imamura, N. Fujimaki, S. Morohashi, H. Tamura, H. Suzuki, H. Hoko, S. Kotani, A. Yoshida, and S. Ohara. I want to express my sincere thanks for their efforts.

The present research effort is part of the National Research and Development Program on "Scientific Computing System", conducted under a program set by the Agency of Industrial Science and Technology, Ministry of International Trade and Industry.

References

[1] S. Kotani, N. Fujimaki, T. Imamura, and S. Hasuo, S., IEEE J. Solid State Circuits, SC-23, 2, pp.591-596 (April 1988).

[2] N. Fujimaki, S. Kotani, T. Imamura, and S. Hasuo, IEEE J. Solid State Circuits, SC-22, 5, pp.886-891 (Oct. 1987).

[3] S. Kotani, N. Fujimaki, T. Imamura, and S. Hasuo, S., Digest of Tech. Papers of 1988 International Solid-Circuit Conf. (ISSCC), San Francisco, 1988, pp.150-151 (Feb. 1988).

[4] M. Gurvitch, M. A. Washington, and H. A. Huggens, Appl. Phys. Lett., 42, pp.472-474 (Mar. 1983).

[5] S. Morohashi, F. Shinoki, A. Shoji, A. Aoyagi, and H. Hayakawa, Appl. Phys. Lett., 46, pp.1179-1181 (June 1985).

[6] S. Morohashi, S. Hasuo, and T. Yamaoka, Appl. Phys. Lett., 48, pp.254-256 (Jan. 1986).

[7] S. Kotani, T. Imamura, and S. Hasuo, Tech. Digest of International Electron Devices Meeting (Washington, D.C., 1987), pp.865-866 (Dec. 1987).

[8] N. Fujimaki, H. Tamura, T. Imamura, and S. Hasuo, S., Digest of Tech. Papers of 1988 International Solid-State Conf. (ISSCC), San Francisco, 1988, pp.40-41 (Feb. 1988).

[9] S. Kotani, N. Fujimaki, S. Morohashi, S. Ohara, and Hasuo, IEEE J. Solid-State Circuits, SC-22, 1, pp.98-103 (Feb. 1987).

[10] N. Fujimaki, T. Imamura, and S. Hasuo, Proc. of the Symposium on Low Temperature Electronics and High Temperature Superconductors, Honolulu, 1987, pp.375-380 (1988).

[11] H. Suzuki, N. Fujimaki, H. Tamura, T. Imamura, and S. Hasuo, IEEE Trans. Magnetics, to be published.

A Superconducting IC Technology Based on Refractory Josephson Tunnel Junctions

SUSUMU TAKADA

Electrotechnical Laboratory, 1-1-4, Umezono, Tsukuba, 305 Japan

ABSTRACT

A superconducting integrated circuit(IC) technology based on refractory Josephson tunnel junctions, which has been developed at ETL over recent years, is presented. Refractory superconducting materials of sputtered Nb and NbN films are employed to integrate Josephson tunnel junctions. In order to make superconducting ICs, a reactive ion etching process including self-alignment insulation has been developed. Logic circuit design, logic simulation and automatic layout of logic cells are performed by the computer aided design(CAD) system for Josephson circuits. A Josephson 1k-bit random access memory(RAM) chip has been developed successfully by introducing a new approach of memory cell including periphery circuits. The refractory tunnel junction IC technology has been used to make progress in other application fields such as fluxon, sampler and so on.

INTRODUCTION

Superconducting electron devices are very attractive for the applications of high speed, low noise, high sensitive, and high accuracy devices, particularly using the integration of Josephson tunnel junctions. Many efforts have been made by an IBM group to realize Josephson computer circuits using a lead-alloy technology [1]. In the lead-alloy technology, a photoresist stencil lift-off method was used to make superconducting electrodes. An rf-oxygen discharge process was invented for tunnel barrier formation. The integration of Josephson tunnel junctions on a chip using the lead-alloy technology has stimulated one to develop Josephson ICs.

Recently, all refractory Josephson tunnel junctions have been developed by employing sputtered Nb and NbN films as superconducting electrodes instead of the lead-alloy ones. These junctions exhibit high reliability and stability for thermal cycling and for long storage due to their mechanical hardness. In these tunnel junctions, very thin insulating films[2,3,4] have been prepared on the superconducting electrode as an artificial barrier. In order to make a uniformity of the Josephson critical current density in the whole wafer, a tri-layer sandwich method has been proposed[2]. The tri-layer sandwich, i.e., counter electrode/tunnel barrier/base electrode, has been first prepared on a substrate. A reactive ion etching(RIE) process with photolithography is used to make small tunnel junctions from the tri-layer. To isolate the small junctions each other self-alignment process such as SNAP[2] and SNEP[5] have been developed. These tunnel junctions fabricated can stand on a process temperature of more than 150 °C. Then, the investigation of the superconducting ICs have been accelerated. At present, many Josephson ICs have been developed using the refractory Josephson tunnel junctions [6,7,8,9,10].

In a superconducting closed loop including Josephson junctions a very small magnetic flux so called single flux quantum(SFQ;$\Phi_o=2.06\times10^{-15}$Wb) can be stored without any power supply. This SFQ has been applied to Josephson memory cells which has extremely low power dissipation and high speed access time. Based on this principle, 1k-bit Josephson memory chips[11,12] have been demonstrated using the refractory tunnel junctions.

In this paper, a superconducting IC technology which has been developed based on the refractory tunnel junctions over recent years at ETL will be described.

REFRACTORY TUNNEL JUNCTION PROCESS

Refractory tunnel junctions have been integrated on a chip using the tri-layer junction sandwich. Figure 1 shows the refractory tunnel junction process of SNEP[5]. First, (a) the tri-layer of the tunnel junction, i. e., counter electrode/barrier/base electrode, is made on a 2 inch whole wafer substrate. The tri-layer is formed successively in the sputtering chamber without any vacuum breaking. In the second step, (b) the tri-layer is patterned by photolithography to make the base electrode area. The tri-layer except the part which forms base electrodes is removed by an reactive ion etching process(RIE) with CF4 gas. In the next step, (c) counter electrodes are formed to make a well defined small junctions area by RIE process. After this process, the insulation layer(SiO) is formed by evaporation on that surface before removing the photoresist. Then, the self-alignment of the insulation layer is achieved completely. Finally, (d) wiring layers are formed by the photolithography and the RIE process.

In Table I, characteristics of refractory tunnel junctions developed for digital applications at ETL are shown. Nb/Aloxide/Nb junction has a gap voltage of 2.8mV. A tunneling parameter of Vm(the product of Josephson critical current and quasi-particle resistance below the gap voltage) is about 100mV. The specific capacitance of the Nb/Aloxide/Nb junction is about 6μF/cm^2. The barrier of Al oxide is formed by thermal oxidation of the surface of the sputtered Al film on the Nb base electrode. The barrier thickness is controlled precisely by oxygen pressure and oxidation time. The thermal oxidation for barrier formation gives a high uniformity of Josephson critical currents in the whole wafer.

NbN/MgO/NbN[4] tunnel junction is very attractive for high speed digital applications due to a large gap voltage of 5mV. Josephson critical current density of the junctions is controlled by the thickness of MgO sputtered film carefully.
It is important to reduce spreads of Josephson critical currents integrated on the chip. Figure 2 shows I-V characteristics of the Nb/Aloxide/Nb tunnel junctions. The I-V characteristics were obtained with 1000 junctions connected in series. The junction size is 10μm x 10μm. The spread is

Table I Characteristics of the refractory junctions.

Tunnel junction	Nb/AlOx/Nb	NbN/MgO/NbN
Gap voltage	2.8mV	5.0mV
Vm(at 2mV)	~ 100mV	~ 100mV
Capacitance	6μF/cm^2	10μF/cm^2

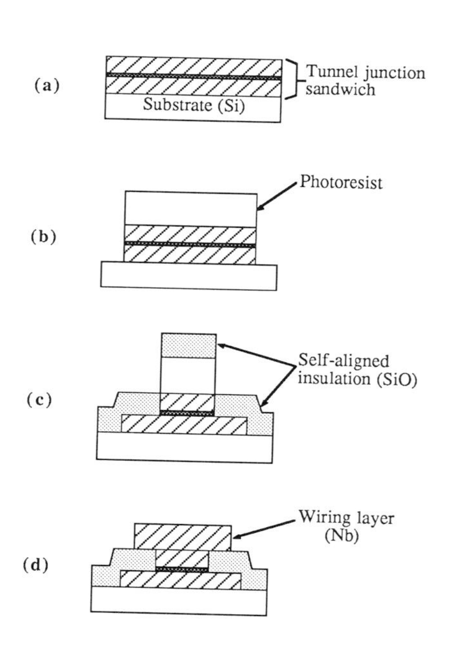

Fig.1. Refractory Josephson tunnel junction process.

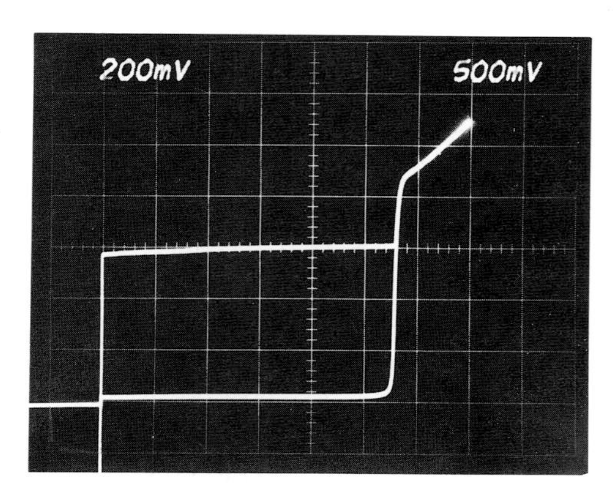

Fig.2. I-V characteristics of 1000 Nb/Aloxide/Nb tunnel junctions connected in series.

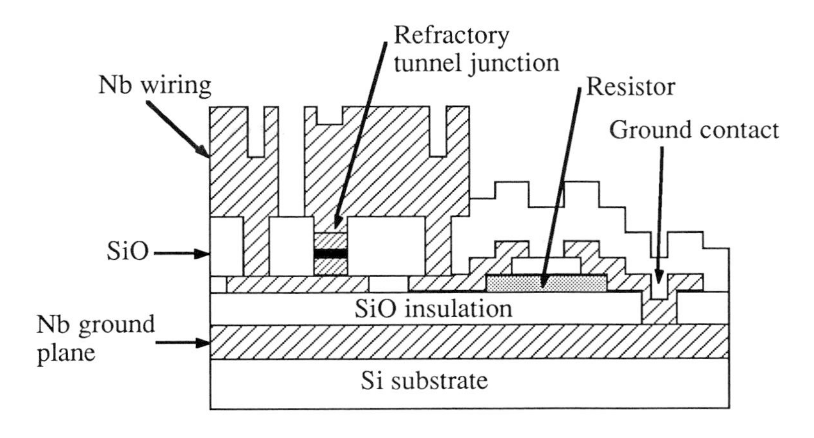

Fig.3. Cross section view of Josephson IC circuits.

less than 3% in the Josephson critical currents in the figure. In order to obtain a high Vm of the junctions, Nb-underlayer method[13] is introduced beneath the base electrodes.
A cross-section view of Josephson logic circuits is shown in Fig.3. On a 2 inch silicon wafer substrate, an Nb ground plane film is first layered using sputtering deposition. In order to cover the ground plane and to isolate the circuits constructed on that plane, SiO film is prepared. The thickness of this insulation layer is very important to determine the electronic parameters of the inductance elements and the strip lines. Ground contacts are obtained by making holes in the SiO insulation layer. Instead of evaporated SiO film, chemical vapor deposited SiO_2[6] is used to reduce defects which causes short in the circuits. Resistors are prepared with sputtered Mo film[14] or evaporated Pd film[15] using the photolithography. Then, refractory tunnel junctions are integrated by the fabrication process shown in Fig.1.

In the RIE process, sharp steps are formed at the film edges. In order to obtain a good coverage at the edges by other layers a planarization method[6] is employed. In order to finish a complete etching, a thin MgO film[16] is inserted on the layers which should be protected from RIE because of a small etching rate of MgO.

CAD SYSTEM FOR JOSEPHSON IC

In order to make Josephson LSI circuits, the CAD system which performs logic circuit design, logic simulation and automatic layout of logic cells has been developed. Figure 4 shows a block diagram of the Josephson CAD system. First, function cells are designed using OR-, AND-, INVERT-, LATCH-gates . By taking account of fabrication process, the cell patterns are designed in the pattern CAD. In the logic CAD, the logic cell models are designed based on the function cells which have specified parameters of fan-in, fan-out, logic delay, and powering. Logic simulations are performed for full logic circuits using the logic cell models. Then, automatic cell layout is performed by the logic CAD. Position of each cell is determined to minimize wiring length among cells. Layout informations of the cells are finally combined with the cell patterns in the pattern CAD. In this way a complete set of photomask patterns is obtained to make LSI chips.

A photograph of the chip[17] which has been fabricated using the Josephson CAD system is shown in Fig.5(a). Figure 5(b) shows a part of the chip enlarged.

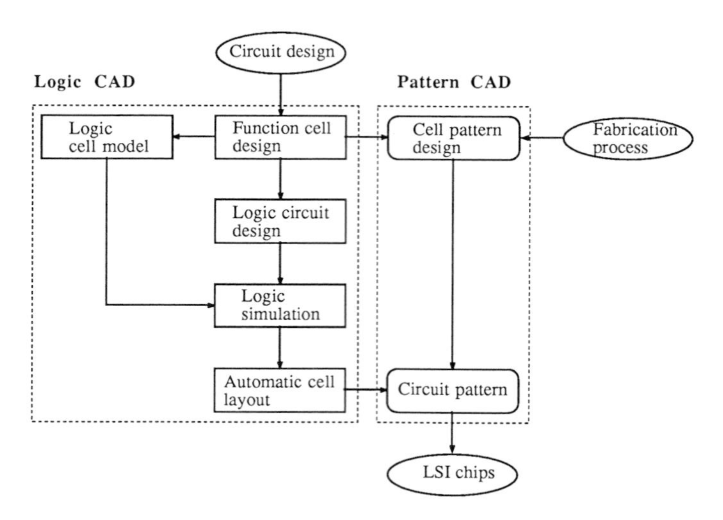

Fig.4. Block diagram of CAD system for Josephson ICs design.

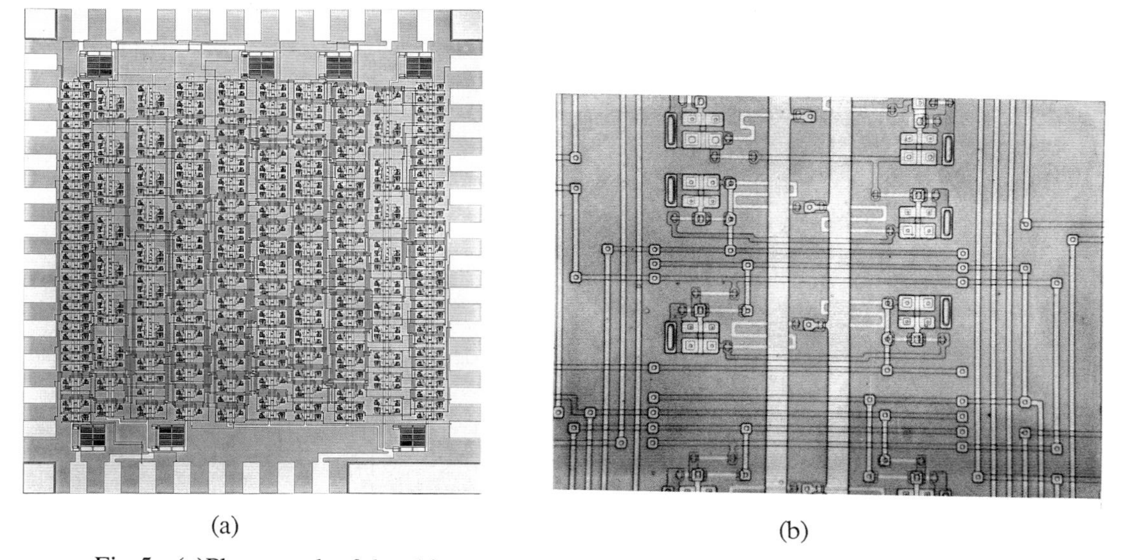

(a) (b)

Fig.5. (a)Photograph of the chip made by refractory junctions using the CAD system, and (b)a part of the circuits in the chip.

JOSEPHSON RAM TECHNOLOGY

There are many efforts to develop Josephson random access memory(RAM) chips. A new type Josephson RAM cell which has been proposed and investigated at ETL is shown in Fig.6. The memory cell has an asymmetric dc SQUID structure and has no sense gate in it as shown in Fig.6(a). Principle of the operation is based on a variable threshold curve of the cell due to the existence of a magnetic single flux quantum(SFQ) in the cell loop. A driving circuit for the variable threshold memory is shown in Fig.6(b). SET gates and RESET gates are used to write and read-out nondestructively with no conventional return lines. Sense gates are placed on each column. Figure 7 shows a block diagram of the 1k-bit Josephson RAM chip[12]. In the 1k-bit memory plane, 4-bit memory cells for 1 word are arranged to 4 columns x 64 lines. Address data of A0-A5 for X lines and A6-A7 for Y columns are used to select the word data. WE enables writing operations. CS makes the chip active for memory operations. Decoder circuits consist of OR-gates and INVERT-gates instead of conventional OR-AND gates. Figure 8 shows the Josephson RAM chip fabricated with 3μm-Nb/Aloxide/Nb tunnel junction process. 1024 memory cells and 1028 logic gates are integrated on the 3.3 x $3.3mm^2$ chip. Stable full operations have been confirmed completely on the 1k-bit memory chip. The chip has a power dissipation of 1.9mW.

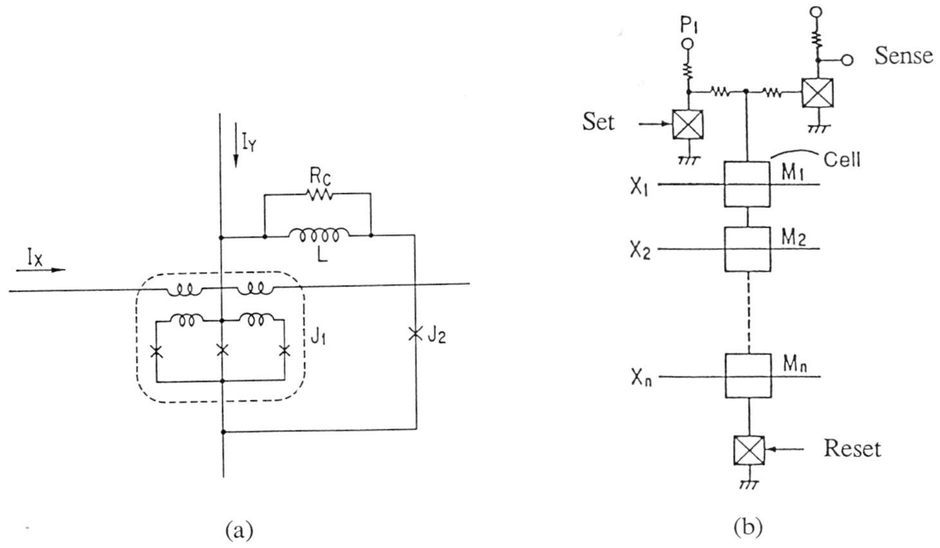

Fig.6. (a)Josephson RAM cell of variable threshold memory and (b)driving circuit of the cells.

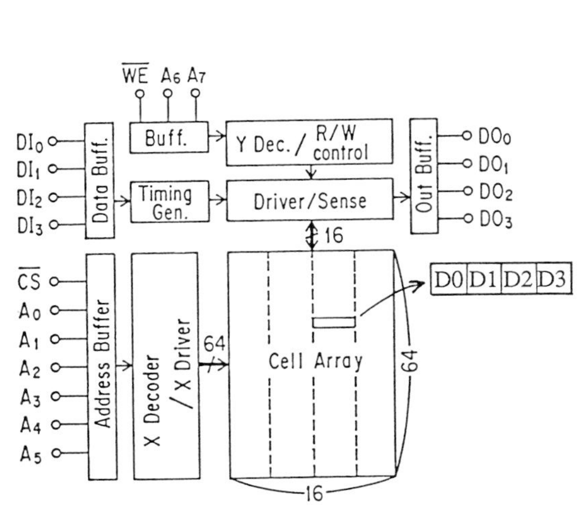

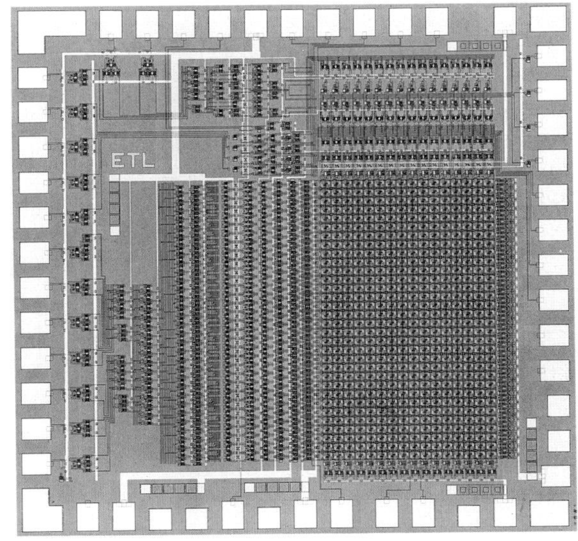

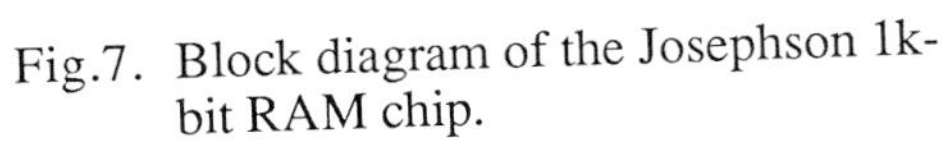

Fig.7. Block diagram of the Josephson 1k-bit RAM chip.

Fig.8. Photograph of the RAM chip.

OTHER APPLICATIONS

The refractory tunnel junctions described above show high reliability and stability. Then, the refractory junction technology has been used to make progress in other superconducting devices such as planar dc SQUIDs[18], 1V Josephson precise voltage generators[19], Josephson samplers[20], fluxon devices[21,22], SIS X-ray detectors[23] and so on. The fluxon propagates in the long tunnel junction transmission line and reflected at the end of the line. Figure 9(a) is an experimental chip of the fluxon. The Josephson sampler detects the waveform of the fluxon in real time. In Fig.9(b), the reflected fluxons are shown as a function of bias current Ie.

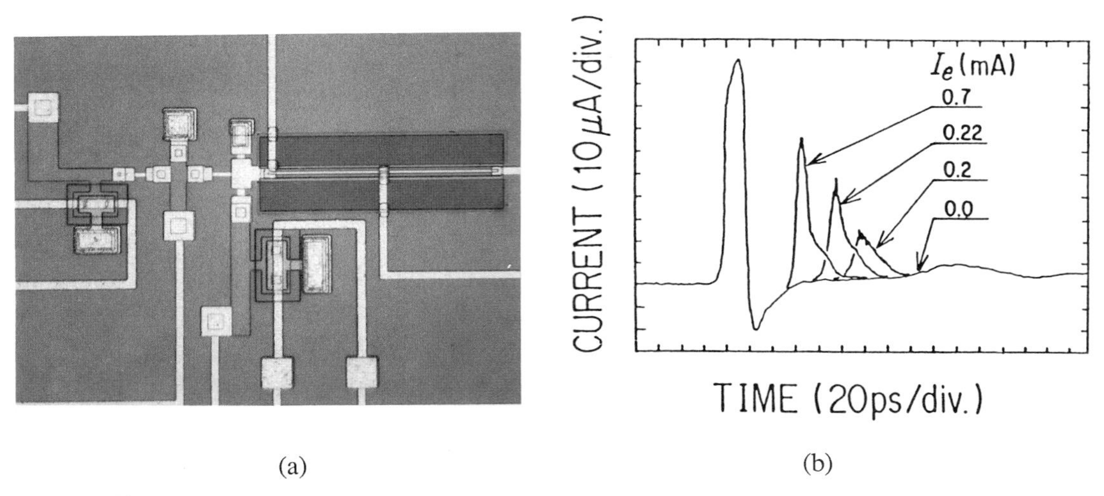

(a) (b)

Fig.9. (a)Experimental chip of the fluxon, and (b)reflected fluxon observed by Josephson sampler.

SUMMARY

A superconducting IC technology based on the refractory tunnel junction process at ETL is described. The IC technology for superconducting digital applications has been achieved to investigate real LSI chips. The present technology is now promising one to develop other new superconductive device applications.

ACKNOWLEDGMENT

The author would like to express his thanks to the members of special section on Josephson computer technology for their valuable discussions and to T.Tsurushima for his encouragement at ETL. The author also thanks to H.Nakagawa for his help during the preparation of this paper.

REFFERENCES

[1]IBM J. Res. Develop., 24, (1980)(Special issue on Josephson Computer Technology). [2]H.Kroger et al., Appl. Phys. Lett., 39, 280(1981). [3]M.Gurvitch et al., Appl. Phys. Lett. 42, 472(1983). [4]A.Shoji et al., Appl. Phys. Lett. 46, 1098(1985). [5]A.Shoji et al., Appl. Phys. Lett. 41, 1097(1982). [6]S.Kosaka et al., IEEE Trans. Magn. MAG-21, 102(1985).[7]H.Nakagawa et al., IEEE Trans. C.S. CAS-34, 1123(1987). [8]N.Fujimaki et al., IEEE J.Solid-State circuits SC-23, 852(1988). [9]Y.Hatano et al., Ext. Abst. of 1987 ISEC 239(1987). [10]S.Kotani et al., IEICE Tech. Rep. Vol.88, 73(1988). [11]Y.Wada et al., 1988 ISSCC Digest Tech.Papers, 84(1988). [12]I.Kurosawa et al., Ext. Abst. 20th Conf. on SSDM 605(1988). [13]H.Nakagawa et al., Jpn. J. Appl. Phys. 25, L70(1985). [14]D.Jillie et al., IEEE, J. Solid State Circuits, SC-18, 173(1983). [15]K.Kuroda et al., Electronics Lett. Vol.23, 163(1987). [16]H.Nakagawa et al., IEEE Trans. Magn. IEEE Trans. Magn., MAG-23, 739(1987). [17]S.Kosaka et al., presented in 1988 Applied Superconductivity Conference. [18]M.Koyanagi et al., Ext. Abst. of 1987 ISEC, 33(1987). [19]Y.Sakamoto et al., Ext. Abst. of 1987 ISEC, 84(1987). [20]H.Akoh et al., Jpn. J. Appl. Phys. 22, L435(1983). [21]S.Sakai et al., Ext. Abst. of 1987 ISEC, 118(1987). [22]H.Akoh et al., Ext. Abst. of 1987 ISEC, 122(1987). [23]K.Ishibasi et al., presented in 1988 Applied Superconductivity Conference.

Josephson Parametric Amplifiers: Low Noise at 9GHz

H.K. OLSSON and T. CLAESON

Physics Department, Chalmers University of Technology, S-412 96 Göteborg, Sweden

ABSTRACT

The performance of a thin film, (Nb/PbBi), resistive rf-SQUID as a parametric amplifier was investigated at 9 GHz. It was matched to a 50Ω coaxial cable via a low impedance (Z_L) microstrip transformer. This allowed relatively high Josephson currents, I_o, to be used. Stable Josephson oscillations were achieved for $\beta_L Q=2eI_oZ_L/hf<1.1-1.8$. Best amplifier performance was obtained in the 3 and 4 photon, externally pumped modes. Both these two types gave signal gains more than 10 dB and negligible noise power contributions. A hot/cold measurement at a bath temperature of 1.2K gave G_{dsb}=13 dB and T_{dsb}=3±4K over a bandwidth of 310 MHz for the 4 photon amplifier (dsb = double side band). Internally pumped amplifiers were noisier, e.g., a 4 photon one gave single side band noise of the order of 50 K.

INTRODUCTION

The non-linear inductive response of a Josephson junction is suitable for parametric amplification. A negative effective resistance cancels the load resistance at the signal frequency (f_s) giving rise to a substantial signal gain transfering power from a pump (p). The amplifier is operated close to bifurcation points where parametrically excited oscillations arise. This can easily lead to a very noisy state, possibly due to operating close to a chaotic regime. This may occur due to thermally induced hopping between a "stable" and a bifurcated state.

An inductive shunt resistor may be used to stabilize a single tunnel junction - i.e. a resistive rf-SQUID. If the inductance is large, the circuit will run into an unwanted relaxation oscillation mode. However, Calander et al [1] managed to obtain a stable self-pumped amplifier with T_n<30K by reducing the inductance of the shunt. Better optimized circuits, used in the three ($f_s \approx f_p/2$ and four photon ($f_s \approx f_p$) modes also gave low values of T_n: 6+15/-7K [2] and 18±35K [3].

We have recently reported [4] very low noise at 9 GHz for a four photon, externally pumped amplifier. The improved circuit is stabilized by a low $\beta_L Q$-value. ($\beta_L=2eI_oL/h$, I_o is the critical current of the tunnel junction, L the shunt inductance. The Q-value of the LC resonance, $\omega^2=1/LC$, is given by $Q=\omega Z_L C$ if circuit losses are assumed negligible, C is the tunnel junction resistance and Z_L the load resistance at the resonance frequency $\omega/2\pi$). In particular, a novel transformer [5], that is used to match the low impedance, shunted junction to the external world, allows the use of a sufficiently high I_o to avoid thermal noise induced hopping.

This report will touch upon some of the aspects of the low noise amplifier, describe its operation also in the three photon and the internally pumped modes, and give an improved value of T_n=3±4K for the externally pumped, four photon mode.

CIRCUIT CRITERIA, DESIGN AND FABRICATION

As an example, let us regard an internally pumped, three photon amplifier. For this we get large gain [6] when $Z_L \approx hf_p/2eI_o$, Z_L being the load impedance at the resonance frequency (we are choosing the LC resonance frequency about equal to the signal frequency and can neglect reactive circuit components). Similar expessions exist for the other modes of operation.

Small critical currents are very sensitive to noise suppression. Therefore we need a large current, I_o>>2ekT/h. However, a critical current of the order of 50μA results in a very low load impedance, about 0.5Ω, at 10GHz. The connecting coaxial cable has an impedance of 50Ω and a matching is needed. This is realized by a thin film transformer of a novel design [5].

By incorporating an inductance L parallel to the tunnel junction and in series with a small shunt resistor, it is possible to resonate out the junction capacitance within a frequency range covering the signal frequency. Both high and low frequencies are effectively short circuited. The transformer also acts as a filter, reducing unwanted thermal noise outside the bandwidth.

Low resistive losses are achieved by choosing the quasi-particle resistance of the junction, R, large and the shunt resistor, R_s, small. To keep the oscillator linewidth small, in the internally pumped mode, R_s should be of the order of 10 mΩ. It is possible to tune the amplifier by varying I_o via a magnetic field, a most useful property.

Nb/oxide/PbBi (or Pb) tunnel junctions were fabricated by evaporating PbBi strips across ion beam oxidized Nb films via holes in SiO. The junction size was 30x19 μm. The resistive shunt was a Cr-Au layer. The circuit layout and fabrication is described elsewhere [7].

MEASUREMENTS AND CIRCUIT PERFORMANCE

The chip microstrip was connected to a "microstrip launcher" of coaxial type. This was connected to a three port cryogenic circulator. By using a 20 dB attenuator (also helium cooled) at the circulator input, it was possible to limit the input noise temperature to 5K (using a value of 21 dB for the circulator isolation). Either a broad band solid state noise source or a Gunn oscillator was used as a signal source. The pump at 9 or 18 GHz was generated by a broad band microwave sweeper.

Reflected signals were amplified by an FET amplifier (+19 dB, T_n=200K), converted to 1 GHz with the aid of a balanced mixer, amplified by a second FET amplifier and detected by a spectrum analyzer.

The measurements were made at 4.2 or 1.2 K. The amplifier results reported here were taken at the lower temperature.

The Josephson oscillations that occur at finite bias voltage can be used as a diagnostic tool, e.g., to determine the resonance frequency, the bandwidth of the transformer, and the stability of the oscillations.

The oscillation linewidth results from thermal voltage noise in the shunt resistor. It was determined to be 830 kHz for R_s=17.5mΩ and T=1.22K. This is close to the calculated value, 890 kHz [8]. The linewidth did not change appreciably with bias voltage or critical current (I_o<65μA).

The impedance seen by the junction was calculated using measured values of the transformer microstrip dimensions [5]. A plot of this impedance from 0 to 40 GHz is given in the inset of Fig. 1. The low frequency behavior is dominated by the inductive shunt while frequencies above the resonance frequency are short circuited by the capacitance. The measured output power is shown in Fig. 1. High output power (more

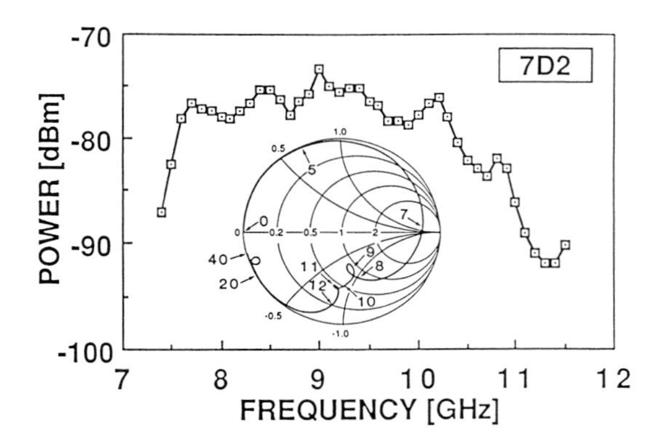

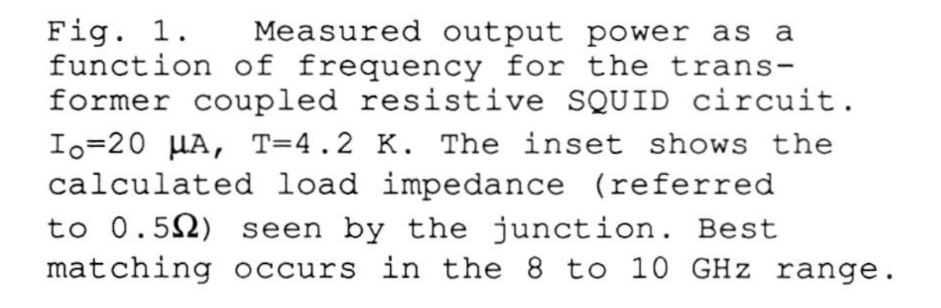
Fig. 1. Measured output power as a function of frequency for the transformer coupled resistive SQUID circuit. I_o=20 μA, T=4.2 K. The inset shows the calculated load impedance (referred to 0.5Ω) seen by the junction. Best matching occurs in the 8 to 10 GHz range.

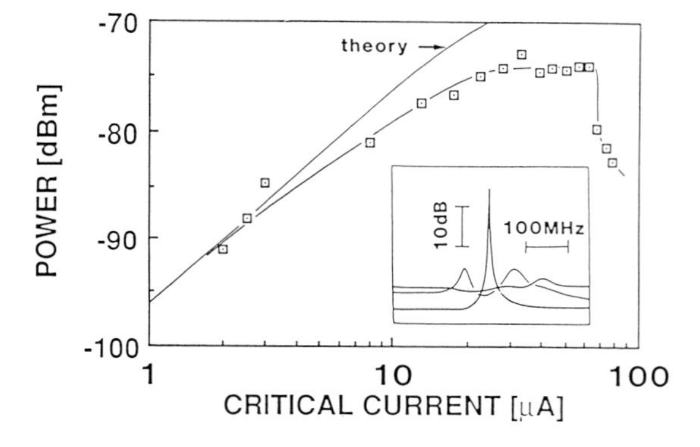

Fig. 2. The output power from the Josephson oscillator for different values of the critical current. The calculated curve (ref [9]) uses Z_L=0.5 Ω. The linewidth is constant up to 65 μA after which it increases, splits into two peaks and transforms into white noise, R_s=18 mΩ.

than -78 dBm) was available between 7.6 and 10.3 GHz in qualitative agreement with calculations. The latter gave two maxima within the passband while there are several peaks in the experimental curve (possibly due to mismatches in the connectors).

The critical current could be tuned from 230 μA down to a few μA for the junction we are discussing here. The oscillator output within most of this range is shown in Fig. 2. A sharp drop of maximum output power and an increase in the linewidth occurred at 65 μA. At 110 μA, the oscillation split into two peaks and at still higher values, the oscillation resembled white noise (T=9000 K), see the inset of Fig. 2.

The tests hence showed that the transformer operated as expected and that stable oscillations occurred at relatively small critical currents with a value of β_{LQ}<1.1-1.8. This is important for the parametric amplification that will be discussed next.

PARAMETRIC AMPLIFICATION

Externally Pumped Modes

The three photon amplifier ($f_s+f_i=f_p$, i=idler) has to be current biased at zero voltage. An example is given in Fig. 3a where a Josephson current of 39 μA (tuned by the magnet) is suppressed to 14.5 μA by the pump power and biased at 10 μA. By comparing the reflected signal from the amplifier (biased as described) with the one from the same junction with no bias, pump, or magnetic field, a signal gain of 11.6 dB is calculated, compare with Fig. 3a.

The externally pumped four photon mode ($f_s+f_i=2f_p$) performs similarly. It is unbiased and has an I_o of 49 μA without pump power. At 9.00 GHz, a signal gain of 11.8 dB is evaluated and it varies smoothly over the pass band. Essentially no extra noise is added in the amplifying state, see Fig. 3b.

Hot/Cold Noise Measurements

The noise temperature estimate of the last section (T_n≈0K for both modes) has a limited accuracy. A better way is to use the hot/cold method. Here, the signal is a broad band noise source and its temperature, T_{in}, is varied. The output noise power is related to T_{in} by $P_n=k_B\Delta f G_{dsb}(T_{dsb}+T_{in})$ where Δf is the bandwidth of the measurement and G_{dsb} and T_{dsb} the double side band gain and noise temperature. (The linear relationship is valid if $T_{in}>>hf/k_B$, which is about 0.5K in the X band). As we use a cold attenuator and circulator, T_{in} can be varied from 5 to 13K and an accurate estimate of a small T_{dsb} can be made.

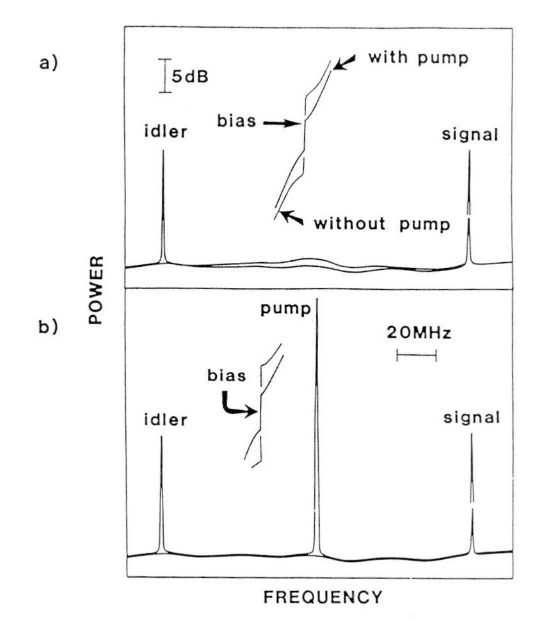

Fig. 3. Externally pumped amplifiers. Reflected signals from a short circuited junction and amplified signals and idlers are shown. The scales are the same in both plots.

(a) The 3 photon amplifier with f_s=9.00 GHz, f_p=18.176 GHz, and T=1.25K. The inset shows IV-curves with and without pump power, I_o=38 μA.

(b) The 4 photon amplifier with fs=9.00 GHz, fp=9.07 GHz, P_p=-60 dBm, and T=1.25 K. The pumped and unpumped IV curves are shown in the inset, I_o=49 μA.

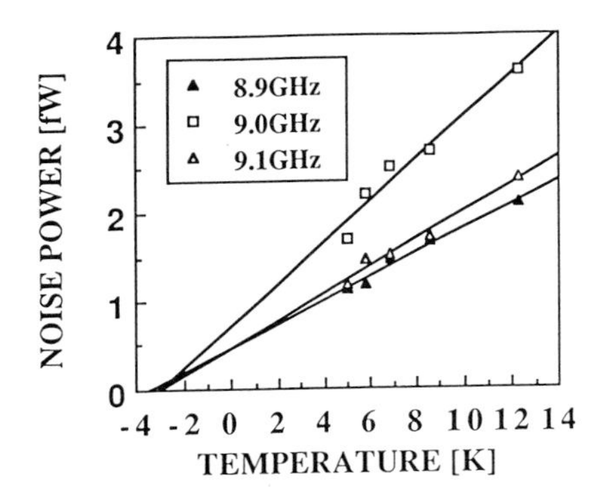

Fig. 4. The noise power output from a 4 photon, externally pumped amplifier vs input temperature. At 9.00 GHz, the plot gives G_{dsb}=12.9 dB, T_{dsb}=3±4K. I_o=48 μA, T=1.27 K.

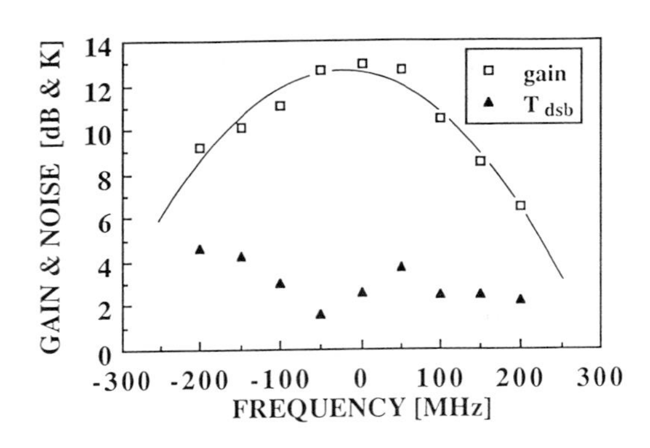

Fig. 5. The frequency dependence of G_{dsb}, and T_{dsb} of the 4 photon, externally pumped amplifier. The zero of the frequency scale is set to the pump frequency, fp=9.00 GHz. I_o=48μA, P_p=-59.1 dBm, T=1.27K.

Fig. 4 shows the linear relation between P_n and T_{in} for the four photon amplifier. A least square fit gives G_{dsb}= 12.9 dB and T_{dsb}=3±4K at 9.00 GHz. The relationships for two other frequencies within the instantaneous band width of 310 MHz are also shown in the figure. A plot of the gain and the noise temperature as a function of frequency is shown in Fig. 5. The values are less accurate for frequencies outside the calibration frequency (9.0 GHz).

Internal Pump

An example of an internally pumped, four photon amplifier is shown in Fig. 6. A critical current of 60 μA gave stable Josephson oscillations with an output power of -73 dBm at 9.07 GHz. The signal gain was 10 dB as compared with the shorted junction. The disadvantage of this mode is evident from the figure. The pump is broad and its flank adds noise to the signal. At a signal-to-pump separation of 70 MHz, this corresponded to a noise temperature of about 50K. The noise adds non-linearly with frequency separation in the limited bandwidth.

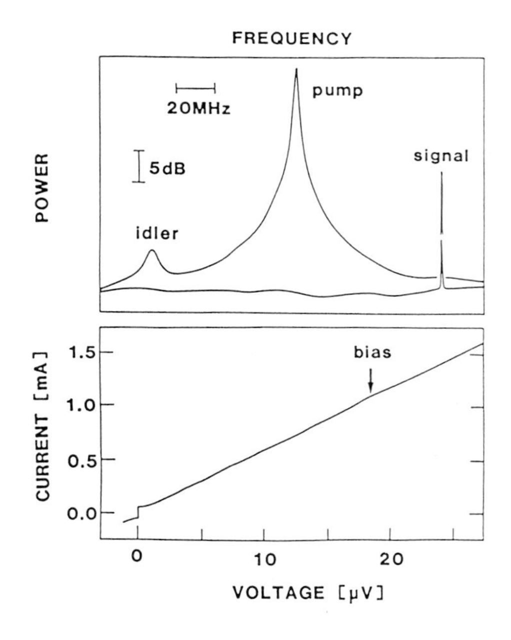

Fig. 6. The upper diagram shows the frequency spectra of an internally pumped 4 photon amplifier: the lower curve is the reflected signal from a short circuited junction and the upper one with voltage bias. The IV curve is given in the lower diagram. I_o=60 μA, f_s=9.00 GHz, f_p=9.07 GHz, T=1.25 K.

The tested circuit is not suited for three photon operation as it lacks a resonance at twice the signal frequency. Oscillations can be excited, however, at subharmonics or non-integer ratios of the bias frequency at large critical current. An operation close to the onset of an oscillation does amplify the signal. With a bias corresponding to 15.9 GHz ($1.7xf_s$), the signal gain increased from 8 to 28 dB as the critical current went from 26 to 34 μA. T_{ssb} remained roughly constant around 20K until I_o exceeded 34μA at which it increased sharply at the same time as the gain saturated.

CONCLUSIONS

We conclude that the externally pumped amplifier has a very low noise. For successful operation, it is desirable to use a low impedance external load at the resonance frequency and important that the Josephson oscillations are stable. The low impedance requirement can be fullfilled using a microstrip transformer. Our design calculations gave a Z_L of 0.5Ω and this value was verified both by measuring the output oscillator power and from the amplifier parameters giving high gain.

The oscillator stability can be controlled by the Josephson critical current. The stability criterion is governed by the $\beta_L Q$ factor (or alternatively the ratio of Z_L and the characteristic Josephson impedance $hf/2eI_o$) and stable oscillations were obtained for $\beta_L Q<1.1$ to 1.8.

The tuning ability given by the magnetic field dependence of the critical current is very useful. Stable oscillations and a control of the effective negative resistance can be realized.

A double sideband measurement gave a gain of 13 dB and a noise temperature as low as 3±4K for the externally pumped four photon mode at 9 GHz. Internally pumped modes gave less favourable noise performance (the three photon mode might be better than the four photon one but it may not be practical to operate; small variations in parameters may lead to instabilities and excessive noise).

The disadvantage, however, of our amplifier is that there exist low noise semiconductor transistors in the X-band. At higher frequencies, say 70 GHz, there are no competitors. In principle, there seem to be no obstacles in extending the Josephson amplifier to that region. A higher freuency would permit a higher critical current, a larger range of stable operation, and a higher saturation level. Smaller junctions and higher current densities can be obtained with modern technology. The present problem is to find a cooled circulator at this frequency.

Acknowledgements

The work was supported by the Swedish Natural Science Research Council and the Board of Technical Development.

REFERENCES

1. N. Calander, T. Claeson, and S. Rudner, Appl. Phys. Lett.*39*, 650 (1981); J. Appl. Phys. *53*, 5093 (1982).

2. A.D. Smith, R.D. Sandell, J.F. Burch, and A.H. Silver, IEEE Trans. Magn. *MAG-21*, 1022 (1987).

3. L.S. Kuzmin, K.K. Likharev, V.V. Migulin, E.A. Polunin, and N.A. Simonov, in *SQUID'85* (ed.by H.D.Hahlbohm and H.Lübbig), Walter de Gruyter, Berlin 1985, p. 1029.

4. H.K. Olsson and T. Claeson, Jpn. J. Appl. Phys. *26*, 1547 (1987).

5. H.K. Olsson, Electron Lett. *23*, 1152 (1987).

6. P. Russer, Archiv der Elektrischen Übertragung *23*, 417 (1979).

7. H.K. Olsson and T. Claeson, J. Appl. Phys, in print, H.K. Olsson, PhD thesis, Chalmers Univ. of Techn., Göteborg, 1988.

8. D. Rogovin and D.J. Scalapino, Ann. Phys. *86*, 1 (1974).

9. K.K. Likharev, *Dynamics of Josephson Junctions and Circuits*, Gordon and Breach Sci. Publ., N.Y. 1986, p. 392.

Losses in AC Superconducting Coils

H. KASAHARA, S. AKITA, T. ISHIKAWA, and T. TANAKA

Central Research Institute of Electric Power Industry (CRIEPI), Advanced Technology Department, 2-11-1, Iwatokita, Komae-shi, Tokyo, 201 Japan

ABSTRACT

AC superconducting wires are being developed so as to be applied to electrical machineries. The main concerns in developing the AC superconducting wires are AC losses. In order to assess AC losses in the form of winding, we made two AC superconducting coils having different rated capacities, 500kVA and 20kVA. The AC losses are measured by a calorimetric method.

Wire AC losses in the function of magnetic field strength and frequency are calculated from coil losses for each coil by a least square estimation method. In this process, we consideration for the magnetic field distribution in the coil windings were given.

The wire losses estimated from the measured coil losses were compared with the AC losses calculated by a set of usual theoretical formulations. In theory, AC loss components in proportion to magnetic field strength and frequency are regarded as hysteresis losses and components in proportion to the square of them are regarded as coupling losses. We divided the estimated AC losses into hysteresis loss components and coupling loss components.

The estimated hysteresis losses were about four times higher than the corresponding theoretical values and the coupling losses were 1.7 to 3 times higher than the corresponding theoretical values.

INTRODUCTION

Development of an ac superconducting wire to be applied in superconducting technology to electrical power apparatus has been intensively desired. The main concerns in such development are ac Losses in an ac superconductor with respect to the efficiency and the stability of a superconducting machine. In order to assess ac losses in the form of winding, we made two ac superconducting coils using NbTi ac superconductors. Their respective rated capacities are 500kVA and 20kVA. We used ac superconducting coils because they should give us more accurate ac losses for windings of electrical machines than in a wire form of

Table I. Parameters of AC Superconducting Coils.

	Coil A	Coil B
Rated Capacity [kVA]	500	20
Max Capacity [kVA]	713	74.4
Voltage [kV_{rms}]	5.8	1.92
Current [A_{rms}]	122	38.8
Frequency [Hz]	50.2	49.0
Winding I.D. [cmφ]	4.0	4.5
O.D. [cmφ]	14.5	11.3
Height [cm]	19.5	6.0
Layer Numbers	22	20
Total Turn No. [Turns]	2645	1986
Conductor Length [m]	758	491
Stand		
Filament Diameter [μmφ]	0.49	0.5
Numbers	14478	14280
Twisting Pitch [mm]	0.98	1.9
Cu Ratio NbTi/Cu/CuNi	1 /0.1/2.5	1/1/3.5
Insulation Thickness[μm]	0	10
Overall Diameter [mmφ]	0.112	0.16
1st Cable		
Number of Strands	7	7
Center Strand	SC Strand	SUS
Cabling Pitch [mm]	3.05	3.5
Insulation Thickness[μm]	0	50
Overall Diameter [mmφ]	0.34	0.6
2nd Cable		
Number of 1st Cable	6	
Center Cable	SUS	
Cabling Pitch [mm]	6.7	
Insulation Thickness[μm]	0	
Overall Diameter [mmφ]	1.04	
Winding Method		
Spacers Inter-Layers	Y	Y
Spacers Inter-Turns	N	N
Impregnation	N	Y

superconductor, since the ac current and acmagnetic field should be applied to the wires simultaneously.

EXPERIMENTS

Coils

Specifications of coils A and B are shown in Table I. Both coils were wound with CuNi matrix wire having 0.5μmφ diameters of NbTi filaments. For ac coils, the mechanical stiffness of the windings is intrinsically important to prevent wires from moving.

Coil A employs 22 layer cylindrical spacers on which the conductor wound. Each spacer has spiral and axial grooves on the surface. The axial grooves are deeper than the spiral grooves, and form channels for liquid helium. The conductor was wound in the spiral grooves to prevent axral wire motions. Over 10 kg/mm^2 tension was applied to suppress wire motions,due to hoop stress.

Coil B was wound with 2.5mm wide and 1mm thick spacers in each of 20 layers. After completion of winding, Coil B was impregnated with epoxy resin. The excess resin was cleaned off to have enough cooling channels. Poyester fibers used for inter turn insulation had absorbed epoxy resin, and gave good mechanical strength like fiber reinforced plastics.

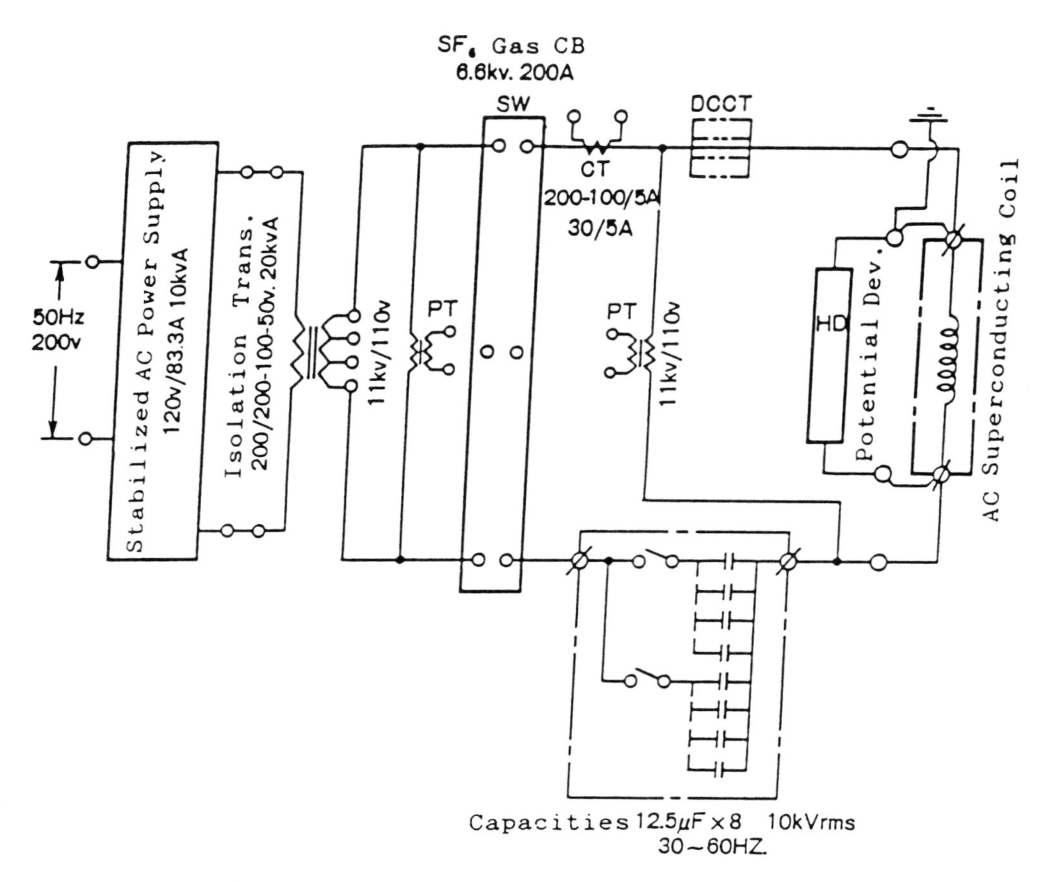

Fig. 1. Power supply for ac superconducting coils.

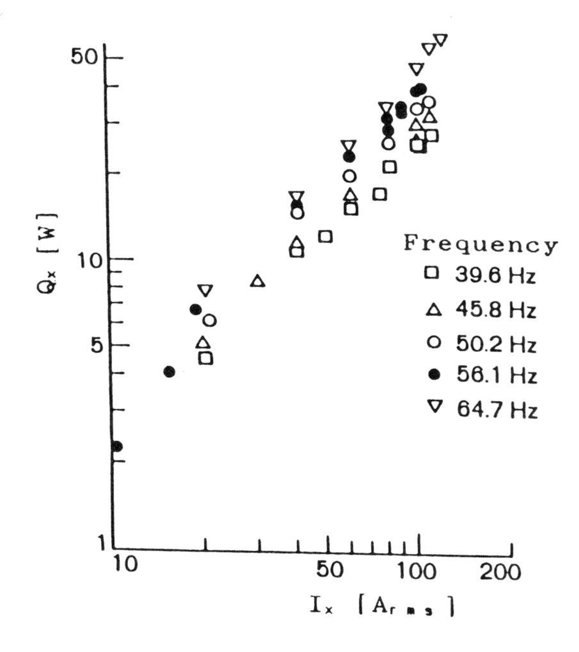

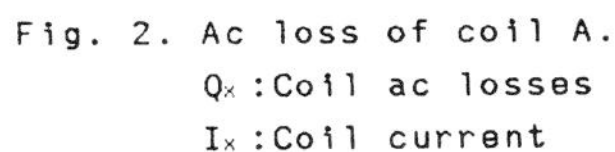
Fig. 2. Ac loss of coil A.
Q_x :Coil ac losses
I_x :Coil current

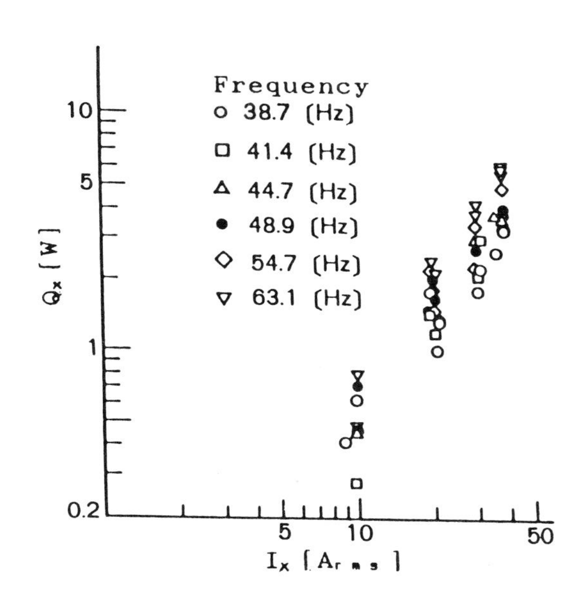

Fig. 3. Ac loss of coil B.
Q_x :Coil ac losses
I_x :Coil current

Power Supply

Superconducting coils are energized by a resonating circuit with capacitors as shown on Fig. 1. The resonant frequency is given by the inductance of the coil and the capacitance of the capacitors. and can be changed by parallel connected numbers of the capacitors. The circuit is driven by transistorized ac an amplifier with a constant current control means. Sine wave signal to the amplifier comes from a frequency synthesizer with resolution of 0.1mHz. Measurement of high accuracy of frequency and amplitude was thus assured.

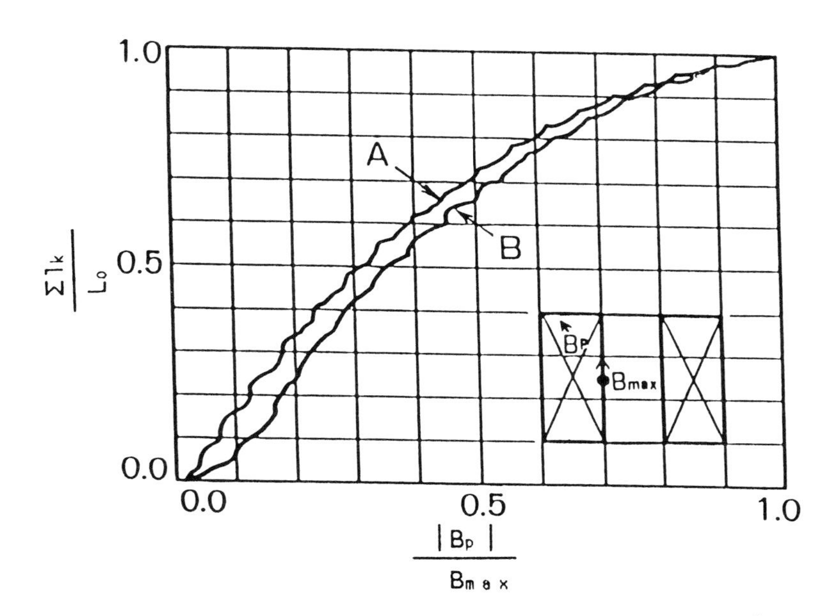

Fig. 4. Magnetic flux density applied to the windings of coil A and B.
$\frac{\Sigma l_k}{L_o}$:Wire length bellow field of $|B_p|$
L_o :Total Length, l_k :Each turn length
$\frac{|B_p|}{B_{max}}$:Applied magnetic field

	Turn. No.	layers	B_{max}
Coil A	2645	22	0.01478[T/A]
Coil B	1985.5	20	0.02705[T/A]

Ac Loss Measurement

Ac losses in coils were measured by a calorimetric method from the consumption rate of liquid helium when heat inleak to the cryostat was between 3W and 4W: from each value there of and the apparent loss was subtracted to get an actual ac loss. The accuracy of measurement by this method was ±1W according to the calibration thanks to utilization of an electric heater.

RESULTS

The measured ac losses for coils A and B are respectively shown in Fig. 2 and 3, respectively. Ac losses were measured on several various values of frequency and amplitude of current.

DISCUSSIONS

Magnetic Field in the Coil

Magnetic field strength of air core solenoid coils like Coil A and Coil B, varies with the location of the winding position. Since the ac loss in a superconducting wire depends on the magnetic field strength, to assess ac losses in objective superconducting coils we should know the exact magnetic field shape . The distributions of field strength for Coil A and Coil B are shown in Fig. 4, as the integrated wire length normalized by the total wire length in the function of normalized field strength by the maximum field. Fig. 4 shows that about 70% of the total wire length of each coil lies below the magnetic field strength of the half of maximum field.

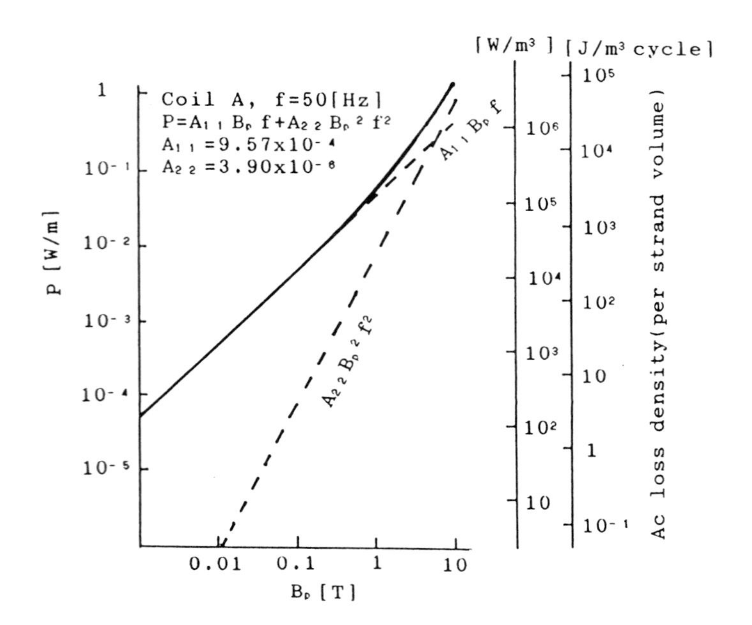

Fig. 5. Estimated ac loss of the wire used for coil A
P:Wire ac loss,
B_p :Magnetic flux density(peak value)

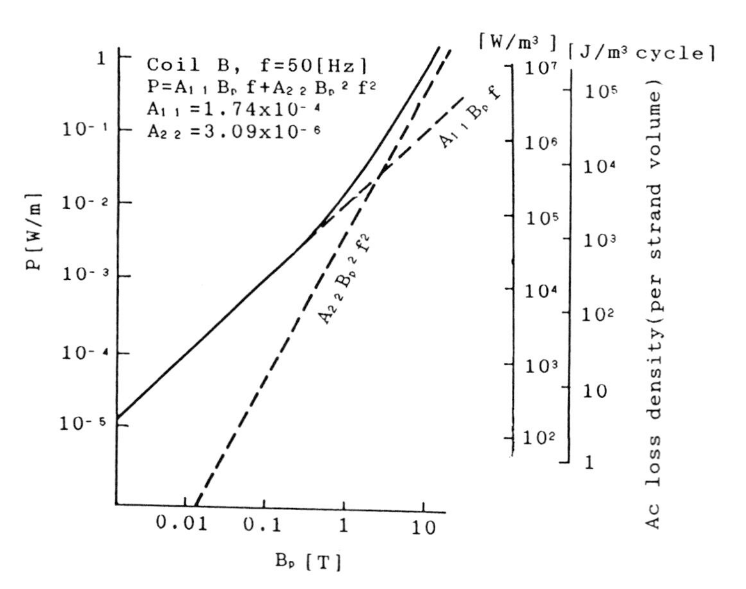

Fig. 6. Estimated ac loss of the wire used for coil B
P:Wire ac loss,
B_p :Magnetic flux density(peak value)

Ac Loss of Wires

We assumed the ac loss in wires, P, to be the function of the frequency f and the peak value of the magnetic field strength B_p, as Equation (1).

$$P=A_{11}B_p f+A_{22}B_p^2 f^2 \quad (1)$$

The coefficients A_{11} and A_{22} in Equation (1) can be known by a least squares method to be used for the comparison of the measured ac losses shown in Fig. 2 and 3 and the calculated values from Equation (1) according to the field distribution shown in Fig. 4. The estimated wire ac losses for Coil A and Coil B are shown on Fig. 5 and Fig. 6, respectively. We assumed $A_{11}B_p f$ corresponds to hysteresis loss and $A_{22}B_p^2 f^2$ to coupling loss. Further, we compared the estimate hysteresis and the coupling losses with the calculated values using following Equations (2) for the hysteresis loss and (3) and (4) for the coupling one.

$$P_h\,[W/m^3]=\frac{8}{3\pi}f d_f \lambda J_c B_p \quad (2)$$

$$P_e\,[W/m^3]=\frac{B_p^2(2\pi f)^2\tau}{\mu_0} \quad (3)$$

$$\tau[sec]=\frac{1}{2}\mu_0\sigma_\perp\left(\frac{l_p}{2\pi}\right)^2 \quad (4)$$

Where Equation (2), (3), and (4), nomenclatures are as follows; d_f [m]: diameter of NbTi filaments, J_c [A/m^2]: critical current density of NbTi filaments, μ_0 [H/m]:magnetic susceptibility of vacuum, τ[sec]:decay time constant for coupling current, l_p [m]:twisting pitch of superconducting strand, $\sigma_\perp$ [S/m]:effective conductivity across strand. The estimated hysteresis losses, $A_{11}B_p f$, for Coil A and Coil B are about 4 times higher than the corresponding calculated values from the equation (2). Also, the estimated coupling loss, $A_{22}B_p^2 f^2$, for Coil A is 1.7 times higher than the calculated value for it, and the estimated coupling loss for Coil B is 3 times higher than the calculated value for it.

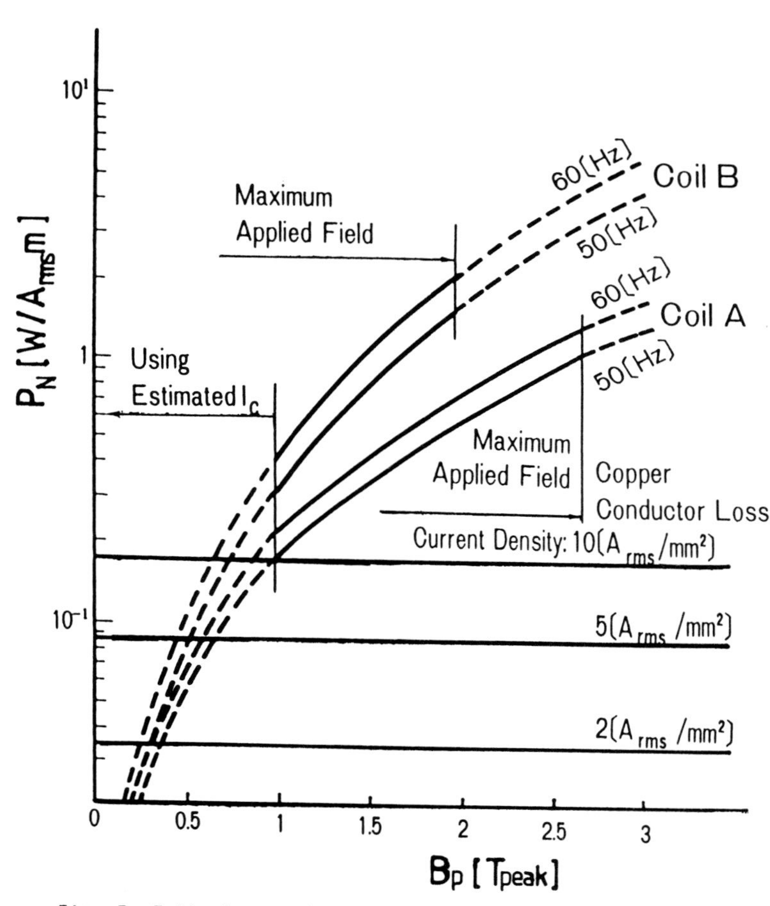

Fig. 7. Estimated ac loss of superconducting wires of coil A and coil B compared with Joule loss of copper conductor.

P_N :Wire AC loss per unit length and per unit current at room temp.(η=1/1000)

$P_N = P/I_c$ [A_{rms}], I_c :Critical current at B_p

B_p :Magnetic flux density(peak value)

Ac Loss Compared to Copper Conductor Loss

An ac superconducting wire would be used for an electric apparatus instead of a copper conductor. Therefore the ac loss in a superconducting wire should be compared with the Joule loss in a copper conductor. The Joule loss varies with current density. In Fig. 7, the Joule losses in a copper conductor at room temperature are shown for current densities of 10 A_{rms}/mm^2, 5 A_{rms}/mm^2, and 2 A_{rms}/mm^2. Each value of Loss is shown as the value per unit length and unit current flow. As a copper conductor has insignificant magnetcresistance effect at room temperature, the Joule losses are constant even when exposed to different values of magnetic field strength. The ac losses in superconducting wires for Coil A and B are also shown in Fig. 7 as the function of the magnetic field. The ac loss values are divided by the critical current at each magnetic field and multiplied by one thousand to get a loss level including the liquid Helium refrigerating power. The ac loss in Coil B wire is higher than that in Coil A wire because of the high coupling loss due to the high copper ratio and the long twisting pitch of the former when compared with those of the latter. By comparing the ac losses in such superconducting wires with the losses of the copper conductor as shown in Fig. 7, we can conclude that a superconductor would have a lower ac loss at a low magnetic field around 0.5T, and a higher ac loss at a high magnetic field around 2T. In other words, ac superconducting wires having 0.5µm diameter class of superconducting filaments have low ac losses suitable to be applied to transformers with magnetic cores, but they are not appropriate to be applied to an air gap armature winding of a superconducting generator because of the loss reduction effect. Of course superconducting wires have several merits other than low ac loss, including the ability of getting high current density in the winding. Winding current densities for Coil A and Coil B at quenchings including cooling channels are 32 A_{rms}/mm^2 and 38 A_{rms}/mm^2, respectively. These values are higher than that for water cooling copper conductors.

CONCLUSIONS

We have made two ac superconducting coils with different NbTi superconducting wires and estimated the ac losses for these wires from the ac loss in the coils by considering the magnetic field distribution in the solenoid coil. Further, we have compared the ac loss in the superconducting wire including refrigerating power with the Joule loss of a copper conductor . Based on, our test results we concluded that an ac superconducting wire having 0.5µm diameter of filaments has lower ac loss than a copper conductor could have at a magnetic field of around 0.5T.

ACKNOWLEDGMENT

We would like to express our appreciation to Prof. T. Ogasawara of Nihon Univ. and Prof. F. Sumiyoshi of Kagoshima Univ. for their meaningful discussions.

3
Organic Superconductors

Chemical and Physical Properties of Organic Superconductor κ-(BEDT-TTF)$_2$Cu(NCS)$_2$

G. SAITO[1], H. URAYAMA[1], H. YAMOCHI[1], and K. OSHIMA[2]

[1] The University of Tokyo, Institute for Solid State Physics, 7-22-1, Roppongi, Minato-ku, Tokyo, 106 Japan
[2] The University of Tokyo, Cryogenic Center, 2-11-16, Yayoi, Bunkyo-ku, Tokyo, 113 Japan

ABSTRACT

An overview of the chemical and physical properties of a new organic superconductor κ-(BEDT-TTF)$_2$Cu(NCS)$_2$, of which Tc (10.4-11.0K) is the highest among those of organic superconductors so far known, is presented. Distorted-hexagon-shaped crystals of the Cu(NCS)$_2$ salt were prepared by the electrochemical oxidation of BEDT-TTF in 1,1,2-trichloroethane in the presence of CuSCN, KSCN, and 18-crown-6 ether (or K (18-crown-6 ether)Cu(NCS)$_2$ complex, or CuSCN and TBA·SCN, as a supporting electrolyte) under a constant current of 1-5 μA. The black shinny crystals were stable against air and moisture and start to decompose above 190 °C.

The crystal structures indicate that both lattice parameters and packing pattern of BEDT-TTF molecules are almost the same to those of κ-(BEDT-TTF)$_2$I$_3$ salt (Tc=3.6K) despite the considerable difference of Tc between them. Two BEDT-TTF molecules are dimerized and the dimers are linked one another by short sulfur..sulfur atomic contacts to construct two-dimensional conducting BEDT-TTF sheets in the bc-plane. The anions, Cu(NCS)$_2^-$, form one-dimensional zigzag polymer along the b-axis and the polymers construct insulating sheets in the bc-plane. Every conducting layer is sandwiched by the insulating layers along the a-axis. Two kinds of layers are linked by hydrogen bonds between terminal ethylene groups of BEDT-TTF and nitrogen and sulfur atoms of the anions. Due to the lack of the inversion center of the crystal, the space group is lowered from P2$_1$/c of κ-(BEDT-TTF)$_2$I$_3$ to P2$_1$. Owing to this symmetry the Cu(NCS)$_2$ salt is optically active (specific rotatory power is about 230° at 632.8nm) and the Fermi surface based on extended Huckel MO is composed of both a cylindrical closed surface (18% of the first Brillouin zone) and a modulated quasi-one-dimensional open surface.

Tc of the BEDT-TTF-h$_8$ salt is 10.2-10.4K by four-probe d.c. resistivity measurements and Tc decreases with increasing pressure. An inverse isotope effect was observed in our samples so far measured (more than seven each samples). The deuterated samples showed higher Tc by 0.5K measured by the RF penetration depth measurements. Magnetic susceptibility measurements indicate that the salt is non-ideal class II superconductor and almost 100% of the perfect diamagnetism was observed below 7K. Upper critical field measurements showed a dimensional cross-over-like behavior (3D to 2D) when the external field is parallel to the 2D plane. Hc$_2$ values at 6K were about 13T and 0.5T in the 2D plane and normal to it, respectively. The estimated anisotropy in the GL coherence length from the critical field near Tc is $\xi_{bc}(0)$:$\xi_{a^*}(0)$=182Å:9.6Å=19:1. The Shubnikov-de Haas signal was observed below 1K and above 8T. The period of 0.0015T^{-1} corresponds to the area of extremal orbit of 18% of the first Brillouin zone. The anisotropic nature of the thermoelectric power of the crystal in the 2D plane can be explained on the basis of the complicated Fermi surface. No EPR signal ascribed to Cu^{2+} was observed. The linewidth of the EPR signal of BEDT-TTF increased monotonically with decreasing temperature in contrast to the predictions of the Elliot formula for the spin relaxation in metals. A Korringa relation was observed in ^{1}H NMR measurements between 77K and 10K. Below 10K a big enhancement of the relaxation rate was observed with a peak at considerably lower temperature than Tc. Anisotropic superconducting gaps were detected by tunneling spectroscopic work. Tc of several BEDT-TTF superconductors were discussed on the basis of the effective volume of one electron for the molecular design of new organic superconductors.

INTRODUCTION

Thirty one organic superconductors so far prepared are classified into five groups depending on the organic component: TMTSF(7 superconductors), BEDT-TTF(13), DMET(7), MDT-TTF(1), and M(dmit)$_2$ molecule(3). A family of BEDT-TTF superconductors can be classified into three groups in terms of the shape of the counter anions, namely tetrahedral (ReO$_4$)[1], linear (I$_3$, IBr$_2$[2], AuI$_2$[3]), and cluster (Cl·H$_2$O[4], Hg$_{2.78}$Cl$_8$[5], Hg$_{2.89}$Br$_8$[6]). Majority of the members are with symmetric linear anions especially I$_3$. This anion gives six different superconductors; iodine doped α-[7] , low Tc β-[8], high Tc β-[9], γ-[10], θ-[11], and κ-I$_3$ salt[12] of BEDT-TTF, with a variety of chemical and physical properties. For example Tc of low Tc β-I$_3$ salt increased from 1.4K to 7.5K by an application of a moderate pressure. Though a further application of pressure depressed Tc, a release of pressure provided a high Tc β-I$_3$ salt in addition to the low Tc β-I$_3$ salt (Fig. 1). Tc of the former one (8.1K) was the highest among those of organic superconductors, but this salt was metastable and can be isolated by some special treatment [13]. Very recently, a stable 8K

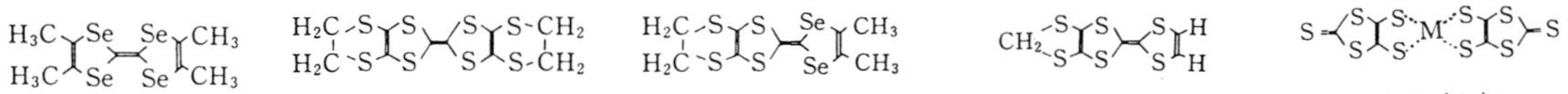

TMTSF | BEDT-TTF-h$_8$ | DMET | MDT-TTF | M(dmit)$_2$

superconductor was prepared by tempering α-I_3 salt [14]. The linear anions IBr_2 and AuI_2 gave β-phase superconductors with Tc lower than that of high Tc β-I_3 salt. To explain the size effect of the anion on Tc of β-phase salt, an idea of "lattice pressure" has been proposed [15]. A use of small anion will cause lattice compression which corresponds the lattice pressure. β-IBr_2 salt (Tc 2.7K) and β-AuI_2 salt (Tc 3.4-4.9K) correspond a lattice pressure of about 4kbar and 3-3.7kbar, respectively. A plot of these lattice pressures and Tcs on the Tc vs pressure curve of β-I_3 salt (Fig. 1) showed good agreement with the curve of high Tc β-I_3 salt. β-I_2Br salt corresponds a lattice pressure of 2-2.5kbar and may have Tc of about 6K. But this salt was not a superconductor due to the lack of symmetry in the anion. On the basis of the above consideration, one might expect a β-phase salt of Tc higher than 8K with a negative lattice pressure. What does the negative lattice pressure mean in terms of molecular design?

In the family of β-phase salts Tc increases linearly with increasing the length of the symmetric linear anion as discussed by Williams et al. and Kistenmacher [16, 17]. From the crystal structures of β-phase salts the following sequence was postulated that a use of small anion, which lies along the a-b axis (Fig. 2), will cause the interplanar distance of BEDT-TTF molecules decrease, which will increase the transfer integral and band width of the salt resulted in a decreased density of states. As a consequence Tc decreased by using small anion. Then a longer anion than I_3, which corresponds the "negative lattice pressure", might yield a β-phase salt with Tc higher than 8K. Above sequence is quite attractive but still need experimental evidences concerning about the quantities of these physical parameters to confirm this sequence. Nevertheless using of a symmetric linear anion which is longer than I_3 is an important working hypothesis for molecular design to get Tc higher than 8K. We started the search of long symmetric linear anion almost two years ago. Out of many anions examined we have obtained a superconductor of $Cu(NCS)_2$ anion with Tc higher than 10K. It turned out that the salt is not β-phase, moreover the anion is neither symmetric nor linear but forms a polymeric cluster.

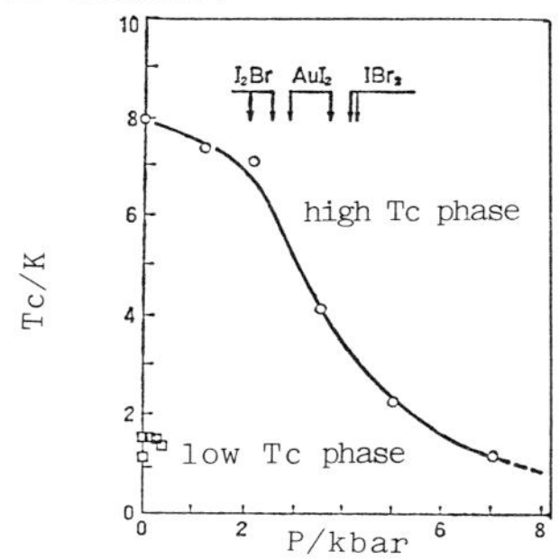

Fig.1. Tc vs pressure phase diagram of β-$(BEDT\text{-}TTF)_2I_3$.

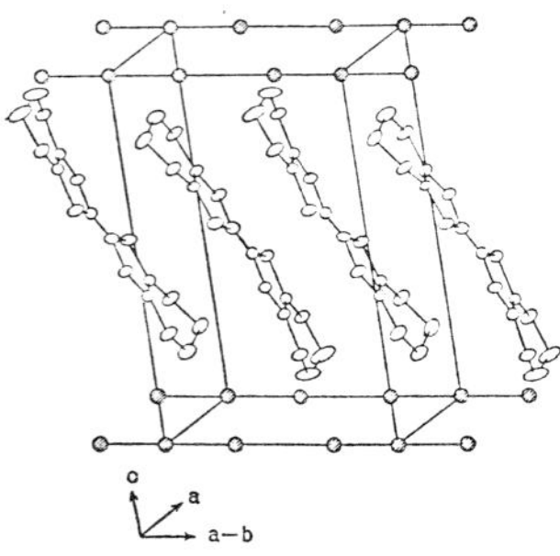

Fig.2. Crystal structure of β-$(BEDT\text{-}TTF)_2I_3$.

CRYSTAL GROWTH

Distorted-hexagon-shaped crystals of $Cu(NCS)_2$ salt ($3x2x0.05mm^3$) were prepared by the electrochemical oxidation of BEDT-TTF. BEDT-TTF was purified by recrystallization from chloroform, THF, then monochlorobenzene (mp. 256-258 °C). 18-Crown-6 ether was recrystallized from CH_3CN. KSCN was recrystallized from abs. EtOH. Crude CuSCN showed a EPR signal due to Cu^{2+} which was eliminated by treatment with aq. KSCN solution, but the $Cu(NCS)_2$ salt of BEDT-TTF prepared using either crude or purified CuSCN gave almost the same EPR and conductivity results. TBA·SCN was prepared from TBA·Br and KSCN and dried. K^+(18-crown-6 ether)$Cu(NSC)_2^-$ complex (abbreviated to K(crown)$Cu(NCS)_2$) was recrystallized from acetone-water, washed with acetone and dried.

We used three kinds of supporting electrolytes to prepare single crystals, and whichever we took the same kind of crystals was obtained so far in our group [18]; 1. K(crown)Cu(NCS) , 2. CuSCN +KSCN+crown, and 3. CuSCN+TBA·SCN. The last one was developed by Carlson et al.[18b] and Ugawa et al.[19], independently. The solvents were 1,1,2-trichloroethane, 1,2-dichloroethane, monochlorobenzene, THF, etc. and they gave the same kind of crystals. In the cases of electrolyte 2 or 3, undissolved materials remained on the bottom of the cell during the course of the electrocrystallization but the precipitation did not affect the crystal growth.

The single crystals were harvested within 1-3 weeks, washed with appropriate solvent and dried. The crystals were stable against air and moisture. No change in DSC, TG, and DTG was observed between -130 °C-190 °C. Above 190 °C the crystals started to loose weight and decomposed at around 230 °C with gas evolution.

CRYSTAL AND BAND STRUCTURES [20-22]

The crystal structures of this salt [20,21] are very analogous to that of κ-phase of I_3 salt. Two BEDT-TTF molecules form a dimerized pair and the pairs are linked one another almost perpendicularly to construct a two-dimensional (2D) conducting sheets in the bc-plane (Fig. 3). Every conducting layer is sandwiched by the insulating layers of anion along the a-axis. Although the packing patterns of BEDT-TTF molecules and the lattice parameters except the space group of the crystals are almost the same between the $Cu(NCS)_2$ salt and κ-I_3 salt, Tc of them are quite different. This is in contrast with the linear correlation between Tc and unit cell volume for β-phase salts [16]. One of the different structural points between two κ-phase salts is that both terminal ethylene groups of one BEDT-TTF molecule participate the formation of hydrogen bonds with nitrogen and sulfur atoms of anion in the $Cu(NCS)_2$ salt at 104K [21]. So every conducting layer is connected by hydrogen bonds to the insulating anion layers. The other different point is that the anion $Cu(NCS)_2$ is neither symmetric nor linear but asymmetrically bent like a boomerang. $Cu(NCS)_2$ instead of $Cu(SCN)_2$ is termed by the structural similarity of the known anion $Cu(CN)_2$. The repeating unit; boomerang, is arranged one after the other along the b-axis to form a zigzag 1D flat polymer. In one polymer, -SCN(I)-Cu-NCS(I)- form an infinite chain and other kind of ligand NCS(II) coordinates to the chain as a pendant. Sulfur and nitrogen atoms within the chain participate in the hydrogen bonding. Every polymers lie in the same direction to form an insulating layer in the bc-plane. Therefore the crystal does not have inversion center and the space group is lowered from $P2_1/c$ of κ-I_3 salt to $P2_1$.

One of the consequences of this symmetry is that the crystal is optically active and the specific rotatory power of the crystal is about 230° (25 °C, 632.8nm). Since these two κ-salts have different space groups, they show different band structures and Fermi surfaces (Fig. 4). There is no gap along the line ZM for the κ-I_3 salt and the Fermi surface is composed of one big closed circle [12]. On the other hand the calculated Fermi surface of the $Cu(NCS)_2$ salt indicates a gap along the line ZM [22]. As a consequence the Fermi surface of the $Cu(NCS)_2$ salt is composed of two different ones; one is a closed cylindrical 2D surface and the other is a modulated 1D open surface.

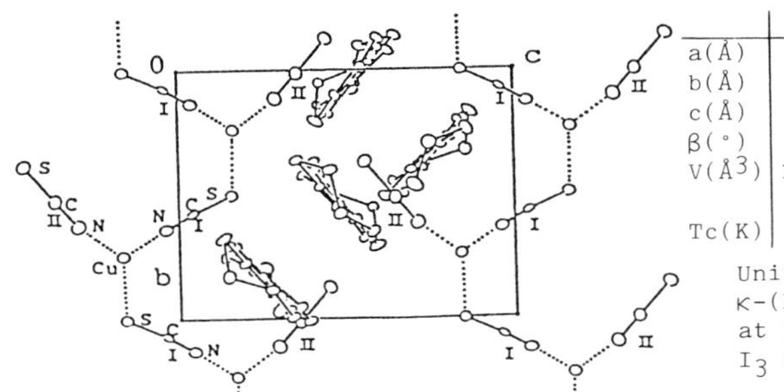

	H(104K)	κ-I_3(room)
a(Å)	16.382(4)	16.387(4)
b(Å)	8.402(2)	8.466(2)
c(Å)	12.833(4)	12.832(8)
β(°)	111.33(2)	108.56(3)
V(Å^3)	1645.3	1687.6
	$P2_1$	$P2_1/c$
Tc(K)	10.4	3.6

Unit cell parameters of κ-$(BEDT\text{-}TTF\text{-}h_8)_2Cu(NCS)_2$ at 104K and κ-$(BEDT\text{-}TTF)_2I_3$ at room temperature.

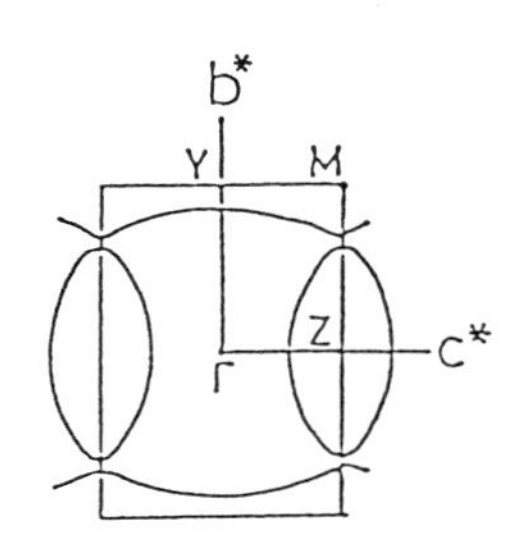

Crystal structure (104K, Fig.3) and Fermi surface (Fig. 4) of κ-$(BEDT\text{-}TTF\text{-}h_8)_2Cu(NCS)_2$.

TRANSPORT PROPERTIES [20,22-25]

The electrical conductivity along the b-axis at room temperature is 10-37Scm^{-1}. σc is a little higher than σb but almost isotropic in the bc-plane. σa* is about 600 times less than σb. The resistivity decreases down to 270K then it increases with a maximum at around 90K by cooling the sample (four probe method[20], Fig. 5). All our samples showed this behavior though the ratio of resistivity at the peak to that at room temperature varies from 2 to 6 depending on samples. The on-set of the superconduction is 11K and Tc is 10.2-10.4K which decreases with increasing pressure (-1.3K/kbar). An isotope effect on Tc was studied by using deuterated BEDT-TTF-d_8 (99.8%) [23-25]. D-salt is structurally isomorphous to the H-salt. By four probe method Tc of the D-salt was observed at 10.8-11.0K. Since four probe method needs electrical contacts which might cause pressure effect on Tc, we have compared the superconducting transition (Tc') of H- and D-salts by RF penetration depth measurements where Tc' is defined by the cross point of the base line and the straight line of the slope as is depicted in Fig. 6. Therefore Tc' is significantly smaller than Tc obtained by the resistive transition. The average Tc' of nine H-salts was 9.3K and that of seven

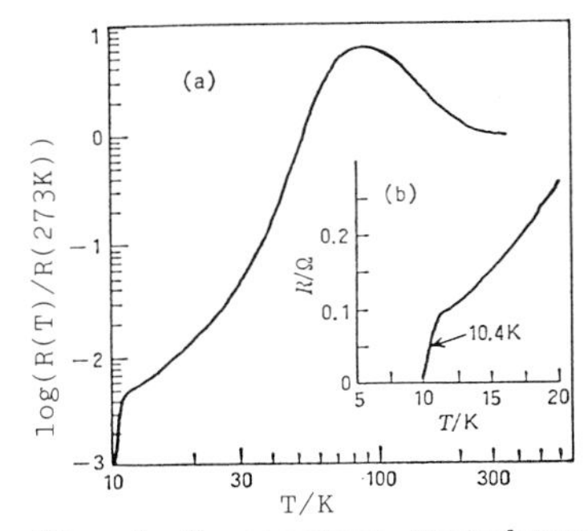

Fig. 5. Temperature dependence of resistivity and superconducting transition at ambient pressure in κ-$(BEDT\text{-}TTF\text{-}h_8)_2Cu(NCS)_2$.

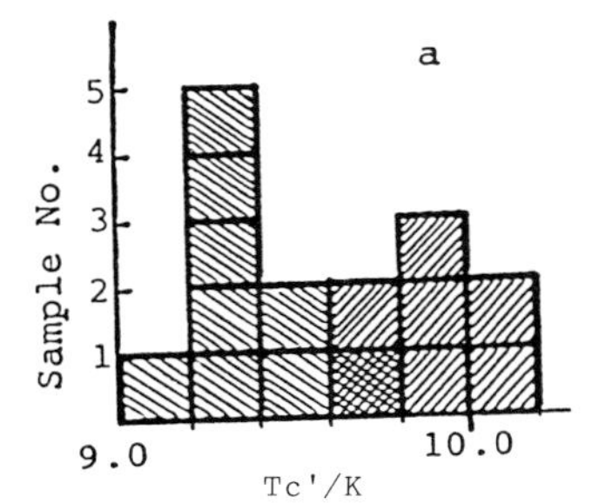

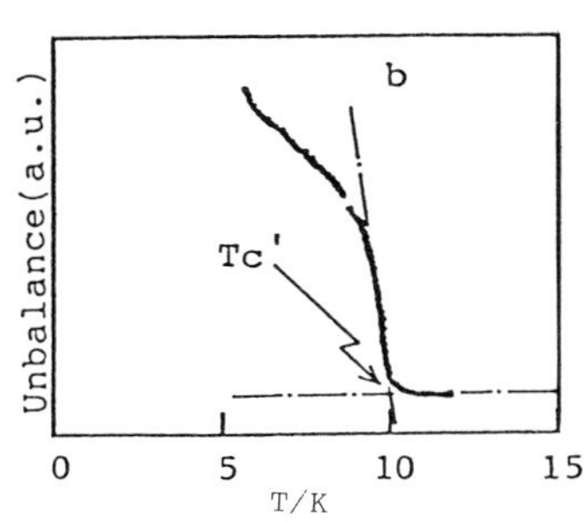

Fig. 6. Scatter in T_c' for the deuterated(▨) and undeuterated (▧) samples of κ-$(BEDT\text{-}TTF)_2Cu(NCS)_2$ (a) determined by RF penetration depth measurements (b).

D-salts was 9.9K. As far as our samples concern an inverse isotope effect by 0.5K on Tc was observed so far measured. One of the explanations of this isotope effect is that the superconduction of this salt is associated with the increase of the electron-phonon coupling constant in the D-salt compared to the H-salt, as has been observed in the system of palladium-hydrogen and palladium-deuterium [26], through the decreased phonon frequency of the carbon-deuterium bond or anion-deuterium bonds. We need more conclusive evidence to confirm this.

The temperature dependence of upper critical field of the H-salt (Fig. 7) shows an inflection at around 9K and Hc_2 exceeds 13 Tesla at 5K within the 2D plane [24]. This kind of inflection is very reminiscent of the 3D to 2D crossover in the layered structure compounds. Along the a*-axis Hc_2 increases steadily to about 0.8 Tesla at 5K then increases rather sharply at low temperatures (Hc_2=5.5T at 0.5K). We do not have any reasonable explanation for this behavior. The estimated anisotropy in the GL coherence length from the critical field near Tc is 182Å(//bc):9.6Å(//a*) =19:1. It is noteworthy that the coherence length along the a* axis is less than the length of a*. The quasi-isotropic nature in the bc plane is also observed in the reflection spectra. The calculated effective masses from the optical data at room temperature were m^*/m_0=5.5 and 4.1 [19] or 4.0 and 3.0 [27] with the polarized light along the b- and c-axes, respectively.

Below 1K and above 8 Tesla Shubnikov-de Haas (SdH) oscillation was observed in the transverse magnetoresistance curve [22, Fig. 8]. This is the first observation of SdH signal in the organic superconductors. The oscillation is $0.0015T^{-1}$ (H and D-salt) and the temperature dependence of the oscillation gave an effective mass at the Fermi level as m^*/m_0=3.5 in good agreement with those derived from optical data. The SdH signal is the conclusive evidence of the existence of the 2D closed Fermi surface. The area of the extremal orbit ($6.37\times10^{14}cm^{-2}$) is 18% of the first Brillouin zone ($3.56\times10^{15}cm^{-2}$), and this rate is equal to that of the calculated cylindrical Fermi surface. This cylindrical Fermi surface is hole-like and the modulated open Fermi surface is electron-like (Fig. 4). These complicated nature of the Fermi surface gave a good accordance with the observed anisotropic thermopower which is positive along the c-axis and negative along the b-axis [28].

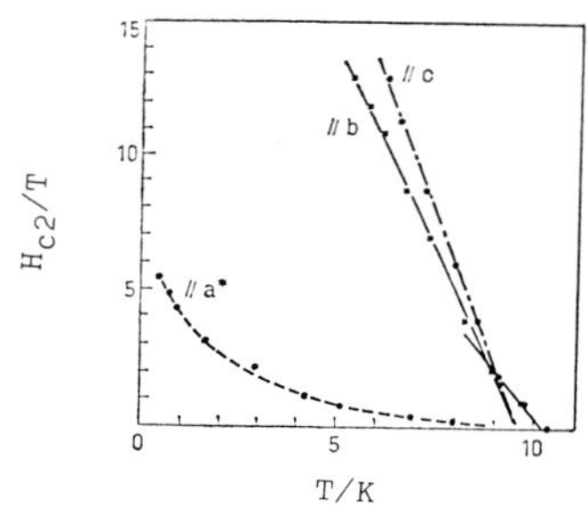

Fig. 7. Temperature dependence of upper critical field of κ-(BEDT-TTF-h_8)$_2$Cu(NCS)$_2$.

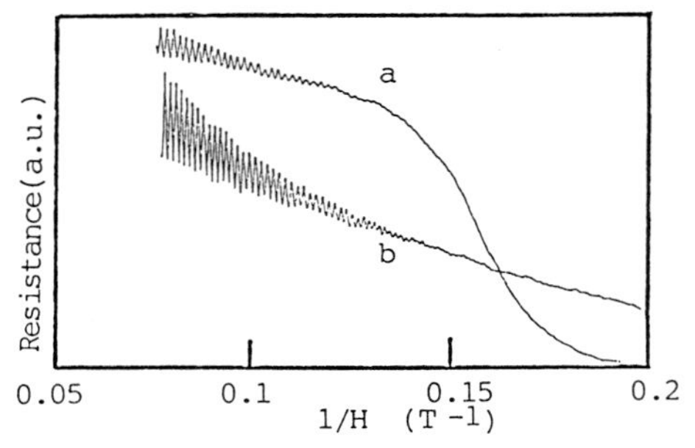

Fig. 8. Shubnikov-de Haas signals (a:1 bar, b:7 kbar) of κ-(BEDT-TTF-d_8)$_2$Cu(NCS)$_2$.

MAGNETIC AND OTHER PROPERTIES [21,28-32]

The d.c. magnetic susceptibility on polycrystalline sample showed that almost constant(with slight decrease at ca. 90K) Pauli paramagnetism of $4.1\text{-}4.6\times10^{-4}$ emu/mol between room temperature to about 10K [29, Fig. 9]. From the room temperature value the density of states per formula unit for a single spin is calculated as $N(E_F)=7.1eV^{-1}$, which is comparable to that of low Tc β-I_3 salt ($6.5eV^{-1}$) and α-I_3 salt ($8.6eV^{-1}$). The critical current density Jc of $1060A/cm^2$ was estimated at 4.9K and 50 Oe. The a.c. susceptibility on polycrystalline sample showed that Tc is 10.3±0.4K and almost 100% of the perfect diamagnetism was observed below 7K at 0.3 Oe [30]. So this salt is a bulk superconductor. The lower critical fields were estimated as 3.0 and 0.5 mTesla for Hc_1//a* and Hc_1//bc, respectively, at 7.3K. We need works on single crystals to discuss in more detail.

No signal ascribed to Cu^{2+} was observed in ESCA at room temperature and EPR down to 4K [28]. A broad Lorentzian signal due to BEDT-TTF$^+$ was observed in EPR with a cylindrical cavity (TE011). Additional sharp signal (ΔH=10-20G, g=2.0075) appeared at the center of the broad signal below 30K. This sharp signal may be due to defect or contamination of another phase of $Cu(NCS)_2$ [18]. Due to the broadening of the main signal and to the appearance of the sharp signal, accurate estimations of g values and line width (ΔH) were not possible below 20K. The constant g-values (2.0078-2.0070) down to 20K where the external magnetic field (H_0) is normal to the bc-plane indicate that the molecular orientation of BEDT-TTF does not change substantially with respect to the crystal axis. The magnitude of the g-values is quite reasonable since H_0 is nearly parallel to the long molecular axis of BEDT-TTF. The line width at room temperature was 61G and increased to about 80G at 110K smoothly. Then ΔH increased more pronouncedly to 100G down to 30K followed by a rapid increase to 120G (20K)(Fig. 10). Since the electrical resistivity increases down to 90K, the increase of the line width can be explained by the increase of scattering rate of conduction electrons in this region. The pronounced increase of ΔH below 90K, especially below 30K, cannot be explained in terms of scattering rate since ΔH does not follow the temperature dependence of the resistivity. So the EPR data are in contrast to the predictions of the Elliot formula for the spin relaxation in metals. We need precise knowledge of temperature dependence of scattering rate (// and ⊥) to know the mechanism of the broadening of ΔH. The EPR intensity obtained with polycrystals is almost constant (3.2×10^{-4}emu/mol at RT) down to 90K in good agreement with the d.c. susceptibility measurements.

In the ^1H NMR measurements on polycrystals [31, Fig. 11] a Korringa relation was observed between 77K and 10K. The T_1T value of this salt (about 1050 secK) is less than 2/3 of that of β-I_3 salt (1730 secK) indicating that density of state of this salt is higher than that of β-I_3 salt. A big enhancement of relaxation rate T_1^{-1} was noticed with a peak at considerably lower temperature than Tc. The peak height is about 30 times than that of the normal state at 3.28KOe. The peak height as well as its position are dependent of the applied magnetic field. Such pronounced enhancement of T_1^{-1} is not explained by known theory [33] and is unexpected for usual superconductors to date.

The superconducting gaps have been measured with a tunneling spectroscopic method on $Cu(NCS)_2$ salt/Al_2O_3/Au junctions [32]. No reliable spectra was obtained at temperatures above 4.2K. Well below Tc the superconducting energy gap somewhat comparable to the BCS theory was detected (2Δ=4meV, $2\Delta/kTc$=4.5 vs. 3.52 for BCS ratio) for one sample. The other sample showed much smaller gap less than 1meV. Therefore at this stage we may postulate that the superconducting gaps of this salt are anisotropic.

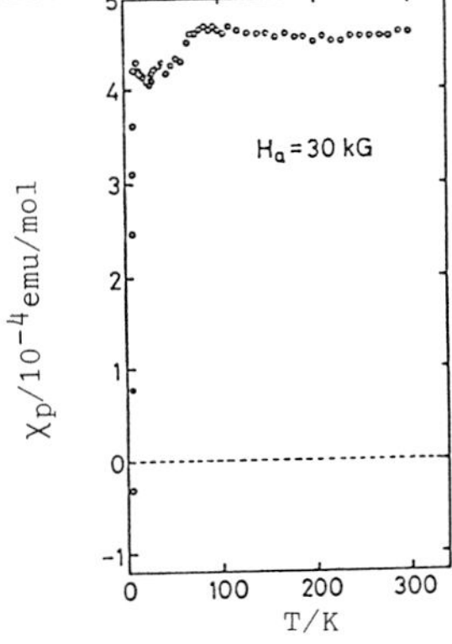

Fig. 9. Temperature dependence of the molar paramagnetic susceptibility at 30kG of κ-(BEDT-TTF-h_8)$_2$Cu(NCS)$_2$ polycrystals.

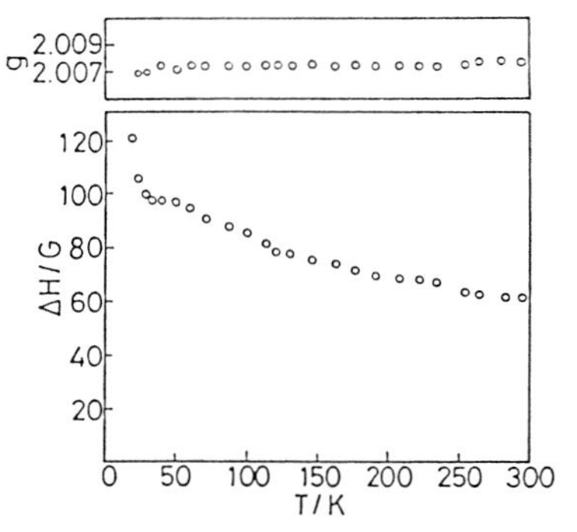

Fig. 10. Temperature dependence of g value and linewidth of κ-(BEDT-TTF-h_8)$_2$Cu(NCS)$_2$ single crystal.

Tc

In Fig. 12, Tc of several BEDT-TTF superconductors are plotted against the effective volume for one electron (Veff). Veff was calculated using unit cell volume (at RT), anion volume (anion is roughly approximated as a cylinder with the radius of the biggest atom in the anion), and numbers of conduction electrons (n) in the unit cell;Veff=(Vunit cell-Vanion)/n. Though Fig. 12 does not include every BEDT-TTF superconductors and the calculated Veff may contain uncertainty, there seems a linear relationship between Tc and Veff even among different phases. It is seen that β-I3 salt has bigger Veff by about 10Å^3 than κ-I_3 salt. Therefore if one could obtain β-phase of $Cu(NCS)_2$ anion, Tc of it might be about 13K as long as the relation is held. Also by choosing an appropriate anion one might extend the line to more than Tc=20K within the family of BEDT-TTF salts. Though there should be certain limit of Tc in this treatment since the larger the effective volume is, the more dispersed BEDT-TTF molecules in a unit cell are, which makes crystal unstable. The manifested feature of the BEDT-TTF superconductors is the strong 2D in the electronic and structural properties. So it is important to search inorganic and organic anions which provide large Veff and layered space where BEDT-TTF molecules can intercalate in a several packing manners.

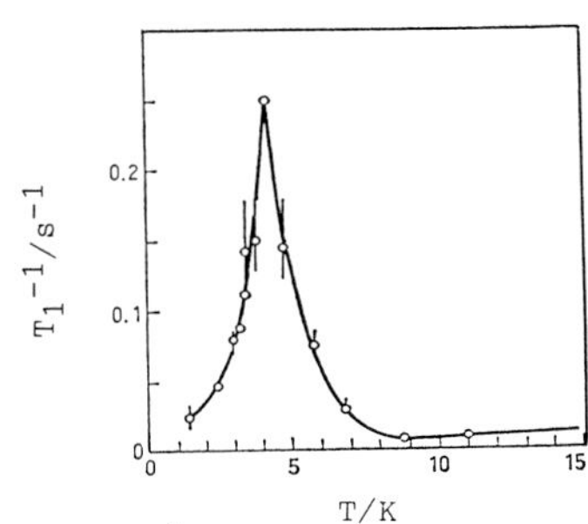

Fig. 11. ^{1}H-NMR relaxation rate $1/T_1$ at 3.28 kOe of κ-(BEDT-TTF-h_8)$_2$Cu(NCS)$_2$ at low temperatures.

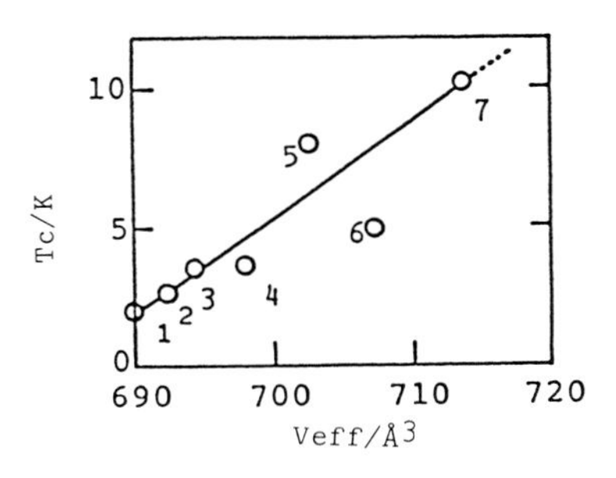

Fig. 12. Plot of Tc vs effective volume Veff. Veff was calculated based on the crystal structures. 1:ReO_4, 2:β-IBr_2, 3:κ-I_3, 4:θ-I_3, 5:high Tc β-I_3, 6:β-AuI_2, 7:κ-$Cu(NCS)_2$ salt of BEDT-TTF.

ACKNOWLEDGEMENTS

We would like to thank all our colleagues: K.Omichi, T.Tsuji, K.Nozawa, T.Sugano, M.Kinoshita, S.Sato, Y.Ishida, T.Yajima (ISSP), T.Mori, H.Inokuchi, T.Inabe, Y.Maruyama (IMS), S.Kagoshima (Univ. Tokyo), T.Takahashi (Gakushuin Univ.), J.Tanaka (Nagoya Univ.)

REFERENCES

1 S.S.P.Parkin, E.M.Engler, R.R.Schumaker, R.Lagier, V.Y.Lee, J.C.Scott and R.L.Greene, Phys. Rev. Lett., 50 (1983) 270.
2 J.M.Williams, H.H.Wang, M.A.Beno, T.J.Emge, L.M.Sowa, P.T.Copps, F.Behroozi, L.N.Hall, K.D.Carlson and G.W.Crabtree, Inorg. Chem., 23 (1984) 3839.
3 H.H.Wang, M.A.Beno, V.Geiser, M.A.Firestone, K.S.Webb, L.Nunez, G.W.Crabtree, K.D.Carlson, J.M.Williams, L.J.Azevado, J.F.Kwak and J.E.Schirber, Inorg. Chem., 24 (1985) 2466.
4 T.Mori and H.Inokuchi, Solid State Commun., 64 (1987) 335.
5 R.N.Lyubovskaya, R.B.Lyubovskii, R.P.Shibaeva, M.Z.Aldoshina, L.M.Gol'denberg, L.P.Rozenber, M.L.Khidekel and Yu.F.Shul'pyakov, Pis'ma Zh. Eksp. Teor. Fiz., 42 (1985) 380. 6 ;R.N.Lyubovskaya, E.A.Zhilyaeva, A.V.Zvarykina, V.N.Laukhin, R.B.Lyubovskii and S.I.Pesotskii, ibid, 45 (1987) 416.
7 E.B.Yagubskii, I.F.Shchegolev, V.N.Laukhin, R.P.Shibaeva, E.E.Kostyuchenko, A.G.Khomenko, Yu.V.Sushko and A.V.Zvarykina, ibid, 40 (1984) 387.
8 E.B.Yagubskii, I.F.Shchegolev, V.N.Laukhin, P.A.Kononovich, M.V.Karatsovnik, A.V.Zvarykina and L.I.Buravov, ibid, 39 (1984) 12.
9 V.N.Laukhin, E.E.Kostyuchenko, Yu.V.Sushko, I.F.Shchegolev and E.B.Yagubskii, ibid, 41 (1985) 68; K.Murata, M.Tokumoto, H.Anzai, H.Bando, G.Saito, K.Kajimura and T.Ishiguro, Solid State Commun., 54 (1985) 1236.
10 V.F.Kaminskii, T.G.Prokhoroea, R.P.Shibaeva and E.B.Yagubskii, Pis'ma Zh. Eksp. Teor. Fiz., 41 (1984) 15.
11 H.Kobayashi, R.Kato, A.Kobayashi, Y.Nishio, K.Kajita and W.Sasaki, Chem. Lett., 1986, 789, 833.
12 A.Kobayashi, R.Kato, H.Kobayashi, S.Moriyama, Y.Nishio, K.Kajita and W.Sasaki, ibid, 1987 459.
13 F.Creuzet, G.Creuzet, D.Jerome, P.Schweitzer and H.J.Keller, J. de Phys. Lett. 46 (1985) L1079; V.B.Ginodman, A.V.Gudenko, I.I.Zasavitskii and E.B.Yagubskii, Pis'ma Zh. Eksp. Teor. Fiz., 42 (1985) 384; W.Kang, G.Creuzet, D.Jerome and C.Lenoir, J. de Phys., 48 (1987) 1035.
14 D.Schweitzer, P.Bele, H.Brunner, E.Gogu, U.Haeberlen, I.Hennig, I.Klutz, P.Swietlik and H.J.Keller, Z. Phys. B-Condensed Matter, 67 (1987) 489.
15 M.Tokumoto, H.Bando, K.Murata, H.Anzai, N.Kinoshita, K.Kajimura, T.Ishiguro and G.Saito, Synth. Met., 13 (1986) 9.
16 J.M.Williams, M.A.Beno, H.H.Wang, U.W.Geiser, T.J.Emge, P.C.W.Leung, G.W.Crabtree, K.D.Carlson, L.J.Azevedo, E.L.Venturini, J.E.Schirber J.F.Kwak and M-H. Whangbo, Physica 136B (1986) 371.
17 T.J.Kistenmacher, Solid State Commun., 63 (1987) 977.
18 Very recent studies of the following groups indicate the existence of other phases of $Cu(NCS)_2$ salt of BEDT-TTF than ours; a) S.Gartner, E.Gogu, I.Heinen, H.J.Keller, T.Klutz and D.Schwietzer, Solid State Commun., submitted, where the magnitude and temperature dependence of EPR line width (alignment of the measurements is not specified) are completely different from ours; b) K.D.Carlson, U.Geiser, A.M.Kini, H.H.Wang, L.K.Montgomery, W.K.Kwok, M.A.Beno, J.M.Williams, C.S.Cariss, G.W.Crabtree, M-H.Whangbo and M.Evain, Inorg. Chem., 27 (1988) 965, where three kinds of EPR line width (15-20, 33-35 & 39, 60-70G; the sample alignment and temperature dependence are not described) are reported. Maybe the last one corresponds to ours but the Fermi surface proposed by them is considerably different from ours; c) N.Kinoshita, K.Takahashi, K.Murata, M.Tokumoto, and H.Anzai, Solid State Commun., submitted, they obtained a salt with metal-insulator transition at around 200K with different crystal structure from ours, see also references in the proceedings of ICSM'88 (Synth. Met. to be published).
19 A.Ugawa, G.Ojima, K.Yakushi and H.Kuroda, Phys. Rev. B, in press.
20 H.Urayama, H.Yamochi, G.Saito, K.Nozawa, T.Sugano, M.Kinoshita, S.Sato, K.Oshima, A.Kawamoto and J.Tanaka, Chem. Lett., 1988 55.
21 H.Urayama, H.Yamochi, G.Saito, S.Sato, A.Kawamoto, J.Tanaka, T.Mori, Y.Maruyama and H.Inokuchi, ibid, 1988 463.
22 K.Oshima, T.Mori, H.Inokuchi, H.Urayama, H.Yamochi and G.Saito, Phys. Rev. B, 37 (1988).
23 K.Oshima, H.Urayama, H.Yamochi and G.Saito, Physica B+C, in press.
24 K.Oshima, H.Urayama, H.Yamochi and G.Saito, J. Phys. Soc. Jpn., 57 (1988) 730.
25 G.Saito, J. J. Appl. Phys., Series 1 Superconducting Materials, (1988) 165.
26 J.E.Schirber and C.J.M.Northrup, Jr., Phys. Rev. B, 10 (1974) 3818.
27 T.Sugano et al., private communication.
28 H.Urayama, H.Yamochi, G.Saito, T.Sugano, M.Kinoshita, T.Inabe, T.Mori, Y.Maruyama and H.Inokuchi, Chem. Lett., 1988 1057.
29 K.Nozawa, T.Sugano, H.Urayama, H.Yamochi, G.saito and M.Kinoshita, ibid, 1988 617.
30 T.Sugano, K.Terui, S.Mino, K.Nozawa, H.Urayama, H.Yamochi, G.Saito and M.Kinoshita, ibid, 1988 1171
31 T.Takahashi, T.Tokiwa, K.Kanoda, H.Urayama, H.Yamochi and G.Saito, Phys. Rev. Lett., submitted.
32 Y.Maruyama, T.Inabe, H.Urayama, H.Yamochi and G.Saito, Solid State Commun., 67 (1988) 35.
33 Y.Hasegawa and H.Fukuyama, J. Phys. Soc. Jpn., 56 (1987) 877.

Organic Metals and Superconductors Based on BEDT-TTF

HATSUMI URAYAMA[1], HIDEKI YAMOCHI[1], GUNZI SAITO[1], and KOKICHI OSHIMA[2]

[1] The University of Tokyo, Institute for Solid State Physics, 7-22-1, Roppongi, Minato-ku, Tokyo, 106 Japan
[2] The University of Tokyo, Cryogenic Center, 2-11-6, Yayoi, Bunkyo-ku, Tokyo, 113 Japan

ABSTRACT

A metal, $(BEDT\text{-}TTF)_3Br_2(H_2O)_2$, and a superconductor, $(BEDT\text{-}TTF)_2Cu(NCS)_2$, based on BEDT-TTF (bis(ethylenedithiolo)tetrathiafulvalene, Fig. 1) are presented. The BEDT-TTF complex with a small anion, Br^-, has been synthesized with the accidental inclusion of H_2O which may stabilize the crystallization. The given black plate crystal, $(BEDT\text{-}TTF)_3Br_2(H_2O)_2$, is isostructual to $(BEDT\text{-}TTF)_3(ClO_4)_2$. Two bromide anions and two water molecules, $Br^-_2(H_2O)_2$, form a planar square cluster with hydrogen bonds. This cluster constructs a insulating sheet which is sandwiched by two-dimensional BEDT-TTF sheets. The electrical resistivity shows the metallic behavior down to 185 K and transforms to semiconducting system, which is also confirmed by the ESR measurements. $(BEDT\text{-}TTF)_2Cu(NCS)_2$ was found to be the first ambient pressure organic superconductor with the Tc higher than 10 K. The salt has a layered structure which is composed of donor and anion sheets. The packing pattern of donors of the $Cu(NCS)_2$ salt is nearly analogous to $\kappa\text{-}(BEDT\text{-}TTF)_2I_3$. In anion layers Cu^I is trigonal coordinated to two N atoms and one S atom to construct a one-dimensional planar polymer. The valence state of Cu^I was observed by ESCA at 298 K and ESR at 298-4 K. One broad Lorentzian ESR signal ascribed to a BEDT-TTF cation radical was obtained. The g-values are independent of temperature, while the linewidth shows broadening with lowering temperature even though the resistivity decreases below 90 K. The ESR magnetic susceptibility is constant down to 90 K, which is suggesting Pauli-like paramagnetism. The thermopower is anisotropic in the two-dimensional bc plane, which is originated from the anisotropy of the band structure.

INTRODUCTION

BEDT-TTF is a talented organic donor to give a variety of charge transfer complexes from semiconductors and metals to superconductors [1]. Contrary to the one-dimensional character of usual organic compounds, pseudo one-dimensional or even two-dimensional character which can prevail over the one-dimensional instability such as Peierls distortion appears in organic conductors of BEDT-TTF. The packing of conducting donors is controllable to some extent with respect to anions of supporting electrolytes by changing the shape and size including the length of linear anion, employing mixed electrolytes, etc. In this paper a organic metal having a small anion Br^-, $(BEDT\text{-}TTF)_3Br_2(H_2O)_2$ [2], and a superconductor containing a polymer anion, $(BEDT\text{-}TTF)_2Cu(NCS)_2$ (having the highest Tc(=10.4 K) among organic superconductors) [3], based upon BEDT-TTF which we have obtained so far are presented.

$(BEDT\text{-}TTF)_3Br_2(H_2O)_2$.

EXPERIMENTAL

Black plate-crystals were grown by electrochemical crystallization of BEDT-TTF in chlorobenzene and tetrahydrofuran by using tetra-n-butylammonium bromide as a supporting electrolyte during 1-2 months. The lattice parameters of $(BEDT\text{-}TTF)_3Br_2(H_2O)_2$ are; triclinic, $P\bar{1}$, a=16.167(2), b=18.125(3), c=7.718(1) Å, α=91.05(2), β=94.09(2), γ=148.18(1)°, V=1176.0(4) $Å^3$ and Z=1. The

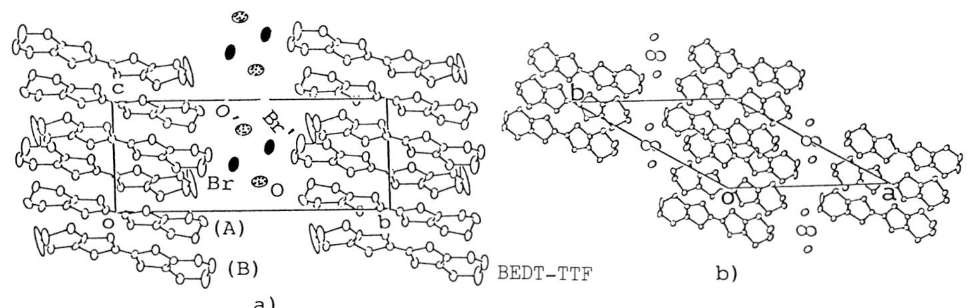

Fig. 1. Crystal structure of $(BEDT\text{-}TTF)_3Br_2(H_2O)_2$.

crystal structure was solved by the heavy atom method and refined by the block-diagonal least-squares method by using 2574 independent reflections (Cu Kα, $2\theta<120°$, $|F_0|>3\sigma(|F_0|)$). The final R is 0.082. The temperature dependence of electrical resistivity was measured by a standard d.c. four-probe technique with a gold paint as an electrode. ESR studies were performed by utilizing a JEOL JES-FE1XG (9.2 GHz) spectrometer over 300-10 K. The sample crystal was placed at the center of cylindrical cavity in the TE_{011} mode.

RESULTS and DISCUSSION

[Crystal structure]

The crystal structure of (BEDT-TTF)$_3$Br$_2$(H$_2$O)$_2$ is illustrated in Fig. 1. The unit cell is composed of three BEDT-TTF molecules, two bromide anions, and two H_2O molecules. The inclusion of H_2O was confirmed by the elemental analysis; Found: C, 26.69; H, 2.09 %. Calcd for $Br_2S_{24}O_2C_{30}H_{28}$: C, 26.40; H, 1.97 %. As the starting materials were purified with the exclusion of H_2O, air humidity had been included into this complex to form the stable single crystals during the course of electrocrystallization for 1-2 months.

(BEDT-TTF)$_3$Br$_2$(H$_2$O)$_2$ has a layered structure: the organic conducting sheets and the inorganic insulator ones stack alternatingly along the [110] axis. In the latter sheets, Br, Br', O, and O' of Br$_2$(H$_2$O)$_2$ form a rectangular unit (Fig. 1a). The distances of Br••O and O••Br' (3.36 and 3.26 Å, respectively) are a little bit longer than the van der Waals contact (3.22 Å) and the angle of Br-O-Br' (111°) is close to that of H-O-H (104°), which suggests the hydrogen contact of Br••O-H-O around an inversion center. This part nearly resembles to (BEDT-TTF)$_3$Cl$_2$(H$_2$O)$_2$ [4]. A little different point between them is that Cl$_4$(H$_2$O)$_4$ is one unit which is composed of a rectangular Cl$_2$(H$_2$O)$_2$ and the additional hydrogen bonded 2(Cl(H$_2$O)). Therefore the unit cell volume of the bromide complex is a half as large as that of the chloride one.

In the donor layers two BEDT-TTF molecules are crystallographycally independent. One(A) is on the center of symmetry and the other(B) is on the general position as shown in Fig. 1. The similar intramolecular bond lengths and angles of two independent donors indicate no charge separation which leads the donor to be BEDT-TTF$^{+1/2}$. As shown in Fig. 2, there are three interacting chains of donors: a stacking along the [111] axis, two side-by-side ones along the [112] and [221] axes. Among them short S••S contacts (<3.6 Å) are only found in the [221] direction. However, this does not mean that the strongest interaction is observed along [221] direction since an extended Hückel calculation and reflectance spectra in the (BEDT-TTF)$_3$(ClO$_4$)$_2$, which is isostructual to the bromide complex show that the strongest interaction [5] and dispersion [6] are detected along another side-by-side direction [112] but not along the direction [221] with short contacts. The inclusion of H_2O made the complex of a small bromide anion belong to structurally a tetrahedral anion system in BEDT-TTF salts.

[Electrical properties]

The electrical resistivity versus temperature on the (110) plane of a typical crystal is plotted in Fig. 3. Above 185 K the salt is weakly metallic with a room-temperature resistivity of 0.013 Ω•cm. The material undergoes a transition to a semiconducting system below 185 K. The resistivity at this temperature is still below (0.05 Ωcm). The activation energy just below 185 K is obtained as 0.021 eV. In order to confirm the metallic state and observe the transition property, ESR measurements were performed.

[Electron spin resonance studies]

As shown in Fig. 1, the two-dimensional sheets parallel to the (110) plane are sandwiched by the anion sheets (2(H$_2$O)..2Br$_2$). The Lorentzian ESR spectra were recorded with the applied magnetic field normal to the [110] axis. The angular dependence of g-values and peak-to-peak linewidths shows that when the magnetic field is applied along the a' axis (⊥[110]) which is closely parallel to the long axis of BEDT-TTF molecules, $g_{a'}$(=2.014) shows the largest g value. On the other hand, with the applied field along the c' axis (⊥[110]) which is nearly parallel to the donor's stacking axis the smallest value ($g_{c'}$=2.003) was obtained. These g-values indicate that these signals are ascribed to the BEDT-TTF cation radicals [7].

The temperature dependence of the g-values, the peak-to-peak linewidths, and the spin susceptibility from 300 K to 10 K were observed when the magnetic field is applied along the a'

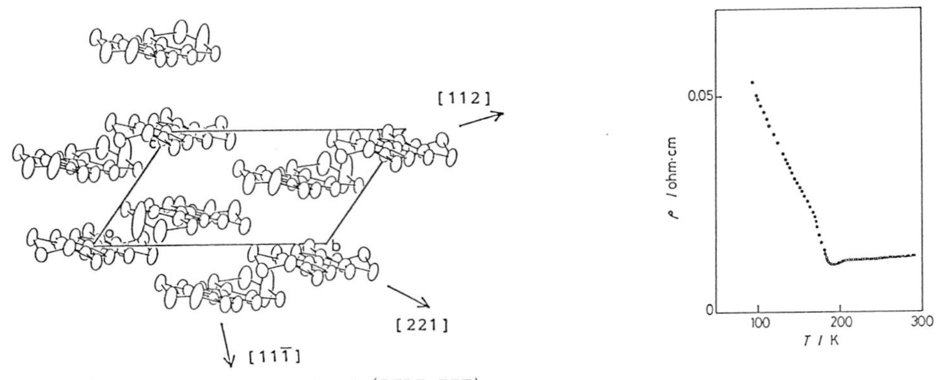

Fig. 2. Molecular arrangement of (BEDT-TTF)$_3$Br$_2$(H$_2$O)$_2$.

Fig. 3. Electrical resistivity of (BEDT-TTF)$_3$Br$_2$(H$_2$O)$_2$.

axis. Over the entire temperature range the g-value (2.0013) does not change significantly. In the high temperature region above around 200 K no change of the linewidth (50 G) is observed. Below it the linewidth gradually narrows with decreasing temperature to 3.5 G at 12 K. The spin susceptibility above 200 K is relatively large (9.7×10^{-4} emu/mol) and independent of temperature, which is consistent with the Pauli-paramagnetism expected for a metal. This behavior is agreeable to the metallic behavior of electrical resistivity above 185 K. Assuming the one-dimensional tight-binding model, magnetic susceptibility can be described as

$$\chi = \mu_B^2 N_A / \pi t \sin(\pi N'/2)$$

where N_A is Avogadoro's number, μ_B is Bohr magneton, N' is the amount of charge transfer per donor molecule. From the above equation, the transfer integral t is estimated as 0.038 eV. In the low temperature region below around 200 K, the spin susceptibility decreases with exponential equation of the form,

$$\chi = C/T \cdot \exp(-J/k_B T)$$

in which χ is the spin susceptibility, C is the Curie constant, J is the magnetic activation energy, and k_B is the Boltzman constant. The calculated fitting gives J of 0.017 eV. The decrease in both the spin susceptibility and the linewidths below around 200 K is consistent with the metal-semiconductor transition of electrical resistivity. The behavior of the linewidth and the magnetic susceptibility is similar to that of $(BEDT\text{-}TTF)_3(ClO_4)_2$ [8]. However, the g-value of the ClO_4 salt changes clearly at the transition temperature from metal to semiconductor, whereas that of the bromide salt is constant. Therefore the further study is necessary to clarify whether the transition is originated from Peierls instability or not.

$(BEDT\text{-}TTF)_2Cu(NCS)_2$

EXPERIMENTAL

We have tried various methods to synthesize the $(BEDT\text{-}TTF)_2Cu(NCS)_2$ salt: in situ electrocrystallization of (A)CuSCN, KSCN, 18-crown-6 ether, (B)CuSCN, TBA^+•SCN, or (C)(K^+18-crown-6)$Cu(NCS)_2$ in 1,1,2-trichloroethane, chlorobenzene, 1,2-dichloroethane, THF, or EtOH under 1.5 - 5µ A. So far, only one phase, κ, which gives black thin plate crystals (2 x 1 x 0.04 mm^3) has been studied in our group.

The crystal structure of this complex was solved with utilizing X-ray reflections collected by the Rigaku automated four-circle diffractometer (Mo Kα, $2\theta<60°$, $|F_o|<3\sigma|F_o|$) and refined by the block-diagonal least-squares method. The crystal data of $(BEDT\text{-}TTF)_2Cu(NCS)_2$ are shown in Table 1. The final R values are 0.0394 measured at 104 K and 0.0837 at 298 K. The isotropic temperature factors were used for the hydrogen atoms.

The electrical resistivity was determined by the d.c. four-probe method. A gold wire (10 µm diameter) was used for an electrical lead with a gold paint as an electrode. The ESCA spectra were recorded on a VG-1000 spectrometer with monochromatic Mg Kα radiation under 5×10^{-8} Torr below. ESR measurements were performed at X-band (9.2 GHz) by a JEOL JES-FE1XG spectrometer by utilizing a cylindrical cavity (TE_{011}) in the temperature range of 4-300 K with an Air Products LTR-3-110 He gas flow cryostat and a Scientific Instruments Series 3700 digital temperature controller. The

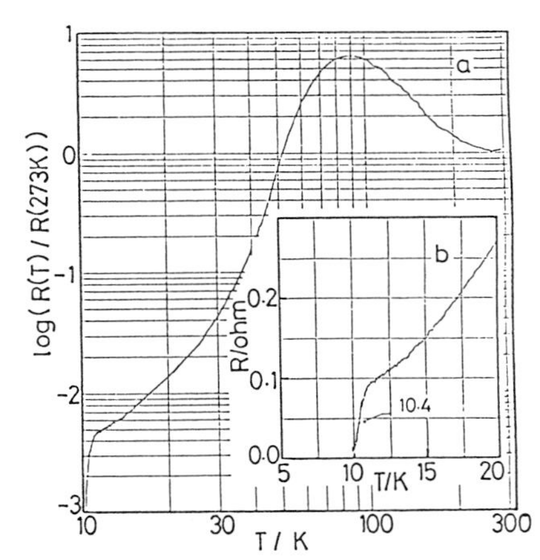

Fig. 4. Electrical resistivity of $(BEDT\text{-}TTF)_2Cu(NCS)_2$.

TABLE 1

Crystal data of $(BEDT\text{-}TTF)_2Cu(NCS)_2$

	298 K	104 K
a /Å	16.248(5)	16.382(4)
b	8.440(2)	8.402(2)
c	13.124(5)	12.833(4)
β /°	110.30(3)	111.33(2)
V /Å^3	1688.0(9)	1645.3(7)
	monoclinic, $P2_1$, and Z=2	

temperature dependence of thermoelectric power was measured by attaching single crystals with 1 μm thick gold foils to two boron nitride heat sinks [9]. The thermopower derived Au foils is corrected from the data.

RESULTS and DISCUSSION

[Electrical resistivity]

The temperature dependence of the electrical resistivity in $(BEDT\text{-}TTF)_2Cu(NCS)_2$ is depicted in Fig. 4. The room temperature conductivity along the crystal long axis, the b-axis, was 14-37 S·cm^{-1}. On the bc plane the resistivities are almost isotropic, while along the a* axis, the resistivity is 600-1000 times higher than that on the bc plane. Down to 270 K the weak metallic behavior was observed. Thereafter the resistivity increases with lowering temperature and the maximum value at around 90-100 K is 4-6 times as large as that at room temperature. As the purified material affords the same curious behavior, the origin of the resistivity increase is still in question. Below 90 K a metallic behavior appeared again following by a superconducting transition at 10.4 K. Addition to the $Cu(NCS)_2$ salt, we have synthesized the $Ag_x(SCN)_y$ (or $Ag_{x'}(SeCN)_{y'}$) salt [10] with electrochemical oxidation of BEDT-TTF, AgSCN (or AgSeCN), KSCN (or KSeCN), and 18-crown-6 ether in benzonitrile (or 1,1,2-trichloroethane). The former salt is a black rhombic plate which is a semiconductor of Ea (=0.12 eV) with a room-temperature resistivity (ρ_{RT}) of 0.4 Ωcm. The latter one is a black needle crystal of a semiconducting behavior (Ea=0.08 eV) with ρ_{RT} of 0.7 Ωcm.

[Crystal Structure]

The crystal structure at 104 K is shown in Fig. 5. This salt has a layered structure: a conducting BEDT-TTF sheet stacks on an insulating $[Cu(NCS)_2]$ sheet alternatingly along the a-axis. In donor sheet the packing pattern of BEDT-TTF moloecules is quite similar to that of κ-$(BEDT\text{-}TTF)_2I_3$. However, the space group is lowered from $P2_1/c$ to $P2_1$ by judging from the strong reflections of ((h0l),l=2n). Two BEDT-TTF molecules are crystallographically independent. These two donors form a pair and each pair places perpendicularly (84-87°) to one another. Moreover there are short S••S contacts between intra- and interpair to construct the two-dimensional layer as shown in Fig. 3.

The bond lengths of the central C=C double bond of two independent BEDT-TTF molecules are the same (1.36 Å), suggesting BEDT-TTF$^{+1/2}$. The BEDT-TTF molecules are slightly bent with the dihedral angles between the least square planes of three tetrathioethylene moieties of 9.4, 3.2° (donorA), 3.1, 10.1° (donorB). Though the κ-$(BEDT\text{-}TTF)_2I_3$ salt has a conformational order of ethylene group at room temperature, one side of the ethylene conformations for the $Cu(NCS)_2$ salt is disordered at 298 K and the both sides are ordered at 104 K.

The crystal structure at 298 K is closely analogous to that at 104 K. The slight differences between the structure at 298 K and that at 104 K are that a) the conformational disorder of the ethylene group exists at 298 K (vide supra), b) the intrapair distance becomes a little wider (3.38 Å at 298 K and 3.30 Å at 104 K), and c) the number of S••S contacts is smaller (Fig. 3).

Based on the crystal structure the extended Huckel band calculation was performed. The band structure are analogous to those of κ-$(BEDT\text{-}TTF)_2I_3$ to show the two-dimensional feature. However, since there is no inversion center because of the space group $P2_1$, the degeneracy on MZ is dissolved and both open and closed Fermi surfaces were obtained (Fig. 9). The observation of Shubnikov-de Haas signal strongly indicates the existence of the closed Fermi surface [2]. An optical isomerism was observed which is also a result of the crystal symmetry. The specific rotatory power of $[\alpha]^{25}_{632.8nm}$=230° was obtained.

Another peculiar feature of the crystal structure is an anion arrangement (Fig. 5). Two crystallographically independent N-C-S groups (I,II) and Cu^I construct a V-shape unit. As Cu^I is coordinated by two nitrogen atoms in a unit and one sulfur atom in a neighboring unit, V-shape unit

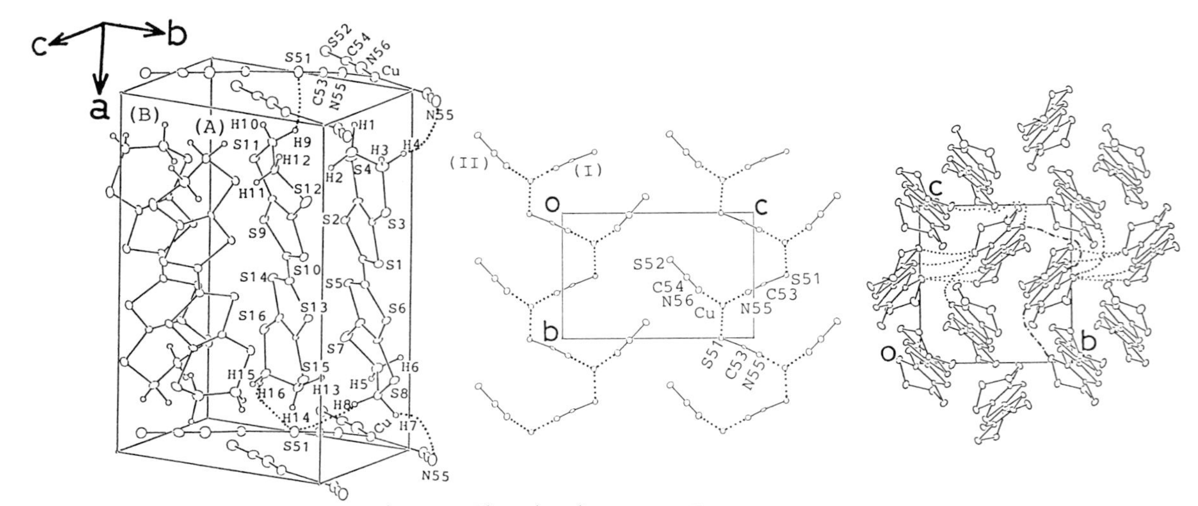

Fig. 5. Crystal structures of $(BEDT\text{-}TTF)_2Cu(NCS)_2$ at 104 K.

was linked by the next one to construct a planar one-dimensional zig-zag polmer along the b-axis. This trigonal coordination of Cu^{I} is very unique. Another novel example is $K[Cu(CN)_2]$, where Cu^{I} is coordinated by two carbon and one nitrogen atoms and they form a not planar but a spiral one-dimensional polymer [11]. Since there is no trigonal coordination of Cu^{I} , the copper anion should be Cu^{I}, which is proved in ESCA and ESR measurements (vide infra).

Some specific cation-anion contacts in the $Cu(NCS)_2$ salt are observed at 104 K. The atomic contacts of N(anion)••H(donor)(<2.6 Å) and S••H(<2.9 Å) within the van der Waals radii are shown in Fig. 5). Both sides of ethylene group of BEDT-TTF molecules are linked by the anion chain, whereas in κ-$(BEDT\text{-}TTF)_2I_3$ such specific contacts were observed only one side of ethylene groups. Whether the key of the high Tc is in hydrogen-bond or not is an interesting subject.

The temperature dependences of the lattice parameters shows that the thermal compressibility of the c-axis decreases monotonically with lowering temperature and five times larger than that of the b-axis. On the other hand, the length of the a-axis is almost constant down to 270 K, and then gradually increases. However, the distance between two anion layers, $a\cdot\sin\beta=1/a^*$, is constant on an entire temperature range, which shows a slight parallel shift of layers due to the subtle movement of BEDT-TTF molecules. The behavior of the a-axis is reminiscent of the temperature dependence of the electrical resistivity of this salt. The relation between them has been further investigated.

[ESCA and ESR measurements]

Figure 6 shows the ESCA spectrum of $(BEDT\text{-}TTF)_2Cu(NCS)_2$ at room temperature. The Cu $2p_{1/2}$ and Cu $2p_{3/2}$ electron peaks were recorded at 951.9 and 931.9 eV, respectively. These binding energies are reasonable for Cu^{I} specimen. The similar spectra of CuSCN was detected under the same conditions. Neither characteristic shake-up satellites nor shoulder peaks of Cu^{II} were present in these spectra.

No Cu^{II} peak was also observed in the ESR measurements. Figure 7 shows the typical signals of $Cu(NCS)_2$ salt. At 295 K one broad Lorentzian signal was recorded. With decreasing temperature the broad one decreased, whereas an additional sharp signal (g=2.0075) appeared. As a typical g-values of Cu^{II} is 2.05-2.50 [12], the sharp one may not Cu^{II} but some imperfections of the crystal. The angular dependences of g-values and linewidths for a single crystal at 298 K shows that the maximum (2.0095) and the minimum (2.0057) of g-values were obtained when the magnetic field was applied nearly parallel to the molecular long axis and tilted to the molecular short axis by 50°, respectively, in the crystal. The broad signal can be attributed to the BEDT-TTF cation radical [7].

The g-values and linewidths (ΔH) versus temperature are plotted in Fig. 8. Below 20 K reliable g-values were not available because of the broadening of the spectra. The g-values are independent (2.0078-2.0070) of temperature over entire measured temperature range. The data suggest nearly no spin-orbit coupling with Cu^{II} if any. Therefore the ESR data strongly confirm that the $Cu(NCS)_2$ salt is cuprous material. The specific feature was that the linewidth showed broadening from 61 G at 295 K to 100 G at 30 K even though the resistivity decreased from around 90 K to 10.4 K. This is the exception of the Elliot rule ($\Delta H \propto \rho$) [13].

The ESR magnetic susceptibilities (χ_s) were measured on single crystals with random orientation to have a satisfactory signal-to-noise ratio. As shown in Fig. 8 χ_s is nearly temperature independent down to 90 K which indicates Pauli-like paramagnetism expected for a metal though the electrical resistivity was quite likely semiconducting. Below 60 K, ESR signal diminished maybe due to skin effect. The intensity at room temperature, 3.2×10^{-4} emu/mol, is consistent to that of the d. c. susceptibility, 4.6×10^{-4} emu/mol. Assuming the tight-binding approximation for the two-dimensional quarter filled band, χ_s is written as,

$$\chi s = \mu_B^2 N_A \cdot \pi^{1/2} \cdot \{\pi t \cdot \sin(\pi^{1/2})\}^{-1}$$

where μ_B is the Bohr magneton and N_A is Avogadro's constant and transfer integral is estimated as t=0.06 eV.

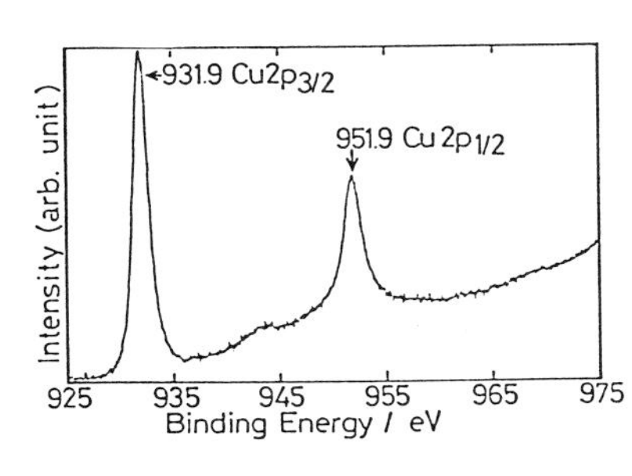

Fig. 6. ESCA spectrum of $(BEDT\text{-}TTF)_2Cu(NCS)_2$.

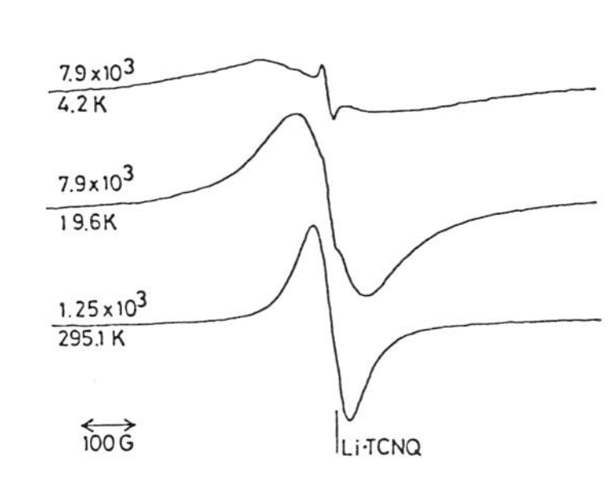

Fig. 7. ESR signals of $(BEDT\text{-}TTF)_2Cu(NCS)_2$.

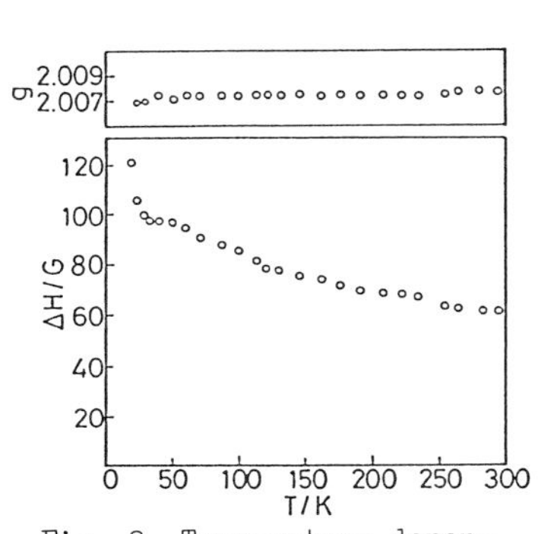

Fig. 8. Temperature dependence of g-values and line width for $(BEDT\text{-}TTF)_2Cu(NCS)_2$.

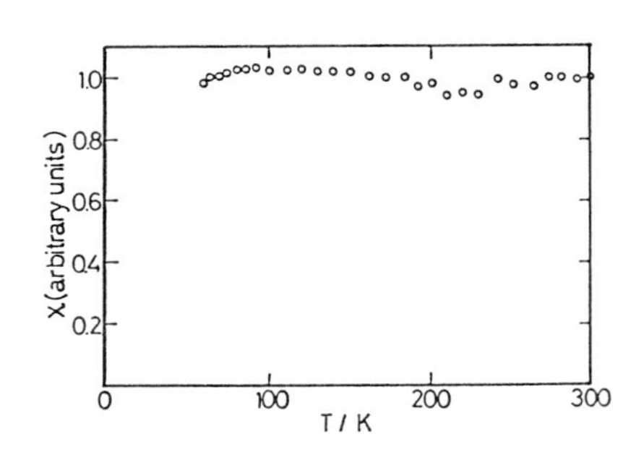

Fig. 9. Temperature dependence of ESR magnetic susceptibility for $(BEDT\text{-}TTF)_2Cu(NCS)_2$.

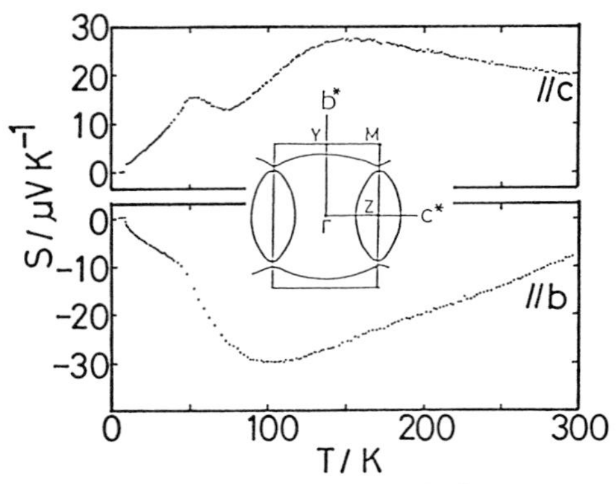

Fig. 10. Temperature dependence of thermopower for $(BEDT\text{-}TTF)_2Cu(NCS)_2$.

[Thermoelectric power]

Temperature dependences of thermopower (S) along the b and c axes are shown in Fig. 9. Since conducting carriers run only in donor layers because of no mixed valance state of $[Cu(NCS)_2]$ in anion layers, the thermopower should be positive assuming the simple tight-binding model. Nevertheless, S along the c axis is positive, while S along the b axis is negative on an entire temperature range measured above 10 K. Unexpectedly the negative S on a certain temperature range can be often observed in BEDT-TTF salts, BF_4, I_3(β), CuI_6, and Hg_3Cl_8 salts [14]. The behavior of S for $(BEDT\text{-}TTF)_2Cu(NCS)_2$ was quantitatively explained on the basis of the band structure calculation on the extended Huckel method [15]. The band calculation shows the open Fermi surface near YM is electron-like which leads S negative and the closed Fermi surface around the Z point is hole-like which leads S positive (Fig. 10). Therefore the specific behavior of S may be the intrinsic feature of the $Cu(NCS)_2$ salt. The additional hump at 50 K can not be explained by the band calculation, but can be attributed to the phonon-drag. The S(=0) along both the b and c axes below 10 K exhibits the superconducting state.

REFERENCES

1 E. B. Yagubskii, I. F. Shchegolev, V. N. Laukhin, P. A. Kononovich, M. V. Karatsovnic, A. V. Zvarykiha, and L. I. Buravov, Pis'ma Zh. Eksp. Theor. Fiz. 39 (1984) 12; H. H. Wang, H. A. Beno, V. Geiser, M. A. Firestone, K. S. Webb, L. Nunez, G. W. Grabtree, K. D. Carlson, J. M. Williams, L. J. Azevedo, J. F. Kwak, J. E. Schirber, Inorg. Chem., 24 (1985) 2466; J. M. Williams, H. H. Wang, M. A. Beno, T. J. Emge, L. M. Sowa, P. T. Copps, F. Behroozi, L. N. Hall, K. D. Carlson, G. W. Crabtree, Inorg. Chem., 23 (1984) 3839; R. N. Lynbovskaya, R. B. Lyubovskii, R. P. Shibaeva, M. Z. Aldoshina, L. M. Gol'denberg, L. P. Rozenber, M. L. Khidekel, and Yu. F. Shul'pyakov, Pis'ma Zh. Eksp. Teor. Fiz., 42 (1985) 380; R. N. Lyubovskaya, E. A. Zhilyaeva, A. V. Zvarykina, V. N. laukhin, R. B. Lyubovskii, and S. I. Pesotskii, ibid., 45 (1987) 416.

2 H . Urayama, G. Saito, A. Kawamoto, and J. Tanaka, Chem. Lett., 1987, 1753; H. Urayama, G. Saito, T. Sugano, M. Kinoshita, A. Kawamoto, and J. Tanaka, Synth. Met., in press.

3 H. Urayama, H. Yamochi, G. Saito, K. Nozawa, T. Sugano, M. Kinoshita, S. Sato, K. Oshima, A. Kawamoto, and J. Tanaka, Chem. Lett., 1988 55; H. Urayama, H. Yamochi, G. Saito, T. Sato, T. Sugano, M. Kinoshita, A. Kawamoto, J. Tanaka, T. Inabe, T. Mori, Y. Maruyama, H. Inokuchi, and K. Oshima, Synthe. Met. in press and references cited in.

4 T. Mori and H. Inokuchi, Chem. Lett., 1987 1657; D. Chasseau, D. Watkin, M. Rosseinsky, M. Kurmoo, and P. Day, Synth. Met., 24 (1988) .

5 H. Kobayashi, R. Kato, T. Mori, A. Kobayashi, Y. Sasaki, G. Saito, T. Enoki, and H. Inokuchi, Chem. Lett., 1984, 179.

6 H. Kuroda, K. Yakushi, H. Tajima, and G. Saito, Mol. Cryst. Liq. Cryst. 125 (1985) 135.

7 T. Sugano, G. Saito, and M. Kinoshita, Phy. Rev. B 34 (1986) 117.

8 T. Enoki, K. Imaeda, M. Kobayashi, H. Inokuchi, and G. Saito, Phys. Rev. B 33 (1986) 1553.

9 P. M. Chaikin and J. F. Kwak, Rev. Sci. Instrum., 46 (1975) 218.

10 The results of $Ag_x(SCN)_y$ and $Ag_{x'}(SeCN)_{y'}$ salts were presented at the symposium of Annual Meeting of Chemical Society of Japan, Tokyo, April 1987. The crystal data of (BEDT-TTF)-$Ag_x(SCN)_y$ are: orthorhombic, a=11.590(1), b=40.132(4), c=4.254(1) Å, and V=1978.9(6) $Å^3$. The diffuse track along the c axis measured by oscillation photographs was detected.

11 D. T. Cromer, J. Phys. Chem., 61 (1957) 1388.

12 A. Abragam and B. Bleaney, Electron Paramagnetic Resonance of Transiiton Ions, (Clawrendon, Oxford, 1970).

13 R. J. Elliot, Phys. Rev. 96 (1954) 266; A. W. Overhauser, ibid., 89 (1953) 689.

14 Int. Conf. on Electronics of Organic Materials, ELORMA 1987 Tashkent; K. Mortensen, J. M. Williams, and H. H. Wang, Solid State Commun., 56 (1985) 105; V. A. Merzhanov, E. E. Kostyuchenko, O. E. Fabrt, I. F. Shchegolv, and E. B. Yagubskii, Zh. Eksp. Teore. Fiz., 89 (1985) 292.

15 T. Mori and H. Inokuchi, submitted to J. Phy. Soc. Jpn.

4
Oxide Superconductors

4.1 Mechanisms of Superconductivity

The Spin Bag Mechanism of High Temperature Superconductivity

J.R. SCHRIEFFER, X.-G. WEN, and S.-C. ZHANG

Institute for Theoretical Physics, University of California, Santa Barbara, CA 93106, USA

ABSTRACT

In oxide superconductors the local suppression of antiferromagnetic correlations in the vicinity of a hole lowers the energy of the system. This quasi two-dimensional bag of weakened spin order follows the hole in its motion. In addition, holes prefer to share a bag, leading to a strong pairing attraction and a high T_c superconductivity. There are many experimental consequences of this mechanism for both the superconducting and normal phases.

INTRODUCTION

In the pairing theory of superconductivity, it is assumed that the elementary excitations (quasi-particles) of the normal phase, like electrons or holes, are charged spin $\frac{1}{2}$ carriers of momentum with decay rate $\frac{1}{\tau}$ small compared to their excitation energy ϵ_k above the Fermi surface. The theory further assumes there exists a net attractive interaction which leads to a cooperative phase transition to the superconducting state, with pairs of quasi-particles having opposite directions of momentum and spin being bound in a pair state $\varphi(r_1 - r_2)$ whose range ξ_0 (the pairing coherence length) is given by $\hbar v_F/\pi\Delta_{SC}$ where v_F is the Fermi velocity of the quasi-particles and Δ_{SC} is the superconductor gap parameter, *i.e.*, the minimum energy to create a quasi-particle. For metallic superconductors $\xi \sim 10^2 - 10^4$ Å while $\xi \sim 12 - 20$ Å for an oxide superconductor.

While there is no firm proof at present, it is widely accepted that the basic aspects of the pairing theory continue to apply to the oxide superconductors [1]. The essential questions are

1) what is the nature of the charged spin $\frac{1}{2}$ quasi-particles in the normal phase and
2) what produces the net attractive interaction $V_{kk'}$ between these excitations which leads to the pairing condensation and high T_c?

Several experimental facts helped to focus theoretical efforts. The discovery of a strongly reduced isotope effect [2] ($\alpha \simeq 0.2$) in La_{2-x} $(Ba,Sr)_x$ CuO_{4-y} (214) with $T_c \sim M^{-\alpha}$ and the (nearly) vanishing α for Y Ba_2 Cu_3O_{7-y} (123) showed that as opposed to conventional materials phonons do not play a dominant role in causing superconductivity in the oxides. This is consistent with theoretical upper limits for phonon mediated T_c which are 30–40° K versus greater than 100° K as observed.

An important experiment was the discovery of commensurate antiferromagnetism [3] in the 214 ($x \rightarrow 0$) and 123 ($y \rightarrow 1$) materials, with strong *ab* plane spin correlations persisting above the Neél temperature T_N. As holes are doped into the antiferromagnet ($x > 0.03$ for 214 or $y > 0.6$ for 123) T_N drops to zero, with strong antiferromagnetic spin correlations remaining with a correlation length ξ_{S-S} large compared to the lattice constant a. With further hole doping, T_c of the superconducting phase rises, coexisting with the antiferromagnetic correlation. A number of investigators suggested that in some fashion antiferromagnetic correlations lead to the required pairing attraction, much as phonons do for conventional superconductors.

Unfortunately, the direct analogy does not work. As shown by Scalapino *et al.* [4], if one starts from the paramagnetic phase of the doped oxide (no long range antiferromagnetic order), the pairing potential given by the exchange of one antiparamagnon (the analog of one phonon) leads to at repulsive $V_{kk'}$. In essence, $V_{kk'}$ is proportional to the wave vector dependent spin susceptibility $\chi(k - k')$ which is necessarily positive if the paramagnetic phase is stable. While non-*s* wave superconductivity can occur in this approach, T_c is found to be very low.

SPIN BAGS

We have taken a complementary starting point [5]. In view of the strong in-plane antiferromagnetic correlation in the superconducting phase with $\xi_{S-S} \sim 30 - 8$ Å, suppose we begin by concentrating on a region where the staggered antiferromagnetic order parameter

$$\vec{M}_n = \vec{S}_n \cos \vec{Q} \cdot \vec{R}_n \tag{1}$$

is roughly constant. Here $\vec{Q}$ is the wavevector of the spin order and $\vec{R}_n$ is the location of the nth site. Of course, the spacial size and location of these regions will fluctuate in time, reflecting the q and ω broadening of the spin-spin correlation function $S(q,\omega)$ around the Bragg peak. However, if the typical length and time scales involved in forming the quasi-particles are short compared to the length and time scales for domain fluctuations, this quasi-static background picture should be a good starting point.

The essential point is the following: if one adds a hole to a region where M is initially uniform M_0, the hole will oppose spin ordering and reduce M to a smaller value. The reason for this reduction depends on whether one is in the spin density wave (SDW) or the Heisenberg (localized electron) limit. In the former, the SDW is formed when the Fermi surface is in contact with the nesting surface for a commensurate SDW, $Q = (\frac{\pi}{a}, \frac{\pi}{a})$, see Fig. 1. This happens for one electron per cell, *i.e.*, $x = 0$. If we add a hole in a localized wavepacket of size L, the local density of electrons is reduced by $(a/L)^2$, pulling the local Fermi surface away from the nesting surface. This reduces the tendency to antiferromagnetism and reduces M_0 to $M_B(L)$ in this vicinity. The smaller L the larger is the reduction.

In the Heisenberg localized electron limit, a hole also reduces the local spin order. In essence, the reduced coordination number due to the hole leads to weaker Neél ordering fields on the spins neighboring the hole. In this case, quantum fluctuations from $S_i^+ S_{i+1}^-$ will be more effective in reducing the ordered moment, reducing the order parameter. Thus, a hole reduces M locally through longitudinal (S_z) fluctuations for the SDW limit and through transverse $(S_\pm)$ fluctuations for the Heisenberg limit. This is supported by explicit calculations.

A second physical effect of the localized hole is that the SDW energy gap Δ_{SDW} is proportional to the local order parameter M. Since the minimum energy to add a hole is Δ_{SDW} if M is reduced to M_B, so is the hole self-energy reduced to Δ_B.

Finally, there remains the kinetic energy required to localize the hole in the region of size L, $h^2/2mL^2$, where m is the SDW band mass. Therefore, the total energy is the sum of the exchange, gap and localization energies

$$E(L, \Delta_B) = 1/2mL^2 + (\alpha/2w)(\Delta_B - \Delta_{\text{SDW}})^2\ L/a + (\Delta_B - \Delta_{\text{SDW}}). \tag{2}$$

where α is of order unity.

Minimizing with respect to Δ_B and L one finds Δ_B is a fraction of Δ_0 and L is of order the SDW electron coherence length $v_F/\pi\Delta_{SDW}$ which is likely of order 3-5 lattice spacings in these materials. We find in more complete calculations the bags are of cigar or star shapes depending on the location of k on the Fermi surface.

While the holes are strongly dressed by the loss of antiferromagnetic order inside the bag in which they live, the entire excitation – bag with its hole – has spin $\frac{1}{2}$ and charge $+e$. The effective mass of the bag is strongly renormalized, as is consistent with the large mass enhancement observed in the heat capacity γ and in the optical properties. In addition, one expects strong dressing effects in the tunneling conductance dI/dV, as have been seen experimentally.

While the above picture is based on a static SDW domain picture, it is evident that the results are insensitive to SDW domain fluctuations of frequency $\omega_0 \ll \Delta_{SDW}$. On the other hand, spin fluctuations about the domain are handled fully regardless of their frequency.

SPIN BAG MECHANISM OF HIGH T_c

Now that we have discussed the nature of the quasi-particles in the normal phase, how do quasi-particles interact? Simply stated, the bag potential established by one hole provides a region of space with a lower gap Δ_B in which a second hole can reside without paying the price to reduce the magnitude of the exchange energy. The above simple model as well as diagram perturbation calculations about the SDW ground state show that the bag-bag potential is attractive, with a depth on the order of a fraction of Δ_{SDW} and a range of order $v_F/\pi\Delta_{SDW}$. In addition there is the strong short range Coulomb potential. Fortunately, the size of the bag is sufficiently large, that it appears that two holes can live within a single bag yet stay further apart than the range of the screened Coulomb repulsion.

An approximate solution of the gap equation based on the spin bag mechanism shows that

$$k_B T_c \sim \Delta_{SDW}\ \exp\left[-\frac{1}{N(E_F)V}\right]. \tag{3}$$

Since Δ_{SDW} is a large energy $\sim 0.1 - 0.5$eV, it appears that a large value of T_c can be obtained for a reasonable range of hopping t and on site Coulomb U energies.

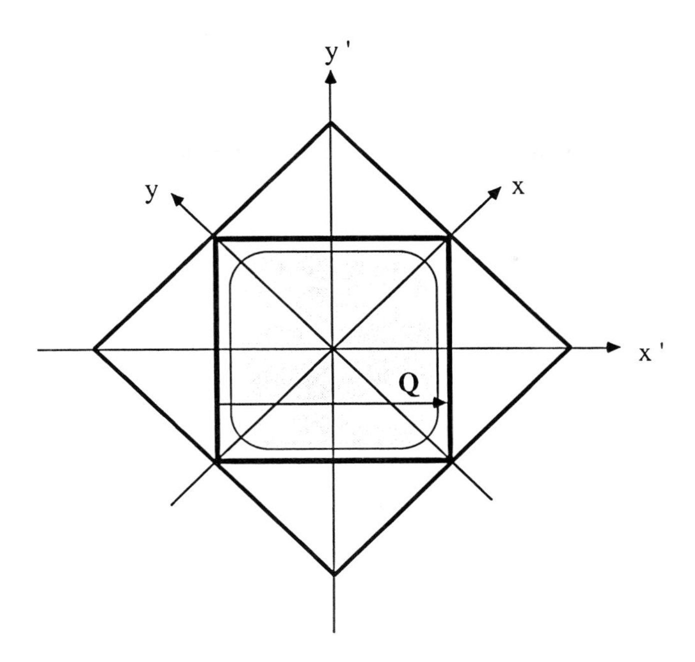

Figure 1: The large square represents the first Brillouin zone. The small square represents the Fermi surface at half fill. Near half filling, the shaded region is occupied by electrons. Q is the nesting vector.

CONCLUSION

We have proposed that a quasi-particle formed as a hole clothed with a region of decreased antiferromagnetism can account for many puzzling features of the normal and superconducting state. These excitations should occur for all ranges of coupling. Our analysis showed that a similar charge bag can occur in CDW systems like Ba (Pb, K) Bi O_3 [6].

This research was supported in part by the National Science Foundation under Grant No. PHY82-17853, supplemented by funds from the National Aeronautics and Space Administration, at the University of California at Santa Barbara.

REFERENCES

1. J.G. Bednorz and K.A. Müller, *Z. Phys.* **B64**, 189 (1986); C.W. Chu, *et al.*, *Phys. Rev. Lett.* **58**, 405 (1987).
2. B. Batlogg, *et al.*, *Phys. Rev. Lett.* **58**, 2333 (1987); L.C. Bourne, *et al.*, *Phys. Rev. Lett.* **58**, 2337 (1987).
3. G. Shirane, *et al.*, *Phys. Rev. Lett.* **59**, 1613 (1987); J. Tranquada, *et al.*, *Phys. Rev. Lett.* **60**, 156 (1988); R.J. Birgeneau, *et al.*, to appear.
4. D.J. Scalapino, E. Loh, Jr. and J.E. Hirsch, *Phys. Rev.* **B34**, 8190 (1986); D.J. Scalapino and E. Loh, Jr., *Phys. Rev.* **B35**, 6694 (1987).
5. J.R. Schrieffer, X.-G. Wen and S.-C. Zhang, *Phys. Rev. Lett.* **60**, 944 (1988).
6. R.J. Cava, *et al.*, *Nature* **332**, 814 (1988).

Electronic States in Oxide Superconductors

S. MAEKAWA

Department of Applied Physics, Nagoya University, Nagoya, 464-01 Japan

Dynamical properties of hole carriers introduced in the strongly correlated electron systems, i.e. the Mott-Hubbard insulators, are theoretically studied by using both the moment method and the operator transformation method. Particular emphasis is put on effects of the quantum spin fluctuation on the dynamics.

1. Introduction

The high T_c Cu-oxides, $La_{2-x}Ba_xCuO_4$, $La_{2-x}Sr_xCuO_4$ and $YBa_2Cu_3O_{7-x}$, show two distinct phases depending on x : antiferromagnetic (AF) insulating and superconducting phases. The parent compounds, La_2CuO_4 and $YBa_2Cu_3O_6$, are considered to be the Mott-Hubbard AF insulators. When hole carriers are introduced in the compounds, the AF states disappear and the compounds become metallic and superconducting. Therefore, it is of crucial importance to study the relation between the dynamics of hole carriers in the insulators and the magnetic ground states.

Soon after the discovery of the high T_c Cu-oxides,[1)] Anderson[2)] has proposed that the essential physics of the oxides is described by a two-dimensional Hubbard model with strong correlation. Although a question as to whether this model is relevant to study the details of the real Cu-oxides has not fully answered yet, it was also noted that the two band Hubbard model, which takes into account the electronic states in both Cu and O ions in the CuO_2 layers, can be reduced to the single band Hubbard model.[3-6)] Here, we take the two-dimensional Hubbard model and examine the dynamics of hole carriers introduced in the Mott-Hubbard insulators.

The half-filled Hubbard model with strong correlation exhibits the insulator with AF exchange interaction between nearest neighbor sites. The magnetic ground state may be the AF Neel state with large quantum spin fluctuation (QAF) or one of the resonating valence bond (RVB) states because of the low dimensionality and the low spin value ($S = 1/2$).

In Sec. 2,[7)] we calculate the Green's function of a hole carrier in the insulator using the moment expansion method,[8-11)] and discuss effects of quantum spin fluctuation on the motion.

One of the interesting observations in the strongly correlated systems is that the degrees of freedom of electron, i.e. charge and spin, can separate.[12-16)] Such observation has naturally led to the formalism of the RVB superconductors. In Sec. 3,[16-17)] we introduce operators for the charge and spin degrees of freedom, and examine the RVB states and the motion of hole carriers in the states in the mean field theory. Finally, we will discuss implications of the theory in some physical properties of the Cu-oxides.

2. Moment Method

Let us consider the single band Hubbard model,

$$H = t \sum_{(ij)\sigma} c_{i\sigma}^{+} c_{j\sigma} + U \sum_{i} n_{i\uparrow} n_{i\downarrow} , \tag{1}$$

where $c_{i\sigma}$ is an annihilation operator of electron with spin σ at site i and $n_{i\sigma} = c_{i\sigma}^{+} c_{i\sigma}$. The summation over (ij) runs over the nearest neighbor sites. We take the correlation energy U to be much larger than the transfer energy t, so that the system is an insulator in the half-filled case. We introduce a hole in the insulator and examine its dynamics.

Taking the transfer term to be a small perturbation, we obtain the effective Hamiltonian in the order of O(t/U),

$$H_{eff} = H_k + H_{ex} ,$$

$$H_k = - t \sum_{(ij)\sigma} (1 - n_{i-\sigma}^{+}) c_{i\sigma}^{+} c_{j\sigma} (1 - n_{j-\sigma})$$

$$H_{ex} = J \sum_{(ij)} (\vec{S}_i \cdot \vec{S}_j - 1/4\, n_i n_j) , \tag{2}$$

where $J = 2t^2/U$, $n_i = \sum_{\sigma} n_{i\sigma}$, and $\vec{S}_i$ is the spin operator with S = 1/2 at site i . Here, we neglect the so-called pair-hopping term, which will be discussed in the next section. We note that in the Hamiltonian (2) all doubly occupied sites are projected out.

The motion of a hole introduced in the insulator in the two-dimensional square lattice depends on the magnetic ground states and the excitations. In the limit of $t/U \to 0$, the hole forces the spins to be aligned ferromagnetically and moves like a free particle.[8] On the other hand, in the classical Neel state with finite J, a hole is confined to a region containing $N \approx \pi\sqrt{t/J}$ spins. The energy for the localization is $\approx 2\pi\sqrt{Jt}$.[9] In order to examine effects of the exchange interaction (J) on the hole in detail, we consider the diagonal component of the Green's function,

$$G_{ii}^{(s)}(\omega) = \sum_{\sigma} \langle S \mid c_{i\sigma}^{+} (\omega - H_{eff})^{-1} c_{i\sigma} \mid S \rangle , \tag{3}$$

where | S > denotes the spin states in the half-filled case. We take the continued fraction representation of the Green's function in the retraceable path approximation neglecting paths with closed loops,[9,18]

$$G_{ii}^{(s)}(\omega) = \cfrac{1}{\omega - \cfrac{b_1^2}{\omega - a_1 - \cfrac{b_2^2}{\omega - a_2 - \ddots}}} \tag{4}$$

where the coefficients a_m and b_m are functions of the moments of the density of states of a hole, and are evaluated by examining the paths of a hole and the exchange energy stored in the paths.

Let us first consider the classical Neel state (CN) at T = 0 . The coefficients b_m are given by $b_1^2 = zt^2$ and $b_{m+1}^2 = (z-1)t^2$ for $m \geq 1$, z being the number of the nearest neighbor sites (z=4). When a hole moves in CN, it creates the "string" of unfavorable spins. Thus, the further the hole goes, the longer the "string" is. Examining the exchange energy in the "string" , we have $a_m = (2m+3)J$ in CN. We solve the Green's function (4) numerically and obtain that all the states are localized unless J = 0 . The energy of the lowest energy state is given by $\omega \approx 2.35\sqrt{Jt}$. Although there are effects coming from paths with closed loops to release a hole from the localization,[21] we neglect the effects since they are weak.

We next consider QAF at T = 0, where the spins fluctuate quantum mechanically around the Neel state. In this case, it is possible for two spins to change the directions simultaneously. Let us operate H_{eff} n times on the AF state with a hole at the origin, and create new states. The states have the following characteristic features depending on n: When n << t/J, the hole travels and leaves behind n overturned spins as in the case of CN. On the other hand, when n >> t/J, the overturned spins flip due to the exchange interaction H_{ex} and "string" is cut up. Then, the most probable state is that of the propagating hole with n_0 overturned spins, n_0 being of the order of t/J. In our continued fraction representation of the Green's function, a_m and b_m are not much different in QAF from those in CN for n << t/J. However, for n >> t/J, a_m and b_m depend on how to relax the overturned spins. Assuming that the spins are completely relaxed when they are cut off from the "string", we find that a_m and b_m are constants for n >> J. We calculate the density of states numerically, and obtain that the states make a band of mobile holes, demonstrating that the quantum spin fluctuation forces holes to be mobile. We also find numerically that there exists a narrow peak at the low energy side in the density of states. This peak may correspond to the coherent motion of holes with the "string". We note, however, that when the spins do not relax well, holes tend to localize.

We have examined effects of the quantum spin fluctuation on a hole in QAF. Another magnetic ground states in the two-dimensional square lattice are the RVB states, where the quantum spin fluctuation exhibits a crucial role. In the next section, we examine the dynamics of a hole in the RVB states by using the operator transformation method.

3. Operator Transformation Method

As discussed in Sec. 1, the degrees of freedom of electron can separate in and near the Mott-Hubbard insulators. Let us transform the electron operator as

$$c_{i\sigma} \rightarrow b_i^+ s_{i\sigma} , \tag{5}$$

where b_i is a boson operator called holon which describes the empty state at site i and $s_{i\sigma}$'s are fermion operators called spinons which describe singly occupied states with spin σ at site i. The transformation (5) is accompanied by the constraint,

$$b_i^+ b_i + \sum_\sigma s_{i\sigma}^+ s_{i\sigma} = 1 , \tag{6}$$

The transformation separates the degrees of freedom of electron as

$$\begin{cases} S_i^z = \frac{1}{2}(s_{i\uparrow}^+ s_{i\uparrow} - s_{i\downarrow}^+ s_{i\downarrow}) , \\ S_i^+ = s_{i\uparrow}^+ s_{i\downarrow} , \quad S_i^- = s_{i\downarrow}^+ s_{i\uparrow} , \end{cases} \tag{7}$$

$$Q_i = e b_i^+ b_i , \tag{8}$$

with e > o and Q_i being the operator of charge at site i. Inserting eq. (5) into eq.(2) and noting eq.(6), we have

$$H_k = - t \sum_{(ij)\sigma} b_i^+ b_j s_{j\sigma}^+ s_{i\sigma} , \tag{9}$$

$$\begin{cases} H_{ex} = J \sum_{(ij)} \hat{\Delta}_{ij}^+ \hat{\Delta}_{ij} , \\ \hat{\Delta}_{ij} = (s_{i\uparrow} s_{i\downarrow} - s_{i\downarrow} s_{j\uparrow}) / \sqrt{2} . \end{cases} \tag{10}$$

We introduce the order parameter for the singlet states,

$$\Delta_\tau = \langle (s_{i\uparrow}s_{i+\tau\downarrow} - s_{i\downarrow}s_{i+\tau\uparrow}) \rangle / \sqrt{2} \quad , \tag{11}$$

where τ shows one of the nearest neighbor sites in the two-dimensional square lattice, and $\langle ... \rangle$ denotes the thermal average. The states with this order parameter are called the RVB states in the mean field theory. In the two-dimensional square lattice, the RVB states depend on the relation between Δ_x and Δ_y. It is also important to note that in the half-filled case, the particle-hole symmetry, i.e. $s^+_{i\uparrow}s_{i\uparrow} = s_{i\downarrow}s^+_{i\downarrow}$, induces the parameter,

$$\xi_\tau = \sum_\sigma \langle s^+_{i\sigma}s_{i\pm\tau\sigma} \rangle / \sqrt{2} \quad , \tag{12}$$

which is not independent of Δ_τ. The kinetic energy of holons is written in the mean field theory as,

$$H_k = - \sqrt{2}t \sum_{i\tau} \xi_\tau b^+_i b_{i+\tau} \quad . \tag{13}$$

Let us consider the so-called d and s+id states.[23-29] These states have the parameters,

$$\begin{cases} \Delta_x = - \Delta_y \\ \quad = \xi_x = \xi_y = C / 2 \ , \text{ for d-state} \ , \end{cases}$$

$$\begin{cases} \Delta_x = - i\Delta_y = C \ , \\ \text{and } \xi_x = \xi_y = 0 \ , \text{ for s+id state} \ , \end{cases}$$

where $C = 0.339$ at $T = 0$. Thus, the effective mass of a holon in the d state is given by $m^* = \hbar^2/2tCa^2$ with a being the lattice constant. On the other hand, in the s+id state, ξ_τ vanishes in the mean field theory so that we must consider the higher order term, i.e. the pair-hopping term,

$$H_p = - J \sum_{(ilj)} \hat{\Delta}^+_{jl} \Delta_{li} \, b^+_i \, b_j \quad , \tag{14}$$

where (ijl) implies the sum over the nearest (l) and next nearest (j) neighbors of site i with j being the nearest neighbors of l. Then, the mass of a holon is written in the mean field theory as $m^* = \hbar^2/8JC^2a^2$. We also calculated[17] the relaxation time of a holon which is also dependent on the magnetic states. Recently, Kane et al.[22] have studied effects of the virtual spinon excitation on the holon in the d state, and obtained that m^* is enhanced by ~ t/J.

We have found that hole carriers are mobile in the RVB states in the mean field theory. The motion, however, is strongly dependent on the states.

4. Discussion

We have studied the dynamics of hole carriers in the Mott-Hubbard insulators. The motion of carriers is determined by the quantum spin fluctuation so that one of the characteristic energies of the transport properties is the exchange energy of the order of J. The optical conductivity in $YBa_2Cu_3O_{7-x}$ shows a structure around 0.2 eV. Rice and Zhang[18] and Kane et al.[22] have noted that this structure may be an indication of the interplay between the hole carriers and the spin states. Following Rice and Zhang,[18] the optical conductivity is expressed by the density of states of a hole. Using the results in Sec.2, we obtain the structure.

In conclusion, we have found that the quantum spin fluctuation may cause a hole to be mobile in the Mott-Hubbard insulators. The present theory suggests that the magnetic properties determine the dynamics of carriers and, thus, the superconductivity in the Cu-oxides.

Acknowledgements

The work reported in this paper has been done in collaboration with J. Inoue, A. Oguri and M. Miyazaki. The detailed results of Sec. 2 and Sec. 3 are given in refs. 7 and 17, respectively. This work has been supported by Grant-in Aid for Scientific Research on Priority Areas "Mechanism of Superconductivity".

References

1) J. G. Bednorz and K. A. Muller : Z. Phys. B64, 189(1986).
2) P. W. Anderson : Science 235, 1196(1987).
3) F. C. Zhang and T. M. Rice : Phys. Rev. B37, 3759(1988).
4) S. Maekawa, T. Matsuura, Y. Isawa and H. Ebisawa : Physica C152, 133(1988).
5) N. Andrei and P. Coleman : to be published
6) F. Mila : to be published
7) J. Inoue, M. Miyazaki, and S. Maekawa : to be published
8) Y. Nagaoka : Solid State Commun. 3, 409(1965) : Phys. Rev. 147, 392(1966).
9) F. Brinkman and T. M. Rice : Phys. Rev. B2, 1324(1970)
10) P. Lederer and Y. Takahashi : to be published in Proc. of 1st Asia-Pacific Conf. on Condensed Matter Physics ; Y. Takahashi: to be published.
11) R. Joynt : Phys. Rev. B37, 7979(1988).
12) S. Kivelson, D. Rokshar and J. Sethna : Phys. Rev. B35, 8865(1987).
13) Y. Isawa, S. Maekawa and H. Ebisawa : Physica B148, 391(1987).
14) P. W. Anderson, G. Baskaran, Z. Zou and T. Hsu : Phys. Rev. Lett. 58, 2790(1988).
15) Z. Zou and P. W. Anderson : Phys. Rev. B37, 627(1988).
16) P. W. Anderson and Z. Zou : Phys. Rev. Lett. 60, 132(1988).
17) S. Maekawa and A. Oguri :to be published in Proc. of 1st Asia-Pacific Conf. on Condensed Matter Physics; A. Oguri and S. Maekawa : to be published.
18) T. M. Rice and F. C. Zhang : to be published
19) S. Schmitt-Rink, C. M. Varma, and A. E. Ruckenstein : Phys. Rev. Lett. 60, 2793(1988).
20) B. I. Shraiman and E. D. Siggia : Phys. Rev. Lett. 60, 740(1988).
21) S. A. Trugman : Phys. Rev. B37, 1597(1988).
22) C. L. Kane, P. A. Lee, and N. Read : to be published
23) G. Kotliar : Phys. Rev. B37, 3664(1988).
24) H. Fukuyama, Y. Hasegawa and Y. Suzumura : Physica C153-155, 1630(1988); Y.Suzumura, H. Hasegawa and H. Fukuyama : to be published.
25) K. Kotliar and J. Liu : to be published.
26) F. C. Zhang, C. Gros, T. M. Rice and H. Shiba : Supercond. Sci. Technol. 1, 6(1988).
27) E. Muller-Hartman, M. Drzazga, A. Kampt and H. Wischmann : Physica C153-155, 1261(1988).
28) G. Baskaran : to be published.
29) I. Affleck and J. B. Marston : to be published.
30) G. A. Thomas, J. Orenstein, D. H. Rapkine, M. Capizzi, A. J. Millis, R. N. Bhatt, L. F. Schneemeyer and J. V. Waszezak : to be published.

Are Metal-Oxide Superconductors Charged Bosonic Superfluids?

L.J. DE JONGH

Kamerlingh Onnes Laboratory, State University Leiden, P.O. Box 9506, 2300 RA Leiden, The Netherlands

Introduction. Where are the holes? Polaronic conductivity and its solitonic aspects.

Prior to a discussion of the superconductivity mechanisms in the high-T_c metal oxides, the nature of the conductivity mechanism itself, and in particular the type of charge carriers should be considered. As advocated by the present author [1-3], the materials should be viewed as doped, large-gap semiconductors, where the excess charges introduced in the lattice by the doping form mixed-valence small-polarons (highly local, charged defects). At sufficiently high concentration, the mobility of the defects is guaranteed by quantum-mechanical tunneling in a band motion for temperatures smaller than the polaron bandwidth, and by thermally activated hopping in the higher temperature range [4]. Since the moving polarons will see a random electric potential from a.o. the impurity atoms (Sr^{2+}, oxygen deficiency), localization will occur at too low carrier concentrations, leading to a metal-insulator transition of the Anderson type. This behaviour is clearly observable in the experimental resistivity curves as a function of doping level [2]. In the semiconducting samples, thermally activated hopping has indeed been observed.

As to the precise nature of the small-polarons, one has to address the question whether the excess charges will be found primarily on the copper or on the oxygen sites, in other words "where are the holes". In our view there exists considerable confusion on this subject, mainly arising from the neglect of polarization effects in most treatments we have seen so far. Also, arguments explaining the nature of the intrinsic gap of these semiconductors are used to define the nature of the defect states that form the narrow polaronic impurity band). Let us consider the material $La_{2-x}Sr_xCuO_4$ as the example.

Pure La_2CuO_4 is a semiconductor with a relatively large intrinsic gap of at least 2-3 eV. There is a general consensus that the origin of the gap will be associated with metal-ligand charge transfer [5]. For intrinsic conduction one needs charge separation, and it has been found that the metal-ligand charge transfer energy i.e. $Cu^{2+} + O^{2-} \rightarrow Cu^{+} + O^{-}$, will be considerably less than the energy of disproportionation of the metal atoms, i.e. $2Cu^{2+} \rightarrow Cu^{3+} + Cu^{+}$. (The latter energy corresponds to the "Hubbard U" in the theory of electron correlation effects in metals).

Next then, the problem arises whether the extra hole introduced in the (basically insulating) Cu-O network upon replacing a La^{3+} by a Sr^{2+} atom will be found primarily on a copper or on an oxygen atom, that is will there be formed a "Cu^{3+}" or a "O^{-}" defect. Here we have put the asterisks to stipulate that these defects do not refer to bare Cu^{3+} or O^{-}, but to the ions dressed with their polarization clouds. One should not forget that oxides are highly polar ionic lattices, for instance most ferroelectrics are oxides. Thus, when the extra charge is put in the lattice both the electronic polarization of the surrounding atoms, and the ionic polarization due to the displacements of these atoms should be taken into account (cf. fig.1). The correct approach thus appears to be a calculation of the formation energies of the "Cu^{3+}" and "O^{-}" defects including the lattice deformation energy and the polarization energies. Such calculations have recently been performed by the group of Catlow [6]. They find in fact that the formation of the dressed "O^{-}" costs about twice as much energy (5.4 eV) as that of the "Cu^{3+}" defect (2.9 eV). This confirms our earlier proposal [1] that the Cu^{3+} small-polaron defects are the charge carriers,

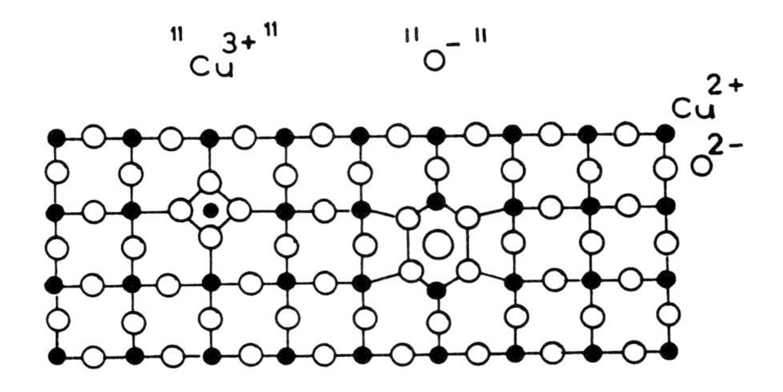

Figure 1: Sketch of the "Cu^{3+}" and "O^{-}" defect states in the 2-dimensional Cu/O network.

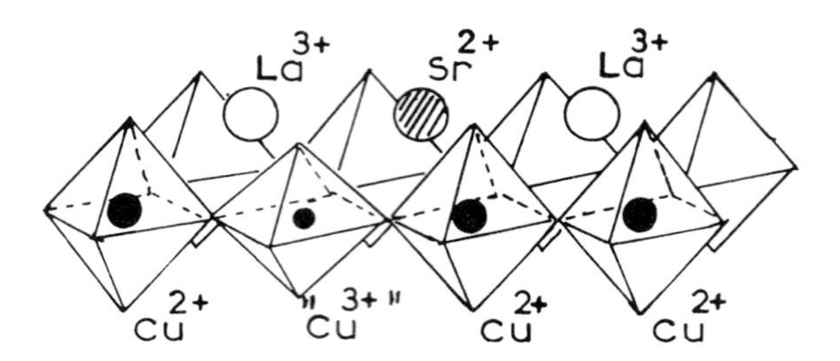

Figure 2: Sketch of the "Cu^{3+}" polaron in the Cu/O layer of $La_{2-x}Sr_xCuO_4$.

as sketched in fig.2. The defect state is the distorted CuO_6 molecule, where the oxygens surrounding the copper are contracted due to the extra charge on the copper atom. The contraction is due to the decrease of the ionic radius of Cu^{3+} with respect to Cu^{2+}, and to the increase in the copper-oxygen Coulombic interactions. In addition to this isotropic distortion of the defect, involving an ionic polarization energy of the order of 2.7 eV, there will also be an anisotropic deformation with respect to the host lattice due to the Jahn-Teller distortions of the latter. However, the Jahn-Teller deformation energies will typically be an order of magnitude smaller. Needless to say that there is a strong hybridization of the copper and oxygen orbitals in this model. It is also important to note that the extra hole will pair its spin to the Cu^{2+} spin that was originally there, leaving a spinless CuO_6 molecular unit [7]. The point is that any theory placing the excess holes on the oxygens must explain why in all the experiments (notably magnetic susceptibility and ESR) so far the introduction of holes in the copper-oxide matrices is found not to be accompanied by the appearance of unpaired spins. It is not easy to account for this with holes on oxygen, whereas with the "Cu^{3+}" defects this difficulty does not arise. In particular we mention that many nominally Cu^{3+} oxide compounds do exist and are known to be diamagnetic. (Some references are given in [2]).

Obviously, we have to reconcile the above "Cu^{3+}" defect state with the results of the high-energy spectroscopic methods, in which so far the presence of excess charge carriers in the lattice has only been detected on the oxygen sites. In our opinion there is still room for the above interpretation, since there will certainly be an extra signal associated with the oxygens that form the distorted octahedral environment of the Cu^{3+}, with which they are highly hybridized. That so far no feature of the extra hole has been seen on the Cu 3d positions may find its explanation in the neglect of the polarization effects in the abnalyses given so far. When the excess charge is taken off the polaronic "Cu^{3+}" defect in the high-energy spectroscopic experiment, the system is left behind in an excited state, since the polarization cloud of the defect can not relax in the time of the experiment. The excitation energy will correspond to the binding energy of the polaron, which is several eV. Thus, it may well be that the dressed "Cu^{3+}" signal should be sought at a different position in the spectra.

Other experimental proof for the validity of the small-polaron model has recently been obtained by Kim et al. [8] from photoinduced infrared absorption measurements on both La_2CuO_4 and $YBa_2Cu_3O_{6.25}$, in which the excess charges are generated by photo-excitation. From the appearance of extra infrared active vibrational modes of the oxygen octahedra in the optical spectra, it may be inferred that the carriers lead to local structural distortions, i.e. they form small-polaronic defects. The dynamic mass of the defects is deduced to be of the order of 10-20 m_e. This is in fact a relatively small effective polaron mass considering the strength of the electron-lattice interaction (the above-mentioned polarization energies and defect formation energies are calculated [6] to be a few eV). Below we shall argue that such strong electron-lattice interactions need not immobilize the defect states because of their solitonic properties, that is they should be considered as nonlinear excitations.

The polarons considered can be called mixed-valence polarons [1], since their formation is to be associated with the ability of the metal atoms to assume different oxidation states just by slight modifications of the ligand coordination. In most cases the metal atom forming the polaron will only differ in valence from the host by one unit of charge. However, as we have pointed out previously, [2,3] there are several metallic elements in the periodic table for which it is well known that they exist primarily in valences differing by a factor of two, the reason being that both oxidation states correspond to closed-shell structures. As an example consider the Bi-atom which has the electronic configuration $6s^2 6p^3$, and thus will exist as Bi^{3+} or Bi^{5+}, but not as Bi^{4+}. Put in another way, the disproportionation energy associated with the reaction $2Bi^{4+} \rightarrow Bi^{3+} + Bi^{5+}$, is greatly reduced by intra- atomic electron correlation effects, and the associated polarization effects in ionic compounds. In the scheme of the just-described polaron formation, one may expect the introduction of extra holes in a Bi^{3+} compound to lead to a Bi^{5+} polaron, whereas the doping with excess electrons of a nominally Bi^{5+} compound should lead to Bi^{3+} polarons. Similar arguments apply to e.g. $Sb(Sb^{3+}/Sb^{5+})$, $Pb(Pb^{2+}/Pb^{4+})$, $Sn(Sn^{2+}/Sn^{4+})$, $Tl(Tl^{+}/Tl^{3+})$, etc. Of importance here is that in case the defect states differ in valence by two charges with respect to the host lattice, the so-formed polaron can be considered as an <u>on-site</u> bipolaron, as it corresponds to an extra pair of charges located on a single site. In case the polaron carries only a single charge, a bipolaron can be formed by pairing two polarons on adjacent sites, corresponding to the <u>intersite</u> bipolaron. Since bipolarons are bosonic real-space pairs, they may account for the observed superconductivity.

We now discuss the mobility of the mixed-valence (bi)polarons, and note that these defect states (e.g. the deformed CuO_6 molecular unit) can be regarded as substitutional impurities in the crystal. As already remarked by Andreev and Lifshitz [9] because of quantum-tunneling effects the local defects in crystals will turn into mobile quasi-particles (defectons, impuritons) at sufficiently low temperatures. Because of the translational invariance, depicted in fig.3a, they will move essentially free through the crystal if the latter is perfect. The defectons should obey Bose statistics if the crystal atoms and the impurities follow the same statistics, and obey Fermi statistics in the opposite case [9]. Thus the mobile bipolarons mentioned in the above can be considered as bosonic quasi-particles. As mentioned by Andreev and Lifshitz the defectons may constitute a strongly degenerate Fermi- or Bose gas of quasiparticles moving on a lattice, and may undergo Bose-condensation in the latter case.

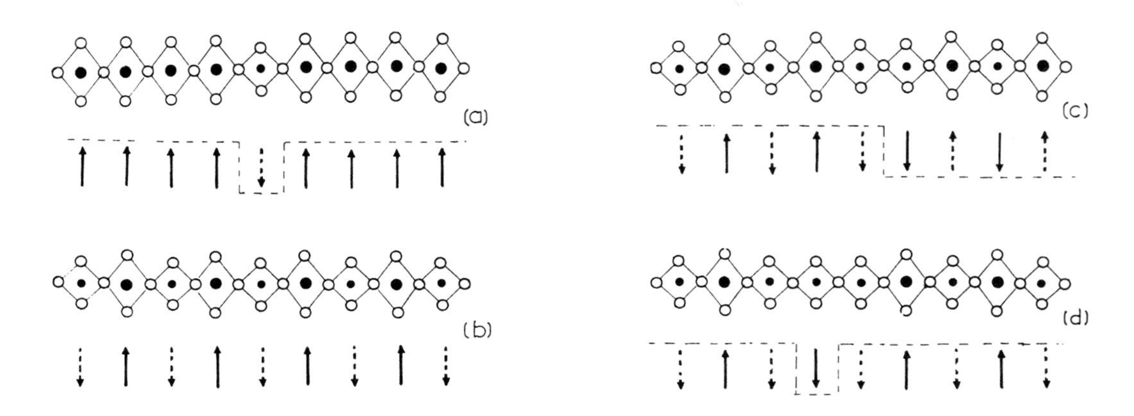

Figure 3: a) The small-polaron in a chain of metal-oxygen molecules viewed as a nonlinear excitation. b) The disproportionated chain of metal-oxygen molecules. c) Topological kink in the chain of fig.3b. d) Nontopological kink in the chain of fig.3b. (In all cases the pseudo-spin analogues have also been drawn).

Evidently, in case the lattice is not perfect, Anderson-type localization [4] may occur when the defect density is too low, as already mentioned. Also, for temperatures comparable to or larger than the quantum mechanical transfer integrals the motion of the defect states will change from a (basically free) band motion into thermally activated variable range hopping. However, for the metal oxides, in particular for Cu($3d^9$)l and Ti($3d^1$), the transfer integrals are so large that the system behaves quantum mechanically even at room temperature [1] and higher! We mention that a decrease in the apparent activation energy with lowering temperature, as expected from the transition to quantum-mechanical tunneling, appears to have been observed by the group of Livage [10] in (amorphous) mixed-valent transition metal oxides (V_2O_5 gels, $W_6O_{19}^{3-}$ polyanions). It also agrees with small-polaron theories [4].

Thus, in spite of the extremely strong electron-lattice interactions that characterize the polaronic defect states, their mobility can be substantially enough [10] to lead to metallic-like conductivity, although their velocity is of the order of the speed of sound. In this respect it is of importance to realize that the defect states can be seen as nonlinear excitations, i.e. as nontopological solitons [11,12]. For instance, the polaron in an ionic crystal can be treated as a nontopological 3-dimensional soliton, as discussed extensively by Davydov [12]. In the model the crystal is approximated by a continuum in which the extra charge interacts with the polarization field of the crystal. With increasing velocity the radius (i.e. the extent of the polarization cloud) of the polaron goes through a minimum, to which corresponds a maximum in the dynamical mass of about 10-20 m_0, where m_0 is the effective mass of the polaron at rest.

As discussed by the present author [2,3], the various solitonic aspects of the problem become quite transparent when 1-dimensional model approximations are used. For instance, in fig.3a a chain of metal-oxygen octahedra with a single defect molecule (a site with an extra charge) is shown. The translational invariance of the defect is obvious. In a pseudospin representation we may represent the defect site by a down-spin and the "unoccupied" sites by an up-spin. Then the polaron corresponds to the spin- flip in a ferromagnetic Ising-type chain [13]. This excitation can be seen as a combination of a kink and an antikink, that is as a bound state of a soliton and an antisoliton [14].

In a similar way the problem of a fully disproportionated lattice, as occurs for example in $BaBiO_3$ which has been shown to be an alternating structure of Bi^{3+} and Bi^{5+} ions, can be modelled by the chain in fig.3b. In the pseudospin picture the analogue of this chain would be the antiferromagnetic Ising-type chain, as shown in the same figure. Another type of domain wall excitations may then occur, namely defects in the alternating sequence of the two different metal-oxygen octahedra, as shown in figure 3c,d. Again, such defects can be viewed as kinks or kink-antikink pairs, and in the latter case correspond to (bound-)magnon states of the Ising problem [14].

The basic features of the polaron-soliton model are thus the following. Becoming self-trapped by the lattice deformation that it itself creates, the extra charge forms a mixed-valence small-polaron that can be regarded as a nontopological soliton. The defect states are not thermally excited but are formed during the synthesis of the compound. They are extremely stable nonlinear excitations, since an energy of the order of the formation energy (several eV) must be supplied to separate the polaron into its constituents, namely the bare charge and its polarization cloud. Although the electron-lattice interaction is thus extremely strong, the phonon energy enters essentially into the formation energy of the defects. They can move about by quantum-mechanical tunneling and the corresponding displacements of the atoms from the equilibrium positions can be described in terms of virtual phonons. The envelope of the defect (the kink-antikink structure) may preserve its shape since it moves with a speed less than the sound velocity, so that no kinetic energy can be dissipated into heat (acoustical phonons). These are the characteristic features of nonlinear excitations [11,12].

In the above we have considered very narrow walls (kinks) for simplicity of discussion. However, depending on the parameters one may consider as well the polaron to extend over more than a single lattice site, that is distribute the extra charge over more than one metal site. As shown by Onodera [15], in the 1-dimensional approximation the problem then becomes equivalent to the charge transport in conjugated polymeric chains, notably to the well-known example of all-trans polyacethylene [16,17]. This illustrates the close resemblance of the superconducting metal oxides to the molecular (synthetic) conductors. As an example we shall use this analogy to estimate the effective mass of the soliton in the dimerized metal oxide chain shown in fig.3c, using the formula derived for the soliton in polyacethylene [16]. We follow the treatment of Onodera [15] for the mixed-valence platinum-halogen chain, and apply it to the metal- oxygen- chain. In fig.4 the structures of the all-trans polyacethylene chain in the hypothetical undimerized and

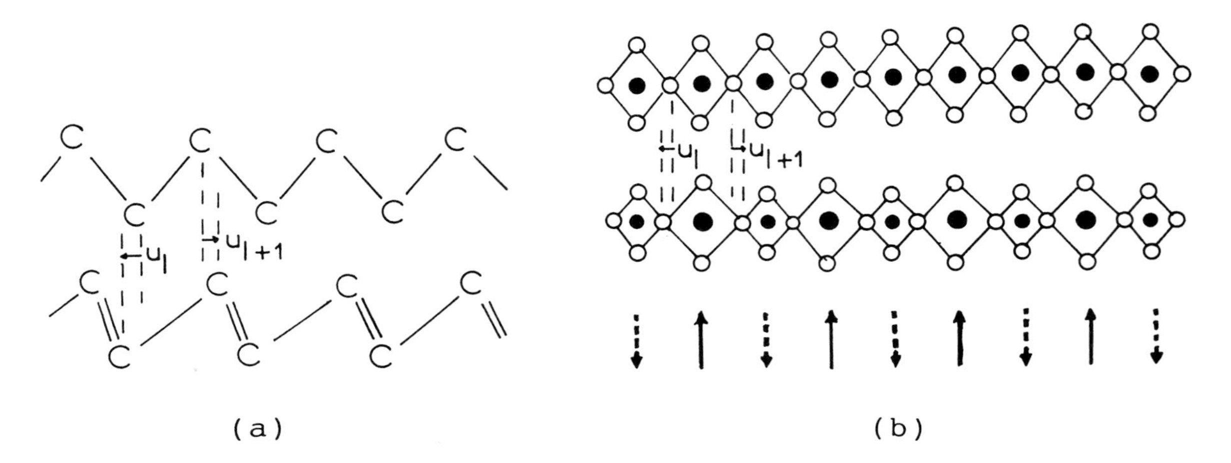

Figure 4: a) Structure of the (hypothetical) undimerized $(CH)_x$ chain, together with the dimerized form in all-trans polyacethylene. b) Structure of the regular and the dimerized metal-oxygen chain.

in the dimerized form are sketched. They are compared to the undimerized metal-ligand chain and the fully dimerized mixed valence chain. Instead of the alternating bond problem we then have the alternating valence problem. In both cases, however, the ground state has the same two-fold degeneracy, leading to the possibility of domainwall formation just as for the antiferromagnetic chain in figs.3b,c,d, to which they are both analogous. The corresponding solitons are the same as those appearing in the ϕ^4 field theory, as recognized by Su, Schrieffer and Heeger (SSH), who laid down the basis for the soliton dynamics in polyacethylene by considering the hamiltonian:

$$\mathcal{H}_{SSH} = \frac{1}{2}K\sum_{\ell}(u_{\ell+1}-u_\ell)^2 - \sum_{\ell,\sigma} t_0(c^+_{\ell,\sigma}c_{\ell+1,\sigma} + c^+_{\ell+1,\sigma}c_{\ell,\sigma}) + \sum_{\ell,\sigma}\alpha(u_{\ell+1}-u_\ell)(c^+_{\ell,\sigma}c_{\ell+1,\sigma} + c^+_{\ell+1,\sigma}c_{\ell,\sigma}) \quad (1)$$

Here the first term accounts for the elastic deformation energy associated with the dimerization. The second term leads to the hopping of the π- electrons along the $(CH)_x$ chain. The third term accounts for the electron- phonon interaction. In the SSH model the latter enters through the modulation of the transfer integral t by the lattice displacements, that is: $t_{\ell+1,\ell} = t = t_0 - \alpha(u_{\ell+1} - u_\ell)$. We note that in hamiltonian (1) the electron lattice interaction is taken explicitly into account, whereas the U-term is discarded, as discussed by Schrieffer [17].

The problem of the ligand-bridged mixed-valence metal chain has been discussed by Ichinose [18] and by Onodera [15]. The corresponding hamiltonian considered by Onodera is written as:

$$\mathcal{H}_O = \frac{1}{2}K\sum_{\ell}(u_{\ell+1}-u_\ell)^2 - \sum_{\ell,\sigma} t_0(c^+_{\ell,\sigma}c_{\ell+1,\sigma} + c^+_{\ell+1,\sigma}c_{\ell,\sigma}) + \sum_{\ell}\alpha(u_{\ell+1}-u_\ell)c^+_\ell c_\ell \quad (2)$$

The first term is again the elastic deformation energy, whereas the second now describes the charge transfer of d-electrons between adjacent metal sites. The difference lies in the third term, which again represents the electron-phonon coupling but in this case the displacements u_ℓ produce a variation in the site-energies for the d-electrons on the metal sites (when the negative ligand anions approach a given metal site, the energy for a d-electron to reside on that site will increase correspondingly). Although the displacements u_ℓ may also produce some variation in the transfer integral, that effect is of second order and is neglected in this model.

As discussed by Onodera, when taking the continuum limit both the hamiltonian $\mathcal{H}_{SSH}$ and $\mathcal{H}_O$ can be described by the same mathematical expression derived by Takayama, Lin-Liu and Maki. Consequently, the nonlinear excitations must be quite similar for both physical problems (and both are analogues of the antiferromagnetic Heisenberg chain problem). The difference of course is that for the $(CH)_x$ chain the staggered charge distribution refers to the bonds (the carbon sites retain equal charge density also for the dimerized case), whereas for the metal chain it refers to the metal sites themselves.

Interestingly, the above analogy leads to an estimate of the mass of the soliton in the mixed-valence chain [15]. One may use the same expression for the translational mass M_s of the soliton as in the SHH-model [16], $M_s = \frac{4}{3}\frac{a}{\xi_0}(\frac{u_0}{a})^2 M$. Here ξ_0 is the soliton width, u_0 the displacement of the ligands and M the ligand mass. For narrow solitons we may put $\frac{4}{3}\frac{a}{\xi_0} \simeq 1$, so that $M_s/M \simeq (u_0/a)^2 \simeq 10^{-3}$. For oxygen $M \approx 3 \times 10^4 m_e$, so that $M_s \simeq 30 m_e$, where m_e is the mass of the electron. This compares well with estimates of the effective masses for the small polarons in the metal oxides [2,8], in particular considering the crudeness of the model used.

As already in the above, to ensure the mobility of charged polarons in real systems they must occur in the form of nontopological defects. Also, for polyacethylene this has been explicitly discussed [17]. Neutral, single soliton states probably only arise as defects created during cis-trans isomerization. At low doping levels, it is found that the extra charges enter on the chain essentially in the form of polarons, that is form states of soliton-antisoliton pairs. In fact the decay of a polaron into such a pair is also the mechanism proposed for the photogeneration of solitons in $(CH)_x$. Fundamental arguments as to why solitons may only be generated in the form of soliton-antisoliton pairs can also be given. We finally note that in Ichinoses treatment [18] the U-term is included. He considers narrow solitons and makes a transformation to the uniaxially anisotropic antiferromagnetic Heisenberg hamiltonian, similar as discussed in the next section. This hamiltonian is known to lead to a rich variety of nonlinear excitations.

Superconductivity by Bose-condensation of real space pairs

As regards the superconductivity in the metal oxides, the major question is whether it is due to the usual BCS-type pairing of fermions in $\vec{k}$- space, or to real-space pairs, preexisting at temperatures well above T_c. The author has become convinced that the latter is the case. One of the strongest experimental indications is the coherence length, which even inside the superconducting Cu-O planes is found to be equal to the average distance between the charge carriers. This would be incompatible with BCS, since it proves that all charge carriers take part in the superconductivity. Then the problem should be treated by Bose-statistics, rather then Fermi-statistics, and the superconductivity becomes analogous to the superfluidity of ^{4}He (Bose-liquid, $T_c \simeq 2.15$ K), instead of the superfluidity of ^{3}He (Fermi-liquid, $T_c \simeq 2$ mK). In fact, many experimental results when interpreted in terms of BCS lead to quite inconsistent results, whereas they appear to be well interpretable in terms of real-space pairing, if only one would care to consider this as an alternative possibility.

Obviously, on basis of these considerations the author was led to the above polaron model. As explained one may form charged bosons either by considering doubly-charged polarons, which then are onsite bipolarons, or by combining two polarons on adjacent sites into an intersite bipolaron. The former case applies [3] to the mixed-valence systems like Bi^{3+}/Bi^{5+}, occurring in $Ba(Bi/Pb)O_3$ and $(Ba/K)BiO_3$ and probably also to the Cu-1 chains in $YBa_2Cu_3O_7$. The latter case applies e.g. to the Cu-O planes in the superconducting oxides [1-3]. The problem then is to find an attractive mechanism able to overcome the repulsive Coulomb interaction between two adjacent polarons, which in most cases will exceed the gain in the total deformation energy that is obtained in the pair by sharing the deformations of the individual polarons. The argument is that the Coulomb repulsion is heavily screened by the polarization, since the dielectric constants are large ($\epsilon_\infty \simeq 5-10$; $\epsilon_s \simeq 20-50$) so that a reduction of the Coulomb repulsion energy to a value of order 0.1 eV or lower appears to be likely. Then, one may find several arguments [1,19,20] favouring a pairing of the nonmagnetic "Cu^{3+}" polarons on basis of the antiferromagnetic structure of the underlying Cu^{2+} magnetic lattice of spins $S=\frac{1}{2}$. One may then also understand [1,2] why the pairing only occurs in the copper oxides and not in the analogous oxides of the other 3d-metals (Ni, Mn, Cr, etc.), the reason being that only for $3d^9$ (Cu^{2+}), and possibly also for $3d^1$ (Ti^{3+}), the magnetic interaction constants arising from the metal-oxygen-metal exchange paths are strong enough to yield a sufficient magnetic energy gain to overcome the Coulomb repulsion.

Once bosonic quasi particles are formed, the problem is solved, since it is a well-established theoretical result that a gas of hard-core bosons on a lattice shows superfluidity, as first pointed out by Matsubura and Matsuda in their classical paper [21]. When the bosons carry a charge, the superfluidity may also correspond to a superconducting state, as shown by Schafroth [22]. In constructing their quantum-lattice-gas model for superfluid ^{4}He, Matsubara and Matsuda based themselves on the hamiltonian:

$$\mathcal{H}_b = \sum_{<i,j>} v_{ij}n_i n_j + \frac{1}{2}\sum_{<i,j>} t_{ij}(b_i^+ b_j + b_i b_j^+) + \mu\sum_i b_i^+ b_i \quad (3)$$

describing a system of bosons with infinite onsite repulsion, weak intersite interaction v_{ij} for nearest-neighbour sites and a kinetic energy term t_{ij} (transfer). The boson operators b_i^+, b_i can then be shown to be replacable by Pauli spin-operators S_i^+, S_i^-, where the pseudo-spins $S_i^z = \pm\frac{1}{2}$ correspond to an occupied and unoccupied lattice site, respectively. The equivalent pseudo-spin hamiltonian is:

$$\mathcal{H}_S = \sum_{<i,j>} [J_z S_i^z S_j^z - J_{xy}(S_i^x S_j^x + S_i^y S_j^y)] + H \sum_i S_i^z \tag{4}$$

where the field term is related to the chemical potential ($\mu = H$, apart from an additive constant). This hamiltonian is well-known in magnetism, since it describes the Heisenberg model with uniaxial (Ising-type) anisotropy. The parallel exchange, $J_z = v_{ij}$, is related to the interaction between bosons on neighbouring sites, and can be ferromagnetic (> 0) or antiferromagnetic (< 0) according to the sign of v_{ij}. The transverse exchange is related to the kinetic energy, $J_{xy} = t_{ij}$, and thus the establishment of ferromagnetic order in the XY plane corresponds to the condensation of bosons in the quantum-lattice-gas model. The excitations in the Bose-condensed phase correspond to the spin-waves in the XY magnet, that is they are collective coherent motions of the particles. The transition temperature T_c is characterized by the disappearance of such collective excitations.

It has already been proven theoretically that the above type of description is indeed applicable to bipolaronic systems. For intersite bipolarons it has been shown [23] that the hamiltonian describing the bipolaronic system may indeed be transformed into one that is analogous to (3), and thus to (4). Furthermore, in treating intrasite fermion pairs an extended Hubbard type hamiltonian for fermions on a lattice with attractive onsite interaction $U < 0$, and a weak intersite interaction [24,25] has been used. The hamiltonian is:

$$\mathcal{H}_f = \sum_{i,j,\sigma,\sigma'} v_{ij} n_{i\sigma} n_{j\sigma'} + \sum_{i,j,\sigma} t_{ij}(a_{i\sigma}^+ a_{j\sigma} + a_{i\sigma} a_{j\sigma}^+) + \sum_i U n_{i\uparrow} n_{i\downarrow} - \sum_{i,\sigma} \mu n_{i\sigma} \tag{5}$$

where the $a_{i\sigma}^+$, $a_{i\sigma}$ are fermion operators. In the limit of large U the fermions will form local pairs, which will behave as hard-core bosons. Thus each site becomes either occupied or unoccupied by a fermion pair and one may introduce pair operators, $b_i^+ = c_{i\uparrow}^+ c_{i\downarrow}^+$, that may be interpreted as bose operators and can thus also be transformed into the spinoperators, as before. The pseudo-spin hamiltonian becomes:

$$\mathcal{H}_S = \sum_{i,j} [(\frac{2t_{ij}^2}{|U|} + v_{ij}) S_i^z S_j^z - \frac{2t_{ij}^2}{U}(S_i^x S_j^x + S_i^y S_j^y)] - H \sum_i S_i^z \tag{6}$$

Again the ferromagnetic order of the XY components corresponds to superconductivity, whereas the antiferromagnetic order of the z-components corresponds to CDW order in the electron system (and $H \leftrightarrow \mu$).

Finally, we mention that, although in the hamiltonian (5) the electron-phonon coupling is not present, it was shown for the case of the mixed-valence chain by Ichinose [18] that when including this coupling one may again arrive at a hamiltonian of the form given in eqs.(4) and (6). Alternatively, one may associate the attractive U-term in (5) with one or the other of the mechanisms leading to onsite bipolaron formation mentioned in the above.

The general conclusion from these results is that, once it would be fully established that the superconductivity in the metal oxides is due to pre- existing bosonic real-space pairs, the transformation to the Heisenberg $S = \frac{1}{2}$ antiferromagnet with uniaxial anisotropy is possible. This relationship can be a very fruitful one, since the corresponding magnetic problems have been so extensively studied in the literature. For instance, the whole field of spinwave theory in magnetic systems is based on the transformation of spin operators into boson operators. The author is presently engaged in trying to understand the consequences of these analogies. The first items that come into mind are the following.

Firstly, it is quite interesting to note that on basis of this analogy one may directly explain the strong fluctuation (precursor) effects observed in e.g. the resistivity above T_c of the superconducting copper oxides. For low-dimensional magnets it has been well-established that spinwave-like excitations may persist in the paramagnetic region up to temperatures quite high compared to T_c. The criterion basically is that the wavelength of the spinwave is smaller than the magnetic correlation length. In low-d magnets the latter can remain quite large over a considerable temperature ranges above T_c [27]. The copper-oxide superconductors have very pronounced 2-d (or 1-d) features. Consequently, collective excitations corresponding to superconducting fluctuations should be present far (up to 20-40%) above T_c. These effects are seen particularly clearly [26] in the new Bi-Cu-O and Tl-Cu-O systems, probably because these materials are less complicated by twinning processes. Secondly, the $H-T$ phase diagram of the uniaxially anisotropic Heisenberg antiferromagnet, shown in the example in fig.5, is the analogue of the phase diagram of the bosonic (bipolaronic) superconductor given in the same figure. Decreasing the field from the saturation value at $T = 0$ corresponds to increasing the density of bipolarons (n_b) from zero. An intuitively obvious result of the bipolaronic theory is that in the presence of interactions v between the bipolarons, a static charge-ordered state should occur at sufficiently high density n_b. In the magnetic problem this corresponds to the competition between the J_z and the J_{xy} coupling terms. Thus the superconducting (BS) state and the CDW state correspond to the spin-flopped (SF) and the magnetically ordered (AF) state, respectively. Under certain conditions the first-order spinflop line that usually separates the AF and the SF state may be split into two second-order lines, encompassing an intermediate (mixed) state as shown in the example [23] in the fig.5. What is of interest here is that the antiferromagnetic phase diagrams in general show a very rich series of phase transitions as a function of the magnetic field. In particular for assemblies of weakly coupled magnetic chains or weakly coupled magnetic layers, it can be shown [28] that the field- induced phases may be in the form of soliton-lattices, the problems being quite analogous to the commensurate-

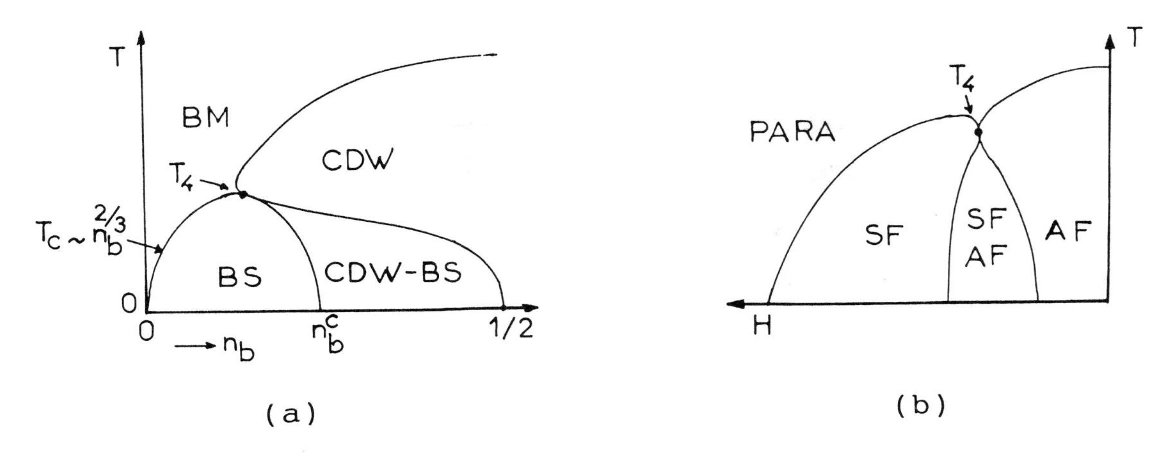

Figure 5: a) Phase-diagram of the bipolaronic superconductor [23]. BM-bipolaronic metal, BS-bipolaronic superconductor, CDW-charge density wave. n_b-density of bipolarons. b) Phase-diagram of the uniaxial Heisenberg antiferromagnet. SF-spinflopped phase, AF: antiferromagnetically ordered phase. H is the magnetic field.

incommensurate transitions. Since the copper- oxides are quasi 2-dimensional superconductors, they can be modelled by a magnetic system of weakly coupled layers, so that they can be expected to show the same nonlinear properties.

Thirdly, and by the same token, the 2-dimensional nature of the copper-oxide superconductors would make the superconducting order extremely sensitive to (randomly) distributed impurities, which will certainly be present. In the magnetic analogue, this is immediately clear from the wealth of literature on the random-field problem for systems with Ising-type anisotropy [29]. Basically, the argument is that systems of lattice dimensionality $d \leq 2$ are unstable for domain formation triggered by random impurities, inhibiting long-range order. Since the PARA-AF transition, and thus also the BM-CDW transition, is of the Ising-type, the latter could well be absent in the quasi-2-dimensional oxides, in contrast with the 3-dimensional ones. As regards the transition to the superconducting state in these materials, the phase boundary surrounding the BS phase in fig.5a is quite similar to the experimental variation of T_c in $La_{2-x}Sr_xCuO_4$ with hole concentration. Interesting in this respect is also the fact that the resistivity curves of the high-T_c oxides become extremely smeared out as soon as magnetic fields (even a few tesla) are applied [26]. At first sight one would conclude from these curves that the phase transition is destroyed in such magnetic fields. Following the above, one could speculate that in one way or the other the vortex lattice induced by the magnetic field leads to a destruction of the long-range-superconducting order, e.g. by acting as a source of impurities. However, the action of the magnetic field on the superconducting system of charged bosons is a much more complicated problem, so that further investigations are needed to disclose a possible difference between the quasi 2-d and the 3-d systems.

Fourthly and finally, one may use the above analogy to discuss a possible increase of the superconducting T_c with respect to the presently attained values of about 100 K. The point is that in the magnetic analogues of a system of weakly coupled magnetic layers, is well-known that the magnetic ordering temperature (T_N) can be lower than a factor of 2 with respect to a 3-dimensional magnetic system with the same magnetic exchange constant J. For instance, consider the mean-field prediction θ for the ordering temperature, given by $\theta = 2zS(S+1)J/k_B$, where z is the number of nearest interacting neighbours. For 3-dimensional magnets, the actual T_N values are in the range of 0.6-0.8 θ, whereas for quasi 2-dimensional magnets they are in the range 0.2-0.5 θ [27]. On this basis one would conclude that for a 3-dimensional copper-oxygen-copper network with an equivalent transfer integral t in all three directions, the superconducting T_c could very well be a factor of two or three higher than for the present layer-type superconducting oxides. Unfortunately, one has to add that in practice, it is very difficult to realize such a 3-dimensionally isotropic t for a copper-oxide network due to the same type of cooperative Jahn-Teller ordering of the copper-oxygen octahedra that gives rise to the 2-dimensional antiferromagnetic properties of the Cu^{2+} spins in the copper-oxides [1]. It can be argued on rather general grounds that for any cubic arrangement of CuO_6 octahedra, the cooperative Jahn-Teller ordering will lead to a system of weakly coupled layers, or to a system of weakly coupled chains, but never to an isotropically coupled 3-dimensional system [1,30]. Therefore, the best way to get a higher T_c at present would seem to be an increase of the number of Cu-O sheets in the multiple layers that occur in e.g. the recent Tl-Cu-O compounds. Probably the transfer between Cu-O sheets in the same multiple layer is not too small compared to the transfer within each Cu-O sheet. Then by increasing the number of Cu-O sheets in the multiple layer, a 3-dimensional structure can be approximated. Already there is evidence for a sizable increase of T_c with increasing number of Cu-O sheets, notably in the Tl-Cu-O compounds. Besides the possible effect of an increase in the density of holes per cm^3, the higher T_c might also be related to the enhanced 3-dimensionality, as the thickness of the multilayer becomes large compared to the coherence lengths.

Acknowledgements

The continuous interaction with the colleagues of the Leiden Materials Science Centre working on the high-T_c superconductors is gratefully acknowledged, as well as the support for this research by the "Stichting FOM" (Foundation for Fundamental Research on Matter), which is sponsored by "ZWO" (Netherlands Organization for the Advancement of Pure Research).

References

[1] L.J. de Jongh, Solid State Commun. 65 (1988) 963.

[2] L.J. de Jongh, Physica C 152 (1988) 171.

[3] L.J. de Jongh, Proceed. NATO Adv. Res. Workshop on Condensed Systems of Low Dimensionality, Val de Courcelles, Paris, June 1988, to appear in J. de Chimie Physique.

[4] For reviews see N.F. Mott, "Conduction in non-crystalline materials", Clarendon Press, Oxford, 1987; I.G. Austin and N.F. Mott, Adv. Phys. 18 (1969) 41; and N.F. Mott and M. Kaveh, Adv. Phys. 34 (1985) 329.

[5] J. Zaanen, G.A. Sawatzky and J.W. Allen, Phys. Rev. Lett.55 (1985) 418.

[6] M.S. Islam, M. Leslie, S.M. Tomlinson and C.R.A. Catlow, J. Phys. C. 21 (1988) L109.

[7] See e.g. F. Mila, F.C. Zhang and T.M. Rice, Physica C 153-155 (1988) 1221, and references cited therein.

[8] Y.H. Kim, C.M. Foster, A.J. Heeger, S. Cox, L. Acedo and G. Stucky, preprint; Y.H. Kim et al., submitted to Phys. Rev. Lett.; Y.H. Kim, A.J. Heeger, L. Acedo, G. Stucky and F. Wudl, Phys. Rev. B. 36 (1987) 7252.

[9] A.F. Andreev and L.M. Lifshitz, Sov. Phys. J.E.T.P. 29 (1969) 1107.

[10] J. Livage, J. Phys. (Paris) 42, Coll. C4 (1981) 981; C. Sanchez, F. Babonneau, R. Morineau and J. Livage, Phil. Mag. B47 (1983) 279; F. Babonneau and J. Livage, Nouveau J. de Chimie 10 (1986) 191.

[11] For reviews see e.g. A.R. Bishop, J.A. Krumhansl, and S.E. Trullinger, Physica 1D (1980) 1; "Solitons and Condensed Matter Physics", eds. A.R. Bishop and T. Schneider, Springer Series in Solid State Sciences vol.8, 1978; or Tarik Ö. Ogurtani, "Solitons in solids", Ann. Rev. Mater. Sci. 13 (1983) 67; Solitons in magnetic systems are reviewed e.g. by A.M. Kosevich, B.A. Ivanov and A.S. Kovalev, Sov. Sci. Rev. A. Phys. 6 (1985) 161; L.J. de Jongh, J. Appl. Phys. 53 (1982) 8018.

[12] A.S. Davydov, "Solitons in Molecular Systems", D. Reidel, Publishing Company, Dordrecht 1985.

[13] J.B. Torrance and M. Tinkham, Phys. Rev. 187 (1969) 5587; ibid 187 (1969) 595; M. Date and M. Motokawa, Phys. Rev. Lett. 16 (1966) 1111; J. Phys. Soc. Jpn. 24 (1968) 41.

[14] N. Ishimura and H. Shiba, Progr. Theor. Phys. 63 (1980) 743; H.J.M. de Groot, L.J. de Jongh, M. ElMassalami, H.H.A. Smit and R.C. Thiel, Hyp. Int. 27 (1986) 93.

[15] Y. Onodera, J. Phys. Soc. Jpn. 56 (1987) 250.

[16] W.P. Su, J.R. Schrieffer and A.J. Heeger, Phys. Rev. Lett. 42 (1979) 1698, and Phys. Rev. B 22 (1980) 2099.

[17] J.R. Schrieffer, Int. School of Physics "Enrico Fermi", Course 89 (1983) 300.

[18] S. Ichinose, Solid State Commun. 50 (1984) 137.

[19] N. Kumar, Physica C 153-155 (1988) 1227; M.M. Mohan and N. Kumar, J. Phys. C. 20 (1987) L527.

[20] T.V. Ramakrishnan, Physica C 153-155 (1988) 555.

[21] T. Matsubara and H. Matsuda, Progr. Theor. Phys. 16 (1956) 569; ibid. 17 (1957) 19.

[22] M.R. Schafroth, Phys. Rev. 100 (1955) 463.

[23] A. Alexandrov and J. Ranninger, Phys. Rev. B. 24 (1981) 1164; A.S. Alexandrov, J. Ranninger and S. Robaszkiewicz, Phys. Rev. B. 33 (1986) 4526 and references cited.

[24] V.J. Emery, Phys. Rev. B 17 (1976) 2989; M. Fowler, Phys. Rev. B 17 (1978) 2989.

[25] P. Nozières and S. Schmitt-Rink, J. Low Temp. Phys. 59 (1985) 195.

[26] See e.g. T.T.M. Palstra, B. Batlogg, L.F. Schneemeyer, and R.B. van Dover, Phys. Rev. B 38 (1988) to appear; T.T.M. Palstra, B. Batlogg, L.F. Schneemeyer, R.B. van Dover and J.V. Waszczak, preprints.

[27] See e.g. L.J. de Jongh and A.R. Miedema, "Experiments on simple magnetic model systems", Taylor and Francis (Monographs in Physics), London, 1974; also published in Adv. Phys. 23 (1974) 1, and references cited in this work.

[28] For a discussion of the magnetic analogues, see e.g. H.J.M. de Groot and L.J. de Jongh, Physica 141B (1986) 1; Physica Scripta T13 (1986) 219.

[29] For a recent review of theory, see e.g. T. Nattermann and J. Villain in "Phase Transitions", 1988, vol.11, p.5, Gordon and Breach Science Publishers Inc., England (reprints available from the publishers).

[30] See e.g. R.L. Carlin and L.J. de Jongh, Chem. Rev. 86 (1986) 675; L.J. de Jongh in "Recent Developments in Condensed Matter Physics" ed. J.T. Devreese, Plenum Press, 1981, vol.1, p.359.

4.2 Crystal Chemistry and Electronic Structure

Structure-Property Relations in High-Temperature Oxide Superconductors: Nature of Oxygen and Copper*

C. N. R. Rao

Solid State and Structural Chemistry Unit, Indian Institute of Science, Bangalore, 560012, India

ABSTRACT

Important structural aspects of the following families of perovskite-related superconductors are presented: $La_{2-x}Sr_x(Ba_x)CuO_4$, $LnBa_2Cu_3O_7$, $A_mCa_{n-x}B_xCu_{n-1}O_{2n+4}$ (A=Tl, m=1,2; A=Bi, m=2; B=Ba,Sr). The nature of copper and oxygen in these oxide superconductors is discussed and it will be shown that oxygen holes (O^{1-} or O_2^{2-}) and Cu^{1+} ions play a crucial role in the superconductivity of the cuprates. Oxygen holes are likely to be present in many other oxides purported to contain metal ions such as Cu, Ni and Bi in high oxidation states; this seems to be the case with copper-free bismuth oxide superconductors as well.

INTRODUCTION

$La_{2-x}Sr_x(Ca_x,Ba_x)CuO_4$, $Ln(Y)Ba_2Cu_3O_{7-\delta}$, $Bi_2(Ca,Sr)_{n+1}Cu_nO_{2n+4}$, $Tl_2Ca_{n-1}Ba_2Cu_nO_{2n+4}$ and $TlCa_{n-1}Ba_2Cu_nO_{2n+3}$ systems of high-temperature superconductors exhibit certain important common features [1-3]. Besides examining these structural commonalities, we shall discuss the nature of Cu and oxygen in these cuprates. We show that holes residing on the oxygen hold the key to the superconductivity possibly through the formation of peroxitons (O^{1-} - O^{1-} bound pairs next to Cu^{1+}). The presence of Bi and Tl in the 3+ state in the Bi and Tl cuprate systems and the observation of fairly high T_c ($\sim$ 30K) superconductivity in copper-free $Ba_{1-x}K_xBiO_3$ [4] also reinforce this view.

COMMONALITIES IN THE PROPERTIES OF CUPRATE SUPERCONDUCTORS

All the superconducting cuprates possess perovskite-related structures. They all have low dimensional characteristics, the two-dimensional Cu-O sheets being common to them. The 123 compounds have, in addition, one-dimensional Cu-O chains. They crystallize in orthorhombic or tetragonal structures and the former is more common. The Cu-O bonds in the superconducting cuprates are highly covalent and the distances are generally in the 1.90 ± 0.5Å range. The coordination of Cu is essentially square-planar. These cuprates could be considered as belonging to the same structural family involving the intergrowth of oxygen-deficient perovskite layers with rock-salt layers [5]. The 123 compounds do not have rock-salt layers and only contain defect perovskite layers. In Table I, the common features of the cuprate superconductors are summarized.

In the $Bi_2(Ca,Sr)_{n+1}Cu_nO_{2n+4}$, $Tl_2Ca_{n-1}Ba_2Cu_nO_{2n+4}$ and $TlCa_{n-1}Ba_2Cu_nO_{2n+3}$ series, T_c increases with the number of Cu-O sheets, n, as shown in Fig. 1. It seems unrealistic to increase T_c to very high values by increasing n. While intergrowths seem to give rise to high T_cs [6],it has not yet been possible to obtain recurrent intergrowths. In $La_{2-x}(Sr,Ba)_xCuO_4$ and $YBa_2Cu_3O_{7-\delta}$, T_c increases proportionally with the nominal concentration of Cu^{3+}, but this can not be the real picture since some of the cuprates which are not expected to contain much Cu^{3+}, show high T_cs.

*Contribution No. 561 from the Solid State and Structural Chemistry Unit.

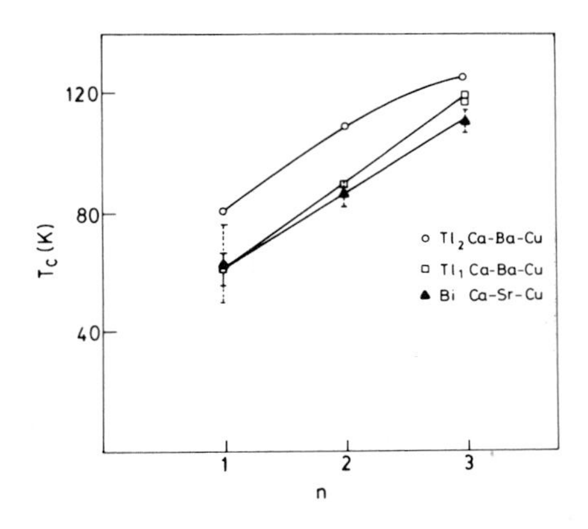

Fig. 1. Variation of T_c with number of Cu-O sheets.

STATES OF COPPER AND OXYGEN

Superconducting $La_{2-x}M_xCuO_4$ and $LnBa_2Cu_3O_7$ are nominally expected to contain a significant proportion of Cu^{3+} ions. Such a nominal Cu^{2+} - Cu^{3+} mixed valence has been considered to play an important role in the superconductivity of these oxides. Superconductivity in $YBa_2Cu_3O_{7-\delta}$ cannot be explained by taking into account the charge balance based on elementary chemical considerations alone. Thus, $YBa_2Cu_3O_{6.5}$ which should formally contain only Cu^{2+} should not be superconducting, but it shows a T_c of $\sim$45K; one possibility is that intergrowth of O_7 and O_6 domains is responsible for this superconductivity. Considerable effort has been made to understand the nature of copper and oxygen ions in the cuprates since all the essential phenomena (magnetic properties and superconductivity) are confined to the CuO_2 planes.

Stoichiometric La_2CuO_4 is not a superconductor and should have every copper atom in the 2+ state ($\underline{d}^9$ electron configuration or 1 hole in the $\underline{d}$-band) and oxygen in the 2-state ($\underline{p}^6$ electron configuration or no hole in the $\underline{p}$-band). Chemical or stoichiometry change such as introduction of 15% Sr^{2+} in place of La^{3+}, we essentially introduce 15% holes or excess positive charge in the basal plane and obtain a $T_c \sim 35K$. In the 123 oxide ($YBa_2Cu_3O_7$), the mobile hole concentration is $\sim$33%, with a $T_c \sim 90K$. An important question that arises is whether the holes introduced result in a Cu^{3+} state ($\underline{d}^8$ electron configuration) or whether the holes appear elsewhere.

X-ray absorption (Cu K or L edge) spectroscopy of $YBa_2Cu_3O_{6.9}$ and $La_{1.85}Sr_{0.15}CuO_4$ has shown no evidence for Cu^{3+}. A 1s ⟶ 2p transition found in the oxygen K-edge spectrum suggests

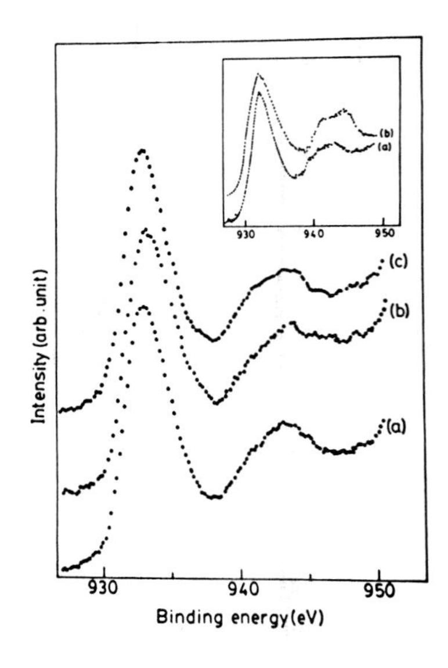

Fig. 2. $Cu(2p_{3/2})$ region in the XPS of $YBa_2Cu_3O_{6.9}$ (a) 300K, (b) after cooling to 80K and (c) after warming (b) to 300K. Inset shows $Cu(2p_{3/2})$ spectra of (a) $Tl_2Ca_2Ba_2Cu_3O_{10}$ and (b) $Bi_2(Ca,Sr)_3Cu_2O_8$ at 300K.

TABLE I SALIENT FEATURES OF SUPERCONDUCTING CUPRATES

	$La_{2-x}(Ca,Sr,Ba)_xCuO_4$	$Y(Ln)Ba_2Cu_3O_7$	$(BiO)_2(Ca,Sr)_{n+1}Cu_nO_{2n+2}$	$(TlO)_m(Ca,Ba)_{n+1}Cu_nO_{2n+2}$
T_c	30-40K	90K	60-110K (n=1-3)[a]	m=2, 90-125K (n=1-3)[a] m=1, 60-120K(n=1-3)
Dimensionality	2(Cu-O sheets)	2(Cu-O sheets) +1(Cu-O chains)	2(Cu-O sheets)	2(Cu-O sheets)
Normal state	Marginally metallic	Marginally metallic	Marginally metallic	Marginally metallic
Crystal structure	Orthorhombic	Orthorhombic[b]	Orthorhombic	Tetragonal
Cu-coordination	Square-planar	Square-planar	Square-planar	Square-planar
Short Cu-O distance, Å	1.90	1.90 ± 0.05	1.90	1.93
Oxidation state of Cu as found	1+, 2+	1+, 2+	1+, 2+	1+, 2+

[a] The n = 1 member with only a single Cu-O layer is comparable to $La_{2-x}Sr_x(Ba_x)CuO_4$ of K_2NiF_4 structure.
[b] Some high T_c tetragonal 123 compounds have been made.

the presence of holes in the oxygen band [7]. We have investigated the nature of holes in the superconducting cuprates for sometime by employing x-ray photoelectron (XP) and Auger electron spectroscopies [8]. Cu(2p) XP spectra as well as L_3VV Auger spectra of $La_{1.85}Sr_{0.15}CuO_4$ and $YBa_2Cu_3O_{6.9}$ show the presence of only the Cu^{1+} and Cu^{2+} species in the ground state of these oxides, with no detectable amount of Cu^{3+}. Thus, the Cu(2p) spectra show an intense peak due to Cu^{1+} ($3\underline{d}^{10}$-final state) at 933 eV and a weaker feature due to Cu^{2+} ($3\underline{d}^9$-final state) at 942 eV (Fig. 2). $Bi_2(Ca,Sr)_3Cu_2O_8$ and $Tl_2CaBa_2Cu_2O_8$ show Cu(2p) and Cu(LVV) features similar to $YBa_2Cu_3O_{6.9}$. Furthermore, both Bi and Tl are essentially in the 3+ state in these cuprates as found from XP spectra (Figs. 3 and 4). The proportion of the Cu^{1+} state seems to increase slightly on lowering the temperature to the T_c region. Since none of the superconducting cuprates

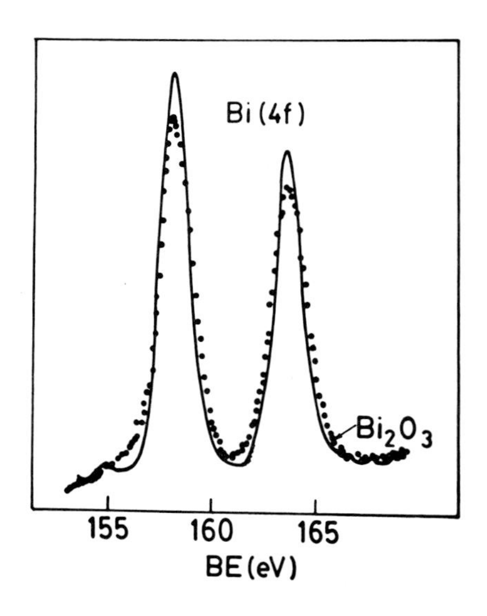

Fig. 3. X-ray photoelectron spectra of $Bi_2(Ca,Sr)_3Cu_2O_8$ and Bi_2O_3 in the Bi(4f) region.

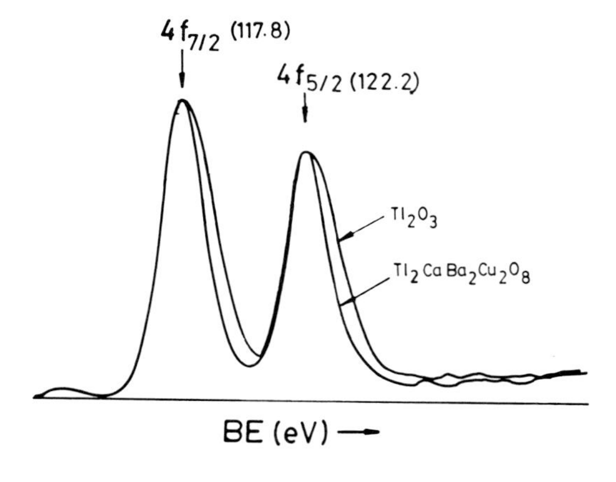

Fig. 4. X-ray photoelectron spectra of 2122 and Tl_2O_3 in the Tl(4f) region.

shows evidence for the presence of Cu^{3+} ions, the holes should reside on the oxygens. In other words, we would expect Cu^{2+} (d^9 or one $\underline{d}$-hole)-O^{2-} ($\underline{p}^6$ or no $\underline{p}$-hole) and Cu^{1+} ($\underline{d}^{10}$ or no $\underline{d}$-hole) - O^{1-} ($\underline{p}^5$ or one $\underline{p}$-hole) type of configurations to be present, with the latter being predominant.

O(1s) core level spectra of the cuprates show the presence of features with binding energies of around 529, 531 and 533 eV respectively. The normally expected oxide species, O^{2-}, with the filled $2\underline{p}^6$ configuration is associated with the 529 eV feature. The O^{1-} species (corresponding to the presence of a hole in the 2p band) is expected to have a higher binding energy than O^{2-}, but the position of this O(1s) feature is likely to be in the 529-531 eV region. Unfortunately CO_3^{2-} and OH^- which are almost always present in such oxides give an intense peak around 531 eV. We, however, find an oxygen species with a binding energy of 533 eV which we consider to be due to the dimerization of O^{1-} holes. The assignment of the 533 eV O(1s) feature to dimerized oxygen holes, O_2^{2-}, is based on the fact that ordinary peroxides and oxygen adsorbed on metal surfaces in the peroxo form show such a feature. The O_2^{2-} species may not be "static" species but could be formed within the vibrational times of the O^{1-} ions. The proportion of the dimerized holes increases substantially with decreasing temperature and the phenomenon is reversible. Furthermore, non-superconducting $YBa_2Cu_3O_6$ does not show the 533 eV feature. The dimerized hole species are likely to play an important role in the mechanism of superconductivity of these oxides. The preponderant presence of Cu^{1+} in the superconducting state dictates that a considerable proportion of the oxygens be present in the hole state (O^{1-} or O_2^{2-}). The presence of holes on oxygens should not surprise chemists since it would be unrealistic to assign a high formal charge to copper (+2, +3) when the Cu-oxygen bonds are so highly covalent.

The holes on oxygen can be visualised in terms of real-space pairings. We can have two possible bound states with O^{1-} - O^{1-} hole pairs (Fig. 5) which we call hole polaron (or hole

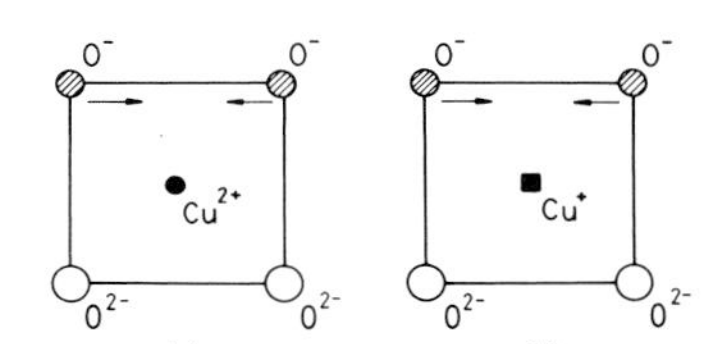

Fig. 5. (a) hole bipolaron; (b) hole peroxiton.

bipolaron) and hole peroxiton [9]. Because of the large inter-site Coulomb repulsion, the bipolaron would have difficulty in forming the bound state while the peroxiton (O^1 - Cu^{1+} - O^{1-}) would have much lower energy. Around a critical value of the parameter λ/V_{pd} (λ is the coupling energy gained as the pair localises due to deformation), the peroxiton binding energy goes to zero. At or above the critical value, peroxiton formation can occur spontaneously.

We believe that oxygen holes are likely to be responsible for the superconductivity of the cuprates. Oxygen holes seem to be present even in oxides such as $Ba_2Cu_2O_5$ and $LaCuO_3$ (rather than Cu^{3+}). Oxygen holes are probably responsible for the superconductivity of $Ba(Bi,Pb)O_3$. The recent discovery of relatively high T_c (~ 30K) in copper-less $Ba_{1-x}K_xBiO_3$ [4] underscores the crucial role of oxygen holes (probably as O^{1-} - Bi^{3+} - O^{1-}). Oxygen holes would be expected to be more favoured in the presence of Cu^{1+} ($\underline{d}^{10}$) ions just as in the case of cuprous chalcogenides where S-S, Se-Se and Te-Te bonds are formed are due to hole dimerization. The oxygen hole dimers would, however, be much less stable. It is interesting that some of the oxygen in $YBa_{2-x}La_xCu_3O_{7+\delta}$ is expected to be in the peroxide form [10]. In Bi and Tl cuprates, there is likely to be hole-transfer from Bi(Tl) to Cu. Since Cu is present in 1+ and 2+ states only, the holes would have to reside on the oxygen in the Bi and Tl cuprates. The hole concentration will increase with the number of Bi(Tl) layers; accordingly, T_cs of $Tl_2Ca_{n-1}Ba_2Cu_nO_{2n+4}$ are always higher than those of the corresponding $TlCa_{n-1}Ba_2Cu_nO_{2n+3}$ (Fig. 1).

Tokura et al [11] have pointed out that the [Cu-O] charge in the sheets largely determines T_c. The charge p of $[Cu-O]^p$ is determined by the relative proportions of the species, a, $(Cu^{3+} - O^{2-})^{1+}$; b, $(Cu^{2+} - O^{1-})^{1+}$; c, $(Cu^{2+} - O^{2-})^0$ and d, $(Cu^{1+} - O^{1-})^0$. Since a is essentially absent, $p = \beta b + \gamma c + \delta d$; $p = 0.33$ if b, c and d are equally important. Since $p \approx 0.2$ is closer to the boundary between superconducting and insulating behaviour, the relative values of γ and δ will determine the value of p as well as the properties. It is possible that $\delta > \gamma$ favours superconductivity.

ACKNOWLEDGEMENT: The author is thankful to the Department of Science and Technology and the University Grants Commission for support of this research.

References

1. C.N.R. Rao, J. Solid State Chem., 74, 147 (1988)
2. C.N.R. Rao, (ed), The Chemistry of Oxide Superconductors, (Blackwell, Oxford, 1988).
3. C.N.R. Rao (ed), Progress in High-Temperature Superconductivity, Vol. VII, (World Scientific, Singapore, 1988).
4. R.J. Cava et al, Nature, 332, 814 (1988).
5. C.N.R. Rao and B. Raveau, to be published.
6. C.N.R. Rao et al, Pramana-J. Phys. 30, 495 (1988)
7. J.A. Yarmoff et al, Phys. Rev. B36, 3967 (1987).
8. D.D. Sarma and C.N.R. Rao, Solid State Commun., 65, 47 (1988).
9. B.K. Chakraverty, D.D. Sarma and C.N.R. Rao, to be published.
10. A. Manthiram, X.X. Tang and J.B. Goodenough, Phys. Rev. B37, 3734 (1988).
11. Y. Tokura, J.B. Torrance, T.C. Huang and A.I. Nazzal, to be published.

Crystal Chemistry of Superconductive Bismuth and Thallium Cuprates

B. RAVEAU, C. MICHEL, and M. HERVIEU

CRISMAT — ISMRa, Université de Caen, Bd du Maréchal Juin, 14032 Caen Cedex, France

The superconductive rush has led to a considerable number of publications, which one often rather similar and are note always of high interest. Nevertheless, many results have been gathered about the 40K-La_2CuO_4 type and 92K-$YBa_2Cu_3O_7$ type superconductors which allow superconductivity in those phases to be better undertood. These two families have been the purpose of numerous papers and several reviews and for this reason will not be discussed here. On the opposite, the bismuth and thallium cuprates which are born recently, have been less studied, perhaps owing to their more difficult chemistry. Thus the present review is mainly focussed on the synthesis, structure, non stoichiometry and extended defects of those latter phases in connection with their superconducting properties.

The idea that trivalent bismuth was a potential cation [1] was founded on the following points :

- Bi^{3+} has a very similar size to lanthanide Ln^{3+} ions,
- Bi^{3+} exhibits a $6s^2$ lone pair, whose stereoactivity may influence the oxygen framework and is susceptible to lead to layer structures like Aurivillius phases.

The ability of bismuth to induce a bidimensional character of the structure was considered as an important factor for the generation of new high Tc superconductors. The validity of this idea was confirmed by the discovery of the first signs of superconductivity at 22K in the Sr-Bi-Cu-O system [2], and subsequently at 90K in the composition $Bi_2SrCa_2CuO_x$ [3]. Up to now three superconductors have been observed in the bismuth cuprates : $Bi_2Sr_2CuO_6$ [2, 4-5], $Bi_2Sr_2CaCu_2O_8$ [4, 6-12] which superconduct at 22K and 80K respectively and are easily synthesized in air, and the 110K-superconductor $Bi_2Sr_2Ca_2Cu_3O_{10}$ [13-16] which has so far not been isolated in the form of bulk samples but appears must of the time as mixed with $Bi_2Sr_2CaCu_2O_8$.

Attempts to substitute thallium (III) for yttrium in $YBa_2Cu_3O_7$, has led to the discovery of superconducctivity at 85K in the system Tl-Ba-Cu-O [17], and subsequently at 125K in the system Tl-Ba-Ca-Cu-O [18]. However synthesis in air led to multiphase samples due to thallium oxide volatility, so that the 108K and 125K-superconductors were first identified [19-22] as $Tl_2Ba_2CaCu_2O_8$ and $Tl_2Ba_2Ca_2Cu_3O_{10}$ respectively, and could be isolated [22] subsequently by reaction of BaO_2, Tl_2O_3, CaO and CuO in evacuated ampoules. Single crystal and electron microscopy studies of $Tl_2Ba_2CuO_6$ [5], $Tl_2Ba_2CaCu_2O_8$ [21-23] and $Tl_2Ba_2Ca_2CuO_{10}$ [20, 21, 24] allowed the structure of those three superconductors to be established. Besides, those three first superconductors, three other superconductive phases were isolated. The 120K-superconductor $TlBa_2Ca_2Cu_3O_9$ was isolated by Martin et al. [25-26] and obtained simultaneously in multiphase samples by Parkin et al. [27], $TlBa_2CaCu_2O_7$ [28] which superconducts at 65K can also easily be isolated. On the opposite the 104K-superconductor $Tl_2Ba_2Ca_3Cu_4O_{12}$" [29] was not isolated as a pure phase : starting from the nominal composition "$TlBaCa_3Cu_3O_x$" one obtains about 90 % of $Tl_2Ba_2Ca_3Cu_4O_{12}$ mixed with $BaCuO_2$.

I - STRUCTURAL PRINCIPLES AND RELATIONSHIPS

The examination of the subcell parameters of bismuth compounds as well as thallium cuprates clearly shows that all those superconductors exhibit a bidimensional accord with the perovskite structure : $a \simeq a_p\sqrt{2}$, $b \simeq a_p\sqrt{2}$ for the orthorhombic bismuth cuprates, and $a \simeq a_p$ for the tetragonal thallium cuprates. The X-ray single crystal and electron microscopy studies show that the stacking sequence of the metallic layers is similar for bismuth and thallium. The high

Fig. 1 - High resolution electron micrographs recorded for a focus value close to -650Å and calculated corresponding images:
a) $Bi_2Sr_2CaCu_2O_8$ and b) $Tl_2Ba_2CaCu_2O_8$.

resolution micrographs allow indeed the stacking sequence to be identified. In that way, the rows of big bright dots in figure 1a and b, recorded for a focus value close to -650 Å, can be correlated to the bismuth and strontium (or thallium and barium) positions whereas the rows of smaller white spots correspond to copper and calcium ones. The comparison of these experimental images with the calculated ones is in agreement with this interpretation. Thus two sorts of stacking sequences are observed :

$$Ba[Tl_2]Ba[Ca_{m-1}Cu_m] \text{ or } Sr[Bi_2]Sr[Ca_{m-1}Cu_m]$$

and $$Ba\text{-}Tl\text{-}Ba[Ca_{m-1}Cu_m]$$

with m ranging from 1 to 4.

The X-ray structure determinations of the thallium compounds, easier than that of bismuth oxides as discussed further, establish that all those oxides are closely related to each other and to the La_2CuO_4-type structure. They can indeed be described (Fig. 2) as intergrowths of multiple oxygen deficient perovskite layers, $ACuO_{3-x}$ with rock salt-type layers AO. Thus they belong to the same structural family $[ACuO_{3-x}]_m[AO]_n$.

Three oxides correspond to the member m = 1 of this series, and are thus formed of single perovskite layers : La_2CuO_4 (Fig. 2a) which exhibits the classical K_2NiF_4-structure, i.e. single rock salt type layers (n = 1) $[LaO]_\infty$, and $Bi_2Sr_2CuO_6$ and $Tl_2Ba_2CuO_6$ (Fig. 2b) in which one observes triple rock salt type layers (n = 3) formed of a thallium or bismuth bilayer sandwidched by barium or strontium single layers. Thus these oxides can be formulated $[LaCuO_3][LaO]$ and $[SrCuO3][(BiO)_2SrO]$ or $[BaCuO_3][(BaO)(TlO)_2]$ respectively.

The members m = 2, are represented by the oxides $Bi_2Sr_2CaCu_2O_8$ and $Tl_2Ba_2CaCu_2O_8$ (Fig. 2c) and the oxide $TlBa_2CaCu_2O_7$ (Fig. 2d) : they all exhibit a double oxygen deficient perovskite layer in which a plane of oxygens has been removed at the some level of the calcium ions, so that such layers can be described as formed of $[CuO_{2.5}]_\infty$ sheets of corner sharing pyramids interleaved by a plane of calcium ions. $Tl_2Ba_2CaCu_2O_8$ and $Bi_2Sr_2Ca_2CuO_8$ exhibit a triple rock salt type layer (n = 3) $[BaO(TlO)_2]_\infty$ and $[SrO(BiO)_2]_\infty$ respectively, whereas a double rock salt-type layer $[(TlO)(BaO)]$is observed for $TlBa_2CaCu_2O_7$. The formulae of these oxides can be represented as follows :
$[SrCa(CuO_{2.5}\square_{0.5})_2]$ $[(BiO)_2(SrO)]$, $[BaCa(CuO_{2.5}\square_{0.5})_2][(TlO)_2(BaO)]$ and $[BaCa(CuO_{2.5}\square_{0.5})_2]$ $[(TlO)(BaO)]$.

Triple oxygen deficient perovskite layers (m = 3) are also observed for $Tl_2Ba_2Ca_2Cu_3O_{10}$, $Bi_2Sr_2Ca_2Cu_3O_{10}$ (Fig. 2e) and for $TlBa_2Ca_2Cu_3O_9$ (Fig. 2f). Such layers are built up from two pyramidal sheets $[CuO_{2.5}]_\infty$ and one $[CuO_2]_\infty$ sheet of corner-sharing CuO_4 square planar groups interleaved with planes of calcium ions. It results in the formulae $[BaCa_2(CuO_{2.5}\square_{0.5})_2(CuO_2\square)][(TlO)_2BaO]$ $[SrCa_2(CuO_{2.5}\square_{0.5})_2(CuO_2\square)][(BiO)_2SrO]$ and $[BaCa_2(CuO_{2.5}\square_{0.5})_2(CuO_2\square)]$ $[(TlO)(BaO)]$ in which triple rock salt-type layers (n = 3) and double rock salt-type sheets (n = 2) are involved.

$Tl_2Ba_2Ca_3Cu_4O_{12}$ (Fig. 2g) is the only member containing quadruple oxygen deficient perovskite layers (m = 4) $[BaCa_3(CuO_{2.5}\square_{0.5})_2(CuO_2\square)]_\infty$. The latter are formed of two $[CuO_{2.5}]_\infty$ pyramidal sheets and two $[CuO_2]_\infty$ sheets of square planar groups interleaved with planes of calcium ions. The rock salt-type layers are double (n = 2) formed of a thallium monolayer sandwidched by barium single layers, according to the formula $[(TlO)(BaO)]_\infty$. In a conclusion the

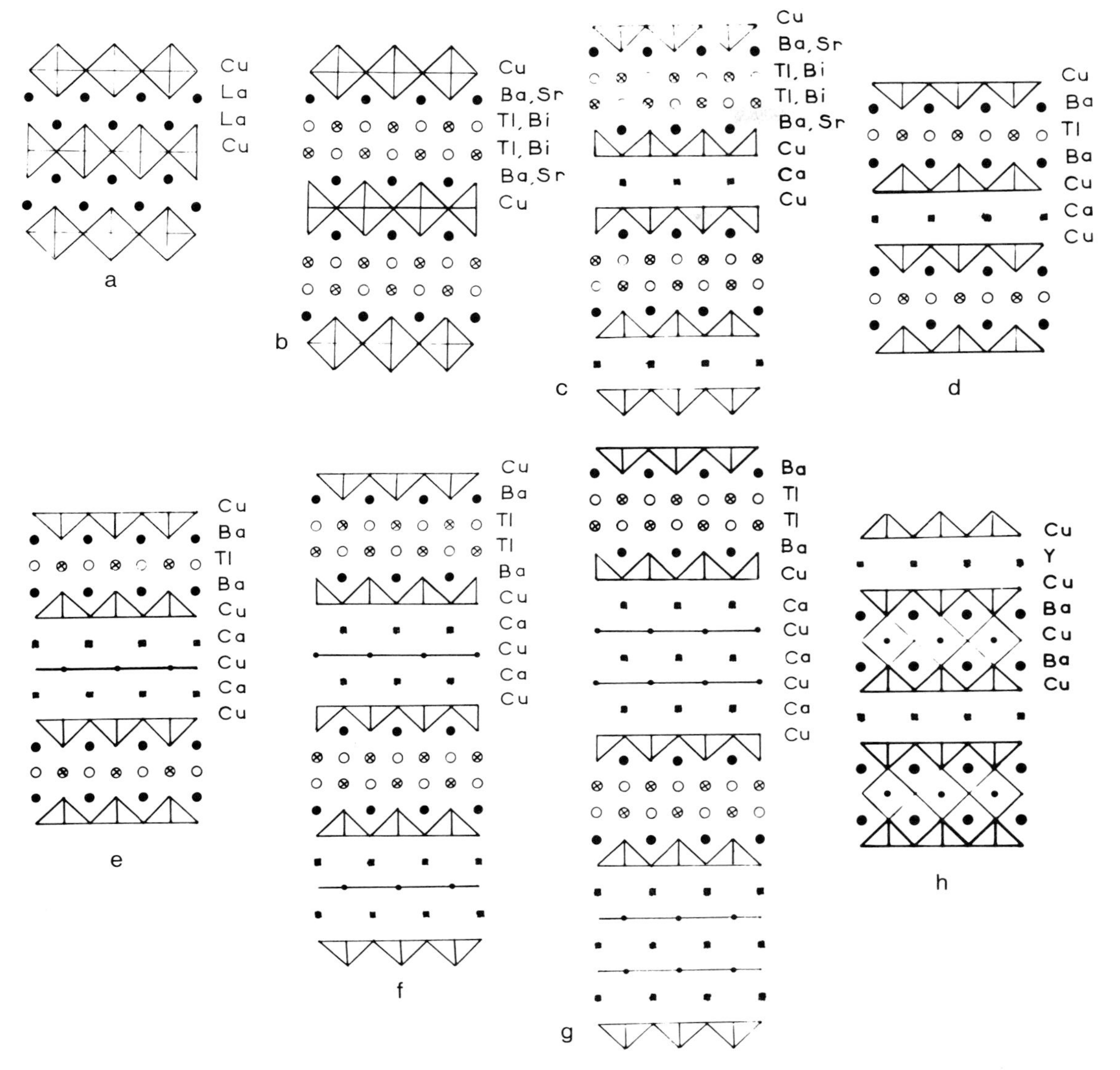

Fig. 2 - Structure of the oxides : a) La_2CuO_4, b) $Bi_2Sr_2CuO_6$ or $Tl_2Ba_2CuO_6$ c) $Bi_2Sr_2CaCu_2O_8$ or $Tl_2Ba_2CaCu_2O_8$, d) $TlBa_2CaCu_2O_7$, e) $Tl_2Ba_2Ca_2Cu_3O_{10}$ or $Bi_2Sr_2Ca_2Cu_3O_{10}$, f) $TlBa_2Ca_2Cu_3O_9$, g) $Tl_2Ba_2Ca_3Cu_4O_{12}$ and h) $YBa_2Cu_3O_7$.

thallium and bismuth compounds can thus be represented by the general formula :

$$[BaCa_{m-1}Cu_mO_{2m+1}]^P \; [BaTl_{n-1}O_n]^{R.S.} \text{ and } [SrCa_{m-1}Cu_mO_{2m+1}]^P \; [SrBi_{n-1}O_n]^{R.S.}$$

respectively. It is also worth pointing out that $YBa_2Cu_3O_7$ belongs to this series and corresponds to n = 0 : one indeed recognizes in this structure (Fig. 2h) pyramidal layers connected through yttrium planes instead of calcium planes.

II - STRUCTURAL PROBLEMS : INCOMMENSURABILITY, NON-STOICHIOMETRY, EXTENDED DEFECTS

The things are not so simple as it could be expected from the above description.

In the case of the bismuth oxides the crystals are lamellar leading to poor crystallographic data. Moreover one observes systematically extra spots in an incommensurate position along [100]* as shown from the E.D. patterns of

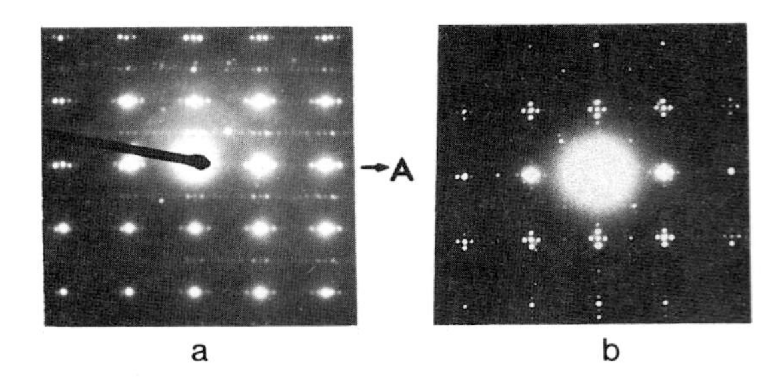

Fig. 3 - [001] E.D. patterns of $Bi_2Sr_2CaCu_2O_8$.

$Bi_2Sr_2CaCu_2O_8$ (Fig. 3a), the commensurate position corresponding to a superstructure of 10 x $a_p \sqrt{2}$. This "superstructure" appears sometimes along two perpendicular directions (Fig. 3b). Thus, if the HREM image does not exhibit any anomaly along [110], it is not the case for the [001] image (Fig. 4) where a modulation of the contrast appears systematically along $\vec{a}$ and in local areas along $\vec{b}$. From these observations it is clear that the structure is so far not completely understood, and that the resolutions performed as well on single crystals in the subcell as by neutron diffraction converge for the positions of the heavy atoms but lead to controversies for the positions of the oxygen atoms. Up to now, it must be stated that the positions of the oxygen atoms which form the $(BiO)_2$ layers are most likely. Nevertheless, localization of oxygen between the bismuth layers like in Aurivillius phases, or at the same level as bismuth but turned of 45° cannot be ruled out. Besides this main structural feature, one observes rarely extended defects which result from the intergrowths of the different structures as shown for instance in the HREM micrograph of $Bi_2Sr_2CaCu_2O_8$ (Fig. 5) which exhibits a "$Bi_2Sr_2CuO_6$" defects (periodicity of 24 Å) in the $Bi_2Sr_2CaCu_2O_8$ matrix (periodicity of 30.7 Å). Another interesting feature concerns the possibility of replacement of calcium by strontium pointed out by Sleight et al. leading to the formulation $Bi_2Sr_{3-x}Ca_xCu_2O_8$ with $0 \leq x \leq 0.9$, so that the exact formula of this 80K-superconductor is not yet absolutely known.

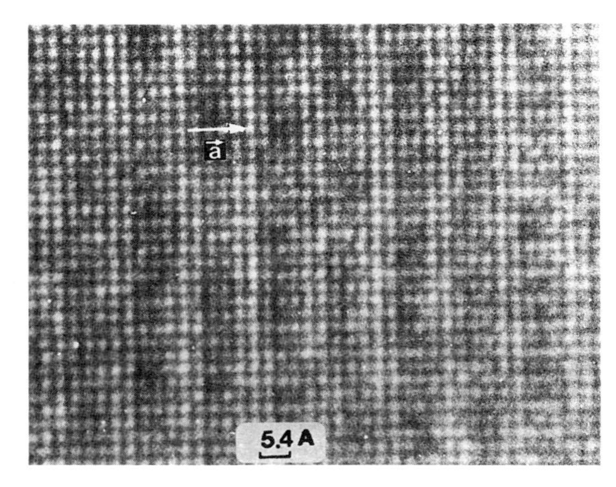

Fig. 4 - Systematical modulation of the contrast along a in the [001] high resolution image ; local modulation are observed along b.

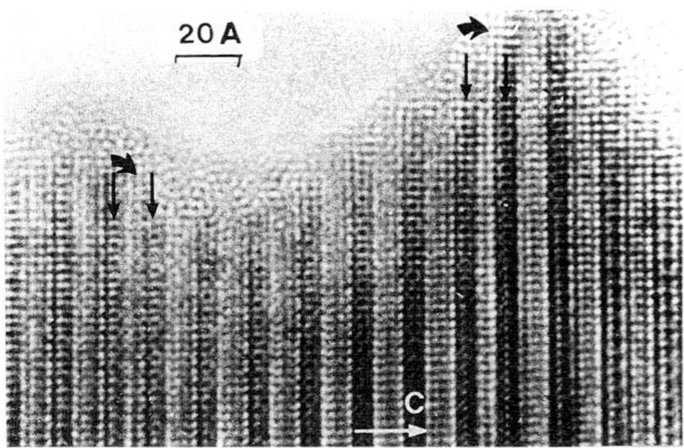

Fig. 5 - $Bi_2Sr_2CuO_6$ layer appears as a defect in $Bi_2Sr_2CaCu_2O_8$ matrix.

Contrary to the bismuth cuprates, the thallium compounds do not exhibit any systematical incommensurability, and their cell is much closer to that of the cubic perovskite, so that there is no ambiguity concerning their atomic positions, i.e. concerning the nature of the different layers of perovskite and rock salt type. Nevertheless there appears a problem with the possible non-stoichiometry or "solid solution" in the thallium and calcium distributions. From the refinements of the B factors during the structure determinations one observes that the thallium sites of the rock salt type layers are deficient whereas the calcium sites of the perovskite layers are partly occupied by thallium. This suggests that the compositions of the superconducting phases are not exactly those given above for $Tl_2Ba_2CaCu_2O_8$ and $Tl_2Ba_2Ca_2Cu_3O_{10}$ but can be formulated $Tl_{2-x}Ca_xBa_2Ca_{1-y}Tl_yCu_2O_8$ and $Tl_{2-x}Ca_xBa_2Ca_{2-y}Tl_yCu_3O_{10}$ respectively with x and y ranging from 0.1 to 0.2. An alternative to this hypothesis can be the presence of cationic vacancies in the thallium rock salt layers leading to the formulae $Tl_{2-x}Ba_2Ca_{1-y}Tl_yCu_2O_8$ and $Tl_{2-x}Ba_2Ca_{2-y}Tl_yCu_3O_{10}$ with $x < 0.10$ and $y \simeq 0.10$. The electron diffraction and high resolution electron microscopy study show that many crystals exhibit stacking faults. E.D. patterns show stacking along the c* axis, witness of a disorder in the layer stacking. Such intergrowth defects, which correspond to a local variation of composition can be easily identified from the HREM images when the contrast was correlated to the structure, through the comparison of experimental and calculated through-focus series. They are related as well to the perovskite-type slabs as to the rock salt-type layers. The variation of the perovskite slab thickness, i.e. of m, is a common feature of such intergrowth structure ; in a general way, the defectuous layers exhibit a n value close to the nominal value in the first members (low m value) but can differ when m increases : an example of the occurence of m = 2 and 4 layers in a nominal m = 3 matrix is shown in Fig. 6a whereas m = 7 is observed in the m = 4 matrix (Fig. 6b).

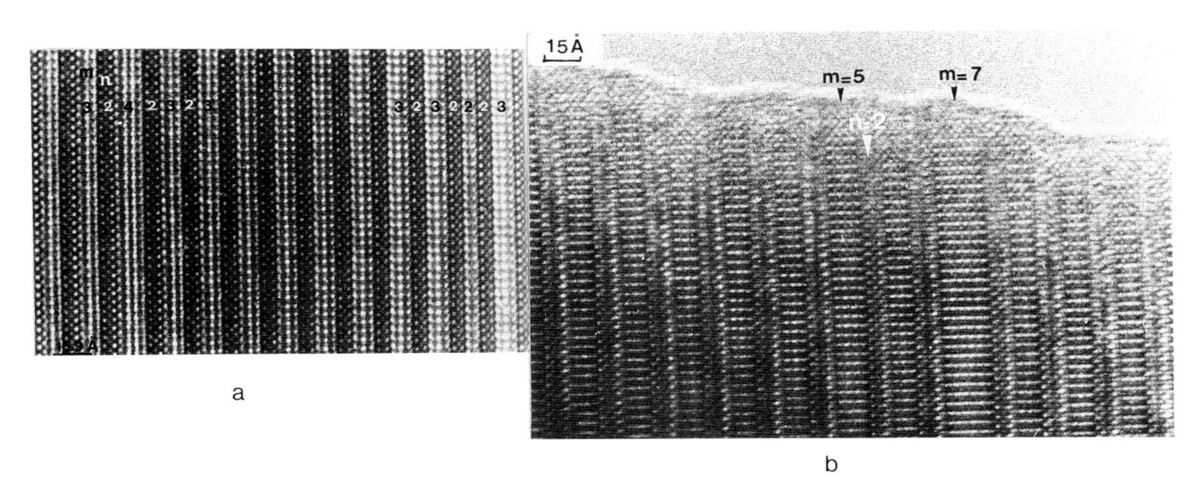

Fig. 6 - Defectuous perovskite layers : a) m = 2 and 4 layers in a nominal m = 3 matrix, b) m = 7 layer in a m = 4 matrix.

The existence of variations in the stacking of the rock salt-type layers is a unique phenomenon. Two types of stacking defects are observed. The first one corresponds to the occurence of defectuous rock salt-type slabs : n = 2 members in a n = 3 matrix is an example (Fig. 7). They can be interpreted as the intergrowth of $TlBa_2CaCu_2O_7$ members with $Tl_2Ba_2CaCu_2O_8$ members. The second one corresponds to the intercalation of additional AO rock salt-type layers between two TlO layers, leading to a stacking sequence : [BaO-TlO-AO-AO-TlO-BaO]. The weak contrast of these additional layers suggests they are CaO layers. This hypothesis is in agreement with the deficient character of the thallium layers.

The number of extended defects tends to increase as the thickness of the oxygen deficient perovskite layers increases.

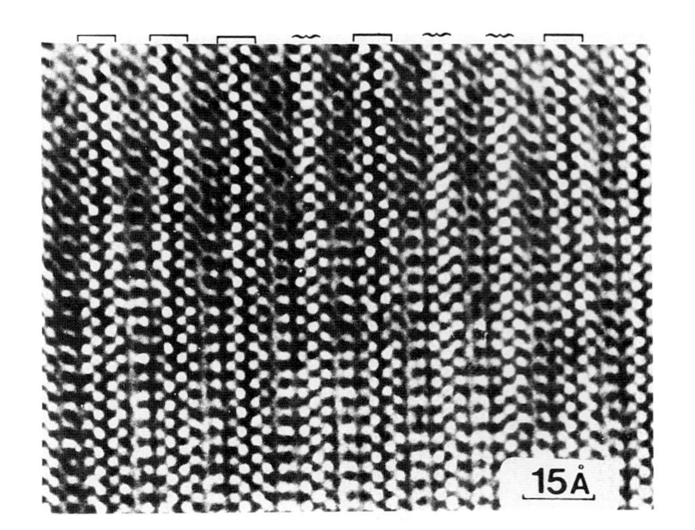

Fig. 7 - Variation in the stacking in the rock salt-type layer : n = 2 in a n = 3 matrix.

In spite of all these structural problems it must be outlined that the copper-oxygen framework is rather well established as well for bismuth compounds as for thallium oxides which exhibit common features : the CuO_5 pyramids in both families are characterized by a long Cu-O distance along $\vec{c}$ (greater than 2.4 Å) and four short distances in their basal plane (close to 1.95-1.98 Å) showing a strong tendancy towards a square planar coordination in agreement with the Jahn Teller effect of copper. In the same way, the Cu-Cu distances between copper sheets, along $\vec{c}$ are very similar for all these oxides and close to 3.2 Å, like in $YBa_2Cu_3O_7$, leading to a very similar cubic coordination for calcium and yttrium. The main difference between bismuth and thallium compounds concern the much greater distance between two successive bismuth layers (3. Å) compared to those observed in thallium bilayers (2.2 Å) and up to now the absence of bismuth cuprates involving bismuth monolayers.

III - MIXED VALENCE AND SUPERCONDUCTIVITY

The influence of the mixed valence Cu(II)-Cu(III) upon the evolution of Tc has been previously discussed [30] for $YBa_2Cu_3O_{7-\delta}$ which exhibits a static disproportionation of Cu(II) into Cu(III) and Cu(I). There is no doubt that Cu(III) does not exist as a cation but that the holes are delocalized over the copper oxygen framework and appear preferentially in the oxygen band, leading to the configuration $3d^9\underline{L}$ ($\underline{L}$ = ligand hole) or $\langle Cu^{II}-O^- \rangle$.

The formulations obtained for the bismuth cuprates from the structural studies suggest that no trivalent copper would be present in those oxides, in contrast with the chemical analysis of the bulk samples performed by several authors. Several hypothesis can be considered to explain the possible mixed valence in these oxides. One possibility deals with the presence of additional oxygen in the bismuth layers, which cannot be detected from the structural determination. A second hypothesis would correspond to a bismuth deficiency, with creation of cationic vacancies or substitution of strontium or calcium for bismuth. The possibility of formation of cationic vacancies on the strontium sites is more attractive since it would not be detected from the structure determination [31]. A very difficult structural investigation remains to be done to understand this problem.

Mixed valence Cu(III)-Cu(II) appears in the thallium cuprates involving thallium monolayers, but a similar problem is observed for the Tl rich oxides - $Tl_2Ba_2CuO_6$, $Tl_2Ba_2CaCu_2O_8$, $Tl_2Ba_2Ca_2Cu_3O_{10}$ and $Tl_2Ba_2Ca_3Cu_4O_{12}$. Nevertheless, the mixed valence in those oxides is explained by the observation of thallium deficiency. But an alternative, concerns the possible dynamical exchange of holes between thallium and copper, so that the thallium layers play the role of a reservoir according to the equation :

$$Tl(III) + 2\ Cu(II) \rightleftharpoons Tl(I) + 2\ Cu(III).$$

As a conclusion, superconductivity in copper oxides is so far not understood, and many investigations remain to be done before giving up the idea superconductivity at room temperature.

REFERENCES

1. B. RAVEAU, R.C.P.-meeting, April 1987, Bordeaux.
2. C. MICHEL et al., Z. Phys., B68, 421 (1987).
3. H. MAEDA et al., Jpn J. Appl. Phys., 27, L209 and L548 (1988).
4. M. HERVIEU et al., Modern Phys. Lett., 2, 491 (1988).
5. C.C. TORARDI et al., Phys. Rev., B38, 225 (1988)
6. J.M. TARASCON et al., Phys. Rev., in press.
7. M.A. SUBRAMANIAN et al., Science, 239, 1015 (1988).
8. M. HERVIEU et al., Modern Phys. Lett., 2, 835 (1988).
9. D. BORDET et al., Physica C, 153-155 619 and 623 (1988).
10. H.G. VON SCNERING et al., Angew Chem., 27, 574 (1988).
11. T. KASITANI et al., Jpn J. Appl. Phys., 27, L587 (1988).
12. S.A. SUNSHINE et al., Phys. Rev., B38, 893 (1988).
13. J.M. TARASCON et al., in press.
14. N. KIJIMA et al., Jpn, J. Appl. Phys., 27, L821 (1988).
15. C.C. TORARDI et al., Science, 240, 631 (1988).
16. S. AMELINCKX et al., Nature, 332, 620 (1988).
17. Z. SHENG et al., Phys. Rev. Lett., 60, 937 (1988).
18. Z. SHEN et al; Nature, 332, 138 (1988).
19. R.M. HAZEN et al., Phys. Rev. lett., 60, 1657 (1988).
20. S.S.P. PARKIN et al., Phys. Rev. Lett., 60, 2539 (1988).
21. C. POLITIS and H. LUO, Modern Physics Lett. B., 2, 793 (1988).
22. A. MAIGNAN et al., Modern Physics Lett. B., 5, 681 (1988).
23. M.A. SUBRAMANIAN et al. Nature, 332, 420 (1988).
24. M. HERVIEU et al., J. Solid State Chem., 74, 428 (1988).
25. C. MARTIN et al., C.R. Acad. Sci., 307, 27 (1988).
26. B. DOMENGES et al., Solid State Comm., in press.
27. S.S.P. PARKIN et al., Phys. Rev. lett., in press.
28. M. HERVIEU et al., Solid State Comm., J. Solid State Chem., 75, 212 (1988)
29. M. HERVIEU et al., Modern Physics Lett., in press.
30. B. RAVEAU et al., International Meeting on High T_c Superconductors, Schloss Mauterndorf, Austria, Feb. 1988.
31. A.K. CHEETHAM et al., Nature, 333, 21 (1988).

Crystal Chemistry of the Copper Oxide Based High Temperature Superconductors

R.J. CAVA

AT&T Bell Laboratories, 600 Mountain Avenue, Murray Hill, NJ 07974, USA

ABSTRACT

The copper oxide based superconductors all display crystal structures which are derivatives of the perovskite and sodium chloride structures. The special geometric characteristics of copper-oxygen bonding result in the formation of many complex ternary and quaternary compounds. The superconductors fall into several broad classes of structures, called homologous series: which are described here. The general formulae derived allow one to predict the crystal structures and compositions of presently unknown materials which would be related to known superconductors.

CRYSTAL CHEMISTRY

The discovery of high temperature superconductivity in oxides based on copper and rare and alkaline earths first caught the solid state physics and materials science communities completely by surprise.[1] In the two years since then, however, much has been learned about these materials, in particular concerning their crystal structures and electronic properties. The theoretical community has not yet, it seems, converged to a universally accepted model for the origin of high temperature superconductivity in these materials. However, a clear set of crystal-chemical requirements can now be empirically devised that point the materials scientist searching for new superconducting copper based compounds towards possible routes of success. This empirical view has evolved over time and is further refined with each new class of materials discovered.

Fortunately, copper based oxides form easily under standard ceramic processing conditions and therefore many compounds are known, and many new compounds are likely to be discovered. The size of the Cu^{2+} ion in oxides is such that when in combination with alkaline and rare earth ions, the perovskite family of compounds and their derivatives are easily accessible. Critical also to the ease of compound formation are the variable formal valence of copper (between +1 and +3) and the variety of coordination polyhedra available for copper and oxygen. Unlike many transition metal oxides the various formal valences of copper are easily accessible under straightforward temperature and oxygen partial pressure conditions: variability between "+2.5" and +1 is possible at temperatures between 500 and 1200C in oxygen partial pressures attainable between bottled O_2 and N_2 gas. Copper oxygen coordination can be linear (2 fold) for Cu +1, and square planar (4 fold), pyramidal (5 fold) and octahedral (6 fold) for valence states of +2 and higher. The "octahedra" for Cu +2 are of course Jahn-Teller distorted. This variability in possible coordination polyhedra increases the variety of geometries that Cu based compounds can have, but has also turned out to be critical in allowing the accommodation of nonideal oxygen stoichiometries. Finally, the coordination requirements of copper ions in the formal valence range of interest, between 2+ and 3+, are similar enough that oxidation/reduction can occur within a single basic coordination polyhedron (with addition or subtraction of oxygen) without serious structural rearrangement. The same seems to be true for Bi^{3+}-Bi^{5+} another cation of interest in superconductivity, but does not seem to be true for Sb^{3+}-Sb^{5+}, for example.

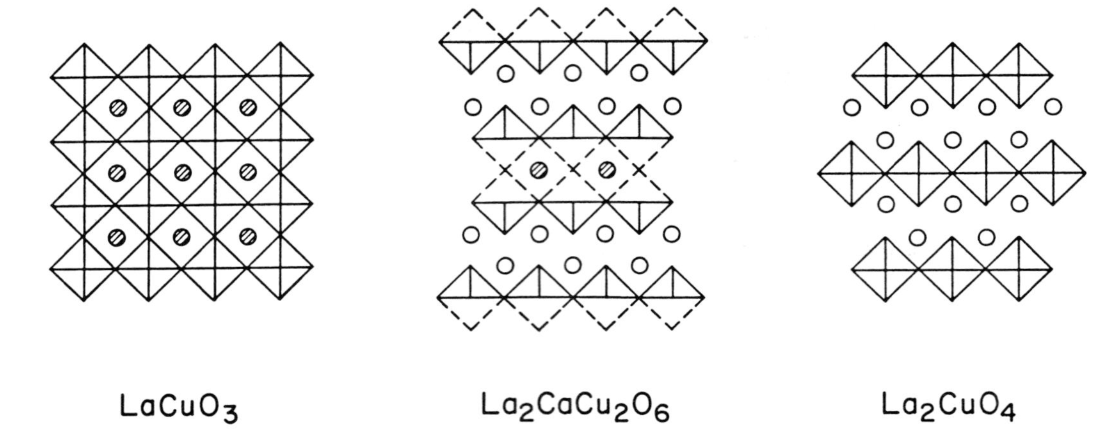

Figure 1: Idealized crystal structures of rare and alkaline earth cuprates. Copper oxygen polyhedra shown as octahedra and square pyramids, rare and alkaline earths shown as clear and shaded circles.

The most productive starting point to begin discussions of the detailed crystal structures of the copper oxide superconductors is the Perovskite structure, shown in an idealized representation in figure 1. $LaCuO_3$, which must be prepared at high oxygen pressures, has a distorted version of this basic structure in which Cu-O-Cu bond angles are not 180°. Angular and dimensional distortions are common in all kinds of perovskites. The 12K and 30K superconductors $BaPb_{.75}Bi_{.25}O_3$[2] and $Ba_{.6}K_{.4}BiO_3$[3] are true perovskites with full compliments of all atoms. For copper based materials however it is best to consider the perovskite structure as made ideally of alternating CuO_2 and AO layers where A represents the large (in our case alkaline earth, or rare earth) perovskite ion. The CuO_2 layers have the basic geometry of a corner shared array of square planar (CuO_4) copper oxygen polyhedra (each oxygen bonded to 2 coppers) with approximately 180° Cu-O-Cu bond angles. The AO layer is also comprised of square planar coordination polyhedra but is completely edge shared such that each oxygen is bonded to 4 A atoms resulting in an AO stoichiometry: the A-O bondlength is approximately $\sqrt{2}$ times the Cu-O bondlength for successful stacking of the AO-CuO_2 arrays, in which the O from the AO layer stacks directly above the Cu of the CuO_2 layer. The stacking sequence is CuO_2-AO-CuO_2-AO-.

The most common copper-oxygen coordinations are those less than 6 fold, and rare and alkaline earth ions are sometimes small, so most often the ideal perovskite coordination numbers (12 for the A ion, 6 for the B ion) are not attained in the compounds of interest. This is accommodated in the known superconducting compounds either through stacking sequence differences or the presence of ordered vacant oxygen sites in either the AO or CuO_2 layers. The recently discovered compound $Ca_{.84}Sr_{.16}CuO_2$[4] for instance is the simplest oxygen deficient variation of the perovskite structure, with a 1:1 A to B ratio but with 1/3 of the oxygens missing in an ordered manner. The stacking sequence is CuO_2-A-CuO_2-A- where A=$Ca_{.84}Sr_{.16}$. The infinite CuO_2 planes are complete, with all oxygens missing from the A layer, resulting in 4 fold Cu-O coordination and 8 fold (square prismatic) A-O coordination. The stacking sequence CuO_2-A-CuO_2 as described is the structural kernel of all of the known superconductors with T_c's greater than 40K, and seems empirically to be clearly indicated as a necessary (but not sufficient) structural component. With that importance in mind, it is of particular interest to note that this crystalline structure although existing for $Ca_{.84}Sr_{.16}CuO_2$ does *not* exist for either pure endmember $CaCuO_2$ or $SrCuO_2$ suggesting that its stability is very critically dependent on the size of the atomic array, with the 8 coordinate radii of Ca^{2+} (1.12) and Sr^{2+}(1.25) differing by only 10%! This suggests in fact that the 123 type materials $Ba_2RECu_3O_7$ with small RE (e.g. Y, Yb, Lu) must be at the borderline of stability for this kind of layer, and may explain why "$Bi_2Sr_2CaCu_2O_8$" appears to form with some Sr on the A type layer between the CuO_2 sheets.

Figure 1 illustrates the manner in which a structural family can be derived from the basic perovskite structure type. On the far left an idealized $LaCuO_3$ perovskite is shown, and on the far right, La_2CuO_4, which is the host compound for the 20-40K superconductors. The La_2CuO_4[5] structure is easily derived from that of $LaCuO_3$ by the insertion of one additional AO layer. The new AO layer is added such that its oxygens are directly above the A ions of the AO layer below. The stacking sequence is now (CuO_2)-(AO)-$(AO)_c$-$(CuO_2)_c$-$(AO)_c$-(AO)- where the subscript distinguishes the shifts of these layers as described above. The (AO)-$(AO)_c$ sequence is exactly the arrangement of atoms found in NaCl, and so one can refer to the known materials as mixtures of perovskite and sodium chloride structure types. This is especially important for the Bi and Tl based cuprates described below. In La_2CuO_4, Cu is surrounded by 6 oxygens, as in the simple perovskites, but the large A atoms now are only 9 coordinate with oxygen as their neighboring layers are CuO_2 and centered AO. The principle of layer insertion is central to the formation of homologous series in oxides and is illustrated in this chemical system in the center of figure 2 which represents the structures of $La_2CaCu_2O_6$ and $La_2SrCu_2O_6$.[6] There are potentially a very large number of compounds (Ruddleston-Popper phases) of the type $A_{n+1}Cu_nO_{3n+1-\delta}$ for which La_2CuO_4 and $LaCuO_3$ are the endmembers n=1 and n=∞. The A ions, which may be mixed, and δ are chosen for charge balance. $La_2CaCu_2O_6$ is the n=2 member of the series. The A ions are ordered such that the small Ca ion takes 8 oxygen neighbors, and we have the stacking sequence: (CuO_2)-(AO)-(AO_c)-$(CuO_2)_c$-$(A')_c(CuO_2)_c$-$(AO)_c$-(AO)-(CuO_2)-(A')-. The structure type that results is a mixture of that of La_2CuO_4 and that of $Ca_{.84}Sr_{.16}CuO_2$. This compound has not to date been successfully doped to give it metallic conductivity, but it clearly satisfies the structural criteria for high T_c superconductivity. As n increases in homologous series of this type (excluding n=∞) the thermodynamic stability of the phases decreases, and in fact compounds in the rare-earth alkaline earth based cuprates with n > 2 are not known.

After the determination of the crystal structure of the 90K T_c compound $Ba_2YCu_3O_7$, with its infinite sheets of CuO_5 square pyramids sandwiching square planar CuO_4 chains, it was natural to attribute the factor of 2 increase in T_c (over doped La_2CuO_4) intimately to the chains, or to a combination of chains and planes. In fact after initial elation over the beauty and uniqueness of a crystal structure which combined 1-d and 2-d structural elements so remarkably, the prospects for finding wholly new superconducting compounds with similarly unique structural characteristics seemed remote: even though copper chains may be found in some compounds, and planes in others, nowhere else but in "123" did both occur together. It is really only after the determinations of the crystal structures of the Bi and Tl based cuprates that the role of the chains in $Ba_2YCu_3O_7$ can be understood. It is now clear that the chains and other secondary structural elements act as charge reservoirs which control the charge concentration on the copper-oxygen planes which are the critical structural component involved in superconductivity. We have observed, for instance, the occurrence of a dramatic structural anomaly accompanying the loss of superconductivity in $Ba_2YCu_3O_{7-x}$ which is due to the internal redistribution of electronic charge from the chains to the planes on oxygen removal. Those experiments also have shown that "orthorhombicity" in $Ba_2YCu_3O_{7-x}$ is at best a tertiary structural parameter, reflecting only the presence or absence of long range orientational ordering of copper-oxygen chain fragments.[7]

Much has been written about the crystal structure of $Ba_2YCu_3O_7$, so I will not discuss that structure in detail here except for the manner in which it forms one member of a homologous series of compounds which are to date considerably less well understood than those based on La, Cu, O. Figure 2 schematically represents the presently known compounds in the series, beginning with $BaYCuFeO_5$[8] which consists of infinite planes of copper/iron oxygen square pyramids stacked through sharing their apical oxygen atoms. The stacking sequence for this compound is then: - Y - $(Cu/Fe)O_2$ - BaO - $(Cu/Fe)O_2$ -. The homologous series is formed by the insertion of copper-oxygen chains between the apices of the pyramids. In this sense, $Ba_2YCu_3O_7$ is the simplest member of the series containing chains, with *1* chain (and a BaO plane) inserted between each set of square pyramids (figure 2). The stacking sequence is: - Y - CuO_2 - BaO - CuO - BaO - CuO_2 -. One way the series can be expanded is through the addition of more layers of chains sharing only corners with each other with a BaO layer accompanying each, with the formulae $Ba_{2+n}Y_2Cu_{4+n}O_{10+2n}$. Only the n=o and n=2 compounds are presently known ($BaYCuFeO_5$, $Ba_2YCu_3O_7$). Another way the series can be expanded is through the addition of edge shared "double chains" of CuO_4 squares. These are really

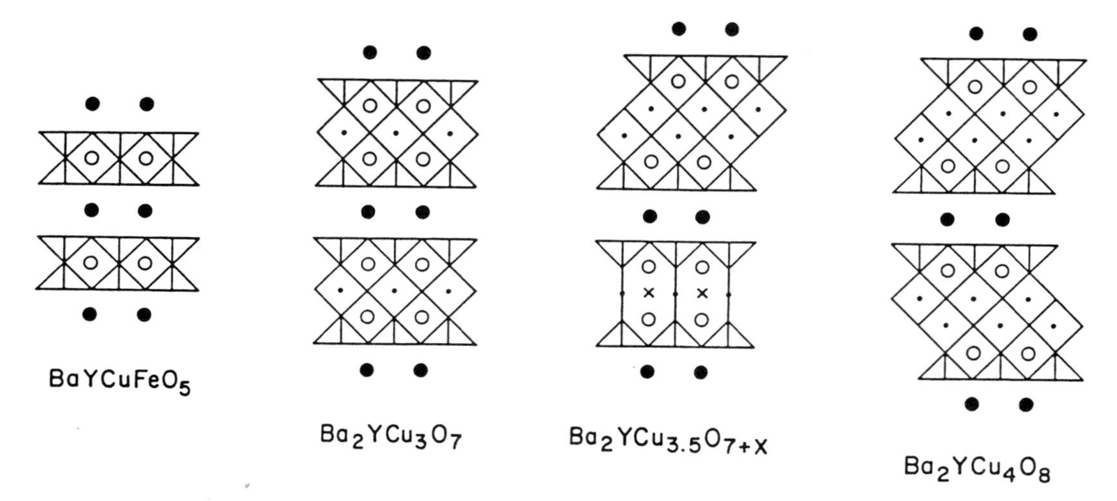

Figure 2: Idealized crystal structures of Ba-Y-Cu-oxides, showing Cu-O coordination polyhedra, Y (black circles) and Ba (open circles).

crystallographic shear derived variations of the first structural series. The crystallographic shear eliminates the otherwise present interleaving "A" atom-oxygen layer, such that only copper-oxygen layers are added, resulting in the ideal formulae $Ba_4Y_2Cu_{6+n}O_{14+n}$. The first 3 members of this series are in fact known (n=0, 1, 2) and are shown in figure 2, the latter two having been synthesized under relatively unconventional synthetic conditions.[9,10] It would in fact be surprising if members with n>2 would be found as there do not seem to be any examples of simple "triple chain" copper oxides known, but we have certainly been surprised before.

From a crystal-chemical point of view, the new Tl and Bi based copper oxides form the most straightforward homologous series of compounds, though their synthesis and characterization is certainly far from straightforward.[11,12] These compounds advanced the field not only by their very high T_c's, but also through the fact that they made it clear that only copper-oxygen pyramidal planes are necessary for superconducting T_c's near 100K. In addition to the classical structure/composition relations found in these materials they display other characteristics of homologous series' which make them of significant interest (and frustration) to the crystal chemist: (1): deviations from ideal stoichiometry are generally accommodated by the formation of extended and not point defects, primarily through intergrowth of different members of the series on a microscopic scale and (2) as the complexity (number of stacks) increases, the compounds have smaller and smaller temperature/composition regions in which they can be successfully formed. In fact the short synthesis times necessary to make some of the Tl based compounds suggests that they may actually be metastable, e.g. that is that there are actually *no* conditions under which they exist as equilibrium compounds.[13] Both of these crystal chemical problems make the formation of single phase, well characterized materials in these chemical systems very difficult, and the unsuspecting person trying to piece together a uniform picture of the structure-property relationships in these materials based on currently available data must certainly beware!

Perhaps the most interesting crystal chemical aspect of these materials is that for those based on Tl there exist not only variable numbers of Cu-O layers but also variable numbers of Tl-O layers, which makes for the possibility of many compounds, in what is really an infinitely adaptive series. The whole series can be written as: $Tl_mBa_2Ca_{n-1}Cu_nO_{m+2n+2}$. The subscript m defines how many rocksalt layers are present, and to date examples have been found for m=1, and 2. The number of copper layers is remarkably variable, with n=1, 2, 3, 4 presently claimed to be prepared as bulk materials. (see figure 3 for the m=2, n=1-3 members). Even more remarkably, all compounds thus far synthesized in this series are claimed to be superconducting with T_c's greater than 80K. Unfortunately, compounds based on Bi-Sr-Ca-Cu-O have proven to be much more difficult to make in anything other than the m=2, n=1, 2 structures. The m=2 n=3 variant has been claimed to be responsible for superconductivity at 100K in the pure Bi system, but it has only been reproducibly and systematically prepared as single phase material in the Bi-Pb-Sr-Ca-Cu-O chemical system.[14]

HOMOLOGOUS SERIES:

$(Bi, Tl)_2 (Sr, Ca, Ba)_{n+1} Cu_n O_{2(n+2)\pm\delta}$

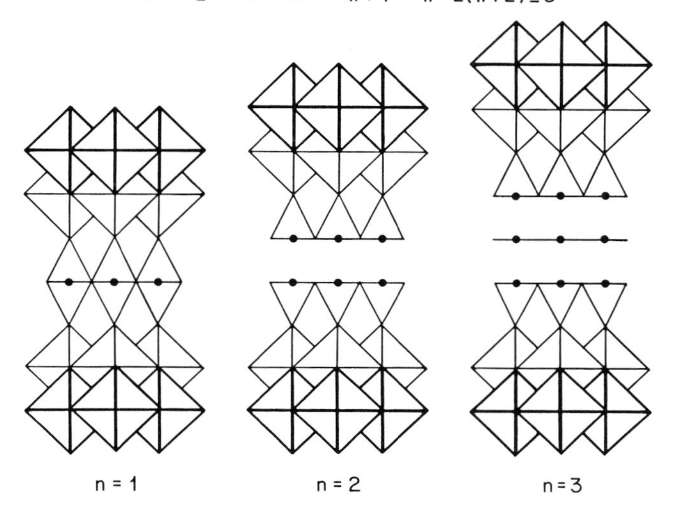

Figure 3: Idealized crystal structures of the m=2, n=1 to 3 compounds $Tl_mBa_2Ca_{n+1}Cu_nO_{m+2n+2}$, in the order n=1, n=2, n=3.

Comparison of the crystal structures of oxygen deficient $Ba_2YCu_3O_{7-\delta}$, and the new Tl and Bi based cuprate superconductors allows one to make general conclusions about the roles that various structural components play in the attainment of high T_c's. The primary structural elements are CuO_2 planes in which CuO_4 squares share corners *exclusively,* with O-Cu-O bond angles near 180° and CuO bond lengths near 1.9Å. However, at this time, it also seems that planes alone are not enough: the Cu in the plane must also be bonded to at least one additional oxygen forming the apex of a square pyramid, or two additional apices of an elongated octahedron. The bond length to the apical oxygens must be near 2.3-2.4 angstroms, indicating a considerably weaker Cu-O bond in that direction. These Cu-O planes then sandwich secondary structural elements such as the Cu-O chains in $Ba_2YCu_3O_{7-\delta}$ or the Tl-O, Bi-O planes in the new cuprates. The role of the secondary structural elements is to regulate the charge (hole concentration) on the Cu-O planes by acting as a charge reservoir which "dopes" the planes. The presence of both kinds of structural components is necessary to attain high T_c. When new high T_c copper oxide based superconductors are found, we can almost be sure they will include both Cu-O planes and M-O charge reservoir layers.

The 30K superconductor $Ba_{.6}K_{.4}BiO_3$ is of interest from a structural point of view in that it displays a simple cubic perovskite crystal structure to temperatures below T_c. Thus it appears to be a high T_c material without the 2-dimensional metal-oxygen arrays so critical to superconductivity in the copper oxides. It certainly suggests the question as to whether 2-d Bi-O superconductors might have still higher T_c's, or similarly, whether 3-d Cu-O superconductors will be found. Unfortunately, in both respects, crystal chemistry plays a central, and not a positive role. Relatively few compounds with Bi valence greater than +3 are known, making the possible high T_c candidates many fewer than for copper oxides. Similarly, 3-dimensional copper oxides might need very highly oxidized copper to form, perhaps near Cu +3, limiting the accessibility of those materials by conventional synthetic techniques. Nevertheless it is clear at the present time that the events of the past few years represent only the beginning for oxide superconductors. I suspect that many new materials and insights are due in the years to come.

ACKNOWLEDGEMENT

Many of the ideas described in this paper are the result of a collaboration of A. Santoro, F. Beech, M. Marezio and R. J. Cava. The detailed report will be published elsewhere.

REFERENCES

[1] J. G. Bednorz and K. A. Muller, Z. Phys. *B64,* 189 (1986).
[2] A. Sleight, J. L. Gillson and P. E. Bierstedt, Sol. St. Comm. *17,* 27 (1975).
[3] R. J. Cava, B. Batlogg, J. J. Krajewski, R. Farrow, L. W. Rupp, Jr., A. E. White, K. Short, W. F. Peck and T. Kometani, Nature *332,* 814 (1988).
[4] T. Siegrist, S. M. Zahurak, D. W. Murphy and R. S. Roth, Nature *334,* 231 (1988).
[5] N. Nguyen, J. Choisnet, M. Hervieu, and B. Raveau, J. Sol. St. Chem. *39,* 120 (1981).
[6] N. Nguyen, L. Er-Rakho, C. Michel, J. Choisnet and B. Raveau, Mat. Res. Bull. *15,* 891 (1980).
[7] R. J. Cava, B. Batlogg, K. M. Rabe, E. A. Rietman, P. K. Gallagher and L. W. Rupp, Jr., to be published.
[8] L. Er-Rakho, C. Michel, Ph. Lacorre, and B. Raveau, J. Sol. St. Chem., *73,* 531 (1988).
[9] P. Marsh, R. M. Fleming, M. L. Mandich, A. M. DeSantolo, J. Kwo, M. Hong and L. J. Martinez-Miranda, Nature *334,* 141 (1988).
[10] P. Bordet, C. Chaillout, J. Chenavas, J. L. Hodeau, M. Marezio, J. Karpanski, E. Kladis, Nature *334,* 596 (1988).
[11] H. Maeda, Y. Tanaka, M. Fukutomi, and T. Asano, Jpn. J. Appl. Phys. Lett. *27,* L209 (1988).
[12] Z. Z. Sheng and A. M. Herman, Nature *332,* 138 (1988).
[13] A. Sleight, private communication.
[14] U. Endo, et. al. this conference and, M. Takano et al, private communication.

The Important Role of High Oxidation States in Cuprate Superconductors

S.X. Dou[1,2], H.K. Liu[1,2], J.P. Zhou[1], A.J. Bourdillon[1], and C.C. Sorrell[1]

[1] School of Materials Science and Engineering, University of New South Wales, P.O. Box 1, Kensington, NSW 2033, Australia
[2] Visiting Professor, Northeast University of Technology, Shenyang, People's Republic of China

ABSTRACT

In cuprate superconductors the presence and concentration of the high oxidation state, Cu^{3+} or $(Cu\text{-}O)^{+}$, are crucial to and correlate with the superconducting properties, such as transition temperature and critical current density. Likewise the stability of the materials in corrosive environments depends on the oxidation state.

INTRODUCTION

The mixed valence condition is crucial to sustaining high T_c superconductivity in the cuprate superconductors: $La_{2-x}Sr_xCuO_{4-y}$ (LSCO), $YBa_2Cu_3O_{7-y}$ (YBCO), Bi-Sr-Ca-Cu-O (BSCCO) and Tl-Ba-Ca-Cu-O (TBCCO). In the latter three of these compounds the high oxidation state copper, Cu^{3+} or $(Cu\text{-}O)^{+}$ complex, has been detected by the volumetric technique (1-4). It was found that the concentration of the Cu^{3+} ions correlates both with superconducting properties, such as transition temperature (T_c) and critical current density (J_c), and with the stability of these materials in corrosive environments. In this paper we review our volumetric measurements on these materials.

Specimens were prepared by solid state reaction or coprecipitation as previously described (5-9). They were characterised in a variety of ways including scanning electron microscopy in a JEOL JSM840 (SEM) with Link Systems energy-dispersive detector, X-ray powder diffraction with a Philips PW1140/00 diffractometer, electrical resistivity and critical current density by the four-probe technique, and Cu^{3+} ionic concentrations from the evolution of oxygen from weighed specimens in dilute hydrochloric acid (2).

CORRELATION OF Cu^{3+} CONCENTRATION WITH T_c IN YBCO

Table I shows the correlation between labile ion concentration, Cu^{3+}, with atomic oxygen concentration, crystal structure, T_c, ΔT_c (10-90%, width of transition sigmoid). For oxygen concentrations $7-x<6.75$, T_c falls to 60 K and below. Our neutron diffraction data show that in samples quenched from 600°C (T_c = 66 K in table I) the oxygen vacancies are disordered. In the absence of Cu^{3+} the tetragonal structure was observed together with semiconducting behaviour, while the oxygen concentration was measured by the weight gained after annealing at 600° in oxygen (as in the top row of data in table I). Dopants such as Fe and Al tend to promote the tetragonal structure (7,10); but this phase can superconduct provided Cu^{3+} ions are present. Thus the superconductivity depends primarily not on the orthorhombic structure; but on the high oxidation state of copper.

TABLE I. CORRELATION OF Cu^{3+} WITH CRYSTAL STRUCTURE AND T_c IN $YBa_2Cu_3O_{7-x}$

Cu^{3+} in total Cu (%)	7-x	Crystal structure	T_c	ΔT_c
29.3	6.94	O	92.5	1.5
26.0	6.88	O	92.1	2.4
22.5	6.86	O	91.0	3.0
22.3	6.83	O	89.3	4.8
16.4	6.75	O	92.3	5.0
13.3	6.71	O	83.0	20.0
6.0	6.60	O	66.0	22.0
0	6.40	T	33.0	44.0
0	6.31	T	-	-
0	6.22	T	-	-

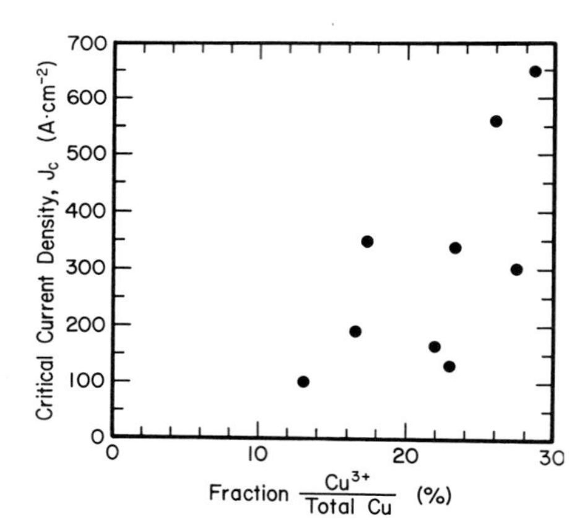

Figure 1. Dependence of J_c on Cu^{3+} concentration

CORRELATION OF Cu^{3+} CONCENTRATION WITH J_c IN YBCO

Figure 1 shows the dependence of the critical current density, J_c, on the Cu^{3+} concentration in sintered specimens. According to recent non-phonon mixed valence theories (11-13) superconductivity is related to the carrier hole concentration which depends on the concentration of Cu^{3+} ions (or Cu-O complexes). Thus J_c depends on the hole pair concentration and therefore on the Cu^{3+} concentration. It has been found that J_c generally increases with increase of physical density due presumably to improved connectivity of grains in dense specimens. However the Cu^{3+} concentration in a specimen more than 99 % dense (figure 2a), made by controlled heat treatment, is only 13 % of total Cu present in YBCO, i.e. much lower than in materials only 95 % dense (29 % Cu^{3+} per total Cu). The decrease in labile ion concentration is attributed chiefly to obstructions to oxygen diffusion for the tetragonal to (superconducting) orthorhombic phase transformation, though heavy twinning observed in the denser material may contribute. Since each twin boundary is shared by two crystals, a row of oxygen atoms is easily eliminated, driving Cu^{3+} to Cu^{2+} by stoichiometry, with consequent low J_c. It is interesting to note that melt textured material (figure 2b) showed few twins in its microstructure, and had a J_c of 2400 A/cm^2.

CORRELATION OF Cu^{3+} CONCENTRATON WITH T_c IN BSCCO

More than fifty different compositions and heat treatment conditions in Bi-Sr-Ca-Cu-O systems were studied. While there is some difficulty involved in preparing single phase superconducting Bi-Sr-Ca-Cu-O, the multiphase assemblage shows evidence of two transitions at temperatures of 110 K and 84 K. Figure 3 shows the effect of the initial composition on the superconducting transition at 110 K. Specimens made from initial compositions with high Cu concentration have a Cu-enriched superconducting phase with strong transitions at 110 K.

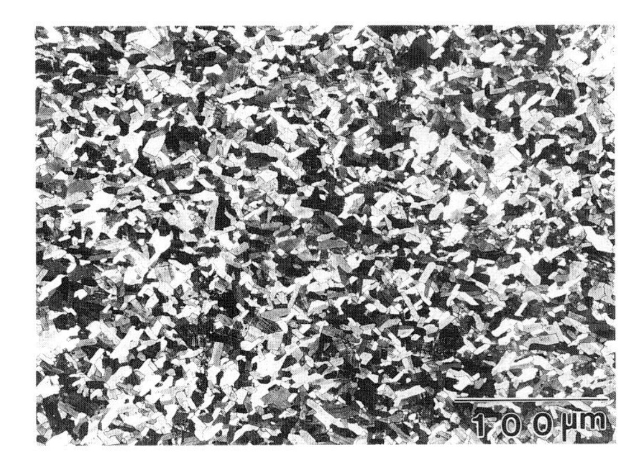

Figure 2a. Secondary electron image of fully dense YBCO

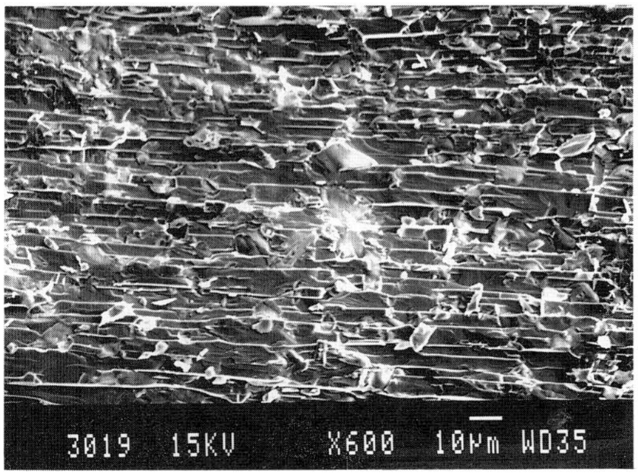

Figure 2b. Secondary electron image of melt textured YBCO

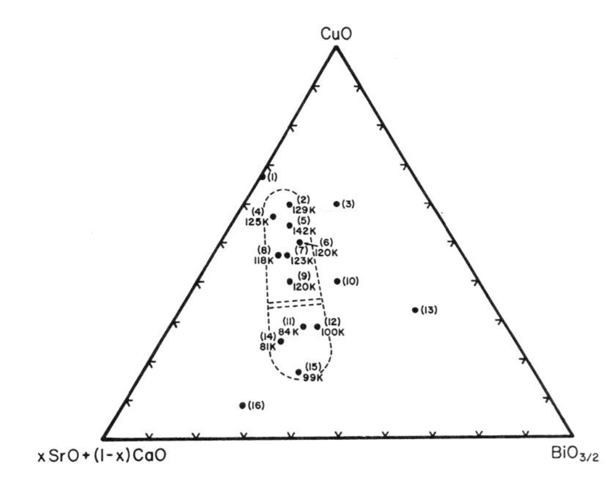

Figure 3. Dependence of superconducting transitions on initial composition

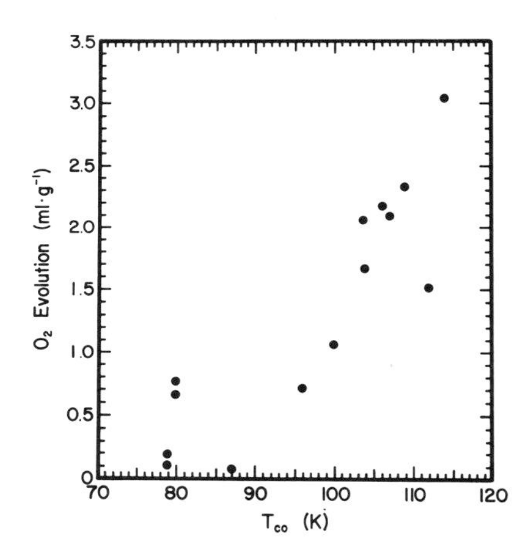

Figure 4. Correlation between T_{co} and volume of O_2 evolution in dilute HCl

Conversely, specimens with initial composition close to that of the superconducting phase $Bi_2Sr_2CaCu_2O_{8+y}$ (2212) give rise to a superconducting phase depleted in Cu and Ca with a transition onset (T_{co}, at 90 % of sigmoid) at only 90 K. The transition at 110 K can be enhanced by selective heat treatment, by which the concentration of Cu^{3+} can be varied and this relates to the superconducting transitions in BSCCO (figure 4). T_{co} increases with the volume of O_2 evolved after dissolution in dilute HCl. At least 1.5 ml/g is evolved from specimens that show evidence of transitions at 110 K. This volume corresponds to a value for y of 12 % Cu^{3+} per total Cu. This demonstates, as with YBCO, that Cu^{3+} ions are crucial to superconductivity in BSCCO.

CORRELATION OF Cu^{3+} CONCENTRATION WITH ELECTROCHEMICAL CORROSION RATE

The high valence of Cu^{3+} is a powerful oxidising agent. In reducing environments superconducting properties are lost as this state reduces to Cu^{2+} or Cu^{+}.

Electrochemical cells have been constructed by ultrasonically bonding (with indium) pairs of tinned copper wires to the silver-painted ends of rectangular bars of YBCO and of BSCCO. To reduce electrochemical effects at the contacts, these were coated with epoxy resin so as to protect them from corrosion failure during tests. The two leads were connected to terminals for room temperature resistance measurement while the samples and leads were immersed in distilled water. The resistance of the sample was recorded as a function of time.

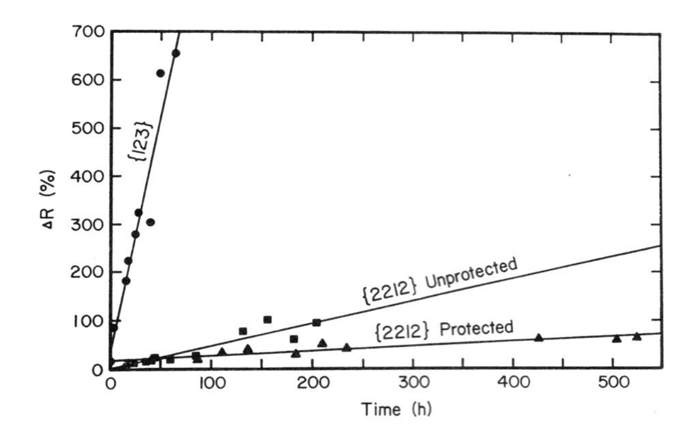

Figure 5. Change in resistance, ΔR, with time for YBCO (●), BSCCO (■) and protected BSCCO (▲).

Figure 5 shows relative resistance plotted against time. The resistance increased much faster in YBCO than in BSCCO, correlating with respective ratios of Cu^{3+} ionic concentrations. This is a consequence of Cu^{3+} being the driving force for the electrochemical cell reactions.

Additional support for this driving force comes from observations on the two leads. In the case of the YBCO cell the 0.3 mm diameter leads were completely corroded in 55 h, while for the BSCCO cell the time taken was four times greater, i.e. 220 h. This correlates again with relative Cu^{3+} concentrations.

Further corrosion protection is obtained if the whole YBCO or BSCCO specimen is covered with epoxy resin. Corrosion was unobservable even after 45 days. This electrochemical behaviour is of technical importance since corrosion around metal-superconductor contacts requires prevention, and optimised coatings should be used to diminish corrosion rates.

ACKNOWLEDGEMENTS: We are grateful to Metal Manufactures Ltd for support (S.X.D), and to the Department of Industry, Technology and Commerce for support (H.K.L) under a general technology grant.

REFERENCES

1. S.X.Dou, H.K.Liu, A.J.Bourdillon, N.X.Tan, J.P.Zhou, C.C.Sorrell and K.E.Easterling. (1988) Mod. Phys. Lett. B 1, 363.
2. S.X.Dou, H.K.Liu, A.J.Bourdillon, J.P.Zhou, N.Savvides and C.C.Sorrell. (1988) Sol. State. Comm. in press.
3. H.K.Liu, S.X.Dou, A.J.Bourdillon, N.Savvides and C.C.Sorrell. (1988) Supercond. Sci. and Tech., in press.
4. S.X.Dou, H.K.Liu, A.J.Bourdillon, N.X.Tan, N.Savvides C.Andrikidis, R.B.Roberts and C.C.Sorrell. (1988) Supercond. Sci and Tech. in press.
5. S.X.Dou, A.J.Bourdillon, C.C.Sorrell, K.E.Easterling, S.Ringer, N.Savvides, J.B.Dunlop and R.B.Roberts. (1987) Appl. Phys. Lett. 51 535.
6. S.X.Dou, N.Savvides, X.Y.Sun, A.J.Bourdillon, C.C.Sorrell, J.P.Zhou and K.E.Easterling. (1988) J.Phys. C 20 L1003.
7. S.X.Dou, A.J.Bourdillon, X.Y.Sun, J.P.Zhou, H.K.Liu, N.Savvides, D.Haneman, C.C.Sorrell and K.E.Easterling. (1988) J. Phys. C 21 L127.
8. S.X.Dou, H.K.Liu, A.J.Bourdillon, N.X.Tan, J.P.Zhou and C.C.Sorrell. (1988) Mod. Phys. Lett., in press.
9. H.K.Liu, S.X.Dou, N.Savvides, J.P.Zhou, N.X.Tan, A.J.Bourdillon and C.C.Sorrell. (1988), submitted to Physica C
10. S.X.Dou, H.K.Liu, J.P.Zhou, A.J.Bourdillon, X.Y.Sun, N.X.Tan and C.C.Sorrell. (1988) J. Am. Ceram. Soc. 71, in press.
11. M.W.Shafer, T.Penney and B.L.Olsen. (1988) Novel Supercondivity, Eds S.A.Wolf and V.Z.Kresig, p 771.
12. L.J.deJongh. (1988) Physica C 152 171.
13. J.A.Wilson. (1988) J. Phys. C 21 2067.

XANES Study of Copper Valence and Structure in High Tc Superconducting Oxides

K.B. GARG, K.V.R. RAO, N.L. SAINI, D.C. JAIN, H.S. CHAUHAN, U. CHANDRA, and J. SINGH

Department of Physics, University of Rajasthan, Jaipur, India

ABSTRACT

The question of Cu^{3+} and role it may play in the mechanism of the HTSC in the perovskites recently has been examined in depth employing XPS and XANES techniques [1-5]. The Cu 2p XPS and valence band emission from CuO and $YBa_2Cu_3O_{7-x}$ when discussed along with the CuL_3 XANES in these clearly ruled out the presence of Cu^{3+} and established that the holes are instead situated in the oxygen band. This is further reinforced by the presence of additional peak in the O 1s XPS emission. Measurements and one-electron full-multiple scattering calculations on the CuK XANES establish that the best fits to experimental data are obtained employing the $3d^{10}$ configuration for copper [6]. The question of alternative structure proposed by Reller, Bednorz and Muller is also examined and rejected through these calculations.

INTRODUCTION

Combination of XPS and XANES data for the same core level provides a complete description of the final state and ground state. Experimental investigations of single phase $YBa_2Cu_3O_7$ by Bianconi et al [1-5] and XANES calculations by Garg et al [6] have been done to ascertain (1) the formation of ground state due to a many-body configuration with itinerant holes in oxygen valence band coupled with localized 3d electrons on Cu sites (2) absence of Cu $3d^8$ configuration and (3) high electronic correlation on Cu 3d sites.

RESULTS AND DISCUSSION

Cu $2p_{3/2}$ XANES and XPS spectra for CuO and $YBa_2Cu_3O_{6.5+x}$ from Bianconi et al [1-5] are shown with the same energy scale in figures 1 and 2 respectively. In figure 1, the peak A is broad and its fine structure is due to multiplet splitting of the 2p $3d^9$ final states, while the peak B appearing at 933.6 eV is due to a final state which involves a charge transfer from the oxygen 2p derived valence band to the Cu 3d derived orbitals i.e. 2p $3d^{10}\underline{L}$ ($\underline{L}$ indicates a ligand hole). White line at the L_3 XANES spectrum (with a 0.95 eV FWHM), like that of other Cu^{2+} compounds [7], is due to an atomic like transition from the ground state Cu $2p^6 3d^9$ to Cu $2p^5 3d^{10}$ final state. The energy scale has been fixed with the maximum of the white line at 931.2 eV according to Koster [7].

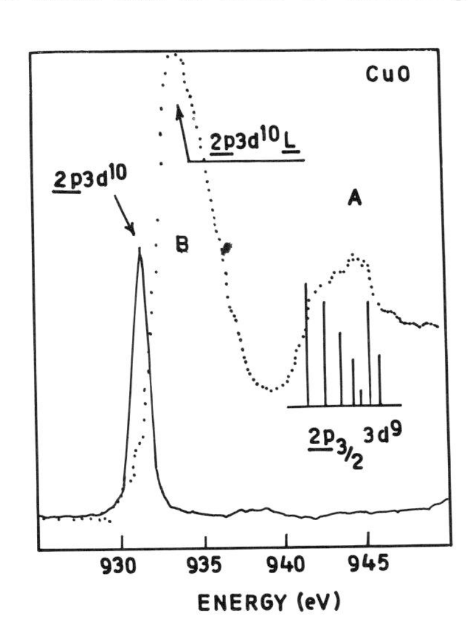

Fig. 1. Cu $2p_{3/2}$ XPS spectrum of CuO (dotted line). The Cu $2p_{3/2}$ XANES spectrum of CuO (solid line) exhibits a single white line due to the $2p3d^{10}$ final state configuration (Ref. 4).

Similar to Cu $2p_{3/2}$ XPS spectrum for CuO, peaks A and B which are separated by 8.5±0.2 eV have been observed in $YBa_2Cu_3O_7$ and are all identically assigned. There is an additional feature appearing at 933 eV as a broad shoulder of the white line in the superconductor which is completely absent in CuO. This new feature is at about the same energy as the peak B in Cu XPS spectrum due to $2p\,3d^{10}\underline{L}$ final states and is therefore assigned to the transition from $2p3d^9\underline{L}$ to final state $\underline{2p}\ 3d^{10}\underline{L}$.

Figure 3(a) shows oxygen 1s XPS spectrum in a freshly scraped sample $YBa_2Cu_3O_7$ measured by Bianconi et al [4]. The stronger peak B is at 530.9 eV and a weaker peak A at 528.9 eV and the intensity ratio B/A is about 4/1. Lowering the O_2 concentration at surface by

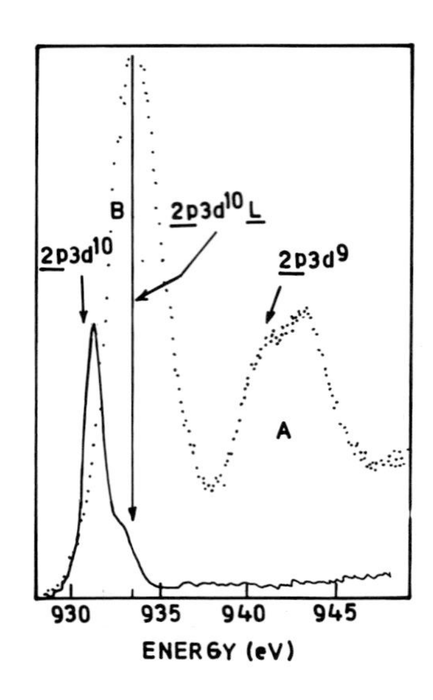

Fig. 2. Cu $2p_{3/2}$ XPS spectrum of $YBa_2Cu_3O_7$ (dotted line) and the Cu $2p_{3/2}$ XANES spectrum of $YBa_2Cu_3O_7$ (solid line) (Ref.4)

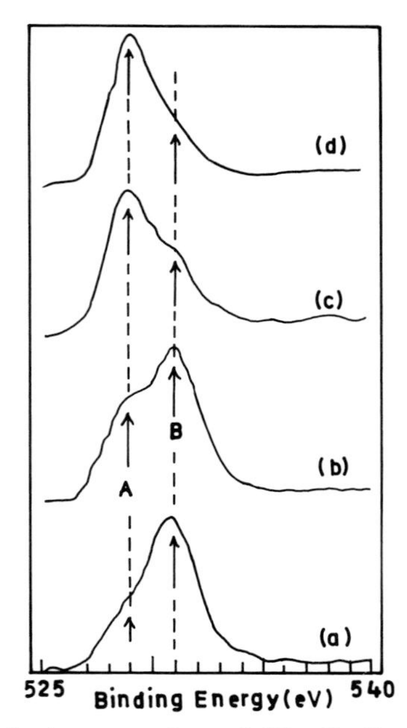

Fig. 3. Oxygen 1s X-ray photoelectron spectra of $YBa_2Cu_3O_{6.5+x}$ obtained by X-ray photoemission using Mg K_α radiation $h\nu$ = 1253.6 eV. The spectra of the freshly scraped sample lower curve(a), after two days in vacuum at 8×10^{-9} torr(b) and after Ar ion bombardment for 10 min (c) and 30 min (d) are shown.(Ref.4).

keeping the sample in vacuum of experimental chamber for 48 hrs (curve b), or by gentle Ar^+ ion bombardment for 10 minutes (curve b) or 30 minutes (curve c), the peak B at 530.9 eV decreases and peak A at 528.9 eV increases. At the lower oxygen concentration (curve d) the Cu 2p XPS spectrum shows only Cu^+ ions and only the 529 eV line is observed. It is well known that upon decomposition of CuO to Cu_2O by argon ion bombardment the O 1s line is not shifted, therefore peak A is assigned to O^{2-} ions in CuO and Cu_2O and the peak B which is quenched by sample reduction is due to the oxygen atoms with holes induced by $3d^9\underline{L}$ configuration. The O 1s energy shift of about 2 eV is small of the same order of magnitude as the shift observed between different chemical compounds. Moreover, the intensity of 530.9 eV peak is much larger than that expected for one hole per oxygen atom. These two points indicate that $\underline{L}$ hole is distributed over 4 or more oxygen atoms probably in the CuO_2 layers [4,5].

Garg et al [6] have studied the Cu K XANES in Y-B-C system and made one electron multiple scattering calculations to interpret the experiment. To see the effect of size of the clusters, XANES calculations for sites Cu 1 and Cu 2 according to the coordinates of Capponi et al [8] have been done and it is found that the characteristic XANES spectrum of CuO_4 square plane [6] strongly modifies by including multiple scattering pathways with the second shell containing 8 Ba atoms, determined by the fact that Ba atoms have large scattering amplitudes for low energy electrons. Inclusion of the third shell containing 6 Cu ions, does not induce relevant changes of the spectrum, since the Cu ions have a low backscattering amplitude for low energy electrons. Only in the low energy side of the main absorption peak Cu ion induces a weak shoulder in the spectrum. Inclusion of 4th shell containing 20 oxygen atoms induces an important modification of the spectrum showing the important role of multiple scattering with these atoms on the XANES spectrum. A new spectral feature appears at about 5 eV beyond the main peak corresponding to peak C_2 in the experimental spectrum. Similarly the effect of size of the clusters in XANES spectra of CuO square pyramid, second shell with 4Ba and 4Y ions, third shell of 6 Cu ions and fourth shell of 18 oxygen atoms respectively have been performed and large changes in spectrum are noticed on considering the fourth shell. In figure 5 the XANES calculations, for largest clusters of Cu 1 and Cu 2 sites and their weighted sum are shown. Because there are two Cu2 sites in the unit cell, the sum reflects mostly the structures of the spectrum of the Cu2 sites in the CuO_2 layer. Comparing with the experimental spectrum in figure 4, all experimental features $P, A_1, A_2, C_1, C_2, D_1$ and D_2 are reproduced in the calculated spectrum.

The XANES spectra for different configuration of Cu 3d orbitals Cu $3d^9$ (or $3d^9$) and Cu $3d^8\ 4s^2p^1(3d^8)$ were also calculated by Garg et al [6] and compared with the results discussed above for the Cu $3d^{10}4s^1$ configuration. The muffin-tin potential and phase shifts have been calculated for each configuration and the XANES calculations have been performed using the same atomic coordinates of the cluster. The effect of changing the electronic configuration induces changes in the lineshape of XANES spectrum within 20 eV but no characteristic spectral feature can be associated with each electronic configuration except for the peak C_2 which is well resolved only for Cu $3d^{10}$ configuration. The pre-edge peak P found in the spectrum of Cu $3d^9$ configuration is due to the partially allowed transition in the distorted square pyramid of the dipole forbidden 1s-3d transition.

The experimental spectrum in high energy range was shown to be in better agreement with the calculated spectrum of Cu $3d^{10}$ configuration than with that of the Cu $3d^9$ configuration. The peak C_2 is obtained in the calculation only for the Cu $3d^{10}$ configuration. This result is in agreement with the fact that increasing the kinetic energy of photoelectrons we reach the sudden approximation, like in the XPS final state where the fully relaxed configuration is the $3d^{10}$ configuration [6].

Calculations have now been made for the unit cell having oxygens; situated out of the plane of Cu site 2. The results are shown in figure 6. The peak corresponding to C_2 in experiment (Fig. 4) now shows much more clearly and is in better agreement with the experiment.

Later on, Reller et al [7] had suggested an alternate structure for the Y-B-C system in which oxygen was present at all the sites around the Y atom and missing from all the sites around Cu site 1. Results of this are shown in figure 7. The peak C_2 is now conspicuous by its absence. As is now well known, this model was however, rejected on basis of other experiments too. In fact the results recently obtained by Durham et al [10] using self consistent calculation for the $(La_{1-x}\ Sr_x)_2\ CuO_4$ indicate that this peak can be obtained for large clusters and are thus in agreement with our findings. We therefore, conclude that there is no Cu^{3+} in these system and that the alternative structure proposed by Reller et al [9] does not

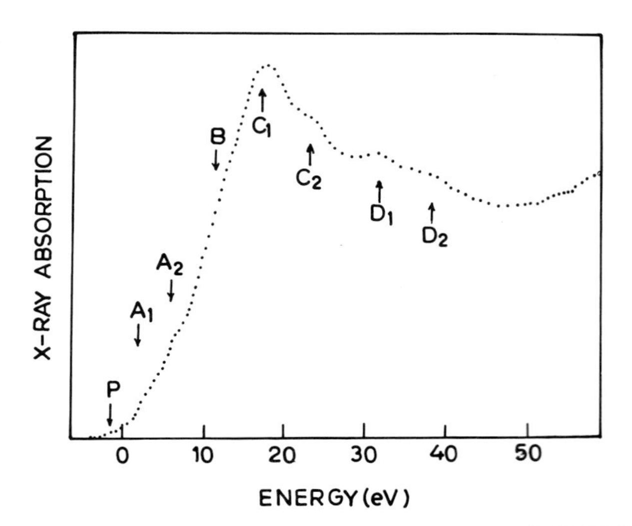

Fig. 4. Cu K-edge XANES of superconducting $YBa_2Cu_3O_{6.9}$ (Ref. 4)

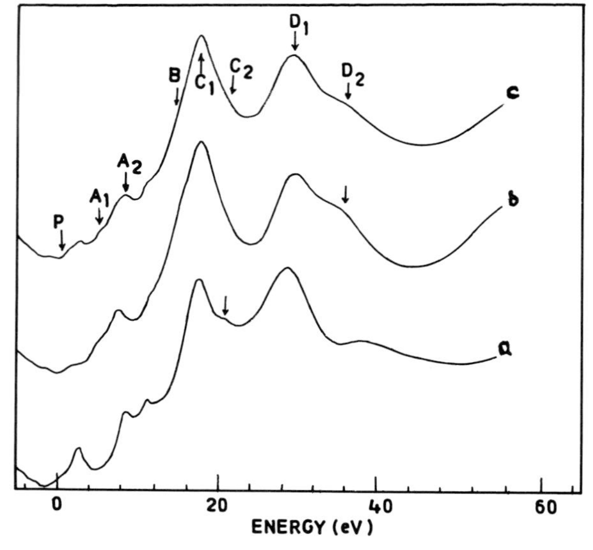

Fig. 5. Cu K-edge XANES spectrum of $YBa_2Cu_3O_7$ upper curve and calculated XANES spectrum using the multiple scattering approach from the two clusters with 4 shells, having as central atom the four fold coordinated Cu 1 (curve a) and the five fold coordinated Cu2 in the CuO_2 layers (curve b) and their weighted sum (curve c) (Ref. 6).

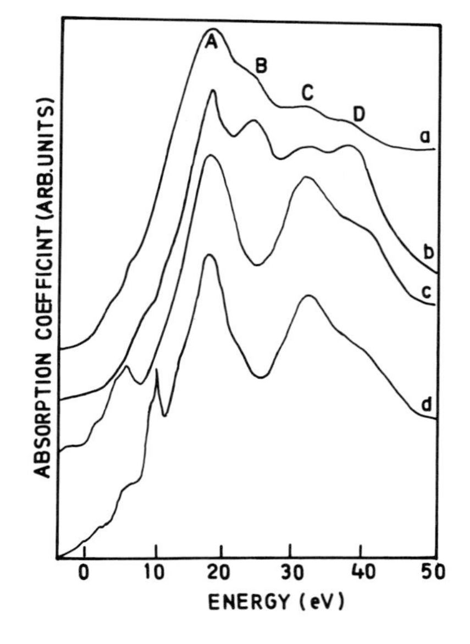

Fig. 6. Experimental and calculated Cu K XANES spectra in case of Orthorhombic $YBa_2Cu_3O_{7-x}$ system, when unit cell having oxygen situated out of the plane of Cu site 2. Curve a - experimental spectrum, Curve b - Cu^{1+} configuration, Curve c- Cu^{2+} configuration, and Curve d - Cu^{3+} configuration.

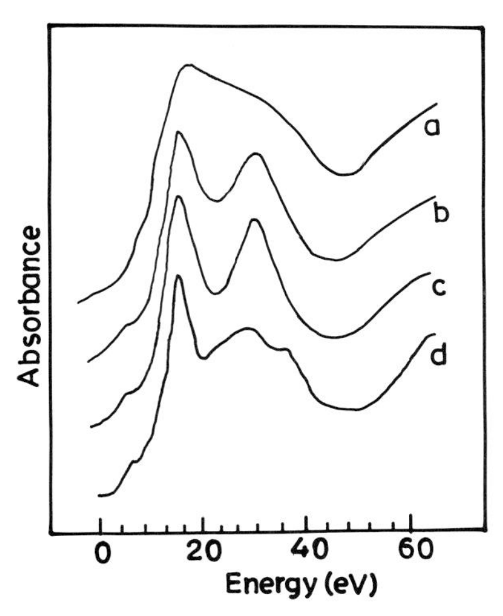

Fig. 7. Calculated Cu K XANES spectra considering a structure for YBC suggested by Reller et al for first (a), second (b), third (c) and fourth (d) shell.

stand the test of our calculations also and is untenable. The earlier conclusions drawn by Oyanagi et al [11] and Alp et al [12] linking the presence of the peak C_2 observed in experimental spectrum to presence of Cu^{3+} is likewise also untenable.

REFERENCES

1. A. Bianconi, A. Congiu Castellano, M. De Santis, C. Politis, A. Marcelli, S. Mobilio, A. Savoia, Z. Physik B Condensed Matter 67, 307 (1987).

2. A. Bianconi, A. Congiu Castellano, M. De Santis, P. Rudolf, P. Lagarde and A.M. Flank, Solid State Commun. 63,1009 (1987).

3. A. Bianconi, A. Congiu Castellano, M. De Santis, P. Delogu, A. Gargano, R. Giorgi, Solid State Commun. 63,1135 (1987).

4. A. Bianconi, A. Clozza, A. Congiu, Castellano, S. Della Longa, M. De Santis, A. Di Cicco, K. Garg, P. Delogu, A. Gargano, R. Giorgi, P. Lagarde, A.M. Flank and A. Marcelli, Int. J. Modern Phys. B. 1,853 (1987).

5. A. Bianconi, A. Clozza, A. Congiu Castellano, S. Della Longa, M. De Santis, A. Di Cicco, K.B. Garg, P. Delogu, A. Gargano, R. Giorgi, P. Lagarde, A.M. Flank and A. Marcelli, Proc. 14th Int. Conf. on X-ray and Inner-shell Processes, Paris Sept. 14-18, 1987 (France).

6. K.B. Garg, A. Bianconi, S. Della Longa, A. Clozza, M. De Santis and A. Marcelli, Phys. Rev. B 38, 244 (1988).

7. A.S. Koster, Molecular Physics 26, 625 (1973)

8. J.J. Capponi, C. Chaillout, A.W. Hewat, P. Lejay, M. Marezio, N. Nguyen, B Raveau, J.L. Soubeyroux, J.L. Tholence, and R. Tournier Europhysics Lett. 3, 1301 (1987) and A.W. Hewat, J.J. Capponi, C. Chaillout, M. Marezio and E.A. Hewat, Sol. State Commun. 64, 301 (1987).

9. A. Reller, J.G. Bednorz, K.A. Muller, Z. Physik B. Condensed Matter 67, 285 (1987).

10. P.J. Durham, W.M. Temmerman, and G.M. Stocks, Proceeding of the 14th International Conference on X-ray and Inner-shell Processes, Paris Sept. 14-18, 1987 (France).

11. H. Oyanagi, H. Ihara, T. Matsubara, T. Matsushita, T. Tokumoto, M. Hirabayashi, N. Terada, K. Senzaki, Y. Kimura and T. Yao, Jpn. J. Appl. Phys. 26, L488 (1987).

12. E.E. Alp, G.K. Shenoy, D.G. Hinks, D.W. Capone II, L. Soderholm and M. Ramanathan, Phys. Rev. B35, 7199 (1987).

Photoemission Study of Single Crystal $Bi_2Sr_2CaCu_2O_8$

TAKASHI TAKAHASHI[1], HIROYOSHI MATSUYAMA[1], HIROSHI KATAYAMA-YOSHIDA[1], YUTAKA OKABE[1], SHOICHI HOSOYA[2], KAZUHIRO SEKI[3], HITOSHI FUJIMOTO[4], MASATOSHI SATO[4], and HIROO INOKUCHI[4]

[1] Department of Physics, Tohoku University, Sendai, 980 Japan
[2] Institute for Materials Research, Tohoku University, Sendai, 980 Japan
[3] Department of Materials Science, Hiroshima University, Hiroshima, 730 Japan
[4] Institute for Molecular Science, Okazaki, 444 Japan

ABSTRACT

Photoemission measurements with synchrotron radiation have been performed on single crystal $Bi_2Sr_2CaCu_2O_8$. Two energy bands with dispersion of 0.2-0.5 eV were observed in the vicinity of the Fermi level and one of them crosses the Fermi level midway between the center and boundary of the Brillouin zone, giving a clear evidence for existence of a Fermi surface. The Fermi-edge peak exhibits a pronounced enhancement at photon energy of the O 2s core threshold, meaning a dominant O 2p nature of the Fermi-edge states. These results indicate existence of the Fermi-liquid states with dominant O 2p nature in the high-Tc superconductor. The superconductivity could be driven by formation of Cooper-pairs of the O 2p holes in the Fermi-liquid states, probably through the spin or charge fluctuation.

INTRODUCTION

The mechanism of high-Tc superconductivity has not been well established although numerious experimental and theoretical studies have been done. It is certain that understanding the electronic structure is the first step to establish the high-Tc mechanism. In this paper, we present a comprehensive result of our synchrotron-radiation photoemission study [1,2] on single crystal $Bi_2Sr_2CaCu_2O_8$. We have performed two types of photoemission measurements, one is an angle-resolved photoemission to determine the band structure and the other is a resonant photoemission to study the atomic-orbital nature of the bands.

We found two electronic bands in the vicinity of the Fermi level and one of them crosses the Fermi level midway between the center and boundary of the Brillouin zone, giving a clear evidence for existence of a Fermi surface in the high-Tc superconductor. We also found that the Fermi-edge peak exhibits a pronounced resonance at photon energy of the O 2s core threshold but not at the Cu 3p, meaning a dominant O 2p nature of the electronic states at the Fermi level. From these experimental results, we discuss the electronic structure of the high-Tc superconductor and the mechanism of the high-Tc superconductivity.

EXPERIMENTAL

A single crystal of $Bi_2Sr_2CaCu_2O_8$, typically 5x5x0.5 mm^3, was grown with KCl flux [3]. The stoichiometric melt of Bi_2O_3, $SrCO_3$, $CaCO_3$ and CuO powders, 99.9 % pure, with KCl of about 20 weight % was slowly cooled down from 890°C to 850 °C by the rate of 0.5°/hr. and then quenched to room temperature. The crystal has a mirror-like plane and can be cleaved very easily along the plane. The single-crystallinity was confirmed by an x-ray diffraction. A magnetic-susceptibility measurement showed that this crystal becomes superconductive at 85 K. Photoemission measurement was

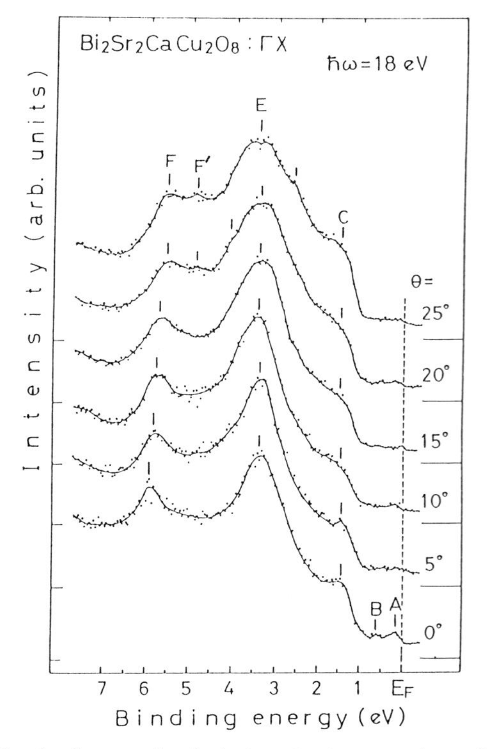

Fig. 1 Angle-resolved photoemission spectra of $Bi_2Sr_2CaCu_2O_8$ measured in the direction of ΓX using ħω=18 eV. Polar angle referred to the surface normal is indicated.

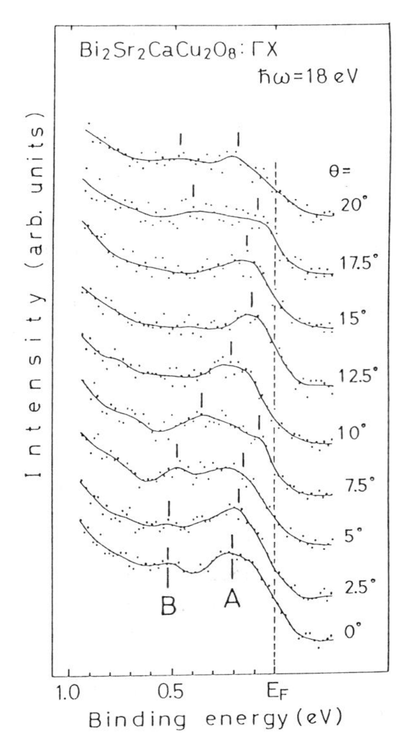

Fig. 2 Angle-resolved photoemission spectra of $Bi_2Sr_2CaCu_2O_8$ in the vicinity of the Fermi level measured in the direction of ΓX using ħω=18 eV.

performed with an angle-resolved photoemission spectrometer at the UVSOR, Institute for Molecular Science, Japan. The energy resolution was 0.2 eV as estimated from the Fermi-edge of a gold film deposited in the spectrometer. The angular resolution was about 2°. The single crystal was cleaved in the spectrometer to obtain a clean and fresh surface and kept at room temperature during the measurements. No surface degradiation was detected throughout the measurement. The Fermi level was referred to that of a gold film.

RESULTS AND DISCUSSION

Angle-resolved photoemission

Figure 1 shows angle-resolved photoemission spectra measured in the high-symmetry direction ΓX. The spectra have several structures denoted by tich marks and letters A-F. Small structures in the vicinity of the Fermi level, A and B, seem to have some angular dependence. Figure 2 shows photoemission spectra in the Fermi-level region obtained by long-time data aquisition, where bands A and B exhibit a substantial energy dispersion. Figure 3 shows the band structure of $Bi_2Sr_2CaCu_2O_8$ determined from the photoemission spectra using the convensional method. Experimental results obtained with ħω = 40 eV and for another high-symmetry direction $\Gamma\bar{M}$ are also included. Four band structure calculations [4-7] for $Bi_2Sr_2CaCu_2O_8$ are available at present and they are essentially the same. One of them [6] is shown in Fig. 3 for comparison. As shown in Fig. 3, band A crosses the Fermi level midway in both ΓX and $\Gamma\bar{M}$, giving a strong evidence for existence of a Fermi surface. Band B seems to slightly touch the Fermi level

The other experimental bands except for band F show no remarkable energy dispersion while band F has a small upward dispersion from the Γ point to the zone boundary.

We compare the present experimental result with a band structure calculation. Band A seems to correspond to the calculated Cu3d-O2p antibonding states located at 0.7 eV at the Γ point. Comparison of the two bands gives a rough estimation of the mass-enhancenent factor of four, which should be checked by other experiments. Band B can also find its theoretical counterpart about 0.3 eV below the experimental points. Thus, the band structure calculation is qualitatively in agreement with the angle-resolved photoemission result in the vicinity of the Fermi level. However, we could not find the electron pocket at the $\bar{M}$ point formed from the Bi6p-O2p antibonding states. In the higher-binding energy region, the observed bands except for band F are almost dispersionless in sharp contrast to the highly dispersive feature of the calculated bands. As for band F, the direction of energy dispersion is just opposite to that of the calculated Cu3d-O2p bonding states, suggesting a substantial renormalization by the strong electron correlation. These results provide a direct experimental evidence for the electronic structure of $Bi_2Sr_2CaCu_2O_8$ consisting of the Fermi-liquid states in the vicinity of the Fermi level and localized states with strong electron correlation in the higher energy region. The qualitative agreement between the experiment and the band calculation near the Fermi level implies that the one-electron approximation may be qualitatively correct at least near the Fermi level. Alternative explanation is that doped O 2p hole states should have a similar band dispersion to that of the band calculation because they should also reflect the symmetry of the crystal. This point will be given a clear answer by the resonant photoemission measurement described below.

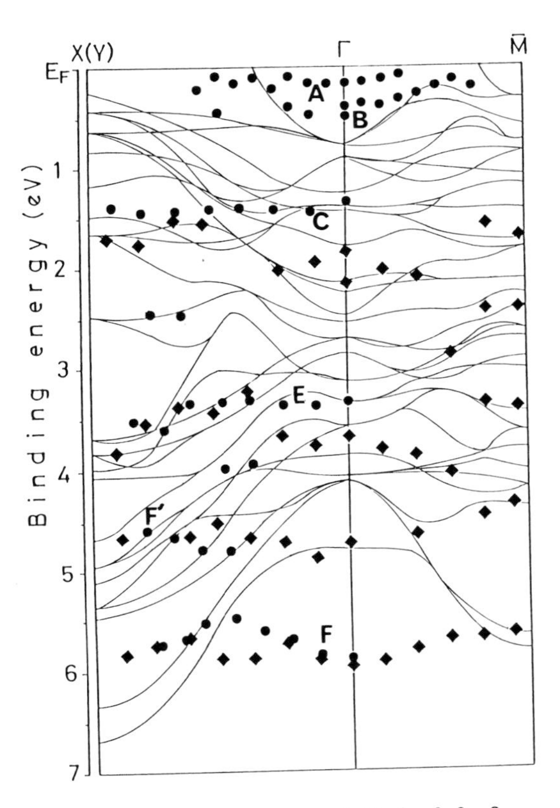

Fig. 3 Band structure of $Bi_2Sr_2CaCu_2O_8$ determined by angle-resolved photoemission with $\hbar\omega$=18 eV (circles) and $\hbar\omega$=40 eV (squares). A representative band structure calculation (Ref. 6) is shown by thin solid lines for comparison.

Resonant photoemission

Figures 4 and 5 show valence-band photoemission spectra of $Bi_2Sr_2CaCu_2O_8$ measured with several photon energies at and near the O 2s and Cu 3p core thresholds, respectively. Figure 6 shows photoemission spectra in the vicinity of the Fermi level taken at both the photon-energy regions by a long-time data aquisition. The photoemission intensity is normalized with respect to the incident photon-flux and the experimental points are smoothed with a least-square-method.

At first we should remark the pronounced enhancement of the Fermi-edge peak (band A in Fig. 6) at photon energy of the O 2s core threshold (18 eV) and its negligible change when passing the Cu 3p core threshold (74 eV). The observed clear resonance at the O 2s core threshold presents a direct evidence for the dominant O 2p character in the electronic states at the Fermi level. This result is quite consistent with the recently discovered BCS-like enhancement of enriched ^{17}O nuclear-relaxation rate just below Tc [8]. In the higher-binding-energy region, some photoemission peaks also show a resonant behavior. In order to make the resonant effect more clearly seen, we plotted

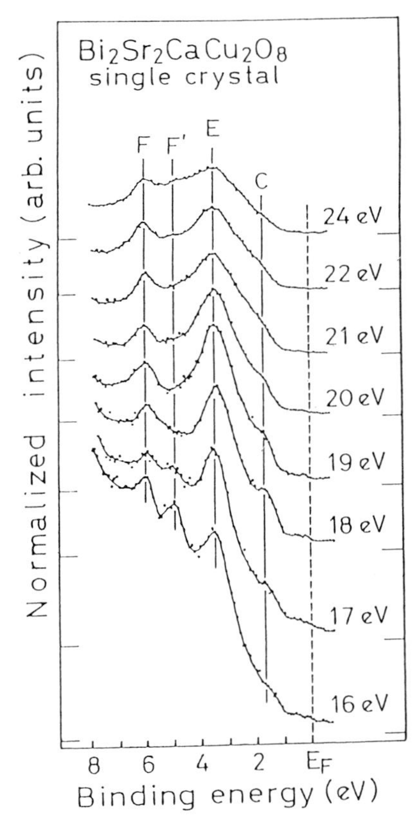

Fig. 4 Valence-band photoemission spectra of $Bi_2Sr_2CaCu_2O_8$ taken at several photon energies at and near the O 2s core threshold (18 eV).

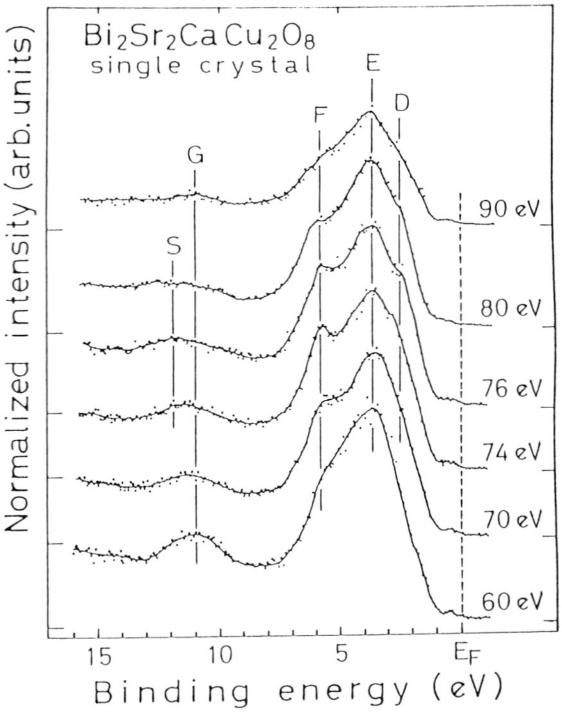

Fig. 5 Valence-band photoemission spectra of $Bi_2Sr_2CaCu_2O_8$ taken at several photon energies at and near the Cu 3p core threshold (74 eV).

the photoemission-peak intensity of each band against the photon energy in Fig. 7. First we remark band S since this band has been already observed in the La and Y systems and discussed as an evidence for divalent copper atoms (Cu^{2+}) and a strong electron correlation between d electrons (U_{dd}=6-7 eV). Band S is ascribed to a valence-band satellite due to the two-hole-bound state and is resonantly enhanced at the Cu 3p core threshold through the super Coster-Kronig process. The observed binding energy of band S (12 eV) a little smaller than those in La and Y systems (12.5 eV) may be due to an accidental overlapping from band G which is attributed to the Bi 6s state. Band C shows a slight resonant enhancement similar to that of band A and disappears at the higher photon-energies where the photoionization cross-section of the O 2p state is much smaller than that of the Cu 3d. This indicates dominant O 2p nature in band C. The opposite case is band D, which appears only at photon energies around the Cu 3p core threshold, meaning a substantial contribution from the Cu 3d state. Band F also exhibits a remarkable resonant enhancement at the Cu 3p core threshold while band E does not show a clear resonant behavior either at the O 2s or Cu 3p core threshold. Table I summarizes the atomic-orbital nature of each photoemission band (A-G and S) determined from the present resonant photoemission. We find in Table I a rough tendency that the O 2p electrons occupy the electronic states near the Fermi level (bands A, B, and C) while the Cu 3d electrons distribute in the higher-binding-energy region.

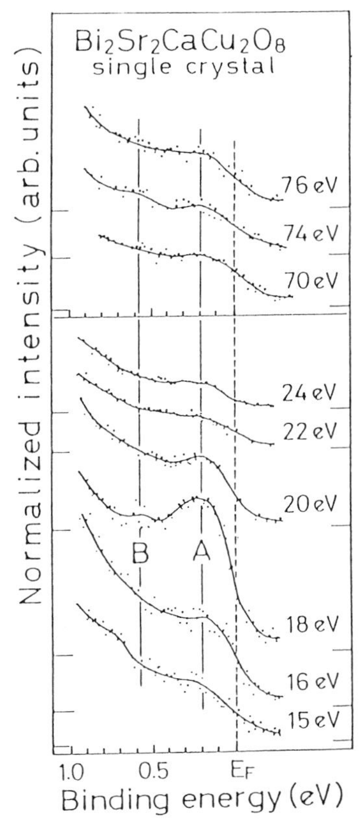

Fig. 6 Photoemission spectra in the vicinity of the Fermi level taken at photon energies of both the O 2s and Cu 3p core threshold.

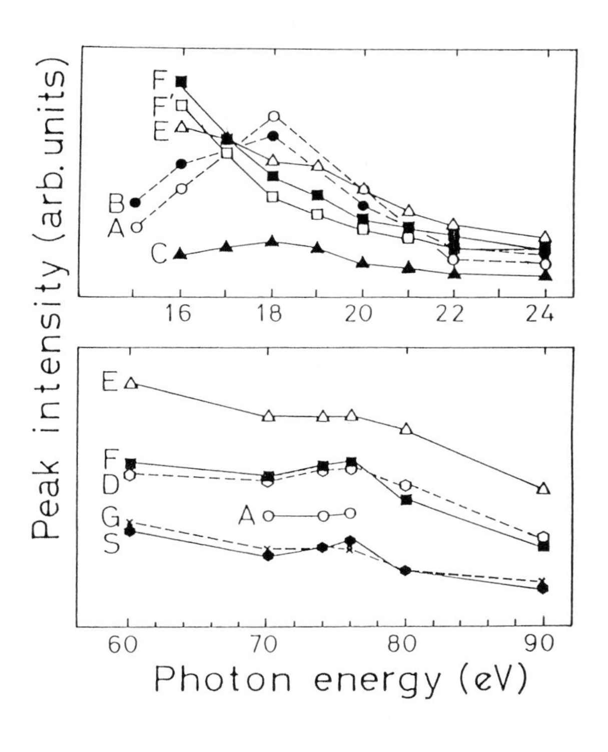

Fig. 7 Peak intensities of each photoemission band (bands A-G and S in Fig. 4-6) plotted against photon energies near the O 2s (upper) and Cu 3p (lower) core threshold.

CONCLUSION

We have performed photoemission measurements on single crystal $Bi_2Sr_2CaCu_2O_8$ with synchrotron radiation. The angle-resolved measurement revealed that (1) there are two bands with substantial energy dispersion (0.2-0.5 eV) in the vicinity of the Fermi level, (2) one of them crosses the Fermi level midway between the center and boundary of the Brillouin zone, giving a clear evidence for existence of a Fermi surface, (3) energy bands in the higher-binding-energy region are almost dispersionless in sharp contrast to the highly dispersive features in the band structure calculation. The resonant photoemission measurement found that (1) the Fermi-edge peak has a dominant O-2p atomic-orbital nature but not of the Cu 3d and (2) the O 2p electrons occupy the electronic states near the Fermi level while the Cu 3d electrons distribute in the higher-binding-energy region of the valence band. Thus, the present synchrotron-radiation photoemission study provides a clear picture for the electronic structure of the high-Tc superconductor, which consists of two parts as shown in Fig. 8; one is the Fermi-liquid states originating mainly from the O 2p states and the other is the strongly localized states with both the Cu 3d and O 2p natures away from the Fermi level. The superconductivity could be driven by formation of Cooper-pairs of the O 2p holes in this Fermi-liquid states, probably through the spin or charge fluctuation.

Table I

Atomic-orbital nature of each photoemission band (A–G and S in Figs. 4–6) in the valence band of $Bi_2Sr_2CaCu_2O_8$ determined by resonant photoemission.

Band	Binding energy (eV)	Nature
A	0.2	dominant O 2p
B	0.6	dominant O 2p
C	1.6	dominant O 2p
D	2.5	dominant Cu 3d
E	3.4–3.6	O 2p + Cu 3d
F'	4.9	O 2p + Cu 3d
F	5.7–5.8	dominant Cu 3d
G	11	Bi 6s
S	12	Cu 3d satellite

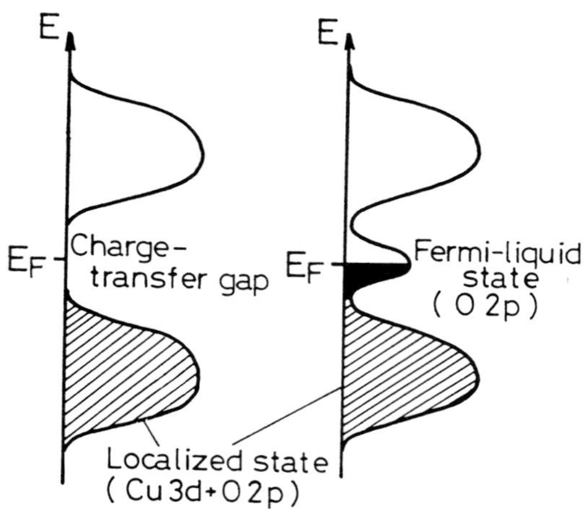

Fig. 8 Schematic diagrams of electronic structure of high-Tc superconductor; left: charge-transfer semiconductor right: high-Tc superconductor.

ACKNOWLEDGEMENTS-We thank Profs. M. Tachiki and Y. Kuramoto and Dr. K. Okada for their useful discussion. This work was supported by a Grant-in-Aid for Scientific Research on Priority Area from the Ministry of Education, Science, and Culture of Japan, and by Nippon Sheet Glass Foundation for Materials Science.

REFERENCES

1. T. Takahashi, H. Matsuyama, H. Katayama-Yoshida, Y. Okabe, S. Hosoya, K. Seki, H. Fujimoto, M. Sato, and H. Inokuchi, Nature 334, 691 (1988).
2. T. Takahashi, H. Matsuyama, H. Katayama-Yoshida, Y. Okabe, S. Hosoya, K. Seki, H. Fujimoto, M. Sato, and H. Inokuchi, unpublished.
3. H. Katayama-Yoshida, Y. Okabe, T. Takahashi and S. Hosoya, unpublished.
4. M.S. Hybertsen and L.F. Mattheiss, Phys. Rev. Lett. 60, 1661 (1988).
5. H. Krakauer and W.E. Pickett, Phys. Rev. Lett. 60, 1665 (1988).
6. S. Massidda, J. Yu, and A.J. Freeman, Physica C 52, 251 (1988).
7. F. Herman, R.V. Kasowski, and W.Y. Hsu, Phys. Rev. B 38, 204 (1988).
8. Y. Kitaoka, K. Ishida, K. Asayama, H. Katayama-Yoshida, Y. Okabe and T. Takahashi, Nature in press.

X-Ray Photoelectron Spectra of $La_{1+x}Ba_{2-x}Cu_3O_y$ (x=0.2, 0.3, 0.4, 0.5)

AKIHIKO SAITO[1], SIN-ICHIRO YAHAGI[1], SHU YAMAGUCHI[2], HIROSI HAYAKAWA[2], and YOSHIAKI IGUCHI[2]

[1] Central Research Laboratory, Daido Steel Co., 2-30, Daido-cho, Minami-ku, Nagoya, 457 Japan
[2] Department of Materials Science and Engineering, Nagoya Institute of Technology, Gokiso-cho, Showa-ku, Nagoya, 466 Japan

ABSTRACT

The X-ray photoelectron spectra were measured on the single phase $La_{1+x}Ba_{2-x}Cu_3O_y$ (x=0.2, 0.3, 0.4, 0.5).

Oxygen content y was be increased by the increase of La content x. The critical temperature Tc was decreased by the increase of La content x. The binding energy of Cu 2p states was closely related with Tc. Superconductivity seemed to occur below the critical value of the Cu binding energy of about 934.25eV. Below the critical value of the Cu binding energy, Tc could be increased by decreasing the binding energy of Cu.

Remarkable differences in O 1s and Ba 3d states were observed between the superconductor and non-superconductor. The spectrum of the non-superconductor was similar to that of La_2O_3 indicating strong La-O bond. In the superconductor the full width of half maximum intensity of Ba 3d 5/2 was lower than that of the non-superconductor.

These results might suggest that the free holes around Cu-O bonds contribute to the superconductivity.

1. INTRODUCTION

Since the discoveries of high Tc superconductors in La-Ba-Cu-O (1) and Y-Ba-Cu-O (2), a number of studies have been made to find out higher Tc materials (3) - (6). In recent reports on La-Ba-Cu-O system, the crystal structure of $La_{1+x}Ba_{2-x}Cu_3O_y$ is essentially the same with $YBa_2Cu_3O_y$ (7) - (9). And Tc decreases with increasing La content x and at the composition of x=0.4 superconductivity was not observed. This La-Ba-Cu-O system has been studied for its interesting crystal chemical aspects. (10) - (12)

However the electronic bonding states of this La-Ba-Cu-O system are not so obvious. Therefore we investigated the X-ray photoelectron spectra of various compositions to clarify the differences of the electronic bonding states between the superconductor and the non-superconductor.

2. EXPERIMENTAL PROCEDURE

2.1 Samples

The samples were synthesised from a stoichiometric mixture of La_2O_3, $BaCO_3$ and metallic copper. The mixture of the starting materials was heated at 930°C, then calcined at 950°C and calcined again at 970°C. Each process was performed for 24 hr in pure oxygen gas atmosphere of one atmospheric pressure followed by slow cooling in a furnace. The products obtained through the process were confirmed as single phases of $La_{1+x}Ba_{2-x}Cu_3O_y$ by X-ray diffraction. The powder was pressed to pallets of 13mm diameter and 1 - 2 mm thick. The pellets were then sintered at 1050°C for 24 hr in an oxygen atmosphere and slowly cooled. The sintered pellets were also single phases.

Finally the pellets were sintered at 950°C for 24 hr followed by step slow cooling holding at 600°C for 24 hr, 500°C for 24 hr, 400°C for 12 hr and 300°C for 6 hr. The final sintering process was performed in an oxygen atmosphere. The final products were confirmed as single phases. The halves of the pellets were used for oxygen content analysis, and the other halves were used for the X-ray photoelectron spectroscopic investigation.

La_2O_3 and $LaCuO_3$ were prepared for comparison which were heated at 1300°C for 40 hr. and 950°C for 23 hr. respectively. The heat treated powder was pressed to pellets and sintered at 1020°C for 16 hr and slowly step cooled holding at 600°C for 4 hr, 500°C for 6 hr, 400°C for 12 hr and 300°C for 12 hr in an oxygen atmosphere. The pellets were also confirmed as single phases.

2.2 Oxygen analysis

The prepared samples were pulverized in an argon atmosphere. The powder of about 35 mg was placed in a Ni capsule which was set in a graphite crucible together with flux of 0.5g Ni and 0.5g Sm. The graphite crucible was degassed in He gas at about 3000°C for 30 seconds. Sample resolution was performed at about 2800°C for 27 seconds. After filtering the impurities through the dust filter of silica wool, CO_2 gas was analyzed by infra-red detector.

2.3 X-ray photoelectron spectroscopic investigation

The surfaces of the pellets tested were chiseled 0.1 - 0.5 mm with pure Al_2O_3. Sputtering procedure was not adopted to avoid oxygen leaking out of the pellets. The X-ray photoelectron spectroscope used was made by PHI operated at output of 150W, 15kV with Mg X-ray gun filament. The degree of vacuum during measurement was 10^{-7} Pa. The analysis line electron multiplier supply was 3000V with pass energy of 25 - 100eV. The binding energy was measured at the range of 0 - 1000eV; O 1s states (522 - 542eV), Cu 2p states (925 - 975eV), Ba 3d states (775 - 800eV) and La 3d states (820 - 870eV).

The binding energy was also measured on the samples of La_2O_3, $LaCuO_3$, CuO and Al_2O_3 of states of O 1s, Cu 2p and La 3d. Before and after the measurement, binding energy of carbon 1s was checked to calibrate the peak point.

3. RESULTS AND DISCUSSIONS

3.1 Composition and lattice parameters

The chemical compositions of the samples tested were determined as $La_{1.2}Ba_{1.8}Cu_3O_{6.8}$, $La_{1.3}Ba_{1.7}Cu_3O_{6.95}$, $La_{1.4}Ba_{1.6}Cu_3O_{7.1}$ and $La_{1.5}Ba_{1.5}Cu_3O_{7.25}$. Fig. 1 shows the relationship between La content x and oxygen content y. The value y can be increased by the increase of x. The data obtained agree well at x=0.5 and disagree at x=0.1 with the previous report (11). The lattice parameters are shown in Fig. 2. Single phases were confirmed between x=0.15 to 0.5. It is reported that two phases of $BaCuO_2$ and $La_{1-x}Ba_{2-x}Cu_3O_y$ coexist at the composition range of $x<0.1$, and two phases of $La_{1+x}Ba_{2-x}Cu_3O_y$ and $La_4BaCu_5O_{13.5}$ coexist for $x>0.5$ (12).

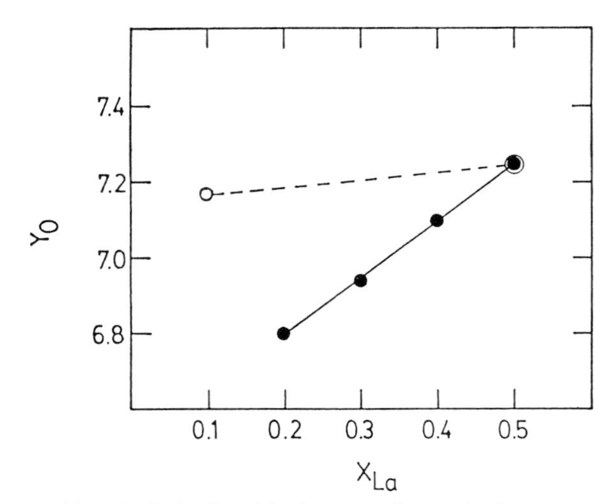

Fig. 1 Relationship between La content x and oxygen content Y

● : this work

○ : Muromachi et al

3.2 Superconductivity and X-ray photoelectron spectra investigation

Critical temperature Tc was determined as the offset temperature at which the electrical resistivity became zero. Tc can be increased by decreasing La content x (Fig. 3). The samples of x=0.4 and 0.5 are non-superconductivity. These results agree well with the report by C.U. Segre et al (8).

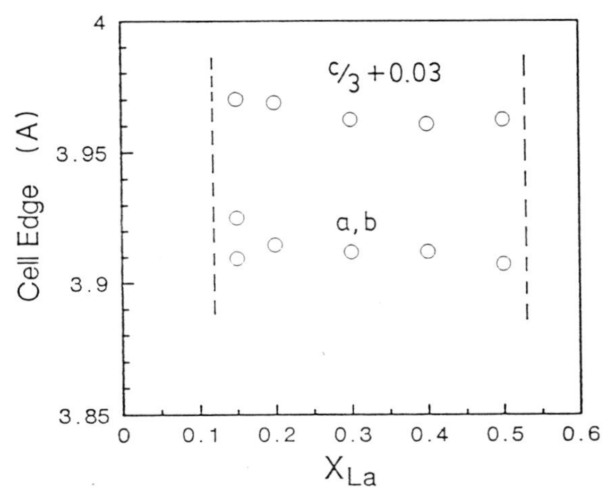

Fig. 2 Dependence of lattic parameters on La content x in $La_{1+x}Ba_{2-x}Cu_3O_y$

3.2.1 Binding energy of Cu 2p 3/2

The binding energy of Cu 2p 3/2 is decreased by the decrease of La content x (Fig. 4). Fig. 5 shows the relationship between Tc and the binding energy of Cu 2p 3/2. A critical value seems to be exist at about 934.25eV. Below the value, Tc can be increased by the decrease of Cu binding energy. Above the value, samples are no more superconductive.

It is reported that in $YBa_2Cu_3O_y$ Tc is about 90K and the binding energy of Cu 2p 3/2 is near 932eV (4). The binding energy of Cu seems to be closely related with Tc.

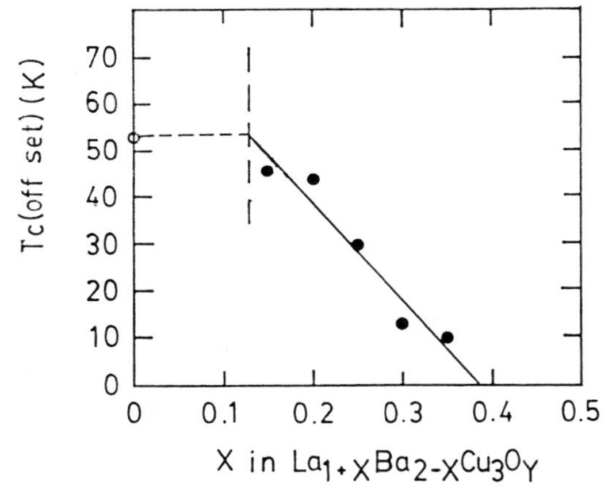

Fig. 3 Dependence of Tc on La content x

3.2.2 Full width of half maximum intensity of Ba 3d 5/2 states

A large difference was observed between superconductor and non-superconductor in the Ba 3d 5/2 peak's full width of half maximum intensity (F.W.H.M.). The superconductive samples with small La content x have sharp peaks and small F.W.H.M. of Ba 3d 5/2 (Fig. 6). It might be thought that Ba-O bond suddenly weakens at about x=0.4.

3.2.3 Binding energy of O 1s states

By increasing La content x the shoulder peak appears at lower energy side near 529eV(Fig. 7). The shoulder peak becomes obvious at higher La content, and the peak energy becomes lower towards those of the lower energy peak of $LaCuO_3$ and La_2O_3. La-O bond may be strengthened like La_2O_3 by increasing La content x.

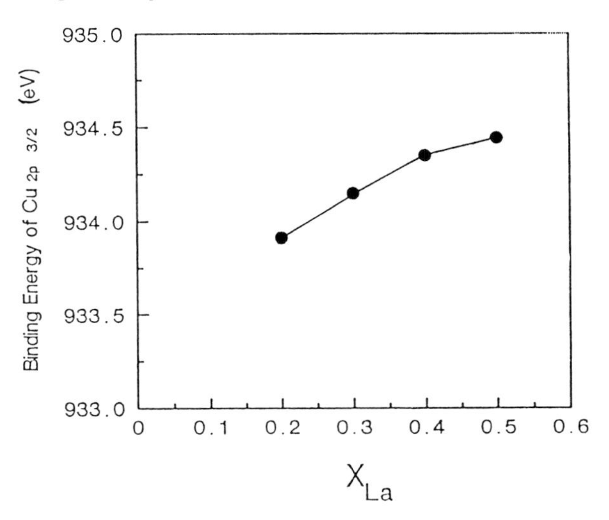

Fig. 4 Dependence of the binding energy of Cu 2p 3/2 on La content x

4. SUMMARY

The results obtained by this experiment on $La_{1+x}Ba_{2-x}Cu_3O_y$ are as follows:

(1) Oxygen content y can be increased by increasing La content x.

(2) The binding energy E_B of Cu 2p 3/2 is closely related Tc. Superconductivity was observed at $E_B < 934.25$eV.

(3) In non-superconductors, weak Ba-O bonds were observed.

(4) In non-superconductors, strong La-O bonds similar to La_2O_3 were observed.

From these results, the reason why Tc is low at high La content might be thought as follows: If La is added to the superconductor, the calculated number of holes must increase. La-O bonds become strong and Ba-O bonds weaken. (Fig. 6, 7). Therefore Cu-O bonds might be unbalanced by adding La. And the free holes might localize to Cu atoms, which might be assisted by the result of the shift of Cu binding energy to higher energy level by increasing La content (Fig. 4).

In the superconductor of La-Ba-Cu-O System, free holes around Cu-O bonds may contribute to the superconductivity.

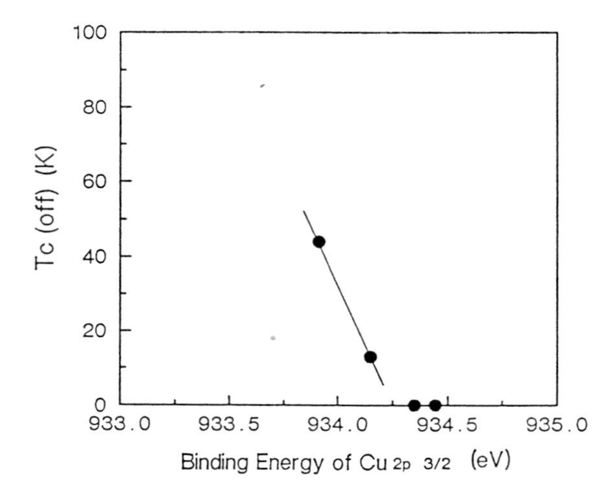

Fig. 5 Dependence of Tc on the binding energy of Cu 2p 3/2

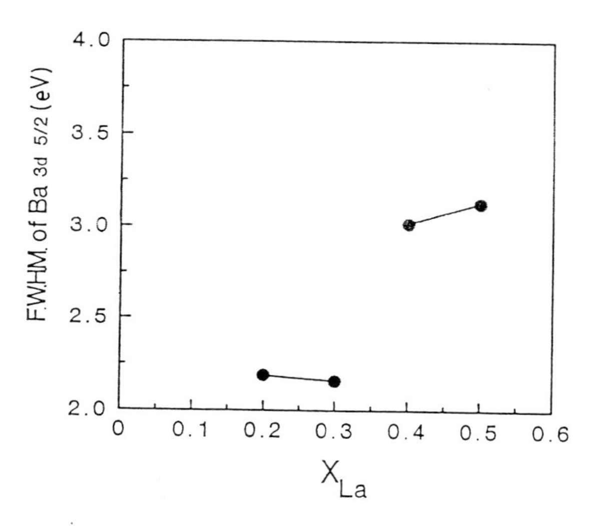

Fig. 6 Dependence of Ba 3d 5/2 peak's full width of half maximum intensity

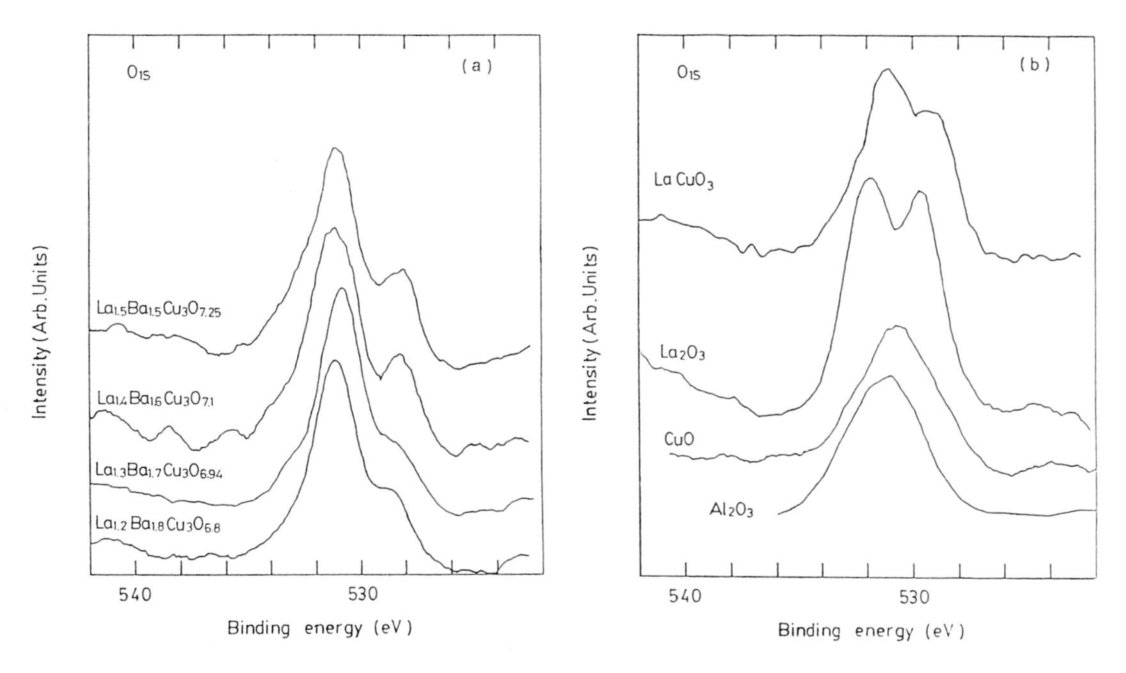

Fig. 7 The XPS spectra of O 1s level

(a) $La_{1+x}Ba_{2-x}Cu_3O_y$ (x=0.2, 0.3, 0.4, 0.5)

(b) $LaCuO_3$, La_2O_3, CuO and Al_2O_3

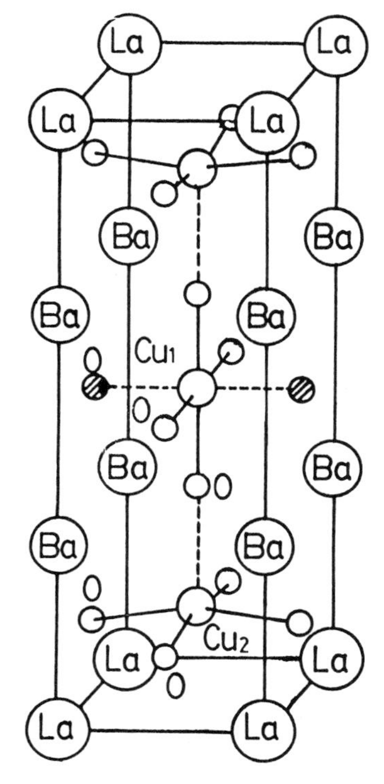

Fig. 8 A view of the structure of $La_{1+x}Ba_2Cu_3O_y$, (9)

REFERENCE

1) J.G. Bednorz and K.A. Müller: Z. Phys. B64 (1986) 189.
2) M.K. Wu, J.R. Ashburm, C.J. Torng, P.H. Hor, R.L. Meng, L. Gao, Z.I. Huang and C.W. Chu: Phys Rev. Lett. 58 (1987) 908.
3) N. Fukushima, H. Yoshino, H. Niu, M. Hayashi, H. Sasaki, Y. Yamada and S. Murase: Jpn. J. Appl. Phys. 26 (1987) L719.
4) H. Ihara, M. Hirabayashi, N. Terada, Y. Kimura, K. Senzaki, M. Akimoto, K. Bushida, F. Kawashima and R. Uzuka: Jpn. J. Appl. Phys. 26 (1987) L460.
5) S. Yamaguchi, K. Terabe, A. Saito, S. Yahagi and Y. Iguchi: Jpn. J. Appl. Phys. 27 (1988) L179.
6) Y. Tokura: Parity 3. No. 6 (1988) 60.
7) W.I.F. David, W.T.A. Harrison, R.M. Ibberson, M.T. Weller, J.R. Grasmeder and P. Lanchester: Nature, 328 (1987) 328.
8) C.U. Segre, B. Dabrowski, D.G. Hinks, K. Zhang, J.D. Jorgensen, M.A. Beno and I.K. Schuller: Nature, 329 (1987) 227.
9) I. Nakai, K. Imai, T. Kawashima and R. Yoshizaki: Jpn. J. Appl. Phys. 26 (1987) L1244.
10) E. Takayama-Muromachi, Y. Uchida, A. Fujimori and K. Kato: Jpn. J. Appl. Phys. 27 (1988) L223.
11) E. Takayama-Muromachi, Y. Uchida, A. Fujimori and K. Kato: Jpn. J. Appl. Phys. 26 (1987) L1546.
12) S. Yamaguchi et al.: to be published.

X-Ray Photoemission Study of High-Tc Superconducting Thin Film

TAKATO MACHI, ATSUSHI TANAKA, NOBUO KAMEHARA, and KOUICHI NIWA

Fujitsu Laboratories Ltd., 10-1, Morinosato-Wakamiya, Atsugi, 243-01 Japan

ABSTRACT

We examined Bi-Sr-Ca-Cu-O superconducting thin film with XPS, and the observed satellite peaks were similar to those of $YBa_2Cu_3O_{7-d}$. The cluster model, which predicts the localization of the Cu 3d electrons and holes in the O 2p, can be applied to the Bi-Sr-Ca-Cu-O superconductor.

INTRODUCTION

The superconducting carrier is the hole, as can be seen from measurements of the hole coefficient[2]. To understand the superconducting mechanism, we must know the location of the hole. In La-Sr(Ba)-Cu-O and Y-Ba-Cu-O, it is reported that the superconducting current flows in the CuO_2 plane. Several recent studies proposed that the Cu 3d electrons were localized and the hole, which doped into CuO_2 plane, was in O 2p[1]. To examine whether this model can be applied to the Bi-Sr-Ca-Cu-O superconductor, we studied a thin film with XPS (X-ray photoemission spectroscopy).

EXPERIMENTAL

The Bi-Sr-Ca-Cu-O film was prepared by RF magnetron sputtering and post-annealing. The sputtering target was $BiSrCaCu_2O_x$. A thin film was sputtered for 4 hours at 600°C on a MgO (100) single crystal. The film was about 1 μm thick. It was post-annealed in air at 890°C for 30 minutes and 866°C for 12 hours.

The film was evaluated by resistivity measurement, X-ray diffraction, and X-ray photoemission spectroscopy. $YBa_2Cu_3O_{7-d}$ bulk and Bi-Sr-Cu-O bulk were also measured by XPS.

RESULTS AND DISCUSSION

Figure 1 shows the X-ray diffraction patterns of the as-deposited and annealed Bi-Sr-Ca-Cu-O thin films. Deposition provided an amorphous film, which was crystallized during post-annealing. The peaks of the annealed film correspond to those of $Bi_2Sr_2CaCu_2O_8$ which is an 80 K superconductor[3]. Since the film shows strong (00n) peaks, the annealed film has c-axis orienta-perpendicular to the substrate.

Figure 2 shows resistivity versus temperature for the annealed thin film. Tco is about 80 K and Tce is 69 K. This is consistent with the results of X-ray diffraction. The change of resistivity in the normal state behaves like a metal. The as-deposited film was an insulator.

Figure 3 shows the photoemission spectra in the Cu core-level region for Cu, $YBa_2Cu_3O_{7-d}$, Bi-Sr-Cu-O, and $Bi_2Sr_2CaCu_2O_8$. Two peaks are observed at 951 eV and 931 eV, which correspond to j=1/2 and j=3/2 in all samples. Two satellite peaks are observed at about 960 eV and 940 eV in the superconducting samples. The main and satellite peaks arose from $2p3d^{10}$ and $2p3d^9$ configurations. The energy separation between these peaks is about 8 eV, which is the intra-atomic Coulomb energy between the 2p hole and the 3d electron.

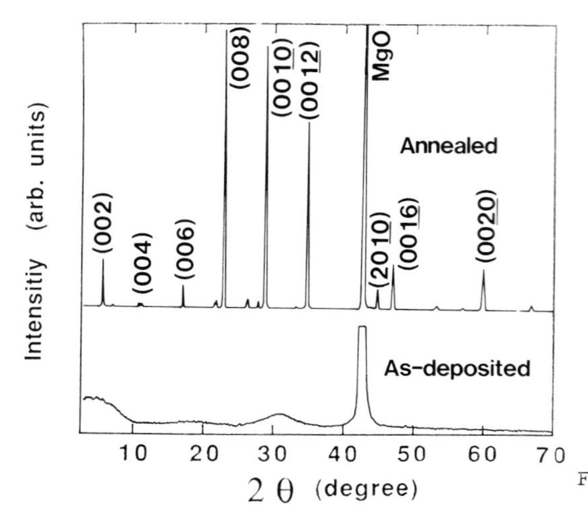

Fig.1. X-ray diffraction patterns of the as-deposited and the annealed thin films of Bi-Sr-Ca-Cu-O.

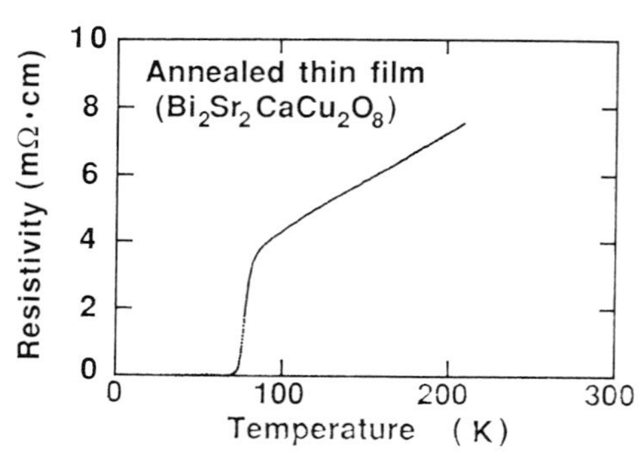

Fig.2. Resistivity versus temperature of the annealed thin film.

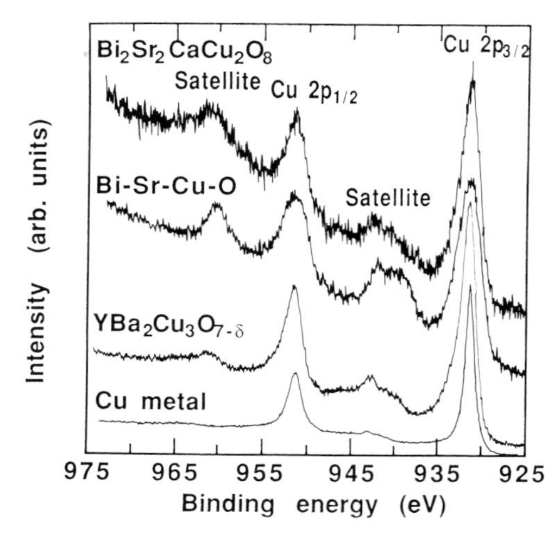

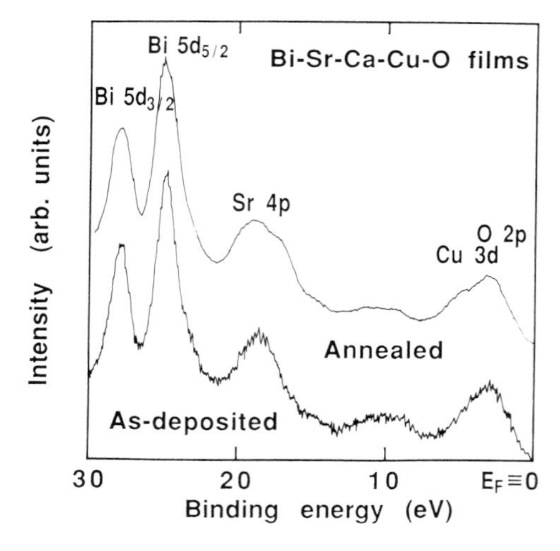

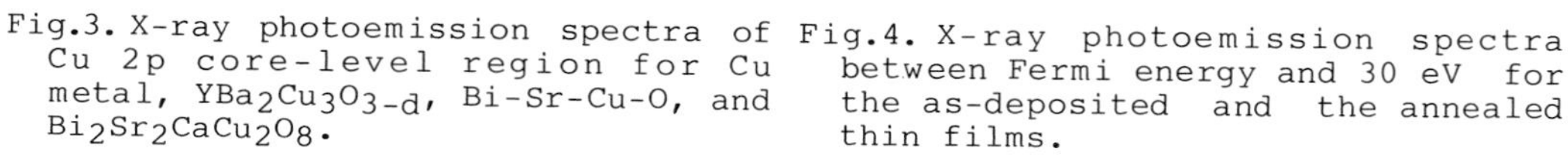

Fig.3. X-ray photoemission spectra of Cu 2p core-level region for Cu metal, $YBa_2Cu_3O_{3-d}$, Bi-Sr-Cu-O, and $Bi_2Sr_2CaCu_2O_8$.

Fig.4. X-ray photoemission spectra between Fermi energy and 30 eV for the as-deposited and the annealed thin films.

Figure 4 shows the photoemission spectra between the Fermi energy and 30 eV for as-deposited and annealed films. The Fermi level was determined from Bi $5d_{5/2}$ which is 25 eV. The peak around 4 eV represents the hybridization between Cu 3d and O 2p. This hybridization peak is far from the Fermi energy for both spectra.

Figure 5 shows the photoemission spectra around Cu 3d for $YBa_2Cu_3O_{7-d}$, Bi-Sr-Cu-O, and $Bi_2Sr_2CaCu_2O_8$. Since the intensity of the spectra around the Fermi level are nearly zero, these materials can not be explained by the band theory. Two satellite peaks are observed at around 12 eV and 10 eV. Because the same peaks are observed in $CuCl_2$[4], which has localized 3d electrons, it is assumed that the Cu 3d electrons in these superconductors are localized. The main peak arises from $3d^9$ and $3d^{10}$ ground states, and the satellite peaks arise from $3d^8$ final state. This interpretation is given by the cluster model[1]. Our photoemission spectra in Bi-Sr-(Ca)-Cu-O support this model.

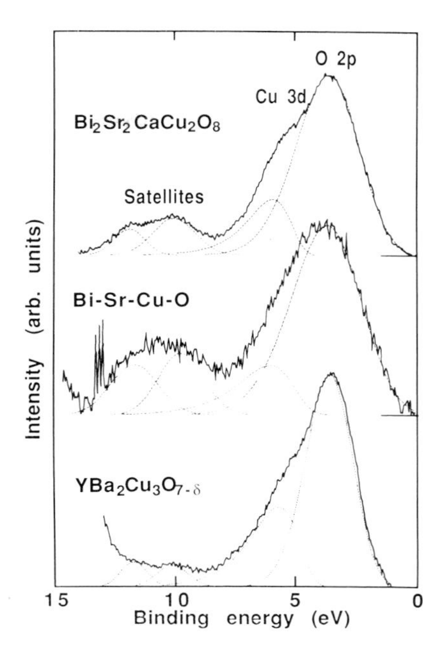

Fig.5. X-ray photoemission spectra between Fermi energy and 15 eV for $YBa_2Cu_3O_{7-d}$, Bi-Sr-Cu-O, and $Bi_2Sr_2CaCu_2O_8$.

CONCLUSION

$Bi_2Sr_2CaCu_2O_8$ superconducting thin film was prepared by RF magnetron sputtering and post-annealing. The film has a c-axis orientation perpendicular to substrate. Satellite peaks were observed in the photoemission spectrum of Cu 2p core-level region. The energy separation between the main and satellite peaks was about 8 eV, which represents the intra-atomic Coulomb energy between the hole in O 2p and the electron in Cu 3d. Two satellite peaks were observed in the photoemission spectra around Cu 3d for Bi-Sr-Cu-O and $Bi_2Sr_2CaCu_2O_8$ superconductors. The satellite peaks were similar to those of $YBa_2Cu_3O_{7-d}$. X-ray photoemission spectroscopy in Bi-Sr-(Ca)-Cu-O supports the cluster model which states that the Cu 3d electrons are localized, and the holes are in O 2p.

REFERENCE

[1] A. Fujimori, E. Takayama-Muromachi, Y. Uchida: Solid State Commun. 63, 857(1987).
[2] J. B. Torrance, Y. Tokura, A. I. Nazzal, S. S. P. Parkin: Phys. Rev. Lett. 60, 542(1988).
[3] M. Onoda, A, Yamamoto, E. Takayama-Muromachi, and S. Takekawa: Jpn. J. Appl. Phys. 27, L833(1988).
[4] G. van der Laan, C. Westra, C. Haas, and G. A. Sawatzky: Phys. Rev. B23, 4369(1981).

Electronic and Structural Properties of $Ba_2Y_{1-x}Pr_xCu_3O_{7-\delta}$

AZUSA MATSUDA, KYOICHI KINOSHITA, TAKAO ISHII, HIROYUKI SHIBATA, TAKAO WATANABE, and TOMOAKI YAMADA

NTT Basic Research Laboratories, 3-9-11, Midori-cho, Musashino, Tokyo, 180 Japan

ABSTRACT

Measurements of the Hall coefficient in the $Ba_2Y_{1-x}Pr_xCu_3O_{7-\delta}$ system show that an increase in Pr concentration reduces the Hall carrier number ($1/eR_H$). The Cu formal valence, calculated from the Pr valence, which has been determined to be about +3.5 from its magnetic moment, slightly decreases as the Pr concentration increases. The transition temperature shows a stronger dependence on Hall carrier number and Cu-O formal valence than in the oxygen depleted $Ba_2YCu_3O_{7-\delta}$ system. X-ray diffraction measurements show that the structure remains almost unchanged throughout the series. The persistence of the chain structure explains the major difference in Cu-O valence dependence on T_c between the present system and the oxygen depleted $Ba_2YCu_3O_{7-\delta}$ system on the basis of the two band model. Measurements of the magnetoresistance show that the pure Y compound has a large positive magnetoresistance up to 250 K, while the Pr substitution drives it negative. The temperature dependence of the positive magnetoresistance indicates that it is caused by the superconducting fluctuation. The experiment suggests that there is strong correlation between the appearance of the negative magnetoresistance and the suppression of superconductivity. Rietveld analysis reveals that the CuO_2 plane becomes less flat as a result of the oxygen displacements, and the oxygen in the BaO plane moves toward the CuO_2 plane. These displacements also possibly cause the Tc reduction.

INTRODUCTION

Of known Ba_2[lanthanoide]$Cu_3O_{7-\delta}$ compounds, only the Pr compound shows neither superconductivity nor metallic conduction, although it has an orthorhombic structure isomorphic to that of $Ba_2YCu_3O_{7-\delta}$.[1),2)] Thus, the differences in the electronic and crystal structure of these compounds might be a key parameter controlling superconductivity as well as normal state properties of the high Tc copper oxides. It is known that a series of single phase $Ba_2Y_{1-x}Pr_xCu_3O_{7-\delta}$ can be synthesized.[3),4)] They range continuously from superconductors to insulators. It is generally accepted that in high Tc copper oxides, the total carrier density is a crucial parameter which determines the conduction properties. There are several systems, for example, $[La,Sr]_2CuO_4$,[5)] $Ba_2YCu_3O_{7-\delta}$,[6)] and $Ba_{2-x}La_{1+x}Cu_3O_{7-\delta}$,[7)] which show a strong correlation between Tc and the formal valence of Cu-O. These relations have also been confirmed by Hall effect measurements.[5),8)] It is pointed out that the present $Ba_2Y_{1-x}Pr_xCu_3O_{7-\delta}$ solid solution system cannot be understood on the same basis as a valence changing system.[4),9)] Therefore, the present understanding is far from complete. From the structural point of view, it is pointed out that local structure is important in determining superconductivity.[10)] However, all systems investigated generally accompany a structural transformation such as an ortho-tetra transition, which may mask intrinsic information. Therefore, it is important to choose a system which doesn't show any structural change to extract local structural information. The present system is suitable for this purpose. The system is investigated from both the structural and the electronic points of view. The crystal structure is analyzed through X-ray diffraction measurements with the Rietveld refinement. The Cu valence is estimated by the measurements of the Pr moment and the results are compared with the Hall effect measurements and the results of the structural analysis. The Hall effect provides a more direct and physical way of investigating the carrier density and the band structure of the system.[11)] Magnetoresistance is measured to obtain further information about the electronic properties of the present system.

BASIC CHARACTERISTICS

(a) Sample Preparation

A series of $Ba_2Y_{1-x}Pr_xCu_3O_{7-\delta}$ samples was synthesized from 4N $BaCO_3$, 4N Y_2O_3, and 4N CuO powders. The powders were mixed, ground, calcined, reground, pressed into pellets, and sintered at 920°C for 50 hours in an oxygen atmosphere. Oxygen concentration analysis showed that the oxygen content was almost constant ($0.03<\delta<0.12$) throughout the series. X-ray diffraction analysis showed that all the samples had a single-phase oxygen deficient perovskite structure, except for a small trace of $BaCuO_2$ observed in the samples with $x\neq0$. The existence of $BaCuO_2$ indicates that a small amount of Pr is substituted into the Ba site as pointed out by Okai et al.[4)] The amount of substitution was estimated by the peak intensity ratio to that of the pure $BaCuO_2$, showing a mono-

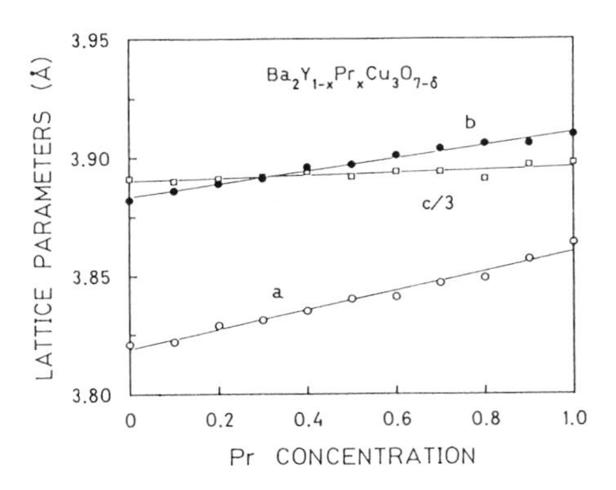

Fig.1 Lattice parameters a, b, and c as a function of Pr concentration (x).

tonic increase with increasing x. At x=1, about 2% of the Ba was substituted by [Y,Pr]. The obtained occupation factor of Pr (or Y) at Ba sites was used for the Rietveld analysis.

The relationship between Pr concentration and lattice parameters is shown in Fig.1. The lattice parameters a and b slightly increase with Pr concentration. However the orthorhombic distortion remains almost constant for the whole range of x. This result, together with the constancy of the oxygen content, indicates that the ordering of oxygen in the CuO_3 chain sites remains unchanged. Thus the present system provides a suitable model system for investigating the electronic effect of the Pr substitution. The lattice parameter c stays almost constant throughout the series. This indicates the ionic radius of Pr is about the same as that of Y, suggesting a Pr valence higher than +3.

(b) Resistivity

The resistivities were measured by a standard four probe configuration with three additional electrodes for the Hall voltage measurement. Rectangular bars 8x2.5x0.3mm were sliced from the as-grown sintered pellets. In Fig.2, the temperature dependence of the resistivity is shown as a function of x. The system is metallic down to x<0.6 as shown in Fig.2a. For x>0.7, the system becomes insulating. In Fig.2b, the resistivities are plotted with three dimensional variable range hopping (VRH) parameterization. Resistivity does not show simple behavior and neither VRH nor activation type T dependence fits the data in the insulating region. In Fig.3a, Tc is plotted as a function of x, where error bars indicate the transition width. Figure 3b shows superconducting volume fraction, which is obtained as a strength of a diamagnetic signal normalized at x=0 at 5 K in a 100 G magnetic field. Tc and the volume fraction decrease with increasing x, while transition widths remain rather narrow.

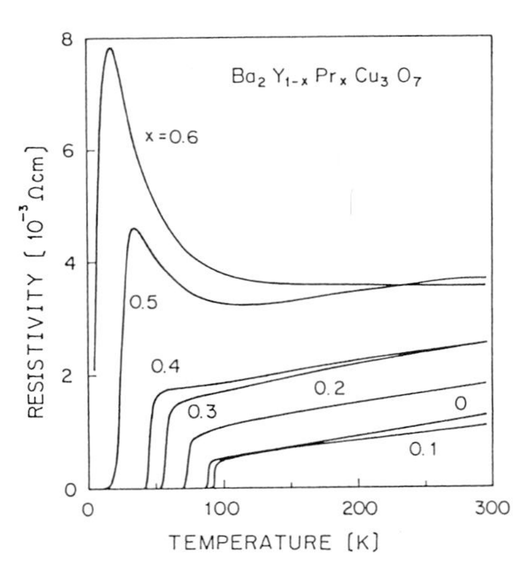

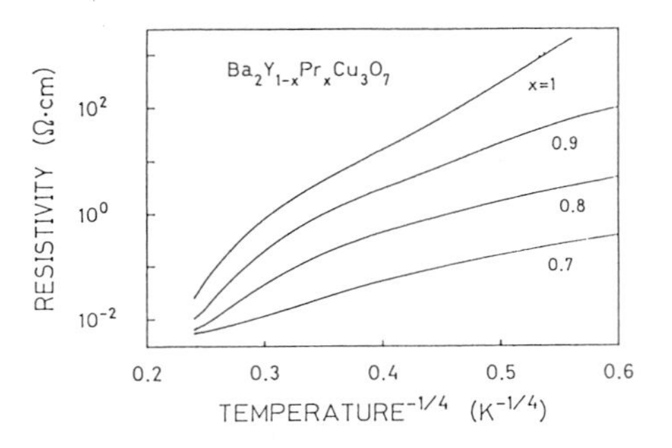

Fig.2 Temperature dependence of resistivity. (a) For samples with x<0.6. (b)For samples with x>0.7. The temperature scale is plotted as $T^{-1/4}$.

ELECTRONIC PROPERTIES

(a) Magnetic Properties

Magnetic susceptibility (χ) was measured by a SQUID magnetometer in a magnetic field of 1KG in a temperature range of 5 to 300K. Since Pr is a major magnetic atom in the system and the moment is a function of the valence, the Pr valence can be estimated from the susceptibility data. Pr^{+3} and Pr^{+4} have moments of 3.6μ_B and 2.5μ_B, respectively. The temperature dependence of χ was expressed as χ_o+C/(T-θ), where χ_o is a temperature-independent component of the susceptibility, and C/(T-θ) is the Curie-Weiss contribution. This temperature dependence well reproduces experimental data for T>50K with θ≈0. For T<50K and x>0.5, deviation from this form becomes evident, indicating the domination of the singlet ground state. The hatched area in Fig.4a, represents the estimated effective Pr moments. Ambiguity comes from the Cu moment (μ_{Cu}) estimation. The solid line represents the eye guide for μ_{Cu}=0μ_B and the dashed line for 0.3μ_B. In either way, the obtained Pr moment is about 3μ_B. This means that Pr valence is about +3.5. The valence obtained was different from the +3.9 obtained by Okai et al. The difference may be due to their neglect of the χ_o term. The Cu valence calculated from the Pr valence and the oxygen content is displayed in Fig.4b. This valence is consistent with the small change in lattice parameters. Since the average ionic radius of $Pr^{+3.5}$ is about 1.04 Å, the complete substitution of the Y ion only produces a 0.02Å increase in the average ionic radius.

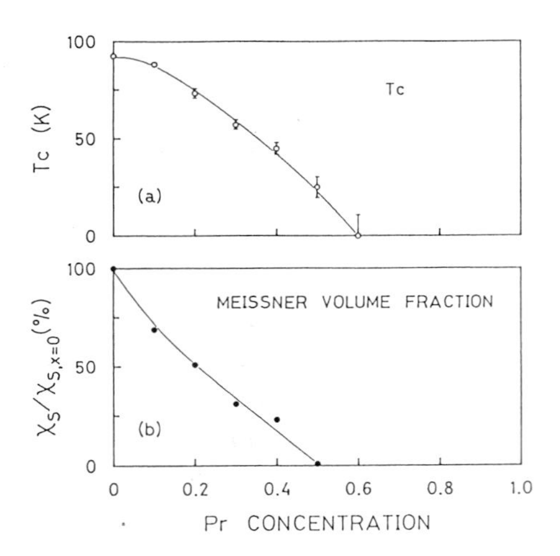

Fig.3(a) Superconducting transition temperatures as a function of Pr concentration (x). Error bars show the transition width. (b) The Meissner volume fraction normalized at x=0.

The temperature independent part of susceptibility χ_o increases rapidly with x, as shown in Fig.4a. Since it is not proportional to x, the major contribution to χ_o might come from Cu electrons. This is in contrast to the oxygen depleted $Ba_2Y_{1-x}Pr_xCu_3O_{7-\delta}$ system where χ_o decreases rapidly with increasing oxygen deficiency.[12]

(b) Hall Effect

The Hall effect was measured in a magnetic field of 5T with a current of 10 to 50mA. Figure 5 shows V/eR_H as a function of temperature. Here V and R_H are unit cell volume and the Hall coefficient. V/eR_H means carrier number within a unit cell, if the system has a single flat band. As reported by many authors,[5,6,8] at x=0, V/eR_H is positive and shows remarkably linear temperature dependence. With increasing x, the dependence of T rapidly becomes less linear, and overall magnitudes decrease monotonically.

Since V/eR_H has strong temperature dependence, the simple interpretation of this quantity, like carrier number, is impossible. Wang et al have proposed a two-carrier model.[8] It consists of a two-dimensional (2D) CuO_2 plane band together with a one-dimensional (1D) CuO_3 chain band, with carrier densities and mobilities of n_{2d}, μ_{2d} and n_{1d}, μ_{1d}, respectively. However, this model seems inconsistent. If the simple two-carrier expression for the Hall coefficient is used and the 1/T temperature dependence of μ_{2d}, which is thought to be realized in the doped La_2CuO_4 system, are adopted, a calculation shows the Hall coefficient has 1/T temperature dependence only if $\mu_{1d} >> \mu_{2d}$. It is hard to beleive that this condition is realized in the actual system. Recent experiments show that the system should be understood as a highly correlated Fermion system having a large on-site coulomb repulsion.[13] In such a system, the Hall coefficient may be temperature dependent because of a strong renormalization effect in a nearly half-filled Hubbard band.

Since overall magnitudes of V/eR_H change monotonically, and temperature dependence weakens due to Pr substitution, the value at a fixed temperature can be used as a measure of carrier density. Figure 6a shows a Hall carrier number at 100K as a function of formal Cu-O valence, which is defined as [calculated Cu valence]-2. There is a

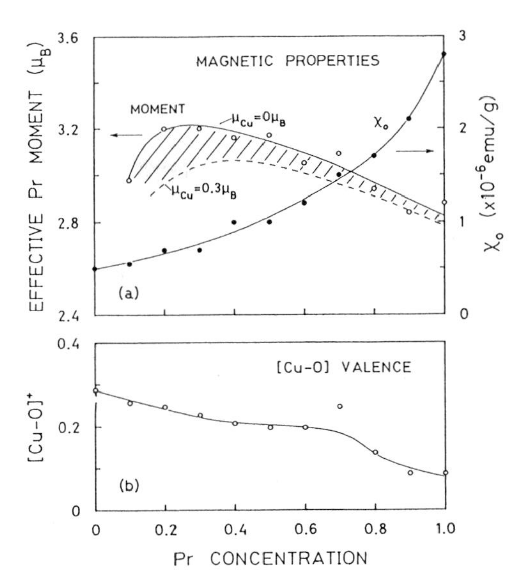

Fig.4 (a) Magnetic properties of $Ba_2Y_{1-x}Pr_xCu_3O_{7-\delta}$. Open circles represent the Pr effective moment obtained by assuming $\mu_{Cu}=0\mu_B$. The dashed line shows the Pr moment obtained by assuming $\mu_{Cu}=0.3\mu_B$. Solid circles represent the temperature independent component of susceptibility. (b) Cu valence calculated from the Pr moment and the oxygen content 7-δ.

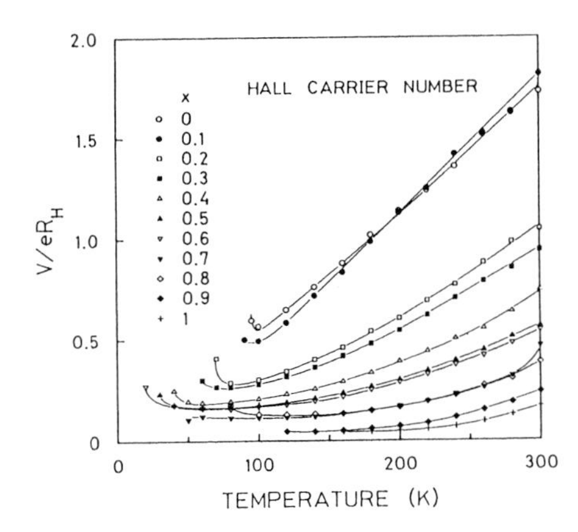

Fig.5 Temperature dependence of Hall carrier number, V/eR_H. V indicates the unit cell volume.

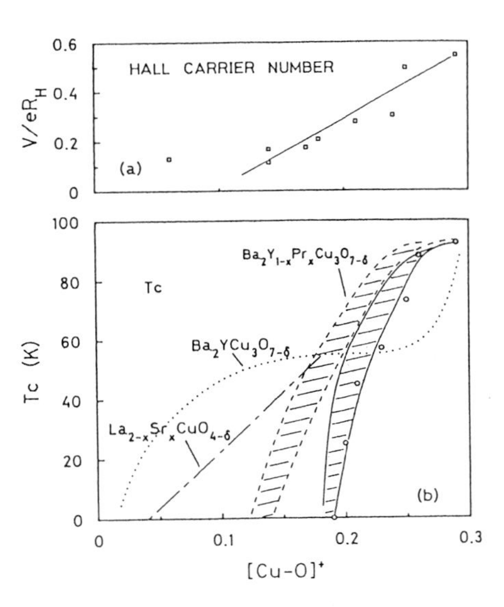

Fig.6 (a) Hall carrier number at 100K as a function of Cu-O valence. (b) Tc as a function of Cu-O valence. The hatched area surrounded by the solid lines represents the ambiguity caused by the Cu moment estimation. The dotted line indicates the relation in oxygen depleted $Ba_2YCu_3O_{7-\delta}$ after ref.8. The dot-dashed line represents the relation held in $La_{2-x}Sr_xCuO_{4-\delta}$ after ref.5. Dashed lines indicate the relation obtained by assuming that the 1D carrier remains constant in the $Ba_2Y_{1-x}Pr_xCu_3O_{7-\delta}$ system.

linear relation between these two quantities, as expected. In Fig.6b, Tc is plotted as a function of Cu-O valence. The hatched area surrounded by solid lines indicates ambiguity stemming from the estimation of the Cu moment. The dotted line and dash-dotted line indicate the relation for oxygen depleted $Ba_2YCu_3O_{7-\delta}$[8] and doped La_2CuO_4,[5] respectively. Figure 7 shows Tc as a function of the Hall carrier number at 100K. The dashed line indicates the result obtained by Wang et al for oxygen depleted $Ba_2YCu_3O_{7-\delta}$.[8] The clear difference between present results and other systems can be seen both in Fig.6b and Fig.7.

(c) Model Consideration

In the present system, the oxygen vacancy concentration and degree of oxygen ordering remains unchanged. This means that a one-dimensional CuO_3 band exists throughout the series. The $Pr^{+3.5}$ substitution into Y^{+3} sites simply compensates the Cu valence and thus the hole concentration.

Figure 8 shows the simplified band picture of the present system and that of the oxygen depleted Y compound. There exist one and two dimensional bands both of which have a Mott-Hubbard gap above the Fermi level (E_F). Considering that the Cu in the chain site tends to be trivalent, the gap edge of the 1D band might be higher than that of the 2D band. In the present system, E_F increases monotonically with x. At about x=0.5, E_F reaches the gap edge of the 2D band, making Tc zero. There, the 1D carrier still exists. However, as is well known, the 1D carrier is easily localized by a small imperfection, resulting in VRH-type conduction. The complex temperature dependence of the resistivity for x>0.6 (Fig.2b), can be understood as the behavior of the almost localized 1D electron system. On the other hand, in the oxygen depleted system, the removal of oxygen from a chain site strongly affects the 1D band. At $\delta\approx1$, chain site Cu is thought to adopt a monovalent state. Therefore, oxygen removal not only elevates E_F but also lowers the 1D band position, which will be reduced to a localized atomic level in a $\delta\approx1$ limit. These movements explain the complex Hall carrier behavior obtained by Wang et al.[8]

Now, the differences appearing in Figs.6b and 7 can be easily understood on the basis of the band picture. The apparent difference in the Hall carrier concentration at which Tc≈0 shows the existence of 1D chain electrons in the present system. This indicates, as a rough estimate, that $n_{1d}\approx0.12$ at Tc=0. Since n_{2d} is estimated to be about 0.18 at Tc=60K, it can be said that the chain band carries a substantial number of carriers. The stronger carrier number dependence on Tc can be attributed to the fact that fewer electrons are needed to fill the fixed 1D band. It is difficult to know the actual carrier concentration n_{2d} from the Hall coefficient and even from the Cu formal valence. However, the Cu-O valence can be estimated in the extremum case where n_{1d} stays constant. This simply means all the substituted Pr ions compensate holes in CuO_2 plane. Even in such an extremum case, which is displayed as the hatched area surrounded by dashed lines in Fig.6b, the result shows stronger dependence than that of the La system. There are two possibilities to be considered: (1) true n_{2d} really has a stronger dependence on Tc in this system, and (2) there is some other pair-breaking mechanism. The latter possibility will be considered in the subsequent sections.

The observed increase in temperature-independent susceptibility χ_o can be understood as a result of the 1D band filling, which provides a divergent state density and Pauli susceptibility.

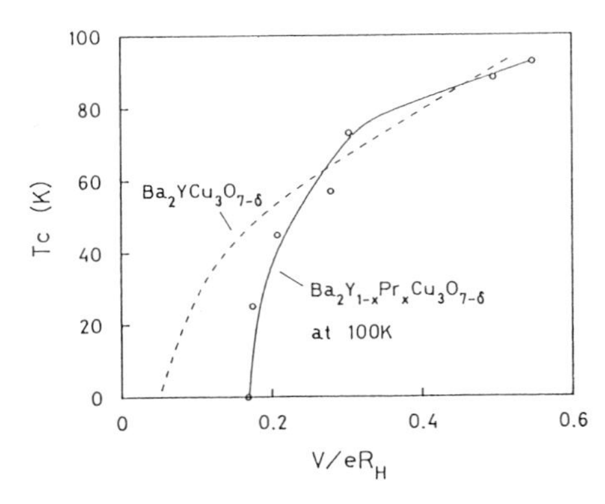

Fig.7 Tc as a function of Hall carrier number at 100K. The dashed line represents the result of the oxygen depleted $Ba_2YCu_3O_{7-\delta}$ after ref. 8.

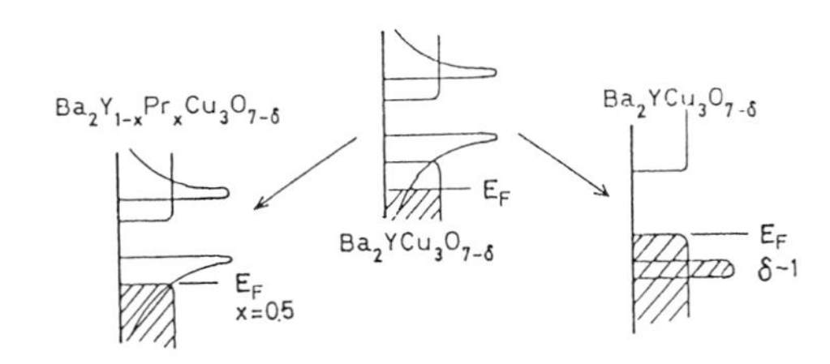

Fig.8 Proposed band picture of $Ba_2Y_{1-x}Pr_xCu_3O_{7-\delta}$ compared with that of $Ba_2YCu_3O_{7-\delta}$.

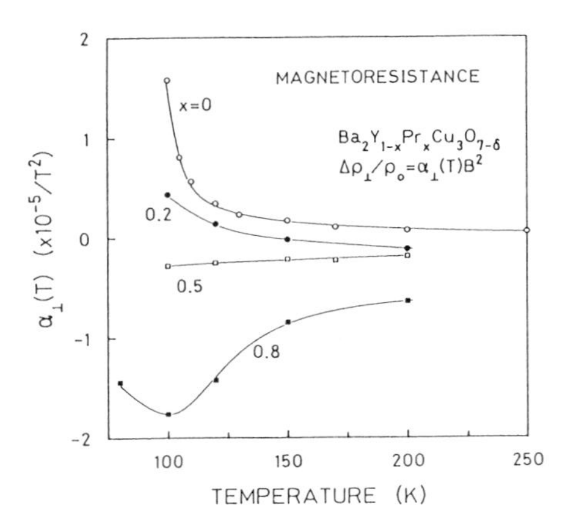

Fig.9 Temperature dependence of the transverse magnetoresistance coefficient. Open circles, filled circles, open squares, and filled squares represent the data for the samples of x=0, 0.2, 0.5, and 0.8, respectively. $\alpha_\perp$ means the proportional coefficient of B^2.

(d) Magnetoresistance

The magnetoresistance of the present system was measured in a temperature range from 50K to 250K and in a magnetic field of up to 7T. Magnetoresistance is generally small in this temperature range. Therefore, the stability of the temperature under the magnetic field is very important for accurate measurements. In the present experiment, a capacitance temperature sensor was used for the temperature control, and great care was taken to minimize the effect of the intrinsic capacitance drift. Main results are discussed briefly here, and the details will be published elsewhere.

Except for the narrow temperature range around Tc, the magnetoresistance $\Delta\rho(T)/\rho(T)$ is expressed as $\alpha(T)B^2$, where $\Delta\rho(T)/\rho(T)=[\rho(B,T)-\rho(0,T)]/\rho(0,T)$. The present discussion is limitéd to the low field criteria where $\Delta\rho$ is proportional to B^2. Figure 9 summarizes the temperature dependence of the proportional coefficient $\alpha_\perp(T)$ for several x values in a transverse magnetic field, where the field is perpendicular to the current. The magnetoresistances observed change the sign from positive to negative with increasing x. Both positive and negative sides are much larger than that of the normal one, where α is proportional to $(\text{mobility})^2$.

In Fig.10, $1/\alpha_\perp(T)$ of $Ba_2YCu_3O_{7-\delta}$ is plotted as a function of temperature. The figure shows that $\alpha_\perp(T) \propto Tc/(T-Tc)$, indicating that the origin of this magnetoresistance comes from the superconducting fluctuation. However, the result of the AL theory under the magnetic field[14),15)] shows that $\alpha_\perp(T) \propto [Tc/(T-Tc)]^{5/2}$ in 3D and $[Tc/(T-Tc)]^3$ in 2D. Therefore, the AL fluctuation theory doesn't explain the present simple behavior. It is known that the fluctuation conductivity decreases more rapidly than the AL prediction in the case of the normal superconductor[16)] and also in the oxide superconductor.[17)] Therefore, it is an open question whether the AL type theory based on the Ginzburg-Landau theory can explain the present simple result or not.

To see the characteristics of the negative magnetoresistance, the transverse ($\Delta\rho_\perp$) and longitudinal ($\Delta\rho_{//}$) magnetoresistances are plotted as a function of B^2 in Fig.11. The small dependence on the direction of magnetic field indicates that the negative magnetoresistance has a magnetic origin. Since the $Ba_2GdCu_3O_{7-\delta}$ shows a positive magnetoresistance like the Y compound, the magnetic ion in the usual lanthanoide superconductors doesn't cause the effect, as expected. Therefore the result from the present system strongly suggests that the reduction of superconductivity is related to the appearance of the negative magnetoresistance which may be caused by the spin scattering.

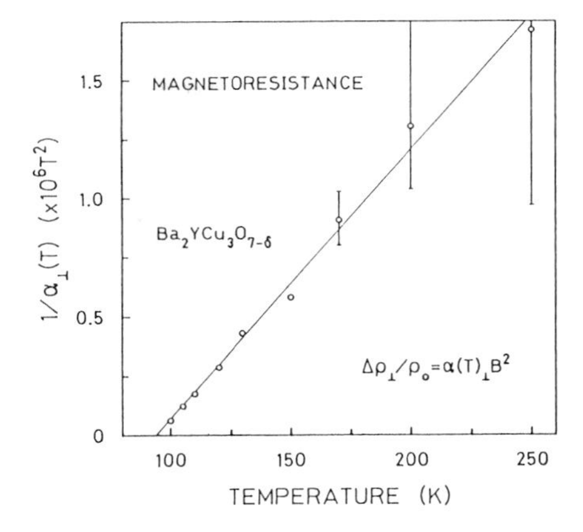

Fig.10 Temperature dependence of $1/\alpha_\perp$ in $Ba_2YCu_3O_{7-\delta}$

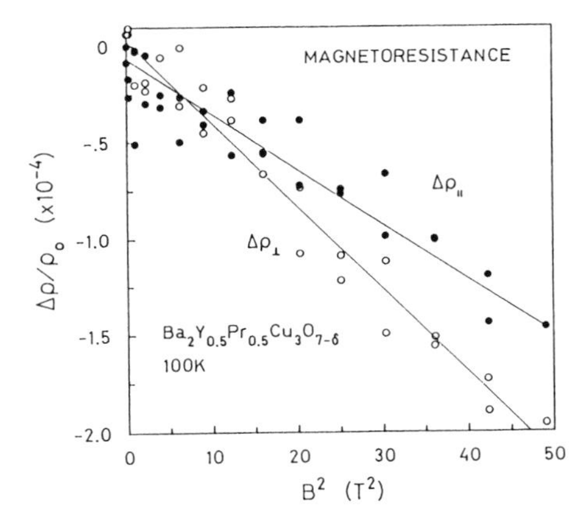

Fig.11 Magnetoresistance of $Ba_2Y_{0.5}Pr_{0.5}Cu_3O_{7-\delta}$ as a function of a magnetic field squared at 100K. Open circles represent transverse magnetoresistance and filled circles represent transverse magnetoresistance.

STRUCTURAL ANALYSIS

For the Rietveld analysis, step-scanned intensity data were collected at 0.006° intervals over a 2θ range from 25° to 70°. The obtained diffraction data were analyzed by the Rietveld program RIETAN[18)]. The refinements was made on the orthorhombic $Ba_2YCu_3O_{7-\delta}$ with a space group Pmmm. The resulting R factors are as good as that of a pure Y compound for all Pr concentrations. The details will be published in a separate paper.[19)] Reflecting the small changes in the lattice parameters, the atomic displacements are, in general, small throughout the series. Ba-Ba, Ba-[Y,Pr], and Cu-Cu lengths are almost constant. However, there are three notable displacements in the oxygen atoms: two in the CuO_2 plane, O(4) along the b axis and O(5) along the a axis, and one [O(3)] between the Cu in the chain [Cu(1)] and the Cu in the plane [Cu(2)]. Since the atomic scattering factor of oxygen is small compared to the other constituent elements, the results always have some ambiguity. However, the constancy of the heavy element frame work and the systematic changes observed make the results reliable. At x=0, both O(4) and O(5) atoms are located at almost the same z-coordinates. With increasing x, O(4) atoms shift toward the BaO plane, while O(5) atoms shifts in the opposite direction. At x=1, the z-coordinate difference of these atoms reaches about 0.9Å. These displacements cause the roughening of the $Cu(2)O_2$ plane. Besides these displacements, O(3) atoms move toward Cu(2). Thus, Cu(1)-O(3) bond length increases. In Fig.12, the bond lengths of Ba-O(4), Ba-O(5), and Cu(1)-O(3) are plotted as a function of x.

The displacement of the O(3) atom can be naturally explained as a result of the charge transfer from [Y, Pr] to the Cu-O complex, although the shift observed is opposite compared to the case of the oxygen depletion and the Co substitution.[10] If the excess electron reduces the Cu(1) valence which is thought to be higher than +2, the Cu(1)-O(3) bond length increases. This supports the proposed band model. On the contrary, it is difficult to explain the displacements in O(4) and O(5) in a simple charge transfer argument. However, the roughening reduces transfer integrals of the Cu-O network and thus reduces the band width. Therefore, the roughness together with the randomness provided by the random site occupation of Pr at the Y site and/or the randomness in the Pr valency (+3 or +4) may change the electronic system so that it is easily localized. This change may also provide the substantial increase in the spin scattering needed to provide the negative magnetoresistance.

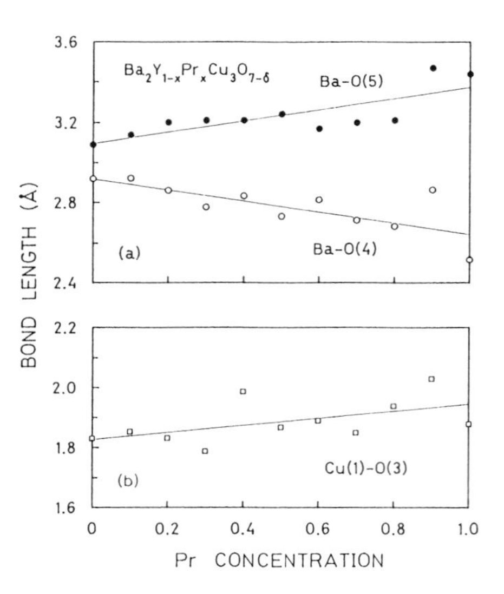

Fig.12 Bond-length changes (a) in Ba-O(4), Ba-O(5), and (b) in Cu(1)-O(3).

CONCLUSION

In conclusion, the magnetic susceptibility measurement of the $Ba_2Y_{1-x}Pr_xCu_3O_{7-\delta}$ system revealed that the Pr valence is about +3.5. The Cu formal valence, calculated from the Pr valence and the oxygen content, showed a slowly decreasing dependence on Pr concentration. Measurements of the Hall coefficient of the system showed that increased Pr concentration reduces the Hall carrier number and strong temperature dependence. Tc showed a stronger dependence on the Hall carrier number and Cu-O formal valence than in the oxygen-depleted $Ba_2YCu_3O_{7-\delta}$ system. X-ray diffraction measurements showed that the structure remains almost unchanged throughout the series. The persistence of the chain structure, which can be derived from the constancy of the basic lattice, explains the major difference in Cu-O valence dependence on Tc between the present system and the oxygen-depleted $Ba_2YCu_3O_{7-\delta}$ system on the basis of the two band model. Measurements of magnetoresistance showed that the pure Y compound has a large positive magnetoresistance up to 250K, while the Pr substitution drives it negative. The divergent temperature dependence of the positive magneto-resistance at Tc indicates that it is caused by the superconducting fluctuation. The appearance of the negative magnetoresistance is related to the suppression of super-conductivity in the present system. Detailed Rietveld analysis revealed that the CuO_2 plane becomes less flat as a result of the oxygen displacements, and the oxygen in the BaO plane moves toward the CuO_2 plane, suggesting a charge transfer from the chain Cu to [Y,Pr].

The results of the magnetoresistance measurement and the structure analysis suggest the existence of some qualitative change in the present electronic system. However, it is very difficult to explain the overall behavior presented here, without understanding the electronic system of the pure Y compound, that is characterized by the three linear temperature dependences: resistivity, Hall carrier number, and the coefficient of magnetoresistance.

ACKNOWLEDGEMENT

The authors would like to thank Dr. H. Koizumi for his Rietveld analysis and Dr. M. Naito for his helpful discussion. The authors are also indebted to Dr. Y. Kato and Dr. T. Izawa for their support and encouragement during the course of this study.

REFERENCES

1) M.K.Wu, J.R.Ashburn, C.J.Torng, P.H.Hor, R.L.Meng, L.Gao, Z.J.Huang, Y.Q.Wang and C.W.Chu; Phys.Rev.Lett 58 (1987) 908.
2) T.Yamada, K.Kinoshita, A.Matsuda, T.Watanabe, and Y.Asano; Jpn.J.Appl.Phys. 26 (1987) L633.
3) L.Soderholm, K.Zhang, D.H.Hinks, M.A.Beno, J.D.Jorgensen, C.U.Segre, and I.K.Schuller; Nature 328 (1987) 604.
4) B.Okai, M.Kosuge, H.Nozaki, K.Takahashi, and M.Ohta; Jpn.J.Appl.Phys. 27 (1988) 41.
5) M.W.Shafer, T.Penny and B.L.Olson; Phys.Rev. B36 (1987) 4047.
6) H.Takagi, S.Uchida, H.Iwabuchi, S.Tajima, and S.Tanaka; Jpn.J.Appl.Phys.Series1 (1988) 6.
7) K.Takita, H.Akinaga, H.Katoh, H.Asano, and K.Masuda; Jpn.J.Appl.Phys. 27 (1988) 67.
8) Z.Z.Wang, J.Clayhold, N.P.Ong, J.M.Tarascon, L.H.Greene, W.R.McKinnon, and G.H.Hull; Phys.Rev. B36 (1987) 7222.
9) A.P.Gonçalves, I.C.Santos, E.B.Lopes, R.T.Henriques, and M.Almeida; Phys.Rev.B37 (1988) 7476.
10) P.F.Miceli, J.M.Tarascon, L.H.Green, and P.Barboux; Phys.Rev. B37 (1988) 5932.
11) A.Matsuda, K.Kinoshita, T.Ishii, H.Shibata, T.Watanabe, and T.Yamada; Phys.Rev.B38 (1988) 2910.
12) T.Takabatake, M.Ishikawa, and T.Sugano; Jpn.J.Appl.Phys. 26 (1987) 1859.
13) T.Takahashi, F.Maeda, H.Arai, H.Katayama-Yoshida, Y.Okabe, T.Suzuki, S.Hosoya, A.Fujimori, T.Shidara, T.Koide, T.Miyahara, M.Onoda, S.Shamoto, and M.Sato; Phys.Rev.Lett. 36 (1987) 5686.
14) K.D.Usadel; Z.Physik 227 (1969) 260.
15) K.Maki; J.Low.Temp.Phys. 1 (1969) 513.
16) W.L.Johson, C.C.Tsuei and P.Chaudhari; Phys.Rev. B17 (1978) 2884.
17) J.S.Moodera, P.M.Tedrow, and J.E.Tkaczk; Phys.Rev. B36 (1987) 8329.
18) F.Izumi; J.Crystallogr. Soc. Jpn. 27 (1985) 23 in Japanese.
19) K.Kinoshita, A.Matsuda, H.Shibata, T.Ishii, T.Watanabe, and T.Yamada; Jpn.J.Appl.Phys (to be published).

New Experimental Results Concerning the Difference Between Various T_cs of $LnBa_2Cu_3O_y$ (Ln=Lanthanide)

TSUNEKAZU IWATA, MAKOTO HIKITA, and SHIGEYUKI TSURUMI†

NTT Opto-Electronics Laboratories, Tokai, Ibaraki, 319-11 Japan

ABSTRACT

The superconducting properties of a sample with large ion radius lanthanide is improved by high temperature annealing. This result is considered to be due to improved atomic ordering in the sample. However, high temperature annealing causes the production of impurity phases such as $BaCuO_2$. Annealing in Ar atmosphere was studied to obtain single phase $LnBa_2Cu_3O_y$.

It is found that the superconducting critical temperature (T_c) of $LnBa_2Cu_3O_y$ (Ln = Lanthanide) changes systematically from 88 K to 96 K depending on the melting temperature of the substance. The small changes in T_c for $LnBa_2Cu_3O_y$ are considered to be due to changes in its Debye temperature.

INTRODUCTION

90 K class superconductivity has been studied in the general formula $LnBa_2Cu_3O_y$ (Ln: Y and lanthanide), and the structure of these compounds has been identified as an oxygen-deficient perovskite structure.[1] We have studied $LnBa_2Cu_3O_y$, and in particular materials with large ion radius lanthanide atoms which show a higher T_c than those with small ion radius lanthanide atoms.[2-6]

In this paper, we describe the procedure used to obtain good quality $LnBa_2Cu_3O_y$ with large ion radius lanthanide atoms.

We show that the T_c is proportional to the melting temperature.

EXPERIMENTAL

Samples were prepared as follows: appropriate amounts of LnO_x, $BaCO_3$ and CuO were mixed and heated at 885 - 935 °C for 12 h in an O_2 atmosphere. The resultant samples were pulverized, pressed into disks, sintered at 885 - 1045 °C for 12 h in O_2 and annealed at 400 °C for 24 h and then cooled slowly to room temperature. For large ion radius lanthanide, annealing in Ar atmosphere was also examined. First, samples were annealed in Ar atmosphere at 880 - 950 °C for 10 h and then annealed at 800 - 900 °C for 10 min. in O_2 atmosphere. The temperature was then reduced to 450 °C and maintained for 20 h and finally the sample was cooled slowly to room temperature.

Powder diffraction data were obtained by X-ray diffractometry. The data were analyzed by the Rietveld method using the RIETAN program written by Izumi [7]. Electrical resistance of the samples was measured by the AC (73 Hz) four-terminal method with Au films and wires as terminals. The composition of the sample was measured by an energy dispersion type electron probe micro analyzer (EPMA). Melting temperatures were determined by thermogravimetry (TG) measurements and differential thermal analysis (DTA).

† present address: NTT R & D Information, Patent and Licensing Center, Midori-cho, Musasino-shi, Tokyo, 180, Japan

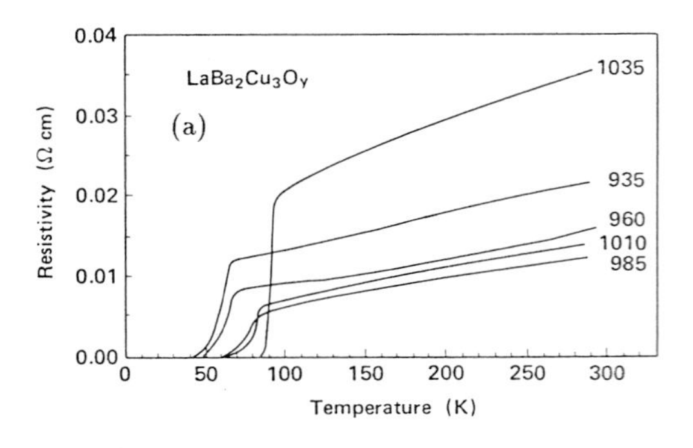

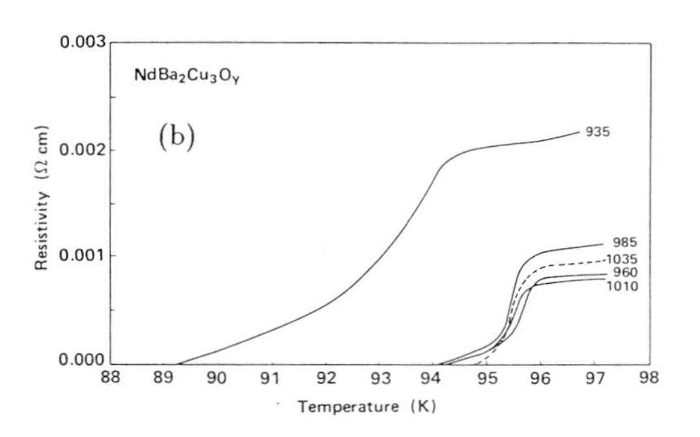

Fig. 1: Temperature dependent resistivity of the sample with various annealing temperatures for $LaBa_2Cu_3O_y$ (a) and $NdBa_2Cu_3O_y$ (b).

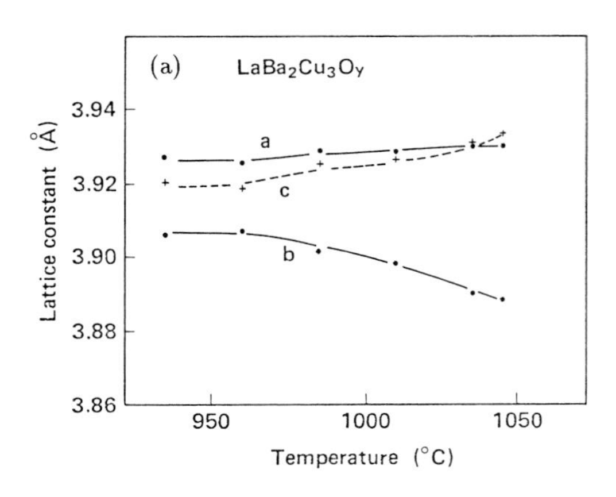

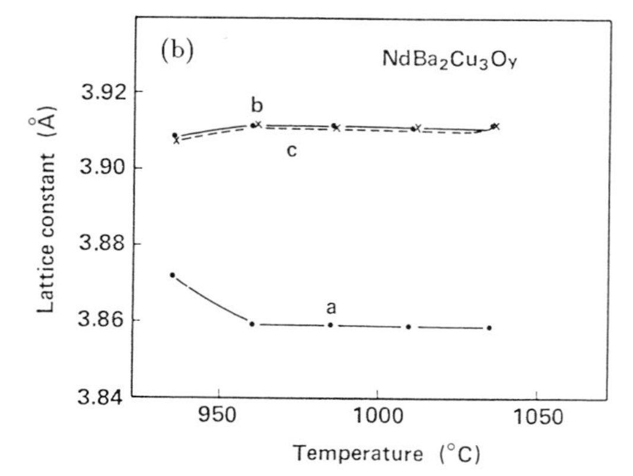

Fig. 2: Lattice constants of the sample with various annealing temperature for $LaBa_2Cu_3O_y$ (a) and $NdBa_2Cu_3O_y$ (b).

RESULTS and DISCUSSION

High temperature annealing

Figure 1 shows the temperature dependent resistivity of the sample of various annealing temperatures for $LaBa_2Cu_3O_y$ (Fig.1(a)) and $NdBa_2Cu_3O_y$ (Fig.1(b)). In the case of La compound, T_c increases gradually as the annealing temperature increases. In contrast, Nd compounds show a rapid increase in T_c as the annealing temperature increases.

Figure 2 shows lattice constants of the sample of various annealing temperatures for $LaBa_2Cu_3O_y$ (Fig.2(a)) and $NdBa_2Cu_3O_y$ (Fig.2(b)), the values were obtained by Rietveld analysis. Lattice constants a and c increase and b decreases as the annealing temperature increases. La compounds require a high temperature to reach saturated values.

These results indicate that the atomic ordering of Ln and Ba in the sample was improved by high temperature annealing. In the case of La compounds, a higher annealing temperature is required to obtain high T_c samples. Because the ion radius of La is nearer to the value of Ba than other lanthanide atoms, it is difficult to site La atoms.

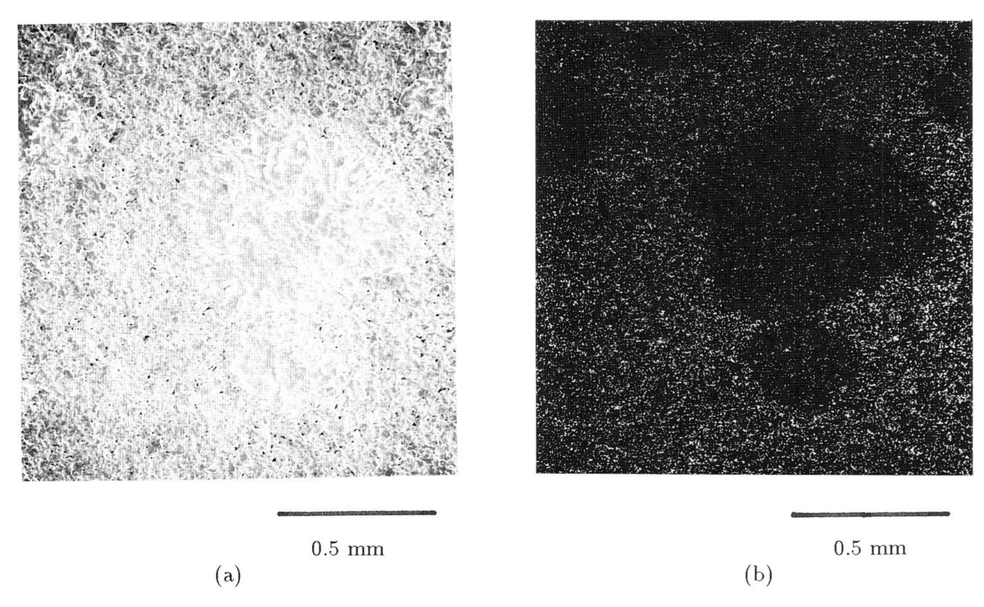

Fig. 3: A SEM photograph (a) and an X-ray (Nd Lα_1) image (b) of $NdBa_2Cu_3O_y$ annealed in O_2 atmosphere at 1035 °C.

The high temperature annealing process makes it possible to obtain high T_c samples, however, it was found that an impurity phase was produced. Figure 3 shows a SEM photograph and an X-ray (Nd Lα_1) image of $NdBa_2Cu_3O_y$ annealed at 1035 °C in O_2 atmosphere. A circular impurity phase is precipitated by high temperature annealing. The impurity phase was identified as $BaCuO_2$ by EPMA. The high resistivity of $LaBa_2Cu_3O_y$ sample annealed at 1035 °C shown in Fig.1(a) is believed to be due to the existence of the impurity phase.

Annealing in Ar atmosphere

To reduce the impurity phase, annealing in Ar atmosphere was studied. Three samples were prepared to study the effect of annealing in Ar atmosphere. The first sample was obtained by annealing at 960 °C in O_2 atmosphere (O_2). The second sample was obtained by annealing at 950 °C in Ar atmosphere after O_2 annealing (Ar). The third sample was obtained by annealing below 850 °C in O_2 atmosphere following Ar annealing described above (Ar/O_2). Figure 4 shows X-ray diffraction patterns of $LaBa_2Cu_3O_y$ annealed under the conditions described above. The sample annealed at high temperature in O_2 atmosphere (O_2), contains a large amount of $BaCuO_2$. Arrows in Fig. 4 show $BaCuO_2$ diffraction peaks. Ar annealing causes these peaks to disappear, and they do not reappear after subsequent low temperature annealing below 850 °C in O_2 atmosphere. The T_c of the sample prepared by Ar annealing followed by low temperature O_2 annealing (Ar/O_2) is as high as that of the sample prepared by high temperature annealing. This result is in good agreement with that for $LaBa_2Cu_3O_y$ annealed in N_2 atmosphere reported by T. Wada *et al.*[8] This is an effective procedure for obtaining high quality samples without an impurity phase, not only for La compounds but also for other compounds containing large ion radius lanthanide such as Nd, Sm and Eu. And the same result was obtained for those compounds.

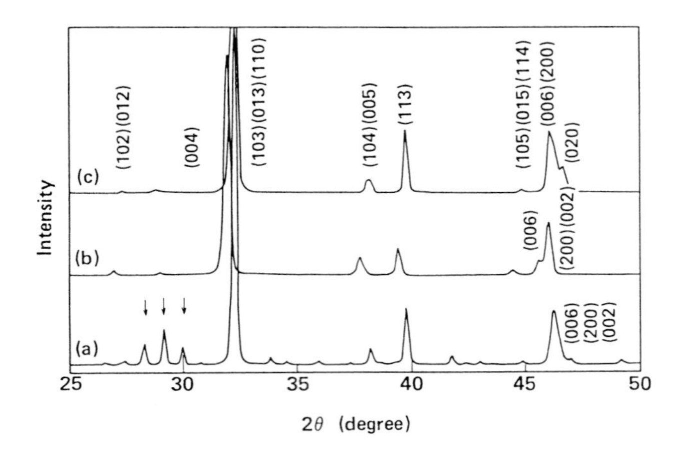

Fig. 4: X-ray diffraction patterns of $LaBa_2Cu_3O_y$ annealed in O_2, Ar, and Ar and O_2 indicated by (a), (b) and (c), respectively.

Melting temperature

Figure 5 shows melting temperature *vs.* ion radius of the lanthanide atoms of $LnBa_2Cu_3O_y$, where the melting temperature was determined by TG and DTA data. The melting temperature increases as the ion radius of the lanthanide atoms increases with the exception of $LaBa_2Cu_3O_y$. Figure 6 shows T_c *vs.* melting temperature of $LnBa_2Cu_3O_y$, where the data of Tamegai *et al.*[9] and J. M. Tarascon *et al.*[10] are also plotted with the data of the present work. It is noted that the T_c increases linearly with the increase in the melting temperature of the samples. The T_c of $LnBa_2Cu_3O_y$ (Ln = Lanthanide) changes systematically from 88 K to 96 K as the melting temperature of the sample increases. It shows that the T_c has a strong correlation with the melting temperature.

BCS theory suggests that the T_c correlates with the Debye temperature. The Debye temperature is expressed by the following equation formulated by Lindemann:[11]

$$\Theta_D \cong C \left[\frac{\kappa T_m}{A V_0^{2/3}} \right]^{1/2}$$

where Θ_D is the Debye temperature, C is a constant, A is the atomic weight, and V_0 is the atomic volume, T_m is the melting temperature.

In this equation, the atomic volume (V_0) also effects the Debye temperature. It is not appropriate to express V_0 in terms of simple cell volume in this oxygen-deficient perovskite structure. We believe that the sub-cell volume of the structure is more suitable for expressing the V_0 value.

This correlation of T_c and melting temperature indicates that the T_c is influenced by the phonon mechanism. This result qualitatively agrees with the result of the isotope effect. The authors believe that the appearance of superconductivity is at least partially due to the phonon mechanism, even if other mechanisms play a more important role.

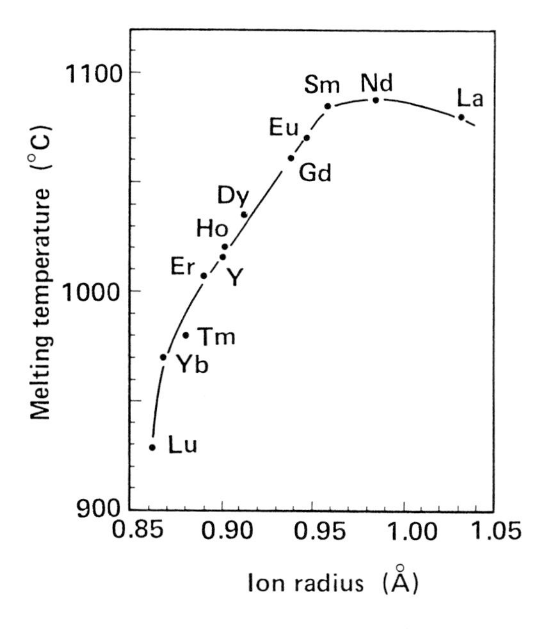

Fig. 5: Melting temperature *vs.* ion radius of the lanthanide atoms of $LnBa_2Cu_3O_y$.

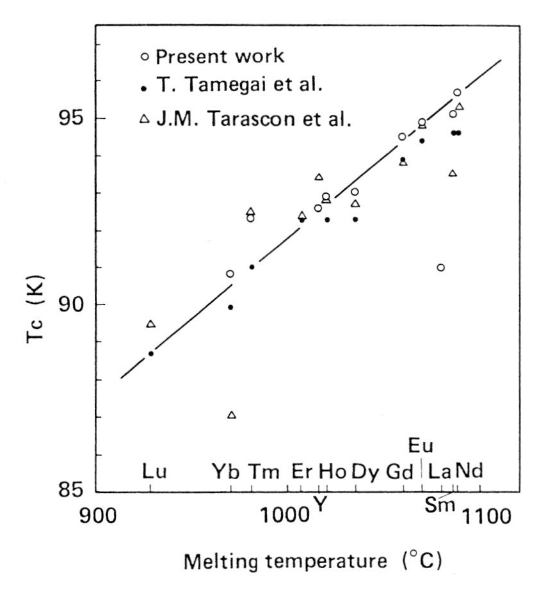

Fig. 6: T_c *vs.* melting temperature of $LnBa_2Cu_3O_y$, where the data of Tamegai *et al.* ref.[9] and J. M. Tarascon *et al.* ref.[10] are plotted with the data of the present work.

CONCLUSION

The superconducting properties of the sample with large ion radius lanthanide were improved by high temperature annealing. This result is considered to be due to improved atomic ordering in the sample. However, high temperature annealing causes the production of impurity phases such as $BaCuO_2$. Annealing in Ar atmosphere was studied to obtain single phase $LnBa_2Cu_3O_y$.

The T_c of $LnBa_2Cu_3O_y$ changes systematically from 88 K to 96 K depending on the melting temperature of the substance. The small changes of T_cs for $LnBa_2Cu_3O_y$ depend on its Debye temperatures. This result indicates that the superconductivity was partially influenced by the phonon mechanism.

The authors would like to thank Y.Katayama, A.Yamaji, T.Yamada and T.Murakami for their valuable advice and encouragement. Thanks are also due to Y.Tajima for T_c measurements.

REFERENCES

[1] for example, S.Tsurumi, T.Iwata, Y.Tajima, and M.Hikita, Jpn. J. Appl. Phys. **26**, L1865 (1987).
[2] T. Iwata, M. Hikita, Y. Tajima, and S. Tsurumi, J. Cryst. Growth **85**, 661 (1987).
[3] T. Iwata, M. Hikita, Y. Tajima, and S. Tsurumi, Jpn. J. Appl. Phys., **26**, L2049 (1987).
[4] T. Iwata, M. Hikita, and S. Tsurumi, Proc. Int. Conf. Erectorical Materials (Tokyo) 1988, to be published.
[5] S. Tsurumi, T. Iwata, Y. Tajima, and M. Hikita, Jpn. J. Appl. Phys., **27**, L397 (1988).
[6] M. Hikita, Y. Tajima, A. Katsui, Y. Hidaka, T. Iwata, and S. Tsurumi, Phys. Rev. **B 36**, 7199 (1987).
[7] F.Izumi, nihon kessyou gakkai shi, **27**, 23 (1985) [in japanese].
[8] T. Wada, N. Suzuki, T. Maeda, A. Maeda, S. Uchida, K. Uchinokura, and S. Tanaka, Appl. Phys. Lett. **52**, 1989 (1988).
[9] T. Tamegai, A. Watanabe, I. Oguro, and Y. Iye, Jpn. J. Appl. Phys. **26**, L1304 (1987)
[10] J. M. Tarascon, W. R. McKinnon, L. H. Greene, G. W. Hull, and E. M. Vogel, Phys Rev. **B 36**, 226 (1987).
[11] F. Lindemann, Phys. z., **11**, 609 (1910).

Raman Study of High-Tc Superconductor $Bi_{0.7}Pb_{0.3}SrCaCu_{1.8}O_y$

T. SUZUKI, K. HOTTA, Y. KOIKE, and H. HIROSE

Nippon Institute of Technology, 4-1, Gakuendai, Miyashiro, Minami-Saitama, Saitama, 345 Japan

ABSTRACT

High-Tc superconductor $Bi_{0.7}Pb_{0.3}SrCaCu_{1.8}O_y$ with the zero resistivity temperature Tc(end) of 110K has been synthesized by sintering method and characterized by Raman spectroscopy in the frequency range between 200cm^{-1} and 1000cm^{-1}. Two or more broad scattering have been observed at 450-700 cm^{-1} region. Wavelength-dependence of these scattering have been measured. Polarization-dependent Raman spectra of 2μmΦ area of one platelike crystal have also been measured with Raman microscope in the frequency range between 800cm^{-1} and 950cm^{-1}.

INTRODUCTION

Many studies have recently been reported on oxide superconducting materials. Maeda *et al.*(1) have discovered a new superconducting ceramic with the composition of $BiSrCaCu_2O_x$. This material was reported to have two different superconducting transition temperatures : one is about 105K and the other is about 80K. Takano *et al.*(2) have found that partial substitution of Pb for Bi in the Bi-Sr-Ca-Cu-O system increases sharply the volume fraction of the high-Tc phase. The dominance of the high-Tc phase over the low-Tc phase has been estimated at 9/1 in volume from the susceptibility data. Yamada *et al.*(3) reported the Pb substitution increases the zero resistivity temperature Tc(end) of Bi-Sr-Ca-Cu-O system. They obtained a maximum Tc(end) value of 85.5K for the composition of $Bi_{0.5}Pb_{0.5}SrCaCu_2O_y$.

Raman spectroscopy is effective to obtain information about the phonon structure, that may lead to the understanding of the mechanism of the high-Tc superconductivity. The Raman spectra of various oxide superconductors have been measured by many groups in order to study the origin of the high-Tc superconductivity(4-6). Polarization-dependence of $YBa_2Cu_3O_x$(7-9) have appeared in order to make Raman mode assignments. Sugai *et al.*(10) have investigated the dynamical spin fluctuation and lattice vibration in Bi-Sr-Ca-Cu-O single crystal with Tc=80K. They observed more than ten phonon peaks and very broad scattering which is assigned to two-magnon scattering. If micro-Raman spectroscopy is used to characterize polycrystal samples, spatial distribution of high-Tc or low-Tc phases and impurity phase might be identified. However it is well known that when measuring sintered samples, signals from the impurity phases at the grain boundaries often appear and disturb the reliability of the data obtained by macroscopic Raman spectroscopy.

In this study, high-Tc oxide superconductor $Bi_{0.7}Pb_{0.3}SrCaCu_{1.8}O_y$ with Tc(end)=110K was synthesized by sintering method and characterized by micro-Raman spectroscopy in the frequency range between 200 and 1000 cm^{-1} in order to identify high-Tc and low-Tc phases spatially.

EXPERIMENTAL

Samples were synthesized by sintering method. Fine powders of Bi_2O_3, PbO, $SrCO_3$, $CaCO_3$ and CuO (all of them with 99.9% purity) were mixed in the metal atom ratio of Bi : Pb : Sr : Ca : Cu = 0.7 : 0.3 : 1 : 1 : 1.8 . The mixture was pressed into 5g pellets with 24mm diameter and 3mm thickness. The pellets were first calcinated at 800℃ for 12 hours in air. Then the specimens were reground, pulverized and pressed into pellets with 10mm diameter and 2mm thickness and sintered at 830℃ for 244 hours in air. Resistivity vs. temperature curves were measured by a standard four-probe method.

Raman scattering was measured with excitation wavelength of 488nm and 514.5 nm of Ar^+ laser (SP:168-09) and 647.1nm of Kr^+ laser (SP:168-11). Measurements were done both with the Raman microscope (backscattering) and in the macrochamber (grazing incidence) at room temperature. The laser power was less than 10mW in the macrochamber and less than 0.5 mW(2μmΦ) in the microscope. A double-monochromator (J/Y:U-1000) with a conventional photon counting system (Hamamatsu: R943-02 + C-1230) was used for the analysis and detection of the scattering light.

RESULTS AND DISCUSSION

Figure 1 shows the temperature dependence of resistivity of $Bi_{0.7}Pb_{0.3}SrCaCu_{1.8}O_y$. Resistivity decreases almost linearly with decreasing of temperature but deviates below 145K, dropping to zero at 110K. This curve indicates that slight higher-Tc phase whose onset temperature (Tc(onset)) is about 145K and dominant lower-Tc phase of Tc(onset)~125K are involved in the sample. But the Tc(onset) of the lower-Tc phase (125K) obtained here is almost the same as that of high-Tc phase that Takano *et al.*(2) reported. So, the higher-Tc phase of Tc(onset)=145K might be a new phase.

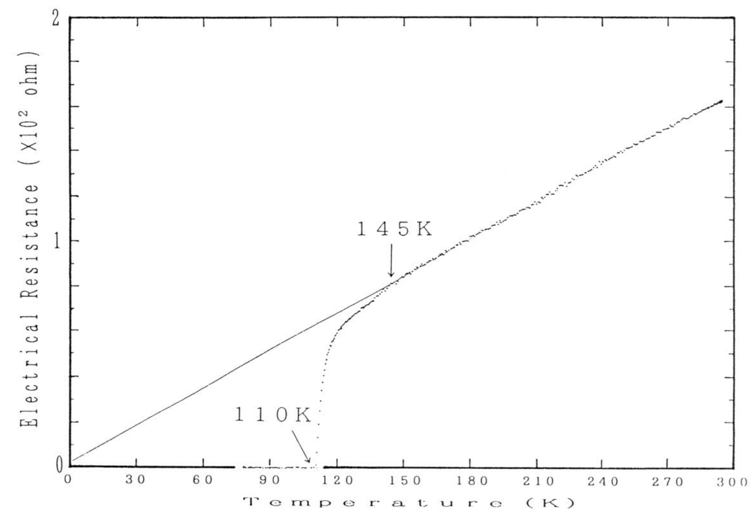

Fig. 1 Temperature dependence of the electrical resistance of $Bi_{0.7}Pb_{0.3}SrCaCu_{1.8}O_y$.

Figure 2 shows the Raman spectra measured in the macrochamber in the 90° scattering configuration with excitation wavelength of 488, 514.5 and 647.1 nm. No sharp phonon peak is observed in the frequency range between 200 cm^{-1} and 1000 cm^{-1} different from the Bi-Sr-Ca-Cu-O (10). Two or more broad scattering are observed in the range between 450 cm^{-1} and 700 cm^{-1}, showing the wavelength-dependence of the incident laser light. Mode of these scattering is now under examination.

Then the sample was characterized by Raman microscope. The Raman system used here is equipped with a CCD TV camera to monitor the measuring position of the sample surface and the condition of incident laser light. Figure 3 shows the photograph of the monitor screen showing the surface of sintered $Bi_{0.7}Pb_{0.3}SrCaCu_{1.8}O_y$ sample. Platelike crystals same as in (Bi,Pb)-Sr-Ca-Cu-O system (2) and Bi-Sr-Ca-Cu-O system (11) are observed. Measured area is the center of the cross

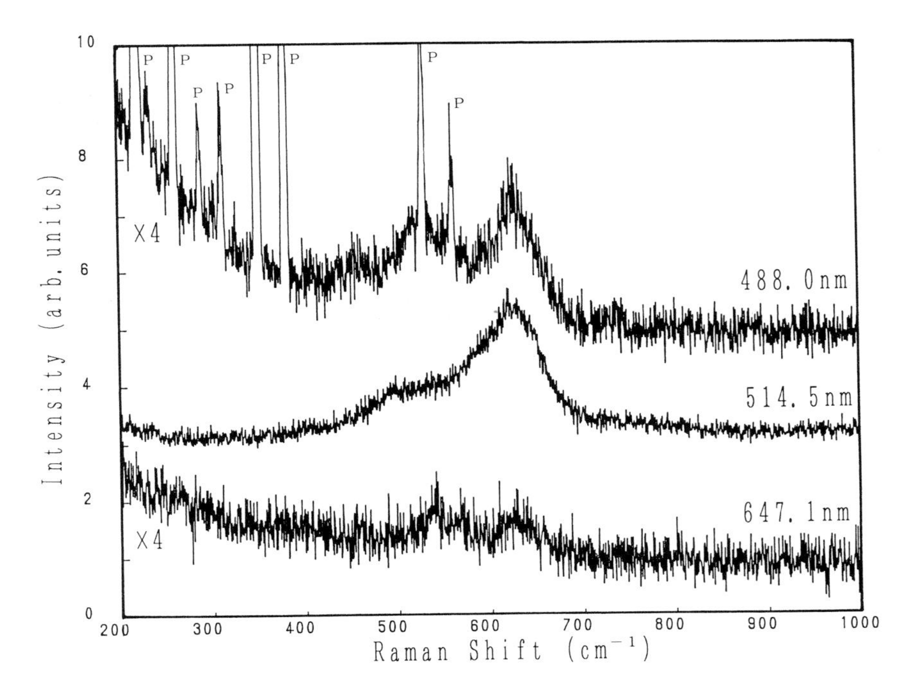

Fig. 2 Raman spectra of $Bi_{0.7}Pb_{0.3}SrCaCu_{1.8}O_y$ in the range between $200cm^{-1}$ and $1000cm^{-1}$ measured with different wavelength of incident laser light. The intense and sharp lines indicated with figure "P" are plasma lines of laser light.

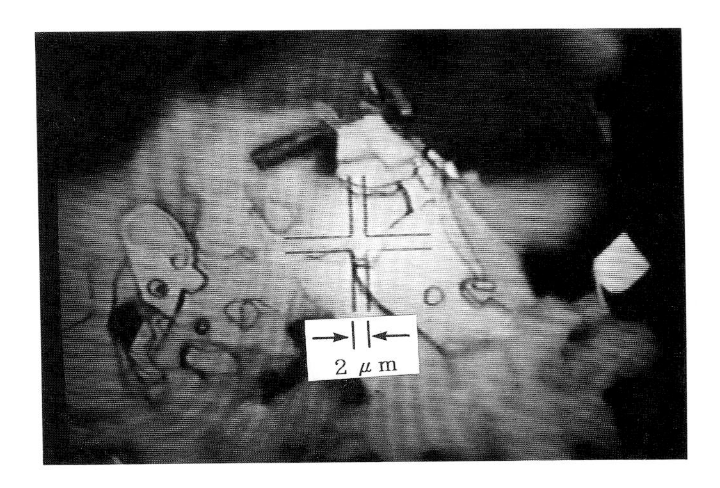

Fig. 3 An optical microscope image of $Bi_{0.7}Pb_{0.3}SrCaCu_{1.8}O_y$ observed by monitor TV camera equipped in the Raman microscope.

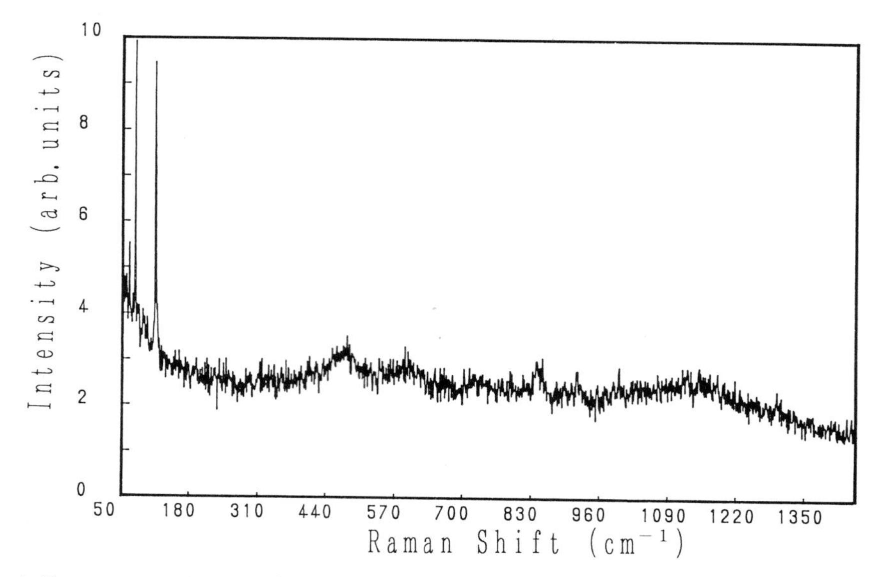

Fig. 4 Raman spectrum of $Bi_{0.7}Pb_{0.3}SrCaCu_{1.8}O_y$ measured with Raman microscope in the backscattering configuration.
Analyzer was not used and the wavelength of incident laser light was 514.5nm.

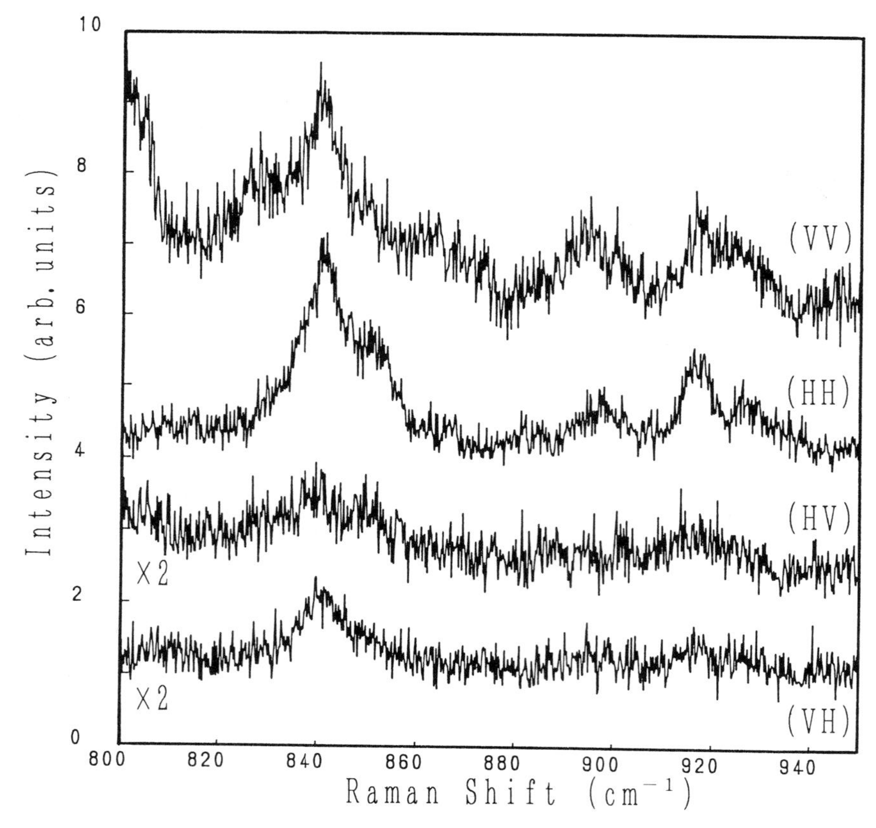

Fig. 5 Polarization-dependent Raman spectra of $Bi_{0.7}Pb_{0.3}SrCaCu_{1.8}O_y$ measured with 514.5nm line.
Insert is an arrangement of the measurement.

slit indicated in the center of the photograph. In this case objective lens of X50 was used, so the width of the two lines of the cross slit is 2 μm and the laser light was focused on the sample surface with a spot size of 2μm diameter. Various surface positions of the sample shown in Fig. 3 such as at grain boundaries or at each platelike crystals were characterized by micro-Raman spectroscopy. Shapes of spectra changed position to position same as in $YBa_2Cu_3O_{7+y}$ case(12). But to date the identification of high-Tc and low-Tc phases from micro-Raman spectroscopy was not obtained clearly. Typical microscopic Raman spectrum measured with 514.5nm line is shown in Fig. 4. This spectrum is rather different from that obtained in the macrochamber (Fig. 2). We think this difference is due to the presence of the effect of impurity phase and boundaries.

Polarization-dependent Raman spectra of the crystal shown in Fig. 3 were measured in the range between 800cm^{-1} and 950cm^{-1} as shown in Fig. 5. Here, insert is an arrangement of the measurement, (VH) for example, denotes that the direction of incident beam is polarized in the vertical plane and that of scattered beam is polarized in the horizontal plane. Two peaks in the 840 cm^{-1} and 918cm^{-1} regions are observed for all geometries. A peak in the 850 cm^{-1} is observed for the (HH) and (HV) geometries. A peak in the 892 cm^{-1} is observed for the (VV) and (HH) geometries. Further information will be obtained from the more careful and elaborate measurements using single crystals.

SUMMARY

$Bi_{0.7}Pb_{0.3}SrCaCu_{1.8}O_y$:a partially substituted $BiSrCaCu_2O_x$ by Pb for Bi was synthesized by sintering method and characterized by Raman spectroscopy. Temperature dependence of resistivity showed that the Tc(end) is 110K. It also indicated the possibility of new higher-Tc phase of Tc(onset)=145 K in its behavior. Two or more broad scattering observed at 600 cm^{-1} region showed the wavelength-dependence of the incident laser light. Polarization-dependent Raman spectra from one platelike crystal showed that there are several kinds of polarization-dependence. Two peaks in the 840 cm^{-1} and 918 cm^{-1} show no polarization-dependence. But peaks in the 850 cm^{-1} and 892 cm^{-1} show different kind of polarization-dependence each other.

Micro-Raman measurement was attempted to identify high-Tc and low-Tc phases but clear results could not be obtained so far. We think micro-Raman spectroscopy will be effective for understanding physics of superconductivity if the related and necessary information be integrated to some extent.

ACKNOWLEDGMENT

The authors wish to thank Dr. K. Endo of Electrotechnical Lab. (ETL) for the measurement of resistivity.

REFERENCES

1) H. Maeda, Y. Tanaka, M. Fukutomi and T. Asano: Jpn. J. Appl. Phys. 27 (1988) L209.
2) M. Takano, J. Takada, K. Oda, H. Kitaguchi, Y. Miura, Y. Ikeda, Y. Tomii and Mazaki: Jpn. J. Appl. Phys. 27 (1988) L1041.
3) Y. Yamada and S. Murase: Jpn. J. Appl. Phys. 27 (1988) L996.
4) (for Y-Ba-Cu-O:see for example)
H. Rosen. E. M. Engler, T. C. Strand, V. Y. Lee and D. Bethune: Phys. Rev. B **36** (1987) 726.
M. Stavola, D. M. Krol, W. Weber, S. A. Sunshine, A. Jayaraman, G. K. Kourouklis, R. J. Cava and E. A. Rietman: Phys. Rev. B **36** (1987) 850.
S. Nakashima, M. Hangyo, K. Mizoguchi, A. Fujii, A. Mitsuishi and Yotsuya: Jpn. J. Appl. Phys. **26** (1987) L1794.

5) (for La-Cu-O:see for example)
G. A. Kourouklis, A. J. Jayaraman, W. Weber, J. P. Remika, G. P. Espinosa, A. S. Cooper and R. G. Maines, Sr. : Phys. Rev. B **36** (1987) 7218.

6) (for La-Sr-Cu-O:see for example)
S. Sugai, M. Satou and S. Hosoya: Jpn. J. Appl. Phys. **26** (1987) L495.
S. Blumenroeder, E. Zirngiebl, J. D. Thompson, P. Killough, J. L. Smith and Z. Fisk: Phys. Rev. B **35** (1987) 8840.

7) R. Liu, C. Thomsen, W. Kress, M. Cardona, B. Gegenheimer, F. W. de Wette, J. Prade, A. D. Kulkarni and U. Schroder: Phys. Rev. B **37** (1988) 7971.

8) C. Thomsen, M. Cardona, B. Gegenheimer, R. Liu and A. Simon: Phys. Rev. **37** (1988) 9860.

9) R. Nishitani, N. Yoshida, Y. Sasaki and Y. Nishina: Jpn. J. Appl. Phys. **27** (1988) L1284.

10) S. Sugai, H. Takagi, S. Uchida and S. Tanaka: Jpn. J. Appl. Phys. **27** (1988) L1290.

11) S. Adachi, O. Inoue and S. Kawashima: Jpn. J. Appl. Phys. **27** (1988) L344.

12) M. P. Fontana, B. Rosi, D. H. Shen, T. S. Ning and C. X. Liu: Phys. Rev. B **38** (1988) 780.

Luminescence Study of New High T_c Oxide Superconductors

YASUFUMI FUJIWARA, MASAYOSHI TONOUCHI, and TAKESHI KOBAYASHI

Faculty of Engineering Science, Osaka University, 1-1, Machikaneyama, Toyonaka, Osaka, 560 Japan

ABSTRACT

We have succeeded in the observation of a characteristic x-ray excited thermally stimulated luminescence (TSL) from new high T_c superconducting systems including no rare-earth elements, e.g. Tl-Ba-Ca-Cu-O and Bi-Sr-Ca-Cu-O. It has been found that the TSL glow curve of the Tl system significantly differs from that of the Bi system, though they share crystallographic properties in common. In TlBaCaCuO with the nominal composition of Tl:Ba:Ca:Cu = 2:2:2:3, exhibiting superconducting transition at 97 K, distinct four glow peaks have been observed, which suggests that some traps exist in this system. Their peak wavelength ranges from ultraviolet to visible region. The thermal activation energies of traps related to the glow peaks are from 0.1 eV and 0.3 eV. The cumulative x-ray irradiation time dependence of their glow peaks has been systematically investigated together with the nominal composition dependence. On the other hand, in $BiSrCaCu_2O$, which consists of high and low T_c phases, one broad and fragile TSL glow peak has been observed around 115 K.

INTRODUCTION

Since the discovery of superconductivity in $YBa_2Cu_3O_7$, the superconducting transition temperature T_c being above liquid-nitrogen temperature, [1] much attention has been paid to the physics and application. In searching other new oxide superconductors exhibiting an even higher T_c, following the recent discovery of high T_c superconductivity in the Bi-Sr-Ca-Cu-O system by Maeda et al [2], Hermann et al have reported superconducting transitions in a Tl-Ba-Ca-Cu oxide with onset temperature of up to 120 K and with zero temperature below 107 K [3]. Quite recently, Parkin et al have obtained bulk superconductivity with a transition to zero resistivity at 125 K in $Tl_2Ba_2Ca_2Cu_3O$, depending on preparation conditions. [4]

We have systematically investigated high T_c superconductors by photo- (PL) and thermally stimulated luminescence (TSL) techniques. [5-7] Previously, we reported the first observation of characteristic PLs from the Er-Ba-Cu-O and Nd-Ba-Cu-O systems, which consisted of a series of sharp emission lines. [5,6] They were tentatively attributed to electronic transitions between weakly crystal-field split spin-orbit levels of the trivalent lanthanide ions, reflecting a local environment of the ion site. TSL measurements were also performed with a particular emphasis on the Gd-Ba-Cu-O system. [7] In this system, some kinds of visible TSL glow curves were obtained, which significantly depended on Gd composition, sample preparation conditions and cumulative x-ray irradiation time. The thermal activation energy of the related traps ranged from 0.1 eV to 0.3 eV. Successive luminescence studies including those from various laboratories in the world [8-14] suggest us the following two feasibilities. One is we can use the luminescence signal as a powerful characterization probe for a high T_c superconductor, which has been successfully performed in a semiconductor field. The other is we can ultimately develop a new-concept device with both of superconducting and light-emitting functions.

In this paper, we report the first observation of characteristic TSLs from the Tl-Ba-Ca-Cu-O and Bi-Sr-Ca-Cu-O systems. These preliminary results suggest the effectiveness of this luminescence characterization technique for the new high T_c superconductors without any rare-earth elements.

Table I. Specifications and superconducting properties of the Tl-Ba-Ca-Cu-O samples used in this study.

Sample	Initial ratio of Tl:Ba:Ca:Cu	Zero resistivity temperature (K)	Volume fraction of superconducting phase at 50 K (%)
A	2:2:2:3	97	50
B	3:2:2:3	80	5

EXPERIMENTAL

Sample Preparation

The samples used in the present work were prepared by the conventional solid state reaction method with appropriate mixtures of standard starting powders. The mixture was pressed into 0.5-inch pellets and processed under proper annealing conditions. The details will be published elsewhere. Table I shows the specifications and superconducting properties of the Tl-Ba-Ca-Cu-O samples used in this work.

TSL Measurement

Prior to TSL measurements, the sample was mounted on a cold finger of a cryostat by a silver paste and cooled down to liquid-nitrogen temperature. The x-ray from a Cu target was used as an excitation source. The x-ray exposure to the sample was performed by a commercially available x-ray diffraction equipment operating at 30 kV and 10 mA. After the x-ray irradiation, the sample was heated by a neighboring heater and the temperature was monitored by a chart recorder. A typical heating rate was about 0.1 K/sec in this work. The luminescence from the sample was detected by a photomultiplier with a detectivity of a visible region and processed by a lock-in system. The block diagram of the TSL measurement system is shown in Fig. 1.

RESULTS AND DISCUSSION

Tl-Ba-Ca-Cu-O System

We have observed characteristic TSLs from two samples studied in the present work. Figure 2 shows a typical TSL glow curve obtained in sample A. A similar glow curve was also obtained in sample B, though the intensity was about 35 times lower than that of sample A, which might be related to a volume fraction of the superconducting phase listed in Tab. I. Distinct four glow peaks, being denoted from A to D in order from the lowest temperature in the figure, can be observed, which suggests that some traps exist in this system. Furthermore, it should be noticed that the TSL glow curve quite differs from those previously reported in the Gd-Ba-Cu-O system [7].

These TSL glow curves were analyzed by a well-known general-order kinetic equation describing the TSL intensity I as a function of temperature T as follows [15];

$$I = sn_0\exp(-E/k_BT)\left[[(l-1)s/\beta]\int_{T_0}^{T}\exp(-E/k_BT')dT'+1\right]^{-l/(l-1)} \quad (1)$$

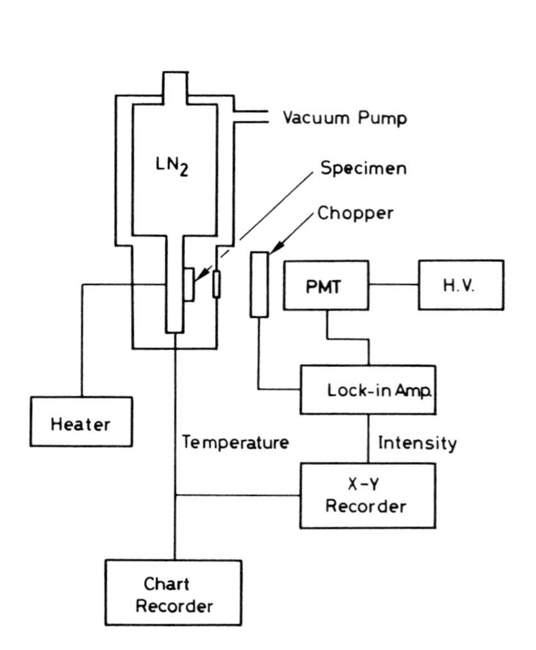

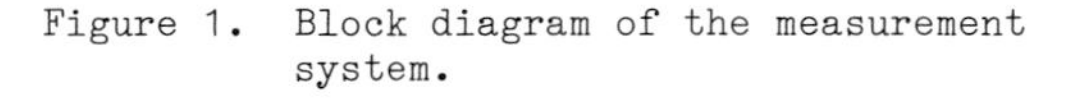
Figure 1. Block diagram of the measurement system.

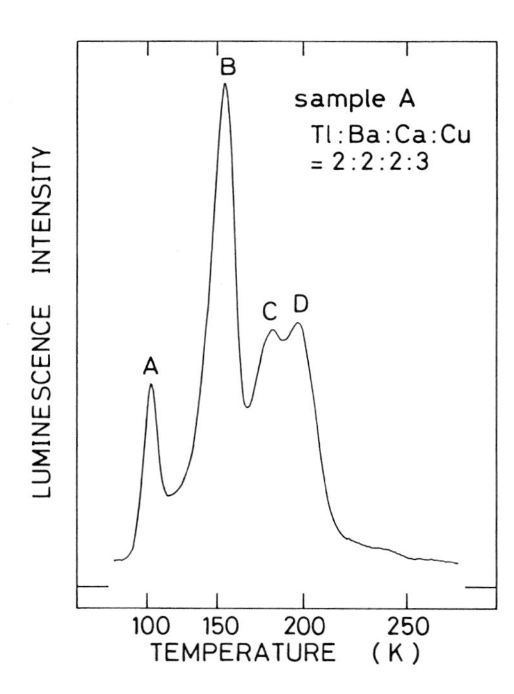

Figure 2. Typical TSL glow curve from the Tl-Ba-Ca-Cu-O system.

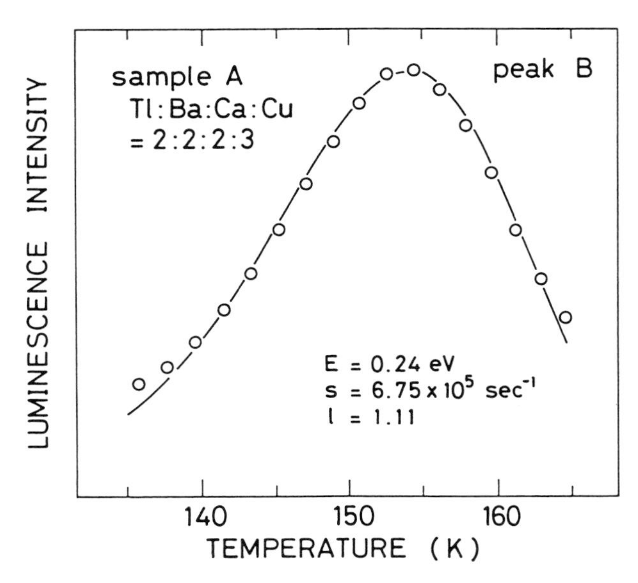

Figure 3. Least-squares fit of eq. (1) to glow peak B observed in sample A. Open circles show the experimental results and the solid line the calculated one.

Here K_B is Bolzmann's factor, n_0 the concentration of trapped charges at $T = T_0$ and $t = 0$.   is the heating rate, which is experimentally obtained and is about 0.1 K/sec in this work. E, s and l are parameters characterizing a related trap and are called thermal activation energy, frequency factor and kinetic order, respectively. These three parameters can be estimated by fitting eq. (1) to an experimentally available glow peak. An example of the least-squares fitting to peak B obtained in sample A is shown in Fig. 3. Open circles display the experimental result and the solid line the calculated one. The resultant E, s and l are 0.24 eV, 6.75×10^5 sec^{-1} and 1.11, respectively. These parameters for traps related to other glow peaks can be determined by a similar procedure. The results are listed in Table II. As peaks C and D superimposed each other and, therefore, peak C could never been separated from peak D by any techniques, the estimation for the three parameters for peak C was not successfully performed. As can be seen in the table, the thermal activation energies of the traps are from 0.1 eV to 0.3 eV. It is interesting to note that the activation energies are comparable to those previously obtained in the Gd-Ba-Cu-O system [7].

The wavelength dependence of the TSL glow peaks has been investigated to obtain further informations about the recombination mechanism, e.g. position of the recombination center, recombination path and so on. Figure 4 shows the preliminary result obtained by using appropriate band-pass and high-pass optical filters. The vertical axis implies the intensities of peak A, C and D normalized by that of peak B to avoid extrinsic effects such as transmittance characteristics of the used filter and optical path. An error bar displays the wavelength at the transmittance of 50 % against the peak transmittance of the filter. Taking it into consideration that the intensity of peak D measured by using the high-pass filter with the cut-off wavelength of 0.62 μm was about three times lower than that with the cut-off wavelength of 0.58 μm, the peak wavelength of these peaks ranges from ultraviolet to visible region. Furthermore, the normalized intensity individually depends on the wavelength, suggesting that their luminescence spectra differ from one another. In addition, the spectral regions of peak B (C), A and D are arranged in order from short wavelength. These observations might reflect a difference in the recombination process and/or a multi-phase property.

Table II. Features of traps related to the glow peaks shown in Fig. 2.

peak	Activation energy (eV)	Frequency factor (sec^{-1})	kinetic order
A	0.12	1.12×10^4	1.21
B	0.24	6.75×10^5	1.11
C	-	-	-
D	0.27	5.91×10^4	1.11

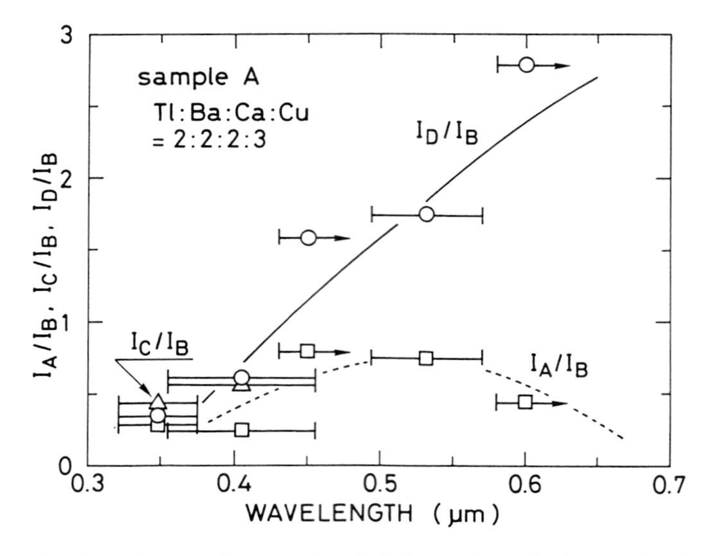

Figure 4. Wavelength dependence of peak intensities of glow peaks A, C and D normalized by that of glow peak B observed in sample A.

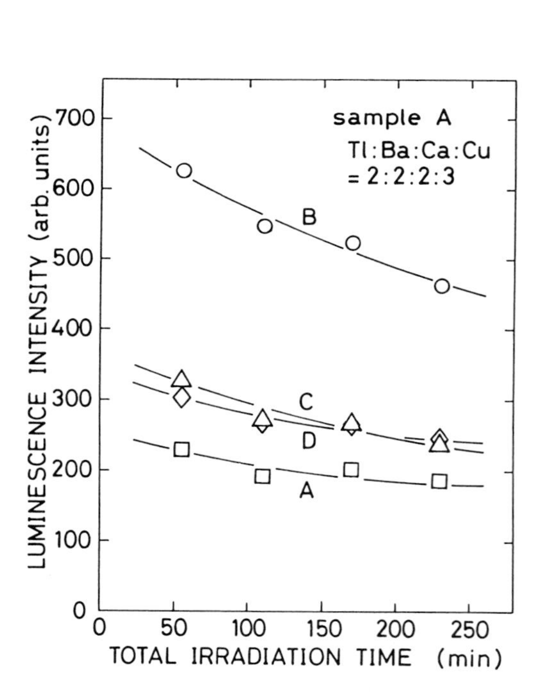

Figure 5. Cumulative x-ray irradiation time dependence of the peak intensities of glow peaks observed in sample A.

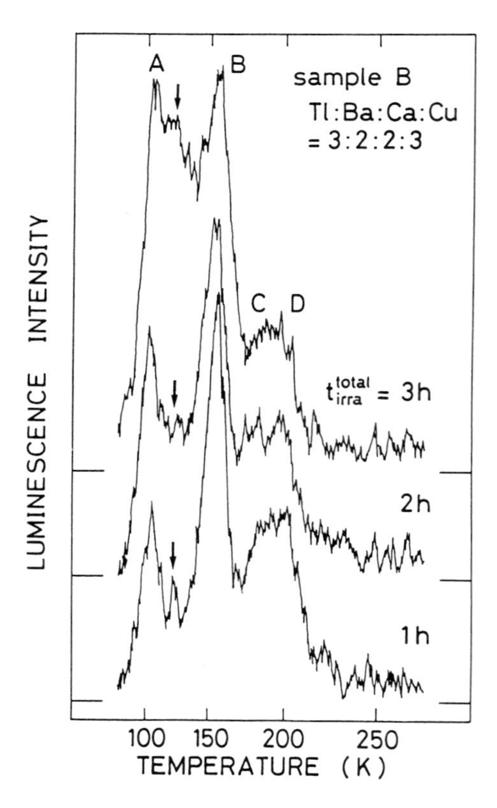

Figure 6. Cumulative x-ray irradiation time dependence of the glow curve observed in sample B. With increasing the irradiation time, one new glow peak appears and grows, which is shown by an arrow.

As previously reported, The glow curves from the Gd-Ba-Cu-O system significantly depended on the cumulative x-ray irradiation time. [7] Though the origin is not clear at present, the investigation will give us a clue to clarify the mechanism of annihilation and creation of the traps. In sample A, with the increase in cumulative x-ray irradiation time, the intensities of the TSL glow peaks gradually decrease without the appearance of any new glow peaks. The behavior is shown in Fig. 5. On the other hand, one new glow peak appears and grows around 120 K, which is shown by an arrow in Fig. 6, with increasing cumulative x-ray irradiation time, though other glow peaks similarly behave as described in sample A. Furthermore, if we store the long irradiated sample B for about 15 hours at room temperature, the new glow peak completely disappears and displays a similar behavior as mentioned before for the successive x-ray irradiation, suggesting that though the trap related to the glow peak is induced by x-ray irradiation, it is metastable and is easily relaxed by a room-temperature annealing. A similar cumulative x-ray irradiation dependence of a glow peak was observed in the Gd-Ba-Cu-O system with a certain Gd composition. [7]

Bi-Sr-Ca-Cu-O System

A characteristic TSL glow curve has also been obtained in the Bi-Sr-Ca-Cu-O system. Figure 7 shows the typical TSL glow curve. The inset implies the temperature dependence of sample resistance, indicating that the sample consists of well-known high and low T_c phases [2]. A broad and fragile glow peak can be observed around 115 K, which is quite different from that in the Tl-Ba-Ca-Cu-O system, though they partly share crystallographic properties in common. The least-squares fit of eq. (1) to the glow peak has not successfully performed at present, which suggests that the glow peak might not be due to a single trap but to a few of traps. Further study is now in progress and the details will be described elsewhere.

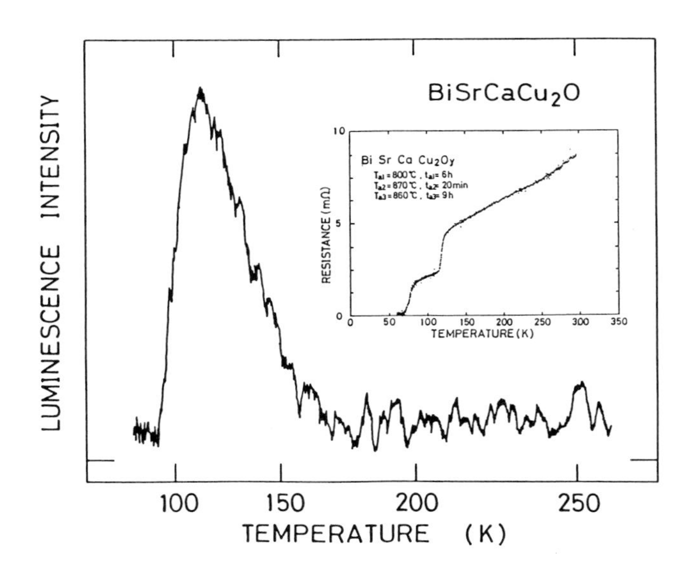

Figure 7. Typical TSL glow curve from the Bi-Sr-Ca-Cu-O system.

CONCLUSION

A characteristic TSL has been successfully observed in new high T_c superconducting systems without any rare-earth elements, e.g. Tl-Ba-Ca-Cu-O and Bi-Sr-Ca-Cu-O. In the Tl-Ba-Ca-Cu-O system, distinct four glow peaks have been obtained, whose spectral region ranges from ultraviolet to visible one. The thermal activation energies of the related traps are from 0.1 eV to 0.3 eV. The cumulative x-ray irradiation time dependence of the glow peaks has been investigated and discussed. In the Bi-Sr-Ca-Cu-O system, one broad and fragile glow peak has been obtained.

ACKNOWLEDGMENTS

This work was partly supported by Science Research Grant-in-Aid from the Ministry of Education, Science and Culture of Japan.

REFERENCES

[1]M.K.Wu,J.R. Ashburn, C. J. Torng, P. H. Hor, R. L. Meng, L. Gao, Z. J. Huang, Y. Q. Wang, and C. W. Chu, Phys. Rev. Lett. **58**, 908 (1987).
[2]H. Maeda, Y. Tanaka, M. Fukutomi, and T. Asano, Jpn. J. Appl. Phys. **27**, L209 (1988).
[3]Z. Z. Sheng and A. M. Hermann, Nature **332**, 138 (1988).
[4]S. S. P. Parkin, V. Y. Lee, E. M. Engler, A. I. Nazzal, T. C. Huang, G. Gorman, R. Savoy, and R. Beyers, Phys. Rev. Lett. **60**, 2539 (1988).
[5]Y. Fujiwara, T. Takahashi, Y. Fukumoto, M. Tonouchi, and T. Kobayashi, Proceedings of the 18th International Conference on Low Temperature Physics, August 20-26, 1987. Jpn. J. Appl. Phys. **26**, Suppl. 26-3, 2123 (1987).
[6]Y. Fujiwara, T. Takahashi, Y. Fukumoto, and T. Kobayashi, Extended Abstracts of 1987 International Superconductivity Electronics Conference, August 28-29, 1987, 401 (Japan Society of Applied Physics, 1987).
[7]Y. Fujiwara, T. Nishino and T. Kobayashi, Proceedings of 1988 MRS International Meeting on Advanced Materials, May 30 - June 3, 1988 (in press).
[8]B. M. Tissue and J. C. Wright, J. Luminescence **37**, 117 (1987).
[9]B. M. Tissue and J. C. Wright, J. Luminescence **40&41**, 313 (1988).
[10]D. W. Cooke, H. Rempp, Z. Fisk, and J. L. Smith, Phys. Rev. Lett. **36**, 2287 (1987).
[11]M. Roth, A. Halperin, and S. Katz, Solid State Commun. **67**, 105 (1988).
[12]V. N. Andreev, B. P. Zakharchenya, S. E. Nikitin, F. A. Chudnovskii, E. B. Shadin, and E. M. Sher, JEPT Lett. **46**, 492 (1987).
[13]B. J. Luff, P. D. Townsend, and J. Osborne, J. Phys. D: Appl. Phys. **21**, 663 (1988).
[14]S. H. Pawar, H. T. Lokhande, C. D. Lokhande, R. N. Patil, B. Jayaram, S. K. Agarwal, A. Gupta, and A. V. Narlikar, Solid State Commun. **67**, 47 (1988).
[15]R. Chen and Y. Kirsh, Analysis of Thermally Stimulated Processes (Pergamon, Oxford, 1981) p. 10.

Consideration of Interaction Between Carriers and Magnetic Rare-Earth Ions in $R_1Ba_2Cu_3O_y$ (R=Rare-Earth Ions) Based on ESR and Resistivity Studies

KAZUSHI SUGAWARA

Central Research Laboratory, Sharp Corporation, Ichinomoto-cho, Tenri, Nara, 632 Japan

ABSTRACT

With regard to the magnetic interaction between magnetic rare-earth ion and conduction carriers in $R_1Ba_2Cu_3O_y$, theoretical and experimental aspects of ESR , inelastic neutron scattering cross-section and normal-state resistivity of RBaCuO system have been examined. ESR of Gd diluted in YBaCuO gave $J_{ex}N_F \simeq 0.001$, which is much smaller than the value obtained by inelastic neutron scattering cross-section for ErBaCuO. Some comments are made on the interpretation of neutron scattering cross-section. The magnitude of magnetic resistivity, R_m, due to magnetic rare-earth ions was theoretically estimated using the value of $J_{ex}N_F$, giving $R_m \simeq 10^{-4}$ μΩ-cm. This indicates that the normal-state resistivity of RBaCuO is dominated by phonon contribution.

INTRODUCTION

It is generally accepted that the Cu-O planes and /or Cu-O linear chains play an important role for the superconductivity of $R_1Ba_2Cu_3O_y$, where R is a rare-earth element. As is well known, the R ions are crystallographically sandwiched between Cu-O planes (see Fig.1)[1]. It is well known that magnetic moment of the R site does not change the superconducting phase-transition temperature T_c considerably, suggesting a weak (probably very weak) coupling between R ions and superconducting electrons in these systems of compounds. From the temperature dependence of ESR line-width of Gd partially substituted for Y in YBaCuO, Sugawara et al.[2] concluded that the magnetic interaction of the type of $J_{ex}\vec{J}\cdot\vec{s}$, is negligibly small in the YBaCuO system, where $\vec{J}$ and $\vec{s}$ are the total angular momentum quantum number of R ion and spin of conduction carrier, respectively, and Jex is the coupling coefficient between R ion and carrier. Contrary to this, inelastic neutron scattering experiments for ErBaCuO done by Walter et al. [3] revealed that there exists considerable magnetic interaction between Er ion and conduction carrier, giving unexpectedly high value for $J_{ex}N_F$ ($\simeq$0.025) Here, N_F is the density of states at the Fermi level in the units of eV^{-1} $atom^{-1}$ $spin^{-1}$. For these reasons, it may be valuable to reexamine the magnetic interaction between R ion and conduction carrier. For doing this, in this study, ESR and inelastic neutron scattering cross-section have been examined on the basis of theory and experiments. As will be discussed later, other useful probe to test the magnetic interaction, like s-f exchange coupling, in the RBaCuO system is the temperature dependent behavior of resistivity in the normal-state. This is, in principle, particularly useful for RBaCuO in which magnetic R ion has a substantial crystal-field splitting.

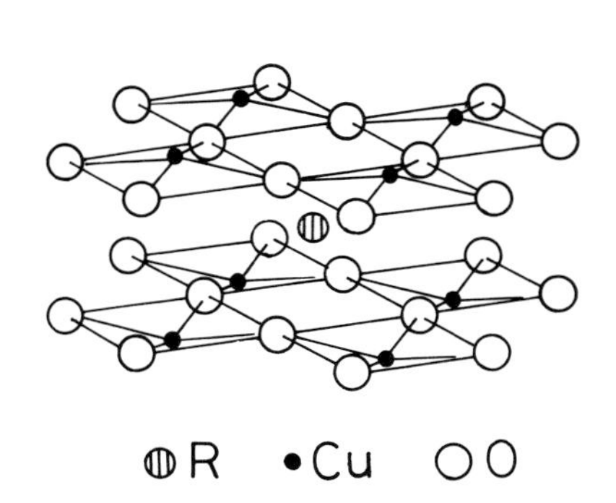

Fig.1 Part of crystal structure of RBaCuO system.

EXPERIMENTAL

Following the conventional powder technique, ceramic samples of $Y_{0.9}Gd_{0.1}Ba_2Cu_3O_y$ were made. Samples with nominal composition $Y_{0.9}Gd_{0.1}Ba_2Cu_3O_y$ were prepared by sintering mixed powders of Y_2O_3, Gd_2O_3, $BaCO_3$ and CuO in appropriate concentrations. The powders were first heated for about 10h in air at about 900°C. The thoroughly reacted samples were then ground, pressed into pellets, heated in an oxygen atmosphere at about 950°C for 10h and slowly cooled to room temperature. The onset and zero-resistance temperatures were about 90-95K and 85-87K, respectively. Using a conventional ESR spectrometer operating at about 9GHz, the peak-to-peak ESR linewidth of Gd was studied at temperatures between about 100K and room temperature. Several samples of $Er_1Ba_2Cu_3O_y$ and $Nd_1Ba_2Cu_3O_y$ were also made in a similar manner, and their resistivities were measured against temperature with particular care.

RESULTS AND DISCUSSION

ESR studies

A. Review of ESR linewidth theory

ESR linewidth ΔH of localized magnetic moment in metal is, in general, proportional to temperature, T, known as the so-called Korringa broadening. This broadening originates from magnetic interaction, of the type $J_{ex}\vec{J}\cdot\vec{s}$ for example, between localized moment and conduction electrons and is simplified as

$$\Delta H \simeq 9.3 \times 10^4 \, [J_{ex} N_F]^2 \cdot T \qquad (1)$$

Here, physical quantities ΔH, J_{ex} and T are measured in the units of Gauss, eV and Kelvin, respectively. It is important to note that Eq.(1) is valid only for magnetic system in which the so-called bottlenecking factor $B \equiv T_{sd}/T_{sL}$ is much greater than unity [4, 5]. The temperature gradient $\partial(\Delta H)/\partial T$ is dependent on B: progressively depressed with decreasing B. In the adiabatic limit,in which the factor B satisfies $B \ll 1$, $\partial(\Delta H)/\partial T$ is vanishingly small [4,5]. The limiting case, in which $B \gg 1$, is called as " isothermal limit ". In order to obtain information about the magnetic interaction between localized moment and conduction carriers based on Eq.(1), it is needed to know as to whether a magnetic system is in the isothermal limit or in the adiabatic limit. The difference between isothermal and adiabatic limits can be qualitatively understood from Fig. 2, in which the vertical axis represents the spectral density of localized magnetic moments and conduction electrons.

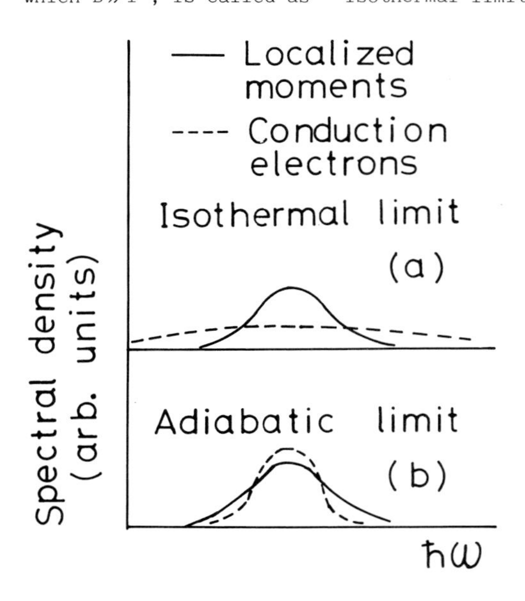

Fig.2 Qualitative representation of magnetic fluctuation spectra in the (a) isothermal limit and in the (b) adiabatic limit. The horizontal axis represents energy.

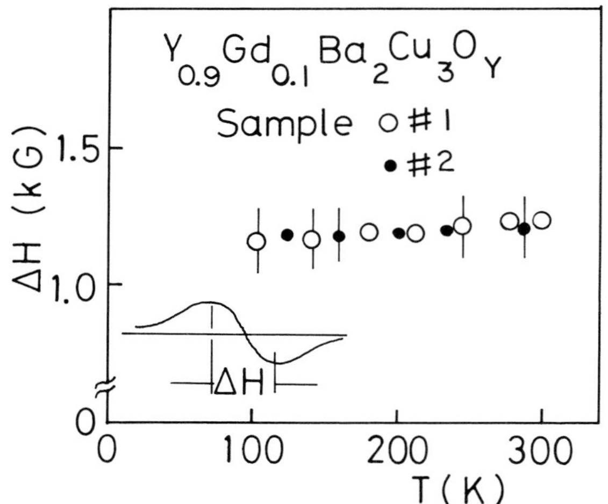

Fig.3 ESR linewidth of Gd in $Y_{0.9}Gd_{0.1}Ba_2Cu_3O_y$.

B. ESR of Gd in YGdBaCuO

There are several reports on the ESR of Gd^{3+} in RBaCuO system. The temperature dependence of Gd ESR linewidth in GdBaCuO done by Mehran et al. [6] gives $\partial(\Delta H)/\partial T \simeq 1$ Gauss/K above 100 K. Similar result has been also obtained by Causa et al. [7] for Gd ESR in $Gd_xEu_{1-x}Ba_2Cu_3O_y$. To confirm this magnitude, we have measured the ESR linewidth of Gd in $Y_{0.9}Gd_{0.1}Ba_2Cu_3O_y$ above 100K. Our results are shown in Fig.3. In this study, we could not obtain accurate value for the temperature gradient of the linewidth because of broad linewidth. However, our rough estimate gives $\partial(\Delta H)/\partial T \simeq 0.5 \pm 0.5$ Gauss/K.

Next, based on the existing papers, the spin fluctuation spectra of conduction carriers in RBaCuO system will be considered, somehow qualitatively. Firstly, according to Chen et al. [8] , no localized moment is present on the Cu atoms in YBaCuO. Secondly, Szpunar et al.[9] theoretically demonstrated that the charge fluctuations in the Cu-O planes force the spin fluctuations on the copper site. Thirdly, up to present, no conduction electron spin resonance has been detected for any RBaCuO system, presumably due to "very" broad linewidth. From these reasons, it may be reasonable to assume that the conduction carriers in RBaCuO are in the isothermal limit in the normal-state. If this is correct, the Gd ESR linewidth vs. temperature relation together with Eq.(1) gives $J_{ex}N_F \simeq 0.001$. This value is substantially smaller than the value (0.025) obtained for ErBaCuO by inelastic neutron scattering experiments [3]. If the neutron data (0.025) is accepted for YBaCuO and conduction carriers are in the isothermal limit in YBaCuO, $\partial(\Delta H)/\partial T$ of Gd diluted in YBaCuO should be as large as ≃60Gauss/K. This is nearly two orders of magnitude much greater than experimental results. In the following section some comments are made on the inelastic neutron scattering cross-section.

Comments on the inelastic neutron scattering cross-section

In general, the energy width of inelastic neutron scattering arising from transition between two energy levels may contain contributions from several sources including magnetic interaction of the type, $J_{ex}\vec{J}\cdot\vec{s}$, [10] and interaction between localized magnetic spins of the type, $J'_{ex}\ \vec{J}_i\cdot\vec{J}_j$, where $\vec{J}_i$ and $\vec{J}_j$,respectively, represent the total angular momentum of i-th and j-th magnetic ions [11]. It is believed that ErBaCuO orders antiferromagnetically below about 1K [12,13,14] . This indicates that there exists considerably strong magnetic interaction between Er ions. The Er-Er interaction may affect the inelastic neutron scattering cross-section [11]. A general trend of the linewidth of inelastic neutron scattering cross-section arising from the magnetic interaction between R ions with substantial crystal-field splitting is its increase with temperature. The situation was theoretically studied for Pr-V (V=group-V element) [11]. For ErBaCuO, a possible contribution of Er-Er interaction to the cross-section could not be overlooked. We feel that inelastic neutron scattering experiments for Er^{3+} diluted in YBaCuO may yield more accurate value for $J_{ex}N_F$ of Er.

Resistivity as a probe of magnetic interaction

A careful analysis of normal-state resistivity of RBaCuO can, in principle, give information about the magnetic interaction between R ion and conduction carriers. This will be discussed in this section. Heremans et al. [15] theoretically investigated the normal-state resistivity of RBaCuO, and concluded that the predominant mechanism for the resistivity is due to electron-phonon interaction. The resistivity due to phonon is proportional to temperature. However, strictly speaking,if R ion is a magnetic ion with substantial crystal-field splitting and if there exists considerable interaction between R ion and conduction carriers, the normal-state resistivity is not exactly proportional to temperature [16,17]. In what follows this will be discussed in more detail. Figure 4 represents a qualitative behavior of electrical resistivity of RBaCuO. The symbols, R_p and R_m, respectively, stand for the resistivity due to phonon and localized magnetic R ions. As mentioned above R_p is proportional to temperature. The magnetic resistivity, R_m, due to magnetic R ions with substantial crystal-field splitting was first studied theoretically by Hirst[16] and his theory was confirmed experimentally for many rare-earth compounds [17]. Following Hirst[16], R_m is given as

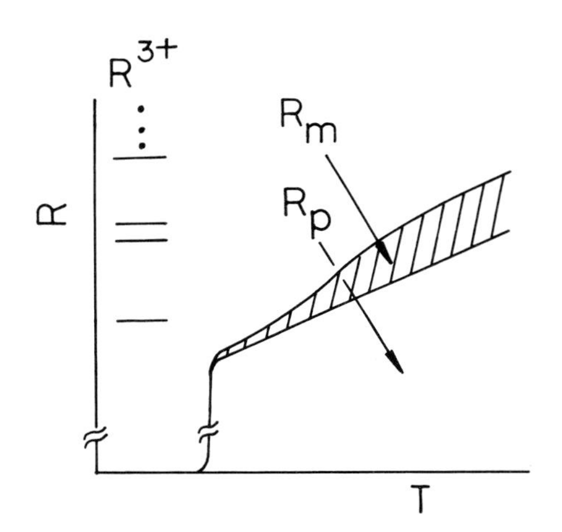

Fig.4 Qualitative temperature dependence of resistivity,$R=R_p+R_m$, of RBaCuO system. R_p and R_m, respectively, represent the resistivity due to phonons and magnetic rare-earth ions with substantial crystal field splitting like the one shown in the inset. (The crystal-field splitting of Er^{3+} is given in ref.3.)

$$R_m = CD \sum_{I,I'} P(I)\,(E_{II'}/kT)[\exp(E_{II'}/kT)-1]^{-1} \times |\langle I'|\vec{J}\cdot\vec{s}|I\rangle|^2 \qquad (2)$$

where C is the concentration of magnetic ions;D is a constant dependent on parameters describing the conduction electron system and the magnetic interaction constant; P(I) is the Boltzmann factor giving occupation of the crystal-field energy level $|I\rangle$, $E_{II'}$ is the energy difference between crystal-field states. There are several reports on the resistivities of RBaCuO system [18-21], in which the temperature dependence of normal-state resistivity has a behavior similar to that shown in Fig.4, a possible indication of the crystal-field effect. To confirm the normal-state restivity, we have examined experimentally and theoretically the resistive behavior of ErBaCuO and NdBaCuO. ErBaCuO may be a useful material for the magnetic interaction study, since its crystal-field splitting scheme is roughly known [3]. We have done the resistivity measurement with particular care. However, our measurements revealed linear temperature dependence for both compounds between about 100K and 300K. No anomaly like the one shown in Fig.4 was identified within experimental accuracy ($\sim 10\mu\Omega$). Using Eq.(2) together with approximate value of J_{ex}, R_m will be estimated.(As a matter of course, accurate value of R_m can not be obtained presently, since detailed information about crystal-field splitting and wave function of each energy level of R^{3+} are not available.) Very crude calculation gives $R_m \simeq (J_{ex})^2 \times 10^2\ \mu\Omega$-cm for RBaCuO. Approximate values of $J_{ex}N_F$ were already obtained from ESR and neutron scattering experiments. However, J_{ex} is not well known. Thus, we simply assume that J_{ex} is of the order of 10^{-2} eV-10^{-3} eV. Under this assumption, the magnitude of R_m is estimated to be $\sim 10^{-2}$- 10^{-4} $\mu\Omega$-cm. This value was too small to detect in our resistivity measurements.

The experimental and theoretical studies of R_m just mentioned above imply that small deviations of resistivity from its linear temperature dependence observed by several authors [18-21] may not be correlated with magnetic resistivity, R_m. In order to obtain J_{ex} accurately , it is needed to perform very precise resistivity measurements. We feel that resistivity measurements using single-crystals samples may be hopeful for this purpose.

CONCLUSION

With regard to the magnetic interaction between R ion and conduction carriers in RBaCuO, ESR linewidth, inelastic neutron scattering cross-section and electrical resistivity of RBaCuO have been examined based on theory and experiments. The Gd ESR in YGdBaCuO gives $J_{ex}N_F \simeq 0.001$, substantially smaller than the value obtained from inelastic neutron scattering cross-section. Some comments are made on the interpretation of the cross-section. As a comparison, $J_{ex}N_F$ of Ce diluted in LaAs will be reviewed. According to the ESR studies on Ce^{3+} in LaAs [10,22] , $J_{ex}N_F$ of Ce is of the order of 0.02. These results together with the Gd ESR in YGdBaCuO indicate that the coupling between R and conduction carriers in RBaCuO is weaker than in rare-earth group V compounds [22,23] . Theoretical estimate of the magnetic resistivity R_m in RBaCuO revealed that the magnitude of R_m is of the order of 10^{-2}-10^{-4} μΩ-cm.

REFERENCES

1. T. Ishigaki, H. Asano, K. Takita, H. Katoh, H. Akinaga, F. Izumi and N. Watanabe, Jpn. J. Appl. Phys. 26, L1681(1987).
2. K. Sugawara, R. Kita, Y. Akagi, H. Taniguchi, Y. Nakajima and S. Kataoka; Proc. 18th Int. Conf. on Low Temp. Physics, p2119, 1987; Jpn. J. Appl. Phys. 26(1987) Supplement 26-3.
3. U. Walter, S. Fahy, A. Zettl, S. G. Louie, M.L. Cohen, P. Tejedor and A.M. Stacy, Phys. Rev. B 36, 8899(1987).
4. H. Hasegawa, Progr. theor. Phys. 21 483(1959).
5. K. Sugawara, phys. stat. solidi (b)84,709(1977).
6. F. Mehran, S.E. Barnes, C.C. Tsuei and T.R. McGuire, Phys. Rev. B36,7266(1987).
7. M.T. Causa, C. Fainstein, G. Nieva, R. Sanchez, L.B. Steren, M. Tovar, R. Zysler, D.C. Vier, S. Schultz, S.B. Oseroff, Z. Fisk and J.L. Smith, Phys. Rev. B38,257(1988).
8. H. Chen, J. Callaway and P. K. Misra, Phys. Rev. B38,195(1988).
9. B. Szpunar and V.H. Smith,Jr, Phys. Rev. B37,2338(1988).
10. K. Sugawara, phys. stat. solidi (b)92,317(1977).
11. K. Sugawara, phys. stat. solidi (b)81,313(1977).
12. S.H. Liu, Phys. Rev. B37,7470(1988).
13. J.W. Lynn, W-H Li, Q. Li, H.C. Ku, H.D. Yang and R.N. Shelton, Phys. Rev. B36,2374(1987).
14. B.W. Lee, J.M. Ferreira, Y. Dalichaouch, M.S. Torikachvili, K.N. Yang and M.B. Maple, Phys. Rev. B37,2368(1988).
15. J. Heremans, D.T. Morelli, G.W. Smith and S.C. Strite III, Phys. Rev. B37,1604(1988).
16. L.L. Hirst, Solid State. Commun. 5,751(1967).
17. K. Sugawara,C.Y. Huang, C.W. Chu and B.R. Cooper, Solid State Commun. 21,189(1977).
18. T. Kobayashi, T. Takahashi, M. Tonouchi, Y. Fujiwara and S. Kita, Jpn. J. Appl. Phys. 26,L1381 (1987).
19. A. Oota, Y. Kiyoshima, A. Shimono, K. Koyama, N. Kamegashira and S. Noguchi, Jpn. J. Appl. Phys. 26,L1543(1987).
20. S. Chittipeddi, Y. Song, D.L. Cox, J.R. Gaines, J.P. Golben and A.J. Epstein, Phys. Rev. B37, 7454(1988).
21. M. Mukaida, M. Yamamoto, Y. Tazoh, K. Kuroda and K. Hohkawa, Jpn. J. Appl. Phys. 27, L211(1988).
22. K. Sugawara and C.Y. Huang, J. Phys. Soc. Japan 40,295(1976).
23. K. Sugawara and C.Y. Huang, J. Phys. Soc. Japan 39,643(1975).

The Effect of the Addition of S, Se and Te in Ba-Y-Cu-O on Its Superconductive Properties

SHIRO KAMBE[1] and MAKI KAWAI[2]

[1] The Institute of Physical and Chemical Research, 2-1, Hirosawa, Wako, Saitama, 351-01 Japan
[2] Research Laboratory of Engineering Materials, Tokyo Institute of Technology, 4259 Nagatsuta-cho, Midori-ku, Yokohama, 227 Japan

ABSTRACT

An addition of small amount of sulfur or selenium in $Ba_2YCu_3O_{7-y}$ gave raise in Tc(ZERO). The lattice constant of the c-axis in the selenium contained $Ba_2YCu_3O_{7-y}$ with highest Tc(zero) was 11.55 Å, which was shorter than those observed in pure $Ba_2YCu_3O_{7-y}$ system.

INTRODUCTION

Since the discovery of the La-Ba-Cu-O superconductor by Müller and Bednorz[1], attempts have been made to increase Tc in the related perovskites; Tc of 90K was found in Ba-Y-Cu-O compound by Wu et al[2]. These works have led to the studies to modify the chemical composition by isomorphous replacement of the elements in order to raise the transition temperature and to examine the high Tc superconductive mechanism[3-9].

In the $Ba_2YCu_3O_{7-y}$ system, the superconductive properties are strongly correlated with the content of oxygen[13]. Substitution for oxygen with other elements should change the superconducting property. Especially, the chalcogens such as S, Se and Te, have similar valence state to oxygen and are expected to form similar binding orbitals with Cu. Sulfur substitution may cause some change in the superconductive properties[10-12,14-16].

In this paper, it is shown that the addition of small amount of sulfur or selenium to Ba-Y-Cu-O system gives sharp transition in the composition of $Ba_2YCu_3O_{7-y}S_{0.05}$ and $Ba_2YCu_3O_{7-y}Se_{0.5}$. The effect of S or Se addition on the structure and electronic property is discussed.

EXPERIMENTAL

$Ba_2YCu_3O_{7-y}$ was prepared by mixing Y_2O_3, $BaCO_3$ and CuO and grinding in an agate mortor, followed by the calcination in air at 950C for 8 hours. The mixture was then ground, pressed into pellets, sintered in flowing O_2 for 5 hours at 950C followed by calcination for 5 hours at 550C, and cooled to room temperature.

Sulfur, selenium or tellurium containing compounds were prepared in the same manner. Either sulfur, selenium or tellurium powder was added in the starting materials.

The composition of barium, yttrium, copper, selenium and tellurium in the sample was observed by the induction coupled plasma atomic emission spectroscopy, ICP-AES (ICPS-50A,Shimadzu Co.). Sulfur composition was determined by measuring the amount of barium by ICP-AES after sulfur was precipitated as barium sulfate.

Resistivity measurement was carried out by dc four-probe method, with copper electrodes attached with 99.99% indium by the supersonic solder. Morphology of the materials thus prepared was observed by a scanning electron microscope (JSM-840A, JEOL). The lattice constant was determined by X-ray powder diffraction (RAD-IIIB, Rigaku-denki). The electronic states were measured by XPS (ESCALAB, VG scientific Co.).

RESULTS AND DISCUSSION

Substitution of selenium for oxygen in $Ba_2YCu_3O_{7-y}$ influenced the superconductive properties. Substituted $Ba_2YCu_3O_{7-y}Se_x$ showed superconductive property with $x \leq 0.5$, while it turned to be an insulator with $x \geq 1.5$. Fig.1

shows the dependence of the resistance on the temperature in $Ba_2YCu_3O_{7-y}Se_x$ with x=0 and 0.5. These two samples were calcinated with the same condition. As shown in the figure, Tc onset of x=0 and 0.5 was in the similar temperature range of 93.0-93.5K, while Tc(zero) of that range was influenced by Se content. As the content of Se increased from 0 to 0.5, Tc(zero) raised from 89.5K to 91.0K, where the highest Tc(zero) was achieved in the composition of $Ba_2YCu_3O_{7-y}Se_{0.5}$. This composition gave the sharpest transition of 2.5K. Further addition of Se changed it into an insulator.

X-ray diffraction patterns of $Ba_2YCu_3O_{7-y}Se_x$ are shown in Fig.2. As the content of Se increases, the insulator phases of $BaCuY_2O_5$, $BaSeO_4$ and CuO_x begin to appear besides the main peaks of $Ba_2YCu_3O_{7-y}$. Further selenium addition to x>1.5 lead to the disappearance of $Ba_2YCu_3O_{7-y}$ and the formation of mainly $BaSeO_4$, CuO_x.

The observed XRD pattern shows that the lattice constants a, b and c of the $Ba_2YCu_3O_{7-y}$ phase in the selenium-substituted sample (x=0.5) varied from that of pure orthorhombic phase of $Ba_2YCu_3O_7$. Changes in the lattice constants of $Ba_2YCu_3O_{7-y}$ with selenium contents are shown in Fig.3. As is shown here, a- and b-axes were almost constant with the increase in the selenium concentration. However the lattice constant for the c axis decreased significantly at the selenium concentration of x=0.5, where the highest Tc(zero) was observed. It is well known that the content of oxygen in $Ba_2YCu_3O_{7-y}$ gives change in the length of the axis of the unit cell[13]. As the oxygen constant decreases, the lattice for c-axis increases. The smallest value for the c-axis is observed with $Ba_2YCu_3O_7$ and is 11.68Å. This value shown by Cava et al. is achieved with the sample of x=0. The value of the lattice constant of c-axis with x=0.5 was 11.55A and was much smaller than that of 11.68Å for $Ba_2YCu_3O_7$[13]. Such a small length of c axis might be explained by the partial substitution of selenium for the oxygen in $Ba_2YCu_3O_7$.

Secondary electron image (SEI) of the selenium-added superconductive sample (x=0.5) shows that they consist of particles, the diameter of which is mainly between 1 and 10 micrometers, and some over 30 micrometers.

The addition of sulfur also raised Tc(ZERO)[16]. In the composition of $Ba_2YCu_3O_{7-y}S_{0.05}$, the highest Tc of 93.0K was achieved. More addition of sulfur lead to the formation of insulator phase as in the case of selenium. The lattice constant for the a-, b- and c- axes varied with S composition. As in the case of

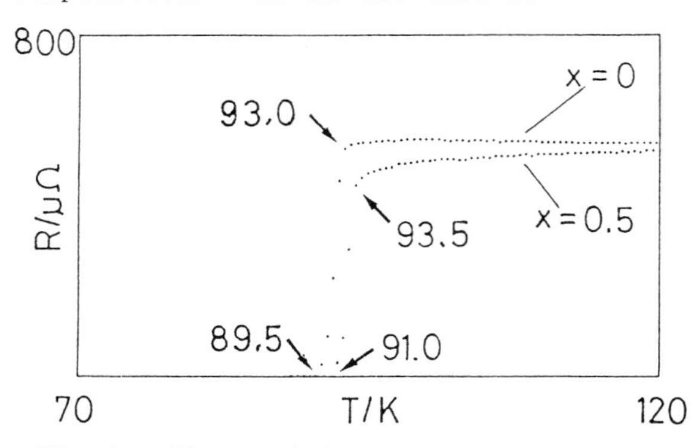

Fig.1 The resistance-temperature curves of $Ba_2YCu_3O_{7-y}Se_x$. In the content of x=0.5, Tc(ZERO) raises.

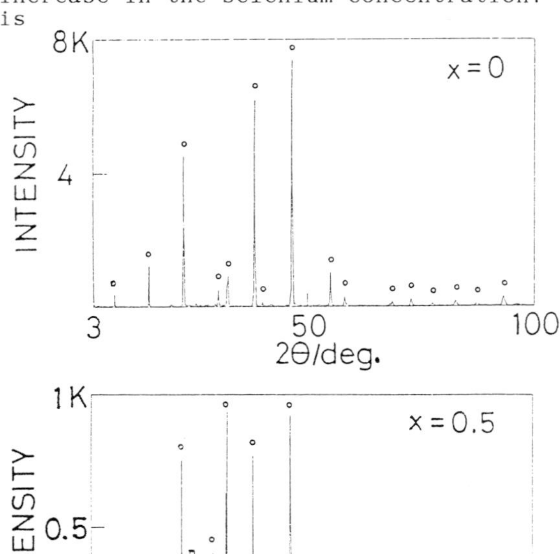

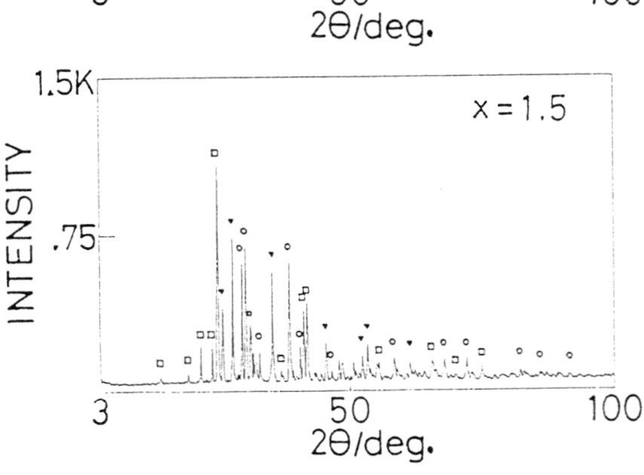

Fig.2 The XRD patterns of $Ba_2YCu_3O_{7-y}Se_x$.

○ : $Ba_2YCu_3O_{7-y}$ or $Ba_2YCu_3O_{7-y}Se_x$,
▽ : $BaCuY_2O_5$,
□ : $BaSeO_4$ or CuO_x.

selenium, the composition of x=0.05 gave the minimum length of c-axis. This may also been explained as an effect of substituting sulfur for the oxygen in the Cu-O chain or plane. The addition of Te resulted in formation of insulator in all cases with Te ratio of 0.7 to 5.

The addition of small amount of sulfur or selenium in $Ba_2YCu_3O_{7-y}$ gave raise in Tc(ZERO). The structural analysis by XRD suggested that the lattice constant of c-axis of these samples was smaller than those observed in $Ba_2YCu_3O_{7-y}$ system without S or Se. These results may suggest that the substitution of S or Se for the chain or plane oxygen in $Ba_2YCu_3O_{7-y}$ causes sharp transition in the R-T curve.

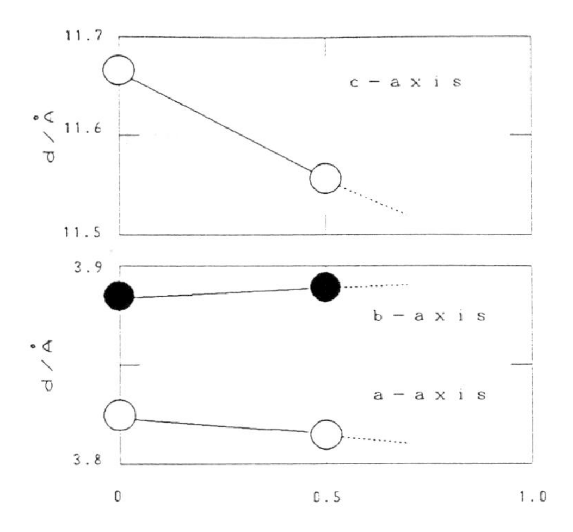

Fig.3
The dependence of the lattice constants on the selenium content. The composition of x=0.05 gave the minimum length for the c-axis.

ACKNOWLEDGMENT

We are grateful to Dr. Takahashi in the Institute of Physical and Chemical Research for ICP-AES and inorganic measurement. This work was carried out in the Frontier Research Program in the Institute of Physical and Chemical Research.

REFERENCES

1)J.G.Bednorz and K.A.Muller: Z.Phys., B64 (1986) 189.
2)M.K.Wu,J.R.Ashburn, C.J.Torng, P.H.Hor, R.L.Meng, L.Gao, Z.J.Huang,Y.Q.Wang and C.W.Chu: Phys.Rev.Lett., 58 (1987) 908.
3)Y.Hakuraku, F.Sumiyoshi and T.Ogushi: Appl.Phys.Lett.,52 (1988) 1528.
4)J.Jung, J.P.Franck, W.A.Miner and M.A.-K.Mohamed: Phy.Rev.B, 37 (1988) 7510.
5)J.S.Kim, J.S.S Swinnea, A.Manthiran and H.Steinfink: Solid State Comm., 66 (1988) 287.
6)D.N.Mathews, A.Bailey, R.A.Vaile, G.J.Russell and K.N.R.Taylor: Nature 328 (1987) 786.
7)In-Gann Chen, S.Sen and D.M.Stefanescu: Appl.Phys.Lett., 52 (1988) 1355.
8)A.K.Tyagi, S.J.Patwe, U.R.K.Rao and R.M.Iyer: Solid State Comm., 65 (1988) 1149.
9)K.Okura, K. Ohmatsu, H.Takei, H.Hitotsuyanagi and T.Hakahara: Jpn.J.Appl.Phys., 27 (1988) L655.
10)K.N.R.Taylor, D.N.Mathews and G.J.Russell: J.Cryst.Growth, 85 (1987) 628.
11)I.Felner and I.Nowik: Phys.Rev.B, 36 (1988) 7496.
12)B.A.Richert and R.E.Allen: Phys.Rev.B, 37 (1988) 7496.
13)R.J.Cava, B.Batlogg, C.H.Chen, E.A.Rietman, S.M.Zahurok and D.Werder: PHys.Rev.B, 36 (1987) 5719.
14)I.Matsubara, H.Tanigawa, T.Ogura, and S.Kose: Jpn.J.Appl.Phys., 27, (1988) L1080.
15)I.Felner and B.Barbara: Phys.Rev.B, 37 (1988) 5820.
16)S.Kambe and M.Kawai: submitted to Jap.J.Appl.Phys.

4.3 Phase Diagram and Crystal Growth

Phase Diagram and Crystal Growth of $RBa_2Cu_3O_{7-y}$ System

KUNIHIKO OKA, MASATOSHI SAITO*, MASAHIRO ITO*, KENJI NAKANE**, and HIROMI UNOKI

Electrotechnical Laboratory, 1-1-4, Umezono, Tsukuba, 305 Japan

ABSTRACT

Phase diagrams of R_2O_3 (rare earth oxide)-BaO-CuO system have been investigated. The temperature versus concentration diagrams are given on the four binary systems ranging from $RBa_2Cu_3O_{7-y}$ to CuO, to $BaCuO_2 \cdot 3CuO$, to $BaCuO_2 \cdot 4CuO$ and to $3BaCuO_2 \cdot 2CuO$ mixture. Crystals of $YBa_2Cu_3O_{7-y}$ and $NdBa_2Cu_3O_{7-y}$ have been grown by the slow-cooling method in accordance with the phase diagrams of R_2O_3-BaO-CuO system.

INTRODUCTION

The $YBa_2Cu_3O_{7-y}$ has been known to be superconductive at around 90 K (1). The interest in $YBa_2Cu_3O_{7-y}$ for studies of the superconducting mechanism have resulted in the requirement for a large and high-quality single crystals. The first step to grow the single crystal is to prepare the phase-equilibrium diagram. Thus we have tried to make up phase diagrams of Y_2O_3-BaO-CuO system. An area surrounded by the CuO-$BaCuO_2$-$YBa_2Cu_3O_{7-y}$ triangle is particularly examined by dividing the area into four binary systems ranging from $YBa_2Cu_3O_{7-y}$ to CuO, to $BaCuO_2 \cdot 3CuO$, to $BaCuO_2 \cdot 4CuO$ and to $3BaCuO_2 \cdot 2CuO$ mixture.

EXPERIMENTAL PROCESS

The phase diagram of Y_2O_3-BaO-CuO system has been determined using differential thermal analysis, quenching technique and X-ray diffraction measurements. The samples for measuring the $YBa_2Cu_3O_{7-y}$ system were prepared by mixing the powders of 99.99 % pure Y_2O_3, $BaCO_3$ and 99.9 % pure CuO. About 60 sort of samples were prepared with every 5-10 mol% mixtures between $YBa_2Cu_3O_{7-y}$, and CuO, $Ba_3Cu_{12}O_{15}$, $Ba_3Cu_7O_{10}$, $Ba_3Cu_5O_8$ as shown in Fig. 1. The analysis was carried out with a Defferential Thermal Analyzer with a halogen lamp as the heating element at a heating and cooling rate of 10 ℃/min in air. Two to three handredth of a gram of

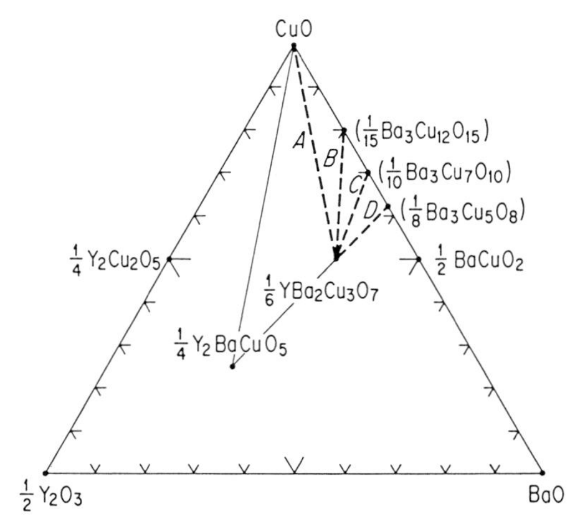

Fig. 1 Y_2O_3-BaO-CuO ternary phase diagram.

---

* Permanent address: Sumitomo Metal Mining Co., Ltd.
** Permanent address: Sumitomo Chemical Co., Ltd.

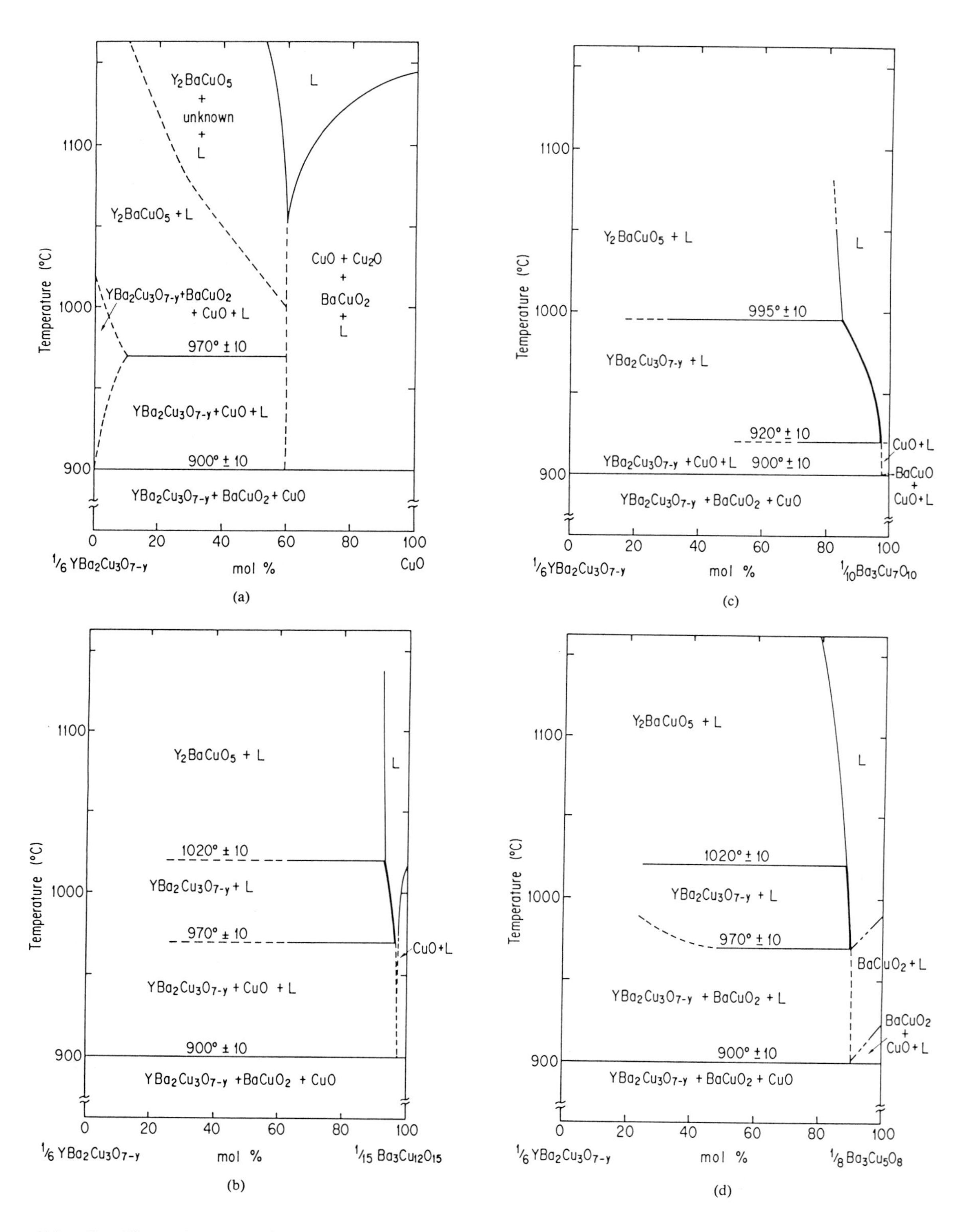

Fig. 2. The temperature vs composition phase diagrams obtained in the present study ; (a): along *A*-, (b): along *B*-, (c): along *C*-, (d): along *D*-directions shown in Fig. 1. The liquidus lines required for the solution growth of $YBa_2Cu_3O_{7-y}$ single crystal are drawn by bold lines.

the mixture was placed in a platinum crucible of 5 mm in diameter and 5 mm in length. The liquidus, eutectic and peritectic temperatures in the phase diagram were determined by melting a few milligram of the sample on a small platinum sheet (20X20 mm^2) which was put into a muffle furnace. The furnace was heated stepwise by 10 degree increment by direct observation through the furnace inlet. After 20 minutes, the sample was removed from the furnace and quenched, then the furnace temperature was raised by 10 degrees and another sample of the adjcent composition was put into the furnace. The quenched specimens were powdered, and phase identification was made at room temperature by X-ray diffraction.

RESULTS AND DISCUSSION

The phase diagrams thus determined are shown in Fig. 2 (2). The eutectic temperature throughout these four figures is common and is 900±10℃. A quenched product of $YBa_2Cu_3O_{7-y}$ compound at temperature from 1020 to 1300℃ was found to contain green material of mainly Y_2BaCuO_5, while in a quenched product from 1350℃ the brown material of mainly Y_2O_3, was found.

For the growth of $YBa_2Cu_3O_{7-y}$ single crystal from solution, the liquidus line has been found to be extremely narrow in concentration range. But it was found that the liquidus line in the $NdBa_2Cu_3O_{7-y}$ phase diagram expands in rather wider range than that in the $YBa_2Cu_3O_{7-y}$ system.

We have tried to grow single crystals of $YBa_2Cu_3O_{7-y}$ and $NdBa_2Cu_3O_{7-y}$ (3) by slow-cooling method based on the phase diagram show in Fig. 2 (C) and (D). The starting material consisted of 15% $YBa_2Cu_3O_{7-y}$ and 85% $Ba_3Cu_7O_{10}$ for $YBa_2Cu_3O_{7-y}$, and 25% $NdBa_2Cu_3O_{7-y}$ and 75% ($Ba_3Cu_7O_{10}$ or $Ba_3Cu_5O_8$) for $NdBa_2Cu_3O_{7-y}$, were mixed and put into 100 mm^3 platinum crucible. The crucible was covered with a platinum lid and put into a muffle furnace. The $YBa_2Cu_3O_{7-y}$ system heated at 1050 ℃ and $NdBa_2Cu_3O_{7-y}$ system heated at 1100℃, held at constant temperature a few hours, and then cooled slowly at a rate of 10 ℃/hr to 900 ℃. The $YBa_2Cu_3O_{7-y}$ crystal dimensions in the c-plane plate are 1 X 1 mm^2 with a thickness of 0.2 mm. The $NdBa_2Cu_3O_{7-y}$ crystal dimensions are 3 X 3 mm^2 with a thickness of 0.5 mm (Fig. 3). The large crystal of $NdBa_2Cu_3O_{7-y}$ suggests that wide liquidus line really yields the favorable effect for crystal growth.

Fig. 3. $NdBa_2Cu_3O_{7-y}$ single crystal grown by the slow-cooling method.

REFERENCES

(1) M.K.Wu et al., Phys. Rev. Lett. 58 908 (1987).
(2) K.Oka et al., Jpn. J. Appl. Phys . 27 L1065 (1988).
(3) A.Katsui et al., Jpn. J. Appl. Phys. 26 L1521 (1987).

Superconductivity on Single Crystals of Tl-Ca-Ba-Cu-O System

HIROMI TAKEI, TOSHIHIRO KOTANI, TETSUYUKI KANEKO, and KOJI TADA

Basic High-Technology Laboratories, Sumitomo Electric Industries, Ltd.
1-1-3, Shimaya, Konohana-ku, Osaka, 554 Japan

ABSTRACT

The investigation on the growth of single crystals of the Tl-Ca-Ba-Cu-O superconducting system was carried out using the flux growth method. The crystal structure and superconducting properties of mm-sized single crystals were measured using X-ray diffraction and the SQUID magnetometer. The grown crystals are typically in platelet form ; 0.5 to 2.5 mm wide and about 0.2 mm thick. The crystals exhibit a mirror-like surface and some growth steps can be observed through a scanning electron microscope. The superconducting onset transition temperature of 108K to 118K was determined by DC magnetic susceptibility obtained for as-grown single crystals. The transition temperature has a tendency to increase as the amount of CaO-CuO in the starting nominal composition rises. According to the X-ray diffraction measurement, the c-axis of the grown crystals is 35.8Å for those with Tc of 118K, and 29.3Å for those with Tc of 108K.

INTRODUCTION

Since the discovery of high-Tc superconducting oxides, preparation of single crystals of the high-Tc oxides has been underway to determine the precise crystal structures and superconducting properties, including anisotropic. The growth of single crystals in La_2CuO_4, $Ba_2YCu_3O_x$, and the low-Tc phase of the $Bi_2(Sr, Ca)_3Cu_2O_x$ system is performed mainly by the flux method. Large single crystals of the high-Tc phases in the Tl-Ca-Ba-Cu-O and Bi-Ca-Sr-Cu-O systems have not yet been obtained [1-4]. Recently, Torardi et al. determined the crystal structure of the high-Tc phase of the $Tl_2Ca_2Ba_2Cu_3O_{10}$ system from single crystal X-ray diffraction data for very small crystals, $0.08 \times 0.09 \times 0.01$ mm^3 in size [5]. The growth of mm-sized crystals in the Tl-Ca-Ba-Cu-O system was carried out by Morosin et al., and three distinct phases with superconducting transition at 103K, 106-108K, and 112-114K, (which are lower than the highest transition temperature of sintered specimen), were reported [4].

In this paper we will report on the growing of large crystals in the high-Tc phase of the Tl-Ca-Ba-Cu-O system using flux method, and the relationship between the crystal structure and the superconducting transition temperature.

EXPERIMENTAL

Crystal growth experiments were performed using CaO-CuO self-flux. The starting materials are shown in Fig. 1. Each charge inserted into a Au tube was placed in a 40-cm^3 Pt crucible covered with a Pt lid. The crucible was rapidly heated to the temperature range of 900°C to 950°C, and held for 1 hour under O_2 gas flow. The temperature was then decreased to 740°C at rate of about 10°C/hour, and afterwards further decreased to room temperature in 6 hours. Some charges were quenched during the decrease from the holding temperature to room temperature in order to examine the phases formed at the holding temperature.

The morphology of the obtained crystals were observed through scanning electron microscopy (SEM). The chemical compositions of the crystals were analyzed through energy dispersive X-rays (EDX), and inductively coupled argon plasma atomic emission spectroscopy (ICP-AES). The crystal structure was analysed through the Debye-Shereer X-ray diffraction (XRD) method. Several

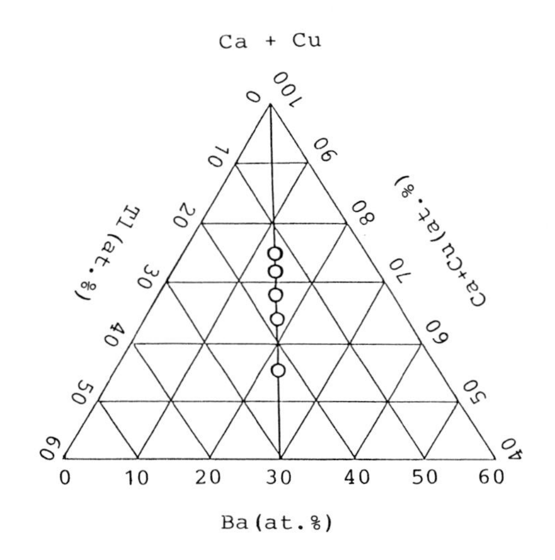

Fig.1. Nominal compositions used in the Tl-Ca-Ba-Cu-O system.

single crystals selected from the crushed specimens were examined through X-ray micro-diffraction. The temperature dependence of DC magnetic susceptibility determined the superconducting transition temperature, as measured by a SQUID magnetometer.

RESULTS AND DISCUSSION

Slow-cooled samples exhibit a superconducting transition temperature of 108K to 120K, as listed in Table I. In the composition range of this study, the transition temperature tends to increase monotonously with the content of CaO-CuO in the starting nominal composition. Fig.2 shows the temperature dependence of DC magnetic susceptibility of a single crystal obtained from specimen No.1 and No.5. The dimensions of the crystals are about 2.5 × 2 × 0.1 - 0.2 mm^3 for the crystal with Tc of 110K, and 0.5 - 1 × 0.5 - 1 × 0.2 mm^3 for the crystal with Tc of 118K. In a typical run, a large crystal from the low-Tc phase was obtained more easily than that of the high-Tc phase. The SEM images of these crystals are shown in Fig.3. These crystals have uniform surface composition, and some growth steps can be seen. According to the X-ray diffraction measurement of the crystals, the c-axis of a unit cell with a transition temperature of 110K is 29.3Å, and 35.8Å for the crystal with a transition temperature of 118K. The crystal with a transition temperature of 115K shows an intergrowth structure containing two phases with a c-axis of 29.3Å and 35.8Å.

In the case of quenched samples, the superconducting transition temperatures are lower than those of the slow-cooled samples as shown in Table I. These samples, like the slow-cooled samples, show an increase in transition temperature as the content of CaO-CuO in the starting nominal composition increase up to n = 4.

Table I. Nominal composition and measured properties of Tl-Ca-Ba-Cu-O system.

Specimen						Heat Treatment		
No.	TlCaBaCu [2223] + n [CaO + 0.75CuO]					Slow cooling		Quenching
	n	Tl	Ca	Ba	Cu	Tc χ (K)	Phases	Tc χ (K)
①	0	2	2	2	3	108~110	2122 2223	80
②	1	2	3	2	3.75	115	2223 2122	78
③	2	2	4	2	4.5	115	2223	92
④	3	2	5	2	5.25	118	2223	95
⑤	4	2	6	2	6	118~120	2223	108

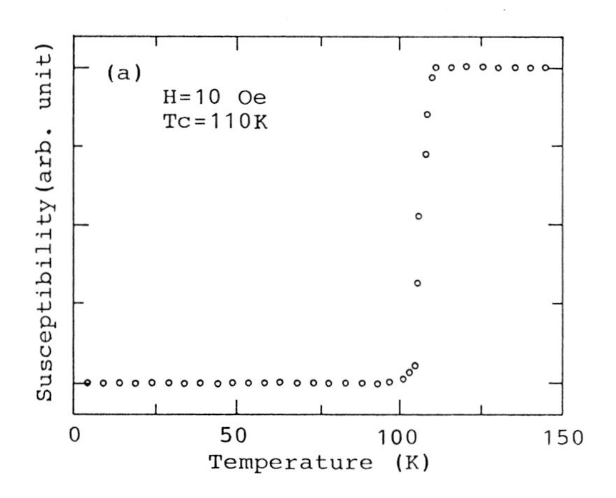

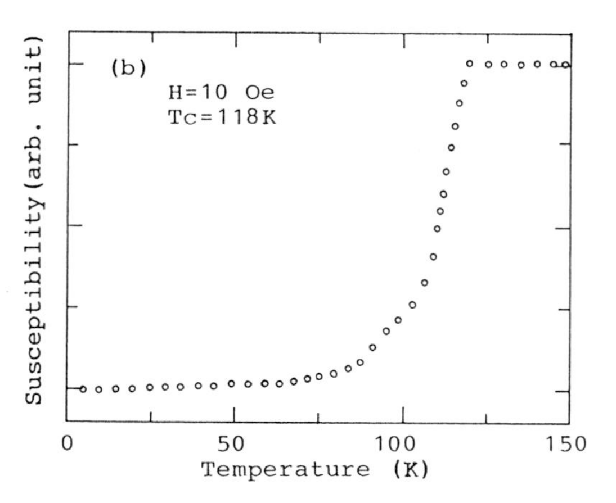

Fig. 2. Temperature dependence of DC magnetic susceptibility for a single crystal obtained from (a) specimen No.1, and (b) specimen No.5.

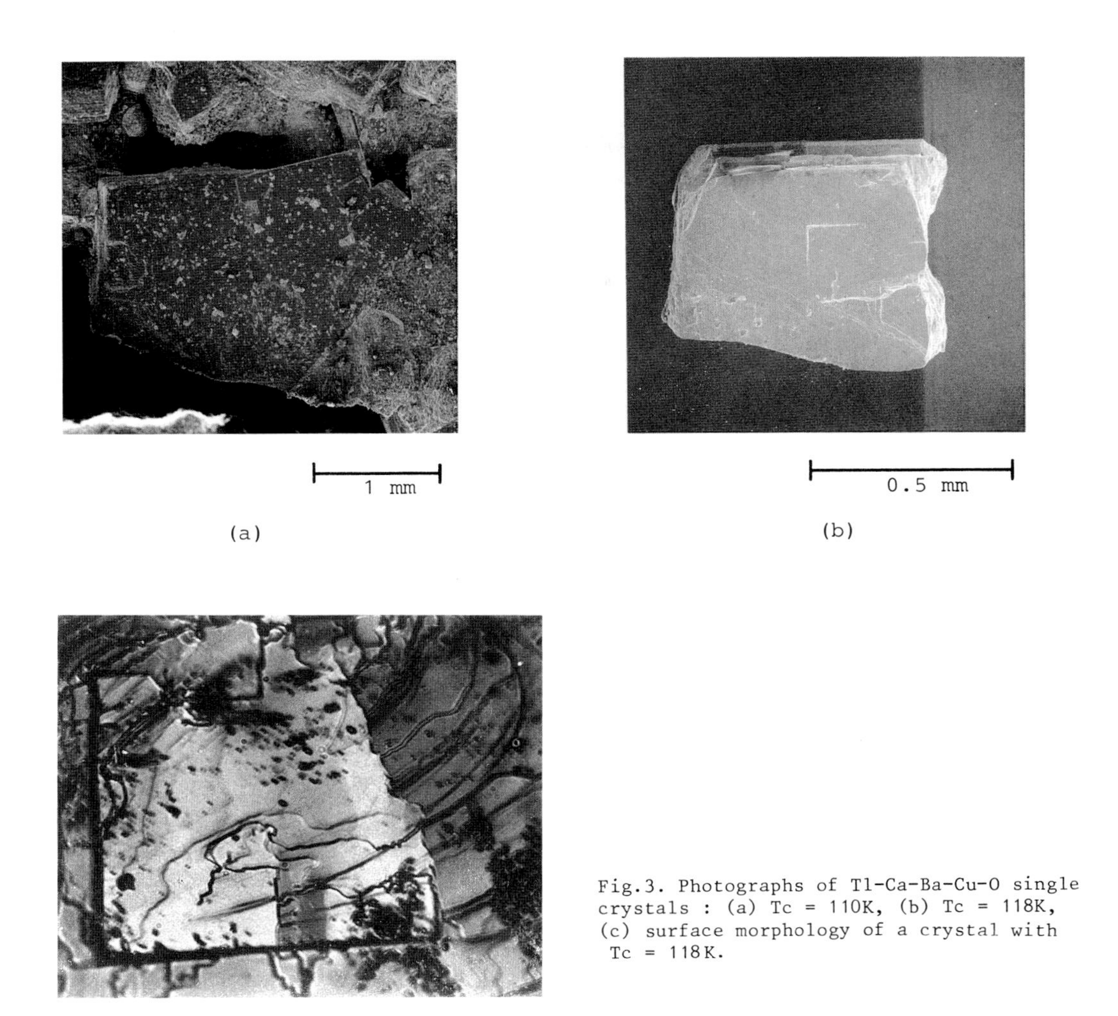

Fig.3. Photographs of Tl-Ca-Ba-Cu-O single crystals : (a) Tc = 110K, (b) Tc = 118K, (c) surface morphology of a crystal with Tc = 118K.

Those results suggest that the high-Tc phase of the Tl-Ca-Ba-Cu-O system changed to the phases with lower transition temperatures, by melting. A higher content of CaO-CuO in the starting materials may be effective in forming a crystal with a high transition temperature. Although the optimum content of CaO-CuO is not clear, a more detailed study is in progress.

We thank Mr. H. Kusuhara for the X-ray diffraction measurements, and for useful discussions.

REFERENCES

1. Y. Hidaka, Y. Enomoto, M. Suzuki, M. Oda and T. Murakami, J. Cryst. Growth 85, 581 (1987).
2. H. Takei, H. Takeya, Y. Iye, T. Tamegai and F. Sakai, Jpn. J. Appl. Phys. 27, Suppl. 2, 51 (1988).
3. L. F. Schneemeyer, R. B. van Dover, S. H. Glarum, S. A. Sunshine, R. M. Fleming, B. Batlogg, T. Siegrist, J. H. Marshall, J. V. Waszczak and L. W. Rupp, Nature 332, 31 March (1988).
4. B. Morosin, D. S. Ginley, P. F. Hlava, M. J. Carr, R. J. Baughman, J. E. Schirber, E. L. Venturini and J. F. Kwak (preprint).
5. C. C. Torardi, M. A. Subramanian, J. C. Calabrese, J. Gopalakrishnan, K. J. Morrissey, T. R. Askew, R. B. Flippen, U. Chowdhry and A. W. Sleight, Science 240, 631 April (1988).

Single-Crystal Growth of Bi-Sr-Ca-Cu Oxides and Its Superconductivity

SHUNJI NOMURA, TOMOHISA YAMASHITA, HISASHI YOSHINO, and KEN ANDO

Advanced Research Laboratory, Research and Development Center, Toshiba Corporation, Kawasaki, 210 Japan

ABSTRACT

Single crystals of Bi-Sr-Ca-Cu Oxide were grown by the self-flux method. Relation between the single crystals produced and the starting compositions in the Bi-Sr-Ca-Cu Oxide system was obtained for $Bi_2(Sr,Ca)_2CuO_y$, $Bi_2(Sr,Ca)_{3-x}Cu_2O_y$ and $(Sr,Ca)BiO_3$. The dimensions of the single crystal of the superconductor obtained were up to $7X5X1mm^3$.

The effect of cation deficiency of $Bi_2(Sr,Ca)_{3-x}Cu_2O_y$ on its superconductivity have been studied. This system forms a solid solution within the x range of -0.2 to 0.3. The lattice parameter of the c axis decreases with the increase of x. The strong sattelite peak in the X-ray diffraction patterns originating from the incommensurate modulated structure are observed for the samples with large cation deficiency.

The critical temperature of the single crystal was 84K which was determined by both magnetization and electrical resistivity measurements. The resistivity ratio in the c plane and along the c axis for $Bi_2(Sr,Ca)_{3-x}Cu_2O_y$ was found to be approximately 35 at Tc onset.

1. *Introduction*

The discovery of the superconductor in Bi-Sr-Cu-O system by Mitchel et al.(1) and by Akimitsu et al.(2) and the superconductors in Bi-Sr-Ca-Cu-O system by Maeda et al.(3) reveal that there are at least three kinds of the superconductors . These new oxide superconductors have been recently developed and intensive studies have been on the way. The oxides show the superconducting transition temperatures (Tc) of 110K, 80K and 6~30K, and are thought to be expressed as $Bi_2Sr_2Ca_2Cu_3O_y$, $Bi_2(Sr,Ca)_3Cu_2O_y$ and $Bi_2Sr_2CuO_y$, respectively.

Among them, the compound with Tc of 110K in a single phase was not obtained. The structure and the chemical composition of this compound, therefore, have not been precisely determined yet. A lot of studies on the aim at determination in the chemical composition and at the isolation of the single phase of the 110K superconductor have been performed.

$Bi_2(Sr,Ca)_3Cu_2O_y$ is a superconductor with Tc of 80K and its structure is determined by X-ray diffraction(4~6), electron diffraction(5), high resolution TEM(5,6) and neutron diffraction(8). It has an orthorhombic symmetry with a=5.4, b=5.4 and c=30.6A and shows a modulation structure(7). In particular, electron diffraction, microscopic observation and X-ray diffraction revealed the presence of strong incommensurate modulation along the b axis, resulting that satellite lines appear in the powder X-ray diffraction pattern. All reflections appeared in the powder X-ray diffraction pattern have been precisely indexable with taking account of the incommensurate modulated structure.

Most of the studies for the compound, $Bi_2(Sr,Ca)_3Cu_2O_y$ were done using polycrystalline materials because of the difficulty of the preparation of large-size single crystals. Recently, single crystals of $Bi_2(Sr,Ca)_3Cu_2O_y$ have been grown and the characteristics of the compound have also been unveiled(9~12). Takagi et al.(10) measured the anisotropy of the upper critical field in the single crystals with typical dimensions of $1.5 \times 0.7 \times 0.1 mm^3$. Schneemeyer et al.(11) found the anisotropy of resistivity dependency on temperature along the a axis and the b axis for the single crystal.

Some crystals grown by the flux method have the stoichiometric composition, $Bi_2(Sr,Ca)_3Cu_2O_y$, whereas some show nonstoichiometry, namely, cation deficiency in alkaline earth elements determined by EPMA and EDX.

In this study, we report on the growth of single crystals of the superconducting Bi-Sr-Ca-Cu Oxide and their related compounds by the self-flux method and the characteristics of the grown crystals. We also report on the cation deficiency in $Bi_2(Sr,Ca)_{3-x}Cu_2O_y$ and on the effects on their superconductivity and the incommensurate modulated structure.

2. *Experimental Procedure*

The singles crystal were grown by a self-flux method. The mixtures of Bi_2O_3, $CaCO_3$, $SrCO_3$ and CuO powders of analytical grade were prepared in the various cation ratios. The mixtures of 50g were ground with a mortor and pestle, placed in alumina crucibles, heated in a furnace at 1000°C for 24 hours, cooled to 800°C at a rate of 10°C/h, and finally cooled to room temperature in the furnace. The reaction product was a solid mass containing single crystals preferentially grown from the bottom to the surface in the solidified matrix Single crystals were mechanically isolated.

The powders of $Bi_2(Sr,Ca)_{3-x}Cu_2O_y$ were prepared by the conventional solid-state reaction. The mixtures having the nominal composition $Bi_4Sr_{3-x}Ca_{3-x}Cu_4O_y$ ($-0.2 \leq x \leq 1.0$) were calcined at 750°C for 8h in air. The powders were reground and pressed into pellets without adding any binder under 1 ton/cm^2 pressure. These pellets were sintered at 850 C for 8h in air.

The powder X-ray diffraction by Cu Kα radiation was used to confirm the phase of the samples and to measure the lattice parameters of the c axis with Si as an internal standard. Magnetization was measured by SQUID in 0.001T magnetic field. Electrical resistivity measurement was carried out by a conventional four probe method with 10mA dc current.

The oxygen content in the samples was determined by thermal decomposition using EMGA-2800 (HORIBA Ltd.). In this system samples were melted in a graphite crucible with Ni-Sn as a flux in He atmosphare and the oxygen content was measured by the nondispersive IR detection as the outgoing CO quantity.

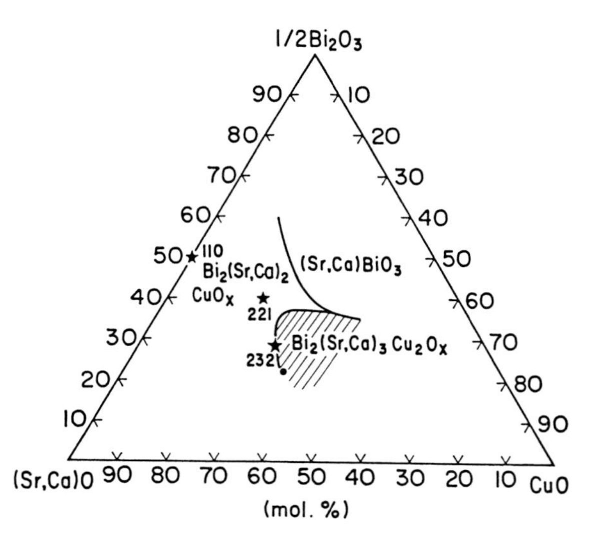

Fig.1 Relation between the the single crystals and the starting compositions

3. *Results and Discussion*

3.1 Single Crystal Growth and Their Characteristics

The single crystals grown by the self-flux method cooling from 1000°C to 800°C are shown in Fig.1. All the growth experiments were carried out in the mixture of Sr/Ca=1.0. Superconducting $Bi_2(Sr,Ca)_{3-x}Cu_2O_y$ crystals were obtained from the solutions with the hatched region. These crystals were black, shiny and irregularly shaped and showed the facet. The single crystals of $Bi_2(Sr,Ca)_2CuO_y$ could be grown from the solutions with less CuO content and were also black and shiny and showed the facet. From the solutions with more Bi_2O_3 content, the single crystals of $(Sr,Ca)BiO_3$ appeared. These crystals were yellowish black and irregularly shaped and could be easily cleaved on the facet.

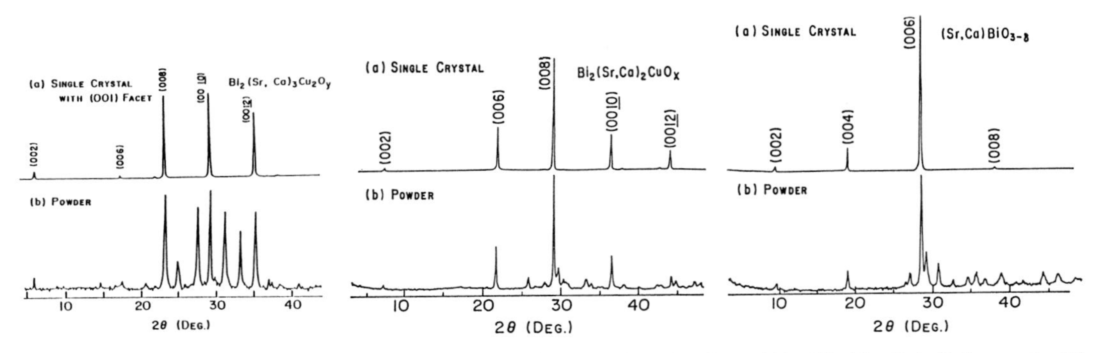

Fig.2 X-ray diffraction patterns of (1) $Bi_2(Sr,Ca)_{3-x}Cu_2O_y$, (2) $Bi_2(Sr,Ca)_2CuO_y$ and (3) $(Sr,Ca)BiO_3$ crystals and powder X-ray patterns

The X-ray diffraction patterns of (1) $Bi_2(Sr,Ca)_{3-x}Cu_2O_y$, (2) $Bi_2(Sr,Ca)_2CuO_y$ and (3) $(Sr,Ca)BiO_3$ crystals with such facets and the powder X-ray patterns were shown in Fig.2. These patterns indicate the plane of the facet to be (001), namely the c plane. The powder X-ray diffraction patern in Fig.2(1) indicates that the crystal is composed of $Bi_2(Sr,Ca)_{3-x}Cu_2O_y$. The pattern in Fig.2(2) shows that the compound has the c axis of 24A reported by Mitchel(1) and Akimitsu(2). Fig.2(3) shows that the crystal has the c axis of 18A reported by Hazen(6). The chemical compositions of the single crystals were analyzed with energy dispersive X-ray analyser. The composition of the $Bi_2(Sr,Ca)_{3-x}Cu_2O_y$ crystal was determined to be $Bi_{2.2}(Sr,Ca)_{2.6}Cu_2O_y$. The cation ratio of Bi, alkaline earth elements and Cu did not change in any part of the single crystal. However the Sr/Ca ratio was found slightly varying from 1.05 to 1.13 in the crystal.

The temperature dependence of the resistivity of the single crystal with the composition of $Bi_{2.2}(Sr,Ca)_{2.6}Cu_2O_x$ measured perpendicular to the c-axis is shown in Fig. 3(a). The resistivity was 1 mΩ cm at 300K for all of the crystals measured. This value is a little larger than those reported by Sunshine et al. (9) and Takagi et al.(10). The temperature dependence of the resistivity was almost linear. The resistive Tc onset and the zero resistance Tc were determined to be 84.0 and 81.5K, respectively.

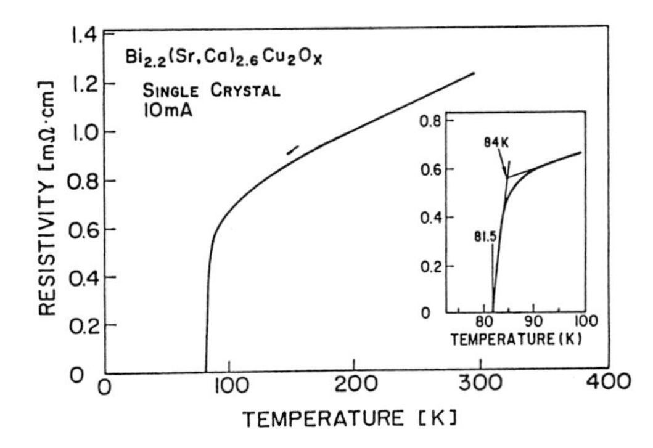

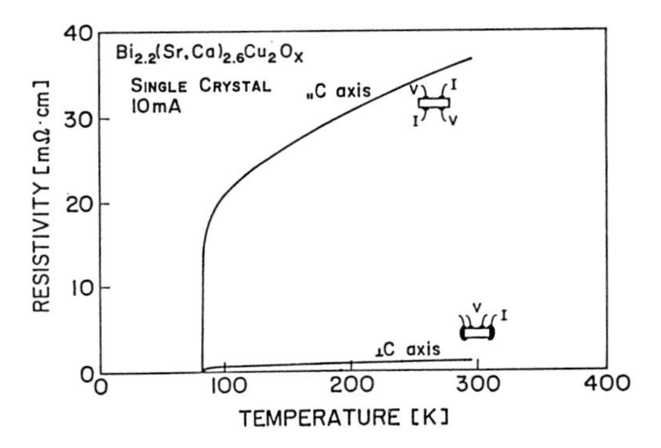

Fig.3 Temperature dependence of the resistivity for single crystal of $Bi_{2.2}(Sr,Ca)_{2.6}Cu_2O_x$ measured (a) in the c plane and (b) along the c axis

The temperature dependence of the resistivity for the single crystal with electrical current parallel to the c axis is shown in Fig. 3(b). The curve measured with the current perpendicular to the c axis is also shown in this figure for comparison. The temperature dependences of the resistivities were very similar to each other and the resistivity showed the linear dependence on temperature. The Tc onset and the zero resistance Tc were identical for the both directions. At Tc onset the resistivity parallel to the c axis, however, was 35 times larger than the resistivity perpendicular to the c axis. The observed anisotropy was considered to be intrinsic, resulting from the anisotropy of the crystal structure, although the absolute value of the resistivity ratio may change due to the fluctuation of the composition in the crystal, especially of the oxygen content.

The temperature dependence of the magnetization of the single crystal was measured by SQUID. The value of the magnetization increases with the increase of temperature and the zero magnetization is observed at 84K, which is coincident with the resistive Tc.

The temperature dependence of the resistivity in the single crystal of $Bi_2(Sr,Ca)_2CuO_y$ is shown in Fig.4. The resistivity at room temperature was several mΩcm, however, the temperature dependence showed that the crystal is semiconductive. Although $Bi_2Sr_2CuO_y$ has the superconductivity, substitution of Sr by Ca may cause the change in its electronic band structure. The superconductivity in this structure, therefore, does not exist in $Bi_2(Sr,Ca)_2CuO_y$

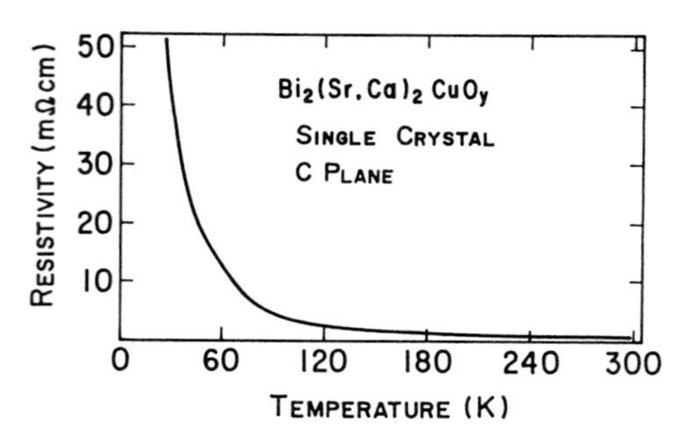

Fig.4 Temperature dependence of resistivity for $Bi_2(Sr,Ca)_2CuO_y$ single crystal

The single crystal of $(Sr,Ca)BiO_3$ was insulative and the resistivity was more than MΩcm at room temperature.

3.2 Cation Deficiency

X-ray diffraction patterns of the sintered samples with various amount of cation deficiency ($-0.2<x<1.0$) are shown in Fig.5. In the samples with x=-0.2 to x=0.3, the X-ray diffraction patterns show that these samples consist of nearly single phase, $Bi_2(Sr,Ca)_3Cu_2O_y$. Whereas, the samples with x>0.4 in $Bi_2(Sr,Ca)_{3-x}Cu_2O_y$ are composed of both $Bi_2(Sr,Ca)_3Cu_2O_y$ and $Bi_2(Sr,Ca)_2CuO_y$. The existence of $Bi_2(Sr,Ca)_2CuO_y$ is comfirmed by the arrow peak in the X-ray diffraction patterns in Fig.5. The samples of $Bi_2(Sr,Ca)_{3-x}Cu_2O_y$, therefore, remain of the same structure of $Bi_2(Sr,Ca)_3Cu_2O_y$ in the range of $-0.2<x<0.3$.

Many satellite reflections are observed in the powder X-ray diffraction pattern as shown in Fig.5. These reflections are presicely coincident with those in ref.(4). The intensities of the reflections of (0071), (021$\overline{1}$) and (0211) are observed to increase with the increase of the cation deficiency. This means that the cation deficiency should release some forces by producing the incommensurate modulated structure of Bi atoms.

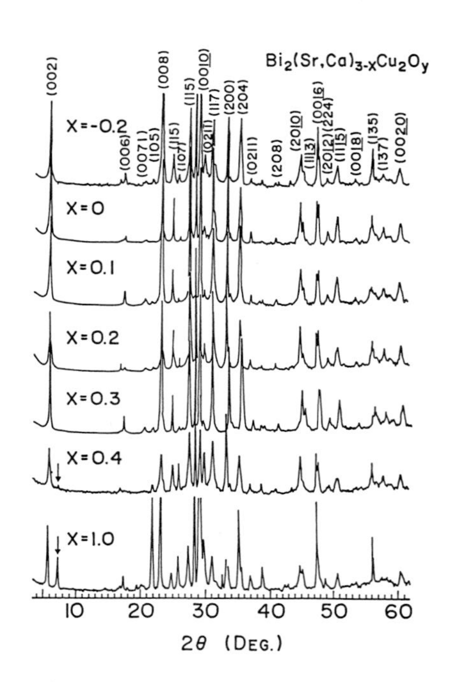

Fig.5 X-ray diffraction patterns for $Bi_2(Sr,Ca)_{3-x}Cu_2O_y$

The lattice parameters of the c axis and the a, b axis measured from the 2θ of (001) (l=2n,n;integer), (200) and (020) peaks in X-ray diffraction patterns are shown in Fig.6. The lattice parameter of the c axis decreases with the increase of the cation deficiency from x=-0.2 to x=0.2. The lattice parameter does not change with x more than 0.3 within the experimantal errors. This results in a solid solution ranging over $-0.2 \leq x \leq 0.3$. The lattice parameter of the a and b axis, however, does not change in this region.

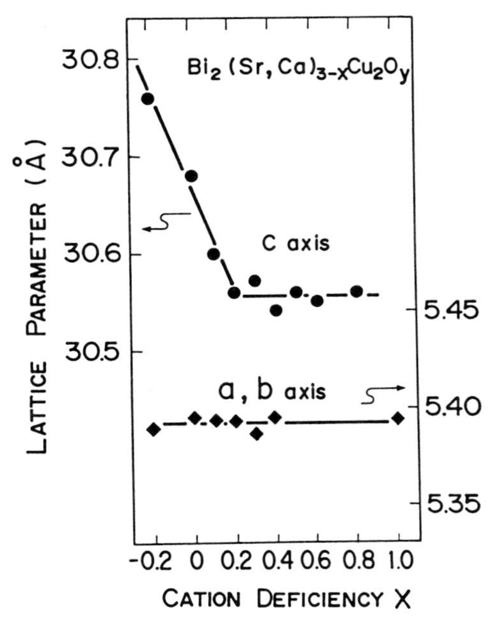

Fig.6 Relation between lattice parameters and cation deficiency

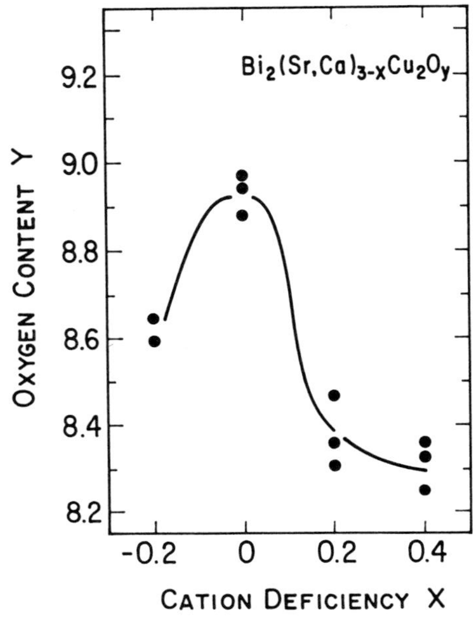

Fig.7 Relation between oxygen content and cation deficiency

The relation between the oxygen content, y, in the samples of $Bi_2(Sr,Ca)_{3-x}Cu_2O_y$ and the cation deficiency, x, is shown in Fig.7. The density measurement shows that there are vacancies in Bi and Cu sites in the sample with x=-0.2. The oxygen content, therefore, has the peak at the stoichiometric $Bi_2(Sr,Ca)_3Cu_2O_y$.(14)

The relation between the zero resistance Tc and the cation deficiency, x, is shown in Fig.8. The zero resistance Tc once increases and decreases with increasing the cation deficiency and shows the peak temperature of 80K in the sample with x=0.1. This curve is similar to the relation between the oxygen content and the cation deficiency. The zero resistance Tc in the single crystal of $Bi_{2.2}(Sr,Ca)_{2.6}Cu_2O_y$ is plotted in the same figure. Bi ions can be substituted for the alkaline earth element sites(13). The chemical composition of this single crystal shows that excess Bi ions are partially substituted for the vacant alkaline earth sites. The chemical composition of the single crystal is thought to be

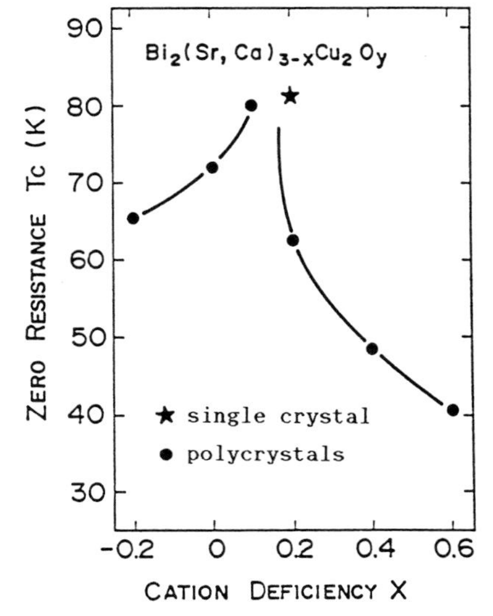

Fig.8 Relation between zero resistance Tc and cation deficiency

$Bi_2(Bi,Sr,Ca)_{2.8}Cu_2O_y$. The zero resistance Tc of the single crystal is, therefore, plotted at the cation deficiency in the alkaline earth site of x=0.2. This value is coincident with those in the polycrystalline samples.

4. *CONCLUSION*

Single crystals of Bi-Sr-Ca-Cu Oxide were grown by the self-flux method. Relation between the single crystals produced and the starting compositions in the Bi-Sr-Ca-Cu Oxide system was obtained for $Bi_2(Sr,Ca)_2CuO_y$, $Bi_2(Sr,Ca)_{3-x}Cu_2O_y$ and $(Sr,Ca)BiO_3$. The dimensions of the single crystal of the superconductor obtained were up to $7X5X1mm^3$.

The effect of cation deficiency of $Bi_2(Sr,Ca)_{3-x}Cu_2O_y$ on its superconductivity have been studied. This system forms a solid solution within the x range of -0.2 to 0.3. The lattice parameter of the c axis decreases with the increase of x. The strong sattelite peak in the X-ray diffraction patterns originating from the incommensurate modulated structure are observed for the samples with large cation deficiency.

The critical temperature of the single crystal was 84K which was determined by both magnetization and electrical resistivity measurements. The resistivity ratio in the c plane and along the c axis for $Bi_2(Sr,Ca)_{3-x}Cu_2O_y$ was found to be approximately 35 at Tc onset.

REFERENCE

1) C.Mitchel, M.Herview, M.M.Borel, A.Gradin, F.Deslandes, J.Provost and B.Raveau:Z.Phys.,B68,(1987)421

2) J.Akimitsu, A.Yamazaki, H.Sawa and H.Fujiki:Jpn.J.Appl.Phys.,26,(1987)L2080

3) H.Maeda, Y.Tanaka, M.Fukutomi and T.Asano:Jpn.J.Appl.Phys.,27,(1988)L209

4) M.Onoda, A.Yamamoto, E.Takayama-Muromachi and S.Takekawa:Jpn.J.Appl.Phys.,27,(1988)L833

5) Y.Syono, K.Hiraga, N.Kobayashi, M.Kikuchi, K.Kusada, T.Kajitani, D.Shindo, S.Hosoya, A.Tokiwa, S.Terada and Y.Muto:Jpn.J.Appl.Phys.,27,(1988)L569

6) R.M.Hazen, C.T.Prewitt, R.J.Angel, N.L.Ross, L.W.Finger, C.G.Hadidiacos, D.R.Veblen, P.J.Heaney, P.H.Hor, R.J.Meng, Y.Y.Sun, Y.Q.Wang, Y.Y.Xue, Z.J.Huang, L.Gao, J.Bechtold and C.W.Chu:Phys.Rev.Lett.,60,(1988)1174

7) M.A.Subramanian, C.C.Torardi, J.C.Calabrese, J.Gopalakrishnan,K.J.Morrissey, T.R.Askew, R.S.Flippen, U.Chowdhry and A.W.Sleight:Science,239,(1988)1015

8) T.Kajitani, K.Kusaba, M.Kikuchi, N.Kobayashi, Y.Syono, T.B.Williams and M.Hirabayashi:Jpn.J.Appl.Phys.,27,(1988)L587

9) S.A.Sunshine, T.Siegrist, L.F.Schneemeyer, D.W.Murphy, R.J.Cava, B.Batlogg, R.B.van Dover, R.M.Fleming, S.H.Glarum, S.Nakahara, R.Farrow, J.J.Krajewski, S.M.Zahurak, J.V.Waszczak, J.H.Marshall, P.Marsh, L.W.Rupp Jr. and W.F.Peck:preprint

10) H.Takagi, H.Eisaki, S.Uchida, A.Maeda, S.Tajima, K.Uchinokura and S.Tanaka:Nature,332,17 March (1988)

11) L.F.Schneemeyer, R.B.van Dover, S.H.Glarum, S.A.Sunshine, R.M.Fleming, B.Batlogg, T.Siegrist, J.H.Marshall, J.V.Waszczak and L.W.Rupp:Nature,332,31 March (1988)

12) S.Nomura, T.Yamashita, H.Yoshino and K.Ando:Jpn.J.Appl.Phys.,27,(1988)L1251

13) K.D.Mackay, M.L.Allan and R.H.Friend:J.Phys.C:Solid State Phys.,21,(1988)L529

14) S.Nomura, T.Yamashita, H.Yoshino and K.Ando:to be published

4.4 Processing and Microstructure

Progress in Processings of Oxide Superconductor

Y. TANAKA

Yokohama R&D Laboratories, The Furukawa Electric Co., Ltd., 2-4-3, Okano, Nishi-ku, Yokohama, 220 Japan

ABSTRACT

The fabrication processes of oxide superconductors have been surveyed on a category of bulk or wire materials. The conventional powder process has been developed and modified considerably in recent years. Very recently grain-oriented processes were realized for enhancing the transport current density of the oxide superconductors. Magnetic properties of the grain-oriented materials are significantly improved and show a similar charcteristic to that of the c-axis oriented thin film. This grain-oriented process is eminently suitable for use in manufacturing oxide superconducting wires. But this process is faced not only in rare earth elements but also in bismuth systems and thallium systems for lengthened wires for keeping good magnetic properties.

INTRODUCTON

Since the discovery of superconducting materials with high critical tempreature, Tc,(1)(2) many researchers around the world have been trying both to accomplish higher Tc and to develop the practical application potential of these new materials. Among the copper oxide-based superconducting materials, $YBa2Cu3O7-x$ compounds, have recived considerable attention because of their high upper ciritical magnetic field, Hc2, and considely high ccritical current density, Jc. Most applications require both desirble superconducting proerties (Tc, Hc2 and Jc) and machanical ones (strengh and ductility). It has been observed that although rare earth elements compounds have high Tc, the value of Jc is limited in polycrystalline mateirals(3). In addition, these materials have been observed to be brittle with very low ductility and fracture toughness(4).

In view of these problems, efforts have been made to improve both the Jc and the mechanical properties of the polycrystalline oxide superconducting materials. For Jc enhancement in bulk or wire materials, these efforts have focused on improveing density, oxygen content, grain-orientation and minimizing weak links at grain boundaries. Reported zero-field transport Jc at 77 K for these materials is rather low in the polycrystalline state, 10^2-10^3 A/cm^2 being ususal(5). However, the high Jc values reported in literature (10^6 A/cm^2) are magnetization Jc's obtained from magnetization hysteresis loops. The measured transport Jc values are low presumably due to intergrain impedance because the grains are weakly conpled, and are significantly affected by processing techniques.

For commercially useful applications, many processing techniques are under consideration. These include preform processes, modified sintering processes, melt processes and a mushy process. In the following, a frief survey of these progressing in process is carried out and some potential processing for bulk or wire materials are pointed out.

PREFORM PROCESS

The oxide superconductors are usually brittle (maximum strain 0.04-0.1 %) and difficult to fabricate into tapes or wires, which are necessary for power applications. For these materials, it is useful to adopt from the conventional fabrication methods, such for polycrystalline $Al2O3$, in which the slurry-type precursor to the oxide is preformed as a desired shape, dried and subjected to heat treatment to remove volatile components and to generate the oxide. These preform processes include tape casting method(6), die-free spinning method(7), sol-gel method(8) and extruding method(9).

In these methods it is rather easy to generate the oxide superconducting phase having higher Tc than the liquid nitrogen temperature (6)(7)(8)(9). But reported density for these methods is rather low in the sintered state, less than 90 %, then it is important to make a powder of highest content in order to improve the mechanical properties and the transport current. Thus the particle size in the spin mixture and the viscosity of the slurry have to control carefully. Recently the sticky and dense precursor has been performed by the extruding mathod(9), in which the oxide powder was suspended in an organic polymer of polyvinyl alcohol (PVA) and an organic solvent. An extruded wire was wound on a mandrel and heated to dry any to sinter. For the coil obtained magnetic fields, 21.1 gauss and 350 gauss were successfully generated by dc operations at 77 K and 4.2 K, respectively. A critical current density of the wire was of over 10^3 A/cm^2 at 77 K in zero-magnetic field, but it was dropped down steeply with increasing magnetic fields.

MODIFIED SINTERING PROCESS

For Jc enhancement in the sintering method, many efforts have focused on improving homogeneity of the oxide powder, density and grain alignment, and minimizing impurities at grain boundaries.

For synthesized or sintered materials it is important to make more homegeneous powder or grain in composition, in order to transform sharply into the superconductive state. We have reported a newly developed processing technique that has greatly improved the homogeneity of compositon and the grain size(10). In this method an aqueous solution of Y, Ba and Cu acetate with a nominal composition of YBa2Cu3O7-x was initially prepared as a precursor to the oxide. Reported Jc of a pellet using this homogeneous powder was of over 300 A/cm² at 77°K in zero-magnetic field.

Relative density of the sintered oxide superconductors prepared by conventional powder solid state reactions was rather low, less than 90 % of the theoretical one, because the particle size was large and its size distribution was broad. In a modified sintering process the YBa2Cu3O7-x was made by the wet-milled slurry. the slurry was dried in air. The dried powder was calcined at 950 C for 6 hr. The calcined powder was pulverized and sized into four average ranges of 10 μm, 5 μm, 1 μm and 0.5 μm. The sized powder was pressed to pellets (1x2x30 mm³) and sintered at 850°C for 100 hr in oxygen. High dense materials (over 95%) have been performed for the pellets having fine particle size of less than 5 μm. As well as the high Jc values have been achieved in the same particle range, in which the highest value was 800 A/cm² at 77°K in zero-magnetic field as shown in Fig. 1. The Jc values were increased with volume fraction of the oxide superconductor which was estimated by the susceptibility measurement as shown in Fig.2. The sintered materials however was observed in random orientation from the results of the X-ray measurement.

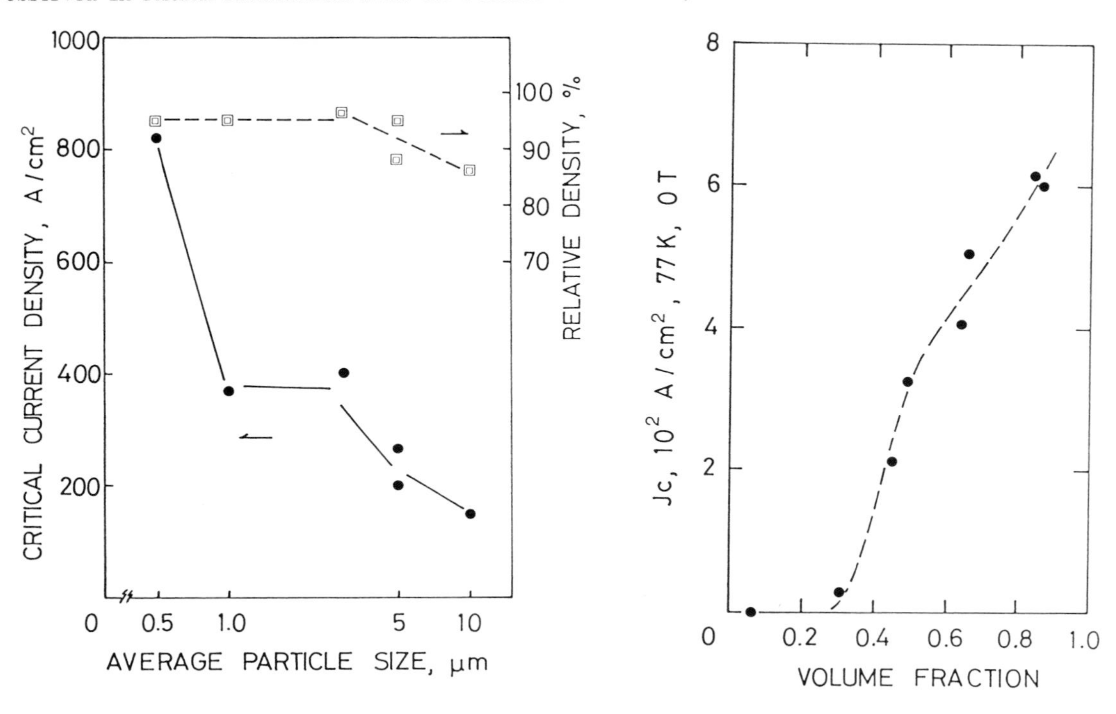

Fig.1 Relationships between particle size and critical current density of the YBCO oxide superconductor prepared by the sintering process.

Fig.2 An effect of the volume fractions on the Jc values of the YBCO oxide superconductors

In the conventional sintering process the sequence of events during calcination and subsequent sintering has not always understood. This is likely to be carried out on the two stages as shown in Fig. 3-1, where it is important to control mixtures of Y2BaCuO5 and BaCu2O2 as a precursor to the final oxide superconductor, YBa2Cu3O7-x(Fig. 3-2). This sequence has been confirmed in a newly developed process(11), in which two reactions was contralled both in the calcination and the sintering process. This is useful for practical applications to make a composite oxide wire which composes of Y2BaCuO5 and BaCuxOy(12), and for understanding a diffusion reaction between Ba-Ca-Cu oxide and Tl2O3 in the thallium system. The doping effect of Ag for the yttrium system(13) and of Pb for the bismuth system(14) on the formation of the high-Tc phases will closely relate with the simmilar reactions.

Recently the grain-oriented techniques have been reported. These techniques include a powder-in-tubing method(5) and a hot-forging method(15). As a result the Jc(10^3-10^4 A/cm² at 77°K in zero-magnetic field) of the heavily deformed and subsequent sintered samlples was eleveated to about ten times that of the conventional samples. Reported results was explained by the grain-orientation factor(15) and somewhat oriented structures(16). The exact mechanism by which grain-orientation and high local stresses promote the acheivement of the high Jc is not known.

MELT PROCESS

Fabrication of high Tc superconductors by an oxide melting method in place of the conventional sintering method will be suitable for incresing the relative density of the obtained materials. Using the melt porcesses, i.e., melt drawing(17), melt spinning(17), preform-wire melting(16), or melt quenching(18), the fabrication of the oxide superconducting wire has been attempted. The density of the melt-processed compound was measured as high as 98% of the theoretical density as shown in Fig.4 and as compared to the value of 80-85% density for the conventional sintered samples. In the melt quenching method, as-quenched meterials of the yttrium system is rather low dense, 5.7 g/cm³ (19), and after post-annealing in oxygen the density of the samples was incresed with the annealing duration, and the transport current density was also incressed during annealing. But the Jc is rather low, 650 A/cm² at 77 °K in zero-magnetic field, because of the random orientation and the appearane of the second non-superconductive phase between the superconductive grains.

(1) Calcination

(a) $Y_2O_3 + 4BaCO_3 + 6CuO \rightarrow 2YBa_2Cu_3O_{7-x} + 4CO_2$

(b) $2Y_2O_3 + 8BaCO_3 + 12CuO$

$\rightarrow 2Y_2BaCuO_5 + 5BaCu_2O_2 + BaO + 8COx + 3O_2$

(2) Phase fransformation

RT $Y_2BaCuO_5 + BaCu_2O_2 + BaO$

400°C $Y_2BaCuO_5 + BaCu_2O_2 + BaO + O_2$

800°C $Y_2BaCuO_5 + BaCuO_2 + CuO + O_2$

900°C $YBaCu_3O_{7-x}$

Fig.3 Reaction sequences of the YBCO oxide superconductor

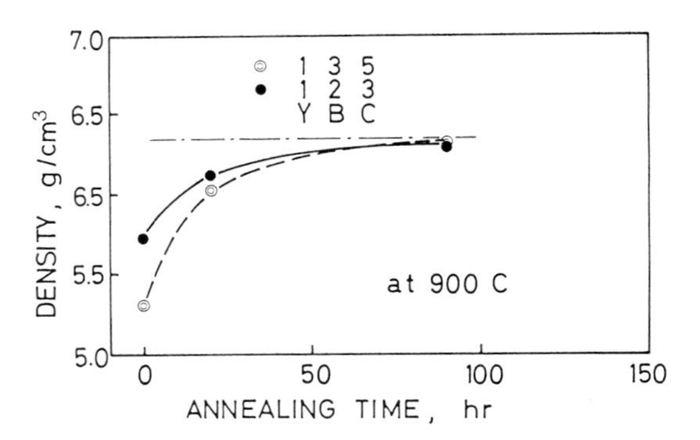

Fig.4 Effects of the annealing on densities of the YBCO oxide superconductors prepared by the melt process.

On the other hand, in the bismuth system the melt quenching method leads to produce amorphous phase during cooling from the melten state, depending on the colling rate from the high temperature(18). Fig.5 shows the sensitivity of the formation of the high-Tc phase due to the cooling rate in some bismuth systems as compared to the sintered materials. Results suggest that the slow cooling or the themally equilibrium process is suitable for the formation of the high-Tc phase.

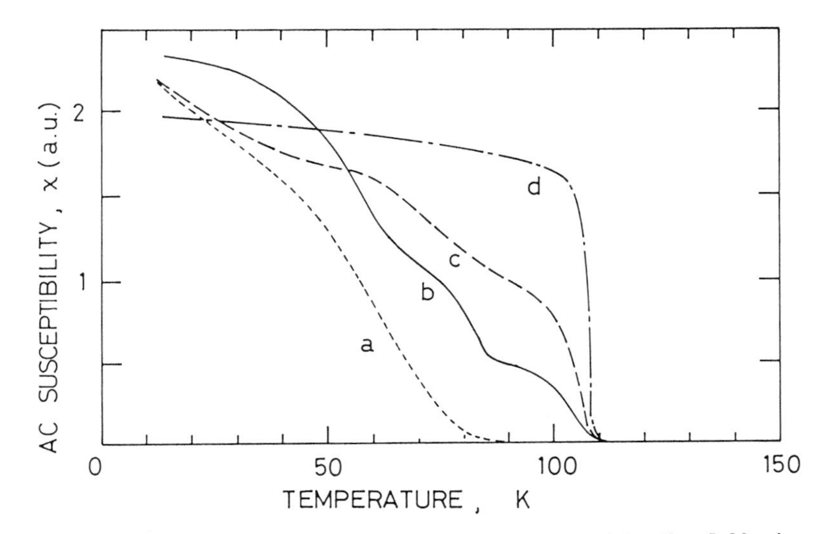

Fig.5 AC susceptivility of the BSCCO oxide superconductors prepared by the following conductions.
(a) BiSrCaCu2Ox melt quenching (10^3-10^4 C/sec)→870°C x10h→750°C x10h →furnace cooling
(b) BiSrCaCu2Ox melt cooling (-10^2 C/sec)→870°C x10h→750°C x10h →furnce cooling
(c) BiSrCaCu2Ox calcination (850 C x5h) sintered (865 C x10h) furnace cooling
(d) (Bi0.7 Pb0.3)SrCaCu1.8Ox→calcination(800°C x12h)→sintered(845°C x270h)→furnace cooling

MUSHY PROCESS

The discrepancy between the magnetizatiion and transport Jc values and severe field dependence of Jc led to the suggestion of weak coupling by tunneling of intrinsically high Jc grains in sintered YBa2Cu3O7-x. A number of possibilities exist for the source of the weak links which are most likely to be located at or near the grain boundaries; severe anisotropy in conductivity, the presence of impurity layers, the presence of porosity due to microcracks/stress-concentration, or other variation in chemistry or crystal structure at grain boundaries(20). Recently it has been demonstrated that the weak-link problem can be greatly reduced by a new "melt-textured growth" (MTG) processing(20)(21), instead of the conventional melt or sintering process.

In the MTG processing, the sintered samples of YBa2Cu3O7-x were heated to a temperature range of 1,050-1,200°C at which the samples were in a mushy state of solid and liquid, and then cooled in a temperature gradient followed by a long oxygenation heat treatment. This method can adopt into a composition region of the phase-equilibrium diagram proposed by D.G.Hinks et.al(22), as shown in Fig.6. The figure suggests that the nominal compositions for the method cannot be always the congruent one of YBa2Cu3O7-x. But the deviation from the congruent composition leads to the formation of the non-superconductive pbases such as Y2BaCuOx, BaCuxOY or CuOx. Fig.7 shows a transmission electron micrograph of the mushy processed sample of which the nominal compositon was

Fig.6 Phase diagram for the Y2O3-BaO-CuO system at around 950°C. Hinks et al's superconducting conmposition and non-superconducting composition are represented by triangles and circles, respectively. A hatched area is showning possible compositon for orented materials prepared by the mushy process.

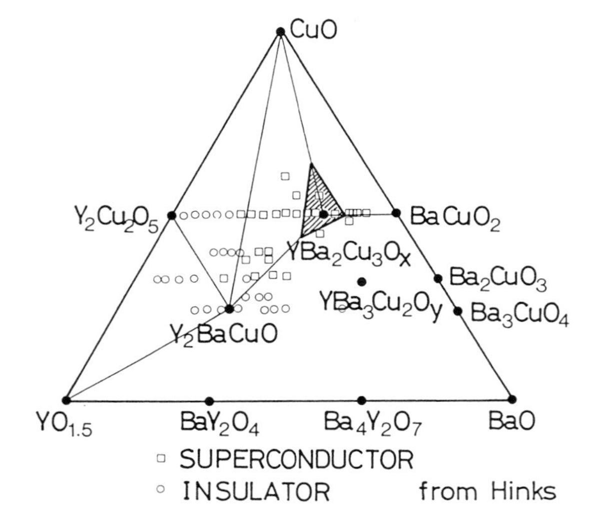

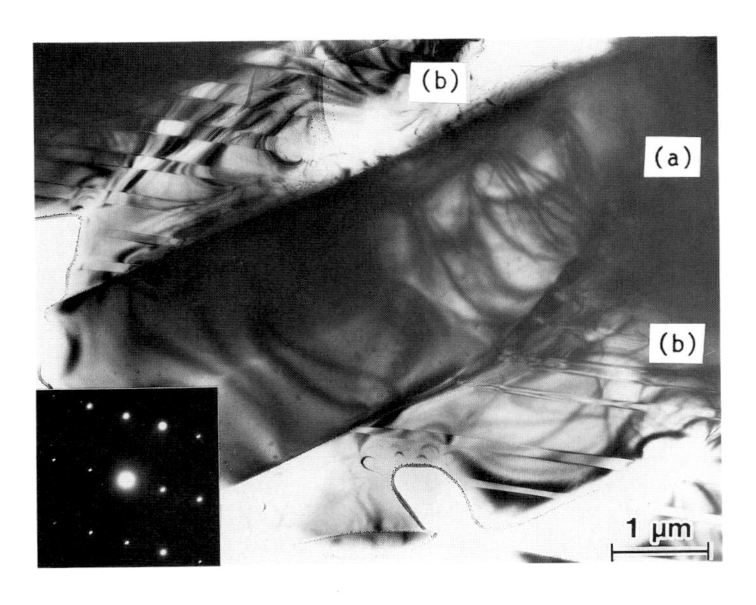

Fig.7 TEM micrograph of a mushy processed YBCO oxide superconductor with a beam along the [001], showing the coexistence of orthorhombic twins labelled (b) and an unknown phase (a).

shifted by Y:Ba:Cu=1:2:3.52. The figure exhibits well-aligned structure consiting of long, needle-shape grains. The needles have been identified to be the superconducting $YBa_2Cu_3O_{7-x}$ phase(b) and a somewhat twin-free phase (a) which may exhibit non-superconductive. The superconducting phase is usually characterized by twin structures, and also determined by X-ray measurements as shown in Fig. 8 , where the oxygen contents are linerly related with the lattice constants of the c-axis(23).

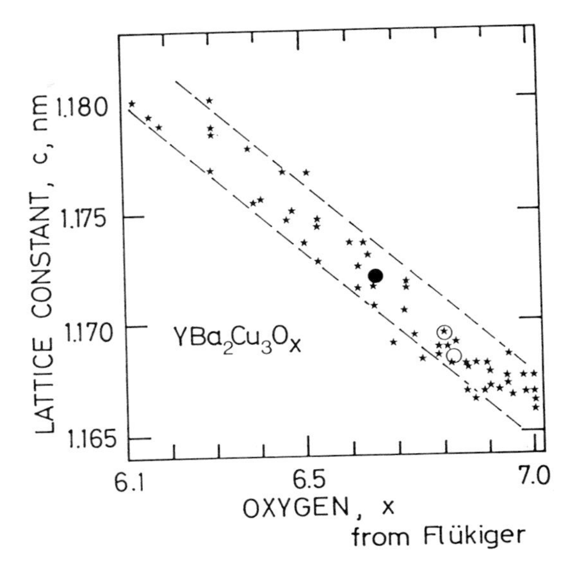

Fig. 8 Relationship between oxygen contents and the lattice constants of c-axis for the YBCO oxide superconductors surveyed by Flükiger et al. Superconductong and non-superconducting material prepared by the mushy process are represented by an open circle and clased circles, respectively.

While the transport critical current density Jc in the mushy-processed samples dramatically improved (7,400 A/cm² at 77°K in zero magnetic field)(20) and revealed considerably good field dependence of Jc, the Jc should be defined by an exact crierion, voltage or resistivity. Fig. 9 shows a relationship between the Jc values and the related criteria in some grain-oriented samples of the $YBa_2Cu_3O_{7-x}$. It is suggested that the Jc values of the non-oriented samples are significantly affected by the voltage criteria. For instance, when the Ic criterion of the non-oriented samples is changed from 1 μV/cm to 10 μV/cm, the Jc values will be increased by about ten times.

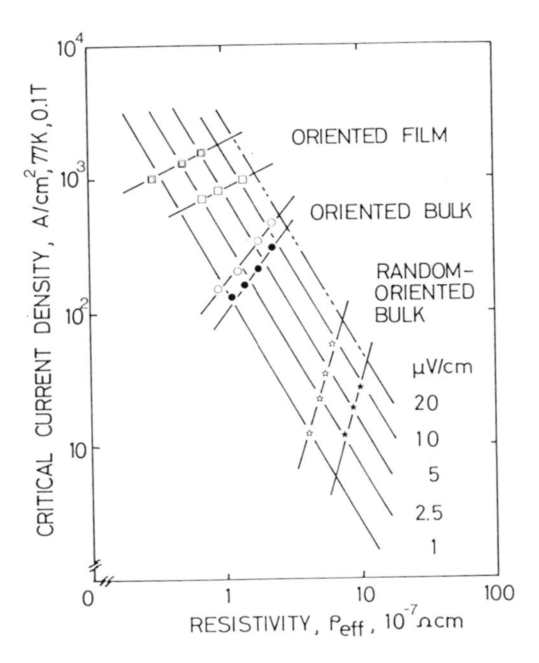

Fig. 9 Effects of the Ic criteria on the Jc values of YBCO oxide superconductors

SUMMARY

On the fabrication process of the oxide superconducting bulks or wires the following progresses are summarized.

(1) In the preform precess with precursor materials to the final oxide superconductor, the Jc values are affected by the density.

(2) The development of the sintering process is significant for the realization of the practical superconducting wires, in which to understand the pricipal method of the grain orientation is important as well as to enhance bulk density and fraction of the superconducting phase or to contral the diffusion processes.

(3) The melt process and the mushy process will give rise to improve the Jc value of bulk materials, and also give valuable knowledges for developing the modified sintering processes.

ACKNOWLEDGEMENTS

This work cooperated with Tokyo Electric Power Co., Inc., Hokkaido Electric Power Co., Inc., Tohoku Electric Power Co., Inc., Electric Power Development Co., Ltd., and The Furukawa Electric Co., Ltd.

The author is also deeply grateful to the whole member of our Superconducting Research Development Department and Material Analysis & Testing Section in Yokohama R&D Laboratories.

REFERENCES

(1) J.G.Bednorz and K.A.Müller, Z.Phys. B 64, 189 (1986).
(2) C.W.Chu, P.H.Hor, R.L.Meng, L.Gao, Z.J.Hung and Y.Q.Wang, Phys. Rev. Lett. 58, 405(1987).
(3) J.W.Ekin, Adv.Ceram.Mater. 2(3B), 586, Special Issue (1987).
(4) R.F.Cook, T.M.Shaw and P.R.Duncombe, Adv. Ceram. Mater. 2(3B),606, Special Issue(1987).
(5) Y.Tanaka, K.Yamada and K.Matsumoto, MRS Symposium, May 30-June 3, (1988) Tokyo.
(6) M.Ishii, T.Maeda, M.Matsuda, M.Takata and T.Yamashita, Jpn. J. Appl. Phys, 26, L1959(1987).
(7) T.Goto and I.Horiba, Jpn. J. Appl. Phys, 26, L1970(1987)
(8) F.Uchikawa, H.Zheng, K.C.Chen and J.D.Mackenzie, MRS Spring Meeting, Nevada(1988).
(9) Y.Tanaka, K.Yamada and T.Sano, Jpn. Appl. Phys, 27,L362(1988)
(10) T.Shibata, R.Setaka, W.Komatsu and Minami, MRS Symposium, May 30-June 3, (1988) Tokyo.
(11) N.Uno, N.Enomoto, Y.tanaka and H.Takami, Jpn. J. Appl. Phys, 27, L1003 (1988).
(12) Y.Ikeno, N.Sadakata, A.Kume, K.Gotho and O.Kohno, MRS Symposium, May 30-June 3, (1988) Tokyo.
(13) J.P.Sing, D.Shiand D.W.Capone II, Appl. Phys. Lett, 53(3), 237(1988).
(14) D.R.Dietderich, H.Kumakura and K.Togano, MRS Sysmpsium, May 30-June 3, (1988) Tokyo.
(15) T.Takenaka, H.Noda, A.Yoneda and K.Sakata, Jpn. J. Appl. Phys, 27, L1209 (1988).
(16) M.Okada, A.Okayama, T.Morimoto, K.Aihara and S.Matsuda, Jpn. J. Appl. Phys, 27, 1185 (1988).
(17) S.Jin, T.H.Tiefel, R.C.Sherwood, G.W.Kammlott and S.M.Zahurak, Appl. Phys. Lett. 51(12), 943(1987).
(18) T.Komatsu, K.Imai, O.Tanaka, K.Matsuita, M.Takata and T.Yamashita, Nippon-Seramikusu-Kyokai-Gakujutsu-Ronbunshi, 94(4), 367(1988).
(19) T.Komatsu, K.Sato, K.Imai, K.Matsusita and T.Yamashita, Jpn. J. Appl. Phys, 27, L550 (1988)
(20) S.Jin, T.H.Tiefel, R.C.Sherwood, M.E.Davis, R. B. van Dover, G.W.Kammlott, R.A.Fastnacht and H.D.Keith, Appl. Phys, Lett. 52(24), 2074(1988).
(21) J.S.Zhang, X.P.Jiang, J.G.Huang, Y.Yu, M.Jiang, Y.L.Ge, G.W.Qiao, Z.Q.Hu, Y.H.Zhao and Y.J.Wang, ICMC-88, GC-9(1988), Shenyang.
(22) D.G.Hinks, L.Sodrholm, D.W.Capone II, J.D.Jorgensen, Ivan K.Schuller, C.U.Segre, K.Zhang and J.D.Grace, submitted to Appl. Phys. Lett.
(23) T.Wolf, I.Apfelstedt, W.Goldacker, H.Küpfer and R.Flükiger, presented at "High-Tc Superconductors Materials and Mechanisms of Superconductivity, (1988), Interlarken.

Microstructure and Transport Properties of Oxide Superconductors

M. Murakami, S. Matsuda, K. Sawano, K. Miyamoto, A. Hayashi, M. Morita, K. Doi, H. Teshima, M. Sugiyama, M. Kimura, M. Fujinami, M. Saga, M. Matsuo, and H. Hamada

R & D Laboratories-I, Nippon Steel Corporation, 1618 Ida, Nakahara-ku, Kawasaki, 211 Japan

ABSTRACT

Microstructural control is crucial to improve transport critical current density in oxide superconductors. Bulk sintered materials contain a number of defects such as porosities, second phases, cracks, impurities, variation in chemical composition and other grain boundary defects. All these defects reduce critical current density of oxide superconductors. The number of these defects can be minimized by controlling the processing conditions. However we have found that the weak-link effects cannot be completely eliminated from the material as far as we employ the solid state reaction. This is confirmed by a sharp degradation of Jc in magnetic field. No sample size dependence in magnetization hysteresis between the increasing and decreasing magnetic field process also indicates that supercurrent localizes intragranularly.

The weak-link problems can be overcome by employing the carefully controlled solidification process. In such solidified samples, magnetic field dependence of transport critical current density was quite small compared with that of bulk sintered material and Jc value exceeding 1000 A/cm^2 was obtained at 77K in 10 T.

INTRODUCTION

After the discovery of LaBaCuO by Bednorz and Muller[1] extensive study has been conducted to search for higher Tc material culminating in the discovery of YBaCuO with Tc exceeding 90K by Wu et al.[2]. This breakthrough has stimulated enormous research to apply new materials for practical applications which require critical current density(Jc) of the order of 10^5 - 10^6 A/cm^2 often in the presence of magnetic field. At first our work had been directed toward producing superconducting wires and tapes. However, we soon realized that we have to clarify the dominant factors controlling Jc prior to the fabrication of commercial products since transport Jc was only a few hundred A/cm^2 which is three orders of magnitude lower than the required value. It is generally known that Jc is sensitive to microstructure, therefore we started detailed microstructural study with a special reference to Jc in the bulk sintered oxide superconductors[3]. It has been found through microstructural investigation[3,4,5] that the possible sources of reduced Jc are i) inhomogeneity ii) the presence of porosity iii) the presence of second phases along grain boundaries iv) the presence of microcracks resulting from the anisotropic thermal expansion along different crystallographic direction and tetragonal to orthorhombic transition. All these defects reduce Jc of oxide superconductors. We have tried to minimize the number of these defects and found that sintering the calcined powders as fine as possible at relatively low temperature can provide the bulk sintered material with Jc of 1200 A/cm^2 at 77K in zero field[3]. However, this value is not sufficient for the practical applications. Furthermore a sharp drop in Jc took place when magnetic field was applied. According to magnetization measurement[3] Jc of three orders of magnitude higher is obtainable intragranularly, indicating that there still exist weak-coupling between grains. This result may imply that severe anisotropy in conductivity causes weak connection at high angle grain boundaries. We also encounter some frustration with powder metallurgical process. That is the optimum sintering temperature for Y based superconductors lies in the region where a liquid phase intrudes, which limits controllable processing variables.

The growth of the superconducting phase from the molten oxides is another promising process. In the course of the attempts to grow single crystals with good quality, we have found[6] that the number of weak-links is extremely small in the solidified samples as compared with the sintered material. However, transport Jc was less than 100 A/cm^2 at 77K. Microstructural observation revealed that nonsuperconducting phases such as $BaCuO_2$ and CuO separate the superconducting phases in such samples. Therefore much higher Jc will be attainable through the microstructural control. This idea has been confirmed by the success of so-called "Melt-Textured-Growth" processing to achieve Jc of 1000A/cm^2 in 1T by Bell group[7], although their design concept was rather different from ours. We have also accomplished Jc of exceeding 1000 A/cm^2 in 10 T by using the melt process.

In view of these results, we first summarize microstructural study in the bulk sintered material and then report microstructure and improvement in Jc using the melt process.

EXPERIMENTAL

The samples were prepared by the conventional solid state reaction using appropriate amounts of oxides and carbonates[3,4,5]. Electrical resistivity was measured by a standard four terminal method. Transport critical currents were obtained from V-I characteristic curves using 1 μV/cm criterion. Then the samples were subjected to microscopic observation, computer aided electron probe microanalysis, Auger electron spectroscopic analysis, X-ray diffraction analysis and magnetization measurement.

As for the melt process, the precursor sintered samples with various chemical compositions(123 stoichiometry and modified compositions) were heated at 1300 to 1500°C for 1 min to 1 h and then quenched into disks about 100 mm in diameter and 2 to 5 mm in thickness. Then the disks were reheated to 1200 to 1300°C for partial melting. The melted samples were then solidified by cooling at -1.5°C/h to 900°C and subsequently cooled at -100°C/h to room temperature. All the treatment was conducted in flowing oxygen. Then the samples were submitted to the same analyses as those of bulk sintered materials.

RESULTS AND DISCUSSION

Microstructure and Jc of bulk sintered oxide superconductors

Jc of bulk sintered material is sensitive to microstructure. In an early stage chemical inhomogeneity was the major problem to reduce Jc of bulk sintered materials. The homogeneity can be forced by controlled calcination and repeated grinding, which lead to the improvement of Jc by two orders of magnitude[3,4].

However, we encounter the basic frustrations at sintering stage. Although the bulk density of oxide superconductors increases with increasing sintering temperature and time as shown in Fig. 1, Jc deteriorates sharply when sintering temperature exceeds 925 °C. Microstructural observation revealed that a liquid phase intrudes on sintering at higher than 950°C as shown in Fig. 2. According to EDS analysis the liquid phase was identified as admixture of $BaCuO_2$ and CuO. Although the presence of the liquid phase promotes densification of oxides, grain boundaries are covered with nonsuperconducting phases. In a dense material microcracks are also formed along grain boundaries as shown in Fig. 3. Stress concentration resulting from anisotropy in thermal expansion in crystallographic orientation and tetragonal to orthorhombic transition is considered to cause cracking at grain boundaries. Incorporation of oxygen is also difficult in a dense material, which leads to the degradation of Jc. The intrusion of microcracks and grain boundary liquid phases can be suppressed by employing lower sintering temperature and clean grain boundaries can be obtained as shown in Fig. 4. However, in that case bulk density is low and a number of porosities remain inside the oxides as shown in Fig. 5. Consequently the effective path for supercurrent is reduced and thereby local critical current density becomes fairly large, which also limits overall Jc. As far as we employ the conventional sintering process, this frustration seems inevitable.

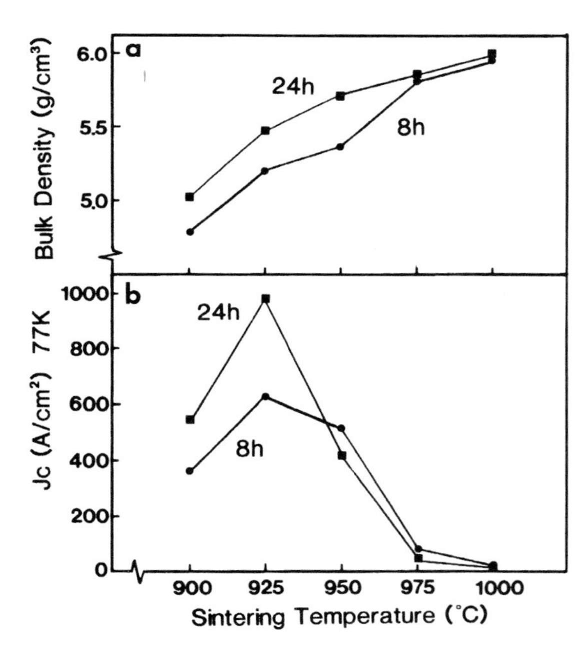

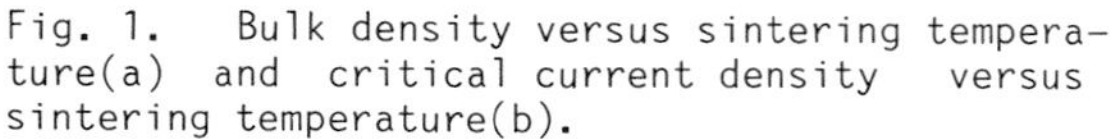
Fig. 1. Bulk density versus sintering temperature(a) and critical current density versus sintering temperature(b).

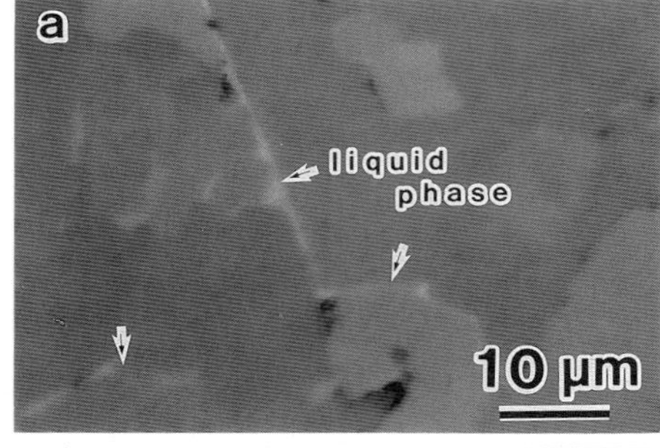

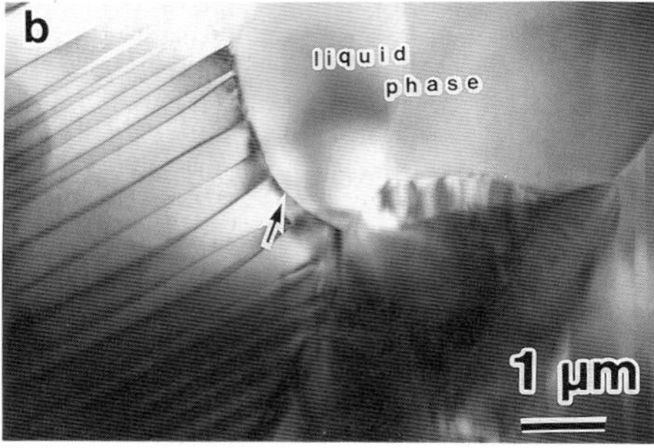

Fig. 2. Optical micrograph(a) and transmission electron micrograph(b) for the oxide superconductor sintered at 950°C.

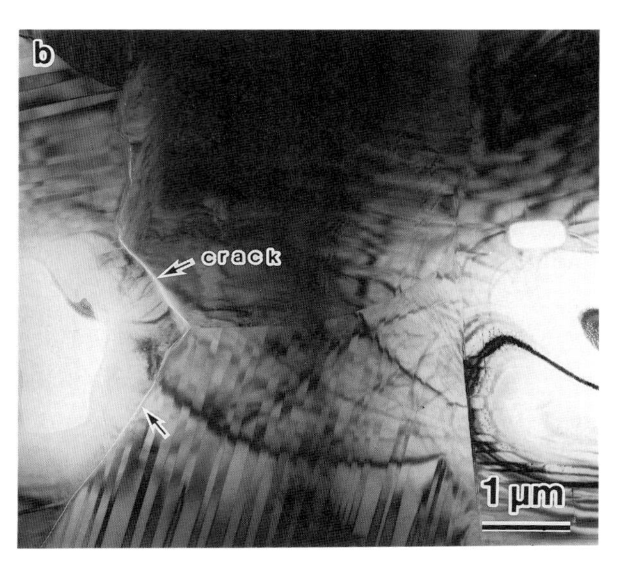

Fig. 3. Microcracks found in the bulk sintered material. (a) Optical micrograph of the sample sintered at 950°C for 8h. (b) Transmission electron micrograph for the sample sintered at 925 °C for 24h.

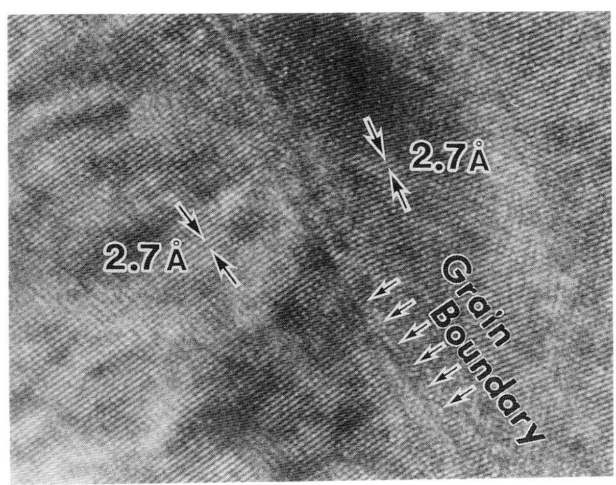

Fig. 4. Transmission electron micrographs for the sample with Jc of 1200 A/cm². Most grain boundaries are free of second phases and microcracks.

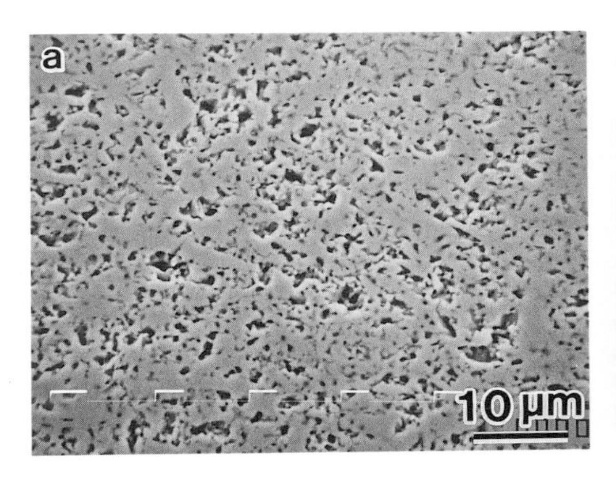

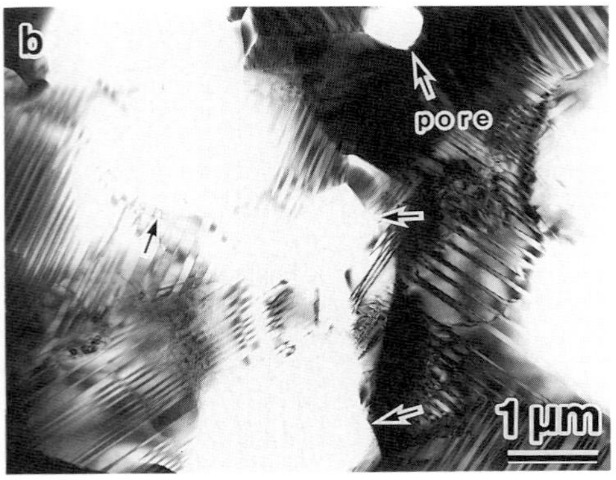

Fig. 5. Scanning electron micrograph(a) and transmission electron micrograph(b) for the sample sintered at 900 °C for 24h. The sample has Jc value of 1050 A/cm² at 77K. Note a number of porosities exist in the sintered body.

The melt process

In the bulk sintered samples magnetization difference between increasing and decreasing the field process is independent of the sample size[3,8]. This indicates supercurrent localizes intragranularly. A sharp drop of Jc in magnetic field also indicates the presence of a weak link network along grain boundaries. However, we have found that a significant sample size dependence in magnetization hysteresis occurs in the samples grown from the melt, which poses the hope that higher Jc may be attainable through the melt process. However, transport Jc was less than 100 A/cm² in such samples. Microstructural observation revealed that the solidified sample comprises admixture of 123 superconducting phase, $BaCuO_2$, CuO and 211 phase as shown in Fig. 6 and this multiphased structure causes the weak connection between superconducting phases leading to low transport Jc. It is known that the superconducting phase forms according to the following reaction[6]:

$$Y_2BaCuO_5 + 3BaO + 5CuO \rightarrow 2YBa_2Cu_3O_{6.5}$$

Admixture of BaO and CuO are forming liquid at this stage. So the 211 phase is surrounded by liquid and the 123 phase grows by the peritectic reaction. Consequently the 211 phase and the liquid phase must coexist for the growth of the 123 phase. Once the 211 phase is trapped inside the 123 phase and separated from the liquid phase no more reaction proceeds. Finally the microstructure as previously shown in Fig. 6 is obtained.

In order to improve transport Jc, the 123 phases must be connected well to each other. For the realization of this the 211 phases must be dispersed finely. At higher temperature the 211 phase also decomposes into Y_2O_3, BaO and CuO. Therefore Y_2O_3 acts as a nucleation site for the 211 phase. The 211 phase grows on cooling and at a certain temperature the peritectic reaction takes place and the 123 phase nucleates at the interface between the 211 and liquid phases. The 123 phase grows gradually, however it stops when the 211 and liquid phases are separated. The 211 phases remain down to room temperature and the liquid phases become $BaCuO_2$ and CuO.

The starting microstructure with finely dispersed 211 phases can be obtained by quenching the molten oxide from fairly high temperature(1300 to 1500°C) and reheating the sample again to high temperature(1200 to 1300°C) and then cooling it to the temperature just above the point where the peritectic reaction takes place. The microstructure with well connected superconducting phases can be attained by slow cooling the material in oxygen atmosphere to incorporate oxygen into the superconducting phase. Fig. 7 shows the typical microstructure of the melt processed sample. Compared with the microstructure of the uncontrolled solidified sample(Fig. 6) the 123 phases are well developed and well connected to each other. Fig. 8 demonstrates the temperature dependence of resistivity and ac susceptibility of the melt processed sample. Zero resistance temperature of 93K has been achieved and transition width is only 3K. It is also noticeable that magnetization hysteresis is quite large at 77K. It is also found that the sample size dependence of magnetization hysteresis for the melt processed material can be described by the critical state model[8], indicating the weak-links are reduced. Fig. 9 shows the magnetic field dependence of transport Jc. Jc value exceeding 1000 A/cm² was achieved at 77K in 10T which is dramatically high value as compared to less than 0.1A/cm² in the sintered material.

The possible mechanisms for the suppression of the weak-link problem in the melt processed samples are i) the formation of dense structure which enables the well connection between superconducting phases, ii) the improved homogeneity and superconducting properties in the superconducting phase, iii) the elimination of high angle grain boundaries, iv) the finely dispersed 211 phases which work as pinning points.

Fig. 6. Optical micrograph of the microstructure for the sample solidified from the molten oxides. The sample comprises of four phases: 211, 123, $BaCuO_2$, and CuO. The superconducting 123 phases are separated by the other phases.

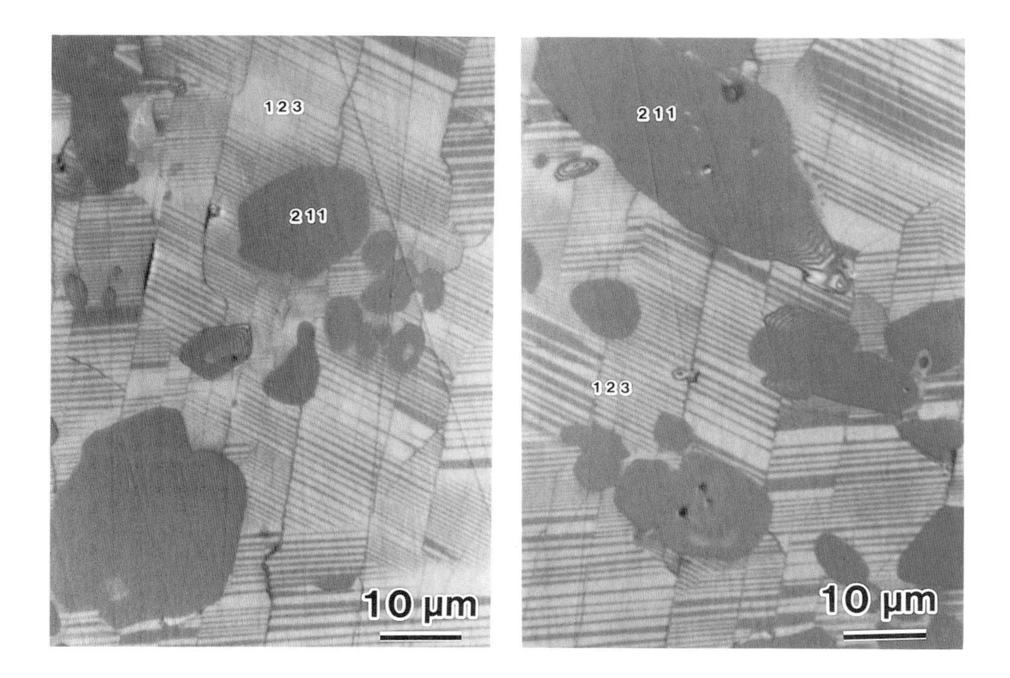

Fig. 7. Optical micrograph of the microstructure of the melt processed sample. Note the superconducting phases are well developed and the connectivity is much improved

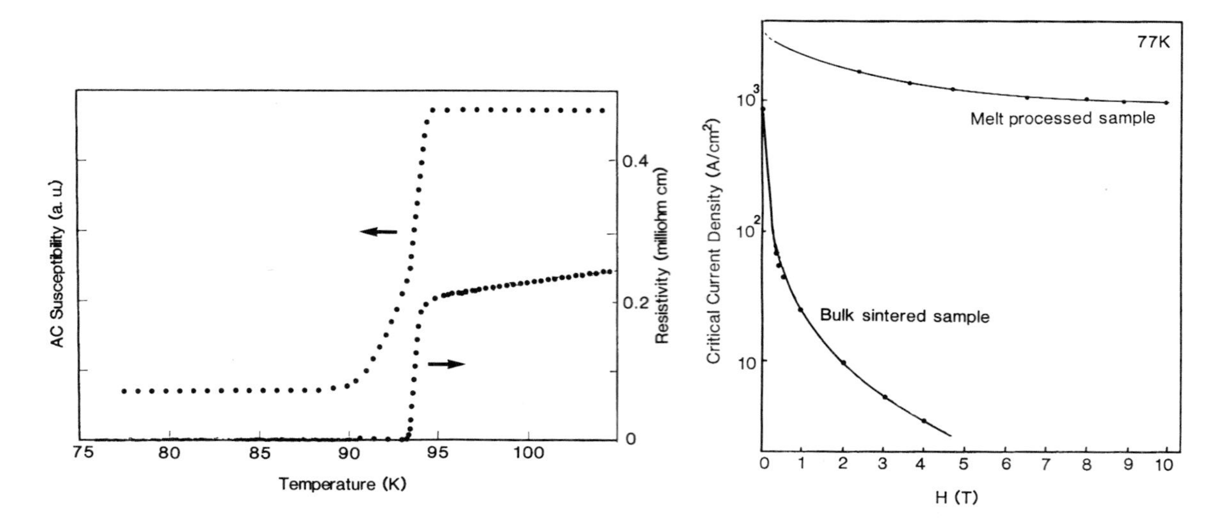

Fig. 8. Temperature dependence of resistivity and AC susceptibility for the melt processed sample.

Fig. 9. Magnetic dependence of transport Jc for the bulk sintered material and the melt processed sample.

Finally we would like to point out that the structure containing finely dispersed 211 phases may provide the improved fracture toughness, which is also beneficial for practical applications.

CONCLUSIONS

In bulk sintered oxide superconductors, there exist the weak-links which reduces transport critical current density and therefore hinders their application. The weak-link problems are difficult to be overcome in the sintered materials. However, the melt process enables the suppression of the weak-links through microstructural control.

REFERENCES

[1] J. G. Bednorz and K. A. Muller, Z. Phys., B64, 189 (1986).
[2] M. K. Wu, J. P. Ashburn, C. J. Thorng. P. H. Hor, R. L. Meng, L. Gao, Z. J. Huang, Y. Q. Wang and C. W. Chu: Phys. Rev. Lett., 58, 908 (1987).
[3] M. Murakami, Paper presented at Fifth Japan-US Workshop on High-field suerconducting material and standard procedures for high field superconducting materials testing held in Fukuoka, 165 (1987).
[4] M. Murakami, M. Morita, K. Sawano, T. Inuzuka, S. Matsuda and H. Kubo, Proceedings of SINTERING '87 held in Tokyo in 1987, ed. by S. Somiya.
[5] M. Murakami, H. Teshima, M. Morita, K. Doi, S. Matsuda, H. Hamada, M Fujinami, M. Saga, M. Sugiyama and M. Nagumo, Proceedings of MRS Meeting held in Tokyo (1988).
[6] M. Murakami and M. Morita, Proceedings of MRS Meeting held in Tokyo (1988).
[7] S. Jin, T. H. Tiefel, R. C. Sherwood, R. B. van Dover, M. E. Davis, G.W. Kammlott and R. A. Fastnacht, Phys. Rev. B, 37, 7850 (1988).
[8] K. Funaki, M. Iwakuma, Y. Sudo, B. Ni, T. Kisu, T. Matsushita, M. Takeo and K. Yamafuji, Jpn. J. Appl. Phys., 26, L1445 (1987).
[9] W. A. Fietz, M. R. Beasley, J. Silcox, and W. W. Webb, Phys, Rev., 136, A335 (1964).

Completely Textured Yttrium Barium Cuprate Superconductors

D.N. MATTHEWS[1], A. BAILEY[1], S. TOWN[1], G. ALVAREZ[1], G.J. RUSSELL[1], K.N.R. TAYLOR[1], F. SCOTT[2], M. McGIRR[2], and D.J.H. CORDEROY[2]

Advanced Electronic Materials and Technology Group, School of Physics[1], and School of Materials Science and Engineering[2], University of New South Wales, P.O. Box 1, Kensington, NSW 2033, Australia

ABSTRACT

It is now generally recognized that if the full potential of the new class of superconducting oxides is to be achieved in bulk samples, then extensive grain orientation (texturing) must be carried out.

We have recently extended our early studies of pressure texturing by studies of materials which have been deliberately doped to be copper oxide rich. The resultant samples show complete surface texturing,with the (00*l*) axis normal to the specimen and with a spread of only ±1° in the axis orientation.

The properties of these materials are discussed.

INTRODUCTION

Many practical applications of the new high T_c bulk Y-Ba-cuprate superconductors will require a transport critical current density, J_c, of approximately 10^4 A cm^{-2} in a magnetic field of 2 to 3T at 77K. Published values of J_c for bulk, randomly oriented polycrystalline material of grain size 5 to 20 μm, are rather scattered but the best materials have a value of ≈ 10^3 A cm^{-2} at 77K in zero magnetic field. This value decreases by an order of magnitude for an applied field of ≈ 0.3T[1]. There would appear to be two main factors causing these low values: 'weak link' coupling between the individual grains, which have the potential to carry large currents, and the high anisotropy of the orthorhombic lattice with respect to current flow. This results in critical currents in the **a-b** plane being an order of magnitude greater than that in the **c**-direction[2].

Therefore, texturing of bulk polycrystalline material,so that the **c**-direction is perpendicular to the transport current flow direction,should provide a significant increase in J_c. This value could be further increased if the coupling between the textured grains is improved. In fact, Jin et al[3] have reported a melt-textured growth process for $YBa_2Cu_3O_{7-\delta}$ (123) material which has raised J_c to 1.75×10^4 A cm^{-2} in zero magnetic field at 77K. In this report, a range of Y-Ba-Cu-O compounds are textured, without material melting and then characterised with emphasis on transport critical current variations.

EXPERIMENTAL

Each sample was prepared by the solid-state reaction method using an appropriate mixture of the starting powders Y_2O_3, $BaCO_3$ and CuO. After mixing and grinding the materials, the powder was axially pressed at 3.9 MPa into pellets (30 mm diameter by 2mm thick) that were calcined at 930°C for 12 hours. The calcined pellets were then re-ground and the powder, suspended in a solution of ethanol, was axially pressed at 463 MPa into pellets that were then sintered at 930°C for 12 hours in flowing oxygen. The pellets were cooled at 1°C per minute to room temperature.

As reported previously[4] the above technique results in almost complete texturing of the top and bottom surface regions of (123) samples. In this work crystallographic texturing of samples, with compositions across the range (123) to (235), has been achieved by the introduction of additional CuO to the sintered powder system and their properties have been studied. Thus fifteen samples with the following compositions: $Y_1Ba_2Cu_3(Cu_x)O_{7-\delta}$, $Y_{1.125}Ba_{1.875}Cu_3(Cu_x)O_{7-\delta}$ and $Y_{1.2}Ba_{1.8}Cu_3(Cu_x)O_{7-\delta}$ where x = 0, 0.25, 0.5, 0.75 and 1.0, were prepared simultaneously using the method described above:

RESULTS AND DISCUSSION

The quality of the texturing for each sample was measured using both x-ray and neutron diffraction techniques. X-ray diffraction spectra from the large area surfaces of all fifteen samples showed enhanced intensity of the (00*l*) peaks compared to those from a statistically oriented (123) powder, Figure 1. The degree of surface texturing across the matrix of compositions is shown in Figure 2 where the normalized intensity of the (006) line [$I_{norm} = I_{(006)}/I_{(110)}$] is used to characterise the degree of texturing. For the statistically oriented (123) powder $I_{norm} \cong 0.4$, for the surface textured (123) sample $I_{norm} \cong 6$ and for the highly textured (124) sample $I_{norm} = 40$, an increase of 100 over randomly oriented samples.

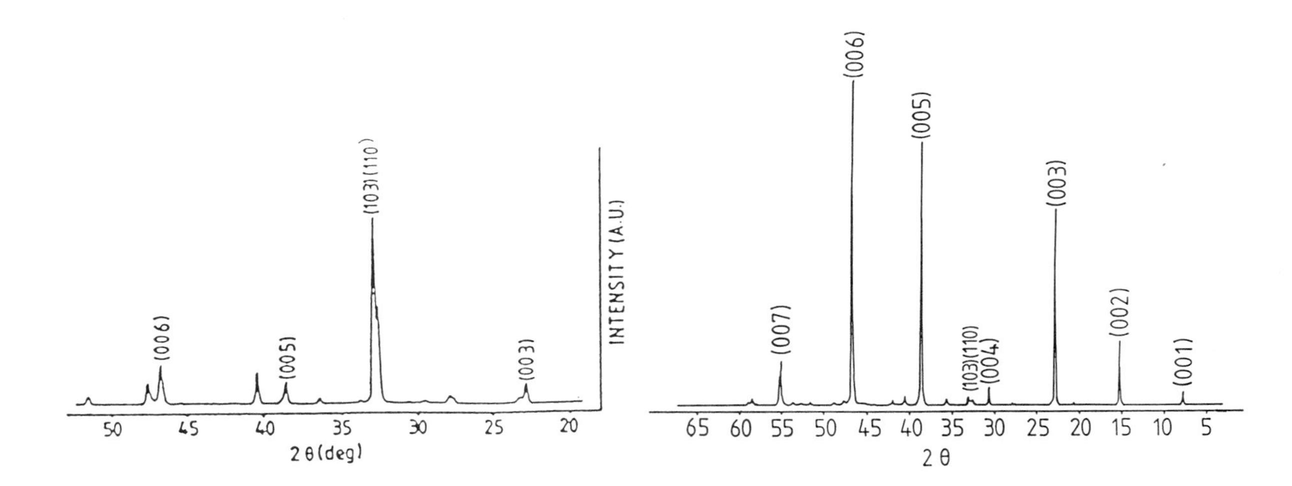

Figure 1: (a) X-ray diffraction spectrum for a powdered $YBa_2Cu_3O_{7-\delta}$ sample.
(b) X-ray diffraction spectrum from the large area surface of the highly textured $YBa_2Cu_3O_{7-\delta}$ sample.

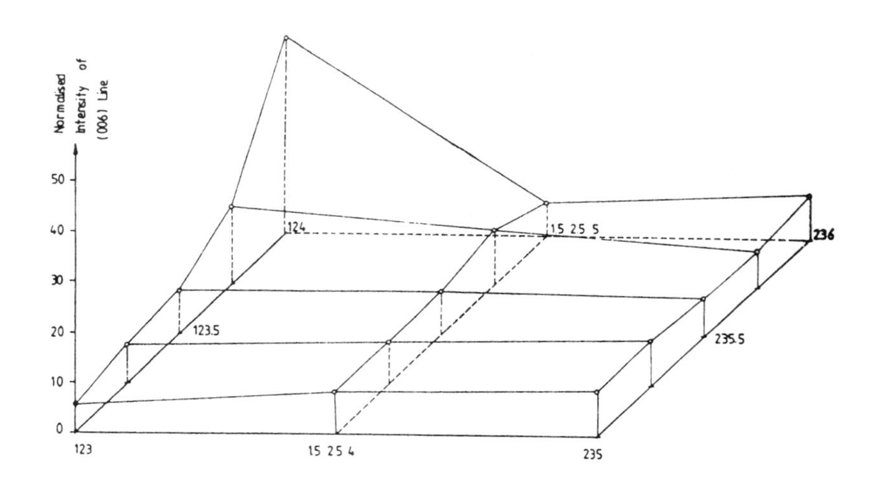

Figure 2: Variation of the normalized intensity of the (006) line with sample composition.

X-ray pole figure analysis, using the Schulz reflection method, further confirmed the preferred alignment of the **c**-axis of the orthorhombic unit cell perpendicular to each of the sample surfaces. In particular, Figure 3 shows pole figure results for the highly textured $Y_1Ba_2Cu_3(Cu_1)O_{7-\delta}$ (124) sample surfaces. Figure 3(a) was derived from the high intensity (006) line and shows **c**-axis alignment to be better than ±1° with respect to the surface perpendicular. The result shown in Figure 3(b) was obtained from the (013) line and clearly indicates that there is no preferred orientation in the **a-b** plane, this process producing a fibrous texture rather than complete alignment.

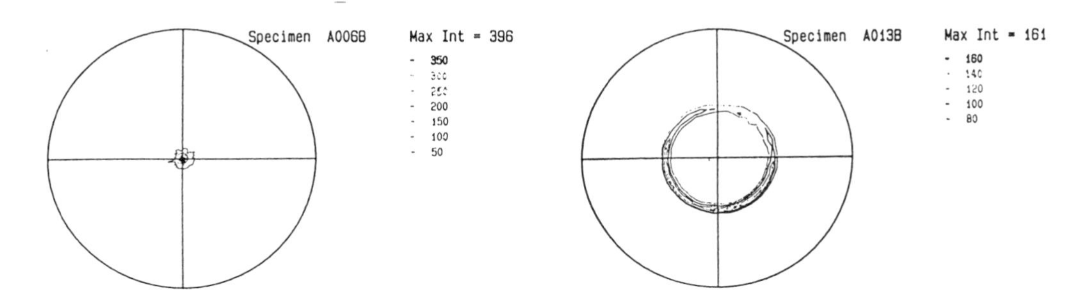

Figure 3: (a) X-ray polar figure, using the (006) line, for the $YBa_2Cu_4O_{7-\delta}$ sample. The almost complete alignment of the **c**-axis perpendicular to the surface is evident.

(b) X-ray pole figure, using the (013) line, for the $YBa_2Cu_4O_{7-\delta}$ sample. This figure shows no preferred orientation in the **a-b** plane.

Neutron diffraction analysis of the (124) and (123.25) samples was undertaken using a four circle single crystal diffractometer and the high intensity (006) line. Figure 4 shows the neutron spectrum for the highly surface textured (124) sample. The 18° spread in the (006) line for this sample should be compared to the 140° spread of the same line for the comparatively poorly surface textured (123.25) sample. This result clearly shows that the texturing extends into the bulk material of the (124) sample, however, the degree of alignment being less than that for the surface region which appears to have a thickness ≈50 μm.

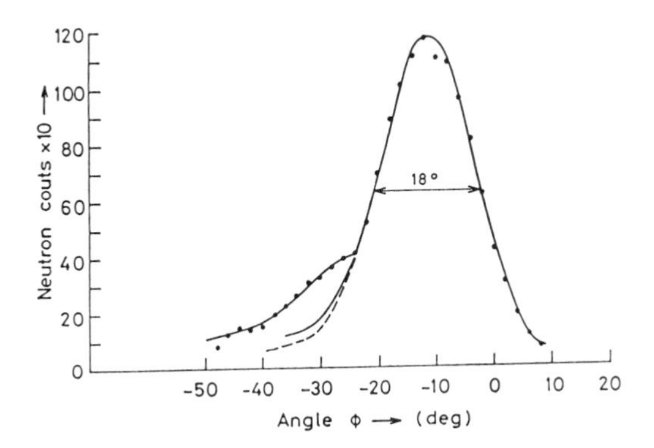

Figure 4: Neutron diffraction spectrum for the $YBa_2Cu_4O_{7-\delta}$ sample with $2\theta = 46.6°$ [(006) line].

The critical current density, J_c, was measured at 77K for all samples using both the four point transport technique in zero magnetic field and the magnetization technique. Transport measurements showed that for all compositions J_c was less than 100 A cm^{-2}, with only a small decrease in value in going from the (123) sample to the (124) sample, as shown in Figure 5. These values should be compared with an average value of 5 A cm^{-2} for our untextured material. This result is rather surprising in view of the high texturing achieved, however, considering the amount of excess CuO introduced into the matrix of some of the compositions it is likely that

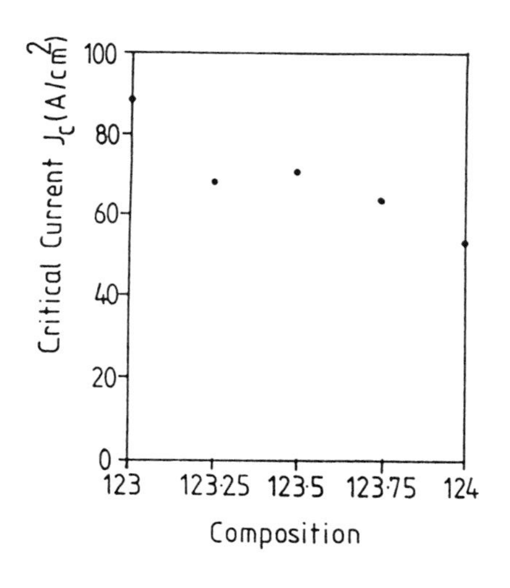

Figure 5: The variation of J_c (transport) along the compositional line from $YBa_2Cu_4O_{7-\delta}$ to $YBa_2Cu_4O_{7-\delta}$.

significant impurity phase precipitation is occuring at the grain boundary. It should also be noted that all calculations have been made on the premise that the superconducting current passes through the bulk of the material. If it is assumed that the current passes basically through the textured surfaces then J_c would be in excess of 5×10^3 A cm^{-2}. Critical magnetization current density measurements, using a conventional double coil arrangement with a ramped applied magnetic field perpendicular to the **c**-axis, gave J_c values of approximately 1.6×10^3 A cm^{-2} (using the Bean model). Thus, critical current densities should be > 10^4 A cm^{-2} for current flow in the **a-b** plane. Transport J_c for the compositions studied here are significantly less than this value and the problem would appear to be directly connected to 'weak link' coupling between the oriented grains of the textured material. However, in view of the strong mis-orientation dependence of J_c reported by Dimos et al[5] which shows a drop of J_c by a factor of 30 for a tilt angle of 10°, the low values of transport J_c may simply be associated with the incompletely textured interior of the mateial (±9°). Work is continuing on this problem in order to understand the nature of these 'weak-links' and to establish processing methods through which the second phase materials can be minimised.

Transport critical current densities were also determined from Josephson supercurrent measurements which used the ceramic bridge method[6]. For this technique, a very small bridge of triangular cross-section was formed by asymmetrical cutting of the bulk sample. This bridge was cut so that the surface textured layer was

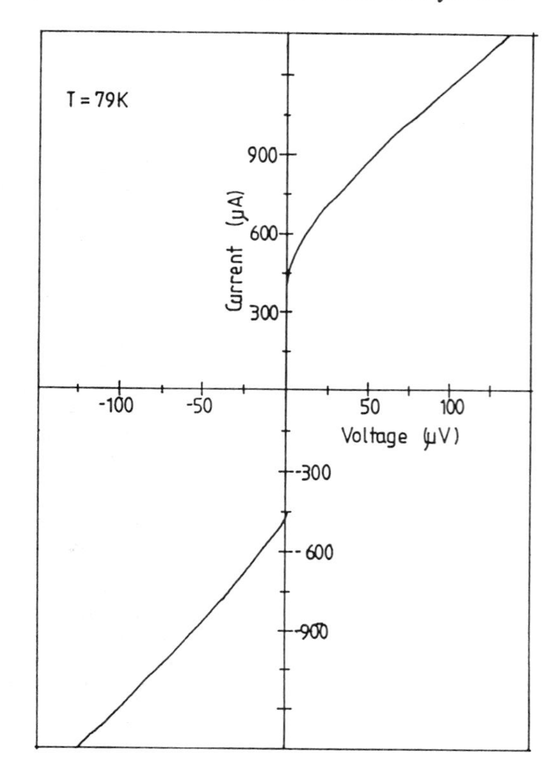

Figure 6: Typical Josephson i-V response for a textured (123) material cut into the form of a triangular ceramic bridge.

undisturbed and formed the base of the triangle. For this type of bridge and a textured (123) sample, Figure 6 shows the typical Josephson i-V response, while Figure 7 shows novel current-voltage characteristics that appear at higher voltages. Figure 7(a) is believed to show tunneling in SIS structures[7,8], the observed steps having an integer relationship. Figure 7(b) is a frequently found characteristic with a very interesting negative resistance region. The shape of the curve is similar to that predicted for SIS and SNS junctions[9]. J_c determined from this technique was, on the average, > 10^4 A cm^{-2} at 77K and zero magnetic field.

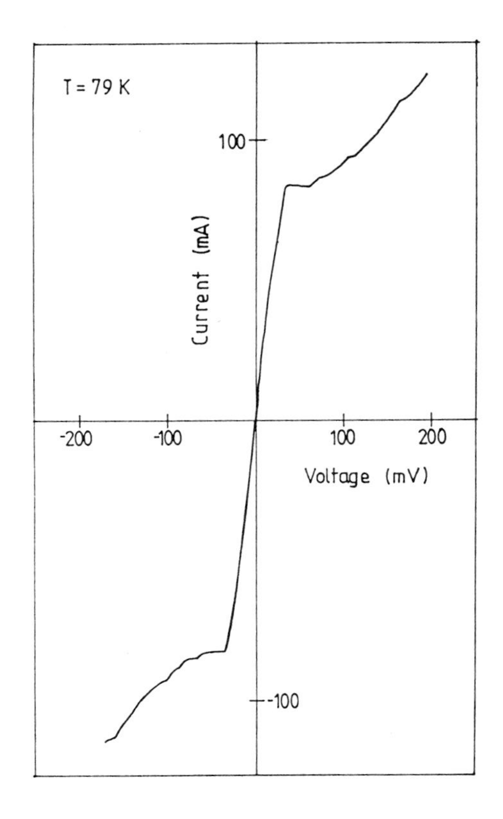

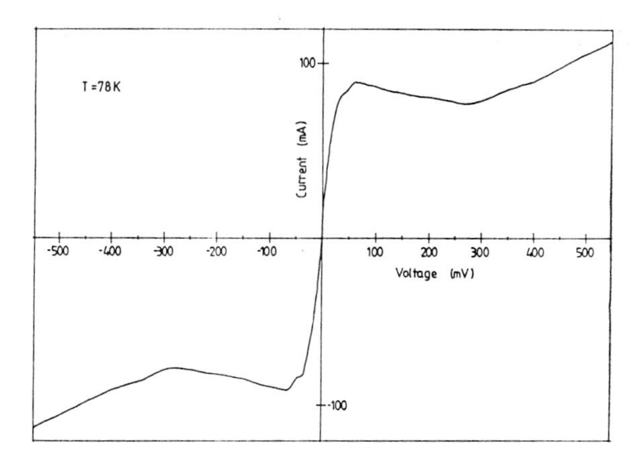

Figure 7: (a) I-V characteristic for the bridge at larger voltages. A number of distinct steps are observed which have an integer relationship.

(b) Another type of i-V characteristic frequently found using the bridge method. Note the negative resistance region.

In summary, the (124) material shows almost complete surface texturing with the bulk material also significantly textured. Further, for the compositions studied a significant variation in J_c was not observed, even though there was a large change in excess CuO associated with the starting materials. Finally, the transport J_c's were lower than the magnetization determined J_c's and intergrain coupling would appear to explain this difference and techniques must be found to strength the coupling between the textured grains.

REFERENCES

1. W.J. Gallagher, J. Appl. Phys. 63, (8), (1988).
2. T.R. Dinger, T.K. Worthington, W.J. Gallagher, R.L. Sandstorm, Phys. Rev. Lett. 58, 2687 (1987).
3. S. Jin, T.H. Tiefel, R.C. Sherwood, R.B. Vandover, M.E. Davis, G.W. Kammlott, R.A. Fastnach, Phys. Rev. B37, 7850 (1980).
4. D.N. Matthews, A. Bailey, T. Puzzer, N. Mondinos, G. Alvarez, G.J. Russell, K.N.R. Taylor, J. Crystal Growth, (in press).
5. D. Dimos, P. Chaudhari, J. Mannhart and F.K. LeGoues, Phys. Rev. Lett., 61, 219 (1988).
6. P.H. Wu, Q.H. Cheng, S.Z. Yang, J. Chen, Y. Li, M. Ji, J.M. Song, H.X. Lu, X.K. Gao, J. Wu and X.Y. Zhang, Jap. J. Appl. Phys., 26, L1579 (1987).
7. J. Rantaszkiewicz, A. Reich, P. Przystupski, A. Pajaczkowska and J. Igalson, Physica C, 153-155, 1391 (1988).
8. G.A. Alvarez, K.N.R. Taylor and G.J. Russell (to be published).
9. R. Kümmel, B. Huckestein, R. Nicolsky, Sol. State Communications, 65, 1567 (1988).

Densification and Critical Current Density of $YBa_2Cu_3O_x$ Ceramics

SHIGEKI KOBAYASHI, TOSHIO KANDORI, YASUYOSHI SAITO, and SHIGETAKA WADA

Toyota Cental Research and Development Laboratories, Inc., Nagakute, Aichi, 480-11 Japan

ABSTRACT

The improvement of Jc was explored from the aspect of densification of sintered bodies in $YBa_2Cu_3O_x$. Two methods were employed to increase the bulk density of sintered bodies; (a) Calcined powder was ball-milled to obtain fine particles and increase the sinterability. (b) Specimens were formed by swaging to get powder compacts of high density.

Sintered bodies of over 6.0 g/cm^3 were easily obtained but they did not necessarily show a high Jc value. Several problems such as the reaction of powder with water during ball milling, the formation of microcracks on sintering and low oxygen contents of specimens were identified. Reducing these problems improved the Jc value up to 600 - 800 A/cm^2.

INTRODUCTION

The discovery of $YBa_2Cu_3O_x$ superconductors [1,2] drew attention from both the scientific and industrial groups because of their high Tc. Many works on this new class of superconductors have revealed that their superconductivity is sensitive to the oxygen content [3,4] and their critical current density (Jc) is generally low, especially in bulk specimens [5-7]. The majority of specimens reported in literature have had densities less than 90% of theoretical one. Since higher density is desirable in terms of current path and specimen integrity, the densification of Y-Ba-Cu-O system was explored in this study through two ways; i) ball-milling to reduce the particle size and to increase the sinterability of powders, ii) swaging to increase the density of powder compacts.

POWDER PREPARATION AND SPECIMEN FABRICATION

Powder Preparation

The starting powder of $YBa_2Cu_3O_x$ was prepared by the solid-state reaction. The appropriate amounts of Y_2O_3, $BaCO_3$, CuO of reagent grade were mixed with ethanol in a polyethylene jar for 24 hours. The mixed powder was calcined at 900 ℃ for 12 hours in air, followed by grounding in a mortar and pestle. This process was repeated seven times to ensure the decomposition of $BaCO_3$ and the reaction among three powders. The calcined powder was sieved to a -145 mesh size (<100 μm). A typical SEM picture of the calcined powder is shown in Fig. 1a. The powder consists of agglomerates of 20-50 μm and the size of individual particles is 5-10 μm.

Ball Milling

In order to obtain powders of fine particle size and increase the sinterability, the calcined powder was ball-milled in ethanol for 1,3 and 5 days using a Si_3N_4 pot and balls. By ball milling, the average particle size was reduced from 50 μm to 3 μm after 1 day but leveled off to 0.5 μm over 3 days. As seen in Fig. 1, agglomerates in calcined powder was first crushed to individual particles and then particles were further pulverized to finer particles. This was also confirmed by the fact that the distribution of particle size changed from the bimodal (1 day) to the normal distribution (3 and 5 days).

The prepared powders were die-pressed into bars of approximately 3 mm thickness and 4 mm width at 20-30 MPa, followed by cold-isostatic-pressing at 200 MPa. The green densities of pressed bars were 4.00, 3.75, 3.61 g/cm^3 for ball-milled powders for 1,3 and 5 days (denoted as BM1D, BM3D and BM5D), compared with 4.57 g/cm^3 for the calcined powder of -145 mesh size (CP). The sintering of specimens

Fig. 1. SEM pictures of (a) the calcined powder (-145 mesh size; d_{50}=50μm), (b),(c) the powders ball-milled for 1 day (d_{50}=3 μm) and 5days (d_{50}=0.5 μm), and (d) the interior of swaged wire.

was performed in flowing oxygen by heating with a rate of 5 ℃/min., holding at 900-980 ℃ for 6 hours and furnace-cooling. Several large cracks were formed for the ball-milled sample BM3D, and in the ball-milled sample BM5D, extensive crack formation destroyed specimen integrity. Therefore, the property measurement was performed only for the ball-milled sample BM1D. The reason for this crack formation is discussed later.

Swaging

The calcined powder of -145 mesh size was packed into copper tubes of 10 mm diameter (1 mm thickness) and 200 mm length, which was swaged down to a diameter of 4 mm with reduction steps of 1 mm. The packing density of powder in Cu tube was 3.0 g/cm^3. After swaging, swaged wires were cut into specimens of an appropriate length and then the copper sheath was carefully removed by mechanical polishing. The density of swaged rods was 5.2 g/cm^3, which is very high compared with the green density of die-pressed samples (3.8-4.6 g/cm^3). Actually, this is higher than the bulk density of sintered bodies using the calcined powder (CP), as shown in Fig. 2. The SEM picture of swaged rod (Fig. 1d) clearly indicates that the packed powder was crushed into finer particles whose size is comparable to that of the ball-milled powder. The sintering was done in a similar way to the ball-milled and die-pressed samples, as described above.

CHARACTERIZATION

Structural Analysis

The variation in the bulk density of sintered bodies with sintering temperature is shown in Fig. 2. The bulk density of the ball-milled (BM) and swaged (SW) samples shows a similar behavior: increasing in the temperature range of 900 to 980 ℃ and decreasing at higher temperatures. The highest densities obtained are 6.27, 6.15 g/cm^3 for the swaged (SW) and ball-milled samples (BM), respectively and these are 99, 97% of theoretical density: 6.36 g/cm^3 [8]. Thus, the densities of samples using fine particles are much higher than those of samples using the calcined powder (CP), indicating that the sinterability is facilitated by reducing the particle size. The weight loss during sintering was about 2% and 1% for the ball-milled and swaged samples, respectively, while the weight loss of the calcined sample was almost negligible.

For all the samples, the grain growth was found to start at 940 - 960 ℃ and the amount of open porosity decreased with increasing the sintering temperature. XRD analyses indicated that the structure of sintered bodies was orthorhombic, but the tetragonal phase existed in the interior of the ball-milled (BM) and the swaged samples(SW).

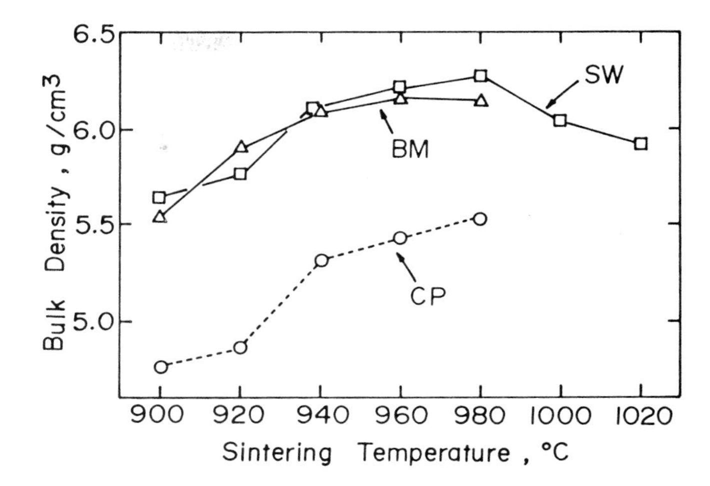

Fig.2. Variation in the bulk density of sintered bodies for the calcined (CP), ball-milled (BM) and swaged samples (SW).

Superconducting Properties

The electrical resistivity was measured by a conventional four-prove dc technique. At first, samples were coated with gold to reduce the contact resistance and copper wires were tightly attached to the surface, which was further secured with silver paint. The measurement was done in a cooling cycle to liquid nitrogen temperature. AC magnetic susceptibility was measured at a frequency of 1kHz.

i) Ball-milled Samples

Superconductivity was confirmed for all the ball-milled samples(BM) except for one sintered at 940 ℃, which did not show zero resistivity even at 77K. All other ball-milled samples show zero resistivity at about 90K. The AC susceptibility and Jc value at 77K for as-sintered samples are plotted in Fig. 3. The Jc values for samples at 940 and 960 ℃ were zero, but in Fig.3, the data were plotted at the bottom of the figure for convenience. The AC susceptibility shows a maximum at 900 ℃ and a minimum at 940 ℃. A similar variation is seen in Jc values, but a maximum is observed at 980 ℃. Since the bulk density is still low at 900 ℃, the current path is not complete. At 980 ℃, the volume fraction of superconducting materials would be lower, but this would be compensated by an increase in current path and higher Jc could be obtained. The minimum of Jc at about 940 - 960 ℃ is common for the relatively high density samples fabricated by die-pressing. Several reasons can be considered. As the bulk density of samples becomes high, the

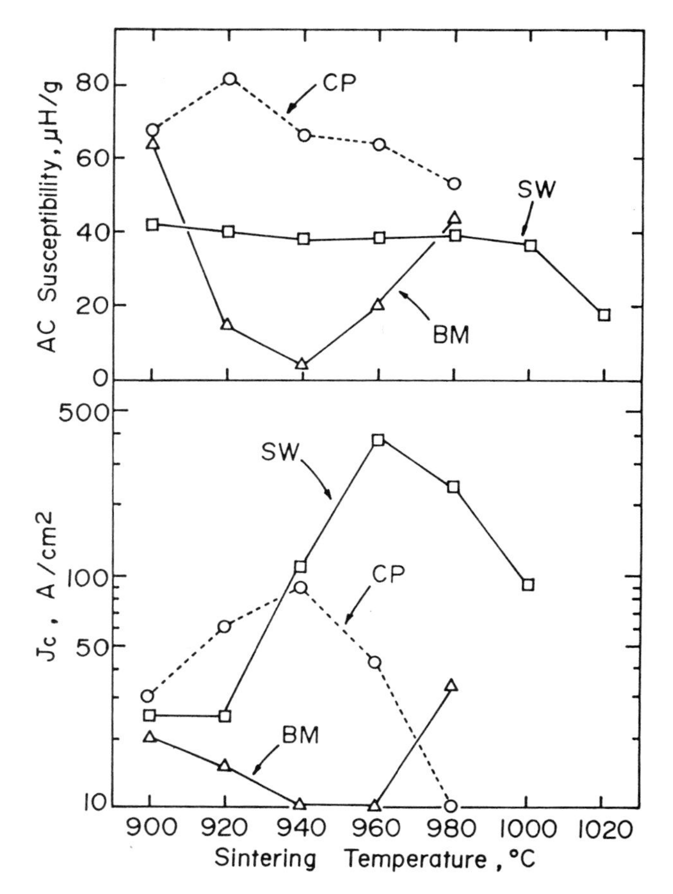

Fig.3. AC susceptibility and critical current density (Jc) at 77k for the calcined (CP), ball-milled (BM) and swaged samples (SW).

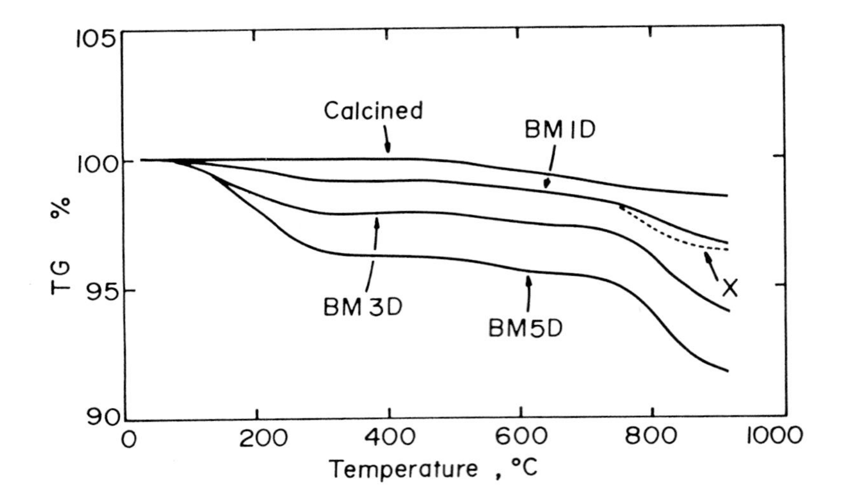

Fig. 4. Thermal gravimetric (TG) analyses of the calcined powder (-145 mesh size), powders ball-milled in ethanol for 1 day (BM1D), 3 days (BM3D) and 5 days (BM5D). The curve for the powder ball-milled in xylene for 5 days is shown by the dotted line (X).

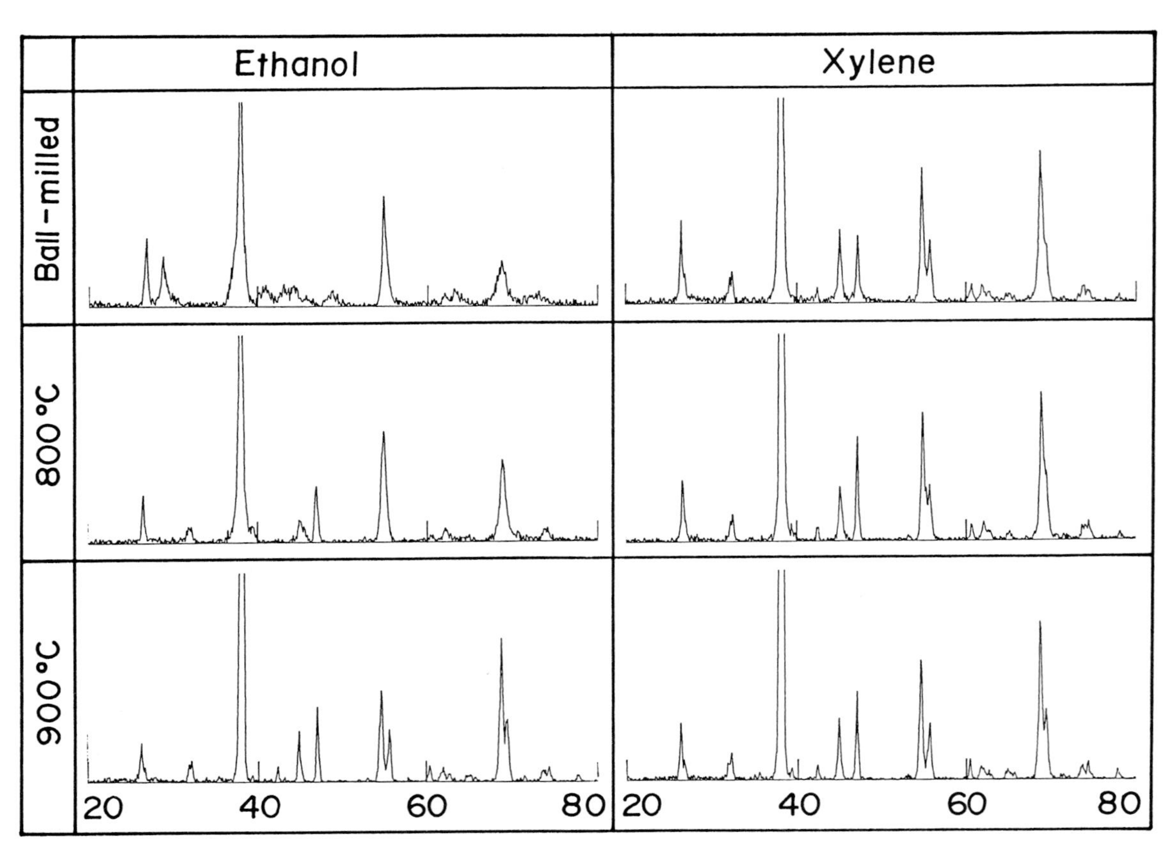

Fig. 5. X-ray diffraction patterns of the ball-milled powders in ethanol and xylene. Powders were also heat-treated at 800 and 900 ℃. (Co-Kα, 30 kV 40 mA)

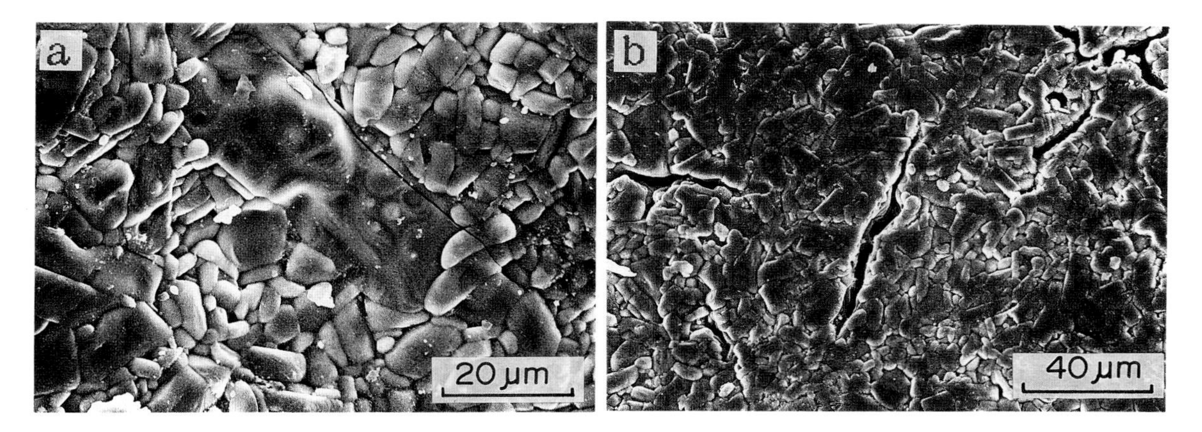

Fig. 6. SEM pictures of cracks formed in the swaged samples. (a) Cracks formed during sintering; similar cracks were also observed in the ball-milled samples. (b) Cracks formed during the swaging process.

amount of open porosity decreases and oxygen should diffuse in the matrix, meaning that the oxygen level in the specimen becomes low, especially inside the specimen [9]. This was confirmed by the fact that polishing the surface reduced the transition temperature (Tc) and the XRD peaks of tetragonal phase became more evident. Since the liquid phase exists in this temperature range [10], the second phases will be formed at the boundaries.

These ideas can also explain the behavior of the sample CP. Since the densities of the calcined samples are low, oxygen can be transported through pores and diffuse easily into the interior of specimens. The Jc value will be determined by the balance between the increase in current path and the formation of second phases.

ii) Swaged Wires

Although the onset of superconductivity was always observed at about 90K, the swaged wires of as-sintered state did not show zero resistivity above 77K. The data shown in Fig. 3 were taken from the samples annealed at 600 ℃ in oxygen for 100 hours after sintering. After annealing, all the samples showed zero resistivity at about 90K, but the data of AC susceptibility did not change substantially. This means that the volume fraction of superconducting materials does not change largely by annealing. Therefore, the surface of as-sintered samples would be covered with some films of high resistance.

The variation in Jc for the swaged samples is quite different from that for the ball-milled samples, as shown in Fig. 3. The contribution of current path to Jc will increase with the sintering temperature. But this is intervened by the formation of second phases, which is closely related to the existence of liquid phase. In the case of the swaged samples, this second phase formation would be somehow deterred and this would contribute to the higher Jc value than the ball-milled (BM) and calcined samples (CP).

LIMITING FACTORS OF Jc

Reaction of Powder with Water

As mentioned before, cracks were frequently found in the ball-milled samples. TG analyses indicated that the ball-milled powders lost more weight than the calcined powder, and the weight loss increased with the ball-milling time (Fig. 4). Ball-milling also affects the x-ray diffraction patterns as shown in Fig. 5. The peaks of superconducting phase became broad and some new peaks appeared. These peaks could not be identified [11,12], but they should be related to the compounds containing Ba, because the lines of $BaCO_3$ appeared during the heat-treatment of the ball-milled powders. The complete recovery of the superconducting phase was attained by heating over 900 ℃. Changing the solvent from ethanol to xylene drastically improved the problem (Figs. 4 and 5). The water content in xylene was found to be 0.015%, which was much lower than that in ethanol (0.2%).

From the above discussions, the crack formation in the ball-milled samples can be explained as follows. During ball-milling, the calcined powder reacts with water in ethanol. This is further facilitated by making fresh surfaces and reducing the particle size during milling. The reaction will lead to the decomposition of powder and the formation of hydrate. The decomposition of hydrate and carbonate, which is formed by the reaction of hydrate and CO_2, causes the evolution of gases during heating. As far as the amount of gases is small, they can escape easily through many pores. When the amount increases, they will expand the specimen and lead to the crack formation.

Microcrack Formation

The crack formation in the ball-milled samples was caused by the reaction of powder with water, as described above, and these cracks were formed mainly below 600 ℃. Milling in xylene could reduce this problem, but fine cracks were still observed on the sintered bodies. Quenching experiments revealed that these cracks were formed during cooling to room temperature. This is probably due to the volume expansion caused by the transformation from orthorhombic to tetragonal phase.

In the swaged samples, two types of cracks were observed, as shown in Fig. 6. The cracks shown in Fig. 6a were similar to those in the ball-milled samples. But the cracks shown in Fig. 6b were unique to the swaged samples and found to be formed during swaging. The formation of these cracks could be reduced by changing the swaging conditions. Since cracks shown in Fig. 6a are formed by the transformation, slow cooling was effective to suppress their formation and cooling in nitrogen could completely remove cracks from the specimen.

Low Oxygen Content

The densifiction of sintered bodies is beneficial to Jc in terms of the increase in current path, but it retards the oxygen diffusion into specimens severely. In the dense samples, the superconducting phase is formed only near the surface. The problem can be overcome by long-time annealing, but required time will be an order of weeks [13]. Another solution for this problem is reducing the specimen size. Actually we have succeeded in the improvement of Jc by reducing the specimen size to about 1 mm thickness or diameter and at the same time, employing the method to reduce the other Jc limiting factors. The maximum Jc values obtained were about 600 A/cm^2 for the pressed samples [14] and 800 A/cm^2 for wires [15]. Details of these results will be published elsewhere.

REFERENCES

1. M. K. Wu, J. R. Ashburn, C. T. Torng, P. H. Hor, R. L. Meng, L. Gao, Z. J. Huang, Y. Q. Wang and C. W. Chu, Phys. Rev. Lett. 58 (1987), 908.
2. S. Hikami, T. Hirai and S. Kagoshima, Jpn. J. Appl. Phys. 26 (1987), L314.
3. R. J. Cava, B. Batlogg, C. H. Chen, E. A. Rietman, S. M. Zahyrak, D. Werder, Nature 129 (1987),423.
4. M. Tokumoto, H. Ihara, T. Matsubara, M. Hirabayashi, N. Terada, H. Oyanagi, K. Murata and Y. Kimura, Jpn. J. Appl. Phys. 26 (1987), L1565.
5. Y. Yamada, N. Fukushima, S. Nakayama, H. Yoshino and S. Murase, Jpn. J. Appl. Phys. 26 (1987), L865.
6. T. Hioki, A. Oota, M. Ohkubo, T. Noritake, J. Kawamoto and O. Kamigaito, Jpn. J. Appl. Phys. 26 (1987), L873.
7. J. W. Ekin, Adv. Ceram. Mater. 2 [3B] (1987), 586.
8. D. W. Johnson, Jr., E. M. Gyorgy, W. W. Rhodes, R. J. Cava, L. C. Feldman and R. B. van Dover, Adv. Ceram. Mater. 2 [3B] (1987), 364.
9. T. Motai and N. Ichinose, J. Ceram. Soc. Jpn. 96 (1988), 373.
10. R. S. Roth, K. L. Davis and J. R. Dennis, Adv. Ceram. Mater. 2 [3B] (1987), 303.
11. M. F. Yan, R. L. Barns, H. M. O'Bryan,Jr., P. K. Gallagher, R. C. Sherwood and S. Jin, Appl. Phys. Lett. 51 (1987), 532.
12. M. Yoshimura, S. Inoue, N. Ogasawara, Y. Ishikawa, T. Nakamura, Y. Takagi, R. Liang and S. Somiya, J. Jpn. Soc. Powder & Powder Met. 34 (1987), 659.
13. T. Yamamoto, T. Furusawa, H. Seto, K. Park, K. Kuwabara, T. Hasegawa, K. Kishio, K. Kitazawa and K. Fueki, J. Jpn. Soc. Powder & Powder Met. 35 (1988), 333.
14. Y. Saito, unpublished data.
15. T. Kandori, unpublished data.

Processing of Dense, Fine Grained $YBa_2Cu_3O_x$ Ceramics Using a Spray-Drying Technique

NAOMICHI NAKAMURA, MASAYOSHI ISHIDA, YOSHIHIRO KOSEKI, and MICHIO SHIMOTOMAI

New Materials Research Center, High-Technology Research Laboratories, Kawasaki Steel Corporation, 1, Kawasaki-cho, Chiba, 260 Japan

ABSTRACT

High-Tc superconductor $YBa_2Cu_3O_x$ has been synthesized using a spray-drying technique with special reference to the influence of the purity of the starting reagents on microstructures and superconducting properties. Sintering is enhanced and high density is easily obtained with ultrapure reagents of acetic salts. Doping with Sr to examine the impurity effects of reagents of conventional grade has resulted in porous microstructures. Magnetic field dependences of magnetization and critical current density of the ceramics are significantly dependent on the grade of the starting reagents.

INTRODUCTION

The discovery of the superconductivity in $YBa_2Cu_3O_x$ above liquid-nitrogen temperature (1) has brought a great impact in the field of application of superconductors. One of the most important properties required for its applications is undoubtedly the critical current density (Jc). Two characteristics of Jc in bulk polycrystalline $YBa_2Cu_3O_x$ have been recognized. Firstly, Jc measured by transport current decreases by an order of magnitude by application of weak magnetic field (2,3). Secondly, the transport Jc is about two orders of magnitude lower than the Jc estimated from magnetization data which do not involve electron transport through the bulk (4). It has been suggested that grain boundaries are responsible for such low values of Jc (5,6). Therefore, it is important to find out the factors affecting the geometry and chemistry of grain boundaries to improve the superconducting properties.

We have already developed a spray-drying technique to minimize contamination during processing (7). Next step is purification of the starting reagents. Among impurities, Sr is chemically very similar to Ba and not easily separated from the latter. Reagent of Ba salt of conventional grade usually contains about a few thousands ppm of Sr. Present paper is concerned with the effects of Sr on microstructures and superconducting properties of the sintered materials.

EXPERIMENTAL

Acetic salts of component metals were chosen as starting reagents. Two grades of reagents were prepared to examine the impurity effects: One is the conventional grade and the other ultrapure grade. Sintered bulk samples were processed as follows. Aqueous solution of acetic salts was sprayed and injected into a stream of hot air. The temperature of the air was about 150-200°C. The dried particles are then cyclonically separated at a collection point. The diameter of the dried particle was about 1μm. These spray-dried powders were heated in a crucible at 900°C for 12hrs in air. After calcination, they were pressed into pellets of 20mm in diameter and 1mm in thickness. Sintering was carried out in flowing oxygen for 12hrs at

table I. Composition of the samples.

sample No.	Y : Ba : Cu	Sr/Ba (appm)	reagents
#1	1 : 2 : 3	< 10	ultrapure grade
#2	1 : 2 : 3	2000	ultrapure grade doped with Sr
#3	1 : 2 : 3	10000	ultrapure grade doped with Sr
#4	1 : 2 : 3	2500	conventional grade

950°C, followed by cooling to room temperature at a rate of about 1°C/min. Four kinds of samples were prepared with these reagents as shown in table I. Samples #2 and #3 were prepared to discriminate the influence of Sr.

Crystallographic phases were identified by X-ray diffraction. Resistivity and Jc measurements were made on bar-shaped samples cut from sintered pellets by a four-probe method using indium and gallium as contact metals. Magnetization measurements were performed with a vibrating sample magnetometer on rod-shaped samples cooled in zero magnetic field. The size of the samples was about 1mm in diameter and 5mm in length. No corrections were made for the demagnetizing factor, which is estimated to be about 0.05 in the present case.

RESULTS AND DISCUSSION

The properties of the samples are summarized in table II. Tc's in all the samples are about 90 K, suggesting same lack of oxygen intake. However, their apparent densities are significantly different to each other: #1 - #3 (made with ultrapure reagents) have much higher densities than that of #4 (made with conventional reagents). SEM-micrographs of these samples are shown in Fig.1-(a-d). #1 has microstructure with exaggerated grain growth. As Sr concentration increases, the grain size becomes smaller and the density of intragranular pores increases. Thus, it is concluded that Sr affects the grain growth in $YBa_2Cu_3O_x$ although the mechanism is still an open question. #4 shows a porous microstructure with the finest grains among the samples in spite of lower Sr concentration than #3, which suggests the presence of other factors affecting sinterability of $YBa_2Cu_3O_x$ in conventional grade reagents. The other impurities such as Fe or Na are suspected.

As is shown in table II, transport Jc's in zero field at 77 K in the samples #1 - #3 are smaller than that in the sample #4, although the former samples have higher densities. Presumably, intergranular microcracks resulted from anisotropic constriction of the large grains govern the transport Jc in the samples #1 - #3.

Magnetization curves of the samples at 77 K in low applied fields are shown in Fig.2. Each magnetization curve has an shouldered structure and its slope changes arround 10 Oe. The shoulder of #1 is faint. However, the shoulder becomes distinct with the increase of the Sr concentration. It is noted that #4 has the most distinct one.

Similar shouldered structures in the magnetization curves for the polycrystalline high-Tc ceramics have been also reported and explained in terms of the weak-coupling between superconducting grains as follows (8-10). In low magnetic field, the grains are coupled by

table II. Summary of measurements.

sample No.	density (%)	Tc (K)	Jc (A/cm^2) at 77 K in zero field
#1	98	90	205
#2	94	90	76
#3	92	92	180
#4	80	92	480

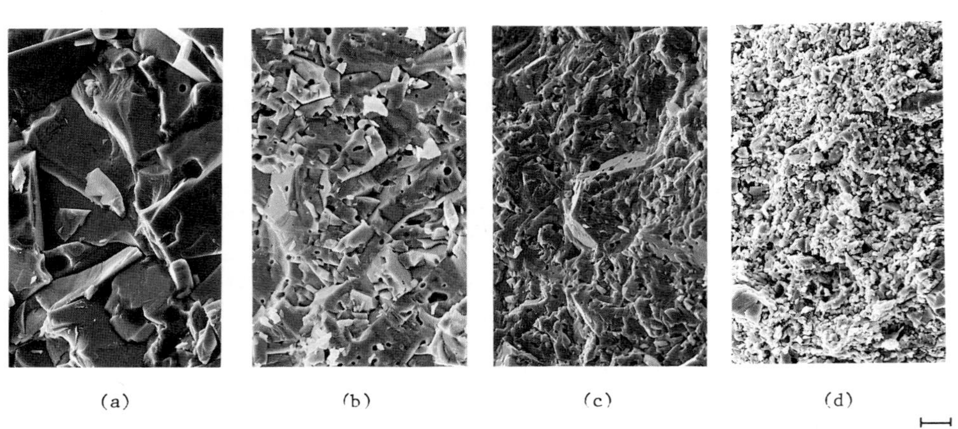

Fig.1. SEM-micrographs of the fractured surfaces of sintered samples; (a) for #1, (b) for #2, (c) for #3 and (d) for #4.

intergranular shielding currents which reject flux penetration into the bulk. With increasing field to the shoulder region, magnetic flux begins to penetrate into the grain boundaries and the grains are decoupled. In fields beyond the shoulder, the remaining superconducting volume is supposed to be unchanged until critical field Hc_1 is attained. In this field range, the material behaves as an aggregation of superconducting grains.

The relatively small shoulders in the magnetization curves of #1 - #3 compared with #4 may be explained as follows. The former samples have dense and large grained microstructures, so that the fraction of grain-boundary region is small. Consequently, the amount of flux penetrating into sample body should be little. Flux penetrates more easily to the samples #2 and #3 than to the sample #1 because of the porous microstructures caused by the Sr doping. This would have resulted in progressive development of the shoulders in #2 and #3. It is unclear at present which is more responsible for the flux penetration characteristics, the microstructure affected by doped Sr or the segragation of them to grain boundaries. Our preliminary study by Auger-electron spectroscopy suggests that the former is the case.

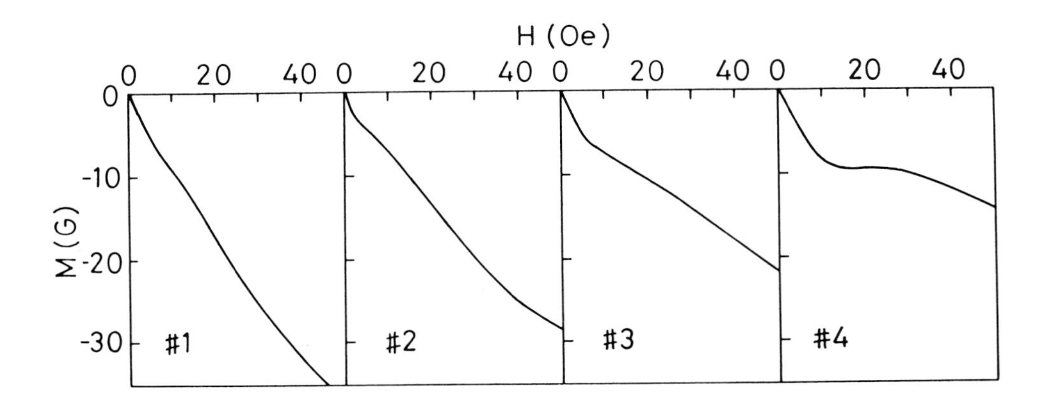

Fig.2. Low field magnetization curves of the samples measured at 77K.

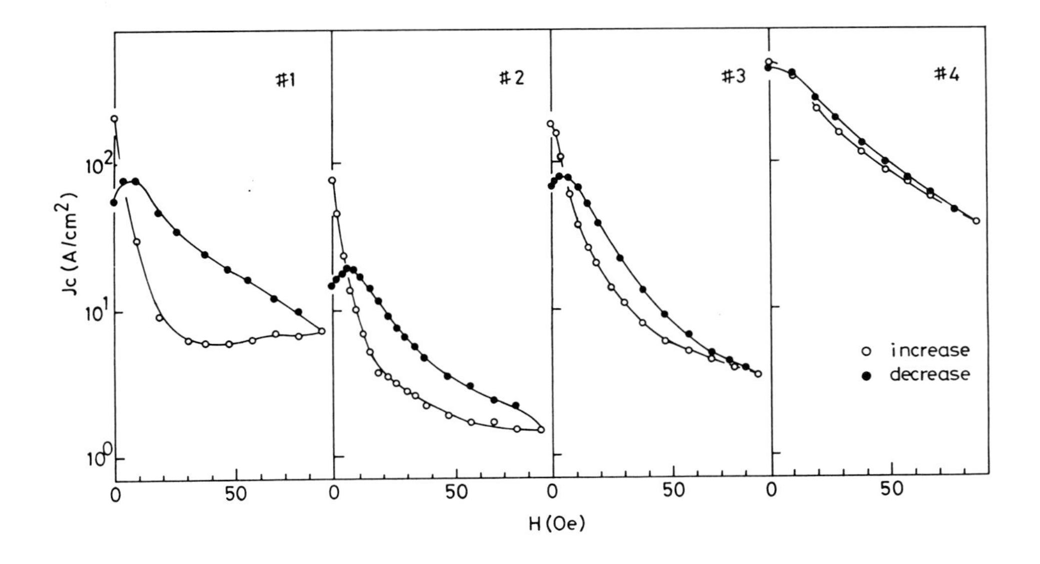

Fig.3. Magnetic field dependence of the critical current density in the samples at 77 K.

Magnetic field dependence of transport Jc (Jc-H curves) at 77 K in low applied fields are shown in Fig.3. It is easily see that they have good correspondence with the magnetization curves. As the shoulder becomes weaker, the degradation of Jc with increasing field and hysteresis in Jc-H curve become more significant. The correlation between the characteristics of the Jc-H curves and the extent of the shoulder of magnetization curves indicates that flux penetration characteristics have influence on the field dependence of Jc-H curves. Pinning characteristics and the distribution of the pinned flux in a sample are currently under study.

Summarizing the present study, 1. sintering is enhanced and high density is easily obtained with ultrapure reagents, 2. addition of Sr yields microstructures with small grains and intragranular pores, 3. a shouldered structure appears in the magnetization curve and develops by doping of Sr, 4. magnitude of the degradation of Jc by applying field and the hysteresis in Jc-H curves are correlated with the extent of the shoulder of magnetization curves.

REFERENCES

1. C.W.Chu, P.H.Hor, R.L.Meng, L.Gao, Z.J.Huang and Y.Q.Wang: Phys. Rev. Lett. 58, 405 (1987).

2. U.Dai, G.Deutscher and R.Rosenbaum: Appl. Phys. Lett. 51, 460 (1987).

3. M.Okada, A.Okayama, T.Morimoto, T.Matsumoto, K.Aihara and S.Matsuda: Jpn. J. Appl. Phys. 27, L185 (1988).

4. K.Funaki, M.Iwakuma, Y.Sudo, B.Ni, T.Kisu, T.Matsushita, M.Takeo and K.Yamafuji: Jpn. J. Appl. Phys. 26, L1445 (1987).

5. J.E.Blendell, C.A.Handwerker, M.D.Vaudin and E.R.Fuller,Jr.: J. Crystal Growth 89, 93 (1988).

6. R.L.Peterson and J.W.Ekin: Phys. Rev. B37, 9848 (1988).

7. N.Nakamura, T.Nakano, S.Gotoh and M.Shimotomai: TMS Topical Meeting on Processing and Application of High-Tc Superconductors, New Brunswick, 5/9-11 (1988).

8. S.Gotoh, N.Nakamura, M.Ishida, H.Shishido and M.Shimotomai: The 90th Annual Meeting of the American Ceramic Society, Cincinnati, 5/2-5 (1988).

9. H.Zhang, S.S.Yan, H.Ma, J.L.Peng, Y.X.Sun, G.Z.Li, Q.Z.Wen and W.B.Zhang: Solid State Commun. 65, 1125 (1988).

10. D.Wong, A.K.Stamper, D.D.Stancil and T.E.Schlesinger: Appl.Phys.Lett. 53, 240 (1988).

Effect of Cu and Sources on Powder Processing and Densification of Superconducting $YBa_2Cu_3O_x$ Ceramic

HUNG C. LING

Engineering Research Center, AT&T Bell Laboratories, P.O. BOX 900, Princeton, NJ 08540, USA

ABSTRACT

The solid state reaction method to form the superconducting $YBa_2Cu_3O_x$ oxide was studied. It was found that the starting cupric and yttrium components accelerated the decomposition of the $BaCO_3$ component. At a constant heating rate of 10°C per minute in the TGA analysis, the temperature of complete decomposition, T_f, was lowered from greater than 1000°C in pure $BaCO_3$ to between 915 and 985°C. The effectiveness in decreasing T_f can be ranked in the order of oxalate, carbonate and oxide. The highest sintered density achieved in this study was 6.03 gm/cm^3 (ρ/ρ_{th}=95%) at 990°C and 5.85 gm/cm^3 (ρ/ρ_{th}=92%) at 960°C. The source of cupric ion had the largest effect on densification. The use of cupric carbonate resulted in a consistently high Archimedes density of about 6.00 gm/cm^3 and large dimensional shrinkage of about 20% at 990°C for 12h. In contrast, the use of cupric oxide gave the lowest density and smallest shrinkage. Within the same powder lot, higher sintered density and smaller dimensional shrinkage were observed in samples with higher initial green density and compaction pressure. However, the data suggested that the enhanced densification and higher density achieved by the use of cupric carbonate and oxalate cannot be accounted for by the different physical characteristics of the powders and the mechanics of powder compaction, measured collectively by the green density.

INTRODUCTION

The solid state reaction method to form the superconducting $YBa_2Cu_3O_x$ ceramic has generally involved the use of barium carbonate ($BaCO_3$), yttrium oxide (Y_2O_3) and cupric oxide (CuO) [1-11]. The sintered densities of compacted samples from these powders were generally quite low, varying between 60 to 85% of the theoretical density. The actual sintered density achieved appeared to depend on the powder processing method used by the different groups. For example, by using repeated calcination and milling steps to ensure complete decomposition of raw materials and small particle size after calcination, the sintered density may be increased to over 90% of the theoretical density[1]. Alternatively, hot deformation techniques[2,4,12], such as forging or hot isostatic pressing, were employed to produce dense superconducting ceramics. There were also reports on chemically prepared powders with smaller particle size, which sintered to densities between 75-95% of theoretical density[13,14].

It is believed that one main reason for the poor sintering characteristic of the powders derived from the solid state reaction method is the slow and incomplete decomposition of $BaCO_3$ below 1000°C. Since sintering of the calcined powder must be below the incongruent melting temperature of $YBa_2Cu_3O_x$ (-1025°C), calcination is typically performed in the temperature range between 900 and 1000°C for a long period of time. As a result, hard and coarse powders are formed after calcination, requiring a secondary "high energy" milling step to produce an active powder with reduced particle size for sintering. In this paper, the impact of starting copper and yttrium sources on the decomposition of $BaCO_3$ and the subsequent densification of $YBa_2Cu_3O_x$ ceramic during sintering will be examined.

EXPERIMENTAL PROCEDURES

In this study, $BaCO_3$ was used solely as the source of barium so that water could be used as the mixing and milling medium. For the other metal ions, Cu and Y, the oxide, carbonate or oxalate of these metals was chosen as the source and constituted one major variable in the study.

A standard powder precessing procedure was adopted. Appropriate weight ratios of $BaCO_3$ and the cupric and yttrium ion sources were mixed in 1000 ml plastic bottle and ball-milled for 4 hours in de-ionized water using ZrO_2 milling media. The slurry was then filtered and dried at 125°C for 16 hours. Prior to calcination, the dried mix was granulated through a 20 mesh screen. Approximately 100 gram batches were calcined at 900°C for 2 hours, removed and allowed to cool outside the furnace, granulated and re-calcined at 900°C for another 2 hours. The calcined powder was granulated using mortar and pestle prior to compaction. Some powders made from a combination of barium carbonate, cupric oxide and yttrium oxide (lot E1305) were also calcined at 950°C for 16 hours. These powders were quite hard and were subjected to a 4-hour secondary ball-milling using zirconia balls and isopropanol.

Samples were dry pressed from the powders in the form of 1.27 cm (0.5 inch) diameter discs, 0.64 cm (0.25 inch) diameter discs or rectangular bars measuring 0.32 cm (0.125 inch) in width and 1.27 cm (0.5 inch) in length under an average pressure of 165.4 MN (24,000 psi) or 103.4 MN (15,000 psi). The samples were heated at a rate of 400°C per hour to a temperature between 960 or 990°C and sintered for 5 minutes, 1h, 3h or 12h in oxygen. All the sintering runs were followed by annealing at 550°C for 2h and slow cooling to room temperature. The variation in sintering temperature and time constituted the second major variable, in addition to the variable of starting Cu and Y sources.

The geometrical densities were determined from the weight and the physical dimensions of the samples after sintering. For dense samples, the Archimedes densities were determined by the submersion method, using carbon tetrachloride as the liquid medium. Dimensional shrinkage was calculated from the change in radius of the disc samples or from the average change in the length and the width of the bar samples before and after sintering. Electrical resistance was measured using the four-point technique from room temperature to 77 K. X-ray diffraction patterns were obtained from the calcined powders as well as the sintered discs using filtered Cu radiation. Thermogravimetric analysis (TGA) of the starting materials and the powder lots before calcination were performed to determine the temperature range of decomposition. TGA was done in flowing air at a constant heating rate of 10°C per minute.

RESULTS AND DISCUSSION

A. Thermogravimetric Analysis (TGA)

In the ball-milled powders prior to calcination, the decomposition characteristics of the individual components were preserved in their respective temperature ranges. As an example, Figure 1 shows the TGA curves of three powder lots: E1314, E1306 and E1329. In this series, $BaCO_3$ and Y oxalate were used in combination with cupric oxide (E1314), cupric carbonate (E1306) or cupric oxalate (E1329). The cupric sources were decomposed by 350°C, the Y oxalate by 700°C while the temperature at which $BaCO_3$ was completely decomposed, T_f, was at 955°C for E1314, 930°C for E1306 and 915°C for E1329. In contrast, there was very little decomposition of pure $BaCO_3$ at 1000°C, with a measured weight loss of only 1.3%.

In all the powder lots, the decomposition of the $BaCO_3$ component started around 800°C, but the temperature at which the decomposition was completed, T_f, was affected by the starting Y and Cu sources. Table I lists T_f and the measured weight loss due to the decomposition of $BaCO_3$ component, together with the calculated weight loss based on the percentage of $BaCO_3$ in the powder lots. The measured weight loss agreed very well with the calculated weight loss to within 0.5%, except for lot E1305 where 1.3% was retained. This indicated that at temperatures above 955°C, the decomposition of the $BaCO_3$ component was completed in most of the powder lots in this study. A small amount of $BaCO_3$ remained above 985°C in lot E1305, which consisted of the most widely used components of cupric oxide and yttrium oxide in addition to $BaCO_3$.

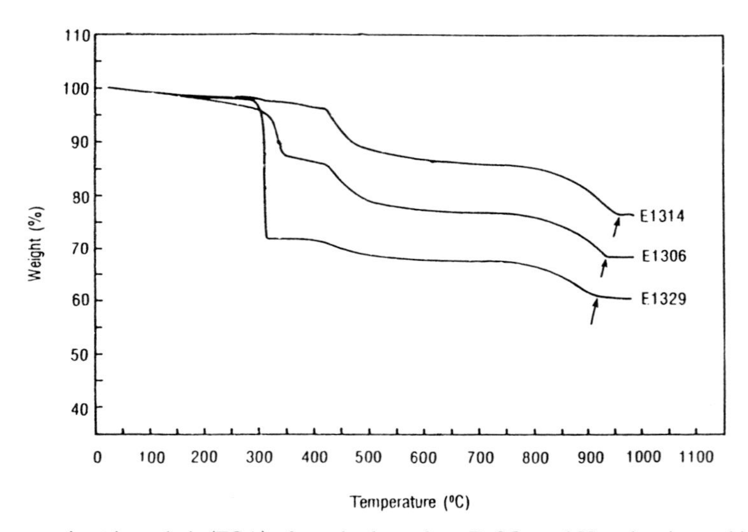

Figure 1 Thermogravimetric analysis (TGA) of as-mixed powders: $BaCO_3$ and Y oxalate in combination with cupric oxide (E1314), cupric carbonate (E1306) or cupric oxalate (E1329).

Table I

Weight Loss due to Decomposition of Barium Carbonate ($BaCO_3$)

lot	T_f(°C)$^+$	measured loss	calculated loss	Δ^{++}	source
E1305	985	10.5%	11.8%	1.3%	Y oxide, Cu oxide
E1309	955	9.9%	10.4%	0.5%	Y oxide, Cu carbonate
E1328	955	8.0%	8.3%	0.3%	Y oxide, Cu oxalate
E1314	955	9.1%	9.4%	0.3%	Y oxalate, Cu oxide
E1306	930	8.3%	8.5%	0.2%	Y oxalate, Cu carbonate
E1329	915	7.0%	7.0%	0%	Y oxalate, Cu oxalate
E1328	955	8.0%	8.3%	0.3%	Y oxide, Cu oxalate
E1331	925	7.3%	7.6%	0.3%	Y carbonate, Cu oxalate
E1329	915	7.0%	7.0%	0	Y oxalate, Cu oxalate
E1309	955	9.9%	10.4%	0.5%	Y oxide, Cu carbonate
E1330	955	9.5%	9.4%	-0.1%	Y carbonate, Cu carbonate
E1306	930	8.3%	8.5%	0.2%	Y oxalate, Cu carbonate
pure $BaCO_3$	1000*	1.5%	22.3%	20.7%	

$^+$ T_f is the temperature at which decomposition of $BaCO_3$ is completed

$^{++}$ Δ - calculated loss - measured loss

* Maximum temperature reached in the TGA analysis

The weight loss of pure $BaCO_3$ at 1000°C was only 1.3% at a constant heating rate of 10°C per minute. Even when a small quantity (~0.5 gm) of pure $BaCO_3$ was calcined at 1000°C for 2 hours, the total weight loss was only 12.7%. This was smaller than the calculated weight loss of 22.3% if the decomposition was complete. Thus, the accelerated decomposition of $BaCO_3$ in the presence of the other components is a significant result. The effectiveness in lowering the decomposition temperature of $BaCO_3$ can be ranked in the order of oxalate, carbonate and oxide regardless whether Y or Cu source was involved. The lowest T_f of 915°C was attained in lot E1329, which was comprised of $BaCO_3$, Y oxalate and Cu oxalate.

For the sintering study, the calcination temperature of all powder lots were fixed at 900°C for 4 hours with the exception of a small batch of E1305 that was calcined at 950°c for 16h. Based on the continuous TGA analysis, a small amount of $BaCO_3$ may have been retained in the calcined powders. The retained amount would be smaller in powders with a lower T_f determined in the TGA. Furthermore, the exact amount would depend on the kinetics of decomposition at 900°C and would be affected by the atmosphere, layer thickness in the calcination boats, etc. In powders using $BaCO_3$, cupric oxide and yttrium oxide as the ingredients, a higher calcination temperature and longer calcination time will be necessary to completely decompose the $BaCO_3$. Otherwise, the remnant $BaCO_3$ could cause poor sintering due to evolution of CO_2 or carbon entrappment during sintering.

B. Sintered Density

Figure 2 shows the normalized electrical resistance of samples sintered at 990°C for 12h. Among the four lots, E1305, E1306, E1309 and E1314, the onset temperature for the superconducting transition was between 90 and 95 K, while zero resistance was reached at temperatures between 87 and 92 K. Thus, all the samples were superconducting above the liquid nitrogen temperature. However, there was a substantial difference in the densification behavior and the highest density reached among the powders prepared using different starting cupric and yttrium sources.

In bulk ceramics, the sintered density has a tremendous influence on the mechanical properties. The electrical, magnetic or superconducting properties may also be affected to different degrees by the bulk density. An accurate way of measuring the bulk density is by the submersion method based on the Archimedes Principle. However, it is only applicable to samples with no open porosity. The alternative, and less accurate, parameter is the geometrical density determined by the weight and the physical dimensions of the sample. In studying the superconductivity phenomena, a general parameter is required to describe the quality of the sintered ceramic, from porous to dense samples. Thus, the geometrical density was chosen as the parameter to correlate the densification characteristics, measured by the linear shrinkage of samples undergoing different heat treatments.

Figure 3 shows the correlation between the Archimedes and the geometrical densities for relatively dense samples. The plot represents the aggregate data from four different powder lots and two different sample geometries (discs and rectangular bars). Samples with higher densities were sintered at the higher temperatures and longer times. As one might expect, an approximately linear relation existed between the geometrical and Archimedes densities. Using a theoretical density of 6.35 gm/cm^3, the Archimedes density as a percentage of the theoretical density (ρ/ρ_{th}) is also labelled on the right abscissa. For ρ/ρ_{th} greater than 87% (or ρ>5.5 gm/cm^3), the Archimedes density gave a good measure of the bulk density. If the geometrical density was smaller than ~ 5.0 gm/cm^3, bubbling was observed when the sample was immersed in the fluid and the Archimedes density remained at about 5.5 gm/cm^3. The highest density achieved in this study was 95% of theoretical density. This is higher than the typical values of 60-85% reported in the literature for powders made by the solid-state reaction route.

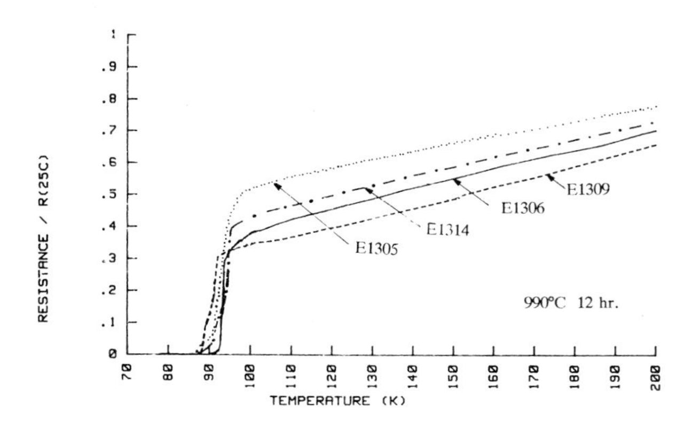

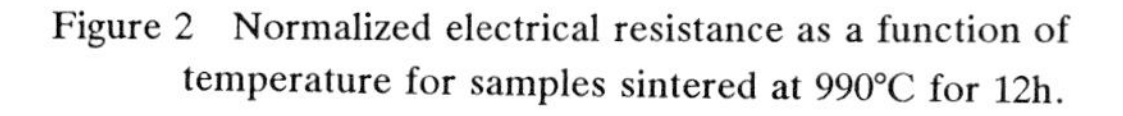

Figure 2 Normalized electrical resistance as a function of temperature for samples sintered at 990°C for 12h.

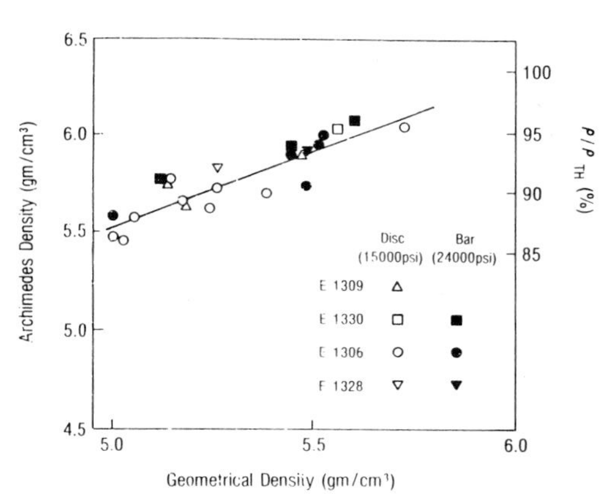

Figure 3 A plot of the Archimedes density versus the corresponding geometrical density on samples of two different geometry and from four different powder lots.

Conventional sintering theory of ceramics suggests that there is no open porosity when the bulk density reaches about 95% of the theoretical density. In the case of $YBa_2Cu_3O_x$, the lower limit shifted to 87% in this study. There are two possible causes. It may indicate that sintering on the surfaces of the compacted samples was more rapid than the interior so that internal pores were closed off during the sintering process. Or it may mean that gas evolution prevents the complete sintering internally while, on the surface, gas can escape to allow for full densification. Figure 4 shows the SEM micrographs of (a) an as-sintered surface and (b) a fracture surface in a dense sample (lot E1306) sintered at 990°C for 12h with an Archimedes density of 6.03 gm/cm^3, and (c) a fracture surface of a less dense sample (lot E1306) sintered at 960°C for 12h with an Archimedes density of 5.85 gm/cm^3. The as-sintered surface contained a mixture of equiaxed grains and elongated grains, which had the c-axis parallel to the short dimension. There was no apparent open porosity. However, approximately hemispherical pores scattered in the fracture

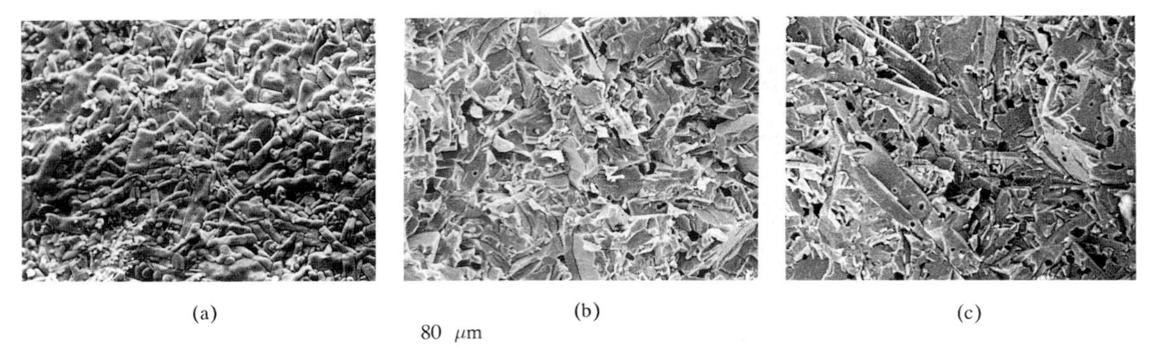

Figure 4 SEM micrographs of (a) as-sintered and (b) fractured surfaces of a E1306 sample sintered at 990°C for 12h and (c) fracture surface of another E1306 sample sintered at 960°C for 12h (samples were 0.5 inch disc pressed at 15000 psi).

Table II

Maximum Density Achieved using Different Cu and Y Sources
(990°C, 12 hours, O_2)

powder lot	density (gm/cm^3)	ρ/ρ_{th} (%)	ion source
E1309	6.01*	94.6	Y oxide, Cu carbonate
E1330	5.92*	93.2	Y carbonate, Cu carbonate
E1306	6.03*	95.0	Y oxalate, Cu carbonate
E1328	5.93*	93.4	Y oxide, Cu oxalate
E1331	4.23**	66.6	Y carbonate, Cu oxalate
E1329	5.73*	90.2	Y oxalate, Cu oxalate
E1305	3.70**	58.3	Y oxide, Cu oxide
E1314	4.53**	71.3	Y oxalate, Cu oxide

* Archimedes density
** Geometrical density

surfaces, with a greater amount of porosity in the less dense sample. This would suggest that gas (probably CO_2) entrappment was mainly responsible for the internal porosity. There was no apparent enhancement in the grain size at the higher sintering temperature of 990°C. Thus, liquid phase assisted sintering was not observed.

The maximum density achieved by the different powder lots were listed in Table II for samples sintered for 12 hours at 990°C. The use of Cu carbonate as the cupric source resulted in consistently high Archimedes densities of about 6.00 gm/cm^3, or greater than 93% of the theoretical density. The use of Cu oxalate in combination with Y_2O_3 or Y oxalate also gave high Archimedes densities between 5.70 and 5.93 gm/cm^3, or between 90 and 93% of the theoretical density. In comparison, the use of CuO generally gave poor densities in the range of 60 to 70% of the theoretical density. These values were on the low end of the reported densities using CuO as the cupric source, probably due to the relatively simple powder processing method that was used in this evaluation.

C. Linear Shrinkage

A second useful parameter to monitor the densification behavior is the linear shrinkage after sintering. Figures 5 and 6 show the linear shrinkage of disc samples as a function of sintering time for two sintering temperatures: 960 and 990°C. For all the powder lots, the shrinkage increased with the sintering time during the first 2 hours, reaching a saturated value after about 4-6 hours at the sintering temperature.

The figures illustrate the dramatic effect on densification by varying the starting cupric and yttrium sources. The use of cupric carbonate as the cupric source (lots E1309, 1306 and 1330) resulted in a rapid shrinkage in the first one hour of sintering, reaching a saturated value of 16% at 960°C and 20% at 990°C. The maximum shrinkage corresponded to samples with the highest Archimedes density of 6.03 gm/cm^3 (ρ/ρ_{th}=95%) achieved in lot E1306. There was a small effect exhibited by the use of different yttrium sources. In combination with cupric carbonate, yttrium oxalate and carbonate behaved similarily while yttrium oxide showed a slower densification rate at the shorter sintering times (<3 hours). However, the final shrinkage achieved was the same at long sintering times greater than 6 hours.

The use of cupric oxalate as the cupric source resulted in a final shrinkage between 9 and 12% at 960°C and between 14 to 19% at 990°C. The largest shrinkage in this group was achieved by the combined use of yttrium oxide and cupric oxalate (lot E1328), corresponding to an Archimedes density of 5.93 gm/cm^3 (ρ/ρ_{th}=93.4%). At 990°C, lot E1328 showed similar shrinkage behavior as the powders lots using cupric carbonate as the cupric source.

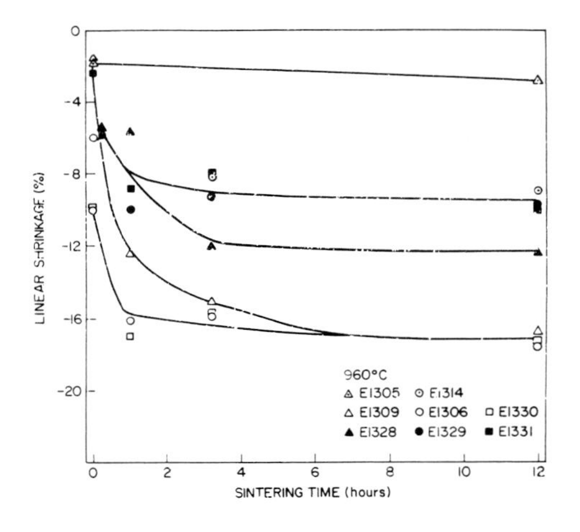

Figure 5 The linear sintering shrinkage of disc samples with a green diameter of 0.5 inch and compacted at 15000 psi as a function of sintering time at the sintering temperature of 960°C.

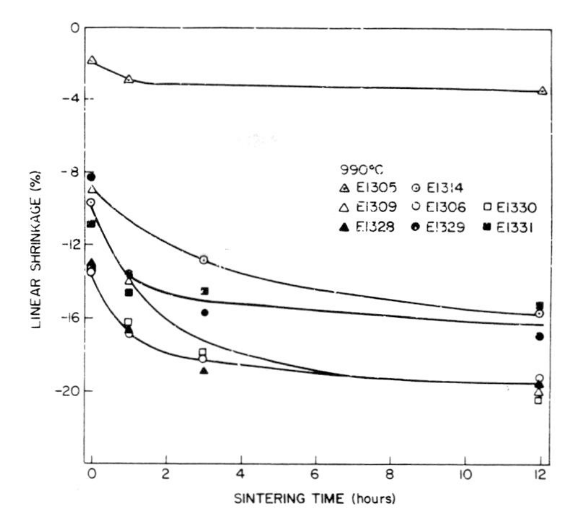

Figure 6 The linear sintering shrinkage of disc samples with a green diameter of 0.5 inch and compacted at 15000 psi as a function of sintering time at the sintering temperature of 990°C.

The combined use of cupric oxide, yttrium oxide with barium carbonate (E1305) showed the least amount of densification. The linear shrinkage ranged from 0 to 5% for all sintering temperatures and times. This behavior was similar for the two calcination conditions investigated: 900°C for 4h followed by grinding with mortar and pestle, and 950°C for 16h followed by 4h ball-milling in isopropanol. Thus, the powder that was prepared from the solid state reaction of $BaCO_3$, CuO and Y_2O_3 exhibited the lowest goemetrical density of about 3.7 gm/cm^3, or about 58% of theoretical density. X-ray diffraction analysis of the calcined powders showed a residue of $BaCO_3$ of 43.9% in the 900°C calcined powder versus 2.4% in the 950°C calcined powder*. This ruled out the incomplete decomposition of $BaCO_3$ as the primary cause of the extremely low density in powder lot E1305. So far, the cause(s) have not been identified.

The use of yttrium oxalate with cupric oxide (E1314) gave a slightly better densification, with a total shrinkage of 8% at 960°C and 15% at 990°C. The highest geometrical density was 4.53 gm/cm^3, or 71% of the theoretical density.

C. General Discussion

An important parameter that determines the final sintered density and shrinkage of compacted samples is the green density after compaction. It is anticipated that a higher green density will result in a smaller shrinkage during sintering and a higher sintered density. Table III lists the green densities measured for the three types of sample geometry and two compaction pressures. A comparison between the green and sintered densities between the four lots (E1305, E1306, E1309 and E1328) suggested that other differences in the powders constituted a more important effect on the densification and sintered density. For example, E1328 had the highest green density among the four lots; but the shrinkage and sintered density at 960°C were smaller than lots E1306 and 1309. On the other hand, the green density of lot E1305 was only about 5% smaller than E1306, but the sintered density was significantly lower. Thus, the physical powder characteristics and the mechanics of compaction, measured collectively by the green density, were insufficient to account for the enhanced densification observed in powder lots using cupric carbonate as the cupric source. One possibility was enhanced surface diffusion during sintering, which would be affected to different degrees by the surface chemistry of powders prepared from different starting raw materials.

Table III

Green Density (gm/cm^3) of Compacted Samples of Different Geometries

sample	E1305	E1306	E1309	E1328
0.50 inch disc (15000 psi)	3.070±0.02	3.185±0.01	3.192±0.01	3.481±0.01
0.25 inch disc (15000 psi)	3.220±0.01	3.410±0.02	3.432±0.02	3.768±0.005
0.25 inch disc (24000 psi)	3.454±0.02	3.609±0.03	3.620±0.03	3.890±0.005
rectangular bar (24000 psi)	3.318±0.01	3.448±0.02	3.445±0.01	3.828±0.01

* Percentage of $BaCO_3$ approximated by the ratio of the strongest $BaCO_3$ peak to the strongest combined $YBa_2Cu_3O_x$ peaks of {(013), (103) and (110)}.

CONCLUSION

The solid state reaction method to form the superconducting $YBa_2Cu_3O_x$ oxide was studied. It was found that the addition of the cupric and yttrium components accelerated the decomposition of the $BaCO_3$ component. At a constant heating rate of 10°c per minute in the TGA analysis, the temperature of complete decomposition, T_f, was lowered from greater than 1000°C in pure $BaCO_3$ to between 915 and 985°C. The effectiveness in decreasing T_f can be ranked in the order of oxalate, carbonate and oxide. The effect of starting Cu and Y sources on the densification of the superconducting $YBa_2Cu_3O_x$ ceramics was studied with respect to the variables of sintering temperature, sintering time, sample geometry and compaction pressure. The highest density achieved in this study was 6.03 gm/cm^3 (ρ/ρ_{th}=95%) at 990°C and 5.85 gm/cm^3 (ρ/ρ_{th}=92%) at 960°C. The source of cupric ion had the largest effect on densification. The use of cupric carbonate resulted in a consistently high Archimedes density of about 6.00 gm/cm^3 and large dimensional shrinkage of about 20% at 990°C for 12h. Using cupric oxalate also gave a high Archimedes density between 5.70 and 5.93 gm/cm^3 at 990°C. The use of cupric oxide gave the lowest density and smallest shrinkage, consistent with the results in the literature which generally use the solid state reacted powder of $BaCO_3$, Y_2O_3 and CuO. Within the same powder lot, higher sintered density and smaller dimensional shrinkage were observed in samples with higher initial green density and compaction pressure. However, a comparison between the different powders lots indicated that the enhanced densification and higher density achieved by the use of cupric carbonate cannot be accounted for by the different physical characteristics of the powders and the mechanics of powder compaction, measured collectively by the green density. One possible mechanism was enhanced surface diffusion during sintering, which would be affected by the surface chemistry of powders prepared from different starting raw materials.

REFERENCES

1. D.W. Johnson, Jr., E.M. Gyorgy, W.W. Rhodes, R.J. Cava, L.C. Feldman and R.B. van Dover, Adv. Ceram. Matls. vol. 2, 329 (1987).
2. Q. Robinson, P. Georgopoulos, D.L. Johnson, H.O. Marcy, C.R. Kannewurf, S.J. Hwu, T.J. Marks, K.R. Poeppelmeier, S.N. Song and J.B. Ketterson, Adv. Ceram. Matls. vol. 2, 386 (1987).
3. J.B. Blendell, C.K. Chang, D.C. Cranmer, S.W. Freiman, E.R. Fuller, Jr., E. Drescher-Krasicka, W.L. Johnson, H.M. Ledbetter, L.H. Bennett, L.J. Swarttzendruber, R.B. Marinenko, R.L. Myklebust, D.S. Bright and D.E. Newbury, Adv. Ceram. Matls. vol. 2, 512 (1987).
4. I. Wei Chen, X. Wu, S.J. Keating, C.Y. Keating, P.A. Johnson and T.Y. Tien, J. Am. Ceram. Soc., vol. 70, C388 (1987).
5. P.K. Gallagher, H.M. O'Bryan, S.A. Sunshine and D.W. Murphy, Mat. Res. Bull., vol. 22, 995 (1987).
6. H. Takagi, S. Uchida, H. Sato, H. Ishii, K. Kishio, K. Kitazawa, K. Fueki and S. Tanaka, Jpn. J. Appl. Phys., vol. 26, L601 (1987).
7. S. Ohshima and T. Wakiyama, Jpn. J. Appl. Phys., vol. 26, L812 (1987).
8. T. Kawai and M. Kanai, Jpn. J. Appl. Phys., vol. 26, L736 (1987).
9. H. Watanabe, Y. Kasai, T. Mochiku, A. Sugishita, I. Iguchi and E. Yamaka, Jpn. J. Appl. Phys., vol. 26, L657 (1987).
10. D.G. Hinks, L. Soderholm, D.W. Capone, J.D. Jorgensen and I.K. Schuller, Appl. Phys. Lett., vol. 50, 1688 (1987).
11. R.C. Budhani, S.H. Tzeng, H.J. Doerr and R.F. Bunshah, Appl. Phys. Lett., vol. 51, 1277 (1987).
12. H.M. O'Bryan and J. Thompson, unpublished data, AT&T Bell Laboratories, MUrray Hill, NJ 07974 (1987).
13. B. Bunker and J. Voigt, Presented at the Spring Meeting of the Materials Research Society, Reno, Nevada (1988).
14. A. Safari, H.G.K. Sundar, A.S. Rao, C. Wilson, V. Parkhe, R. Caracciolo and J.B. Wachtman, Jr., Proceedings of 38th Electronic Components Conference, pp 181-187, Los Angeles (1988).

$YBa_2Cu_3O_{7-\delta}$ Superconductor-Insulator Composites — The Percolation Limit and Device Potential

K.N.R. Taylor, D.N. Matthews, G.J. Russell, M. Shepherd, G. Alvarez, and K. Sealey

Advanced Electronic Materials and Technology Group, School of Physics, University of New South Wales, P.O. Box 1, Kensington, NSW 2033, Australia

ABSTRACT

The superconducting properties of composites formed in the series $YBa_2Cu_3O_x$-Y_2BaCuO_y have been extensively investigated. These materials, formed along a tie-line in the phase diagram are intrinsically two-phase, with the ratios of the two phases established by the working composition.

Since the (211) compound is intrinsically an insulator, by varying the composition we have been able to reach the percolation limit and establish the behaviour of both the inter- and intra-granular behaviour independently. The dependence of the V-i characteristics have been used to establish a number of devices and the tunnelling spectroscopy observations show clear evidence for an unusual series of transitions at high voltages.

INTRODUCTION

The superconducting properties of bulk, sintered samples of yttrium barium cuprate materials are dominated by their granular structure through the inherently weak coupling which occurs between the grains of the solid. Measurements of both the electrical and magnetic behaviour of $YBa_2Cu_3O_{7-\delta}$ (hereafter (123)) in its sintered form reveal very low critical current densities and critical field strengths in comparison with the results for both single crystals [1] and thin films [2]. These effects are particularly noticeable in the changes they lead to in the form of the resistive transition [3] and the complex susceptibility transition [4] at the critical temperature.

The weak-link structure is also generally taken as being responsible for the Josephson junction and SQUID like behaviour observed in narrow bridges [5,6] and for the detailed variation of the critical current with applied magnetic field strength [7].

In attempting to understand these observations it is necessary to understand the nature of the weak-link contacts in detail, and a number of aspects of these materials have been identified as contributing to the weak-link performance. The highly anisotropic conductivity of single crystal (123) must obviously play a significant role in determining the current carrying capacity of sintered discs of randomly oriented grains. Studies of fibre textured materials [8,9] have shown improvements in the critical current densities (j_c) and more recent observations of orientation effects have shown the importance of intergrain misalignment [10] and twin boundaries [11] in establishing j_c. The inherent sensitivity of this material to the appearance of impurity phases during processing is also of obvious importance, since their precipitation at the grain boundaries will provide natural SIS junctions whose thickness may lead to Josephson junction behaviour or more conventional tunnelling processes. In a recent preliminary study of two phase materials formed as composites of $YBa_2Cu_3O_{7-\delta}$ and Y_2BaCuO_5, we have shown that the observed properties vary rapidly with composition close to the percolation limit. We have now extended these observations to include the entire series of materials over the range from the (123) to the (211) composition. The results of these measurements are described in the following.

EXPERIMENTAL

Samples were prepared by our normal sintering route [12] at compositions between the two terminal compounds on the tie line joining them. This was done to ensure that only a single impurity phase would be present, namely the insulating (211) phase, along with the superconducting material. To avoid effects arising from small differences in processing conditions all samples were processed simultaneously in the furnace. In the following a specific composition in the general range $Y_{1+x}Ba_{(2-x)}Cu_{3-2x}O_y$ for $0 < x < 1$ will be identified by its yttrium composition.

The final samples consisted of homogeneous mixtures of the two phases with optical and SEM micrographs showing grain sizes of ≅2.5μm for the (123) phase and 0.2 - 0.5μm for the (211) phase. Beyond $Y_{1.5}$, the distribution of the superconducting phase became inhomogeneous and consisted of long chains of (100μm x 5-10μm) and roughly spherical clusters (40μm radus) of the small grains embedded in the insulating material.

Xray diffraction confirmed that the materials were two phase, and had compositional ratios close to those predicted by the initial stoichiometries of the starting mixtures. The lattice parameters showed no significant changes across the series and were typically

a = 3.82Å   b = 3.88Å   c = 11.67Å   for the (123) component

and a = 12.22Å   b = 5.61Å   c = 7.15Å   for the (211) phase

All the conventional electrical measurements were made using a four-point probe technique with the samples mounted in a continuous flow cryostat and contacts established with silver paste and copper wire. The critical temperatures were determined from the resistance-temperature variation of each sample with a measuring current of 1mA and a limiting voltage sensitivity of 0.1μV. A typical variation for a sample with significant second phase concentration is given in Fig. 1, with an expanded transition region shown below. As may be seen, there are a number of characteristic features in this transition, all of which are indicated on the diagram. The compositional dependence of these parameters is shown in Fig. 2,

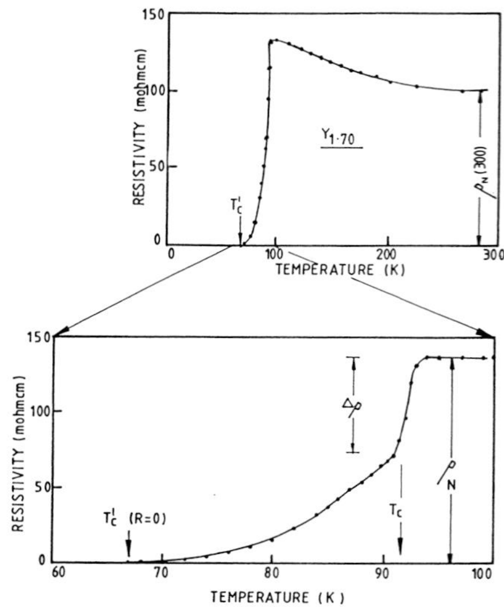

Figure 1. The resistive transition of the $Y_{1.7}$ sample showing the details of the extended foot associated with the weak-link structure of these materials. Various features used in the text are shown on the diagram

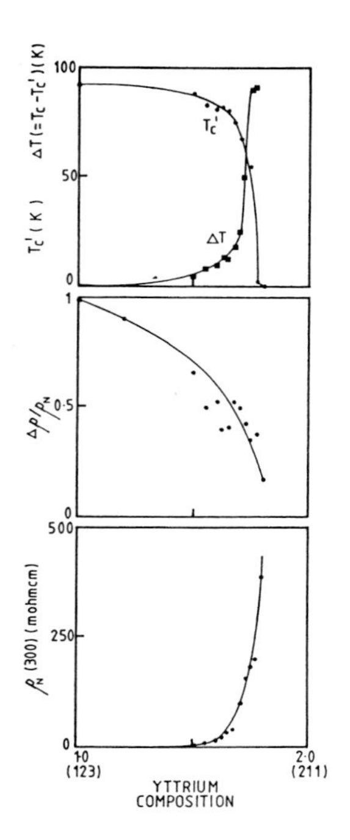

Figure 2. The composition dependence of T_c', (T_c-T_c'), $\Delta\rho/\rho_N$ and the normal state resistivity showing the clear approach to the percolation limit at $Y_{1.8}$.

from which it is clear that a critical composition exists at approximately $Y_{1.8}$. Beyond this, there is no transition to zero resistivity and the normal state resistivity increases rapidly, both of which suggest that there is no longer full connectivity throughout the superconducting granular components.

Critical current densities (j_c) for each of the samples were determined as a function of temperature in zero applied magnetic field strength. This was achieved by continuously recording the sample voltage as the transport current was increased from zero. Typical V-i characteristics obtained this way are shown in Fig. 3. Since the normal practical definition of j_c under these conditions is associated with a detectable voltage appearing above some minimum value, these curves, which are essentially asymptotic to the current axis, are unsuitable for the determination of accurate j_c values. To improve this situation the V-i characteristics have been replotted as logV-logi relations as shown in Fig. 4 for a typical sample. As may be seen, at a single temperature, the data can be represented by one or two linear regions which intercept the 1μV axis (our limiting voltage sensitivity) at a well defined current density, which we take to be the critical current density. The variation of this critical current density with temperature at a fixed composition, or with composition at a fixed temperature, is readily obtained, and an example of a j_c-T variation is shown in Fig. 5. A second,

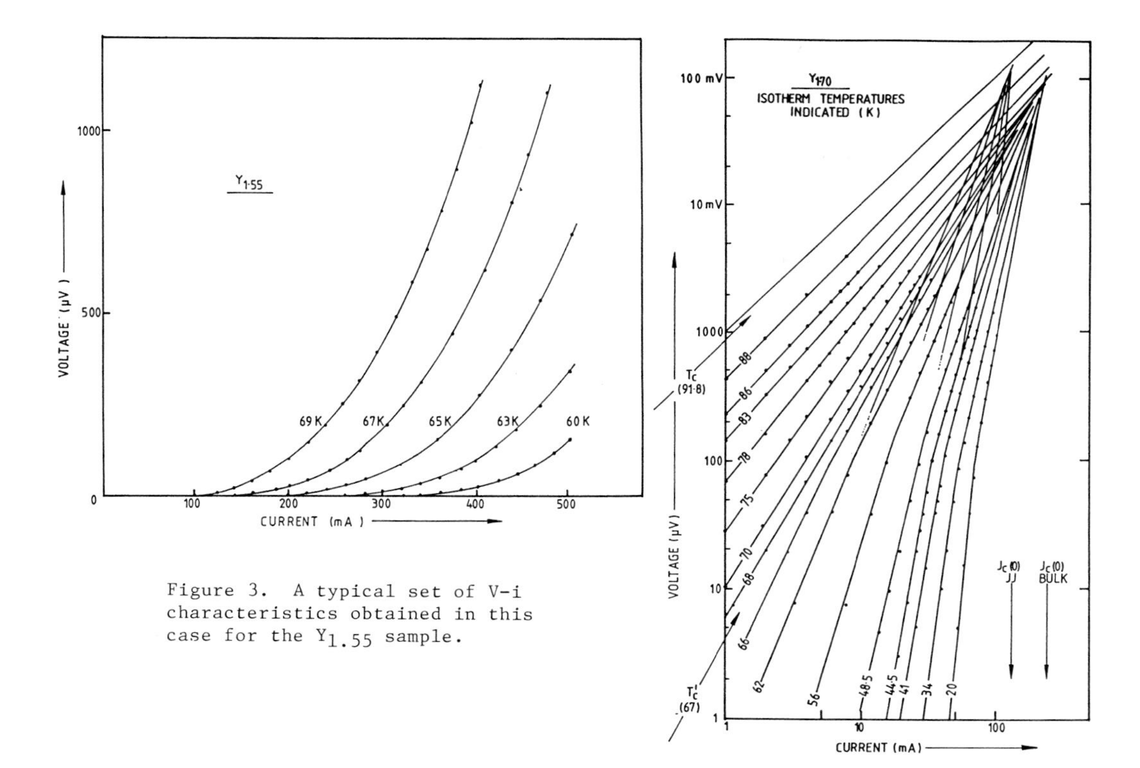

Figure 3. A typical set of V-i characteristics obtained in this case for the $Y_{1.55}$ sample.

Figure 4. A logV-logi isotherms for $Y_{1.70}$ showing the two linear regimes below T_c' and the convergence points $J_c(o)_{JJ}$ and $J_c(o)_{BULK}$ referred to in the text. The upper voltage boundary of the steepest sections are indicated(*).

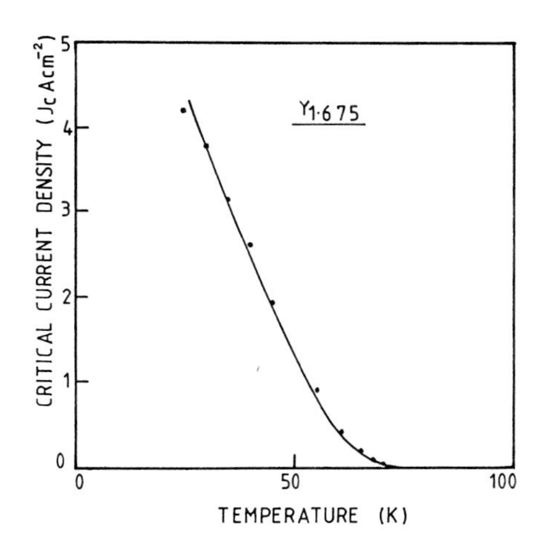

Figure 5. The variation of j_c with temperature for a typical sample.

obvious feature of Fig. 4 is the change from the double linear variation at low temperatures, to the single linear variation at higher temperatures, with the changeover occurring at $T \cong T_c'$ the lower limit transition temperature. It is also clear that the low temperature linear regime of highest slope is bounded by a constant sample-voltage line (as indicated) and above this voltage, the slopes of the log V - logi relationships vary continuously with those found for $T > T_c'$. These linear regions, of course, define power law relationships between V and i of the form $V = i^{a(T)}$ which is similar to the behaviour reported in other granular superconductors [13], with the index a(T) tending to unity at high temperatures where we can anticipate normal ohmic behaviour. The temperature dependence of a(T) for several of the samples is shown in Fig.6 from which it is clear that the general behaviour may be described by $a(T) \propto T^{-n}$ with a strong composition dependence of this power law. Assuming that the straight line variations shown can adequately describe the data it is evident that the strength of the a(T) variation is extremely composition dependent and as Fig. 7 shows is given by $n = \beta(p-p_c)^{1.7}$ where $p_c \cong 1.8$ (corresponding to the $Y_{1.8}$ material). If we take p_c as the percolation limit for this particular series of granular materials, this behaviour is close to that predicted theoretically for a randomly connected 3D system.

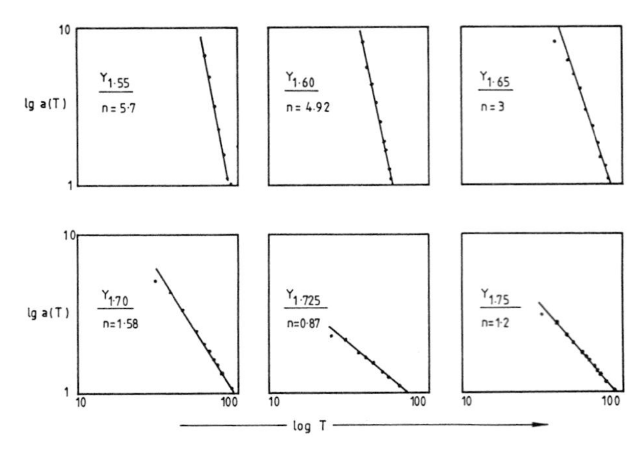

Figure 6. The temperature dependence of the index a(T) which describes the V-i characteristics for a range of compositions approaching the percolation limit. The negative gradient n of these individual graphs is shown.

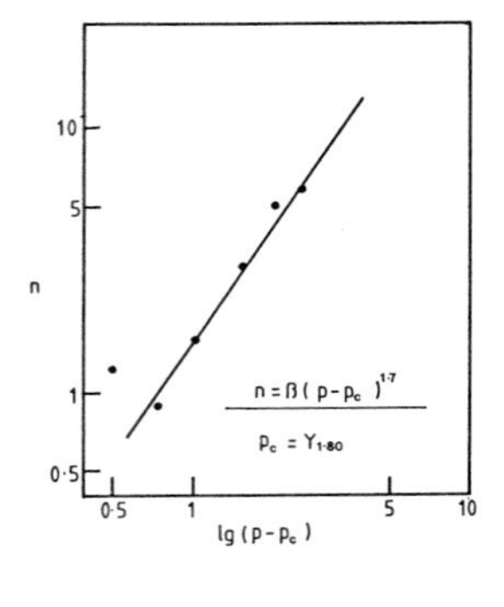

Figure 7. The variation of the index n (Figure 6) with composition p relative to the percolation limit p_c.

DISCUSSION

The form of the resistive transition shown in Fig. 1 is typical of that to be expected from a granular composite superconductor in which the superconducting path between grains is maintained by weak-link junctions. On cooling, the initial steep decrease in resistivity can be associated with the bulk properties of the grains while the extended tail reflects the intergrain coupling characteristics. This tail is very sensitive to the transport current since locally this can exceed the current carrying capacity of individual junctions and establish a resistive component in the conduction path. At our particular measuring current, only below T_c' do we have full connectivity throughout the sample and clearly this value of T_c' is a function of both the current and the voltage sensitivity.

For sample temperatures in this region below T_c' the two stage characteristics of Fig.3 can be interpreted as the gradual breaking of the weak-link structure at the lower currents and voltages and the response of a system of disconnected superconducting grains at the higher currents. Above T_c' the measuring current is already in excess of j_c and we only observe the return of the granular system to its fully normal state with increasing current. It is very noticeable that the two systems of linear logV-logi characteristics in Fig. 4 converge to different limiting (V,i) points. Since the gradients of each isotherm vary inversely as a high power of temperature, the T = OK characteristics should lie perpendicular to the current axis at some critical current value which corresponds to these convergence points (V,i). In that case, these convergence points can be taken as the T = OK limit of the critical currents in the two sample regimes. These values are indicated in Fig. 4 and as would be expected, $J(O)_{JJ}$, the weak-link critical current is always less than $J(O)_{BULK}$ although not by as large a margin as would be expected if the latter value was associated with the intragranular superconductivity.

While there seems little doubt that the low current regime below T_c' corresponds to the weak-link system it is necessary to look for an alternative explanation for the broader, high current behaviour characterized by the convergence point $J_c(o)_{BULK}$ and shown in Figs. 5,7. One possible origin of the two regimes lies in the observed microstructure of the materials which was mentioned in the previous section. This ball and chain morphology consisting of both long, narrow strings and spherical aggregates of grains, offers two types of situation for weak-link effects. Conduction along the chains will be highly sensitive to the current density since they represent essentially 1D current paths. In contrast, the granular clusters, although poorly connected, are much nearer to the 3D limit. This interpretation of the data in terms of closely related mechanisms is supported by the observed compositonal dependence of both critical current values shown in Fig. 8, from which it is clear that both tend to zero at the percolation limit.

In addition to the fundamental value of these composite materials in providing an insight into the properties of the terminal (123) compound, they also provide the basis for improved device application. Several specific areas can be identified in which the combination of superconducting percolation paths and weak-link coupling provide unique opportunities to optimize performance.

In one area of such applications, the high sensitivity of the transition to the resistive state can be used as the basis of a range of versatile sensors. In the form of bolometric detectors, we have already demonstrated sensitivities of 1-10V/watt with reasonably high response rates.

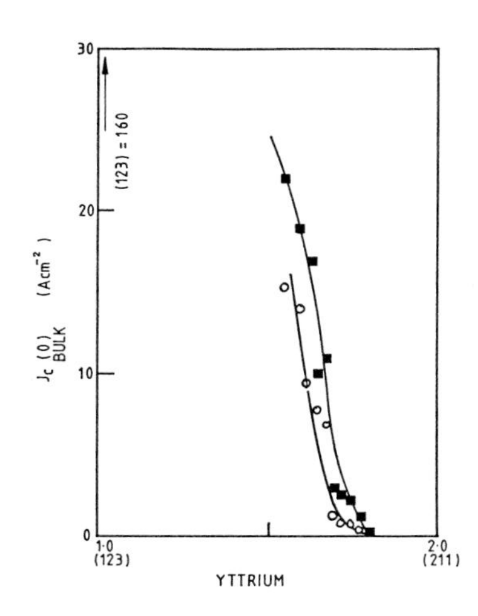

Figure 8. The variation of the critical convergence currents from the logV-logi relationships, as a function of composition The closed square symbols relate to $J(o)_{BULK}$ while the open circles are for $J(o)_{JJ}$.

In a second application field which is attracting considerable attention, the reduction of the number of superconducting percolation paths through the material as the percolation limit is approached can restrict the number of competing junctions in narrow bridge SQUID systems. In the limit, it should be possible to observe both single and double junction effects similar to those more commonly found in low temperature devices operating at 4.2K.

In preliminary observations of the tunnelling characteristics of these composites at 77K we have observed well defined Josephson behaviour at low voltages and clear evidence for tunnelling processes which probably involve the SIS jsunctions formed between the (123) grains by the (211) intergranular material.

REFERENCES

1. W.J. Gallagher. J. Appl. Phys. 63 4216 198
2. P. Chaudhari, R.H. Koch, R.B. Laibowitz, T.R. McGuire and R.J. Gambino Phys. Rev. Letts. 58 2684 1987
3. I. Kostadinov, M. Mateev, I.V. Petrov, P. Vassilev and J. Tihov, Physica C. 153-155 320 1988.
4. R.B. Goldfarb, A.F. Clark, A.I. Braginski and A.J. Pauson. Cryogenics 27 475 1987
5. C.E. Gough, M.S. Colclough, E.M. Forgan, R.G. Jordan, M. Keene, C.M. Muirhead, A.I.M. Rae, N. Thomas, J.S. Abell and S. Sutton. Nature 326 855 1987.
6. C.M.Pegrum,G.B. Donaldson, A.H. Carr and A. Hendry. App. Phys. Lett. 51 1364 1987.
7. R.L. Peterson and J.W. Ekin. Phys. Rev. B. 37 9848 1988.
8. J. Huang, T.W. Li, X.M. Xie, J.H. Zhang, T.G. Chen and T. Wu. Materials Letters 6 222 1988.
9. D.N.Matthews, A. Bailey, G.J. Russell, G. Alvarez, S. Town, K.N.R. Taylor, F. Scott, M. McGirr and D.J.H. Corderoy. These Proceedings.
10. D. Dimos, P. Chaudhari, J. Mannhart and F.K. LeGoues. Phys. Rev. Letts. 61 219 1988.
11. G. Deutscher Physica C. 153-155 15 1988.
12. K.N.R. Taylor, D.N. Matthews, J. Cochrane, G.J. Russell, S. Bosi, H.B. Sun, G. Alvarez, N. Mondinos, B. Hunter, R.A. Vaile, A. Bailey and T. Puzzer. Physica C. 153-155 818 1988.
13. K. Epstein, A.M. Goldman and A.M. Kadin. Phys. Rev. Letts. 47 534 1981.

observation [4], where any diffraction peaks for 10h MA powders disappeared below the level of the resolution. The fact that we observed a broad peak indicates the coexistence of an amorphous phase. Indeed, a sharp diffraction line emerges in place of the halo upon heating above 473 K. We take this as evidence for the crystallization of the amorphous phase.

The mechanically alloyed powders are generally 0.1-1 μm in diameter and thus made the direct TEM observation difficult. Instead, we analysed the composition of three constituent elements Y, Ba and Cu at various places selected randomly in a particle and also in different particles. The results for the 36h MA powders are shown in Fig.2. The spatial resolution of EDX was limited to about 0.1-0.3 μm because of the dissipation of the incident electron beam in a particle with an average diameter of 1 μm. It can be seen that, within this limited spatial resolution, the three elements are more or less homogeneously distributed after 36h MA. The yet measurable composition fluctuations may be associated with the presence of crystalline phases in the amorphous matrix. We, therefore, conclude that MA treatment for 36h is reasonably satisfactory in synthesizing a homogeneous master alloy YBa_2Cu_3, though efforts for the further refinement, particularly the atomic-level homogenization, should be continued.

Chemical analysis revealed the presence of Fe, Cr and Ni impurities originating from the stainless steel balls and the wall of the vial. The total content turns out to be less than 1 % in the 36h MA powders.

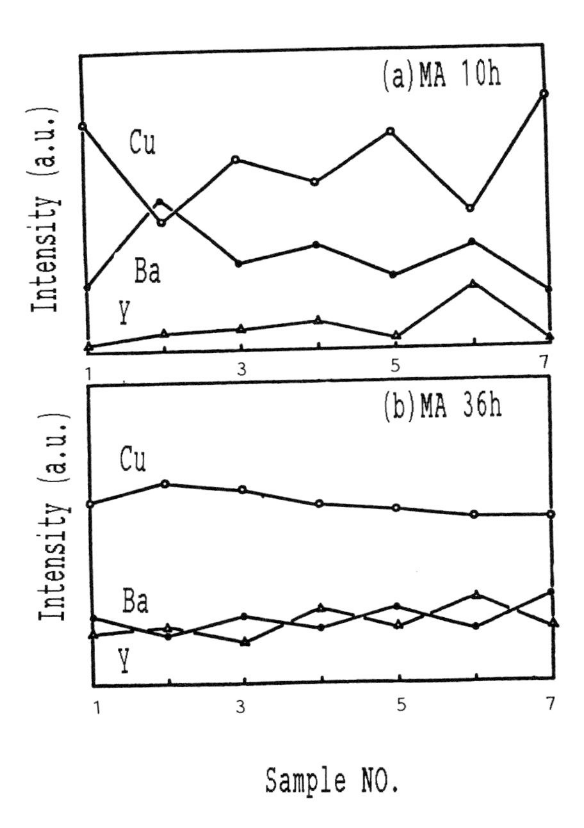

Fig.2 EDX analysis for YBa_2Cu_3 powders after (a) 10h and (b) 36h MA treatments. The abscissa refers to the randomly selected position of the spot where the analysis made.

Formation of $YBa_2Cu_3O_y$ compound through oxidation

The TG/DTA analysis was performed for the 36h MA powders under two different conditions; one in air and the other in Ar gas atmosphere. As shown in Fig.3, the TG curve remains unchanged in the Ar gas whereas the DTA curve exhibits a clear exothermic peak followed by a small endothermic peak in the temperature range 473-573 K. This can be attributed to the onset of the crystallization and is consistent with the X-ray diffraction data in Fig.1. On the other hand, the TG curve clearly shows a peak above about 473 K when measured in air. This implies that the oxidation becomes significant in the temperature range of 473-573 K where the crystallization takes place.

Realizing the TG/DTA results shown in Fig.3, we attempted to oxidize the MA powders at 623 K for 24h in air. Fig.4 shows the X-ray diffraction patterns of the powders thus prepared. Unfortunately, no orthorhombic superconducting phase has been identified. Instead, the formation of $BaCO_3$ is evident. The electron diffraction experiment revealed the presence of $BaCO_3$ and its dark-field image showed that $BaCO_3$ is exclusively formed over the surface of a YBa_2Cu_3 particle. We realized that air-exposed YBa_2Cu_3 powders have to be oxidized at temperatures high enough to decompose $BaCO_3$. Otherwise, the existence of stable $BaCO_3$ on the surface of each particle hampers the formation of the perovskite phase. It is, therefore, critically important to oxidize the mechanically alloyed powders without exposing air or CO_2 in order to transform them directly to the orthorhombic phase at low temperatures below about 873 K, where the orthorhombic phase is most stable [6].

It may be worthy mentioning that the bulk sample prepared by sintering the mechanically alloyed powders at 1213 K can be well characterized by the orthorhombic single phase and undergoes a sharp superconducting transition at 90 K with its width of approximately 2 K. The volume fraction of the superconducting phase was estimated to be about 50 % from the field dependence of magnetization at 4.2 K [4]. The value is higher than that of 20-30 % obtained by the conventional sintering technique based on solid state reaction of oxides.

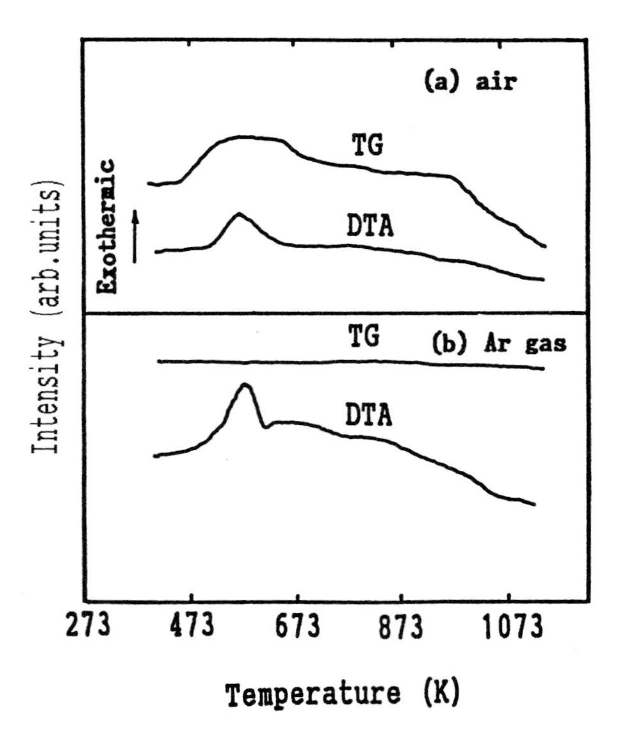

Fig.3 TG/DTA curves for YBa_2Cu_3 powders after 36h MA treatment. Heating was made in (a) air and in (b) Ar gas atmosphere with a heating rate of 10 K/min.

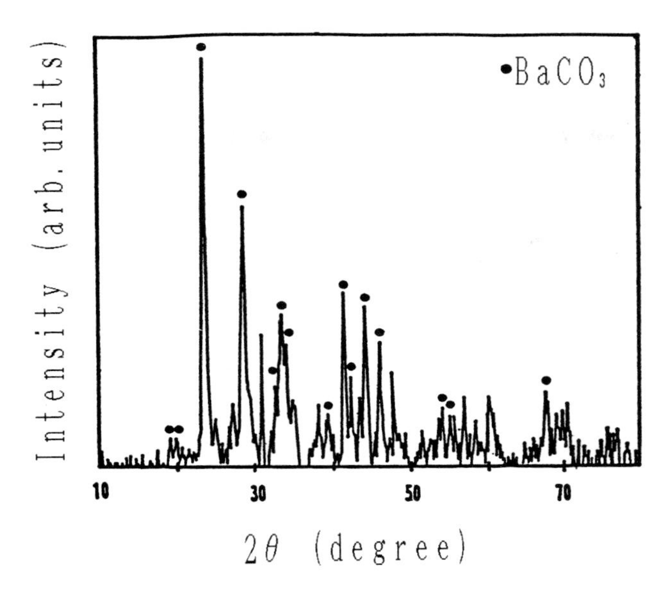

Fig.4 The X-ray diffraction patterns for YBa_2Cu_3 powders oxidized in air at 623 K for 24h.

$BiSrCaCu_2O_y$ produced by MA and subsequent oxidation

The X-ray diffraction pattern for the $BiSrCaCu_2O_y$ sample prepared by oxidizing and sintering the mechanically alloyed powders is shown in Fig.5. The lines can be essentially identified as the low-temperature phase of the $Bi_2Sr_2CaCu_2O_y$ structure. As shown in its inset, the superconducting transition is found to take place at 110 and 75 K, indicating the presence of two superconducting phases. The results are essentially identical to those obtained for samples prepared by the ordinary sintering technique.

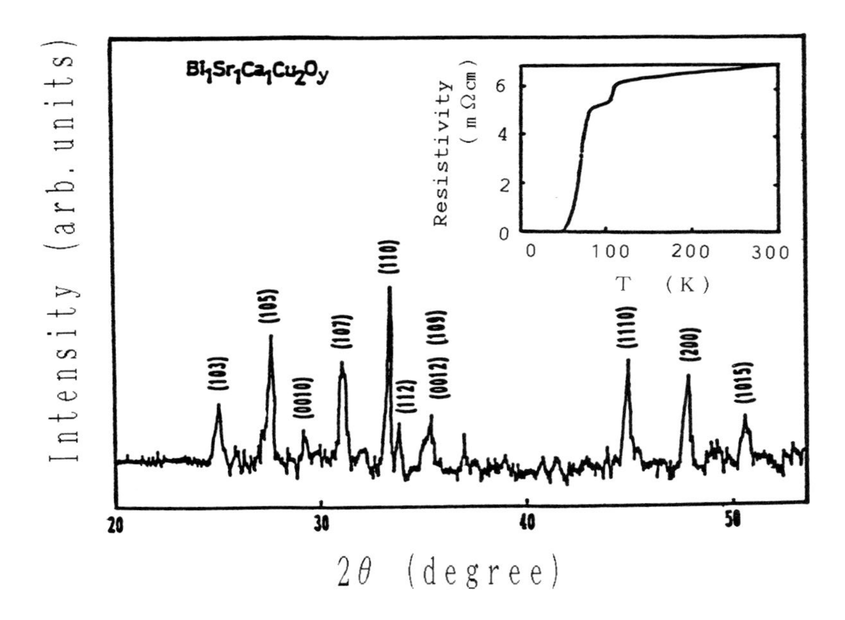

Fig.5 The X-ray diffraction patterns for $BiSrCaCu_2$ mechanically alloyed powders oxidized in air at 1073 K for 24h. The lines are indexed in terms of the low-temperature phase of $Bi_2Sr_2CaCu_2O_y$. Its inset shows the resistivity versus temperature curve for the sample after sintering.

ACKNOWLEDGMENT

The authors wish to acknowledge K.Kotani, Industrial Research Institute of Aichi Prefectural Government, for her assistance with the electron microscope observations.

REFERENCES

[1] K.Matsuzaki, A.Inoue, H.Kimura, K.Moroishi and T.Masomoto, Jpn.J.Appl.Phys. **26** (1987) L334.
[2] R.Haldar, Y.Z.Lu and B.C.Giessen, Appl.Phys.Lett. **51** (1987) 538.
[3] R.B.Schwarz, Mat.Sci.Eng. **97** (1988) 71
[4] Y.Yamada, S.Murasaki, M.Suganuma and U.Mizutani, Jpn.J.Appl.Phys. **27** (1988) L802.
[5] K.Matsuzaki, A.Inoue and T.Masumoto, Jpn.J.Appl.Phys. **27** (1988) L779.
[6] K.Nakamura and K.Ogawa, Jpn.J.Appl.Phys. **27** (1988) 577.

Densification Behavior of $YBa_2Cu_3O_{7-\delta}$ Sintering

H. SAKAI, M. YOSHIDA, and K. MATSUHIRO

Materials Research Laboratory, NGK Insulators, Ltd., 2-56, Suda-cho, Mizuho-ku, Nagoya, 467 Japan

ABSTRACT

The densification of $YBa_2Cu_3O_{7-\delta}$ to 5.9g/cm^3 was achieved by the reaction sintering at 950°C for 3h with using the mixture of Y_2BaCuO_5 and the unknown X phase as a starting powder under the controlled oxygen partial pressure ranging 0.1 to 0.2atm. The better sinterability was caused by the liquid phase formation during sintering. The X phase decomposed to $BaCuO_2$ and the CuO-rich unknown phase around 600°C in oxygen partial pressure ranging 0.1 to 0.2atm. Y_2BaCuO_5 reacted with $BaCuO_2$ in the liquid phase formed from the CuO-rich phase above 820°C. The formation of $YBa_2Cu_3O_{7-\delta}$ and the densification proceeded simultaneously, hence the high density was achieved. On the other hand the X phase oxidized to $Ba_3Cu_5O_8$ in oxygen partial pressure ranging 0.5 to 1.0atm. Y_2BaCuO_5 reacted with $Ba_3Cu_5O_8$ above 900°C with small amount of liquid phase. $YBa_2Cu_3O_{7-\delta}$ densified insufficiently because the densification started after the formation of $YBa_2Cu_3O_{7-\delta}$.

INTRODUCTION

It has been studied intensively to increase the critical current density, Jc, of the high Tc superconducting oxide since the high Tc material is a must for the practical application. One of the methods to increase Jc of the bulk material may be the densification of the sintered body because high Tc superconducting oxides such as $YBa_2Cu_3O_{7-\delta}$ possess the property of bad sinterability and are usually very porous.

The high density oxide can be obtained by using the fine and homogeneous powder, for example, produced by the co-precipitation method[1]. However, the precise stoichiometric composition is very difficult to be controlled and the production yield is often limited to be small.

The process in reaction sintering is very similar to the usual solid state reaction method. Therefore the composition can be controlled precisely and the process is very simple. In particular the powder with coarse grains can be possibly densified in reaction sintering. $YBa_2Cu_3O_{7-\delta}$ was reported to be densified to 5.5g/cm^3 by the reaction sintering with using the mixture of $BaCuO_2$ and $Y_2Cu_2O_5$ as a starting powder[2].

In this study the reaction sintering was applied to $YBa_2Cu_3O_{7-\delta}$ with using the powder calcined in vacuum as a starting powder which consisted of two phases, Y_2BaCuO_5 and the unknown phase. The sintering behavior depending on the oxygen partial pressure was analysed to clarify the mechanism of densification.

EXPERIMENTAL

Starting powder preparation

Y_2O_3, $BaCO_3$ and CuO powders of 99.9% purity were mixed so that the cation composition should be Y:Ba:Cu=1.0:2.0:3.0. The mixture was milled with Zr balls in ethanol solution for 100h and calcined at 900°C for 10h in vacuum of 1-10Pa. The calcined powder was pulverized in an agate mortar for 2h and the larger particles than 100μm was screened. The mean diameter of the calcined powder was 8μm and the cation composition was estimated to be Y:Ba:Cu=1.0:2.0:2.9 by chemical analysis. The X-ray diffraction pattern of the calcined powder is shown in Fig. 1. The powder consists of Y_2BaCuO_5 and the unknown phase (called X phase). This X phase was identified as the same crystalline phase as reported previously[3].

Sample preparation and measurement

The starting powder was pressed to 7ton/cm^2 isostatically into the pellet with the diameter of 17mm and the thickness of 4mm and then sintered in three different atmospheres, oxygen, air and nitrogen at the heating rates of 5 to 30°C/min up to 920°C. The samples were quenched out of the furnace at 300, 400, 600, 820, 860, 890 and 920°C in the midway of heating process and after keeping 920°C for up to 9h.

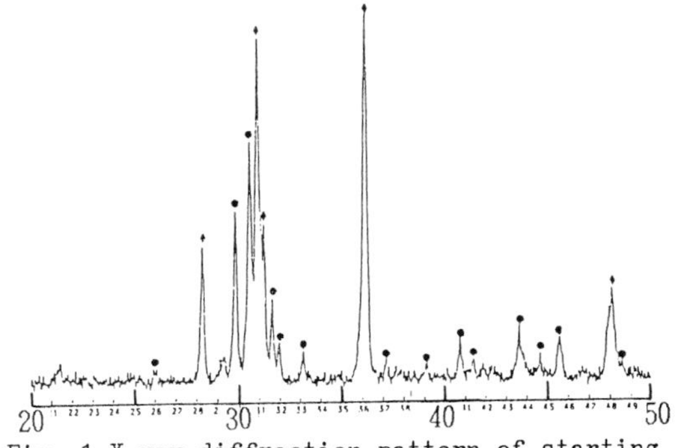

Fig. 1 X-ray diffraction pattern of starting powder. ● Y_2BaCuO_5 , ♦ unknown X phase.

Another set of samples were sintered at 920 and 950°C for 3h in the flowing gas of 100ml/min under the controlled oxygen partial pressure ranging 0.1 to 1.0atm by mixing the oxygen gas with the nitrogen gas using the flow meter and by monitoring with the calcia-stabilized zirconia solid electrolyte oxygen sensor.

The changes in diameter and weight of the pellet before and after the sintering were measured by the calliper and the electric balance, respectively. The crystalline phase was detected by X-ray diffractometry and the microstructure was observed by scanning electron microscopy(SEM). The starting powder was analyzed by thermogravimetry and differential thermal analysis(TG-DTA). The bulk density was measured by means of Archimedes' method using kerosene.

RESULTS

Changes in dimension and weight

The changes in dimension Δl and weight ΔW in the course of sintering process under three different atmospheres, oxygen, air and nitrogen, are shown in Figs. 2(a) and (b).

The sintering in oxygen atmosphere resulted in the expansion of diameter of the pellet by 2% up to 900 °C and the weight gain by 4% up to 600 °C. The shrinkage and the weight loss began simultaneously when the temperature was maintained at 920°C. The sintering at 920°C for 6h led to the shrinkage by 2% (by 4% during keeping at 920°C).

The sintering in air at the heating rate of 30°C/min showed the expansion by 1% and the weight gain by 3% up to 800°C. The rapid shrinkage by 4% was observed from 820°C through 920°C accompanied with the weight loss, while a little shrinkage proceeded after keeping at 920°C. The similar behavior was observed in the case of the sintering in air at the heating rate of 5°C/min except that the final shrinkage reached more than 7%.

Nitrogen atmosphere caused little changes in dimension and weight up to 920°C.

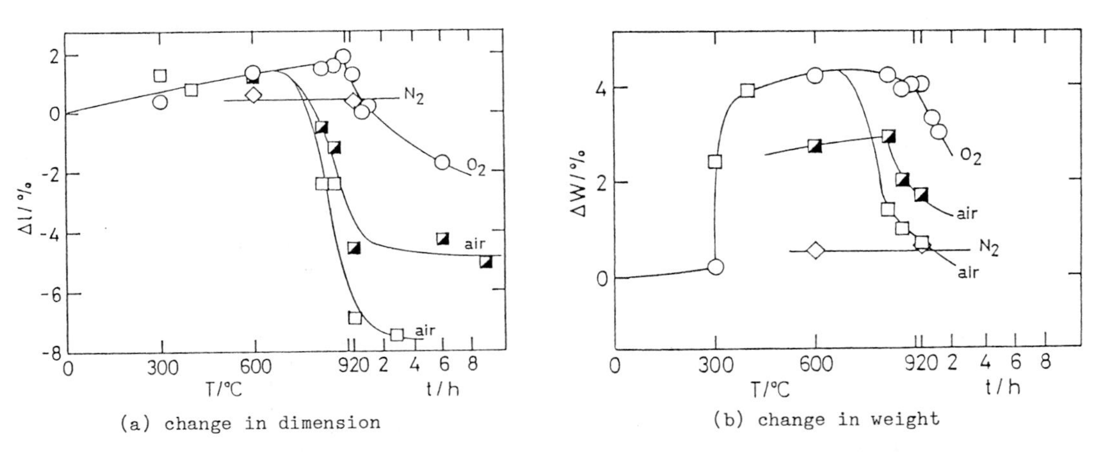

(a) change in dimension (b) change in weight

Fig. 2 Changes in dimension Δl and weight ΔW of quenched samples under three different atmospheres. ○ in oxygen at the heating rate of 20°C/min, ◪ in air, 30°C/min, □ in air, 5°C/min, ◇ in nitrogen, 20°C/min.

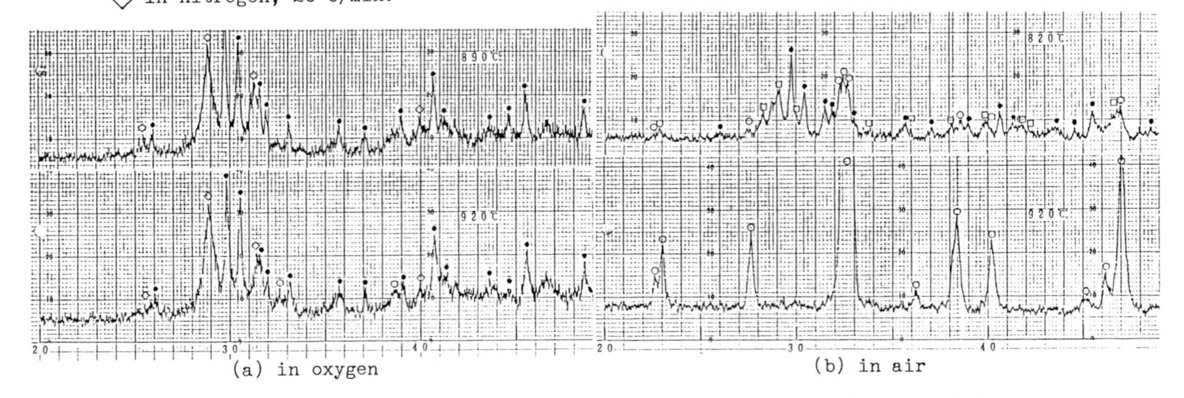

(a) in oxygen (b) in air

Fig.3 Typical X-ray diffraction patterns of quenched samples in oxygen(a) and air(b).
● Y_2BaCuO_5 , ○ $YBa_2Cu_3O_{7-\delta}$, □ $BaCuO_2$, ◇ $Ba_3Cu_5O_8$.

Crystalline phase

The crystalline phases produced under three different atmospheres are partly shown in Figs. 3(a) and (b), and summarized in Table I.

The unknown X phase contained in the starting powder changed into $Ba_3Cu_5O_8$ phase, which will be clarified in the discussion section, between 300 and 600°C in oxygen atmosphere. $Ba_3Cu_5O_8$ was stable until the production of $YBa_2Cu_3O_{7-\delta}$ phase by the reaction between Y_2BaCuO_5 and $Ba_3Cu_5O_8$ at 920°C.

On the other hand the unknown X phase changed into $BaCuO_2$ phase by way of $Ba_3Cu_5O_8$ up to 600°C in air. Above 820°C the disappearance of Y_2BaCuO_5 and $BaCuO_2$ phases followed by the appearance of $YBa_2Cu_3O_{7-\delta}$ phase and the only $YBa_2Cu_3O_{7-\delta}$ phase was present when the temperature reached 920°C.

Little change in crystalline phases was observed in nitrogen atmosphere, although small amount of $BaCuO_2$ phase was detected at 920°C.

Table I Crystalline phases produced during sintering.

quenched temp.(°C)	atmosphere/heating rate(°C/min) O_2 20	air 30	air 5	N_2 20
300	G,X	-	G,X	G,X
400	-	-	G,B,(X)	-
600	G,B,(X)	G,C,(B)	-	-
820	G,B	G,C,(S)	S,(G,C)	-
860	G,B	S,(G,C)	S,(G,C)	-
890	G,B	-	-	-
920	G,B,(S)	S	S	G,X,(C)
920°Cx0.5h	S,(C)	S	-	-
920°Cx1h	S,(C)	-	-	-
920°Cx3h	-	-	S	-

G:Y_2BaCuO_5, S:$YBa_2Cu_3O_{7-\delta}$, C:$BaCuO_2$, B:$Ba_3Cu_5O_8$
X:unknown X phase.
( ) means weak peak(s).

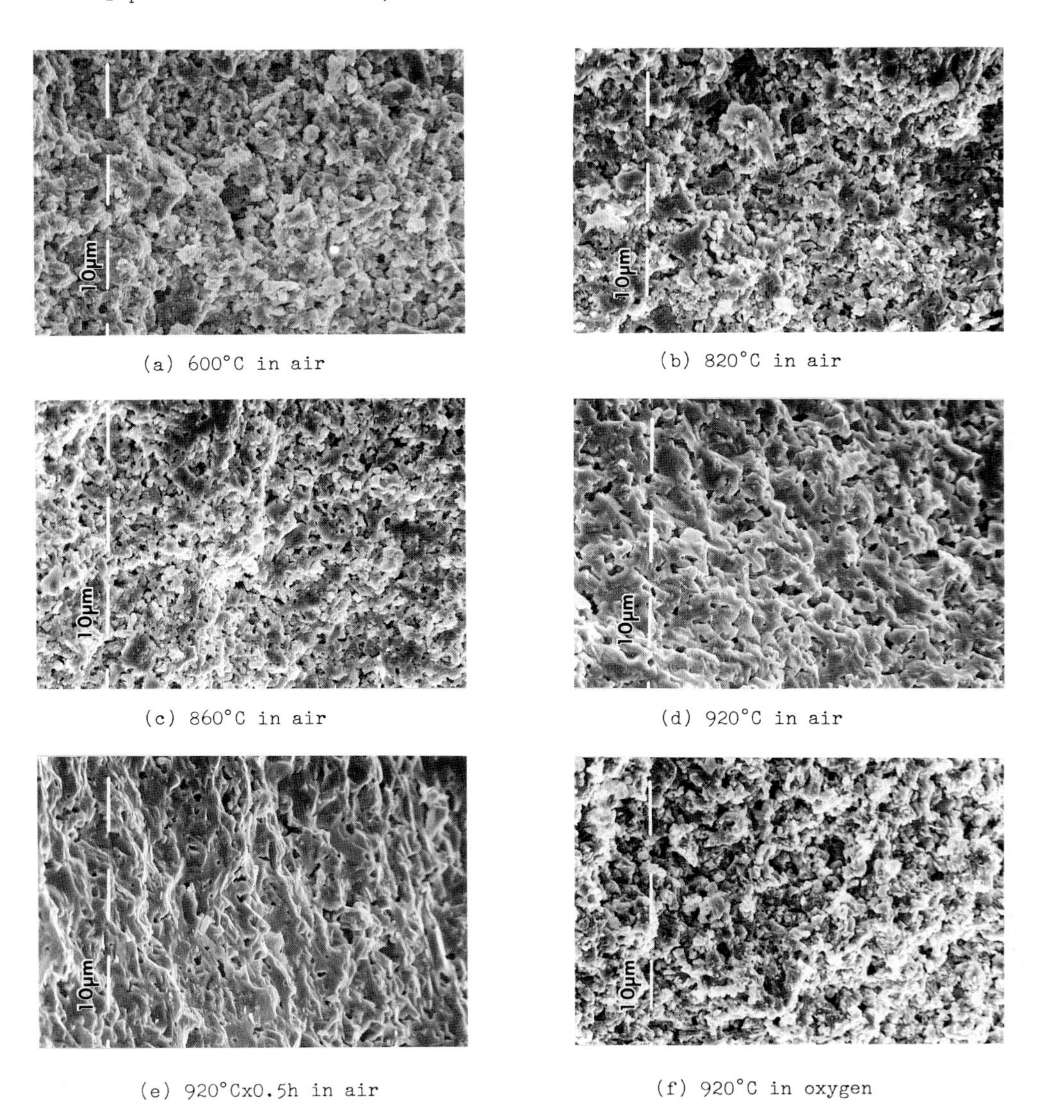

(a) 600°C in air (b) 820°C in air
(c) 860°C in air (d) 920°C in air
(e) 920°Cx0.5h in air (f) 920°C in oxygen

Fig.4 Microstructural changes during sintering observed by SEM.

Microstructure

The microstructural change in the course of sintering process in air at the heating rate of 30 °C/min is shown in Figs. 4(a) through (e).

The onset of densification (indicated by the appearance of the neck between grains) was observed at 820 °C and the densification proceeded with increasing temperature from 820 °C to 920 °C.

The microstructure at 920 °C in oxygen atmosphere is also shown in Fig. 4(f). The comparison of Fig. 4(d) with Fig. 4(f) clearly indicates that little densification occurred in oxygen atmosphere even at 920 °C.

TG-DTA analyses

The results of the TG-DTA analyses of the starting powder in air and oxygen atmosphere are shown in Figs. 5(a) and (b).

The group of exothermic peaks were observed around 315 °C in air and around 345° C in oxygen atmosphere accompanied with the rapid gain of the weight. The sharp endothermic peaks were observed at 1014° C in air and at 1033 °C in oxygen atmosphere. The endothermic peak at 835 °C in air was rather sharp, while in oxygen atmosphere the endothermic peak was broad and small around 900 °C.

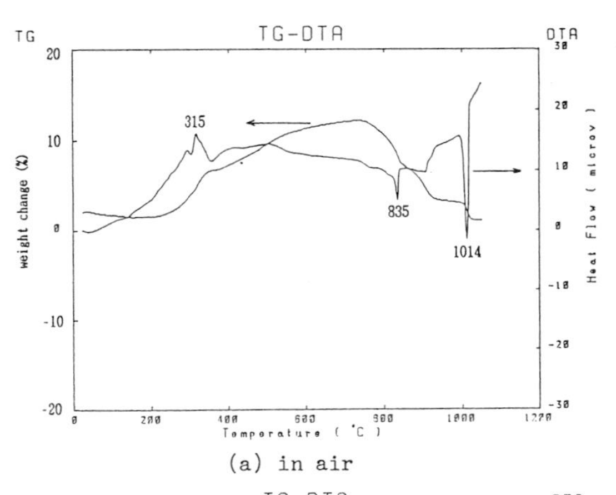

(a) in air

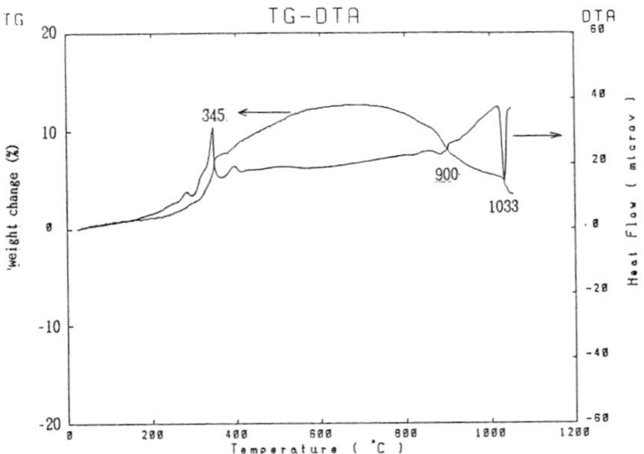

(b) in oxygen

Fig. 5 TG-DTA analyses of starting powder.

Bulk density

The bulk densities of the samples sintered at 920 and 950 °C for 3h under various oxygen partial pressures are plotted in Fig. 6.

At 920°C the bulk density reached more than 5.6g/cm^3 under the oxygen partial pressure lower than 0.2atm, while about 5.0g/cm^3 under oxygen partial pressure higher than 0.5atm. At 950 °C the bulk density became large even under the high oxygen partial pressure, however was still larger than 5.9g/cm^3 under the oxygen partial pressure lower than 0.2atm.

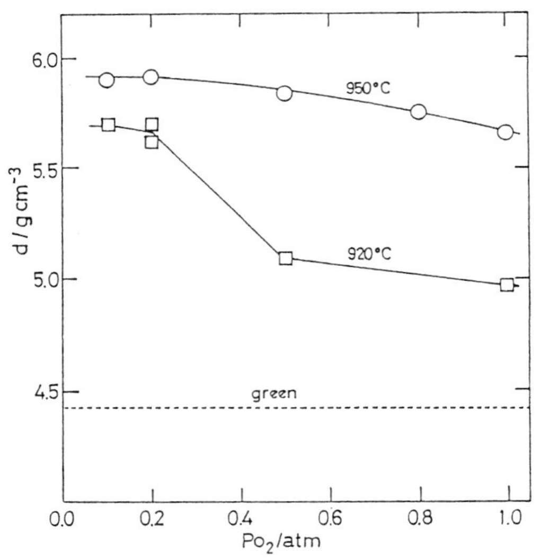

Fig. 6 Bulk density against oxygen partial pressure P_{O_2} at 920 □ and 950 °C ○ .

DISCUSSION

Densification mechanism

The drastic change in densification behavior depending on the oxygen partial pressure can be explained as follows.

The unknown X phase changed to $Ba_3Cu_5O_8$ by oxidation around 345 °C in oxygen atmosphere since the oxidation reaction is usually exothermic and the sample is able to gain weight only by oxygen absorption from the atmosphere. Between 900 and 920 °C Y_2BaCuO_5 reacted with $Ba_3Cu_5O_8$ into $YBa_2Cu_3O_{7-\delta}$ with endothermic reaction. The weight loss may be partly due to the oxygen release by the following reaction,

$$Y_2BaCuO_5 + Ba_3Cu_5O_8 \longrightarrow 2YBa_2Cu_3O_{7-\delta} + (\delta-0.5)O_2, \tag{1}$$

since the nonstoichiometry δ is reported to be about 0.6 at 920 °C in oxygen[4], which corresponds to the weight loss of 0.5%. As the densification proceeded after the completion of the reaction to $YBa_2Cu_3O_{7-\delta}$, the bulk density reached only 5.0g/cm^3 at 920°C under the oxygen partial pressure ranging 0.5 to 1.0atm.

In oxygen partial pressure ranging 0.1 to 0.2atm, the unknown X phase oxidized to $Ba_3Cu_5O_8$ around 315°C and decomposed into $BaCuO_2$ around 600°C. Considering the composition balance before and after the decomposition, CuO-rich phase in addition to $BaCuO_2$ phase should be present as indicated by the following reaction,

$$Ba_3Cu_5O_8 \longrightarrow 3BaCuO_2 + \text{CuO-rich phase}. \quad (2)$$

No peaks from the CuO-rich phase were detected by X-ray diffractometry probably because the peaks from the CuO-rich phase were very weak and superimposed by the peaks from other phases. The partial melting is reported to be present near CuO-rich region in $YO_{1.5}$-BaO-CuO system[5]. Therefore it is suggested that the CuO-rich phase became liquid at 820°C to promote the densification as well as the reaction between Y_2BaCuO_5 and $BaCuO_2$. In the case of the slower heating rate such as 5°C/min, the liquid phase may be formed more uniformly, hence more densified sample is considered to be obtained.

$Ba_3Cu_5O_8$ phase

In BaO-CuO system, $BaCuO_2$ phase and possibly Ba_2CuO_3 phase are reported to be present[5] and no compound have been confirmed in the composition between $BaCuO_2$ and CuO. In $YO_{1.5}$-BaO-SrO-CuO system, it is recently reported[6] that $Sr_3Cu_5O_x$ phase is present and one third of the Sr sites can be substituted by Y or Ba ions, which constitutes the solid-solution region in phase diagram. The crystal structures of $Sr_3Cu_5O_x$, $Sr_2BaCu_5O_y$ and $Sr_2YCu_5O_z$ phases are very similar and possess the orthorhombic symmetry. The lattice volume of $Sr_3Cu_5O_x$ expands a little in the case of Ba substitution and shrinks in the case of Y substitution.

$Ba_3Cu_5O_8$ phase, which was newly found in this work, was able to be identified from the following two results. One is the similarity of X-ray diffraction pattern to $Sr_2BaCu_5O_y$ phase. The other is that the composition should be Ba:Cu=3:5 if $YBa_2Cu_3O_{7-\delta}$ phase is formed by the reaction with Y_2BaCuO_5 phase. The oxygen number was determined assuming that the valences of Ba ion and Cu ion are two. $Ba_3Cu_5O_8$ may be stable phase at reduced oxygen partial pressure or the metastable phase formed by oxidation of the unknown X phase.

SUMMARY

(1) $YBa_2Cu_3O_{7-\delta}$ densified to 5.9g/cm^3 at 950°C for 3h in the reaction sintering with using the mixture of Y_2BaCuO_5 and the unknown X phase under the controlled oxygen partial pressure ranging 0.1 to 0.2atm.

(2) In oxygen partial pressure ranging 0.1 to 0.2atm, the sintering proceeded by the following reaction.

$$X \longrightarrow Ba_3Cu_5O_8 \longrightarrow 3BaCuO_2 + \text{CuO-rich phase}. \quad (3)$$

$$BaCuO_2 + \text{CuO-rich phase} + Y_2BaCuO_5 \longrightarrow 2YBa_2Cu_3O_{7-\delta}. \quad (4)$$

(3) In oxygen partial pressure ranging 0.5 to 1.0atm, the sintering proceeded by the following reaction.

$$X \longrightarrow Ba_3Cu_5O_8. \quad (5)$$

$$Ba_3Cu_5O_8 + Y_2BaCuO_5 \longrightarrow 2YBa_2Cu_3O_{7-\delta}. \quad (6)$$

(4) The better sinterability in oxygen partial pressure ranging 0.1 to 0.2atm may be due to the presence of liquid phase formed from CuO-rich phase above 820°C.

REFERENCES

[1] T.Yamamoto, T.Furusawa, H.Seto, K.Park, K.Kuwahara, T.Hasegawa, K.Kishio, K.Kitazawa and F.Fueki: J. Jpn. Soc. Powder and Powder Metall. 35(5)(1988)333.
[2] T.Kasakoshi, Y.Ogata, Y.Iyori, K.Maruta and T.Iimura: Proc. Fall Meeting of Jpn. Soc. Powder and Powder Metallurgy 1987, 3-10.
[3] N.Uno, N.Enomoto, Y.Tanaka and H.Takami: Jpn. J. Appl. Phys. 27(6)(1988)L1003.
[4] K.Kishio, J.Shimoyama, T.Hasegawa, K.Kitazawa and K.Fueki: Jpn. J. Appl. Phys. 26(7)(1987)L1228.
[5] R.S.Roth, K.L.Davis and J.R.Dennis: Adv. Ceram. Mater. 2(3B)(1987)303.
[6] Y.Ikeda, Y.Oue, K.Inaba, M.Takano, Y.Bando, Y.Takeda, R.Kanno, H.Kitaguchi and J.Takeda: J. Jpn. Soc. Powder and Powder Metall. 35(5)(1988)329.

Preparation and Superconducting Properties of Y-Ba-Cu and Y-Ba-Cu-F Oxides

I. NAKAGAWA[1], K. YASUDA[1], H. MATSUI[2], and S. ITO[2]

[1] Government Industrial Research Institute, Nagoya, 1-1, Hirate-cho, Kita-ku, Nagoya, 462 Japan
[2] Department of Nuclear Engineering, Nagoya University, Furo-cho, Chikusa-ku, Nagoya, 464 Japan

ABSTRACT

Preparation of high-T_c superconductors, Y-Ba-Cu and Y-Ba-Cu-F oxides, were investigated by chemical processing methods. The electrical resistivity and magnetic susceptibility were measured for the resultant powders and sintered pellets treated at 900-950°C in air or in flowing O_2. The $T_{c(offst-onset)}$ were mostly found to be around 91-95 K. A slightly positive effect of fluorine-doping was observed in the superconducting of Y-Ba-Cu(1:1:2) oxide, while an electron irradiation showed a negative effect. Deterioration of the superconductive characters in time duration was also examined.

INTRODUCTION

Since the discovery of a high-T_c(30 K class) superconductor in the La-Ba-Cu-O system[1,2], a number of studies have been focussed to develop higher T_c superconductors. Up to now, the Y-Ba-Cu [3], Bi-Sr-ca-Cu[4] and Tl-Ba-Ca-Cu[5] oxides have been found. Among the multiphases of $Y_{1-x}Ba_xCuO_{3-y}$ compound, an orthorhombic $YBa_2Cu_3O_{7-\delta}$ has been found most important in the manifestation of superconductivity. At present, to make the single orthorhombic phase of $YBa_2Cu_3O_{7-\delta}$ with adequate oxygen content is desired to increase the critical current density for practical use. Y-Ba-Cu oxide has been prepared by two processes. One is a solid state reaction using metal oxides and carbonate as starting materials. Another is a chemical process, containing copresipitation[6], spray drying[7] and colloidal or sol-gel methods[8]. Submicrom particles were synthesized by these mehtods. The bulk materials of Y-Ba-Cu oxide have been prepared by sintering at an appropriate temperature in air or in O_2 atmosphere. Annealing effect of the bulk materials has been intensively examined for an improvement of the superconductive characters. Studies of irradiation effects by high energy particles(electron[8], proton[9] and neutron[10,11]) have been carried out. However, except [10] were negative effects to the superconductive characters.

The present work is a summary of our studies of the preparation methods of powder and bulk materials, measurements of electrical resistivity and magnetic susceptibility, effect of electron irradiation, and deterioration of superconductive characters of Y-Ba-Cu oxides.

EXPERIMENTAL

Preparation of powder materials

(1) Y-Ba-Cu oxide : Powder $Y_{1-x}Ba_xCu_{3-y}$(x=0.4, 0.5, 0.6) and Y-Ba-Cu(1:2:3) oxides were synthesized by a procedure as shown in Fig. 1. Y, Ba and Cu nitrates as starting agents were separately dissolved in distilled water with the concentrations of 0.25 mol. The 0.5M oxalic acid solution was used as precipitant. The pH of the solution was adjusted at 1-4 with NH_4OH. The solution containing Y^{3+}, Ba^{2+} and Cu^{2+} ions was dropped in the 0.5M oxalic solution to form oxalates. After that, hydrazine monohydrate was added to precipitate the remained Cu^{2+} ions in the solution. The dryed oxalates were crushed and heated to decompose to oxides at 600°C in air.

(2) Y-Ba-Cu-F oxides : Powders of Y-Ba-Cu(1:1:2)-F and Y-Ba-Cu(1:2:3)-F oxides were prepared from a procedure as shown in Fig. 2. Except HF as a fluorinating agent, the other agents and processing were similar to to that of the procedure (1). Y^{3+} and Cu^{2+} ions were selectively fluorinated.

Powder samples obtained from the process (1) and (2) were examined by X-ray diffraction, chemical analyses(ICP, photospectrometry and thermal analysis).

Bulk Materials

Bulk samples of $Y_{1-x}Ba_xCuO_{3-y}$, $YBa_2Cu_3O_{7-\delta}$ with the composition of Y:Ba:Cu=1:2:3, and $YBaCu_2(F,O)_y$ (1:1:2) and $YBa_2Cu_3(F,O)_{7-\delta}$ (1:2:3) were prepared by sintering. The powders obtained in the process (1) or process (2) were pressed and sintered at 900-950°C in air. The sintered pellets were annealed at 300°C in air or in flowing O_2. The sintered samples thus obtained were examined by X-ray diffraction and chemical analysis by EPMA and ICP.

Measurements of electrical resistivity and magnetic susceptibility

The electrical resistivity was measured by dc four-probe method. Whereas the magnetic susceptibility was measured by the Faraday method. Both measurements were carried out between liquid nitrogen temperature(LNT) and room temperature(RT).

Electron irradiation

The sintered sample of $YBaCu_2O_y$ (1:1:2) was irradiated with 7 MeV electron ($1x10^{17}e/cm^2$) at LNT. $Y_{0.6}Ba_{0.4}CuO_y$ sample was irradiated with electrons($1x10^{18}e/cm^2$) at RT. The electrical resistivity and magnetic susceptibility of both samples were measured before and after irradiation.

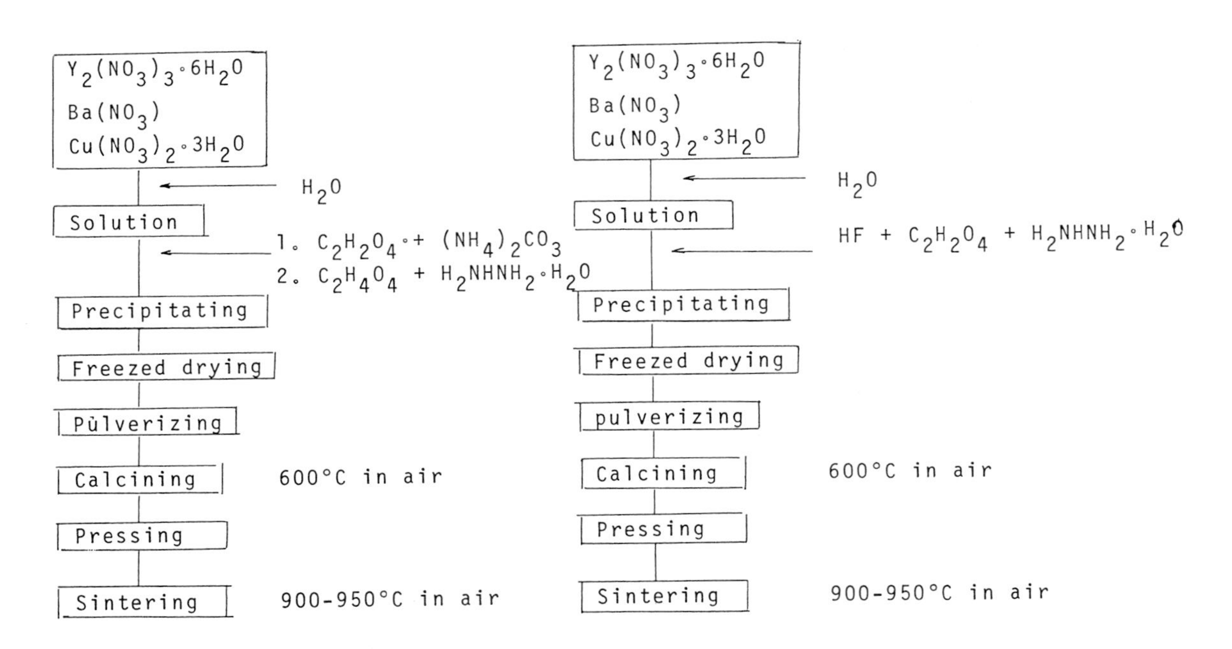

Fig. 1 Preparation process (1) of Y-Ba-Cu oxide samples.

Fig. 2 Preparation process (2) of Y-Ba-Cu-F oxide samples.

RESULTS AND DISCUSSION

Powder materials

Examples of SEM observation synthesized powders are shown in Photo. 1. Fine parricles(0.2-0.3 μm) were obtained with greater than 80% meissuner effect. No grain growth was observed up to 950°C in heat-treatment. Above 950°C, grain growth was observed and the decomposition occurred to to yield a tetragonal phase and other oxides.

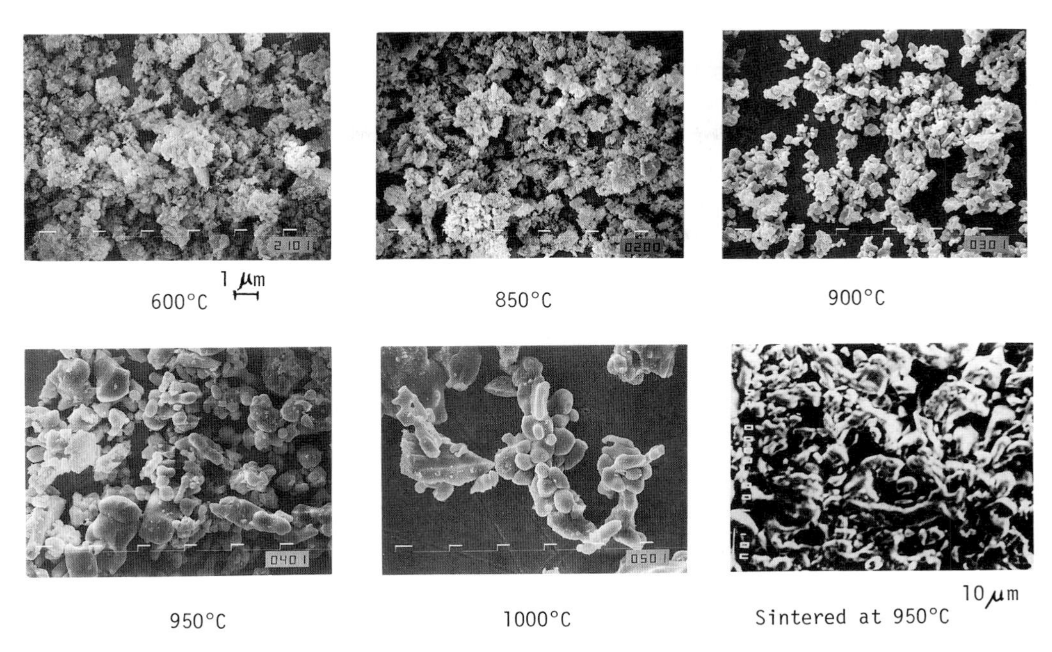

Photo. 1 SEM figures for Y-Ba-Cu (1:1:2) powder treated in air for 4 hrs.

A single orthorhombic phase of $YBa_2Cu_3O_{7-\delta}$ was synthesized in an optimum condition as illustrated in Photo. 2 and in Fig. 3. The lattice parameters of this compound were a=3.729, b=3.849, c=11.681 Å.

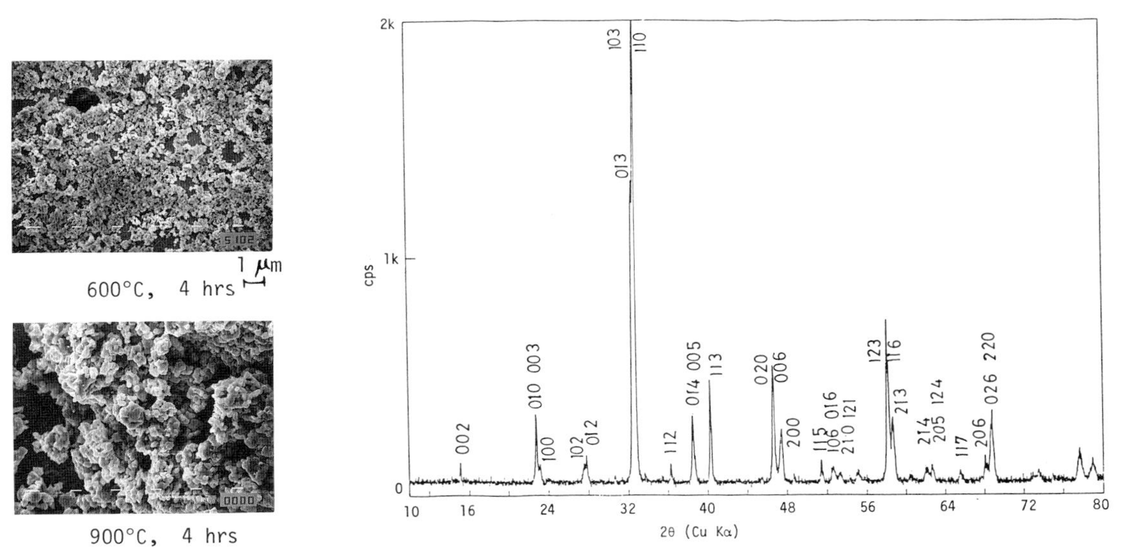

Photo. 2 SEM figures of Y-Ba-Cu (1:2:3) oxide.

Fig. 3 X-ray diffraction profile for Y-Ba-Cu (1:2:3) oxide powder treated at 900°C 4 hrs in air.

Fluorine content in F-doped powders of the Y-Ba-Cu-F (1:1:2:X) and Y-Ba-Cu-F (1:2:3:X) was analyzed to be X=1.0 and X=0.6 wt%, respectively. Both oxides were unstable to a successive heat-treatment. Large amount of BaF_2 was observed in X-ray diffraction after the heat-treatment.

Sintered materials

Sintered samples were prepared from the resultant powders by the process (1) and (2) shown in Figs. 1 and 2. Pressed pellets were sintered at 900-950°C for 4-46 hr in air. In addition, they were further annealed at 300°C for several cycles of 10 hour's annealing in air and in O_2 atmosphere. A typical X-ray profile is illustrated in Fig. 4. Change of the lattice parameters on temperature dependency is shown in Fig. 5. The oxygen content in the Y-Ba-Cu (1:2:3) compound was determined to be 6.6 from a relationship between the c-axis and oxygen concentration[13]. The differences of the chemical components of Y, Ba and Cu between the calculated and analytical results were 4%, 6% and 5%, respectively.

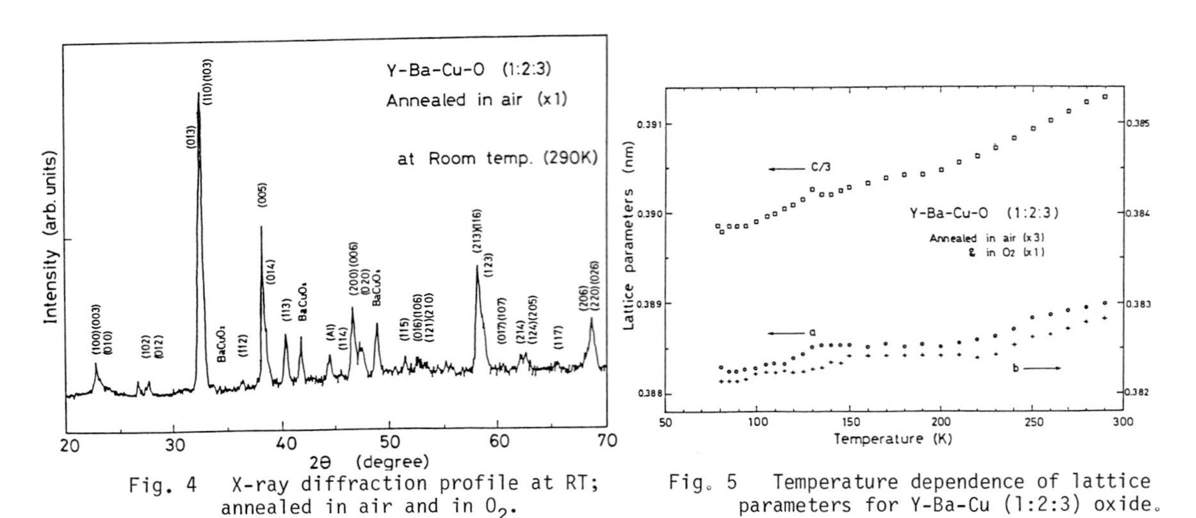

Fig. 4 X-ray diffraction profile at RT; annealed in air and in O_2.

Fig. 5 Temperature dependence of lattice parameters for Y-Ba-Cu (1:2:3) oxide.

Superconducting characters (resistivity and magnetic susceptibility)

The resistivity and magnetic susceptibility of the sintered $YBaCu_2O_y$ (1:1:2) are represented in Figs. 6 and 7, respectively. x1, x2 and x3 in the figures denote the times of annealing in terms of x=10 hrs/one time. Those for $YBa_2Cu_3O_{7-\delta}$(1:2:3) are illustrated in Figs. 8 and 9, respectively. As seen in Figs. 6 and 8, the superconducting characters of as-prepared samples were bad. However, after a couple annealing, the superconducting characters became good ($T_{c(onset)}$= 98.7 K, $T_{c(offset)}$=92.4 K, $\Delta T_{(onset-offset)}$=1.3 K for the $YBaCu_2O_y$ (1:1:2)). The results of $YBa_2Cu_3O_{7-\delta}$ (1:2:3) were similar to that of the $YBaCu_2O_y$ (1:1:2).

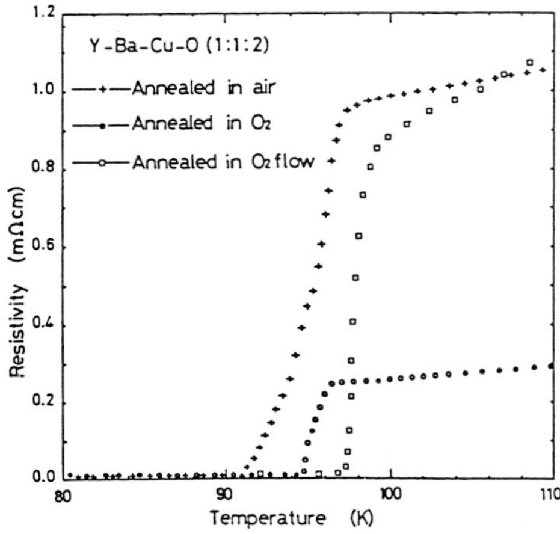

Fig. 6 Resistivity of annealed $YBaCu_2O_y$(1:1:2) at 300°C.

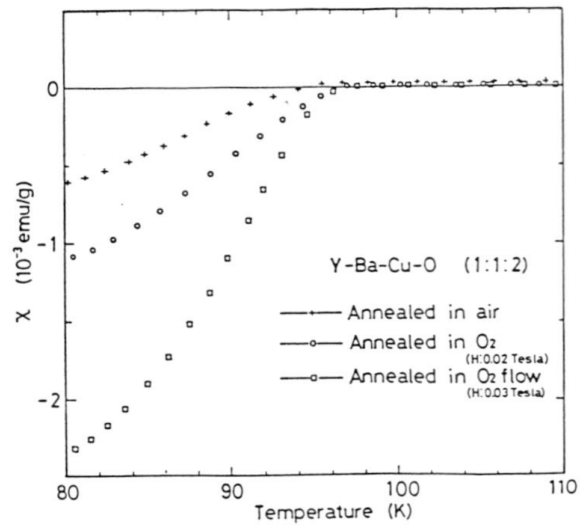

Fig. 7 Magnetic susceptibility of annealed $YBaCu_2O_y$(1:1:2) at 300°C.

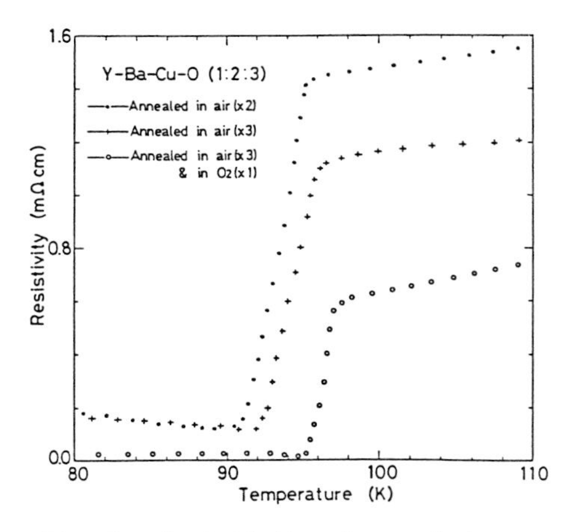

Fig. 8 Resistivity of annealed $YBa_2Cu_3O_{7-\delta}$(1:2:3) at 300°C.

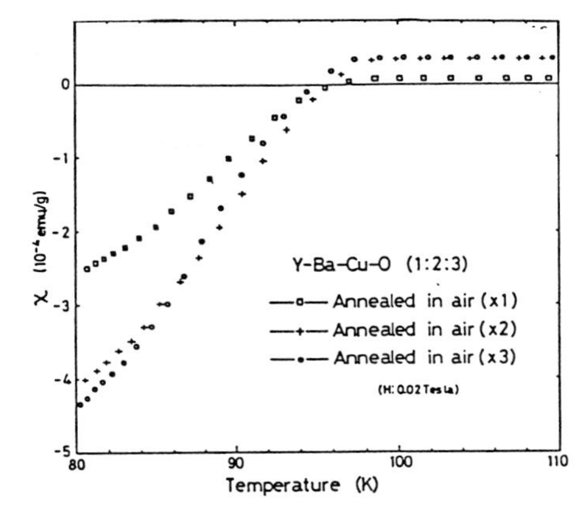

Fig. 9 Magnetic susceptibility of annealed $YBa_2Cu_3O_{7-\delta}$(1:2:3) at 300°C.

F-doping effect

The magnetic susceptibility of the F-doped YBa_2CU_3-$(F,O)_{7-\delta}$ is shown in Fig.10 compared with that of non-doped one. T_c of the former was higher about 2 K than the later, but the susceptibility showed a diamanetism even at room temperature. Those phenomena could not be fully understood in the present stage.

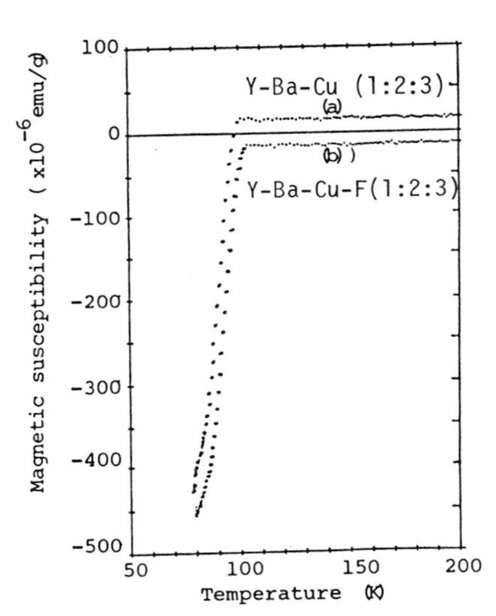

Fig. 10 Magnetic susceptibility for Y-Ba-Cu (1:2:3) and Y-Ba-Cu-F (1:2:3) oxides.

Electron irradiation

Before irradiation, the specimen ($YBaCu_2O_y$(1:1:2)) showed a zero-resistance(offset) at 91.1 K and at 95.7 K with ΔT_c=0.7 K(transition width). Those temperatures shifted to slightly lower temperatures in the irradiation specimen. On the other hand, the magnetic susceptibility decreased and an appearance of the diamagnetism shifted to a lower temperature with electron irradiation. At a mabnetic field of 0.5 T, only the paramagnetic suscept-ibility was observed for the irradiated specimen.

Deterioration of superconductive characters[13]

The stability of the supercondutive characters were examined for the $YBa_2Cu_2O_y$(1:2:2) and an electron irradiated $YBaCu_2O_y$(1:1:2). The changes of the electrical resistivity and the magnetic susceptibility were examined for both oxides in time duration for about one year. Within a few month, there was no remarkable change in the superconductive characters. After a half year, however, an increase of resistivity and decrease of the diamagnetic susceptibility were observed together with a slight sift of the transition temperature T_c to lower temperatures. Deterioration of the superconductive character was further accerelated after a half year, for both oxides.

REFERENCES

1) J.G. Bednorz and K.A. Müller: Z. Phys. B64 (1986) 189-193.
2) S. Uchida, H. Takagi, K. Kitazawa and S. Tanaka: Jpn. J. Appl. Phys.:26 (1987) L1-L2.
3) C.W. Chu, et al.: Phys. Rev. Lett.: 58 (1987) 405.
4) H. Maeda, Y. Tanaka, M. Fukutomi and T. Asano: J. Appl. Phys. 27 (1988) L209.
5) Z.Z. Sheng, et al.: Phys. Rev. Lett. 60 (1988) 937.
6) M. Hirabayashi: J. Metals Jpn 26 (1987) 943.
7) M. Awano, et al.: Nippon Seramikkusu-Kyokai-akujutsu-Ronbunshi 96[4] (1988) 407.
8) S. Hirano, et al.: Chem. Let.(Chem. Soc. Jpn) (1988) 665.
9) H. Matsui, S. ITO, I. Nakagawa, K. Yasuda and M. Takeda: Jpn. J. Appl. Phys. 27[7] (1988)L1281.
10) K. Atobe and H. Yoshida: Phys. Rev. B. 36 (1987) 7194).
11) G. Samann-Ishenko: Int. Conf. Low. Temp. Phys. (LT-18), Kyoto, 1987.
12) S. Sueno, I. Nakai, F.P. Okamura and A. Ono: Jpn. J. Appl. Phys. 26 (1987) L309.
13) H. Matsui, S. Ito, I. Nakagawa, M. Takeda and K. Yasuda: Jpn. J. Appl. Phys., Submitted.

4.5 Tapes and Thick Films

Preparation of High Tc Superconductive Thin Tape from Acid Salts

TAKAO NAKAMOTO[1], MASATADA FUKUSHIMA[1], HIROKI KOBAYASHI[1], TAKEO SHIONO[1], ETUO HOSOKAWA[1], HIROYUKI NASU[2], TAKESHI IMURA[2], and YUKIO OSAKA[2]

[1] Chemical Material R&D Department, Showa Electric Wire & Cable Co. Ltd., 2-1, Odasakae, Kawasaki-ku, Kawasaki, 210 Japan
[2] Faculty of Engineering, Hiroshima University, Saijo, Higashi-Hiroshima, 724 Japan

ABSTRACT

High-Tc superconducting thin tapes were prepared by pyrolysis using organic acid salts. 2-ethylhexanoates of Y, Ba, and Cu etc. were used as the starting materials. The organic acids were dissolved in a mixed solvent at suitable metal concentration, and mixed at suitable metal ratio.

The mixed organic acid was dropped on substrates and heated. Several kinds of substrates, yttria-stabilized zirconia, MgO, and metal foils such as Ag, stainless steel, invar etc. were used. We obtained superconducting thin films of about 5μm thick on metal foil.

INTRODUCTION

Since Bednorz and Müller,[1)] researchers at the IBM Zurich Laboratories, discovered the high critical temperature (Tc) La-Ba-Cu-O superconductors, many other oxide superconductors have been found one after another. In particular, Y-Ba-Cu and Bi-Sr-Ca-Cu oxide superconductors which have critical temperatures (end) above 77 k have been the focus of attention, and much work has been done in many respects including thier applications.

Since application of high Tc superconducting materials requires the development of thin-film forming techniques, attempts have been made on a variety of techniques such as rf sputtering, vapor deposition, screen printing, and thermal spray coating.

We used each 2-ethylhexanoate of (Y, Ba, Cu) or (Bi, Sr, Ca, and Cu) as starting materials, and ceramics and metal tapes as substrates.

EXPERIMENTAL

Preparation of Specimens

2-ethylhexanoates of Y, Ba, Cu, Bi, Sr, and Ca ($M(C_7H_{15}COO)n$: M = Y, Ba, Cu, Bi, Sr, Ca and n = 2 or 3) were used in powder or liquid forms as raw materials. The raw materials were mixed in a mole ratio of Y : Ba : Cu = 1 : 2 : 3, to obtain a Y-Ba-Cu-O product, and Bi : Sr : Ca : Cu = 1 : 1 : 1 : 2 to obtain a Bi-Sr-Ca-Cu-O product. Subsequently, they were dissolved in a solvent at a given concentration to produce a metal salt.[2~5)]

Using a syringe, a starting material of the composition above was added dropwise to MgO, Ag, Cu, invar, and stainless steel substrates, and dried at room temperature. Then, it was calcined in an electric furnance at 500°C. The cycle of its dropwise addition and calcining was repeated to produce a calcined film. The resulting Y-Ba-Cu-O film was sintered at 800 or 900°C in the presence of oxygen, and the Bi-Sr-Ca-Cu-O film was sintered at 950°C in the air.

Measuring Method

The resulting specimens were investigated for temperature dependency of electrical resistance using the four-probe method. During measurement current was applied, both forwardly and reversely, and the average values were obtained.

The films obtained were examined for appearance and fracture surface under a scanning electron microscope (SEM). They were also analyzed for crystal structure by means of X-ray diffraction over the range of 20 to 70°.

RESULTS AND DISCUSSION

Surface Analysis

Fig.1 and 2 show the SEM-photos of the film deposited on Ag and MgO substrates. This film is $YBa_2Cu_3O_{7-x}$ sintered at 900°C. As is evident from Fig.1, there were neither cracks nor delaminations throughout the surface of the MgO and Ag substrates. The area shown in Fig.1 may be concidered to represent the overall film condition.

When the test specimen was compared with the bulk prepared by the solid-phase method, it appears that the grains are finer with the former.

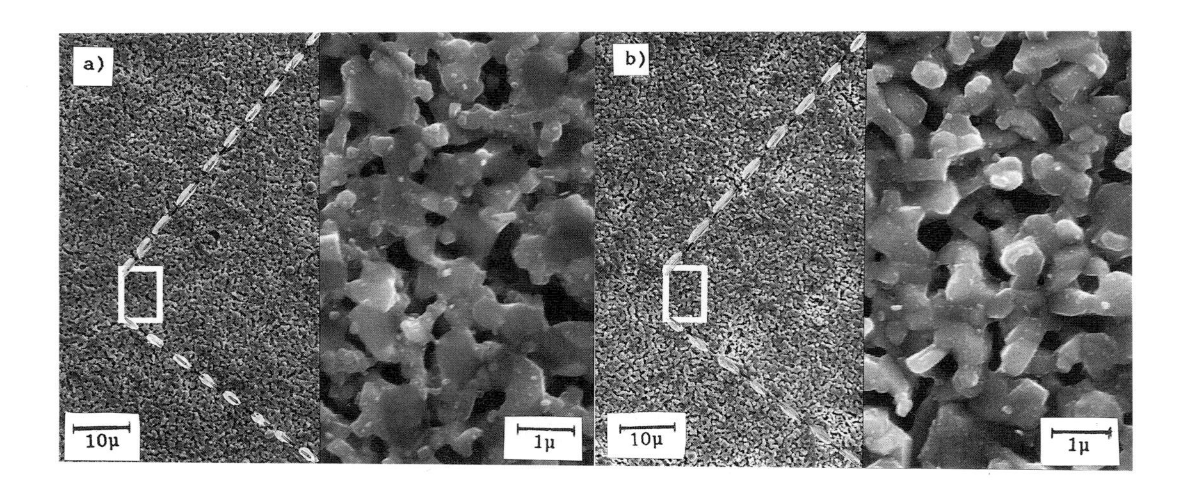

Fig.1. SEM photographs of free surfase on a) Ag substrate b) MgO substrate

On the fracture surface, the deposited film does not appear to have reacted with either Ag or MgO substrate.(see Fig.2) Furthermore, an attempt was made to form films on other metal substrates. It was found that they were delaminated from the Cu tape during calcining and delaminations seem to have occured between the Cu oxide formed on Cu and the Cu substrate. When films were formed on the invar substrate, they curled inward during sintering at 900°C.

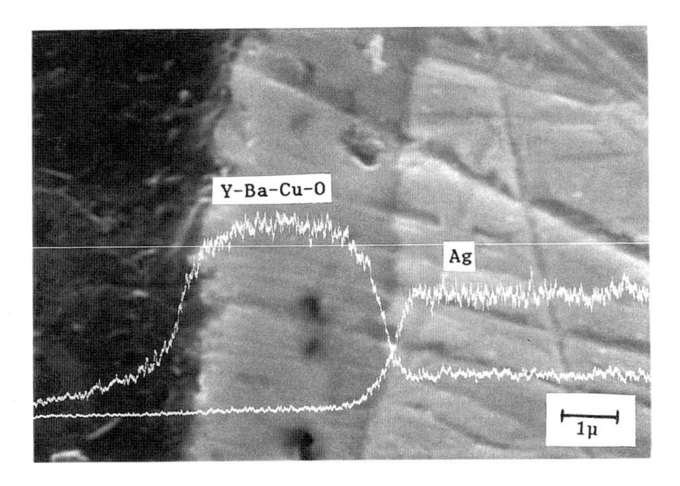

Fig.2. X-ray micro analyzing of film on Ag substrate

Crystal Structure

Fig.3 shows the X-ray diffraction (XRD) patterns in the film discussed above. As is evident from this figure, the film has a $YBa_2Cu_3O_{7-x}$ structure of the perovskite type. As is clear from peak distribution shown in the XRD pattern, no orientation was observed and the MgO(100) surface of the substrate seems to have exerted no adverse effect on orientation.

With the film on the Ag substrate, the tetragonal structure seems to be a little stronger than other structures. This fact suggests that oxygen deficiency is more frequent on the Ag substrate than on MgO.

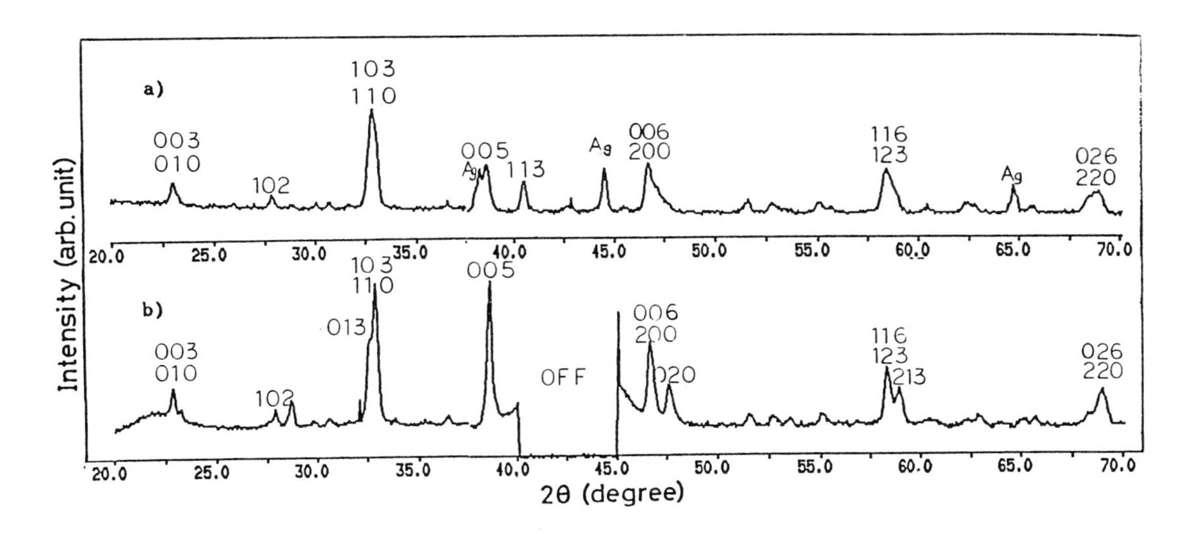

Fig.3. X-ray diffraction pattern of the film on a) Ag substrate b) MgO substrate

Electrical Characteristics

Fig.4 shows the temperature dependency of electrical resistance for the film deposited on the MgO substrate.

When sintered at 900°C, the specimen showed Tc (onset) and Tc (end) of 95 K and 80K, respectively.

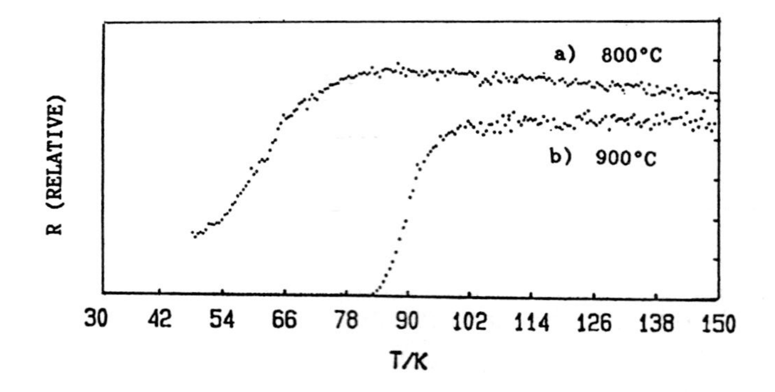

Fig.4. Temperature dependence of resistivity of the film on the MgO substrate

Fig.5 indicates the temperature dependency of magnetic susceptibility for the film formed on the Ag substrate. This specimen, a Bi-Sr-Ca-Cu-O film, was found by X-ray diffraction to have low-Tc phase as its main constituent.

The susceptibility is in the positive area when the temperature is 75K or over, but decreases sharply below 75K, as can be known from the kink in the curve. The diamagnetism is detected below 73K, indicating that superconduction takes place in the films. The susceptibility at 8K is approximately -5.0emu/g at 5000Oe.

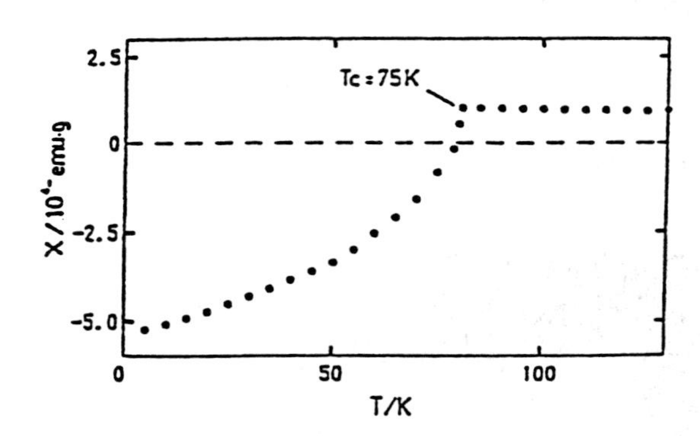

Fig.5. Temperature dependence of magnetic susceptivility for Bi-Sr-Ca-Cu-O film on Ag substrate

CONCLUSION

We obtained High-Tc superconducting thin film on metal foil by pylorisis using 2-ethyhexanoates of (Y,Ba,Cu) and (Bi,Sr,Ca,Cu).

REFERENCES

1) J.G. Bednorz and K.A. Müller, Z. Phys., B64, 189 (1986).
2) H.Nasu, S.Makida, T.Kato, Y.Ibara, T.Imura and Y.Osaka, Chem. Lett., 2403 (1987)
3) H.Nasu, S.Makida, T.Imura and Y.Osaka, J. Mat. Sci. Lett., 7, (1988) in press
4) H.Nasu, S.Makida, Y.Ibara, T.Kato, T. Imura and Y.Osaka, Jpn. J. Appl. Phys., 27, L536 (1988)
5) Y.Ibara, H.Nasu, S.Makida, T.Imura, Y.Osaka, T.Shiono, and T.Nakamoto, Chem. Lett.,(1988) to be published.

Superconducting Properties of Y-Ba-Cu-O Sheet by a Tape Casting Method

SHOZO YAMANA[1], HIDEJI KUWAJIMA[1], TORANOSUKE ASHIZAWA[1], SHUICHIRO SHIMODA[1], KEIJI SUMIYA[1], TETSUO KOSUGI[1], and TOMOICHI KAMO[2]

[1] Ibaraki Research Laboratory, Hitachi Chemical Co., Ltd., Katsuta, Ibaraki, 312 Japan
[2] Hitachi Research Laboratory of Hitachi Ltd., Hitachi, Ibaraki, 316 Japan

ABSTRACT

$YBa_2Cu_3O_{7-\delta}$ (YBCO) superconducting sheets were successfully fabricated by a tape casting method. Critical temperature (Tc^{zero}) was 92.1K with a tape sintered at 940°C for 10 hours in oxygen. The (001) planes of YBCO particles were oriented parallel to the casting plane of the green sheet. The highest Jc was 1,510A/cm^2 at 77K in zero magnetic field. It is supposed that the grain orientation contributed principally to the enhancement of Jc of the YBCO sheet.

INTRODUCTION

Since the discovery of high-Tc superconductors with zero resistance above liquid nitrogen temperature (77K)[1], numerous studies on new high-Tc superconductors, physical and chemical properties, and improvement of critical current have been widely undertaken.

There are a number of reports on YBCO thin films formed on MgO and $SrTiO_3$ single crystal substrates and the critical current density (Jc) is found to be 10^6A/cm^2 [2](at 77K under zero magnetic field). On the other hand Jc of bulk materials is reported to be 100 [3]to 950A/cm^2 [4], which is two to three orders of magnitude lower than those reported on thin films. This fact suggests that a weak link is formed at grain boundaries which is supposed to result from, for example secondary insulation phase oxygen deficient layer at the YBCO surface and misorientation of anisotropic YBCO.

YBCO crystals have a tendency that the (001) plane develops preferentially[5], where Jc in the (001) plane is 10 to 100 times larger than those perpendicular to (001) plane[6].

In a tape casting method, it is expected that the plate like particles will be oriented with it's larger surface parallel to green sheet surface due to the anisotropic particle shape and shearing force between doctor blade and carrier sheet.

We expected that a tape casting method was useful to improve Jc of YBCO. In this paper, we report superconducting properties of YBCO sheet by a tape casting method.

EXPERIMENTAL

(1) Sample preparation : The nominal composition of YBCO was prepared from Y_2O_3, $BaCO_3$, and CuO powders. These were mixed in a ball mill and calcined at 960°C for 10 hours in oxygen. The calcined powder was thoroughly milled with zirconia media in a zirconia pot. YBCO slurry was prepared by mixing the YBCO powder with mean particle size 1.5μm, organic binder, plasticizer, and organic solvent for 24 hours. The slurry was cast under the doctor blade into a 500mm wide and 0.5mm thick green sheet on the carrier sheet. The green sheets were rolled into a thickness of 0.1mm to 0.3mm. The rolling ratio is calculated from thickness of green sheet before and after the rolling.

The orientation factor of YBCO particles which is defined later is calculated from X-ray diffraction intensities. Specimens 5mm in width and 40mm in length were cut out from the rolled sheet. The specimens were sintered at 900∼940°C for 10 hours on an alumina plate in oxygen and cooled down slowly in a furnace at a rate of 1°C/min. Then silver electrodes were painted and fired at 800°C for 1 hour in oxygen and cooled slowly (1°C/ min.) in a furnace.

(2) Measurement of properties : Tc was measured with a current of 50mA by the standard four-probe method in a cryostat, where the temperature was measured by a AuFe-chromel thermocouple.

We define that Tc^{onset} is the starting temperature of superconducting transition and Tc^{zero} is the temperature of zero resistance state.

Jc was measured resistively in liquid nitrogen (at 77K). Jc was defined as the current density where the voltage rose 1μV across 10mm length of the tape.

Bulk densities of sintered bodies were measured by Archimedes' method using ethyl alcohol.

RESULTS AND DISCUSSION

We examined the orientation of YBCO particles in the green sheet and sintered bodies. The orientation was calculated by Lotgering method[7]. Lotgering method can be applied to the case that all of diffraction peaks are separated each other. Although in the case of YBCO, 001 and hkl diffractions often overlap together, we have chosen the peaks at $2\theta=15°$ and $33°$ (Cu $K\alpha$) which are indexed as 002 and 013,103 and 110 respectively, because they were clearly separated. In this paper, the calculation of orientation factor F is defined as follows.

$$F=P_{obs}/P_{calc} \cdots (1)$$

P_{obs} : $I_{obs}(002)/I_{obs}(013),(103),(100)$
P_{calc} : $I_{calc}(002)/I_{calc}(013),(103),(100)$
I_{obs} : observed intensity
I_{calc} : calculated intensity

When the green sheet is non-oriented, F equals to 1.

The reduction ratio R is defined in formula (2)

$$R=(t_0-t_1)/t_0 \cdots (2)$$

t_0:thickness of sheet before rolling
t_1:thickness of sheet after rolling

Fig.1 shows the relationship between the reduction ratio and the orientation factor of green sheets and sintered bodies. The orientation factor of the unrolled sheet (reduction ratio=0%) was about 5, while that of the isostatic pressed sample was about 1. The orientation factor increased by rolling of the green sheets. And the orientation factor of sintered body was almost the same as that of the green sheet. In this study, the orientation factor was 17 at the reduction ratio of 70%. These results show that the combination of tape casting and rolling of green sheet are effective to improve of grain orientation.

Fig.2 shows the effect of the reduction ratio of the green sheet on the density of sintered bodies. In this figure, as the reduction ratio increased, the density of the sintered body increased. In the region of reduction ratio about 60% to 70%, the density of the sintered body was about 6.26g/cm^3. This value was 98.4% of the theoretical density 6.36g/cm^3 which was calculated by quantitative analysis of the chemical composition and the lattice constants of the sample. The chemical composition of samples were determined as $YBa_2Cu_3O_{6.82}$ by an iodometry method and ICP-emission spectroscopy. Thus the rolling of YBCO sheet is an effective method for densification. As shown in Fig.3, the density of the green sheet was

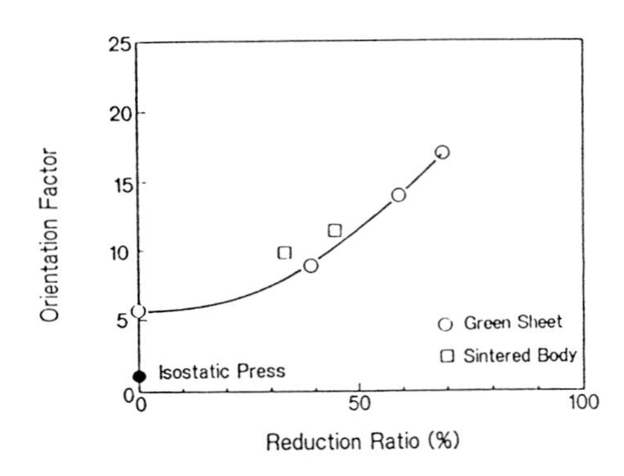

Fig.1 The Relationship between Reduction Ratio and Orientation Factor

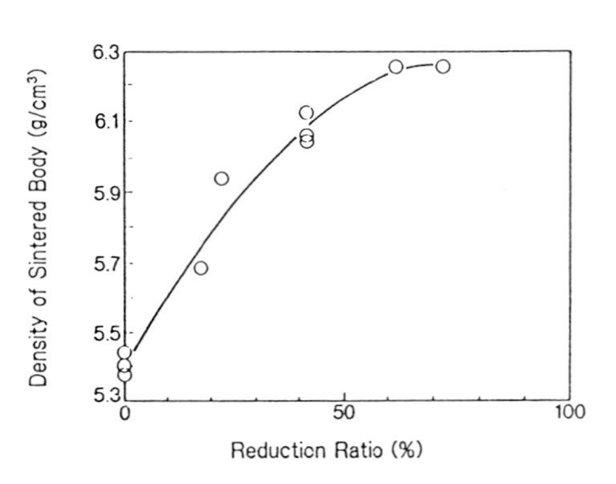

Fig. 2 The Effect of Reduction Ratio of Green Sheet on the Density of Sintered body

increased by rolling. The improved sintering resulted from density increase after rolling of green sheet and very fine particles which were used in the experiment (mean particle size equals to 1.5μm).

Fig.4 shows SEM micrographs of a sintered sample with the reduction ratio of 60%. These micrographs show a pore free texture in surface and cross section. Some of grains were grown to 20 to 40μm.

Superconducting properties of sintered sample were examined. Fig.5 shows the relationship between the density of sample and critical temperature (Tc^{onset}, Tc^{zero}). Both Tc^{onset} and Tc^{zero} did not change with the density of samples from 5.37 to 6.26g/cm^3.

Fig.6 shows the relationship between orientation factor of green sheet and Jc. Jc increased, as the orientation factor increased. When the factor increased to 17, Jc was 520A/cm^2.

Fig.7 shows the relationship between Jc and density of samples. Increasing the density of samples by a tape casting method, Jc increased. Open circles show tape cast samples and a closed circle shows isostatically pressed sample in which was not texturing observed. When compared Jc between the tape cast and isostatic pressed sample, a higher Jc was obtained in the tape casting by about 4 times than in the isostatic pressing. This means that Jc is not always high even if the density of sintered body is high. When the density of sintered body is 6.26g/cm^3, Jc is 520A/cm^2. We considered that the grain alignment contributed principally to the enhancement of Jc of the YBCO sheets.

In a tape casting method , plate like particles were oriented by the influence of shearing force and the anisotropy of particle shape. Thickness of green sheet is controlled by the distance between a doctor blade and a carrier sheet. When that distance becomes narrower, shearing force become larger, which results in higher orientation factor and Jc. Then we prepared thinner green sheets to examine these effects. Fig.8 shows the dependence of Jc on the thickness of the sintered body. In Fig.8, as the thickness of the sintered body decreased, Jc increased. At the thickness of sintered body of 0.07mm, Jc was 1,510A/cm^2, but the orientation factor was 13. Now the reason for the dependence of Jc on the thickness of sample is not so clear, further investigations should be needed.

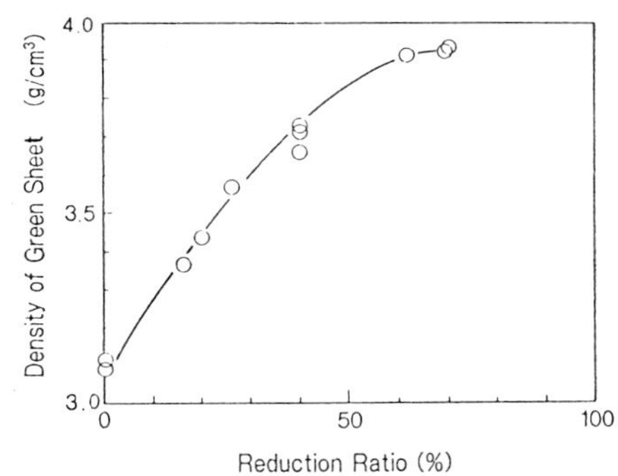

Fig. 3 The Effect of Reduction Ratio of Green Sheet on the Density of Green Sheet

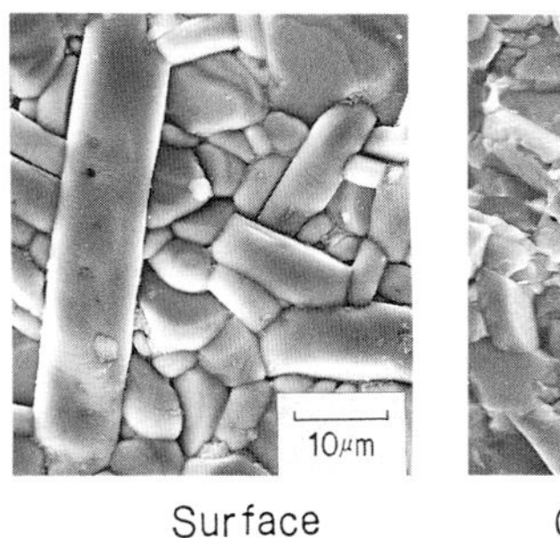

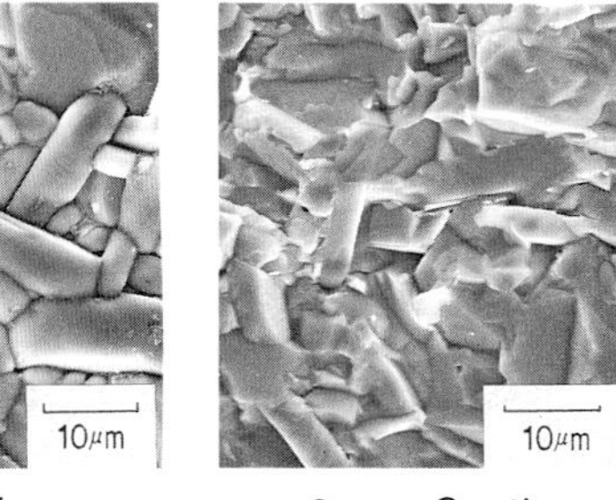

Fig.4 SEM Micrographs of Sintered Sheet

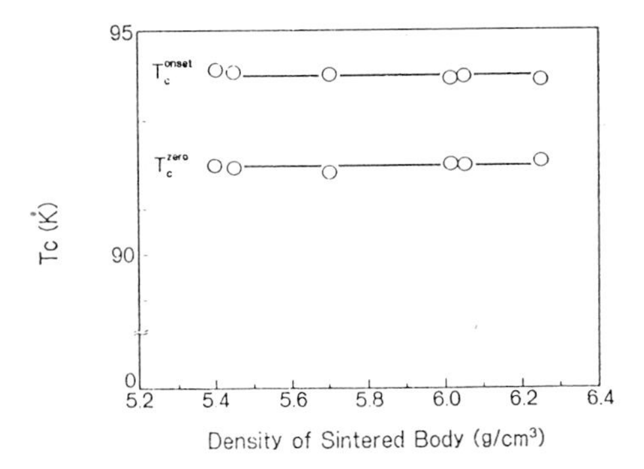

Fig. 5 The Relationship between Density of Sintered Body and Critical Temperature Tc

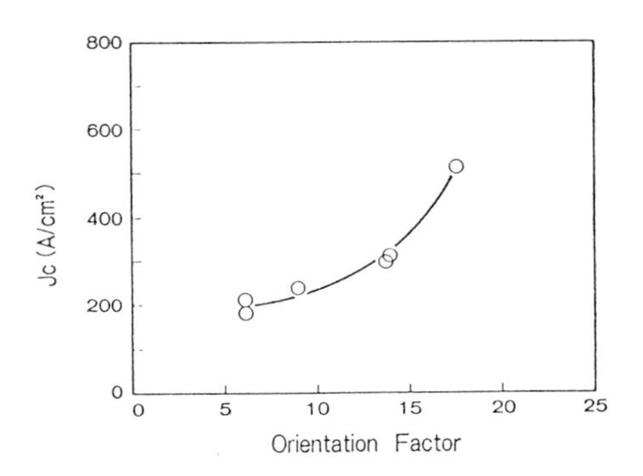

Fig. 6 The Relationship between Orientation Factor of Green Sheet and Critical Current Density

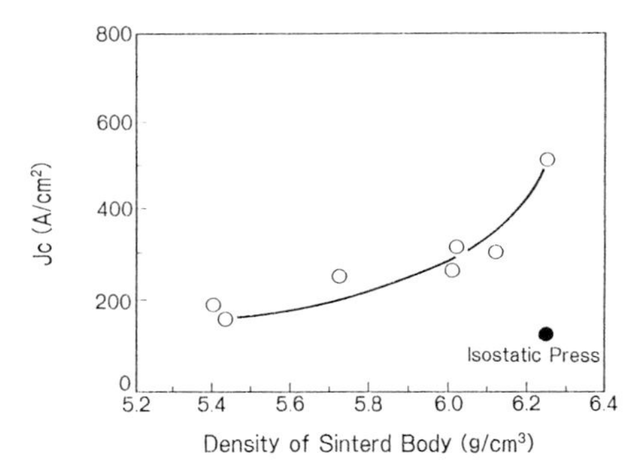

Fig. 7 The Relationship between Critical Current Density and Density of Sintered Body

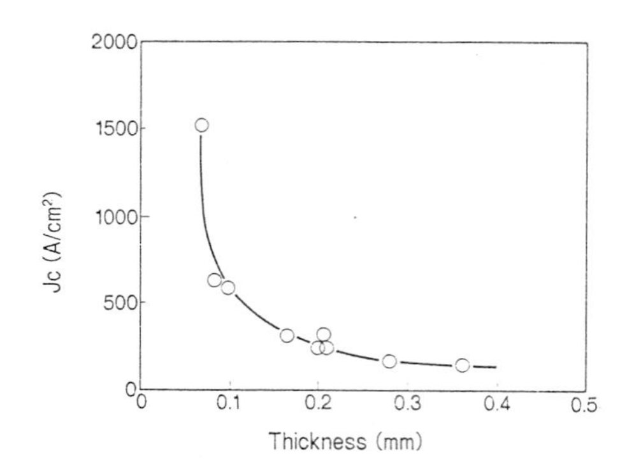

Fig. 8 Thickness Dependence of Critical Current Density of Sintered Body

CONCLUSIONS

We have examined superconductivity of $YBa_2Cu_3O_{7-\delta}$ sheets prepared by a tape casting method. Combination of tape casting and rolling of green sheet is effective to improve the grain orientation. The relative density of sintered body was 98.4% by this method. This combination is effective for preparing a high Jc YBCO. It is considered that grain alignment is a principal contribution to the enhancement of Jc of YBCO sheet. In this experiment, the highest Jc was 1,510A/cm^2 at 77K in zero magnetic field.

REFERENCES

1) M.K.Wu,J.R.Ashburn,C.J.Torng,P.H.Hor,R.L.Meng,L.Gao,Z.J.Huang,Y.Q.Wang and C.W.Chu:Phys. Rev.Lett.**58**(1987)908
2) T.Terashima,K.Iijima,K.Yamamoto,Y.Bando and H.Mazaki:Jpn.J.Appl.Phys.**27**(1988)L91
3) K.Togano,H.Kumakura,H.Shimizu,N.Irisawa and T.Morimoto:ibid.**27**(1988)L45-L47
4) D.W.Johonson,Jr.Warren,W.Rhodes and G.S.Grader:International Meeting on Advanced Materials Final Program and Abstracts of the 1988 MRS,PD-26
5) H.Takei,H.Takeya,Y.Iye,T.Tamegai and F.Sakai:Jpn.J.Appl.Phys.**26**(1987)L1425-L1427
6) Y.Enomoto,T.Murakami,M.Suzuki and K.Moriwaki:ibid.**26**(1987)L1248-L1250
7) T.Takenaka,H.Noda,A.Yoneda and K.Sakata:ibid.**27**(1988)L1209
8) M.Ishii,T.Maeda,M.Matsuda,M.Takata and T.Yamashita:ibid.**26**(1987)L1959-L1961

Critical Current Densities of Silver Sheathed Tl-Ba-Ca-Cu Oxide Superconducting Tapes

M. SEIDO[1], F. HOSONO[1], Y. ISHIGAMI[1], T. KAMO[2], K. AIHARA[2], and S. MATSUDA[2]

[1] Metal Research Laboratory of Hitachi Cable Ltd., Tsuchiura, Ibaraki, 300 Japan
[2] Hitachi Research Laboratory of Hitachi Ltd., Hitachi, Ibaraki, 319-12 Japan

ABSTRACT

The Tl-Ba-Ca-Cu oxide (TBCCO) superconducting material was formed into tapes by means of the SSRP (Silver Sheath & Rolling Process). The superconducting properties of these tapes were measured and compared with those of the YBCO tapes fabricated by the same method.

The critical temperature Tc (zero resistivity) of the TBCCO tape was 107 K, and the maximum critical current density Jc of 3500 A/cm^2 at 77 K was achieved in a tape with 0.05 mm thickness. It was observed that the Jc apparently depended on the thickness in the TBCCO as well as YBCO tape. The Jc of the TBCCO tapes was higher than that of YBCO tapes, which came from the thermal margine due to the higher Tc of the material.

INTRODUCTION

Recently various high-Tc superconducting oxides have been discovered. The $Y_1Ba_2Cu_3O_y$ (YBCO) [1], Bi-Sr-Ca-Cu-O (BSCCO) [2] and Tl-Ba-Ca-Cu-O (TBCCO) [3] are already well-known superconducting oxides with the critical temperature Tc exceeding the nitrogen boiling temperature (77 K). In particular, the TBCCO has a high-Tc of 105 - 125 K. On the other hand, the critical current density Jc of the wire type superconducting oxides is too low to use in the power apparatus applications; such as, the superconducting magnets with the liquid nitrogen cooling. The Jc of the YBCO wires which have been reported so far is in a level of 10^2 - 10^3 A/cm^2 at 77 K, and the Jc of the TBCCO is in a level of 10^2 A/cm^2. For a practical use, a critical current density of at least 10^4 A/cm^2 is necessary. The research to improve the Jc of silver sheathed YBCO tapes fabricated by the SSRP has been already reported [4].

In this paper, we report the properties of silver sheathed TBCCO tapes fabricated by the SSRP. The purpose of this research is to make clear the relation between fabrication conditions and the Jc of TBCCO and YBCO tapes.

EXPERIMENTAL

The TBCCO powder was prepared from Tl_2O_3, $BaCO_3$, $CaCO_3$ and CuO powder (Tl:Ba:Ca:Cu=2:2:2:3) by mixing and sintering under oxygen flowing gas at 1073 - 1223 K. We formed the superconducting powder to a tape by means of the SSRP (Silver Sheath & Rolling Process) - the superconducting powder was packed into a silver tube and this clad rod was drawn to a smaller diameter, and rolled into a thin tape, as described in the previous paper [4]. The formed tapes were heat-treated under oxygen flowing gas at 1073 - 1173 K in order to sinter oxide particles in the silver sheath.

After the procedure, we estimated the superconductivities (Tc and Jc) of the tapes by the usual four-probe technique. The Tc of the tape was measured at 100 mA current, monitoring the temperature by a platinum-cobalt resistance thermometer. The Jc was also measured in a magnetic field under 0.1 T at nitrogen boiling temperature (77 K). The magnetic field was applied perpendicularly to the tape surface. Criterion for the critical currents was 1 μV/cm. The microstructures were observed by SEM and a polarized optical microscope.

RESULTS AND DISCUSSION

By the SSRP, long superconducting tapes can be rolled without difficulties. Figure 1 shows the appearance of the rolled silver sheathed TBCCO tape (0.1 mm^t x 4 mm^w x 3 m^l). Figure 2 shows the cross sectional view of the silver sheathed TBCCO tape after sintering. The clad tape has a good contact between the TBCCO and silver sheath, and no large voids in the cross-section.

Fig.1 Appearance of the Rolled TBCCO/Ag tape

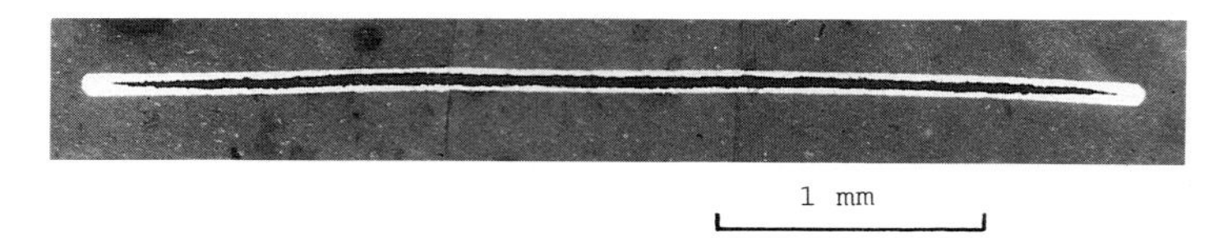

1 mm

Fig.2 Cross-sectional view of the sintered TBCCO/Ag tape

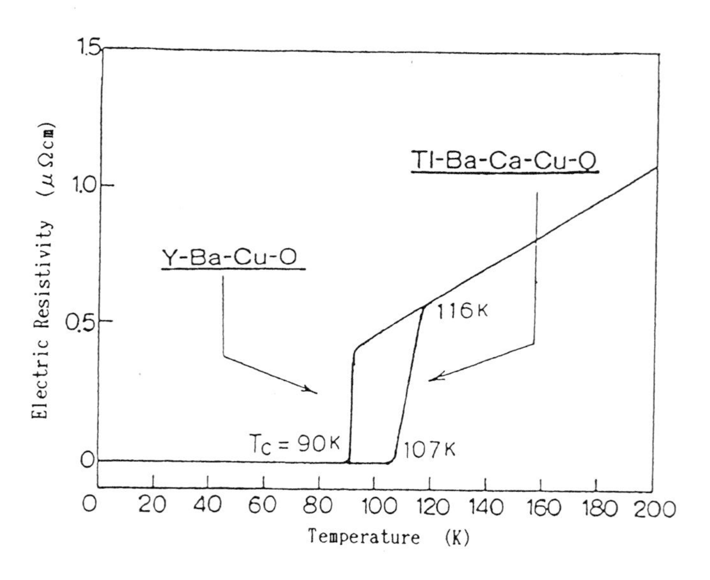

Fig.3 Temperature dependence of the electrical resistivity of TBCCO/Ag and YBCO/Ag tapes

Figure 3 shows the temperature dependence of the electrical resistivity of the silver sheathed TBCCO tape as well as YBCO tape. The electrical resistivity was estimated by the cross-sectional area of the silver sheath. The electrical resistivity of the tape at the temperature above Tc was almost same with that of the pure silver wire. In the case of TBCCO tape, Tc(onset) = 116 K, Tc(zero resistivity) = 107 K, while in the case of YBCO tape, Tc(onset) = 91 K, Tc(zero resistivity) = 90 K. The Tc(zero resistivity) of TBCCO tape was higher than that of YBCO tape by 17 K. Besides, The difference between the Tc(onset) and the Tc(zero resistivity) of the TBCCO tape is much larger than that of the YBCO. It seems that this broad transition was caused by the difference of the superconducting phases in the two oxide tapes. X-ray diffraction analysis was performed to confirm it. It was shown that the YBCO tape consisted of a single phase $Y_1Ba_2Cu_3O_y$ and that the TBCCO tape consisted of two phases (the high-Tc phase $Tl_2Ba_2Ca_2Cu_3O_y$ and the low-Tc phase $Tl_2Ba_2Ca_1Cu_2O_y$ [5]). Those mixed phases induced the broad super-normal transition of the TBCCO tape.

Figure 4 shows the relation between the tape thickness and the Jc at 77 K under zero magnetic field. It was clearly observed that the Jc sharply increased with the tape thickness less than 0.2mm. The highest Jc values of TBCCO and YBCO were 3500 A/cm^2 and 3300 A/cm^2, respectively, at the tape thickness of 0.05 mm. Moreover, the Jc of the TBCCO tape was 3 times higher than that of the YBCO tapes when the tape thickness was more than 0.2 mm. It seems that the merits of the TBCCO were brought by the thermal margine due to the high-Tc of the material. And, it is thought that the high Jc of TBCCO thin tape was attained by the same reason as the YBCO thin tape [4] - (a) densification of the oxide superconducting core ,(b) uniformity of the shape of the oxide core and (c) good contact with sheath. Besides, the orientation of oxide grains in sheath during SSRP caused the Jc-increase of the superconducting tape.

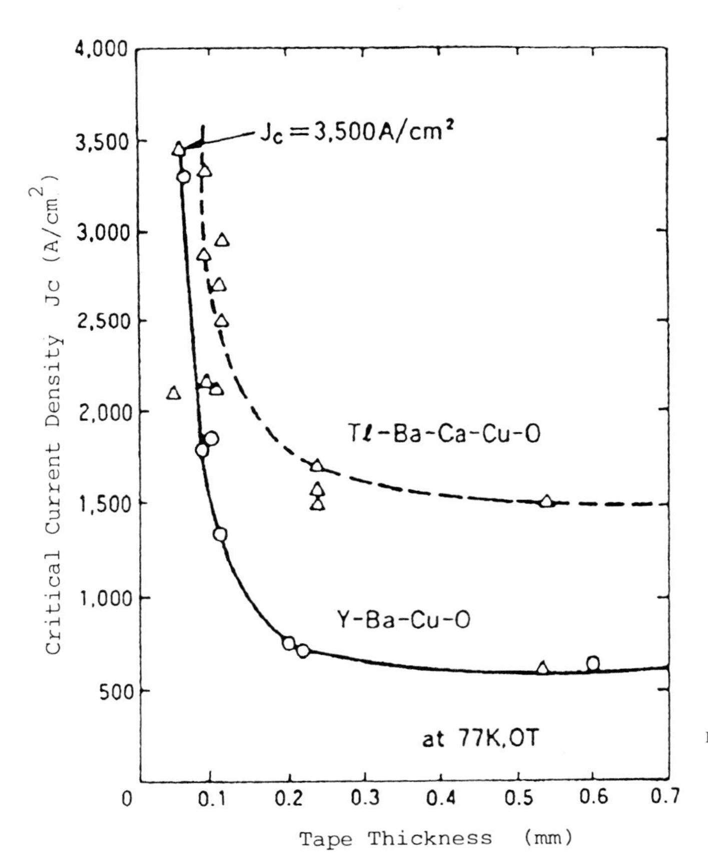

Fig.4 Critical current density of TBCCO/Ag and YBCO/Ag tapes with various thickness at 77 k, 0 T

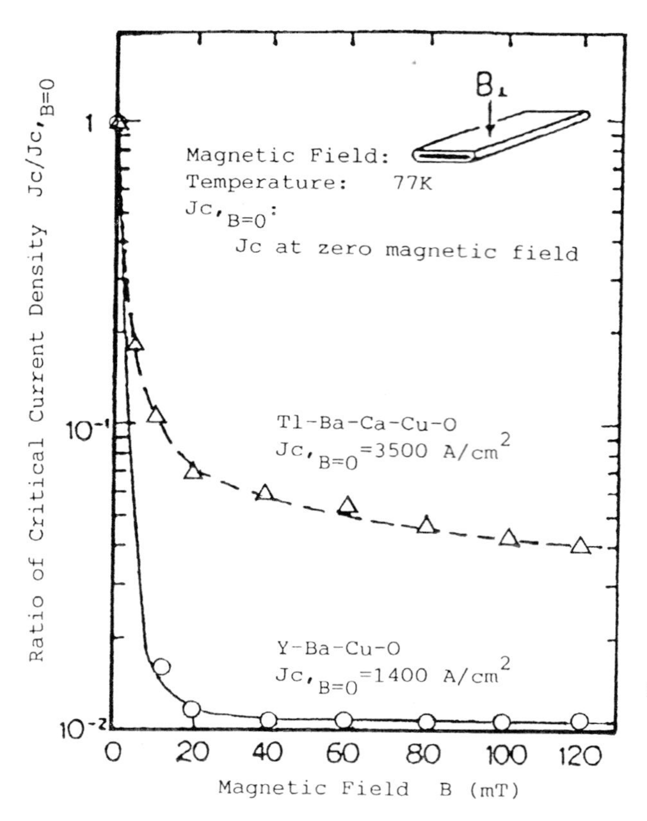

Fig.5 Critical current density of TBCCO/Ag and YBCO/Ag tapes in magnetic field at 77 K

Figure 5 shows the magnetic field dependence of Jc at 77 K for a TBCCO tape with 0.05 mm thickness and a YBCO tape with 0.1 mm thickness. At zero magnetic field, Jc values of the TBCCO tape and the YBCO tape were 3500 and 1400 A/cm^2, respectively. But Jc/Jc(B=0) of the TBCCO and YBCO decreased to 1/25 and 1/100 at 120 mT, respectively. The Jc reduction by the magnetic field of the TBCCO is smaller than that of the YBCO.

REFERENCES

[1] M. K. Wu, J. R. Ashburn, C. J. Torng, P. H. Hor, R. L. Meng, L. Gao, Z. J. Huang, Y. Z. Wang and C. W. Chu, Phys. Rev. Lett., 58, 908 (1987)

[2] H. Maeda, Y. Tanaka, M. Fukutomi and T. Asano, Jpn. J. Appl. Phys., 27, L209 (1988)

[3] Z. Z. Sheng, A. M. Hermann, A. El Ali, C. Almasan, J. Estrada, T. Datta and R. J. Matson, Phys. Rev. Lett. 60, 937 (1988)

[4] M. Okada, A. Okayama, T. Morimoto, T. Matsumoto, K. Aihara and S. Matsuda, Jpn. J. Appl. Phys., 27, L185 (1988)

[5] R.M. Hazen, L.W. Finger, R.J. Angel, C.T. Prewitt, N.L. Ross, C.G. Hadidiacos, P.J. Heaney, D.R. Veblen, Z.Z. Sheng, A.El Ali and A.M. Hermann, Phys. Rev. Lett. 60, 1657 (1988)

Fabrication and Properties of Superconducting Ba-Y-Cu-O and Bi-Ca-Sr-Cu-O Tapes

TOSHIYA MATSUBARA, YASUJI YAMADA, JUN-ICHIROU KASE, NAOSHI IRISAWA, JUN-ICHI SHIMOYAMA, MICHIO MITSUHASHI, MIKIO SASAKI, HIROSHI ABE, and TAKESHI MORIMOTO

Research Center, Asahi Glass Co., Ltd., Hazawa-cho, Kanagawa-ku, Yokohama, 221 Japan

ABSTRACT

$Ba_2YCu_3O_y$ green tapes were fabricated by a doctor blade method and sintered at temperatures ranging from 910 to 1010°C to examine the effect of sintering temperature on their properties. The tape sintered at 930 °C showed the largest critical current density (Jc) of 1500 A/cm^2 at 77 K. It was concluded that the Jc of our tapes are determined by the compromise of porosity, portion of superconducting phase and the loss of contact at the grain boundary.

Bi-Ca-Sr-Cu-O green tapes were also fabricated by a doctor blade method from powders with nominal composition of $Bi_{1.4}Pb_{0.6}Ca_2Sr_2Cu_{3.6}O_y$. The tape with the critical temperature (Tc) of 106K and Jc of 100 A/cm^2 at 77K was obtained by sintering green tape made of the powder containing high critical temperature phase.

INTRODUCTION

Layered-structure perovskite of the formula $Ba_2YCu_3O_y$, layered-structure oxides of Bi-Ca-Sr-Cu-O system and Tl-Ca-Ba-Cu-O system have been found to show superconductivity above liquid nitrogen temperature. Because of the high superconducting critical temperature and critical magnetic field, many studies dealing with the processing of the new materials either in bulk or thin films have been carried out. The application of these compounds to a superconducting magnet requires the winding of tape or wire conductors with sufficient flexibility into a solenoid. Since oxide superconductors are intrinsically brittle, special technique must be developed for preparing tape or wire conductors. [1,2,3] In this paper, fabrication process of Ba-Y-Cu-O and Bi-Ca-Sr-Cu-O tape by a doctor blade process and their superconducting properties are described.

EXPERIMENTAL

A powder specimen with a nominal composition of $Ba_2YCu_3O_y$ was prepared by mixing purified fine Y_2O_3, $BaCO_3$ and CuO powders, grinding and calcining the mixture at 850 °C for 10 hours in a flowing oxygen atmosphere. The obtained $Ba_2YCu_3O_y$ powder was suspended in trichloroethylene containing dispersant, binder and plasticizer, and further milled to get viscous slurry. The resulting slurry was casted through a doctor blade into a 125 mm wide and 100 μm thick green tape on a carrier sheet made of polyethylene terephthalate film. Rectangular sheets of the size 5 mm x 25 mm cut from the green tape were heated at 500 °C for 3 hours to decompose and remove the binder and sintered at temperatures 910 - 1010 °C for 10 hours on magnesium oxide plates in a flowing oxygen atmosphere.

In the case of Bi-Ca-Sr-Cu-O system, powder with a nominal composition of $Bi_{1.4}Pb_{0.6}Ca_2Sr_2Cu_{3.6}O_y$ was prepared by mixing purified fine Bi_2O_3, PbO, $CaCO_3$, $SrCO_3$ and CuO powders, grinding and calcining the mixture at 800 °C for 12 hours or at 845 °C for 100 hours in air. A green tape was fabricated by the same process as described above. Further, it was cut into a rectangular sheet and sintered at 845°C in air.

Tc was measured by dc four-probe resistive method in a cryostat. Current and voltage lead wires were attached to the surface of the sample using conductive silver paste. The temperature was measured by a thermocouple. Critical current (Ic) of the tape was also measured in liquid nitrogen by resistive method. Ic was defined as the current which generated voltage of 1 μV across a 10 mm length of the sample. Jc was calculated dividing Ic by the cross-sectional area of the sample. Oxygen contents of the tapes were determined by iodometric titration method. Porosities of the tapes were obtained using mercury porosimeter. The magnetization measurements were carried out in liquid nitrogen using a vibrating sample magnetometer.

Unsintered green tape was flexible enough to be wound into a roll of less than 1 cm diameter. After the sintering, the tape became brittle and shrinked to a certain extent maintaining the original shape.

RESULTS AND DISCUSSION

Ba-Y-Cu-O tape

Figure 1 shows the X-ray diffraction (XRD) patterns of the powders obtained from grinding the tapes sintered at temperatures ranging from 910 °C to 970 °C. No crystal except orthorhombic $Ba_2YCu_3O_y$ is observed in the samples. The relative intensities of the (0,0,n) peaks of the samples sintered above 950 °C are stronger than those of the samples sintered below 940 °C, which means that the grain growth occurred preferentially along the ab plane.

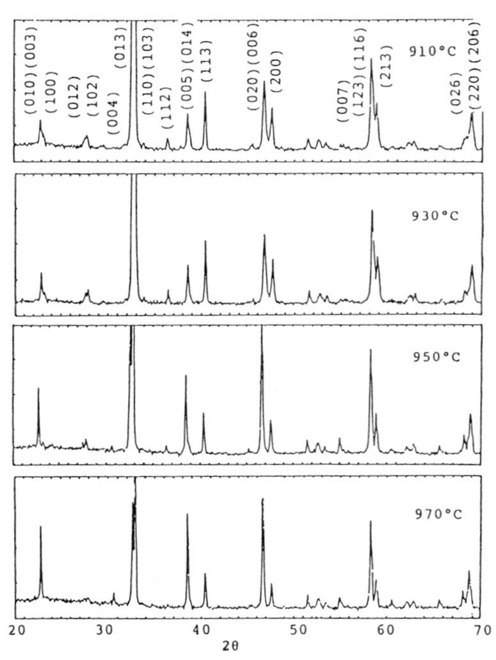

Fig. 1 X-ray diffraction patterns of powders obtained by grinding the sintered tapes

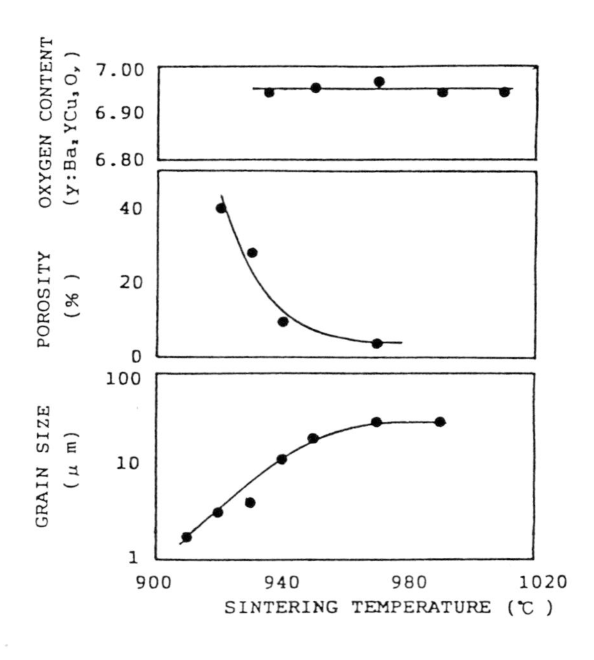

Fig. 2 Effect of sintering temperature on oxygen content, porosity and grain size of the tape

Figure 2 shows the oxygen content, porosity and grain size of the tapes as a function of sintering temperature. Oxygen content of all the $Ba_2YCu_3O_y$ tapes is more than 6.9 and independent of sintering temperature. Porosity of the tapes decreases with the elevation of the sintering temperature. On the other hand, the grain size increases with the elevation of the sintering temperature.

Figure 3 shows Tc, Jc and magnetic susceptibility of the tapes as a function of sintering temperature. The magnetic susceptibility increases with elevation of sintering temperature and kept constant above the sintering temperature of 970°C. Portion of superconducting phase is considered to be proportional to magnetic susceptibility. Therefore, Jc is expected to increase with the decrease in porosity and increase in magnetic susceptibility. However, the Jc increases as the sintering temperature elevates and decreases above the sintering temperature of 940 °C, showing the highest critical current density of 1500 A/cm^2 at a sintering temperature of 930 °C.

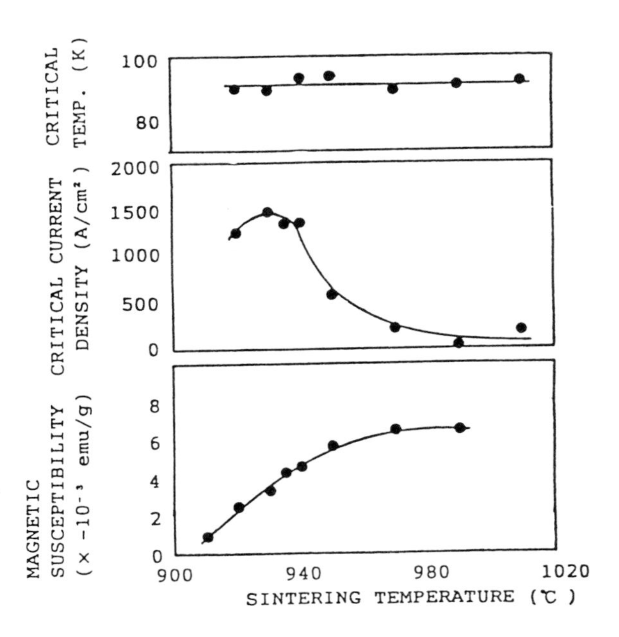

Fig. 3 Effect of sintering temperature on critical tempearture, critical current density and magnetic susceptibility of the tape

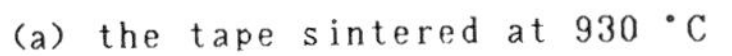

(a) the tape sintered at 930 °C

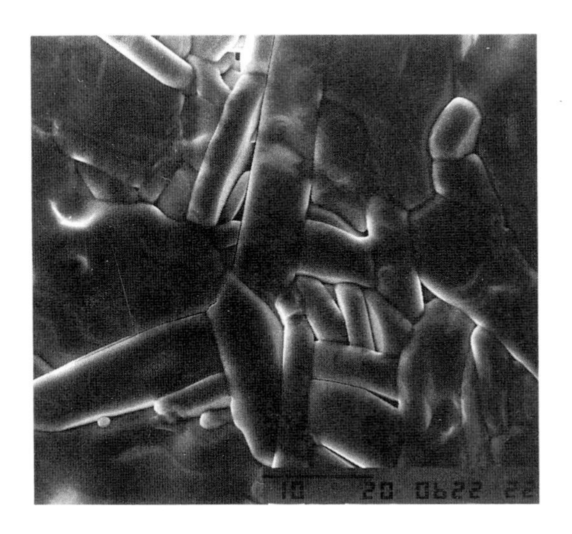

(b) the tape sintered at 970 °C

Fig. 4 SEM micrographs of surface of the tape sintered at 930 °C and 970 °C

One of the reasons of the sharp decrease in Jc above the sintering temperature of 940°C is considered to be the formation of cracks along the grain boundaries. Figure 4 shows SEM micrographs of the surface of the tapes sintered at 930 °C and 970 °C. In the tape sintered at 930 °C, grains have spherical shape of a few microns in diameter. As the boundaries appear to be obscure, grains seem to be intimately contacted with each other. On the other hand, grains in the tape sintered at 970 °C are plate in shape of more than 10 microns in length. As the boundaries are observed clearly, cracks seem to be formed along the boundaries. It is well known that the cell-volume contracts on cooling from sintering temperature to room temperature with phase transition from tetragonal to orthorhombic. This anisotropic thermal contractions are considered to be the cause of formation of cracks along the grain boundaries particularly in large-grained and dense sample.

Another reason of low Jc value is considered to be due to the presence of a second phase at the grain boundaries. Figure 5 shows the Auger electron spectra of the fractured surface of the tape sintered at 970 °C obtained by scanning Auger microscopic study. The ratio of the peak height of each element in the spectrum obtained at transgranular area was slightly different from that obtained at intergranular area. The average ratios of the peak heights of barium to copper are summarized in Table 1. As the peak height is proportional to an amount of a element, the atomic ratio of barium to copper at intergranular area is larger than that at transgranular area. The difference in atomic ratio of barium to copper between intergranular area and intergranular area suggests the presence of a second phase at the grain boundaries.

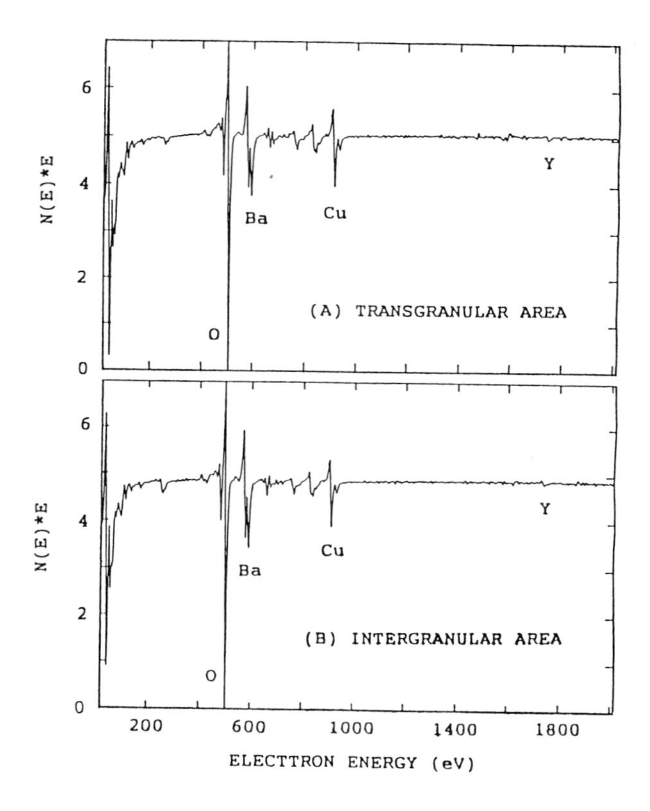

Fig. 5 Auger electron spectra of the fractured surface of the tape sintered at 970 °C

Table 1 The peak height ratios of barium to copper of the Auger electron spectra of the transgranular area and the intergranular area of the fractured surface of the tape sintered at 970 °C

Site	Peak height ratio (Ba/Cu)
Transgranular area	1.30
Intergranular area	1.63

Therefore, it is concluded that Jc in our tapes are determined by the compromise of porosity, portion of superconducting phase and the weak junction due to the presence of a second phase and cracks at the grain boundary.

Bi-Ca-Sr-Cu-O tape

Figure 6 shows the temperature dependence of the resistivity of the tapes and the bulk sample with a nominal composition of $Bi_{1.4}Pb_{0.6}Ca_2Sr_2Cu_{3.6}O_y$. Starting powder for the bulk sample and the green tape was obtained by calcining raw materials at 800 °C for 12 hours in air. The bulk sample was prepared by pressing the powder into a pellet and sintering it at 845 °C for 105 hours in air. The tape (a) was prepared by sintering the green tape at 845 °C for 160 hours. The bulk sample showed zero-resistivity at 108 K. However, despite the tape (a) showed a small resistive transition at 100 K, it did not show zero-resistivity at 77 K. Figure 7A and 7B show the XRD patterns of these samples. As shown in these figure, although the high critical temperature phase is observed in the bulk sample, it is not detected in the tape (a). To acertain the reason, metal composition of the bulk sample and the tape (a) was analyzed by an inductively coupled plasma - atomic emission spectroscopy. The results reveals that originally charged amount of lead almost remains in the bulk sample, while no lead remains in the tape (a). Thus it is possible that the absence of high critical temperature phase in the tape (a) is attributed to the evaporation of lead during sintering at 845 °C.

Therefor to minimize the evaporation of lead and to prepare powders which is mainly composed of high critical temperature phase, starting powder was pressed into a pellet, sintered at 845°C for 100 hours in air and pulverized into powder. The green tape was fabricated using this powder and sintered at 845 °C for 62 hours in air. The tape (b) thus obtained showed zero-resistivity at 106 K and Jc of 100 A/cm2 at 77 K. It was confirmed by XRD that the tape (b) was mainly composed of high critical temperature phase. These results suggest that the lead atoms contained in high critical temperature phase are hard to evaporate during sintering at 845 °C.

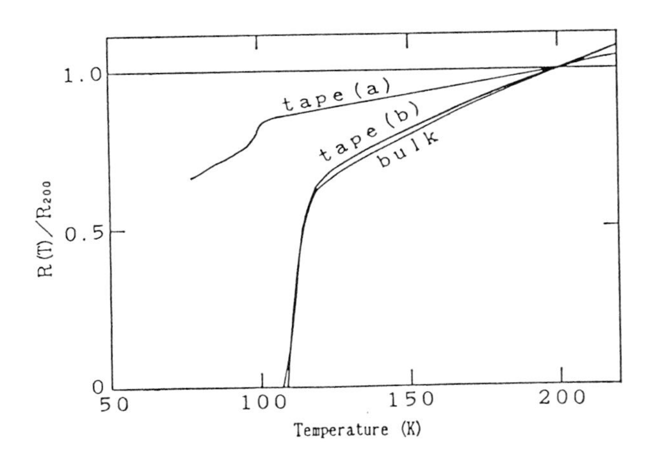

Fig. 6 Temperature dependence of the resistivity of the tapes and the bulk sample of Bi-Ca-Sr-Cu-O system

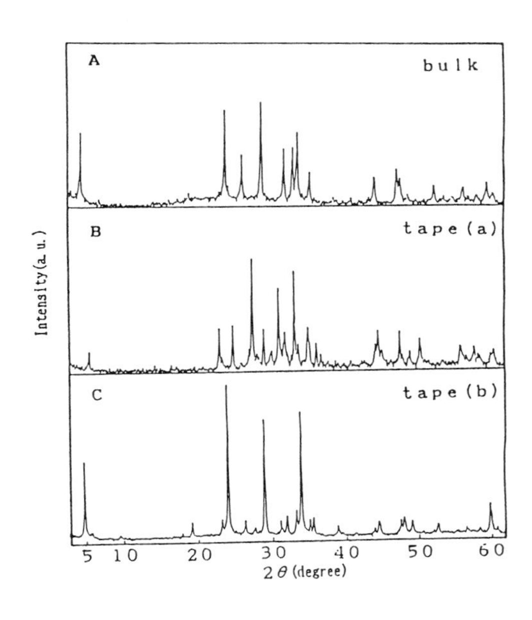

Fig. 7 X-ray diffraction patterns of the tapes and the bulk sample of Bi-Ca-Sr-Cu-O system

REFERENCES

1) T. GOTO, et. al., Jpn. J. Appl. Phys. 26 (1987) L1527-L1528

2) M. ISHII, et. al., ibid 26 (1987) L1959-L1960

3) K. TOGONO, et. al., ibid 27 (1988) L45-L47

Multi-Layered Superconducting $YBa_2Cu_3O_y$ Thick Films on Y_2BaCuO_5 Substrate

KIICHI YOSHIARA[1], KENJI KAGATA[1], TSUTOMU HIROKI[1], and KIYOTAKA NAKAHIGASHI[2]

[1] Materials and Devices Laboratory, Mitsubishi Electric Corporation, 8-1-1, Tsukaguchi-Honmachi, Amagasaki, 661 Japan
[2] Department of Materials Science, University of Osaka Prefecture, Mozu-Umemachi, Sakai, 591 Japan

ABSTRACT

Multi-layered superconducting $YBa_2Cu_3O_y$ thick films have been prepared on Y_2BaCuO_5 substrate by the thick film method. The superconducting films were tightly bound with insulating Y_2BaCuO_5, Dy_2BaCuO_5 or Yb_2BaCuO_5 films and Y_2BaCuO_5 substrate by an anchor effect. The superconducting critical temperature (Tc) of the multi-layered films were 70K and 77K for the first and second layer respectively. The electric resistivity of the Y_2BaCuO_5, Dy_2BaCuO_5 and Yb_2BaCuO_5 insulating films sandwitched by the superconducting $YBa_2Cu_3O_y$ films were 10^4-$10^7\,\Omega\cdot cm$.

The insulating materials used in the present studies seems to be preferable to the $YBa_2Cu_3O_y$ thick film formations because of less interactions between the superconducting and the insulating films (or the substrate).

INTRODUCTION

Since the discovery of high Tc superconducting oxides in the La-Ba-Cu-O and Y-Ba-Cu-O system[1,2], many efforts have been paid to research for more high Tc materials. Recently, new materials with higher Tc of 110K-125K has also been reported in the Bi-Sr-Ca-Cu-O and Tl-Ca-Ba-Cu-O system[3,4]. Although the new superconductors have better resistivity against humidity and the impacts to application have been stronger than the YBCO system, an exact chemical compositions with higher Tc phase have not been hitherto determined. Therefore, at this stage, the study for application still concentrates mainly on the Y-Ba-Cu-O system. Their applications to the superconducting electronic devices require preparations of the material in thick and thin films. In the preparation of thin films, many studies[5-8] by sputtering method have been carried out and high critical temperature Tc(92K) and high critical current density Jc(1.8×10^6 A/cm^2) have been achieved. On the other hand, in the thick films prepared by screen printing method, the reported values of Tc and Jc were very low[9-10]. However, thick film process has some advantage over the thin films because it is a lower cost process and easy to be used for fabrication of the small integrated circuit where the relatively coarse line resolution is acceptable. Then, it is important to enhance the superconductivity of thick film. Recently, throughout our successful preparation of superconducting thick films on Y_2BaCuO_5 substrate by screen printing and sintering method[12], we have revealed that the films on Y_2BaCuO_5 substrate indicate high Tc and Jc and important factor on the fabrication of the thick films is interactions between films and substrate. The high temperature annealing causes interdiffusion between films and substrate, and degradation of the superconductivity of the films. These points are serious problems for electronic device applications. The advantage of choosing the Y_2BaCuO_5 substrate is its thermal stability and possible compensating effect for the $YBa_2Cu_3O_y$.

Futhermore, the multi-layered films to integrat circuits become an important problem for the practical use. The fabrication of multi-layered film contains the same or more complications as described the above.

In this paper, we report the preparations of the multi-layered $YBa_2Cu_3O_y$ thick films on the substrate Y_2BaCuO_5 by the screen printing and sintering method and their superconducting properties. The multi-layered $YBa_2Cu_3O_y$ thick films were made by formating an insulating films such as Y_2BaCuO_5, Dy_2BaCuO_5 and Yb_2BaCuO_5 between two $YBa_2Cu_3O_y$ films. These insulating materials have nearly the same physical properties to each other and are stable in thermally even at 1150℃ ~ 1250℃ in air[13].

EXPERIMENTAL

The ceramic powders $YBa_2Cu_3O_y$, Y_2BaCuO_5, Dy_2BaCuO_5 and Yb_2BaCuO_5 were prepared by calcining the mixtures of prescribed amounts of 4N Y_2O_3, Yb_2O_3, Dy_2O_3, $BaCO_3$ and CuO at 950℃ in air for 12 hours, and then pulverizing. The calcining and pulverizing were repeated until a single phase of each powder sample was confirmed by X-ray diffractometry.

The Y_2BaCuO_5 substrate was prepared from the powder by pressing into a disc pellet with 50mm in diameter and 4mm in thickness, and then firing at 1100℃ in air for 24 hours. For the practical use, a surface of the Y_2BaCuO_5 substrate thus obtained was polished with #1500 abrasive paper. The superconducting paste was prepared by mixing the powder $YBa_2Cu_3O_y$ and organic agents (diethylene glycol mono-N-butyl etheracetate, ethylcellulose, terpineol and 2-propanol) in an agate mortar. The insulating paste was also prepared in the same method by using the powder of Y_2BaCuO_5, Dy_2BaCuO_5 and Yb_2BaCuO_5.

The fabrication of multi-layered thick film was achieved by repeating the printing and the sintering of the superconducting and the insulating paste alternately on the substrate, where the sintering process was done after every stage of printing. Namely the superconducting paste was printed on the Y_2BaCuO_5 substrate through a 325 mesh screen and dried at 120℃ for 30 min. The film (1st-layer) was sintered at 960℃ in air for 1 hour and then the insulating paste was printed and sintered at the same condition. The multi-layered thick films were obtained by repeating the above processes untill the number of the desired layeres was achieved. The thickness and some morphological properties of the sintered films were clarified based on the observations of film cross sections by scanning electron microscope. The crystal structure of the films was investigated by X-ray diffraction spectrometry with a monochromatized CuKα radiations. The electrical resistivity and its temperature dependences of the each film in the multi-layered films were measured, independently, on the fabricated line by the dc/four-probe method. The pattern of fabrication was shown in Fig.1 and the length and the width of fabrication were 20 mm and 5 mm,respectively.

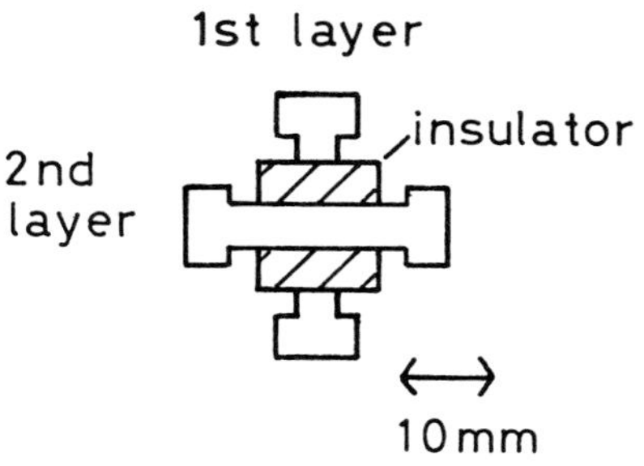

1st layer
2nd layer
insulator
10mm

Fig.1 Fabrication pattern of multi-layered superconducting films

Thermal and mechanical properties of the substrate Y_2BaCuO_5 were also measured for the purpose of a practical use. The thermal expansion ($\Delta l/l$) of the $YBa_2Cu_3O_y$, Y_2BaCuO_5 and some other ceramic substrates were measured, in the bulk state, over the temperature range from R.T. to 900℃ by Thermal Mechanical Analysis (TMA). The flexural strength was measured by three-point bending test technique under the conditions that sample size was 5 mm×20 mm×2 mm, span distance and loading speed were 15 mm and 1 mm/min respectively. The dielectric constant was measured at the frequencies 1 MHz and 10 MHz by Q-meter method. The shrinkage was estimated from the changes in diameter of the disk pellets in before and after sintering.

RESULTS AND DISCUSSION

The mechanical, electrical and thermal properties of the Y_2BaCuO_5 substrate used in this experiment are shown in Table Ⅰ. The sintered density and shrinkage were 5.04 g/cm^3 and 16 % respectively, and the flexural strength was 500 kg/cm^2 - 1000 kg/cm^2. The substrate has a relatively low dielectric constant such as a value of 8.29 at 1 MHz and 10 MHz, which is closed to that of alumina, and a high volume resistivity (>10^8 Ω·cm). The thermal expansion

Table Ⅰ The properties of Y_2BaCuO_5 substrate

Sintered Density		5.04 g/cm^3
Shrinkage		16%
Flexural Strength		1000 kg/cm^2
Dielectric Constant	(1MHz)	8.29
	(10MHz)	8.29
Volume Resistivity	(100V DC)	>10^8 Ω·cm
Thermal Expansion Coefficient		110×10^{-7}/℃

coefficient is 110×10^{-7}/℃ and this is nearly to the value 126×10^{-7}/℃ of $YBa_2Cu_3O_y$. These substrate properties are suggesting that Y_2BaCuO_5 substrate is adequate for $YBa_2Cu_3O_y$ film formations.

Figure 2 shows a scanning electron micrograph of the cross section of the multi-layered $YBa_2Cu_3O_y$ thick films with insulating layer Dy_2BaCuO_5. From the Figure 2, the microstructure of $YBa_2Cu_3O_y$ thick films are in a porous state for both of 1-st and 2-nd layer. On the other hand, the Y_2BaCuO_5 substrate and the insulating layer Dy_2BaCuO_5 are in a dense state with the microstructure being composed of nearly spherical particles with diameters of about 1 μm -20 μm due to the grain growth during the high-temperature annealing. The thicknesses of the films were estimated as 50μm for 1-st layer, 50μm for Dy_2BaCuO_5 layer and 50μm for 2-nd layer, respectively.

Fig.2 Scanning electron micrograph of the cross section of the multi-layered films

Figure 3(a) and (b) show temperature dependence of an electrical resistivity of the $YBa_2Cu_3O_y$ thick films of, (a) first layer, and of (b) second layer for multi-layered thick film. The onset temperature and zero resistivity temperature of the thick film were 80K and 70K in the first layer and were 85K and 77K in the second layer.

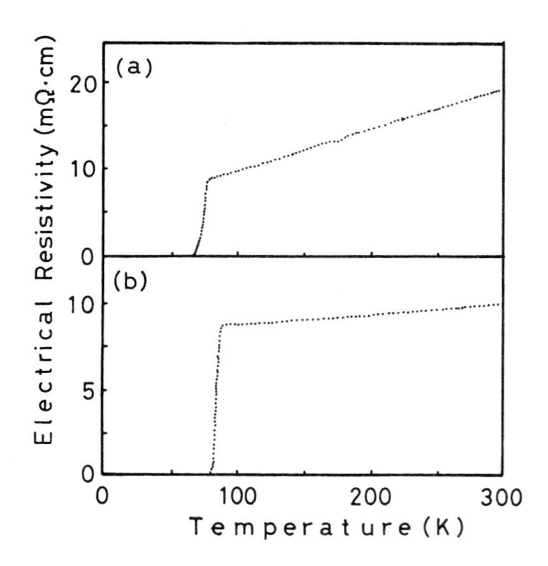

Fig.3 Temperature dependence of a electrical resistivity of $YBa_2Cu_3O_y$ films in multi-layered films

Figure 4(a), (d) and (e) shows the X-ray diffraction spectrums of $YBa_2Cu_3O_y$, Dy_2BaCuO_5 and Yb_2BaCuO_5 film sintered only one cycle on the Y_2BaCuO_5 substrate. As is clear from the figures, these spectrums come from a single phase of the respective materials. However, increasing the cycle of the sintering, unexpected Y_2BaCuO_5 spectrums together with $YBa_2Cu_3O_y$

spectrums were observed as shown in the figure 4(b) and (c). From the above, the sintering of multi-times at high-temperature in superconducting thick film prompt to some appearances in thick film. So, the conditions of multi-sintering must be considered very carefully.

Figure 5 shows the relations between thermal expansion and temperature with heating of the $YBa_2Cu_3O_y$, Y_2BaCuO_5, Dy_2BaCuO_5, Yb_2BaCuO_5, Al_2O_3, MgO, $SrTiO_3$, YSZ(Yittrium Stabilized Zirconia). Thermal expansion of the $YBa_2Cu_3O_y$ is the highest value at any temperateres observed and increases abruptly at near 460℃. This abrupt increase is caused by a change in the oxygen content. The thermal expansion of Y_2BaCuO_5, Dy_2BaCuO_5 and Yb_2BaCuO_5 are the closest to that of the $YBa_2Cu_3O_y$ except for the MgO.

The Y_2BaCuO_5, Dy_2BaCuO_5 and Yb_2BaCuO_5 are stable insulator and are able to coexist with $YBa_2Cu_3O_y$. Moreover, if these materials diffuse to $YBa_2Cu_3O_y$ materials, the impurity phase would appear the $YBa_2Cu_3O_y$, $DyBa_2Cu_3O_y$ and $YbBa_2Cu_3O_y$ phase and the other. these impurity are the superconductor in itself, and damaged only slightly to the superconductivity of $YBa_2Cu_3O_y$ film. From the above, we are thinking that these materials are adequate for the multi-layered superconducting thick film.

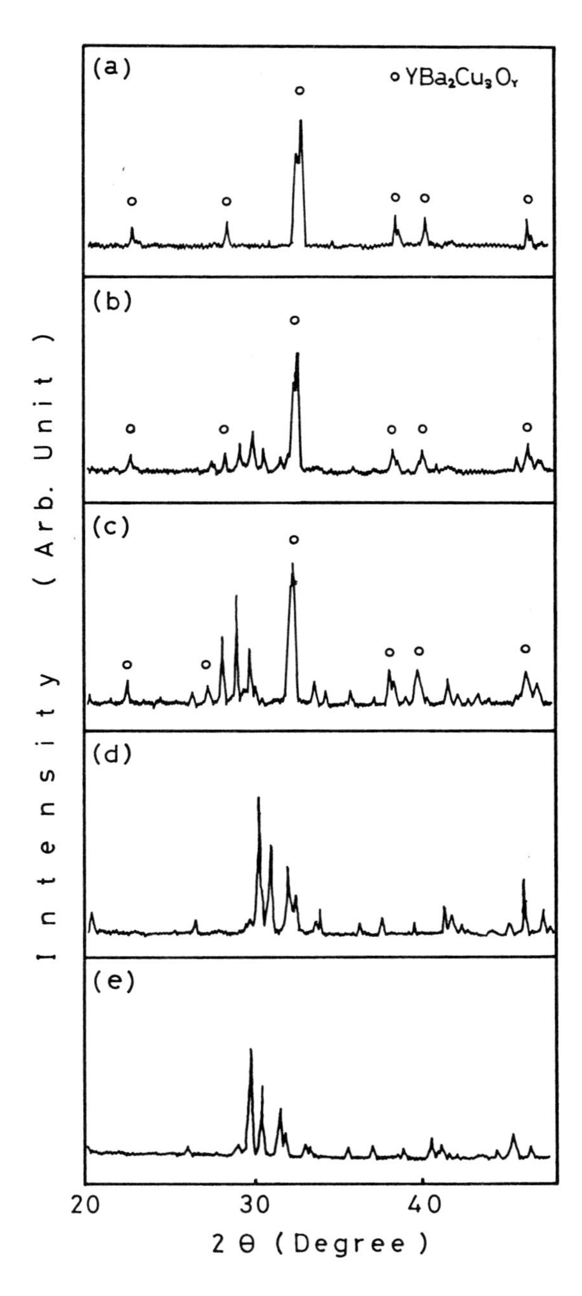

Fig.4 X-ray diffraction spectra of $YBa_2Cu_3O_y$ film sintered (a)1 time, (b)2 times, (c)3 times, and the insulating film on the 1-st superconducting film (d) Dy_2BaCuO_5, (e) Yb_2BaCuO_5.

CONCLUSION

The multi-layered $YBa_2Cu_3O_y$ thick films were obtained successfully by using the Y_2BaCuO_5, Dy_2BaCuO_5 and Yb_2BaCuO_5 as an insulating interlayer substances and exhibited fairly high Tc of 70K for the 1-st layer and 80K for the 2-nd layer, respectively. The thermal expansion coefficient of the $YBa_2Cu_3O_y$ was fairly agreed with those of the insulating materials used in present studies, and chemical reactions by thermal diffusion between the superconducting films and insulating films (or the substrate) were very small. Therefore, it is suggesting that Y_2BaCuO_5, Dy_2BaCuO_5 and Yb_2BaCuO_5 films are promising materials for multi-layered superconducting thick films.

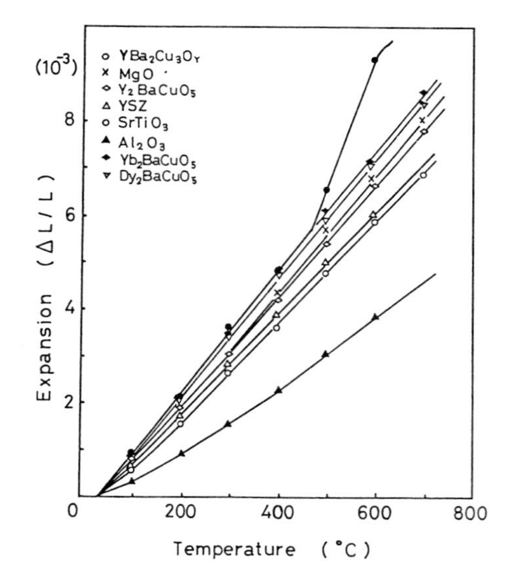

Fig.5 Temperature vs thermal expansion relationship of Al_2O_3, MgO, Y_2BaCuO_5, $YBa_2Cu_3O_y$, $SrTiO_3$, YSZ, Dy_2BaCuO_5 and Yb_2BaCuO_5.

REFERENCE

[1] J.G.Bedonrz and K.A.Muller, Z.Phys.B64,189(1986).

[2] M.K.Wu, J.R.Ashburn, C.J.Torng, P.H.Hor, R.L.Meng, L.Gao, J.Huang, Y.Q.Wang and C.W.Chu, Phys.Rev.Lett.58,908(1987).

[3] H.Maeda, Y.Tanaka, Y.Fukutomi and T.Asano, Jpn.J.Appl.Phys.27,L209(1988).

[4] Z.Z.Sheng, A.M.Hermann, Nature 332,55(1988).

[5] T.Aida, T.Fukazawa, K.Kanke, Z.Wen, S.Yokoyama, H.Asano, Iguchi and E.Yamaka, Jpn.J.Appl.Phys.26, L1483(1987).

[6] Y.Enomoto, T.Murakami, M.Suzuki and K.Moriwaki, Jpn.J.Appl.Phys.26,L1248 (1987).

[7] M.Futamoto and Y.Honda, Jpn.J.Appl.Phys.27,L73(1988).

[8] A.Nakayama, A.Inoue, K.Takeuchi and Y.Okabe, Jpn.J.Appl.Phys.26,L2055 (1987).

[9] H.Koinuma, T.Hashimoto, T.Nakanuma, K.Kishio, T.Kitazawa and K.FueKi, Jpn. J.Appl.Phys. 26,L761(1987).

[10] M.Itoh and H.Ishigaki, Jpn.J.Appl.Phys. 27,L420(1988).

[11] R.C.Budhani, S.H.Tzeng, H.J.Doerr and R.F.Bunshah, Appl.Phys.Lett. 51, 1277(1987).

[12] K.Yoshiara, K.Kagata, S.Yokoyama, T.Hiroki, H.Higuma, T.Yamazaki and K.Nakahigashi, Jpn.J.Appl.Phys.29,L1492(1988).

[13] K.Nakahigashi, K.Yoshiara, M.Kogachi, S.Nakanishi, H.Sasakura, S.Minamigawa, N.Fukuoka and A.Yanase, Jpn.J.Appl.Phys.27,L378(1988).

Preparation of Superconducting Printed Thick Films of Bi-Sr-Ca-Cu-O and Bi-Pb-Sr-Ca-Cu-O Systems

K. HOSHINO[1], H. TAKAHARA[1], and M. FUKUTOMI[2]

[1] Central Research Laboratory, Mitsui Mining & Smelting Co, Ltd., 1333-2 Haraichi, Ageo, 362 Japan
[2] National Research Institute for Metals, Tsukuba Laboratories, Tsukuba, 305 Japan

ABSTRACT

The thick films of Bi-Sr-Ca-Cu-O and Bi-Pb-Sr-Ca-Cu-O systems were prepared on various substrates by a screen printing method. Superconducting properties in these thick films were examined as a function of powder preparations for pastes and postprinting annealing conditions. By appropriate heat treatments after printing for Bi systems, highly c-axis oriented thick films on (100) MgO substrate were obtained, which showed zero resistance at 107 K. For Bi-Pb systems, the powders contained the high-Tc phase of Bi systems in a large volume fraction were prepared for a printing paste. The films were screen printed using this paste, and then annealed. These films deposited on (100) MgO and Ag metal substrates showed zero resistance at 106 and 100 K, respectively.

1. INTRODUCTION

Following epockmaking discoveries of La-Sr-Cu-O[1] and Y-Ba-Cu-O[2] superconductors, the new superconducting oxides of Bi-Sr-Ca-Cu-O and Tl-Ba-Ca-Cu-O systems with transition temperature, Tc above 100 K have been recently reported by Maeda et al.[3] and Sheng et al.[4], respectively. Although Tl compounds show the highest Tc of 125 K among the series of high-Tc superconducting oxides at this time, detailed studies on industrial applications have been prevented by the toxic nature of Tl element. While, Bi systems have two superconducting phases with Tc of 85 K (the low-Tc phase) and 110 K (the high-Tc phase)[3], and that the syntheses of the single high-Tc phase are quite difficult in their systems. Very recently, Takano et al.[5] reported that the partial substitution of Pb for Bi systems increases drastically the volume fraction of the high-Tc phase. The Bi-Pb systems, therefore, have become the very attractive materials on developing the technologies for practical applications. The thick film technologies for these superconducting materials are expected to have a wide variety of applications including printed thick film circuits, superconducting tapes formed by thick films and so on. Thus, it is essential to develop superconducting thick films onto various substrates such as Alumina ceramics or metal tapes.

In this paper, we report the preparation of Bi-Sr-Ca-Cu-O (BSCCO) and Bi-Pb-Sr-Ca-Cu-O (BPSCCO) thick films by a screen printing method. Firstly, the BSCCO thick films were deposited on (100) MgO substrates. Secondly, the printing paste were prepared from the powders containing the high-Tc phase, and then thick films were deposited on various substrates using those pastes. As substrates, (100) MgO, MgO polycrystal, commercial Al_2O_3, and Ag materials were used. Superconducting properties of these films as well as the adhesion and interaction between the films and substrates were investigated.

2. EXPERIMENTAL

The BSCCO or BPSCCO powders with nominal compositions listed in Table 1 were prepared from high purity powders of Bi_2O_3, PbO, $SrCO_3$, $CaCO_3$ and CuO. The powders were produced by

Table 1. Nominal compositions and heat treatment conditions for BSCCO and BPSCCO powders.

Powder (Paste)	Nominal compositions Bi:Pb:Sr:Ca:Cu	Heat treatment condition (in air)
(I)	1 : 0 : 1 : 1 : 2	800 °C x 12 hr
(II)	0.7: 0.3 : 1 : 1 : 1.8	800°C x 12 hr + 845°C x 100 hr
(III)	0.7: 0.3: 1 : 1.5 : 2	800°C x 12 hr + 845°C x 100 hr

a conventional solid state reaction including mixing, grinding and calcining. The heat treatment conditions for each powders are also listed in Table 1. The resultant compound was pulverized to fine powders past through a 280 mesh screen. The powders were subsequently converted into a paste by thoroughly mixing and grinding them with an acrylic resin. The substrates used in this experiment were cleaved (100) MgO single crystal, MgO and Al_2O_3 polycrystals with the purity of 99.9% and 96%, respectively, and Ag tapes of 0.01 mm in thickness.

The films dried after printing were heated in a furnace at the rate of 300°C/hr, and then annealed in the temperature range of 800-890°C followed by cooling to room temperature at the rate of 100-200°C/hr. All the heat treatments in this experiment were carried out in air atmosphere. The film thickness varied from 5 to 50 μm, which was measured by examining the cross section under the scanning electron microscopy(SEM). The electrical resistivity was measured using a standard dc four-probe method with silver paste contacts. The ac magnetic susceptibility was measured using a Hartshorn bridge. The temperature was measured by a well-calibrated Pt-Co thermometer. The crystal structure and surface morphology of the annealed films were studied by X-ray diffraction and SEM techniques, respectively. The composition of the films was examined by energy dispersive X-ray microanalysis(EDX), and the reaction at the interface between the films and substrates was examined by electron probe microanalysis(EPMA).

3. RESULTS AND DISCUSSION

The experimental data of printed thick films deposited on various substrates are listed in Table 2. The data on Tc, annealing conditions, the reactivity and adhesion between films and substrates are included in this table. The detailed results are described in the following sections.

Table 2. Tc and annealing conditions of printed thick films deposited on various substrates. The data on the reactivity and adhesion between films and substrates are also included.

Film	Substrate	Paste	Annealing condition (in air)	Tc (K)	Reactivity	Adhesion
(a)	(100) MgO	(I)	860°C x 0.5 hr	41	no	good
(b)	(100) MgO	(I)	880°C x 0.5 hr	80	no	excellent
(c)	(100) MgO	(I)	885°Cx1 hr + 872°C x72 hr	107	no	excellent
(d)	(100) MgO	(II)	845°C x 24 hr	49	no	good
(e)	(100) MgO	(III)	845°C x 24 hr	106	no	good
(f)	Ag metal	(III)	845°C x 24 hr	100	little	good
(g)	MgO poly	(III)	845°C x 24 hr	103	no	good
(h)	Al_2O_3 poly	(III)	845°C x 12 hr	92	reactive	poor

(1) BSCCO thick films

Figure 1 shows typical curves of resistivity vs temperature for BSCCO films deposited on (100) MgO substrate. The thickness of films (a) and (b) is about 10 μm. Film (c), which was a much thicker film of 50 μm than films (a) and (b), prepared in order to minimize the effect of the interaction between the film and substrate on the

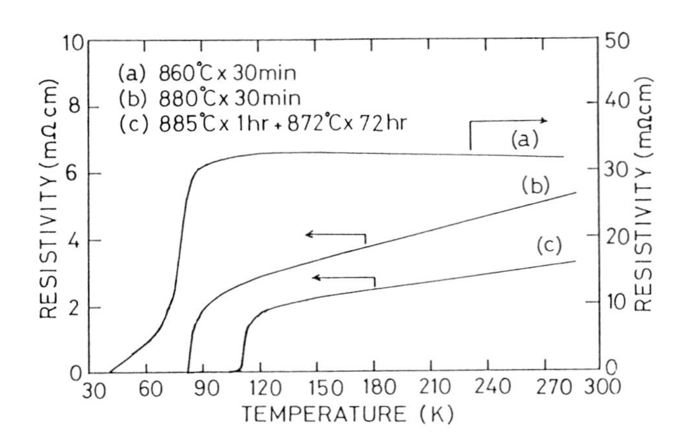

Fig. 1 Resistivity transition of BSCCO thick films (a) and (b) annealed at 860 and 880°C for 30 min, respectively. Film (c) was annealed at 885°C for 1 hr, followed by annealing at 872°C for 72 hr.

Tc behavior. As shown in Fig. 1, film (c) shows zero resistance Tc(end) of 107 K, which corresponds to the superconducting transition of the high-Tc phase in BSCCO materials. Figure 2 shows the X-ray diffraction patterns for films (a), (b) and (c). The peaks of these diffraction patterns were indexed using the results by Tarascon et al.[6] and Takayama-Muromachi et al.[7]. The highly oriented films with the c-axis perpendicular to the substrate were obtained from the annealing above 880°C (films (b) and (c)), but not for the annealing at 860°C (film (a)). In film (c), both the peaks of 2θ = 4.7° and 5.6° for the high-Tc and the low-Tc phases[5], respectively, were found as shown in Fig. 2. It is noted that the diffraction patterns of film (c) are very similar to that of the thin films with the Tc(end) of 103 K reported by Yoshitake et al.[8]. The main compositions of films (b) and (c) was found to be those of the low-Tc phase (Bi:Sr:Ca:Cu=2:2:1:2) and the high-Tc phase (Bi:Sr:Ca:Cu=2:2:2:3), respectively, by EDX examination. However, the process to obtain film (c) is very sensitive to the heat treatment condition and film thickness. Figure 3 shows the surface morphology for films (a), (b) and (c) by SEM. The rough surface with many voids was observed in film (a). On the other hand, the surface of films (b) and (c) were very smooth and densified. This suggests that the annealing condition of films (b) and (c) included a melting process, which was probably responsible for the formation of the highly oriented films.

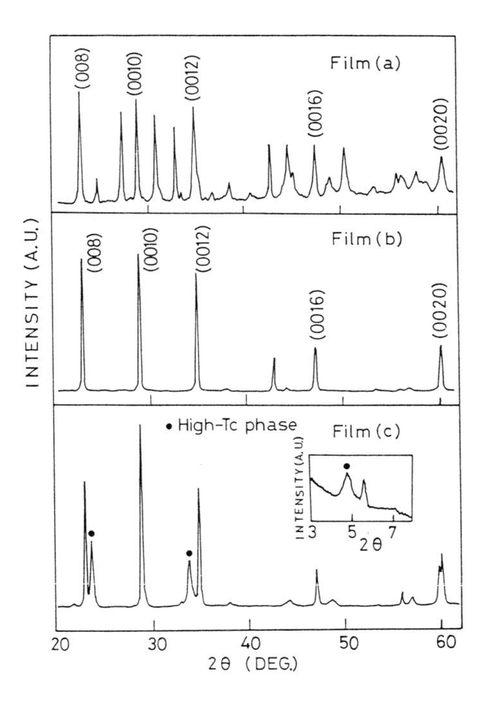

Fig. 2 X-ray diffraction patterns of BSCCO thick films on (100) MgO. Films (a), (b) and (c) correspond to those shown in Fig. 1.

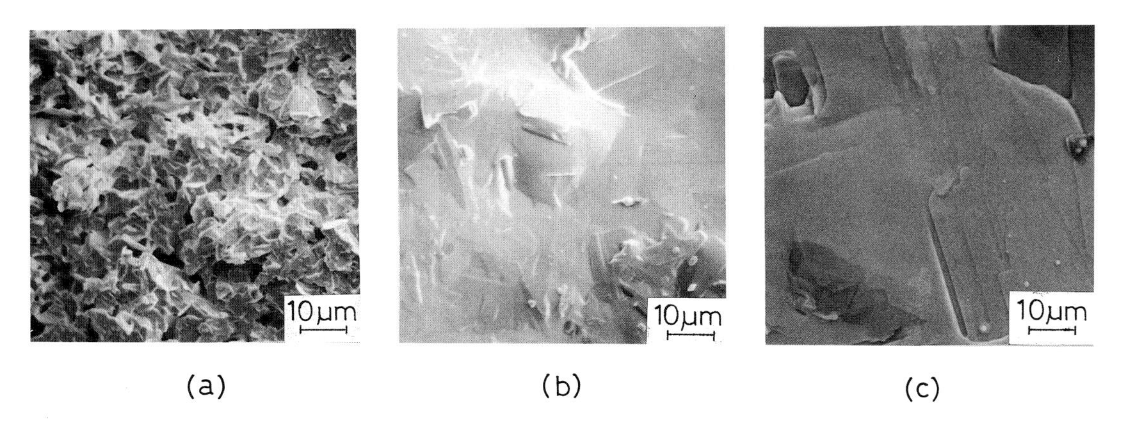

Fig. 3 SEM micrographs of BSCCO thick films (a), (b) and (c) on (100) MgO. Films (a), (b) and (c) correspond to those shown in Fig. 1.

(2) BPSCCO thick films

Superconducting powders

The X-ray diffraction patterns for powders (II) and (III) are shown in Fig. 4. In powder (III), the intensity for the (002) peak of the high-Tc phase at 2θ = 4.7° is much higher than that of the low-Tc phase at 2θ = 5.6°. While, in powder (II), the peaks of the low-Tc phase still exist in a high intensity. As shown in Fig. 5, the relative volume fraction of the high-Tc phase in these powders can be estimated from the ac magnetic susceptibility measurements. The onset temperatures equal to 107 K are almost the same for both powders (II) and (III). However, Meissner effect for powder (III) at 107 K is about five times larger than that of powder (II). Moreover, two step Meissner effects were observed for powder (II), which suggests coexistence of the low- and high-Tc phase. The difference in the volume fraction of the high-Tc phase for these powders may be due to their calcium contents. The excess amount of calcium in powder (III) seems to enhance the formation of the high-Tc phase.

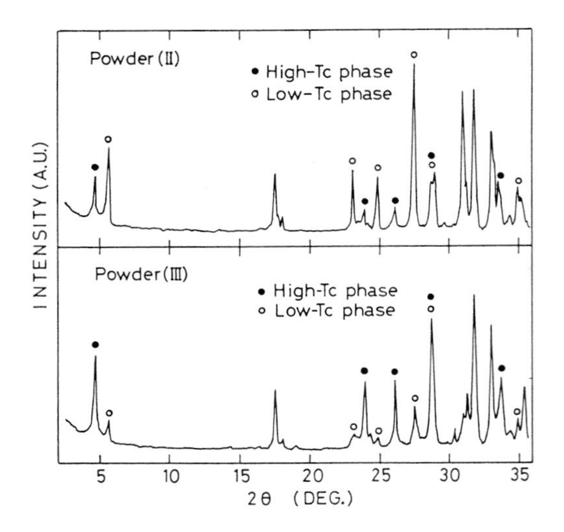

Fig. 4 X-ray diffraction patterns of BPSCCO powders. Powders (II) and (III) correspond to those listed in Table 1.

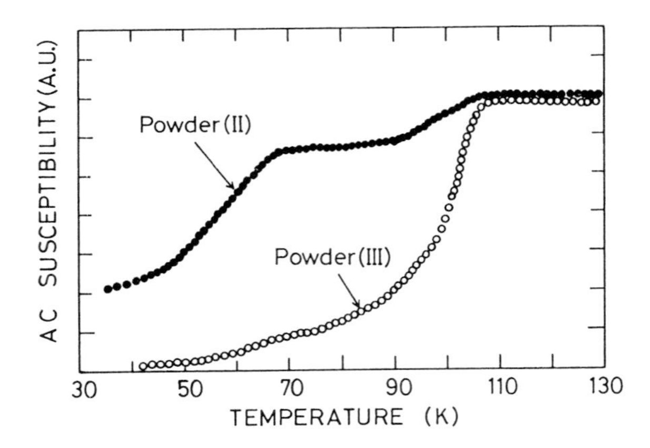

Fig. 5 Temperature dependence of ac magnetic susceptibility for powders (II) and (III). Powders (II) and (III) correspond to those listed in Table 1.

Thick films on (100) MgO and Ag substrates

Figure 6 shows typical superconducting transition curves for BPSCCO thick films on (100) MgO substrates prepared using paste (II) and (III), where the number of paste is labelled as same as that of powders. These films with 10-20 μm in thickness were annealed at 845°C for 24 hr. Film (e) prepared from paste (III) shows sharp superconducting transition with Tc(end) of 106 K. The film (d) from paste (II), however, shows the two step decreasing on resistivity and zero resistance around 50 K. Figure 7 shows the temperature dependence of resistance for the film (f) deposited on the Ag substrate using paste (III). Film (f) was annealed at 845°C for 24 hr. The zero resistance was achieved at 100 K. Although the critical current density, Jc, at 77 K was not measured, the values of Jc for these films on both MgO and Ag substrates seemed to be very small (∿30 A/cm^2), which was estimated from the applied constant current and the cross-sectional area of the films in the Tc measurements. The X-ray diffraction patterns for films (e) and (f) are shown in Fig. 8. The c-axis orientation perpendicular to the substrate for film (f) seems to be much better than film (e). As shown in Fig. 9, the surface of the film (e) on the MgO substrate is very rough and porous. On the contrary, a plate like grain growth morphology can be seen for the film on the Ag substrate. Such the difference on the morphology for the films annealed by the same heat treatment is probably due to the substrate characteristics, which also affect the preferencial orientation behavior of the films. It seems that the annealing process of BPSCCO films does not include a melting process, so the surface morphology is quite different from those of films (b) and (c) described in the previous section. In order to obtain a molten surface, the BPSCCO films were annealed above 860°C. However, the annealing process lowered the Tc of the films below to 90 K.

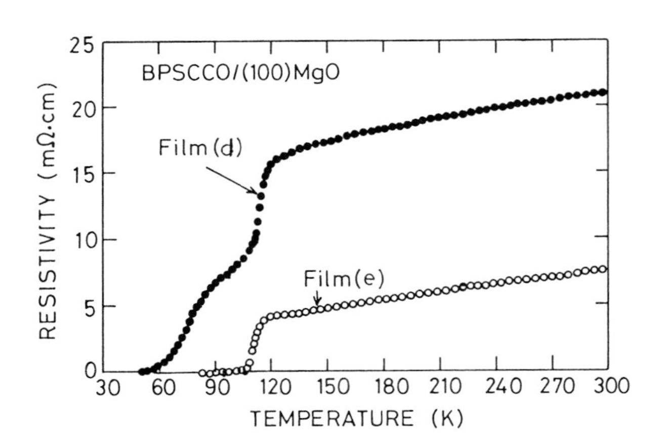

Fig. 6 Resistivity transition of BPSCCO thick films (d) and (e) annealed at 845°C for 24 hr. Films (d) and (e) were prepared from paste (II) and (III), respectively.

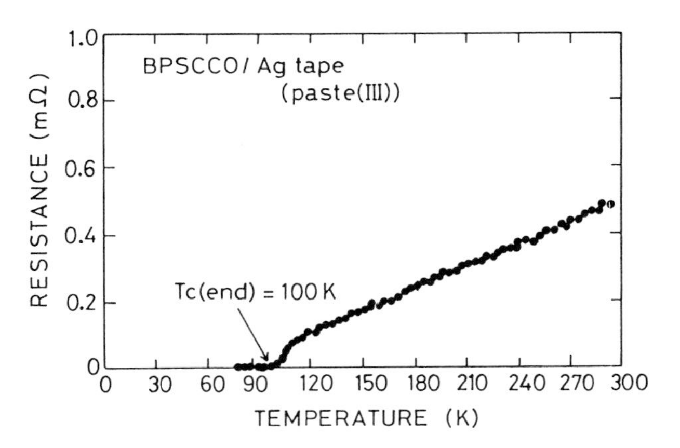

Fig. 7 Temperature dependence of resistance for BPSCCO thick film on Ag substrate annealed at 845°C for 24 hr. The film was prepared from paste (III).

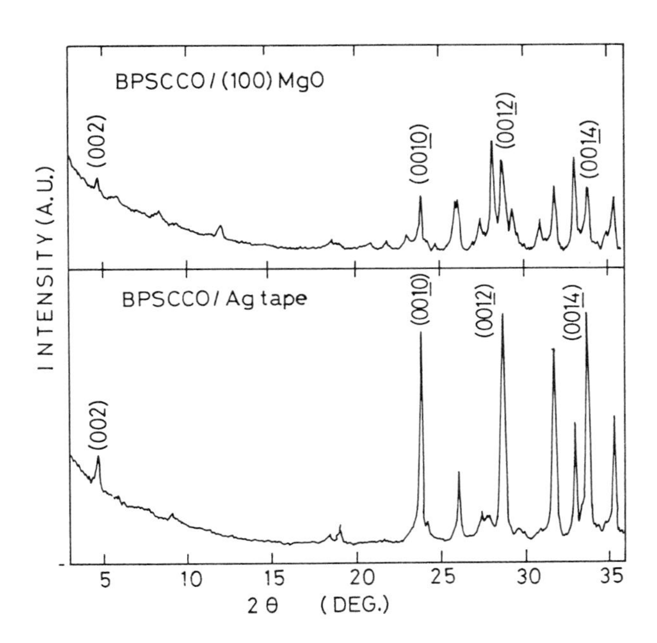

Fig. 8 X-ray diffraction patterns of BPSCCO thick films (e) and (f) deposited on (100) MgO and Ag substrates, respectively. These films were annealed at 845°C for 24 hr.

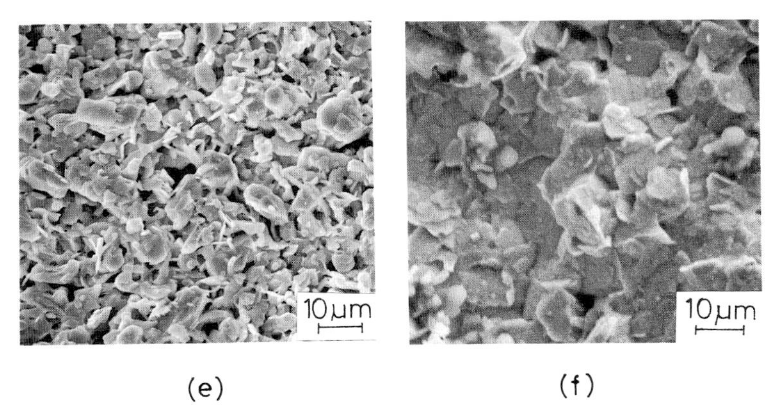

Fig. 9 SEM micrographs of BPSCCO thick films (e) and (f), which correspond to those shown in Fig. 8.

MgO and Al_2O_3 polycrystal substrates

The thick films prepared from paste (III) were deposited on MgO (99.9% purity) and Al_2O_3 (96% purity) polycrystal substrates. The experimental results are listed in Table 2. Superconducting behavior for the film (g) on the MgO polycrystal substrate is almost the same as the film (e) on the (100) MgO substrate. Although zero resistance for the film (h) on the Al_2O_3 substrate was obtained at 92 K, the reaction at the interface between the film and substrate was rather severe compared with the other substrates used in this experiment.

4. CONCLUSIONS

The high-Tc superconducting thick films were prepared on various substrates by a screen printing technique. In Bi systems, highly c-axis oriented thick films on (100) MgO substrates were successfully obtained, which showed zero resistance at 107 K. For Bi-Pb systems, the powders for printing paste were prepared, which contained the high-Tc phase in a large volume fraction. Using these pastes, the thick films were screen printed and then annealed. These films deposited on (100) MgO, MgO polycrystal, commercial Al_2O_3 substrates and Ag tapes showed zero resistance at 106, 103, 92 and 100 K, respectively. At the present stage, the particles with the high-Tc phase in these films were not well sintered and densified. However, this will be improved by optimizing the powder processing as well as annealing conditions.

REFERENCES

1. J. G. Bednorz and K. A. Muller, Z. Phys. B64, 189 (1986).
2. M. K. Wu, J. R. Ashburn, C. T. Torng, P. H. Hor, R. L. Meng, L. Gao, Z. J. Huang, Y. Q. Wang and C. W. Chu, Phs. Rev. Lett. 58, 908 (1987).
3. H. Maeda, Y. Tanaka, M. Fukutomi and T. Asano, Jpn. J. Appl. Phys. 27, L209 (1988).
4. Z. Z. Sheng and A. M. Hermann, Nature, 333, 138 (1988).
5. M. Takano, J. Takada, K. Oda, H. Kitaguchi, Y. Miura, Y. Ikeda, Y. Tomii and H. Mazaki, Jpn. J. Appl. Phys. 27, L1041 (1988).
6. J. M. Tarascon, Y. LePage, P. Barboux, B. G. Bagley, L. H. Green, W. R. McKinnon, G. W. Hull, M. Giroud and D. M. Hwang, Phys. Rev. B (submitted).
7. E. Takayama-Muromachi, Y. Uchida, A. Ono, F. Izumi, M. Onoda, Y. Matsui, K. Kosuda, S. Takayama and K. Kato, Jpn. J. Appl. Phys. 27, L365 (1988).
8. T. Yoshitake, T. Satoh, Y. Kubo and H. Igarashi, Jpn. J. Appl, Phys. 27, L1089 (1988).

Formation of Y-Ba-Cu-O Superconducting Films from Chloroform-Formalin Solution of Acetyl Aceton Complexes

NUKIO NISHIYAMA[1], JUNZOH FUJIOKA[1], OSAMU MURATA[1], KOHJI NISHIO[1], HITOSHI TABATA[1], and TAKEO YOSHIHARA[2]

Technical Institute[1] and Steel Structure & Industrial Equipment Division 2, Kawasaki Heavy Industries, Ltd., 1-1, Kawasaki-cho, Akashi, 673 Japan

ABSTRACT

A thick film of $YBa_2Cu_3O_{7-x}$ with T_c=73K is obtained by a spray pyrolysis method using mixed solutions of Ba- and Y-acetyl aceton complexes in formalin and a Cu-acetyl aceton complex in chloroform. This film is dense and oriented with c-axis perpendicular to substrate surface. The composition of $Ba(ClO_3)_2$ which lowers T_c is often observed in the film prepared by this method.

INTRODUCTION

Since Bednorz and Muller reported the possibility of high T_c superconductivity in La-Ba-Cu-O system〔1〕, various researches have been made ,and critical temperatures (T_c) exceeded the liquid nitrogen temperature in the Y-Ba-Cu-O system. Ceramic superconductors are, however, hard to be fabricated ,as is often the case in other ceramic materials. This is one of the most difficult but challenging subjects for oxide superconductors to be put into commercial use. For this purpose, various studies have been made on the development of processing techniques. Among them ,a spray pyrolysis method using organometallic compounds is considered to be one of the most effective methods because of its exellence of uniform composition in the film ,easiness of producing films on complex shaped parts and unnecessity of expensive equipments.

Several papers have appeared reporting successful preparations of Y-Ba-Cu-O films by a painting pyrolysis technique using solutions of organometallic compounds. The reagents used in the methods were organic acid salt in n-butanol or toluene〔2〕 ,propoxide or acetyl aceton complex for yttrium, etoxy ethyl alkoxide or acetyl aceton complex for copper and metallic barium in 2-metoxy ethanol or 2-etoxy ethanol〔3〕,and yttrium butoxide in xylene and Cu-,Ba-methoxides in triethanol and methanol〔4〕.

We report here that a thick film of $YBa_2Cu_3O_{7-x}$ showing superconducting properties can be successfully formed by the spray pyrolysis method using the solution of Ba- and Y-acetyl aceton complexes in formalin and a Cu-acetyl aceton complex in chloroform.

EXPERIMENTAL

We have investigated various kinds of solvents for acetyl aceton complexes of yttrium, barium and copper, and consequently found a new solution ; Ba- and Y-acetyl aceton complexes in formalin and Cu-acetyl aceton complex in chloroform. Using the process shown in Fig. 1, the atomic ratio of yttrium, barium and copper in the solution was adjusted to give a compound with ratio of 1:2:3 respectively after sintering.

Before making films, differential thermal analysis (DTA) and thermogravimetric analysis (TG) were performed in order to understand the behavior of the thermal decomposition process. The original solution(10mg) was analyzed at a heating rate of 5°C/min from room temperature to 1000°C.

The starting solutions were sprayed on the yttria stabilized zirconia (YSZ) substrates heated at 200 to 400·C for 6min, decomposed at 500·C for 2h , preheat treated at 850·C for 0.5h , sintered at 880 to 960·C for 1 to 15mins and finally annealed at 550·C forl to 8.5h. The temperature of decomposition was determined from results of thermal analysis. The optimum substrste temperature was determined by observation on the surface morophology after decomposition. After annealing , films were investigated by T_c measurement , x-ray diffraction analysis , surface morphology observation and chemical analysis.

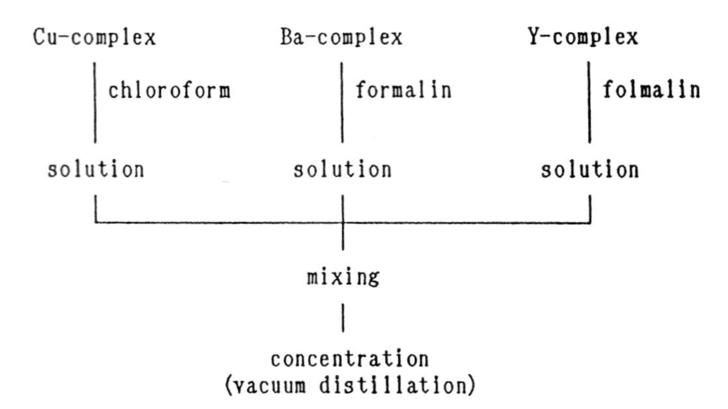

Fig. 1 Process of preparation of the starting solution from acetyle aceton complexes.

RESULTS AND DISCUSSION

In various kinds of solvents , formalin was found to be a good solvent for Ba-and Y-acetyl aceton complexes. A solution of the Ba-acetyl aceton complex was transparent and an emulsive solution of the Y-acetyl aceton complex was homogeneous and stable. Chloroform was found to be a good solvent for Cu-acetyl aceton complex and the solution exhibited blue color of Cu ion. The mixture of these three solutions separated into two layers, and Cu ions moved from the under layer of chloroform to the upper layer of formalin. This mixed solution lost the chlorofolm layer by vacuum distillation , resulting in formation of sol in the upper layer. As the formaldehyde in formalin consists of hydromonomer-methyleneglycol : $CH_2(OH)_2$ and hydropolymer-polyoxy methylene glycol : $(CH_2O)_n \cdot H_2O$ [5] , these glycols seems to be associated with stabilization and gelling of Cu-complex.

Fig. 2 shows the results of the thermal analysis on the solutions. Endothermic reactions occur at 50 and 85·C seemingly due to evaporation of the solvent and a large amount of weight loss occurs until 150·C. An exothermic peak appers at 393·C resulting supposedly from thermal decomposition of the complexes and a small amount of weight losss occurs at 390 to 400·C. The endothermic peak is observed around 870·C probably due to decomposition of $BaCO_3$. It is considered from the results of thermal analysis and IR absorption spectra that the solvents are evaporated until about 150·C , acetyl aceton complexes are decomposed thermally until about 400·C and the decomposition of $BaCO_3$ occurs around 870·C.

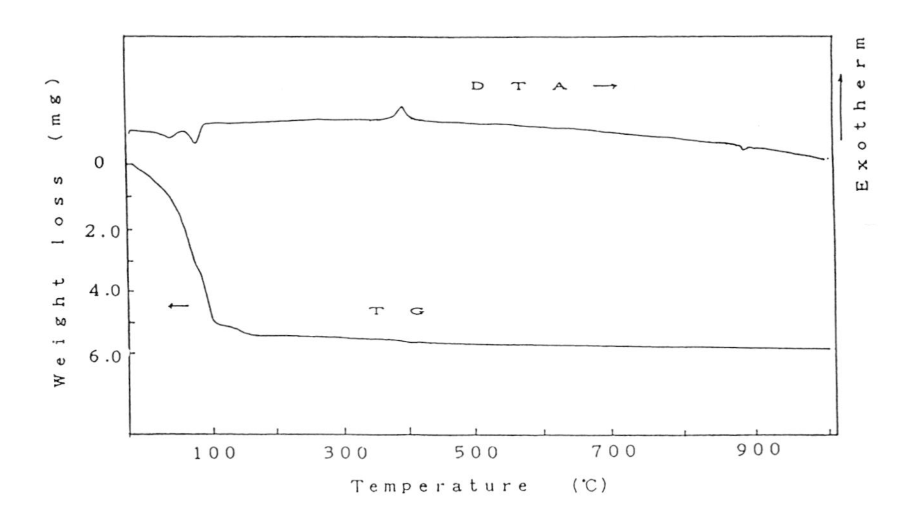

Fig. 2 DTA and TG curves of the starting solution measured at heating rate of 5·C/min

Fig.3 Optical micrscope photographs of surface structures of films sprayed for 2 min. at temperatures of of 200(a), 250(b) and 350~400·C(c) and decomposited at 500·C.

Fig.3 shows the optical micrscope photographs of surface structures of about 1μm thick films sprayed at various substrate temperatures after decomposition at 500·C. The most favorable films are obtained at the substrate temperature of 250·C. Cracking occurs at 200·C and decomposed materials agglomerate at 350 to 400·C.

Fig.4 shows the dependency of resistivity on temperature. The $YBa_2Cu_3O_{7-x}$ films with about 3μm thickness was sprayed on the YSZ substrate at 250·C, decomposed at 500·C for 2h, preheattreated at 850·C for 0.5h, sintered at 880·C(sample A) or 920·C(sample B) for 5min and annealed at 550·C for 2h (sample A) or 5h(sample B). The T_c of the samples A and B are 73K and 45K, respectively. It is presumed from the resistivity curves that semiconducting phases exist in both samples and the amount of their phases in sample A is smaller than that in sample B.

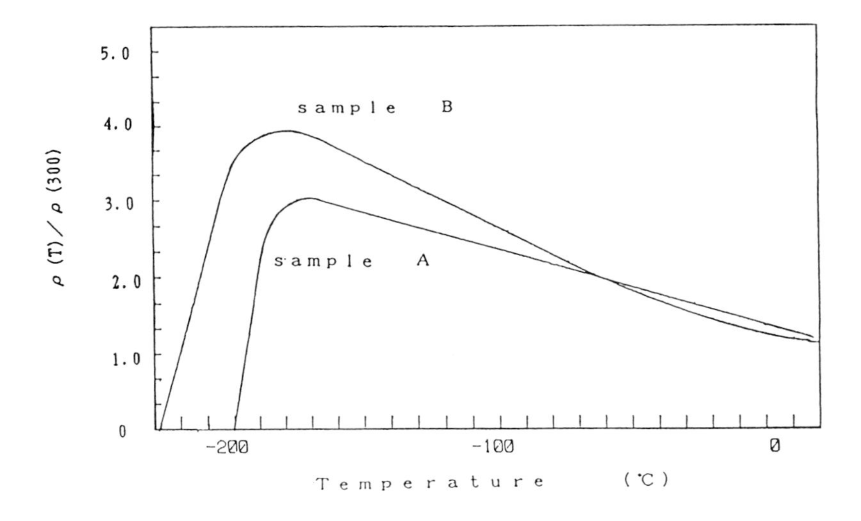

Fig.4 Temperature dependence of the resistivity of $YBa_2Cu_3O_{7-x}$ films.

The x-ray diffraction patterns of samples A and B in Fig.5 show the presence of $YBa_2Cu_3O_{7-x}$ and $Ba(ClO_3)_2$. The peak of $Ba(ClO_3)_2$ is lower in sample A than that in sample B. Except for the peak of $Ba(ClO_3)_2$, the diffraction lines of the (001) reflections of $YBa_2Cu_3O_{7-x}$ are dominantly observed in sample A. Thus, sample A has a preferred orientation with c axis perpendicular to substrate surface. In sample B, the lines of $YBa_2Cu_3O_{7-x}$ except for (001) orientation are observed. It is expected that Cl is mixed in the form of $Ba(ClO_3)_2$ from the chloroform used as the solvent.

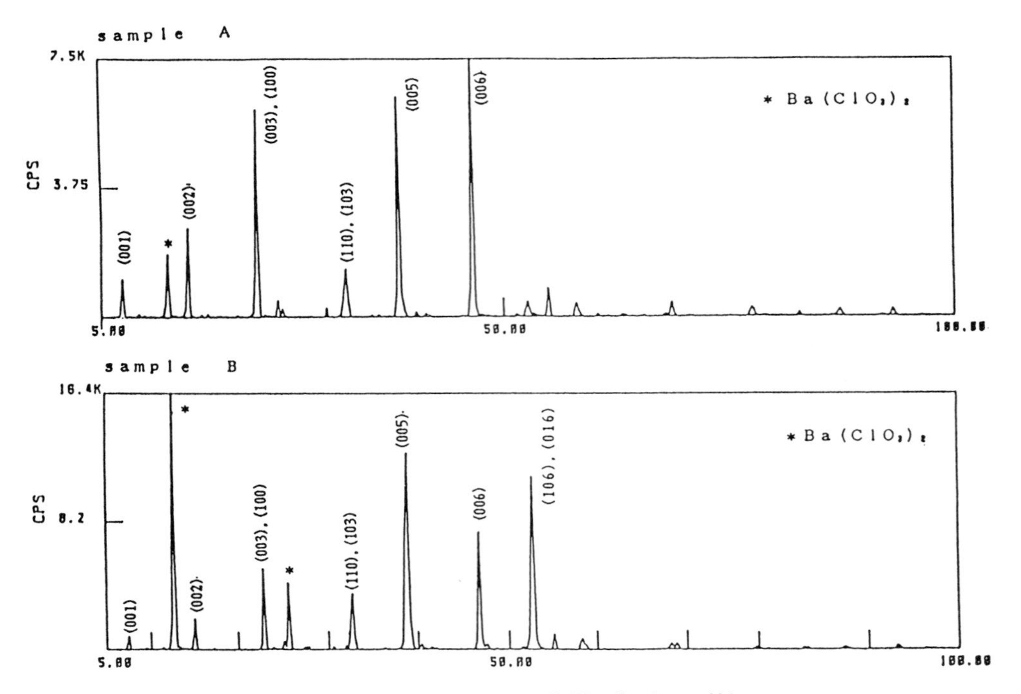

Fig. 5 X-ray diffraction patterns of $YBa_2Cu_3O_{7-x}$ films.

Fig. 6 shows the secondary electron image and the distribution of yttrium and copper on sample A. The dense film with a collection of uniform, planar grains, most of which lie approximately parallel to the surface are observed. However, grains standing in slightly different directions are also observed, which shows that (001) alignment is not perfect. The dimension of the grains is about 5μm, and a few small particles less than 1μm are also seen. The morphology of sample B was almost same as that of sample A, but grains of the (001) direction in Sample B was outnumbered those in sample A. The grain size of sample B was about 10μm with a few less than 2μm. Difference in grain size between the two samples appears to result from difference in sintering temperatures.

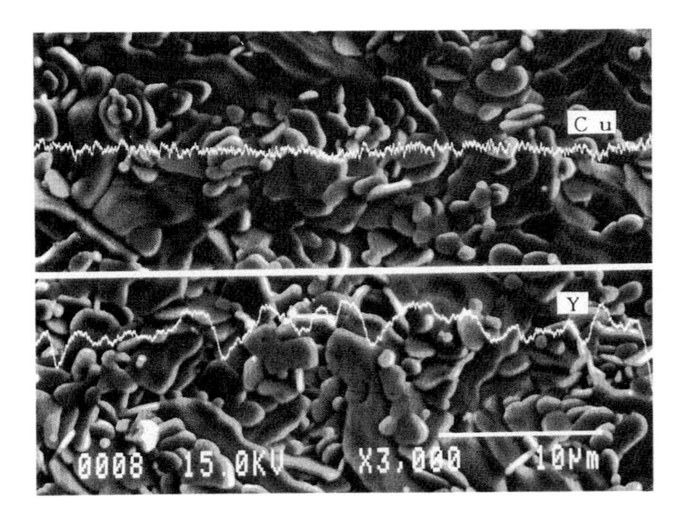

Fig. 6 Secondary electron image of sampl A.

From the results of these investigations, one of the most probable reasons for difference of T_c between the two samples is considered to be attributed to the difference in the amount of $Ba(ClO_3)_2$.

We wish to thank T. Kawai (Osaka Univ.) for helpful advice, and H. Saito for experiments

[1] J. G. Bednorz and K. A. Muller: Z. Phys., B64 (1986), 139. [2] T. Kumagai et al: J. Japan Ceramics Soc., 98 [4] (1988), 417. [3] S. Hirano et al : Proc. Symp. Japan Soc. Powder and Powder Metallurgy, (1988. 4), p16. [4] S. Sakuhana and H. kouzuka: J. Japan Soc. Powder and Powder Metallurgy, 98 (1988), 339. [5] M. Imoto et al : Formaldehide, ('65. 4), Asakura Syoten, p12.

Fabrication of High Tc Superconducting Films by Spray Pyrolysis

HITOSHI NOBUMASA[1], KAZUHARU SHIMIZU[1], TAKAHISA ARIMA[1], KIICHIRO MATSUMURA[1], YUKISHIGE KITANO[2], and TOMOJI KAWAI[3]

[1] Composite Material Laboratory, Toray Ind. Inc., 3-2-1, Sonoyama, Otsu, Shiga, 520 Japan
[2] Toray Research Center Inc., 1-1-1, Sonoyama, Otsu, Shiga, 520 Japan
[3] The Institute of Scientific and Industrial Research, Osaka University, Ibaraki, Osaka, 567 Japan

ABSTRACT

A 100K superconducting Bi(Pb)-Sr-Ca-Cu-O film was formed on a MgO (100) single crystal by a spray pyrolysis method. Fifteen hours heating of the as-sprayed film at 845°C in air was enough to give the superconducting film with a Tc_{zero} higher than 100K. An X-ray diffraction pattern showed that this film mainly consisted of the high-Tc phase with the orientation of the c-axis perpendicular to the surface. Furthermore a microstructure having triple Cu-O layers without intergrowth was directly proved by high resolution TEM.

INTRODUCTION

The Bi-Sr-Ca-Cu-O superconductor which was found by Maeda et al.[1] has always consisted of the mixture of high-Tc , lower-Tc and other phases. For practical uses , the high-Tc phase should be increased , and some attempts have been made[2,3,4,5,6]. In these experiments, a prolonged heat treatment with a narrow temperature range was needed.

When Bi is partially substituted by Pb [Bi(Pb)-Sr-Ca-Cu-O] in the bulk ceramic samples[7,8,9,10,], the high-Tc phase can be increased.

For the practical application of this Bi-Sr-Ca-Cu-O superconductor, film formation is essential, and the method should be speedy, simple and low in cost. A spray pyrolysis method [11] has the following advantages and is suitable for such film formation : the film is homogeneous and stoichiometric and the method is applicable in air and can produce an oriented film.

In this letter, we report the formation of a Bi(Pb)-Sr-Ca-Cu-O superconducting film made by the spray pyrolysis method. The film consists of the high-Tc phase with the Tc_{zero} higher than 100 K and the c-axis is oriented perpendicular to the substrate surface. The ratio of the high-Tc phase to the low-Tc one is 9.6 . The microstructure of this film has triple Cu-O layers without intergrowth.

EXPERIMENTAL

An aqueous solution of Bi,Pb,Sr,Ca and Cu nitrates was prepared from $Bi(NO_3)_3 5H_2O$, $Pb(NO_3)_2$, $Sr(NO_3)_2$, $Ca(NO_3)_2 4H_2O$ and $Cu(NO_3)_2 3H_2O$ by dissolving them into triply distilled water (25mmol/100cc) to give the atomic ratio of Bi, Pb, Sr, Ca and Cu to be 0.8:0.2:0.8:1.7:1.6. The aqueous solution was sprayed over a single crystal of MgO(100) substrate which was kept at 400°C on a hot plate. The distance between the substrate and the spray nozzle(0.3mmφ) was about 20 cm. The carrier gas for the spray was O_2 with the compressive pressure of 1 kg/cm^2. The solution of 20ml - 50ml was sprayed for one run. Finally, the hot plate was turned off, and the substrate was cooled to room temperature, taking 1 - 2 hours. These samples were piled up and heated at 845°C for 2h, 5h, 10h, 15h, 20h, 40h or 80h in air, and then cooled to room temperature in a furnace.

The resistances (R) of the Bi(Pb)-Sr-Ca-Cu-O films were measured versus temperature (T) with a standard four-point method at a current density of 100 mA/cm^2. Temperature was measured by using a calibrated Au-Fe/chromel thermocouple. The X-ray diffraction was carried out with Cu-Kα radiation. The atomic ratio of Bi, Pb, Sr, Ca and Cu in the films was determined by an inductively coupled plasma atomic emission spectrometry (ICP). The microstructure was proved by high resolution TEM. The film thickness was measured by SEM to be 40μm - 50μm.

RESULTS AND DISCUSSION

The electrical resistance ratios (R.R) of these samples are shown in Fig.1. The high-Tc phase above100 K appeared more clearly with a longer heating time. When the heating time was 15h, thelow-Tc phase in the R.R-T curve was not observed, and the Tc_{zero} at 100 K

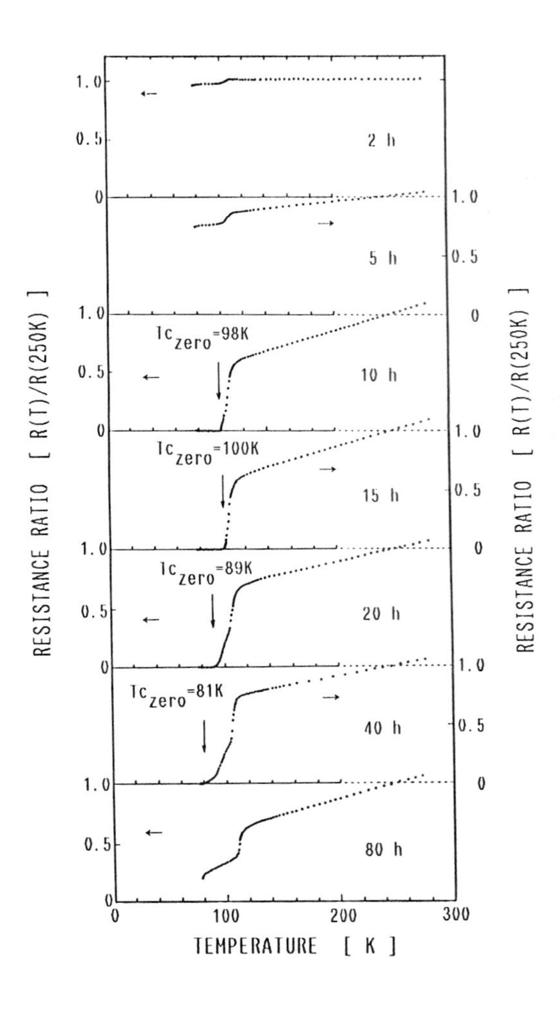

Fig.1

Temperature dependences of the electric resistance ratios of Bi(Pb)-Sr-Ca-Cu-O films on MgO heated at 845°C in air for seven different periods;2h, 5h, 10h, 15h, 20h, 40h and 80h.

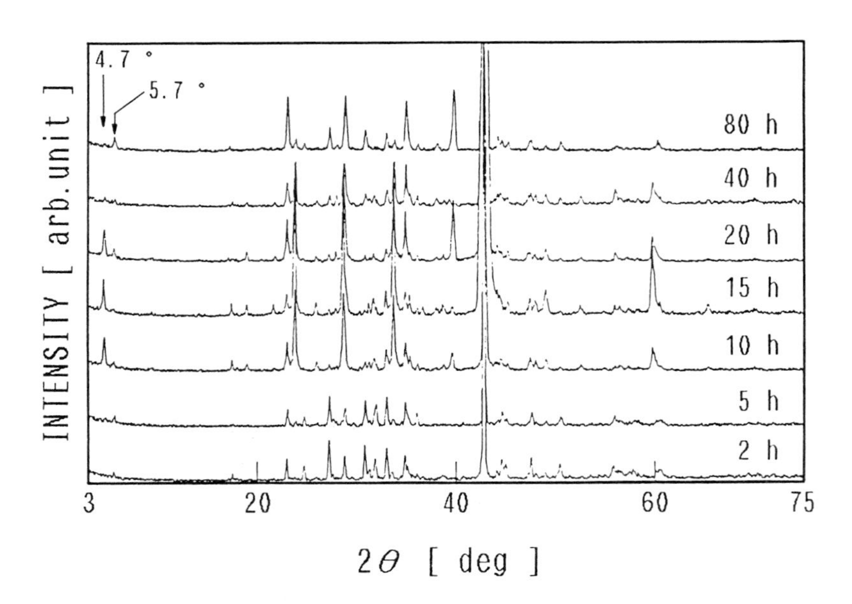

Fig.2 X-ray diffraction patterns of the Bi(Pb)-Sr-Ca-Cu-O films samples shown in Fig.1.

was obtained. When the heating time became longer than 15h, the lower-Tc phase appeared again, and Tc_{zero} became lower.

The X-ray diffraction patterns of these samples (Fig.2) showed a clear correlation with the R.R-T curves. As it was already known, the peak of 2θ=5.7° corresponded to the low-Tc phase with double Cu-O layers[12], the 7.2° peak to a 7 K (or semiconducting) phase with single Cu-O layer and the 4.7° peak to the high-Tc phase with triple Cu-O layers [2,3,4]. The 7.2° peak was not found in any diffraction pattern, so the relative intensity of the 4.7° peak to the 5.7° peak ($I_{4.7°}/I_{5.7°}$) was examined. The $I_{4.7°}/I_{5.7°}$ value became higher with the longer heating time, as seen in Fig.3, and took the highest value ($I_{4.7°}/I_{5.7°}$ = 9.6) when the heating time was 15h. However, a heating time longer than 15 h lowered the $I_{4.7°}/I_{5.7°}$ value. These results correspond very well to the relationship between the R-T curve and the heating time.

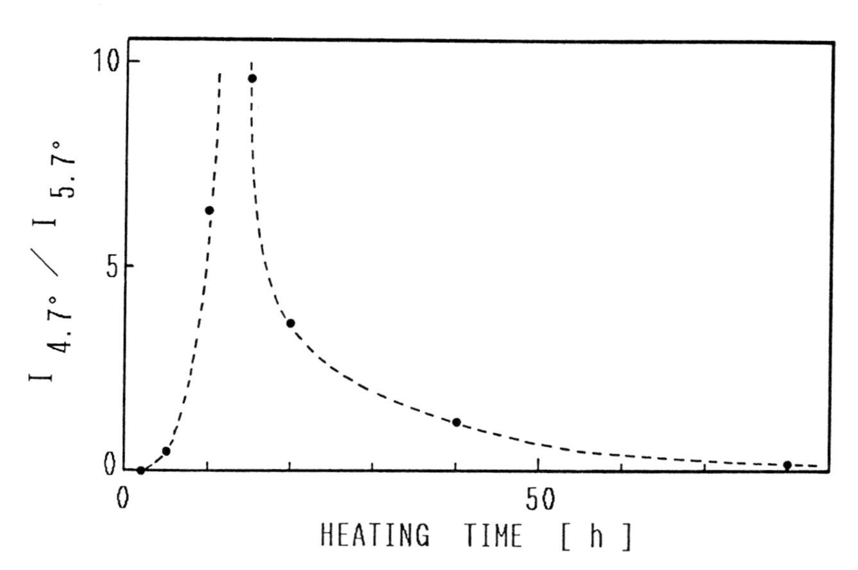

Fig.3 The ratio of the peak intensity of 2θ=4.7° to 2θ=5.7° X-ray diffraction ($I_{4.7°}/I_{5.7°}$) versus the heating time.

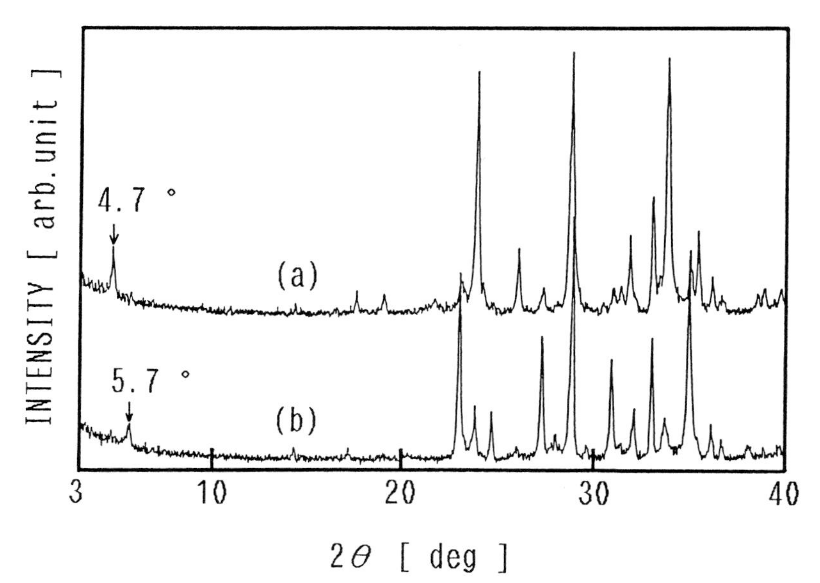

Fig.4 X-ray diffraction patterns of the Bi(Pb)-Sr-Ca-Cu-O films
(a) The sample was piled up during the heating.
(b) The sample was heated without piling up.

Besides the heating time and the heating temperature, it was important for increasing the high-Tc phase that the samples were piled up during the heating. When the films were exposed to air directly and heated, the high-Tc phase was not so dominant. This seems to indicate that the oxygen-deficient condition is good for making the high-Tc phase of Bi(Pb)-Sr-Ca-Cu-O superconductor. Fig.4 shows the effectiveness of heating in piles.

In Fig.5 is shown in detail the X-ray diffraction pattern for the sample of 15 hours heating, in which the proportion of the high-Tc phase was the highest. The peaks of the high-Tc phase were indexed by the calculation assuming that this phase had a tetragonal unit cell with a=b=5.396 Å and c=37.140 Å with the P4/mmm space group. The diffraction lines due to the (00l) index were strongly detected. This result indicated the formation of an oriented high-Tc superconducting film on the substrate. The formation of this c-axis orientation is very sensitive to the heating temperature and profiles. The main X-ray diffraction pattern of this film corresponded to the high-Tc phase, although there were a few other peaks. These other peaks showed that this sample included a small amount of the low-Tc phases , Ca_2PbO_4 and CuO.

The atomic ratio of Bi, Pb, Sr, Ca and Cu in the film was analysed by ICP, and was assigned to 0.8:0.09:0.73:1.32:1.7. Comparing this value with the starting composition in the aqueous solution, the percentage of Pb, Sr and Ca decreased. Especially the decrease of Pb was remarkable. After 80 hours of heating, the ratio of Pb to Bi became only 0.01. If the high-Tc phase contained triple Cu-O layers between Bi_2O_2 layers and Pb substitutes for Bi, the atomic ratio of Bi(Pb), Sr, Ca and Cu would be 2:2:2:3. Accordingly the amount of Bi(Pb), Ca and Cu were too much, although Sr might have been partially substituted by Ca. This result corresponded to the existence of the peaks of Ca_2PbO_4 and CuO compounds in the X-ray diffraction pattern.

There have been four reports in which a Tc_{zero} of more than 100 K was achieved on the form of films. Two of these were accomplished by using rf magnetron sputtering[13,14]. The film showed a broad peak at around $2\theta=4.7°$ in the X-ray diffraction pattern. This indicated that many phases were mixed on a microscale. The other films were made by screen printing[15] or electron-beam deposition[16]. The X-ray diffraction patterns of the films indicated that the volume fraction of the high-Tc phase was less than 50%.

On the contrary , the film made in this study had a high and sharp peak at $2\theta=4.7°$ with a very small peak at 5.7°. The ratio of the peak of the high-Tc phase to that of low-Tc one was 9.6 . Fig.6 shows the lattice image of this film which was proved directly by high resolution TEM. The microstructure has triple Cu-O layers with c/2=18 Å . Any intergrowths has not been found all over the sample. The partial substitution of Pb for Bi promoted the formation of the high-Tc phase.

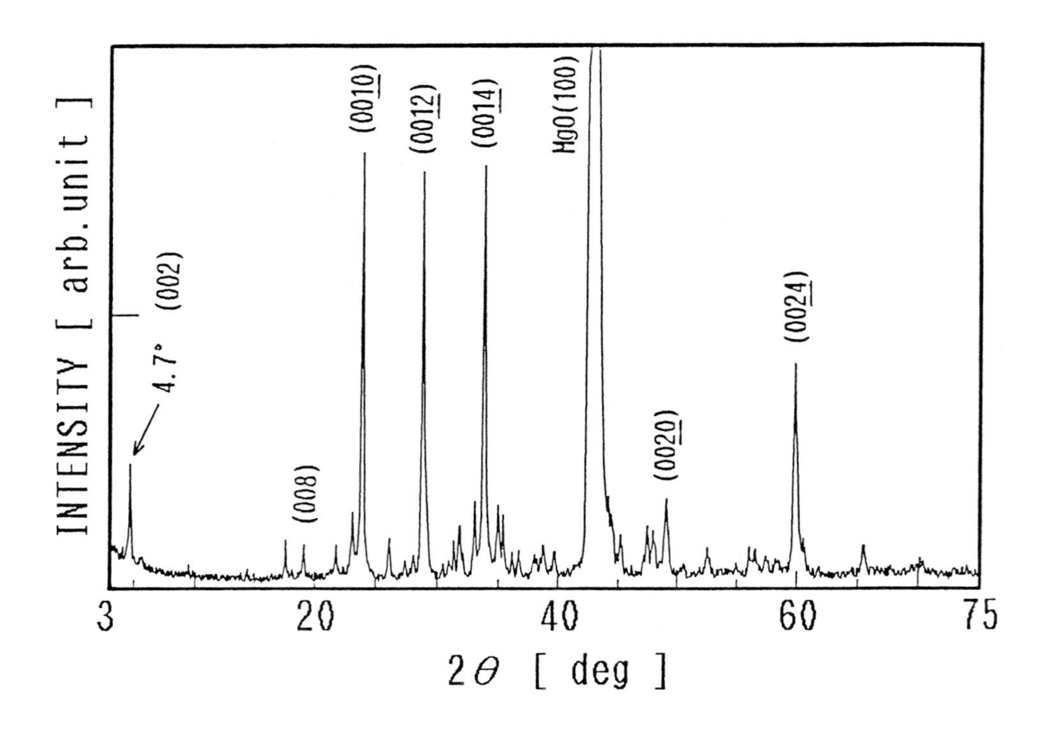

Fig.5 X-ray diffraction pattern of the sample heated for 15 h.

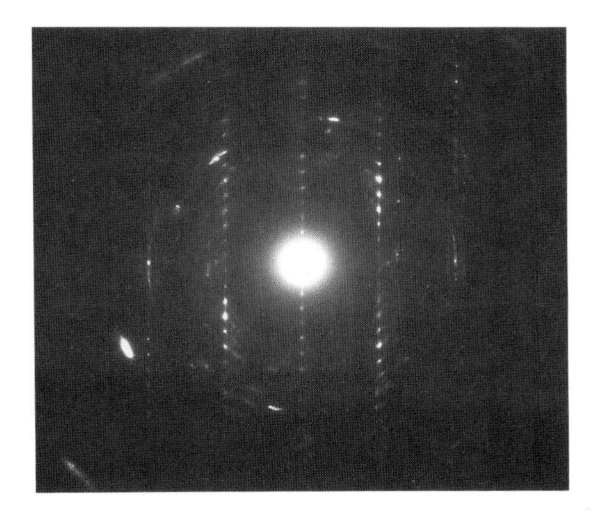

Fig.6 High-resolution electron microscope images and electron diffraction pattern of the Bi(Pb)-Sr-Ca-Cu-O films.

CONCLUSION

We obtained a film mainly consisting of the high-Tc phase in as short a heating time as 15 hours by the spray pyrolysis method. When the heating time became longer than that, the high-Tc phase began to decrease and the low-Tc phase appeared again. When the heating time and the starting composition in the aqueous solution are controlled more strictly, a perfect single high-Tc phase will be obtained, and Tc_{zero} will get to 110K**

**Most recently, by controlling the atomic ratio of Bi, Pb, Sr, Ca, and Cu in the aqueous solution, Tc_{zero} has got to 103 K as shown in Fig.7.

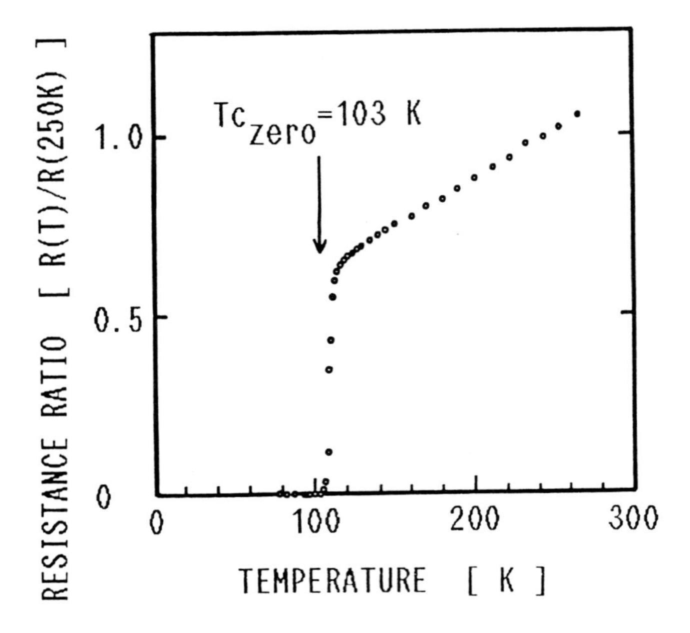

Fig.7 Temperature dependence of the electric resistance ratio of the highest Tc_{zero} Bi(Pb)-Sr-Ca-Cu-O film formed by a spray pyrolysis method.

ACKNOWLEDGMENT

The authors wish to thank M.Nishina at Toray Ind. Inc. and Y.Murata at Toray Research Center Inc. for their support in the experiments.

REFERENCES

[1] H.Maeda,Y.Tanaka,M.Fukutomi and T.Asano : Jpn. J. Appl.Phys. 27 (1988) L209.
[2] H.Nobumasa,K.Shimizu,Y.Kitano and T.Kawai : Jpn. J. Appl.Phys. 27 (1988) L846.
[3] K.Kitazawa,S.Yaegashi,K.Kishio,T.Hasegawa,N.Kanazawa,K.Park and K.Fueki : to be published in Adv. Ceram. Mater. .
[4] K.Kuwahara,S.Yaegashi,K.Kishio,T.Hasegawa and K.Kitazawa : to be published in Proc. Latin-American Conf. on High Temperature Superconductivity.
[5] A.Sumiyama,T.Yoshitomi,H.Endo,J.Tsuchiya,N.Kijima,M.Mizuno and Y.Oguri: Jpn. J. Appl.Phys. 27 (1988) L542.
[6] N.Kijima,H.Endo,J.Tsuchiya,A.Sumiyama,M.Mizuno and Y.Oguri : Jpn. J. Appl.Phys. 27 (1988) L821.
[7] R.J.Cava:1988 Spring Meeting of the American Physical Society,March (1988).
[8] C.Politis:MRS 1988 Spring Meeting, Reno, April 6 (1988).
[9] M.Takano,J.Takada,K.Oda,H.Kitaguchi,Y.Miura,Y.Ikeda,Y.Tomii and H.Mazaki : Jpn. J. Appl. Phys. 27 (1988) L1041.
[10] H.Nobumasa,K.Shimizu,Y.Kitano and T.Kawai : MRS Int. Meeting on Advanced Materials,Tokyo, June 2 ,1988,(Materials Research Society,Pittsburgh,in press).
[11] M.Kawai,T.Kawai,H.Masuhira and M.Takahasi : Jpn. J. Appl.Phys. 26 (1987) L1740.
[12] M.Onoda,A.Yamamoto,E.Muromachi and S.Takekawa : Jpn. J. Appl.Phys. 27 (1988) L833.
[13] Y.Ichikawa,H.Adachi,K.Hirochi,K.Setsune and K.Wasa:MRS Int. Meeting on Advanced Materials, Tokyo,June 2 ,1988, (Materials Research Society,Pittsburgh,in press).
[14] M.Fukutomi,Y.Tanaka,T.Asano,H.Wada and H.Maeda : MRS Int. Meeting on Advanced Materials, Tokyo,May 31 ,1988, (Materials Research Society,Pittsburgh,in press).
[15] K.Hoshino,H.Takahara and M.Fukutomi: MRS Int. Meeting on Advanced Materials,Tokyo,June 2 , 1988,(Materials Research Society,Pittsburgh,in press).
[16] T.Yoshitake,T.Satoh,Y.Kubo and H.Igarashi : Jpn. J. Appl.Phys. 27 (1988) L1089.

Preparation of Superconducting Thick Films of Y-Ba-Cu-O by Gas Deposition of Ultrafine Powder

K. HATANAKA[1], M. KAITOU[2], M. UMEHARA[2], S. KASHU[2], and C. HAYASHI[2]

[1] ULVAC JAPAN, Ltd., 2500 Hagisono, Chigasaki, 253 Japan
[2] Vacuum Metallurgical Co.,Ltd., 516 Yokota, Sanbu-machi, Chiba, 289-12 Japan

ABSTRACT

Superconducting thick($\sim$20 μm) films of $YBa_2Cu_3O_x$ have been successfully made by gas deposition of ultrafine(<0.1 μm) powder of Y, Ba and Cu on MgO substrates. After high temperature anneal, films are superconducting with $T_c \sim 76$ K and $\Delta T_c \sim 5$ K. Because of excellent sintering properties of ultrafine powder, critical current densities are relatively large (in excess of 10000A/cm^2 at 4.2 K) and resistivities at room temperature are about 0.3 mΩcm.

INTRODUCTION

Since the discovery of high T_c oxide superconductors[1], a great deal of activity has been made to apply these oxides to various technical applications. Although much effort has been devoted to fabricate epitaxial thin films by sputtering[2] or electron beam evaporation[3], high quality superconducting thick(10 to 100 μm) films[4,5,6] are also needed for some technical applications such as grain boundary device, magnetic shield, and superconducting wire or tape fabrications.

In this report, we describe the fabrications of superconducting thick films of $YBa_2Cu_3O_x$ by gas deposition[7] of ultrafine powder(UFP)[8]. This process will probably produce denser, more uniform thick films than those obtained from other methods.

EXPERIMENT

A. Preparation of films

Figure 1 shows the gas deposition system schematically. UFP of Y, Ba and Cu were generated by gas evaporation method[8] from metal sources and transported to the deposition chamber by means of gas flow caused by a pressure difference. Oxygen gas was charged to improve the oxidation of deposited films. Mixed UFP of Y, Ba, and Cu with the carrier Ar, He, and O_2 gases were sprayed from 0.8 mm i.d. nozzle onto a (100) MgO substrate, which was fixed on a moving XY table. The pressure of the deposition chamber was typically 2 Torr(0.5 Torr O_2) and the substrate temperature was maintained at 400°C. A line shape film of 20μm x 0.8 mm x 20 mm was deposited in about 5 min.

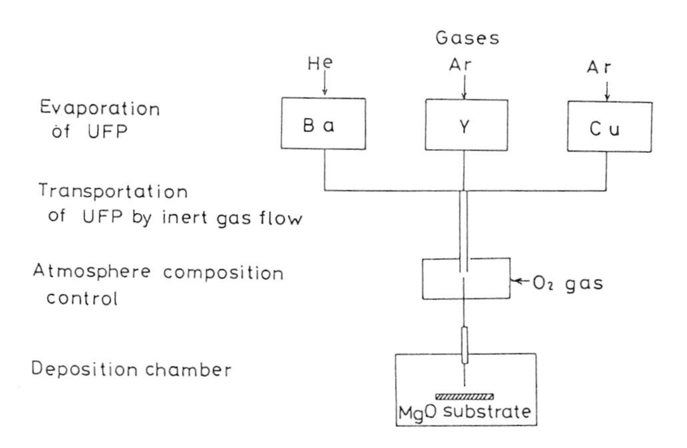

Fig. 1. Schematic diagram of the UFP gas deposition system.

B. Heat treatment

As-deposited films were set into a furnace and heat treated to obtain the superconducting characteristics. Sintering were done between 800°C and 940°C for 1 to 3 hours and cooling rate was about 100°C/h. Oxygen gas was flowed in all heat treatment periods.

C. Transition temperature and normal state resistivity

The transition temperatures and the normal state resistances were determined on all samples by a standard dc four probe methods in a special cryostat modified from cryopump. Temperature were determined by calibrated platinum thermometer[9]. Distance between each probe is 5 mm. Cross-sectional area of films were calculated by monitoring film thickness. Transition temperatures were also determined inductively using Lindsay bridge[10] operating at 20 kHz.

D. Critical current densities

Critical current data were obtained using four probe transport measurement. The probes were the same as for the resistance measurements. We defined critical current as the current necessary to produce an electric field equal to one micro volt/cm. All measurements were carried out in zero magnetic field.

RESULTS AND DISCUSSION

A. Film characterization

Figure 2 shows typical X-ray diffraction patterns of a film before (a) and after (b) heat treatment at 930°C for 1 hour. Strong diffraction peaks corresponding to (00ℓ) indices shows the formation of a c-axis oriented film of $YBa_2Cu_3O_x$.

Figure 3 shows an example of the back-scattered electron image of a heat treated sample. On the left side of the photograph, stratums caused by grain growth were observed. Wide white stripes are the surface steps of unpolished MgO substrate.

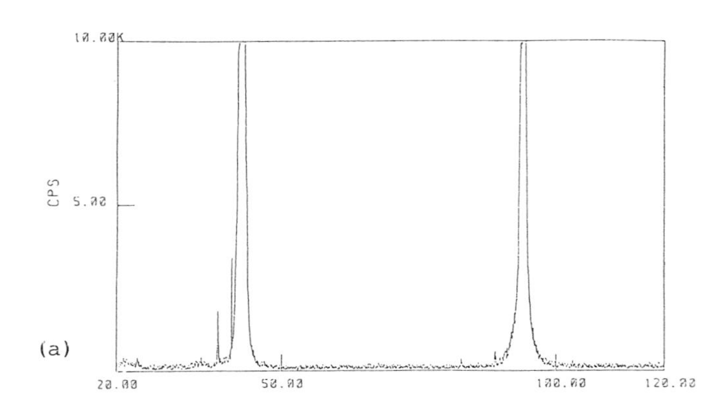

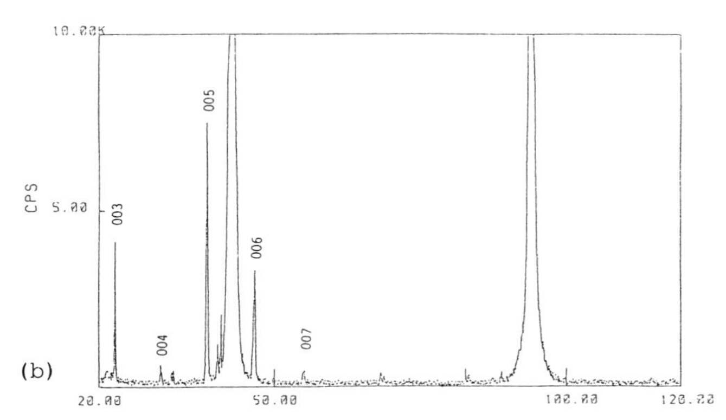

Fig. 2. X-ray diffraction pattern of before(a) and after(b) 930°C x 1h heat treatment for sample No.202-2.

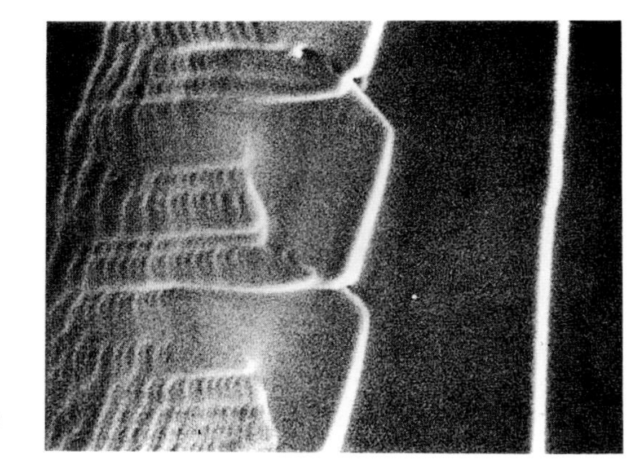

Fig. 3. The back-scattered electron image of heat treated sample No.125-2.

B. Transition temperature and normal state resistivity

Figure 4 shows the temperature dependence of the resistivity of the heat treated films. There is a nearly linear temperature dependence between T_c and room temperature. The resistivities at 300 K is 1-5 mΩcm and for some samples about 0.3 mΩcm. These films were stable, and we found that T_c had decreased only slightly in two months. The Meissner effect was observed for all samples shown in Figure 4.

The temperature dependence of the resistivity and the susceptibility of the sample 125-2 is compared in Figure 5. We notice that the onset temperature of the susceptibility measurement is in good agreement with the endpoint of the resistivity measurement.

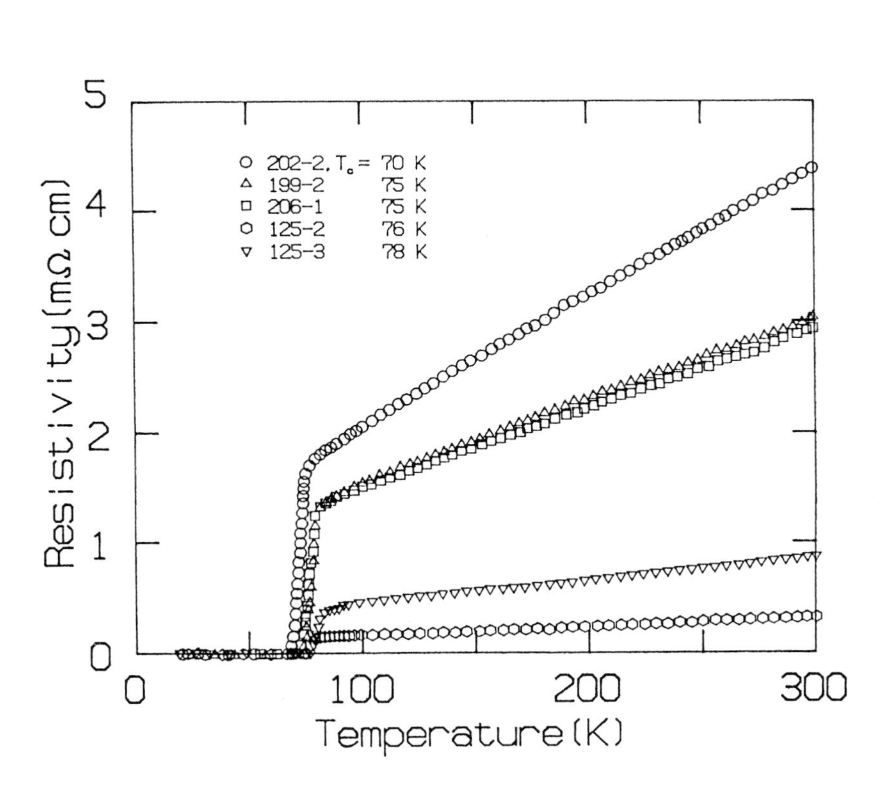

Fig. 4. Temperature dependence of the resistivity of Y-Ba-Cu-O thick films.

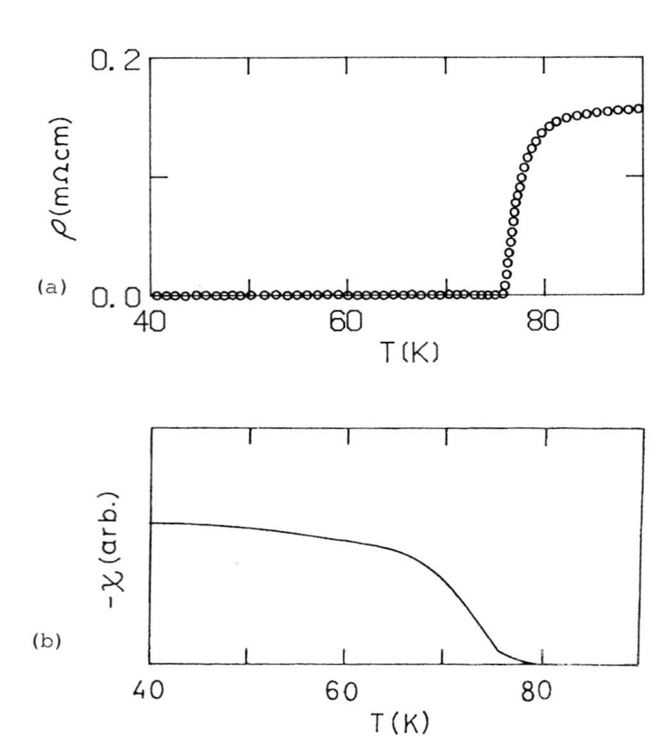

Fig. 5. The temperature dependence of the resistivity (a) and susceptibility (b) of sample No. 125-2.

C. Critical current densities

The temperature dependence of critical current density is shown in Fig 6. In one of these samples(No.162-3) the critical current density at liquid helium temperature was in excess of 10000 A/cm^2. Although critical current densities at liquid nitrogen temperature is zero or very low at present, we believe the gas deposition of UFP can be a useful technique to increase the critical current density of the oxide superconductors.

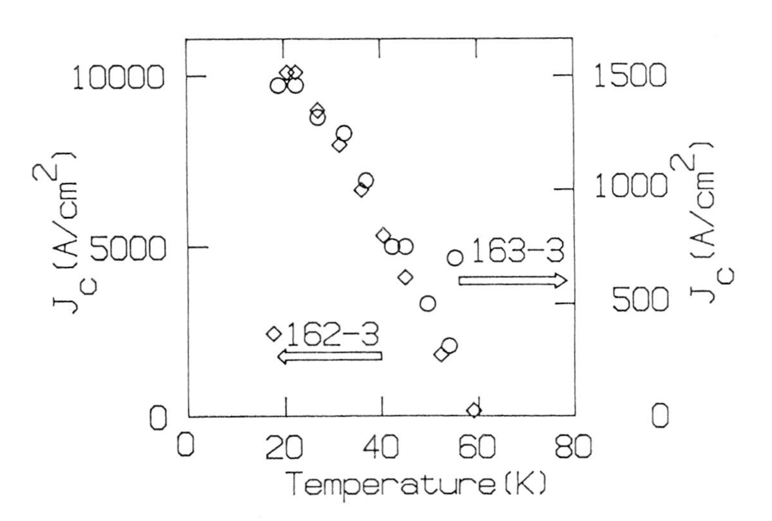

Fig. 6. The critical current density vs temperature for two samples measured directly.

CONCLUSIONS

In conclusion, we have grown preferentially oriented $Y_1Ba_2Cu_3O_x$ films of $T_c \sim 76K$ on a (100) MgO substrate using the gas deposition method. Although critical temperatures are slightly lower than bulk sample or carefully deposited thin films, stability of the film and the normal state resistivity measurements showed the favorable characteristics of our films. Such results make gas deposition of UFP extremely promising for future applications of high-temperature superconducting oxides.

ACKNOWLEDGMENTS

The authors would like to thank T. Ohtsuka, H. Yamakawa and H. Naruse for their valuable discussions, K. Onoue for EPMA analysis, K. Sekino for X ray diffraction, and S. Hukui for susceptibility measurements. Part of this work was an extension of Hayashi UFP Project of Exploratory Research for Advanced Technology organized by Research and Development Corporation of Japan(RDCJ).

REFERENCES

[1] J. B. Bedonorz and K. A. Muller: Z. Phys. B64 (1986) 189
[2] Y. Enomoto, T. Murakami, M. Suzuki and K. Moriwaki: Jpn. J. Appl. Phys. 26 (1987) L1248
[3] T. Terashima, K. Iijima, K. Yamamoto, Y. Bando and H. Mazaki: Jpn. J. Appl. Phys. 27(1988) L91
[4] H. Koinuma, T. Hashimoto, T. Nakamura, K. Kishio, K. Kitazawa and K. Fueki: Jpn. J. Appl. Phys. 26 (1987) L761
[5] K. Terashima, K. Eguchi, T. Yoshida and K. Akashi: Appl. Phys. Lett. 52(1988) 1274
[6] S. Shibata, T. Kitazawa, H. Okazaki and T. Kimura: Jpn. J. Appl. Phys. 27 (1988) L646
[7] S. Kashu, E. Fuchita, T. Manabe and C. Hayashi: Jpn. J. Appl. Phys. 23 (1984) L910
[8] C. Hayashi: PHYSICS TODAY, December 1987, 1
[9] LAKE SHORE CRYOTRONICS, INC., Model PT-103
[10] VACUUM METALLURGICAL CO.LTD., Model TC-201M

4.6 Wires and Coils

Stabilization of a Superconducting Magnet Wound by a High Tc Oxide Conductor

HIDEAKI MAEDA

Toshiba Research and Development Center, 4-1, Ukishima, Kawasaki, 210 Japan

ABSTRACT

Potential problems of stabilization and protection for a high T_c oxide superconducting magnet are examined. Firstly, both magnetic and mechanical disturbances, appearing during magnet charging, are discussed; allowable conductor dimension is given based on the adiabatic stability criterion. Secondly, effect of cryogenic stabilization is compared with the case of conventional superconducting magnet at 4.2 K. Finally, magnet protection during coil quenching is examined.

INTRODUCTION

After discovery of type II superconductors in 1960s, some magnets were wound by the superconductors which revealed degraded behaviors at low fields; i.e., the magnets became normal (quenched) and burned out at far below the critical current. The degradation is due to "flux jumping", thermal instability of the magnetic fluxes threading through the superconductor. The flux jumping was removed by using a multi- filamentary composite conductor [1].

Large magnets such as those for MHD or fusion are further stabilized by cryogenic stabilization by liquid helium[2]. Smaller high current- density magnets for industrial or medical use are still suffering from premature quenchings at high fields, due to mechanical disturbances by the electromagnetic force[3][4].

The present paper examines potential problems of stabilization and protection of superconducting magnets wound by high T_c superconductors.

PHYSICAL PARAMETERS

The superconductor considered in this paper is $Y_1Ba_2Cu_3O_{7-x}$. The critical current density, J_c, for a bulk sintered sample is less than 10^8 A/ m^2, if J_c is directly measured by the transport current, while Jc measured by magnetization reaches as high as 7 X 10^8 A/m^2.[5] Thus, Jc of 1 X 10^9 A/m^2 is assumed in this paper, which might be attained in the near future.

Physical parameters of NbTi at 4.2 K and $Y_1Ba_2Cu_3O_{7-x}$ at 77 K are listed in Table 1; those for copper stabilizer are also included in the table. Difference between 4.2 K and 77 K is summarized as follows: (a) A volumetric specific heat for $Y_1Ba_2Cu_3O_{7-x}$ is 209 times as high as that for NbTi. (b) Copper volumetric specific heat at 77K is 2000 times of that at 4.2 K.

The thermal and magnetic diffusivities for NbTi, $Y_1Ba_2Cu_3O_{7-x}$ and copper are listed in Table 1. The thermal diffusivity, D_t, for copper at 77 K is 1/ 2391 of that at 4.2 K.

FLUX JUMPING

Stored energy by the shielding current in the superconductor is dissipated into heat by a flux jumping. The D_t/ D_m value, 1.1 X 10^{-6}, for $Y_1Ba_2Cu_3O_{7-x}$ ensures the local adiabaticity during flux jumping; thus, an adiabatic criterion for removing a flux jumping on $Y_1Ba_2Cu_3O_{7-x}$ is[6]

$$\mu_0 Jc^2 w^2/ (4 \gamma C (\theta_c - \theta_0)) < 3 \qquad (1)$$

where μ_0 is the permeability of free space, w is the slab thickness, γ C is the volumetric specific heat, θ_c is a critical temperature and θ_0 is a bath temperature.

Table 1 Physical parameters.

	Thermal conductivity (W/mK)	Density (Kg/m^3)	Specific heat (J/Kg K)	Resistivity (Ω m)
4.2 K[6]				
NbTi	0.11	6.2×10^3	0.87	6×10^{-7}
Copper	1000	8.9×10^3	0.1	1×10^{-10}
77 K[7]				
$YBa_2Cu_3O_{7-x}$	7	6.4×10^3	176	8×10^{-6}
Copper	820	8.9×10^3	200	2.3×10^{-9}

Table 2 Thermal and magnetic diffusivities.

	Thermal diffusivity (m^2/s)	Magnetic diffusivity (m^2/s)
4.2 K		
NbTi	2.0×10^{-5}	0.5
Copper	1.1	8.0×10^{-5}
77 K		
$Y_1Ba_2Cu_3O_{7-x}$	6.7×10^{-5}	6.4
Copper	4.6×10^{-4}	1.8×10^{-3}

Table 3 Critical conductor width for removing flux jumpings.

	NbTi (4.2 K)	$Y_1Ba_2Cu_3O_{7-x}$ (77 K)
Jc (A/m2)	1.0×10^{10}	1.0×10^9
θc(K)	9.2	95
θo (K)	4.2	77
The conductor width	< 50 μm	< 14 mm

Permissible conductor thickness for removing the flux jumping is given in Table 3 for Nbti and $Y_1Ba_2Cu_3O_{7-x}$: the acceptable $Y_1Ba_2Cu_3O_{7-x}$ thickness is 14 mm, much larger than that for Nbti at 4.2 K, 50 μm. Therefore, it is unnecessary to fabricate a multi- filamentary type of conductor.

A solenoid wound by a Nb_3Sn tape conductor tends to be quenched at the coil top or bottom end, where the flux is bent outwards in the radial direction, threading through the broad surface of the conductor. Copper foils are interleaved between Nb_3Sn tapes to damp the flux motion and to cool the conductor (dynamic stabilization); such a flux jumping does not occur in a $Y_1Ba_2Cu_3O_{7-x}$ tape wound magnet, if the tape width is less than 14 mm.

A transport current, supplied by an outer power supply, generates magnetic field in the circumferential direction inside the round conductor. The shielding effect pushes the current to the conductor outer surface. Instability of such a circumferential magnetic flux is called as a self field flux jumping. An adiabatic criterion for removing the self field flux jumping is given as[6]

$$\mu_o Jc^2 d^2/ (4 \gamma C (\theta_c - \theta_o))$$

$$< (-1/2 \ln \varepsilon -3/8 + \varepsilon^2/2 - \varepsilon^4/8)^{-1} \quad (2)$$

where d is the conductor diameter and $\varepsilon = (1 - \text{transport current}/ \text{critical current})^{0.5}$. A maximum permissible *conductor* diameter is given by this criterion. Lower Jc and larger heat capacity at 77 K permit a larger conductor diameter ; e.g. an $Y_1Ba_2Cu_3O_{7-x}$ round conductor with a diameter of 10 mm does not show self field flux jumpings, while an NbTi conductor with a diameter of 1 mm shows a self field flux jumping at 20 % of the critical current at 0 Tesla. Thus, the self field flux jumping is nearly harmless for a high Tc magnet.

MECHANICAL DISTURBANCES

While a superconducting magnet is energized, an electromagnetic force causes conductor motion or epoxy debonding. Though a released energy is smaller than that by a flux jumping, the coil is quenched if an energy margin becomes scarce at high fields.

The maximum mechanical energy stored in an epoxy impregnant is expressed as $\sigma^2 / 2 Y$,where σ is the epoxy yield strength, ~ 100 MPa, and Y is the epoxy Young's modulus,~ 10 GPa[6]. The maximum available heat energy in case of cracking or debonding is 5×10^5 J/m^3, corresponding to the conductor temperature rise of less than 1 K (see Table 1). Thus, it is suggested that the mechanical disturbance does not induce coil quenching.

RECOVERY

The heat generated in the normal zone is removed through conduction along the conductor. If the heat removal rate is more than the generation, the conductor recovers to the superconducting state. The normal zone propagates if its length exceeds the minimum propagation zone (MPZ) given as

$$\pi /2 (k (\theta_c-\theta_g)/ G_c)^{0.5}, \quad (3)$$

$$G_c = J^2 \rho \quad (4)$$

where k is the average thermal conductivity, θ_g is a heat generation temperature, J is the overall current density and ρ is the average resistivity of the conductor. The MPZ at J = 1.0×10^8 A/ m^2 is only 50 μm for the $Y_1Ba_2Cu_3O_{7-x}$, while it is as large as 2 mm for the copper clad $Y_1Ba_2Cu_3O_{7-x}$ conductor with a copper/ superconductivity ratio, ν , Of 2.0. Therefore, recovery by thermal conduction is nearly negligible for the magnet.

If the available cooling by cryogen is much more than the heat generation in the normal zone, the conductor recovers to the superconducting state after any large disturbances. The generated heat in a copper clad $Y_1Ba_2Cu_3O_{7-x}$ at 77 K is 23 times as large as that of a copper clad NbTi, if the copper ratio and the current densities for both magnets are assumed to be the same: On the other hand the minimum heat flux for liquid nitrogen is 3- 6 times as high as that for liquid helium at 4.2 K [8]. Therefore, the current density for the high T_c magnet, operated at 77 K, is less than that for NbTi or Nb_3Sn at 4.2 K, if it is stabilized by cryogenic stabilization. Considering a harmlessness of the magnetic and mechanical disturbances in the magnet winding, cryogenic stabilization may be an unnecessary procedure for the high T_c magnet.

PROTECTION

A normal section is heated during the current decay after coil quenching. The adiabatic temperature rise for the section is calculated as[6]

$$\int J^2\ dt = \int \gamma C/\rho\ dT, \qquad (5)$$

The integrals for $Y_1Ba_2Cu_3O_{7-x}$ at 1.0 X10^8 A/ m^2 shows that the conductor temperature reaches room temperature within 24 μsec after quench initiation, while that for a copper clad conductor, with ν =2, is as long as 7 sec, which is sufficiently long if the normal zone propagates rapidly.

A normal zone propagation velocity under adiabatic condition is[6]

$$(J/\gamma C)(\rho k/(\theta_s-\theta_o))^{0.5}, \qquad (6)$$

where $\theta_s = (\theta_g+\theta_s)/2$. The velocity for the copper clad $Y_1Ba_2Cu_3O_{7-x}$ at 77 K with a current density of 1.0X 10^8 A/m2 is 0.02 m/sec, if ν = 2. The propagation is much slower than that for the NbTi conductor with the same J and ν, 9- 10 m/sec, due to larger heat capacity at 77 K; i.e. the magnet energy is dissipated in a small quenched section, which is gradually overheated, even if $Y_1Ba_2Cu_3O_{7-x}$ is clad by copper. Thus, an active protection circuit, with a switch and an external dump resistor R_d[6], should be used to decay the coil current as fast as possible after the coil quenching with a time constant of L/ R_d.

The higher a dumping voltage, the quicker the coil current decays. As the electrical breakdown voltage for nitrogen gas is several times higher than that for helium gas, the stored energy in liquid nitrogen can be dumped more quickly than in liquid helium.[9]

CONCLUSIONS

(i) The high T_c magnet is operated stably at 77 K, as the magnetic or mechanical disturbances are harmless for coil operation.
(ii) From a view point of coil protection, the conductor should be clad by a stabilizer such as copper or silver. Furthermore, the magnet should be actively protected by a switch and dump resistor.

REFERENCES

(1) P. F. Smith, M. N. Wilson and A. H. Spurway, J. Phys. D, Appl. Phys.3,1561 (1970).
(2) Z. J. J. Stekly and J. L. Zar, IEEE Trans. NS-12, 367 (1965).
(3) H. Maeda, O. Tsukamoto and Y. Iwasa, Cryogenics, vol 22, 287 (1982).
(4) Y. Iwasa, Cryogenics, vol 25, 304 (1985).
(5) E. Shimizu and D. Ito, presented at this conference.
(6) M. Wilson, " *Superconducting magnets*", Clarendon Press Oxford (1983).
(7) M. Ishikawa, Parity, an extra issue No 4, 135, in Japanese (1988).
(8) J. M. Robertson, " *Cryogenic Engineering*", edited by B. A. Hands, Academic Press (1986).
(9) Y. Iwasa, IEEE Trans. MAG-24, 1211(1988).

Preparation of High Tc Oxide Superconducting Filaments by Suspension Spinning Method

TOMOKO GOTO, HIROYUKI INAJI, KATSUHIKO TAKEUCHI, and KEISAKU YAMADA

Department of Materials Science and Engineering, Nagoya Institute of Technology, Gokiso-cho, Showa-ku, Nagoya, 466 Japan

ABSTRACT

The suspension spinning of Ln-Ba-Cu-O (Ln=Dy, Y, Ho and Er), Bi-Sr-Ca-Cu-O and Tl-Ba-Ca-Cu-O was examined to prepare a long filament with high Jc. A fine powder of the oxide was suspended in PVA solution containing a dispersant. The viscous suspension was extruded as a filament into precipitating medium and the filament was coiled on a winding drum. The obtained filament about 200 μm in diameter was heated to remove volatile components and to generate the superconducting phase.

The Ln-Ba-Cu-O filament was a superconductor at 84 K and The highest Jc value at 77 K, 0 field, of 1128 A/cm^2 was attained for the $Ho_1Ba_2Cu_3Ox$ filaments. The Jc increased at 2500 A/cm^2 by one directional solidification treatment.

The Bi-Sr-Ca-Cu-O filament was found to consist of two different superconducting phases with Tc values of 110 K and 80 K. The maximum Jc at 77 K was 2 A/cm^2. Since Tl_2O_3 is chemically instable at the synthesis temperature, the preparation of Tl-Ba-Ca-Cu-O superconducting filament by this method was not easy because of the side reaction with PVA and volatility from large surface area of the filament. The Tl-Ba-Cu-O filament exhibited the best superconducting properties with an onset temperature as high as 90 K and zero resistivity at 40 K.

INTRODUCTION

The discovery of a oxide superconductor with Tc exceeding the liquid nitrogen temperature promised wide application in future technology [1-4]. The progress toward major application of high Tc oxide superconductor has been hindered by the difficulty in fabrication of the brittle ceramic materials and by the weak-link problem in the sintered materials, which is typified by very low transport critical current densities and their severe degradation in weak magnetic fields.

We have studied the preparation of oxide superconducting long filament using a textile fiber spinning technology for the precursor of the oxide[5-10]. In this paper the suspension spinning of Ln-Ba-Cu-O (Ln=Dy, Y, Ho and Er), Bi-Sr-Ca-Cu-O and Tl-Ba-Ca-Cu-O was examined to prepare a long filament with high Jc.

EXPERIMENTAL

It is important in this technique how to enhance densification of the oxide superconducting phase in the filament and the Jc of the filament was strongly dependent on the fineness of the starting powder. Then the various starting oxides were initially prepared.

Two types of fine mixed powders with nominal composition of $Ln_1Ba_2Cu_3Ox$ were employed in our experiment. One of them was prepared by coprecipitating the carbonates of Ln, Ba and Cu, and the other was by firing a gel of the citrates of Ln, Ba and Cu. The gel was formed by evaporating of an aqueous solution of the salts. Both powders were annealed at 1123 K for 7.2 ks and the particle size of the heated powder was up to 500 nm.

A mixed powder specimen with nominal composition of $Bi_1Sr_1Ca_1Cu_2Ox$ was prepared by a standard solid states reaction technique from Bi_2O_3, $SrCO_3$, $CaCO_3$ and CuO. The powder was throughly mixed, heated at 1093 K for 2 hr, ground and filtered through 846 mesh sieves to make a fine mixed powder less than 15 μm in diameter. Two types of fine mixed powder were prepared by firing a gel of the citrates and oxalates of Bi, Sr, Ca and Cu. The other types of fine mixed powder were also prepared by coprecipitating the carbonates and oxalates of Bi, Sr, Ca and Cu. These powders were calcinated at 1023 K for 2 hr.

Various compositions of Tl-Ba-Ca-Cu-O powder used in the present experiment were prepared as follow: A fine mixed powder of Ba-Ca-Cu-O, or Ba-Cu-O and Ca-Cu-O was prepared by coprecipitating the carbonates of Ba, Ca, and Cu. Appropriate amounts of Tl_2O_3 and Ba-Ca-Cu-O, or Ba-Cu-O and Ca-Cu-O were completely mixed, ground and filtered through 300 mesh sieves.

The powder produced by these methods was suspended in PVA solution containing a dispersant. The suspension dope was extruded as a filament into a precipitating medium and was coiled on a winding drum. The filament was heated under various heating conditions.

The resistivity (ρ) of the filament heated was measured by a standard four probe method. A silver paint was used to contact Ag electrodes 50 μm in diameter. A sample current density was of the order of 0.01 A/cm^2. Specimens temperature was measured using a chromel-gold + 0.007 % iron thermocouple. The crystal structure was measured by X-ray diffractometer and the tensile strength of the filament heated was measured with an Instron type machine.

RESULTS AND DISCUSSION

Ln-Ba-Cu-O filament

Previous studies had shown that Jc of $Y_1Ba_2Cu_3$ oxide filament was strongly dependent on the spinning conditions and high Jc was observed for the filament spun by using PVA with optimum degrees of polymerization, dispersant of mixture of anionic and nonionic surfactants and powder content of 93 wt % [6]. The suspension spinning of $Ln_1Ba_2Cu_3$ oxide was examined in aqueous solution and nonaqueous solution. Fine powder was initially suspended in PVA aqueous solution containing a mixture of sodium dodecyl sulfate and polyoxyethylene (10) octyl phenyl ether. The spinning dope was extruded as a filament into a mixed aqueous solution of NaOH and Na_2SO_4 at 313 K and was coiled on a winding drum at a low winding speed of 1 m/s. The spinning of the powder was continuously performed. The filament about 200 μm in diameter was heated under various heating conditions such as 1223 K for 5 hr, 1253 K for 300 s and 1303 K for 60 s in O_2 followed by furnace cooling at a cooling rate of 100 K/hr. The resistivity of the filament heated was measured. A rapid resistivity drop is observed around 90 K and a zero resistivity state at 83 K.

It is well known that the superconducting $Y_1Ba_2Cu_3O_x$ perovskite phase breaks up in the aqueous solution. Therefore PVA dimethyl sulfoxide (DMSO) solution was used for a spinning dope and methyl alcohol for precipitating medium. The spinning was easily performed and the resistivity of the filament heated was measured. The maximum Jc at 77 K, the value of ρ at 100 K and the tensile strength of the filament spun in both aqueous and nonaqueous systems are listed in Table I with Ln^{+3} ionic radius. As the present spinning conditions is suitable for obtaining Y oxide filament with high Jc, the filament of the Dy, Ho and Er oxides which are close to the ionic radius of the Y oxide exhibits a high Jc. As the Jc increases, ρ value decreases and tensile strength increases. The filament spun in nonaqueous solution has higher Jc and lower ρ value than that for the filament spun in aqueous solution. The highest Jc of 1128 A/cm² was attained for the filament of $Ho_1Ba_2Cu_3$ oxide with tensile strength of 57 MPa. The filament had a single layered perovskite phase. On the observation of scanning elecreon micrographs, the filament had smooth surface and grain boundary of the oxide was disappeared, while the blade like grains with grain size of 3μm x 5 μm x 1 μm was observed on the surface of the $Y_1Ba_2Cu_3O_x$ filament exhibiting Jc of 680 A/cm². The $Ho_1Ba_2Cu_3$ filament was melted by heat treatment due to the low melting point.

S.Jin et.al. studied the melt textured growth of polycrystalline $Y_1Ba_2Cu_3O_x$ superconductor using directional solidification and reported the new microstructure exhibited dramatically improved transport Jc values at 77 K of about 17000 A/cm² in zero field and 4000 A/cm² at H=1 T [11]. We have studied the one directional solidification processing of the $Ho_1Ba_2Cu_3$ oxide filament produced by suspension spinning. The as-drawn filament was heated at more than 1223 K and one directional solidification was then performed by removing the filament from the center to one side of the tube furnace which had a temperature gradient of 25 K/cm to about 1173 K in flowing O_2 atomosphere followed furnace cooling at a cooling rate of 100 K/hr. A high Jc was observed for the filament heated at more than 1273 K in flowing O_2 of 3-7 l/min. The maximum Jc of the one directionally solidified filaments removed various speeds is shown in Fig. 1. The Jc of the filament removed less than 5 mm/min is superior to that of the melted filament. The highest Jc of 2100 A/cm² was attained for the as-drawn filament after one directional solidification. A scanning electron microscopy photograph of the filament is shown in Fig.2. The filament consists of a fine needlelike crystals with 0.2-0.4 μm in short and 2-5 μm in long dimensions. The crystalline was packed densely in the filament. The microstructure of the filament was quite different from that of the melt textured growth of $Y_1Ba_2Cu_3O_x$ polycrystalline consisting of locally aligned, long, needle-shaped grains (typically 40-600 μm in length) [11].

The sintered filament was also treated by the one directional solidification processing. A high Jc was observed for the filament heated at more than 1300 K in flowing O_2. The effect of sample removing is also shown in Fig.1. The Jc of the filament was enhanced by removing less than 5 mm/min and the highest Jc of 2500 A/cm² was attained in this case. The texture of the filament was same as for the filament without sintering. The prefered orientation of the crystal was not observed on the measurement of X-ray transmission Laue method.

We are making effort to prepare a long filament with higher Jc by one directional solidification processing.

Table.I. Jc at 77 K, ρ at 100 K and tensile strength of the $Ln_1Ba_2Cu_3$ oxide filament produced by suspension spinning comparing with Ln^{+3} ionic radius.

Ln	Ionic radius (nm)	Jc at 77 K (A/cm²)	ρ at 100 K (mΩ · cm)	Tensile strength (MPa)	Elongation (%)
Dy	0.092	17.8*	1.0*	31.1*	2.3*
		648**	0.9**	33.9**	2.2**
Y	0.092	230*	1.9*	16.0*	2.0*
		680**	0.3**	37.0**	1.2**
Ho	0.091	268*	0.8*	48.5*	0.8*
		1128**	0.4**	57.0**	1.0**
Er	0.089	160*	1.8*	46.6*	2.4*
		228**	0.6**	37.0**	1.0**

* aqueous system, ** nonaqueous system

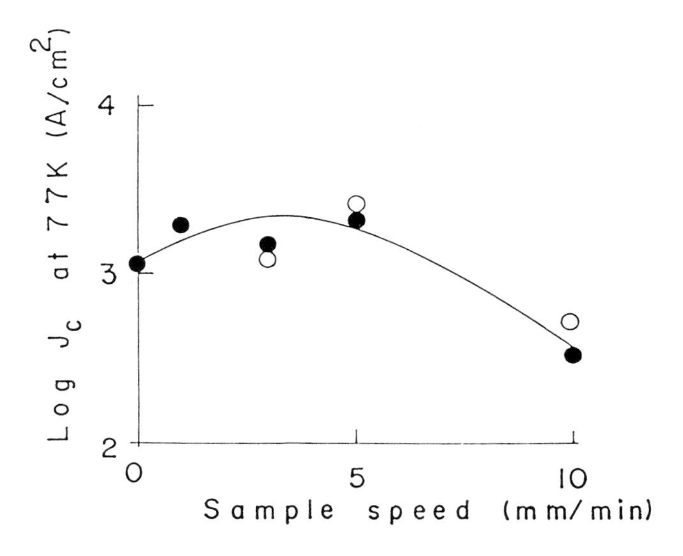

Fig.1 Maximum Jc of the one directionally solidified $Ho_1Ba_2Cu_3Ox$ filament removed at various speeds
● as-drawn filament
○ sintered filament

Bi-Sr-Ca-Cu-O filament

The Bi-Sr-Ca-Cu powder was suspended in PVA DMSO solution. The viscous suspension was extruded as a filament into a precipitating medium of methyl alcohol and coiled on a winding drum. The as-drawn filament was heated to remove volatile components and to generate the superconducting phase. A long filament of 300 μm in diameter was spun through various starting oxides and heated at 1148 k for various hours in air followed by furnace cooling at a cooling rate of 100 K/hr. The resistivity of the filament heated was measured by a standard four probe method. The Tc, Jc at 77 K, values of ρ at 150 K and tensile strength of the filament spun under optimum spinning condition to exhibit the best superconductivity are listed in Table II. As the starting powder particle size decreases, optimum heating time is shorter and tensile strength of the filament obtained increases. The filament spun from the powder prepared by firing the citrates had the highest ρ value at normal state. The superconductivity exceeding liquid nitrogen temperature was detected for the filament from the powder of coprecipitating method. The highest Jc at 77 K and 4.2 K of the filament was 2 A/cm^2 and 800 A/cm^2, respectively. Figure 3 shows the temperature dependence of the electrical resistance for the filament spun through the powder by coprecipitating method. There are two distinct superconducting transitions; one centered near 110 K and the other at 78 K. The proportion of high Tc transition to low Tc transition is larger for the filament from oxalates than

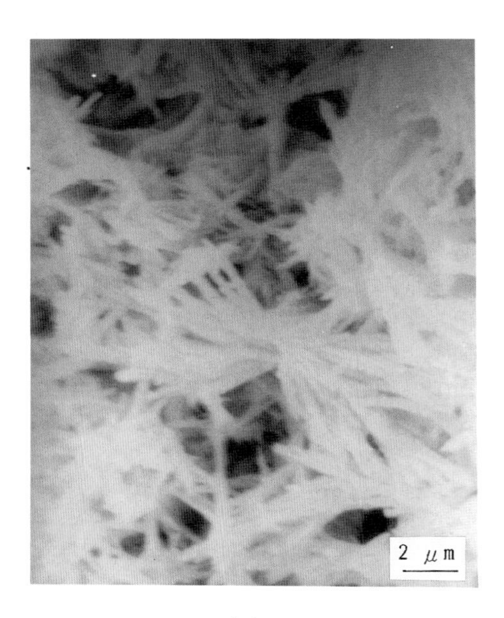

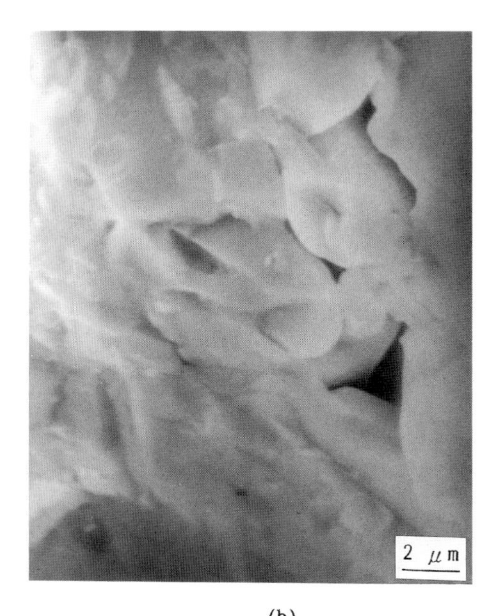

(a) (b)

Fig.2. Surface and cross section of the one directionally solidified filament exhibiting Jc = 2100 A/cm^2. (a) surface, (b) cross section

Table II Tc, Jc at 77 K, ρ at 150 K and tensile strength of the $Bi_1Sr_1Ca_1Cu_2Ox$ filament spun through various kinds of oxide powder and heated at 1148 K.

Powder method	Particle size (μm)	Heating time (hr)	Tc onset (K)	offset (K)	Jc at 77 K (A/cm²)	ρ at 150 K (mΩ · cm)	Tensile strength (MPa)	Elongation (%)
A	10	2	110	60	-	0.66	1	0.4
B-citrate	1	1	115	50	-	2.17	1.3	0.6
B-oxalate	0.2	1	110	68	-	1.94	9.4	0.8
C-carbonate	0.2	1	115	77.5	2	1.02	10.6	0.6
C-oxalate	0.2	1	115	77.5	1.6	0.91	8.7	0.8

A:Solid state reaction B:Firing C:Coprecipitating

that for the filament from carbonates. The powder X-ray patterns of the filaments were measured. All main peaks of the filaments from carbonates were assigned for the low Tc phase of $Bi_2(Sr,Ca)_3Cu_2Oy$, a=0.5396 nm, b=0.5395 nm c=30.643 nm [12]. On the other hand the filaments from oxalates had a mixed structure of the low Tc and high Tc phases. The surface and cross-section of the filament from oxalates were observed by SEM and it was found that the oxide powder was packed densely in the filament and flake-like grains, which are characteristic of the Bi-Sr-Ca-Cu-O superconductore were observed in the cross-section of the filament.

As the Jc of $Y_1Ba_2Cu_3$ oxide filament depended on the spinning condition, the spinning condition of Bi-Sr-Ca-Cu oxide was studied. Various kinds of dispersant such as anionic, nonionic and cationic surfactants and their mixture were examined, but no surfactant was effective for the Bi-Sr-Ca-Cu oxide filament. The effect of the oxide powder content in the filament on the Jc of the filament obtained was also measured and it was found out maximum Jc was observed at powder content of 95 wt %. Because the low Tc phase still remained in the present filaments, high Jc at 77 K was not attained.

Since the discovery of superconducting at onset temperature above 100 K in the Bi-Sr-Ca-Cu-O system, much effort has been made to single out the high Tc phase. Partial substitution of Pb for Bi has been reported to be effective in increasing of the proportion of the high Tc phase [13]. A mixed fine powder of nominal composition of $Bi_{0.7}Pb_{0.3}Sr_1Ca_1Cu_{1.8}Ox$ was prepared by coprecipitating the carbonates. The filament was made by suspension spinning as same as for the $Bi_1Sr_1Ca_1Cu_2Ox$ filament. The as-drawn filament was heated at 1118 K for various hours followed by furnace cooling at a cooling rate of 200 K/hr. The resistivity of the heated filament was measured and data are shown in Fig.4. A remarkable resistance drop at 110 K is observed but a complete zero resistance temperature is 70 K. The resistivity value at normal state decreases with increasing heating time. The highest proportion of the high Tc phase was observed for the filament heated at 1118 K for 17 hr but a single phase of the high Tc was not yet attained.

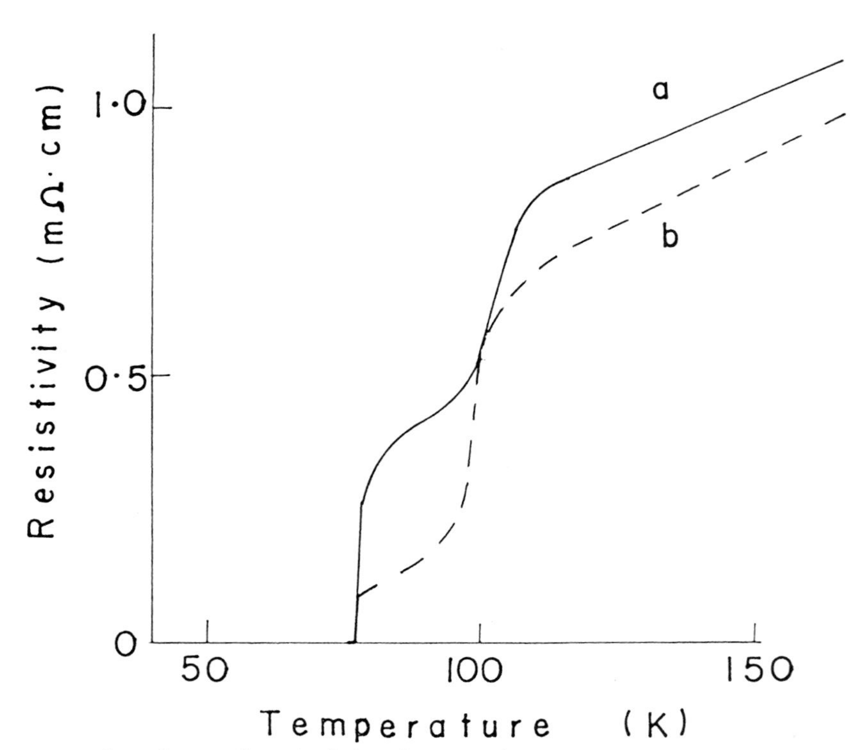

Fig.3 Temperature dependence of resistivity for the $Bi_1Sr_1Ca_1Cu_2Ox$ filaments spun through oxide powder prepared by coprecipitating the carbonates (a) and oxalates (b)

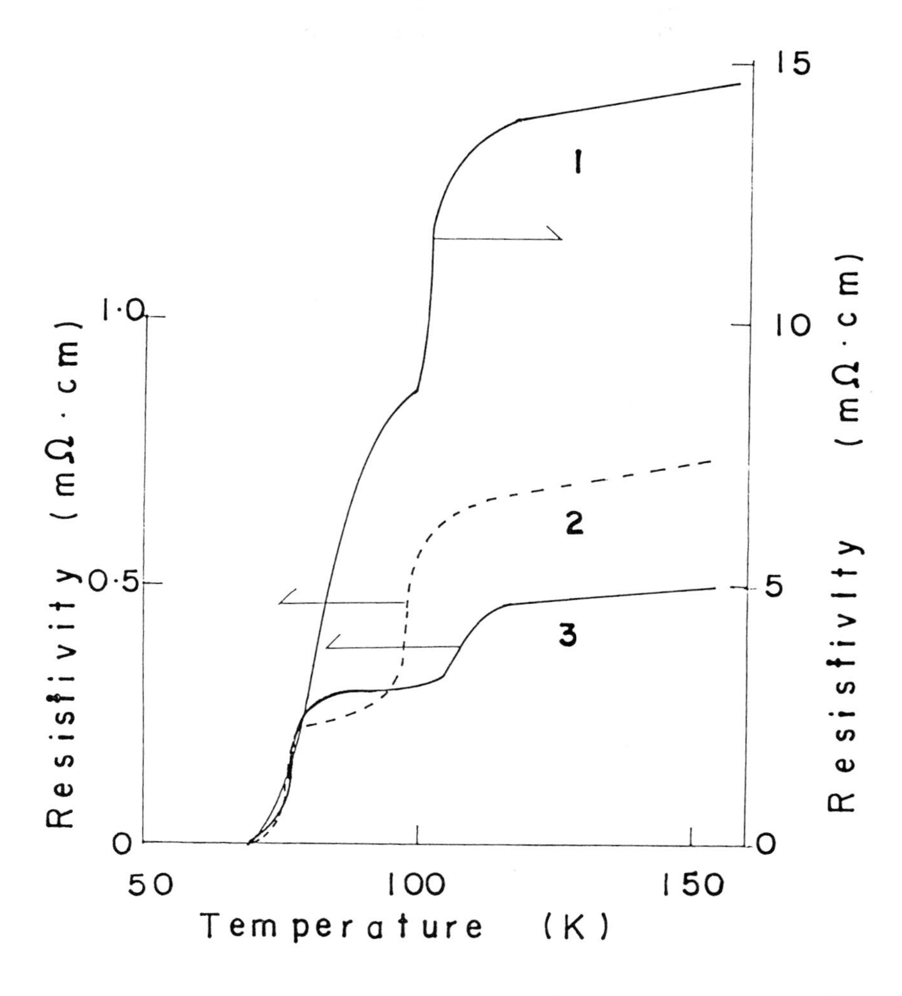

Fig.4 Temperature dependence of resistivity for the $Bi_{0.7}Pb_{0.3}Sr_1Ca_1Cu_{1.8}Ox$ filament heated at 1118 K for various hours. (1) for 5 hr (2) for 17 hr (3) for 24 hr.

Tl-Ba-Ca-Cu-O filament

The chemistry of Tl-Ba-Ca-Cu-O system is complicated by the instability of Tl_2O_3. Tl_2O_3 decomposes at the synthesis temperature to Tl_2O. Tl_2O melts between 573 K and 673 K and is quite volatile. In these ceramic materials, which are typically processed between 1073 K and 1173 K, this necessitates short anneal times as well as excess Tl_2O_3 in order to compensate for Tl loss and to assure formation of the correct superconducting phase [14]. Then various starting compositions of the oxide were prepared for the Tl-Ba-Ca-Cu-O system with atomic ratios of 1:1:1:2, 2:2:2:3, 3:2:2:3, 4:2:2:3, 5:2:2:3, 4.5:2.2:2.5:3.0, 4:1:2:2 and 4:1:4:3. The oxide powder was mixed, ground and sieved through 300 mesh and the continuous filament was obtained as same as for the Bi-Sr-Ca-Cu-O filament. Heat treatment of the filament was carried out using three different methods. 1) flush annealing in open vessels at up to 1173 K for 10 min in flowing O_2. 2) annealing in sealed quartz tubes at 1173 K for up to 30 min. 3) flush annealing in open quartz tube up to 748 K for 5 min in flowing O_2 and then sealed followed annealing at 1173 K for 10 min. It was examined in some cases to fill the crucible or tube with pellets composed of the same materials used for the starting oxide, or pellets of Tl_2O_3 or bulk metals of Tl. A large number of filaments were prepared and annealed using the three methods described above. However, most of the filaments heated were insulator or semiconductor. Only one sample of the $Tl_{4.5}Ba_{2.2}Ca_{2.5}Cu_{3.0}Ox$ filament, which was annealed in sealed quartz tube at 1173 K for 10 min filling the tube with pellets composed of the same material used for the starting oxide, exhibited superconductivity as shown in Fig.5. A pellet of $Tl_2Ba_2Ca_2Cu_3Ox$ was also made by pressing the mixed powder and annealing at 1173 K for 10 min in open vessel in flowing O_2. The resistivity of the pellet was also shown in Fig.5. A sharp resistance drop at 120 K was observed and a complete zero resistance temperature was 83 K for the pellet, whereas the filament showed the zero resistivity state at 40 K.

X-ray diffraction measurements were performed for the filament whose resistance measurement is shown in Fig.5. The filament had a mixed structure of $Ba_2Ca_2Cu_3O_7$ phase and superconducting low Tc phase of $Tl_2Ca_1Ba_2Cu_2Ox$ with a=0.544 nm, b=5.44 nm and c=29.55 nm psedotetragonal subcell[15].

For the present filament, the decomposition of PVA occured from 473 to 733 K [6] and it is considered that some side reaction between Tl_2O_3 and those volatile components proceeds as well as a substancial loss of Tl from the filament with a large surface area. Therefore the generation of the supercondutivity of Tl-Ba-Ca-Cu-O system is very difficult in the present filament.

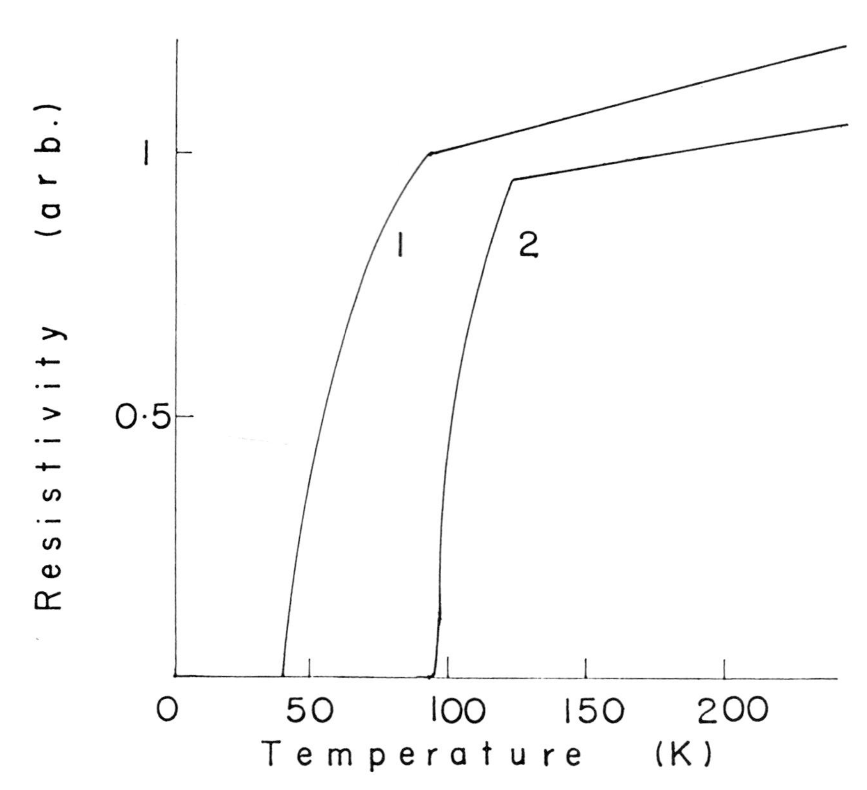

Fig.5 Temperature dependence of reisitivity for the $Tl_{4.5}Ba_{2.2}Ca_{2.5}Cu_3Ox$ filament (1) and the $Tl_2Ba_2Ca_2Cu_3Ox$ pellet (2).

REFERENCES

1. J.Bednorz and A.K.Muller, Z.Phys. B 64, 189 (1986)
2. M.K.Wu, J.R.Ashburn, C.J.Torng, P.H.Hor, R.L.Meng, L.Gao, Z.J.Huang. Y.Q.Wang and C.W.Chu, Phys.Rev.Lett. 58, 908 (1987)
3. H. Maeda, Y. Tanaka, M. Fukutomi and T. Asano, Jpn. J. Appl. Phys. 27, L209 (1988).
4. Z.Z.Sheng and A.M.Hermann, Nature 332, 10 (1988).
5. T. Goto, I. Horiba, M. Kada and M. Tsujihara, Jpn. J. Appl. Phys. 26-3, p-1211 (1987).
6. T.Goto and M.Kada, Jpn. J. Appl. Phys. 26, L1527 (1987).
7. T. Goto, Jpn. J. Appl. Phys. 27 (1988) L680.
8. T. Goto, I. Horiba, M. Kada and M. Tsujihara, Physica C. 153-155, 800 (1988).
9. T. Goto and M. Tsujihara, Proc. Int. Meet. on Advanced Materials, Tokyo 1988.
10. T.Goto and K.Takeuchi, Jpn. J. Appl. Phys. to be published.
11. S.Jin, T.H.Tiefel, R.C.Sherwood, M.E.Davis, R.B.van Dover, G.W.Kammlott, R.A.Fastnacht and H.D.Keith, Appl.Phys.Lett. 52, 2074 (1988)
12. M. Onoda, A. Yamamoto, E. Takayama-Muromachi and S. Takekawa, Jpn. J. Appl. Phys. 27, L-833 (1988).
13. R.J.Cava, B.Batlogg, S.A.Sunshine, T.Siegrist, R.M.Fleming, K.Rabe, L.F.Schneemeyer, D.W.Murphy, R.B.van Daver, P.K.Gallagher, S.H.Glarum, S.Nakahara, R.C.Farrow, J.J.Krajewski, S.M.Zahurak, J.V.Waszczak, J.H.Marshell, P.Marsh, L.W.Rupp.Jr., W.F.Peck and E.A.Rietman, Physica C 153-155, 560 (1988).
14. B.Morosin, D.S.Ginley, P.F.Hlava, M.J.Carr, R.J.Baughman, J.E.Schirber, E.L.Venturini and J.F.Kwak, Physica C, 152 413 (1988)
15. R.M.Hazen, L.W.Finger, R.J.Angel, C.T.Prewitt, N.L.Ross, C.G.Hadidiacos, P.J.Heaney, D.R.Veblen, Z.Z.Sheng, A.El Ali and A.M.Hermann, Phys. Rev. Lett. 60, 1657 (1988)

Superconducting Wire of High Tc Oxide by Diffusion Process

O. KOHNO, Y. IKENO, N. SADAKATA, and A. KUME

Fujikura Ltd., 1-5-1, Kiba, Koto, Tokyo, 135 Japan

ABSTRACT

The diffusion process between Y_2BaCuO_5 and Ba-Cu oxide has been investigated for fabricating a $YBa_2Cu_3O_{7-y}$ wire. At above 926°C, the solid-liquid reaction at the boundary between Y_2BaCuO_5 and Ba-Cu oxide has been observed because of decomposion of Ba-Cu oxide. By X-ray diffraction analysis, it was revealed that Ba-Cu oxide was consisted of $BaCuO_2$ and CuO_2 at the room temperature. The obtained maximum critical current density of a wire was 730A/cm^2 which was not so high comparing with the value of the basic bulk specimen of 1,900A/cm^2. The heat--treating conditions and the microscopic diffused texture will be investigated in detail to improve the critical current density.

INTRODUCTION

Advantages of the new high Tc superconducting materials are the possibility of using the high temperature coolant of liquid nitrogen. In order to realize that, many high Tc superconducting oxides have been studying in the world[1-3]. When they may be applied to the actual electric machine in the future, a high critical current which is desirous to be independent to the magnetic field like the usual metallic superconducting materials will be extremely required. The critical current density of a superconducting oxide film by sputtering process has already reached 10^6A/cm^2 at OT and liquid nitrogen temperature, however, for wire or conductor by the conventional powder process it still wanders about 10^3 to 10^4 A/cm^2[4-6]. It is considered that the low critical current density by the powder process depends on the weak coupling between grains, anisotropy in the critical current density, porosity and so on.

The authors have proposed the diffusion process between Y_2BaCuO_5 and Ba-Cu oxide phases to form the dense $YBa_2Cu_3O_{7-y}$ phases which is expected to be effective to increase the critical current density[7]. As the results, the critical current density of 1,900A/cm^2 has been reported by the basic bulk specimen.

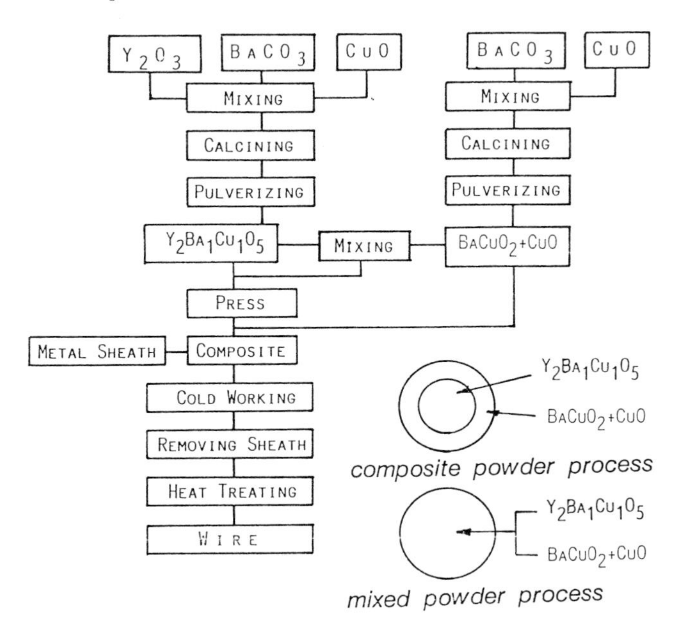

Fig.1 Fabrication flow of the powder diffusion process

In this paper, the fabrication process for a wire and the obtained results including the critical current density and X-ray diffraction patterns are presented.

EXPERIMENTAL

For this diffusion process, the green colored Y_2BaCuO_5 non-superconducting powder and Ba-Cu oxide powder consisted of $BaCuO_2$ and CuO were separately prepared by other routes. In order to prepare Y_2BaCuO_5 powder, a mixture of Y_2O_3, $BaCO_3$ and CuO with mol ratio of 1/1/1 were calcined at 900°C for 24H in air, and then cooled to the room temperature for pulverizing, and again heated at 950°C for 24H in air. Ba-Cu oxide was prepared by mixing appropriate amount of $BaCO_3$ and CuO and heating at 880°C for 24H in air. A mixture of $BaCuO_2$ and CuO were consequently formed after heat-treating.

Two kinds of the fabrication process for a wire, mixed powder diffusion process and composite powder diffusion process, were investsigated. The fabrication flows are shown in Fig.1. For mixed powder process, Y_2BaCuO_5 were mixed with Ba-Cu oxide powder and pressed to a bar shape with rubber shell in a hydrostatic press machine. The bar was sheathed with a silver tube and cold-worked by the swaging and drawing machine. For the composite powder process, Y_2BaCuO_5 powder was pressed to a bar shape and inserted into a silver tube with a little larger diameter than that of the bar. Ba-Cu oxide powder was poured into the gap between a bar and an outer silver sheath. After that, it was cold-worked to a desired diameter with the same process as the mixed one. A wire specimen was heat-treated in oxygen atmosphere at the final diameter of 1.1mm after removing a silver sheath with nitric acid in order not to form the cracks on the core surface.

RESULTS AND DISCUSSION

Measuring of the superconducting properties and observation of the cross section of a wire specimen were carried out. The critical temperature and the critical current were measured resistively at liquid nitrogen temperature by dc four-probe method. Four lead wires were put on the ceramics surface directly with the ultrasonic soldering. For measuring the critical temperature, we adopted the transport current of 1mA and used a Pt-Co thermometer for calibration. The critical current was determined with the criterion of 1uV/cm across the specimen, and the critical current density was calculated with an area of the reacted layer for composite process and an overall area of a wire for mixed process.

The critical currents of the wire specimens are shown in Fig.2 which represents the critical current not the critical current density for evaluating the superconducting current capacity of a wire. The maximum critical currents were 1.54A(=730A/cm^2, calculated from an area of the reacted layer only) for a composite wire, and 3.2A(=340A/cm^2, calculated from an area of the overall wire) for a mixed one. These values are considerably lower than the maximum critical current density of 1,900A/cm^2 for a basic bulk specimen. In order to improve the critical current density, the heat-treating conditions should be further studied in detail. It can be estimated from the results in Fig.2, that the low temperature below 930°C and long time of several hundred hours for a composite powder process and also the high temperature above 970°C and the long time of several hundred hours for the mixed powder process should be investigated.

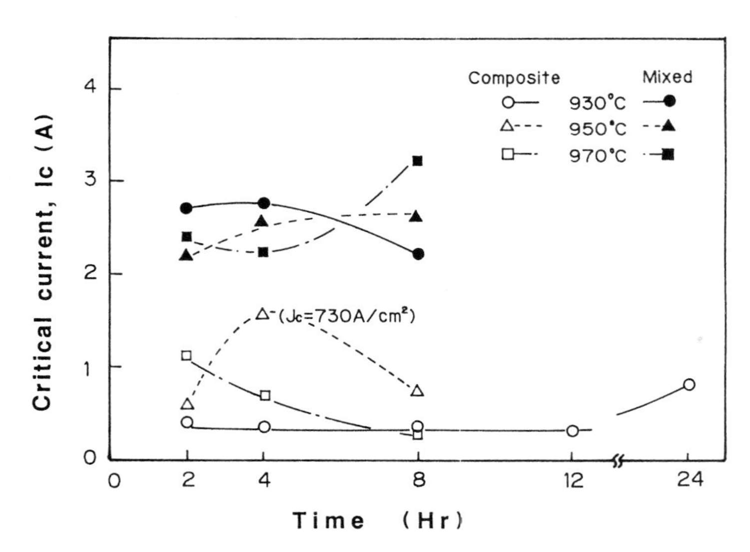

Fig.2 Critical current vs heat-treating time curves of the wire specimens by both composite and mixed powder diffusion processes

The critical current values were not scattered for the mixed powder process, however, they varied widely for the composite powder process shown in Fig.2. The thickness and the shape of reacted layer around Y_2BaCuO_5 core for the composite process are uncertainty because of melting state of Ba-Cu oxide during heat-treating. For the composite process, the shape of a wire was stable even after heat-treating because of enclosing melted Ba-Cu oxide layer by Y_2BaCuO_5 matrix.

The decomposition points of Ba-Cu oxide are shown in Fig.3, and strongly depend on the composition ratio of Ba/Cu and heat-treating atmosphere. The lowest decomposition point in oxygen atmosphere was 926°C at 3/5 for Ba/Cu ratio. This indicates that Ba-Cu oxide in the diffusion couple for this study was liquid phase, probably at the first stage of the heat-treating. Therefore, the reaction at the boundary between Y_2BaCuO_5 and Ba-Cu oxides looked like solid-liquid diffusion.

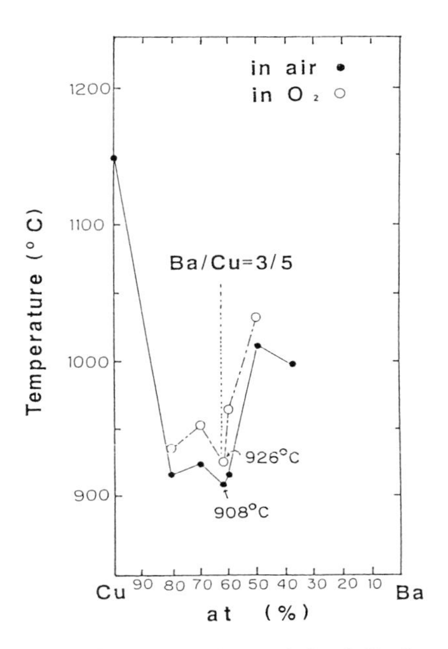

Fig.3 Decomposition point of Cu-Ba oxide

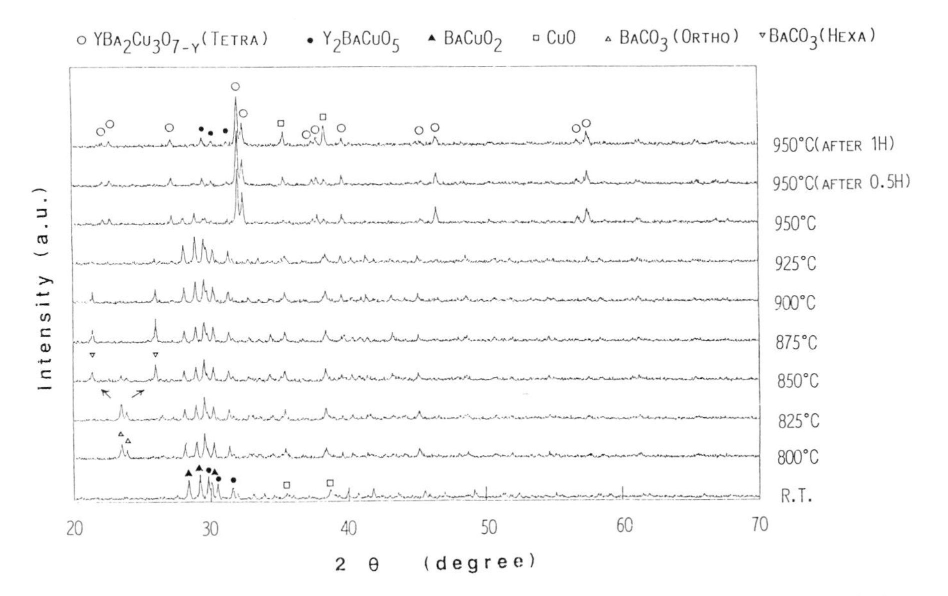

Fig.4 X-ray diffraction patterns of the mixed powder consisted of Y_2BaCuO_5, $BaCuO_2$ and CuO at the high temperatures

In order to reveal the structure at such high temperature as sintering condition, X-ray diffraction pattern analyses of the mixed Y_2BaCuO_5 and Ba-Cu oxide powders were carried out with the heating rate of 5°C/min. X-ray diffraction patterns at 800°C up to 950°C are shown in Fig.4. When the specimen reached 950°C, the peaks of Y_2BaCuO_5 and $BaCuO_2$ disappeared and simultaneously $YBa_2Cu_3O_{7-y}$(tetragonal) appeared.

The microscopic observation of the cross-sections of both specimens was carried out. Fig.5 and 6 show the cross-sectional views of the wire by the composite and mixed powder processes, respectively. They include the overall cross-sections of before and after heat-treating and a part of the reacted area. Fig.7 and 8 show the part of the reacted areas which are heat-treated for 4 Hr at 930°C(a), 950°C(b) and 970°C(c). For the composite powder process, the dense reacted $YBa_2Cu_3O_{7-y}$ layer around the Y_2BaCuO_5 core was clearly observed as shown in Fig.5 and 7 and also Fig.9 of the analyzed profile by an electron probe X-ray micro-analyzer. For the mixed powder process, however, it was not observed clearly so much because of microscopic melting of Ba-Cu oxide in Y_2BaCuO_5 matrix. It is commonly observed for both processes that the heat-treated wire includes fair amount of non-reacted residual Y_2BaCuO_5 green phase, so the optimum heat-treating condition should be investigated furthermore.

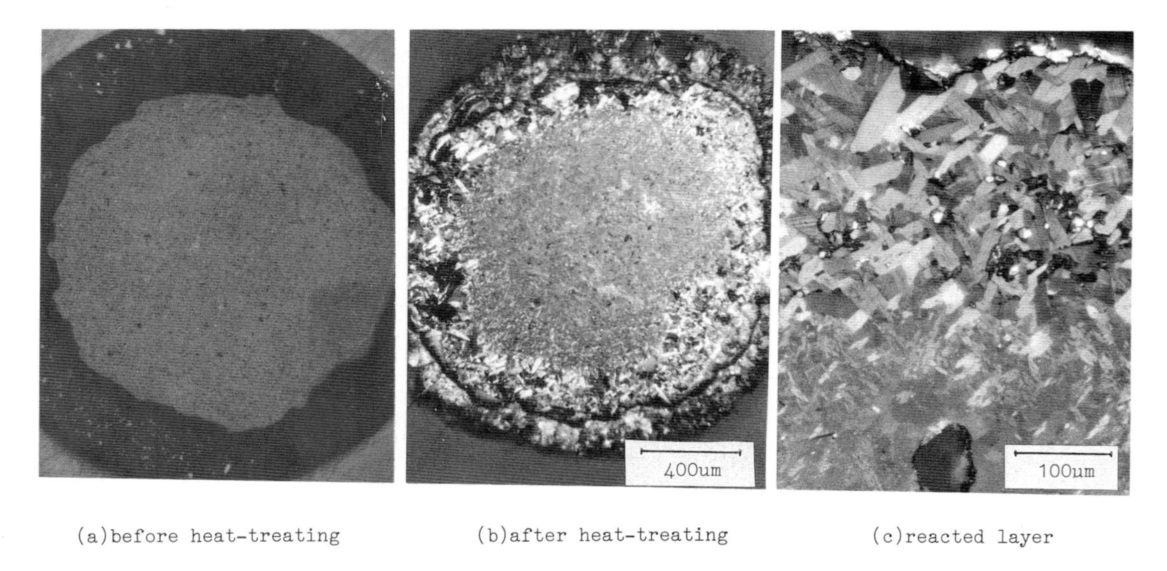

(a)before heat-treating (b)after heat-treating (c)reacted layer

Fig.5 Cross-sections of a wire by composite powder process

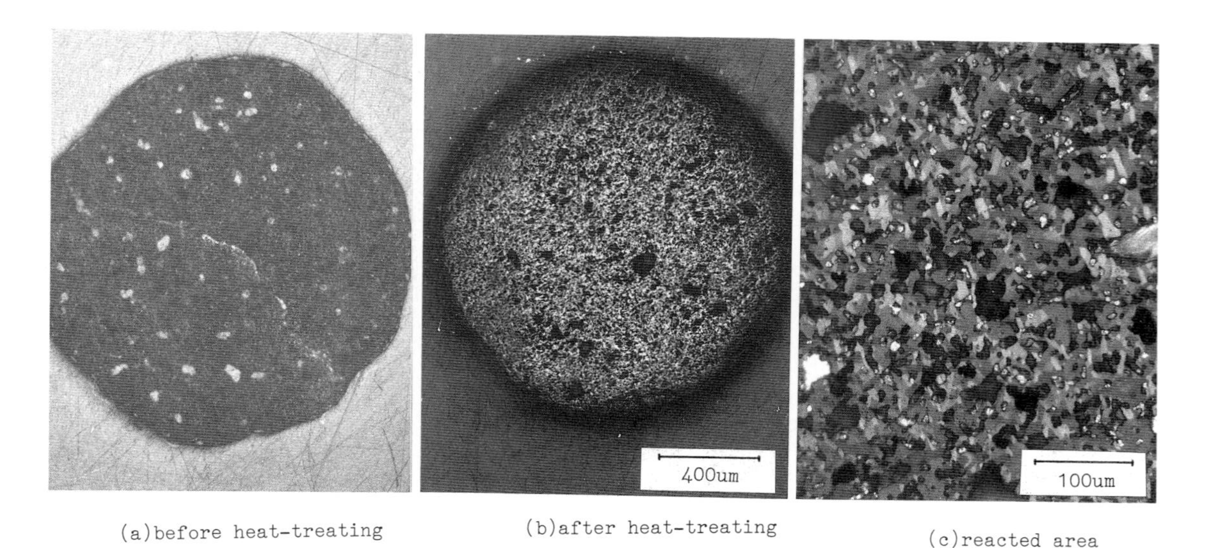

(a)before heat-treating (b)after heat-treating (c)reacted area

Fig.6 Cross-sections of a wire by mixed powder process

930°Cx4H 950°Cx4H 970°Cx4H

Fig.7 Reacted layers of a wire by composite powder process

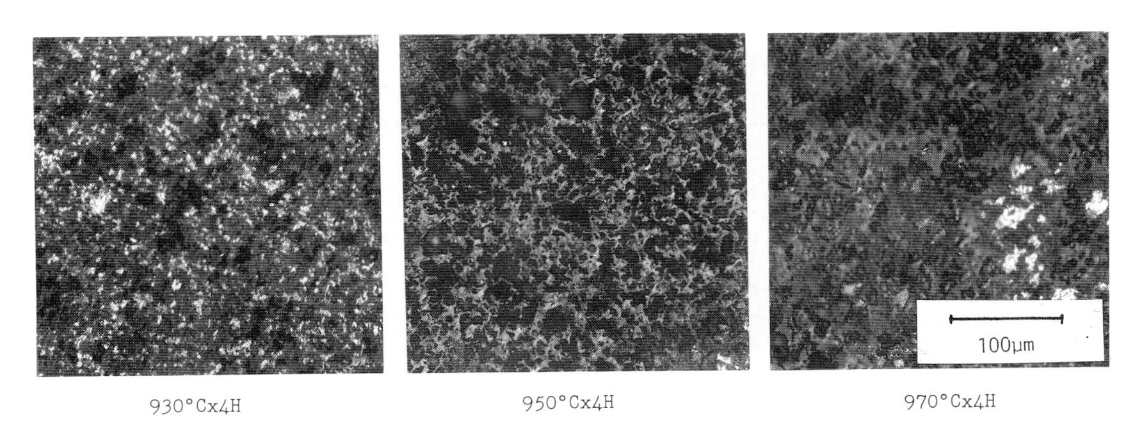

930°Cx4H 950°Cx4H 970°Cx4H

Fig.8 Reacted areas of a wire by mixed powder process

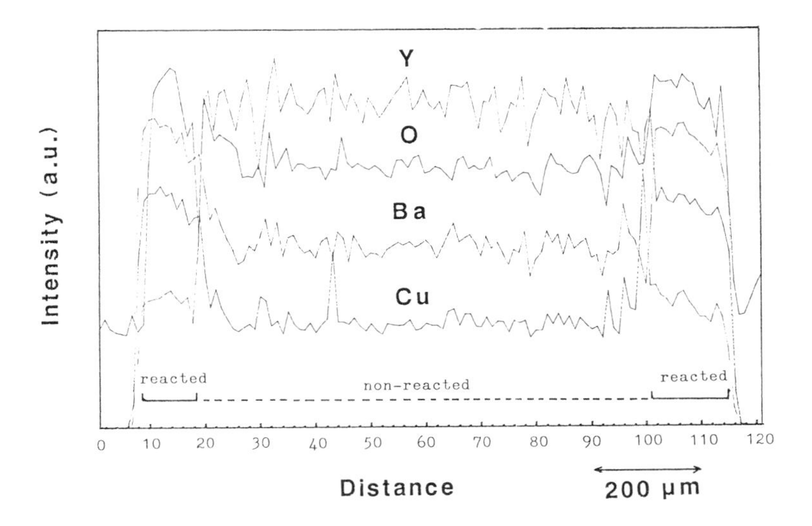

Fig.9 EPMA profile of cross-section of a heat-treated wire at 930°C for 4H for mixed powder process

A $Y_{1.5}$-BaO-CuO phase diagram is shown in Fig.10. The composition of Ba-Cu oxide with mol ratio of Ba/Cu=3/5 is marked as an open circle on the line between CuO and BaO, and also situates on the extension of the line between Y_2BaCuO_5 and $YBa_2Cu_3O_{7-y}$. According to the results of X-ray diffraction patterns shown in Fig.4, Ba-Cu oxide was proved to be mainly consisted of $BaCuO_2$ and CuO at the room temperature.

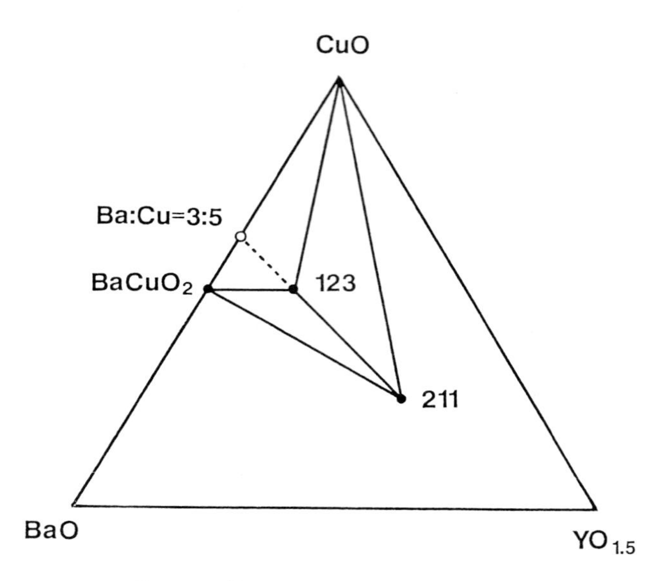

Fig.10 $Y_{1.5}$-BaO-CuO phase diagram

CONCLUSION

Two powder diffusion processes consisted of the reaction between Y_2BaCuO_5 and Ba-Cu oxide powders, the composite and the mixed powder diffusion processes, were proposed as one of the fabrication processes for an oxide wire. The obtained maximum critical current density was 730A/cm^2 calculated by an area of the reacted layer for the composite process. This value is not so high comparing with that by the basic bulk specimen. According to the results of the observation of the reacted layer by microscopy, it was revealed that the diffused $YBa_2Cu_3O_{7-y}$ phase had a dense structure and included fair amount of non-reacted green phase. Therefore it can be estimated that the critical current characteristics of a wire can be expected to be improved by adopting the optimum heat-treating condition and controlling the microscopic diffused texture.

REFERENCES

[1]M.K.Wu, J.R.Ashburn, C.J.Torung, P.H.Hor, R.L.Meng, L.Gao, Z.J.Huang, Y.Q.Wang and C.W.Chu, Phys.Rev.Lett.,58,908(1987)
[2]H.Maeda, Y.Tanaka, M.Fukutomi and T.Asano, Jpn.J.Appl.Phys.,27,L209(1988)
[3]Z.Z.Sheng and A.M.Herman, Naure, 332,138(1988)
[4]S.Tanaka and H.Itozaki, Jpn.J.Appl.Phys., 27,L622(1988)
[5]O.Kohno, Y.Ikeno, N.Sadakata and K.Goto, Jpn.J.Appl.Phys., 27,L77(1988)
[6]S.Jin, T.H.Tiefel, R.C.Sherwood, M.E.Davis, R.B.van Dover, G.W.Kammlott, R.A.Fastnacht and H.D.Keith, Appl.Phys.Lett., 52,2074(1988)
[7]N.Sadakata, M.Sugimoto, O.Kohno and K.Tachikawa, presented at Appl.Super.Conf., Aug(1988)

Microstructures and Superconducting Properties of Y-Ba-Cu Oxide Coils Prepared by the Explosive Compaction Technique

S. HAGINO[1], M. SUZUKI[1], T. TAKESHITA[1], K. TAKASHIMA[2], and H. TONDA[2]

[1] Central Research Institute, Mitsubishi Metal Corporation, 1-297, Kitabukuro, Omiya, Saitama, 330 Japan
[2] Faculty of Engineering, Kumamoto University, 2-39-1, Kurokami, Kumamoto, 860 Japan

ABSTRACT

Explosive compaction technique possesses considerable potential as a means of densifying metal, ceramics and mixtures of these powders. We applied this technique to preparation of silver sheathed coils of Y-Ba-Cu superconducting oxide. The coils to be compacted were placed into metal tube and the tube was filled with SiC powder as a pressure propagating medium and the tube was explosively compacted by a cylindrically axi-symmetric method. The coils explosively compacted were then sintered, and microstructure observations and superconducting property measurements were made. The oxide core was seen to be very dense, and a part of a Y-Ba-Cu oxide coil had a critical current density higher than 13,000A/cm^2 at 77K. This results show that the explosive compaction technique will be an effective means to prepare coils with excellent superconducting properties.

INTRODUCTION

The recent discovery of the superconducting La-Ba-Cu-O system with a critical temperature (Tc) of 33K by Bednorz and Müller[1] has led to the subsequent discovery of the superconducting Y-Ba-Cu-O system (YBCO) with Tc of around 90K by M.K.Wu[2]. The highest critical current density of this YBCO observed is the order of 10^6A/cm^2[3] for thin film at liquid nitrogen temperature. However, the critical current density of bulk materials or of wire-type materials of YBCO reported are in the range of $10^2\sim10^3$A/cm^2 at liquid nitrogen temperature. For practical uses of superconductors in energy areas or in magnets, a critical current density of at least $10^4\sim10^5$A/cm^2 is required. Generally speaking, the critical current density is much dependent on the microstructures of superconducting materials and the microstructures are very much controlled by fabrication processes.

The purpose of this study is to make the wire-type YBCO core dense and to improve its critical current density. The explosive compaction technique possesses a considerable potential as a means of densifying various powder materials including metals, cermets and ceramics[4]. We applied this technique to preparation of silver sheathed coils of YBCO superconducting oxide.

EXPERIMENTAL

The typical experimental procedure is shown in Fig.1. YBCO powders in the stoichiometric ratio of Y:Ba:Cu=1:2:3 were first prepared by the co-precipitation technique. Scanning electron micrographs of calcined and pulverized powders are shown in Fig.2. The average particle size of powders are about 1.8μm. The powder is seen to have the typical oxygen-deficient triperovskite structure by the X-ray powder diffraction analysis.

To prepare a coil, this powder was packed into a silver tube of 6mm in diameter and 100mm in length, and this composite was cold-rolled to a wire with a diameter of 1~2mm without the intermediate annealing. This wire was molded into a coil of 8~15 turns and of a diameter of about 10~15mm. Explosive compaction was applied to this coil as shown in Fig.3. The coil was placed into a steel tube of 30mm diameter and 2mm wall thickness, and the tube was filled with SiC powder as a pressure propagating medium which converts explosive stress to hydrostatic pressure. The tube was then surrounded with an explosive (Asahi Chemical Industry, PAVEX:detonation velocity is 2100~2400m/sec). Compacted samples were sintered at 920°C and 930°C in flowing oxygen gas atmosphere. The critical temperature and the critical current density were measured by the conventional four-probe method. The microstructures were observed by SEM and a polarized optical microscope.

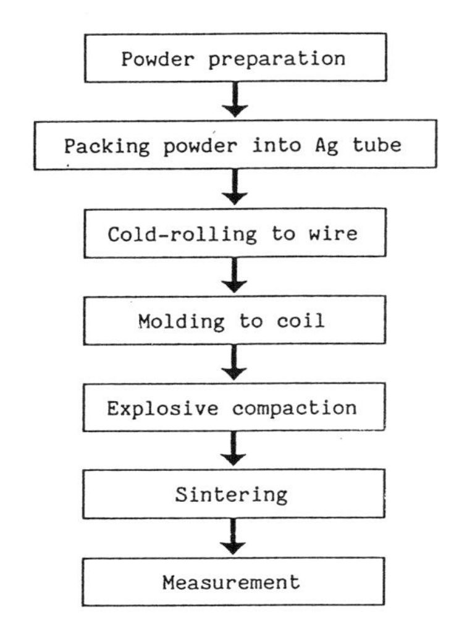

Fig.1 Experimental procedure

Fig.2 Scanning electron micrograh of YBCO powder

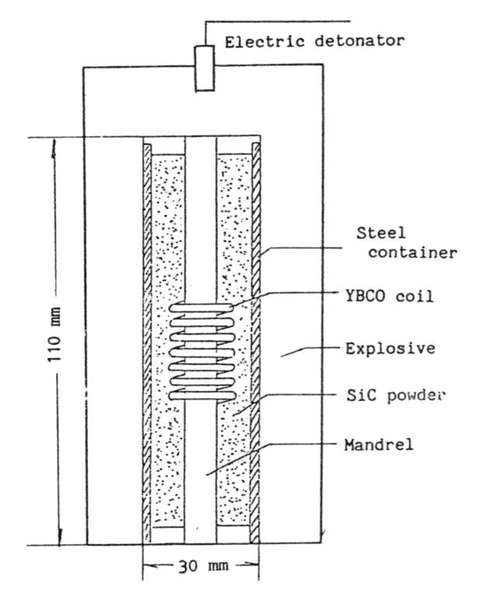

Fig.3 Experimental arrangement for explosive compaction

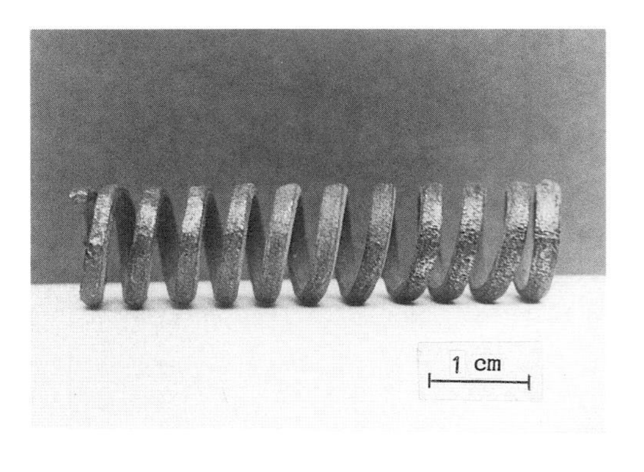

Fig.4 YBCO superconducting coil with silver sheath

RESULTS AND DISCUSSION

Fig.4 shows an explosively compacted silver sheathed coil of 10 turns of diameter of 12mm, of which wire diameter is 1.7mm. Fig.5 shows X-ray diffraction patterns of the explosively compacted YBCO core of a silver sheathed coil. Fig.5(a) and Fig.5(b) represent the X-ray diffraction peaks of as-compacted and subsequently sintered cores respectively. The X-ray diffraction peaks of as-compacted coil are broad. This could be explained by the fine crystallite size and/or the crystal lattice distorsion of cores which was induced by the explosive compaction. Fig.6 shows the microstructure of the longitudinal cross section of as-compacted YBCO silver sheathed coil. One can see that YBCO core is very dense and its grain size is very fine. In some parts of the core, grain growth areas are observed as shown in Fig.6(b). This grain growth is considered to be caused by local heating due to very high explosive compaction stress which is estimated to be $10^9 \sim 10^{10}$Pa.

The explosively compacted coils were heat-treated at temperatures of 550°C, 700°C, 850°C, 920°C and 930°C respectively in the flowing oxygen gas atmosphere. For the coils heat-treated at temperatures below 850°C, we could not observe a finite current density by the four probe method. For the coils sintered at temperatures of 920°C and 930°C, we could observe the current density higher than a few hundreds A/cm^2. It is believed that the penetrative connection between grains by sintering is required to obtain a finite current density. Fig.7 shows the scanning electron micrograph of the YBCO core heat-treated at 850°C. The microstructure consists of fine grains, and neither penetrative connections nor grain growth between grains can be observed. Fig.8 shows the longitudinal cross section view and micrograph of the coil sintered at 930°C in flowing oxygen gas atmosphere. YBCO oxide core is very dense and has clean grain boundaries, and each grain consists of well-grown twins.

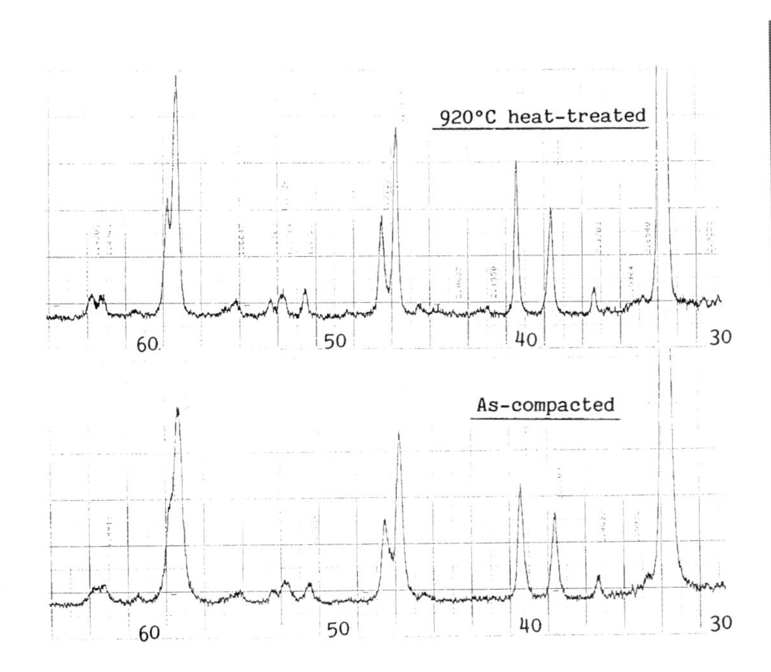

Fig.5 X-ray diffraction pattern (CuKα)

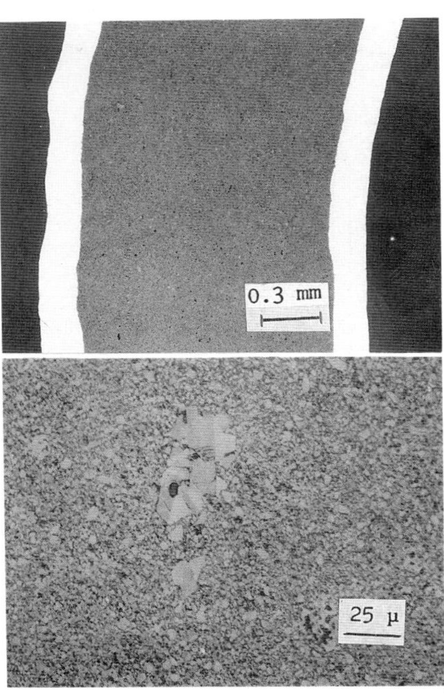

Fig.6 Longitudial cross section view and micrograph of as-compacted YBCO coil.

Fig.9 shows experimental results of critical current density measurements of YBCO coils at liquid nitrogen temperature in zero external magnetic field. Fig.9(a) shows the current vs the induced voltage curve of the coil which undergoes the usual resistive transition as the current approaches the critical value. Fig.9(b) shows that of the coil which exhibits a ohmic behavior at lower current region, and a usual resistive transition at the critical current. Fig.9(a) type curves were usually observed but Fig.9(b) type curves were seldom seen in our measurements. It is considered that a ohmic behavior at lower current region in Fig.9(b) could be explained by the current-transfer from the outer metal sheath to the inner superconducting core. We observe various defects in the some area of oxide core of silver sheathed coil as shown in Fig.10. Fig.10(a) shows the two types of cracks. One is half plane type and another is the complete type. As the current flow in the core will be stopped by the cracks, it will transfer from the core to the metal sheath, and then it will transfer from the metal sheath to the core again. Fig.10(b) shows the cracks at the vicinity of the metal sheath. These types of cracks also decrease the current flow.

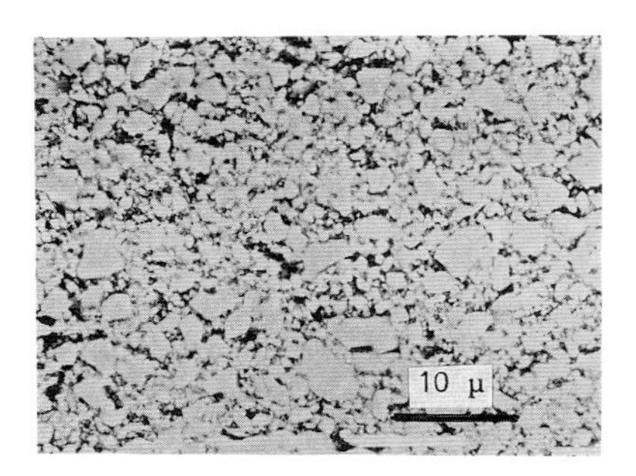

Fig.7 Scanning electron micrograph of YBCO core heat-treated at 850°C.

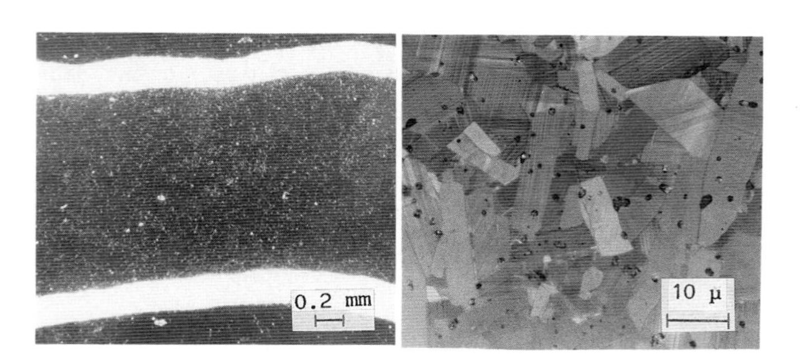

Fig.8 Longitudinal cross section view and micrograph of YBCO core sintered at 930°C.

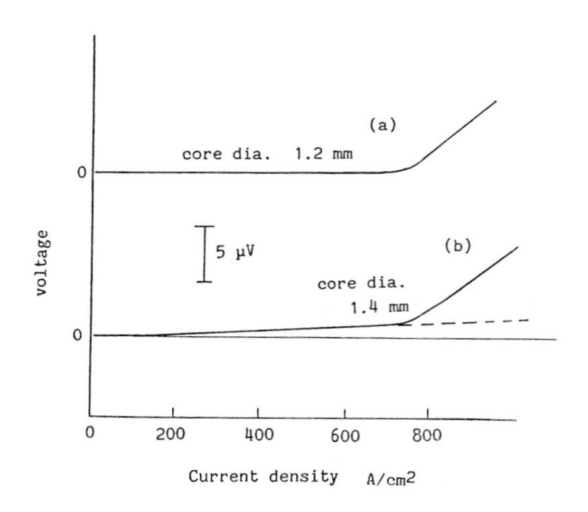

Fig.9 Critical current characteristics of YBCO coils. (a) the usual resistive transition, (b) the ohmic behaviour at lower current region and a usual resistive transition at the critical current.

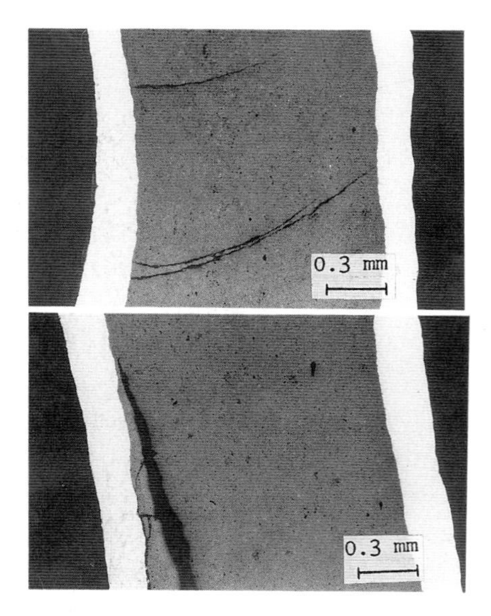

Fig.10 Longitudial cross section of YBCO core sintered at 930°C.

As mentioned above, the phenomena of the current transfer from the metal to the oxide or from the oxide to the metal is so complex, but the appearance of ohmic region at lower current region in Fig.10(b) could be explained by considering some of the current flow in a part of the metal sheath which is a normal conductor. It is also known that an inadequate separation between current and voltage contacts results in a finite linear slope in the measured voltage-current characteristic at lower current range. This has been reported for multifilamentary NbTi and Nb_3Sn composite superconductors.[5][6].

Fig.11 shows the current density vs voltage curve at liquid nitrogen temperature for a YBCO silver sheathed coil. A large critical current density of higher than 13,000A/cm^2 was obtained in this case. In this figure, the upper current was limited by the power of the current supply. To our knowledge, the largest critical current density measured so far is 11,000A/cm^2 for the wire type sample, and it appears that our present data is the highest. However, this measurement shows a finite slight linear slope in the measured voltage-current characteristics, so the critical current density determined by the standard criterion of 1μV/cm is around 8,000A/cm^2.

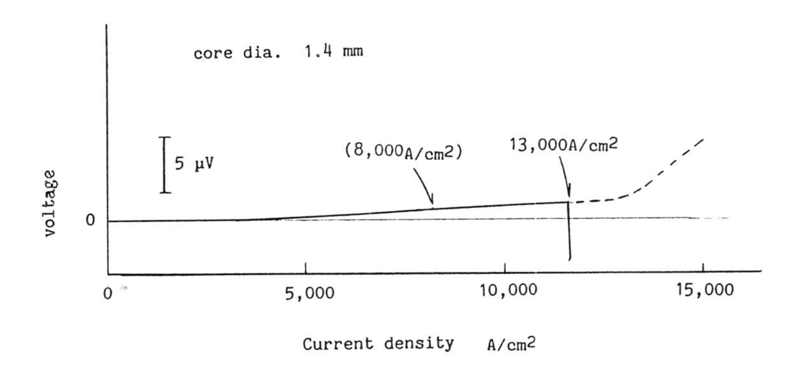

Fig.11 The current density vs the induced voltage curve of the YBCO coil sintered at 930°C.

CONCLUSION

The experimental results are summarized in the following ;

1) Explosive compaction technique is an effective means to make the YBCO core very dense.
2) For preparation of a superconducting coil by the explosive compaction technique, the following procedure is effective : The coils are placed into a metal tube and the tube is filled with SiC powders as a pressure propagating medium and the tube is explosively compacted by a cylindrically axi-symmetric method.
3) The coils after sintering were seen occasionaly to have finite slight linear slopes in the measured V-I curves at low current. This could be explained by existences of various types of cracks which cause the current-transfer between the metal sheath and the superconducting core.
4) The critical current density measurements of the coil were made and we saw that a part of a coil has critical current density higher than 13,000A/cm^2 at 77K, although major portion of the coil has a small critical current density.

REFERENCES

[1] J.G.Bednorz and K.A.Müller : Z. Phys. B64 (1986) 189.

[2] M.K.Wu, J.R.Ashburn, C.J.Torng, P.H.Hor, R.L.Meng, L.GaO, Z.J.Haung, Y.Q.Wang and C.W.Chu : Phys. Rev. Lett. 58 (1987) 908.

[3] Y.Enomoto, T.Murakami, M.Suzuki and K.Moriwaki : Jpn. J. Appl. Phys. 26 (1987) L1248.

[4] K. Takashima, H.Tonda, M.Ueno, T.Toraishi and M.Miyano : J. Iron & Steel Inst. 73 (1987) 2219.

[5] J.W.Ekin : J. Appl. Phys. 49 (1978) 3406.

[6] J.W.Ekin and A.F.Clark : J. Appl. Phys. 49 (1978) 3410.

Fabrication and Characterization of Ceramic Superconducting Composite Wire

MICHAEL R. NOTIS, MIN-SEOK OH, BETZALEL AVITZUR, QIN-FANG LIU, and HIMANSHU JAIN

Department of Material Science and Engineering, Lehigh University, Bethlehem, PA 18015, USA

ABSTRACT

Sheathed high T_c superconducting ceramic wire composites have been fabricated by wire drawing and by hydrostatic extrusion. A number of different superconducting ceramic compositions have been fabricated including $YBa_2Cu_3O_7$ ("1-2-3"), and the Bi-Sr-Ca-Cu-O system. Under certain circumstances, a displacement reaction occurs at the interface between the ceramic superconductor and the containment tube; the nature of this reaction and its prevention are discussed.

INTRODUCTION

One of the most desirable configurations for a useful superconductor is as a wire product [1]. In this form it is useful as a conductor for DC or AC power transmission, and for the generation of high magnetic fields. The production of fine wire is critical to minimize problems with stabilization and to lower the AC or dynamic losses associated with changing field. If ceramic superconductors can be fabricated in a wire configuration, its potential for commercialization is considerable over a broad industrial spectrum, and this commercialization might be accomplished much earlier than if other forms, i.e. thin films, of ceramic superconductors are required. For this reason, much research related to the fabrication of ceramic superconducting wire has already been performed.

The methods used so far for the fabrication of bare ceramic wire require the use of an organic binder [2]; however, the $Y_1Ba_2Cu_3O_{7-x}$ ("1-2-3") superconductors of most common use today are believed to be degraded by carbon contamination [3]. The metal-core-composite wire [4] leaves the ceramic superconductor exposed to reaction with the CO_2 in the atmosphere, and the geometry is such to put the brittle material furthest from the central-axis thus making it more susceptible to cracking due to bending. We have therefore chosen to use the metal-clad composite wire configuration for our experiments. In this configuration, it has already been shown [5] that the superconductor can be fabricated using silver as the tube material. Silver is non-reactive with respect to the cuprate superconductors but allows for rapid oxygen diffusion. Further, silver or silver oxide can be incorporated into the cuprate matrix, possibly enhancing ductility and providing a source of oxygen.

Wire can be fabricated either by wire drawing or by extrusion. Wire drawing is a more economical and commercially desirable process because it can be easily made into a continuous operation. On the other hand hydrostatic extrusion often allows for easy separation of experimental variables and is therefore of great benefit for fundamental studies or for situations where wire drawing is just not suitable. In our research, two different configurations for a composite ceramic wire have been used. In one case, we have used a thick wall silver tube with a variety of cuprate superconducting ceramic powders packed inside; so far we have used this configuration with our experiments on wire fabrication by hydrostatic extrusion.

This composite has been demonstrated to be in the superconducing state after final fabrication by levitation testing of the ceramic powder core over a magnet submersed in liquid nitrogen, by direct transport measurements, and by x-ray diffraction. The other configuration we have used has consisted of a thin wall silver tube containing the superconducting ceramic powder, both of which are placed inside a thick wall tube made of either stainless steel, nickel, or copper. The higher strength outer material makes the assembly more amenable to wire drawing because of the higher wire pulling load that can be applied. Here also, the outer tube may be removed after wire drawing for further processing. However, as will be described, after interim annealing operations, these composites have shown significant reaction between the superconductor and the outer tube material, right through the thin silver interlayer. This has significant implication for the future development of commerical wire products, for fabrication of multifilamentary wires, and for joining technology in general.

EXPERIMENTAL PROCEDURE

The "1-2-3" powder was fabricated by first calcining the mixed component carbonates at about 850°C and then sintering in air at about 920°C for a few hours; this material was then given a low temperature anneal in air at about 550°C and held for eight hours and then slow cooled. The powder was observed to be superconducting prior to wire fabrication, by levitation of a cold-compacted pellet over a magnet held in liquid nitrogen.

Experiments have been performed using both conventional wire drawing and hydrostatic extrusion. The size ratio of the inner core radius to outer tube radius of about 0.5 was found to be the most efficient compromise between the need for compressive pressure to compact the powder, and sleeve strength to carry the tensile load during wire drawing. For hydrostatic extrusion, the sleeve material was a thick-wall (0.25 in OD; 0.1675 in ID) silver tube; for wire drawing, a thin-wall (0.2 in ID) silver tube was placed inside a thick wall (0.4 in OD; 0.25 in ID) 403 stainless steel tube. In one experiment, this stainless steel tube was replaced by a nickel tube, and in another, by a copper tube. Some experiments were also performed without the silver thin tube being present. Most experiments were carried out using a "1-2-3" ceramic superconducting powder packed inside the tube as the core material; some later experiments were performed using a number of Bi-superconductor compositions. Details of the drawing and hydrostatic extrusion schedules and heat treatments are described in another publication [6], but at least one temperature excursion between 600-900°C for 1-8 hours was typical.

RESULTS AND DISCUSSION

After fabrication, the wire products were tested for the presence or absence of superconductivity by noting the presence or absence of levitation of the ceramic core material above a Nd-Fe magnet immersed in liquid nitrogen.
For a few specimens, direct transport current measurements were made using a four point probe method; these are not reported in the present work, as they are tentative. Each of the wire configurations were subjected to detailed metallographic and microchemical study using a combination of light optical microscopy, scanning electron microscopy (SEM) and electron probe microanalysis (EPMA).

Despite the currently accepted paradigm that the superconducting ceramic is not reactive with respect to silver, we have noted that reaction is possible under certain circumstances. In all of the experiments performed using wire drawing with 403SS, Ni, or Cu as the outer tube material, a reaction was noted between the tube material and the ceramic core after annealing. This reaction was found to be present even when a thin wall silver tube was inserted between the outer tube and the ceramic powder core (Figure 1). The reaction was not observed for material fabricated using only a silver tube as the containment material.

The results for the 403SS/Ag/"1-2-3" composite (Figure 2) were generally typical of the reactions observed as well in all other material combinations; therefore, this one example will be described in detail. In this case, after annealing, the silver interlayer appears to have present within it a dark, somewhat spheroid shaped phase (Figure 3). This phase was identified as Cu_2O using EPMA. Further analysis in the region around the Ag/"1-2-3" interface indicated a depletion of copper in the ceramic near the Ag interlayer. The region around the 403SS/Ag interface was also observed to have reacted, forming an iron-rich oxide layer nearest the 403SS outer tube and a copper oxide layer nearest the Ag inner tube material (Figure 4). A similar reaction was noted between Ni-tubing and Bi-based cuprate superconductors. In all cases examined at lower annealing temperatures, the copper oxide formation was limited to the grain boundary regions within the Ag tube wall; at higher temperature the copper oxide particles appeared within the Ag grains. Although based only on qualitative observation, it appears that the extent of reaction is actually greater when the Ag tube interlayer material is present, versus the case when the superconductor is in direct contact with the 403SS, Ni or Cu outer tube. This is indicated by the total loss of superconductivity for the cases when a silver interlayer is present, and may be explained by the dispersion and increased surface area of the copper oxide phase within the silver layer. When the silver is not present, the copper oxide forms as a continuous layer which limits further reaction.

We believe that the microstructure described above is developed due to a displacement reaction between the outer tube material (Fe-Cr, Ni or Cu) and the cuprate superconductor where copper cations are present at least in the divalent state. Rapp, et al. [7], and Yurek, et al. [8], have described similar displacement reactions to occur between Fe, Ni or Co and Cu_2O. More recent work by Vosters, et al. [9] and von Loo, et al. [10] has examined the effect of impurities on the kinetics and morphology of the displacement reaction, as well as to provide a more definitive thermodynamic basis for the occurrence of such reactions in multiphase higher order multicomponent systems. In our particular case, we propose that elements such as Fe, Cr, & Ni, which have greater oxidation potential than copper, give up electrons which travel across the interface or through the Ag interlayer to reduce copper ions in the ceramic from the divalent to monovalent state. These monovalent copper ions together with ionic oxygen then migrate toward the outer tube region (down their own chemical gradient). We presume that when the source of divalent copper would be depleted, the system would move to a lower valence state, i.e., $Fe^{\circ} + Cu_2O$ giving $FeO + Cu^{\circ}$, as is the case described in the literature.

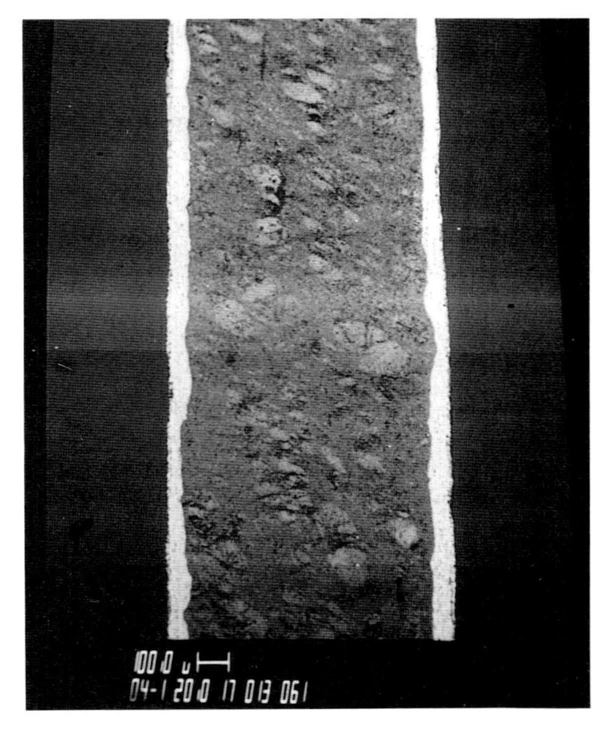

Figure 1. Longitudinal section of composite sheathed wire. Outer tube material is 403SS; interlayer is pure silver; core $YBa_2Cu_3O_{7-x}$ ("1-2-3") ceramic superconductor composite; fabricated by wire drawing.

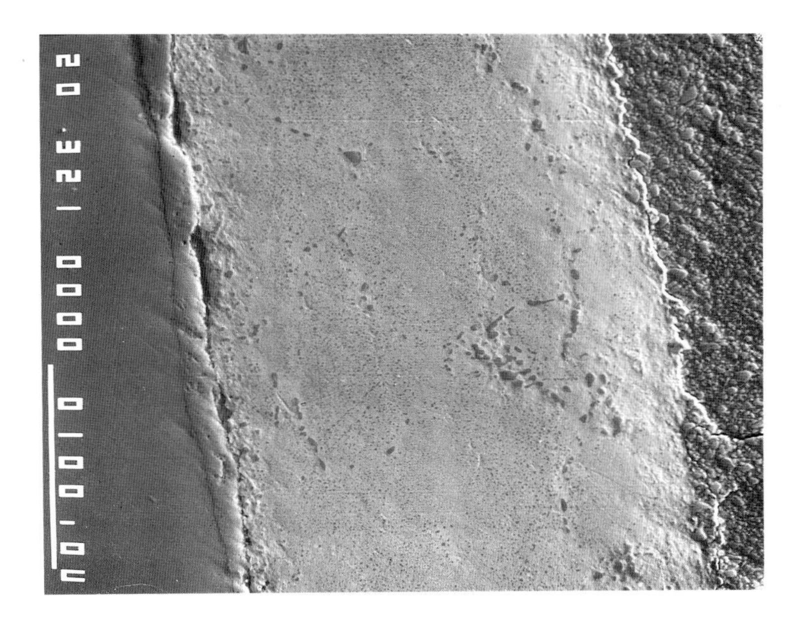

Figure 2. Overall reaction zone for composite sheathed wire shown in Figure 1. 403SS outer tube on left, ceramic core on right, silver thin interlayer at center.

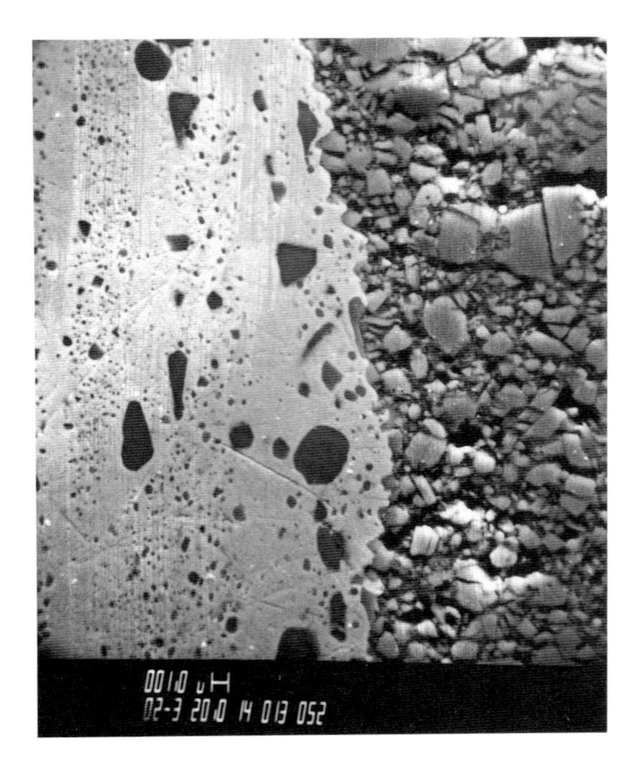

Figure 3. Interface region between thin silver interlayer (left) and ceramic core (right) after annealing. Note the dark spheroid shaped phase within the silver region.

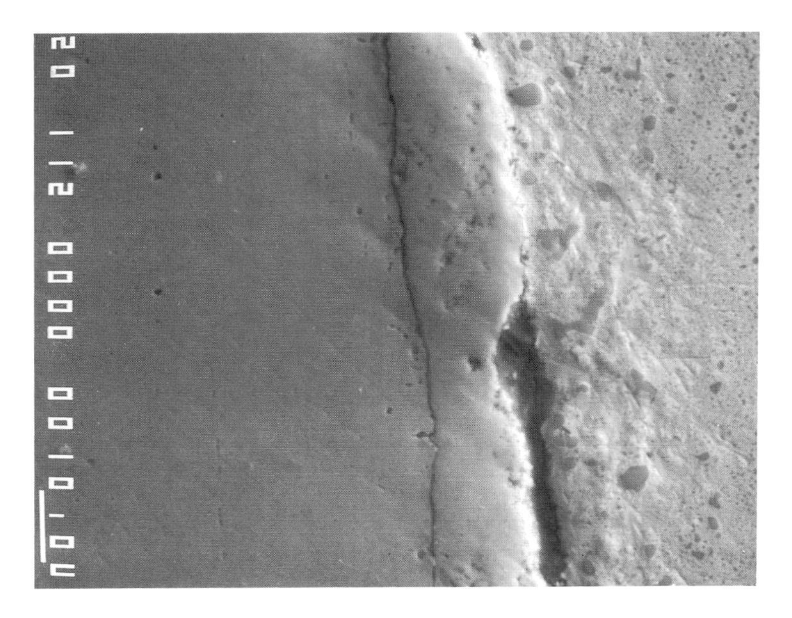

Figure 4. Interface region between the 403SS outer tube (left) and silver thin interlayer (right) showing double oxide layer forming between the two metal tubes.

Superconducting ceramic materials, no matter what their form, size or shape, must eventually make contact with non-superconducting materials in order to accomplish current transfer to other parts of a real operating system, or for testing and measurement of properties. Thus, whether the configuration is a clad wire, a bulk superconducting disc, tape, or a thick or thin superconducting film on a substrate, the physical and mechanical behavior of interfaces (interconnections, joints, etc.) between superconductors and normal conductor materials of all kinds is of extreme importance to the technological development of these systems. Fabrication heat treatments associated with the joining process (such as we have described above) would allow possible reactions between the superconducting ceramic and the contact to occur, and consequently influence properties at the interface region. The nature of these reactions is therefore of great broad interest, as these may be a primary determinant for the real capability of these materials.

CONCLUSIONS

1. Composite wire-form products with ceramic superconductor cores can be successfully drawn or extruded using a variety of composite metal tube combinations.
2. Reaction products which form between the ceramic core and the outer sheath during fabrication annealing are attributed to a displacement reaction related to the oxidation potential development between the metallic sheath and the ceramic core.

ACKNOWLEDGEMENTS

The authors wish to thank the National Science Foundation (Grant No. MSM8715 863) and the Lehigh University Consortium for Superconducting Ceramics (LUCSC) for their support of this research.

REFERENCES

[1] J. K. Hulm, "The Need for Fine Filaments in Superconducting Magnets," Supercurrents, p.24, March, 1988.

[2] J. K. Degener, et al., "Extrusion of Superconducting Ceramic Wires, to be published in Proc. Symposium on High T_c Ceramic Superconductors, Am. Ceram. Soc., Cincinnati, OH, May, 1988.

[3] L. J. Zhang, M. P. Harmer and H. Chan, "Formation of C-Containing Phase at $YBa_2Cu_3O_{7-x}$ Grain Boundaries During Annealing," Fall Meeting, MRS, Boston, MA., 1988, to be published.

[4] S. Jin, et al., Appl. Phys. Lett. 51, 943 (1987).

[5] R. W. McCallum, et al, "Problems in the Production of YBa_2Cu_3Ox, Superconducting Wire," Advances in Ceramics 2, 388-400 (1987).

[6] M. Oh, Q. F. Liu, W. Misiolek, A. Rodrigues, B. Avitzur and M. Notis, "Hydrostatic Extrusion and Drawing of Ceramic Composite Superconducting Wire," submitted for publication J. Am. Ceram. Soc.

[7] R. A. Rapp, A. Ezis and G. J. Yurek, "Displacement Reactions in the Solid State," Met. Trans. 4, 1283-1292 (1973).

[8] G. J. Yurek, R. A. Rapp, and J. P. Hirth, "Kinetics of the Displacement Reaction between Iron and Cu_2O," Met. Trans. 4, 1293-1300 (1973).

[9] P. J. C. Vosters, M. A. J. Th. Laheij, F. J. J. von Loo and R. Metselaar, "The Influence of Impurities on the Kinetics and Morphology of the Displacement Reaction Between Ni or Co and Cu_2O," Oxid. Met. 20, 147-160 (1983).

[10] F. J. J. von Loo, J. A. von Beek, G. F. Bastin, and R. Metselaar, "The Role of Thermodynamics and Kinetics in Multiphase Ternary Diffusion," p. 231 in Atomic Transport in Concentrated Alloys and Intermediate compounds, TMS-AIME (1985).

Development of High Tc Superconducting Wire by Powder Method

MASAYUKI NAGATA, KAZUYA OHMATSU, HIDEHITO MUKAI, TAKESHI HIKATA, YOSHIKADO HOSODA, NOBUHIRO SHIBUTA, KEN-ICHI SATO, HAJIME HITOTSUYANAGI, and MAUMI KAWASHIMA

Sumitomo Electric Industries, Ltd., Shimaya, Konohana-ku, Osaka, 554 Japan

ABSTRACT

Ag-sheathed high Tc superconducting wires of $YBa_2Cu_3O_{7-x}$ and $Bi_{0.8}Pb_{0.2}SrCaCu_{1.5}O_x$ were developed. Improvements in the density and the orientation of crystals brought about by press processing have been investigated, and by adopting adequate processing procedures the critical current densities were increased up to 4140A/ cm² and 4400A/ cm² for Ag-sheathed $YBa_2Cu_3O_{7-x}$ and $Bi_{0.8}Pb_{0.2}SrCaCu_{1.5}O_x$ superconductors, respectively.

After investigating Jc distribution and the bending tolerance of Ag-sheathed $YBa_2Cu_3O_{7-x}$ tapes, we made a coil with an inner diameter of 50 mm. This superconducting coil generated 47gauss at the center and 64gauss on the conductor with a transport current of 35A.

INTRODUCTION

Technology for manufacturing high Tc superconducting wires is vital to the applications of superconductivity. Superconductors are widely applicable to high field magnets composed of long tighly wound superconducting wires.

However, there are many problems involved in fabricating superconducting wires and magnets. Much effort was put forth in developing high Tc superconducting wires. We tried to fabricate wires by the powder method using a silver tube sheath.[1] Through this method we can utilize a conventional metal working process such as drawing, swaging or rolling.[2] We previously found that the Jc of the Ag-sheathed oxide superconductor could be increased through press processing, especially in $YBa_2Cu_3O_{7-x}$ and $Bi_{0.8}Pb_{0.2}SrCaCu_2O_x$ superconductors.[3] On the contrary the Jc of $BiSrCaCu_2O_x$ superconductor with Ag-sheath did not increase through press processing. In this paper we will report the improvement in Jc of short samples and the development of a superconducting coil using a 15m long Ag-sheathed superconducting tape.

EXPERIMENTAL

In order to make wires, the powders of Y_2O_3, $BaCO_3$ and CuO are weighed for 1-2-3 and mixed, then calcined at 900℃ for several hours followed by sintering and oxygen annealing. Then the ground oxide superconducting powder was put into a silver tube, and this composite was reduced to a 0.2 ~ 3.0mm diameter wire over 20m long by a metal drawing machine. The round wires were cold rolled or pressed into tapes in order to increase powder density to more than 90%. These tapes were again sintered at around 950 ℃ in air for five to fifty hours followed by oxygen annealing. For the Bi-system we made both silver sheathed wires of $BiSrCaCu_2O_x$ and $Bi_{0.8}Pb_{0.2}$-$SrCaCu_2O_x$ superconductors. The former was reported by Dr.H.Maeda of NRIM [4], and the latter is the material which is likely to have high Tc phase.[5] As the $BiSrCaCu_2O_x$ superconductor with Ag-sheath did not show any increase in Jc by press processing, we concentrated our efforts on $Bi_{0.8}Pb_{0.2}SrCaCu_2O_x$ and $Bi_{0.8}Pb_{0.2}SrCaCu_{1.5}O_x$ superconductors.

The powders were mixed, calcined and sintered before being put into a silver tube. After the diameter was reduced through drawing, the composite was introduced to a cold roll process or press processing. Finally it was sintered again, followed by annealing.

The critical current densities were measured in liquid nitrogen (77.3K) by the four probe method using DC transport currents supplied from outside of the silver sheath. The critical current density was calculated for the cross section of the oxide area with the silver sheath excluded.

We measured bending tolerances using several mandrels of different diameters.

The Ag-sheathed $YBa_2Cu_3O_{7-x}$ tape was wound around a ceramic bobbin to make a coil and sintered together with the bobbin.

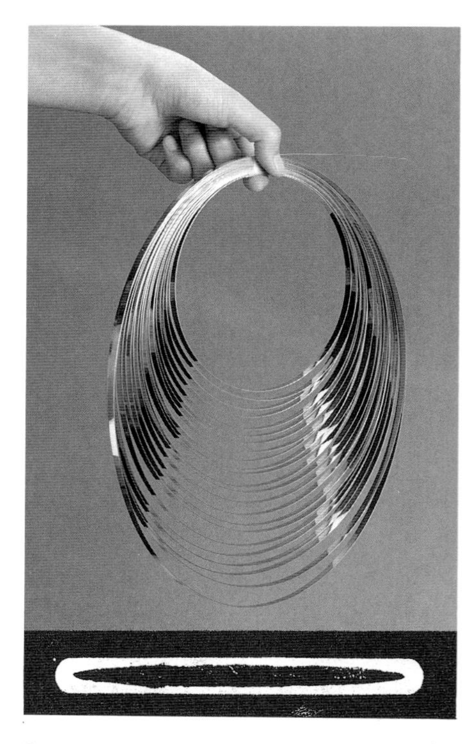

Figure 1. A cross section of the Ag-sheathed $YBa_2Cu_3O_{7-x}$ superconductor and a 20m long tape.

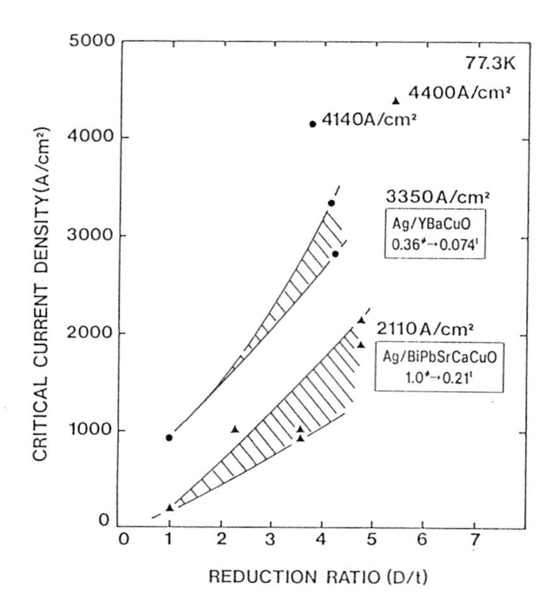

Figure 2. Jc increase by reduction ratio (D/t), where D is wire diameter before press processing and t is the thickness of tape after press processing. Black circles and triangles indicate the data for Ag-sheathed $YBa_2Cu_3O_{7-x}$ and $Bi_{0.8}Pb_{0.2}SrCaCu_2O_x$ tapes respectively. The top data of $4400A/cm^2$ was obtained for $Bi_{0.8}Pb_{0.2}SrCaCu_{1.5}O_x$ superconductor.

RESULTS AND DISCUSSION

The cross section and the appearance of the wire are shown in figure 1. It is well known that the orientations of the crystals of $YBa_2Cu_3O_{7-x}$ powder are aligned by uniaxial pressure especially on the surface.

We expect to increase Jc by alignment of crystal orientations. Therefore, it seems necessary to compress the inner powders from outside the silver sheath by cold pressing or cold rolling.

The alignment of crystal orientations seems to reduce grain boundaries and improve contacts between grains. We expect to increase Jc due to the compaction and orientation of the grains by the press or rolling process. Figure 2 shows this increase in Jc due to press or roll processing. The abscissa is the reduction ratio of D/t where D is the diameter of the wire before press processing and t is the thickness of the tape after press processing.

The data for both Ag-sheathed $YBa_2Cu_3O_{7-x}$ and $Bi_{0.8}Pb_{0.2}SrCaCu_2O_x$ superconducting tapes are shown in this figure. Both materials showed a sharp increase in Jc as the reduction ratio increased.

The area shaded by oblique lines shows the distribution of data when wire size, heat treatments and/or pre-conditioning of the powders were changed. The highest Jc obtained was **$4140A/cm^2$** for Ag-sheathed $YBa_2Cu_3O_{7-x}$ and **$4400A/cm^2$** for Ag-sheathed $Bi_{0.8}Pb_{0.2}SrCaCu_{1.5}O_x$ superconducting tape respectively, both at 77.3K in zero magnetic field.

The $BiSrCaCu_2O_x$ superconductor, which does not include Pb, did not show any increase in Jc after press processing. The maximum Jc of Ag-sheathed $BiSrCaCu_2O_x$ wire was **$426A/cm^2$** at 77.3K and 0T.

Superconducting properties were improved and formation of the high Tc phase was accelerated through press processing. Our Ag-sheathed $Bi_{0.8}Pb_{0.2}SrCaCu_2O_x$ wire did not include a low Tc phase judged by the measurement of DC susceptibility. The critical temperature (R=0) was 104K.

Magnetization of the $Bi_{0.8}Pb_{0.2}SrCaCu_2O_x$ and $YBa_2Cu_3O_{7-x}$ superconductors revealed some hysteresis during measurement by a vibrating sample magnetometer, while the $BiSrCaCu_2O_x$ superconductor showed no hysteresis in a range above 0.1T. Scanning Electron Microscopy (SEM) showed the orientation of the crystal structure of the $Bi_{0.8}Pb_{0.2}SrCaCu_2O_x$ superconductor. The crystals looked like thin flakes stacked tightly in c-direction showing multiple layers.

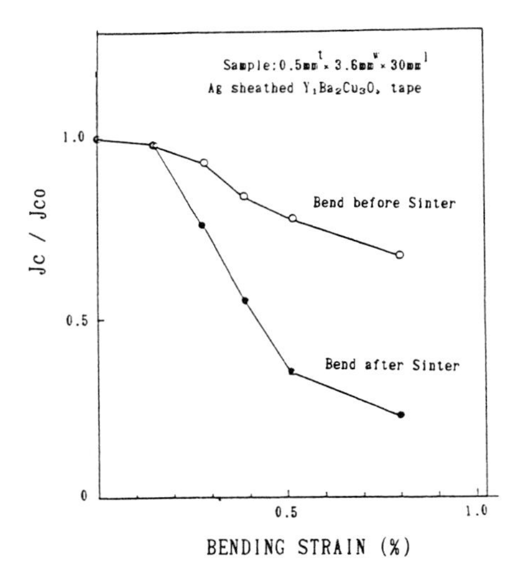

Figure 3. Jc degradation due to bending before and after sintering for Ag-sheathed $YBa_2Cu_3O_{7-x}$superconducting tape.

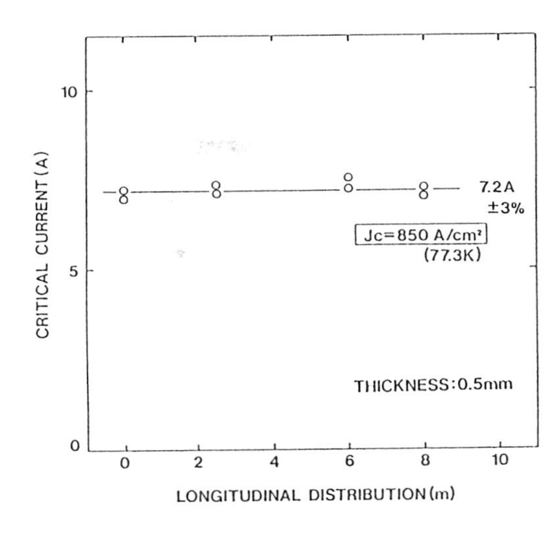

Figure 4. Critical current distribution in long Ag-sheathed $YBa_2Cu_3O_{7-x}$ superconducting tape.

To study the fundamental properties of magnet coils, we investigated the bending properties of the tape, magnetic field dependence of Jc,and Ic distribution in a length of tape.

The critical current densities are shown in figure 3 when 0.5mm thick tape was bent either before or after sintering. The Jc for sintered material rapidly degraded over 50% when the strain exceeded 0.3%, because sintered ceramics are usually brittle. The magnetic field dependence of Jc will be reported elsewhere.[6]

The Ic distribution in a 10m long tape is shown in figure 4. The eight samples were cut from four points of a long $YBa_2Cu_3O_{7-x}$tape with Ag-sheath. They were sintered at once, followed by oxygen annealing. The deviation from the mean value of 7.2A is only 3% or less for eight samples cut from a 10m length.

These deviations seem to be small enough to permit coil manufacture. We then made a coil using the " wind and react " technique, which is commonly applied in making superconducting coils for the Nb_3Sn intermetallic compound superconductor. The coil shown in figure 5 is wound with 7 turns of 12 layered Ag-sheathed YBa_2Cu_3O tape on a ceramic bobbin of 50mm in diameter. Twelve layers of superconducting tape is difficult to join in a series connection, so we connected both ends in parallel. Therefore we can enlarge the total current carrying capacity.

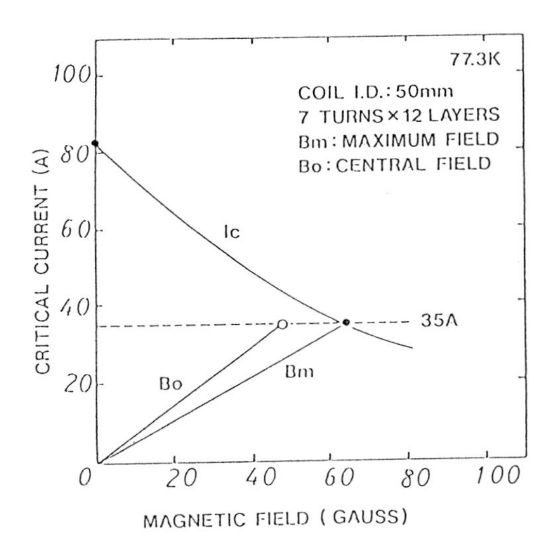

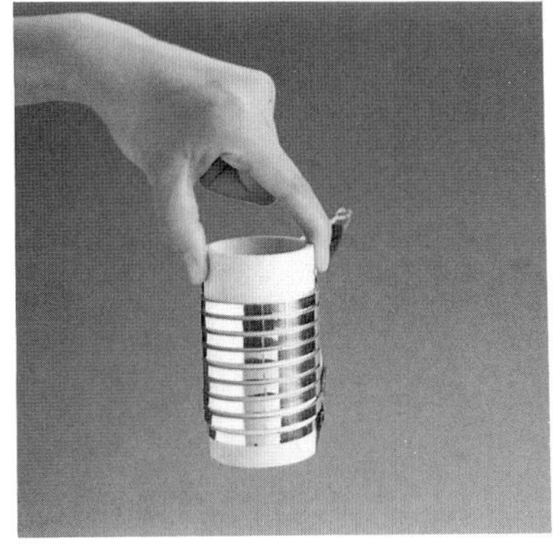

Figure 5. High Tc superconducting coil of Ag-sheathed $YBa_2Cu_3O_{7-x}$tape wound on a ceramic bobbin and coil load line.

Currents as large as 35A were successfully supplied to this coil generating 47gauss at the center of the coil, and 64 gauss on the conductor. The summation of critical currents of each tape at zero field was 81A. From these values we can reasonably anticipate that the coil properties of high Tc superconductors coincide with short sample properties. There remain many problems to making high Tc superconducting magnets, such as insulation, mechanical support, and stability of coil, but we believe that the feasibility of high Tc magnets has been confirmed from these results.

CONCLUSION

We investigated the critical current density of Ag-sheathed $YBa_2Cu_3O_{7-x}$ and $Bi_{0.8}Pb_{0.2}SrCaCu_2O_x$ superconductors and found through press processing from round wire to tape, that their properties are dependent on the reduction ratio. We improved the critical current density of the $YBa_2Cu_3O_{7-x}$ and $Bi_{0.8}Pb_{0.2}SrCaCu_{1.5}O_x$ superconductors to 4140A/ cm^2 and 4400A/ cm^2 respectively, at 77.3K in a zero magnetic field.

We investigated the bending properties of these tapes as well as the Jc distribution in a long conductor. We made a coil of 50mm bore diameter using Ag-sheathed $YBa_2Cu_3O_{7-x}$ tape that could generate a 64gauss magnetic field when a 35A current was transported. Thus, the feasibility of high Tc coil was demonstrated.

ACKNOLEDGEMENT

We greatly appreciate Dr.T.Nakahara and T.Mitsui for thier constant encouragement and advice.

REFERENCES

[1] K. Ohmatsu, K. Ohkura, H. Takei, S. Yazu and H. Hitotsuyanagi: Jpn. J. of Appl. Phys. vol.26, pp.1207-1208, (1987) Suppl.26-3, Proc. 18th Int. Conf. on Low Temperature Phys., Kyoto, 1987.

[2] M. Okada, A. Okayama, T. Morimoto, M. Matsumoto, K. Aihara, and S. Matsuda: Jpn. J . of Appl. Phys. vol.27, pp.L185-L187,(1988)

[3] K. Kawashima, M. Nagata, Y. Hosoda, S. Takano, N. Shibuta, H. Mukai and T. Hikata: '88 Appl. Superconductivity Conference, Sanfrancisco, CA, August 21-25, 1988. paper MI-6

[4] H. Maeda, Y. Tanaka, M. Fukutomi, and T. Asano : Jpn. J. Appl. Phys. vol.27,pp.L209-L212 (1988)

[5] R. J. Cava et al. A paper presented at Int'l Conf on High Tc S.C and Material and Mechanism of Superconductivity, Interlaken, Feb.29-March3, 1988 ; Physica C153-155 (1988) pp.560-565, North-Holland, Amsterdam.

[6] T.Nakahara : " Development of Superconducting Material and Impacts on the Social Economy. " A paper which will appear in this conference (ISS'88)

4.7 Coherence Length, Magnetic Properties, and Critical Current

Properties of High Tc Short Coherence Length Superconductors

GUY DEUTSCHER

School of Physics and Astronomy, Raymond and Beverly Sackler Faculty of Exact Sciences, Tel Aviv University, Ramat Aviv, Tel Aviv, Israel

ABSTRACT

Two fundamental consequences of the very short coherence length of the new high Tc superconducting oxides are discussed. The first one is an enhanced sensitivity to crystallographic defects which results in local depressions of the superconducting order parameter and weak link behavior. This effect becomes very pronounced when the coherence length approaches the lattice parameter, and results in much reduced critical current densities accross boundaries perpendicular to the c axis. The second consequence are enhanced critical fluctuations of the order parameter in the small coherence volume. We show that there exists a direct relationship between the width of the critical region and the weak pinning often observed in the high Tc oxides.Both of the effects that we discuss - local depressions of the order parameter at boundaries and enhanced critical fluctuations - contribute to a reduction of the critical current density, flux motion and glass behavior.

INTRODUCTION

Since the historical discovery of the high Tc oxides (1), their granular behavior has been noticed as one of their prominent features. Low critical current densities, susceptibility and magnetization behavior reminescent of that of spin glasses - hence the term glassy behavior (2) -, anomalous microwave absorption at fields lower than the bulk thermodynamical field for first vortex penetration H_{c1} (3), a.c. Josephson effects occuring at "internal" junctions (4) were noticed very early in polycrystalline samples. Later on, similar effects were also observed in single crystals (5,6,7), showing that they are of a fundamental nature rather than resulting from the poor connectivity of sintered ceramic samples that are always to some extent porous and often contain second phase precipitates. Such defects further enhance the granular behavior of the high Tc oxides, but there is growing evidence that it is already present to some extent in single crystals.

Nevertheless, single crystals (8) and oriented films (9,10) have much higher critical current densities than sintered samples, showing that the high Tc oxides have a high intrinsic critical current density. At low temperatures, it approaches 1.10^8 A/cm^2 in the (a,b) plane and 1.10^7 A/cm2 along the c axis (10) for YBCO. These critical current densities are close to the intrinsic depairing limit.

The general behavior of the critical current in the high Tc oxides - low in polycrystalline samples and high in single crystals and oriented films - is just the opposite of that observed in the conventional Type II superconductors. There, the critical current is enhanced by the presence of such defects as grain boundaries, small second phase precipitates, dislocations, while it is low in high quality single crystals. It is well understood that in conventional superconductors, defects are in fact necessary to generate the so called "pinning centers" that prevent the free motion of vortices. The empirical evidence available today shows that the classical pinning centers are not useful in the high Tc oxides - they rather appear to be detrimental.

As first suggested by Deutscher and Müller (11), the short coherence length of the new oxides appears to be at the origin of this anomalous behavior. When the coherence length is so short as to approach the lattice parameter, a local depression of the order parameter is expected at surfaces and interfaces. Extended defects such as grain boundaries, stacking faults, off-stoechiometry planes and twins can then become weak junctions. At least for the case of grain boundaries, there is now clear direct evidence for such a depression (10).

A small coherence length also results in a condensation energy per coherence volume that may not be much larger than k_BT_c. As pointed out previously (12) this leads to non mean field behavior in a measurable range of temperatures near Tc, another feature that distinguishes the new high Tc oxides from the conventional superconductors.Besides the expected anomalies in the behavior of the electronic heat capacity and related quantities, complete reversibility of the magnetization curves must occur in the critical region (or equivalently zero critical current density). This is because, following Anderson (13), the condensation energy per coherence volume is also the typical (maximum) energy barrier that must be overcome to free vortices from pinning centers. Hence, there can be no effective pinning in the critical region.

The depression of the order parameter predicted at boundaries reduces even further the local condensation energy. Easy flux motion can then occur well beyond the critical region, and at fields much lower than H_{c2}, as recently reported by Yeshurun and Malozemoff (6). In practice, the critical region appears to be less than 1K in width. We shall therefore argue that the absence of effective pinning over a much broader range of temperatures is indeed an extrinsic (defect related) rather than intrinsic phenomenon.

THE COHERENCE LENGTH IN THE HIGH Tc OXIDES

In a BCS superconductor, the coherence length is given in the clean limit by the Fermi velocity divided by the energy gap (in units where h = 1). In most metals, the Fermi velocity $v_F = 1.10^8$ cm/s. For a critical temperature of the order of 10K, one then obtains in the weak coupling limit a coherence length of the order of 1000Å. Applying the same formula to the high Tc oxides, with $v_F = 1.10^7$ cm/s and Tc = 100K, one gets a coherence length of the order of 10 Å. This is close to the value determined from measurements of the upper critical field H_{c2}, as discussed below.

This agreement might well be fortuitous, i.e. it does not imply that the high Tc oxides are BCS superconductors.

The currently accepted value for v_F has been derived from Hall effect, electrical conductivity and heat capacity measurements whose interpretations are not straightforward. In the absence of fundamental experiments such as de Haas van Halfen measurements, we are not even sure that these oxides are Fermu liquids, a point that remains much debated (14).

The value of the energy gap remains also in question, ranging from the weak coupling BCS value 1.75 k_BT_c up to twice this value or more(15). On the other hand, there might not even be an energy gap (14).

The experimental determinations of the coherence length are also questionable. They are mostly based on measurements of the upper critical field H_{c2}. The existence of vortices with the usual flux quantum value having been established (16), we are entitled to use the conventional expression for H_{c2} to calculate the (temperature dependent) coherence length. But the experimental determination of H_{c2} has proven to be quite difficult. The resistive transitions are much broader than in the conventional superconductors. The definition of H_{c2} then becomes a matter of interpretation. If it is determined as the field at which half of the normal state resistance is restored - for instance - an unusual temperature dependence is also observed, different from the linear dependence predicted by the Landau Ginzburg theory.Instead, one obtains a power law dependence, H_{c2} varying as $(T_c - T)^n$, with n in general close to 1.5 (17). It is now understood that this field is related to the irreversibility line first reported by Müller et al. (19) in ceramic materials and more recently in single crystals (6), rather than being a determination of the true upper critical field where the average order parameter goes to zero.

Therefore, values of the coherence length originally obtained from such measurements - about 30 Å in the (a,b) plane and 7 Å along the c axis - are probably only upper bounds. Better values of H_{c2} might rather correspond to the onset of the resistive transitions - giving about 12 Å in the (a,b) pane and 2 Å along the c axis (17,20). The values quoted here are based on the measured slopes (dH_{c2}/dT) at Tc, and assume the conventional Landau-Ginzburg temperature dependence.

This assumption might not be correct. Two groups have actually reported that the correct H_{c2} might actually have an infinite slope at Tc. In one measurement the upper critical field was determined as the field at which the critical current density extrapolates to zero (21), and in the other where the magnetization extrapolates to its normal state value (22). It is clear that more measurements - particularly bulk measurements - are necessary to establish the exact values of the coherence length, as well as the true temperature dependence of H_{c2}.

Values of the coherence length quoted above are for the YBCO compound. The study of the upper critical field in different oxides - particularly those with a lower Tc - is of fundamental interest and could well shed more light on the correct values of the coherence length and on the validity of the BCS expression. The critical temperatures of the new oxides now span a broad rage, from about 10K to more than 100K. This should be sufficient to establish whether the coherence length is roughly inversely proportional to Tc, as predicted by BCS and observed in the conventional superconductors, or whether it follows a different dependence. The variation of the coherence length with Tc is also of great practical importance. As discussed below, its small value is at the origine of the difficulties encountered in the practical applications of these materials. If the coherence length is shorter in the higher Tc compounds, as predicted by BCS, these difficulties cannot be cured by raising Tc. If, on the other hand, the BCS relation does not apply (for instance, if for some reason the coherence length turns out to be only weakly dependent on Tc), higher Tc oxides will be of great practical importance.

This being said, it is sufficient to retain for the following discussion that in the new superconductors the coherence length is probably equal to a few lattice spacings in the (a,b) plane, and significantly less along the c axis.

In the following section we discuss we discuss the influence of this short coherence length on the behavior of the order parameter at defects, particularly boundaries; and in the next one, its impact on the activated motion of vortices.

THE DEPRESSION OF THE ORDER PARAMETER AT BOUNDARIES

In the conventional supersonductors, defects of atomic size act as scattering centers and decrease the coherence length, but do not modify the value of the order parameter at the site of the defect. This is because the coherence length is so much larger than the lattice spacing.
This situation is radically modified in the new oxides where, due to the short coherence length as discussed above, one must expect that local perturbation of the electronic structure will correlate strongly with the order parameter. Since these oxides are in general definite compounds, defects should in general bring about a local depression of the order parameter.
In particular, as shown by Deutscher and Mûller (11), one expects a depression of the order prameter at boundaries between the superconductor and an insulator. This is in contrast with the zero gradient boundary condition that applies to conventional superconductors.
In order to understand the physical origin of this depression, it is worthwhile to remark that the superconductor/insulator interface for the new superconductors is analog to the superconductor/normal metal interface for the conventional superconducors from the standpoint of the relevant length scales.In the later case, the order parameter on the S side is decreased because the superconducting wave function "leaks out" in the normal side N. This leakage occurs in general over a length scale that is comparable to the coherence length of S, and the depression in S is significant. In an insulator, the leakage occurs only on the atomic scale - but this is about the same as the coherence length in the new oxides.

Landau-Ginzburg analysis of the depression

In a Landau Ginzburg analysis, the depression is a function of the ratio r of an "extrapolation length" b to the temperature dependent coherent length of the superconductor. The order parameter at the interface is proportional to r for r much smaller than 1, and regains the bulk value for r much larger than 1. For a superconductor/insulator interface, the length b is temperature independent. Since the superconducting coherence length goes to infinity at Tc, the order parameter at the interface goes to zero in that limit. The length b being roughly given by the square of the zero temperature coherence length divided by an interatomic distance, this depression only occurs extremely close to Tc in a conventional large coherence length superconductor (typically, within 1 microdegree from Tc). In practice, the effect is undetectable and the zero gradient boundary condition always applies. But in the new superconductors, the coherence length is of the order of the lattice parameter as noted, and the depression occurs over a sizable range of temperatures. It should be very pronounced at all temperatures at a boundary perpendicular to the c axis, and less pronounced but still substantial for a boundary parallel to the c axis, at least at high temperatures.

Physical consequences of the depression

The depression immediatly explains the gapless characteristics obtained in most tunneling experiments (23). Since a tunneling experiment probes the superconductor over a coherence length, over which the order parameter recovers its bulk value, the I(V) characteristic is both sensitive to the bulk value of the gap, but also to the significant density of states within the gap that exist at the surface. In practice, a (rough) value for the gap can be given by the energy scale of the characteristic, but at the same time an anomalously large value of the initial slope is measured, even at low temperatures. While the zero temperature value of the energy gap derived from such measurements is probably correct, the determination of its temperature dependence requires a much finer analysis.
Another consequence of the depression is the strong reduction of the critical current density observed in polycrystalline samples. Even a "clean" grain boundary is several interatomic distances in width, with in general a high density of dislocations. It can be metallic, semiconducting or insulating, but in any case non-superconducting for reasons discussed above. A grain boundary will therefore be either an S/N/S or an S/I/S juction. As far as the depression of the order parameter is concerned, there is not much difference between these two types of junctions, as shown above. This depression brings about a reduction of the critical current density accross the boundary. In the case of an S/I/S junction, it is of course further reduced by the transmission coefficient of the barrier.
Analysis of the temperature dependence of the critical current accross grain boundaries (10) clearly indicates that it involves a depression of the order parameter, not just a reduced transmission coefficient as in a conventional tunnel junction. For the boundaries parallel to the c axis investigated in (10), this depression is at least a factor of 2 at low temperatures, consistent with a coherence length in the (a,b) plane of the order of 10 Å (20). We expect a much larger depression at boundaries perpendicular to the c axis, because of the shorter coherence length in that direction.

Internal junctions

There is now convincing experimental evidence that junctions also exist in single crystals, as first suggested in Ref.11. Daeumling et al. have reported size independent magnetization of powders of $DyBa_2Cu_3O_7$ (24,25). After presenting data showing a size dependent magnetization according to the Bean model in powders of $YBa_2Cu_3O_7$ (26), Campbell et al. have now come (27) to conclusions identical to those of Daeumlinget al. In an applied field of 1T, and at 77K and 30K, the magnetization is clearly independent of the particle size between 2 and 30 micron. Hence the conclusion that the grains consist of sub-units, for which 2 micron is an upper limit.

A further proof of the existence of internal junctions in single crystals comes from the experiments of Küpfer et al. who performed a similar experiment by powdering a single crystal(28) into grains of smaller and smaller size, measuring at each stage the magnetization. Again, the later was found to be independent of the particle size, down to 10 micron (the smallest size investigated).

A similar conclusion has also been reached from susceptibility measurements on powders by Polturak et al. (29) and Monod et al. (30).

Practical consequences for the critical current densities

Hence, the existence of intragrain as well as intergrain junctions is now well established, substantiating the predictions of Ref.11. In a polycrystalline sample, the most damaging boundaries are those perpendicular to the c axis. We estimate that in that case the order parameter at low temperatures might be depressed by as much as a factor of 10, and hence the critical current by a facor of 100. If we assume a relatively large transmission coefficient of 0.1 for such boundaries (as seems to be the case for boundaries parallel to the c axis from an analysis of Ref.10), the total reduction is of about three orders of magnitude. With a depairing current of about 1.10^8 A/cm^2, one ends up with a probable upper limit of 1.10^5 A/cm^2 for polycrystalline samples of YBCO at low temperatures, and less than 1.10^4 A/cm^2 at a reduced temperature of 0.9. In the absence of a detailed knowledge of the properties of boundaries, the above are of course no more than educated guesses. Nevertheless, we note that the above figures are higher than measured in sintered samples, but not too far from measurements on good polycrystalline films. Let us emphasize again that these limitations are extrinsic rather than intrinsic. An appropriate treatment of the boundaries, aimed at locally reducing the anisotropy, could well improve the critical current of polycrystalline samples.

CRITICAL BEHAVIOR AND PINNING

In the conventional superconductors, the second order phase transition is perfectly well described mean field theory. Critical thermodynamic fluctuations must exist, but they occur in such a small temperature range that they cannot be detected. The applicability of mean field theory reflects the fact that there are many Cooper pairs within the correlation radius of one pair, or equivalently that the critical temperature is much smaller than the Fermi energy. These two related conditions are clearly violated in the new superconductors, in which the coherence length is small and comparable to the distance between the carriers, the critical temperature is high and the Fermi energy rather low due to the small carrier density.

In terms of directly measurable parameters, the width of the critical region - the range of the critical fluctuations - is determined by the condition that the condensation energy per coherence volume, U(T), is of the order of k_BT_c or smaller. The condensation energy per unit volume being given by $(H_c^2/8\pi)$, where H_c is the thermodynamic critical field, one can then calculate the width of the critical region by using experimentally determined values of H_c and of the coherence length. In doing this, one uses the mean field exponents under the accepted assumption that mean field theory is good enough to tell when it breaks down. With all the uncertainties attached to the determination of H_c and of the coherence length, one obtains for the new superconductors a width that is much larger than for conventional superconductors (31).

It is also useful to remark that the energy U(T) is precisely that which enters into the expressions for flux creep - at least at small fields. Hence, instead of calculating U(T) from poorly known values of H_c and of the coherence length, it is simpler and safer to estimate it from measurements of the width of the critical region. Although the later has not been yet studied in great detail due to experimental difficulties, heat capacity (32,33) as well as very careful conductivity measurements (34) indicate a width that is probably not larger than 0.1K.This immediatly gives us a lower bound for U(0) of about 0.3 eV.

Values of U(0) recently given from flux creep analysis on single crystals (0.01 eV, Ref.35) and on thin films (0.07 eV, Ref.36) are significantly smaller than this lower bound. Hence, these values must be extrinsic rather than intrinsic, i.e. related to the particular defect structure of the samples rather than reflecting the value of the condensation energy per coherence volume.

Glass behavior

The existence of internal junctions having been clearly established by the many experiments cited above (Ref. 24 to 30), we believe that the small values of the activation energy obtained from a flux creep analysis stem from the depression of the order parameter at boundaries, and the correspondingly reduced value of the condensation energy. A description of these boundaries as pinning centers would be misleading. Vortices will actually move easily along the boundaries; this is the relevant situation when a magnetic field is applied parallel to the boundaries, and a current flows accross them.

An analysis of the critical current temperature dependence and of magnetization data in terms of a single activation energy actually only applies at low temperatures, where the experiments naturally "pick-up" the smallest activation energy in the system. At higher temperatures, the more complex structure of the boundaries network becomes detectable. In an applied field, boundaries do not act independently from each other but form a frustrated system of loops that is at the origin of the glassy behavior of the high Tc oxides(11). Aging effects, specific to glasses, have indeed very recently been observed in an YBCO single crystal by Rossel et al. (37).

CONCLUSIONS

Much works remains to be done to determine the exact values of the coherence length in the high Tc oxides, through better measurements of H_{c2} and of its temperature dependence. Such measurements, carried on a variety of oxides covering a broad range of critical temperatures, will be helpful to understand the origin of the high Tc, and to evaluate better the potential for applications of the oxides. There is now considerable experimental evidence for the existence of intergrain as well as intragrain junctions in these materials, basically due to the fact that the coherence length is of the order of the lattice parameter. Boundaries perpendicular to the c axis are predicted to be the most damaging; a local reduction of the anisotropy by an appropriate metallurgical treatment might in part remedy this problem. The short coherence length provides a broad range of activation energies in the oxides, which is at the origin of glassy behavior. The low activation energy determinedby a flux creep analysis of low temperature experiments is significantly smaller than the condensation energy per coherence volume, as determined from the width of the critical region region. It is therefore extrinsic rather than intrinsic, and is not the fundamental limitation to the critical current densities.

We are much indebted to Prof. K.A. Müller for a number of illuminating conversations. This work was supported in part by the US-Israel Binational Science Foundation and by the Oren Family Chair for Experimental Solid State Physics.

REFERENCES

+ Currently on leave at the Ecole Supérieure de Physique et de Chimie Industrielles de la Ville de Paris, 10 rue Vauquelin, 75231 PARIS Cedex05, France.

1) J.G. Bednorz and K.A. Müller, Z.Phys. B64, 189 (1986).
2) K.A. Müller, M. Takashige and J.G. Bednorz, Phys. Rev. Lett. 58, 1143 (1987).
3) K.W. Blazey, K.A. Müller, J.G. Bednorz, W. Berlinger, G. Amoretti, E. Buluggiu, A. Vera and C. Matacotta, Phys. Rev. B36, 7241 (1987).
4) D. Esteve, J.M. Martinis, C. Urbina, M.H. Devoret, G. Collin, P. Monod, M. Ribault and A. Revcolevschi, Europhys. Lett. 3, 1237 (1987).
5) T.R. Dinger, T.K. Worthington, W.J. Gallagher and R.L, Sandstrom, Phys. Rev. Lett. 58, 2687(1987)
6) Y. Yeshurun and A.P. Malozemoff, Phys. Rev. Lett. 60, 2202 (1988).
7) K.W. Blazey, A.M. Portis, K.A. Müller and F.H. Holzberg, to appear in Europhys. Lett.
8) T.R. Dinger, T.R. McGuire, D. Keane and M. Chisholm, to be published (reported at APS Meeting New Orleans, March 1988.
9) A. Kapitulnik et al., Proceedings of the Interlaken Conference, Switzerland, Feb. 29-Mar.4,1988.
10) D. Dimos, P. Chaudari, J. Manuhart and F.K. Le Gones, Phys. Rev. Lett., July 1988.
11) G. Deutscher and K.A. Müller, Phys. Rev. Lett. 59, 1745 (1987).
12) G. Deutscher in "Novel Superconductivity", ed. S.A. Wolf and V.Z. Kresin, Plenum Press 1987, p.p. 293; C.J. Lobb, Phys. Rev. B36, 3930 (1987); A. Kapitulnik, M.R. Beasley, C. Castellani and C. Di Castro, Phys. Rev. B37, 537 (1988).
13) P.W. Anderson, Phys. Rev. Lett. 9, 309 (1962).
14) P.W. Anderson, Proceedings of the Interlaken Conference, Op. cit.
15) A. Kapitulnik, presented at the International Conference on Critical Currents in High Temperature Superconductors, Snowmass Village, Colorado, August 16-19,1988.
16) P.L. Gammel, D.J. Bishop, G.J. Dolan, J.R. Kwo, C.A. Murray, L.F. Schneemeyer and J.V. Waszczak, Phys. Rev. Lett.59, 2592 (1987).
17) B. Oh, K. Char, A.D. Kent, M. Naito, M.R. Beasley, T.H. Geballe, R.H. Hammond and A. Kapitulnik, Phys. Rev. B37, 7861 (1988).
18) M. Tinkham, preprint.
19) K.A. Müller, M. Takashige and J.G. Bednorz, Phys. Rev. Lett. 58, 1143 (1987).
20) G. Deutscher, Proceedings of the Interlaken Conference, Op. Cit.
21) U. Dai and G. Deutscher, Appl. Phys. Lett. 51, 6 (1987).
22) M.M. Fang, V.G. Kogan, D.K. Finnemore, J.R. Clem, L.S. Chumbley and D.E. Farrell, Phys.Rev. B37, 2334 (1988).
23) See for instance L.H. Greene in Proceedings of the Interlaken Conference, Op. Cit.
24) M. Daeumling, J. Seuntjeus, and D.C. Larbalestier, Appl. Phys. Lett. 52, 7 (1988).
25) M.Daeuling, J. Seuntjens, X. Cai and D.C. Larbalestier, presented at the International Conference on Critical Currents, Snwmass Village, Op. Cit.
26) A.M. Campbell et al., presented at the Materials Research Meeting, Boston, December 1987.
27) A.M. Campbell et al. , presented at the International Conference on Critical Currents , Snowmass Village, Op. Cit.

28) H. Küpfer, R. Flükiger, J. Apfelstedt, C. Keller, R. Meier-Hirmer, B. Runtsch, A. Turowski U. Wiech and T. Wolf, presented at the International Conference on Critical Currents, Snowmass Village, Op. Cit.
29) E. Polturak, D. Cohen and A. Brokman, to be published.
30) P. Monod et al., Proceedings of the Interlaken Conference, Op.Cit.
31) G. Deutscher, in "Novel Superconductivity", ed. S.A. Wolf and V.Z. Kresin, Plenum Press 1987, p.p. 293.
32) R.A. Butera, Phys. Rev. B37, 5909 (1988).
33) A. Voronel et al. and D.M. Ginsberg et al., Proceedings of the Interlaken Conference, Op. Cit.
34) S.J. Hagen, Z.Z. Wang and N.P.Ong, to appear in Phys. Rev.B.
35) Y. Yeshurun et al., presented at the INternational Conference on Critical Currents, Snowmass Village, Op. Cit.
36) J. Mannhart et al., presented at the International Conference on Critical Currents, Snowmass Village, Op. Cit.
37) C. Rossel et al., presented at the IBM Workshop on High Temperature Superconductivity, Oberlech, Austria, August 8-12, 1988, to be published in the IBM Journal of Research and Development.

Free Energy Surfaces in the Superconducting Mixed State

D.K. FINNEMORE[1], M.M. FANG[1], N.P. BANSAL[2], and D.E. FARRELL[3]

[1] AMES Laboratory, Iowa State University, Ames, IA 50011, USA
[2] NASA Lewis Research Center, Cleveland, OH 44135, USA
[3] Department of Physics, Case Western Reserve University, Cleveland, OH 44106, USA

ABSTRACT

The free energy surface for $Tl_2Ba_2Ca_2Cu_3O_{10}$ has been measured as a function of temperature and magnetic field to determine the fundamental thermodynamic properties of the mixed state. The change in free energy, G(H)-G(0), is found to be linear in temperature over a wide range indicating that the specific heat is independent of field.

INTRODUCTION

Free energy measurements provide an important first step in understanding the fundamental processes involved in the performance and behavior of high temperature superconductors. Thermodynamic reversibility, of course, is essential to the definition of the Gibbs free energy so it is important to establish the range where magnetic flux moves easily. For most of these materials there is a temperature interval below the transition temperature, T_c, where the magnetization curves are thermodynamically reversible and the free energy can be measured. In this range the vortices and magnetic flux move in and out of the sample reversibly enough to permit measurement of the Gibbs free energy to an accuracy of better than 1%. The $Y_1Ba_2Cu_3O_7$ (123) materials are generally reversible for a 6K interval below T_c. In this work we have found that the $Tl_2Ba_2Ca_2Cu_3O_{10}$ (2223) is reversible for a 30K interval below T_c.

Both the Y-based and Tl-based superconductors are highly anisotropic materials with high conductivity in the a-b plane and relatively lower conductivity along the c-axis. This leads to an anisotropic effective mass tensor and an anisotropic coherence distance. In addition the transition temperature is high with a relatively low Fermi velocity so the magnitude of the coherence distance is small, on the order of 0.7 nm along the c-axis and about 2.0 nm in the a-b plane. This is on the order of a unit cell dimension so the vortex behavior may be quite different from that described by the usual Ginsburg-Landau-Abrikosov-Gorkov [1] theory. In fact, the vortices could show melting, tangles [2] and a variety of unusual behavior.

There are many clues in the literature that the thermodynamic properties of these high T_c materials differ from those of the A-15 materials. The specific heat jump at T_c shows very large fluctuation effects [3]. There is relatively little change in specific heat with magnetic field for (123) in the reduced temperature range T/T_c less than 0.9 [4]. The jump in specific heat occurs at essentially the same temperature as the magnetic field increases from 0 to 7T. From the values of the upper critical field derived from measurements of the onset of vortex motion [5] (e.g., resistivity) the temperature where the jump in specific heat occurs should have been suppressed by about 10K for a field of 7T. This does not occur.

The purpose of this paper is to summarize a series of measurements of the free energy as a function of magnetic field and temperature in the mixed state of the 2223 in order to study these phenomena.

EXPERIMENTAL DETAILS

Measurements of the magnetization are conducted on grain aligned [6] samples that have the c-axis aligned but a random orientation of the a-b planes. X-ray rocking curve show a full width at half maximum spread of less than two degrees. Magnetization measurements were carried out by slowly pulling the sample through a superconducting quantum interference device (SQUID) coil in a Quantum Designs instrument.

RESULTS AND DISCUSSION

Magnetization data in a field of H = 20 Oe show an onset of flux exclusion at 120K and a very sharp drop at 117K as shown in Fig. 1. For fields parallel to the a-b plane the magnitude of the exclusion (M_{11}) is about 30% as large as for fields perpendicular to the a-b plane (M_1). At this low field the reversibility window is only about 1K wide.

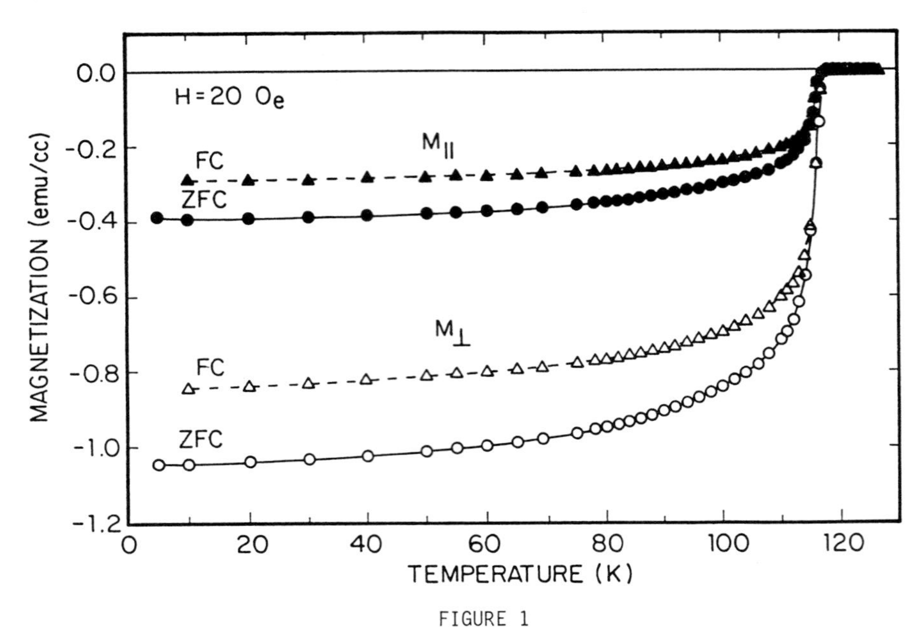

FIGURE 1

Low Field Magnetization Data. M_{11} Indicates H Parallel to a-b Plane.

At fields much larger than the lower critical field (H_{c1}) the magnetization curves are qualitatively similar to that shown for H = 2T in Fig. 2.

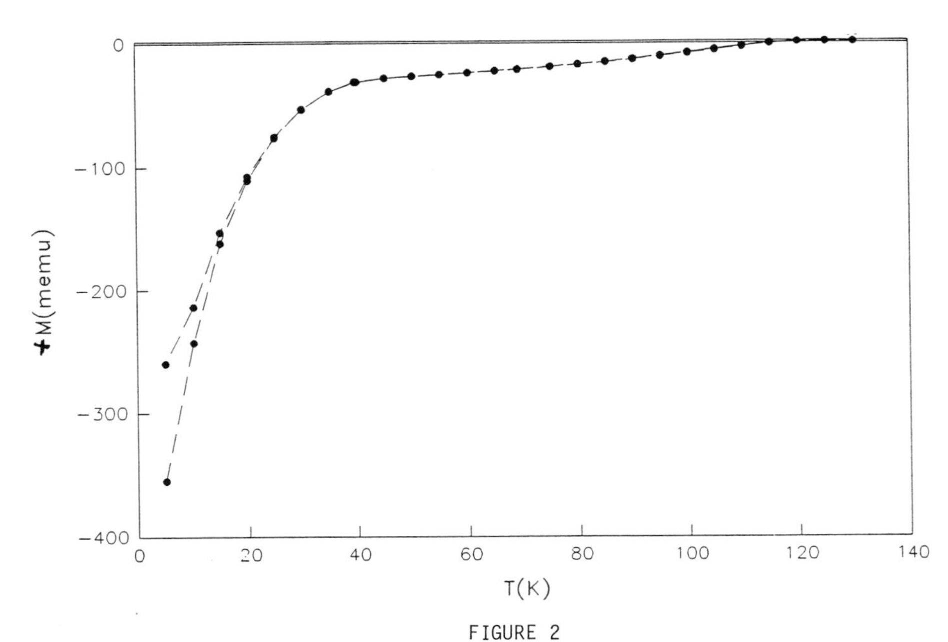

FIGURE 2

Magnetization at H = 5T Showing the Onset of Irreversibility at 25K.

Here the magnetization falls linearly with temperature from 117K to about 40K. At this point M drops sharply and the magnetization becomes irreversible below about 25K. At 0.1T the behavior is similar except that the temperature for the onset of reversibility is 75K. There is, therefore, a very large region of the H-T plane where the magnetization is completely reversibly and thermodynamic quantities can be derived. In the low field region where irreversibility occurs, the critical state model assumptions are used to determine the reversible magnetization to be the average of the field cooled (FC) and zero field cooled (ZFC) values.

The free energy for the (2223) sample evaluated from $G_H - G_O = \int_0^H MdH'$ is shown in Fig. 3 for 5T (solid circles) and 2T (solid squares). With the accuracy of the data,

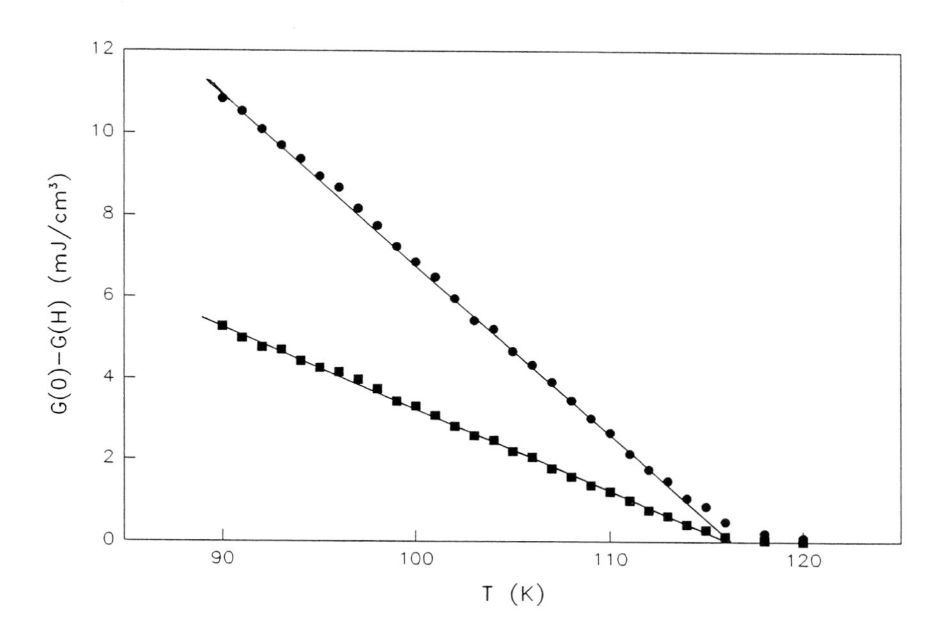

FIGURE 3
Free Energy Curves For 5T (Circles) and 2T (Squares)

G(0) - G(H) is linear in temperature between 90K and 112K. These data are then thermodynamically related to the change in specific heat by

$$C_O - C_H = -T \frac{d^2(G_O - G_H)}{dT^2}$$

Errors in the free energy introduced from this are less than 1%. The linear $G_H - G_O$ curve then means that the specific heat is independent of H. This is shown more explicitly in Fig. 4.

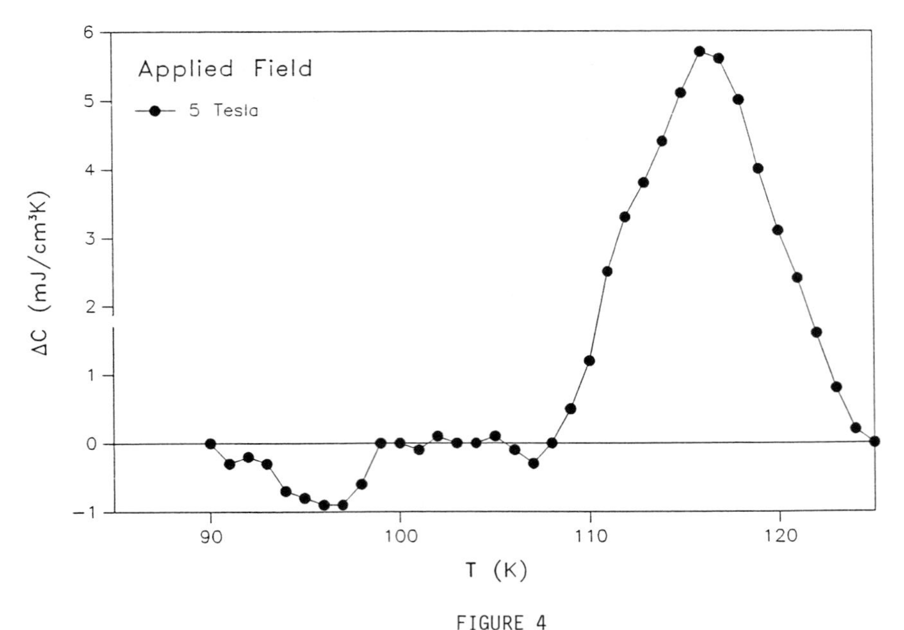

FIGURE 4
Specific Change With Magnetic Field, C_o - C_H.

Here C_O - C_H is calculated by fitting seven consecutive points to a parabola to evaluate the curvature. The data set is then incremented to lower temperature by one point and the process is repeated. The resulting specific heat shown for H = 5T indicates an interval from 110 to 120K where the specific heat changes by approximately 4 mJ/cm^{3}K. Outside this range the specific heat is essentially independent of H.

CONCLUSIONS

There is a very wide temperature interval where the magnetization is thermodynamically reversible, from 90K to T_c. For these grain aligned samples, flux pinning must be very weak in this range. The Gibbs free energy at constant magnetic field is linear in temperature over a 20K temperature interval from 90 to 110K. This, in turn, implies that the specific heat is independent of magnetic field. This behavior is quite different from that observed for the A-15 superconductors.

REFERENCES

1. B. Serin, Superconductivity Vol. 1, p. 925, R. D. Parks, Ed., Marcell Dekker, Inc., New York 1969.

2. David Nelson, Phys. Rev. Lett. 60, 1973 (1988).

3. S. E. Inderhees, M. B. Salamon, N. Goldenfeld, J. P. Rice, P. G. Pazol, D. M. Ginsberg, J. Lin, G. W. Crabtree, Phys. Rev. Lett. 60 1178 (1988).

4. R. A. Fisher, J. E. Gordon, N. E. Phillips and A. M. Stacy, Physica C (Interlaken); N. E. Phillips, R. A. Fisher, S. E. Lacy, C. Marcenat, J. O. Olsen, W. K. Ham, A. M. Stacy, J. E. Gordon and M. L. Tan, Physica 148B 360 (1987).

5. T. K. Worthington, W. J. Gallagher and T. R. Dinger, Phys. Rev. Lett. 59 1160 (1987).

6. D. E. Farrell, B. S. Chandrasekhar, M. R. DeGuire, M. M. Fang, V. G. Kogan, J. R. Clem and D. K. Finnemore, Phys. Rev. B 36 4025 (1987).

History Dependence of Weakly Coupled Intergrain Currents in a Sintered Oxide Superconductor

T. MATSUSHITA[1], B. NI[1], K. YAMAFUJI[1], K. WATANABE[2], K. NOTO[2], H. MORITA[2], H. FUJIMORI[2], and Y. MUTO[2]

[1] Department of Electronics, Kyushu University, 6-10-1, Hakozaki, Higashi-ku, Fukuoka, 812 Japan
[2] Institute for Materials Research, Tohoku University, 2-1-1, Katahira, Sendai, 980 Japan

ABSTRACT

History dependence of the critical current density was measured for a sintered Y-Ba-Cu-O superconductor by resistive and ac inductive methods. The critical current density measured by the two methods took a smaller value in an increasing-field process than in a decreasing one. It was also shown that closed intragrain currents with much larger value were not influenced by the history. This reveals that the history dependence comes only from a weakly coupled intergrain currents. The effect of an excursion of the magnetic field, i.e., addition and removal of a perturbation in the field, on the intergrain current density was examined in detail. A self-field of the closed intragrain currents, which changes its direction depending on the history, is considered to be a cause of the history effect.

INTRODUCTION

Discovery of oxide superconductors with high critical temperatures and high critical magnetic fields gave a large impact to the society not only in the field of physics but also in that of application. In the latter field, especially for power application, the critical current density, i.e., the maximum density of the nondissipative current, is needed to be large enough. Hence, the potentiality of these materials entirely depends on their critical current characteristics. The critical current densities in thin films [1,2] are fairly large and are of a comparable order in magnitude with commercial Nb-Ti (4.2 K) even at the liquid nitrogen temperature. This fact assures the potentiality of these materials.

However, sintered ceramics and powder-processed or diffusion-processed wires, which are suitable for mass-producible cables for large-scale apparatus, can carry only small transport current. It is known [3,4] in Y-Ba-Cu-O that closed shielding currents of fairly high density flow inside grains. Hence, one of the causative factors which depress the critical current density is grain boundaries. Recent studies for bicrystalline films [5,6] elucidated indeed that the grain boundary severely reduces the critical current density. However, the difference between the intergrain current density and the intragrain one observed commonly in polycrystalline specimens are much larger than that observed in bicrystalline films. This indicates that there exist other factors which characterize the critical current density in polycrystalline materials.

Recently it was shown [7] that the critical current density in sintered oxides was not uniquely determined as a function of the magnetic field but depended on its history: the critical current density took a smaller value in the increasing-field process. Such a history effect was observed also in ordinary superconductors [8]. This comes from different arrangements of fluxoids on distributed pinning centers according to the field-application history. It has not yet been clarified, however, if the history effect in high T_c oxides is the same as that in ordinary superconductors. In this paper, the history effect in high T_c oxides was investigated by the resistive and ac inductive measuring methods. It turned out that the observed history effect was different from the ordinary one caused by flux pinning but originated from weak coupling between grains. The cause of the observed history effect is also discussed.

EXPERIMENTS

Specimens were prepared in a usual manner by calcining and sintering powders mixed with desired fractions. The temperatures of calcination and sintering were 900 °C and 930 °C, respectively. Calcination was repeated four times. After the sintering, the temperature was gradually decreased down to 200 °C, and then, quickly reduced to room temperature. The obtained pellets were mechanically cut, and the final sizes are typically 5 mm in width, 0.8 mm in thickness and 20 mm in length.

The critical current density was measured resistively and ac inductively at the liquid nitrogen temperature. The magnetic field was applied normal to a flat surface of the specimen in the resistive measurement. The criterion used for determination of the critical current was 1 μV/cm. In the ac inductive measurement, the dc and superposed ac magnetic fields were applied parallel to a long axis of the specimen. In this case, the shielding current flowing in a plane perpendicular to this axis was measured. The frequency of the ac field was 40.2 Hz. The magnetic flux going in and out of the specimen was measured as a function of the ac field amplitude. According to the analysis described in ref. 4, the inter- and intragrain current densities are obtained.

In order to investigate the history effect in detail, a variation of the critical current density after addition and removal of a perturbation in the field was measured resistively. The patterns of this field excursion is shown in Fig. 1. And the critical current density in process of cooling in the field was also measured for comparison.

RESULTS AND DISCUSSION

Figure 2 represents the critical current density observed by the two methods. The results obtained by the two methods are approximately the same. There is no anisotropy in the transport

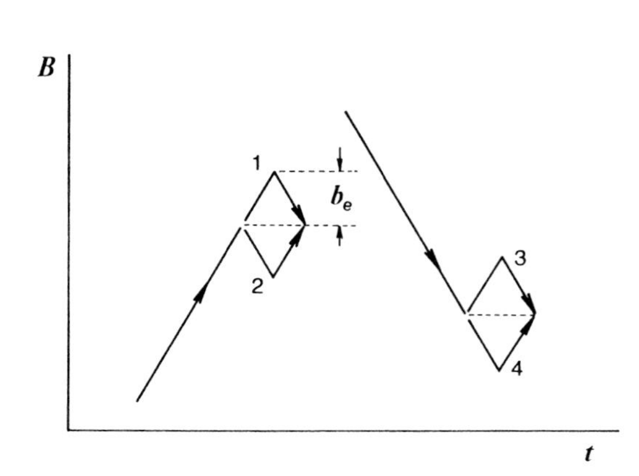

Fig. 1. Patterns of field excursion (b_e; amplitide of excursion, t; time).

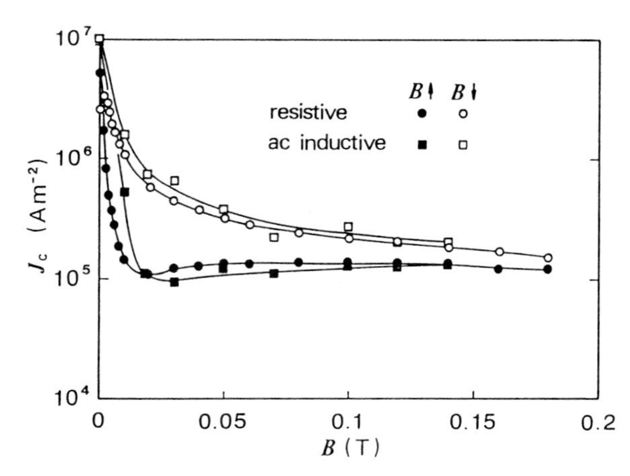

Fig. 2. History dependent critical current densities observed by two methods.

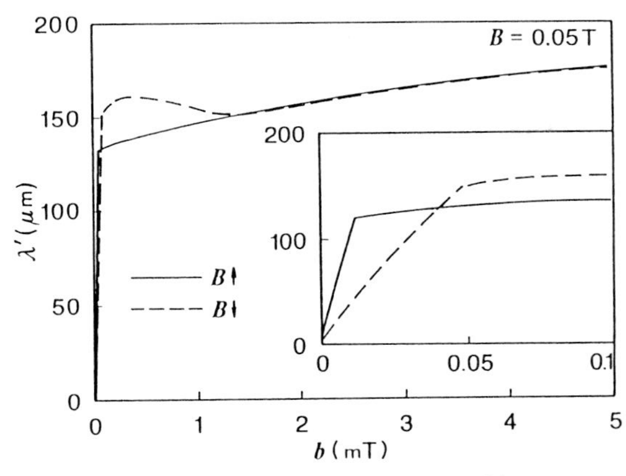

Fig. 3. Ac field penetration depth λ' vs. ac field amplitude b. The insert shows λ' vs. b curve at small b values.

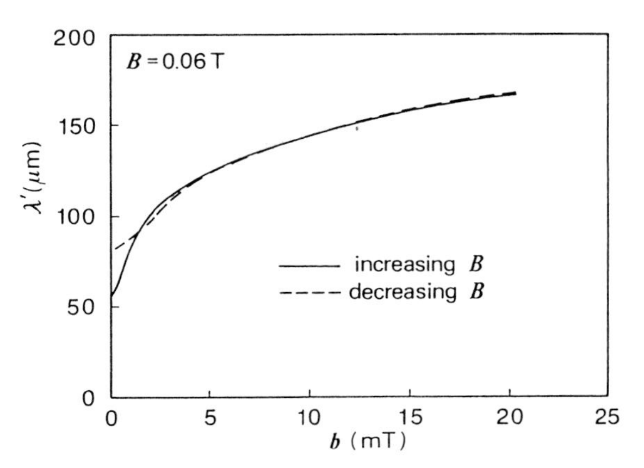

Fig. 4. Ac field penetration depth λ' vs. ac field amplitude curves in powdered specimen.

current density, suggesting a completely random texture in the specimen. The obtained critical current density takes a smaller value in the increasing-field process than the decreasing one. This is qualitatively the same as other experiments [7].

An example of the ac field penetration depth λ' versus the ac field amplitude b curve is depicted in Fig. 3. The characteristic at small field amplitude associated with the small intergrain current density depends on the history as shown in the insert, but that at large amplitude associated with the large intragrain current density does not. That is, the history effect comes only from the weakly coupled characteristic but not from the flux pinning which characterises the intragrain current density. This is certified again from the result on a powdered specimen shown in Fig. 4. Hence, it is concluded that the history effect in oxide superconductors is different from that in ordinary superconductors which is caused by a difference in an arrangement of the fluxoids on randomly distributed pinning centers [8]: If the fluxoids arrange themselves so as to be suitable for pinning, a larger critical current density is obtained.

A similar history effect was observed in a superconducting microbridge [9]. This history effect depends on a shape of the bridge and is considered to be caused by a shielding current flowing in bank regions [9]. Variations of the critical current density due to the field excursion at various values of the bias magnetic field are shown in Fig. 5. The numbers in the figure denote the patterns of the field excursion depicted in Fig. 1. In the cases of patterns 2 and 3, where the final field distribution and the shielding current distribution inside grains do not change from those before the excursion. In patterns 1 and 4, on the other hand, the field distribution of the shielding current changes as illustrated in Fig. 6. The critical current density in these cases changes: When the self field at the surface due to the shielding current is parallel to the external bias field, the critical current density has a smaller value. It takes a larger value when the self field is antiparallel to the bias field. This is qualitatively the same as the history effect in bridge. Figure 7 shows that the critical current density observed inductively in the process of cooling in the field, where the shielding currents are not induced, takes an intermediate value of those in increasing and decreasing field processes. This result also supports the idea that the shielding current inside grains plays some important role in determining the critical current density. This idea is also supported from the fact that the excursion necessary for reaching the opposite branch of the critical current density is of the same order in magnitude with the penetration field of grains, i.e., the variation in the field necessary to invert the direction of the shielding current in grains.

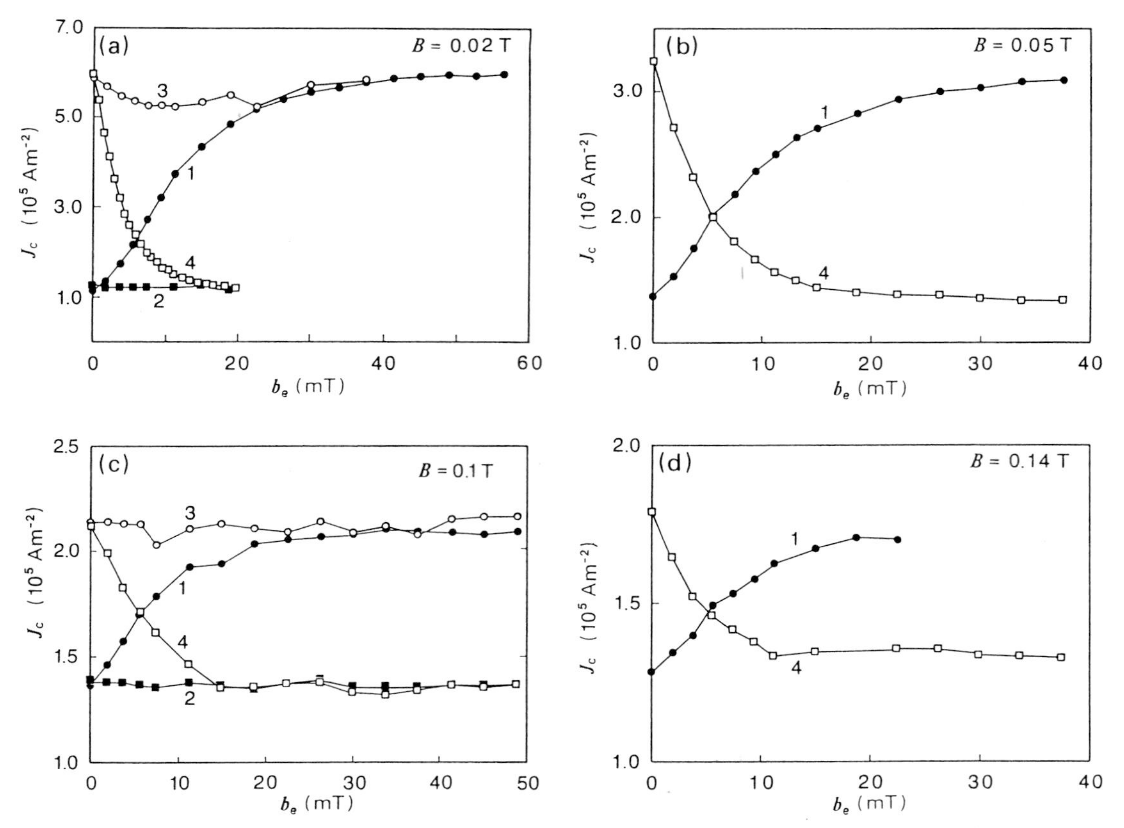

Fig. 5. Variation of the critical current density due to the field excursion at (a) B = 0.02 T, (b) B = 0.05 T, (c) B = 0.10 T and (d) B = 0.14 T.

However, there remains a problem that the self-field due to the shielding current is too small (typically 10 mT) to explain the observed history effect shown in Fig. 2, if the critical current is uniquely determined by the magnetic field which weakly linked regions suffer. Hence, it is more realistic that the weak link current is not uniquely determined by the local magnetic field but directly depends on the history through the shielding currents in the grains. Fundamental investigation in a simple geometry seems to be necessary.

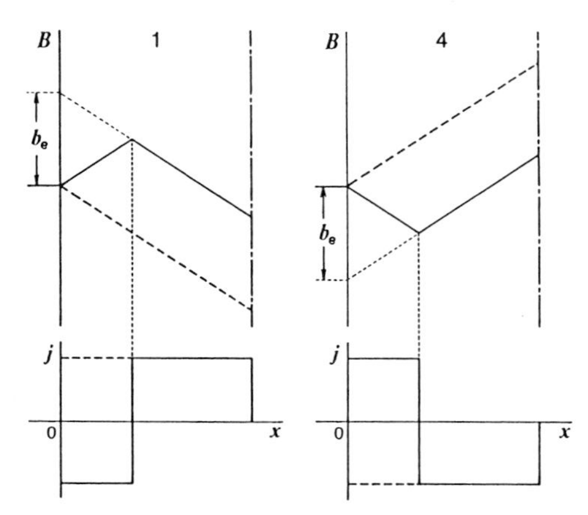

Fig. 6. Distributions of magnetic flux and shielding current in a grain in processes 1 and 4. The broken lines are the ones before the excursion.

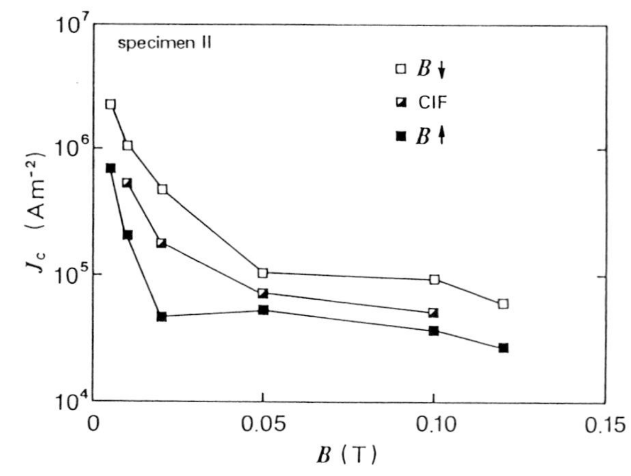

Fig. 7. History dependence of the critical current density. It takes an intermediate value in CIF (cooled in field) process.

REFERENCES

1. Y. Enomoto, T. Murakami, M. Suzuki and K. Moriwaki: Jpn. J. Appl. Phys. 26 (1987) L1248.
2. S. Tanaka and H. Itozaki: Jpn. J. Appl. Phys. 27 (1988) L622.
3. H. Küpfer, I. Apfelstedt, W. Schauer, R. Flükiger, R. Meier-Hirmer and H. Wühl: Z. Phys. B 69 (1987) 159.
4. B. Ni, T. Munakata, T. Matsushita, M. Iwakuma, K. Funaki, M. Takeo and K. Yamafuji: submitted to Jpn. J. Appl. Phys.
5. P. Chaudhari, J. Mannhart, D. Dimos, C. C. Tsuei, J. Chi, M. Oprysko and M. Scheurermann: Phys. Rev. Lett. 60 (1988) 1653.
6. D. Dimos, P. Chaudhari, J. Mannhart and F. K. LeGoues: Phys. Rev. Lett. 61 (1988) 219.
7. K. Noto, H. Morita, K. Watanabe, T. Murakami, Y. Koyanagi, I. Yoshii, I. Sato, H. Sugawara, N. Kobayashi, H. Fujimori and Y. Muto: Physica 148B (1987) 239.
8. H. Küpfer and W. Gey: Philos. Mag. 36 (1977) 859.
9. T. Aomine and A. Yonekura: Phys. Lett. 114A (1986) 16.

Trapped Magnetic Flux in $YBa_2Cu_3O_{7-\delta}$ Superconductors

D.N. MATTHEWS, G.J. RUSSELL, K.N.R. TAYLOR, and B. PURCZUK

Advanced Electronic Materials and Technology Group, School of Physics,
University of New South Wales, P.O. Box 1, Kensington, NSW 2033, Australia

ABSTRACT

Simultaneous measurements of the magnetic field dependence of the critical current density and the **magnetization** of yttrium barium cuprate superconductors have shown the presence of significant residual magnetic fluxes even in zero applied field. These can be sufficient to leave the material in the mixed state and hence can seriously influence the interpretation of the observed electrical and magnetic properties. This trapped flux appears to be associated with a magnetic field component threaded through the intergranular regions, whose motion is inhibited by circulating currents flowing in the granular system surrounding the field and preserved through the weak-link coupling. Once this coupling is destroyed, the field can move more freely.

INTRODUCTION

Since the early publications appeared dealing with the performance of bulk, sintered specimens of the new class of superconductors based on rare earth barium cuprates [1-3] it has been clear that the experimentally available critical current densities are disappointingly low and extremely sensitive to the presence of an applied magnetic field. Both single crystal [4] and epitaxial thin film [5] measurements show that the intrinsic properties of $YBa_2Cu_3O_{7-\delta}$ superconductors are those of a high performance Type II material with observed critical current densities in excess of $10^6 Acm^{-2}$ in magnetic field strengths in excess of 10T.

It was rapidly realised that at least part of the origin of the poor performance of the bulk materials must lie in their granular nature with extensive evidence being produced for the existence of weak-link structures [6,7] between the grains. While the results of measurements of Josephson tunnelling and SQUID like behaviour in macroscopic junction structures have been attractive from the point of view of device development, they have also confirmed the high sensitivity of these granular materials to the combined effects of electrical currents and applied magnetic fields.

The effects of the highly anisotropic electrical properties of crystalline $YBa_2Cu_3O_{7-\delta}$ [8] are also seen as being important in determining the behaviour of granular sintered compacts, in which the microcrystallites are randomly oriented with respect to each other along any conduction path. As a result, considerable effort is now being directed towards textured material [9-11] and the effects of intergrain misalignment [12] and intragrain twinning [13].

Finally, using the normal conditions of material processing followed by the majority of workers in this field, it is known to be extremely difficult to prepare pure, single phase material at the point compound composition $YBa_2Cu_3O_{7-\delta}$. Since all the impurity phases likely to occur at this composition in the ternary phase diagram are known to be insulating, the intergranular regions are likely to be severely affected by their presence through the formation of extreme SIS junctions. Off-stoichiometric effects can also be expected within the grains themselves, where the oxygen deficiency is likely to vary with depth into the individual crystallites as a result of incomplete oxygenation during processing. This will clearly depend on grain size, and may be partially responsible for the decrease in critical current density with increasing grain dimensions [14].

In the course of our recent investigations into the development of crystallographic texture in these materials [11,15] and its effect on the critical superconducting parameters, we have been making simultaneous measurements of both the V-B and the V-i characteristics and also the magnetization, as a function of applied field strength in sintered samples prepared in a variety of ways. In the early stages of this study we have found evidence in the electrical properties for pinning effects which leave a residual flux trapped in the sample after the external field has been removed [16]. The present work extends these early observations to allow a comparison to be made of the magnetic and electrical behaviour in this region.

EXPERIMENTAL

All the measurements described in this work were carried out on samples of $YBa_2Cu_3O_{7-\delta}$ prepared by the standard processing technique of solid state reaction and sintering. Stoichiometric amounts of the dry constituent powders (CuO, Y_2O_3 and $BaCO_3$) were thoroughly mixed by grinding, pressed into a 3cm diameter pellet and fired for 12 hours at 930°C to produce a reacted sample. These discs were then reground and pressed at 300MPa before sintering at 930°C for 12 hours in oxygen. Subsequent cooling to room temperature was also carried out in flowing oxygen at 60°C/hour.

This processing method leads to extremely hard, high density materials with an oxygen stoichiometry close to 7.0. Xray powder diffraction shows no evidence for second phases (ie. < 2%) but clearly shows the distorted orthorhombic lattice parameters a = 3.819, b = 3.883 and c = 11.669. From the c - δ relation of Tarascon et al [17], this suggests a final oxygen composition of $O_{6.98}$ (i.e. δ = 0.02).

Optical polarimetric microscopy shows only small, equiaxed polygonal grains of mean dimensions 5µm x 5µm, with little sign of optical twin structures. This is typical of material processed at these low temperatures and is consistent with our previous studies [14] which showed that smaller grain sizes lead to higher critical current densities.

The electrical measurements were made using a conventional 4-point probe technique. All contacts were made using silver epoxy baked onto the sample at 900°C with final lead attachment achieved using soft soldered copper wires. After baking, the samples were cooled very slowly in flowing oxygen to preserve the high oxygenation value. Subsequent Xray observations showed little or no change in the Xray lattice parameters. The measuring currents in all the electrical observations could be varied up to maximum values of 30A, which with typical specimen cross-sectional dimensions of 3 x 2 mm^2 correspond to transport current densities of up to $500Acm^{-2}$. The magnetization measurements were made using a balanced coil, inductive method [18] with the applied magnetic field lying parallel (or antiparallel) to the current flow direction. This applied field was generated using a liquid nitrogen cooled solenoid and could be increased at a constant rate, to a predetermined maximum value and subsequently decreased to zero at the same fixed rate. The balanced coil system consequently provided a direct output $\frac{dM}{dt}$ with H, or after integration, the M-H hysteresis loop, corresponding to the observational conditions.

In a typical observational run using a fixed transport current, the specimen voltage and the coil voltage ($\frac{dM}{dt}$) were recorded simultaneously as the applied magnetic field was ramped linearly up to its maximum value H_{max} and subsequently ramped down to zero. The ramp is highly linear, and 'no-specimen' observations were taken to establish any coil unbalance.

A typical data set is shown in Fig. 1a for relatively low H_{max} values. As may be seen, $\frac{dM}{dt}$ shows two clear maxima, which on integration can be seen to be associated with two different susceptibilities. The higher susceptibility which occurs at the lowest field strength is similar to that reported by other workers [19,20] and is believed to be associated with bulk superconductivity, while the transition to the lower susceptibility is thought to occur at a transport current-applied field strength combination which allows flux to enter the specimen through the intergranular structure.

Figures 1b to 1d give similar data for increasing values of the maximum field reached during the field ramp cycle.

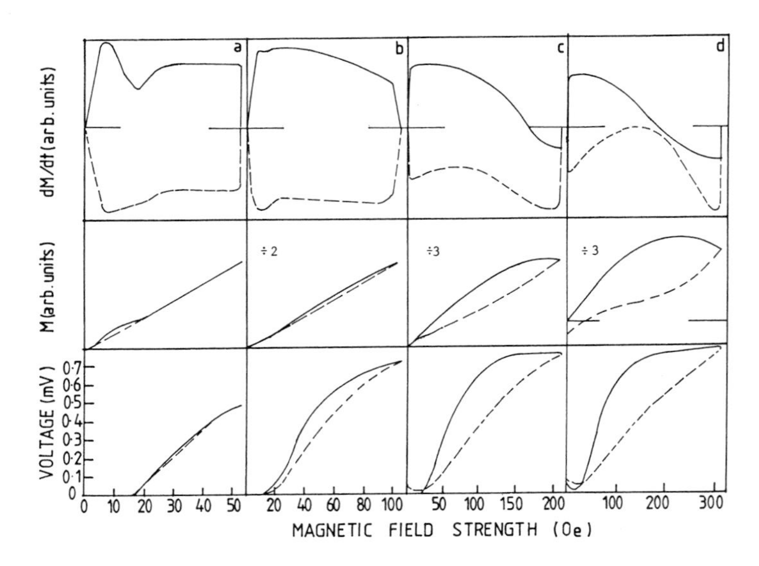

Figure 1. The variation of dM/dt, M and the specimen voltage V as a function of applied magnetic field strength for various peak fields H_{max} achieved during the increasing and decreasing ramp cycle. Note the scale changes for the magnetization data between the four diagrams and the zero offset in d) for the M-H loop. The sample current was maintained at 2.0A for all the observations.

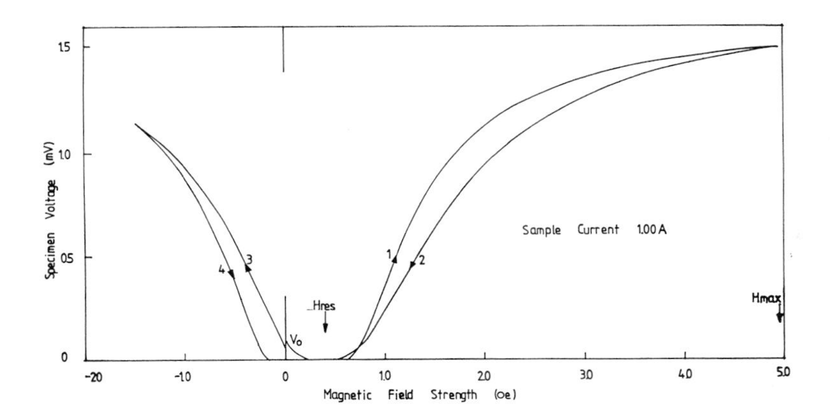

Figure 2. The effect of reversing the magnetic field on the specimen voltage after the residual V_o signal has been established. For larger reverse H_{max} values the voltage minimum in the negative quadrant can reappear along with the associated V_o value.

The striking features of these observations which may be seen in these figures are:

a) the appearance of non-reversible behaviour in the V-i characteristics in the cycle as (H_{max}) is increased (Fig.1b)

b) the appearance of a minimum in the specimen voltage as the field is ramped to zero and an associated residual voltage V_o at zero field (Fig.1c)

and c) at higher H_{max} values this minimum occurs in both increasing and decreasing fields.

If the applied magnetic field is left at zero after a cycle to a high value of H_{max}, the residual voltage decays slowly over a period of minutes to approximately 0.6 of its initial (t = o) value. After this initial decay, there is no clear evidence of further decrease over a period of approximately 1 hour.

When the magnetic field is reversed, the specimen voltage increases from its V_o value as the field strength increases, however if H_{max} is greater than the field strength (H_{res}) at the voltage minimum for the initial field direction the ramp-down results in the appearance of normal superconducting behaviour once more. The form of these results are shown schematically in Fig.2 on which V_o, H_{res} and H_{max} are indicated. If H_{max} for the reverse field is large enough, of course, then both the voltage minimum, and V_o appear as shown. The variation of both V_o and H_{res} with H_{max} are shown in Fig. 3 in which a marked switching behaviour occurs as H_{max} is varied.

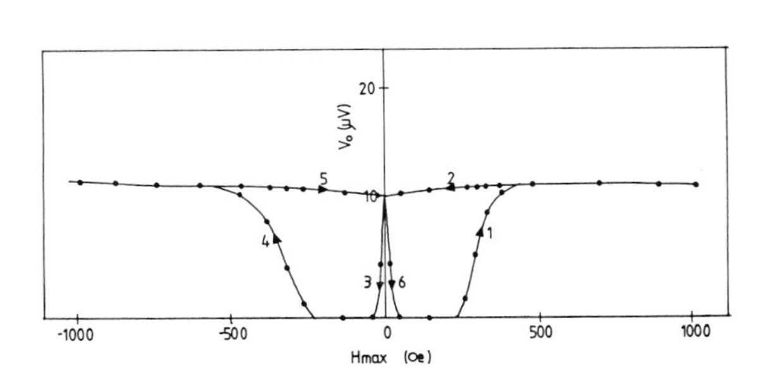

Figure 3a. The variation of V_o during the full field cycle for the sample at a lower current showing a higher onset H_{max} for the appearance of V_o.

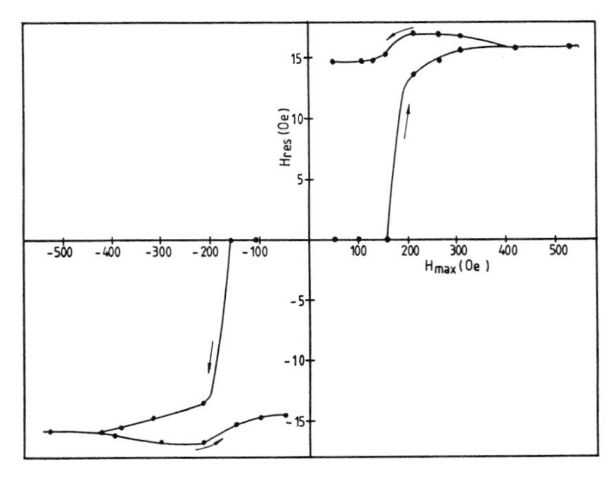

Figure 3b. The hysteresis like behaviour of the magnitude of H_{res} during the full cycle for the same sample at the original specimen current. The detailed shape of H_{res} with H_{max} during the decreasing field ramp is real.

DISCUSSION

The low field behaviour obtained in the initial field cycle shown in Fig. 1a is similar to that to be expected from a normal granular superconductor. The initial sharp peak in $\frac{dM}{dt}$ can be interpreted in terms of a full Meissner effect associated with screening currents circulating in the surface layers of the bulk specimen. The decrease in susceptibility which can be observed at approximately 10 Oe is then associated with the first entry of flux into the network of intergranular boundaries and sample porosity.

Once this has occurred, flux motion driven by the Lorentz force of the transport current is inhibited by the presence of screening currents flowing between the grains surrounding any flux supporting intergranular regions of the specimen. Since flux could only enter the specimen by overcoming the weakest of the weak-link couplings, this flux pinning can only occur for a limited range of applied field strengths before the weak-link system is reduced to the point that these intergranular screening currents are reduced to zero. Flux motion can then occur within the intergranular regions and a specimen voltage will appear comparable to that observed in the mixed state of normal Type II superconductors under comparable observing conditions.

In subsequent cycles the full Meissner susceptibility never reappears,presumably indicating that even at these low fields the flux does not emerge from the granular pores within the time of the experiment. This assumption is supported by the irreversibility of the M-H characteristic in Fig. 1a in the region of 0-15 Oe.

With H_{max} = 100 Oe (Fig.1b) the irreversibility of the M-H loop becomes more obvious as the inertia of the penetrating flux system becomes more important. Finally for H_{max} = 200 Oe a significantly open loop appears and at H_{max} 300 Oe a residual magnetization occurs at H = 0.

As the M-H loop opens, the voltage-field characteristics also develop a highly irreversible behaviour, with the observed voltages during the decreasing field ramp being significantly less than those during the field increase. While the two variations converge at $H \cong H_{c1}$, the decreasing field voltage passes through a minimum before reaching a value of V_o at H = 0. Once a residual magnetization has been established (Fig.1d) this zero field voltage V_o and the minimum in the V-B characteristic occur in both increasing and decreasing fields.

The appearance of these unusual features in the voltage-field characteristics of the current carrying superconductors is clearly associated with the presence of a residual, trapped field in the specimen after magnetization has been carried out to field strengths in excess of some critical value. In Fig. 1 this clearly lies between 100 and 200 Oe and as Fig.3b shows, a critical value of 152 Oe must be exceeded in this sample, at a transport current of 2.0A, in order to establish the observed effects. At 1.0A (Fig.3a) this is increased to 215 Oe.

By comparison with the magnetization data of Fig.1 this critical H_{max} value is close to, but in excess of H_c, occurring at field values which result in the marked deviations from the initial Meissner susceptibility and it is inescapable that we associate the trapped flux with the field which has entered the intergranular regions. Whether or not the field has also entered the body of individual grains will depend on the intrinsic flux pinning and the barriers to flux penetration into the pure (123) material of the grains.

In view of the behaviour observed during both forward and reverse field cycles, and shown in Fig.2 it would appear that the net field seen within the sample may become zero at H_{res}, at which the material appears to return to its full superconducting properties. This would imply that the directions of the trapped and external fields are of equal magnitude and opposite to one another at $H = H_{res}$. Further decrease in H to zero(and to negative values)simply allows the net field to increase from zero at $H = H_{res}$ until the external field is of sufficient strength to reverse the trapped flux. Unfortunately, in the mixed state of Type II superconductors, and under the experimental conditions at the $H = H_{res}$ voltage minimum, both the external and trapped fluxes are parallel and no cancellation is possible.

Despite this, the evidence of both Fig.2 and Fig.3 points clearly towards trapped flux as the mechanism responsible for this phenomenon. In that case, one must look in more detail at the nature of the flux trapping and the origin of the voltages observed in this work.

In relation to the latter, it is well understood [ ] that the voltages appearing in the mixed state of a type II superconductor are associated with the energy loss caused by the motion of fluxons against pinning forces in the sample. These voltages are directly proportional to the fluxon velocity and to the transport current density. In our case, the current is held constant and the changes in observed voltage can be related entirely to the fluxon motion. In other words, the present minimum in V (at $H = H_{res}$) and the appearance of V_o at H = 0 may arise from changes in the fluxon mobility rather from with the magnitude of the trapped flux.

While it is normally assumed that the mobile flux is entirely associated with fluxons within the body of the Type II material, this is not necessary since any flux within the granular sample will be subjected to the same Lorentz force. Only the dynamics of the system will be different since fluxons are associated with a flux quantum and hence are independent of H whereas flux linked with the intergranular structure will depend directly on the applied field.

On this basis, the observations described above can be understood in a relatively straightforward way in terms of the flux states described earlier. It is clear that after the initial magnetization, the full Meissner susceptibility never reappears provided that the sample remains at 77K. Consequently for most of the observations the magnetic field penetrates the sample and is screened internally by currents flowing between the grains which surround a particular field carrying region. Once this local field exceeds the weak-link coupling strength, this screening current disappears and the flux is free to move under the influence of the Lorentz force. This motion may be within the intergranular spaces or from the intergrain regions into the grains themselves provided the barriers to flux entry can be overcome and provided the pinning forces are not excessive. This state now describes the behaviour to $H = H_{max}$ with increasing field.

On decreasing the field, the Lorentz force on flux within the grains (in the form of fluxons) remains constant (and proportional to ϕ_o) while in the intergranular regions it decreases as the applied field decreases since some of this flux must leave the sample. The net motion, and hence the observed voltage, will depend critically upon the detailed properties of the granular structure & on the intrinsic properties of the crystalline material. In view of the nature of the observations it is clear that the rate of flux loss from the specimen is less than that of flux entry. Once the flux density within the intergranular spaces is sufficiently small as to allow the reestablishment of the weak-link coupling and hence the internal screening currents the flux motion will be greatly reduced. Whether or not it becomes zero (and hence $V = 0$) will depend on the internal field gradients and hence on the rate of field change, since the internal energetics will establish a characteristic relaxation time for the system.

As the field decreases below H_{res} , the continuing increase in field gradient will allow the flux to be driven from the sample to reach some equilibrium state. This increase in flux mobility will then lead to the observed V_o.

Since the direction of the voltage depends only on the sign of the transport current, reversing the magnetic field will lead to a continuing increase in V as shown in Fig.2. as the trapped flux is driven out of the sample and as other flux, parallel to the new field direction enters it. Under these conditions it is straight forward to predict the remaining features of this field cycle.

At sufficiently high H_{max} values, the net flux gradients are likely to be large enough to allow flux motion at all external field values and the appearance of the minimum in both increasing and decreasing fields observed in Fig.1d. will result.

CONCLUSIONS

The observed behaviour of the magnetization and apparent resistance of a current carrying $YBa_2Cu_3O_{7-\delta}$ superconductor in an applied field reveals the existence of a number of anomalous properties which may be associated with flux trapped in the sample. The detailed behaviour suggests that this flux is trapped in the intergranular regions by internal screening currents circulating in the granular composite and surrounding the flux area.

Well defined field conditions have been established for the appearance of these anomalies and hysteresis like behaviour of both H_{res} and V_o have been observed.

REFERENCES

1. J.G. Bednorz and K.A. Müller. Z. Physik B. 64 188 1986.
2. U. Dai, G. Deutscher, R. Rosenbaum. Appl. Phys. Lett. 51 460 1987
3. C.W. Hagen, M.R. Bom, R. Griessen, B. Dam and H. Veringa. Physica C. 153-155 322 1988.
4. W.J. Gallagher. J. Appl. Phys. 63 4216 1988.
5. P. Chaudhari, R.H. Koch, R.B. Laibowitz, T.R. McGuire and R.J. Gambino. Phys. Rev. Letts. 58 2684 1987.
6. R.L. Peterson and J.W. Ekin. Phys. Rev. B. 37 9848 1988.
7. C.M. Pegrum, G.B. Donaldson, A.H. Carr and A. Hendry. Appl. Phys. Lett. 51 1364 1987.
8. S.W. Tozer, A.W. Kleinsasser, T. Penney, D. Kaiser and F. Holtzberg. Phys. Rev. Letts. 59 1768 1987.
9. J. Huang, T.W. Li, X.M. Xie, J.H. Zhang, T.G. Chen and T. Wu. Materials Letters 6 222 1988.
10. L. Lynds, F. Galasso, F. Otter, B.R. Weinberger, J. Budnick, D.P. Yang and M. Filipkowski. Comm. Am. Ceram. Soc. 71 C130 1988.
11. D.N. Matthews, T. Puzzer, A. Bailey, N. Mondinos, G. Alvarez, G.R. Russell and K.N.R. Taylor. J. Crystal Growth, 1988. Accepted for publication.
12. D. Dimos, P. Chaudhari, J. Mannhardt and F.K. Goves. Phys. Rev. Letts. 61 219 1988.
13. G. Deutscher. Physica. C. 153-155 15 1988.
14. A. Bailey, D.N. Matthews, K.N.R. Taylor, G.J. Russell and D.R. Misra. J. Mats. Sci. Letts. 1988. Submitted for publication.
15. D.N. Matthews, A. Bailey, G.J. Russell, G. Alvarez, S. Town, K.N.R. Taylor, F. Scott, M. McGirr and D.J.H. Corderoy. These proceedings.
16. D.N. Matthews, G.J. Russell, A. Bailey and K.N.R. Taylor. Nature. 1988. Submitted for publication.

17. J.M. Tarascon, P. Barboux, B.G. Bagley, L.H. Greene, W.R. McKinnon and G.W. Hull. Chemistry of High Temperature Superconductors. Eds. D.L. Nelson, M.S. Whittingham and T.F. George, ACS Symposium Series 351 198 1987.
18. H. Zijlstra. Experimental Methods of Magnetism (NH 1967), Vo. 2, p.103-
19. K.V. Rao, R. Puzniak, D.X. Chen, N. Karpe, M. Baran, A. Wisniewski, K. Pytel, H. Szumczak, K. Dyrbye and J. Bottiger. Physica C. 153-155 347 1988.
20. T. Finlayson. Private Communication.
21. E.J. Thomas, Type II Superconductivity [D. St. James, E.J. Thomas and G. Sarma], Pergamon 1969 pp.209-

Ceramic Problems/Challenges in High Temperature Oxide Superconductors; Hysteretic Force Measurements As a New Analysis Tool

P.E.D. MORGAN, J.J. RATTO, R.M. HOUSLEY, J.R. PORTER, D.B. MARSHALL, and R.E. DE WAMES

Rockwell International Science Center, 1049 Camino Dos Rios, Thousand Oaks, CA 91360, USA

INTRODUCTION

Increasing realism that high temperature oxide superconductors have many unusual and difficult aspects is setting in. It is becoming apparent that many of the problems/challenges are directly dependent upon the ceramic-like properties of these materials, when in polycrystalline form, and that many of these have been encountered before in other ceramic guises. We originally guessed, without making hundreds of specimens, that the peculiar R/T double step (with one drop at ~ 100K, and a second drop at ~ 85K) in the Bi HTSC and the inability initially to get R = 0 at >100K came about through syntactic (coherent) intergrowths [1,2] which are so well known in minerals [2]. We therefore disagree with many ideas that have been put forward to explain this phenomenon [3-12]. We will show that problems with the Bi containing superconductors are extreme to the point where our personal decision is to abandon work on their polycrystalline forms in favor of Tl materials, at least for the time being. We prefer to make bad ceramics of a 125K material than of a 90K variety for obvious reasons. Space applications, where passive cooling, with attendant weight saving and increased robustness, are our major incentive.

In view of these difficulties it is also necessary to devise new techniques for analyzing these materials and here we present a method of using hysteretic magnetic force/distance measurements.

EXPERIMENTAL

Powders of three Bi,Ca,Sr,Cu compositions, 1,1,1,1; 1,1,1,2; and 1,1.5,1,2.5, were prepared by reacting hot Bi and Cu nitrate solutions in a heated blender with $CaCO_3$ and $SrCO_3$ powders, drying the thick slurries at 150°C in air, and ball milling the powders for 1 h. This slurry method ensures that Bi and Cu are finely dispersed as the carbonates react with the nitrates, and upon decomposition, give a very reactive source for production of BCSCO crystals. XRD of the powders showed production of Ca and Sr nitrates and disappearance of the Ca and Sr carbonates.

A series of flash heating/quenching and slow heating/quenching experiments were conducted with the powders. For flash heating/quenching experiments, a small sample was dropped quickly onto a heated alumina plate inside a preheated furnace; the samples were heated at temperature for 15 min, and the alumina plate was removed to ambient (cooldown ~ 200°C/min). For slow heating/quenching experiments, powders were heated in alumina boats at 5°C/h to the desired temperature, followed by quenching in air to ambient.

Slow cooling experiments were conducted by heating the three powders to 800°C at 150°C/h, holding at 800°C for 5 h, slow cooling at 5°C/h to a desired temperature, and quenching in air to ambient.

Two series of flash heating/quenching and slow heating/quenching experiments were also performed using seeded powders, prepared by hand grinding 10 wt% previously flash heated material (seeds) with unreacted powder. Seeds were prepared at temperatures corresponding to maxima of the temperature curves in Fig. 1: 1,1,1,1 (850°C); 1,1,1,2 (790°C); and 1,1.5,1,2.5 (810°C), as the desired product would be the superconducting 30.6Å (2122 type) material. The seeds predominantly contained the 30.6Å phase with lesser amounts of the 24.4Å phase. For the seeded experiments, the heating schedules were the same as those used for corresponding unseeded experiments.

XRD spectra were acquired at 2°/min on a Diano 1057 powder diffractometer with a graphite diffracted beam monochromator, except for a "single crystallite" (Fig. 7) which was run at 0.1°/min, using silicon as an external standard. The relative ratios of the 30.6Å and 24.4Å phases were determined from the (1"5"3) diffraction lines of each phase [2] (the "5" is to approximate the actual incommensurate value of ~ 4.8). These lines were chosen for their ease of identification, lack of overlaps and because, as cross planes, we partially avoid problems that occur due to preferred orientation of the platy structures during preparation of the XRD sample. To obtain the XRD of a single crystallite (~ 0.5 mm in diameter) formed from a 1,1.5,1,2.5 melt cooled at 6°C/h from 900°C, it was mounted and aligned with the basal plane normal to the x-ray beam (at 0° 2θ).

Elemental analysis by energy dispersive x-ray spectroscopy (EDS) in an analytical electron microscope (AEM) was performed on individual particles from the two samples of predominantly 24.4Å material, which had been flash heated/quenched at 768°C and 859°C, respectively (Fig. 1). Powder samples were ground and dispersed in dichloromethane and collected on carbon-covered Be grids for analysis in the AEM. Thin, electron-transparent particles were chosen for EDS analysis. Particles identified as Bi_2CuO_4 and Ca_2CuO_4 were assumed to have a fixed stoichiometry, and were therefore used to calibrate the elemental quantification factor (thin film k-factor) for the Bi/Cu and Ca/Cu ratios. A calculated k-factor was used for the Sr/Cu ratio computation. The results are presented in bar chart form in Fig. 3, organized according to the sum of the Bi and Cu fraction of each particle.

$Bi_2Sr_2CuO_5$ crystals were grown from an oxide melt cooled from 900°C at 6°C/h.

RESULTS AND DISCUSSION - PART I

During sintering of ceramic compositions with complex phases of similar stoichiometry, the product yields can often be substantially changed by varying the heating rate [13]. Through a series of flash heating (~ 400°C/min)/quenching (Fig. 1) and slow heating (5°C/h)/quenching experiments (Fig. 2), this idea was extended to the BCSCO system. The ratio of the BCSCO phases is very sensitive to starting composition and temperature effects.

The flash-heated samples show inconsistent results as a function of Cu content, while all samples show pronounced maxima for the 30.6Å/24.4Å ratio at ~ 800°C. The bell-shaped ratio curves are fairly narrow; for instance, in Fig. 1 for the 1,1.5,1,2.5 composition, the ratio of the 30.6Å phase triples within 30°C. When the reciprocal ratios of Fig. 1 are considered, changes in concentration of the 24.4Å phase are more evident. For instance, in the 1,1.5,1,2.5 system, the concentration of the 24.4Å phase is bimodal from 760 to 860°C with maxima at ~ 768°C and 859°C. As the samples are flashed at higher temperatures, from 770 to ~ 800°C, the 30.6Å phase increases relative to the 24.4Å phase. Above ~ 800°C, the concentration of the 30.6Å phase decreases, while the amount of the 24.4Å phase steadily increases with a maximum at ~ 860°C. Above ~ 850°C, melting begins, as determined by visual examination in an optical microscope.

Included in Fig. 1 are several single data points which indicate operations that increase the concentration of the 30.6Å phase. These include: (1) longer soak times of 15 h at temperature, (2) first flash heating at 768°C followed by flash heating at 826°C, and (3) seeding of the precursor powders with 10 wt% of previously flash heated material followed by flash heating at 828°C.

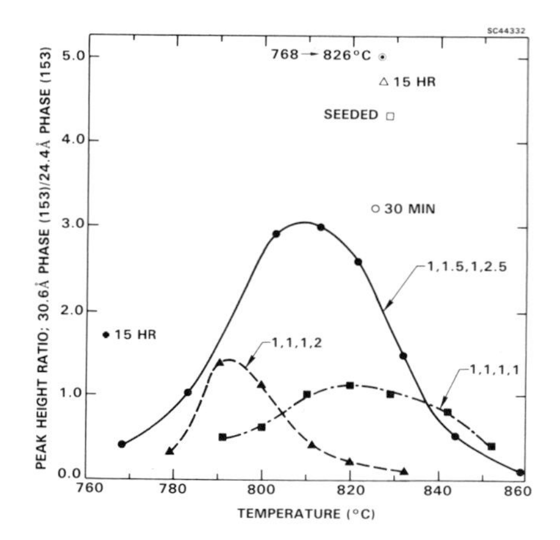

Fig. 1 Flash heat/quench of 1,1,1,1; 1,1,1,2; and 1,1.5,1,2.5 powders with single data points showing pre- and post-conditioned results.

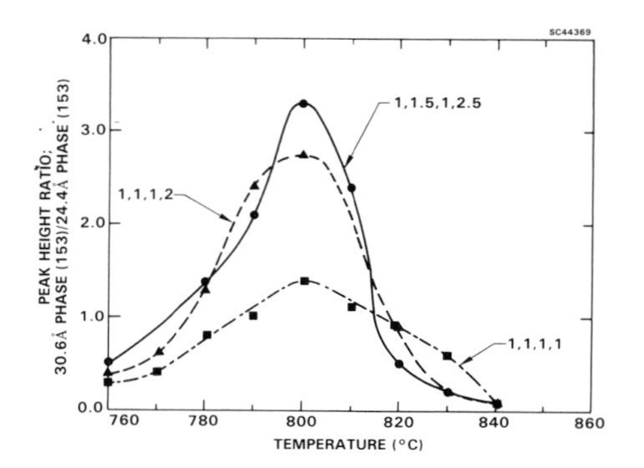

Fig. 2 Slow heat/quench of 1,1,1,1;1,1,1,2; and 1,1.5,1,2.5 powders.

For three compositions, slow heating (Fig. 2) showed no advantage over flash heating in increasing the 30.6Å phase; except for the 1,1,1,2 composition, the ratios are very similar to the flash heating experiments.

Listed in Table I are mole fractions roughly estimated from the major diffraction lines for each phase nearest to maxima of the bell-shaped curves in Figs. 1 and 2. The mole fraction of 30.6Å phase increases with Cu content of the starting powders, and CuO and Ca_2CuO_3 are present as minor phases. As analyzed by EDS/SEM and XRD, the slow heated samples are also seen to contain a noneuhedral Sr-Bi phase believed to be similar to $Sr_{0.9}Bi_{1.1}O_{2.55}$, PDF File 31-1341. This structure is believed to be different from an intergrown Bi-Ca-Sr phase [2].

Table I. Mole Fraction of Phases from Flash and Slow Heating Experiments

Figure	Starting Composition	Temp. (°C)	Mole Fraction 30.6Å	24.4Å	CuO	Ca_2CuO_3	Sr-Bi Phase
Flash Heated	1,1,1,1	820	0.45	0.41	0.11	0.03	--
	1,1,1,2	790	0.50	0.35	0.12	0.03	--
	1,1.5,1,2.5	813	0.60	0.20	0.15	0.05	--
Slow Heated	1,1,1,1	800	0.44	0.31	0.14	0.01	0.10
	1,1,1,2	800	0.55	0.20	0.18	0.01	0.06
	1,1.5,1,2.5	800	0.56	0.17	0.22	0.01	0.04

Previously [2], we compared the results of individual particle analysis (EDS) from the 768°C (predominantly 24.4Å phase) and 813°C (predominantly 30.6Å phase) fractions of the 1,1.5,1,2.5 composition which were flash heated/quenched (Fig. 1). Only a few of the 768°C particles showed the expected Bi/Cu ratio of 2.0 (24.4Å phase), while most of the 813°C particles had Bi/Cu ratios of approximately 1.0 (30.6Å phase). The EDS analyses establish interchange between Ca and Sr and perhaps of Cu with Bi.

In Fig. 3, the results of particle analysis (EDS) of the 768°C and 859°C samples (predominantly 24.4Å phase, formula 2021) from each side of the bell-shaped curve in Fig. 1 are listed. Most of the particles heated at 859°C have a lower Cu content than at 768°C, and the deviation from the average Cu concentration is greater for the 859°C particles. While the analyses of particles in Fig. 3 result from fractions which contain predominantly 24.4Å phase the procedure of analyzing only small particles may have biased the results so that in fact 30.6Å particles were being preferentially selected. In any event, there is little difference at temperatures below and above (where liquid may occur) the peak temperature.

Particle number 13 in Fig. 3 is a Cu-rich Ca compound, is not simply the $CaCu_2O_3$ type, but is a much more complex phase currently under study [14].

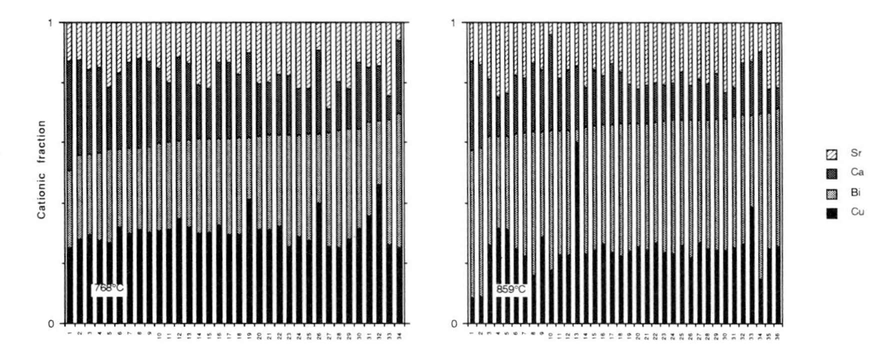

Fig. 3 Bar chart of the EDS results for two 1,1.5,1,2.5 preparations: flash heated/ quenched at (a) 768°C and (b) 859°C, both predominantly 24.4Å by XRD. The five divisions on the y-axis correspond to the five cations in $Bi_2Sr_2CuO_{6-x}$.

To investigate the possible changes occurring at lower temperatures on cooling the sintered phases, samples of powders were slowly cooled from 800°C at 5°C/h to the desired temperature, and quenched in air to ambient (Fig. 4).

In Fig. 4, the 1,1,1,2 and 1,1.5,1,2.5 compositions show a substantial increase in the 30.6Å phase in the cooling range from 800°C to ~ 760°C below which the ratios are almost constant, indicating very slow or no further reaction.

Note that while the 1,1.5,1,2.5 composition resulted in the largest ratios during flash and slow heating experiments, the 1,1,1,2 composition gave the largest ratios in the slow cooling experiment. Below ~ 760°C, the 1,1,1,2 ratios are of similar magnitude to the preconditioned samples (Fig. 1) prepared by long soak times, seeding or double firings. The 1,1,1,1 composition curve is almost flat with a slight downward slope, indicating negligible continuing reaction below 800°C. To investigate reaction at even lower temperatures, samples slow cooled to 735°C were held at 700°C and 600°C for an additional 12 h; insignificant changes in the ratio implied that the reaction had ceased.

Since seeding of the flash-heated 1,1.5,1,2.5 composition resulted in a large increase of the 30.6Å phase (Fig. 1), flash heating/quenching experiments and slow heating/quenching using seeded powders were performed to study the effect of temperature on the polytypoids (Figs. 5 and 6).

In Fig. 5, the bell-shaped curves are very narrow with widths at half height in the range of 25-40°C; the ratios at peak heights are much greater than in the unseeded runs; and the 1,1,1,1 and 1,1,1,2 bell-shaped curves are shifted to higher temperatures with maxima at 830°C.

In unexpected contrast to the seeded flash heated/quenched experiments (Fig. 5), the seeded slow heated/quenched compositions (Fig. 6) show broader bell-shaped curves with ratio maxima at ~ 835°C. Also, the peak height ratios are ~ 2.0, which are significantly lower than the seeded flash heated compositions.

In Fig. 5, we attribute the increased ratios to nucleation by the seeds which are rich in the 30.6Å phase, while in Fig. 6, the reduced ratios may be the result of either the 24.4Å phase being more thermodynamically stable and overgrowing on the 30.6Å phase, or the 24.4Å phase growing much more rapidly on the smaller number of 24.4Å seeds introduced along with the 30.6Å variety.

In an earlier paper [2], we examined intergrowths in large "single crystals" visible by backscattered electron imaging, EDS analysis, and XRD. By positioning a thin crystallite, ~ 0.5 mm in diameter, from a 1,1.5,1,2.5 melt on the diffractometer such that the basal (ab) plane was normal to the x-ray beam-diffracted beam plane, strong, narrow $(0,0,\ell)$, $\ell = 2n$ lines were seen and assigned to both the 24.4Å and 30.6Å phases (Fig. 7). (Even such small specimens gave strong diffraction lines.) Other crystallites showed an ~ 18.3Å repeat assumed to be intergrowths of "2010" Bi-Ca-Sr-Cu oxide, observed also by SEM-backscattered electron imaging. In the SEM, these crystallites showed intergrowths in which large areas of the 30.6Å phase join on a macroscopic scale with the 24.4Å phase with interfaces nearly parallel to the c-axes (as well as with interfaces normal to the c-axes).

The original compound of Michel et al.[3] $Bi_2Sr_2CuO_5$ (i.e., 2021) was also prepared by slow cooling from the melt. The composition appeared to melt congruently, forming good single crystals up to 1 cm in length. Analyses by XRD and EDS/SEM indicated absence of intergrowths. In this one layer type (2021), all cations on each side of the Cu-O layer are IX coordinated, the preference for Sr, and there are no VIII fold sites. If multiple layers cannot form without Ca, then obviously the syntactic problem cannot occur in this structure.

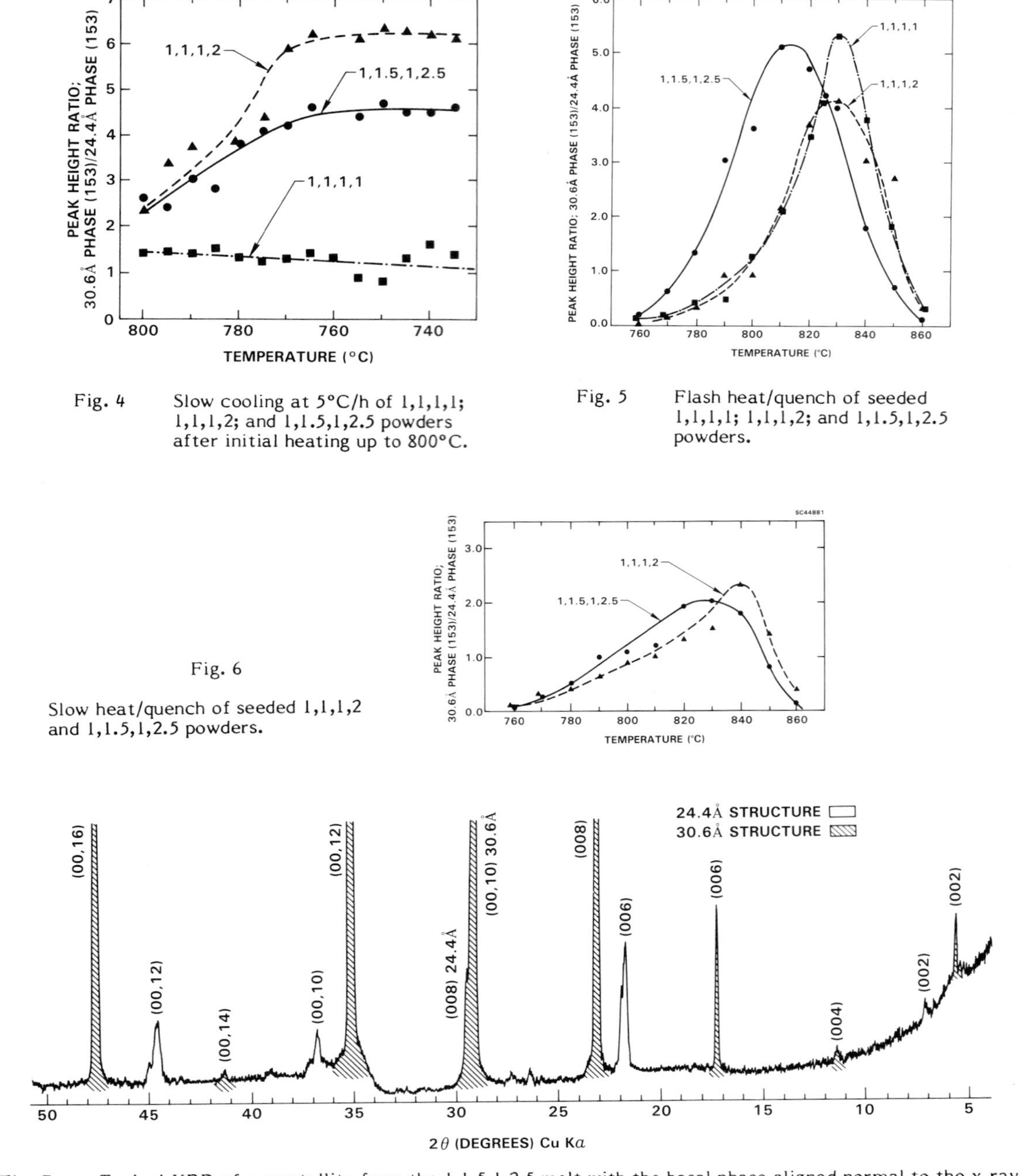

Fig. 4 Slow cooling at 5°C/h of 1,1,1,1; 1,1,1,2; and 1,1.5,1,2.5 powders after initial heating up to 800°C.

Fig. 5 Flash heat/quench of seeded 1,1,1,1; 1,1,1,2; and 1,1.5,1,2.5 powders.

Fig. 6 Slow heat/quench of seeded 1,1,1,2 and 1,1.5,1,2.5 powders.

Fig. 7 Typical XRD of a crystallite from the 1,1.5,1,2.5 melt with the basal phase aligned normal to the x-ray beam (at 0° 2θ).

There are narrow temperature ranges during heating (770-800°C) and cooling (800-760°C) where the 30.6Å polytypoid forms preferably, and a temperature range (820-860°C) where the 24.4Å phase grows increasingly more rapidly at the expense of the 30.6Å phase; this leads to the bell-shaped curves shown for the flash and slow heated powders, but the exact mechanisms of this will require further study. We conclude that the heating data (bell-shaped curves) represent a complicated set of interactions influencing each polytypoid. Differences in kinetics, nucleation variations, melting temperatures and possibly vaporization all play a role in polytypoid formation. The 24.4 and 30.6Å phases show extensive interchange between Ca and Sr, and perhaps between Cu and Bi.

Single crystal XRD work and our previous SEM and EDS analyses [2] show that the BCSCO structures are heavily intergrown on both a micro and macro level (syntactic intergrowths).

We believe then that any analysis of the average electrical and magnetic properties of a polycrystalline ceramic form of these Bi containing materials would have to address the percolative or shielding complexities of a microstructure which would, in a simplified form, look as in Fig. 8.

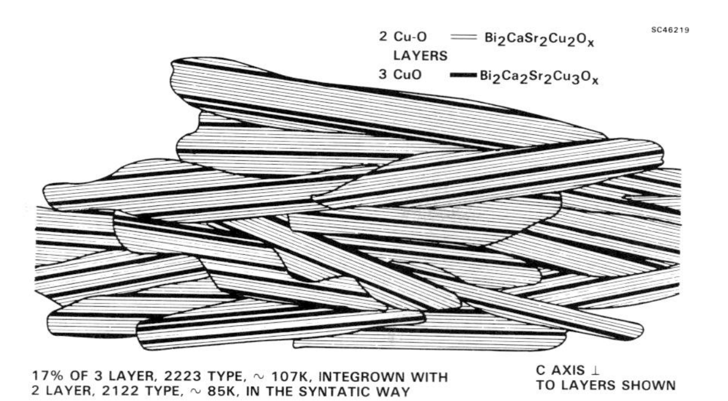

Fig. 8

Syntactically intergrown structure of grains of BiCaSrCuO ceramic.

Crystals of any one pure polytypoid, without intergrowths, may, in fact, turn out to be extremely difficult to achieve. Our initial reaction to these difficulties was to abandon work on the Bi materials and concentrate on the Tl containing materials, which are much simpler in structure [15-17].

Part II: Magnetic Properties - Having detailed one type of problem (of many) with HTSC, we suggest that most standard techniques such as XRD, SEM, TEM, Tc, etc. will be insufficient to characterize HTSC completely. Here we suggest one new technique (of many possible ones) that may help to address the problem.

The technique is based on measurement of the force between a magnet and the superconductor as a function of their separation. An example is shown in Fig. 9. A superconducting disc of $YBa_2Cu_3O_x$ (40 mm diameter, 8 mm thickness) was immersed in liquid N_2 and the magnet (Neodymium-Iron-Boron disc, 13 mm diameter, 5 mm thickness) was attached to the movable crosshead of a commerical mechanical testing machine. Measurements were made with the dipole of the magnet normal to the surface of the superconductor and the disks coaxial [18]. Forces measured as a function of separation, for several load-unload cycles, are shown in Fig. 9. The results are similar to those reported recently by Moon et al [19] for a magnet oriented with its dipole parallel to the surface of the superconductor. There is a large hysteresis in the force-displacement relation, with a distinct change in slope at reversal points. This slope is a measure of the magnetic stiffness.

The hysteresis arises from flux penetration and pinning in regions of the superconductor where the magnetic field exceeds the lower critical field, H_{c1} [20]. The critical field for $YBa_2Cu_3O_x$ has been reported [21] to be ≈ 100 G. From Hall probe measurements of the field along the axis of the magnet (Fig. 10), it is evident that over most of the range of separations in Fig. 9 the field at the upper surface of the superconductor directly beneath the magnet exceeds this value. The magnitudes of the forces in Fig. 9 are also consistent with the superconductor exhibiting type II behavior; the forces are up to several orders of magnitude smaller than expected for complete flux expulsion [18].

The existence of flux pinning and penetration in a type II superconductor is necessary for high critical currents. However, there are also several other important implications that have been demonstrated [18].

1. Since magnetic fields penetrate the type II superconductor, it is not a good shielding material. In fact, the reverse, i.e., focusing of magnetic fields, has been demonstrated [18].

2. The hysteresis allows stable levitation over a range of heights defined by the intersection of the line representing the weight of the magnet with the two branches of the hysteresis loop (at 4 and 10 mm in Fig. 9), in agreement with observations.

3. Hysteretic behavior also allows stable suspension with several combinations of magnets and superconductors.

The measurements in Fig. 9 characterize the magnetic properties of the superconductor averaged over a large volume. It is likely that similar measurements using a smaller magnetic probe could provide a means for local evaluation of microstructural variations, and thereby provide a tool for material evaluation.

CONCLUSION

We have demonstrated that, in detail, the oxide HTSC can be microstructurally, topologically, very complex but, nevertheless, new techniques will arise to address the specific problems. Uses may well, indeed, need to be tailored to the specific actual physical properties that can easily be reached rather than to some desired properties that may be very difficult to achieve.

ACKNOWLEDGEMENTS

Support from the Rockwell Independent Research and Development fund is acknowledged. D.R. Szulc, supported by the Rockwell International Youth Motivation Program, ably prepared samples and conducted some of the heating experiments. The samples slow cooled from the melt were prepared by K.L. Keester with support by the JPL/Caltech Innovative Space Technology Center, which is sponsored by the SDI Innovative Science and Technology Office through an agreement with NASA.

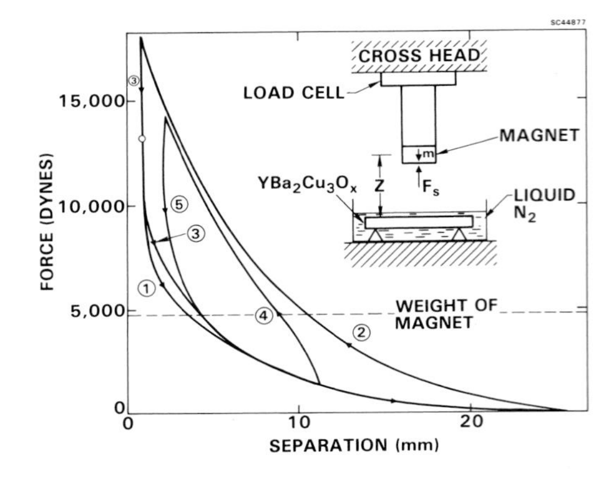

Fig. 9 Repulsive force between magnet and superconductor. Separation was cycled in the order ①②③④⑤.

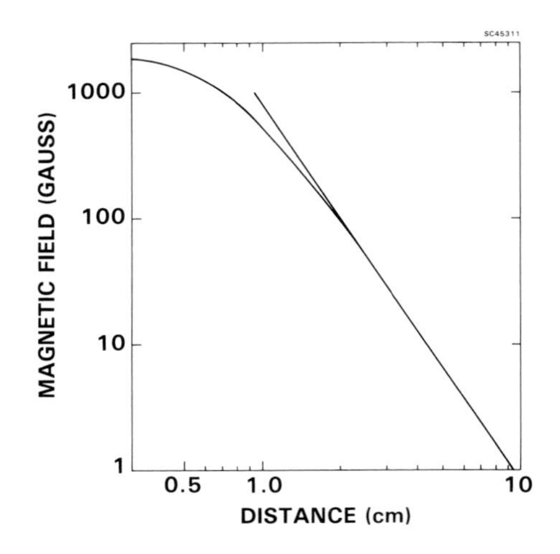

Fig. 10 Magnetic field strength, measured using a Hall probe.

REFERENCES

1. A.B. Harker, P.H. Kobrin, P.E.D. Morgan, J.F. DeNatale, J.J. Ratto, I.S. Gergis and D.G. Howitt, Appl. Phys. Lett. 52, 2186 (1988).
2. P.E.D. Morgan, J.J. Ratto, R.M. Housley and J.R. Porter, MRS Spring Meeting, "Better Ceramics Through Chemistry III," April 5-9, 1988, Reno, NV, in press.
3. C. Michel, M. Hervieu, M.M. Borel, A. Grandin, F. Deslandes, J. Provost and B. Raveau, Z. Phys. B68, 421 (1987).
4. M.A. Subramanian, C.C. Torardi, J.C. Calabrese, J. Gopalakrishnan, K.J. Morrissey, T.R. Askew, R.B. Flippen, U. Chowdhry and A.W. Sleight, Science 239, 1015 (1988).
5. J. Akimitsu, A. Yamazaki, H. Sawa and H. Fujika, Jap. J. Appl. Phys. 26 (12), L2080 (1987).
6. H. Maeda, Y. Tanaka, M. Fukutomi and T. Asano, Jap. J. Appl. Phys. 27 (2), L209 (1988).
7. R.M. Hazen, C.T. Prewitt, R.J. Angel, N.L. Ross, L.W. Finger, C.G. Hadidiacos, D.R. Veblen, P.J. Heaney, P.H. Hor, R.L. Meng, Y.Y. Sun, Y.Q. Wang, Y.Y. Xue, Z.J. Huang, L. Gao, J. Bechtold and C.W. Chu, Phys. Rev. Lett. 60 (12), 1174 (1988).
8. Y. Matsui, H. Maeda, Y. Tanaka and S. Horiuchi, Jap. J. Appl. Phys. 27 (3), L361 (1988).
9. Y. Bando, T. Kijima, Y. Kitami, J. Tanaka, F. Izumi and M. Yokoyama, Jap. J. Appl. Phys. 27 (3), L358 (1988).
10. S. Adachi, O. Inoue and S. Kawashima, Jap. J. Appl. Phys. 27 (3), L344 (1988).
11. S. Ikeda, H. Ichinose, T. Kumura, T. Matsumoto, H. Maeda, Y. Ishida and K. Ogawa, Jap. J. Appl. Phys. 27, L999 (1988).
12. J.M. Tarascon, Y. Le Page, P. Barboux, B.G. Bagley, L.H. Greene, W.R. McKinnon, G.W. Hull, M. Giroud and D.M. Hwang, Phys. Rev. B, 38, 2504 (1988).
13. P.E.D. Morgan, Mat. Res. Bull. 19, 369 (1984).
14. Private communication - David M. Lind.
15. Z.Z. Sheng and A.M. Hermann, Nature 332, 55 (1988).
16. C.C. Torardi, M.A. Subramanian, J.C. Calabrese, J. Gopalakrishnan, K.J. Morrissey, T.R. Askew, R.B. Flippen, V. Chowdhry and A.W. Sleight, Science 240, 631 (1988).
17. M.A. Subramanian, J.C. Calabrese, C.C. Torardi, J. Gopalakrishnan, T.R. Askew, R.B. Flippen, K.J. Morrissey, V. Chowdhry and A.W. Sleight, Nature 332, 420 (1988).
18. D.B. Marshall, R.E. DeWames, P.E.D. Morgan and J.J. Ratto, to be published.
19. F.C. Moon, M.M. Yanoviak and R. Ware, Appl. Phys. Lett. 52, 1534 (1988).
20. F. Hellman, E.M. Gyorgy, D.W. Johnson, Jr., H.M. O'Bryan and R.C. Sherwood, J. Appl. Phys. 63, 447 (1988).
21. R.J. Cava, B. Batlogg, R.B. van Dover, D.W. Murphy, S. Sunshine, T. Siegrist and J.P. Remeika, Phys. Rev. Lett. 58, 1676 (1987).

Magnetization of Various Bulk Oxide Superconductors

SHOICHI YOKOYAMA[1], MASAO MORITA[1], TADATOSHI YAMADA[1], and HIROKO HIGUMA[2]

[1] Central Research Laboratory, Mitsubishi Electric Corp., 8-1-1, Tsukaguchi-Honmachi, Amagasaki, Hyogo, 661 Japan
[2] Materials and Electronic Devices Laboratory, Mitsubishi Electric Corp., 1-1-57, Miyashimo, Sagamihara, Kanagawa, 229 Japan

ABSTRACT

The magnetization measurement at 77K was carried out for various bulk specimens of $YBa_2Cu_3O_{7-x}$ prepared by spray drying process.

The magnetization at low field (< 0.015 T) had hysteresis ΔM. The transport current was applied to specimens and their magnetization was measured. As the transport current was increased, the hysteresis ΔM was decreased. When current was over the critical current Ic of the specimen, ΔM did not changed with the transport current. At high field (> 0.03 T), ΔM was not related to transport current. Therefore ΔM at high field is expected to be caused by shielding current in grains.

The effects of grain size of specimens on magnetization were studied. The hysteresis ΔM at low field of small grain bulk specimen was larger than the large grain specimen. And critical current Ic of small grain specimen was higher. Nevertheless the hysteresis ΔM at high field of the small grain specimen was smaller than the large grain specimen and it depended on magnetic field more strongly.

We prepared the bulk specimen of $YBa_2Cu_3O_{7-x}$ with Ag, using Ag_2O powder. The hysteresis ΔM of this specimen increased by three to eight times .

INTRODUCTION

The high critical current density of a high Tc superconducting oxide system is requisite for power applications. There are a number of measurements on single crystal films of $YBa_2Cu_3O_{7-x}$ that show a value of critical current density Jc in excess of 10^5-10^6 A/cm^2 at 77K [1]. However the value of Jc of bulk specimen of Y-Ba-Cu-O is fairly small in comparison with that of film specimen. This result has been attributed to weak coupling at grain boundaries [2][3]. The magnetization of bulk specimen is due to the sum of bulk transport current and shielding current in grain.

In this paper, we report the effects of grain size of specimen and mixing Ag in $YBa_2Cu_3O_{7-x}$ on magnetization at 77K. In order to distinguish between bulk transport current and shielding current in grain, the magnetization of specimen was measured while transport current was applied to the specimen.

SAMPLE PREPARATION AND EXPERIMENTAL

$YBa_2Cu_3O_{7-x}$ powders were made by thermal decomposition of spray dried powders [4]. This powder was well dispersive and fine (0.1-1μm). Two kinds of final sintered pellets with different microstructures were prepared. The final sintering conditions are 950-1010℃ for 2 hrs in O_2. SEM photographs of fracture surfaces of these specimens are shown in Fig.1. As shown in Fig.1, the microstructure of type A is fabricated with larger grains,which is probably caused by high driving force in sintering. Many grains of type A show rectangular shape ($100\mu m^{\square} \times 10\mu m^{t}$) as elongated to a certain direction. The microstructure of type B is porous and composed of smaller grains ($2\mu m^{\square}\times 2\mu m^{t}$). It was found that such a microstructure of specimens derived from spray drying as type A or B was greatly influenced by process after preparation of oxide powder. The bulk specimen type C of $YBa_2Cu_3O_{7-x}$ with Ag was prepared from mixed powder of $YBa_2Cu_3O_{7-x}$ and Ag_2O. The mixtures were cold-pressed and sintered at 1100℃ for 2hrs in O_2. The microstructure of type C is close and composed of rectangular $YBa_2Cu_3O_{7-x}$ grains ($10\text{-}20\mu m^{\square}\times 5\mu m^{t}$). Ag are separated out on the bulk surface and existed in all voids.

D.c. measurement of superconductivity in the bulk specimen was made using a standard four-terminal method at liquid nitrogen temperature. Critical temperatures Tc of 91K (type A), 86K (type B) and 90K (type C) were obtained. The effect of sintering temperature on the critical current density Jc is shown in Fig.2 for type A and B. These Jc values are increased by increasing the sintering temperature. It is found the effect of sintering temperature on Jc for type A was distinguishable especially. Fig.3 shows the magnetic field dependence of the Jc normalized by that at 0T for samples sintered at 980℃. As shown in Fig.3, the field dependence of the Jc for type A is less than that for type B. In spite of the higher Jc for type B, the magnetic field dependence of the Jc for type B is stronger, which may be because of different behavior of weak links between type A and type B. Jc of type C was about 470-1900A/cm^2 at zero external field. The field dependence of the Jc for type C was same as that for type A.

The magnetization measurement at 77K was carried out for type A-C. External changing magnetic fields, $\pm\mu_0 H$, were applied to these samples in the direction parallel to the longitudinal axis. The frequency of the external field was 0.5-1Hz.

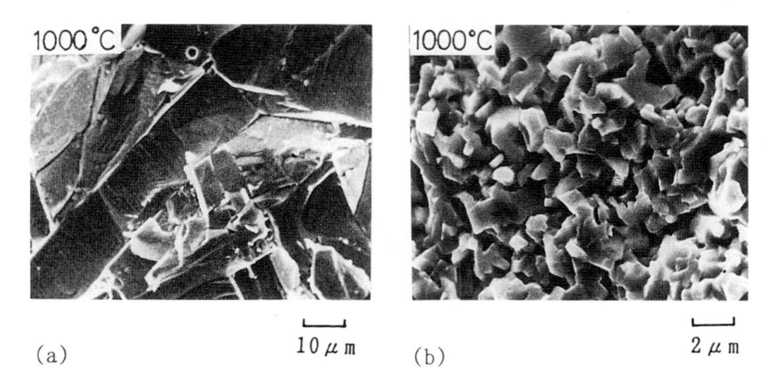

Fig.1 SEM photograph of fracture surface of the bulk specimens.
(a) Type A (1000℃-2hrs)
(b) Type B (1000℃-2hrs)

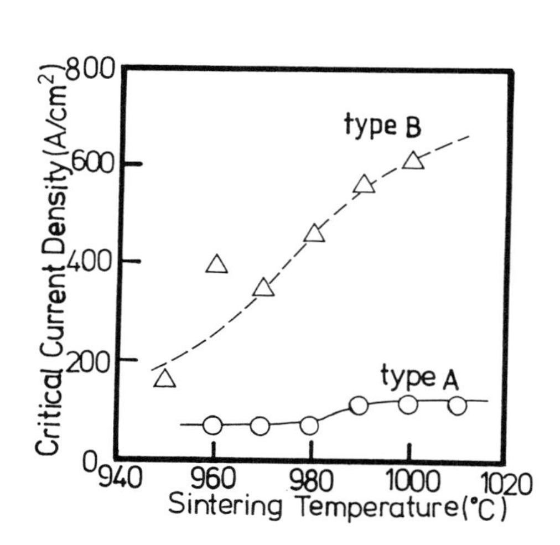

Fig.2 The effect of sintering temperature on Jc for type A and B.

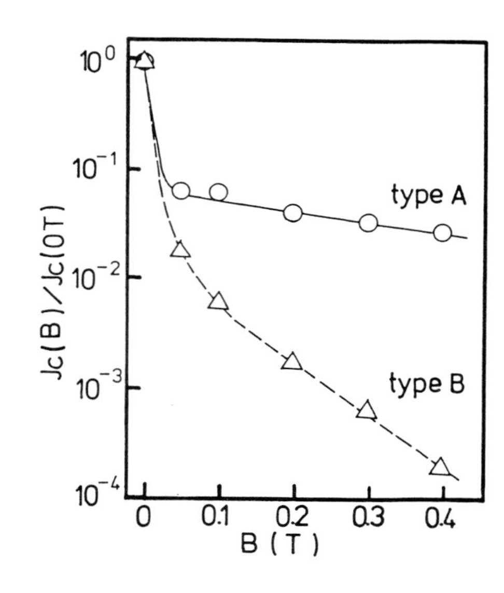

Fig.3 The magnetic field dependence of the Jc normalized by that at 0T.

MAGNETIZATION PROPERTIES

A typical magnetization of $YBa_2Cu_3O_{7-x}$ bulk specimen has hysteresis at low magnetic fields (<150Gauss). However hysteresis of magnetization for the ground specimen at low field was about one order less than of the bulk specimen. It is expected from this result that the hysteresis of magnetization for the bulk was generated with transport current in the bulk. This transport current flows through weak-link regions. In order to distinguish between bulk transport current and shielding current in grain, the magnetization of specimen was measured while transport current was applied to the specimen, as shown in Fig.4-a. These results of the magnetization for the bulk specimen are shown in Fig.4-b ($\mu_0H=\pm0.05T$) and Fig.4-c ($\mu_0H=\pm0.15T$). The hysteresis ΔM was decreased by increasing the applied current. Fig.5 shows that the dependence of the applied current on the hysteresis at 77K and zero field. When the current is over the critical current Ic of the specimen, this hysteresis doesn't change with the current. If the difference, ΔM_T, between hysteresis at zero applied current and one at Ic is equal to the magnetization by transport current, Jc can be calculated from ΔM_T as $\Delta M_T=\mu_0 J_C d/2$ where μ_0 is permeability of the Vacuum ,Jc is a constant critical current density and d is width of the bulk specimen (a 'Bean-London' model). From this equation we have $Jc=145A/cm^2$. This value is almost same as Jc ($=119A/cm^2$) by Four-terminal measurement. As shown in Fig.4-c, the hysteresis of specimen at high field was not related to transport current. Therefore this hysteresis is expected to be caused by shielding current in grains.

The magnetizations of type A-C at low field ($\pm0.015T$) and high field ($\pm0.2T$) are shown in Fig.6. It is found in Fig.6(B-1) that the magnetization property of type B has a peak clearly at 0.002T. The magnetization of type A or C has no-peak at low field (<0.015T). From these results it is expected that the magnetization by transport current in the bulk for type A and C is less than the magnetization by shielding current in grains, or the magnetic field dependence of Ic at weak links is almost same as the dependence of Ic in the grain. At high field (> 0.03T) in Fig.6 (A-2, B-2 and C-2) the magnetization is almost generated by shielding current in grains. The effect of grain size on magnetization is found in Fig.6 (A-2 and B-2). The external field dependence

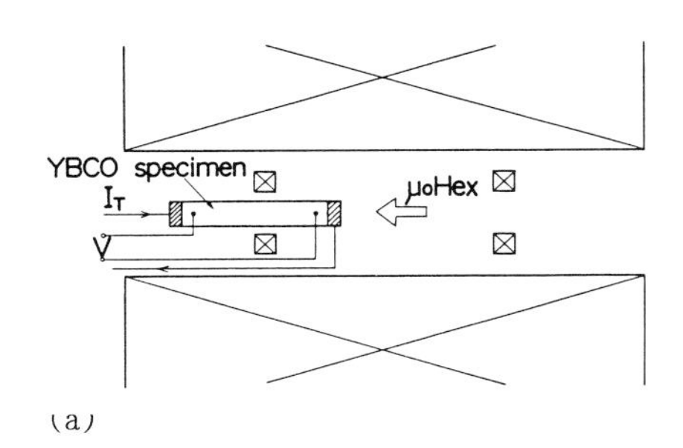

Fig.4 The magnetization of specimen was measured while transport current was applied to specimen.
(a)measurement system (b)magnetization at low field ($\pm0.04T$) (c)magnetization at high field ($\pm0.15T$)

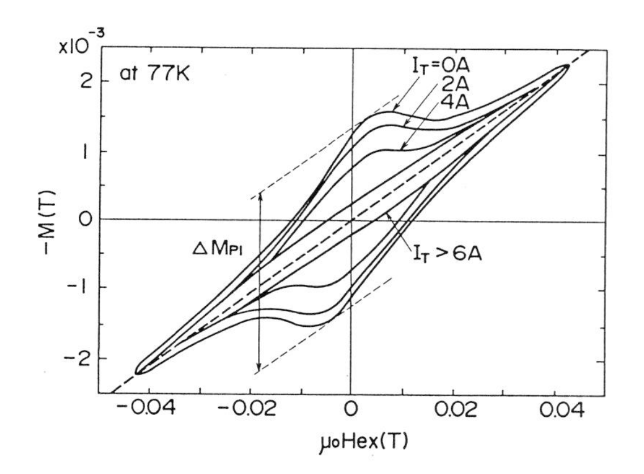

(b)

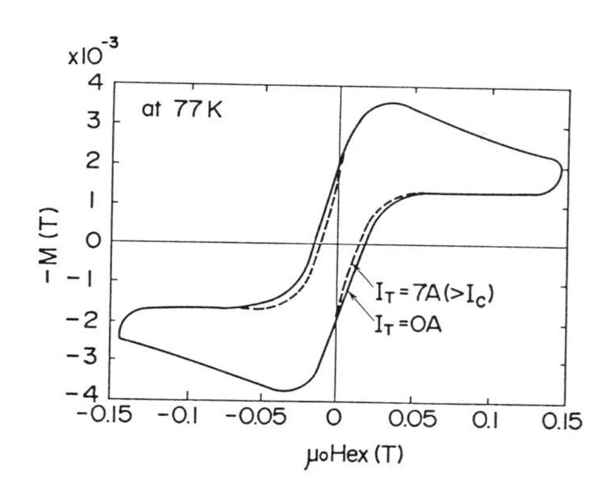

(c)

of hysteresis of magnetization at high field for type A is less than that for type B, that is we expected the field dependence of Jc in the grain for type A is also less than that for type B. The effect of mixing Ag in $YBa_2Cu_3O_{7-x}$ is, found in Fig.6 (C-2). The hysteresis of magnetization for type C is three to six times larger than that for type A or B, that is, Jc of the grains for type C is two to five times larger that for type A or B using a 'Bean-London' model. In this result the grains of type C have strong pinning force.

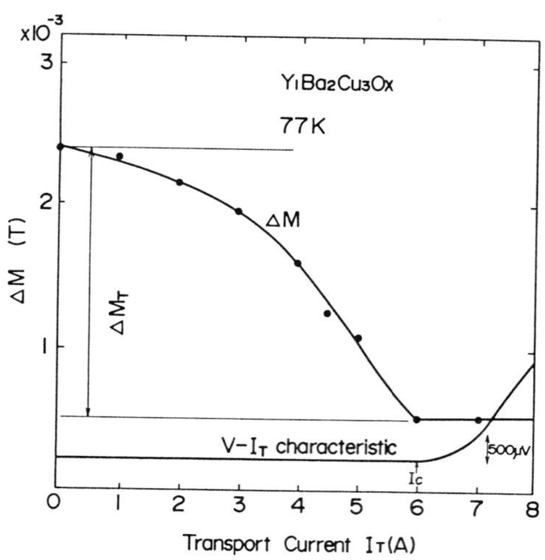

Fig.5 The hysteresis dependence of the applied current at 77K and zero field.

CONCLUSION

We prepared the two types of bulk specimens of $YBa_2Cu_3O_{7-x}$ with different grain sizes using spray drying process. We also prepared bulk specimens from a mixture of Ag_2O and $YBa_2Cu_3O_{7-x}$ powder. The two types of sintered pellets with different microstructures exhibit different superconducting properties. The critical current density Jc and the magnetization of the small grain specimen at 77K and zero field were higher than these of the large grain specimen. The magnetic field dependences of the Jc and the magnetization for the large grain specimen were less than these of the small grain specimen. This result is supposed to be related to weak-links which existed in grain boundaries. In order to distinguish between the magnetization by transport current in bulk and that by shielding currents in grains, the magnetization was measured while transport current was applied to the specimen. From this result, the hysteresis of magnetization at low field

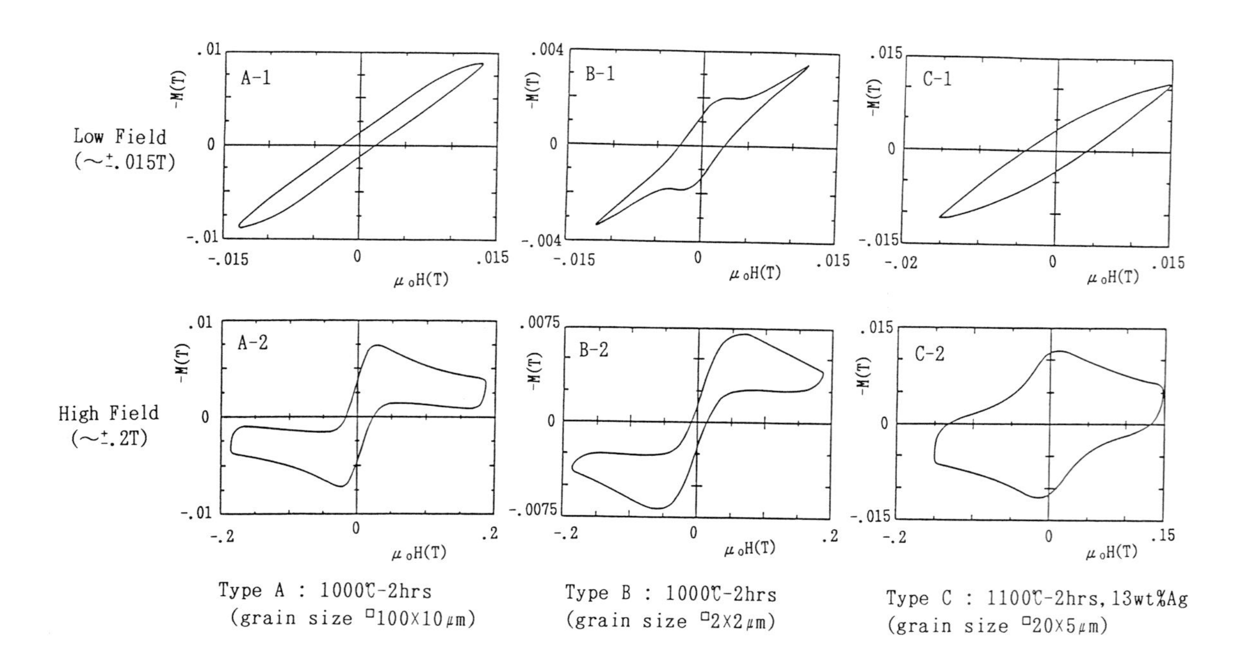

Fig.6 The magnetization of the bulk specimens.

included the magnetization by the transport current in bulk and the hysteresis of magnetization at high field was generated mainly by the shielding current in grains. The hysteresis of magnetization for the specimen containing Ag was larger than that for the pure $YBa_2Cu_3O_{7-x}$ bulk specimen.

REFERENCES

[1] P.Chaudhari, J.Mannhart, D.Dimos, C.C.Tsuei, J.Chi, M.M.Oprysko, and M.Scheuermann, Phys Rev Lett (1988) 60, 1653
[2] J.W.Ekin, A.I.Braginski, A.J. Panson, M.A.Jancko, D.W. Capone II, N.J. Zaluzec, B. Flandermeyer, O.F.de Lima, M.Hong, J.Kwo, and S.H.Liou, J.Appl.Phys. (1987) 62, 4821
[3] V.Ottobini, A.M.Ricca, G.Ripamonti, and S.Zannella, IEEE, Trans., Mag., (1988) 24, 1153
[4] H.Higuma, H.Nakajo, M.Wakata, and K.Egawa, Proceedings MRS, (1988)

Preliminary Results of Tc and Jc Measurements on $YBa_2Cu_3O_x$

Y. KIMURA[1], N. HIGUCHI[1], S. MEGURO[2], K. TAKAHASHI[2], K. UYEDA[2], T. ISHIHARA[3], E. INUKAI[3], and M. UMEDA[4]

Electrotechnical Laboratory, AIST[1], Engineering Research Association for Superconductive Generation Equipment and Materials (Super-GM)[2], Japan Fine Ceramics Center[3], and Moonlight Project Promotion Office, AIST[4],
Umeda UN Bldg., 5-14-10 Nishitenma, Kita-ku, Osaka, 530 Japan

ABSTRACT

In our research and development of oxide superconducting materials, methods for measuring the critical properties such as critical temperature and critical current were investigated and studied by way of round robin tests. The measurement of critical temperature and critical current were made mainly using the 4 terminal method, and partially the magnetization method. The results were found to disperse, depending on the measurement methods and definitions of Tc and Ic. Data analysis with these factors clarified is important.

INTRODUCTION

Oxide superconducting materials are far higher in critical temperature than conventional superconducting materials mainly based on metals, and what critical current density can be achieved in future by oxide superconducting materials at any temperature higher than the liquid nitrogen temperature is surmised to be important in considering the application of oxide superconducting materials.

However, oxide superconducting materials are not so clarified as existing superconducting materials, and in the field of physical property study, high performance single crystals and polycrystals are being synthesized, to obtain precise experimental data, and macroscopic and microscopic material evaluation is being made through crystal structure analysis, chemical composition analysis, electronic structure analysis, etc..

In the present investigation and study, we are mainly examining what subjects should be taken up for research at first in order to develop oxide superconducting materials for application of power apparatuses[1].

Oxide superconducting materials are premature. With attention paid to the fact that superconducting properties are evaluated differently from material to material, it has been attempted to raise the reliability of data used in research and development,for allowing discussion on a common base, by using common samples, with an intention to identify the differences and problems in the measurement methods adopted by the respective member research organs.

INVESTIGATIONS ON THE METHODS FOR MEASURING CRITICAL PROPERTIES OF SUPERCONDUCTIVITY

The Procedure of Round Robin Test concerning the critical properties of superconductivity is shown in Fig.1. As the first step, questionnairing was carried out on the present measurement methods and definitions used for critical temperature and critical current by the member research organs. Common samples were prepared and distributed to the member research organs, and later investigation was made on both the common samples and members own samples concurrently.

With regard to the measurement methods, mainly considering the critical properties of oxide superconducting materials at the present stage, the following problems could be identified.

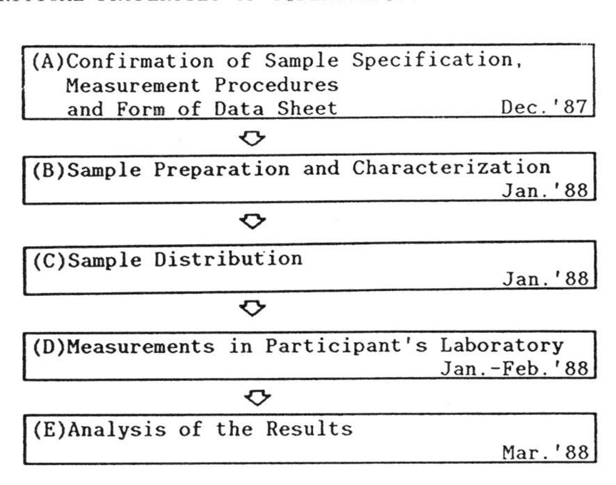

Fig.1 Procedure of Round Robin Tests of Tc and Ic on $YBa_2Cu_3O_x$ Sample

(1)Homogeneity of sample as taken out by bulk material
(2)Past record of superconductor before measurement after its synthesis
(3)Measurement methods and causes of differences between 4 terminal method and magnetization method and(particularly in critical current measurement)
(4)Definitions
(5)Methods for setting the current and voltage terminals in the case of 4 terminal method
(6)Treatment applied to the surface of superconductor when the current and voltage terminals are set
(7)Areas of current and voltage terminals, and spacings between them
(8)Reproducibility of the measurement methods
(9)Critical properties change with the passage of time

Opinions concerning the above were collected, and arranged in reference to the items specified for the critical properties (critical temperature and critical current) measurement of samples.

EXPERIMENTAL METHODS

For the experiments, common samples of $YBa_2Cu_3O_x$ prepared by Japan Fine Ceramics Center were distributed from Super-GM and measured data were compared.
Mainly the 4 terminal method was used for measuring the critical temperature and critical current, but the magnetization method was also partially used for measuring the critical temperature and critical current.

Critical temperature measurement

In the measurement of critical temperature by the 4 terminal method, the sample was mostly fixed to a sample holder made of copper by vacuum grease, etc. with an insulator such as Kapton tape. The copper sample holder is large in thermal capacity and high in thermal conductivity and shows smooth temperature change. However, when the sample holder is moved, the oxide superconducting material which is lower also in thermal conductivity than the metal is liable to undergo stepped temperature change. It was also observed that hysteresis was brought about in the process of temperature rise and fall. It is required to keep as small as possible the temperature difference between both the ends of a sample and the temperature difference between the sample and the thermometer and to lower the temperature change rate, not to bring about the hysteresis in the process of temperature rise and fall. It is necessary to specify the temperature change rate, for obtaining highly reliable data.
When Joule heat is generated at the electrode joints, the sample temperature may rise. Therefore, when the current-voltage characteristic is measured by the 4 terminal method to confirm zero resistance, it is required to examine disturbances such as connection resistances of respective electrode terminals due to the setting of sample holder, sample, etc.[2], and thermo-electromotive force.

In the case of critical temperature measurement by the magnetization method, the critical temperature at which magnetic susceptibility began to change (onset) was mainly measured. The measured values dispersed. It is necessary to examine the shape of pickup coil, method for setting the sample to the holder (coil), and others in the practice of measurement.
Furthermore, in the measurement of critical temperature by the 4 terminal method, when the test current was large even at a low temperature change rate,it was observed that the zero resistance value of the temperature-resistance curve tended to shift toward low temperature side.
Comparing the 4 terminal method with the magnetization method, the 4 terminal method cannot avoid difficulty in identifying the transition onset temperature between superconducting state and normally conducting state but can advantageously decide the transition completion temperature as the temperature of electric resistance zero, while the magnetization method can easily detect the transition onset temperature but cannot avoid difficulty in identifying the transition completion temperature since the magnetization susceptibility keeps changing even if the temperature declines further. For this reason, it is surmised that the superconducting state can be accurately identified by using plural measurement methods such as 4 terminal method and magnetization method in combination.
The research for new materials lays emphasis on the transition onset point, and in the development of wires, etc., the transition completion point is considered important. For respective objects of research and development, the definitions of critical temperature should be selectively used, and in this case, it is surmised to be necessary to clearly state the measurement conditions and definitions to avoid misunderstanding.

Critical current measurement

In the critical current density measurement by the 4 terminal method, the test current applied to the sample is increased, and the current density at which any voltage is generated in the sample with the superconducting state destroyed is defined as the critical current. The critical current measurement by the 4 terminal method of this time used a voltage as adopted in the ASTM standard[3] for the critical current of metal superconducting materials such as NbTi, as a temporary criterion value of voltage generated with the superconducting state destroyed.

Some results showed the decline of critical current density. This may have been caused by the degradation of the superconducting material due to the deterioration of sample surface, etc.[4]. Among them, sample which showed remarkable decline had a varnish used for fixing electrodes and for insulation and was removed by alcohol. Therefore, it can be considered that the sample may have reacted with water contained in alcohol, and water and carbonic dioxide gas in air, to deteriorate the sample surface[5], and to lower the critical current density at the time of remeasurement.

There were also results showing the rise of critical current density on the contrary. As for the cause, it can be considered that since the sample was a polycrystal prepared by sintering, the test current may have taken a different path each time of measurement, due to the influence of electrode contacting method, etc., to change the current density.

The evaluation of critical current density (Jc) by the 4 terminal method is surmised to involve the following problems.

(1)Since the superconducting current flows spatially unevenly in the sections of the sample,the value of Jc varies depending on how to define the effective section in the evaluation of critical current density Jc from the critical current Ic.
(2)If the contact resistance[6] between the current leads and the sample is large, and if the value of Jc is large, the heat generation at the contacts makes the evaluation of Jc inaccurate.
(3)Depending on the method of contacting the leads, oxides may be produced at the contacting portions or the current distribution becomes uneven, to lower Jc.
(4)Microscopically, crystal orientation makes the current flow anisotropic[7], to change the current density of the sample.

On the other hand, the critical current density Jc by the magnetization method is higher than the critical current density obtained by the 4 terminal method.

In the measurement of critical current density by the 4 terminal method, current is fed, to measure the inter-terminal voltage, and the current flowing at the moment when the superconducting condition is destroyed is measured. Therefore, between the voltage terminals of the oxide superconducting material,transport current is required to flow perfectly. If any disturbance exists in the current path, current flow is disturbed, to lower the critical current.

Table I Research Results of Critical Characteristics for Oxide Superconducting Materials

1.Critical Temperature(Tc)		2.Critical Current(Ic)	
A)4 terminal Method Influence of Temperature Change Rate	Higer change rate makes electric resistance hysteresis and the Tc higher Appropriate temperature change rate is necessary	A)4 terminal Method Influence of Connection Resistance	It has been found that measured Ic values are greatly different from kind to kind of electrode materials A higer current applied raises the sample temperature more due to Joule heat generation at the electrode joint
Influence of Applied Current Density and Connection Resistance	A higher current applied raises the sample temperature more due to Joule heat generation at the electrode joint		
B)Magnetization Method Influence of AC Magnetic Field	The transition became unclear	B)Magnetization Method Comparision between 4 terminal Method and Magnetization Method	Jc values are differnt, depending on evaluation method Measurement by magnetization method is surmised to give local information
C)Comparison between 4 terminal method and magnetization method	4 terminal method identify the transition transition temperature of zero resistance, while the magnetization method can easily detect the transition onset temperature It is necessary to clarify the measurement conditions and definitions to avoid misunderstanding		

Various reasons can be macroscopically considered for the disturbance of transport current. One of them is that a sintered $YBa_2Cu_3O_x$ sample has many clearances. It can be considered that sintered grains less contact each other, to decrease the available path area through which current flows across grain boundaries, there by lowering the critical current. Another reason can be surmised to be that the critical current is greatly affected by crystal orientation.

The critical current density obtained by the magnetization method is relatively large probably due to the addition of locally flowing microscopic current.

CONCLUSIONS

Conclusions obtained from the above investigation results are shown in Table I. The methods for measuring the critical properties of oxide superconducting materials, criteria, etc. now used include various methods and temporary definitions. To raise the reliability of measured data of superconductivity properties, and to make discussion based on a common base, common samples were used to identify differences and problems of measurement methods adopted by the respective member research organs. Furthermore, under the present project, it is planned to promote the research and development on the process for synthesizing oxide superconducting materials. It is surmised to be more important in future, to clarify the reliability of data by establishing methods for measuring the properties of oxide superconducting materials, criteria, etc.

ACKNOWLEDGEMENT

This investigation and study were carried out as part of FY 1987 Feasibility Study on Superconducting Machinery and Materials Technology Related to Electric Power Generation (Moonlight Project Promotion Office, Agency of Industrial Science and Technology). The authors are much indebted to the members of Superconducting Materials Subcommittee shown in the following list and other people concerned for their assistance and cooperation.

Subcommittee of superconducting materials in Super-GM

Y.Kimura(Chairman)	Electrotechnical Laboratory,AIST
T.Okada	Osaka University
K.Osamura	Kyoto University
K.Kitazawa	The University of Tokyo
K.Noto	Tohoku University
M.Kosaki	Toyohashi University of Technology
T.Matsushita	Kyushu University
N.Higuchi	Electrotechnical Laboratory,AIST
T.Inoue	National Research Insitutute for Metals
M.Nakamura	The Tokyo Electric Power Co.,Inc.
N.Haruki	The Kansai Electric Power Co.,Inc.
S.Oi	Chubu Electric Power Co.,Inc.
S.Akita	Central Resaeach Institute of Electric Power Industry
N.Tada	Hitachi,Ltd.
K.Tanaka	Hitachi,Ltd.
H.Yoshino	Toshiba Corp.
M.Yamaguchi	Toshiba Corp.
K.Yoshizaki	Mitsubishi Electric Corp.
T.Yamada	Mitsubishi Electric Corp.
H.Takei	Sumitomo Electric Industries,Ltd.
Y.Tanaka	The Furukawa Electric Co.
O.Kono	Fujikura,Ltd.
Y.Kubo	Japan Fine Ceramics Center
K.Uyeda	Super-GM
S.Meguro	Super-GM
T.Saito	Super-GM
K.Takahashi	Super-GM
H.Kubokawa	Super-GM

REFERENCE

[1]Feasibility study on superconducting machinary and materials technology related to electrical power generation (Superconducting material)(1987,march), p110.
[2]J.W.Ekin,A.J.Panson and B.A.Blankenship:Appl.Phys.Lett., **52**,331(1988).
[3]ASTM,Designation B714-82:Standard Test Method for DC Critical Current of Composite Superconductors
[4]R.L.Barns and R.A.Laudise:Appl.Phys.Lett., **51**,26(1987).
[5]S.L.Qiu,M.W.Ruckman,N.B.Brookes,P.D.Johnson,J.Chen,C.L.Lin,M.Strong,B.Sinkovic J.E.Crow and Cham-Soo Jee:Phys. Rev., **37**,3747(1988).
[6]I.Nakai,S.Sueno,F.P.Okamura and A.Ono:Jpn.J.Appl.Phys., **26**,L682(1987).
[7]Y.Iye,T.Tamegai,H.Takeya and H.Takei:Jpn.J.Appl. Phys., **26**,L1052(1987).

A Deterioration Mechanism on Critical Current Densities of High Temperature Superconductors

KUMIKO IMAI and HIRONORI MATSUBA

The Furukawa Electric Co.,Ltd., Yokohama Laboratories, 2-4-3, Okano, Nishi-ku, Yokohama, 220 Japan

ABSTRACT

Deterioration of oxide superconductors has been investigated by evaluating the critical current densities at 77K and the resistivity at room temperature. The oxide superconductors were exposed to moisture or heat cycle between 77K and room temperature. The experimental results indicate that the external substance penetrates into the superconductors and brings about chemical reactions to form non-superconductive products so that it involves volume expansion at the site to cause micro-cracks. An analysis of the relation between the critical current density and the conductivity has been carried out and we have estimated the mechanism of the deterioration by the analysis.

INTRODUCTION

Since oxide superconductors are discovered, much investigation on its applications has been carried out from the viewpoint that the superconductors could be used in liquid nitrogen.
When we intend to put these to practical use, the stability and the life of the superconductors might become an essential item.
Oxide superconductors have been said to be very sensitive to environmental substances such as water, alcohol or some other chemicals. [1]~[4]
Quantitative investigations or the mechanism how to influence the superconductive property like critical current density or critical temperature have not yet been investigated.
We have estimated the deterioration of oxide superconductors by evaluating the transition of the critical current densities and of the conductivities accompanied by the deteriorating process.

EXPERIMENTS ON DETERIORATION BY MOISTURE

Samples are prepared by sintering method. Y_2O_3 $BaCO_3$ and CuO were mixed as for Y:Ba:Cu=1:2:3, then calcined and sintered.
Those were polycristalline substance, the typical size of samples was 2mm X 1mm X 20mm. The critical currents (Jc's) and resistivities were measured by the

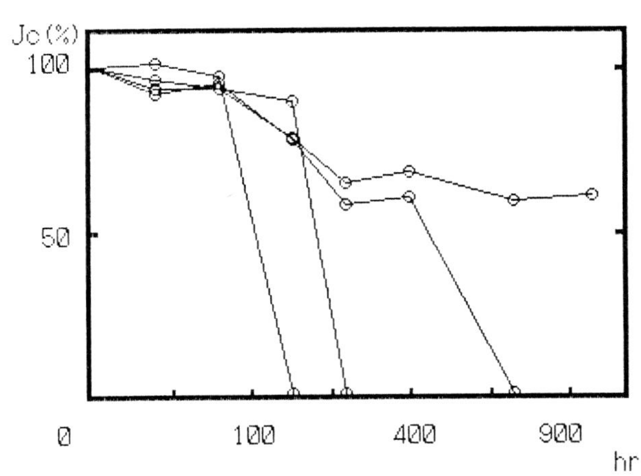

Fig.1 Jc transition of YBCO samples exposed to the air at 25°C and R.H. 95%

four probe method. Four YBCO samples were tested under the aqueous vapor. Initial values of Jc's were around $200A/cm^2$. These specific gravities were about 90 % of the theoretical value. Fig.1 shows the percent transition of Jc's tested in the air at 25°C and R.H.95%. The Jc's initially decrease slowly at a fixed rate. The decreases seem to depend on the diffusion rate of water because the rate is proportional to the square root of the elapsed time. The fast decreases of Jc were observed to follow after the slow decreases. The critical temperature (Tc) was not changed during the deteriorating process. However, the conductivities of the samples at room temperature decreased. YBCO samples are deteriorated in the air at 150°C. The results are shown in Fig.2. Jc's dropped greatly for the first measurement. Thereafter, Jc decreased at a fixed rate.

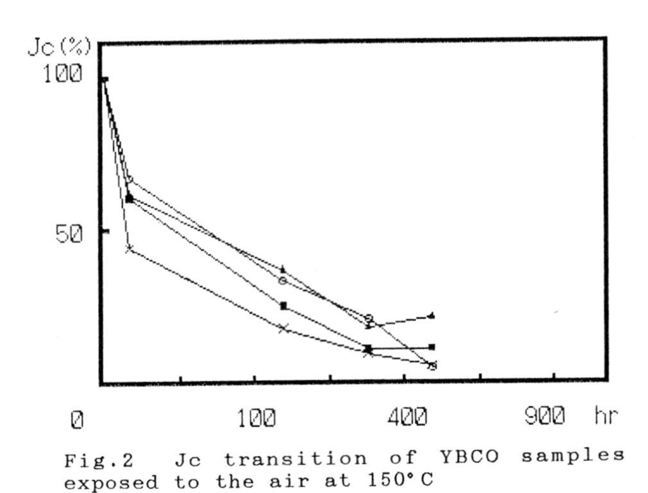

Fig.2 Jc transition of YBCO samples exposed to the air at 150°C

EXPERIMENTS ON DETERIORATION BY HEAT CYCLES

Heat cycle tests were carried out by cycling the samples between liquid nitrogen and ambient temperature. The transition of the critical current densities at 77K and conductivities at room temperature were measured. Fig.3 shows the transitions corresponding to repeated numbers of heat cycles. the conductivities of two samples are plotted against number of heat cycles N, for the lower curves, and the critical current densities are plotted in the upper curves. In these plots, the conductivity measurements start from N=0, which represents the initial state before cooling and the Jc measurements start from N=1, which represents the initial start at 77K. The critical current densities and the conductivities of the same samples are plotted with the same plotted styles respectively.

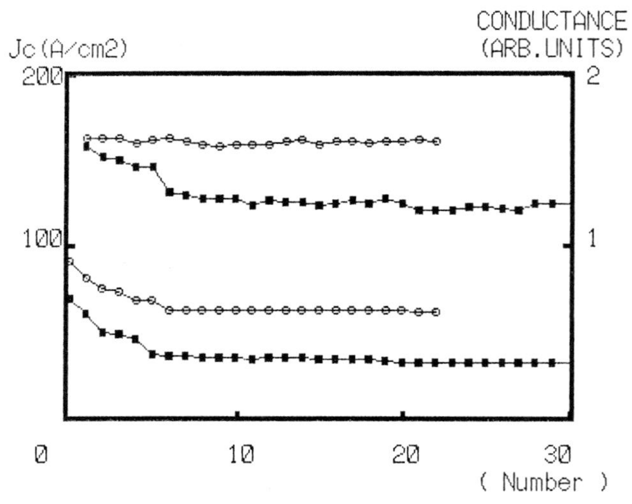

Fig.3 Transitions of Jc and conductance of YBCO samples under heat cycles between room temperature and 77K

The critical current density of the sample plotted with open symbols changes little during the repetition of heat cycles, but the conductivity decreases during the initial several cycles. For the sample plotted with closed symbols, the current densities decreased, following the decrease of the conductivity.
The decrease rate of conductivities is larger in the initial several cycles, especially so for the first cooling process.

CONSIDERATION

The experimental results suggest that the deterioration of the oxide superconductors is caused by some physical transition accompanied by the chemical reactions or heat distortion. Because the fast decrease of Jc after the slow decrease in moisture or the initial decrease of the conductivity by heat cycle is difficult to be explained only by chemical reactions.

We have estimated the physical transition to be the generation of cracks and have made a model of a superconductor containing small cracks in it. Using this model the relation between the conductivity and the critical current density has been introduced. The procedure to introduce the relation is described in APPENDIX. The relation are illustrated in Fig.4 with bold line for equation (13) and with broken bold line for equation (14).

The conductivities and the critical current densities measured at the experiment on deterioration by moisture are also plotted in Fig.4. These plots accord substantially with the calculated relation. However the relation of some samples such as a sample exposed to heat cycle does not accord with the calculated relation. The reason could be that the critical current density is decided only by the minimum critical current path, that is to say, only one cross-sectional area of the sample is dominant; on the other side, every part of the sample affects the conductivity. This relation could be used for evaluating the deterioration of Jc by the first cooling to boiling liquid nitrogen temperature.

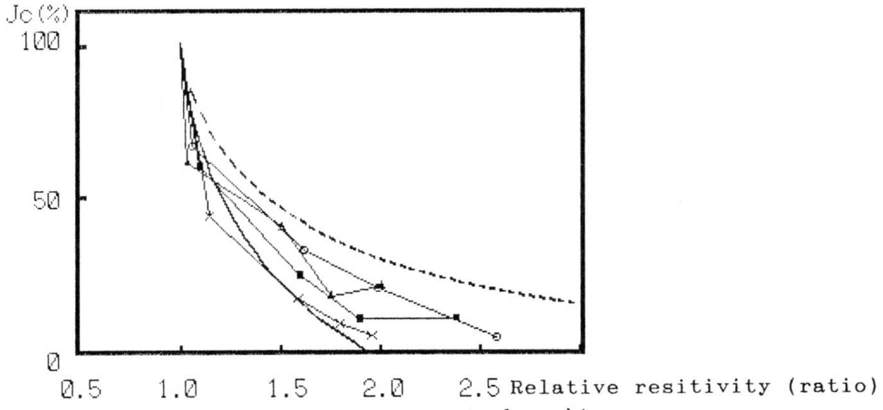

Fig.4 The critical current density versus conductivity of the deteriorated samples and calculated curves

CONCLUSION

Penetration of the external substances into the superconductors brings about chemical reactions with the superconductor at the inter-granular surfaces and forms non-superconductive products. The chemical reactions involve volume expansion by the chemical products and cause micro-cracks at the inter-granular surfaces. The heat distortion also causes micro-cracks or macro-cracks. The critical current density and conductivity of superconductors change with the amounts of non-superconductive products and cracks. We have developed a model which explains the relation of non-superconductive contents to the resistivity or to the critical current density.

REFERENCES

(1) M.F.Yan,R.L.Sarns,H.M.O'Bryan etal Appl. Phys. Lett. 51(7)17 Aug.1987
(2) I.Nakada,S.Sato,Y.Oda and T.Kohara Jpn. J. Appl. Phys. 26 (1987) L697
(3) K.Kitazawa,K.Kishio,T.Hasegawa,O.Nakamura,J.Shimoyama, N.Sugii, and K.Fueki Jpn.J.Appl.Phys.26(1987) L1979
(4) K.Imai and H.Matsuba Preprint for Applied Super. Conf. ME-3

APPENDIX

Relation between Conductivity and Critical Current Density of Superconductive Substance Containing Dispersed Non-conductive Particles in it

1. Derivation of an equation to give the conductivity

As illustrated in Fig.5, Conductivity is calculated on a model that small particles with conductivity of σ_2 are dispersed homogeneously in a medium of which conductivity is σ_1. When the shape of the small particles is ellipsoid, and both the medium and the particles are conductive dielectrics of ε_1 and ε_2 in their complex dielectric constants ε^* respectively, The complex dielectric constant of the composition has been known as:

$$\varepsilon^* = \varepsilon_1^* \{1+ q \frac{n(\varepsilon_2^* - \varepsilon_1^*)}{(n-1)\varepsilon_1^* + \varepsilon_2^*}\} \quad --- (1)$$

where q is volume ratio of small particles to the medium, and n is a function of eccentricity e of the ellipsoid. The eccentricity is defined by the axes a, b, c of the ellipsoid:

$$e = \sqrt{1-a^2/b^2} \quad --- (2)$$

When a<b=c, n is expressed by the equation:

$$n = \frac{e^3}{e - \sqrt{1-e^2}\sin^{-1}e} \quad --- (3)$$

Since the complex dielectric constant is defined by:

$$\varepsilon^* = \varepsilon - j\sigma/\omega \quad --- (4)$$

we can calculate the destined equation of the conductivity by setting $\varepsilon_1 = \sigma_0$ and $\varepsilon_2 = 0$ in the equation (1) as:

$$\frac{\sigma}{\sigma_0} = 1 - q \cdot \frac{n}{n-1} \quad --- (5)$$

While q is expressed as:

$$q = \frac{N}{V} * \frac{4}{3}\pi abc = \frac{N}{V} * \frac{4}{3}\pi b^3\sqrt{1-e^2} \quad --- (6)$$

where N is the number of the particles and V denotes the volume of the composition. Substituting this relation in equation (4), the conductivity can be expressed by the shape of the small particles in the form:

$$\frac{\sigma}{\sigma_0} = 1 - \frac{N}{V} * \frac{4}{3}\pi b^3 \frac{e^3}{\sin^{-1}e - e\sqrt{1-e^2}} \quad --- (7)$$

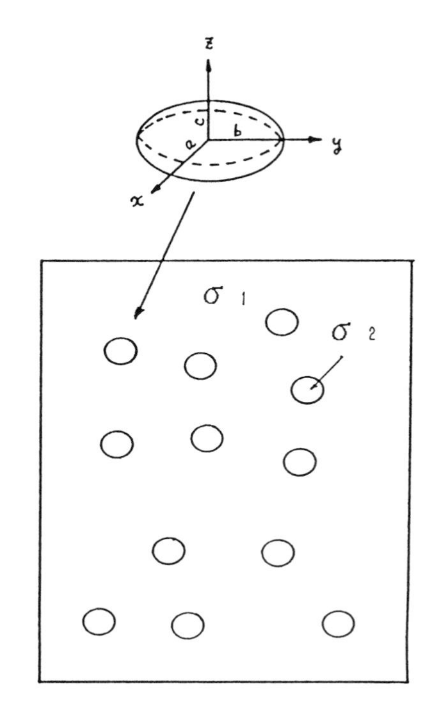

Fig.5 A Model for calculation on conductivity

When we regard the small particles as very thin disks with diameter b, we introduce the following equation in the limit e → 1 from the equation (7):

$$\frac{\sigma}{\sigma_0} = 1 - \frac{N}{V} * \frac{8}{3} b^3 \quad --- (8)$$

If we regard these as spherical particles with diameters b, then the conductivity of the composition is given in the limit e → 0 as:

$$\frac{\sigma}{\sigma_0} = 1 - \frac{N}{V} * 4 * b^3 \qquad --- (9)$$

This equation only differ from equation (8) in the coefficients.

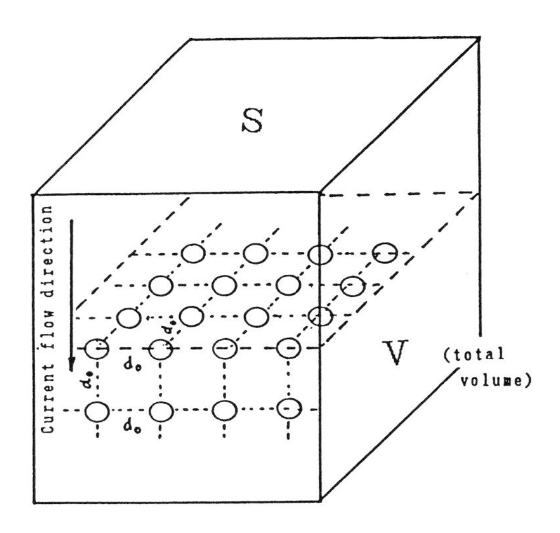

Fig.6 A Model for calculation on critical current density

2 Derivation of an equation to give critical current density

Let us consider the critical current density of the model shown in Fig.5 where the small particles are non-superconductive and the medium is a superconductor of which critical current density is J_{c0}.
The critical current may be determined by the minimum cross-sectional area of the superconductor perpendicular to the current flow direction. Approximate calculation on the cross-sectional area can be made by a simple model that the small particles are arranged regularly as shown in Fig.6. Fig.7 illustrates a cut-out with thickness d_0 and cross-sectional area S, taken out from the composition shown in Fig.6.
The most simple estimation to give the critical current density could be obtained if we assume that the critical current density is considered to be proportional to the projected area Sc of the medium on cut plane,i.e. the superconductive current only flows in the same direction. Thus we obtain:

$$\frac{Jc}{J_0} = \frac{Sc}{S} = 1 - \frac{Ns}{S} \cdot \pi b^2 \qquad --- (10)$$

where Ns is the number of the particles in the cut-out. By the total number N of the particles and the total volume V of the system, this equation can be rewritten as:

$$\frac{Jc}{J_0} = \frac{Sc}{S} = 1 - \pi \left(\frac{N}{V} b^3\right)^{2/3} \qquad --- (11)$$

If we consider that the particles are arranged regularly like closest packing, a little increased critical current density is obtained in the form:

$$\frac{Jc}{J_0} = \frac{Sc}{S} = 1 - 0.687\pi \left(\frac{N}{V} b^3\right)^{2/3} \qquad -(12)$$

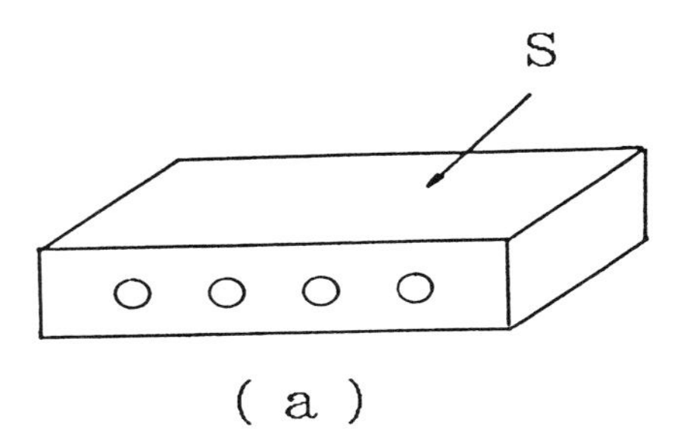

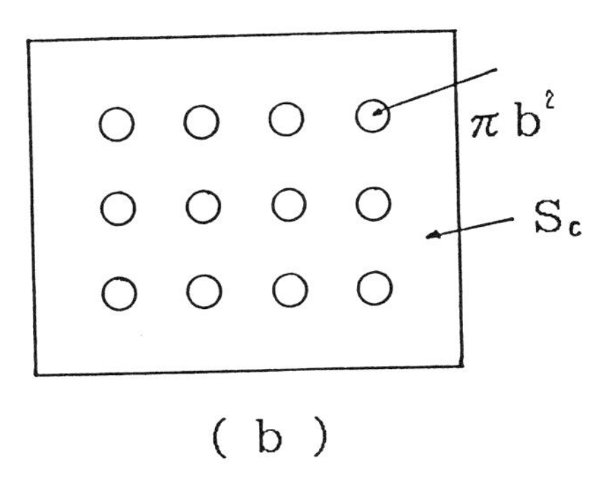

Fig.7 A cut-out from the composition shown in Fig.2

3 Relation between conductivity and critical current density.

By eliminating V,N and b from equations (8) and (10), we can obtain the relation

$$Jr = 1 - 0.52\pi (1 - 1/\rho_r)^{2/3} \qquad --- (13)$$

where $Jr = Jc/J_0$ and $\rho_r = \sigma_0/\sigma$.
From equations (8) and (11), we can also introduce the similar relation:

$$Jr = 1 - 0.357\pi (1 - 1/\rho_r)^{2/3} \qquad --- (14)$$

Effect of Chemical Non-Stoichiometry on Critical Current Density of Oxide Superconductors

KENJI DOI[1], MASATO MURAKAMI[1], SHOICHI MATSUDA[1], and AKIO TAKAYAMA[2]

[1] R & D Laboratories-I, Nippon Steel Corporation, 1618 Ida, Nakahara-ku, Kawasaki, 211 Japan
[2] Muroran Steel Works, Nippon Steel Corporation, 12 Nakamachi, Muroran, 050 Japan

ABSTRACT

Critical current density of oxide superconductors is strongly dependent on microstructure. In the bulk sintered material Jc is remarkably reduced by the formation of a liquid phase. It is possible to suppress the intrusion of the liquid phase by sintering at relatively low temperatures. However, the bulk density cannot be increased by low temperature sintering. We have found that modification of chemical composition from the 123 stoichiometry enables sintering at higher temperatures without forming the liquid phase and Jc was increased compared with that of the 123 composition. Microstructural observation revealed that grain sizes were fairly small in good Jc samples, which is the indirect evidence for the suppression of the liquid phase formation.

INTRODUCTION

After the discovery of YBaCuO with Tc exceeding 90K[1], extensive study has been conducted to apply the material for practical applications which require critical current density(Jc) of 10^5 A/cm^2 or higher often in the presence of magnetic field. However, Jc of bulk sintered materials is quite low and typically a few hundred A/cm^2 in zero magnetic field[2], which hinders the application of high Tc oxides. One of the major problems in the powder metallurgical process is the intrusion of a liquid phase during sintering[3]. Although the bulk density can be increased by sintering the sample at higher temperature, grain boundaries are wet with nonsuperconducting liquid phases: admixture of $BaCuO_2$ and CuO[4]. Because of this Jc is extremely reduced. We have found that the liquid phase formation can be suppressed by modifying the chemical composition slightly off the 123 stoichiometry. In this paper, the effects of chemical variation from the stoichiometric 123 composition on superconducting properties are reported.

EXPERIMENTAL

The samples were prepared by the conventional sintering method. Appropriate mixtures of Y_2O_3, $BaCO_3$, and CuO powders were calcined at 850 to 900℃ for 24 hours. After grinding to powders of 1 to 2 μm and compressing, the pellets were sintered at 900 to 1000℃ for 8 to 24 hours. Electric resistivity was measured by a standard four probe method using ultrasonic solder for the contact to electric lead. Transport critical current density was obtained from V-I characteristic curve using 1×10^{-6} V/cm criterion. Microstructures were observed with an optical microscope, an analytical transmission electron microscope. Fracture surfaces were also observed using a scanning electron microscope.

RESULTS

Effect of chemical non-stoichiometry on resistivity

Figure 1 shows resistivity versus temperature relationships for bulk sintered samples with various chemical compositions. It is found there is a tendency that normal state resistivity increases and zero resistance temperature decreases with deviating chemical composition from the 123 stoichiometry except a slight increase in Cu content.

Critical current density

Figure 2 shows the effect of compositional variation from the 123 stoichiometry on Jc's of the bulk sintered samples. It is noticeable that Jc can be improved by varying cation ratio slightly off the 123 stoichiometry. Slight increases in Y and Ba contents and a decrease in Cu content were effective in increasing Jc. It's also notable that relatively high Jc is preserved even decreasing Cu content down to 2.5. The volume fraction of the superconducting phases must be largely reduced in the sample with 2.5 Cu. This result suggests that superconducting paths are not running through the whole sample.

On the other hand, a slight decrease in Y content or a slight increase in Cu content reduced Jc markedly. However, as the deviation from the stoichiometric composition becomes larger, Jc tends to deteriorate rapidly due to the increase in the volume fraction of second phases, except for Cu rich region as mentioned above.

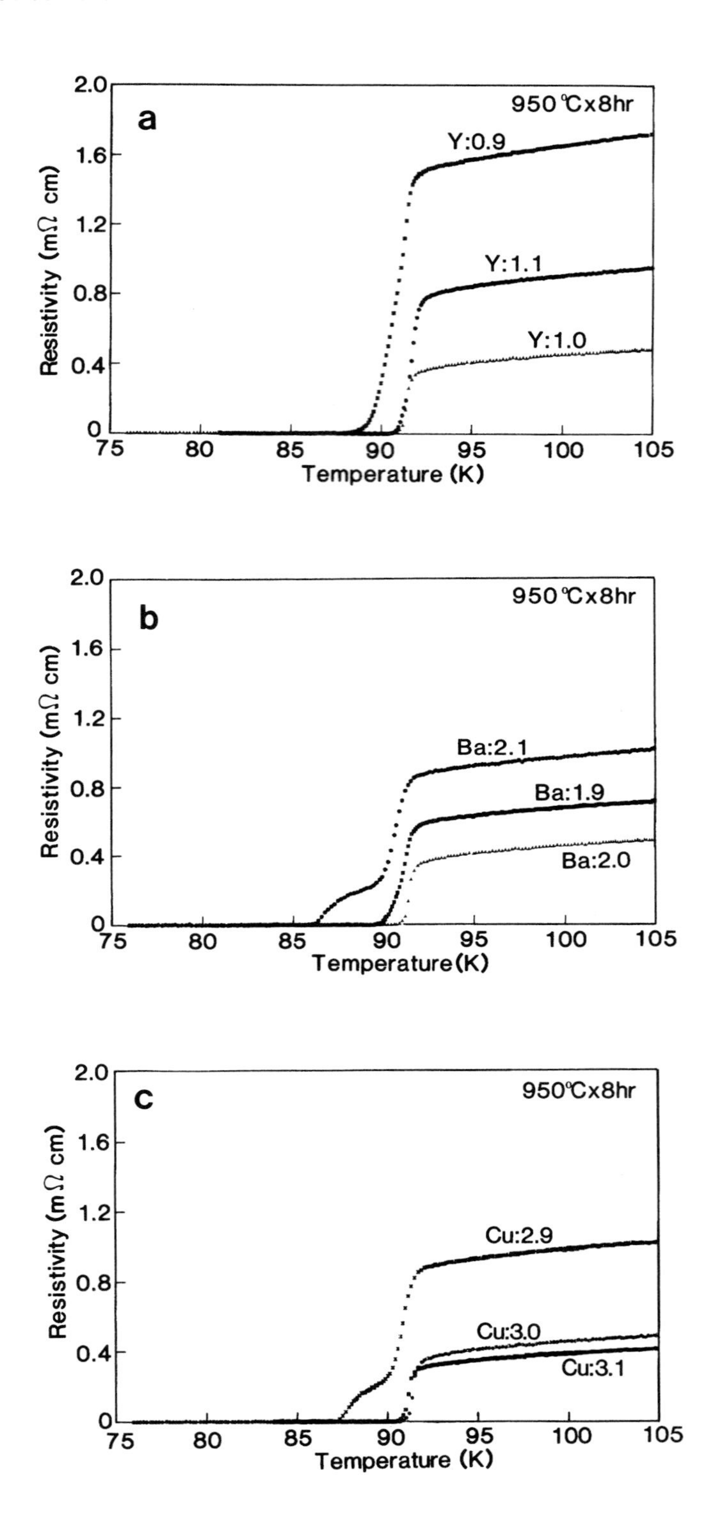

Fig. 1 Temperature versus resistivity curves for the samples deviating the composition of : a) Y; b) Ba; c) Cu.

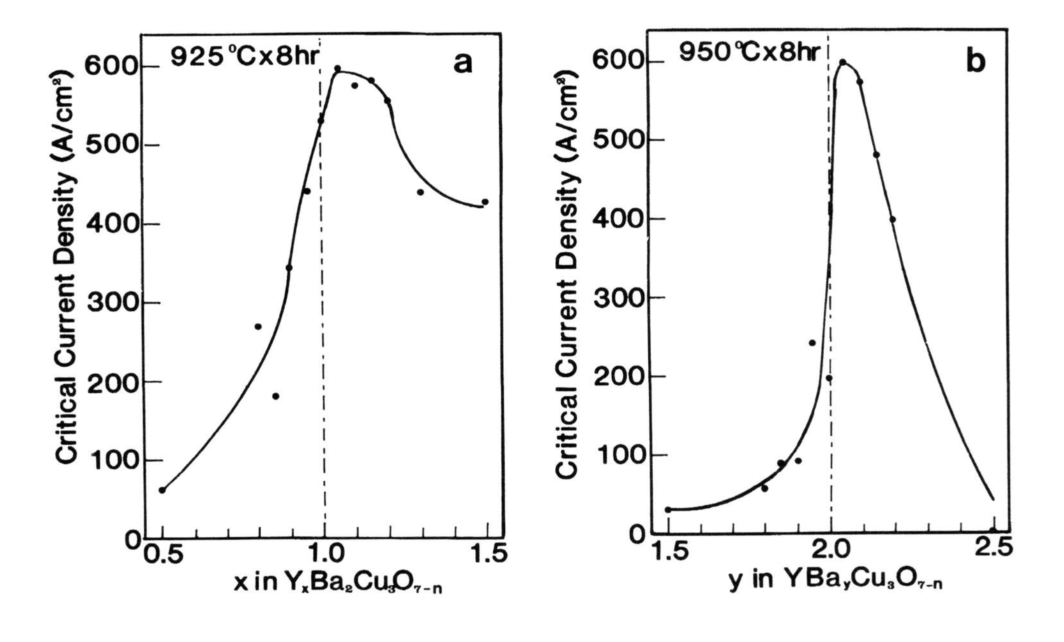

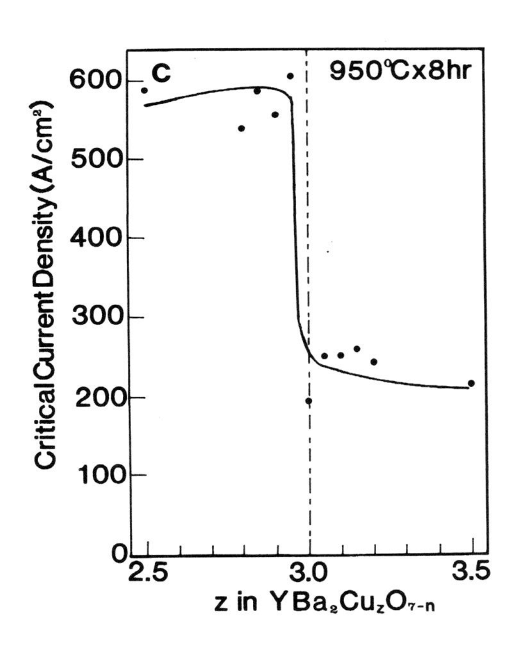

Fig. 2 Critical current densities for the samples deviating the concentration of : a) Y; b) Ba; c) Cu.

Microstructure

Figures 3 to 8 show the effect of deviation of chemical composition on fracture surface and microstructure. With either increase in Y and Ba contents or a slight decrease in Cu content, grain sizes are reduced extremely. This seems to be associated with the suppression of the liquid phase formation, since the presence of liquid phase contributes to the grain growth. Transmission electron microscopy confirmed this and no $BaCuO_2$ was observable, instead the 211 phase was found as a second phase.

A slight decrease in Y content or a slight increase in Cu content promoted grain growth and the appearance of fracture surface clearly indicates the presence of liquid phases along grain boundaries. The network of these liquid phases along grain boundaries reduces transport Jc. In the case of a slight decrease in Ba, grain growth was suppressed, but Jc was poor. According to the observation by a computer aided microanalyser, excessive Y and Cu caused the formation of significant amount of second phases, thereby Jc was reduced.

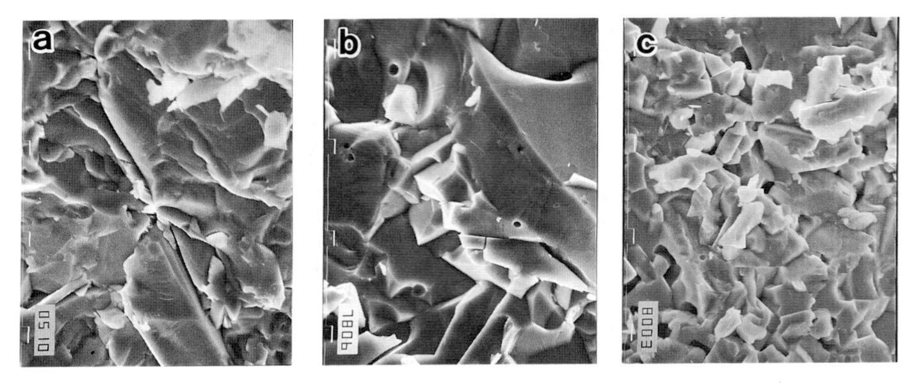

Fig. 3 Effect of Y deviation on fracture surface.
a) Y:0.9, b) Y:1.0, c) Y:1.1(950°Cx8hr).

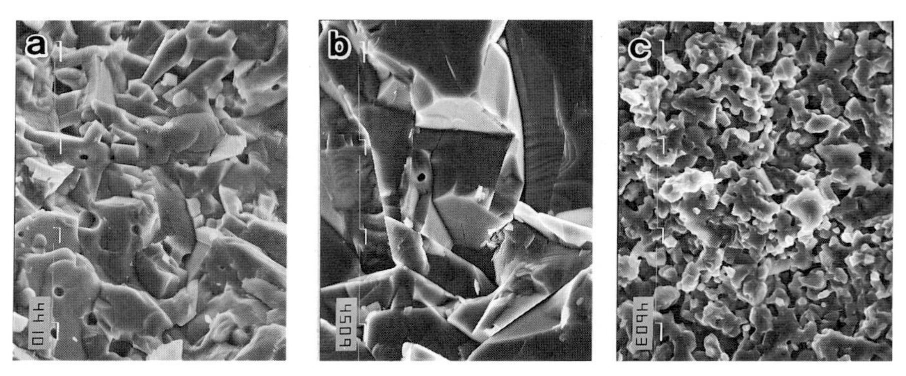

Fig. 4 Effect of Ba deviation on fracture surface.
a) Ba:1.9, b) Ba:2.0, c) Ba:2.1(950°Cx8hr).

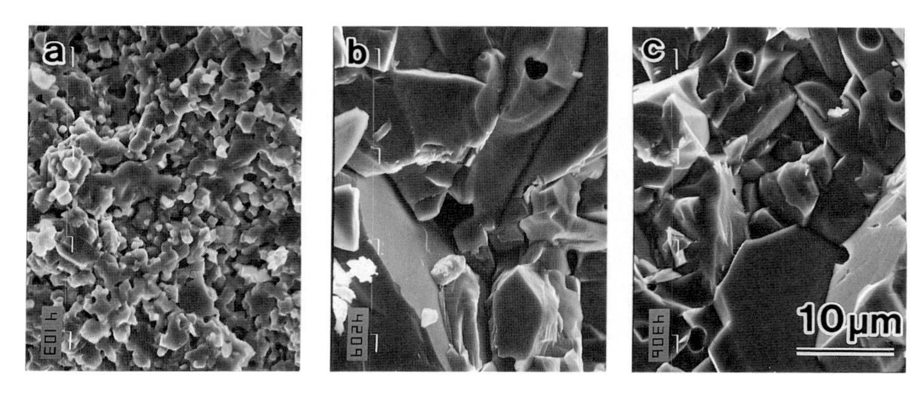

Fig. 5 Effect of Cu deviation on fracture surface.
a) Cu:2.9, b) Cu:3.0, c) Cu:3.1(950°Cx8hr).

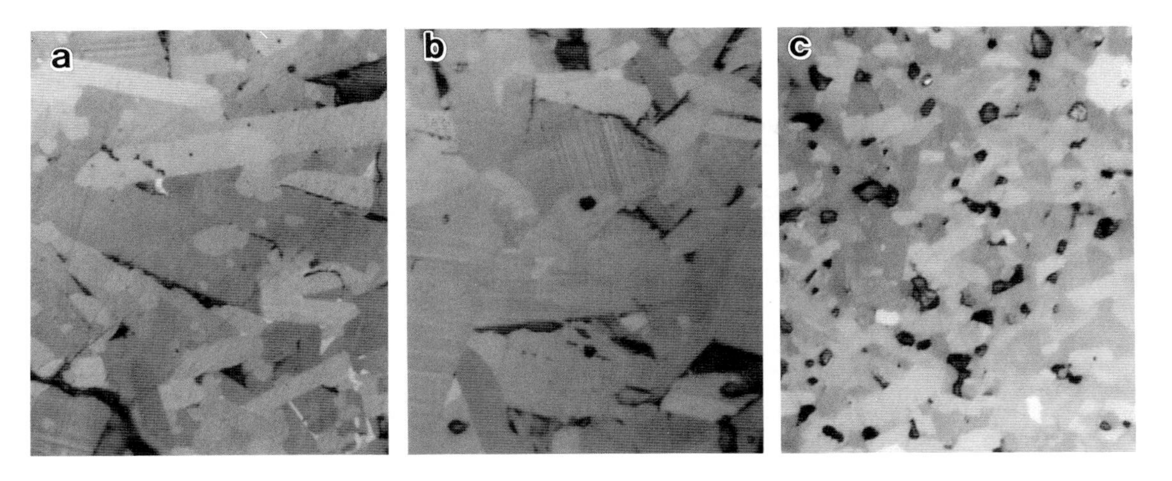

Fig. 6 Effect of Y deviation on microstructure.
a) Y:0.9, b) Y:1.0, c) Y:1.1(950°Cx8hr).

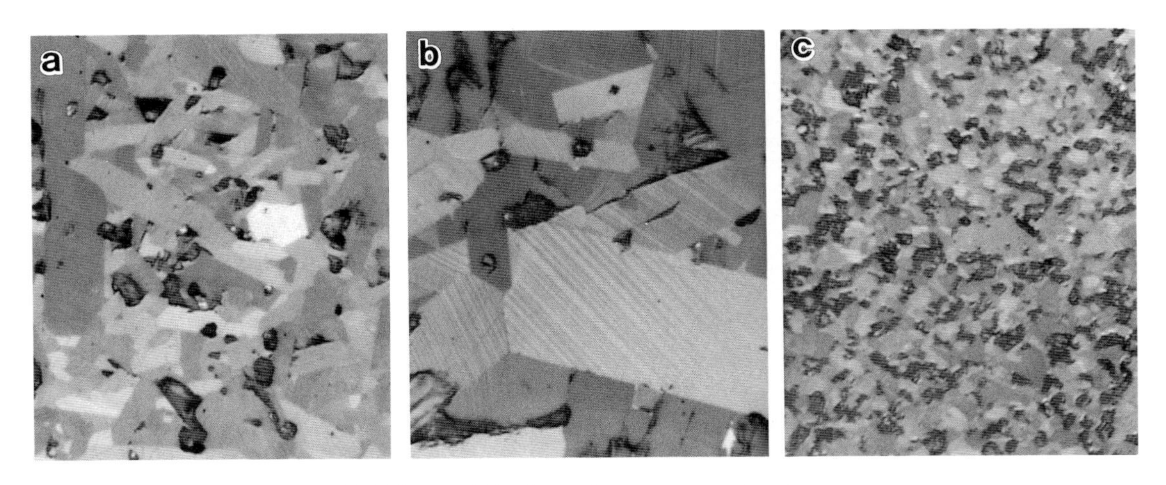

Fig. 7 Effect of Ba deviation on microstructure.
a) Ba:1.9, b) Ba:2.0, c) Ba:2.1(950°Cx8hr).

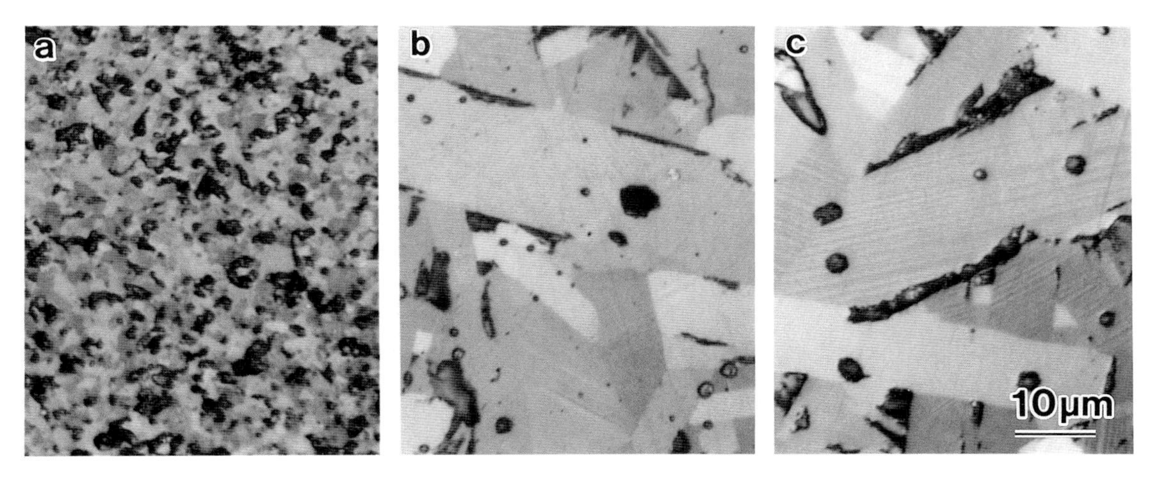

Fig. 8 Effect of Cu deviation on microstructure.
a) Cu:2.9, b) Cu:3.0, c) Cu:3.1(950°Cx8hr).

DISCUSSIONS

The effects of variations in Y or Cu content are understandable by considering the intrusion of a liquid phase. Either an increase in Y content or a decrease in Cu content shifts the composition toward the region where the liquid phase formation is suppressed. This enables sintering at relatively high temperature without introducing the liquid phase, leading to the improved Jc. While either a decrease in Y content or an increase in Cu content promotes the formation of liquid phases, thereby Jc is reduced.

On the other hand, the effect of Ba content is rather complicated. Since the liquid phase is admixture of $BaCuO_2$ and CuO, an increase in Ba content seems to promote the liquid phase formation. However, microstructural study revealed that an increase in Ba content reduced grain growth. According to the phase diagram[4] it's difficult to determine how liquid formation is affected by Ba content. But our result suggests the liquid phase formation is suppressed by varying Ba content from the stoichiometric ratio.

It's also interesting that critical current density can be improved by deviating chemical composition from the 123 stoichiometry in spite of the degradation of normal state resistivity and zero resistance temperature. Electrical resistivity is measured by using a fairly low current. Because of this the resistivity data only reflect the information from the percolative path. While transport Jc reflects the overall superconducting path at 77K. Therefore the connectivity between superconducting phases is the key factor to determine Jc. Since normal state resistivity reflects the connectivity between normal conducting phases, low electrical resistivity is not necessarily the sign of a good connectivity between the superconducting phases. Consequently the samples with low normal state resistivity and high zero resistance temperature sometimes have low Jc values, and vice versa.

CONCLUSIONS

It has been found that the liquid phase formation can be suppressed by the adequate control of the chemical compositions. This enables the sintering at higher temperature without the intrusion of liquid phases. But the attainable Jc is 600 A/cm^2, which is still two to three orders of magnitude lower than the value required for practical applications.

REFERENCES

[1] M. K. Wu, J. P. Ashburn, C. J. Thorng, P. H. Hor, R. L. Meng, L. Gao, Z. J. Huang, Y. Q. Wang and C. W. Chu, Phys. Rev. Lett., 58, 908 (1987).
[2] J. W. Ekin, Adv. Ceram. Mat., 2, 586 (1987).
[3] M. Murakami, M. Morita, K. Sawano, T. Inuzuka, S. Matsuda and H. Kubo, Proceedings of SINTERING '87 held in Tokyo (1987).
[4] e. g. R. S. Roth, K. L. Davis and J. R. Dennis, Adv. Ceram. Mat., 2, 303 (1987).

Temperature and Magnetic Field Dependence of the Critical Current in Polycrystalline $Ba_2YCu_3O_y$

H. OBARA[1], H. YAMASAKI[1], Y. KIMURA[1], Y. HIGASHIDA[2], and T. ISHIHARA[2]

[1] Electrotechnical Laboratory, 1-1-4, Umezono, Tsukuba, 305 Japan
[2] Japan Fine Ceramics Center, 2-4-1, Mutsuno, Atsuta-ku, Nagoya, 456 Japan

ABSTRACT

Temperature and magnetic field dependence on the critical current density J_c of polycrystalline high-T_c oxide superconductor, $Ba_2YCu_3O_y$, have been measured. In the low magnetic field range, 0.6 $\sim$ 7 kOe, the J_c behavior changed at around 70 K. Below 70 K, J_c showed different temperature dependence between field cooling and zero-field cooling, that is, the J_c value measured when the sample was cooled in a fixed magnetic field, was different from that measured when the sample was cooled in zero magnetic field and then a magnetic field was applied. Above 70 K, however, such different temperature dependence on J_c was not observed. These experimental results can be attributed to the effects of anisotropy, and a crossover between the two- and three-dimensional superconductivity is considered to occur.

INTRODUCTION

The discovery of high temperature oxide superconductors [1-4] has brought a great impact on basic science and technology, and has stimulated a remarkably wide range of research activity. One of the purposes of these researches is to produce new compounds which have higher T_c and another is to clarify the mechanism of these high-T_c superconductivity. In regard to technological aspects, there is a need to prepare specimens which have higher critical current density, J_c. Transport critical current density of polycrystalline bulk samples, however, have been very low up to now, while considerably high transport J_c has been reported for thin films [5] and the magnetization measurement has indicated that J_c is very high inside each grain [6]. It is very important to investigate the J_c behavior in polycrystalline bulk samples and to find out the factor which determines J_c.

High-T_c oxide superconductors have some different natures, compared with the conventional superconducting materials, such as Nb_3Sn and $NbTi$, which have been applied for high current conductors. One of the remarkable properties of high-T_c oxide superconductors is its anisotropy. Anisotropic behavior of J_c is observed in good quality thin films and the anisotropy is considered to play an important role for J_c [7], also for polycrystalline samples. Another remarkable nature is that these systems are multinary compounds, and inhomogeneity at grain boundaries is a severe problem for the improvement of J_c [8], probably because of unstable oxygen stoichiometry. For such complicated natures of oxide superconductor, it is very difficult to distinguish one effect on J_c from another, and to clarify the mechanism of J_c.

In the present work, we have measured temperature and magnetic field dependence of J_c of polycrystalline $Ba_2YCu_3O_y$ very carefully, and discussed the observed J_c behavior in correlation with the anisotropy of this system.

EXPERIMENTAL

$Ba_2YCu_3O_y$ samples were prepared by sintering $BaCO_3$, Y_2O_3 and CuO mixed powders of an appropriate composition. Starting materials were mixed in alcohol and heated at 930°C for 12 h in air. Such calcining processes were repeated twice, and the calcined powder was finally pressed and heated at 950°C for 24 h and at 500°C for 24 h in oxygen. Samples were in the form of a bar, 2 cm long, with a cross-sectional area of about $2mm^2$. Critical current was determined by the 1 μ/cm criterion.

The critical temperature of samples was determined by the ac magnetization measurement, which was done in an atmosphere of helium and by using an operating frequency of 930 Hz and an ac field of 1 Oe.

The measurement of J_c was performed using a four-terminal technique in magnetic fields ranging from 0.05 T to 1.5 T, which is calibrated by an NMR measurement. Temperature was controlled by a closed-cycle refrigerator system capable of covering the temperature range from 20 K to 300 K and monitored by Pt resistance thermometer in zero magnetic field and by Au/Au-Fe thermocouple in nonzero magnetic field. Samples were mounted on a copper block with Apiezon grease for thermal contact and with epoxy plates for electrical insulation. Moreover, the sample holder was in an atmosphere of helium which made the temperature of the sample holder uniform. Before making the electrical contacts, samples were mechanically polished and gold was deposited on the contact area. After the deposition of gold, samples were heated at $550°C$ and slowly cooled in an oxygen flow. Finally the wires were attached on gold electrode using silver paint. These processes are very important in obtaining low resistance contacts to measure J_c [9]. If electric contacts are not good, samples are heated at the current contacts and the voltage vs current (V-I) curves become quite different from those measured with a sample in liquid N_2. In the present work, we have obtained good electric contacts ($< 10^{-4}\Omega\ cm^2$) and ensured that V-I curves were almost the same as those measured in liquid N_2.

RESULTS AND DISCUSSION

Figure 1 shows the temperature dependence of ac magnetization of the sample. The onset temperature, at which temperature 10 % of the magnetization at sufficiently low temperature was observed, was 92 K and the midpoint temperature of transition curve was 91 K. Such a sharp transition indicates that the sample was a homogeneous superconductor.

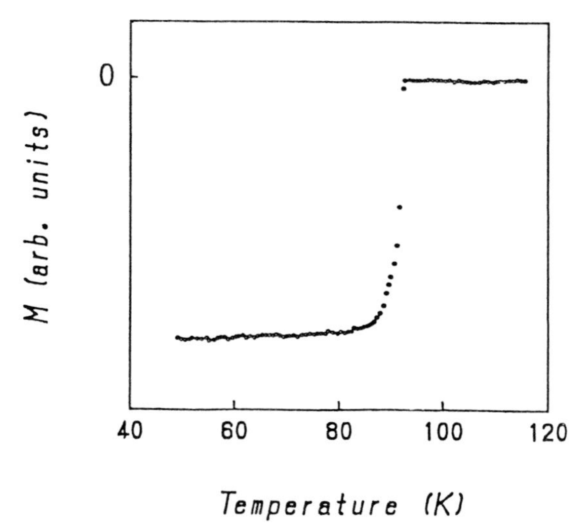

Fig. 1. Temperature dependence of ac magnetization of the sample. The onset temperature of the transition curve is 92 K.

Typical V-I characteristics at several temperatures are shown in Fig. 2. Considerably low resistance contact ($< 10^{-4}\Omega\ cm^2$) enables us to measure J_c of samples without heating at the current contacts. Using only silver paint for electrical contacts or without Apiezon grease for thermal contact, the apparent resistance above J_c was quite larger than that measured in liquid N_2. Therefore, the good electrical contacts are indispensable to the measurement of J_c in a helium gas atmosphere. In our samples the resistance above J_c increased with increasing temperature and showed almost no magnetic field dependence in the measured magnetic field range, $< 1.5\ T$.

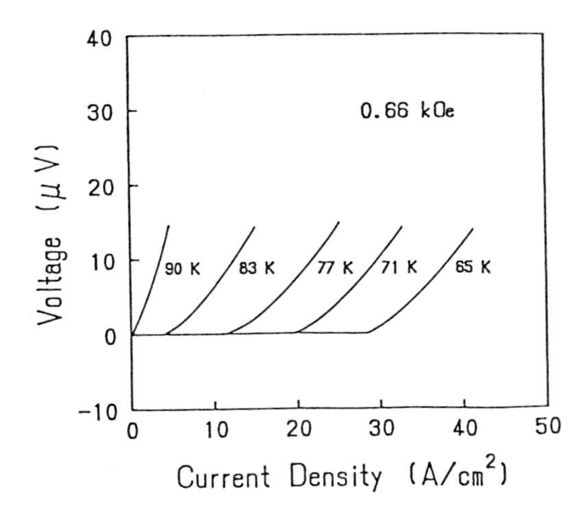

Fig. 2. Typical voltage-current (V-I) characteristics at several temperatures.

Magnetic field dependence of J_c at several temperatures is plotted in Fig. 3. As the temperature decreased, the value of J_c increased and the magnetic field dependence on J_c became smaller. In low magnetic fields a small hysteresis was observed when the field was increased (■) and when it was decreased (●) and it became larger as temperature decreases. At present, the mechanism of such hysteresis is not clear. Although J_c decreases rapidly in low magnetic field range ($< 100\ Oe$), the small field dependence at low temperature indicates that high-T_c oxide superconductors have the possibility of high current conductor in high magnetic field at low temperatures.

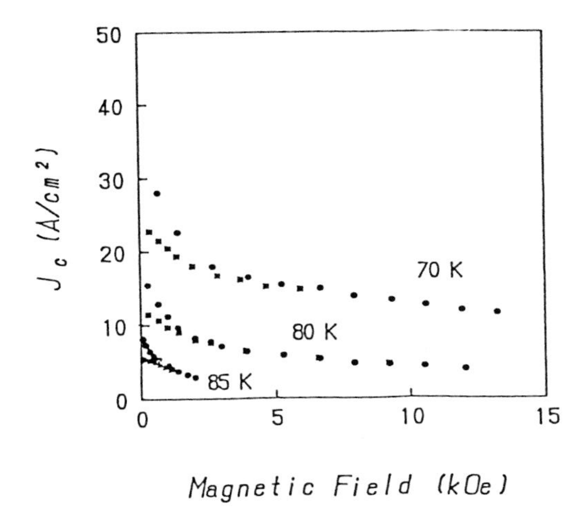

Fig. 3. Magnetic field dependence of J_c, when the field increases (■) and when it decreases (●).

Figure 4 shows the temperature dependences of J_c when samples were cooled in the magnetic field (●) and when samples were cooled in zero magnetic fields and the field was applied at low temperature ($\leq 30K$) (○). The different temperature dependence between field cooling and zero-field cooling was apparently observed below 70 K. Above 70 K, such a difference was not observed. Although the mechanism of such a different temperature dependence is unknown, this phenomenon may be attributable to trapped fluxes in samples because these compounds do not show the perfect Meissner effect [10].

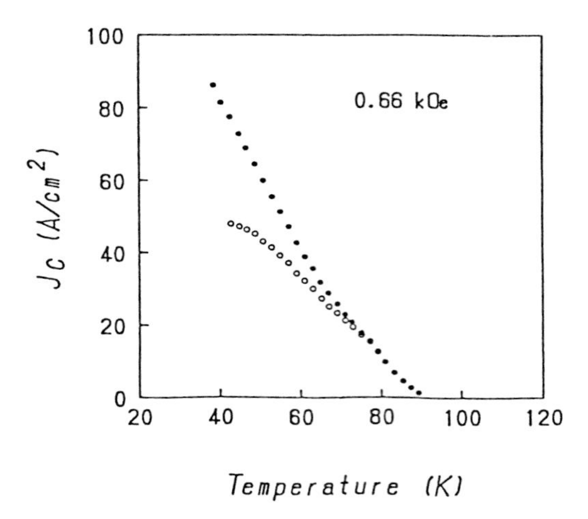

Fig. 4. Temperature dependence of J_c when the magnetic field applied in normal states (●) and when the field was applied after sample was cooled in zero magnetic field (○). A large difference was observed below $\sim 70K$.

Most remarkable thing is that the difference of temperature dependence of J_c appears between field cooling and zero-field cooling less than 70 K and the curvature of the temperature dependence on J_c changes at the same temperature. From the magnetization measurement, no anomaly was observed about 70 K and sample inhomogeneity was not the origin of this phenomenon. From the experimental results, we may propose two possible explanations about this different temperature dependence of J_c.

One possible explanation is that there are low T_c region with very small volume such as grain boundaries. If the weak link at the grain boundaries is important to J_c [11], J_c may be very sensitive to the state of grain boundaries. The other explanation is that the anisotropy is the origin of this phenomenon. In a layered superconductor, when the temperature dependent (GL) coherence length normal to the layers, $\xi_z(T)$, is shorter than the layer distance s, the system is considered to obey the two-dimensional GL equations, and when the temperature is close to T_c and GL coherence length becomes longer than the layer distance, the system is considered as anisotropic three-dimensional superconductor. Therefore, at the temperature when $\xi_z(T)$ becomes equal to $\frac{S}{2}$, dimensional crossover occurs [12,13]. If we can consider that oxide superconductors are like layered superconductors, such a dimensional crossover behavior is expected in these compounds, and several groups have discussed dimensionality of high-T_c oxide superconductors [14-16]. It is quite reasonable to consider that two dimensional behavior, which makes oxide superconductors largely anisotropic, affects the temperature dependence of J_c in low temperature and is the origin of the anomalous temperature dependence of J_c. Such anisotropy may become smaller near T_c when the system is considered as an anisotropic three-dimensional superconductor. Taking $\xi_z(T) = \xi_z(0)\left(\frac{T_c}{T_c - T}\right)^{\frac{1}{2}}$, $s = 11.7$ Å and $\xi_z(0) = 3.5$ Å [16], a dimensional crossover is expected to occur at 67 K in this compound, and this value is consistent with our experimental results. At the present stage we can not throughly explain the temperature dependence of J_c and further study is necessary to clarify the mechanism of J_c of high-T_c oxide superconductors.

CONCLUSION

In the present work, we have measured temperature and magnetic field dependence of J_c of polycrystalline $Ba_2YCu_3O_y$ very carefully and observed that the J_c behavior changes at about 70 K. This phenomenon can be considered as the dimensional crossover effects which originates in anisotropy of the superconductor. Still there is possibility of the effect by grain boundaries and mechanism of J_c is not clear, our experimental results will become a hint to clarify the mechanism of J_c in high-T_c oxide superconductors.

ACKNOWLEDGEMENT

The authors would like to express their thanks to S. Maekawa for his encouragement throughout this research and Chichibu Cement Ltd. for sample preparation.

REFFERENCES

[1] J. G. Bednortz and K. A. Müller: Z. Phys. **B64** (1986) 189.

[2] K. Kishio, K. Kitazawa, S. Kanbe, I. Yasuda, N. Sugii, H. Takagi, S. Uchida, K. Fueki and S. Tanaka: Chem. Lett. **182** (1987) 429.

[3] M. K. Wu, J. R. Ashburn, C. J. Torng, P. H. Hor, R. L. Hor, R. L. Meng, L. Gao, Z. J. Huang, Y. Q. Wang and C. W. Chu: Phys. Rev. Lett. **58** (1987) 908.

[4] H. Maeda, Y. Tanaka, M. Fukutomi and T. Asano: Jpn. J. Appl. Phys. **27** (1988) L209.

[5] Y. Enomoto, T. Murakami, M. Suzuki and K. Moriwaki: Jpn. J. Appl. Phys. **26** (1987) L1248.

[6] H. Kumakura, M. Uehara, Y. Yoshida and K. Togano: Phys. Lett. A **124** (1987) 367.

[7] T. Matsushita, M. Iwakuma, Y. Sudo, B. Ni, T. Kisu, K. Funaki, M. Takeo and K. Yamafuji: Jpn. J. Appl. Phys. **26** (1987) L1524.

[8] R. A. Camps, J. E. Evetts, B. A. Glowacki, S. B. Nencowb and W. M. Stobbs: J. Mater. Res. **2** (1987) 750.

[9] J. W. Ekin, A. J. Panson and B. A. Blankenship: Appl. Phys. Lett. **52** (1988) 33.

[10] M. Tokumoto, M. Hirabayashi, H. Ihara, K. Murata, N. Terada, K. Senzaki and Y. Kimura: Jpn. J. Appl. Phys. **26** (1987) L517.

[11] J. W. Ekin, A. I. Braginski, A. J. Panson, M. A. Janosko, D. W. Caponell, N. J. Zaluzec, B. Flandermeyer, O. F. de Lima, M. Hong, J. Kwo and S. H. Jiou: J. Appl. Phys. **62** (1987) 4821.

[12] W. Lawrence and S. Doniach: *Proc. Int. Conf. on Low-Temperature Physics., Kyoto, 1971* (Academic Press of Japan, Kyoto, 1971) p.361.

[13] R. A. Klemm, A. Luther and M. R. Beasley: Phys. Rev. **B12** (1975) 877.

[14] P. P. Freitas, C. C. Tsuei and T. S. Plaskett: Phys. Rev. **B36** (1987) 833.

[15] A. Kapitulnik, M. R. Beasley, C. Custellani and C. D. Castro: Phys. Rev. **B37** (1988) 537.

[16] K. Kanoda, T. Kawagoe, M. Hasumi, T. Takahashi, S. Kagoshima and T. Mizoguchi: preprint.

Critical Current Density of Y-Ba-Cu-O Ceramics Prepared by the Coprecipitation Method and their Application to Superconducting Devices

TAKAAKI IKEMACHI, MINORU TAKAI, TERUHIKO IENAGA, TOSHIAKI YOKOO, YORINOBU YOSHISATO, SHOICHI NAKANO, and YUKINORI KUWANO

Functional Materials Research Center, Sanyo Electric Co., Ltd., 1-18-13, Hashiridani, Hirakata, Osaka, 573 Japan

ABSTRACT

The effect of the grain boundary formed in the sintering process on characteristics of critical current density (Jc) was investigated. The experiments were performed using samples prepared with a coprecipitated Y-Ba-Cu-O compound at a sintering temperature of 920 °C for 3 ~ 100 hrs in O_2. Though the packing density and superconducting volume fractions of the sample increased with increasing sintering time, Jc had a maximum value at a sintering time of around 12 hrs. A maximum Jc of 540 A/cm^2 was obtained at 77K in zero field. Moreover, it was found that the magnetic field dependence of Jc which is indicative of "Josephson weak-link current" changed clearly with the sintering time. The temperature denendence of Jc showed that the grain boundary having S-N-S junctions at the initial stage of the sintering process, turned into S-I-S junction during prolonged sintering. Using this Y-Ba-Cu-O ceramics, granular superconducting devices for optical detection were examined.

INTRODUCTION

The discovery of high-Tc Y-Ba-Cu-O system superconductors has brought the possibility of superconductor application at liquid nitrogen temperature.[1] However the critical current density (Jc) of Y-Ba-Cu-O ceramics have been much smaller than the significant application value. It is generally said that Jc carried in the superconducting phase of a ceramic depends on the local micro-structure, particle size and grain-boundary.[2] We have already reported on the preparation of $YBa_2Cu_3O_{7-x}$ ceramics with small particles with less than 0.5 μm diameters by a coprecipitation method using water-soluble metal nitrates.[3] We have found that a calcination process preceeding the sintering process is important for synthesizing superconductors and affects Jc characteristics. Moreover limitation of Jc by the grain-boundary effect was less predominant in a sample prepared by the coprecipitation method than that prepared by the conventional solid state reaction method.

Here we examined the effect of the sintering process on characteristics of $YBa_2Cu_3O_{7-x}$ ceramics prepared by the coprecipitation method. The changes of Jc and its magnetic field dependence in the sintering process are presented. An explanation based on a model of the Josephson weak-link network is given. We also describe preliminary experiments on granular superconducting devices using these samples.

SAMPLE PREPARATIONS

Sample powders with a nominal composition of $YBa_2Cu_3O_{7-x}$ were prepared by the coprecipitation method, where $Y(NO_3)_3 \cdot 3.5H_2O$, $Ba(NO_3)_2$ and $Cu(NO_3)_2 \cdot 3H_2O$ were mixed in an aqueous solution and oxalic acid was added to coprecipitate. Detailed procedures have been described in a previous report.[3) Sample powders were pressed into pellets and sintered at 920 °C in flowing oxygen for 3 ~ 100 hrs. From the X-ray diffraction patterns, it was confirmed that the samples were single phase $YBa_2Cu_3O_{7-x}$ with orthorhombic structures.

The magnetic field dependence of Jc was measured by the standard four-probe method using long-bar-type samples with a dimension of 1.0 x 1.0 x 6.0 mm^3 under a magnetic field of up to 600 gauss in liquid N_2. The temperature dependence of Jc was measured using a bridge-type sample whose cross section and length were 0.1 x 0.05 mm^2 and 0.5 mm respectively, which removed the influence of contact resistance by low measuring current.

To examine the granular superconducting effect, a device with a zigzag pattern having a 3.6mm-length, 0.09mm - width and 0.05mm-thickness was fabricated. Optical responses of I-V characteristics were measured.

RESULTS & DISCUSSION

Fig. 1 shows the sintering time dependence of Jc, magnetization(σ), superconducting volume fraction (S.C) and bulk density (ρ), where the magnetization (σ) and the superconducting volume fraction (S.C) were measured by a vibrational sample magnetometer at 77K and bulk density (ρ) was measured by the Archimedes method. It was shown that the bulk density, magnetization and superconducting volume fraction increased with sintering time, however Jc had a peak at a sintering time of around 12 hrs and then decreased gradually on prolonged sintering. The maximum value of Jc was 540 A/cm^2 at 77K in zero field in the long-bar-type sample. Fig. 2 shows SEM photographs of samples with sintering time of 6, 12, 48 and 100 hrs. It was shown that the average grain size of particle size was about 0.5 μm and the particles were packed together at a higher density with increasing sintering time. This is a typical feature of samples prepared by the coprecipitation method.

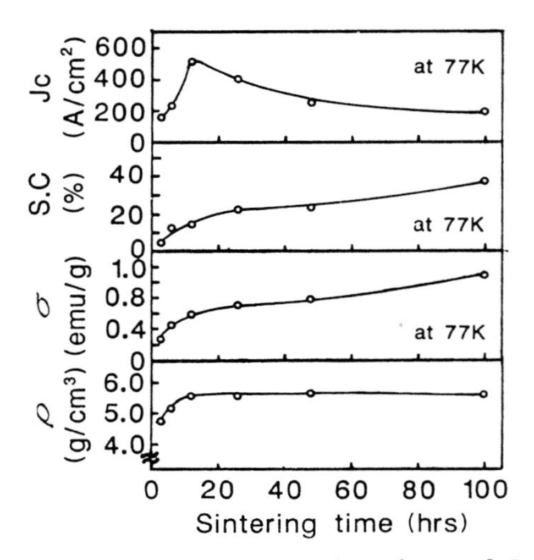

Fig. 1 Sintering time dependence of Jc, S.C. σ and ρ for YBCO superconductors prepared by the coprecipitation method at a sintering temperature of 920 °C.

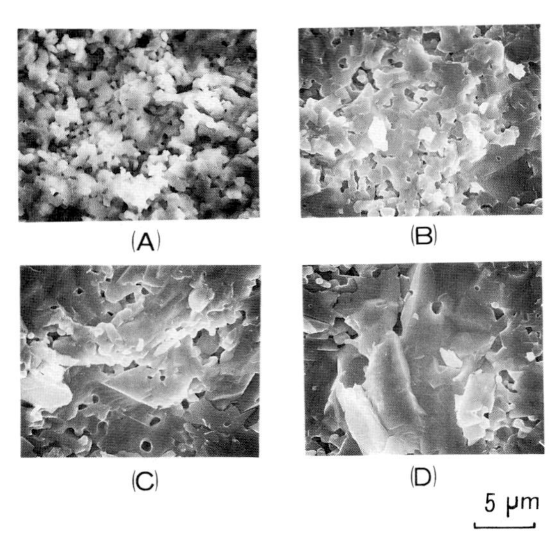

Fig. 2 SEM photographs of YBCO superconductors prepared by the coprecipitation method with sintering time of (A) 6 hrs, (B) 12 hrs, (C) 48 hrs, and (D) 100 hrs.
Sintering temperature is 920 °C.

Fig. 3 shows the magnetic field dependence of Jc in samples with sintering time of 12 and 100 hrs respectively. A drastic decrease in Jc under a low magnetic field was clearly shown in long-sintering-time sample (100 hrs). However, this drastic decrease of Jc under a low magnetic field was not observed for samples with a short sintering time (12 hrs). The reduction of Jc in samples sintered for 6, 12, 48 and 100 hrs under magnetic fields ranging from 0 to 100 G were 70%, 65%, 91% and 98% respectively. Contrary to these results, Jc in short-sintering-time samples showed steeper decreases than that of long-sintering-time samples under a high magnetic field (> 500 G). X-ray diffraction patterns showed no ordering or orienting of the grain even in the long-sintering-time samples. Moreover, second phase segregations can not be seen in microstructure observations using an electron probe micro-analyzer(EPMA), where minimum resolution was about 1μm. From these results, it was confirmed that the difference of the characteristics of Jc in the above experiments was not caused by conduction anisotropy or second phase formation in the grain.

In order to examine the difference of boundary effect between long-time and short-time sintering samples, the temperature dependence of Jc for both samples were measured. Fig. 4 shows a Jc-T curve for 12-hr and 100-hr sintering sample respectively. There was a clear difference between both samples. The 12-hr sintering sample having a large current density of 540 A/cm^2 and showing slow decrease under a low magnetic field seemed to follow the behavior of S-N-S junctions, which satisfies the following equation near critical temperature (Tc):[4]

$$Ic(T) \sim (1-(T/T_0))^2 \qquad (1)$$

where Ic is the critical current of the junction.

On the other hand, the 100-hr sintering sample with a low current density of 180 A/cm^2 and showing a steep decrease under a low magnetic field exhibited behavior similar to S-I-S junctions, which can be described by Ambegaokar-Baratoff theory:[5]

$$Ic(T) = (\pi/2)(\Delta(T)/R_N)\tanh(\Delta(T)/2k_BT) \qquad (2)$$

where R_N is the junction resistance, $\Delta(T)$ is the energy gap.

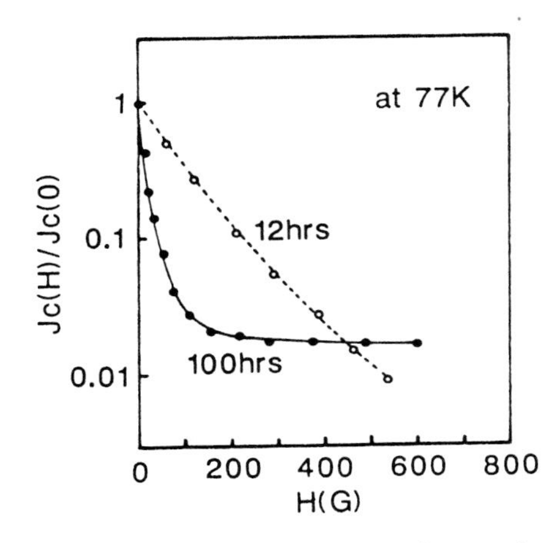

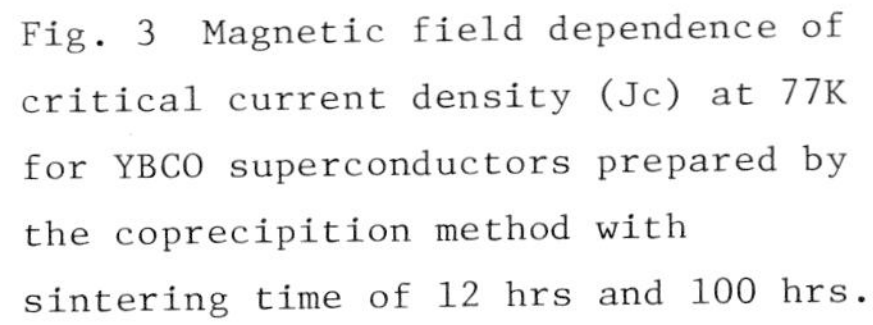
Fig. 3 Magnetic field dependence of critical current density (Jc) at 77K for YBCO superconductors prepared by the coprecipition method with sintering time of 12 hrs and 100 hrs.

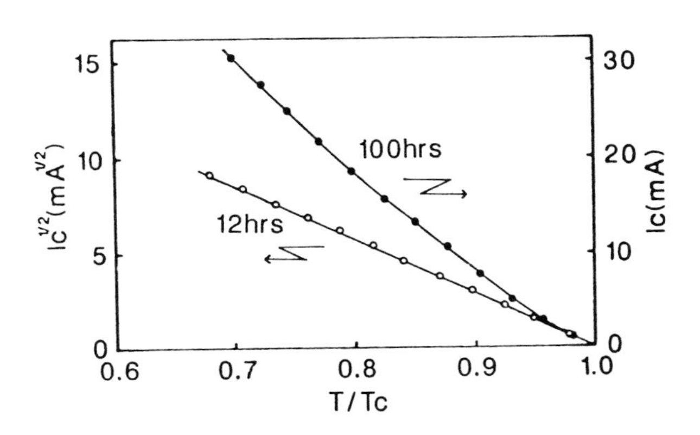

Fig. 4 Plots of critical current(Ic) vs T/Tc for YBCO superconductors prepared by the coprecipitation method. $Ic^{1/2}$ vs T/Tc for the sample with a sintering time of 12 hrs. Ic vs T/Tc for the sample with sintering time of 100 hrs.

It has been said that the relatively low value of Jc in the present $YBa_2Cu_3O_{7-x}$ ceramics and its rapid decrease under a low magnetic field attributes to a "weak-link" region between grains and intrinsic conduction anisotropy.

The increase of Jc in the initial stage of the sintering process seems to correspond to an increase in packing density and formation of S-N-S junctions. The decrease in Jc with a long sintering time seemed to correspond to the formation of S-I-S junctions. Thus Jc had a peak value at an appropriate sintering time while the superconducting phase increased with sintering time. It was suggested that quite thin insulating layers were segregated in long-time-sintering and formed S-I-S junctions in the grain-boundary while the superconducting phase of intra-grains increased.

The magnetic field dependence of Jc was also consistent with a model of Josephson weak-link behavior in a low magnetic field. For a single Josephson junction, Jc in the magnetic field satisfies a Fraunhoufer diffraction pattern:[6].

$$Jc(H) = Jc(0)(\sin(\pi H/H_0)/(\pi H/H_0)) \quad (3)$$

where $H_0 = \Phi_0/\mu dL$, is the superconducting diffraction period, d is the effective junction thickness and L is the junction length. While a network of a large amount of Josephson junctions makes the diffraction period of the network indistinguishable in superconducting ceramics, the sensitivity of Jc to a weak magnetic field was well explained qualitatively. Furthermore, analysis showed that magnetic field dependence of Jc is influenced by the inhomogeneous current density distribution of the junction area.[7] The experimental results seem to show the difference in current density distribution for both samples. It is suggested that segregation of the insulation layer or formation of a non-superconducting phase in the grain-boundary through the sintering process causes the change in current density distribution at the grain-boundary and shows a large change in the magnetic field dependence of Jc. These seem to be also related to the formation of pinning centers which make a difference in the high magnetic field dependence of Jc.

APPLICATION TO GRANULAR SUPERCONDUCTING DEVICES

Using the $YBa_2Cu_3O_{7-x}$ ceramics prepared by the coprecipitation method, granular superconducting devices for optical detection were examined. Though the ceramic is about 90% dense, the particle sizes of the samples are about 0.5 μm in diameter and are homogeneous compared to those of samples prepared by the usual solid state reaction method using oxide powders. This is suitable for fabrication of a granular superconducting device. Fig. 5 shows a photograph of the device pattern. In Fig. 6, the change in the I-V characteristics of the device by irradiation of light is presented. Clear optical response is observed. This will be the first report on optical detection using bulk superconducting ceramics.

CONCLUSIONS

The effects of the sintering process on the properties of Y-Ba-Cu-O ceramics prepared by the coprecipitation method were examined. It was found that Jc has a maximum value of 540 A/cm^2 for a sintering time of 12 hrs at a temperature 920 °C in O_2, where the grain-boundary was S-N-S junction. The junctions became S-I-S junctions and Jc decreased by increasing sintering time over 12 hrs. The clear difference in the magnetic field dependence of Jc was observed in samples with various sintering times, which also seemed to correspond to differences in junctions with inhomogeneous current density distribution. Granular superconducting devices for optical detection using bulk superconductive ceramics were examined.

Fig. 5 Photograph for a pattern of superconducting device having a 3.6mm-length, 0.09mm-width and 0.05mm-thickness.

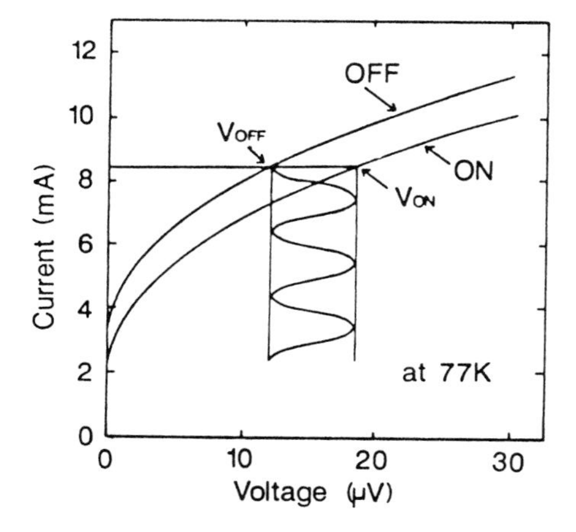

Fig. 6 I-V characteristics by He-Ne laser-irradiation at 77K for a device using YBCO superconductors prepared by the coprecipitation method. Laser power is about 2 mW.

REFERENCES

1) M. K. Wu, et al., Phys. Rev. Lett., 58, 908 (1987)
2) D. K. Finnemore, et al., Phys. Rev., B35, 5319 (1987)
3) Y. Yoshisato, et al., Seramikkusu Ronbunshi, 96, 459 (1988)
4) P. G. de Gennes, Rev. Mod. Phys., 36, 226 (1964)
5) V. Ambegaokar and A. Baratoff, Phys. Rev. Lett., 10, 486 (1963)
6) T. Van Duzer and C. W. Turner, "Principles of Superconductive Devices and Circuits" (Elsevier, New York 1981), Chap. 4.
7) A. Barone and G. Paterno, "Physics and Applications of the Josephson effect" (John Wiley & Sons, New York 1982), Chap. 4.

Magnetization and Critical Current of $Yb_1Ba_2Cu_3O_{7-x}$ Samples Prepared by the Plasma-ark Melting and Rapid Quenching Method

YORINOBU YOSHISATO, TOSHIAKI YOKOO, MARUO KAMINO, HIROSHI SUZUKI, SHOICHI NAKANO, and YUKINORI KAWANO

Functional Materials Research Center, Sanyo Electric Co., Ltd., 1-18-13, Hashiridani, Hirakata, Osaka, 573 Japan

ABSTRACT

The dependence of magnetization (σ) and critical current density (Jc) on annealing temperature was investigated for $Yb_1Ba_2Cu_3O_7$-x superconductor prepared by the plasma-arc melting and rapid quenching method (PMQ-method). It was found that σ and the superconducting volume fraction increased with annealing temperatures up to 980 °C and reached maximum values of 1.98 emu/g and 90% respectively, where particle size also increased to over 25 μm. Futhermore at annealing temperature of more than 980°C, drastic changes in the magnetization curves and an increase in Jc were observed. From X-ray diffraction analysis and EPMA observation, it was found that small $Yb_2Ba_1Cu_1O_5$ particles (1μm diameter) segregated in the large $Yb_1Ba_2Cu_3O_7$-x grain (25 μm diameter). Using the high density and large grain of $Yb_1Ba_2Cu_3O_7$-x ceramics prepared by the PMQ method, grain-boundary Josephson junction effects were examined. Bridges with S-N-I-N-S weak links were fabricated with good reproducibility and showed microwave induced Shapiro steps at 77K.

INTRODUCTION

The discovery of high Tc superconducting ceramics has been followed by intense efforts to fabricate thin or thick films of these copper oxide-based materials for research in electronic device applications. Many attempts have been made to form a constricted Josephson junction from superconducting films.[1] However it was difficult to observe the Josephson effect at higher than 77K which is lower than the Tc of the original films because a degradation was caused in the fabrication process.[2] Furthermore, there were so many grain boundaries in the bridge region that it was hard to observe coherent operation of each junction.[3] Particles in thin films are generally less than 1 μm in diameter.

Using bulk superconducting ceramics, the Josphson effect at temperatures above 77K was reported.[4] However, these Josephson devices were unstable and unreproducible because the Josephson junctions were made with mechanical point contacts or break-contacts.[5] Bulk samples prepared by the sintering method were porous and it was difficult to make the bridge as the same order as the grain size. Moreover, the distribution of grain sizes was quite wide. These ceramics are not appropriate for practical use.

In order to avoid poor density of the bulk sintered samples we have developed the plasma arc melting and rapid quenching method (PMQ-method) to obtain superconducting materials that are highly dense and homogeneous in a very short time.[6]

Here, we examined the grain-boundary Josephon junction of $Yb_1Ba_2Cu_3O_7$-x ceramics prepared by the PMQ-method for the first time. The experimental results of grain-size dependence on annealing temperatures and the microwave induced effect of the bridge sample prepared by a precision mechanical procedure are presented. We also describe drastic changes in the magnetic properties and the critical current density (Jc) in the annealing process, which were found in this study for the first time.

SAMPLE PREPARATIONS

Powders of $BaCO_3$, Yb_2O_3 and CuO with nominal compositio of $YbBa_2Cu_3O_{7-x}$ were mixed and melted in a plasma-arc melting equipment in an Ar gas an quenched rapidly on a copper hearth. Details of the pellet fabrication method (PMQ method) were described in a previous report.[6] The quenched pellets were sliced into wafers with a thickness of 1mm and annealed at temperatures between 920°C and 1020°C for 3 hrs in O_2. The samples were analyzed by X-ray diffraction to identify the crystalline phase.

In order to observe the grain size and its microstructure, an optical micrograph, a scanning electron microscope (SEM) and an electron probe micro-analyzer (EPMA) were used. The magnetization(σ) and superconducting volume fraction (S.C %) were measured by using a vibrational sample magnetometer (VSM). The critical current density and its magnetic field dependence were measured by the usual four probe method. The sample size was 1mm x 1mm x 6mm.

A microbridge-type Josephson device with a cross section of 30 μm x 20 μm was prepared by a mechanical procedure. The grain size was about 25 μm.

RESULTS & DISCUSSION

Fig. 1 shows SEM photographs of the particles in $Yb_1Ba_2Cu_3O_{7-\sigma}$ ceramics whose annealing temperatures were 940°C, 960°C and 980°C respectively. Each particle was obtained by crushing the annealed samples and those samples were pulverized into single grain particles. Observation of the samples by an optical microscope and a SEM showed the same results. It was shown that grain size increased annealing temperatures and that the particles were not ordered and oriented, which was also confirmed

30μm

Fig. 1 SEM micrographs for Yb-Ba-Cu-O ceramics perpared by the PMQ method with annealing (a) at 960°C, (b) at 980°C and (c) at 1000°C for 3 hours in O_2.

by X-ray diffraction patterns of each sample. Fig.2 shows the annealing temperature dependence of the mean particle size. It was shown that grain size increased lineary up to 980°C and reached about 25 μm, while the temperature is higher than 980°C, grain size could not be measured, which seemed to indicate that individual grains cannot be divided at the boundary due to partial melting.

In order to confirm the uniformity of each particle, intra-grain critical current density was estimated from a measurement of magnetization curve of each particle.[7] In Fig. 2, the hysteresis of magnetization ΔM of each particle is presented, where ΔM is defined as a difference in magnetization obtained with increasing and decreasing the magnetic field.[8] Using the critical state model with the simple formula:[9]

$$Jc(A/cm^2) = 30 \times \Delta M / d \qquad (1)$$

where d is the mean size of the particles, Jc in the intra-grain can be calculated and a value of 3×10^4 A/cm^2 is estimated in all particles at annealing temperatures between 920°C and 980°C. The results conform to the intra-grain uniformity of the particles as shown in Fig. 1.

At annealing temperature higher than 980°C, a drastic increase in magnetization and intragrain critical current density was measured by the standard four probe method. Fig. 3 and Fig. 4 show the magnetization curves and the magnetic field dependence of Jc for samples with annealing temperatures of 960°C and 1000°C, respectively. It was shown that maximum magnetization of both samples were the same while the large difference in the hysteresis curve and a correspl nding drastic increase

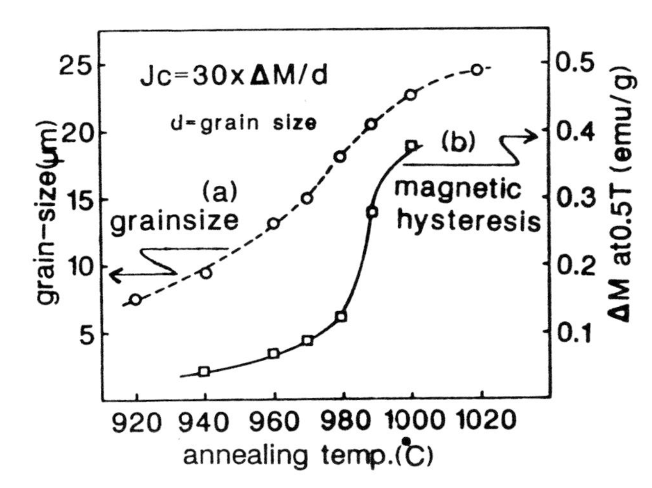

Fig. 2 Grain-size and Magnetic hysteresis M vs annealing temperature. M is the magnetization difference between increasing and decreasing magnetic field.

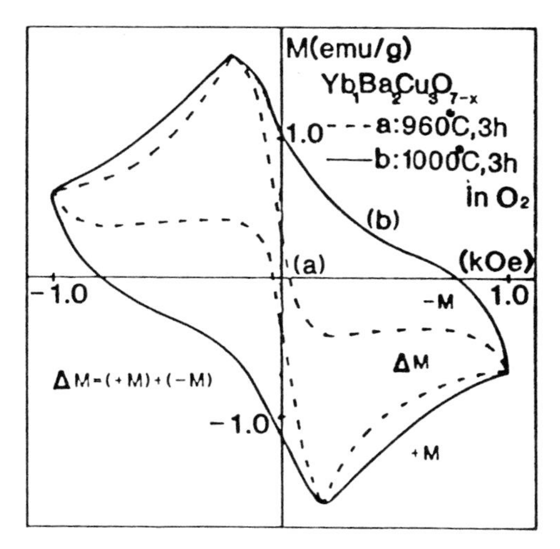

Fig. 3 Magnetization curves for Yb-Ba-Cu-O measured at 77K annealed (a)at 960°C, (b)at 1000°C for 3 hours in O_2

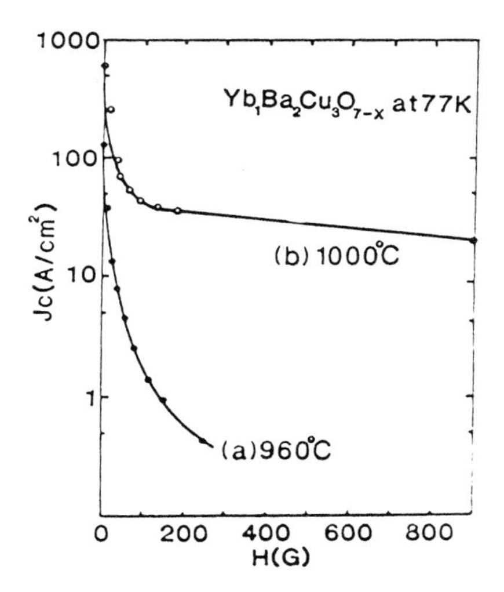

Fig. 4 Magnetic field dependence of the critical current for a sample annealed (a)at 960°C, (b)at 1000°C for 3 hours in O_2

Fig. 5 (a) A micrograph for a bulk annealed at 1000°C for 3 hours in O_2
(b) A SEM micrograph for a bulk annealed at 1000°C for 3 hours in O_2

in Jc were observed. The intra-grain Jc was increased to $1x10^5 A/cm^2$ and the inter-grain Jc was increased from $100A/cm^2$ to $600\ A/cm^2$. The results suggest a new pinning center was formed in the annealing process at a temperature around 980°C.

Fig. 5 shows an optical micrograph and a SEM micrograph of a sample annealed at 1000°C. It was shown that the $Yb_2Ba_1Cu_1O_5$ phase particles of about 1 μm segregated dispersively in large grains (25 μm). Below 980°C annealing, this segregation of the $Yb_2Ba_1Cu_1O_5$ phase was not observed. It is suggested that this segregation of the (211) phase is related to the new pinning effects.

APPLICATION GBJJ DEVICES.

Using $Yb_1Ba_2Cu_3O_{7-x}$ superconductors prepared by the PMQ-method, a microbridge type weak-link Josephson junction device was developed.

Fig. 6 shows a schematic diagram and a photograph of the microbridge, where the mean grain-size is 25μm and the width and thickness are about 30μm and 20μm, respectively. Usual mechanical rapping and grooving procedures were applied. Weak-link devices having a few grain boundaries in the microbridge region were fabricated with good reproducibility.

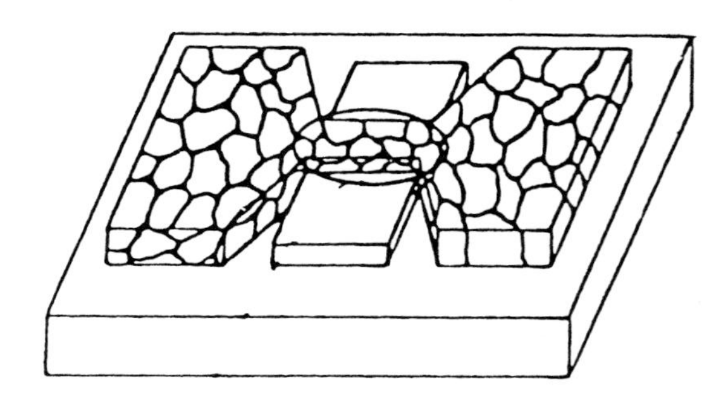

Fig. 6 Schematic of the GBJJ microbridge.

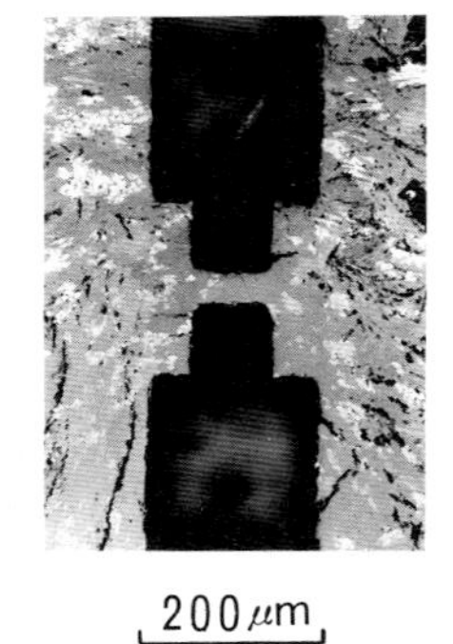

A micrograph for the GBJJ micro-bridge using the sample annealed at 1000°C for 3 hours in O_2.

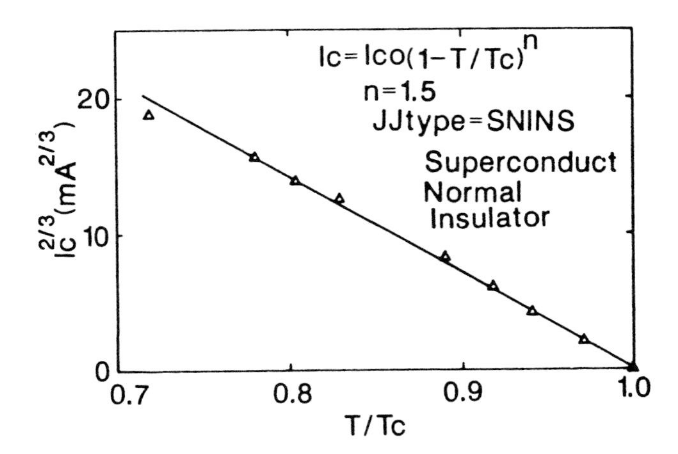

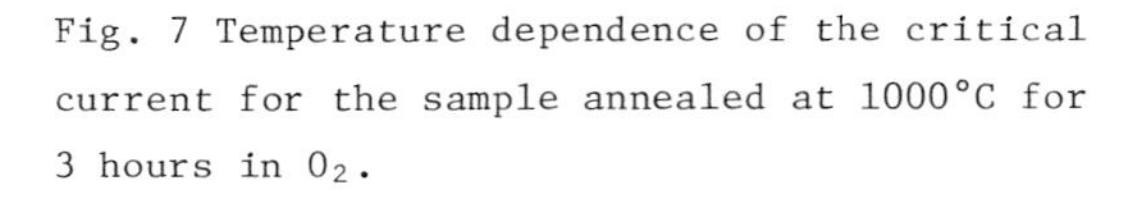

Fig. 7 Temperature dependence of the critical current for the sample annealed at 1000°C for 3 hours in O_2.

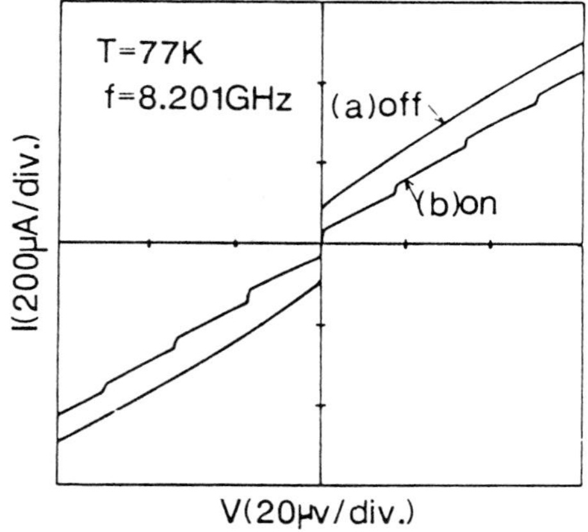

Fig 8 The Shapiro steps observed from the GBJJ micro bridge for the sample annealed at 1000°C for 3 hours in O_2.

Fig. 7 shows the temperature dependence of critical current (Ic(T)) of the microbridge. The results seem to depend on the following relation

$$Ic(T) \sim (1 - T/Tc)^{1.5} \quad (2)$$

near the critical temperature (Tc).[10] This suggests that the grain boundary junction is a S-N-I-N-S junction. In Fig. 8, the microwave induced Shapiro steps which is common evidence of the quantum interference effect is observed in the I-V curve of the bridge at 77K. While Ic is about 20 μA and Jc is several A/cm^2 which is much smaller than that of the sample before fabrication, the characteristics of the device is reproducible and stable. The device may have high potential use as high frequency detectors and flux sensors operative at 77K.

CONCLUSIONS

High-density (more than 98%) and homogeneous superconducting ceramics of $Yb_1Ba_2Cu_3O_{7-x}$ were prepared by the plasma-arc melting and rapid quenching method. It was found that grain size increased with annealing temperature and that grain size of more than 25 μm without orientation can be obtained at 980°C annealing. Using $Yb_1Ba_2Cu_3O_{7-x}$ with high density and large grain size, a GBJJ superconducting device was fabricated. It was confirmed that the device had a S-N-I-N-S Josephson junction and showed microwave induced Shapiro-steps at 77K. A new pinning effect which intra-grain and inter-grain current density was also found at an annealing temperature higher than 980°C.

REFERENCES

1) H. Akoh, et al., Appl. Phys. Lett., 52, 1732 (1988)
2) R. H. Koch, et al., Appl. Phys. Lett., 51. 200 (1987)
3) H. Tanabe, et al, Jpn. J. Appl. Phys. 26 (1987) L1961
4) T. Komatsu, et al, Jpn. J. Appl. Phys. 26 (1987) L1148
5) J. S. Tsai, et al, Jpn. J. Appl. Phys. 26 (1987) L701
6) Y. Yoshisato, et al, FED HiTcSc-ED Work Shop 209 (1988)
7) A. M. Cambell, et al, Mat. Res. Soc. Proc. 99 883 (1988)
8) W. A. Fietz, et al, Phys. Rev, 178 657 (1969)
9) C. P. Bean, Phys. Rev. Lett., 8 250 (1962)
10) H. C. Yang et al, J. Low Temp. Phys. 70 493 (1988)

Critical Current Density Obtained from Particle Size Dependence of Magnetization in $Y_1Ba_2Cu_3O_{7-\delta}$ Powders

ERIKO SHIMIZU and DAISUKE ITO

Toshiba R&D Center, Toshiba Corporation, 4-1, Ukishima-cho, Kawasaki-ku, Kawasaki, 210 Japan

ABSTRACT

$Y_1Ba_2Cu_3O_{7-\delta}$ pellets and wires show a rather small critical current density Jc, determined by the four-probe method, compared with a single crystal. In order to develop wires that have a higher Jc, it is important to determine the reason why Jc degrades to such a small value. The authors estimated Jc from the particle size dependence of magnetization at 4.2 K and 77 K for $Y_1Ba_2Cu_3O_{7-\delta}$ powders calcined in air.

A number of $Y_1Ba_2Cu_3O_{7-\delta}$ powders, with average particle diameters ranging from 3 to 53 μm, were prepared. These powders were sorted by a classifier. Jc values were derived from linear relationship between magnetization and powder size by using the Bean model. Three major results were obtained from these measurements. The first is that the magnetization is in linear proportion to the particle size in the 1-20 μm range and that it saturates above 20 μm. SEM and TEM observations suggest that individual powders, being smaller than 20 μm, seem to a single-crystal. However, each powder larger than 20 μm in size is made from several grains. This result suggests that grain boundaries limit current flow in the $Y_1Ba_2Cu_3O_{7-\delta}$ powder. The second result is that Jc in the $Y_1Ba_2Cu_3O_{7-\delta}$ powder, whose particle diameter is less than 20 μm, is 2×10^6 A/cm^2 for 0.3 T at 4.2 K, and 7×10^4 A/cm^2 for 0.03 T at 77 K. The third result is that the temperature dependence of a magnetization is composed by two curves with different Tc values each other. London penetration depths at 0 K, λ_0, obtained from initial magnetization at 4.2 K were different from the value obtained at 77 K. This result can be explained by the two phase model.

INTRODUCTION

Since the discovery of the 90 K class critical temperature, Tc, superconductor $Y_1Ba_2Cu_3O_{7-\delta}$[1], a number of groups have reported Jc determined by the four-probe method for pellets and wires. These Jc values under the zero magnetic field, approximately 10^4 A/cm^2 at 77 K[2][3][4][5], are lower than the value about 10^7 A/cm^2 at 4.2 K exhibited by conventional materials, such as NbTi and Nb_3Sn. In order to improve the Jc value in wires and pellets, the authors have studied fundamental properties for $Y_1Ba_2Cu_3O_{7-\delta}$ powders.

This paper reports the magnetic properties on many $Y_1Ba_2Cu_3O_{7-\delta}$ powder samples calcined in air, with average particle diameters ranging from 3 to 53 μm. These data provide information about the attainable maximum Jc and the penetration depth at 0 K, λ_0, for this material.

Table 1. Properties for $Y_1Ba_2Cu_3O_{7-\delta}$ powders

Sample No.	Diameters D^{25} (μm)	D^{50}	D^{75}	Packing density (%)	Slope at 4.2K P(4.2)	λ_0 ~λ(4.2) (μm)	Slope at 77K P(77)	λ(77) (μm)	λ_0 from λ(77) (μm)	x_n P(77)/P(4.2)
1	2.6	3.1	3.7	29	0.24	0.67	0.09	1.23	0.89	0.39
2	4.0	4.8	5.8	38	0.55	0.44	0.24	1.03	0.74	0.44
3	6.6	7.7	9.2	44	0.45	0.93	—	—	—	—
4	7.0	8.8	10.5	51	0.82	0.28	0.39	1.25	0.90	0.48
5	7.0	10.0	12.5	49	0.71	0.54	0.39	1.42	1.02	0.55
6	13.0	15.2	17.9	52	0.90	0.26	0.67	0.96	0.69	0.74
7	16.0	19.2	22.1	55	1.02	—	0.62	1.42	1.02	0.61
8	16.5	20.4	24.5	55	1.05	—	0.55	1.88	1.36	0.52
9	16.0	21.5	25.5	55	1.00	—	0.53	2.09	1.51	0.53
10	9.5	23.1	43.0	56	1.11	—	0.64	1.61	1.16	0.58
11	12.4	24.9	45.1	53	0.97	0.13	—	—	—	—
12	18.2	26.7	38.8	57	1.04	—	0.68	1.62	1.16	0.65
13	22.1	32.3	46.0	57	1.04	—	0.65	2.18	1.57	0.63
14	25.0	34.8	48.0	53	1.04	—	0.67	2.19	1.57	0.64
15	18.0	36.0	63.5	57	0.98	0.12	0.68	2.19	1.57	0.69
16	26.5	52.5	74.0	58	1.07	—	—	—	—	—

Figure 1. Scanning electron microscopy, SEM, photomicrographs of $Y_1Ba_2Cu_3O_{7-\delta}$ powder. (a) Sample 1 (D^{50}=3.1 μm), (b)Sample 6 (D^{50}=15.2 μm) (c)Sample 15 (D^{50}=36.0 μm).

EXPERIMENTAL

Powdered samples were made using a coprecipitating method. They were heated in air at a rate of 50℃ per hour up to 950℃, fired at 950℃ for 10 hours and cooled at a rate of 50℃ per hour down to room temperature. The samples for the measurement were prepared simply by passing these powders through a 100-mesh sieve and then separating them into different size particles (diameters ranging from 3 to 53 μm) by a DONALDSON classifier. Table 1 shows the average diameters for 16 samples. The size distribution for each sample was determined with dry sieves and a coulter counter, which is based on the x-ray absorption of the particles during their sedimentation in a NaCl liquid. Figure 1 shows typical examples of scanning electron microscopy, SEM, photomicrographs. Figure 2 shows an example of their X-ray diffraction patterns. This pattern shows that the major part of this sample is attributed to the orthorhombic-I structure of $Y_1Ba_2Cu_3O_{7-\delta}$, but a small amount of impurities was contained therein. The impurity peaks are due both to $BaCuO_2$ and Y_2BaCuO_5.

Figure 3 shows the temperature dependences of magnetization for sample 1. Flux-exclusion (shielding effect) data were obtained in a 20 gauss field. The measurement was carried out with a SQUID magnetometer. The figure indicates that the Tc value for these samples was 93 K.

Most magnetic measurements were carried out with D.C. magnetization measurement systems, which have two coaxial search coils in a backup field coil. The backup coil can generate a magnetic field up to 2 Tesla at 4.2 K and up to 1 Tesla at 77 K. The samples are packed in the form of a hollow cylinder, whose density is given in Table 1. From slopes of the initial magnetization curve, the London penetration depth, λ, can be estimated. From width ΔM value for the magnetization loop, Jc can be estimated.

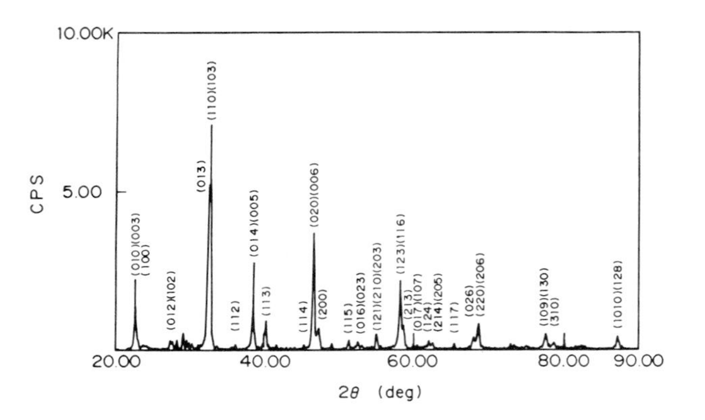

Figure 2. X-ray powder patterns of $Y_1Ba_2Cu_3O_{7-\delta}$ powders for sample 1. Impurity peaks are indicated with arrows, most of which belong to $BaCuO_2$, and Y_2BaCuO_5.

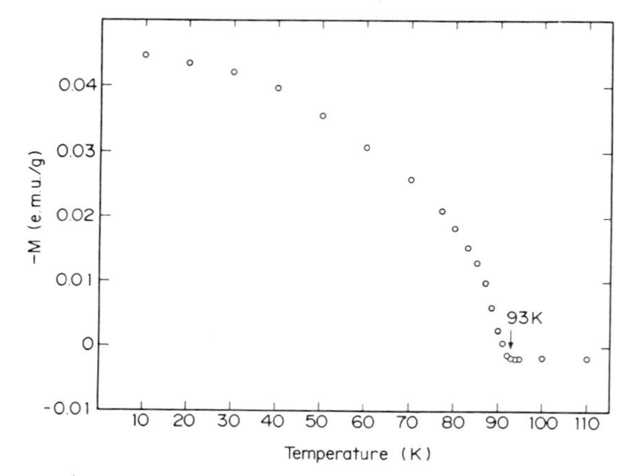

Figure 3. Magnetization temperature dependence 20 gauss for sample 1. These data are from exclusion warming data. This shows that superconducting onset is 93 K.

RESULTS AND DISCUSSION

Penetration depth

From the initial slopes, P(T), of the magnetization curve obtained by zero-field cooling condition, the λ value at T K is estimated. The authors used the formula given by London[6] for P(T) of a spherical superconductor, radius R with $\lambda(T)$.

$$P(T)=1-\frac{3\lambda(T)}{R}\coth\left(\frac{R}{\lambda(T)}\right)+3\left(\frac{\lambda(T)}{R}\right)^2 \qquad (1)$$

$$\lambda(T)=\frac{\lambda_0}{\sqrt{1-(T/Tc)^4}} \qquad (2)$$

where an empirical form was used for $\lambda(T)$. Thus ,using Eq.(1) together with the initial slopes of the magnetization curve and the particle size, the authors have calculated the $\lambda(T)$ values.The penetration depth at 0 K, λ_0, can be obtained by using Eq.(2).

Table 1 lists initial slopes and calculated values for $\lambda(T)$ and λ_0 for both 4.2 K and 77 K. The initial slope decreased with decreasing particle size and increasing temperature, as shown in Table 1. These data suggest that the flux penetrated deeply into the superconducting particle, if the particle size is of the same order as the penetration depth. The penetration depth increases with increasing temperature. The λ_0 values, obtained from $\lambda(4.2)$, are in the range from 0.1 to 1 μm. The reason for the data scattering seems to be according to anisotropy of $Y_1Ba_2Cu_3O_{7-\delta}$ or particle shape deviation from a sphere. The λ_0 values obtained from $\lambda(77)$, are in the range from 0.7 to 1.6 μm, and are several times larger than the λ_0 values from $\lambda(4.2)$.

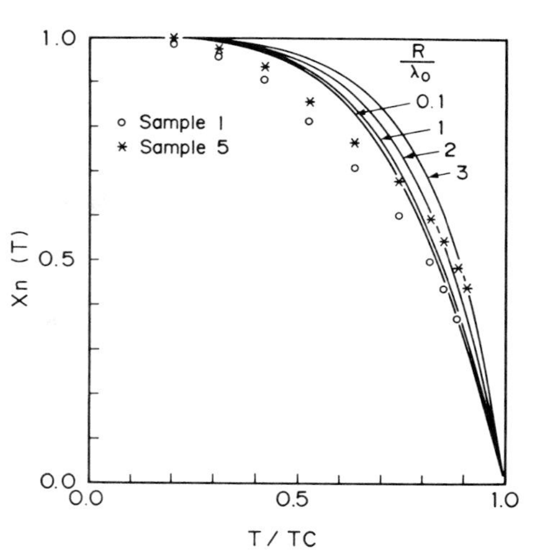

Figure 4. Normalized magnetic susceptibility Xn as a function of reduced temperature T/Tc. Curves 1,2 and 3 are theoretical results. R/λ_0=0.1,1, and 2, respectively. Open circles are data points for sample 1 (D^{50}=3.1 μm) and closed circles are data points for sample 5 (D^{50}=10.0 μm).

In order to study the temperature dependence of λ in detail, the authors compared the shielding effect data with the calculated value. $R/\lambda(T)$ is very close to R/λ_0 at 4.2 K. Hence, a normalized susceptibility of the sample Xn can be explained, using the following equation.

$$Xn=\frac{P(T)}{P(4.2)}=\frac{1-(3\lambda(T)/R)\coth(R/\lambda(T))+3(\lambda(T)/R)^2}{1-(3\lambda_0/R)\coth(R/\lambda_0)+3(\lambda_0/R)^2} \qquad (3)$$

In Fig.4, experimental values and the theoretical curves for Xn are plotted versus T/Tc. The initial slopes of magnetization dependence on particle size and temperature are in qualitative agreement with Eq.(1) and Eq.(2). But entering into detail, the experimental values seem to consist of two different curves with different Tc values. These two components seem to correspond to the orthorhombic I and orthorhombic II phase, whose Tc is approximately 60 K[7]. Therefore, the experimental value deviations from the theoretical curves are subject to this concept. The previously descrived λ difference between that calculated from $\lambda(4.2)$ and $\lambda(77)$ can be explained by the same concept.

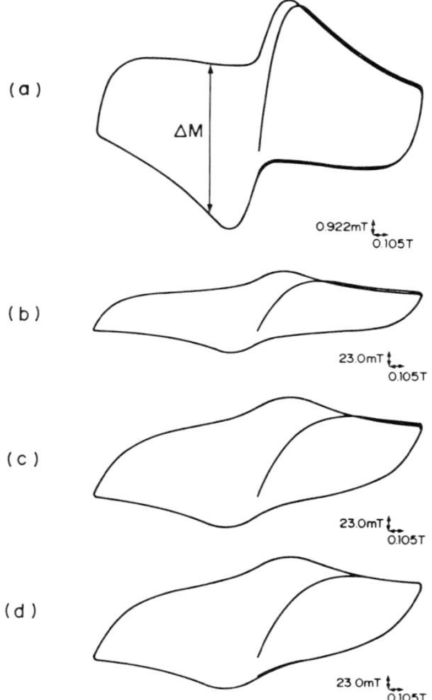

Figure 5. Magnetization curves at 4.2 K (a)Sample 1 (D^{50}=3.1 μm) (b)Sample 6 (D^{50}=15.2 μm) (c) Sample 15 (D^{50}=36.0 μm) (d) Sample 16.(D^{50}=52.5 μm)

Critical density

Typical magnetization curves at 4.2 K are shown in Fig.5, obtained for samples 1,6,15 and 16. Figure 5 shows that the magnetization decreases with decreasing powder size.

The magnetization difference ΔM, as defined in Fig.5, can be explained based on the Bean model, as follows.

$$\Delta M=\frac{\mu_0 dJc}{2} \quad (4)$$

where μ_0 is permeability in a vacuum and d is the mean diameter for each powder particle. Jc values can be derived from the slope of the magnetization (ΔM) and the powder diameter relationship.

Figure 6 shows the ΔM dependence on powder size for 0.3 T at 4.2 K. The magnetization is in linear proportion to particle size in the 1-20 μm range and saturates at above 20 μm. Jc is 2×10^6 A/cm^2 for 0.3 T at 4.2 K, obtained from linear relationship between magnetization and the particle size over the 1-20 μm range. The magnetization saturation suggests that the magnetization current cannot flow through boundaries on 20 μm diameter grains.

Typical magnetization curves at 77 K are shown in Fig.7 for samples 1,2,9 and 15. These magnetizations are smaller than those in Fig.5. Figure 8 shows the ΔM dependence on powder particle size for 0.03 T at 77 K. The magnetization for 0.03 T at 77 K is also in linear proportion to the particle size in the 1-20 μm range and saturates above 20 μm. Jc at 77 K is obtained as 7×10^4 A/cm^2 for 0.03 T. These Jc values are almost the same or somewhat larger compared with that for the single crystal[8]. TEM and SEM observations suggest that individual powders, whose particle sizes are smaller than 20 μm, seem to be made of a single-crystal or its fragment. On the other hand, each particle, whose size is larger than 20 μm, is made from several grains. Therefore, current flow over the 20 μm range is limited by the grain boundaries.

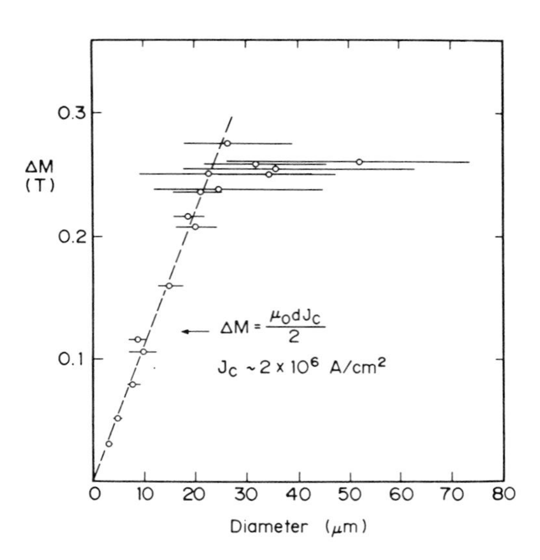

Figure 6. Magnetization powder diameter dependence at 4.2 K, 0.3 T.

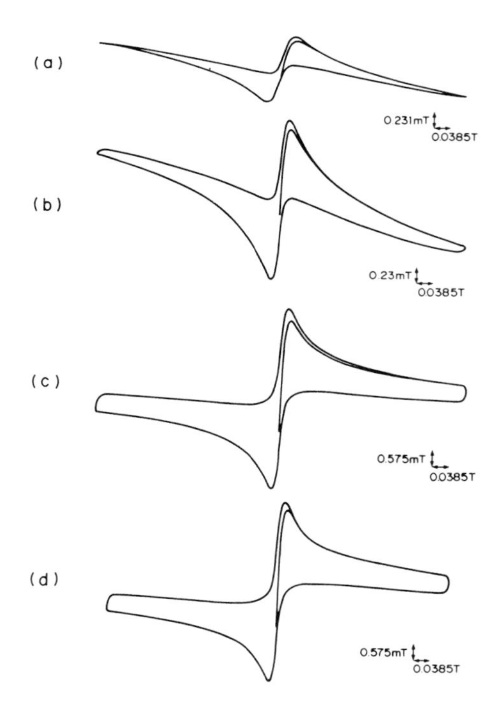

Figure 7. Magnetization curves at 77 K
(a)Sample 1 (D_{50}=3.1 μm),
(b)Sample 2 (D_{50}=4.8 μm),
(c)Sample 9 (D_{50}=21.5 μm)
(d) Sample 15 (D_{50}=36 μm)

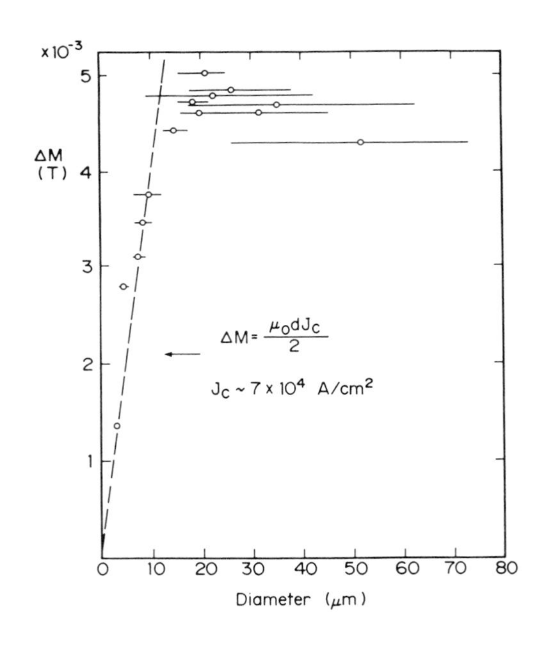

Figure 8. Magnetization powder diameter dependence at 77 K, 0.03 T.

CONCLUSION

Magnetization is in linear proportion to particle size, in 1-20 μm range and it saturates above 20 μm. Jc values, obtained from the linear relationship, are $2x10^6$ A/cm^2 for 0.3 T at 4.2 K, and $7x10^4$ A/cm^2 for 0.03 T at 77 K. The current flow over the 20 μm range are limited by the grain boundaries.

The temperature dependence of the magnetization curve consists of two different curves with different Tc values. London penetration depths at 0 K, λ_0, obtained from the initial magnetization at 4.2 K were different from the value obtained at 77 K. This result can also be explained by the two phase model.

ACKNOWLEDGMENTS

The authors wish to thank S. Nakamura and S. Takeno for their help in TEM and K. Higashibata for his help in SEM.

REFERENCES

1) M.K. Wu, J.R. Ashburn, C.J. Torng, P.H. Hor, R.L. Meng, L. Gao, Z.J. Huang, Y.Q. Wang, and C.W. Chu:Phys. Rev. Lett. **58**(9)(1987) 908.
2) H. Kumakura, M. Uehara, and K. Togano:Appl. Phys. Lett. **51**(19),9 November (1987).
3) Y. Tanaka and K. Matsumoto:Proc. of Japan-US Workshop on High-field Superconducting Materials for Fusion (Fukuoka,1987).
4) Y. Yamada, N. Fukushima, S. Nakayama, H. Yoshino, and S. Murase:JJAP **26**(5)(1987) L865.
5) N. Sadakata, Y. Ikeno, M. Nakagawa, K. Gotoh, and O. Kohno:Mat. Res. Soc. Symp. Proc. (1988) **99** 293.
6) D. Shoenberg:Proc. R. Soc. London, Ser. **A175**(1940) 49.
7) Y. Nakazawa, M. Ishikawa, T. Takabatake, K. Koga, and K. Terakura:JJAP **26**(5)(1987) L796.
8) T.R. Dinger, T.K. Worthington, W.J. Gallagher, and R.J. Sandstorm:Phys. Rev. Lett. **58**(25)(1987) 2687.

4.8 Irradiation Effect

In-Situ Observation of Bubbles in Superconductor $YBa_2Cu_3O_{7-\delta}$ Materials During He Ion Irradiation in an Electron Microscope

SHIGEMI FURUNO[1], KIICHI HOJOU[1], HIROSHI MAETA[2], HITOSHI OTSU[1], and MITSUO WATANABE[2]

Department of Chemistry[1] and Department of Physics[2], Japan Atomic Energy Research Institute Tokai-mura, Ibaraki, 319-11 Japan

ABSTRACT

In-situ observation of formation and growth of bubbles in superconductor $YBa_2Cu_3O_{7-\delta}$ materials during 10 keV He ion irradiation was carried out using a transmission electron microscope. Irradiations were performed inside the electron microscope linked with an ion gun at room temperature. The images on the fluorescent screen were recorded by VTR. The small bubbles were observed at the fluence of $9x10^{16}$ ions/cm^2. These bubbles increased in number and continued to grow, as the irradiation proceeded. At the fluence of $3x10^{17}$ ions/cm^2 bubbles began to coalesce with each other, followed by blisters and exfoliation. New bubbles were formed underneath, grew and coalesced. Preferential formation and growth of He bubbles at crystal grain boundary were found during the irradiations. It was also found that amorphization of $YBa_2Cu_3O_{7-\delta}$ crystals occurred as the irradiation proceeded.

INTRODUCTION

Structural and chemical changes induced by ion irradiation and implantation are an important problem from the up-to-date needs for superconducting magnets in fusion reactors subjected to irradiation damage by 14 MeV neutron and nucleation and formation of helium bubbles by nuclear mutation, because superconducting properties are strongly affected by structural disorders and defects. Our understanding is incomplete on detailed information of structural disorders and defects. Direct observation methods using electron microscope are capable of providing information on atomic scale structure in materials. In-situ observations using an electron microscope linked with an ion gun is most powerful for elucidating the fundamental mechanisms and the processes of structural changes and radiation damage during ion irradiation and implantation[1, 2].

Recently we have developed a new in-situ observation system consisting of 100 keV transmission electron microscope linked with an ion beam of relatively low energy and of high current density[3]. Using this system, we studied dynamic behavior of growth, coalescence, burst and disappearance of helium bubbles and blisters in the aluminum[4] and the ceramic SiC and TiC[3] during helium ion irradiation and subsequent annealing. This paper reports the results of in-situ observation of $YBa_2Cu_3O_{7-\delta}$ superconducting material during 10 keV helium ion irradiation. We have recorded the whole processes from the initial stage to the heavy irradiation up to 10^{17} ions/cm^2 by VTR, including the dynamic processes of bubble growth, coalescence and bursting of helium bubbles in crystal grains and at grain boundaries in $YBa_2Cu_3O_{7-\delta}$ materials during the irradiation at room temperature.

EXPERIMENTAL PROCEDURE

In this experiment, materials were prepared from mixtures of Y_2O_3, $BaCO_3$ and CuO powder by a standard ceramic sintering process. Homogeneous mixtures were calcinated at 900°C for 10 hrs in air. Pressed pellets were sintered at 950°C for 10 hrs in oxygen atmosphere and then slowly cooled (60°C/hr) to room temperature. In the sintered $YBa_2Cu_3O_{7-\delta}$ the transition temperature was observed to be 89.3 K. From measuring c-axis spacing by x-ray diffraction we estimated oxygen contents to be about 6.9[5]. Thin specimens suitable for electron microscope observation were made from discs of 0.2 mm thickness and 3 mm in diameter by Ar ion thinning after dimpling. C-basal plane of used specimen was observed to be nearly parallel to the surface of the specimen by electron diffraction.

Ion irradiations and simultaneous observation were performed by using the in-situ observation system. Fig. 1 shows the in-situ observation system in our laboratory. This system consists of the electron microscope of JEM-100C type and a Duo-plasmatron type ion gun with accelerating voltage up to 10 kV. The mass selected ion beam is incident at an angle of 72° to the surface of the specimen as shown in fig. 2.

In the present experiment, 10 keV He ions were used for the irradiations with the flux of 1.5×10^{14} ions/cm^{2}sec at room temperature. Temperature of specimen during the

Fig. 1 Appearance of in-situ observation system consisting of 100 keV transmission electron microscope linked with ion source.

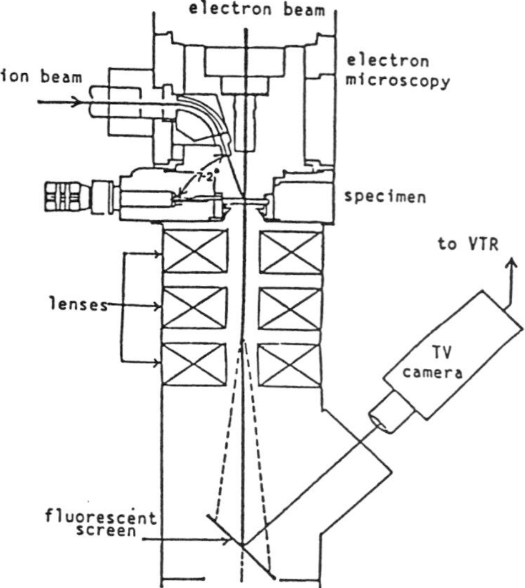

Fig. 2 Schematic diagram of in-situ observation system

irradiation was estimated to be about 50°C[3]. Results of in-situ observation during the ion irradiations were recorded with a VTR through a TV camera and a monitor TV.

RESULTS AND DISCUSSION

Typical examples of observed results are shown in a series of VTR pictures of figs. 3(a) to

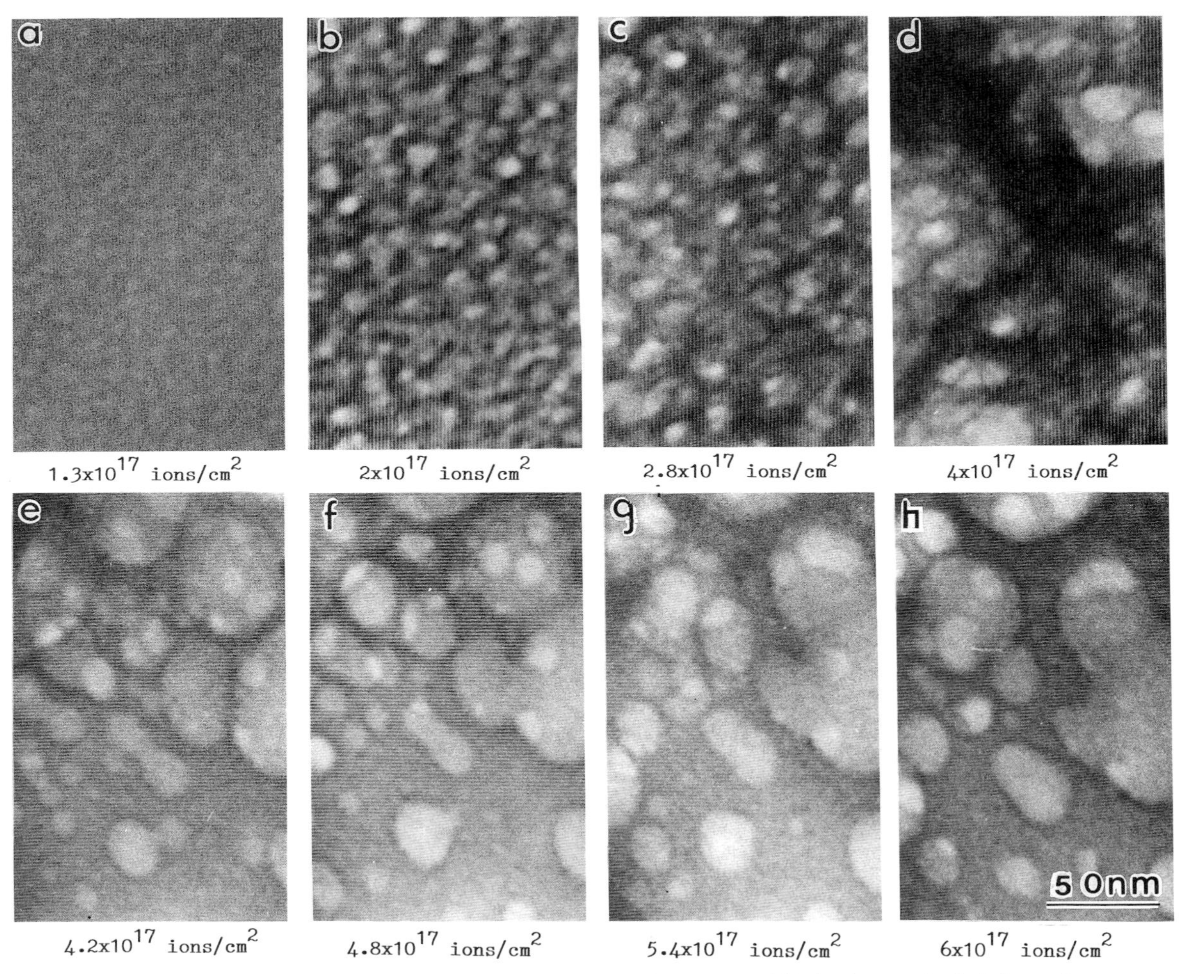

Fig. 3 Dynamic behavior of bubble growth, coalescence, blistering and shrinkage in crystal grains having its c-basal planes parallel to the surface of the specimen during He ion irradiation at room temperature with a flux of 1.5×10^{14} ions/cm^{2}sec.

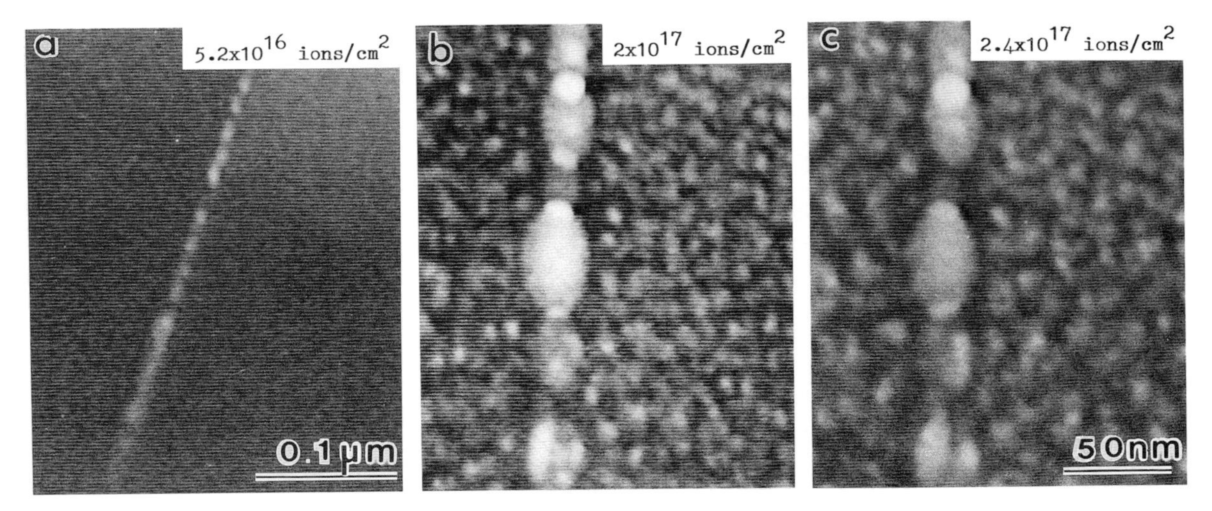

Fig. 4 Preferential growth of bubbles on boundary between two crystal grains having each c-basal planes parallel to the surface of the specimen.

(h). In this case, bubbles are formed at the fluence of about 9×10^{16} ions/cm^2 and then increase in number and size as increasing the fluence as shown in Figs. 3(a) to (c). After the fluence of 3×10^{17} ions/cm^2, bubbles grow largely by coalescence and then at the fluence of 4×10^{17} ions/cm^2 blistering and exfoliation occur as shown in Figs. 3(d) to (e). After the exfoliation, the subsequent behavior of underlying bubbles are shown in figs. 3(e) to (h), where the typical mode of coalescence and shrinkage of bubbles are shown. It is also found that bubbles are formed even in very thin part, which would be associated with anisotropic crystal structure. As the irradiation proceeds to 8×10^{17} ions/cm^2, almost all electron diffraction patterns show halo patterns. This is caused by the amorphization of the $YBa_2Cu_3O_{7-\delta}$ crystal by ion bombardment.

Preferential bubble formation and growth were observed at crystal grain boundary as shown in figs. 4(a) to (c). Prior to bubble formation in the grains, bubbles were formed in the grain boundary at the fluence less than 3×10^{16} ions/cm^2, and then grew more rapidly than in the grains. In spite of the preferential bubble formation and growth at the grain boundary, no denuded zone was observed in both close sides of the grain boundary.

It is also found that bubble formation and growth are retarded in grains not parallel to the surface of the specimen.

CONCLUSIONS

In-situ observation clarified dynamic behavior of helium bubbles and structural changes in $YBa_2Cu_3O_{7-\delta}$ crystals.

(1) Bubbles are formed at the fluence of 9×10^{16} ions/cm^2, grow uniformly and coalesce with each other, followed by blistering or exfoliation.

(2) Preferential formation and growth of He bubbles occur in crystal grain boundary.

(3) Amorphization of the $YBa_2Cu_3O_{7-\delta}$ crystals occurs for the fluence exceeding 3×10^{17} ions/cm^2 and all the crystals become amorphous at the fluence of 8×10^{17} ions/cm^2.

REFERENCES

(1) P. A. Thackery, R. S. Nelson and H. C. Sensom, AERE-R-5817 (1968).

(2) S. Ishino, H. Kawanishi, K. Fukuya and T. Muroga, IEEE Trans. Nucl. Sci. NS-30(1983)1255.

(3) K. Hojou, S. Furuno, H. Otsu, K. Izui and T. Tsukamoto, J. Nucl. Mater., 155-157 (1988)298.

(4) S. Furuno, K. Hojou, K. Izui, N. Kamigaki and T. Kino, J. Nucl. Mater., 155-157(1988) 1149.

(5) M. Watanabe, T. Kato, H. Naramoto, H. Maeta, K. Shiraishi, Y. Kazumata, A. Iwase and T. Iwata, in this proceedings.

Effects of Localisation in Atomic-Disordered High-T_c Superconductors

S.A. Davydov, B.N. Goshchitskii, A.E. Karkin, A.V. Mirmelshtein, V.I. Voronin, M.V. Sadovskii, V.L. Kozhevnikov, S.V. Verkhovskii, and S.M. Cheshnitskii

Institute for Metal Physics, USSR Academy of Sciences, Ural Branch 620219, Sverdlovsk GSP-170, USSR

ABSTRACT

The influence of disordering on the properties of high-T_c superconductors is investigated. The results show that pairing takes place in the systems of localised electrons even in slightly disordered samples. Superconductivity exists till radius of localisation exceeds the typical size of Cooper pairing in highly disordered system.

INTRODUCTION

Fast neutron irradiation is one of the most pure methods to investigate effects of disordering on physical properties of high-temperature superconductors (HTSC). It is known that disorder which is not connected with magnetic impurities according to Anderson theorem should not appreciably affect superconducting transition temperature T_c of the conventional (wide-band) superconductors. However, the properties of such superconducting compounds as A-15, C-15 and Shevrel phases turn out to be highly sensitive to disorder. Thus, response of the system to introduction of defects helps to better understand the peculiarities of its electronic properties in the ordered state characterised in our case by high values of T_c.

Previously [1-3] we have shown that disordering of YBaCuO and LaSrCuO leads to a rapid decrease in T_c. The derivative of the upper critical field H'_{c2} practically does not change. At large fluences instead of linear dependence the Mott-type temperature dependence of $\rho(T)$ is observed in the temperature interval $5 < T < 300$K:

$$\rho(T) = A\exp(Q/T^{1/4}), \quad Q = 2.1[N(E_F)R_{loc}^3]^{-1/4}. \qquad (1)$$

While measuring ρ of both systems at 80 K directly during fast neutron irradiation we found that ρ_{80} grows exponentionally with fluence Φ, i.e. $\rho_{80} \sim \exp(a\Phi)$ beginning with the smallest Φ. The observed variation of ρ depending both on fluence and temperature may be described using the empirical formula [1]:

$$\rho(T) = f(T)\exp(a\Phi/T^{1/4}). \qquad (2)$$

According to experimental data localisation effects are essential in these systems even at the smallest extent of disorder.

The present study is further investigation of localisation effects in high T_c superconductors including BiSrCaCuO (zero resistivity at T=70 K, $\rho_{300} \approx 2$ mohm·cm and linear temperature dependence of ρ (T)), single-crystal sample of YBaCuO ($T_c \approx 80$ K, $\rho_{300} \approx 2$ mohm·cm). Here we investigated also the influence of isochronous low-temperature annealing ($T_{ann} \leq 300$ K) on electrical resistivity of high-

T_c materials disordered under fast neutron irradiation (T_{irr} = 80 K). Besides, we present here the data on the NQR spectrum and spin-lattice relaxation time T_1 for the sample of $YBa_2Cu_3O_{6.95}$ irradiated by a fluence of 5 x 10^{18} cm^{-2}.

EXPERIMENTAL RESULTS

Radiation disordering, deviation from oxygen stoichiometry, introduction of impurities into Cu sublattice suppress superconductivity in $YBa_2Cu_3O_{7-\delta}$. Structural changes are different in all cases. Orthorhombic parameter (b-a) gradually decreases with decreasing oxygen content. In this case oxygen is removed from sites O4 (in chains). As for sites O5, they are filled slightly with increasing ρ just before the tetragonal phase transition [4] . Radiation disordering results in oxygen distribution over sites O4 and O5 and growth of the mean-quadratic atomic displacements. The structure, however, remains orthorhombic. Appearance of oxygen vacancies in sites O4 is well seen in the NQR spectra for the irradiated sample of YBaCuO (Fig.1). The structure of the NQR line on the ^{63}Cu atoms in Cu2 site is associated with vacancy configurations in O4 sites as well as in oxygen-deficient samples [5] . Spin-lattice relaxation time T_1 on ^{63}Cu nuclei in Cu2 sites in normal state ($T > T_c$) increases from 0.5 to 150 msec on changing the oxygen content from $\delta \approx 0$ to $\delta \approx 0.2$. Radiation disordering ($T_c \approx 70$ K, Φ = 5 x $10^{18} cm^{-2}$) leads also to increase in T_1 but to a lesser extent. Assuming that the whole T_1 increase is connected with the change of electron states density on the Fermi level $N(E_F)$ it will decrease more than by order in the first case and by 25% in the second. Distribution of relaxation times over Cu1 sites of the irradiated sample points to appearance of disorder in the electron environment of the Cu atoms in these sites. Measuring T_1 in a superconducting state ($T_c \approx 70$ K) of the irradiated sample yields the value $2\Delta/kT_c \sim 12$ (as in the initial sample). This evidences a very strong coupling regime (bearing in mind the maximum value of a gap). In an oxygen-deficient sample with approximately the same T_c ($\delta \approx 0.2$) the value of a reduced gap decreases to $2\Delta/kT_c \sim 4$-6 for both sites of the Cu atoms. Temperature independent contribution to magnetic susceptibility χ_0 also behaves differently in both cases under consideration, i.e. it drops with increasing δ and grows under irradiation. Growth of χ_0 under irradiation and simultaneous increase of spin-lattice relaxation time shows that an attempt to explain their change only by $N(E_F)$ variations leads to a contradiction.

Variation of the oxygen content directly alters electron structure of the YBaCuO system since the concentration of charge carriers changes. Radiation disordering affects the electron structure only through disorder. Differences in behavior of

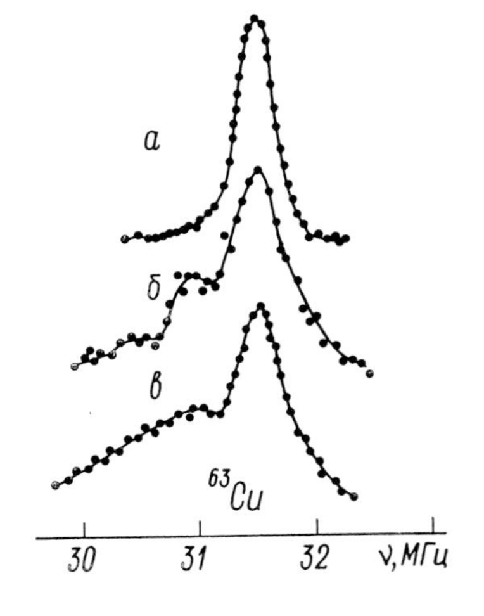

FIGURE 1. Evolution of the NQR spectrum on the ^{63}Cu nuclei in Cu2 site for $YBa_2Cu_3O_{7-}$:a)δ=0.05; б) the sample irradiated by a fluence of 5 x $10^{18}cm^{-2}$($T_c \approx 70$ K); B) δ = 0.2 ($T_c \approx 70$ K). Frequency is given in MHz.

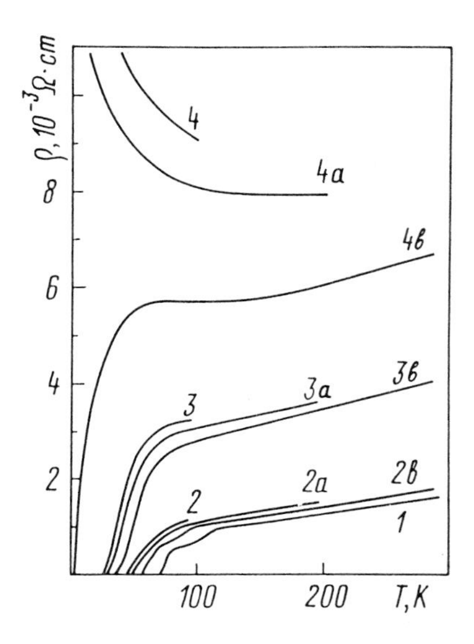

FIGURE 2
Temperature dependence of electrical resistivity for the disordered BiSrCaCuO sample: 1)the initial sample; 2) ϕ = 2 x x 10^{18}; 3) ϕ = 5 x 10^{18}; 4) ϕ =7 x $10^{18}cm^{-2}$. Indexes a and b correspond to annealing at T = 200 K and T = 300 K, respectively.

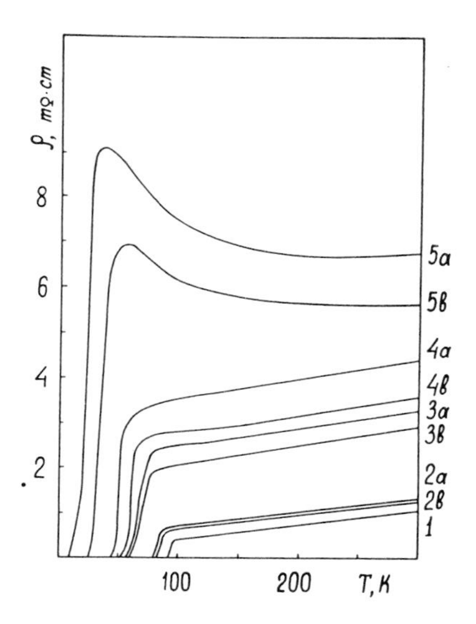

FIGURE 3
Temperature dependences of electrical resistivity for the disordered $YBa_2Cu_3O_{6.95}$ samples: 1)the initial sample; 2) ϕ = 2 x 10^{18}; 3) 5 x 10^{18}; 4) 7 x 10^{18}; 5) 10 x $10^{18}cm^{-2}$. Indexes a and b correspond to annealing at T = 300 K for 20 min and two weeks

YBaCuO on introduction of defects of different types underline, however, the general property: degeneration of T_c always goes with transition to dielectric behavoir of resistivity.

The samples of BiSrCaCuO, as well as YBaCuO, with high enough T_c (relatively small disordering) are characterised by linear growth of electrical resistivity with temperature (Figs.2,3). Increasing extent of disorder leads to exponential temperature dependence of electrical resistivity described by eq.(1). After annealing for 20 min at T = 300 K T_c appears and linear dependence of ρ(T) reestablishes practically within the whole temperature range beginning from T_c to 300 K. For $\phi \geq 10^{19}cm^{-2}$ annealing at T$\leq$300 K leads only to decrease in coefficients A and Q in eq.(1) while T_c does not appear at all. Fig.4 shows the dependence of ρ_{80} on fluence of fast neutrons for materials under study. The value of ρ_{80} grows exponentially with fluence. According to measurements made in the basal plane of a single crystal YBaCuO sample exponential growth of resistivity of ρ_{80} is the property of the material itself and not the consequence of the preparation technique of ceramic samples. The data from Fig.4 as well as the behavior of dH'_{c2}/dT [6] and ρ(T) in the disordered HTSC sample show that localisation effects are likely to appear well before the disappearance of T_c. But the superconducting transition does not allow to observe this experimentally on the temperature dependence of ρ(T).

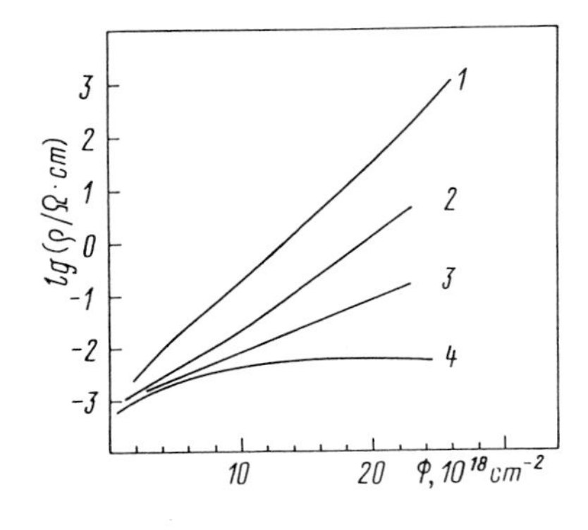

FIGURE 4. Dependences of ρ_{80} on fluence of fast neutrons for ceramic and single-crystal YBaCuO samples (curve 1 and 2, respectively), BiSrCaCuO (3) and $SnMo_6S_8$ (4).

DISCUSSION

It is well known that full enough disorder may result in the metal-insulator transition due to electron localisation (Anderson transition). Anderson showed [7] that in localised electron system Cooper pairing may take place only between the electrons with centres of localisation lying within the sphere of the order of localisation length R_{loc}. The states of such electrons are splitted by energy to a value of the order $[N(E_F)R^3_{loc}]^{-1}$ [8]. Apparently the condition should be held that the value of a superconducting gap Δ (at T=0) is essentially higher than the value of this splitting :

$$\Delta \sim T_c \gg \left[N(E_F)R^3_{loc}\right]^{-1} \tag{3}$$

Eq.(3) is equivalent to the requirement that localisation radius should be essentially higher than the typical size of the Cooper pair in highly disordered systems [6]. Using experimental data on electrical resistivity for the disordered samples of $YBa_2Cu_3O_{7-\delta}$ (for fluences of $\Phi > 5 \times 10^{18} cm^{-2}$) and empirical formula (2) for smaller fluences one may calculate the variation of localisation radius R_{loc} as a function of fluence. On the other hand eq.(3) gives the limited values of R_{loc}, at which superconductivity may still exist in the system of localised electrons. Assuming that for all fluences $N(E_F) = 5 \times 10^{33}\ (erg\cdot cm^3)^{-1}$ (in the model of free electrons this corresponds to one electron per a unit cell, i.e. a carrier concentration of $6 \times 10^{21} cm^{-3}$) and the left-hand side of eq.(3) equals 5 T_c, one will obtain the result shown in Fig.5. From this figure it is seen that eq.(5) is not valid for $\Phi > (5\text{-}7) \times 10^{18} cm^{-2}$. Bearing in mind the qualitative character of this estimate it should be noted that it is in surprisingly good agreement with the experiment. Fig.5 allows to easily interprete the data concerning the influence of low-temperature annealing. In the vicinity of $\Phi \sim (5\text{-}7) \times 10^{18} cm^{-2}$ annealing results in R_{loc} increasing, eq.(3) is valid and superconductivity appears. At high fluences low-temperature annealing is not enough for appearance of superconductivity.

From the given estimates, however, it is not clear why the superconducting transition temperature should drop for high (in comparison with the limited) values of R_{loc}. It is difficult to make a conclusion about the reasons of T_c decreasing with increasing disorder because of lack of theoretical understanding of the nature of T_c in HTSC. However, proceeding from the traditional considerations about the pairing interaction one may expect that one of the reasons for suppression of T_c may be connected with the growth of the Coulomb pseudopotential μ^* ,

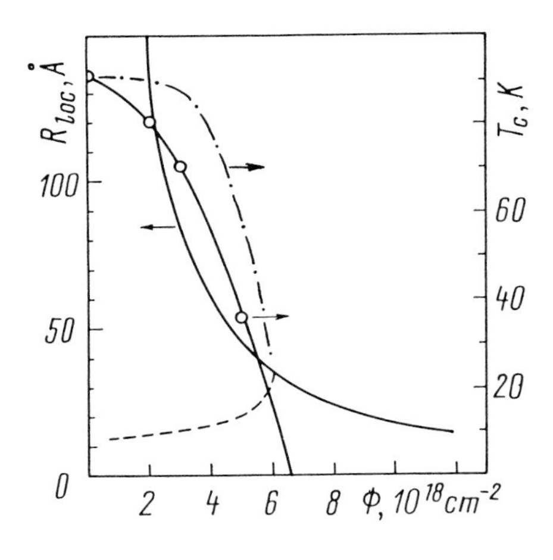

FIGURE 5. Dependence of R_{loc} (solid line) and T_c (open circles) on fluence of fast neutrons for YBaCuO. The dashed line shows the values of R_{loc} calculated by eq.(3). The dash-dot line shows the values of T_c calculated by eq.(5).

which describes Coulomb repulsion of electrons forming the Cooper pair. This effect is connected with growing lag effects of Coulomb repulsion in Cooper pair with decreasing diffusion coefficient during the Anderson transition. According to [6] in localisation region we have

$$\ln \frac{T_{co}}{T_c} = \Psi\left(\frac{1}{2} + \frac{\mu A_{E_F}}{4\pi T_c N(E_F)}\right) - \Psi\left(\frac{1}{2}\right), \quad (4)$$

where Ψ is the digamma function, $\mu = V_o N(E_F)$ is the Coulomb potential, $A_{E_F} \sim R_{loc}^{-3}$. This formula describes T_c suppression due to growing effects of Coulomb repulsion in a single quantum (localised) state. Taking into account eqs.(1) and (2) from eq.(4) we get

$$\ln \frac{T_{co}}{T_c} = \Psi\left(\frac{1}{2} + \frac{\mu T \left(\ln \frac{\rho(T)}{A}\right)^4}{4\pi T_c (2.1)^4}\right) - \Psi\left(\frac{1}{2}\right). \quad (5)$$

Then using the data from Fig.4 and assuming that $\mu \approx 1$ one may easily calculate T_c as a function of disordering extent (Fig.5). As for the functional dependence it slightly differs from the experimental one (note that in the theoretical curve the derivative $dT_c/d\Phi \rightarrow 0$ for $\Phi \rightarrow 0$) however, the qualitative agreement is out of question. Faster suppression of T_c at small extent of disorder observed experimentally may be connected with additional regular contribution of electron densities correlator which was neglected in eq.(4).

This approach helps to elucidate some typical features in behavior to disordered HTSC. The fact that these materials are close to Anderson transition metal-insulator and the existence of superconductivity with high enough T_c in a system of localised electrons is likely to be a specific feature of HTSC. This "nonmetalic" behavior of these compounds becomes evident when comparing, for example, with A-15 superconductors. The latter behave themselves more like metals than HTSC's. As for the superconductors with the structure of Shevrel phase, they present an intermediate case between the A-15 and HTSC.

CONCLUSION

Quasi-two-dimensional systems to which the superconductors under study belong are expected to show strengthening of effects of localisation that occurs at values of conductivity appreciably exceeding typical values of three-dimensional systems The properties of electronic system of HTSC's make them close to Anderson metal-insulator transition even in the ordered state. This feature differs new HTSC's from previously known superconductors of other systems. In the first case pairing is likely to occur in systems of localised electrons even at the smallest disorder. T_c sharply decreases when at low temperatures the dependence ρ (T) of type (1) is observed. Superconductivity is fully suppressed when energetic splitting between the localised states becomes comparable with the value of a superconducting gap.

REFERENCES

1. S.A.Davydov, A.E.Karkin et al. Pis'ma JETP 47, 193 (1988).
2. B.A.Aleksashin, I.F.Berger et al. Effect of disordering on the properties of high-T_c superconductors, in : Proceedings of M^2HTSC (Switzerland, Interlaken, 1988).
3. B.A.Aleksashin, I.F.Berger et al. Effect of disordering on the properties of high-T_c superconductors, in : Series of preprints of scientific reports "Problems of high-temperature superconductivity" (USSR, Sverdlovsk, 1988) N1.
4. V.I.Voronin et al. to be published.
5. V.V.Serikov, Yu.I.Zhdanov, S.V.Verkhovskii. Pis'ma JETP 47, 571 (1988).
6. L.N.Bulaevskii, M.V.Sadovskii. Pis'ma JETP 39, 524 (1984); J.Low Temp.Phys. 59, 89 (1985).
7. P.W.Anderson. J.Phys.Chem.Solids 11, 26 (1959).
8. N.F.Mott, E.A.Davis, Electron Processes in Non-Crystalline Materials (Clarendon Press, Oxford, 1979).

Electron Irradiation Effects on $Y_1Ba_2Cu_3O_y$ Superconductors

M. WATANABE[1], T. KATO[1], H. NARAMOTO[1], H. MAETA[1], K. SHIRAISHI[2],
Y. KAZUMATA[1], A. IWASE[1], and T. IWATA[1]

[1] Department of Physics, Japan Atomic Energy Research Institute, Tokai, 319-11 Japan
[2] Department of Development, Japan Atomic Energy Research Institute, Takasaki, 370-12 Japan

ABSTRACT

3 MeV electron irradiations are performed in the broad range of electron dose, up to 2.0×10^{18} e^-/cm^2, to investigate the effect of irradiation-induced defects on the superconducting transition for oxygen-controlled specimens. The electron irradiations depress Tc at the rate of 0.5 ∿ 3.5 $K/10^{18}e^-/cm^2$ depending on the specimen treatments and the probe currents. A specimen with the lower contents of oxygens is more sensitive to the irradiation. The detailed analysis on the transition region reveals the complicated behaviors. The three typically different phenomena are found in the three kinds of specimens(y=6.6, 6.7 ∿ 6.8, 6.8 ∿ 6.9). In a specimen with y=6.8 ∿ 6.9, the two stage transition is observed, and only the transition at the lower temperature is shifted to the lower temperature region with increasing probe currents. The electron irradiations do not influence on this feature. In a specimen with y=6.7 ∿ 6.8, the single transition is observed. The single transition is shifted to the lower temperature region by the irradiation without changing the sharpness of the transition. Contrary to the above two cases, in a specimen with y=6.6, the two stage broad transition is found, and the transition at the lower temperature depends on the probe current. The irradiation affects both of the peaks. With the increase of the electron dose, the transition at the higher temperature is broadened, but the transition at the lower temperature is sharpened with the shift to the lower temperature region.

INTRODUCTION

Since the discoveries of new oxide superconductors with higher transition temperatures,[1,2] intensive studies have been conducted from both the application and the fundamental points of view. Although the physical mechanism for this class of superconductors has not been understood fully, it is generally recognized that the superconducting characteristics are highly structure-sensitive, which causes the controversial problems among the same kinds of measurements. The transition temperature, T_c, has been found to depend sensitively on the oxygen contents and their atomic arrangement.[3, 4)]

In the particle irradiations, the incident energies of particles can be adjusted precisely and the number of point defects and the depth-distribution are controllable. Thus, the particle irradiation experiment will play an important role to know the fundamental aspects of the superconducting transition even in such a complicated compound. In the ion irradiation experiment on thin film superconductors, the electrical and structural changes are associated with the loss of phase coherence in the grain boundaries.[5, 6)] The electron irradiations have been made to introduce the simpler defects in a bulk specimen.[7 ∿ 10)] The considerably large damage rate was found in Y-Ba-Cu-O superconductors compared with the classical ones, and the irradiation is more influential in the change of T_c than the effect of an oxygen stoichiometry. It is expected that the state of the induced defects depends on the oxygen contents because the structure of Y-Ba-Cu-O is defective and it is easy to stabilize the displaced atoms through the formation of aggregates and/or the charge compensation.

In the present study, four kinds of specimens with the different oxygen contents are prepared through the rapid quenching technique, and a detailed comparison is made among the irradiation effects of 3 MeV electrons on the superconducting transition by measuring the electrical resistivity with the different probe currents. The importance of the inhomogeneous structure associated with the oxygen configurations is described for the study of the irradiation effects.

EXPERIMENTAL PROCEDURE

The specimen preparation process is almost the same as a conventional method for making a sintered pellet. The raw materials of Y_2O_3(99.99 %), $BaCO_3$(99 %) and CuO(98 %) are the reagent grade, and the homogeneous mixtrures were calcinated at 900 °C, 10 hrs in air, and then the pellets pressed at 3 ton/cm^2 were sintered at 900 °C, 10 hrs in air. The platellet specimens, 0.5 x 1.0 x 9.0 mm^3, were cut out from a sintered pellet with a razor blades. These platellets were treated at 950 °C, 10 hrs in flowing oxygen gas, and were rapidly quenched into a liquid nitrogen bath. Five kinds of quenching temperatures(500, 550, 600, 700 and 950 °C) were employed to prepare specimens with the broad range of oxygen contents. For comparison, the slowly cooled specimens were also prepared at the rate of 60 °C/hr.

The x ray diffraction analysis was employed on these quenched specimens to determine the lattice parameters. The data obtained were ordered reasonably depending on the treatments, and the oxygen contents were estimated by fitting our data to the relation determined by Ono.[11)] The lattice parameters in a, b and c axis change in a similar manner to the published results as a function of quenching temperature, and any evidence of a second phase was not detected from this analysis. In Table 1, the estimated oxygen contents are tabulated for four kinds of quenched specimens. The oxygen contents in a slow-cooled and 500 °C quenched specimens are almost the same within an experimental error, but the superconducting characteristics are different with each other as described in the latter part of this paper.

3 MeV electrons from Dynamitron accelerator in JAERI were employed for the irradiation. Scanned electron were taken out into open air through thin aluminum foil, and the irradiation was made on the specimens placed on a water-cooled copper base. 3 MeV energy of electrons is large enough to introduce the uniform distribution of defects along the depth in a bulk specimen. Electron beam intensity was about 3 x 10^{13} e^-/cm^2.sec, and the fluence was increased up to 2.0 x 10^{18} e^-/cm^2 in a step of 0.5 x 10^{18} e^-/cm^2.

A standard four probe resistance method was used to measure the electrical resistivity as a function of temperature. The probe current was changed in the range of 0.5 ∿ 10.0 A/cm^2 to detect the current-dependent behavior of T_c if any. The output voltage in this measurement was recorded changing the direction of probe current for the precise determination of the resistivity, and the data were digitally processed for the later analysis like the calculation of the temperature derivatives of the resistivity, ($d\rho/dT$).

RESULTS AND DISCUSSION

Fig. 1 shows the temperature dependence of the resistivity, measured at 0.5 A/cm^2 in various specimens quenched from high temperatures. Only a specimen quenched from 950 °C is not superconductive, and has the rather high resistivity as illustrated in a different scale. It is observed with the increase of quenching temperature that the resistivity increases in the normal conducting region and that the transition from normal to superconducting (from now on, denoted by n-s transition) shifts to the lower temperature region. The feature of the n-s transition is dependent on the quenching conditions, and in 600 and 700 °C quenched specimens, the two stage n-s transition can be found clearly. As referred in the following figures, T_c(endpoint) corresponds to the temperature where the electrical resistivity becomes lower than 1 μΩ cm.

In the previous figure, the details are not shown, however, the n-s transition is sensitive to the probe current. Fig. 2 illustrates the change of T_c(endpoint) as a function of 3 MeV electron fluence for slow-cooled, 500 °C quenched, 550 °C quenched and 600 °C quenched specimens. Two kinds of probe currents were used to observe the irradiation effect associated with the structural change, and the results at 0.5 and 10.0 A/cm^2 are denoted by solid circles and open triangles, respectively. The values of T_c(endpoint), determined at the different probe currents between the above, are in the shaded area systematically. The data obtained are rather scattered, but the general features of the irradiation effect can be extracted. 3 MeV electron irradiations depress T_c(endpoint) at the rate of 0.5 ∿ 3.5 K/10^{18} e^-/cm^2, depending on the specimen treatment and the probe current. A slow-cooled specimen is less dependent on the probe current, and is the most resistive to the irradiations. It seems that the quenching treatment makes the specimens more sensitive to the irradiations, and the depression rate of T_c(endpoint) is increased. The data connected with dashed lines in this figure indicate the change of the resistivity in the normal state(ρ(100 K, 0.5 A/cm^2) as a function of electron dose. ρ at 100 K reflects the bulk irradiation effect, and increases with the electron dose at almost the same rate among all specimens. This tendency is different from the treatment-sensitive nature of T_c(endpoint), which implies the existence of inhomogeneous structure in the sense of the superconducting properties.

In the previous figure, the width of T_c(endpoint) in the shaded area indicates the difference of T_c(endpoint) determined at two different probe currents(0.5, 10.0 A/cm^2). For the detailed analysis, the electron dose dependence of T_c(endpoint) difference is summarized for the same kind of specimens as in Fig. 3. The T_c(endpoint) difference itself is simply ordered depending on the quenching treatment, but its response to the electron irradiations is not so simple. Both of slow-cooled and 600°C-quenched specimens are not so influenced by the irradiation, but the irradiation effect becomes pronounced in 500°C- and 550°C-quenched specimens.

Fig. 4 shows the probe current dependence of the T_c(endpoint) depression after the 3 MeV electron irradiations up to the dose of 2 x 10^{18} e^-/cm^2. The probe current was changed in the range of 0.5 ∿ 10.0 A/cm^2. The above T_c depression is smaller and larger for slow-cooled and 550°C-quenched specimen, respectively, and the others are in between the above kinds of specimens. The T_c depression in slow-cooled specimen does not depend on the current density. The same kind of the tendency is observed at the probe current of 5.0 A/cm^2, but at 10.0 A/cm^2 the depression is enlarged. On the contrary, the specimens quenched at 500 and 550°C are more sensitive to the probe current.

As shown already, the irradiation effect depends on the preparation process of specimens, but the notable change is found especially in 600°C-quenched specimen. The detailed analysis is shown on this specimen as a representative. Fig. 5 illustrates the change of ρ-T curve in the n-s transition region for the five different doses of 3 MeV electrons. In this figure, one can distinguish the dose-dependent change of the n-s transition. The two stage transition is observed in an as-prepared specimen, but with the increase of electron dose the transition at the higher temperature is smeared out, and the transition at the low temperature lasts with the shift to the lower temperature region.

These results can been seen more clearly through the differentiation of the ρ-T curve as shown in Fig. 6. The temperature derivatives of the resistivity, (dρ/dT), are plotted for two different probe currents(0.5 and 10.0 A/cm^2), and the current dependent nature of this specimen can be recognized. Before the irradiation, (dρ/dT) curve forms two peaks reflecting the two stage n-s transition. The peak at higher temperature(referred to as "H" peak) is not so sensitive to the probe current and, on the contrary, the peak at the lower temperature side(referred to as "L" peak) is influenced by the probe current. When the probe current is increased to 10 A/cm^2, a "L" peak is broadened with the shift to the lower temperature region. The irradiation effect of 3 MeV electrons on these peaks is different. A "H" peak loses its feature with the irradiations and forms a broad plateau after the irradiations. But the insensitiveness to the probe current still lasts up to 2 x 10^{18} e^-/cm^2. A "L" peak with the high sensitiveness to the probe current holds its feature with the irradiations. The peak shift to the lower temperature region is a common feature between these peaks in 600°C-quenched specimen. Different from the above results, in a slow-cooled specimen, the irradiation effect can not be seen in practise(not shown here for simplicity).

Depending on the preparation process of specimens, are observed the two stage n-s transition and/or the n-s transition sensitive to the probe current. The two stage transition reflects the formation of the two phase structure with different T_c's including the grain boundaries. It is considered that the slow-cooling process can produces the resistive nature to the irradiation even in a second phase. Quenching process induces the probe current dependent nature. In a 600°C quenched specimen, the inhomogeneous nature is also expected as in a slow-cooled specimen, but considering the broadening by irradiation and the sensitiveness to the probe current in a "H" peak, the irradiations possibly change the bulk phase into the continuously modified structure in the sense of oxygen configuration. The minor phase may be sensitive to the probe current. Quenching at 500 and 550°C produces the specimens with the oxygen contents of 6.75∿6.85, and these are expected to be highly sensitive to the n-s transition judging from the data by Cava et al.[2)]. Thus, the broad n-s transition and the sensitiveness to the irradiations are reasonably explained by the formation of the substructure where the oxygen contents and/or configurations are varied continuously within a specimen. This kind of the nature is recognized by a comparison between the results from slow-cooled and 500°C-quenched specimens. These specimens have almost the same contents of oxygen atoms, but the T_c(endpoint) and the probe current dependence are considerably different. This result supports the possible existence of the substructure even in the same crystallographic structure with the same average contents of oxygen atoms.

CONCLUSIONS

Quenching process can control the oxygen contents in a specimen fairly well, but the nature of n-s transition seems to be influenced by more other factors such as the distribution of oxygen contents and/or the configuration of oxygen atoms. The sensitiveness to the electron irradiations depends on these microstructural differences. The irradiations depress T_c(endpoint) at the rate of 0.5 ∿ 3.5 $K/10^{18}$ e^-/cm^2 depending on the nature of specimens and the probe current. The slow-cooled specimen with y≃6.9 is most resistive to the irradiations.

The authors would like to thank Dr. N. Shikazono and Dr. K. Ozawa for their valuable suggestions.

Table 1 : Oxygen contents estimated through X ray diffraction analysis on $Y_1Ba_2Cu_3O_y$ specimens quenched from various temperatures.

Specimen	y ($Y_1Ba_2Cu_3O_y$)
Slow-Cooled	6.87
500°C,Quenched	6.85
550°C,Quenched	6.76
600°C,Quenched	6.60
700°C,Quenched	6.35
950°C,Quenched	6.15

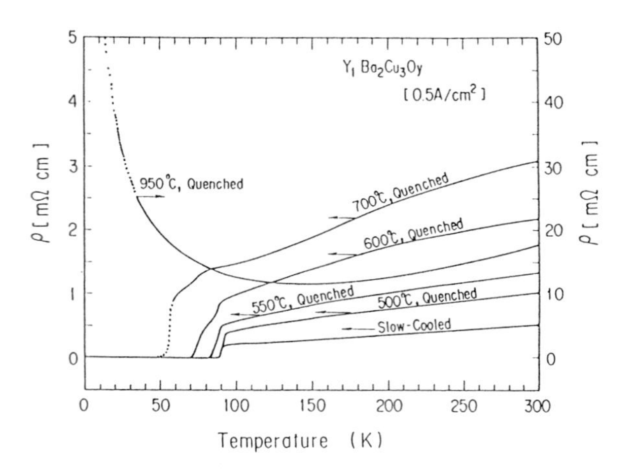

Fig. 1 : Temperature dependence of the electrical resistivity (ρ) for various kinds of $Y_1Ba_2Cu_3O_y$ specimens prepared through various processes.

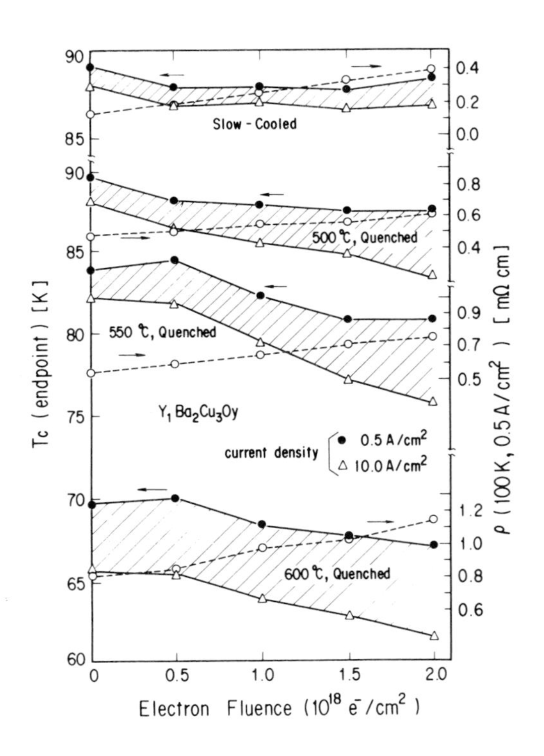

Fig. 2 : T_c's(endpoint) determined at two probe currents (0.5 and 10.0 A/cm^2) are plotted as a function of 3 MeV electron fluence for various quenched $Y_1Ba_2Cu_3O_y$ samples. Change of ρ at 100 K by electron irradiation is shown in a different scale.

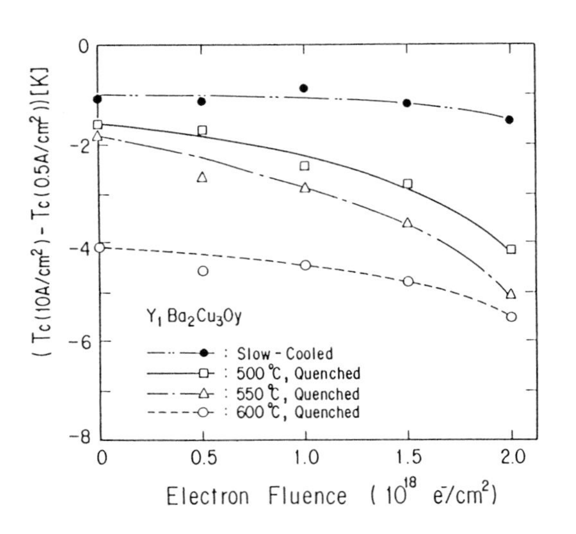

Fig. 3 : Electron fluence dependence of T_c difference at two different probe currents (0.5 and 10.0 A/cm^2) for the same specimens as in Fig. 2.

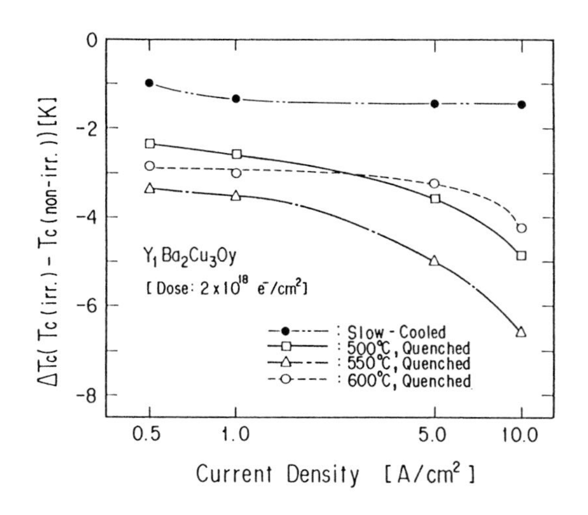

Fig. 4 : Current density dependence of T_c depression, $\Delta T_c = T_c(\mathrm{Irr.}) - T_c(\mathrm{Non\text{-}irr.})$ for various quenched $Y_1Ba_2Cu_3O_y$ specimen irradiated with 3 MeV electrons to $2x10^{18} e^-/cm^2$.

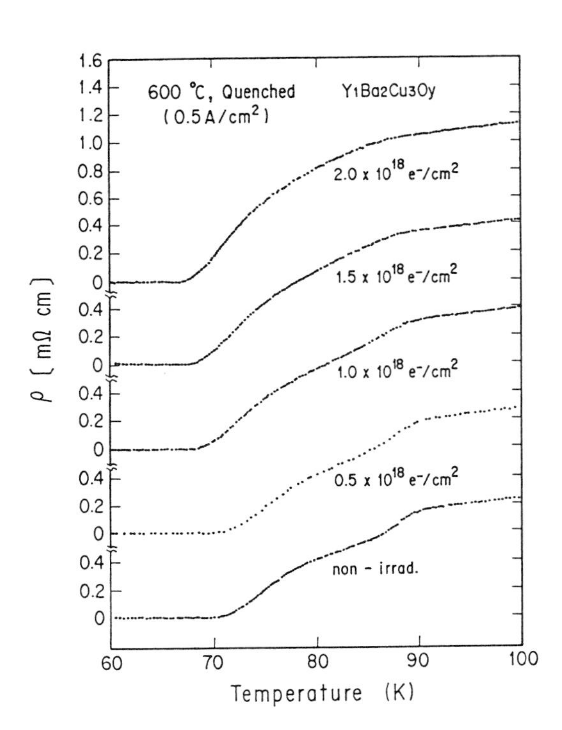

Fig. 5 : Temperature dependence of the electrical resistivity (ρ) for 600 °C-quenched $Y_1Ba_2Cu_3O_y$ specimen irradiated with 3 MeV electrons up to $2x10^{18} e^-/cm^2$ in a step of $0.5x10^{18} e^-/cm^2$.

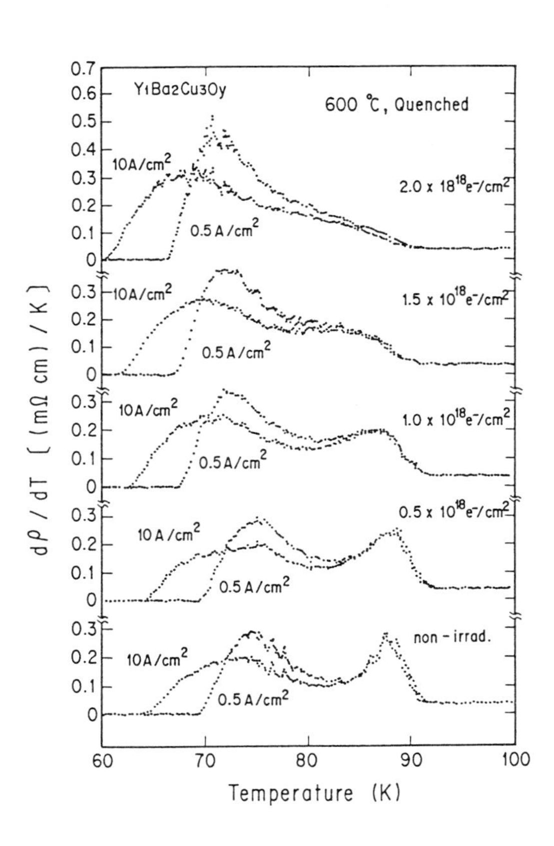

Fig. 6 : Irradiation effect on the temperature dependence of dρ/dT for two different probe currents (0.5 and 10.0 A/cm^2) in 600 °C-quenched specimen.

REFERENCES

1) J. G. Bednortz and K. A. Muller, Z. Phys. B64(1986)189.
2) M. K. Wu, J. R. Ashburn, C. J. Torng, P. H. Hor, R. L. Meng, L. Gao, Z. J. Huang, Y. Q. Wang and C. W. Chu, Phys. Rev. Lett. 58(1987)908.
3) R. J. Cava, B. Batlogg, C. H. Chang, E. A. Rietman, S. M. Zahurak and D. Werder, Phys. Rev. B36(1987)5719.
4) R. J. Cava, B. Batlogg, A. P. Ramirez, D. Weder, C. H. Chen, E. A. Rietman and S. M. Zahurak, Proc. Mat. Res. Soc. Symp. 99(1988)19.
5) G. J. Clark, F. K. LeGoues, A. D. Marwick, R. B. Laibowitz and R. Koch, Appl. Phys. Lett. 51(1987)1462.
6) G. J. Clark, A. D. Marwick, F. LeGoues, R. B. Laibowitz, R. Koch and P. Madakson, Nucl. Inst. Meth. in Phys. Res. B32(1988)405.
7) H. Matsui, S. Ito, I. Nakagawa, K. Yasuda and M. Takeda, J. J. Appl. Phys. 27(1988)L.1281.
8) M. A. Kirk, M. C. Baker, J. Z. Liu, D. J. Lam and H. W. Weber. Proc. Mat. Soc. Symp. 99(1988)209.
9) N. Moser, A. Hofmann, P. Schule, R. Henes and H. Kronmüller, Z. Phys. B71(1988)37.
10) B. Stritzker, W. Zander, F Dworschak, U. Poppe and K. Fischer, Proc. Mat. Res. Soc. Symp. 99(1988)491.
11) A. Ono, J. J. Appl. Phys. 26(1987)L.1223.

4.9 Thin Film Processing and Properties: Part 1

The Preparation, Processing and Properties of Thin and Thick Films for Microelectronic Applications

B.G. Bagley, L.H. Greene, P. Barboux, J.M. Tarascon,
T. Venkatesan, E.W. Chase, Siu-Wai Chan, W.L. Feldmann,
B.J. Wilkins, S.A. Khan, and M. Giroud

BELLCORE, 331 Newman Springs Road, Red Bank, NJ 07701, USA

ABSTRACT

In this overview of Bellcore film research, we present seven points: 1) Crystallization of the $YBa_2Cu_3O_{7-y}$ composition in the "forbidden" temperature region ($\sim 400°C$ to $\sim 650°C$) produces the polyphase microstructure because of the rapid crystallization kinetics of the competing (undesired) phases; 2) In the preparation of $YBa_2Cu_3O_{7-y}$ from BaF_2 sources, the importance of reducing the HF partial pressure and increasing the H_2O partial pressure for the reaction $BaF_2 + H_2O \rightarrow BaO + 2HF$ and the use of SiO_2 as an HF scavenger is shown; 3) Pulsed laser deposition of $YBa_2Cu_3O_{7-y}$ on bare Si at 600°C is shown to produce films with a T_c (R = 0) at 50K, the interposition of a 50nm ZrO_2 diffusion barrier increases T_c to 80K; 4) Our progress in determining the target composition for rf-magnetron and ion beam sputter deposition in order to obtain the $Bi_2Sr_2Ca_2Cu_3O_y$ phase implies a cation concentration of 5.1-1.8-2.2-3; 5) We argue that target-substrate cation differences in magnetron and ion beam sputter depositions arise from loss due to scattering in the target→film flux related to sputtered-ion velocities; 6) An aqueous-based, sol-gel process useful for the preparation of $YBa_2Cu_3O_{7-y}$ thick films and the importance of the age of the $Y(OH)_3$ component is described; 7) A glycerol solution technique produces R = 0 at 75K and 95K for the Bi- and Tl-based superconductors, respectively, and the importance of firing time and temperature in controlling cation concentration is presented.

INTRODUCTION

The discovery by Bednorz and Muller [1] of superconductivity with a transition temperature, T_c, near 40K in the La-Ba-Cu-O system led to the subsequent discoveries of superconductivity in yttrium-based materials [2] with a T_c of 90K, in bismuth-based materials [3,4] having T_c's of 10, 85 and 110K, and in thallium-based materials [5,6] having T_c's as high as 125K. It was clear that, as T_c continued to increase, these materials have a potential application in microelectronics. Two applications were immediately apparent which involved the direct replacement of present technology current-carrying lines with superconductors in order to reduce the heat generated and decrease pulse dispersion thereby resulting in higher packing densities and higher speeds. In present technology the current-carrying lines consist of thin films, which are used for on-chip interconnects, and thick films, which are used for chip-carrier interconnects. Any additional applications of these new high T_c materials as an active device component (e.g., Josephson junction) will also require quality thin films. In this paper we discuss recent Bellcore research on the preparation of thin and thick films of the high T_c oxides.

THIN FILMS

Thin films of $YBa_2Cu_2O_{7-y}$ which are prepared, or heat treated, at temperatures up to $\sim 400°C$ have amorphous structures. Preparation at $\sim 700°C$ results in the proper perovskite structure [7]. When prepared (or if the amorphous material is heat-treated) at temperatures between $\sim 400°C$ to $\sim 650°C$, a polyphase mixture results. This polyphase mixture is not superconducting and often contains, in addition to other phases, CuO, BaO and $BaCuO_2$. A possible explanation for this is that below $\sim 650°C$ the formation of other phases becomes thermodynamically feasible (if the phase diagram were known in detail this point could be examined). In this "forbidden" temperature region, therefore, the phases which form are those that have faster nucleation and growth kinetics. In general, a crystal that is chemically and structurally simple could be expected to nucleate and grow with faster kinetics. Thus crystals of CuO, BaO and $BaCuO_2$ could be expected to nucleate and grow faster than the much more chemically and structurally complex $YBa_2Cu_3O_{7-y}$. Once any undesired crystal is nucleated it drives the composition off stoichiometry making growth of the 123 phase even more unlikely. If the polyphase structure is formed, then growth of the 123 phase requires high temperature processing (e.g., 850°C) analogous to that for preparation involving a solid state reaction. In addition, when BaO is exposed to a laboratory ambient, $BaCO_3$ is formed which also requires high-temperature processing to decompose.

If a deposition is made at $\sim 400°C$ or lower (room temperature is common) then the film must pass through the forbidden region to reach temperatures appropriate for forming the proper perovskite phase. If the incorrect phases are to be avoided, then their crystallization kinetics must be slowed and one way to do this is to impede the reaction of one of the rate controlling species.

Mankiewich et. al., [8] recently discovered that the use of BaF_2 as the barium source in a multiple source thermal co-evaporation yielded more consistent results in deposition behavior. They [9] subsequently discovered that a small amount of water in the high temperature processing ambient enhanced the removal of fluorine. This technique permits traversing the forbidden temperature region without nucleating the incorrect phases, because the Ba is bound as a fluoride. This process also has the advantage that it precludes the formation of $BaCO_3$. The water is then introduced at the desired elevated temperatures. We have recently studied [10] in some detail the effect of the post deposition processing ambient on the preparation of $YBa_2Cu_3O_{7-y}$ thin films using this technique.

In Fig. 1 is shown the results of a calculation of the free energy, ΔG, for the reaction $BaF_2 + H_2O \rightarrow BaO + 2HF$ for two values of the ratio P^2_{HF}/P_{H_2O}. These two values span the processing temperature region (700-950°C) of interest where the single phase perovskite can form and where ΔG goes negative such that the BaF_2 can be converted to the oxide. There are two ways to make ΔG more negative thereby increasing the driving force for this desired reaction to proceed; one is to increase the partial pressure of water and the other is to reduce the partial pressure of HF at the film surface. A particularly convenient way to lower the HF pressure is to have silica present which reacts with the HF to form SiF_4 and H_2O. The ambient effects on the resulting film properties are shown in Fig. 2. In this figure are shown the temperature dependence of the resistance for three pieces of film grown on single crystal $SrTiO_3$ during the same deposition. Curve (a) is the resistance of a film processed at 725°C in 0.3atm H_2O and in the presence of SiO_2. Curve (b) is for a film treated at 690°C in 0.7atm H_2O also with SiO_2 present. Finally, curve (c) is for a film processed the same as that for curve (a) except that there was no SiO_2 present. The anticipated effects of the processing ambient are clearly evident.

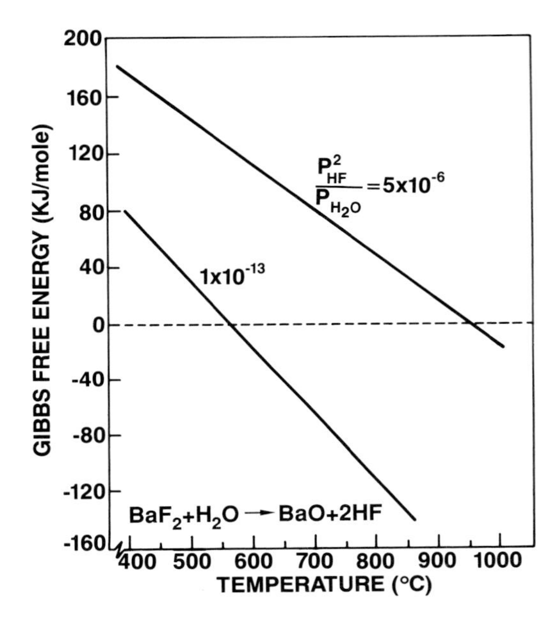

Fig. 1 Gibbs free energy vs. temperature for the reaction $BaF_2 + H_2O \rightarrow BaO + 2HF$

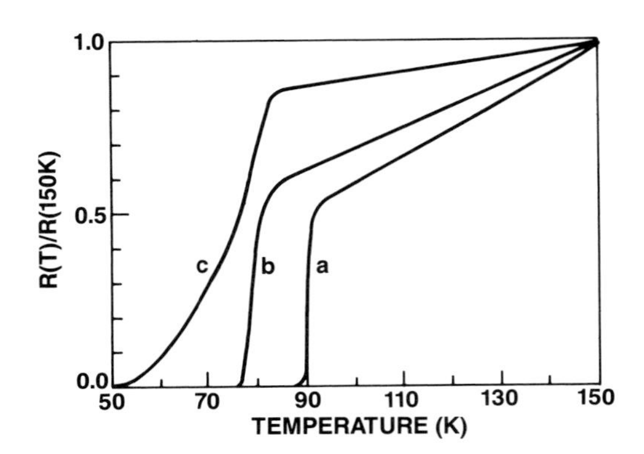

Fig. 2 Temperature dependence of the resistance (normalized to the value at 150K) for three pieces of the same film after post-deposition processing as follows: Curve (a), annealed at 725°C in 0.3atm H_2O in the presence of SiO_2; curve (b), annealed at 690°C in 0.7atm H_2O in the presence of SiO_2; curve (c), processed as in curve (a) but without SiO_2. Ref [10].

Another way to bypass the temperature region where the incorrect polyphase microstructure obtains is to deposit at a temperature above this region, and then cool through it under oxidizing conditions such that the average formal copper valence is maintained at Cu(II) or greater. One of the most stringent tests of a preparation procedure is to deposit on bare silicon, as any reaction with the substrate markedly affects the properties of the superconducting film. In Fig. 3a is shown the superconducting transition of a film of $Y Ba_2Cu_3O_{7-y}$ deposited on bare silicon using pulsed laser deposition [11]. During deposition the measured substrate temperature was 600°C and the oxygen pressure 5mTorr. A lower temperature (450°C) post-deposition anneal in 1 atm oxygen was also done. T_c (zero resistance) is observed to be 50K in this film. The interposition of a 50nm ZrO_2 layer as a diffusion barrier raises T_c ($R=0$) to 80K (Fig. 3b). Recently Berberich et al [12] have achieved a transition temperature ($R=0$) of 85K on bare silicon using thermal co-evaporation in an oxygen atmosphere. Clearly there has been much progress towards the preparation of silicon on-chip interconnects.

The recent discovery of the bismuth-based high T_c materials with transition temperatures of 85 and 110K [4] provided a new system for thin film studies, and this material is particularly attractive because of its water resistance and less sensitivity to oxygen processing [13]. In a preliminary study [14] the RF magnetron sputter deposition from a single Bi-based composite target onto $SrTiO_3$ yielded a film having a 15% resistive drop near 110K (2223 phase) and zero resistance at 80K (4424 or "4334" phase). With a target having a Bi-Sr-Ca-Cu cation ratio of 4-3-3-4, an RBS analysis of the resulting film indicated a 30% Bi deficiency.

We have now undertaken a systematic study of the target to film composition changes with time to obtain film deposition reproducibility with the 2223 composition. These films were prepared by RF magnetron sputtering from a single 5cm diameter composite target having a cation ratio of 4-3-3-5. The power to the sputter-gun was 100W and the pressure during deposition was 100 to 200mTorr using a 2:1 Ar to O_2 mixture. Film composition was determined by RBS analysis. In Fig. 4 are shown the resulting film compositions (normalized to Cu = 3) as a function of sputtering time for depositions at room temperature from the 4-3-3-5 sputter-target. As can be seen, the composition asymptotes to 1 : 1.9 : 1.6 : 3 (errors; ±0.1 : 0.1 : 0.2 : 0.1). In addition, after the target-film composition had stabilized, depositions were then made at 200, 400, 500, 600 and 700°C. We observe a decrease, as compared to room temperature, in the Bi(14%) and Cu(9%) contents at depositions over the temperature range 500 to 700°C indicating a slight decrease in the Bi and Cu sticking coefficients (this assumes that radiation from the substrate to the target produces a negligible effect). An extrapolation of these results (averaging data for all temperatures) suggests that to consistently produce films having a composition of 2-2-2-3 requires a target with a composition 5.1 : 1.8 : 2.2 : 3 (errors given above), and presputtering long enough to achieve composition reproducibility. Experiments with this target composition are now underway.

As to why the target-film compositions are different we offer the following possibility: The initial changes in film composition reflect a changing incident flux because the target surface is changing composition. When the target surface reaches equilibrium (at about 300min) then the flux leaving the target is presumed to be the average target composition, i.e., 4-3-3-5. At low temperatures (e.g., room temperature) the sticking coefficients are likely to be close to one for these oxides (also supported by the small composition change to 700°C). We do not believe that negative ion resputtering from the film is a significant factor in our magnetron-sputter system (this would be a major factor in a diode system) based on; 1) a high deposition rate in our sputter system (400Å/min for the Bi-based materials) and 2) the composition correction (windage) for preparation of $YBa_2Cu_3O_{7-y}$ materials for an ion beam sputter system (B. G. Bagley, unpublished), in which there is no negative ion resputtering, is the same (1 : 2.1 : 4.5) as for many magnetron systems. We are led to the conclusion that the component loss occurs in the path from the target to the substrate. This is related to the kinetic energy with which the atoms (or clusters) leave the target and the number of collisions they undergo (calculable) before reaching the substrate. Unfortunately, we do not know the sputtered kinetic energies for all the atoms and molecules of interest in these new high T_c oxides, but we do know that Cu is low compared to many other atoms [15] and therefore would be more easily scattered out of the target to substrate flux. This is consistent with the need for added Cu in the Y-based, and possibly the need for the extra Bi in the Bi-based, sputter targets.

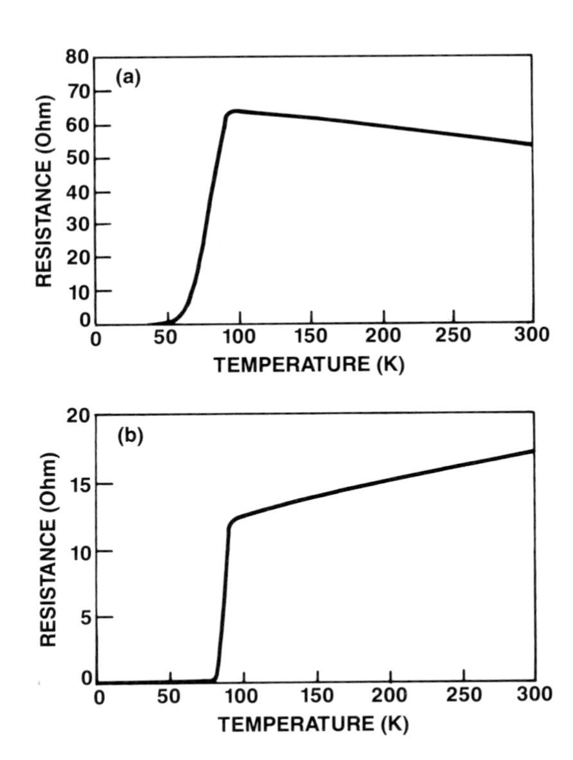

Fig. 3 Resistance vs. Temperature for $YBa_2Cu_3O_{7-y}$ films pulse laser-deposited onto Si substrates at 600°C in 5mTorr oxygen and post-annealed at 450°C for 3hr in 1atm O_2. Curves (a) and (b) are for films deposited directly on Si and on a 50nm ZrO_2 diffusion barrier, respectively. Ref. [11].

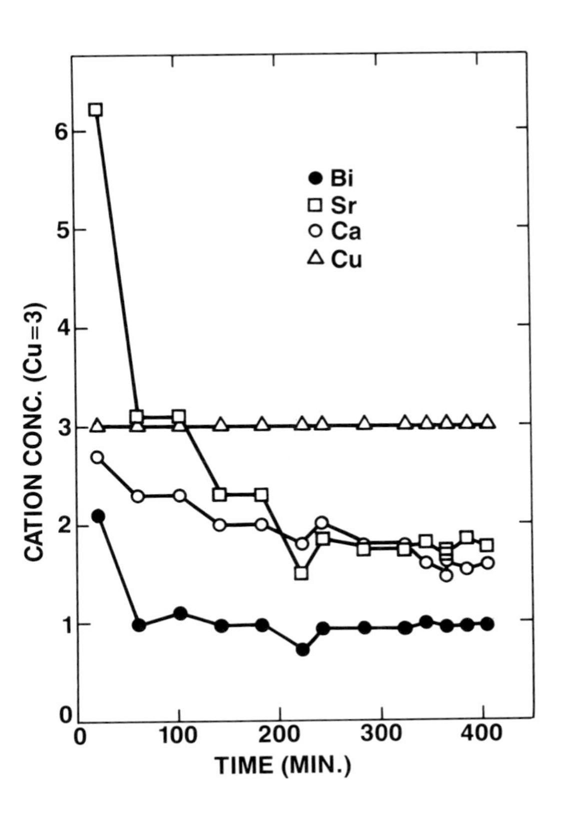

Fig. 4 Sputter-deposited film cation composition vs. sputter time from a $Bi_4Sr_3Ca_3Cu_5O_y$ target as determined by RBS analysis, normalized to Cu = 3.

THICK FILMS

Thick films for chip carrier interconnects (as an example) can be conveniently applied to the dielectric chip carrier by viscous solution techniques such as spin-, spray- or dip-coating and then fired to obtain the desired superconducting properties. For this purpose an aqueous-based solution/sol-gel preparation technique has been developed [16,17] for the Y-based high T_c materials. We believe this aqueous-based process to be generally safer, more convenient and less expensive than an alkoxide-based process.

A solution which is a precursor for the 90K material is prepared [16,17] by mixing a colloidal suspension of yttrium hydroxide with appropriate amounts of barium and copper acetates at a pH close to 7. The yttrium hydroxide is prepared by passing yttrium nitrate through a commercial anionic resin to convert $Y(NO_3)_3$ to colloidal $Y(OH)_3$. The viscosity of the precursor solution increases within a few hours and a blue viscous gel forms which, if dried, produces a blue glassy-like material. This material can then be fired at 850°C to produce small particles ($\sim 1\mu m$) with a narrow size dispersion. Pressing and sintering this powder produces a dense (90%) bulk ceramic having a T_c of 92K (Fig 5a) and a ΔT_c (defined as 10%-90% of the resistive transition) of 0.6K. In these experiments low cost, reagent grade materials were used and yet this is one of the sharpest transitions measured on a bulk material. The depression in T_c (2K) observed in this material, as compared to materials synthesized by the solid state reaction of very high purity starting materials, we attribute to impurities and not to the differences in preparation procedure.

If the viscous precursor solution, instead of being processed to form a bulk sample, is applied to a substrate and fired, then a superconducting thick film is made. The 10K depression in the T_c of this thick film (Fig. 5b) we attribute to an interaction with the substrate, as the properties of the bulk from the same precursor solution are well defined. Because of the interaction with the substrate, films have to be fired a short time at temperatures above 900°C in order to get superconducting behavior in the transport properties. Also, barium carbonate is formed during the heating process which prevents good sintering and thereby good critical currents being achieved, even though there is a preferred orientation of the grains parallel to the substrate within the films.

In the course of this work on the synthesis of the $Yba_2Cu_3O_{7-y}$ materials via sol-gel techniques one processing parameter that we observed to be important was the "age" of the colloidal yttrium hydroxide ; "age" being the length of time between when it is prepared and when it is used in the precursor solution. A rheological study [18] has examined this effect in detail. In Fig. 6 are shown the elastic (G') and loss (G") moduli as a function of time for two different precursor solutions using yttrium hydroxide solutions two weeks (fresh) and ten weeks (aged) old. The viscous contribution, G", is

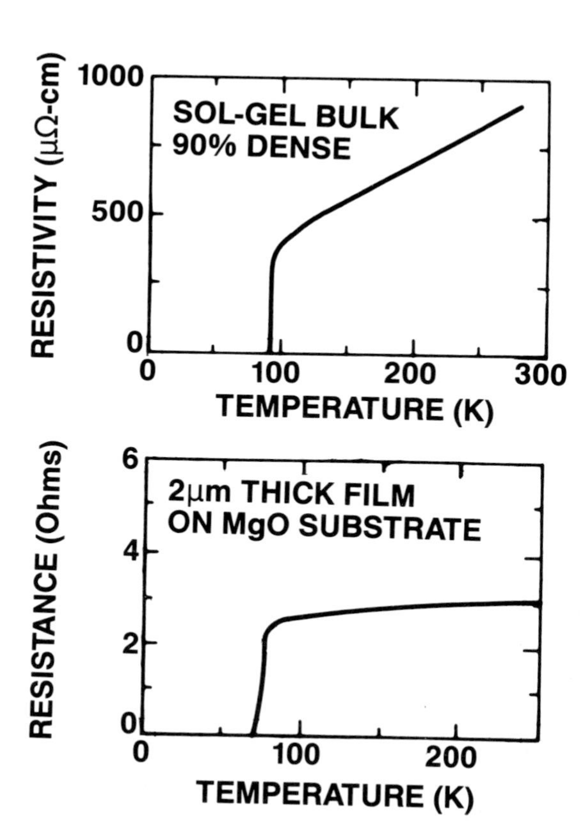

Fig. 5 The resistivity of a bulk $YBa_2Cu_3O_{7-y}$ ceramic and the resistance of a thick film, both prepared by a sol-gel process, are shown as a function of temperature.

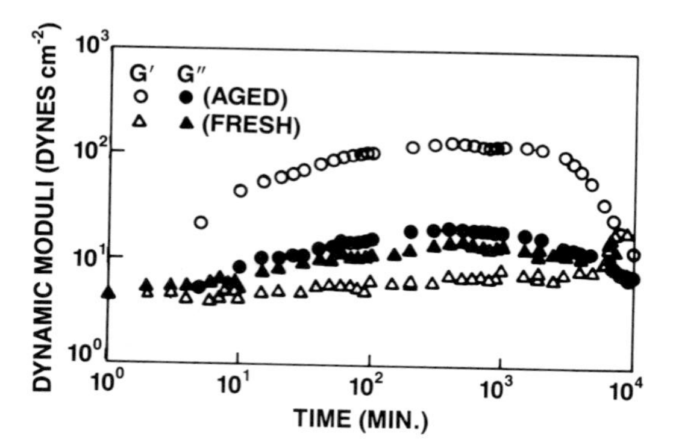

Fig. 6 Elastic (G') and loss (G") moduli of two different precursor solutions (for the preparation of $YBa_2Cu_3O_{7-y}$) as a function of time. Ref. [18].

essentially the same for both solutions but the elastic component, G', is significantly different. It is concluded that the solution made with the "aged" yttrium hydroxide has a more gel-like character and yields better fired thick-film superconducting properties. This solution, however, has a shorter shelf life. The mechanism which has been proposed [19] to account for these differences is that, upon aging, colloidal yttrium hydroxide particles grow due to Ostwald ripening. When used in superconducting precursor gels these particles, depending upon their grain size, yield a different degree of mixing of the copper and yttrium. Older yttrium hydroxide gels produce a microheterogeneity that results in gels that are more elastic and contain more unstable copper hydroxide, but are more reactive upon firing and thus give the correct superconducting phase.

We have also investigated the preparation of thick films in the systems Bi-Sr-Ca-Cu and Tl-Ba-Ca-Cu. Cooper et al.[20] have successfully prepared superconducting thick films of the Bi-based cuprates via the decomposition of aqueous solutions containing nitrates of the elements. We have investigated [21] this preparation procedure and applied it to the thallium system. We found that dissolving the constituent nitrates in a water-glycerol solution increased the reaction rate in comparison to a nitrate-aqueous solution. Substrates are coated prior to heat treating, either by dipping or by spraying surfaces heated to 200°C. In addition to substrate interaction problems which require short firing times, short times are also required to minimize the loss of Bi (Tl). The use of the glycerol-based solution produces the correct superconducting phase within 2-3 minutes before reaction with the substrate occurs, or too much of the constituent Bi (Tl) is lost, during the firing cycle. Fig. 7 illustrates the effect of firing time on the resistivity behavior for a film heat-treated for 10min at 875°C (Curve a) which has a lower onset temperature than the same material after only a 2min annealing (Curve b). This is explained by a loss of bismuth, as was confirmed by Rutherford Back Scattering. However, because of extensive thallium evaporation, the thallium phases cannot be obtained unless the films are fired in the presence of an excess thallium vapor pressure in a sealed capsule. However, thallium loss cannot be completely prevented. Starting with a 2223 mixture and firing for 5min results in only the 1223 phase being observed in the x-ray diffraction pattern (Fig. 8a). A shorter time (2min) is required to keep enough thallium in the film to grow the 2223 phase, as shown in Fig. 8b. An interesting result is that this 1223 phase is easy to obtain when prepared from a fired solution, whereas difficulties are encountered using oxides and a solid state reaction procedure. Since the 1-5 μm thick films are composed of platelets a few microns in diameter, the natural tendency is for the crystals to orient parallel to the substrate with their c axis normal. This explains the textured x-ray patterns as shown in Fig. 8.

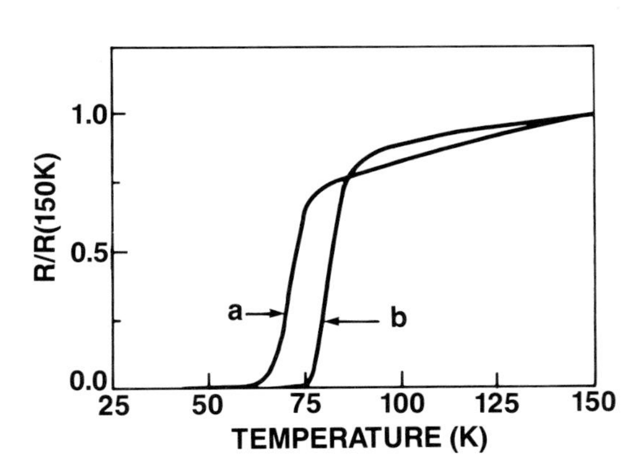

Fig. 7 Temperature dependence of the resistance (normalized to the value at 150K) of a Bi-based thick film on a $SrTiO_3$ substrate heated for 10min at 875°C and air-quenched (Curve a) or 2min at 875°C and air quenched (Curve b). Ref. [21].

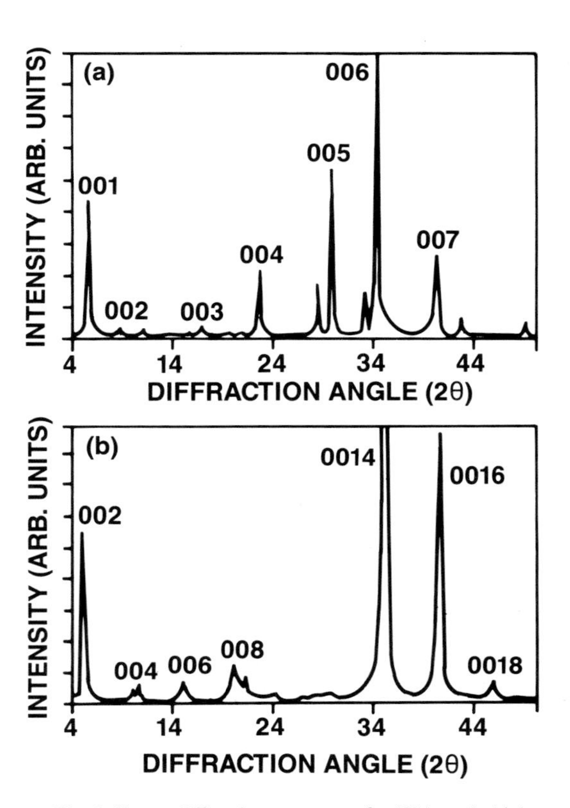

Fig. 8 X-ray diffraction pattern of a Tl-based thick film obtained by spraying a glycerol solution of 2223 starting composition onto a $SrTiO_3$ substrate and firing at 925°C for 5min (Curve a) where only 1223 is observed or 2min where only 2223 is observed. Ref. [21].

Bi-based films show an onset temperature at 85K and zero-resistance at 75K (Fig. 7b) whereas the Tl-films show an onset temperature of 105K and zero-resistance at 95K (Fig. 9). The critical currents obtained to date are quite low ($\sim 50A/cm^2$ at 77K for the thallium phase). Both the 10K depression in T_c (compared to bulk phases obtained by solid state reactions) and the low observed critical currents can be explained by the very short firing times (2min) not allowing enough time for good grain sintering to take place.

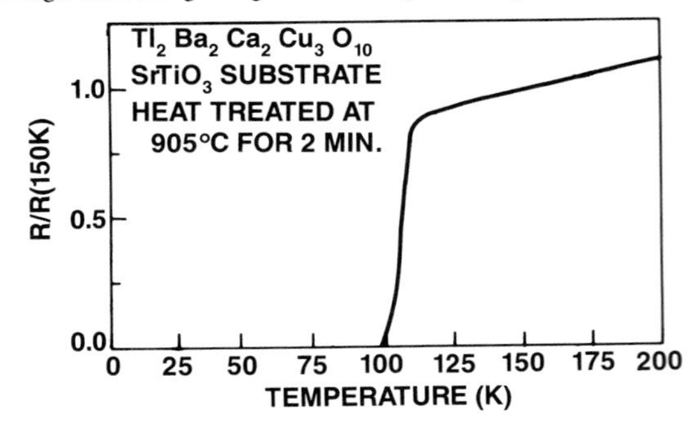

Fig 9 Temperature dependence of the resistance (normalized to the value at 150K) of a Tl-based thick film obtained by a solution technique and fired at 905°C for 2min Ref[21].

ACKNOWLEDGEMENTS

We wish to thank S. Gupta, J. M. Rowell and J. H. Wernick.

REFERENCES

1. J. G. Bednorz and K. A. Muller, Z. Phys. B64, 189 (1986).
2. M. K. Wu, J. R. Ashburn, C. S. Torng, P. H. Hor, R. L. Meng, L. Gao, Z. J. Huang, Y. Q. Wang, and C. W. Chu, Phys. Rev. Lett. 68, 908 (1987).
3. C. Michel, M. Hervieu, M M. Borel, A. Grandin, F. Deslandes, J. Provost and B. Raveau, Z. Phys. B58, 421 (1987).
4. H. Maeda, Y. Tanaka, M. Fuksutumi, and T. Asano, Japan. J. Appl. Phys. 27, L209 (1988).
5. Z. Z. Sheng and A. M. Hermann, Nature 332, 55 (1988); ibid 332, 138 (1988).
6. S. S. P. Parkin, V. Y. Lee, A. I. Nazzal, R. Savoy, R. Beyers and S. J. La Placa, Phys. Rev. Lett. 61, 750 (1988).
7. Y. Enomoto, T. Murakami, M, Suzuki and K. Morawaki, Japan. J. Appl. Phys. 26, L1248 (1987).
8. P. M. Mankiewich, J. H. Scofield, W. J. Skocpol, R. E. Howard, A. H. Dayem and E. Good, Appl. Phys. Lett. 51, 1753 (1987).
9. P. M. Mankiewich, R. E. Howard, W. J. Skocpol, A. H. Dayem, A. Ourmazd, M. G. Young, and E. Good, Mat. Res. Soc. Sym. Proc. Vol. 99, 119, (1988).
10. Siu-Wai Chan, B. G. Bagley, L. H. Greene, M. Giroud, W. L. Feldman, K. R. Jenkin, II and B. J. Wilkins, Appl. Phys. Lett., submitted (1988).
11. T. Venkatesan, E. W. Chase, X. D. Wu, A. Inam, C. C. Chang, and F. K. Shokooki, Appl. Phys. Lett. 53, 243 (1988).
12. P. Berberich, J. Tate, W. Dietsche, and H. Kinder, preprint (1988).
13. J. M. Tarascon, Y. Le Page, P. Barboux, B. G. Bagley, L. H. Greene, W. R. McKinnon, G. W. Hull, M. Giroud and D. M. Hwang, Phys. Rev. B37, 9382 (1988).
14. L. H. Greene, W. L. Feldmann, Siu-Wai Chan, B. G. Bagley, M. Giroud, B. A. Wilkins, L. A. Farrow, D. M. Hwang, J. M. Tarascon and P. Barboux, Materials Research Society Fall Meeting, Boston, Mass., 1988.
15. G. K. Wehrner and G. S. Anderson, in "*Handbook of Thin Film Technology*", edited by L. I. Maissel and R. Glang, (McGraw-Hill, New York, 1970), p. 3-20.
16. P. Barboux, J. M. Tarascon, L. H. Greene, G. W. Hull and B. G. Bagley, J. Appl. Phys. 63, 2725 (1988).
17. P. Barboux, J. M. Tarascon, B. G. Bagley, L. H. Greene, G. W. Hull, B. W. Meagher and C. B. Eom, Mat. Res. Soc. Sym., Proc. Vol. 99, 49 (1988).
18. S. A. Khan, P. Barboux, B. G. Bagley and J. M. Tarascon, Appl. Phys. Lett., in press (1988).
19. S.A. Khan, P. Barboux, B.G. Bagley and F.E. Torres, to be published.
20. E.I. Cooper, E.A. Giess and A. Gupta, Mat. Lett. submitted (1988).
21. P. Barboux, J. M. Tarascon, F. Shokooki, B. J. Wilkins and C. L. Schwartz, J. Appl. Phys., in press (1988).

Basic Thin Film Processing for Rare-Earth-Free High-Tc Superconductors

K. WASA, H. ADACHI, Y. ICHIKAWA, K. HIROCHI, and K. SETSUNE

Central Research Laboratories, Material Science Laboratory, Matsushita Electric Co., Ltd., 3-15, Moriguchi, 570 Japan

ABSTRACT

This paper describes the basic thin film processing for the Bi and/or Tℓ system high-Tc superconductors with controlled numbers of the Cu-O layer, i.e. superconducting phase control. In a conventional deposition process the numbers of the Cu-O layer are controlled by a choice of the deposition condition and/or the postannealing condition. However, there exists a limitation to a fine control of the Cu-O layer numbers. It is found that the layer by layer deposition by a multi-target sputtering achieves the fine control of the Cu-O layer numbers, i.e. the superconducting phase control.

INTRODUCTION

Much attention has been paid to the rare-earth-free high-Tc oxide superconductors such as the Bi-Sr-Ca-Cu-O [1] and Tℓ-Ba-Ca-Cu-O [2]. Recent studies suggest that these rare-earth-free superconductors show a layered structure comprizing Bi-O and Cu-O layer and the critical temperature Tc varies with the numbers of the Cu-O layer. The $Bi_2Sr_2CaCu_2O_x$ comprising two layers of the Cu-O exhibits Tc ≃ 80 K (2-2-1-2 structure) and the $Bi_2Sr_2Ca_2Cu_3O_y$ comprizing three layers of the Cu-O exhibits Tc ≃ 110 K (2-2-2-3 structure). Further increase of the Cu-O layers is expected to increase the Tc.

The thin films of the Bi and/or Tℓ system were prepared by a conventional deposition process including electron beam deposition, sputtering, and chemical vapour deposition [3], [4], [5]. However, there exists a limitation in a fine control of the Cu-O layer. In this paper first we consider the basic thin film processing for controlling the numbers of the Cu-O layers in the conventional deposition process. Secondly we describe a layer by layer deposition for a fine control of the Cu-O layer.

BASIC THIN FILM PROCESSING

Thin film processing for the rare-earth high-Tc superconductors are classified into three processes: (1) deposition at a low substrate temperature followed by a postannealing at around 900°C; (2) deposition at a crystallizing temperature of 600-800°C, followed by the postannealing; (3) deposition at the crystallizing temperature under oxydizing atmosphere [6].

The thin film processes (1) and/or (2) are used for the deposition of the rare-earth-free high-Tc superconductors. Sputtering and electron beam deposition were widely used for the deposition of the thin films. However, the resultant films often showed mixed phases comprizing 2-2-1-2 and 2-2-2-3 structure.

PHASE CONTROL IN A CONVENTIONAL PROCESS

3-1) Bi-Sr-Ca-Cu-O thin films

Thin films of the Bi system are prepared by a conventional rf-planar magnetron sputtering. Typical sputtering conditions are shown in table 1. The target is complex oxides of Bi-Sr-Ca-Cu-O.

Table 1. Sputtering conditions.

Target	Bi:Sr:Ca:Cu:=1-1.7:1:1-1.7:2	100 mm in diameter
Sputtering gas	Ar/O_2=1-1.5	
Gas pressure	0.5 Pa	
rf input power	150 W	
Substrate temperature	200-800 °C	
Growth rate	80 Å/min	

The composition is around 1-1-1-2 ratio of Bi-Sr-Ca-Cu. The processes (1) and/or (2) are used for the deposition. Single crystals of (100) MgO are used as the substrates. The superconducting properties are improved by the postannealing at 850 - 900°C in 5 hr in O_2 [7].

The experiments suggested that the superconducting properties were strongly affected by the substrate temperature during the deposition. Fig. 1 shows typical X-ray diffraction patterns with resistivity-temperature characteristics for the Bi-Sr-Ca-Cu-O thin films of around 0.4 μm thickness deposited at various substrate temperatures. It is seen that the films deposited at 200°C exhibit $Bi_2Sr_2CaCu_2O_x$ structure with the lattice constant C ≃ 30 Å which corresponds to the low Tc phase. The films show the zero resistance temperature of ≃ 70 K [Fig. 1(a)].

When the substrate temperature is raised during the deposition the high Tc phase with Tc ≃ 110 K, the $Bi_2Sr_2Ca_2Cu_3O_x$ structure with the lattice constant C ≃ 36 Å, is superposed on the X-ray diffraction pattern [Fig. 1(b)]. At the substrate temperature of around 800°C a single high-Tc phase is observed. The films show the zero resistance temperature of ≃ 104 K [Fig. 1(c)].

These experiments suggest that the superconducting phases of the Bi-Sr-Ca-Cu-O thin films are controlled by the substrate temperature during the deposition.

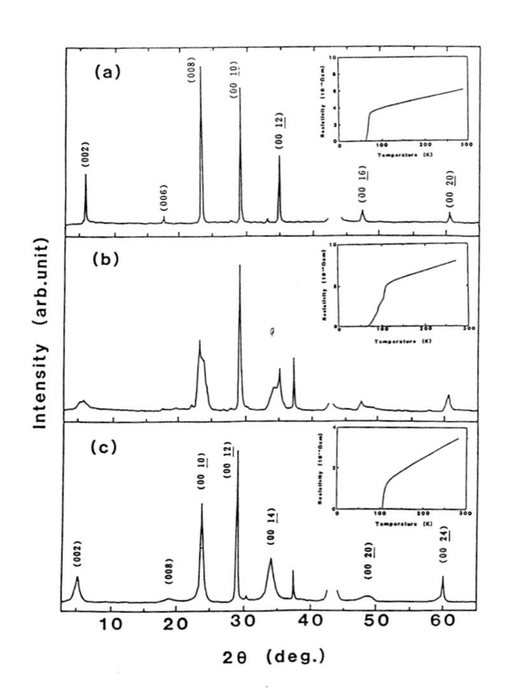

Fig. 1. X-ray diffraction patterns with resistivity versus temperature for the annealed Bi-Sr-Ca-Cu-O films.

3-2) Tℓ-Ba-Ca-Cu-O thin films

Similar to the Bi-Sr-Ca-Cu-O system, thin films of Tℓ-Ba-Ca-Cu-O system are prepared by the rf-magnetron sputtering on MgO substrates. Typical sputtering conditions are shown in Table 2. However, their chemical composition is quite unstable during the deposition and the postannealing process due to the high vapour pressure of Tℓ. Thin films of the Tℓ system are deposited without intentional heating of substrates (∿ 200°C) and annealed at 890 - 900°C in Tℓ vapour [8]. It is found that the superconducting phase of the resultant films strongly depends on the postannealing conditions.

Table 2. Sputtering conditions.

Target	Tl:Ba:Ca:Cu:=2:1-2:2:3	100 mm in diameter
Sputtering gas	$Ar/O_2=1$	
Gas pressure	0.5 Pa	
rf input power	100 W	
Substrate temperature	200 °C	
Growth rate	70 Å/min	

Fig. 2 shows typical X-ray diffraction patterns with resistivity-temperature characteristics for the Tℓ-Ba-Ca-Cu-O thin films annealed at different conditions.

The 0.4 μm thick film exhibits the low temperature phase, $Tℓ_2Ba_2CaCu_2O_x$ structure, with the lattice constant C ≃ 29 Å after slight annealing at 900°C for 1 min [Fig. 2(a)]. The 2 μm thick films heavily annealed at 900°C for 13 min show the high temperature phase, $Tℓ_2Ba_2Ca_2Cu_3O_x$ structure, with the lattice constant C ≃ 36 Å [Fig. 2(b)]. In the specific annealing condition the other superconducting phase $TℓBa_2Ca_3Cu_4O_x$ structure with the lattice constant C ≃ 19 Å is also obtained [Fig. 2(c)] [9]. These experiments suggest that the superconducting phase of the Tℓ-Ba-Ca-Cu-O thin films is controlled by the choice of the annealing conditions.

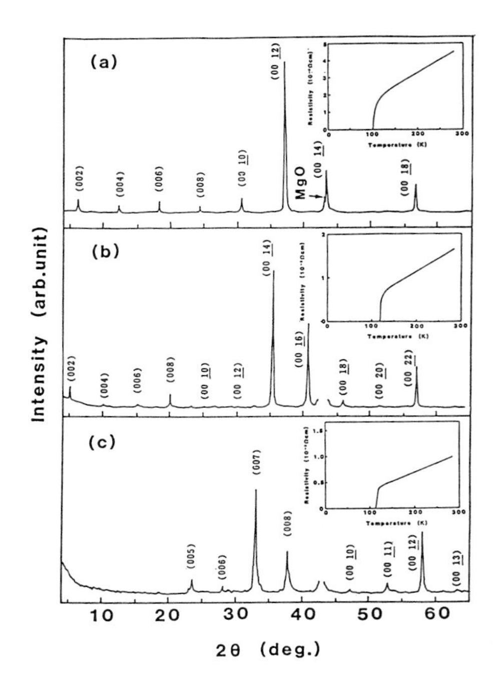

Fig. 2. X-ray diffraction patterns with resistivity versus temperature for the annealed Tl-Ba-Ca-Cu-O films.

DISCUSSIONS AND PHASE CONTROL LAYER-BY-LAYER DEPOSITION

The microstructure of these rare-earth-free superconducting thin films comprises the different superconducting phases, although the X-ray diffraction pattern and/or the resistivity-temperature characteristics may show the single superconducting phase. For instance TEM image suggests the sputtered Bi-Sr-Ca-Cu-O thin films with the zero resistivity temperature of 104 K (which corresponds to the Tc in the 2-2-2-3 structure) comprize 2-2-1-2, 2-2-2-3, 2-2-3-4, and 2-2-4-5 structures. The presence of the mixed phases is also confirmed by the spreading skirt observed in the X-ray diffraction pattern at the low angle peak around $2\theta \simeq 4°$.

It is reasonably considered that the presence of the mixed phases results from the specific growth process of the present rare-earth-free superconducting thin films: The rare-earth-free superconducting thin films may be molten during the annealing process. The superconducting phase will be formed during the cooling cycle.

Table 3. Sputtering conditions.

Target	Bi, two SrCu and CaCu
	60 mm in diameter
Substrate	(100) plane of MgO
	10 mm x 10 mm
Target-substrate spacing	100 mm
Sputtering gas	Ar + O_2 (5:1)
Gas pressure	3 Pa
Input power	Bi: 5-6 W
	SrCu: 9-10 W
	CaCu: 30-60 W

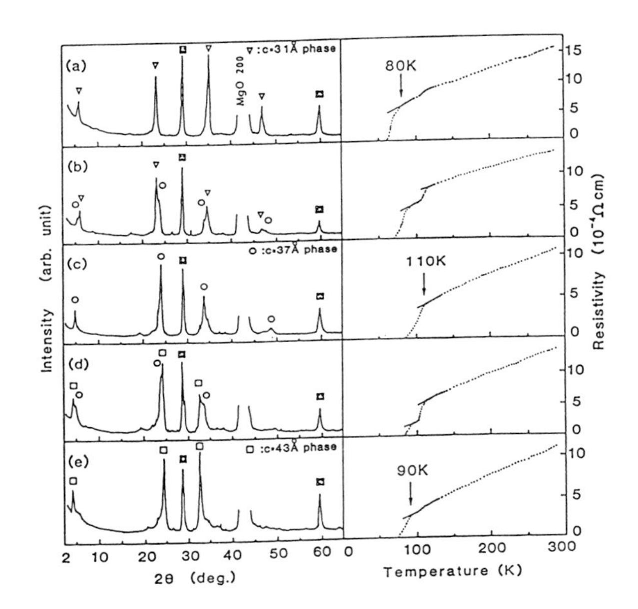

Fig. 3. X-ray diffraction patterns and temperature dependence of resistivity for the annealed Bi-Sr-Ca-Cu-O films with various c-axis lattice spacing.

XPS measurements for the crystallized Bi-Sr-Ca-Cu-O films suggest that the annealing process modifies the crystal structure near the $Cu-O_2$ layer, increase of density of Cu^{3+}. The Bi-O layered structure is stable during the annealing [10]. This implies that the single superconducting phase will be synthesized when the Bi-O basic structure is crystallized and the stoicheometric composition is kept for the unit cell of the Bi-Sr-Ca-Cu-O.

These considerations have been confirmed by the experiments: The layer by layer deposition was conducted by the multi-target sputtering system. Typical sputtering conditions are shown in Table 3. The deposition rate is selected so as to pile up the Bi-O, Sr-O, Cu-O, and Ca layer in an atomic scale range. The substrate temperature was kept around crystallizing temperature of 650°C. Fig. 3 shows the typical results for the layer by layer deposition. It is noted that the phase control is achieved simply by the amounts of Cu-Ca-O during the layer by layer deposition. The experiments show that the Tc does not increase monotonously with the numbers of the Cu-O layer. In the Bi bi-layer system the Tc shows maximum, 110 K, at three layers of Cu-O, $Bi_2Sr_2Ca_2Cu_3O_x$. At the four layers of Cu-O, $Bi_2Sr_2Ca_3Cu_4O_x$, Tc becomes 90 K [11].

CONCLUSIONS

The present layer by layer deposition is one of the most promising process for a fine control of the superconducting phase of the rare-earth-free superconductors. Man-made high-Tc superconductors will be synthesized by the layer by layer deposition.

ACKNOWLEDGEMENTS

The authors thank S. Kohiki for his XPS analyses. They also thank S. Hayakawa and T. Nitta for their continuous encouragements.

REFERENCES

[1] H. Maeda, Y. Tanaka, M. Fukutomi, and T. Asano, Jpn. J. Appl. Phys. 27, L209 (1988).

[2] Z.Z. Sheng and A.M. Hermann, Nature 332, 138 (1988).

[3] H. Adachi, K. Wasa, Y. Ichikawa, K. Hirochi, and K. Setsune, J. Cryst. Growth, 91 (1988) 352.

[4] Y. Ichikawa, H. Adachi, K. Hirochi, K. Setsune, S. Hatta, and K. Wasa, Phys. Rev. B., 38 (1988) 765.

[5] J.H. Kang, R.T. Kampwirth, K.E. Gray, S. Marsh, and E.A. Huff, Physics Lett., 128 (1988) 102.

[6] K. Wasa, M. Kitabatake, H. Adachi, K. Setsune, and K. Hirochi, American Institute of Physics Conference Proceedings No.165, New York 1988, p.38.

[7] Y. Ichikawa, H. Adachi, K. Hirochi, K. Setsune, and K. Wasa: Proc. MRS Int. Meeting on Advanced Materials, June 1988, Tokyo (in press).

[8] Y. Ichikawa, H. Adachi, K. Setsune, S. Hatta, K. Hirochi, and K. Wasa, Appl. Phys. Lett., 53 (1988) 919.

[9] J. Zhou, Y. Ichikawa, H. Adachi, T. Mitsuyu, and K. Wasa, Jpn. J. Appl. Phys., (to be submitted).

[10] S. Kohiki, K. Hirochi, H. Adachi, K. Setsune, and K. Wasa, Phys. Rev. B, 38 (1988) (in press).

[11] H. Adachi, S. Kohiki, K. Setsune, T. Mitsuyu, and K. Wasa, Jpn. J. Appl. Phys., 27 (1988) (in press).

High-Tc Superconductor Prepared by Chemical Vapor Deposition

KAZUHIKO SHINOHARA, FUMIO MUNAKATA, and MITSUGU YAMANAKA

Materials Research Laboratory, Central Engineering Laboratories, Nissan Motor Co., Ltd., 1, Natsushima-cho, Yokosuka, 237 Japan

ABSTRACT

Y-Ba-Cu-O films were prepared by chemical vapor deposition using tris(2,2,6,6-tetramethyl-3,5-heptanedionato)yttrium(III) ($Y(DPM)_3$), bis(1,1,1,5,5,5-hexafluoro-2,4-pentanedionato)barium(II) ($Ba(HFA)_2$) and bis(1,1,1,5,5,5-hexafluoro-2,4-pentanedionato)copper(II) ($Cu(HFA)_2$). A typical Y-Ba-Cu-O film on a (001) $SrTiO_3$ substrate exhibited the presence of the $YBa_2Cu_3O_y$ phase after an annealing process according to X-ray diffraction patterns, and a superconducting transition was obtained with T_c(onset) = 83 K and T_c(zero resistance) = 65 K.

INTRODUCTION

Since the discovery of high-T_c superconductors, unprecedented effort has been made by many researchers to explore both the physical mechanism and its technological potential. The use of superconducting oxide in the development of microelectronic devices requires preparation of these materials in thick and thin film forms. Recently, thin film high-T_c superconductors have been prepared by magnetron sputtering[1], pulse laser evaporation[2], electron beam evaporation[3], and molecular beam epitaxy[4]. However, high temperature treatment of the deposited film is important to obtain high-J_c. Low temperature (< 873 K) process has been established for the preparation of Y-Ba-Cu-O films by activated reactive evaporation[5] and plasma-assisted laser deposition[6].

The growth of superconducting films by chemical vapor deposition (CVD) was also reported[7-10], however, the electrical properties of these films were inferior to those of films prepared by other physical vapor deposition techniques. It is important to determine the thermal properties of the precursors in order to prepare films by CVD. The thermal properties of the precursors have not yet been determined sufficiently for the preparation of high-T_c superconductors. We have reported results of thermogravimetry (TG) and differential thermal analysis (DTA) for tris(2,2,6,6-tetramethyl-3,5-heptanedionato)yttrium(III) ($Y(DPM)_3$; $Y(C_{11}H_{19}O_2)_3$), bis(1,1,1,5,5,5-hexafluoro-2,4-pentanedionato)barium(II) ($Ba(HFA)_2$; $Ba(C_5HF_6O_2)_2$) and bis(1,1,1,5,5,5-hexafluoro-2,4-pentanedionato)copper(II) ($Cu(HFA)_2$; $Cu(C_5HF_6O_2)_2$)[11].

In this paper, we report the preparation of Y-Ba-Cu-O films by CVD using ($Y(DPM)_3$), ($Ba(HFA)_2$), and ($Cu(HFA)_2$) as the sources of Y, Ba, and Cu, respectively, and describe the characteristics of these films.

EXPERIMENTAL PROCEDURE

Figure 1 shows a schematic diagram of the CVD system used in this study. Y-Ba-Cu-O thin films were prepared on (001) $SrTiO_3$ and (001) MgO substrates (10×10 mm^2) by CVD using $Y(DPM)_3$-$Ba(HFA)_2$-$Cu(HFA)_2$-O_2 reaction gas system. Each vaporizer was heated up to an appropriate temperature based on the results of TG and DTA of each precursor [11]. The gas line and the mixing tube were heated up to higher temperatures than that of the vaporizer to prevent condensation of each precursor, but their temperatures were kept lower than the decomposition point of each precursor. Each precursor was carried by Ar into the mixing tube and mixed with O_2. The mixed gas was introduced into the quartz reactor heated by an electric furnace. The substrates were put on a quartz susceptor in the reactor. The substrate temperature was monitored by a CA thermocouple. In order to detect condensation and decomposition of each precursor, the vaporizer, the gas line, and the mixing tube were made of glass.

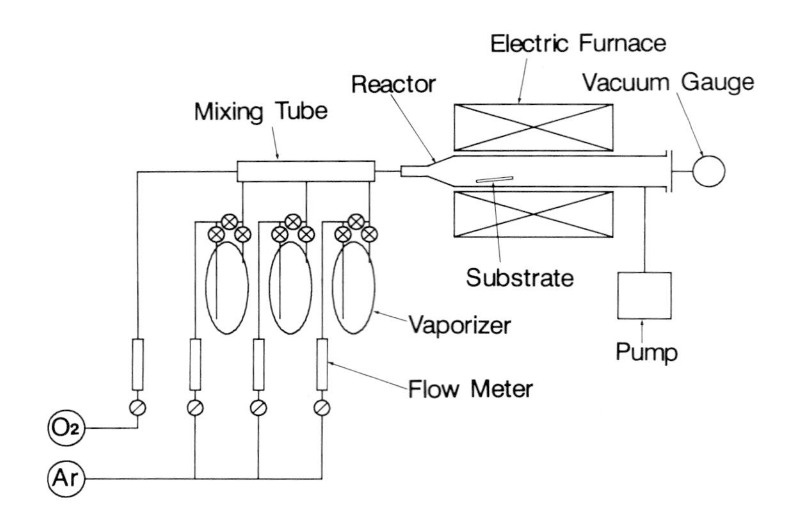

Fig. 1. Schematic diagram of a chemical vapor deposition system.

Deposited films were annealed at 1123 K for 180 min, cooled to 773 K at a rate of 2 K/min, annealed at 773 K for 180 min, and finally cooled to room temperature at the same cooling rate. This annealing was performed in air.

Resistance measurements were carried out for the annealed films using a DC Voltage Current Source/Monitor (ADVANTEST TR6143). Electrical connections to the films were made with four probe silver paint (DUPONT 4929) contacts. In the measurement, the effect of thermoelectric voltage caused by the inhomogeneity of temperature over the sample was eliminated by averaging the data obtained by reversing the current direction. Temperatures below 77 K were obtained by evacuation of liquid nitrogen.

RESULTS AND DISCUSSION

In order to optimize the deposition condition, we prepared the film using each precursor based on the results of TG and DTA of each precursor. According to the analytical results[11], $Y(DPM)_3$ starts to sublime at 416 K and reacts with O_2 above 663 K, $Ba(HFA)_2$ starts to sublime at 493 K and decomposes above 533 K, and $Cu(HFA)_2$ starts to sublime at 330 K and decomposes above 593 K. The deposition condition was summarized in Table I. The X-ray diffraction (XRD) measurement was carried out for the as-deposited films prepared from an individual precursor in order to estimate the structure of the films using CuK_α radiation. Oxides of yttrium (Y_2O_3) and copper (CuO and Cu_2O) were easily deposited. Oxide of barium was not deposited but BaF_2 was made from $Ba(HFA)_2$. The nature of the vapor of each precursor has not sufficiently clarified, however, it is obvious that Y, Ba, and Cu were carried from the vaporizers to the reactor.

Based on the above results, a typical deposition condition of Y-Ba-Cu-O films is listed in Table II. The deposition rate was about 5.5 nm/min and the thicknesses of the samples were about 1000 nm. The as-deposited films were amorphous and insulating, however, they became superconducting after a thermal annealing process.

Table I. Structure of films prepared from individual precursors.

Precursor		$Y(DPM)_3$	$Ba(HFA)_2$	$Cu(HFA)_2$
Temperature of Vaporizer	(K)	423	493	338
Flow Rate of Carrier Gas	(cc/min)	100	500	20
Flow Rate of O_2	(cc/min)	3000	3000	3000
Temperature of Substrate	(K)	873	873	873
Temperature of Gas Line	(K)	513	513	513
Temperature of Mixing Tube	(K)	513	513	513
Gas Pressure	(Pa)	1330	1330	1330
Structure		Y_2O_3	BaF_2	CuO, Cu_2O

Table II. Deposition condition of Y-Ba-Cu-O films.

Precursor		$Y(DPM)_3$	$Ba(HFA)_2$	$Cu(HFA)_2$
Temperature of Vaporizer	(K)	423	493	338
Flow Rate of Carrier Gas	(cc/min)	100	500	20
Flow Rate of O_2	(cc/min)		3000	
Substrate			(001) $SrTiO_3$, (001) MgO	
Temperature of Substrate	(K)		873	
Temperature of Gas Line	(K)		513	
Temperature of Mixing Tube	(K)		513	
Gas Pressure	(Pa)		1330	

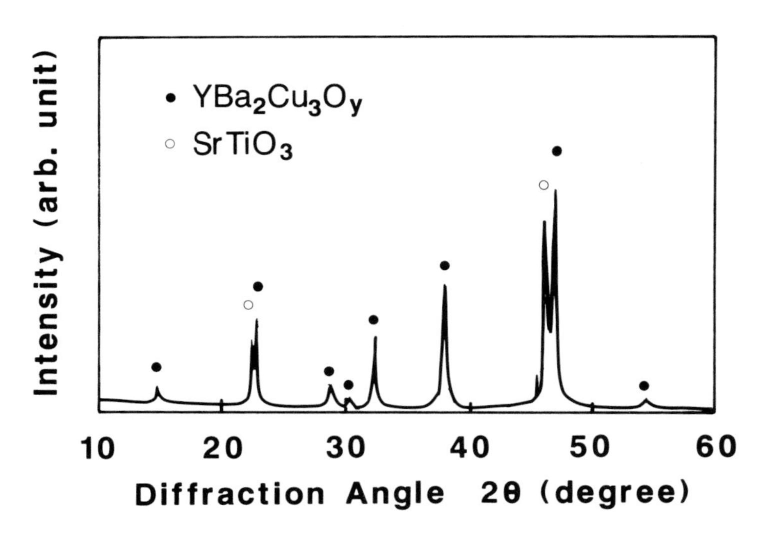

Fig. 2. X-ray diffraction pattern for the annealed Y-Ba-Cu-O film on the (001) $SrTiO_3$ substrate. A few different diffraction peaks are observed from (00n) diffraction lines. Diffraction peaks of the $SrTiO_3$ substrate are also observed.

Figure 2 shows the pattern of the X-ray diffraction (XRD) performed on the film deposited on the (001) $SrTiO_3$ substrate after thermal annealing. The diffraction pattern in Fig. 2 shows the presence of the $YBa_2Cu_3O_y$ phase. The film is not perfectly c-axis-oriented because of observing a few different diffraction peaks from (00n) diffraction lines. BaF_2 is thermally stable, however, the BaF_2 phase was not confirmed in the film.

Temperature dependence of the resistance of the annealed film on the (001) $SrTiO_3$ is shown in Fig. 3. A superconducting transition was obtained with T_c(onset) = 83 K and T_c(zero resistance) = 65 K. The film shows the broad transition to superconductivity, as shown in the inset of Fig. 3. The electrical property of this film is inferior to that of the films prepared by physical vapor deposition such as magnetron sputtering[1].

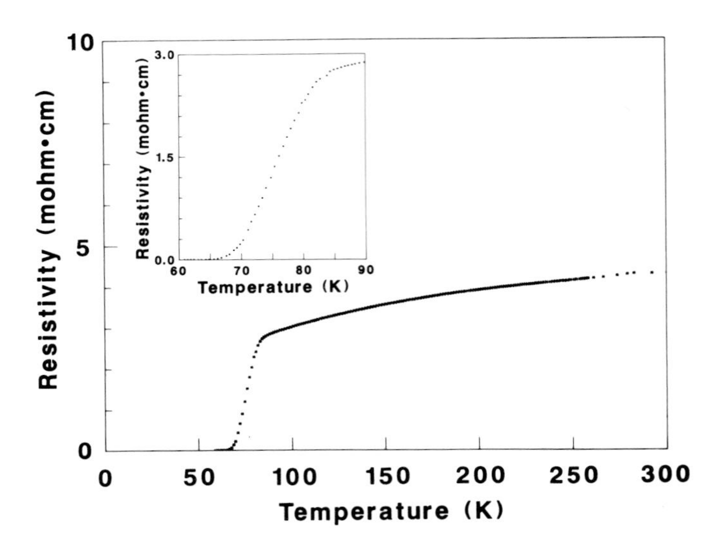

Fig. 3. Temperature dependence of resistivity for the annealed Y-Ba-Cu-O film on the (001) $SrTiO_3$ substrate. A superconducting transition is obtained with T_c(onset) = 83 K and T_c(zero resistance) = 65 K as shown in the inset.

We performed inductively coupled plasma mass spectroscopy (ICPMS) and Auger electron spectroscopy (AES) to know the cause of this broadening of the superconducting transition. ICPMS and AES were carried out for the as-deposited and the annealed films, respectively.

The film composition determined by ICPMS showed the Y-Ba-Cu ratio in the film was 1.0:2.0:4.1; Cu-rich film was prepared. Figure 4 shows the depth profile of the elements (Y, Ba, Cu, Sr, and Ti) in the Y-Ba-Cu-O films on the (001) $SrTiO_3$ obtained by AES. Sr and Ti were confirmed to exist in the film. This is considered to be due to the diffusion of Sr and Ti into the film in the annealing process, even though chemical interaction between $YBa_2Cu_3O_y$ and $SrTiO_3$ has been reported to be low[12].

Therefore, the broadening of the superconducting transition of the Y-Ba-Cu-O film on the (001) $SrTiO_3$ is considered to be due to the following points:

1) Cu-rich film is prepared.
2) Sr and Ti exist as the undesirable elements in the film.
3) The film is not perfectly c-axis-oriented.

And it may be also related to the grain size and/or the oxygen content of this CVD film.

Figure 5 shows temperature dependence of the resistance of the annealed film on the (001) MgO substrate. A superconducting transition was obtained with T_c(onset) = 80 K, but not obtained with T_c(zero resistance) above 60 K, and the resistivity was higher than that of the film on the (001) $SrTiO_3$ substrate. According to the results of ICPMS, the Y-Ba-Cu ratio was 1.0:2.2:3.4. Mg was confirmed to exist in the film by electron prove microanalysis (EPMA). Therefore, the existence of Mg in the film is considered to make the electrical property worse compared to the film on the (001) $SrTiO_3$ substrate.

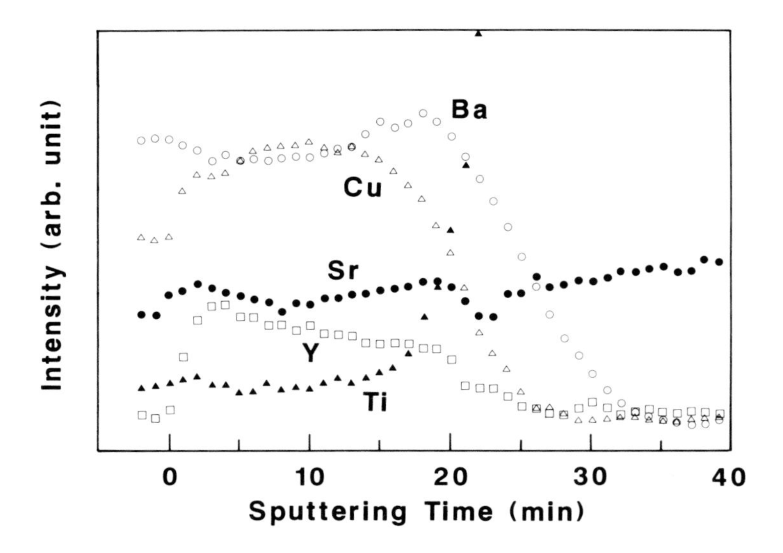

Fig. 4. Result of Auger electron spectroscopy of the annealed Y-Ba-Cu-O film on the (001) $SrTiO_3$ substrate. Sr and Ti are confirmed to exist in the film, and Ba is observed to diffuse into the $SrTiO_3$ substrate.

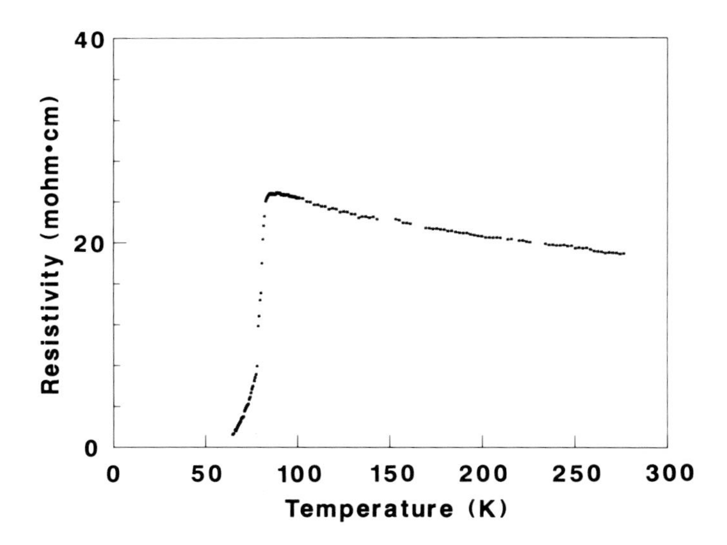

Fig. 5. Temperature dependence of resistivity for the annealed Y-Ba-Cu-O film on the (001) MgO substrate.

In our CVD system, the whole reactor is heated up to the same temperature as the substrate. The precursors begin to decompose and/or react before they reach the substrate. There is a problem of less controllability of the reaction between the precursors and O_2 on the substrate. This causes the film composition to differ from Y:Ba:Cu = 1:2:3. Therefore, it is necessary to keep the reactor at the similar temperature to those of the gas line and the mixing tube and to heat the substrate independently.

As shown in Fig. 4, Ba diffuses into the $SrTiO_3$ substrate and Y concentration at the surface decreases due to the vaporization during the thermal annealing. The diffusion of Ba and the vaporization of Y make the composition of the annealed film differ from that of the as-deposited one, and may lower the electrical property of films. Therefore, as-deposited superconducting films must be prepared by CVD to obtain good reproducibility.

SUMMARY

We have prepared the Y-Ba-Cu-O films by CVD using $Y(DPM)_3$, $Ba(HFA)_2$ and $Cu(HFA)_2$, and obtained the superconducting transition with T_c(onset) = 83 K and T_c(zero resistance) = 65 K for the film on the (001) $SrTiO_3$ substrate. The electrical property of this film is inferior to that of the films prepared by physical vapor deposition at the present time, however, this is considered to be the first step to produce epitaxial films at low temperatures. Further research of the deposition condition will lead to the realization of single-crystalline Y-Ba-Cu-O films by CVD.

ACKNOWLEDGMENTS

The authors wish to acknowledge H.Yamaguchi and I.Ehama for the measurements of XRD and AES and are also grateful to H.Takao, K.Kuno, and Y.Komiya for their continuous encouragement.

REFERENCES

[1] For example; Y.Enomoto, T.Murakami, M.Suzuki and K.Moriwaki, Jpn. J. Appl. Phys. 26, L1248 (1987).
[2] For example; X.D.Wu, D.Dijkkamp, S.B.Ogale, A.Inam, E.W.Chase, P.F.Miceli, C.C.Chang, J.M.Tarascon and T.Venkatesan, Appl. Phys. Lett. 51, 861 (1986).
[3] For example; B.Oh, M.Naito, S.Arnason, P.Rosenthal, R.Barton, M.R.Beasley, T.H.Geballe, R.H.Hammond and A.Kapitulunik, Appl. Phys. Lett. 51, 852 (1987).
[4] For example; J.Kwo, T.C.Hsieh, R.M.Fleming, M.Hong, S.H.Liou, B.A.Davidson and L.C.Feldman, Phys. Rev. B36, 4039 (1987).
[5] T.Terashima, K.Iijima, K.Yamamoto, Y.Bando and H.Mazaki, Jpn. J. Appl. Phys. 27, L91 (1988).
[6] S.Witanachchi, H.S.Kwok, X.W.Wang and D.T.Shaw, Appl. Phys. Lett. 53, 234 (1988).
[7] A.D.Barry, D.K.Gaskill, R.T.Holm, E.J.Cukauskas, R.Kaplan and R.L.Henry, Appl. Phys. Lett. 52, 1743 (1988).
[8] H.Yamane, H.Kurosawa, T.Hirai, H.Iwasaki, N.Kobayashi and Y.Muto, Nippon-Seramikkusu-Kyokai-Gakujyutsu-Ronbunshi 96, 799 (1988).
[9] T.Nakamori, H.Abe, T.Kanamori and S.Shibata, Jpn. J. Appl. Phys. 27, L1265 (1988).
[10] H.Yamane, H.Kurosawa, H.Iwasaki, H.Masumoto, T.Hirai, N.Kobayashi and Y.Muto, Jpn. J. Appl. Phys. 27, L1275 (1988).
[11] K.Shinohara, F.Munakata and M.Yamanaka, to be published in Jpn. J. Appl. Phys.
[12] H.Koinuma, K.Fukuda, T.Hashimoto and K.Fueki, Jpn. J. Appl. Phys. 27, L1216 (1988).

CVD of Bi-Sr-Ca-Cu-O Thin Films

T. KIMURA, M. IHARA, H. YAMAWAKI, K. IKEDA, and M. OZEKI

Fujitsu Laboratories Ltd., 10-1, Morinosato-Wakamiya, Atsugi, 243-01 Japan

ABSTRACT

We developed a chemical vapor deposition (CVD) method for high-Tc superconducting Bi-Sr-Ca-Cu-O thin films, and obtained a high-Tc phase (about 110 K) without postannealing in an atmosphere containing oxygen. Films were deposited on (001) MgO substrates in He in an open-tube reactor with O_2 and/or H_2O oxidizing agents. The source materials -- anhydrous metal halides such as $BiCl_3$, SrI_2, CaI_2, and CuI -- were evaporated in the CVD system to be used as source gases. The deposition temperature was between 700°C and 850°C. The average deposition rate was about 1.5 nm/min and the average film thickness was about 0.1 μm. Films were a mixture of low-Tc (about 80 K) and high-Tc (about 110 K) phases. Their C-axis orientation was perpendicular to the substrate. Their electrical resistance decreased abruptly below 115 K and the zero-resistance temperature was around 98 K. A critical current density of about 1.7×10^4 A/cm^2 was obtained at 4.2 K. The O_2 in the CVD atmosphere played an important role in determining the phase of films at higher temperature depositions.

INTRODUCTION

Since the high-Tc superconducting material La-Ba-Cu-O was discovered [1], other Perovskite superconducting high-Tc materials have been found. Materials used to fabricate electronic devices must be of high quality -- thin single-crystal films in particular. Many ways have developed to fabricate films, e.g., sputtering, electron beam (EB) evaporation, and molecular beam epitaxy (MBE). We have further developed the chemical vapor deposition (CVD) of Bi-Sr-Ca-Cu-O superconductors, discovered by Maeda et al. [2, 3, 4].

We fabricated high-Tc Bi-Sr-Ca-Cu-O superconductors by CVD without postannealing, and discuss the relationship between film characteristics and deposition conditions in this paper.

EXPERIMENT

Figure 1 diagrams the CVD system, which has four source zones and a deposition zone, each heated by a separate furnace. The source materials -- anhydrous metal halides such as $BiCl_3$, SrI_2, CaI_2, and CuI -- are put in boats placed in the source zones and evaporated by heating. The temperature of each zone is controlled independently so that source gas concentrations can be varied. The source gases are transported to the growth zone by the carrier gas (He). The oxidizing agents, O_2 and/or H_2O, are transported to the deposition zone separately from the source gases. (001) MgO substrates 30 mm x 30 mm x 0.5 mm are placed in the deposition zone, where the source gases are mixed and react with O_2 and/or H_2O, and films are deposited on the substrate. The deposition temperature T_{sub} ranged 700°C to 850°C; typical temperatures of each source zone are shown in Fig. 1. After deposition, the substrates were cooled rapidly at about 30°C/min. The oxygen concentration (C_{ox}) ranged from 0.1% to 12.5% of the total gas flow, and the H_2O concentration (C_w) was between 320 ppm and 3200 ppm. The chemical reaction of H_2O in CVD is not understood precisely, but H_2O increases the deposition rate. Film was typically about 0.1 μm thick. The films were not annealed in oxygen or an oxygen mixture after deposition. Film characteristics were determined by X-ray diffraction (XRD) using Cu-Kα radiation, and by electrical resistance measurements. We measured film composition by inductively coupled plasma luminescence spectrometry (ICP) and X-ray fluorescence analysis. The composition was Bi:Sr:Ca:Cu=1:(0.6-1.5):(0.6-1.25):(0.5-3.5). Bumps and segregations in the films prevented the composition estimates from being completely accurate.

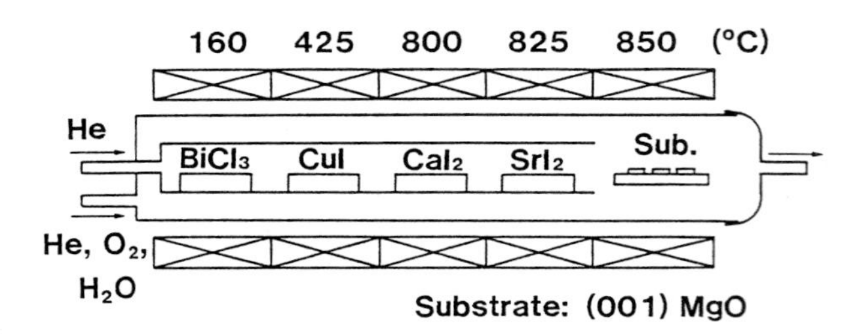

Figure 1 The Bi-Sr-Ca-Cu-O CVD system

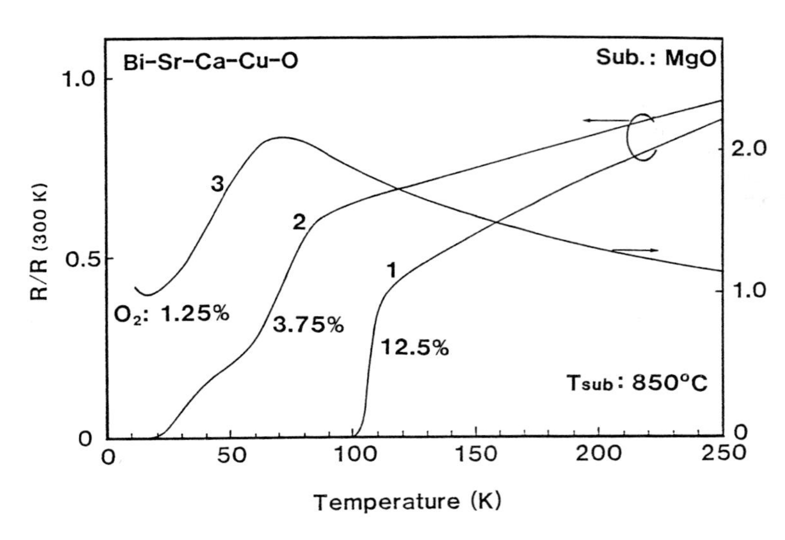

Figure 2 Electrical resistance of films. Samples were deposited on (001) MgO at 850°C in an O_2 concentration of 12.5%, 3.75%, and 1.25%.

RESULTS AND DISCUSSION

Figure 2 shows the electrical resistance of the films versus temperature. The curves are normalized to the resistance at room temperature (300 K). The samples were obtained at T_{sub}=850°C, and at C_{ox}=12.5% for sample 1, 3.75% for sample 2, and 1.25% for sample 3. The resistivity of sample 1 at 300 K is about 1.3 mΩ-cm. The resistance of sample 1 decreases from 300 K to 115 K much like a metal, and drops abruptly below 115 K. The zero-resistance temperature is 98 K and the critical current density is 1.7×10^4 A/cm^2 at 4.2 K. The resistivity of sample 2 at 300 K is about 30 mΩ-cm, about 20 times larger than that of sample 1. The resistance of sample 2, also much like a metal, decreases abruptly below 90 K, and has a shoulder around 50 K. The zero-resistance temperature is 16 K. There is no change in slope at around 100 K. Thus, the resistance characteristics of this sample do not indicate the existence of a high-Tc phase. The resistivity of sample 3 at 300 K is about 82 mΩ-cm, about 60 times larger than that of sample 1. Sample 3 is not superconducting, even at 10 K, but resistance decreases at around 70 K, indicating that this film has a superconducting phase. As these samples show, the superconductivity and high Tc of the films depends on the oxygen concentration during deposition. This concentration plays an important role in deposition at 850°C. Samples 4 and 5 (Fig. 3) were deposited at 700 K. The oxygen concentrations in the atmosphere were 12.5% for sample 4 and 1.25% for sample 5. The resistivity of sample 4 was 1.3 mΩ-cm and that of sample 5 was 0.85 mΩ-cm. The resistance of these samples starts to decrease around 100 K, and drops abruptly at about 90 K. Sample 4 has a shoulder around 100 K. The zero-resistance temperature of samples 4 and 5 is around 77 K. These results show that the oxygen concentration dependence is less in low-temperature than in high-temperature deposition.

Figures 4 and 5 show the X-ray diffraction patterns of the films. In all samples, the C-axis is perpendicular to the surface of the MgO substrate. The X-ray diffraction pattern for sample 1 shows two main kinds of peaks related to the C-axis of the high- and low-Tc phases. This indicates that the film was a mixture of phases, even though this sample is superconducting at 98 K, and that the film has high-Tc phase paths. The pattern for sample 2 shows that the film has only a low-Tc phase. Sample 3 also has a low-Tc phase, but is not superconducting. The resistance characteristics of sample 3 shows that it is a mixture of low-Tc and semiconducting phases. The results for samples 1, 2, and 3 indicate that the concentration of oxidizing agents in the deposition atmosphere determines the phase of the Bi-Sr-Ca-Cu-O system. Samples 4 and 5 also have a small high-Tc phase. The ratio between the peak intensities of high- and low-Tc phases in samples 4 is almost the same as that of the phases in sample 5 in terms of the XRD. This indicates that the deposition of a film with a high-Tc phase may be possible even at a low deposition temperature of 700°C.

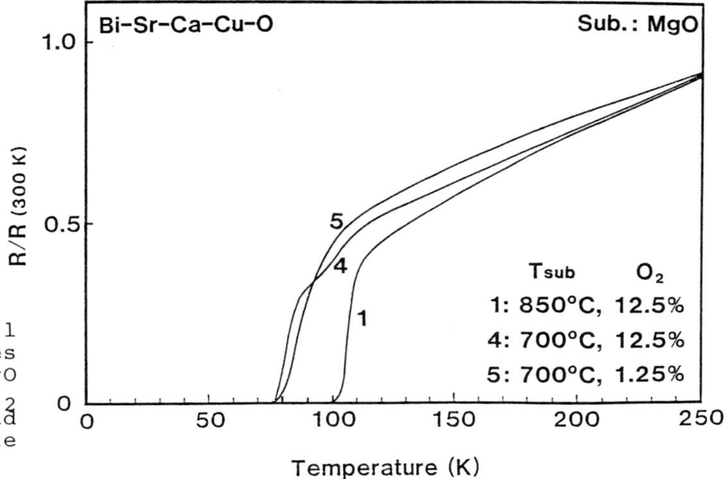

Figure 3 Electrical resistance of films. Samples were deposited on (001) MgO at 700°C in an O_2 concentration of 12.5% and 1.25%. Sample 1 is the same material as that in Fig. 2.

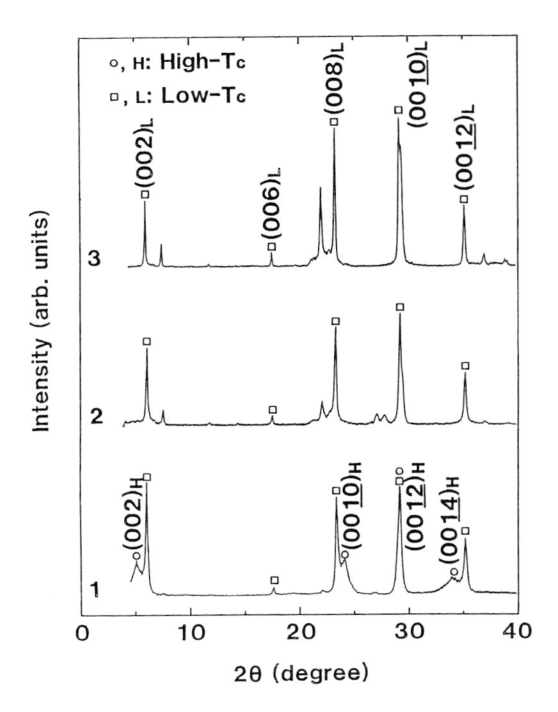

Figure 4 X-ray diffraction patterns of samples 1, 2, and 3 from Fig. 2.

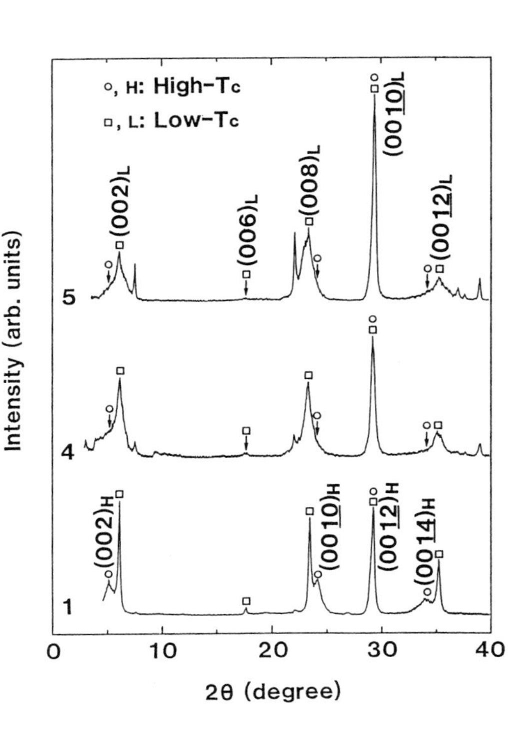

Figure 5 X-ray diffraction patterns of samples 1, 4, and 5 from Fig. 3.

Resistance measurements and X-ray diffraction patterns show the high-Tc phase to be obtained at T_{sub}=700-850°C. At low deposition temperatures, the broad X-ray diffraction peaks indicate that the crystal quality is poor. In 850°C deposition at low O_2 concentration, there is no high-T_c phase. Thus, the high-Tc phase strongly depends on the O_2 concentration in the CVD atmosphere.

A critical current density of $Jc=1.7x10^4$ A/cm^2 at 4.2 K was obtained in sample 1. This value is lower than that of the thin films deposited by sputtering [5]. Film deposited by sputtering had a critical current density of over 10^5 A/cm^2, even at 77 K. Other X-ray diffraction peaks in the patterns differed from superconducting phases, indicating nonsuperconducting phases in the films. These extra phases may be between superconducting grains, and reduce the critical current density.

CONCLUSION

We fabricated a high-Tc phase Bi-Sr-Ca-Cu-O system on a (001) MgO substrate by CVD without postannealing in an oxygen mixture. The source materials -- anhydrous metal halides such as $BiCl_3$, SrI_2, CaI_2, and CuI -- were evaporated in the CVD system to be used as source gases. The source gases reacted with O_2 and/or H_2O oxidizing agents, and the films are deposited on MgO substrates at between 700°C and 850°C. The films were a mixture of low-Tc (about 80 K) and high-Tc (about 110 K) phases, and the C-axis was perpendicular to the substrate. The electrical resistance of the films deposited at 850°C decreased abruptly below 115 K and the zero-resistivity temperature was around 98 K. A critical current density of about 1.7×10^4 A/cm^2 at 4.2 K was obtained.

At higher deposition temperatures, oxygen concentration clearly affects film characteristics. The superconductivity of the films depends on the oxygen concentration of the deposition atmosphere. The high- or low-Tc phase of the films depends on both the deposition temperature and the oxygen concentration. Our results indicate that high-Tc phase films can be deposited for Bi-Sr-Ca-Cu-O superconductors, even at a low temperature of 700°C.

ACKNOWLEDGMENT

We thank Dr. N. Awaji for making the X-ray fluorescence measurements.

REFERENCES

[1] J. G. Bedonorz and K. A. Müller, Z. Phys. B64 189 (1986).
[2] H. Maeda, Y. Tanaka, M. Fukutomi, and T. Asano, Jpn. Appl. Phys. Lett., 27 L209 (1988).
[3] M. Ihara and T. Kimura, Ex. Abstracts of FED HiTcSc-ED Workshop, June, 1988, Miyagi-Zao, Japan, 137.
[4] M. Ihara, T. Kimura, H. Yamawaki, and K. Ikeda, Applied Superconductivity Conference Proceedings, Aug. 21-25, 1988, San Francisco, to be published in IEEE Trans. on Magnetics.
[5] K. Setsune, K. Hirochi, H. Adachi, Y. Ichikawa, and K. Wasa, Appl. Phys. Lett., 53, 600 (1988).

MO-CVD of HTcS

C.I.M.A. Spee[1], A. Mackor[1], and P.P.J. Ramaekers[2]

[1] TNO Institute of Applied Chemistry, P.O. Box 108, 3700 AC, Zeist, The Netherlands
[2] TNO/TUE Center for Technical Ceramics, P.O. Box 513, 5600 MB, Eindhoven, The Netherlands

ABSTRACT

Thin films of HTcS, *e.g.* BiSrCaCu-oxides or $YBa_2Cu_3O_7$, may be prepared on suitable substrates (*e.g.* $SrTiO_3$) by metal-organic chemical vapour deposition MO-CVD, as recently indicated in the literature. However, for optimal process conditions and film properties, a number of variables must be controlled. These include the selection of appropriate precursors with sufficient volatility and stability at a work temperature of $\pm$ 200 °C, like metal β-diketonates. The thermochemistry of these compounds must be such as to provide metal oxide deposits. After these steps (selection, vapour pressure measurements and thermochemistry) have been taken, the actual MO-CVD process will be carried out. The progress in the TNO programme will be reported at the meeting.

INTRODUCTION

For the preparation of thin films of HTcS on substrates (*e.g.* BiSrCaCu-oxides or $YBa_2Cu_3O_7$ on $SrTiO_3$), chemical vapour deposition CVD has an attractive potential. This is due to the well-controlled and mild process conditions and microstructure, composition and product properties of the deposited layer in CVD and its combination of preparation and shaping (non-flat substrates!). Therefore, we have recently undertaken an effort in MO-CVD of HTcS, with financial support from the National Programme on HTcS.

Our research programme consists of four steps. In the first step, a broad range of possibly suitable precursors is selected and synthesized, if necessary. Three classes of precursors are under investigation: a) β-diketonate complexes; b) alkoxides; c) organometallic precursors with ligands like n-alkyl, phenyl, cyclopentadienyl, *etc.* For each of the elements concerned, *i.e.* Cu, Bi, Y, the alkaline earth metals Ca, Sr or Ba, a precursor must have a vapour pressure of at least 1 torr below 200 °C, with sufficient stability.

Presently, we are studying two series of metal β-diketonates : one series with the acac (acetylacetone = 2,4-pentanedione) ligand $M(acac)_n$ (M = Cu, Ca, Sr or Ba, n = 2; Y, n = 3) and the second series with the same metals and the thd (2,2,6,6-tetramethyl-3,5-heptanedione) ligand. For bismuth, an organometallic precursor will be used as for instance $Bi(Ph)_3$ (b.p. 180 °C).

The second step in our programme consists of measurements of vapour pressure *versus* temperature on these precursors.

The third step in our programme consists of studing the thermochemical behaviour of these precursors. In particular, the nature of the metal oxide or other deposit and the volatile residues from the organic ligands are a subject of investigation.

The fourth part of our programme is the actual deposition of pure and mixed metal oxides on the substrate ($SrTiO_3$, MgO or other substrate), for which a CVD reactor has been constructed.

EXPERIMENTAL

Vapour pressure measurements are performed by a static method using an isoteniscope [1,2] (Fig. 1). In this method, the sample is brought into a sample bulb. If necessary, this can be performed under nitrogen. While the sample is cooled by a dry ice/acetone bath (-78.5 °C), the system is brought under a vacuum of less than 0.001 torr. After the system is closed from the

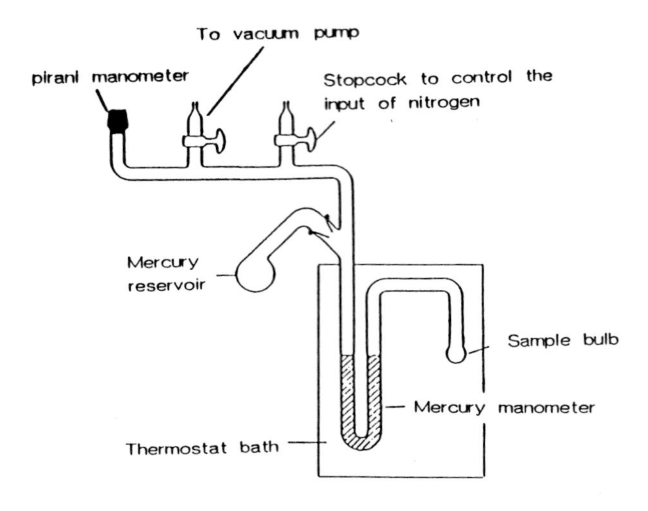

Figure 1: Isoteniscope.

vacuumpump, mercury is introduced from a reservoir into a U-tube closing off a small volume, containing the sample, from the remainder of the system. The mercury lock and sample bulb are then brought to the desired temperature in a thermostat bath. Vapour pressure is measured by a Pirani manometer by introducing the same pressure as in the sample bulb. This is done by the inlet of nitrogen in such a quantity as to keep the mercury level in the lock in the same position at both sides of the U-tube. Limitations of this method are a minimum vapour pressure of 0.1 torr, which can be measured, and a maximum temperature of approximately 180 °C because of increasing p(Hg).

The thermochemical behaviour of the single precursors is studied by thermogravimetric analysis (TGA), differential scanning calorimetry (DSC), and in a small CVD-like apparatus (see Fig. 2).

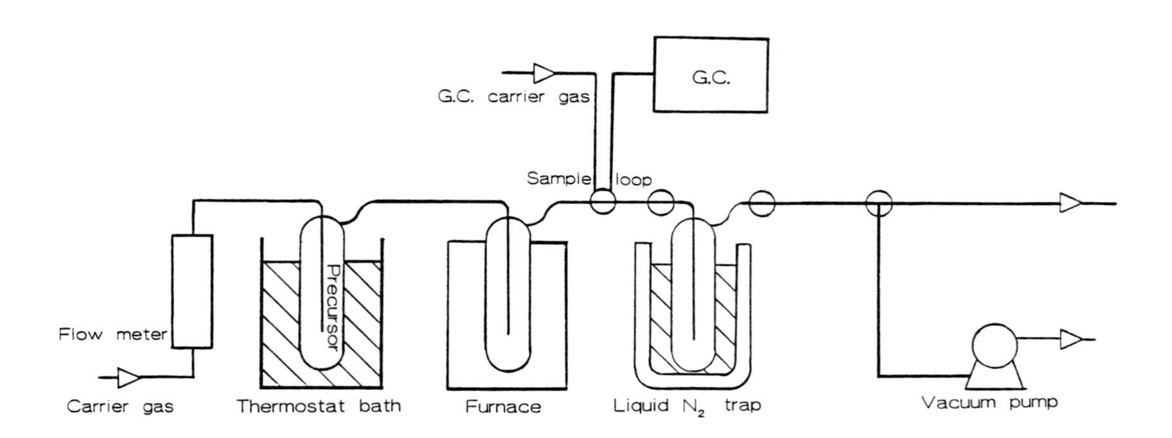

Figure 2: Apparatus for study of thermochemistry of HTcS precursors.

In this apparatus, the precursors are placed in a thermostatted evaporator and thus vapour is drawn along the tube to the furnace by a stream of helium, nitrogen or, if necessary, hydrogen. The gases, resulting from the thermolysis, are analyzed by means of a gas chromatograph or condensed in a trap for further analysis by GC and/or GCMS. Deposits in the reaction tube are examined by elemental analysis techniques (AAS, ICP).

The CVD apparatus is shown in Figures 3 and 4. It consists of a gas-handling system, four evaporators and a mixing chamber, where all gases are mixed prior to entering the reactor. All pipes

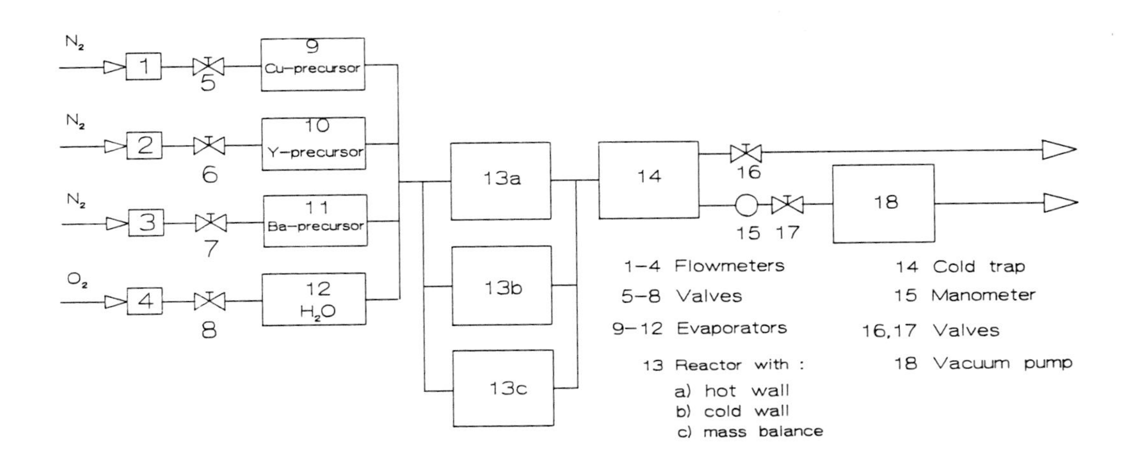

Figure 3: Deposition arrangement shown here for deposition of $YBa_2Cu_3O_{7-x}$.

Figure 4: (LP)MOCVD-apparatus under construction.

from the evaporators to the reactor can be heated in order to hold the vapour pressures set by the evaporators. Reactor configuration can be chosen out of three options, of which the first two have a horizontal reactor arrangement and the third one a vertical arrangement: a) substrate + reaction chamber are heated by a furnace (hot-wall configuration); b) substrate is inductively heated (cold-wall configuration); c) an arrangement, whereby kinetic measurements may be performed (Figure 5) by continuous measurement of mass differences during deposition.

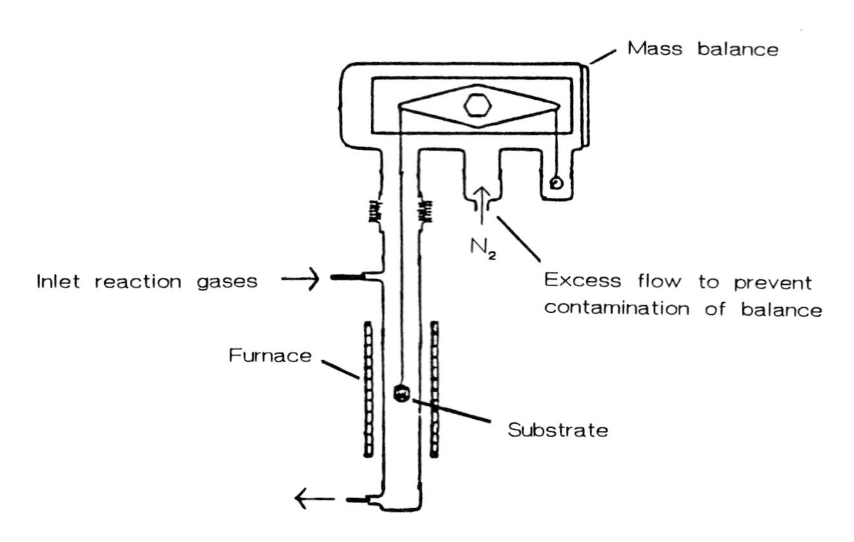

Figure 5: Mass balance arrangement.

PRELIMINARY RESULTS

Since our HTcS programme has started very recently, only a few experimental results are presently available, besides the construction and testing of equipment.

Vapour pressures have been measured for two copper precursors, $Cu(acac)_2$, 1 and $Cu(thd)_2$, 2. Starting with a commercial sample of 1, already at low temperatures a considerable vapour pressure is found, which increases only slightly upon raising the temperature further. We also note that at room temperature 1 has the characteristic smell of the free ligand. Therefore, we conclude from both observations that the sample of 1 as bought contains free acetylacetone as an impurity. This has been removed by heating 1 at 80 °C for 30 minutes at a pressure of 0.0008 torr, prior to the vapour pressure determinations. After this purification, the vapour pressures were determined as shown in Figure 6, series 2, *e.g.* as ± 3 torr at 150 °C (after correction for the vapour pressure of mercury, as also shown in Figure 6).

For a commercial sample of 2, the measurements of vapour pressure do not indicate the presence of the free ligand (Figure 7).

Other groups have also recently reported on similar measurements on precursors for CVD of $YBa_2Cu_3O_7$ [3,4] or on preliminary CVD results [5]. The fact that the vapour pressures of the metal β-diketonates, as determined on practical samples, are in many (all?) cases larger than those of a pure sample, has not yet been noted in the literature.

We are now in the process of investigating purified samples for vapour pressure determinations and their thermochemical behaviour.

Finally, we wish to acknowledge Mr. J.L. Linden for performing the vapour pressure measurements.

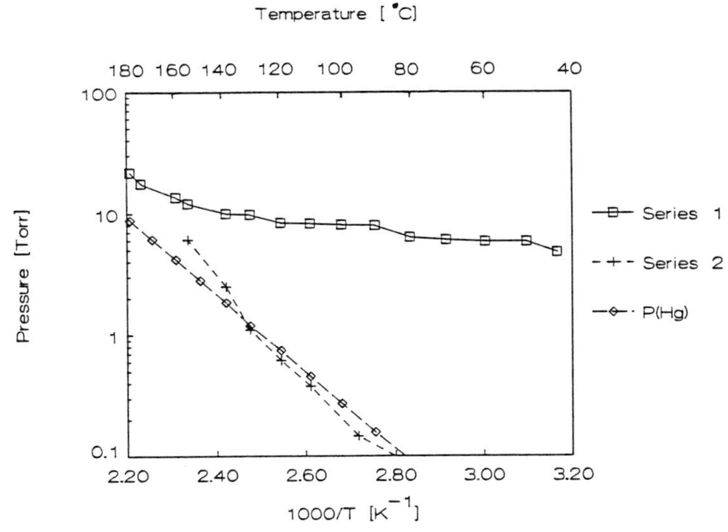

Figure 6: Vapour pressure measurements on $Cu(acac)_2$.

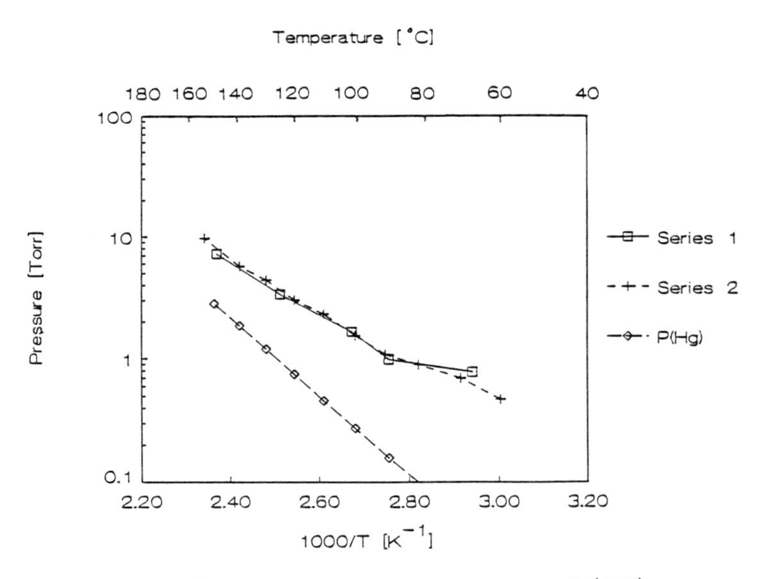

Figure 7: Vapour pressure measurements on $Cu(thd)_2$.

REFERENCES

1. Weissberger, A., Physical Methods, Part 1, Interscience Publishers, New York, 1960, p.434-439.
2. Smith, A., and Menzies, A.W.C., J. Am. Chem. Soc. 32, (1910) 1412-1447.
3. Soer, H., Oehr, Ch., Holzschuh, H., Schmaderer, F., Wahl, G., Kruck, Th., and Kevinen, A., Physica C 153-155, (1988) 784-785.
4. Gonzales-O., C., Schachner, H., Tippmann, H., and Trojer, F.J., Physica C 153-155, (1988) 1042-1043.
5. Berry, A.D., Gaskill, D.K., Holm, R.T., Cukauskas, E.J., Kaplan, R., and Henry, R.L., Appl. Phys. Lett. 52, (1988) 784-785.

High-Tc Superconducting Bi-Ca-Sr-Cu-O Thin Films Prepared by Metalorganic Deposition (MOD)

KAZUHIRO SAKAI, YOSHIHIKO KAN, and TAKUO TAKESHITA

Central Research Institute, Mitsubishi Metal Corporation, 1-297, Kitabukuro-cho, Omiya, 330 Japan

ABSTRACT

High-Tc superconducting Bi-Ca-Sr-Cu-O thin films were prepared by the spin-on pyrolysis technique using a mixed solution of 2-ethylhexanoates of constituent metals dissolved in an organic solvent. Effects of the pyrolyzing temperature were investigated, and it was found that the films prepared by pyrolyzing as-deposited organometals at 1073K showed superconductivity without further heat treatment. X-ray diffraction examinations of these films showed that the crystallographic c-axis was oriented perpendicularly to the film surface. The superconducting transition temperature of these films observed was Tc = 90K(onset) and 77K(end).

INTRODUCTION

The recent discovery of the new superconducting ceramics of the Bi-Ca-Sr-Cu system [1] has been attracting much interest because of its favorable properties; (1) insensitive to water and moisture; (2) easy control of oxygen content; (3) no need of rare earths.

The film preparation techniques used for another superconducting ceramics such as $Ba_2YCu_3O_x$ [2-7] are expected to be utilized for this new ceramic superconductor. And among these techniques, the metalorganic deposition(MOD) method, i.e., the pyrolysis of organic acid salts [8], has the following advantages; (1) it is a very simple process, i.e., no need of the powder making stage; (2) the film composition is easily controlled and it is uniform over the entire film because the solution containing constituent elements is coated uniformly on the substrate; (3) it can produce films of uniform and of desired thickness by choosing number of repetitions of coating processes; and (4) this method requires no vacuum system and does not need expensive equipments.

In this paper, we report results of the investigation of effects of pyrolysis temperature of mixtures of 2-ethylhexanoates of Bi, Ca, Sr and Cu, spun on substrates. We were able to prepare Bi-Ca-Sr-Cu-O thin films with Tc(end) = 77K.

EXPERIMENTAL

Bi- and Ca-2-ethylhexanoates toluene solutions were obtained from Nihon Kagaku Sangyo Co.,Ltd. and Sr- and Cu-2-ethylhexanoates toluene solutions were prepared by reacting metal hydroxides with 2-ethylhexanoic acid in toluene solvent [8,9]:

$$2C_4H_9CH(C_2H_5)COOH + M(OH)_2 \rightarrow M[C_4H_9CH(C_2H_5)COO]_2 + 2H_2O$$
(where M = Sr or Cu)

These solutions were mixed to about 5wt% as total weight of the each metal oxides, Bi_2O_3, CaO, SrO and CuO, and the viscosities of the mixed solutions were about 1.8cps. The starting solutions of nominal composition of Bi:Ca:Sr:Cu = 1:1:1:2 (A) or 2:2:2:3 (B) in molar ratio, were dropped and spun on yttrium-partially stabilized zirconia(Y-PSZ) or single crystal MgO([100]orientation) substrates. The spinning rate was 3000rpm and a period of coating was 10 seconds. The coated films were dried at 423K in air oven for 15 minutes, and were pyrolyzed at several temperatures between 773K and 1073K in air for 15 minutes. This spinning-on, drying and heating process was repeated 5-20 times. The final heat treatment was carried out at 1073K for 60 minutes except the samples which were pyrolyzed at 1073K. These temperatures were selected on the basis of the results of thermogravimetric analyses(TGA) of the decomposition behavior of the starting solutions. TGA of the starting solutions dried to remove the solvent showed that the initial decomposition occurres first at from 473K to 723K and then at from 873K to 1073K. It is likely that the initial decompositions are due to pyrolysis of constituent 2-ethylhexanoates to corresponding oxides and carbonates, i.e., Bi_2O_3, CuO, $CaCO_3$ and $SrCO_3$, and the second to pyrolysis of $CaCO_3$ and $SrCO_3$ to CaO and SrO.

The films were characterized by X-ray diffraction analysis(XRD), scanning electron microscopy(SEM), and the surface resistance by the conventional four probe method with press type contact. Resistivity of the films was measured by the four point method, and electrical contacts were made by silver paste.

RESULTS AND DISCUSSION

XRD patterns of the films on Y-PSZ, which were pyrolyzed at 873 and 973K, were not those of the superconductor of the Bi-system, and SEM observations showed mesh-like structures which may have been created by combustion of organic compounds or decomposition of Ca- and Sr-carbonates, and the XRD patterns can be assigned to those of Bi_2O_3, $CaCO_3$,$SrCO_3$ and CuO. After heat treating them at 1073K, their XRD patterns were those of the superconductor, and SEM observations showed many voids which were probably originated from the mesh-like structures of films.

In the film on Y-PSZ, which were pyrolyzed at 1073K and were not further heat treated, the superconductor of the Bi-system was confirmed by XRD. Although SEM observations showed that the surface of the film was free of voids and cracks, the spots were seen on the films. This is thought to be created by the reaction of the superconductor of the Bi-system with the substrate Y-PSZ. The surface resistance of these films at room temperature by the conventional four probe method with press type contact is shown in Table I. It is seen in this table that the surface resistance decreases as the pyrolysis temperature increases. This tendency is considered to be due to the mesh-like structures observed by SEM, and this mesh-like structure of the films may make the reactivity of solids low and/or cause poor electrical contacts. Consequently, these results indicate the necessity of the substrate selection although the Bi-system superconductor of good superconductivity is prepared by the pyrolysis of dried deposits at 1073K.

Table I. Surface resistance of films.[1)]

Pyrolysis temp.(K)	Surface resistance (Ω)[2)] after pyrolysis	after heating[3)]
873	∞	20
973	0.5M	10
1073	7	----

1) The films on Y-PSZ substrates. 2) Measured by conventional four probe method using press type contact. 3) Heat treated at 1073K for 60 min.

Next we used single crystal MgO ([100] orientation) as the substrate which was considered to be less reactive with the superconductors of the Bi-system, and investigated the relationship of the pyrolysis temperature and the superconductivities of films.

The films, which were prepared from the solution (B) by 10 repetitions of the spinning-on, drying and pyrolysis(at 773, 873 and 973K) and then heat-treated at 1073K for 60 minutes, had Tc(onset) = about 90K and Tc(end) = 50 and 73K. SEM observations and XRD analysis of these samples were similar to those of the films on Y-PSZ substrates. These results are consistent with those of surface resistance measurements of films on Y-PSZ substrates.

The films prepared by the pyrolysis at 1073K for 15 minutes with 5-20 repetitions of the spinning-on, drying, and heating process had Tc(onset) = about 90K and Tc(end) = 77K. Resistivities versus temperature curves of these samples with 10 repetitions using the starting solution (B) are shown in Figure 1.

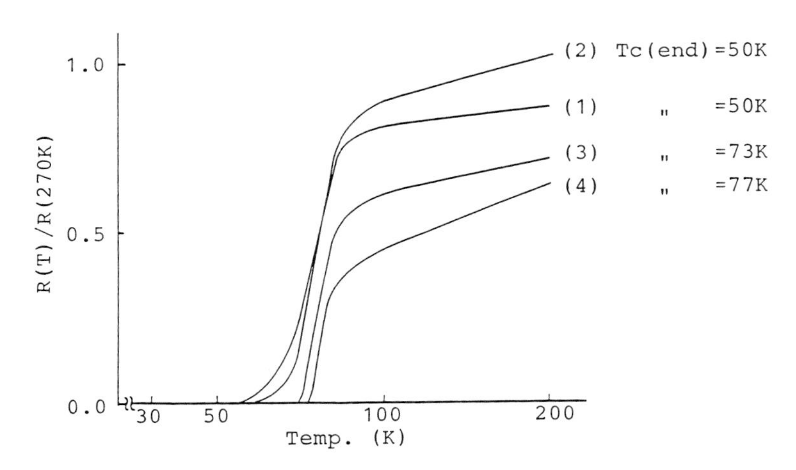

Figure 1. Resistivity vs. temperature curves of films prepared by different pyrolysis temperature;(1) pyrolyzed at 773K, (2) at 873K,(3) at 973K, and (4) at 1073K.

SEM observations of the samples pyrolyzed at 1073K showed that the surface of the films were free of voids and of cracks, and the films were made of grains of 2-3 microns radius together with smaller ones(0.2 microns), and small grains were attached to the large one(Figure 2). Thickness of the films which were prepared by 5 and 20 repetitions of the process were 0.5 and 2 microns respectively. Thus, the films of about 0.1 microns thickness would be formed by the single application of the process.

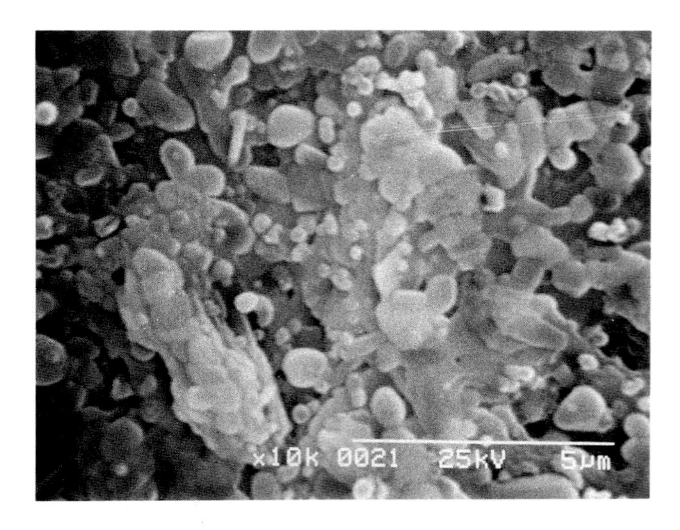

Figure 2. SEM picture of the film pyrolyzed at 1073K.

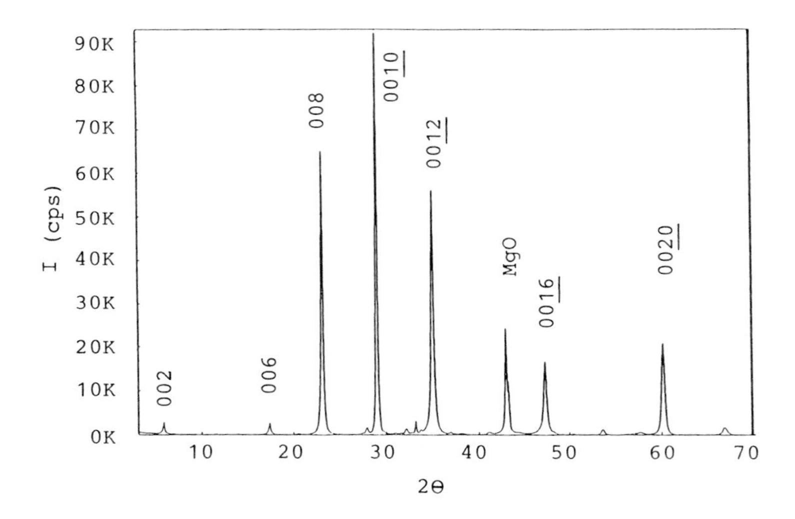

Figure 3. XRD pattern of the film pyrolyzed at 1073K.

From their XRD pattern shown in Figure 3, the intensities of (00n) reflections, where n is an integer, are stronger than those of other reflections. This shows that the c-axis is oriented perpendicularly to the film surface.

The critical current density of the film of 5 repetitions of the process was about 40,000A/cm^2 at 10K.

The film prepared by the pyrolysis at 1073K with 10 repetitions of the process using the starting solution (A) had Tc(onset) = about 90K and Tc(end) = 78K. The difference of the composition of the starting solutions was not obvious in films' characteristics, i.e., in their SEM and XRD analysis, and in superconducting properties.

CONCLUSIONS

(1) The simple preparation technique of high-Tc superconducting thin films by spinning-on and pyrolysis without further heat treatment using organic acid salts, specifically, 2-ethylhexanoates is shown to be a very effective method.

(2) The pyrolysis temperature of this study, i.e., 1073K, is lower than those of the previous investigations [10,11] using organic acid salts solutions or precursors of the Bi-system superconductor.

(3) The films obtained by this method have Tc(onset) of about 90K, Tc(end) of 77K and Jc of 40,000A/cm^2 at 10K.

(4) From XRD analysis of the films, it is shown that the c-axis is oriented perpendicularly to the film surface.

REFERENCES

[1] H.Maeda, Y.Tanaka, M.Fukutomi, and T.Asano, Jpn. J. Appl. Phys. 27, L209(1988).

[2] C.E.Rice, R.B.vanDover, and G.J.Fisanick, Appl. Phys. Lett. 51(22), 1842(1987).

[3] A.H.Hamdi, J.V.Mantese, A.L.Micheli, R.C.O.Laugal, and D.F.Dungan, Appl. Phys. Lett. 51(25), 2152(1987).

[4] H.Nasu, S.Makida, T.Kato, Y.Ibara, T.Imura, Y.Osaka, Chem. Lett. 2403(1987).

[5] T.Kumagai, H.Yokota, K.Kawaguchi, W.Kondo, and S.Mizuta, Chem. Lett. 1645(1987).

[6] M.E.Gross, M.Hong, S.H.Liou, P.K.Gallagher, and J.Kwo, Appl. Phys. Lett. 52(2), 160(1988).

[7] J.V.Mantese, A.H.Hamdi, A.L.Micheli, T.L.Chen, C.A.Wong, J.L.Johnson, M.M.Karmarkar, and K.R.Padmanabhan, Appl. Phy. Lett. 52(19), 1631(1988).

[8] G.B.Vest, and S.Singaram, Mat. Res. Soc. Symp. Proc. Vol.60, 35(1986).

[9] R.C.Mehrotra, and R.Bohra, "Metal Carboxylates." (Academic press, London,1983).

[10] H.Nasu,S.Makida, Y.Ibara, T.Kato, T.Imura, and Y.Osaka, Jpn. J. Appl. Phys. 27, L536(1988).

[11] S.L.Furcone, and Y.-M.Chiang, Appl. Phys. Lett. 52(25), 2180(1988).

Preparation of Superconducting $YBa_2Cu_3O_y$ Films by Thermal Decomposition Technique Using Organometallic Precursors

YUTAKA YAMADA, TOSHIYA MATSUBARA, and TAKESHI MORIMOTO

Research Center, Asahi Glass Co.,Ltd., 1150 Hazawa-Cho, Kanagawa-ku, Yokohama, 221 Japan

ABSTRACT

Thin films of $YBa_2Cu_3O_y$ and $Bi_2Sr_2Ca_2Cu_3O_y$ superconductors were formed by a thermal decomposition technique using organometallic precursor solutions. Dependence of resistivity on temperature showed the Tc(onset) at 93 K and Tc(zero) at 90 K with the $YBa_2Cu_3O_y$ film on yttria stabilized zirconia substrate, and the Tc(onset) at 115 K and Tc(zero) at 80 K with the $Bi_2Sr_2Ca_2Cu_3O_y$ film on MgO(100) substrate. The X-ray diffraction patterns showed that the both $YBa_2Cu_3O_y$ and $Bi_2Sr_2Ca_2Cu_3O_y$ films were strongly oriented along the c-axis perpendicular to the substrate.

INTRODUCTION

Since the discovery of superconductivity at higher temperatures in the Y-Ba-Cu-O system [1], there have been a large number of publications on the preparation of superconducting films using a variety of vacuum and non-vacuum techniques. Non-vacuum technique is simple, and requires only inexpensive, readily available equipment. Typically, a homogeneous coating solution is prepared by dissolving metal compounds in a suitable solvent. This solution is sprayed, dripped or spun-on to a substrate and dried. After drying it is heat-treated in air or oxygen to produce a thin film. Besides process simplicity, the use of homogeneous solution precursors allows accurate control of composition and the molecular level mixing of each element. This enables the production of the film with highly reproducible properties.

Metal compounds of Y, Ba, and Cu, such as nitrates [2], acetates [3], alkyl carboxylates [4], trifluoroacetates [5], alkoxides [6] and acetylacetonates [7] have been used for preparing coating solutions. These solutions have problems such as, low solubility of metal compounds which leads to precipitation of a particular element producing an inhomogeneous film, low wettability of the substrate by the solution and contamination of impurities arising from the synthetic process of metal compounds.

In this study, we report new solution precursors of Y-Ba-Cu-O and Bi-Sr-Ca-Cu-O systems containing a mixture of metal alkoxides and metal 1,3-diketonates in an alcoholic solvents to solve the above mentioned problems and the fabrication of c-axis oriented films of YBa_2Cu_3Oy and $Bi_2Sr_2Ca_2Cu_3O_y$ from these solution precursors.

EXPERIMENTAL

The copper ethyl acetoacetate was prepared by the reaction of cupric acetate with ethyl acetoacetate in ehanol. The yttrium ethyl acetoacetate was prepared by gradually adding the mixture of ethyl acetoacetate and ammonia water to the aqueous solution of yttrium nitrate. The calcium acetylacetonate and the bismuth 2-ethylhexanoate with chemical pure grade were commercially available both from Nihon Kagaku Sangyo Co. The barium and strontium 2-methoxyethoxides were synthesized as 2-methoxyethanol solution by direct reaction of the alcohol with respective metals. The metal content of the β-diketonates were determined by the thermogravimetric analysis in air.

The superconducting Y-Ba-Cu-O films were prepared in the following way. Yttrium ethyl acetoacetate and copper ethyl acetoacetate were added to the 2-methoxyethanol solution of barium 2-methoxyethoxide in a certain mole ratio and stirred at 50 °C for 1 h under nitrogen to give a homogeneous solution. The viscocity was adjusted by varying the concentration. The mixed solution was then applied to a substrate positioned on a photoresist spinner. The substrate was rotated at 2000 rpm for 10 s to produce a uniform film. Then the film was dried in air at 120 °C for 5 min and pyrolyzed in air for 5 min at 500 °C. After this procedure was repeated 10 times to increase film thickness, the film was subjected to final heat-treatment in oxygen at 950 °C for 30 min, and followed by gradual cooling.

In preparing Bi-Sr-Ca-Cu-O films, calcium acetylacetonate was added to the toluene solution of bismuth 2-ethyhexanoate in a given atomic ratio and evaporated to dryness. Then copper ethyl acetoacetate and strontium 2-methoxyethanol solution of Strontium 2-methoxyethoxide were added in a certain mole ratio and stirred at 50 °C under nitrogen. Finally, N,N-dimethylethanolamine was added to obtain the dark blue homogeneous solution. Coating was carried out in a similar process as Y-Ba-Cu-O system except the final heat-treatment condition which was carried out at 870-880 °C for 1 h followed by treatment at 850-860 °C for 6 h. Figure 1 shows the flowsheets for preparation of precursor solutions.

The resistivity of the films was measured by the conventional dc four-probe method. The thickness of the films was measured with a profilometer. Crystal structure of the films were analyzed by an X-ray diffractometer (Rigaku Denki, Cu-Kα). The surface morphology of the films was observed by SEM (JEOL).

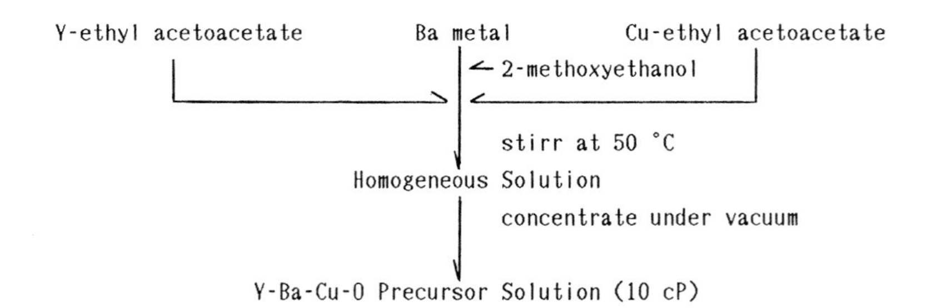

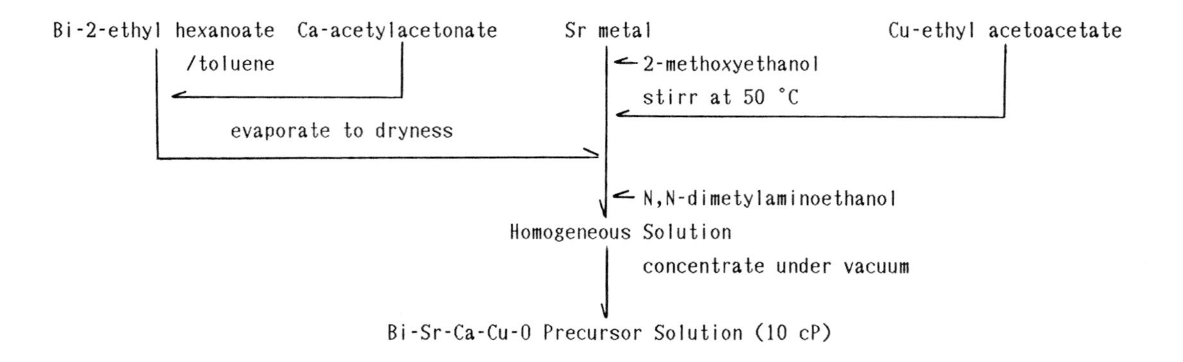

Fig. 1 Flowsheets for preparation of Y-Ba-Cu-O and Bi-Sr-Ca-Cu-O precursor solution

RESULTS AND DISCUSSION

The selection of solvents and metal compounds was based on the following desired properties: (1)metal compounds should be highly soluble in a solvent suitable for coating and decompose cleanly and leave no residual carbon; (2)the number of carbon contained in metal compounds should be small; (3)precursor solution should exhibit appropriate viscocity to perform coating, have good wettability of the substrates and form homogeneous, continuous film; (4)precursor solution should be stable for a certain period. Thus 2-methoxyethanol and N,N-dimethylethanolamine were selected as the solvent for the reasons of (3) and (4). Ethyl acetoacetates and acetylacetonates were selected to fullfil the requirements of (1) and (4). Alkoxides were chosen for the reasons of (1) and (2).

From XRD patterns both Y-Ba-Cu-O and Bi-Sr-Ca-Cu-O films were found to be amorphous after spinning on to the substrate and dried at 120 °C. The films were still mostly amorphous after pyrolyzing at 500 °C. The thickness of the film was typically 3-7 μm after the final heat-treatment process.

Two kinds of precursor solutions were prepared to examine the relation between the starting Y:Ba:Cu composition and film properties for the Y-Ba-Cu-O system. Figure 2 shows the XRD patterns of films on MgO(100) substrates with different starting Y:Ba:Cu mole ratio. Film with the starting mole ratio of 1:2.4:4.2 shows strong diffraction peaks corresponding to (00n) indices suggesting the formation of c-axis oriented structure, while that with 1:2:3 shows no orientation. Figure 3 shows the SEM photographs of the films with both compositions. One may easily find remarkable grain growth on the free surface for the 1:2.4:4.2 film, whereas much smaller grain size assisted with more random orientations for the 1:2:3 film. These facts implies that the preferrable starting composition for preparing the c-axis oriented film on MgO(100) substrate is Y-poor and Cu-rich composition. Probably the excess CuO is acting as a flux to lower the melting point and enhance the crystallization similary to the single crystal preparation [8]. Figure 4 shows dependence of resistivity on temperature for the films with different compositions on MgO(100) and YSZ(polycrystalline) substrates. For

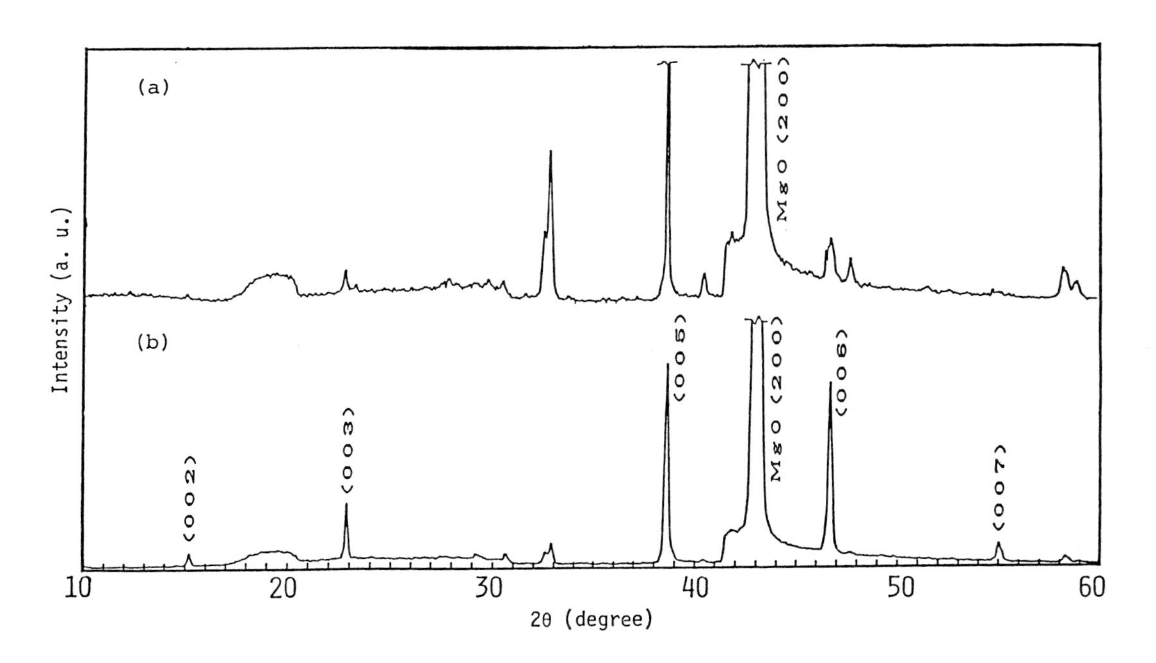

Fig. 2. X-ray diffraction patterns of the films with different starting Y:Ba:Cu composition formed on MgO(100) substrate. (a) Y:Ba:Cu=1:2:3 (b) Y:Ba:Cu=1:2.4:4.2

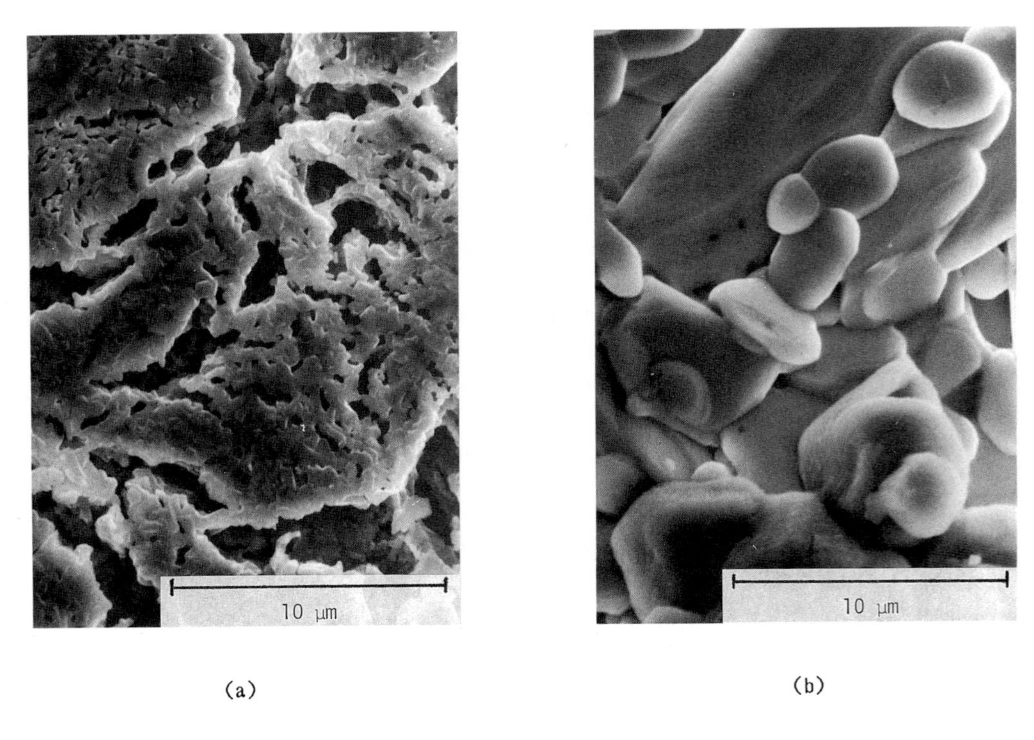

Fig. 3. Scanning electron micrographs of the films with different starting Y:Ba:Cu composition formed on MgO(100) substrate. (a) Y:Ba:Cu=1:2:3 (b) Y:Ba:Cu=1:2.4:4.2

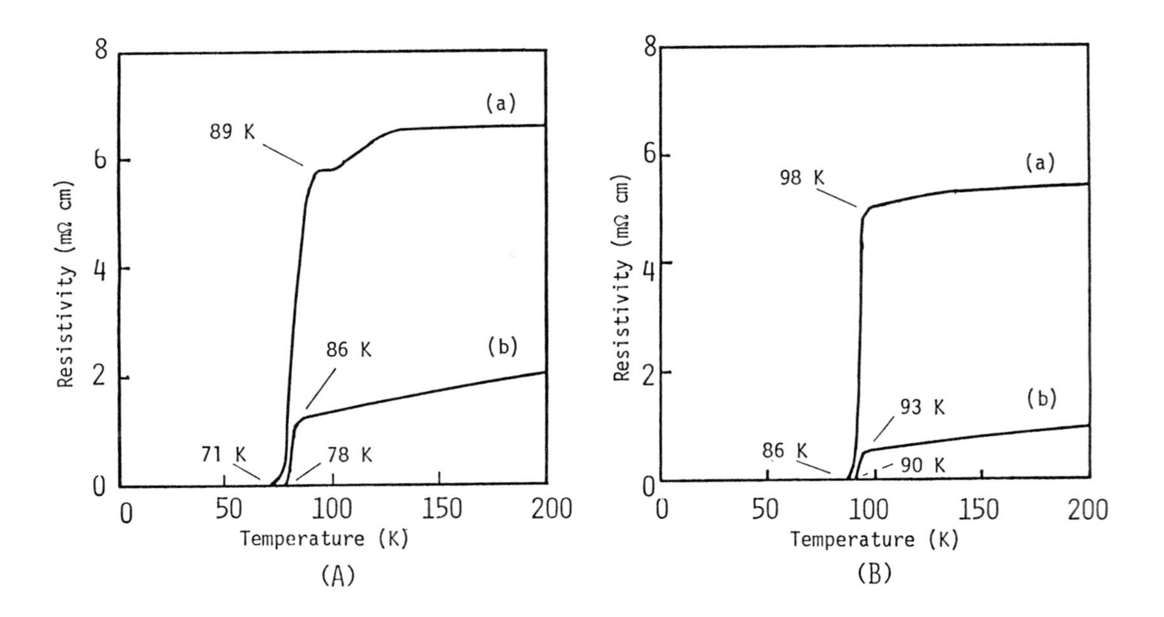

Fig. 4. Temperature dependence of resistivity for different starting Y:Ba:Cu composition films formed on a (A) MgO(100), (B) yttria stabilized zirconia substrate. (a) Y:Ba:Cu=1:2:3 (b) Y:Ba:Cu=1:2.4:4.2

the films on MgO(100) substrate, film with the Y:Ba:Cu=1:2.4:4.2 gave the Tc(zero) of 78 K and the temperature dependence behaviour was metallic, while that with 1:2:3 gave the Tc(zero) of 71 K and indicated semiconducting behaviour. In the case of the films on YSZ substrate, Tc(zero) was 90 K and 86 K for the film of Y:Ba:Cu=1:2.4:4.2 and 1:2:3, respectively . For this substrate, however, XRD patterns showed that the films of both compositions were c-axis oriented. The critical current density of films with Y:Ba:Cu=1:2.4:4.2 on both MgO(100) and YSZ substrates were about 10^3 A/cm^2 at 13 K and 10^1 A/cm^2 at 77 K. That of films with Y:Ba:Cu=1:2:3 on both MgO(100) and YSZ substrates were about 10^2 A/cm^2 at 13 K.

These results suggest that the Tc(zero) of the films relates to the starting composition of the precursor solution and the substrate used. Starting mole ratio of Y:Ba:Cu=1:2.4:4.2 may serve to correct the deviation of the metal composition from 1:2:3 due to the interaction of the film with the substrate during heat-treatment. On the other hand, the factors affecting orientation of the film is much more complicated. Low critical current density may result from the weak linking of superconducting grains and/or from the appearance of the impurity phases.

Two kinds of solutions with different Bi:Sr:Ca:Cu mole ratio were also prepared for the Bi-Sr-Ca-Cu-O system. Figure 5 shows the temperature dependence of resistivity for the films on MgO(100) substrate. The starting compositions of the samples were Bi:Sr:Ca:Cu=2:2:2:3 and 2:2:3:4.5. Although the XRD patterns of both films were quite similar showing highly c-axis oriented structure, temperature dependences of resistivity were different for two films. Almost no drop in resistance was observed at 115 K for the film of 2:2:2:3, but showed a sharp drop to zero at about 74 K. This indicates that there is little high-Tc phase superconductor in this film, but only a low Tc phase superconductor. On the other hand, apparent drop in resistance was observed at 115 K and the Tc(zero) was 80 K for the film of 2:2:3:4.5 suggesting the existence of high-Tc phase superconductor.

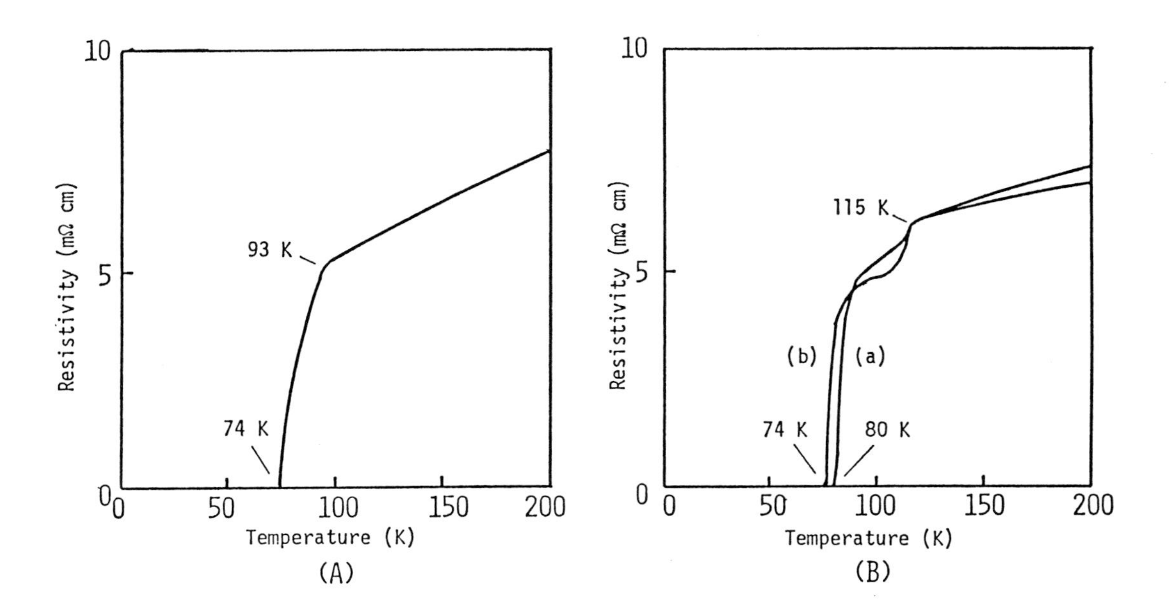

Fig.5. Temperature dependence of resistivity for the films with different starting Bi:Sr:Ca:Cu composition formed on MgO(100) substrate. (A) Bi:Sr:Ca:Cu=2:2:2:3, heat-treated at 880 °C for 1 h followed by 860 °C for 6 h (B) Bi:Sr:Ca:Cu=2:2:3:4.5, (a) heat-treated at 880 °C for 1 h and subsequently kept at 860 °C for 6 h (b) heat-treated at 870 °C for 1 h and subsequently kept at 850 °C for 6 h

The drop became clearer by lowering the final heat-treatment temperature as shown in figure 5. The critical current density of films were about 10^3 A/cm^2 at 13 K. These results implies that Ca-rich and Cu-rich composition are preferred for the preparation of the high-Tc phase. This agrees well with the appropriate compositions for synthesizing a ceramic sample [9].

It is concluded that the existence of a high Tc phase appears to be quite sensitive to such parameters as composition of the starting solution and final heat-treatment temperature for the Bi-Sr-Ca-Cu-O system. Further optimization of these parameters is currently under way to see if higher Tc can be achieved.

REFERENCES

[1] M. K. Wu, J. R. Ashburn, C. J. Torng, P. H. Hor, R. J. Meng, L. Gao, Z. J. Huang, Y. Q. Wang, and C. W. Chu, Phy. Rev. Lett., 58, 908 (1987)

[2] M. Kawai, T. Kawai, H. Masuhira, and M. Takahasi, Jpn. J. Appl. Phys., 26, L1740 (1987); R. L. Henry, H. Lessoff, E. M. Swiggard, and S. B. Qadri, J. Cryst. Growth, 85, 615 (1987); A. Gupta, G. Koren, E. A. Giess, N. R. Moore, E. J. M. O'Sullivan, and E. I. Cooper, Appl. Phys. Lett., 52, 163 (1988); A. K. Saxena, S. P. S. Arya, B. Das, A. K. Singh, R. S. Tiwari, and O. N. Srivasta, Solid State Comm., 66, 1063 (1988)

[3] C. E. Rice, R. B. van Dover, and G. J. Fisanick, Appl. Phys. Lett., 51, 1842 (1987)

[4] T. Kumagai, H. Yokota, K. Kawaguchi, W. Kondo, and S. Mizuta, Chem. Lett., 1987, 1645; A. H. Hamdi, J. V. Mantese, A. L. Micheli, R. C. O. Laugal, D. F. Durgan, Z. H. Zhang, and K. R. Padmanabhan, Appl. Phys. Lett., 51, 2152 (1987); H. Nasu, S. Maeda, T. Kato, Y. Ibara, T. Imura, and Y. Osaka, Chem. Lett., 1987, 2403; M. E. Gross, M. Hong, S. H. Liou, P. K. Gallagher, and J. Kwo, Appl. Phys. Lett., 52, 160 (1988)

[5] A. Gupta, R. Jagannathan, E. I. Cooper, E. A. Giess, J. I. Landman, and B. W. Hussey, Appl. Phys. Lett., 52, 2077 (1987)

[6] T. Monde, H. Kozuka, and S. Sakka, Chem. Lett., 1988, 287; S. Shibata, T. Kitagawa, H. Okazaki, T. Kimura, and T. Murakami, Jpn. J. Appl. Phys., 27, L53 (1988); T. Nonaka, K. Kaneko, T. Hasegawa, K. Kishino, Y. Takahashi, K. Kobayashi, K. Kitazawa, and K. Fueki, Jpn. J. Appl. Phys., 27, L867 (1988); S. A. Kramer, G. Kordas, J. McMillan, G. C. Hilton, and D. J. Van Harligen, Appl. Phys. Lett., 53, 156 (1988)

[7] T. Kumagai, W. Kondo, H. Yokota, H. Minamiue, and S. Mizuta, Chem. Lett., 1988, 551

[8] D. L. Kaiser, F. Holtzberg, B. A. Scott and T. R. McGuire, Appl. Phys. Lett., 51, 1040 (1988)

[9] A. Sumiyama, T. Yoshitomi, H. Endo, J. Tsuchiya, N. Kijima, M. Mizuno, and Y. Oguri, Jpn. J. Appl. Phys., 27, L821 (1988)

Preparation of 107K Superconducting Thin Film in Bi-Sr-Ca-Cu-O System by Coevaporation

T. YOSHITAKE[1], T. SATOH[2], Y. KUBO[1], and H. IGARASHI[1]

Fundamental Research Laboratories[1], and Resources and Environment Protection Research Laboratories[2], NEC Corporation, 4-1-1, Miyazaki, Miyamae-ku, Kawasaki, 213 Japan

ABSTRACT

Thin films of $Bi_{2.0}(Sr_{1-x}Ca_x)_{3.3}Cu_{2.3}O_y$ system were prepared on (100)MgO substrates by coevaporation of Bi_2O_3, Sr-Ca alloy and Cu metal, and effects of film composition and annealing condition on superconducting properties were investigated. High-T_c phase was formed for the films with relatively high Ca composition and with post-deposition annealing at high temperature. The superconducting transition temperatures of both low-T_c phase and high-T_c phase decreased with increasing Ca composition. And the films with $x=0.51$ showed the highest superconducting transition temperature of 107K.

INTRODUCTION

The discovery of new high T_c superconducting materials in the Bi-Sr-Ca-Cu-O system caused a new development in the research of oxide superconducting materials [1]. This Bi system has two superconducting phases with different T_c's of about 85K (low-T_c phase) and about 110K (high-T_c phase). It is reported that the low-T_c phase has a crystal structure with c-axis of about 30.7Å and has an ideal chemical composition of $Bi_2(Sr,Ca)_3Cu_2O_y$ [2-3]. And the high-T_c phase has a crystal structure with c-axis of about 37Å and has an ideal chemical composition of $Bi_2(Sr,Ca)_4Cu_3Oy$ [4-5]. The high-T_c phase has one extra CuO_2 and Ca layer between two BiO layers compared with the low-T_c phase.

Thin film preparation of the high-T_c phase is an attractive target for both fundamental understanding and practical application of this Bi system. Many groups have already prepared the thin films of this Bi system [6-10]. However there has been little information on the thin film with zero resistance T_c beyond 100K [11-12] because of the difficulty in preparing the high-T_c phase.

Recently, we have succeeded in preparing thin films with zero resistance T_c over 100K in Bi-Sr-Ca-Cu-O system by coevaporation [13-15] and indicated the importance of post-deposition annealing at high temperature such as 890°C in order to form the high-T_c phase. In this paper, we will report the dependence of the lattice constant and T_c's on the Ca composition for the $Bi_{2.0}(Sr_{1-x}Ca_x)_{3.3}Cu_{2.3}O_y$ system.

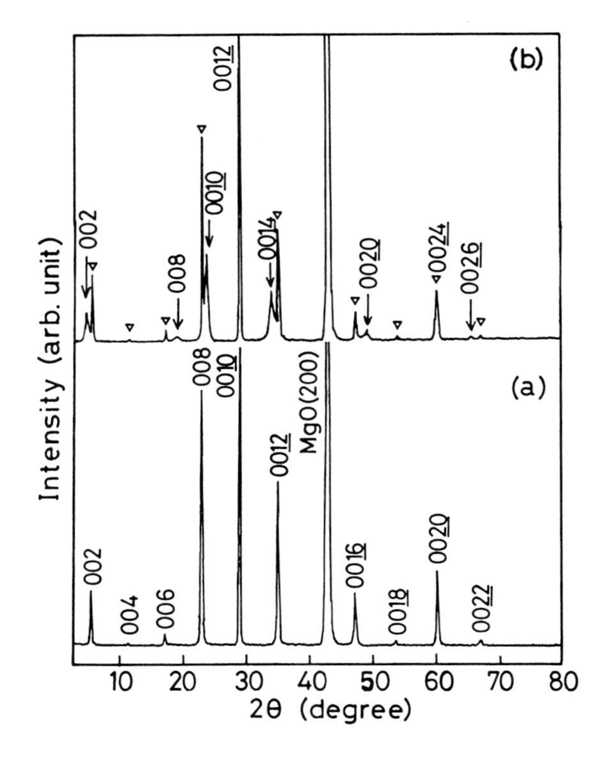

Fig. 1. X-ray diffraction patterns of the thin films with $x=0.52$, (a)annealed at 850°C for 3h, (b)annealed at 890°C for 5h. In (a), indices are for the low-T_c phase. In (b), indices are for the high-T_c phase and the lines of the low-T_c phase are indicated by ∇.

EXPERIMENTAL

Thin films were prepared on (100)MgO substrates by the multiple source evaporation system (ANELVA VD-45S). Bi_2O_3, Sr-Ca alloy and Cu metal were evaporated from three separate electron-beam heating sources. In this experiment, we changed the Ca composition x of the film by changing the composition of the Sr-Ca alloy ingot. The detail conditions of the film preparation were described elsewhere [13].

As-deposited films were amorphous and insulating. Therefore, these films were annealed at 850°C~890°C for 1~5 h in flowing oxygen atmosphere. The chemical compositions of the films were determined by using electron probe microanalysis (EPMA). The structure of the films were studied by X-ray diffraction method with CuKα radiation. The dc resistivity was measured by a conventional four probe method using an current density of $2A/cm^2$. Electrical contacts were made by using gold electrodes sputtered on the film surface and silver paste for outer lead connection.

RESULTS AND DISCUSSION

Figure 1 shows the X-ray diffraction patterns of the films with $x=0.52$. In this figure, patterns (a) and (b) correspond to the films annealed at 850°C for 3h and at 890°C for 5h, respectively. The diffraction lines of the pattern (a) are completely indexed as the (00l) lines of the low-T_c phase. On the other hand, we can observe several somewhat broad diffraction lines in addition to the (00l) lines of the low-T_c phase in the pattern (b). These lines are indexed as the (00l) lines of the high-T_c phase. Thus, the high-T_c phase was observed for the films annealed at relatively high temperature. No lines with indices different from (00l) are seen in both diffraction pattern (a) and (b). This indicates that both the low-T_c phase and the high-T_c phase have strongly preferred orientations with the c-axis perpendicular to the film plane.

Figure 2 shows the formation ranges of phases determined by the X-ray diffraction measurements. In this figure, L, H and M indicates the low-T_c phase, the high-T_c phase and a phase with c-axis of ~24Å, respectively. And N indicates other phase which is insulating.

In the films annealed at 850°C for 3h, only the low-T_c phase was formed over the wide range of $x>0.25$. On the other hand, in the films annealed at 890°C for 1h, the high-T_c phase formed in the composition range of $x>0.6$ in addition to the formation of the phase M in the Ca poor site. Further, by annealing the films at 890°C for 5h, the formation range of the high-T_c phase extended to the composition range of $x>0.5$. From these results, the formation of the high-T_c phase seems to be preceeded by the formation of the low-Tc phase. And the preparation of the high-T_c phase requires the relatively high Ca composition of $x>0.5$ and the annealing condition of high temperatures and long durations.

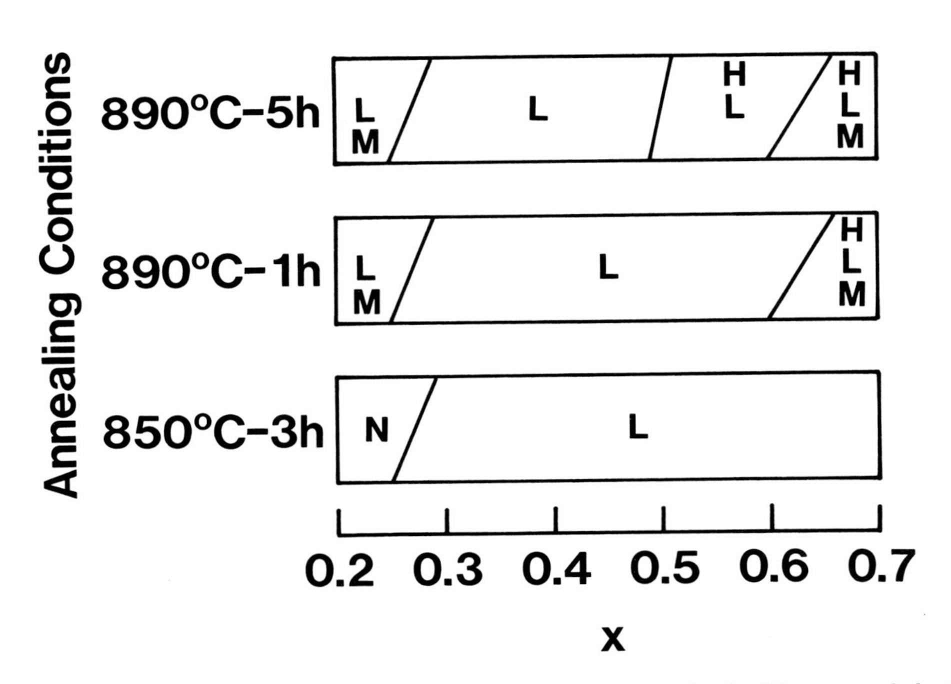

Fig. 2. The formation range of the superconducting phases in the films annealed at 850°C for 3h, at 890°C for 1h and at 890°C for 5h. L: the low-T_c phase, H: the high-T_c phase, M: the phase with c-axis of ~24Å, N: other phase (non-superconducting).

Table I. Typical examples of the compositions of the as-deposited films and both the high-T_c phase and the low-T_c phase in the annealed films. The Ca fraction to the total amount of (Ca+Sr) is shown as x values. The composition of each phase was analyzed separately, and was normalized to the two bismuth atoms in formula. L: the low-T_c phase, H: the high-T_c phase, a: 850°C-3h, b: 890°C-5h.

Film No.	As-deposited		Annealed			
	Bi:Sr:Ca:Cu	x	Bi:Sr:Ca:Cu	x	phase	condition
1	2.0:2.0:0.8:2.1	0.29	2.0:1.8:0.7:1.4	0.28	L	a
2	2.0:1.8:1.3:2.2	0.41	2.0:1.7:1.1:1.5	0.39	L	a
3	2.0:1.8:1.9:2.5	0.51	2.0:1.7:1.9:2.0	0.53	H	b
			2.0:1.6:1.4:1.8	0.47	L	
4	2.0:1.1:2.1:2.2	0.66	2.0:1.2:1.9:1.9	0.61	L	a
			2.0:1.3:2.0:2.0	0.61	H	b
			2.0:1.4:1.6:1.6	0.53	L	

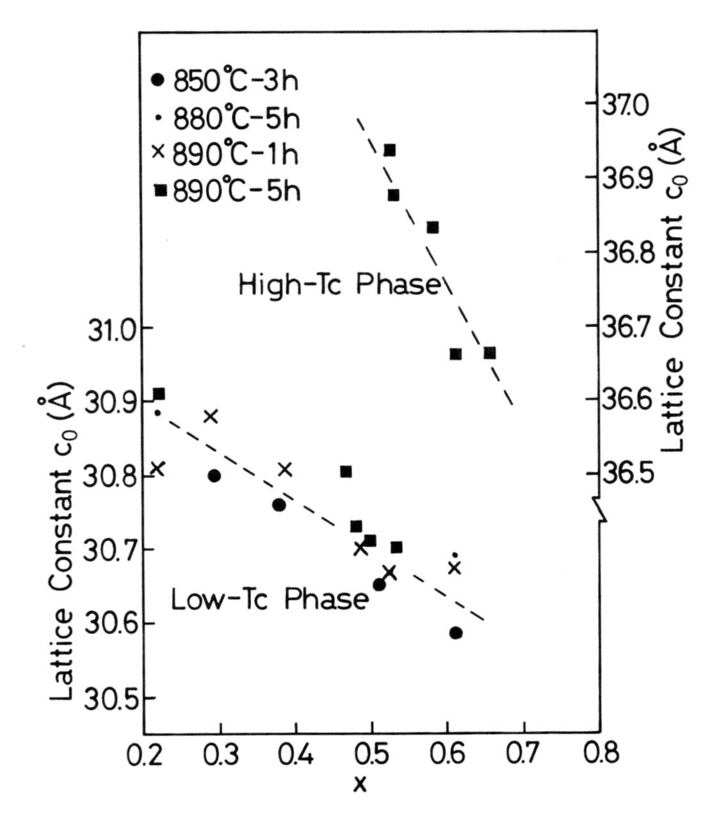

Fig. 3. The dependence of c-axis lattice constant of the low-T_c phase (below) and the high-T_c phase (above) on Ca composition x in the annealed films.

Table I shows the relation between the compositions of the as-deposited films and those of both the high-T_c phase and the low-T_c phase in the annealed thin films. Because these two phases in the annealed films could be distinguished each other in a backscattered electron micrographs as already reported [15], we were able to determine the composition of each phase separately by EPMA. In Film 1 and Film 2, only the low-T_c phase was formed. Therefore, the Ca composition x of the low-T_c phase is nearly equal to that of the as-deposited film. In Film 3, both the high-T_c phase and the low-T_c phase were formed. The analyzed composition of the high-T_c phase contained more Ca and Cu as compared with that of the low-T_c phase. In Film 4, the Ca composition x of the film annealed at 850°C for 3h is nearly the same as that of the as-deposited film. On the other hand, the Ca composition x of the low-T_c phase formed after annealed at 890°C for 5h becomes smaller ($x \sim 0.53$) than that of the as-deposited film. This is caused by the formation of the high-T_c phase which has higher Ca composition of $x \sim 0.61$ than the low-T_c phase. These result confirms us that the compositions of both the low-T_c phase and the high-T_c phase of the annealed films are not always the same as those of the as-deposited films. And both the low-T_c phase and the high-T_c phase have mutual solubilities between Ca and Sr.

In Table I, the analyzed composition of both phases were not equal to their ideal composition. This suggests an existence of some kind of defects such as cation substitutions or phase intergrowths.

Figure 3 shows the dependence of the lattice constant c_0 on the Ca composition x in the annealed film. In this figure, the Ca composition x corresponds to that of each phase appeared in the annealed films. Since these films have a strongly preferred orientation with the c-axis perpendicular to the film plane, we did not obtain any information about the lattice constants of a- and b-axis. And we derived the lattice constant c_0 from only one diffraction peak of (00$\underline{14}$) for the high-T_c phase because of the broadness of diffraction peaks as shown in Fig.1.

The c-axis lattice constant of the low-T_c phase decreases monotonously from~30.9Å to ~30.6Å with increasing Ca composition. The c-axis lattice constant of the high-T_c phase also decreases from ~36.95Å to ~36.6Å with increasing Ca composition. These results show that both the low-T_c phase and the high-T_c phase have mutual solubilities between Ca and Sr, and the change of the c-axis lattice constant of both phases results from the difference of the ionic size between Sr and Ca. That is, the c-axis lattice constants of both phases decrease with increasing Ca composition, which has a smaller ionic size than Sr.

Figure 4 shows some examples of the temperature dependence of the dc resistivities of the films annealed at 850°C for 3h. Three curves ($x=0.41$, 0.52, 0.66) show the superconducting transition at around 85K. These superconducting transitions correspond to those of the low-T_c phase. The T_c's of these films decrease with increasing Ca composition.

Figure 5 shows some examples of the temperature dependence of the dc resistivities of the films annealed at 890°C for 5h. The resistivity curves of the film with $x=0.41$ shows the superconducting transition at around 85K. On the other hand, the resistivity curves of the films with $x=0.52$ and $x=0.66$ show the superconducting transition at around 110K. These superconducting transitions correspond to those of the high-T_c phase. This figure shows that the T_c's of the high-T_c phase also change with x. And the film with $x=0.52$ shows the highest superconducting transition with midpoint T_c of 111K and zero resistance Tc of 107K.

Figure 6 shows the dependence of the midpoint T_c of both the low-T_c phase and the high-T_c phase on Ca composition x. The T_c's of both phases are decreased monotonously with increasing Ca composition x. This tendency corresponds well to the change of the c-axis lattice constant as shown in Fig.3.

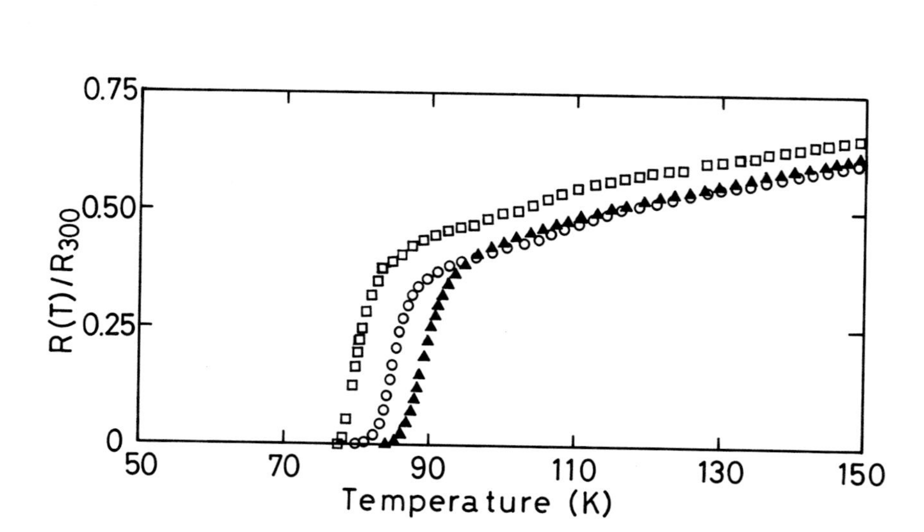

Fig. 4. Temperature dependence of dc resistivity of the thin films annealed at 850°C for 3h, ▲: $x=0.41$, ○: $x=0.52$, □: $x=0.66$.

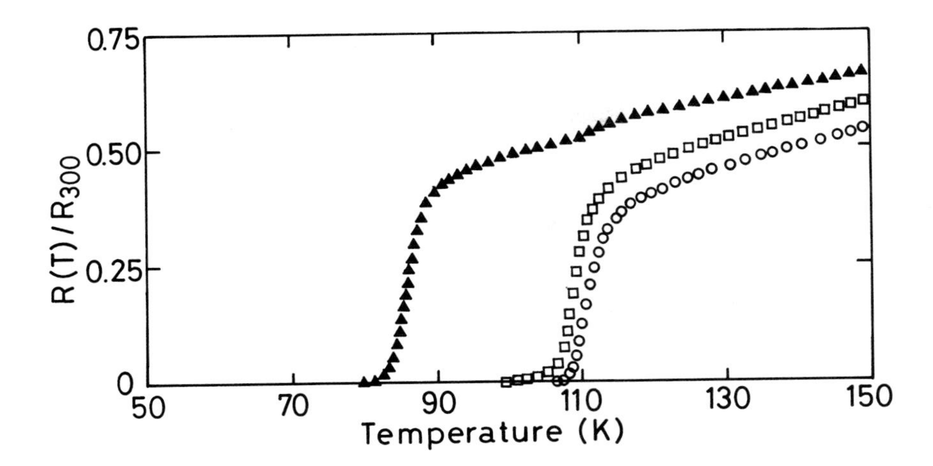

Fig. 5. Temperature dependence of dc resistivity of the thin films annealed at 890°C for 5h, ▲: x=0.41, ○: x=0.52, □: x=0.66.

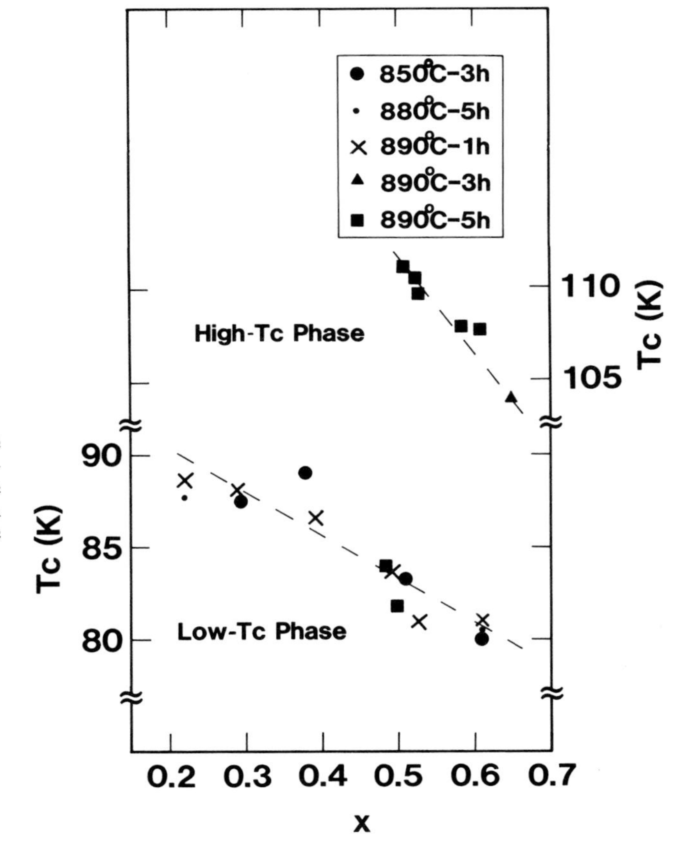

Fig. 6. The dependence of the midpoint T_c's of the low-T_c phase (below) and the high-T_c phase (above) on Ca composition x in the annealed films.

These results suggest that the T_c's of both the low-T_c phase and the high-T_c phase in Bi system are well related to the c-axis lattice constant. Thus, the T_c of the high-T_c phase might become higher if we can obtain the high-T_c phase in the composition range of $x<0.5$ or by the substitution of Ba for Sr.

CONCLUSION

The films of the Bi-Sr-Ca-Cu-O system were prepared on (100)MgO substrates by coevaporation. The high-T_c phase was formed for the film with relatively high Ca composition and with post-deposition annealing at high temperature. Both the low-T_c phase and the high-T_c phase have mutual solubilities between Ca and Sr. The changes of the c-axis lattice constant and T_c's of these two phases showed the similar tendency for the Ca composition x. This tendency suggest that the T_c's are well related to the c-axis lattice constant in the Bi-Sr-Ca-Cu-O system.

We would like to thank M. Yonezawa, T. Kamejima, J. Nakashima, J. Fujita and T. Manako for the supports of this work, and M. Sugimoto for the expert assistance.

References

1. H. Maeda, Y. Tanaka, M. Fukutomi and T. Asano, Jpn. J. Appl. Phys. 27, L209 (1988).
2. M. A. Subramanian, C. C. Torardi, J. C. Calabrese, J. Gopalakrishnan, K. J. Morrissey, T. R. Askew, R. B. Flippen, U. Chowdhry and A. W. Sleight, Science 239, 1015 (1988).
3. J. M. Tarascon, Y. Le Page, P. Barboux, B. G. Babley, L. H. Greene, W. R. McKinnon, G. W. Hull, M. Giroud and D. M. Hwang, Phys. Rev. B37, 9382 (1988).
4. E. Takayama-Muromachi, Y. Uchida, Y. Matsui, M. Onoda and K. Kato, Jpn. J. Appl. Phys. 27, L556 (1988).
5. H. W. Zandbergen, Y. K. Huang, M. J. V. Menken, J. N. Li, K. Kadowaki, A. A. Menovsky, G. van Tendeloo and S. Amelinckx, Nature 332, 620 (1988).
6. M. Hong, J. Kwo and J. J. Yeh, submitted to Appl. Phys. Lett.
7. C. E. Rice, A. F. J. Levi, R. M. Fleming, P. Marsh, K. W. Baldwin, M. Anzlowar, A. E. White, K. T. Short, S. Nakahara and H. L. Stormer, Appl. Phys. Lett. 52, 1828 (1988).
8. M. Nakao, H. Kuwahara, R. Yuasa, H. Mukaida and A. Mizukami, Jpn. J. Appl. Phys. 27, L378 (1988).
9. K. Kuroda, M. Mukaida, M. Yamamoto and S. Miyazawa, Jpn. J. Appl. Phys. 27, L625 (1988).
10. M. Fukutomi, J. Machida, Y. Tanaka, T. Asano, H. Maeda and K. Hoshino, Jpn. J. Appl. Phys. 27, L632 (1988).
11. S. Hatta, Y. Ichikawa, K. Hirochi, K. Setsune, H. Adachi and K. Wasa, Jpn. J. Appl. Phys. 27, L855 (1988).
12. Y. Ichikawa, H. Adachi, K. Hirochi, K. Setsune, S. Hatta and K. Wasa, Phys. Rev. B38, 765 (1988).
13. T. Yoshitake, T. Satoh, Y. Kubo and H. Igarashi, Jpn. J. Appl. Phys. 27, L1089 (1988).
14. T. Yoshitake, T. Satoh, Y. Kubo, T. Manako and H. Igarashi, Jpn. J. Appl. Phys. 27, L1094 (1988).
15. T. Yoshitake, T. Satoh, Y. Kubo and H. Igarashi, Jpn. J. Appl. Phys. 27, L1262 (1988).
16. T. Satoh, T. Yoshitake, Y. Kubo and H. Igarashi, submitted to Appl. Phys. Lett.

Superconducting Thin Films of $YbBa_2Cu_3O_7$ and $ErBa_2Cu_3O_7$ Grown by Molecular Beam Epitaxy: Determination of Epitaxial Conditions

R. CABANEL, J.P. HIRTZ, G. GARRY, F. HOSSEINI TEHERANI, and G. CREUZET

THOMSON-CSF/LCR, B.P. 10, 91401 Orsay, France

ABSTRACT

We have successfully achieved the epitaxial growth of $(RE)_1(BaF_2)_2Cu_3$ (where RE= Yb or Er) on (100) $SrTiO_3$ substrates. The oxygen stoichiometry is obtained by annealing after growth under 1 atm oxygen pressure, leading to the transition from $(RE)_1(BaF_2)_2Cu_3$ to $(RE)_1Ba_2Cu_3O_{7-x}F_y$. The epitaxial conditions are destroyed either for thicknesses above 2000 Å or by introduction of an oxygen partial pressure during growth. The final films are highly oriented with c-axis perpendicular to the substrate. They exhibit sharp transitions observed by transport and magnetic measurements. In addition, the films are very stable without any particular storage conditions. However, RBS studies show diffusion of the species between the film and the substrate, due to the high temperature annealing stage.

INTRODUCTION

Since the discovery of the high temperature superconductivity in copper oxides compounds [1], several laboratories attempt to obtain HTSC (High Tc Superconducting) thin films using various techniques like sputtering [2-4], MOD (Metallorganic Deposition) [5,6], Spray deposition [7], Laser Ablation [8], simultaneous or sequential evaporation [9,10], or Molecular Beam Epitaxy [11-13]. All these techniques demonstrate their ability to lead to thin films with resistive transitions around 90K. However it clearly seems that these techniques are not equivalent with respect to the applications. In fact, due to the large field of microstructure properties, the growth parameters, and the large influence of these characteristics on the final superconducting properties, it appears especially that to obtain controlled and reliable Josephson junctions we need the expertise of atomic layer epitaxy what is specific of a few of them. That is the reason why we have chosen the MBE technique.

ELABORATION

Starting from Cu, BaF_2 and RE which are heated in PBN (Pyrolitic Bore Nitride) crucibles, the evaporated species are growing on (100) $SrTiO_3$ substrates at 130°C under a $\sim 10^{-8}$ Torr vacuum. So the compound is obtained by codeposition of the elements whose relative rate are $V_{BaF2}/V_{Yb}= 1.9$, $V_{Cu}/V_{Yb}= 0.42$, $v_{BaF2}/V_{Er}= 6.2$, $V_{Cu}/V_{Er}= 1.4$ leading to a total rate between 0.1 and 0.3 Å/s. These values demonstrate the influence of the sticking coefficients which particularly depend on the rare earth .

In order to study the epitaxial condition on this substrate, we have looked at the reconstruction of the (100) surface, using RHEED (Reflected High Energy Electron Diffraction). After a 650°C annealing under high vacuum, we can see on fig.1 the characteristic streaks of a highly ordered smooth surface. This pattern is still observed after growth of the first 2000Å of the film. Then a degradation appears due to the constraints in the deposit. Such lines are also sensitive to a 10^{-5}Torr pressure of oxygen in the chamber, which is relevant of a chemical reaction with the species.

After deposition, samples are released to air before the thermal annealing in O_2 at high temperature which leads to the so-called 1,2,3 phase ($(RE)_1Ba_2Cu_3O_7$). At this stage, using Barium fluorine revealed to be essential [14]. In fact with Ba source, films are not stable enough to be processed in an external oven. The superconducting phase is obtained after one hour thermal ramp to 850°C, half-an-hour plateau and six hours cooling ramp to room temperature under flowing oxygen.

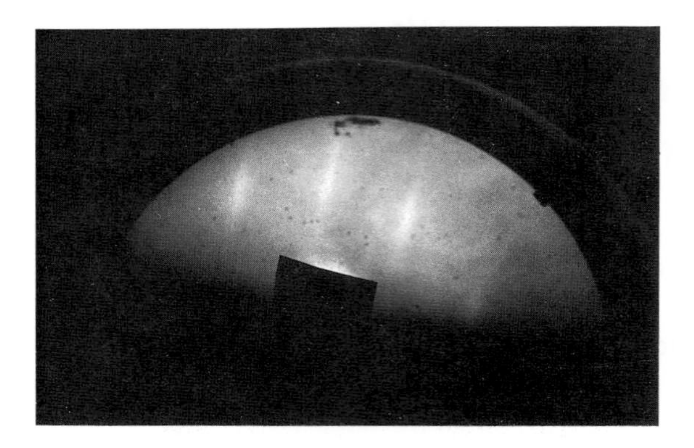

Fig1: Typical RHEED pattern of substrate surface after 650°C thermal annealing.

PROPERTIES

TRANSPORT AND MAGNETIC PROPERTIES [13]

Resistivity versus temperature curves show sharp transition at 87K and 77K (width between 90% and 10% of the onset resistivity is 2K and 8K) respectively for Yb and Er. Low field susceptibility versus temperature curves also exhibit sharp edge at a lower temperature than that was seen in the resistivity measurement (82 K). Using the Bean formula [15], magnetization versus field hysteresis data leads to a critical current of 2 10^4 A/cm^2.

COMPOSITION

The composition of the as-deposited films is systematically checked on a CAMECA EPMA (Electron Probe Micro-Analysis) using Cu Lα, Ba Lβ, F Kα, Er Mα and Yb Lα lines. With thicknesses greater than 4000 Å and electron beam energy less than 10 kV, no substrate species are detected. Moreover, using this energy it is possible to correctly analyze the Fluorine content (30 minutes under e-beam do not change results), which is in agreement with the stoichiometry of the BaF_2 deposit.

After annealing, Fluorine content is between 0.3 and 1.7 depending on the samples. This large range of content could be relied on a variation of a parameter of the thermal treatment. In fact, P.M. Mankiewich [16] and S.W. Chan et al[17] saw that a low partial pressure of H_2O during the high temperature stage, could lead to a strong variation of the fluorine content in thin films. The residual Fluorine content could explain the good stability of the films, while the resistivity transition evolves very little after exposure in air for several months.

Due to the long high temperature plateau, interdiffusion of species occurs at the film/substrate interface. This was verified using RUMP simulation on the RBS spectra [18]: a strong diffusion of Sr and Ba occurs. Such a diffusion could explain the rather low critical temperature value we have noticed on our films compared with other results or bulk samples.

STRUCTURE

Crystalline structure of the film was investigated by X-Rays diffraction (Cu Kα line). Thin films revealed to be granular and crystallized in the 1,2,3 phase (fig. 2). However, some extra phases are present in the sample. Another point is that films are textured with c-axis perpendicular to the substrate.

This texture is remarkably increased by a subsequent short annealing (fig.3). Nevertheless, in the meantime, the critical temperature decreases in large proportion (fig.4). This variation is relied on the evolution of the oxygen content in the film, since the short annealing does not present any slow cooling in the 500-300°C range. Moreover, a subsequent long annealing under oxygen restores the high temperature transition. We also can notice that

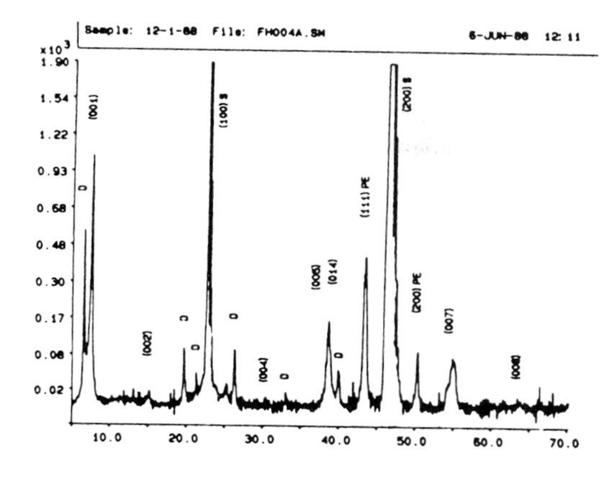

Fig2: X Ray diffraction pattern
after first long annealing.
S = substrate;D = phase which
disappears within short annealing

Fig3: Same sample than in Fig2
after short annealing.
A = phase which appears.

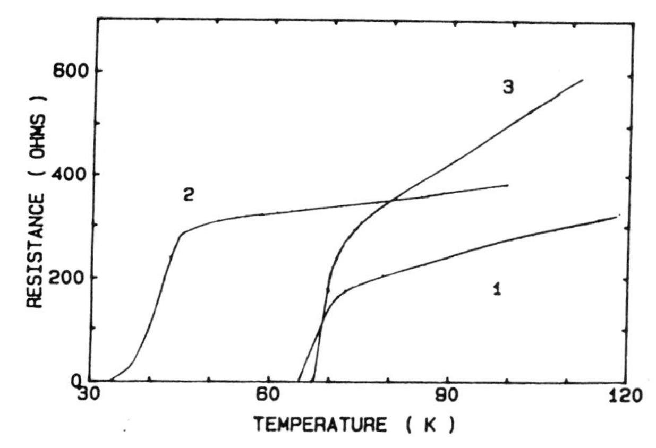

Fig4 Resistance vs Temperature behaviour as a function
of thermal treatments. 1: after first long annealing, 2:after
short annealing, 3: after the last long annealing.

the transition width increases after the rapid annealing, what could be due to an inhomogeneous oxygen content inside the film. Following the last treatment this width reaches 6 K on Er-based sample. These results are in agreement with those of J.V. Manteśe et al[19] on MOD thin films.

CONCLUSION

Epitaxial growth of $(RE)_1(BaF_2)_2Cu_3$ compound on (100) $SrTiO_3$ substrates have been achieved by MBE. After ex-situ thermal treatment under Oxygen, sharp superconducting transitions have been checked by resistivity and susceptibility measurements. The use of BaF_2 instead of pure Ba revealed to be essential to successfully process the films. It points out that this low temperature way of deposition is compatible with lift-off and other technological processes. However, it is worthy to note that the high temperature annealing induces a large diffusion of species (RBS simulation has been roughly confirmed by microanalyses in a Scanning Transmission Electron Microscope [20]). We could expect that Rapid Thermal Annealing will lead to right structure and lower diffusion of Sr, Ba and RE [21].

Nevertheless our main goal is to obtain a one-step epitaxial growth of $(RE)_1Ba_2Cu_3O_{7-x}$ using in-situ process. Optimum one should successively lead to the growth of CuO_2, BaO,

CuO, RE, CuO, BaO, CuO_2 layers. In this field we actually know, that introducing pure molecular oxygen is not sufficient to achieve it. In fact our first results let us think that, like others do [12], the use of activated atomic oxygen is essential to oxidize copper during the growth. This technique yet revealed to be efficient in a low temperature Plasma Assisted Laser Deposition process [22]. In MBE, recent results of R.J. Spah et al [23] are very encouraging. Such a multilayer epitaxy could exhibit less grain boundary defects, what should result in high transport performances.

ACKNOWLEDGMENTS

It is a pleasure to acknowledge magnetic and resistivity measurements by L.Fruchter and C.Giovanella, EPMA by O.Lagorsse, RBS study by J.Siejka and G.Vizkelethy. The authors also are particularly indepted to D.Dubreuil for thermal treatments.

REFERENCES

1 M.K. Wu, J.R. Ashburn, C.J. Torng, P.H. Hor, R.L. Meng, L.Gao, Z.I. Huang, Y.Q. Wang and C.W. Chu, Phys. Rev. Lett. 58 , 908 (1987).
2 H. Adachi, K. Hiroshi, K. Setsune, M. Kitabatake, and K. Wasa, Appl. Phys. Lett. 51, 2263 (1987).
3 R.L. Sandstrom, W.J. Gallagher,T.R. Dinger, R.H. Koch, R.B. Laibowitz, A.W.Kleinsasser, R.J. Gambino, B.Bumble, M.F.Chisholm, Appl. Phys. Lett. 53, 444 (1988)
4 H. Myoren, Y. Nishiyama, H. Nasu, T. Imura, Y. Osaka, S. Yamanaka, M. Hattori, Japn.J.Appl.Phys.27, L 1068 (1988).
5 A.H. Hamdi, J.V. Mantese, A.L. Micheli, R.C.O. Laugal, D.F. Dungan, Z.H. Zhang, K.R. Padmanabhan, Appl.Phys Lett. 51, 2152 (1987).
6 M.E. Gross, M. Hong, S.H. Liou, P.K. Gallagher, and J. Kwo, Appl. Phys Lett. 52, 160 (1988).
7 A. Gupta, G. Koren, E.A. Giess, N.R. Moore, E.J.M. O'Sullivan, E.I. Cooper, Appl. Phys. Lett. 52 (1988).
8 L. Lynds, B.R. Weinberger, G.G. Peterson, H.A. Krasinski, Appl. Phys.Lett. 52,320 (1988).
9 M. Futamoto, Y. Honda, Jpn. J. Appl. Phys. 27, L73 (1988).
10 A. Mogro-Campero, B.D. Hunt, L.G. Turner, M.C. Burnell, W.E. Balz, Appl. Phys. Lett. 52, 584 (1988).
11 J. Kwo, T.C. Hsieh, R.M. Fleming, M. Hong, S.H. Liou, B.A. Davidson, L.C. Feldman, Phys. Rev. B 36, 4039 (1987).
12 E.S. Hellman, D.G. Schlom, N. Missert, K. Char, J.S. Harris Jr, M.R. Beasley, A. Kapitulnik, T.H. Geballe, J.N. Eckstein, S.L. Weng, C. Webb J. Vac. Sci. Technol. B6, 799 (1988).
13 R. Cabanel, J.P. Hirtz, P. Etienne, L. Fruchter, C. Giovanella, G. Creuzet, Proceedings HTSC-M^2S, Interlaken (March 1988).
14 P.M. Mankiewich, J.H. Scofield, W.J. Skocpol, R.E. Howard,A.H. Dayem, E. Good, Appl. Phys. Lett. 51, 1753 (1987).
15 C.P.Bean, Phys.Rev.Lett.8, 250 (1962).
16 P.M. Mankiewich Proceedings T.M.S. New Brunswich 9-11 May 1988.
17 S.W. Chan, B.G. Bagley, L.H. Greene, M. Giroud, W.L. Feldmann, K.R. Jenkin II, B.J. Wilkins, Submitted to Appl. Phys. Lett. June 1988.
18 R.Cabanel, J.P. Hirtz, C. Giovanella, D. Dubreuil, G. Creuzet, J. Siejka, G. Vizkelethy, Proceedings T.M.S. New Brunswich 9-11 May 1988.
19 J.V. Mantese,A.H. Hamdi, A.L. Micheli, Y.L. Chen, C.A. Wong, J.L. Johnson, M.M. Karmarkar, K.R. Padmanabhan, Appl.Phys.Lett. 52, 1631 (1988).
20 J. Chazelas, R. Cabanel, J.P. Hirtz, F. Hosseini Teherani, G. Creuzet, to be submitted to J.Electron Micros.Techn.
21 N.W. Cody, U. Sudarson, R. Solanki, Appl.Phys.Lett. 52, 1531 (1988).
22 S.Witanachchi, H.S. Kwok, X.W.Wang, D.T.Shaw, Appl.Phys.Lett.53, 234 (1988).
23 R.J. Spah, H.F. Hess, H.L. Stormer, A.E. White, K.T. Short, Appl.Phys.Lett. 53, 441 (1988).

Ln-Ba-Cu-O Thin Film Deposited by Ion Beam Sputtering

TSUTOMU YOTSUYA[1], YOSHIHIKO SUZUKI[1], SOICHI OGAWA[1], HAJIME KUWAHARA[2], TETSURO TAJIMA[3], KOHEI OTANI[4], and JUNYA YAMAMOTO[5]

[1] Osaka Prefectural Industrial Technology Research Institute, 2-1-53, Enokojima, Nishi-ku, Osaka, 550 Japan
[2] Nissin Electric Co., Ltd., 47 Umezu-takasecho, Ukyo-ku, Kyoto, 615 Japan
[3] Daikin Industries Ltd., 1304 Kanaokacho, Sakai, 591 Japan
[4] Hitachi-Zosen Co., 1-3-22, Sakurajima, Konohana-ku, Osaka, 554 Japan
[5] Laboratory for Applied Superconductivity, 2-1, Osaka University, Yamadaoka, Suita, 565 Japan

ABSTRACT

We have successfully fabricated the high quality yttrium barium copper oxide (YBCO) thin film with Tc_0=88K and ytterbium barium copper oxide (YbBCO) film with Tc_0=77K by the method of Ion Beam Sputtering. The as-grown films (YBCO and YbBCO) on the MgO substrate showed superconductive transition up to 54 K. To improve the superconductivity, heat treatment was necessary at around 800°C for YbBCO film and around 900°C for YBCO film in the oxygen atmosphere. The opitimized conditions for heat treatment depended on substrate material and film composition.

Three kinds of targets were used, two were oxide targets for the YBCO system and the other was alloy target for the YbBCO system.

The c-axis of the perovskeit structure was perpendicular to a substrate surface for films fabricated on the MgO substrate. On the contrary, for the film deposited on the $SrTiO_3$(100) substrate, the a-b plane was perpendicular to the substrate surface.

INTRODUCTION

After the discovery of the high Tc oxides[1,2,3], many efforts were made to fabricate high quality thin film because of future applications and measurements of physical interests. The high-Tc thin film was formed by various methods such as rf-sputtering[4], electron beam deposition[5], laser ablation [6] and spin coating[7]. We have studied to form the high quality thin films by using the ion beam sputtering method. Because the operation pressure is two order of magnitude lower than that of the ordinary rf-sputtering method and it is expected that the film formed by the ion beam sputtering(IBS) has good quality and good adhesion with a substrate. For the lanthanum-oxide compounds, oxygen deficiency plays the most important roll for the superconducting behavior. Since the operation pressure is low(10^{-4} Torr), low energy oxygen ion beam could be used to realize oxygen treatment during deposition. Considering this feature we adopted the IBS method.

EXPERIMENTS

The sputtering chamber was evacuated by a conventional rotary pump and a cryopump down to $1x10^{-6}$ Torr. After evacuation, oxygen gas of $1x10^{-4}$ Torr and argon gas of $5x10^{-4}$ for ion beam were introduced into the chamber. The target voltage was 400 V and ion current 400mA. As shown in Fig. 1, the target was mounted at 45 degree to the ion beam. Three kind of targets were used. The oxide target compositions were Y:Ba:Cu=1:2:3.6 for the MgO substrate and Y:Ba:Cu=1:2.5:5 for the $SrTiO_3$ substrate and the alloy target of which composition was Yb:Ba:Cu=1:2.5:3.7. The film composition could slightly vary by adjusting deposition conditions such as oxygen partial pressure, substrate temperature and deposition rate. The substrate temperature could be elevated up to 600°C during depositions. However the film composition was almost

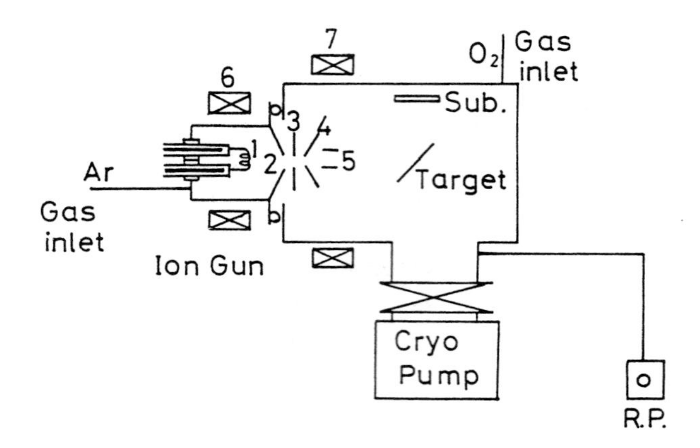

Fig. 1 A schematic diagram of the Ion Beam Sputtering apparatus.

constant over the substrate temperature range between 500 to 600°C.

The film composition was determined by the inductive coupled plasma atomic emission spectroscopy(ICPAES). The accuracy was within 5%.

The structure of the film was examined with the X-ray diffractmeter. The incident beam was CuKα radiations. All the films were heat treated in a tubular furnace under oxygen gas flowed. Typical temperature sequence was as follows. The temperature was increased at the rate of 6°C/min to the highest temperature and was kept for 30 minutes and then decreased to 500°C by 6°C/min. After keeping at 500°C for 4 hours, it was cooled down to room temperature. The highest temperature was changed from 500 to 900°C from annealing to annealing. After annealing, the resistivity and Tc was measured.

The resistivity and superconducting transition of the oxide films were measured with the ordinary four probe method. The electrical contact was made by using silver paint, and the temperature measured with the Pt resistance thermometer.

In order to measure the critical current density, a rectangular sheet of 1mm in width by 2mm in length was formed by chemical etching. The Az-1350J photoresist was used for patterning and 1% diluted HNO_3 acid for etching. To measure the critical current density, electrical contact was made by indium soldering.

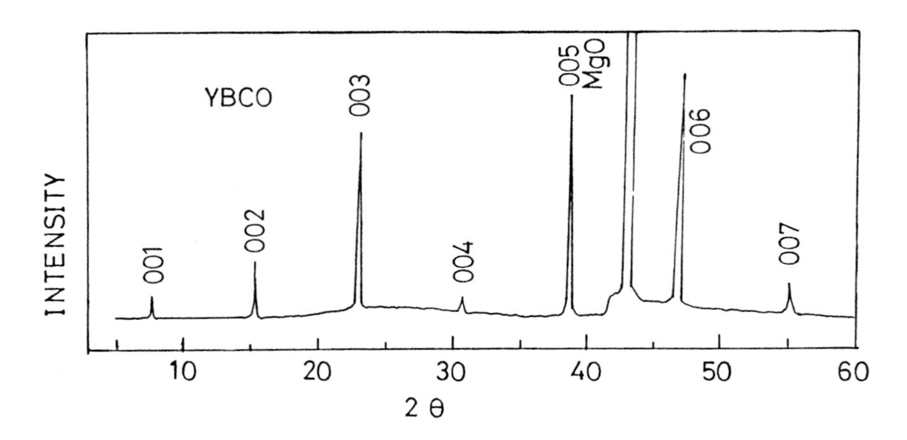

Fig. 2 X-ray diffraction pattern for the YBCO thin film on the MgO(001) substrate. The film thickness was about 1 um.

RESULTS and DISCUSSIONS

The as-grown film had already oxygen-deficient-perovskeit structure. The c-axis of the perovskeit structure of a Ln-Ba-Cu-O films formed on the MgO substrate was considerably oriented normal to the substrate surface. The lattice parameter c was as large as 12 Å for the as-grown film. After annealing, the lattice constant decreased as short as 11.66 Å. It is difficult to explain this large change of the lattice constant c in terms of oxygen deficiency only. We assume that the lattice constant c was enlarged by excess atoms such as Cu atom. These excess atoms were expelled from the crystal by heat treatment. The best c-axis orientation was achieved for YBCO thin film of which composition was Y:Ba:Cu=1:2:3.3 and the lattice constant c 11.66 Å as shown in Fig. 2. However the Tc_0 of this film was only 74 K. On the contrary the film on the MgO substrate, of which composition was very close to 1:2:3 composition, exhibited superconductive transition at 88 K. The critical current density achieved was as large as 10^5 A/cm^2 at 70 K.

The crystal morphology was observed by the SEM after annealing. A very smooth surface was obtained for a $YBa_2Cu_3O_x$ thin film for early stage of annealing. However granular grains became obvious for the film as annealing temperature increased.

On the contrary, the a-b orientation was observed for the YBCO film deposited on the $SrTiO_3$(100) substrate. The best superconductive properties were obtained for the film of which composition was Y:Ba:Cu=1:2.3:4.4. This composition was measured before annealing. Needlelike crystal growth was observed. This needlelike crystalline structure was degraded within a few months[8].

We have fabricated the YbBCO thin film by using an alloy target. In this case, it was necessary to do pre-sputtering for only 10-30 minutes. This was the largest advantage over the oxide target since 5 hours pre-sputtering was necessary for the YBCO oxide target. The deposition rate was measured for the alloy target. The deposition rate was as large as 110 Å for the virgin condition. However as shown in Fig. 3, the deposition rate decreased as the sputtering was carried out. After 100 hours the rate was changed to 40 Å/min, this was similar to the value obtained with the YBCO oxide target. It is supposed that the surface was oxidized during sputtering, so that the surface of the alloy target finally became very similar to that of the oxide target.

The composition of the film was almost independent from deposition conditions such as oxygen partial pressure, deposition rate, and substrate temperature under present experimental conditions. Although the Cu/Ba ratio was constant, Yb/Cu ratio was varied as the deposition was carried out. This was caused by the slight segregation existed in the alloy target.

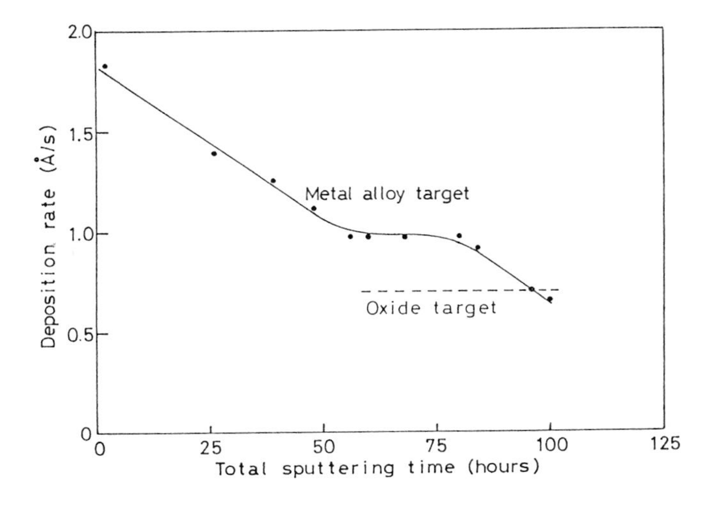

Fig. 3 The deposition rate for the YbBaCu alloy target is shown as a function of accumulated sputtering time.

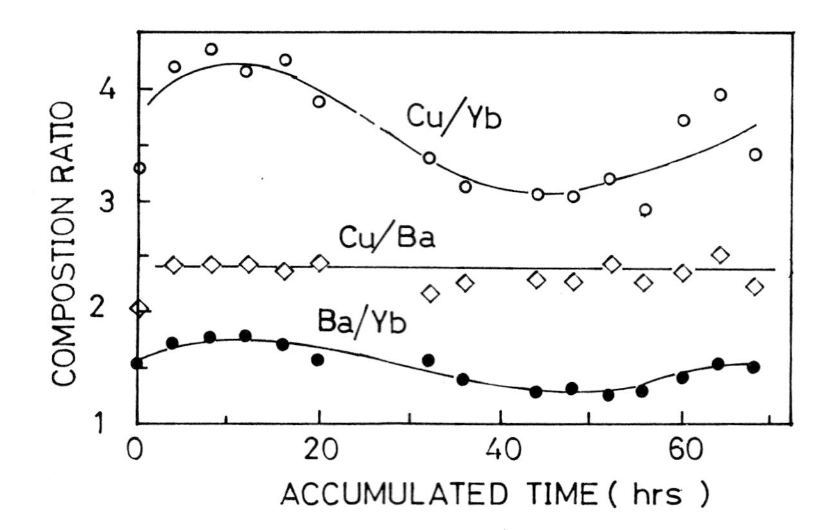

Fig. 4 The variation of the YbBaCuO film's composition is shown as a function of accumulated sputtering time.

The variation of the composition was shown in Fig. 4 and the highest Tc in this system was 77K after annealing at 850°C and the composition of the film was Yb:Ba:Cu = 1:1.7:4.2.

CONCLUSION

We have investigated LnBaCuO thin films by the method of ion beam sputtering using the YBCO oxide targets and the YbBaCu alloy target. The as-grown film showed superconducting transition up to 54 K. To improve the superconductivity, it was annealed above 800°C. The results are summarized as follows.

(1)The film structure was already oxygen deficient perovskeit structure for the as-grown films. The c-axis was perpendicular to the substrate surface for the MgO substrate. On the contrary the c-axis was parallel to the surface for the $SrTiO_3$ substrate. The lattice parameter c was as large as 12 A for the as-grown film, however it was made short by annealing the film.

(2)The YBCO films deposited on the MgO(100) and the $SrTiO_3$(100) substrate showed superconducting transition up to 88 K.

(3)The heat treatment condition depended on the substrate material.

(4)Even though the film composition of YbBCO film was not still close to 123 composition, superconducting transition up to 77 K was obtained after annealing.

(5)Although the alloy target needed very short pre-sputtering, the segregation in the target caused the variation of the film's composition.

ACKNOWLEDGMENT

We would like to thank Dr. T. Hamada for x-ray analysis and useful discussions and Mr. F. Uratani and Mr. M. Yoshikawa for the ICPAES measurements. The Yb-Ba-Cu alloy target was prepared by Dr. T. Dekawa and Mr. T. Kawae of Mitsui Engineering & Ship Building Co. Ltd.. This work was supported by the Osaka Cooperative Research Project for High-Tc Superconductors.

REFERENCES

1. J.G.Bednotz and K.A.Muller; Z.Phys. **B 64**, 189 (1987).
2. M.K.Wu, J.R.Ashburn, C.J.Torng, P.H.Hor, R.L.Meng, L.Gao,Z.J.Huang and C.W.Chu; Phys. Rev. Lett., **58** 908, (1987).
3. H.Maeda, Y.Tanaka, M.Fukutomi and T.Asano; Jpn. J. Appl. Phys., **27** L209 (1988)
4. Y.Emonoto, T.Murakami, M.Suzuki, and K.Moriwaki; Jpn. J. Appl. Phys. Lett. **26** L1845 (1987)
5. P.Chaudrari, R.H.Koch, R.B.Lainbowitz, T.R.McGuire, and R.J.Gambino;Phys. Rev. Lett., **58**, 2684 (1987)
6. J.Narayan, N.Biunno, R.Singh, O.W.Holland, and A.Auciello; Appl. Phys. Lett., **51**, 1845 (1987)
7. M.Kawai, T.Kawai, H.Hashiura, and M.Takahashi; Jpn. J. Appl. Phys. Lett. **26**, L1740 (1987)
8. T.Yotsuya,Y.Suzuki, S.Ogawa, H.Kuwahara, K.Otani, T.Emoto and J.Yamamoto; Procceeding of ICMC-88(Shenyang, China) to be published.

Preparation and Characterization of High Tc Superconductive Thin Film

K. MAEDA[1], N. SAKAMOTO[1], Y. NAMIKI[1], Y. AOKI[1], T. SHIONO[1], H. YAMAMOTO[2], and M. TANAKA[2]

[1] Chemical Material R&D Department, Showa Electric Wire and Cable Co. Ltd., 2-1, Odasakae, Kawasaki-ku, Kawasaki, 210 Japan
[2] College of Science & Technology, Nihon University, 7-24, Narashinodai, Funabashi, 274 Japan

ABSTRACT

Superconductive YBaCuO films were deposited by reactive rf sputtering on MgO and several kinds of metals. The changes of the film composition were studied with respect to sputtering parameters. The as-sputtered film deposited on MgO at about 650 °C revealed a comparatively good superconductivity. A preferential growth of ab planes took place on MgO , while no preferential growth was observed on the metal substrates. The surface morphology was strongly affected mainly by chemical properties of the metals.

INTRODUCTION

Thin film techniques become a key technology for the purposes of electronic and power applications of high temperature superconductors. Many works for syntheses of YBaCuO films have been done concerning with a low temperature process[1,2)], a growth of ultrathin film[3,4)] or an achievement of a high critical current[5)]. In such cases, ceramic substrates, for example MgO, $SrTiO_3$ or ZrO_2 etc, have been adopted. However, the availability of metal substrates has not been confirmed and still unknown, though many groups succeeded in the YBaCuO film preparation at a comparatively low substrate temperature.

In this work, after considerations of sputtering conditions, the YBaCuO film was deposited on several kinds of metals which were closen taking notice the value of thermal expansion coefficient and/or the chemical stability. The crystal structure and surface morphology were studied and compared each other films. As a result, it was found that chemical properties mainly gave influences on the morphology of the film rather than physical properties.

EXPERIMENTAL

Sputtering conditions investigated are shown in Table 1. The target was made by a sintered YBaCuO powder. It was prepared as following process : Y_2O_3, $BaCO_3$, and CuO powder were mixed in a desired ratio, calcined at 930°C for 8 hours, sintered at 950°C for 8 hours, and ground. The product powder was pressed and formed into a disk with diameter of 80 mm and thickess of 1 mm. The sputtering gas was a mixture of Ar and O_2 in a ratio of 1:1. The substrate was heated up to about 650°C. The substrate investigated here were mirror polished MgO(100) planes and metal foils with thickess of about 0.1 mm (invar, sus430, Ag, Ti, Ta, W). The surface of metals was not made particularly clean by chemical treatments. After the film deposition at a fixed temperature, the prepared film was cooled down to room temperature in the atmosphere of 1 atm. oxygen.

The electronic resistivity was measured by the four probe technique using electrodes of silver paste/evaporated Ag film. The crystal structure was investigated by X-ray diffraction patterns. The film composition was measured by inductively coupled plasma (ICP). The morphology of the film was observed by scanning electron microscopy.

RESULTS AND DISCUSSION

The changes of the film composition were studied with respect to typical sputtering parameters. The results are summarized in Fig.1. The concentration of Cu and Ba decreased as increasing the sputtering gas pressure and/or substrate temperature Ts. These results were qualitatively agreement with the results already reported. There was no difference between the results from a conventional sintered bulk target.

Table 1 Sputtering Conditions

Sputtering gas	Ar : O_2 = 1 : 1
Gas pressure	2~9 (x 10^{-2} Torr)
rf power	100 (W)
Deposition rate	20 (Å/min)

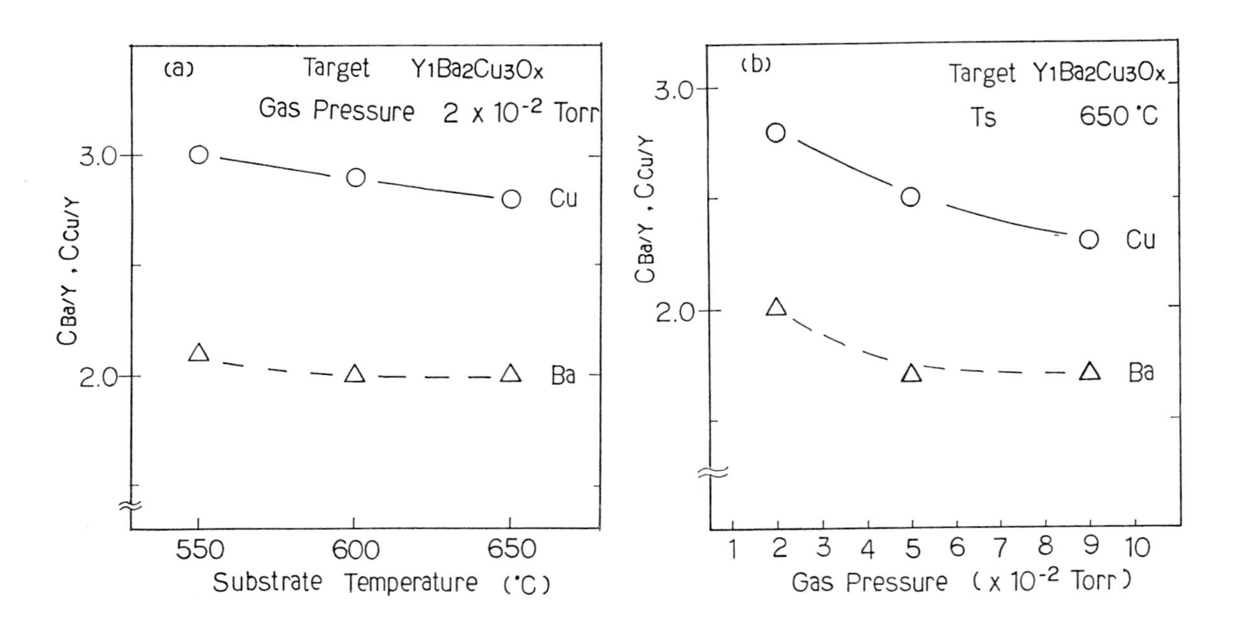

Fig. 1 The changes of the film composition depending on substrate temperature (a), gas pressure (b).

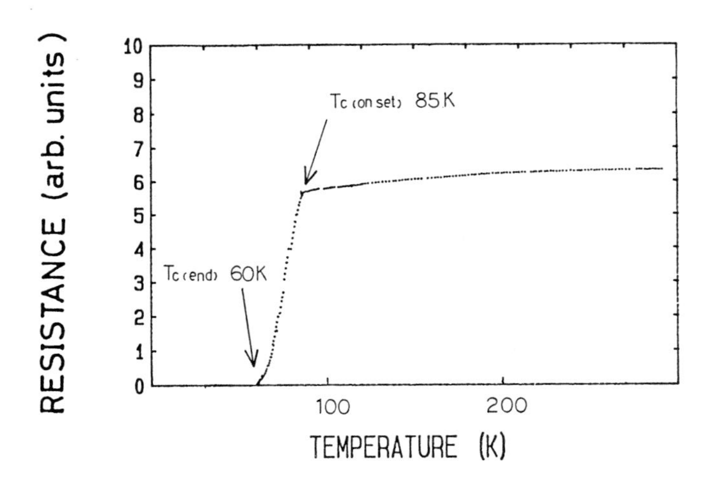

Fig. 2 Temperature dependence of resistance in the film deposited on MgO at 650°C

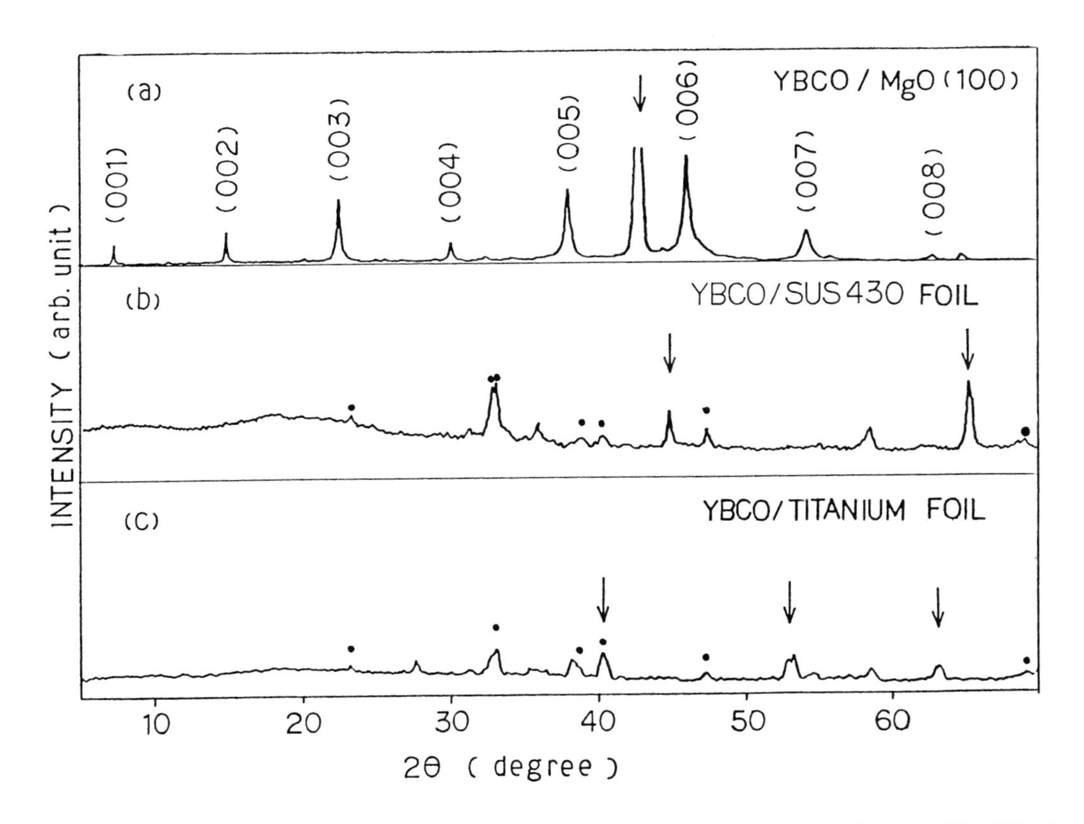

Fig. 3 Typical X-ray diffraction patterns of the film deposited on MgO (a), SUS 430 (b) and Ti (c) at 650°C. (Peak indicated by arrows are due to substrate. Peak indicated by " ● " are due to YBaCuO superconductor.)

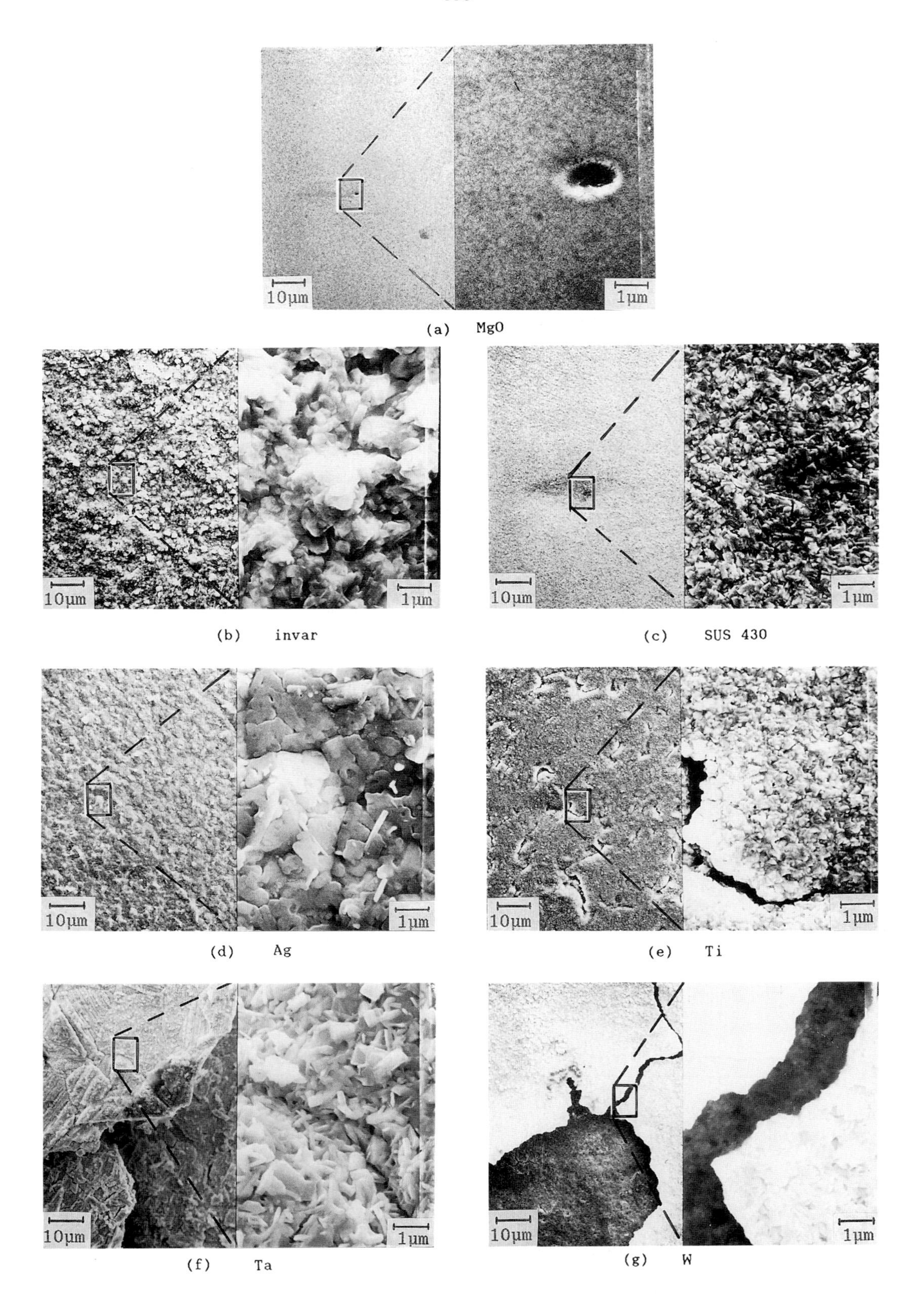

Fig. 4 Scanning electron micrographes of the surfaces of the films deposited on MgO (a), invar (b), SUS 430 (c), Ag (d), Ti (e), Ta (f), and W (g).

The resistive superconducting transition of the film deposited on at Ts of about 650°C is shown in Fig.2. The resistance fell into zero at about 60 K. Though the critical temperature is not so high, the optimum sputtering conditions obtained here are almost satisfied in order to study the following material effects of the substrate.

Typical X-ray diffraction patterns of the films are shown in Fig.3. The film on MgO revealed a strong preferentical growth of ab planes, similar to the results reported by other many groups. On the other hand, the metal substrates resulted in a growth of a randum orientation. Furthermore, the small unknown diffraction peaks also appeared in the cases of invar, Ti, Ta, and W, provably from the result of reactions between the film substance and the metals.

Fig.4 shows the surface morphology of the films which were propared on various substrates at Ts of about 650°C. The film on MgO had a vary smooth surface and revealed a good sticking. In the case of the metal substrates, the morphology of the film was changed obviously depending on the kind of the substrate. Particularly, the films on Ti, Ta, and W had many cracks and delaminations. In the cases of invar, SUS 430 and Ag, the films with minute structure and flatness were obtained, though the size of grains was quite different. The flat and large grains were observed in the film on Ag, while the grains were very fine in the cases of SUS 430 and Ti.

Obtained results were not interpreted well from the point of physical properties. For example, the value of a thermal expansion coefficient of Ag is about twenty times as large as that of invar. The hardness of invar or SUS 430 is not so different with that of Ti or Ta. The appearance of cracks and delaminations had not any relations with physical properties of the metal substrates. The chemical properties of the metal affinity of the metal with the film seem to be important. Then, the fact may be noticed that oxides of Ti, Ta, W are comparatively stable and easily formed in the interface of the film and the substrate. Detailed characterizations of the interface are needed to understand these material effects of the metal substrates.

CONCLUSION

The YBaCuO film was deposited on several metal substrates under the optimized sputtering conditions. The morphology of the film was strongly affected by chemical properties rather than physical properties of the metals. In this stage, invar, SUS 430 or Ag gave a comparatively good morphology and reveald an availability as a metal substrate. Further investigations will be done with respects to the detailed chemical reactions between the metal and the film substance in order to get a suitable metal substrate.

REFERENCES

1) H.Adachi, K.Hirochi, K.Setsune, M.Kitabatake, and K.Wasa, Appl.Phys.Lett., 51(1987)2263.
2) T.Terashima, K.Iijima, K.Yamamoto, Y.Bando and H.Mazaki, Jpn.J.Appl.Phys., 27(1988)L91.
3) T.Miura, Y.Terashima, M.Sugoi and K.Kubo, Ext.Abst. of 5th International Workshop of Future Electron Devices (Togatta-Onsen, Japan 1988) p.75.
4) H.Yamamoto, Y.Morikawa, H.Okukawa and M.Tanata, ibid, p105.
5) Y.Enomoto, T.Murakami, M.Suzuki and K.Moriwaki, Jpn.J.Appl.Phys., 26(1987)L1248.

Epitaxial Y-Ba-Cu-O Films on Si with Intermediate Layer by RF Magnetron Sputtering

S. MIURA[1], H. TSUGE[2], T. YOSHITAKE[1], S. MATSUBARA[1], T. SATOH[3], Y. MIYASAKA[1], and N. SHOHATA[1]

Fundamental Research Laboratories[1], Microelectronics Research Laboratories[2], and Resources and Environment Protection Research Laboratories[3], NEC Corporation, 4-1-1, Miyazaki, Miyamae-ku, Kawasaki, 213 Japan

ABSTRACT

Epitaxial films of Y-Ba-Cu-O were obtained on Si substrate using epitaxial intermediate layer consisting of $SrTiO_3$(or $BaTiO_3$)/$MgAl_2O_4$. $MgAl_2O_4$ was epitaxially grown on Si(100) substrate by chemical vapor deposition, and then $SrTiO_3$ or $BaTiO_3$ was also epitaxially grown on $MgAl_2O_4$ layer by means of RF magnetron sputtering. Y-Ba-Cu-O films were prepared on $SrTiO_3$($BaTiO_3$)/$MgAl_2O_4$/Si substrates by RF magnetron sputtering and their epitaxial growth was confirmed by RHEED observation and X-ray diffraction measurements. Epitaxial orientations of Y-Ba-Cu-O films varied in dependence on RF input power; lower RF power resulted in c-axis oriented film and higher RF power resulted in a-axis oriented film. Preparation of Y-Ba-Cu-O directly on $MgAl_2O_4$/Si was also studied, but only randomly oriented polycrystal film has been obtained so far. In sputter Auger depth measurement, any notable diffusion between Y-Ba-Cu-O film and the substrates was not observed. Resistive superconducting transitions with zero resistance at 65K on $SrTiO_3$/$MgAl_2O_4$/Si and at 70K on $BaTiO_3$/$MgAl_2O_4$/Si were observed.

INTRODUCTION

Since the discovery of the 90K class high T_c superconductor $YBa_2Cu_3O_{7-x}$[1], much effort has been devoted to preparing thin film of this material. As single crystal silicon is the primary electronic material, preparation of Y-Ba-Cu-O thin film on silicon is of great interest for a variety of electronic applications. It has been reported that Y-Ba-Cu-O films were prepared directly on Si[2,3] or on Si with a zirconia buffer layer[3,4] and that those films were polycrystal. $Y_1Ba_2Cu_3O_{7-\delta}$ oxide superconductor has a marked anisotropic crystal structure and notable anisotropy in energy gap, critical current density and coherent length was discovered. From these facts it will be more desirable to grow epitaxial Y-Ba-Cu-O film on silicon for well-controlled device. From fundamental point of view, the authors have examined whether Y-Ba-Cu-O films were epitaxially grown on Si or not, using three kinds of epitaxial intermediate layers.

The epitaxial growth technique for magnesia-spinel ($MgAl_2O_4$) on Si has been well established through an approach to the silicon-on-insulator (SOI) technology[5,6]. Recently, the authors have successfully grown epitaxial films of ABO_3 type perovskite oxides, such as $SrTiO_3$, $BaTiO_3$, $PbTiO_3$ and so on, on the epitaxial $MgAl_2O_4$ layer on Si[7,8]. It is expected that Y-Ba-Cu-O can be epitaxially grown on epitaxial $SrTiO_3$(or $BaTiO_3$)/$MgAl_2O_4$/Si structure. It would be also expected that Y-Ba-Cu-O can be epitaxially grown directly on epitaxial $MgAl_2O_4$/Si structure as in the case of ABO_3 type oxides, because Y-Ba-Cu-O has a perovskite-based crystal structure. This paper presents the authors' preliminary results regarding on the properties of Y-Ba-Cu-O thin films on Si substrate with epitaxial intermediate layers of $MgAl_2O_4$ and $SrTiO_3$($BaTiO_3$)/$MgAl_2O_4$.

EXPERIMENTAL

As shown in Fig. 1, Y-Ba-Cu-O films were deposited on three kinds of structure and their structural and superconducting properties were compared. The 5000 Å thick (100) $MgAl_2O_4$ epitaxial films were grown on (100) Si (n-type) substrate by chemical vapor deposition (CVD) method following the process described in an earlier paper[5,7]. $SrTiO_3$ and $BaTiO_3$ films 3500 Å thick were grown on $MgAl_2O_4$ by RF magnetron sputtering. These films were epitaxially grown and the <100> axes of the films were oriented normal to the substrate surface. This was confirmed by reflection high energy electron diffraction (RHEED) and X-ray diffraction (XRD). Details of preparation condition and film properties for $SrTiO_3$ and $BaTiO_3$ will be described elsewhere[9].

$YBa_2Cu_3O_{7-x}$ films were prepared by RF magnetron sputtering, using a composite oxide target with 1-2.2-5 ratios of Y-Ba-Cu. Sputtering conditions are listed in Table 1. So as to prevent oxygen in the film from being removed, the samples were cooled in an oxygen gas flow. Structural and superconducting properties were measured for as deposited Y-Ba-Cu-O films. The RHEED measurement was carried out under 40 kV incident energy and 1 to 3 degrees incident angle. Resistivity vs. temperature was measured using the conventional four probe technique, where contacts were made by indium solder.

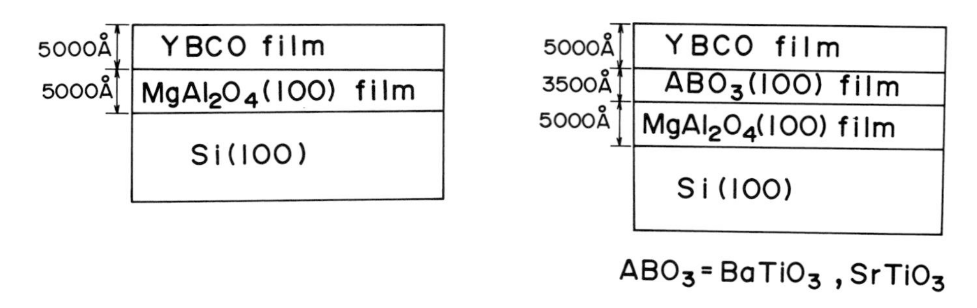

Fig. 1 Schematic of possible substrate structures for epitaxial Y-Ba-Cu-O film

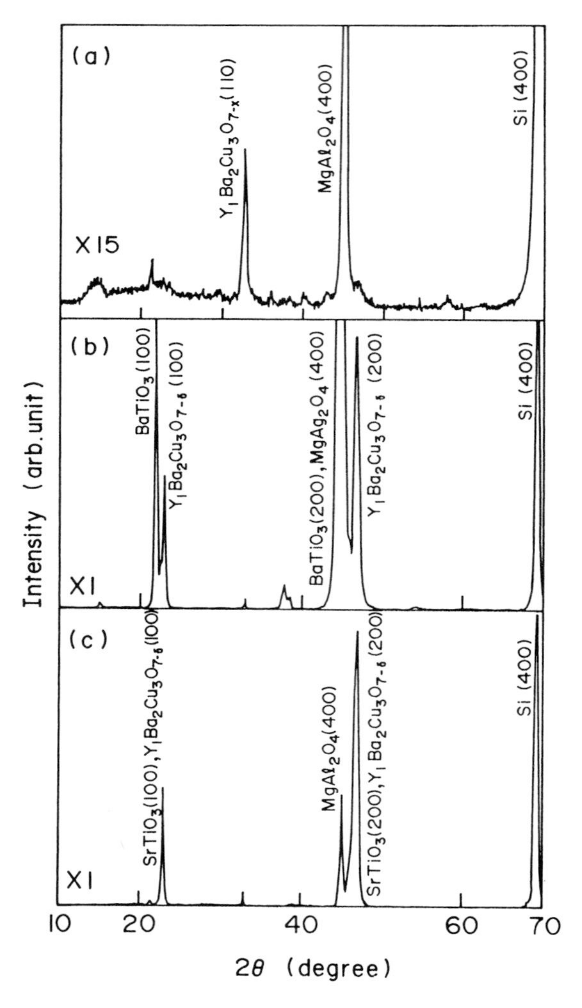

Fig. 2 X-ray diffraction results for Y-Ba-Cu-O films (a) on $MgAl_2O_4$/Si, (b) on $BaTiO_3$/$MgAl_2O_4$/Si and (c) on $SrTiO_3$/$MgAl_2O_4$/Si substrates.

Table I. Sputering conditions for Y-Ba-Cu-O film

Target	$Y_1Ba_{2.5}Cu_5O_x$ (100 ø)
Substrate temperature	640°C
Sputtering gas	Ar(50%) + O2(50%)
Gas pressure	3×10^{-2} Torr
RF power	0.9~2.2 W/cm^2
Substrate-target distance	37 mm

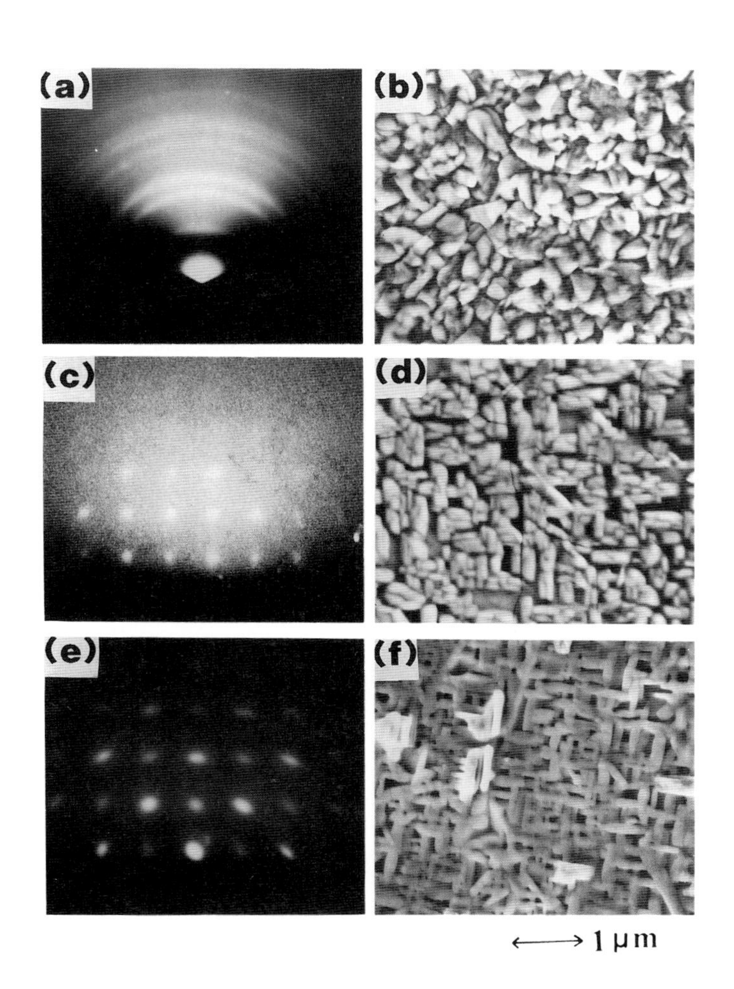

Fig. 3 RHEED patterns and SEM images for Y-Ba-Cu-O films
(a), (b) on $MgAl_2O_4$/Si
(c), (d) on $BaTiO_3/MgAl_2O_4$/Si
(e), (f) on $SrTiO_3/MgAl_2O_4$/Si

RESULT AND DISCUSSION

Figure 2(a) shows the XRD spectrum for Y-Ba-Cu-O film on $MgAl_2O_4/Si$. Only the (110) peak, which is the main peak for polycrystal Y-Ba-Cu-O ceramic, was observed in the spectrum. The RHEED pattern and the scanning electron microscope (SEM)image of the same sample are shown in Figs. 3(a) and (b). There are only rings in the RHEED pattern, and the film consists of many randomly oriented grains in the SEM image. From these results, it was concluded that the Y-Ba-C-O films would not be epitaxially grown on $MgAl_2O_4/Si$. Growth on (100) $MgAl_2O_4$ bulk single crystal substrate was also studied. Only randomly oriented polycrystal film was obtained, confirming the conclusion.

Figure 2(b) shows an XRD spectrum for Y-Ba-Cu-O film on $BaTiO_3/MgAl_2O_4/Si$. Strong (100) and (200) peaks of Y-Ba-Cu-O were observed in the spectrum, which indicates that the a-axis of the film was preferentially oriented normal to the substrate surface. RHEED pattern and SEM image for the same sample are shown in Figs. 3(c) and (d). The RHEED pattern consists of regular spots. Rectangular grains are in sequence in the SEM image. Therefore, it is concluded that the Y-Ba-Cu-O film was epitaxially grown on $BaTiO_3/MgAl_2O_4/Si$ and that the a-axis of the film was preferentially oriented normal to the substrate surface.

Similar XRD spectrum, RHEED pattern and SEM image were obtained for the Y-Ba-Cu-O film on $SrTiO_3/MgAl_2O_4/Si$, as shown in Figs. 2(c), 3(e) and 3(f), respectively. Therefore, it is concluded that the Y-Ba-Cu-O film was epitaxially grown on $SrTiO_3/MgAl_2O_4/Si$ and that the a-axis for the film was preferentially oriented normal to the substrate surface.

In this experiment, whether the film is epitaxially grown on the substrates or not seems to depend mainly on the similarity of crystal structures for two material. Y-Ba-C-O and $SrTiO_3$ have similar crystal structures with relatively small lattice mismatch. The films are epitaxially grown on $SrTiO_3/MgAl_2O_4/Si$. Y-Ba-Cu-O and $BaTiO_3$ have similar crystal structures. The films are epitaxially grown on $BaTiO_3/MgAl_2O_4/Si$, in spite of relatively large lattice mismatch. On the contrary, Y-Ba-Cu-O has a different crystal structure from $MgAl_2O_4$. The films are not epitaxially grown on $MgAl_2O_4/Si$ despite the lattice mismatch between Y-Ba-Cu-O and $MgAl_2O_4$ being nearly equal to that between Y-Ba-Cu-O and $BaTiO_3$. The reason why ABO_3 type perovskite oxides are epitaxially grown on $MgAl_2O_4$, but Y-Ba-Cu-O is not, has not been clarified yet.

In Y-Ba-Cu-O films on $BaTiO_3/MgAl_2O_4/Si$ and on $SrTiO_3/MgAl_2O_4/Si$, epitaxial orientation axis varied with RF input power. The films whose c-axes were preferentially oriented normal to the substrate surface (c-axis oriented films) tended to be formed with low RF input power (<100W). On the contrary, the films whose a-axes were preferentially oriented normal to the substrate surface (a-axis oriented films)were formed with high RF input power (>150W). Figure 4 shows a RHEED pattern and an SEM image of c-axis oriented Y-Ba-Cu-O film on $BaTiO_3/MgAl_2O_4/Si$. In the SEM image small grains and some cracks were observed. In the RHEED pattern streaks normal to the substrate surface and spots corresponding to c axis were observed. Therefore, it is concluded that the c-axis oriented Y-Ba-Cu-O film was epitaxially on $BaTiO/MgAl_2O_4/Si$.

Figure 5 shows the resistivity versus temperature results for the c-axis oriented films on $SrTiO_3/MgAl_2O_4/Si$ and $BaTiO_3/MgAl_2O_4/Si$. The film on $SrTiO_3/MgAl_2O_4/Si$ showed T_c onset at 83K and zero resistance at 65K. The film on $BaTiO_3/MgAl_2O_4/Si$ (same sample that is shown in Fig. 4) exhibited Tc onset at 80K, zero resistance at 70K, and critical current density of 1.8×10^4 (A/cm^2) at 40K. Figure 6 shows the Auger depth profile for c-axis oriented films on $SrTiO_3/MgAl_2O_4/Si$. As seen in Fig. 6, metallic elements are almost constant inside the film and notable interdiffusion was not observed at the interface. The substrate structure for $SrTiO_3/MgAl_2O_4/Si$ is stable against chemical reaction with Y-Ba-Cu O film, under the present deposition condition.

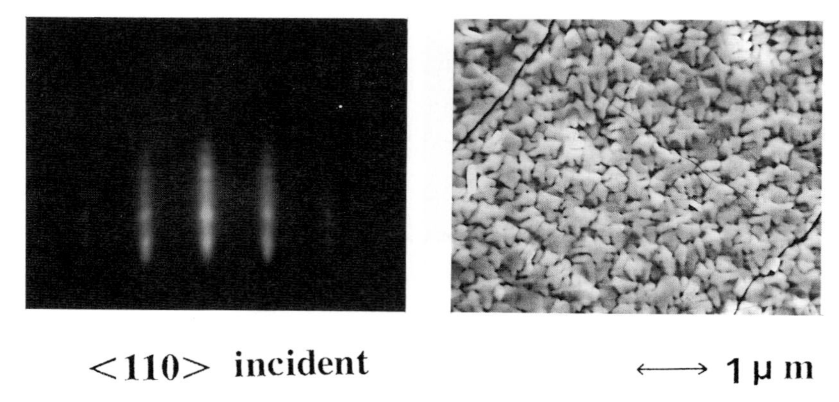

Fig.4 RHEED pattern and SEM image for a c-axis oriented Y-Ba-Cu-O film on $BaTiO_3/MgAl_2O_4/Si$.

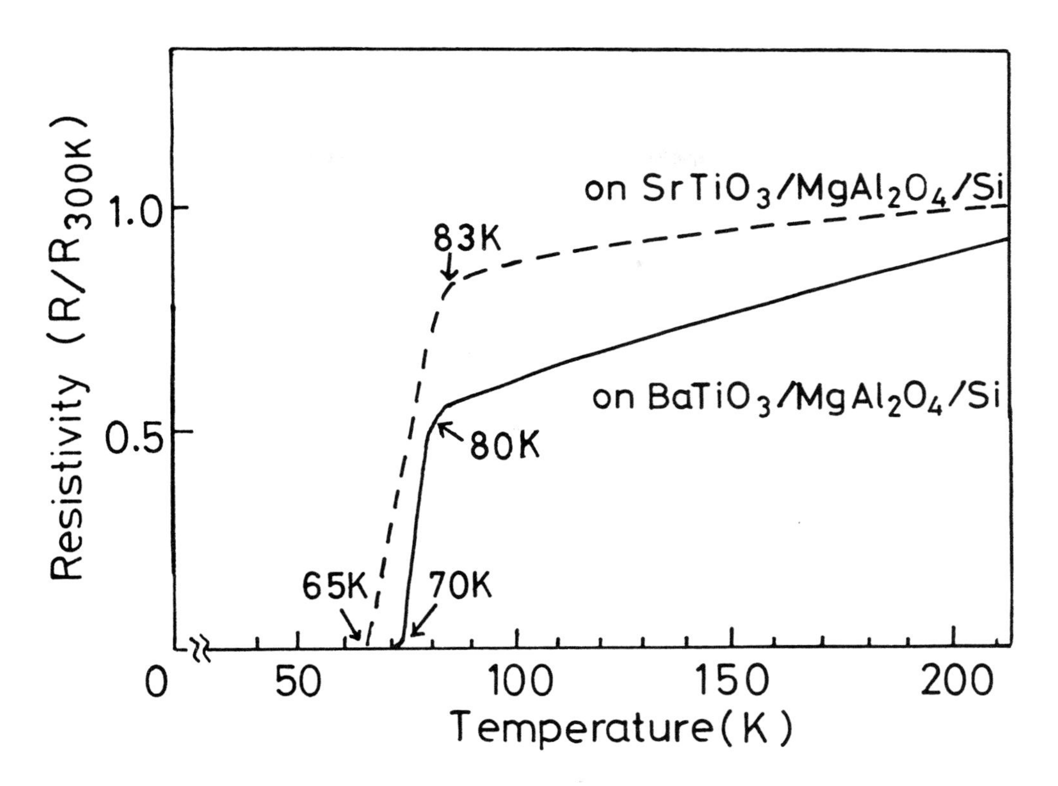

Fig. 5 Temperature dependence of resistivity for Y-Ba-Cu-O films on $SrTiO_3/MgAl_2O_4/Si$ and $BaTiO_3/MgAl_2O_4/Si$ substrates

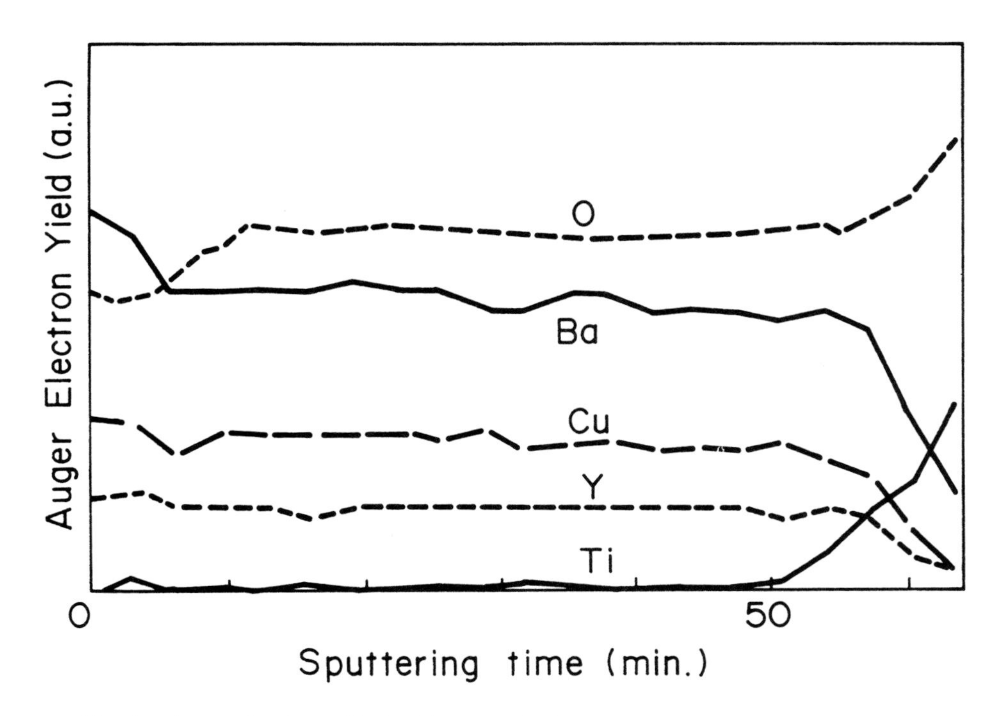

Fig. 6 Auger depth profile for Y-Ba-Cu-O film on $SrTiO_3/MgAl_2O_4/Si$ substrate

In summary, the authors have shown that the a-axis and c-axis oriented epitaxial Y-Ba-Cu-O films can be prepared on both $SrTiO_3/MgAl_2O_4/Si$ and $BaTiO_3/MgAl_2O_4/Si$ substrates by RF magnetron sputtering technique. These structures are stable against chemical reaction with Y-Ba-Cu-O film, under the present experimental conditions. The film on $SrTiO_3/MgAl_2O_4/Si$ showed zero resistance at 65K. The film on $BaTiO_3/MgAl_2O_4/Si$ showed zero resistance at 70K and critical current density of 1.8×10^4 (A/cm^2) at 40K.

The authors would like to thank M. Yonezawa, H. Igarashi, K. Kubo and J. Fujita for their helpful discussions. Thanks are also given to M. Tawarayama, M. Sugimoto and A. Nakai for their experimental assistance throughout this work.

References

1. M.K. Wu, J.R. Ashburn, C.J. Tong, P.H. Hor, R.L. Wong, L. Gao, Z.J. Huang, Y.Q. Wang, and C.W. Chu, Phys. Rev. Lett. 58, 908 (1987)
2. W.Y. Lee, J. Salem, V. Lee, T. Huang, R. Savoy, V. Deline, and J. Duran, Appl. Phys. Lett. 52, 2263 (1988).
3. T. Venkatesan, E.W. Chase, X.D. Wu, A. Inam, C.C. Chang, and F.K. Shokoohi, Appl. Phys. Lett. 53, 243 (1988)
4. A. Mogro-Campero and L.G. Turner, Appl. Phys. Lett. 52, 1185 (1988)
5. Masao Mikami, Yasuaki Hokari, Koji Egami, Hideaki Tsuya and Masaru Kanamori, Extended Abstract of the 15th Conf. Solid State Devices and Materials, Tokyo, 31 (1983) published by the Physical Society of Japan and the Japan society of Applied Physics.
6. Takafumi Kimura, Hideki Yamawaki, Yoshihiro Arimoto, Kazuto Ikeda, Masaru Ihara and Masashi Ozeki, Mat. Res. Soc. Sym. Proc. 53, 143 (1986)
7. Shogo Matsubara, Nobuaki Shohata and Masao Mikami, Jpn. J. Appl. Phys. Suppl. 24-3, 10 (1985)
8. Shogo Matsubara, Yoichi Miyasaka and Nobuaki Shohata, Mat. Res. Soc. Sym. Abst., Boston, 378 (1987)
9. Shogo Matsubara, Yoichi Miyasaka and Nobuaki Shohata (unpublished)

Preparation and Properties of the Y-Ba-Cu-O Thin Films

MICHIHITO MUROI, TOSHIYUKI MATSUI, YUJI KOINUMA, KOICHI TSUDA,
MEGUMI NAGANO, and KAZUO MUKAE

Fuji Electric Corporate Research and Development Ltd., 2-2-1, Nagasaka, Yokosuka,
240-01 Japan

ABSTRACT

We investigated superconducting $YBa_2Cu_3O_y$(YBCO) thin films, which were prepared with two synthetic processes using rf-magnetron sputtering :(a)deposition at low substrate temperature with post-annealing at around 900°C and (b)deposition at high substrate temperature around 650°C with or without post-annealing. We obtained the high Tc superconducting YBCO thin films with each process. The films prepared with process(a) had, however, poor Jc below $10A/cm^2$ at 77K and their surfaces were rough. On the other hand, the films prepared with process(b) had higher Jc and smoother surfaces than those with process(a). Further improvement in surface morphology was made by raising substrate temperature to 700°C at the start of the deposition.

INTRODUCTION

Since the discovery of high Tc superconducting oxides[1,2] such as YBCO, tremendous effort has been made to synthesize thin films of these materials from both scientific and technological viewpoints. Among various techniques such as sputtering, vacuum evaporation, chemical vapor deposition and molecular beam epitaxy, sputtering is the most popular way to fabricate high Tc superconducting thin films.

We tried the fabrication of the YBCO thin films using rf-planar magnetron sputtering, and succeeded in obtaining the high Tc films with either of two synthetic processes:(a)deposition at low substrate temperature(Ts) followed by post annealing and (b)deposition at high Ts above crystallizing temperature with or without post annealing.[3]

YBCO is essentially a line compound with all adjacent phases being either insulating or semiconducting, and it has strong anisotropy due to its crystal structure. Therefore two major problems must be solved to fabricate high quality YBCO thin films. One is to achieve strict stoichiometry in order to avoid the formation of second phases. The other is to make the axis of each grains oriented to the same direction or preferably to realize epitaxial growth in order to increase Jc.

Considering the application to the devices, we must solve the third problem, namely, to make the surface of the film smooth. It is required for the processing such as etching or the formation of a junction.

In this paper, we compare the characteristics of the films prepared with two processes mentioned above in these points of view, and report the relationship between the synthetic conditions and the properties of the films, such as Tc, Jc, crystal structure, surface morphology and chemical composition.

EXPERIMENTAL

Sample Preparation

The films were prepared using rf-planar magnetron sputtering. Targets used were sintered Y-Ba-Cu-O discs with various compositions, which were 6 inches in diameter and 5mm in thickness. MgO(100) or $SrTiO_3$(100),(110) were used as substrates. They were attached to the block heater and could be heated to 700°C or kept at room temperature with water cooling. Input rf-power was kept at 300W during the deposition. The synthetic processes were roughly divided into two processes.

process(a):deposition at room temperature followed by post-annealing at around 900°C in pure oxygen or the mixture of oxygen and nitrogen.

process(b):deposition above crystallizing temperature with or without post-annealing.

Process(a) Since YBCO is a multicomponent material, the components of which have different vapor pressure, deposition at low Ts such as room temperature seems to lead the less difference in composition between the targets and the films. For this reason, we first tried process(a). Typical sputtering and post-annealing conditions are listed in Table I. The target composition was chosen to make up a deficiency of Ba and Cu in the films. The annealing conditions were determined by the optimization concerning temperature, time and atmosphere.

Table I. Sputtering and post-annealing conditions in process(a) and process(b)

	Process(a)	Process(b)
Target Composition	Y:Ba:Cu=1:2.6:4	Y:Ba:Cu=1:2.6:6
Substrate	MgO(100) $SrTiO_3$(100)(110)	MgO(100) $SrTiO_3$(100)(110)
Substrate Temperature	Room Temperature	650°C
Sputtering Gas	Pure Ar	Ar(50%)+O_2(50%)
Gas Pressure	2Pa	10Pa
Growth Rate	250A/min	33A/min
Annealing Temperature	950°C	400-1000°C
Annealing Time	1h	1h
Annealing Atmosphere	O_2(20%)+N_2(80%)	Pure O_2

Process(b) Direct formation of the (123)-phase during deposition is desirable to fabricate high-quality films with high Jc and excellent surface morphology. Crystallized films were obtained when the substrate temperature was raised above 500°C. Higher Ts around 650°C was required for the direct formation of the (123)-phase with no other phases in our experimental conditions. Sputtering and post-annealing conditions are also listed in table I. Rather high gas pressure of 10Pa was chosen to prevent a deposited film from selective resputtering by negative oxygen ions or neutral oxygen atoms, which might cause a serious compositional change.

Characterization of the films

Resistance versus temperature characteristic and Jc were measured using the conventional dc four-probe technique with gold electrodes evaporated at intervals of 4mm. Crystal structure and surface morphology of the films were studied by X-ray diffraction(XRD) and scanning electron microscopy(SEM), respectively. The chemical composition was analyzed by Inductively Coupled Plasma Emission Spectroscopy(ICP) and Auger analysis for the depth profile.

RESULTS AND DISCUSSION

Films Prepared with Process(a)

Though we could obtain the high Tc (75-90K) YBCO thin films with the conditions of process(a) listed in Table I both on MgO and $SrTiO_3$, there were some problems in these films. The first problem is disappearance of superconducting properties in the thinner films. Fig.1 shows the thickness dependence of Tc of the films on MgO and $SrTiO_3$ substrates. As can be seen in Fig.1, the films thinner than 0.5μm on MgO and those thinner than 1.3μm on $SrTiO_3$ did not show superconducting behavior. They had, however, the same crystal structure of (123)-phase as the thicker films according to XRD measurement. We found by SEM that thinner films below 0.5μm had an island structure as shown in Fig.2 in contrast to the thicker ones, which had the ordinary polycrystalline structure with their grains connected each other. This is the reason why

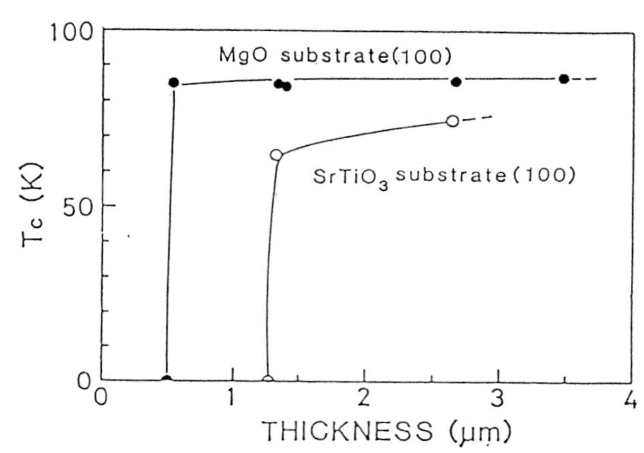

Figure 1. Thickness Dependence of Tc for the Films Prepared with Process(a)

the thinner films did not show the superconducting properties.

The second problem is the lower Tc of the films on $SrTiO_3$ than those on MgO. Especially the films which have a thickness of 0.5-1.3μm showed no superconducting properties, though their grains were connected each other. This phenomenon can be explained by the reaction between $SrTiO_3$ substrates and the films. Fig.3 shows the depth profile of the film on $SrTiO_3$ by Auger analysis. This film was 2μm in thickness. It can be seen that the constituent atoms of the film and the substrate had diffused each other near the interface. Such interdiffusion is particularly marked in the region closer than 1μm from the interface. Above all, Sr was detected even on the surface. We suppose that 1.3μm is a critical thickness below which the films suffer serious compositional change by the interdiffusion of the atoms and lose superconductivity. Fast diffusion of Sr explains that the films on $SrTiO_3$ thicker than 1.3μm still have lower Tc than those on MgO. It is probable that the diffused Sr atoms have replaced the Ba atoms and lowered Tc. Such interdiffusion of atoms was not observed in the case of the film on MgO substrate. After all, annealing temperature of 950°C seems to be too high for the films on $SrTiO_3$ to keep stoichiometric composition, and lower temperature process must be developed when $SrTiO_3$ is used as a substrate.

In addition to these problems, even the thickest film on MgO, which had the best superconducting properties, had poor Jc below $10A/cm^2$ at 77K probably due to the existence of grain boundaries and random orientation of the axis of the grains.

Figure 2. SEM Photographs of the Surfaces of the Films on MgO Prepared with Process(a)

Films Prepared with Process(b)

The films prepared with the conditions listed in Table I were 0.2-0.3μm in thickness after two-hour deposition. The as-deposited films on MgO with smooth surfaces had very low Tc bellow 20K in spite of their nearly stoichiometric composition and the (123)-phase with perfect c-axis orientation. On the other hand, those on $SrTiO_3$ (110) had (110) orientation and rather high Tc of 75K, but their surfaces were rough. Higher Tc was obtained probably because the rough surfaces made it easy for the films to take oxygen while they were cooling.

The superconducting properties were remarkably improved by post-annealing in case of the films on MgO. Fig.4 shows the change in Tc as well as that in the lattice constant along the c-axis with post-annealing temperature. Annealing was done in pure oxygen atmosphere for one hour

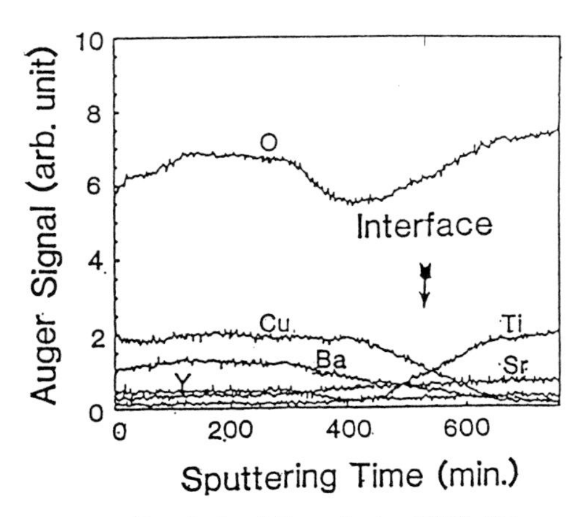

Figure 3. Depth Profile of the YBCO Film on $SrTiO_3$ Prepared with Process(a)

followed by the cooling to room temperature with the rate of 5°C/min. The higher the annealing temperature was up to 950°C, the higher Tc became with gradual shortening of the lattice constant along the c-axis probably caused by oxygen uptake. Tc reached 82K by annealing at 950°C, and typical values of Jc at 77K were around $10^4 A/cm^2$ with the best value of $4x10^4 A/cm^2$. These values are three or four orders higher than those of the films prepared with process(a). Furthermore, steeper increase in Jc with decreasing temperature was observed in the films with process(b) than in those with process(a) as shown in Fig.5. Further increase in annealing temperature up to 1000°C resulted in a drop of Tc presumably owing to the partial dissolution of the (123)-phase or the reaction between the film and the substrate.

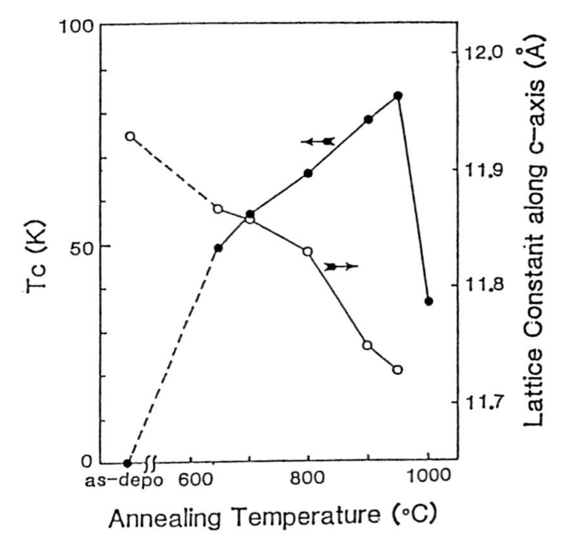

Figure 4.
Change in Tc and Lattice Constant along the c-axis with Annealing Temperature

To examine the mechanism of the rise in Tc with post-annealing temperature, we checked XRD spectra more precisely. (005)peaks in the XRD spectra of the films annealed at various temperature are drawn in Fig.6. It can be seen from this figure that as annealing temperature is raised up to 900°C, the peak moves toward higher angle with its shape unchanged, which indicates the shortening of the lattice constant along the c-axis as described above. Remarkable change suddenly occurs when annealing temperature reaches 950°C. With further shift of the peak toward higher angle, the peak intensity increases markedly and becomes sharp suggesting the improvement in crystallinity. Although it is not clear why the sudden improvement in crystallinity occurs with annealing at 950°C, we suppose it owes to the occurrence of some solid phase epitaxy on the single crystal substrate or the fine adjustment in composition.

The SEM photograph of the surface of the film annealed at 950°C is shown in Fig.7. No grain boundaries are observed even after annealing. It has, however, many small educts like bubbles all over the surface, which were not observed before the annealing. The composition of these educts and the rest of the film were analyzed by EDX, and the Cu content of the former turned out to be less than that of the latter.

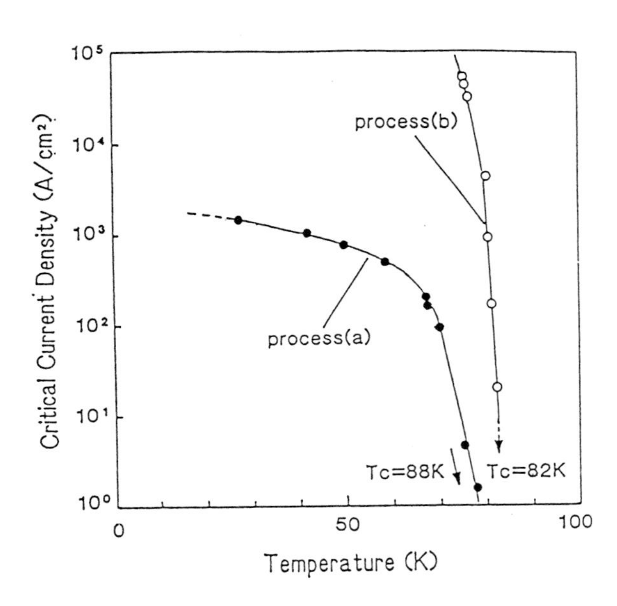

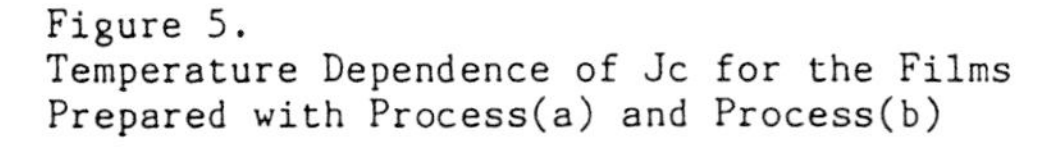
Figure 5.
Temperature Dependence of Jc for the Films Prepared with Process(a) and Process(b)

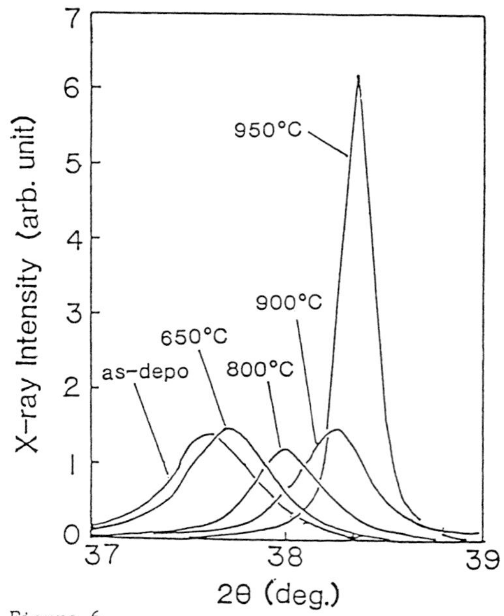

Figure 6.
(005)peaks of the XRD Spectra for the Films Annealed at Various Temperature

On the other hand, the films on $SrTiO_3$ were degraded rather than improved in Tc by post-annealing, probably because of the reaction between the films and the substrates as well as in the case with process(a).

Improvement of Surface Smoothness

The films prepared with process(b) have the crystal structure of the (123)-phase with their c-axis perpendicular to the substrates, the lattice constant of which is close to that along the a-axis or the b-axis of YBCO. Therefore the crystal growth during deposition seems to be epitaxial. Because the crystal of YBCO must grow on the different kind of material in the initial stage of the deposition, the best conditions for the crystal growth may be different from those in the rest stage of the deposition.

From this point of view, we tried the following two processes as the modifications of process(b).

(b-1):deposition at raised Ts of 700°C, at the start with gradual down to 650°C in five minutes followed by that at constant Ts of 650°C.

(b-2):deposition at low Ts below 200°C for first five minutes followed by that at 650°C.

Total deposition time was two hours in both processes. The aims of these processes are to improve the crystallinity of the film formed in the initial stage of the deposition and to enhance the sticking coefficient of atoms arriving to the substrate, respectively. The as-deposited films on MgO(100) substrates with both processes had the same crystal structure of (123)-phase with c-axis orientation as with process(b) and poor superconducting properties. These films were annealed at 950°C for one hour in oxygen atmosphere, which was the same annealing condition as in process(b).

After annealing, the films with process(b-1) had almost the same Tc and Jc as those with process(b), 80-82K and ten thousands level in A/cm^2 at 77K, respectively. The surface morphology was, however, remarkably improved as can be seen in the SEM photograph shown in Fig.8. In addition to the smooth surfaces with no grain boundaries, no educts were observed, which existed on the surfaces of the films prepared with process(b). Such compatibility of superconducting properties with surface smoothness is essential to apply the thin films to the electronic devices.

On the other hand, those with process(b-2) had very low Tc below 20K even after annealing. To make matters worse, they had very rough surfaces as is shown in Fig.9.

These results suggest that the properties of the sputter-deposited films are strongly dependent on the crystallinity of the films formed in the initial stage of the deposition.

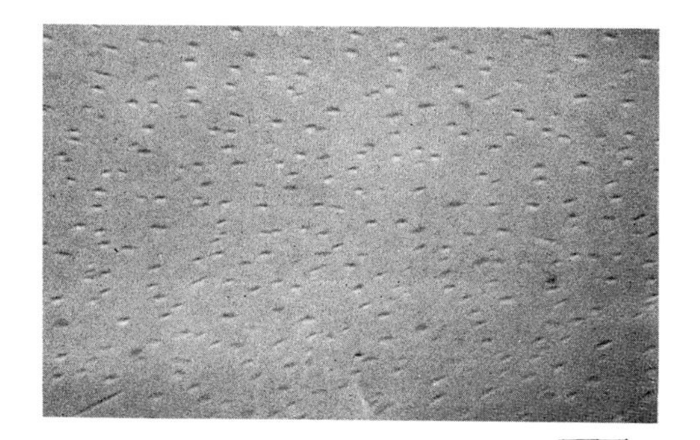

Figure 7.
SEM Photograph of the Surface of the Film Prepared with Process(b)

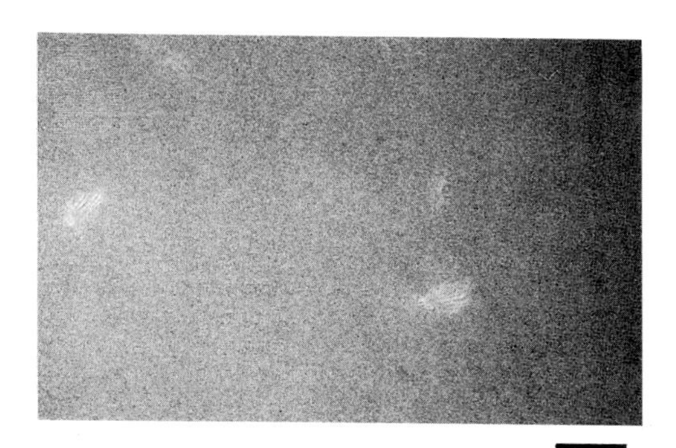

Figure 8.
SEM Photograph of the Surface of the Film Prepared with Process(b-1)

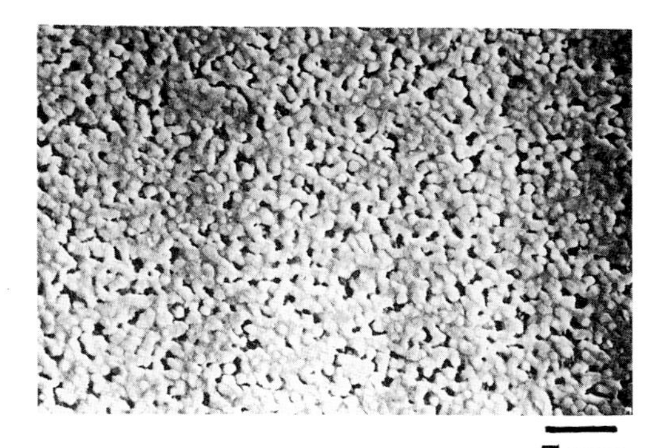

Figure 9.
SEM Photograph of the Surface of the Film Prepared with Process(b-2)

SUMMARY

We fabricated the high Tc superconducting YBCO thin films with two synthetic processes using rf-planar magnetron sputtering and clarified the relationship between the synthetic conditions and the properties of the films.

The films with high Jc, ten thousands level in A/cm^2 at 77K, were fabricated by keeping Ts around 650°C to form the (123)-phase directly during deposition followed by post-annealing.

High-quality films with both high Jc and smooth surfaces were obtained with Ts raised to 700°C only at the start of the deposition, which must have affected the crystallinity of the films formed in the initial stage of the deposition.

REFERENCES

[1]J.Bednortz and K.A.Muller, Z.Phys.B 64,189,1986
[2]M.K.Wu,J.R.Ashburn,C.J.Torng,P.H.Hor,R.L.Meng,L.Gao,Z.J.Huang,Y.Q.Wang and.W.Chu, Phys.Rev.Lett. 58,908,1987
[3]K.Tsuda, M.Muroi, T.Matsui, Y.Koinuma ,M.Nagano and K.Mukae, Proceedings of the International Conference on High Temperature Superconductors and Materials and Mechanisms of Superconductivity 788,1988

Properties of Superconducting Bi-Sr-Ca-Cu-O Thin Films Prepared by Controlling Sputtering Conditions and Annealing Atmospheres

K. OHBAYASHI, T. USHIDA, T. TSUNOOKA, K. OHYA, and H. BANNO

NTK Technical Ceramics Division, NGK Spark Plug Co., Ltd., 14-18, Takatsuji-cho, Mizuho-ku, Nagoya, 467 Japan

ABSTRACT

Bi-Sr-Ca-Cu-O thin films were prepared by using a rf magnetron sputtering. The films were grown on MgO substrates at 300°C cyclically by sputtering through an off-centered shutter hole while rotating the substrate table. Then the films were annealed at 860°C for 10 hours in the mixture of $O_2/N_2=1/10$.

The thin films had no segregation of the constituent elements and showed a high critical temperature of Tc(on)=120K and Tc(zero)=103K.

INTRODUCTION

Since a high Tc oxide superconductor of Bi-Sr-Ca-Cu-O system was discovered by Maeda et al.[1], many efforts have been made to realize a zero resistivity at a temperature above 100K by optimizing a heat treatment.

This Bi system was found to have two phases with a low-Tc phase of 80K and a high-Tc phase of 110K. A number of studies on the low-Tc phase revealed that the 80K phase has a pseudotetragonal crystal structure with a 5.4A x5.4A x31A subcell, which has double perovskite layers sandwiched with Bi_2O_2 layers along the c-axis direction and has an ideal chemical composision of $Bi_2Sr_2Ca_1Cu_2O_x$.[2] Takayama-Muromachi et al.[3] found another tetragonal phase with a larger c-axis of 37A for the high-Tc phase and proposed the crystal structure model which has triple perovskite layers sandwiched with Bi_2O_2 layers, whose ideal chemical composition is $Bi_2Sr_2Ca_2Cu_3O_x$.

It is very important to develope a method for making a single phase material with a high-Tc of 110K or a high-Tc phase dominant material in this Bi-Sr-Ca-Cu-O system.

Several research groups have already reported thin films of the Bi system, which were prepared by sputtering technique on substrates of MgO, $SrTiO_3$, Al_2O_3 and YSZ, at the heating temperature ranging from 200°C to 800°C during sputtering, and annealed at the temperature ranging from 700°C to 900°C for 0.5 to 5 hours in gas atmospheres of O_2 or air.[4]-[9]

This paper describes the superconducting, crystallographic and micrographic properties of the Bi-Sr-Ca-Cu-O thin films prepared by controlling sputtering conditions, annealing temperatures and gas atmospheres.

EXPERIMENTAL

Thin films of Bi-Sr-Ca-Cu-O were prepared on MgO(100) substrates by rf magnetron sputtering from a sintered $Bi_1Sr_1Ca_1Cu_2O_x$ compound target. The sputterings were carried out at the substrate table temperature of 300°C in an Ar atmosphere with a back ground pressure of 5×10^{-6} Torr. The rf input power was 200W on the target of 100 mm diameter set at 55 mm distance from the substrate. In order to control the depositing rate, the films were grown through an off-centered shutter hole cyclically with rotation of the substrate table at 15 r.p.m. or continuously grown without rotation of it. Films were 2μm thick. Then the deposited films were annealed at a temperature of 840-880°C for 10-100 hours in a gas atmosphere of O_2, N_2 or a mixture of O_2/N_2. The resistivities of the films were measured as a function of temperature in a standard four probe method using a pressed indium contact at a constant current of 100 μA from 65K to 300K. The atmosphere in which the films were placed was replaced with a He gas. The crystal structure of the films was examined by the X-ray diffraction (XRD) method with Cu-Ka radiation. The element dispersion of the films was examined by the electron probe micro analysis(EPMA).

RESULTS AND DISCUSSION

The temperature dependeces of resistivity of the thin films deposited at various sputtering conditions and annealed in various gas atmospheres are shown in Fig.1. Thin film of sample A was prepared in continuous depositing on the substrate without rotation of the table and annealed at 860°C for 10 hours in O_2 gas of a flowing rate of 5 l/min. The superconducting onset temperature Tc(on) of this film was approximately 100K and the zero resistivity temperature Tc(zero) was 80K. Thin film of sample B was prepared at the same sputtering condition as the sample A and annealed at 860°C for 10 hours in the mixture gas atmosphere of O_2/N_2 in the flowing rate of 0.5 l/min of O_2 and 5 l/min of N_2. In the thin film of sample B, a clear drop of the resistivity at around 120K was observed and the zero resistivity was achieved at 80K. The phenomena of the resistivity drop at around 120K was observed in the ranges of 1/2 to 1/10 of the O_2/N_2 ratio where the resistivity drop

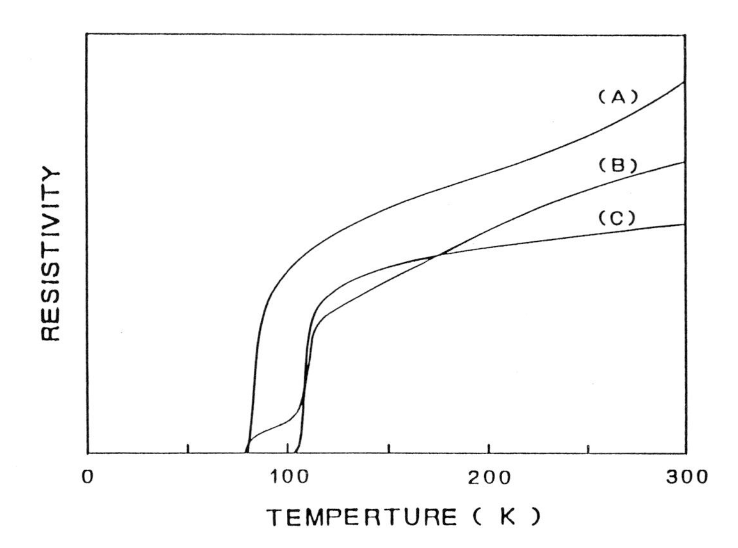

Fig.1 Temperature dependence of the resistivity for Bi-Sr-Ca-Cu-O films on MgO prepared (A) without rotation of the substrate table during sputtering and annealed at 860 °C for 10 hours in O_2 gas with flowing rate of 0.5 ℓ/min, (B) by the same sputtering condition as (A), and annealed in a mixed gas of O_2/N_2 with the flowing rate of 0.5/5 ℓ/min at the same temperature and hour, and (C) with rotation of the table and annealed at the same condition as (B).

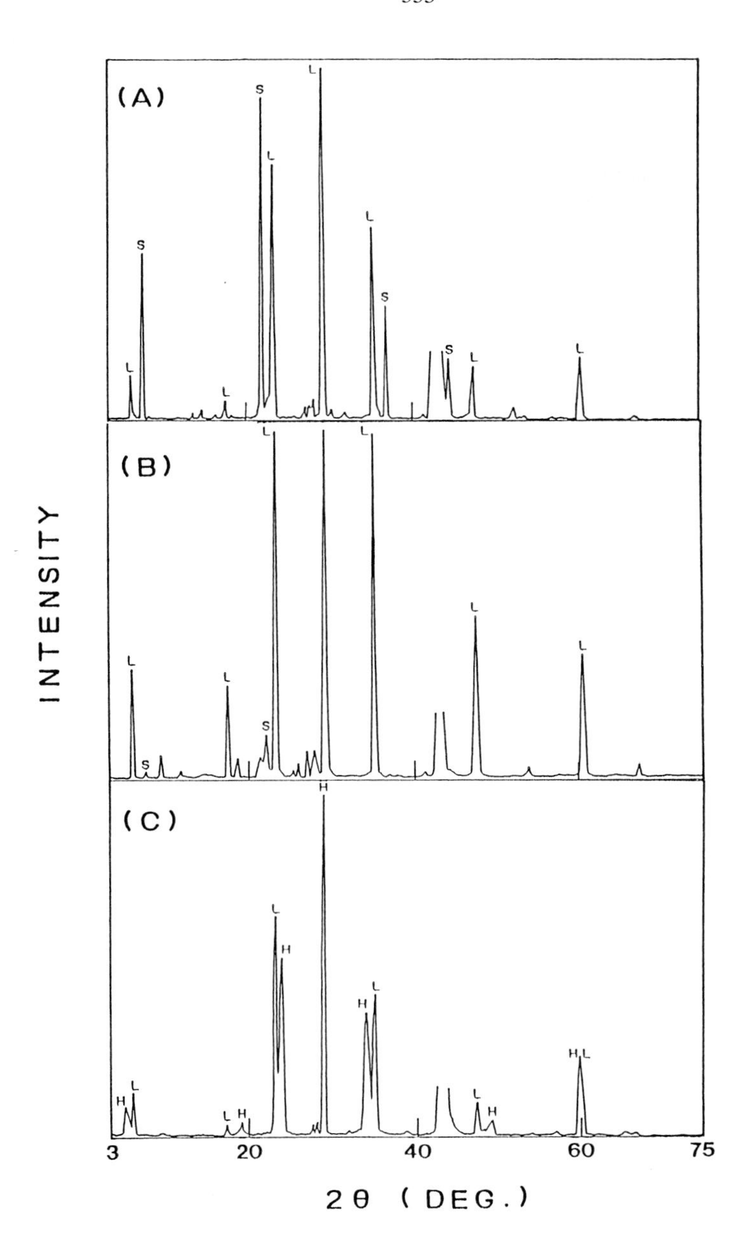

Fig.2 X-ray diffraction patterns of Bi-Sr-Ca-Cu-O thin films on MgO. (A),(B) and (C) correspond to those shown in Fig.1, respectively. (S ; a semiconductor-like phase peak, L ; a low-Tc phase peak, H ; a high-Tc phase peak.)

was clearer in the N_2 rich atmosphere. However, most of the films showed Tc(zero) at around 80K. Thin film of sample C was grown on MgO substrate at 300°C cyclically by sputtering through an off-centered shutter hole while rotating the substrate table at 15 r.p.m. and annealed at 860°C for 10 hours in the mixture of $O_2/N_2=1/10$. It showed high critical temperature of Tc(on)=120K and Tc(zero)=103K.

The X-ray diffraction patterns of the three samples are shown in Fig.2, where peaks S,L and H are a semiconductor-like phase with c-axis of 24 Å of the composition $Bi_2Sr_2Cu_1O_x$, a low-Tc phase with c-axis of 30 Å and a high-Tc phase with c-axis of 37 Å, respectively. Accordingly, the deposited film material of

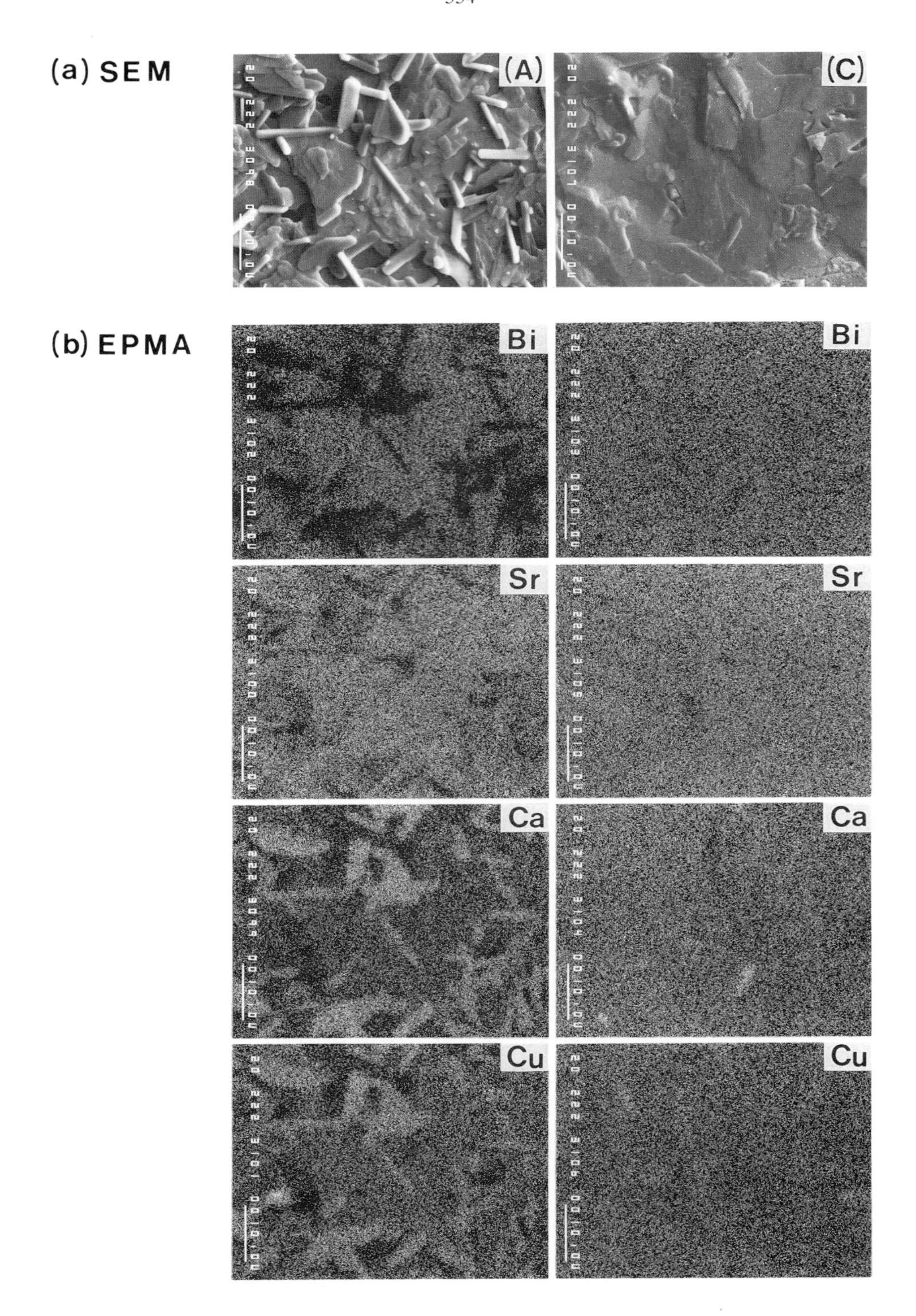

Fig.3 (a) Scanning electron micrograph and (b)electron probe micro analysis of Bi-Sr-Ca-Cu-O thin films on MgO. (A) and (C) correspond to those shown in Fig.1. (A bar indicates 10μm length.)

sample A consists of a semiconductor-like phase and a low-Tc phase. In the sample B, the peaks of semiconductor-like phase are low and the peaks of low-Tc phase are peaks of high-Tc phase appear although the peaks of low-Tc phase are observed. From these results, it shall be considered that when the as-deposited films are annealed, they are affected seriously by the gas atmosphere, especially of the partial pressure of O_2 to change their crystal structures.

The SEM observation shows the difference in surface roughness of the thin films. The surface of sample A is rough and that of sample C is smooth as shown in Figs.3(a)A and C. Investigation results by EPMA on element dispersion are shown in Figs.3(b)A and C. Some segregations of constituent elements of Ca and Cu are observed in the sample A, whereas no segregation is observed in the sample C of the smooth surface. Before annealing the as-deposited films, all the films are amorphous, lustrously smooth in their surfaces, and uniform in element dispersions. By annealing, crystallization and element segregation start in the thin films to change the surface smoothness. The thin film of sample A develops rod-shaped crystallites where the element segregations of Ca and Cu appear during the annealing, whereas the thin film of sample C does not develop any rod-shaped crystallites where no element segregation of the constituents appears. Both the chemical compositions of the sample A and C defined by EPMA were $Bi_{2.0}Sr_{2.1}Ca_{1.2}Cu_{2.4}O_x$. This estimation is correct within the accuracy of 10%. While the superconducting properties of the sample A and C are different each other. This suggests that the high-Tc phase is scarecely formed on the surface or at the grain boundary of the low-Tc phase of the thin films.

The properties of the superconducting Bi-Sr-Ca-Cu-O thin films sputtered on MgO(100) substrate and annealed in O_2 or air atmosphere have been investigated [5]-[8], but the zero resistivity temperature above 100K has not been reported except Ichikawa's.[9]

Ichikawa et al. have obtained the zero resistivity temperature of 100K at the film sputtered by using a target of $Bi_{1.6}Sr_{1.0}Ca_{1.0}Cu_{5.2}O_x$ composition at the substrate temperature of 800°C and annealed at 900°C for 20 min and at 865°C for 5 hours. This thin film reportedly consist of two phases of the low-Tc and the high-Tc. Although there are some differences between results of Ichikawa et al. and those of the present authors in the preparations of the thin films, the superconducting and crystallographic properties are almost the same each other.

Further investigations are needed to realize a single high-Tc phase or high-Tc rich phases in the thin films.

SUMMARY

Superconducting Bi-Sr-Ca-Cu-O thin films were prepared on MgO substrate by controlling sputtering conditions by using $Bi_1Sr_1Ca_1Cu_2O_x$ target, and annealed with changing O_2/N_2 ratio of gas atmosphere. The results obtained in this experiment are summarrized as follows.

1) The highest values of Tc(on)=120K and Tc(zero)=103K were obtained in the films which were deposited at the substrate temperature of 300°C cyclically with rotation of the substrate table and annealed at 860°C in O_2/N_2=1/10 gas atmosphere for 10 hours.

2) The films were confirmed by X-ray diffraction to consist of two phases of a low-Tc and a high-Tc phase. No thin film having a single high-Tc phase was obtained.

3) The results of SEM and EPMA revealed that an as-sputtered film has a smooth surface and changes the surface smoothness due to crystallization from amorphous state according to the annealing conditions. An annealed thin film having the more smooth surface and the less segregation of the constituent elements shows the higher critical temperatures.

REFERENCES

[1]H.Maeda, Y.Tanaka, M.Fukutomi and T.Asano: Jpn.J.Appl.Phys.27(1988)L209
[2]J.M.Tarascon, Y.Le Page, P.Barboux, B.G.Bagley, L.H.Green, W.R.McKinnon, G.W.Hull, M.Giroud and D.M.Hwang: Phys.Rev. B37(1988)9382
[3]E.Takayama-Muromachi, Y.Uchida, Y.Matsui, M.Onoda and K.Kato: Jpn.J.Appl.Phys. 27(1988)L556
[4]H.Koinuma, M.Kawasaki, S.Nagata, K.Takeuchi and K.Fueki: Jpn.J.Appl.Phys.27 (1988)L376
[5]M.Nakao, H.Kuwahara, R.Yuasa, H.Mukaida and A.Mizukami: Jpn.J.Appl.Phys.27 (1988)L378
[6]M.Fukutomi, J.Machida, Y.Tanaka, T.Asano, H.Maeda and K.Hoshino: Jpn.J.Appl. Phys.27(1988)L632
[7]H.Adachi, Y.Ichikawa, K.Setsune, S.Hatta, K.Hirochi and K.Wasa: Jpn.J.Appl. Phys.27(1988)L643
[8]T.Kato, T.Doi, T.Kumagai and S.Matsuda: Jpn.J.Appl.Phys.27(1988)L1097
[9]Y.Ichikawa, H.Adachi, K.Hirochi, K.Setsune, S.Hatta and K.Wasa: Phys.Rev.B38 (1988)765

High T_c Superconductor $YBa_2Cu_3O_{7-x}$ Thin Films Obtained by RF Magnetron Sputtering

Yu.V. GULIAEV, I.M. KOTELIANSKII, V.B. KRAVCHENKO, V.A. LUZANOV, and A.T. SOBOLEV

Institute of Radioengineering and Electronics, Academy of Sciences of the USSR, K. Marx Av. 18, Moscow, 103907, USSR

High temperature superconductor $YBa_2Cu_3O_{7-x}$ and similar rare earth compounds were prepared as thin films with superconducting transitions above 77 K by a number of authors. By that the best results were obtained on single crystal SrTiO substrates where unit cell parameters and crystal structure are close to those for $YBa_2Cu_3O_{7-x}$ compound [1,2]. $SrTiO_3$ crystals are difficult to grow so the price of the crystals is rather high. Besides high dielectric constant ($\varepsilon \sim 300$) of strontium titanate is a great obstacle for ultra-high frequency device applications of the superconducting films.

This paper reports preparation of the films having superconductivity transition temperature T above 77K. Widely used in optics BaF_2 single crystals with $\varepsilon \sim 7$ were applied as substrates.

Two principal ways $YBa_2Cu_3O_{7-x}$ films preparation by magnetron sputtering are known now. The first one includes amorphous film deposition on unheated substrate and subsequent high-temperature annealing in oxygen-containing atmosphere (for instance, [3]). The second one consists in film deposition in oxygen-containing atmosphere on substrate heated up to 600-700 $^\circ$C (for instance, [4]).

Our investigations have shown that the $YBa_2Cu_3O_{7-x}$ films chemical composition practically does not differ from the target composition when the deposition is made in argon atmosphere and substrate temperature T_S does not exceed 400°C. Increase of T_S leads to relative Ba and especially Cu contentration decrease in the films. By that Cu concentration decrease is substatially higher than the calculated one using temperature dependence of Cu condensation coefficient This is possibly connected with formation of some intermediate compounds of the metallic components which results in the change of condensation coefficient temperature dependence for each of the films components. Presence of oxygen in the reactive vapor mixture also decreases considerably relative barium and copper concentrations in the films deposited. This is due to preferential sputtering of some film components by oxygen ions and atoms bombarding its surface. Therefore $YBa_2Cu_3O_{7-x}$ film deposition without subsequent annealing requires careful choice of target composition for each specific magnetron design. That makes the film deposition much more complicated.

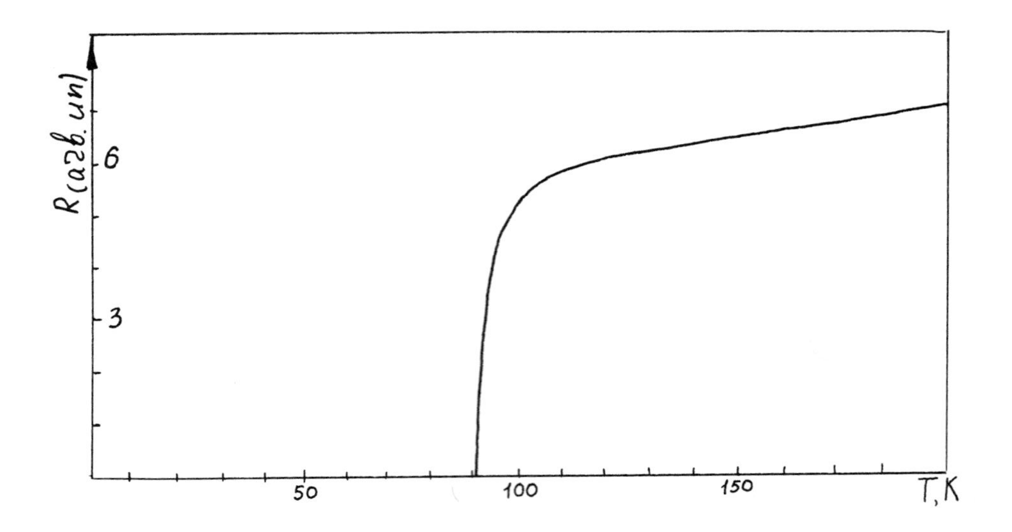

Fig.1 Temperature dependence of film resistance

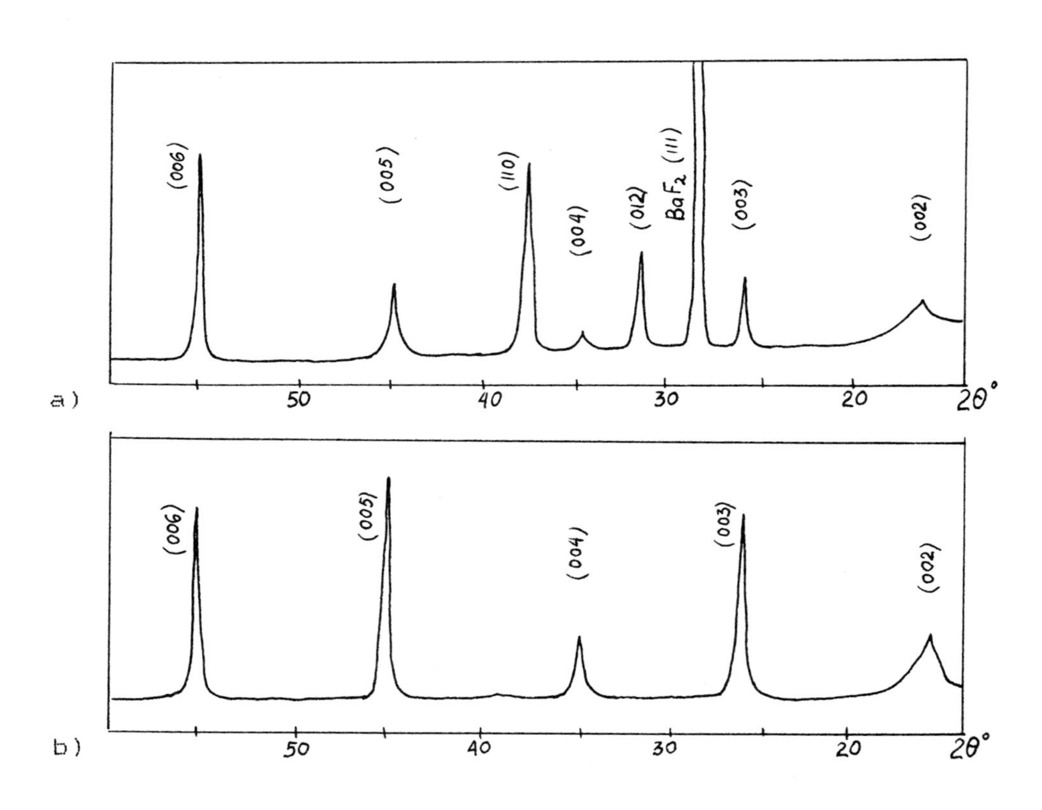

Fig.2 X-ray diagram (λ Co) of polycrystalline (a) and oriented (b) $YBa_2Cu_3O_{7-x}$ films

$YBa_2Cu_3O_{7-x}$ thin films were deposited on unheated BaF_2 substrates with subsequent annealing. The substrates had (111) orientation and were subjected to preliminary chemico-mecanical polishing treatment. Superconducting $YBa_2Cu_3O_{7-x}$ ceramics obtained by usual pressing and high temperature ($\sim 950^{\circ}C$) annealing of Y_2O_3, $BaCO_3$ and CuO powder mixture was used as a target. 13.56 MHz frequency and 100 W power were used for sputtering; deposition velocity was (0.8-1) mcm per hour, thickness of the films deposited was about 1 mcm. The layers had unruffled surfaces, specific resistivity higher than 10 Ohm.cm and did not exhibit superconducting transitions. Annealing at 900-930 C resulted in sharp decrease of films resistivity, down to 10^{-2} Ohm.cm.

The temperature dependence of resistivity is shown in fig.1. Between 300 K and 110 K the resistivity decreased linearly. Then superconducting transition was observed. Characteristic temperatures at which the resistivity was $0.9R_{110}$ (R_{110} is the resistivity at T=110 K), $0.1R_{110}$ and 0 are correspondingly 95 K, 91 K and 89 K. The change of measuring current density in 0.01-1A/cm range did not influence the temperature-resistivity curve.

Two types of films were obtained. For the polycrystalline ones (fig.2) intensity maxima positions correspond to those for $YBa_2Cu_3O_{7-x}$ ceramics which was used as the target. Films with preferred (001) orientation of grains in the film plane were also obtained. X-ray diagram (fig.2b) shows only strong (001) reflections for these films. Grain size was 1-30 mcm in dependence on annealing conditions. Small (up to 0.3%) quantity of F was discovered on the film surface using electron beam X-ray microanalyzer. This is connected with F diffusion out of the substrate.

Strong absorption line was discovered during ESR thin films investigations at 77 K in small ($\sim$20 Oe) magnetic fields using 9.2 GHz frequency.

References.

1. Y.Enomoto, T.Murakami, M.Suzuki, K.Moriwaki. Jap.J.Appl.Phys.,1987, v.28, No.7, p. L 1248.

2. P.P.Chudhart, R.H.Koch, R.P.Laibowitz, T.R.Mc Yire, R.J.Yambino. Phys.Rev.Lett., 1987, v.58, No.25, p.2684.

3. H.Adachi, K.Setsune, T.Mitsuyu, K.Hirochi, Y.Ichikawa, T.Kamada, K.Wasa. Jap.J.Appl.Phys.,1987, v.26, L 709.

4. H.Adachi, K.Hirochi, K.Setsune, M.Kitabatake, K.Wasa. Appl.Phys.Lett.,1987, v.51, No.26, p.2263.

4.10 Thin Film Processing and Properties: Part 2

Some Properties of High-Tc Oxide Superconducting Thin Films

SHUJI YAZU

Sumitomo Electric Industries Ltd., 1-1-1, Koyakita, Itami, Hyogo, 664 Japan

ABSTRACT

High Tc superconducting LnBaCuO (Ln: Ho,Er,Y), BiSrCaCuO and TlBaCaCuO films have been produced by rf magnetron sputtering using single composite targets. X-ray, RHEED and SEM analyses show that LnBaCuO films made on MgO (001) are epitaxially grown single crystals with extremely smooth surface. The critical current densities in the LnBaCuO films at 77.3 K are excess of 3×10^6 A/cm^2. Highly textured c-axis-oriented poly-crystalline BiSrCaCuO films around the cation ratio 2:2:2:3 having Tc (zero-resistance temperature) of 105 K show critical current densities of 1.9×10^6 A/cm^2 at 77.3 K and 2.1×10^7 A/cm^2 at 40 K. In TlBaCaCuO films having a maximum Tc of 109K (zero-resistance temperature), a critical current density of 3.2×10^6 A/cm^2 at 77.3 K was obtained.

INTRODUCTION

There are many problems to be solved for practical application of high-Tc superconducting materials. Especially, it is necessary to obtain the property of high critical current density at liquid nitrogen temperatures considering the requirements in electronic devices and circuits as well as in cables and magnets. World-wide research efforts have brought about the improvement of the property of the oxide thin films using various deposition procedures. As we report here, high quality LnBaCuO (Ln: Ho,Er,Y) thin films can be grown epitaxially on single-crystal MgO (001) and the critical current densities of these films can be quite high, exceeding 3×10^6 A/cm^2 at 77.3 K.

Newly discovered rare-earth-free high-Tc oxide materials of BiSrCaCuO and TlBaCaCuO systems are hopefully expected to have enough temperature margins for operation at liquid nitrogen temperatures. We have recently succeeded in developing BiSrCaCuO and TlBaCaCuO films with high critical current densities exceeding 10^6 A/cm^2 at 77.3 K.

EXPERIMENTAL

LnBaCuO, BiSrCaCuO and TlBaCaCuO films have been prepared by rf magnetron sputtering with a single composite target. The typical experimental conditions are summarized in Table 1. The target was prepared by sintering mixed powder of Ln_2O_3, $BaCO_3$, CaO, Bi_2O_3, Tl_2O_3 and CuO for each system as the cation composition listed in Table 1. In this experiments, we used single-crystal MgO as a substrate material. The glow discharge was radio-frequency excited (13.56 MHz) under 80% Ar and 20% oxygen atmosphere. The substrate temperatures were varied from 350 °C to 800 °C depending on the kind of depositing materials. The deposited films were annealed in the temperature range of 800 to 950 °C under 1 atm. oxygen. In the case of TlBaCaCuO films, the substrate temperatures were kept at 350 °C and the films were post-annealed in a sealed tube containing Tl_2O_3 powder, because of a high vapor-pressure of Tl compound. The temperature dependence of the film resistivity and the critical current density were measured with a four-point probe technique in a cryostat. The composition of the films was determined by inductively coupled plasma spectroscopy (ICP). Microstructure, surface morphology and crystal chemistry of the films were examined by scanning and transmission electron microscopy, reflective high-energy electron diffraction (RHEED) and x-ray diffraction.

RESULTS AND DISCUSSION

LnBaCuO Thin Films

We have investigated surface morphology and crystal chemistry of the HoBaCuO thin films deposited on single-crystal MgO (001) as a function of post-annealing temperatures Ta and substrate temperatures Ts [1]. The results which we obtained are schematically drawn in Figure 1. The map can be divided into three regions of crystallinity of the films; amorphous, poly-crystal and epitaxial growth regions. Ta of more than 900 °C is necessary to crystallize amorphous films, while thin films deposited with Ts > 500 °C are poly-crystalline and epitaxial thin films are grown at Ts > 600 °C. Although, both of poly-crystalline and epitaxial films are superconductive, high critical current densities were obtained in the epitaxial films.

We have grown high-Jc single crystal thin films of LnBaCuO (Ln: Ho, Er, Y) by optimizing the sputtering and post-annealing conditions [2]. In the case of HoBaCuO films, deposition rate was controlled to be 2 - 5 nm/min. and post-annealed at 920 °C. X-ray diffraction patterns of the films indicated that the films had only strong (00n) peaks and that the HoBaCuO (001) planes grew epitaxially parallel to the (001) surface of the MgO substrate. Figure 2 shows RHEED patterns for the post-annealed HoBaCuO film. The streak patterns indicate that this film is a mostly perfect single-crystal and the surface is extremely smooth. We measured the surface roughness of the film by means of SEM with a pair of secondary electron detectors and found that the maximum roughness was less than 5nm (Figure 3). The resistivity-temperature curve of the film showed that superconducting transition temperature (zero resistance) was 84 K and the transition width (10%-90%) was 0.4 K.

Some results of critical current measurements of LnBaCuO films are shown in Figure 4 as a function of temperature. In this experiments, the highest Jc value of 3.53×10^6 A/cm^2 at 77.3 K was obtained for a HoBaCuO film. Recently, the value improved up to 4×10^6 A/cm^2 for a YBaCuO film. The detailed data are reported in this proceedings by T. Nakahara.

Magnetic field dependence of critical current densities were measured for a HoBaCuO film.(Figure 5) Under parallel magnetic field to the current direction in the film, the critical current density was kept constant in this applied field range. This result suggests that the superconductive current pass is straight in the film and JxB forces do not affect the flux movement in the film under the parallel magnetic field. Jc under magnetic field applied perpendicular to the current direction was 1.5×10^6 A/cm^2 at 1.5 tesla. The value is considerably high compared with reported values for sinterd poly-crystalline materials [6].

BiSrCaCuO Thin Films

BiSrCaCuO films have been prepared as listed in Table 1. It is reported that in the BiSrCaCuO system, two phases of of superconductors with different Tc are present; one is the phase of cation ratio 2:2:1:2 with Tc around 80 K and the other is estimated to be 2:2:2:3 phase with Tc around 110 K [3]. To obtain pure higher Tc phase, many efforts have been done. Recently, it was found that partial replacement of Bi by Pb were effective to increase the volume fraction of the higher Tc phase [4]. We have prepared high quality films of simple BiSrCaCuO with Tc 105 K (zero resistance) as shown in Figure 6. XRD patterns of the films revealed that the films were composed of a single phase and had layered structure with c = 37.2 A. We observed the films by transmission electron microscopy (TEM). Figure 7 shows a characteristic [001] electron diffraction patterns of illustrating orthorhombic symmetry and an incommensurate five-times modulation along the b-axis. The films were highly textured with the c-axis oriented normal to the film surface and RHEED patterns indicated that the films were poly-crystalline. The critical current densities of the films were measured as a function of temperature. The typical results are shown in Figure 8. More detailed discussions are reported in this proceedings by H. Itozaki.

TlBaCaCuO Thin Films

TlBaCaCuO films were prepared using the procedures as listed in Table 1. The as-grown films were amorphous state because of the low substrate temperature. After annealing at a higher temperature than 850 °C, the films crystallized and turned to be superconductive as shown in Figure 9. The XRD patterns of the annealed film in the figure indicates that the film is textured with c-axis normal to the film surface and the film is comprised of a single high Tc 2223 compound having tetragonal structure (c = 35.6 A) [5]. RHEED patterns of the film indicated that the film was poly-crystalline.

The typical resistivity versus temperature curve is shown in Figure 10. The zero resistance transition temperature is 109 K. The chemical composition of the film was analyzed by ICP as Tl:Ba:Ca:Cu ratio was 2.1:1.9:2.1:3.3. In Figure 11, temperature dependence of critical current densities of a TlBaCaCuO film is shown. The high critical current density 3.2×10^6 A/cm^2 at 77.3 K was achieved. Applied magnetic field strongly suppressed the critical current density in the film. (Figure 12) We found many tiny pores in the surface of the film probably caused by evaporation of Tl compound. These defects of the films and the poly-crystalline characteristics will be the cause of rapid decrease of the critical current density with increasing applied magnetic field.

CONCLUSION

The experimental results are summarized in Table 2. We have prepared high quality single crystal films of LnBaCuO (Ln: Ho, Er, Y) with Jc more than $3x10^6$ A/cm^2 at 77.3 K. BiSrCaCuO films comprising of single high Tc phase with Tc 105 K (zero resistance) and Jc $1.9x10^6$ A/cm^2 at 77.3 K were obtained. We observed a high critical current density as $3.2x10^6$ A/cm^2 at 77.3 K in poly-crystalline TlBaCaCuO thin films. Applied magnetic field strongly suppress the Jc in poly-crystalline BiSrCaCuO and TlBaCaCuO films compared with in the case of single-crystal LnBaCuO films.

Table 1. Procedures for preparation of high Tc superconducting thin films.

	LnBaCuO	BiSrCaCuO	TlBaCaCuO
METHOD	RF MAGNETRON SPUTTERING		
SUBSTRATE	MAGNESIA SINGLE CRYSTAL		
DEPOSITION CONDITION			
TARGET	Ln:Ba:Cu =1:2.2:3.4	Bi:Sr:Ca:Cu =1.4:1:1:1.5	Tl:Ba:Ca:Cu =0.7:2:2:3
GAS	$Ar+O_2$(20%)	$Ar+O_2$(20%)	$Ar+O_2$(20%)
GAS PRESSURE	0.05Torr	0.02Torr	0.05Torr
SUB. TEMP.	750 °C	600-800 °C	350 °C
THICKNESS	700nm	200nm	700nm
POST ANNEAL			
TEMP.	850-950 °C	800-900 °C	850-900 °C
ATMOSPHERE	(O_2)/(1atm)	(O_2)/(1atm)	(O_2+Tl)/(1atm)

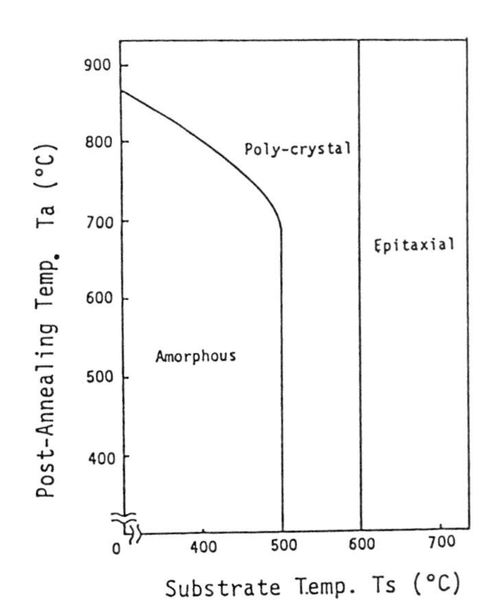

Figure 1. Map of crystallinity for HoBaCuO films as a function of Ta and Ts.

Figure 2. RHEED patterns for an as grown HoBaCuO thin film on a MgO (001) substrate.

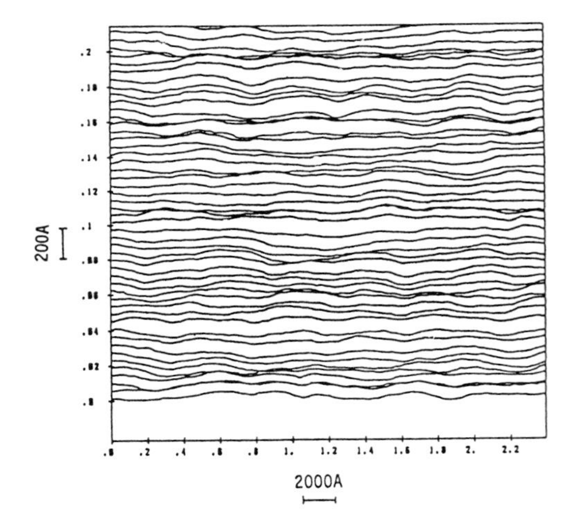

Figure 3. Surface roughness for HoBaCuO thin film observed SEM with a pair of secondary electron detectors.

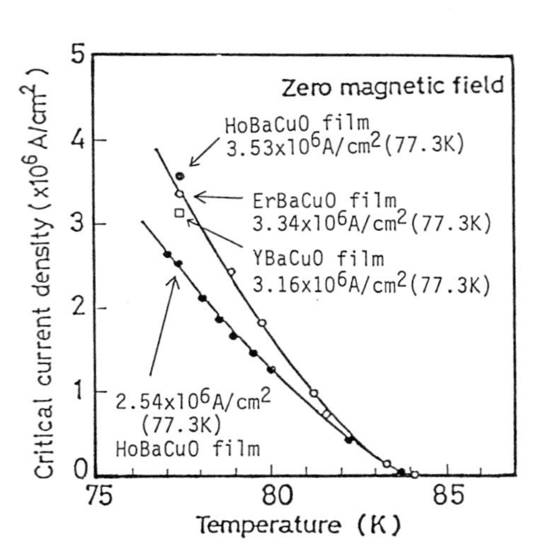

Figure 4. Temperature dependence of critical current density for LnBaCuO thin films.

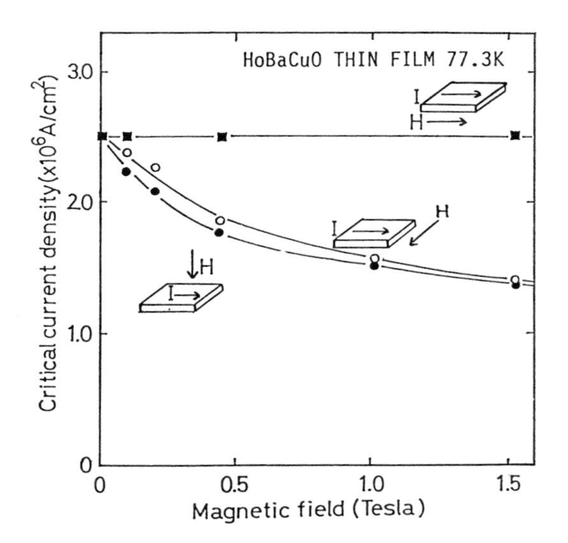

Figure 5. Magnetic field dependence of critical current density for a HoBaCuO thin film.

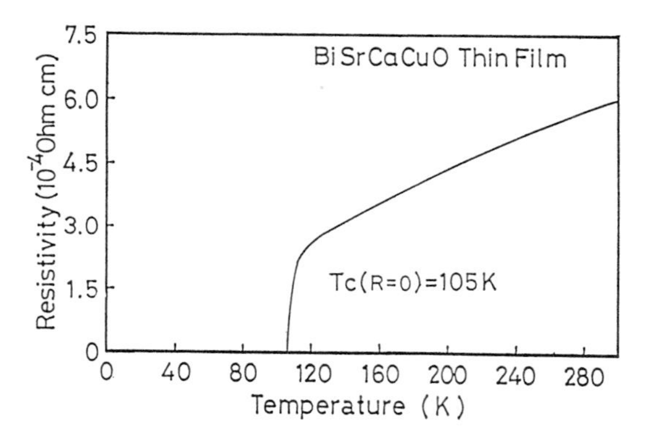

Figure 6. Resistivity versus temperature curve for a BiSrCaCuO thin film.

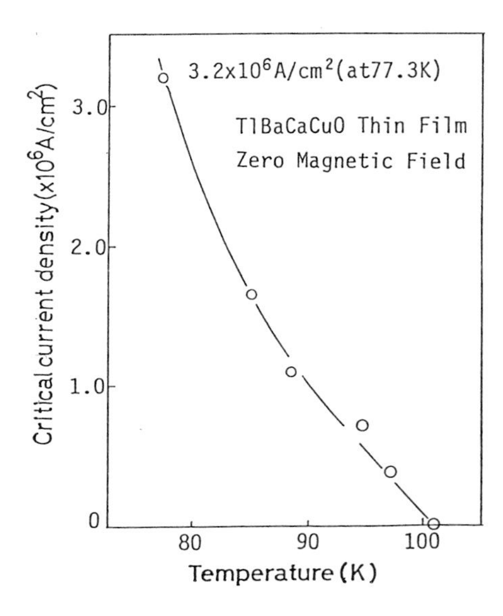

Figure 11. Temperature dependence of critical current density for a TlBaCaCuO thin film.

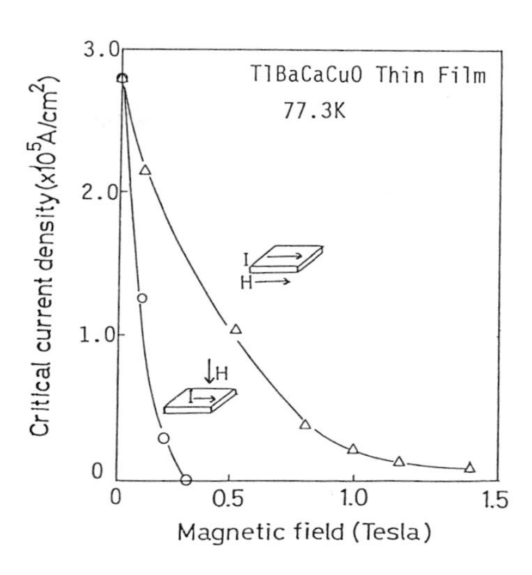

Figure 12. Magnetic field dependence of critical current density for a TlBaCaCuO thin film.

Table 2. Summary of the properties of 3 kinds of high-Tc superconducting thin films.

	LnBaCuO	BiSrCaCuO	TlBaCaCuO
Tc(R=0)	90K	105K	109K
Jc(77.3K)	$4X10^6 A/cm^2$	$1.9X10^6 A/cm^2$	$3.2X10^6 A/cm^2$
CRYSTALINITY	SINGLE	POLY	POLY
MOPHOLOGY	FLAT	ROUGH	ROUGH
Jc UNDER MAG. FIELD	$1.5X10^6 A/cm^2$ (at 1.0T)	$1.0X10^5 A/cm^2$ (at 0.5T)	$1.2X10^5 A/cm^2$ (at 0.1T)

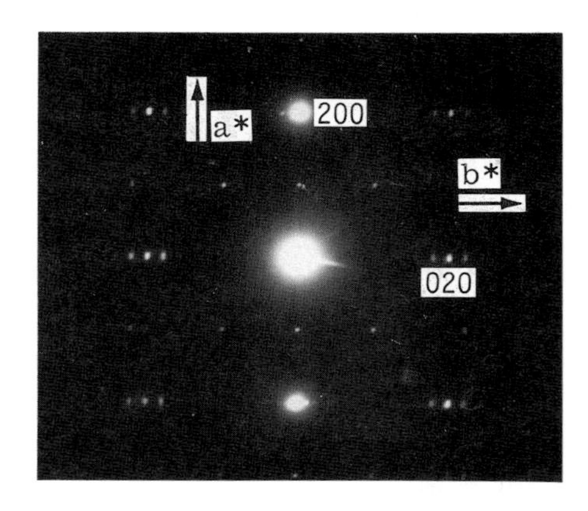

Figure 7. [001] electron diffraction patterns for a BiSrCuCuO thin film.

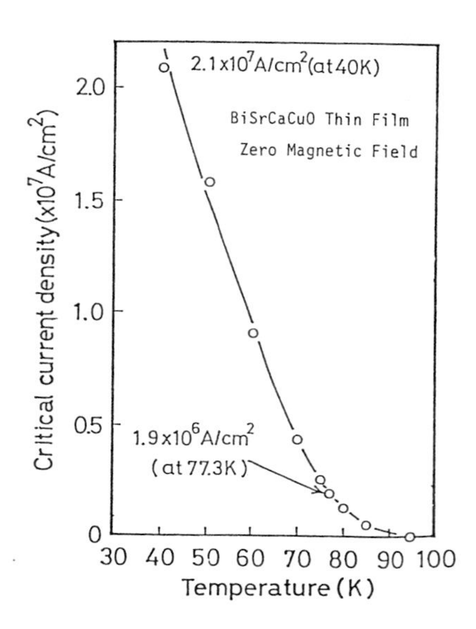

Figure 8. Temperature dependence of critical current density for a BiSrCuCuO thin film.

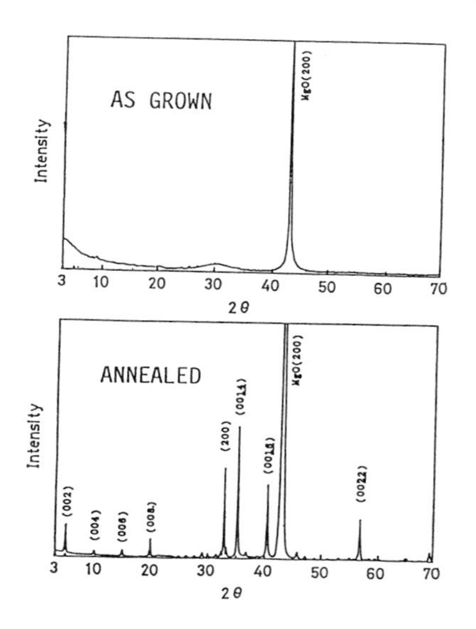

Figure 9. X-ray diffraction patterns for TlBaCaCuO thin films.

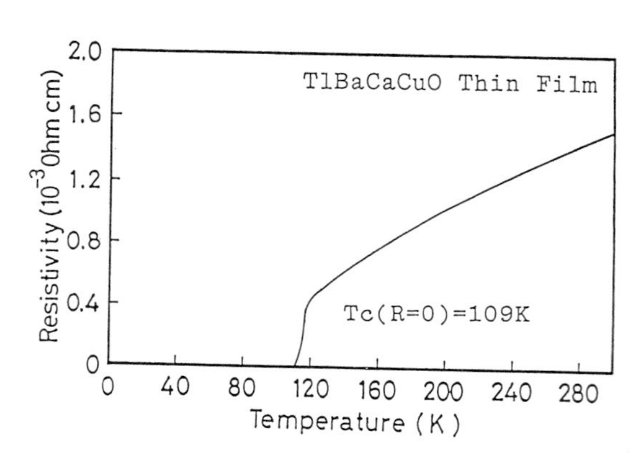

Figure 10. Resistivity versus temperature curve for a TlBaCaCuO thin film.

REFERENCES

[1] S. Tanaka and H. Itozaki: Jpn. J. Appl. Phys. (to be submitted)
[2] S. Tanaka and H. Itozaki: Jpn. J. Appl. Phys. 27(1988)L622
[3] H. Maeda, Y. Tanaka, M. Fukutomi and T. Asano: Jpn J. Appl. Phys. 27(1988)L209
[4] R.J. Cava et al.: Physica C 153-155(1988)560
[5] M.A. Subramanian et al.: Nature, 332(1988)420
[6] J.W. Ekin: Adv. Ceram. Mater. 2(1987)586

Microstructural Development in Ion Beam Deposited Superconducting $YBa_2Cu_3O_{7-x}$ and Bi-Ca-Sr-Cu-O Thin Films

ALAN B. HARKER, J.F. DENATALE, P.H. KOBRIN, and I.S. GERGIS

Rockwell International Science Center, Thousand Oaks, CA 91360, USA

ABSTRACT

The microstructural development of superconducting thin films in the Y-Ba-Cu-O (YBCO) and Bi-Ca-Sr-Cu-O (BCSCO) systems has been observed by transmission electron microscopy as a function of ion beam deposition and annealing conditions. 100 nm thick YBCO samples deposited between 580 and 650°C on $SrTiO_3$ and MgO were found to be single crystal with c-axis normal orientation. Higher temperature deposition or annealing to improve the oxygen stoichiometry and critical properties enhanced microstructure development and produced grain oriented polycrystalline material. In the BCSCO system, the 85K superconducting phase is formed by recrystallization from a non-superconducting phase and exhibits c-axis normal grain orientation even on polycrystalline substrates. Device structures have been prepared from these materials which show the presence of at least two distinct types of weak link behavior associated with the grain boundaries in the superconductors.

INTRODUCTION

The desire to incorporate the new class of ceramic oxide superconductors into hybrid device configurations has led to numerous investigations of techniques for fabricating thin films of these materials. Energetic beam approaches (ion-beam, RF, and DC sputtering) offer the potential advantage of lowering the deposition temperature by enhancing reactivity and mobility of the adatoms on the substrate, thus reducing the need for high temperature post-deposition anneals. Such temperature considerations become of greatest importance when trying to combine the superconducting oxides with semiconductors such as silicon and gallium arsenide. Though reducing the deposition temperature is desirable, the films must also be formed in a manner that minimizes twinning and grain boundary regions. Weak links from non-stoichiometries and secondary phases at grain boundaries inhibit the formation of active devices with Josephson junction behavior [1,2].

In this study, reactive single ion-beam deposition has been investigated as a means of producing films in the $YBa_2Cu_3O_{7-x}$ (YBCO) [3] and Bi-Ca-Sr-Cu-O (BCSCO) [4] systems, both with and without post-deposition annealing, on single crystal $SrTiO_3$, MgO, and Si, as well as polycrystalline Y stabilized cubic ZrO_2. The microstructural development during deposition and progressive annealing in the thin films has been followed by transmission electron microscopy (TEM) using films deposited onto ion-beam-thinned foils of the substrate materials.

EXPERIMENTAL

All deposition experiments were carried out in a copper-lined, ultra-high vacuum chamber with a turbomolecular pumped base vacuum of 10^{-5} Pa without baking. The deposition source was a Kaufman-type ion gun with a 0.5 cm diameter focus, normally operated at 40 mA and 1 KeV with either argon or an argon-oxygen gas mixture, with an oxygen background pressure near 0.01 Pa. Substrates were supported on a stationary resistively heated holder, capable of achieving 800°C. Ceramic oxide targets were fabricated for the system by mixing base oxides, isopressing, and sintering. Thin films were either annealed in-situ in oxygen or removed from the vacuum chamber and annealed in a controlled atmosphere tube furnace. The chemistry of the films was determined by inductively coupled plasma emission spectroscopy.

Transmission electron microscopy (TEM) samples were prepared by direct deposition onto previously dimpled and ion-beam-thinned substrates of single crystal materials as shown in Fig. 1. By this process, thin film microstructure and thin film-substrate orientations can be obtained without any need for sample manipulation or thinning after fabrication [5].

RESULTS

The chemical composition of the multi-cation films was found to be a function of the target stoichiometry, beam focus, and the temperature of the substrate. Compositional control was established by preparing separate ceramic targets for each deposition temperature and by using mechanical apertures in addition to normal electrostatic focusing to attain a reproducible, confined 1 KeV argon-oxygen ion beam.

YBCO depositions were normally carried out with (001) single crystal MgO and $SrTiO_3$ substrates, though superconducting films were also produced on Si and yttrium stabilized cubic ZrO_2. Films 50 to 2,000 nm thick were grown at 580 to 750°C. The as-deposited films with in-situ annealing at the deposition temperature for 1 hour were cosmetically smooth, black, and conductive at room temperature. The crystallography of the YBCO films is highly dependent upon composition and the properties of the single crystal substrates. Films very near to

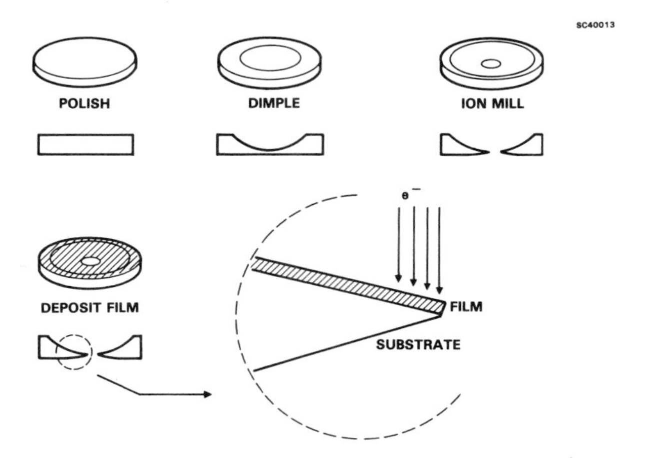

Fig. 1 Schematic of TEM substrate preparation for thin film deposition and study.

the desired 1:2:3 stoichiometry on $SrTiO_3$ substrates with a uniform habit and good polish were typically oriented with the c-axis of the YBCO structure normal to the plane of the substrate. On MgO substrates, films with both a- and c-axis normal orientation could be obtained, with a slight excess in Cu favoring c-axis normal orientation. X-ray diffraction analysis showed the c-axis lattice constant of the as-deposited films to be large, about 11.8 to 11.9Å, indicating insufficient oxidation unless the samples were left in oxygen to anneal after deposition.

The microstructure of a 100 nm thick YBCO film as-deposited at 600°C on (001) SrTiO3 is shown in Fig. 2. The electron diffraction pattern shows the film to be fully oriented to the substrate with the slight lattice mismatch visible in the shift of the diffraction spots. The electron bend contours of the substrate are completely reproduced in the film, with no indication of any discrete grain structure in the as-deposited film. A similar film grown on (001) single crystal MgO also showed lack of defined grain structure and exhibited full orientation to the substrate. The film on MgO did, however, have discrete regions in which individual fine scale twinned grains were present. Annealing of the YBCO films on MgO and $SrTiO_3$ in 1 atmosphere of oxygen for 12 hours at 500°C did produce increased electron contrast in micrographs of discrete regions in the film, but no obvious grain structure. The electron bend contours of the substrate were no longer apparent, though Moire fringes, visible over broad areas, demonstrated the continued orientation of the film to the substrate.

Annealing the films to 675°C for 2.5 hours produced significant coarsening in the films, with individual 0.02-0.05 micron scale grains forming and growing out of the plane of the substrate at the expense of overall film integrity. This process continued with further annealing at 750°C for one hour as can be seen in Fig. 3 for YBCO on $SrTiO_3$. At this point, significant texture has developed in the thinner portions of the film, with the grain development producing porosity. However, even with individual grains being formed, the film retains its c-axis normal orientation to the substrate as shown by the electron diffraction pattern and Moire fringing. Texture development continued with further annealing, and after 20 minutes at 910°C the films were fully polycrystalline

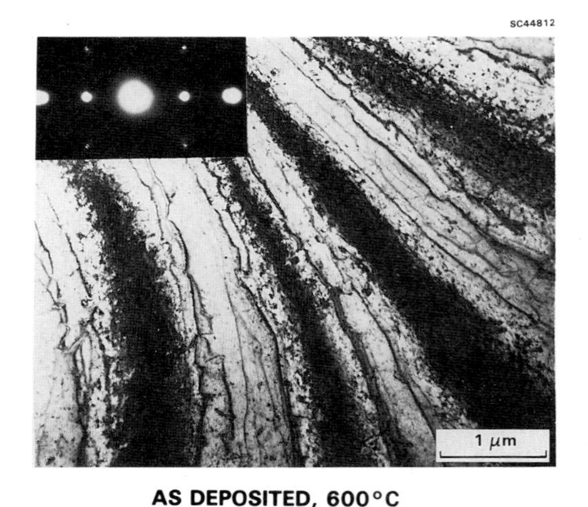

AS DEPOSITED, 600°C

Fig. 2 TEM micrograph and electron diffraction (SAD) pattern of an 100 nm thick YBCO film on (001) $SrTiO_3$, deposited at 600°C.

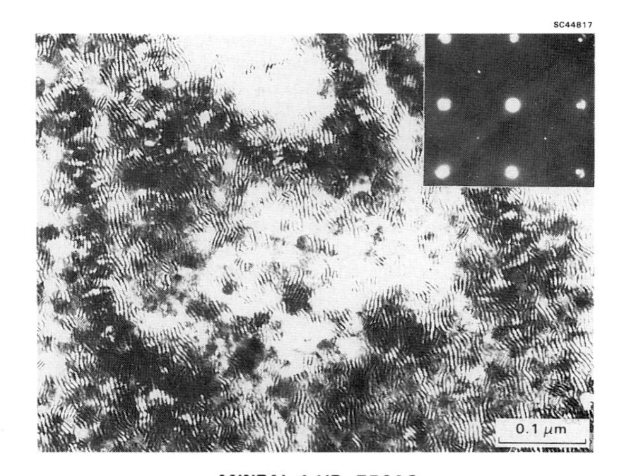

ANNEAL 1 HR, 750°C

Fig. 3 TEM micrograph and SAD pattern of a 100 nm thick YBCO film on $SrTiO_3$ after annealing 1 hour at 750°C.

with broad area fine scale twinning with an orthogonal pattern as shown for a YBCO film on MgO in Fig. 4. This micrograph also shows individual grains nucleating on top of the highly oriented portion of the film in contact with the substrate. Many of these grains retain c-axis orientation though some are oriented with their a-axis normal and others random. Often thicker films show the presence of both a- and c-axis orientation and it is likely that secondary nucleation processes, occurring away from the substrate, are responsible.

Electrical measurements made on thicker films, similar to those deposited on the thin foil substrates, show that the as-deposited films are oxygen poor, achieving a zero dc resistance at 60K or less with low critical currents. Annealing at 750°C and above further improves the critical properties with Jc levels of greater than 10^5 A/cm^2 at 50K. Even with such high critical currents, the current-voltage plots of the annealed films show a characteristic curvature, which is indicative of multiple weak links occurring between the oriented grains [2]. Such behavior limits the device applications of the films and further motivates the optimization of single crystal films.

Films from the Bi-Ca-Sr-Cu-O system, with a stoichiometry of $Bi_{0.8}Ca_{1.0}Sr_{1.1}Cu_{0.9}$ (BCSCO), were also deposited at 650 to 750°C and annealed in oxygen to 880°C. Annealed films were polyphase, containing both the main superconducting phase with a c-axis dimension of 30.6Å and a minor non-superconducting polytypoid with c = 24.4Å [4]. X-ray diffraction showed the material to be highly oriented with the c-axis normal to the MgO substrate. This orientation is developed by recrystallization of the c = 30.6Å BCSCO phase from the non-superconducting c = 24.4Å phase, which appears to be the stable phase at temperatures below 800°C.

As deposited, the films showed fine grained microstructure with the c-axis normal oriented to the MgO substrate and the c = 24.4Å phase dominant. After annealing to 880°C, the film formed the sheet-like microstructure shown in Fig. 5. The very thin micaceous layers form rectangular grains as large as 5 to 10 microns wide which are fully c-axis normal oriented to the MgO substrate. Interestingly, the BCSCO material can also be formed with c-axis normal orientation by annealing of initially amorphous films upon polycrystalline ZrO_2 substrates. This one-dimensional orientation is apparently caused by the preferential formation of the layered grain structure upon recrystallization.

Fig. 4 TEM micrograph and SAD pattern of a 100 nm thick YBCO film on MgO after annealing 20 minutes at 910°C.

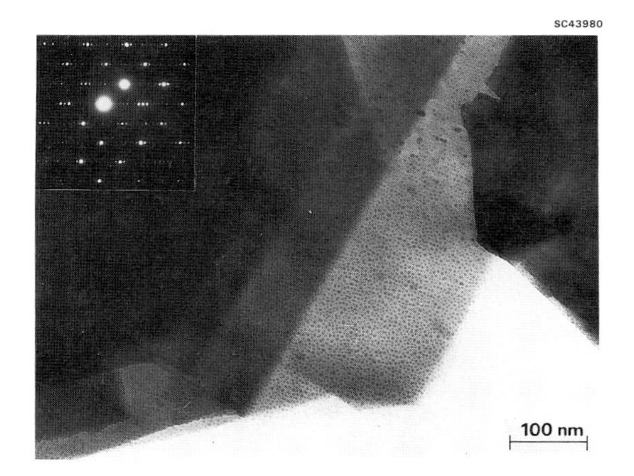

Fig. 5 TEM micrograph and SAD pattern showing the sheet structure of the 85K superconducting phase Bi-Ca-Sr-Cu-O system.

Strip lines patterned on the large grained, oriented BCSCO films produced critical currents of 5×10^4Å/cm^2 at 15K, which is greater than those obtained in typical unoriented YBCO films. The I-V behavior of 10 µm strip lines crossing adjoining grains shows the presence of two types of weak links in the BCSCO films, one type having a significantly higher dynamic resistance than the other [2]. This work supports the observations of Chaudhari et al [1] that intergrannular boundary regions can degrade the electrical properties of the ceramic superconductors.

CONCLUSIONS

Electrical and microstructural analysis of superconducting ceramic oxide thin films in the YBCO and BCSCO systems show that the electrical characteristics of polycrystalline and grain-oriented materials are dominated by grain-boundary effects. Epitaxial growth of the BCSCO materials was not achieved at temperatures as high as 750°C by ion beam deposition, with grain-oriented films being produced only by recrystallization. YBCO, conversely, forms as single crystal 100 nm thick films on $SrTiO_3$ and MgO at temperatures as low as 580°C, but shows coarsening and texture development with oxygen annealing above 650°C.

Further efforts in the development of the ion beam deposited YBCO films are focusing upon techniques of achieving full oxygen stoichiometry at temperatures below 650°C, while efforts with the BCSCO films are directed toward higher temperature deposition and investigating stoichiometry variations in the system to enhance epitaxy at lower temperatures.

REFERENCES

1. P. Chaudhari, J. Mannhart, D. Dimos, C.C. Tsuei, J. Chi, M.M. Oprysko, and M. Scheuermann, Phys. Rev. Let. 60(16), 1653 (1988).
2. I.S. Gergis, J.A. Titus, P.H. Kobrin, and A.B. Harker, submitted Appl. Phys. Lett. (1988).
3. P.H. Kobrin, J.F. DeNatale, R.M. Housley, J.F. Flintoff, and A.B. Harker, Advanced Ceramic Materials, Vol. 2(3B), 430 (1987).
4. A.B. Harker, P.H. Kobrin, P.E.D. Morgan, J.F. DeNatale, J.J. Ratto, I.S. Gergis, and D.G. Howitt, Appl. Phys. Lett. 52, p. 2186 (1988).
5. J.F. DeNatale and A.B. Harker, to be submitted J. Elect. Micro. Tech. (1988).

Transport Properties of High Tc Y-Ba-Cu-O Thin Films Made by RF Magnetron Sputtering

JON-CHI CHANG, SATORU SEO, AKIRA SAYAMA, MASAKAZU MATSUI, KIYOSHI YAMAMOTO, and NAKAHIRO HARADA

Yokohama R&D Laboratories, The Furukawa Electric Co., Ltd., 2-4-3, Okano, Nishi-ku, Yokohama, 220 Japan

ABSTRACT

High Tc YBCO thin films were prepared on MgO(100) single crystal substrates by rf magnetron sputtering. In order to obtain fine control of the composition of thin films, three targets sequential sputtering technique was used. Similar to single target sputtering, sputtered polycrystalline films had completely oriented c-axis. After heat treatment in flowing oxygen, the best measured critical current density at 77K was $10^5 A/cm^2$. In order to investigate the effect of the heat treatment, lattice constant c_o of the film at various temperature was measured by XRD. The Cu-O structure in the film was also studied by using static SIMS for the first time.

INTRODUCTION

Since the discovery of high Tc superconductivity on perovskite-related oxides, various techniques, including sputtering,[1],[2],[3] electron beam evaporation,[4],[5] ion beam sputtering,[6] laser ablation,[7] molecular beam epitaxy,[8] and vapor phase epitaxy,[9] have been applied to prepare these superconducting films. Single target sputtering is likely to be a dominant technique, since its simple process meets the fabrication requirement. However, with this method, compensating the target composition for different sputtering yields of the constructing elements at the substrate surface is always required to obtain the stoichiometric composition of the superconducting films.

In this paper, we describe the usefulness of three targets sequential sputtering technique. The YBCO films were deposited onto MgO(100) single crystal substrates. Similar to the single target sputtering, films obtained were also well c-axis oriented. However, this method allows to obtain fine control of the composition of the films more effectively. In order to improve the superconducting properties of the films, effect of the heat treatment was investigated. Also, for the first time, the static SIMS was applied to observe the Cu-O structure in YBCO films. The experiment and the results will be detailed below.

FILM PREPARATION

The YBCO thin films were deposited onto MgO(100) substrate by using rf magnetron sputtering, either single target sputtering or multi-targets sequential sputtering. For the former method, a conventional rf magnetron sputtering technique from the compensated target, $Y_1Ba_{2\sim2.2}Cu_{3.6\sim4.5}O_x$, was used. For the latter one, three targets of Y, Cu and $BaCuO_2$ were used. In this paper, for the most part, we describe the deposition by using this three targets sequential sputtering technique. Figure 1 shows a schematic diagram of this sequential sputtering system. The rf power applied to the targets is controllable and independent to each other. As shown in the figure, the substrate holder is rotated during deposition. The elements of three targets were deposited onto the substrate in sequence. Sputtering was carried out at 40 to 100mTorr total pressure ($Ar:O_2$=1:1-4:1). The substrate temperature was 600°C to 800 °C. The distance between substrate and target was 70mm. The composition of the films was measured by inductively coupled

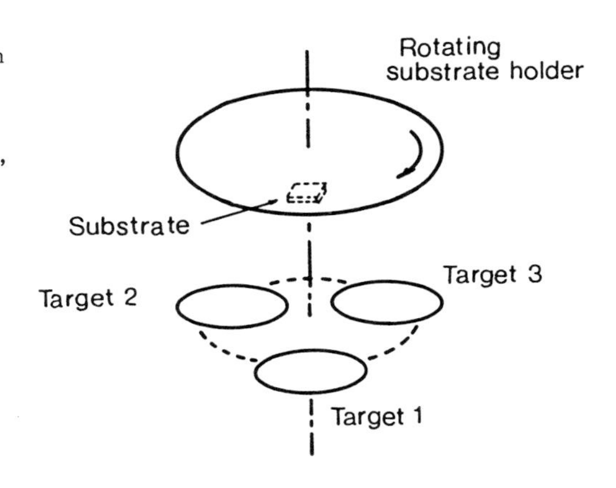

Fig. 1 Schematic diagram of the three targets sequential sputterieng system.

plasma atomic emission spectrometry (ICP).

The structures and morphology were analyzed by X-ray diffraction (XRD) and high resolution scanning electron microscopy (SEM). Figure 2(a) shows the film composition as a function of the rf power applied to $BaCuO_2$ target, where the incident powers of Y and Cu targets were 100watts and 200watts, respectively. The substrate was heated to 750°C at 50mTorr total pressure. By adjusting the incident power of $BaCuO_2$ target, the film composition was close to $Y_1Ba_2Cu_3O_x$. Figure 2(b) shows another approach to reach $Y_1Ba_2Cu_3O_x$ composition. The incident powers of $BaCuO_2$ and Y targets were adjusted to obtain the composition ratio of Y:Ba=1:2, previously. Then, the rf power applied to Cu target was adjusted. As shown in Fig.2(b), the film composition was $Y_1Ba_2Cu_3O_x$ as the incident powers of Y, $BaCuO_2$, and Cu targets were 300watts, 200watts and 80watts, respectively.

Figure 3 shows the structure of the films as a function of the rotating speed of the substrate, where the substrate was heated at 770°C. At low rotating speed (3rpm and 7rpm), a perovskite-like structure with random orientation was obtained. At 15rpm rotating speed, films were c-axis oriented completely.

Figure 4 shows the sputtering conditions to obtain oriented c-axis thin films. In comparison with the conditions by using single target sputtering, sequential sputtering method needs higher substrate temperature to promote the interdiffusion of the sputtered elements. The substrate temperature at 80watts average rf power (average power of the powers apllied to the three targets) is about 50°C lower than that at 170watts average incident power. This indicates that the deposition temperature is lowered by reducing the damages caused by resputtering effects. Besides, the grain size of as-deposited films are about 0.1μm and 1μm for the single target sputtering and three targets sequential sputtering, respectively. A typical X-ray diffraction pattern of as-sputtered films is shown in Fig.5. Only (0, 0, 1) (1=1, 2, 3, ...) reflections of YBCO structure are clearly observed.

EFFECT OF HEAT TREATMENT

The zero resistivity temperature Tc,zero of as-sputtered c-axis oriented thin films is below 77K. In order to provide a proper amount of oxygen into the films, the heat treatment was employed.

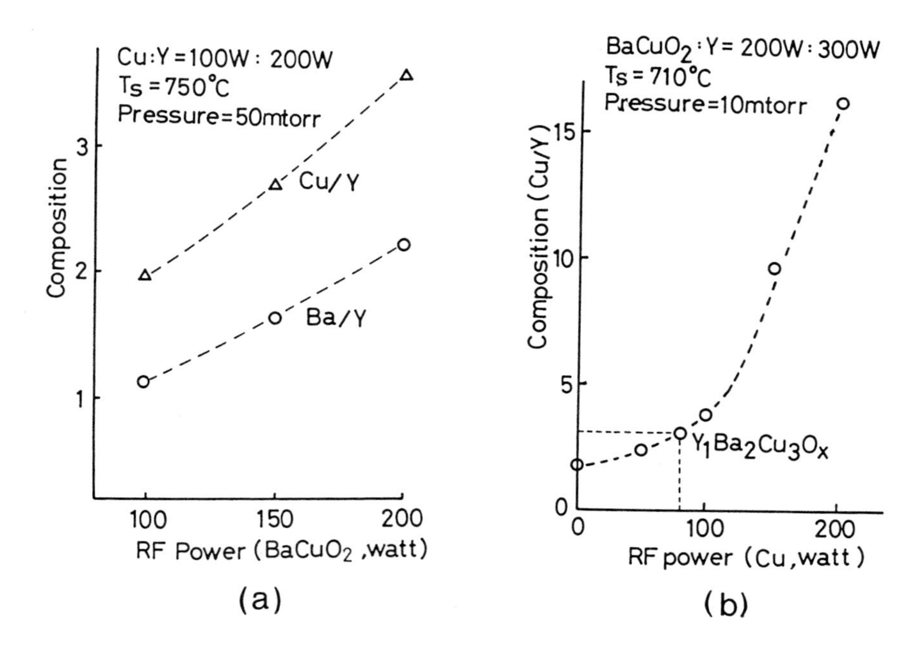

Fig. 2

The composition ratios of the YBCO films prepared by using three targets sequential sputtering technique.

(EFFECT OF O_2 PRESSURE)

There is a correlation among c-axis lattice constant c_o (calculated from the result of X-ray diffraction measurement), oxygen content and Tc,zero.[10] Therefore, we investigated the effect of annealing by measuring the variation of the c_o and Tc,zero of these films due to annealing. Figure 6 shows the variation of c_o of the films which were annealed in a tube furnace with various O_2 pressure (in flowing N_2, O_2 gaseous mixture). Annealing profile is also shown in Fig.6. It shows that O_2 pressure larger than 0.5atm is necessary to increae oxygen content in the films.

(BEHAVIOR OF THE LATTICE CONSTANT Co DURING THE ANNEALING)

In order to study the variation of O_2 content in thin films, c_o of the films was measured at various temperature by using X-ray diffraction analysis. Heating was employed in flowing pure oxygen. Value of 2θ of MgO(200) (about 42.9°) was used as a reference. During annealing the intensity of MgO(200) was confirmed to be constant. Variations of 2θ(006) relative to that of MgO(200) at various temperature were measured. The variation of c_o due to the difference of the coefficients of expansion of MgO substrates and YBCO thin films was so small (about 10^{-5}Å/°C) and was therefore negligible in comparison with the variation of c_o due to the annealing (10^{-2}Å order). Figure 7 shows the variation of c_o at various temperature with the annealing profile shown in Fig.8. The c_o becomes longer as the temperature rises. During keeping at high temperature and cooling down to 400°C, c_o varied rapidly. The variation of c_o during cooling down from 400°C to room temperature was about 0.03Å. This value is

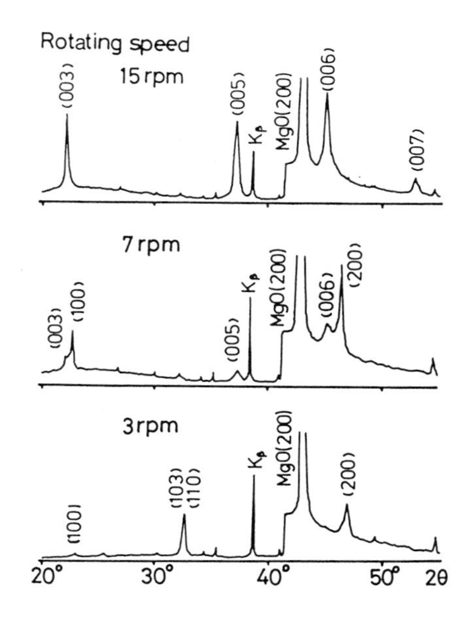

Fig. 3 X-ray patterns as a function of the rotating speed of the substrate during deposition.

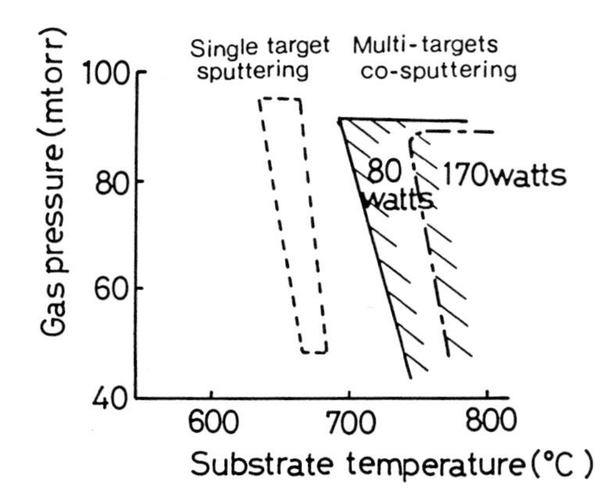

Fig. 4 The sputtering conditions to obtain oriented c-axis thin films.

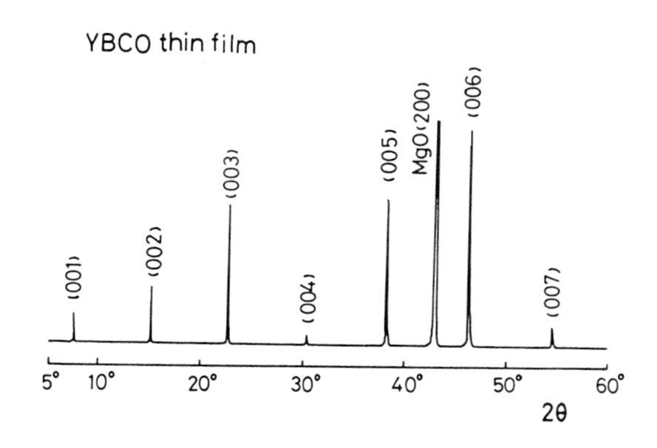

Fig. 5 A typical X-ray pattern of YBCO thin films deposited on MgO(100) substrates.

the same among different samples. During keeping at 400°C, c_o changed slowly but the width of 2θ(006) peak became smaller. In order to fill in sufficient oxygen content into the films, high temperature around 900°C is necessary.

Figure 8 shows the Tc,zero as a function of annealing temperature. It also indicates that high annealing temperature is required for improving Tc,zero of YBCO thin films.

(OBSERVATION OF Cu-O STRUCTURE BY STATIC SIMS)

For orthorhombic $Y_1Ba_2Cu_3O_x$, the oxygen content has been determined corresponds to be 7 from neutron scattering experiments.[11] Therefore, CuO_5 polyhedra and CuO_4 square-planar units exist in it. The X-ray photoelectron spectroscopy (XPS) data for YBCO compounds have been used to study the crystalline structure by measuring the electronic strucutre of Cu[12],[13] Here, the static SIMS[14] (ATOMIKA A-DIDA3000) was used for the first time to study the Cu-O binding situation of YBCO films directly. The measuring conditions were as follows:

primary ion = Ar^+ or O_2^+,
ion energy = 0.5-3.0keV,
primary ion current = about 5nA,
scanning area = about 600μm square.

Before measurement, the surface of the samples was cleaned by sputtering for 1hour (1200μm^2 sputtered area for 30minutes continued with 600μm^2 for 30minutes). As-sputtered and annealed films with 0.3μm film thickness were observed. The c_o for as-sputtered film and that after annealing at 900°C for 2hours were 11.75Å and 11.69Å, respectively.

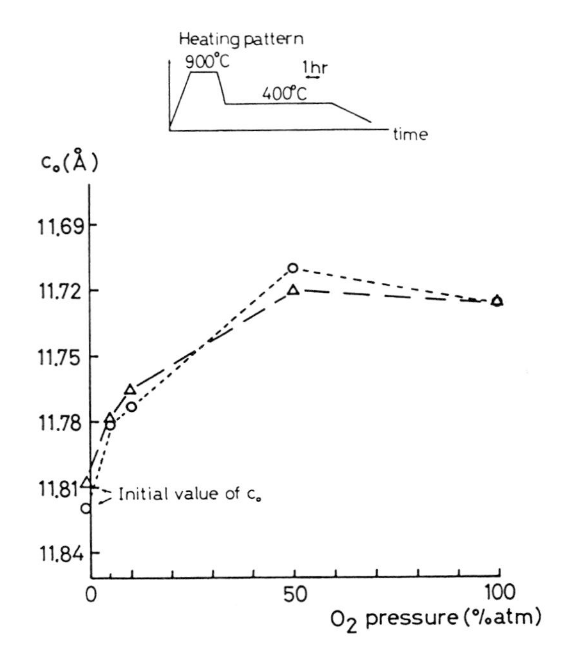

Fig. 6 The variation of the lattice constant c_o for various O_2 pressure in annealing. (○ and △ indicate two different samples used)

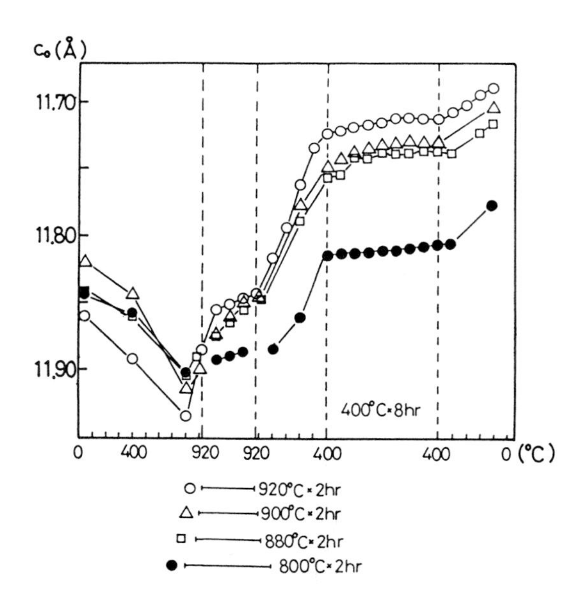

Fig. 7 The lattice constant c_o of YBCO films at various annealing temperature.

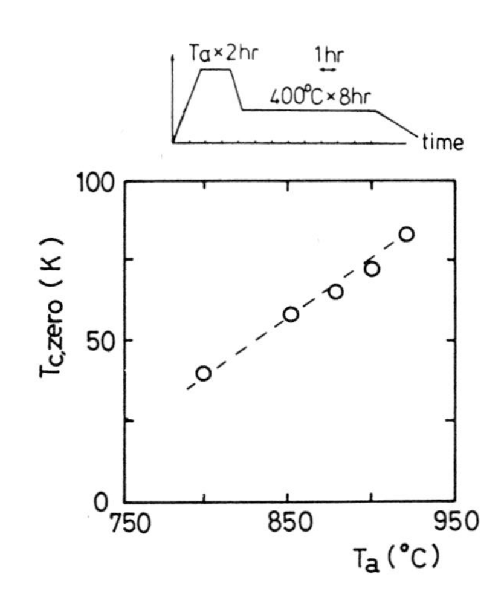

Fig. 8 The zero resistivity temperature of YBCO films as a function of annealing temperature.

Figure 9 shows the mass spectra of these two samples, where 0.5keV Ar^+ primary ions were used. The intensity of CuO_4 and CuO_5 changed much due to annealing. As shown in Fig.10, the intensity decreased as the bombarding ion energy increased. It suggests that CuO_4 and CuO_5 belong to the molecular ions. Figure 11 shows the mass spectra measured by using 0.5keV O_2^{2+} ions. CuO_4 and CuO_5 were not found in the as-sputtered films. The results agree with those shown in Fig.9. The effect that CuO_4 and CuO_5 ions were reduced by Ar^+ ions could be negligible. (For XPS measurement, oxygen reduction occurs by Ar^+ etching for removing the surface contamination.[15])

(Tc, Jc of YBCO FILM)

The zero resistivity temperature Tc,zero, and the critical current density, Jc, were measured by using conventional four-probe transport method. The results of Tc,zero and Jc for a 0.7μm thickness of YBCO film are shown in Fig.12. Tc,zero=82K and $Jc=10^5 A/cm^2$ at 77K were obtained.

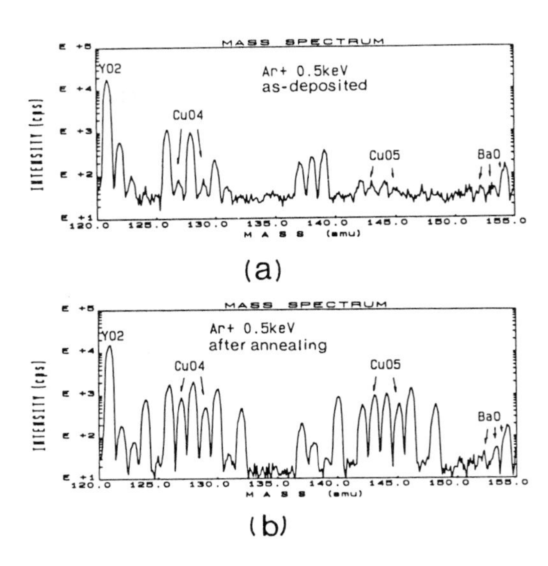

Fig. 9 The mass spectra of YBCO thin films where primary ions of Ar^+ were used for sputtering. (a)as-sputtered,(b) after annealing

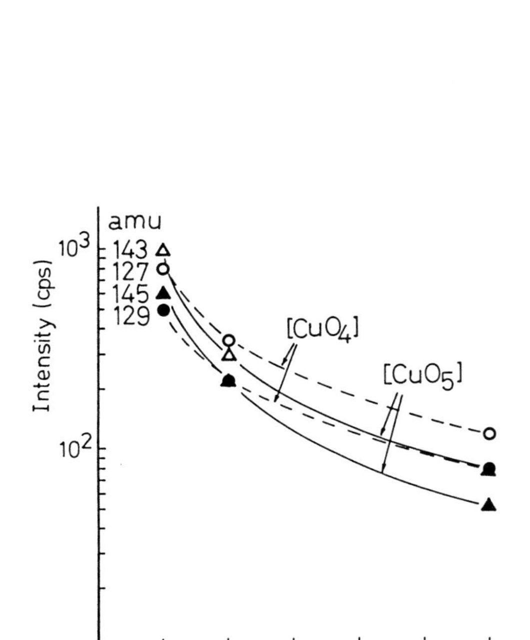

Fig.10 Intensity of CuO_4 and CuO_5 as a function of the Ar^+ ion energy.

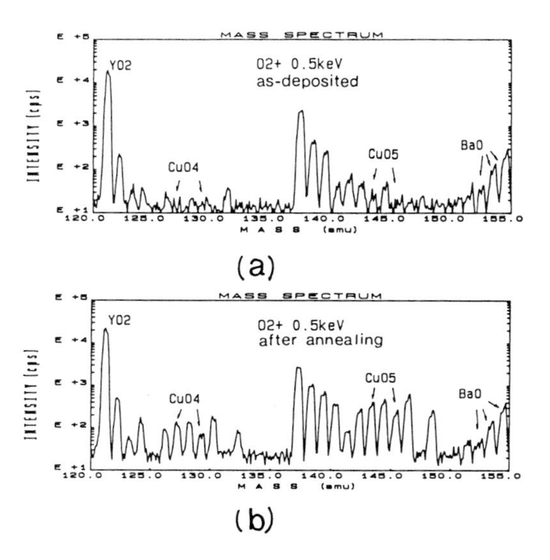

Fig.11 The mass spectra of the YBCO thin films where primary ions of O_2^+ were used for sputtering. (a) as-sputtered, (b) after annealing

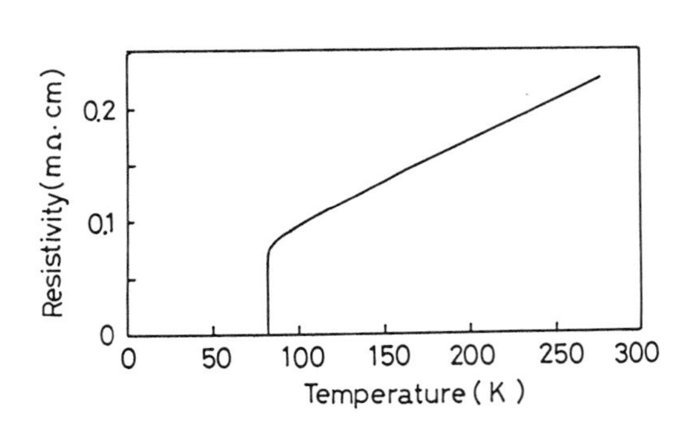

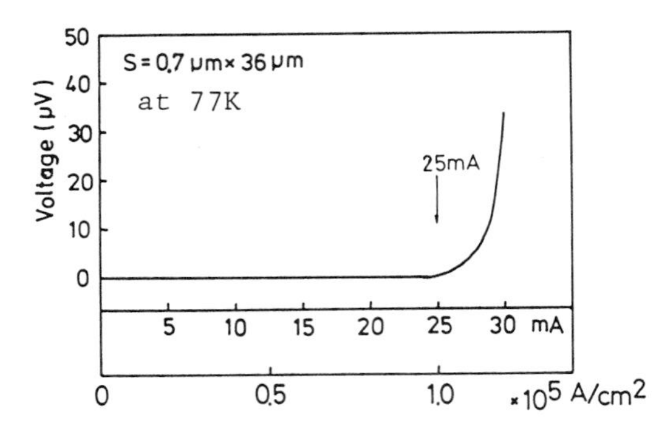

Fig.12 The Tc,zero and J_c for a YBCO thin film.

CONCLUSIONS

In summary, we have obtained good quality YBCO polycrystalline films with completely oriented c-axis by using the three targets sequential sputtering technique. This method is especially attractive because fine control of the composition of the films can be obtained. However, in comparison with the single target sputtering, higher processing temperature which is against the fabrication requirement is needed to obtain well c-axis oriented films. Further studies are needed to solve this demerit. We show that measuring of the c_0 at various temperature by XRD is an effective mean to understand the behavior of the heat treatment. The correlation between the Cu-O structure and the superconducting properties of the films was also studied by using static SIMS method. It shows that this method will be a promissing means to obtain a higher level of structural information in comparison with XPS measurement.

ACKNOWLEDGEMENTS

Authors wish to express great appreciations for fruitful technical discussions and funding bestowed by a group of Japanese electric power companies---Tokyo Electric Power Co., Tohoku Electric Power Co., Hokkaido Electric Power Co., and Electric Power Development Co.

REFERENCES

(1) Y.Enomoto, T.Murakami M.Suzuki and K.Moriwaki, Jpn. J. Appl. Phys. 26, L1248 (1987).
(2) J.L.Makous, L.Maritato and C.M.Falco, Appl. Phys. Lett. 51, 2164 (1987).
(3) K.Char, A.D.Kent, A.Kapitulnik, M.R.Beasley and T.H.Geballe, Appl. Phys. Lett. 51, 1370 (1987).
(4) R.B.Laibowitz, R.H.Koch, P.Chaudhari and R.J.Gambino, Phys. Rev. B, 35, 8821 (1987).
(5) T.Terashima, K.Iijima, K.Yamamoto, Y.Bando and H.Mazaki, Jpn. J. Appl. Phys. 27, L91 (1987).
(6) T.Yotsuya, Y.Suzuki, S.Ogawa, H.Kuwabara, K.Otani, T.Emoto and J.Yamamoto, FED HiTcSc-ED Workshop, 57 (1988).
(7) D.Dijkkamp and T.Venkatesan, Appl. Phys. Lett. 51, 619 (1987).
(8) J.Kwo, T.C.Hsieh, R.M.Fleming, M.Hong, S.H.Liou, B.A.Davidson and L.C.Feldman, Phys. Rev. B, 36, 4039 (1987).
(9) M.Ihara and T.Kimura, FED HiTcSc-ED Workshop, 137 (1988).
(10) A.Ono and Y.Ishizawa, Jpn. J. Appl. Phys. 26, L1043 (1987).
(11) F.Izumi, H.Asano, T.Ishigagi, E.Takayama, Y.Uchida, N.Watanabe and T.Nishikawa, Jpn. J. Appl. Phys. 26, L649 (1987).
(12) H.Watanabe, K.Ikeda, H.Miki and K.Ishida, Jpn. J. Appl. Phys. 27, L783 (1988).
(13) H.Ihara, M.Hirabayashi, N.Terada, Y.Kimura, K.Senzaki, M.Akimoto, K.Bushida, F.Kawashima and R.Uzuka, Jpn. J. Appl. Phys. 26, L460 (1987).
(14) M.J.Hearn and D.Brggs, Surface and interface analysis. 11, 198 (1988).
(15) M.Tonouchi and T.Kobayashi, MRS-Tokyo, D3.10 (1988).

The Stress-Strain Relationship for Multilayers of the High Tc Superconducting Oxides

HIROAKI HIDAKA and HIROSHI YAMAMURA

Tokyo Research Center, Tosoh Corporation, 2743-1, Hayakawa, Ayase, Kanagawa, 252 Japan

ABSTRACT

The calculation of the stress-strain relationship for multilayers of the high Tc superconducting oxides was performed with regard to a wide variety. The elucidation of this relationship is expected quite helpful for the preparation of high-quality multilayers of these materials. This calculation is possible to do in the same way of Timoshanko's bimetal treatment. We did for the first time computation of the residual stress and strain, and the state of stress and strain for these multilayers has been acquired in detail by this calculation.

INTRODUCTION

The investigational work for practical application of high critical temperature superconducting materials has been active recently. Paticularly, device application of thin films of these materials has been tride by many workers[1,2,3]. The applicatin to SQUID,three-terminal device or optical detector has been reported. The preparation of high-quality multilayers using high Tc superconducting oxides is very important for these applications. However, it is well known that these multilayers are highly stressed mostly due to the large discrepancy of the thermal expantion coefficients between the substrate material and grown layers[4]. These stress and strain impaire physical properties such as critical temperature and, occasionally, induce cracking inside the film. Therefore, the elucidation of stress-strain relationship is expected quite helpful for the preparation of high-quality multilayers using superconducting oxides.

In this work, we did for the first time the calculation of the residual stress and strain with regard to a wide variety of multilayers. This calculation is possible to do in the same way of Timoshanko's bimetal treatment[5]. The stress-strain relationship for these maltilayrs were revealed and occasinally, that in a case of using silicon substrate with buffer layer, which is very significant from a practical standpoint, were done in detail.

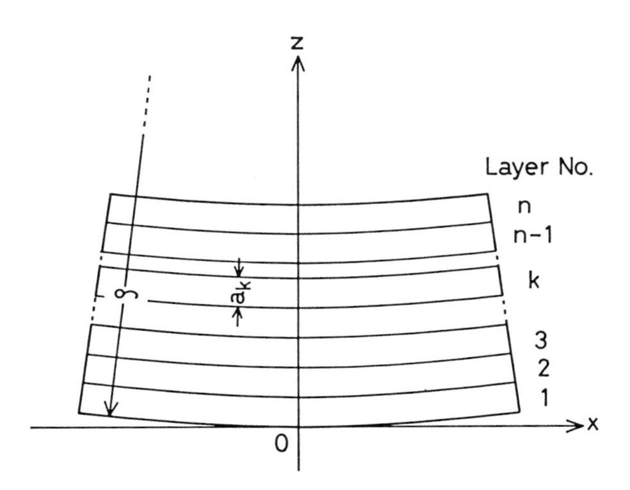

Figure 1. The schematic view of multilayer in the stress-strain calculation.

COMPUTATAITIONAL METHOD

The calculation of the stress-strain relationship for epitaxial multilayers is able to be performed by the same way of Timoshanko's bimetal treatment[5]. Now, that way is briefly introduced as follows. First of all, it is supposed that multilayers overspread infinitely and the influence of gravity is ignored. In this case, the state of stress and strain is exactly alike at each point around the plane of the film. Figure 1 shows epitaxial multilayer which is constructed with n layers. The thickness of the layer k is here indicated as a_k, and the curvature of that is done as ρ . Equations which shall be solved are set up following.

$$\sum_{k=1}^{n} P_k = 0 \qquad (1)$$

$$(1/\rho)\cdot\sum_{k=1}^{n}(I_k/S_k) = \sum_{k=1}^{n}(z_{k-1}+a_k/2)\cdot P_k \qquad (2)$$

$$(S_k/a_k)\cdot P_k + \alpha_k \Delta T_k + a_k/(2\rho) = (S_{k-1}/a_{k-1})\cdot P_{k-1} + \alpha_{k-1}\Delta T_{k-1} - a_{k-1}/(2\rho) \qquad (3)$$

$$(2 \leqq k \leqq n)$$

The equation (1) is to be a balance between forces in layers, and the equation (2) is to be that between moments of layers. Equations (3) are the condition that materials don't at all slip in the interface between layers k and (k+1). In these equations, P_k is the force which acts in the layer k, α_k is the coefficient of thermal expantion of the layer k. ΔT_k is the deviation in temperature for the film preparation. I_k is the moment of inertia;$a_k^3/12$. S_k is the elastic compliance ;$(1-\nu(k))/E(k)$ ($\nu(k)$:Poisson's ratio, $E(k)$:Young's modulus). That is to say, a set of these equations is the eigenvalue problem in (n+1) dimensions. Solving this problem, forces; P1,P2,...,Pn and the curvature; ρ are aquired with regard to a wide variety of multilayers.

According to these values, stress and strain are calculated as follows.
The stress in under-surface of the layer 1; $\sigma_{1s}$is

$$\sigma_{ls} = \frac{1}{S_1}\left(\frac{S_1}{a_1}P_1 + \frac{a_1}{2\rho}\right) \qquad (4)$$

The stress in upper-surface of the layer n; $\sigma_{us}$is

$$\sigma_{us} = \frac{1}{S_n}\left(\frac{S_n}{a_n}P_n + \frac{a_n}{2\rho}\right) \qquad (5)$$

The stress in the neighborhood of the layer k at the interface between layers k and (k+1); $\sigma_{kl}$is

$$\sigma_{kl} = \frac{1}{S_k}\left(\frac{S_k}{a_k}P_k - \frac{a_k}{2\rho}\right) \qquad (6)$$

The stress in the neighborhood of the layer (k+1) at that interface; $\sigma_{ku}$is

$$\sigma_{ku} = \frac{1}{S_{k+1}}\left(\frac{S_{k+1}}{a_{k+1}}P_{k+1} - \frac{a_{k+1}}{2\rho}\right) \qquad (7)$$

In addition, corresponding to each stress, strain; $\varepsilon_i$is

$$\varepsilon_i = \sigma_i \cdot (1-\nu(i))/E(i) \qquad (8)$$

These calculations are performed in practice using a desk computer by the way of numerical program such as Gaussian method.

RESULTS AND DISCUSSION

$Y_1Ba_2Cu_3O_{7-x}$ thin film on various substrates

For preparation of superconducting oxide thin film, various materials have been proposed as substrates by many workers[4]. For example, yttria stabilized zilconia (YSZ), strontium titanate, magunesium oxide or glass has been done up to the present. The physical properties of these materials are shown in table 1[4,6,7,8]. The coefficient of thermal expantion of substrate material is differrent from that of $Y_1Ba_2Cu_3O_{7-x}$ (YBCO) as shown in this table. This

Table 1. Physical Properties of Materials

Material	α ($\times 10^{-6}$)	E ($\times 10^{4}$MPa)	ν
YBCO	16.9	10.4	0.35
MgO	13.8	24.9	0.18
$SrTiO_3$	11.1	30.3	0.23
YSZ	10.3	20.0	0.25
CaF_2	24	7.6	0.28
Si	2.4	13.1	0.22
Pyrex Glass	3	1.17	0.20

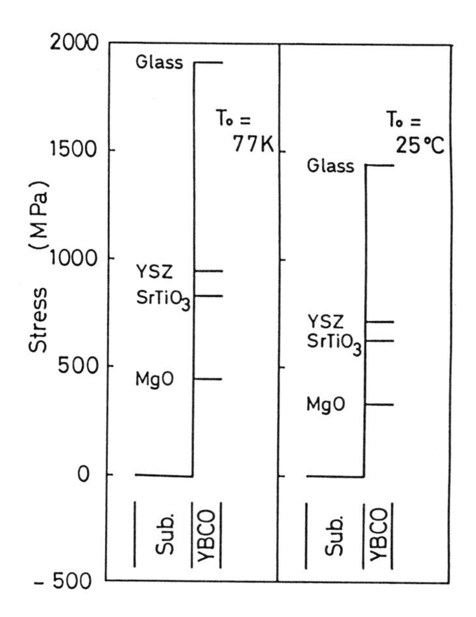

Figure 2. Calculated stresses of YBCO/Substrate structures.

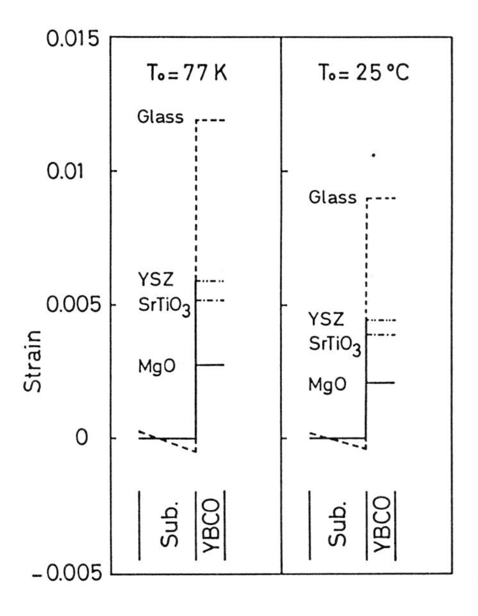

Figure 3. Calculated strains of YBCO/ Substrate structures.

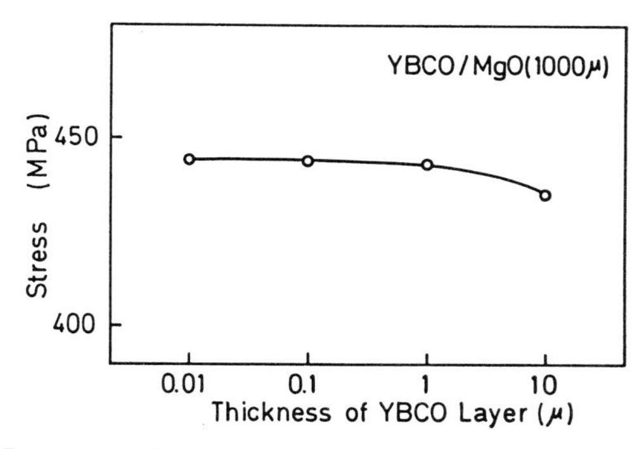

Figure 4. Dependence of YBCO thickness on that stress.

difference is a problem for high temperature preparation of superconducting oxide thin film because of occurrence of residual stress and strain in that film. Figures 2 and 3 show respectively stress and strain which are obtained from the calculation explaned above. In this treatment, the thickness of substrate is 1 mm, and that of thin film YBCO is 1 μm, and the preparation tempertature (T_S) is 700 °C. Values of stress and strain are calculated for room and liquid nitrogen temperatures(TOs) respectively. In the case of a pyrex glass for substrate, large tensile stress, 1910 MPa occurs in YBCO thin film at 77 K, supposing that the deformation is completely elastic between T_S and T_O. Originating in that large stress, this thin film spreads as 1.2 % around the plane. This tensile stress gives rise to crack formation or deterioration of physical properties of the film such as a critical temperature(T_c). The rupture strength, in reference, for sapphire single crystal is approximately 500 or 1500 MPa. On the other hand, in cases of YSZ, $SrTiO_3$ and MgO for substrate, values of the residual stress are 944 MPa, 830 MPa and 443 MPa respectively, and that of strains are 0.6%, 0.5% and 0.3% respectively. In cases mentioned here, tensile stresses occur in thin films, however, those values are smaller than that in a case of pyrex glass. Therefore, the possibility of crack formation or property deterioration is supposed to be comparatively small. According to these results and the report of Koinuma et al.[4], the upper critical stress for practical use of YBCO thin film is estimated to be 1000 or 1500 MPa. That value is reasonable in comparison with the rupture strength of sapphire.

In a case of various thickness of YBCO thin film on MgO substrate, residual stress were calculated. Figure 4 shows the results at 77K. That stress takes somewhat the value 440 MPa over a range from 100 Å to 1 µm of film thickness. That decreases a little at 10 µm of the film thickness, which is about 1 % of that of substrate.

Multilayers on silicon substrate

We have done a trial of usage of silicon wafer for substrate material. In this case, chemical stability of grown film on silicon sabstrate such as unreactivity of film and substrate materials must be under consideration because of high temperature formation of the film about 700 °C. Figures5 and 6 show depth profiles of Auger electon spectroscopy(AES) for superconducting films prepared on Si wafer without and with MgO buffer layer. Thin films were deposited in RF-magnetron sputtering system at the substrate temperature of 600 °C. Figure 5 shows that a reaction layer are made for 1000 or 2000 Å between $Dy_1Ba_2Cu_3O_{7-x}$ thin film and Si substrate. On the other hand, figure 6 show that the superconducting film is quite sound on silicon substrate with MgO buffer layer.

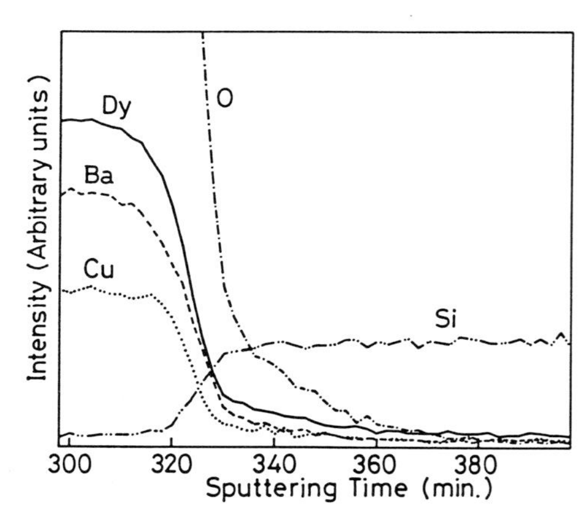

Figure 5. Depth profile of AES for film prepared on Si wafer.

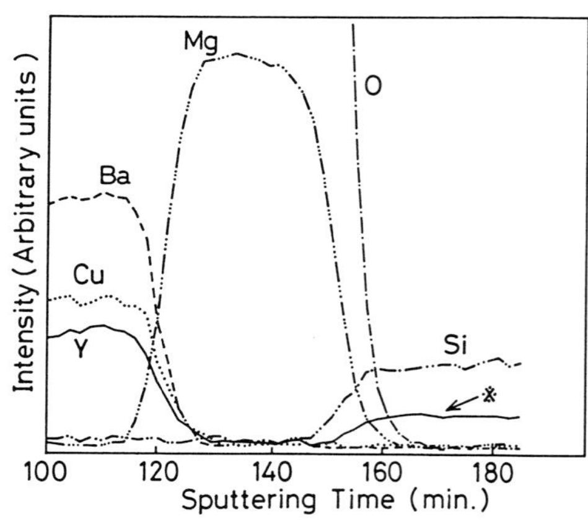

Figure 6. Depth profile of AES for multilayered films prepared on Si wafer.

Figure 7 shows the dependence of preparation temperature at the calculation on residual stress of YBCO thin film at 77K. The solid line and the dashed and dotted curve in this figure are cases of YBCO thin film on Si substrate with and without MgO buffer layer respectively. Si substrate is 500 µm thick and, YBCO and MgO films are 1µm thick. The decrease in residual stress when forming a buffer layer is very small. And, according to the critical stress 1000 or 1500 MPa for cracking of thin film, the preparation temperature shoud be appropriate to be 400°C or less for forming these thin films well. Figure 8 shows the dependence of buffer layer thickness on the residual stress of YBCO thin film prepared on Si substrate at 77K. The decrease in that stress when increasing the thickness of that layer from 0.1 µm to 1 µm is small as shown in this figure. In order to release that stress by the way of forming buffer layer, it shall be necessary to thicken the layer for 10 µm or more. And, the dependence of coefficient of thermal expantion and Young's modulus of buffer layer on the residual stress of YBCO layer at 77K were studied by the calculation. Si substrate is 500 um thick and YBCO and buffer layers are 1 µm thick. The stress of YBCO layer decreases as the coefficient of thermal expantion and the Young's modulus decrease. However, for the perpose of releasing the residual stress, following that result, that values are necessary to be considerably large.

Figure 9 shows the profile of residual stress of multilayer which is constructed with YBCO/MgO/$SrTiO_3$/Si at 77K. Thicknesses of Si substrate and $SrTiO_3$, MgO and YBCO thin films are 500 μm, 1μm, 1μm and 1μm respectively. The residual stress of YBCO thin film is the value of 2018 MPa and this is similar to that of the film without buffer layer. Moreover, buffer layers between YBCO film and Si substrate are very stressed as shown in this figure. Table 2 shows the stress in YBCO thin film in a wide variety of multilayers. That stress scarcely decreases for each combination as shown in this table and the thickness of grown film is too small to release residual stress. That reason may be due to the assumption that the deformation caused by thermal shurinkage is only elastic. Therefore, in practical use, the plastic deformation of buffer layer in the preparation is very important mechanism in order to release residual stresses. This may be, for example, achieved with the adoption of annealing prosess at suitable temperatures. Needless to say, it is very helpful for practical use to lower the preparation temperature or, if possible,to heighten the critical temperature of superconducting thin film.

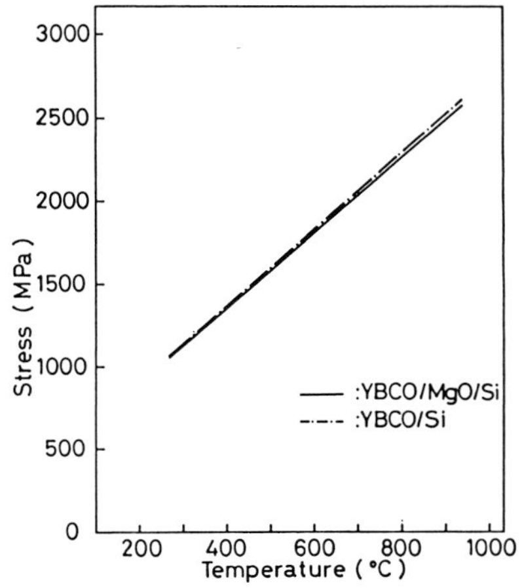

Figure 7. Dependence of YBCO deposition temperature on that stress.

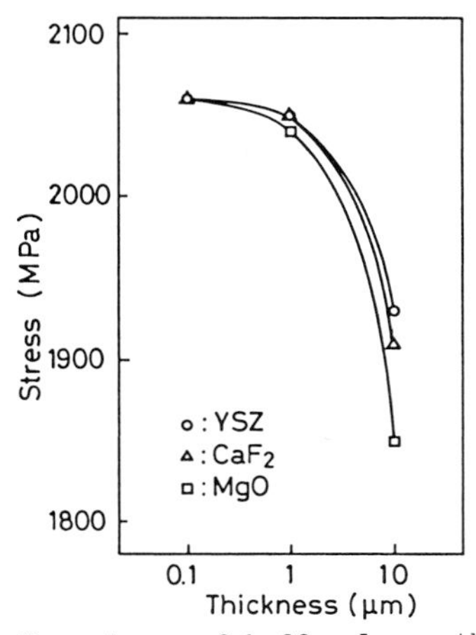

Figure 8. Dependence of buffer layer thickness on the residual stress.

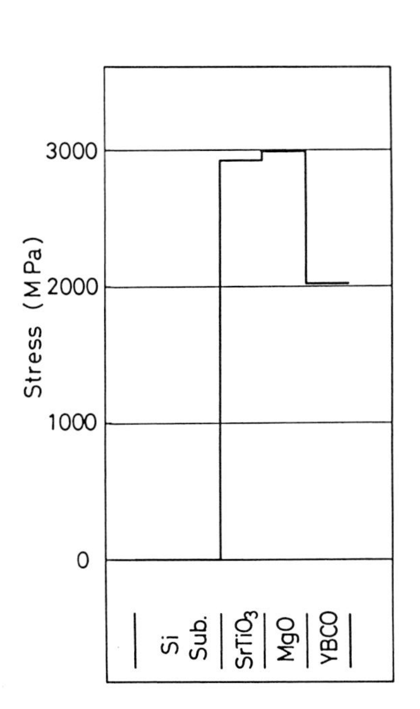

Figure 9. Profile of residual stress of a multilayer.

Table 2. The stress in YBCO film in a wide variety of multilayers

Structure	Stress (MPa)
YBCO/MgO/SrTiO3/Si	2018
YBCO/MgO/CaF2/Si	2025
YBCO/CaF2/MgO/Si	2025
YBCO/CaF2/YSZ/Si	2034
YBCO/SrTiO3/Glass/Si	2040
MgO/YBCO/SrTiO3/Si	2018
CaF2/YBCO/YSZ/Si	2034
YSZ/YBCO/MgO/Si	2026

CONCLUSION

We did the computation of the residual stress and strain for superconducting multilayers, and the state of stress and strain for that layers has been acquired in detail. Dependence of substrate material, preparing temperature or film thickness on residual stress was made clear. And, it is suggested that a multilayer prepared on Si substrate is highly stressed and the intensity of the stress is too large to induce cracking or property deterioration .

ACKNOWLEDGEMENTS

The authors would like to thank Professor T.Kobayashi of Faculty of engineering Science, Osaka University for his constant encouragement and usefull suggestion. We are indebted to Y.Nakanishi and his co-workers for the help in the experiment.

REFERENCES

[1] H.Tanabe,S.Kita,Y.Yoshizako and T.Kobayashi,Jpn.Jppl.Phys.26,1961(1987).
[2] K.Hashimoto,U.Kabasawa,M.Tonouchi and T.Kobayashi, Shingaku-gihou SCE88-25(1988).
[3] Y.Enomoto & T.Murakami,J.Appl.Phys.59,3807(1986).
[4] T.Hashimoto,K.Fueki,A.Kishi,T.Azumi and H.Koinuma,Jpn.J.Appl.Phys.27 L.214(1988).
[5] S.Timoshanko,J.O.S.A.&R.S.I.,11 233(1925).
[6] .S.Kudou,'Bunkougakuteki-seishituwo-shutoshita-kisobusseizuhyou',Kyouritu-Shuppan(1972).
[7] J.L.Tallon,A.H.Shuitema & N.E.Tapp,Appl.Phys.Lett.52 507(1988)
[8] R.O.Bell & G.Rupprecht,Phys.Rev.129 90(1963).

Effects of Heat Treatment on Crystalline Quality of Er-Ba-Cu-O Thin Films

KAZUMASA TAKAGI, TSUTOMU ISHIBA, TOSHIYUKI AIDA,
TOKUUMI FUKAZAWA, and KATSUKI MIYAUCHI

Central Research Laboratory, Hitachi Ltd., Kokubunji, Tokyo, 185 Japan

ABSTRACT

$ErBa_2Cu_3O_7$ epitaxial films are prepared on heated MgO substrates by r.f. magnetron sputtering, and changes in crystallinity during annealing are examined by an X-ray diffraction method. Plane spacing parallel to film plane (001) is as large as that in the oxygen deficient semiconductor phase. However, the film becomes a superconductor at low temperatures. Recrystallization of the amorphous region in the as-sputtered film occurs during the annealing. Plane spacing decreases and crystallinity is improved in this process. It is assumed that plane spacing enlargement is due to stress related to epitaxial growth.

INTRODUCTION

Since the discovery of new high Tc copper perovskite superconductors [1,2], efforts to prepare thin films of these materials have been made by sputtering [3,4], electron beam evaporation [5,6], and pulsed laser evaporation [7]. High current density (~10^6 A/cm²) at 77K has been achieved in thin films deposited on strontium titanate ($SrTiO_3$) [3,8] and magnesia (MgO) substrates [9].

Electrical properties of oxide superconducting films are sensitive to crystalline qualities, including grain boundaries, oxygen content, and impurity concentrations. These are affected by the substrate temperature during film preparation and by the post-annealing temperature. In order to obtain utility superconducting films with high current densities for micro-electronic applications, optimum conditions for deposition and heat treatment have been investigated by many researchers. High temperature annealing is desirable to improve crystallinity, however, it causes an interface reaction between films and substrates.

In this experiment, changes in crystalline quality during annealing are examined mainly by an X-ray diffraction methods.

EXPERIMENTAL

Film Preparation

$ErBa_2Cu_3O_7$ thin films were prepared on an MgO (100) substrate by r.f. magnetron sputtering using a sintered oxide target. Films 0.07-0.7 μm thick were deposited at a rate of 150nm/sec. Sputtering gas was a mixture of argon (50 %) and oxygen (50 %) at 4 Pa. The substrates were heated in order to achieve epitaxy. The substrate holder temperature varied in the range from 730 to 830 ℃. After deposition, oxygen gas was introduced into a chamber followed by cooling to room temperature. Subsequently, the films were annealed at 830 and 870 ℃ for two hours in an oxygen atmosphere, and were slowly cooled to room temperature.

Characterization

Because all films were oriented preferentially along the c-axis, their lattice parameters c_0 were measured by two methods. The diffraction angle 2θ of the 006 refrection was measured by using a θ -2 θ scanning diffractometer with Cu-K α radiation . In the θ -2 θ scan, a narrow slit of 50 μ m, and a singly bent pyrolytic graphite monochrometer were used to improve accuracy. Moreover, an X-ray double crystal method with Cr K α_1 radiation monochronized by an Si (110) wafer, was employed. Plane spacing was calculated from the rotation angle difference between the 006 reflection of a film and the 200 reflection of a substrate. Rocking curve broadness (ω -scan) $\Delta\omega$ consists of deviation both in plane spacing (Δd) and in orientation ($\Delta\phi$). The dispersion angle $\Delta\phi$ of each film was estimated from the difference between $\Delta\omega$ and $\Delta\theta$. Crystallinity of films was also examined by cross-sectional transmission electron microscopy (TEM), and by an RHEED method.

RESULTS AND DISCUSSION

Effects of Heat Treatment on Plane Spacing and Dispersion Angle

As-deposited films were crystalline single phase and were oriented preferentially to the c-axis as shown in the RHEED pattern in Fig. 1(a). Diffraction angles 2 θ of the 006 reflection were small and ranged from 45.6 to 45.9°. The representative lattice constant c_0 was 1.192 nm. Measurement by the double crystal method also showed large lattice constants which correspond to the oxygen deficient semiconductor phase [4,10]. However, these as -sputtered films became superconductors at 40 ~ 60 K. These results indicate that causes other than oxygen defficiency were responsible for the large plane spacing.

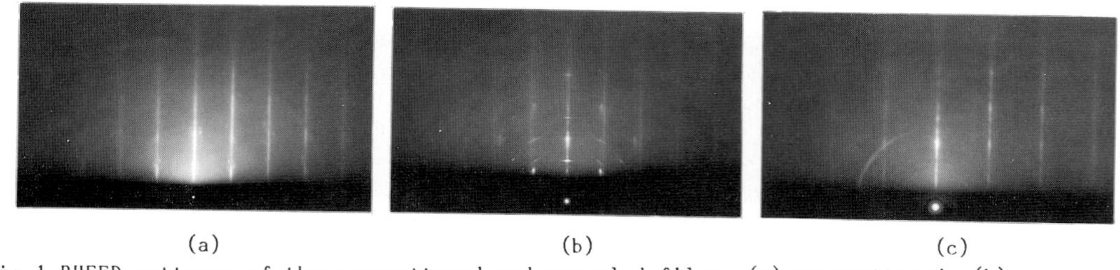

(a) (b) (c)

Fig.1 RHEED patterns of the as-sputtered and annealed films. (a) as-sputtered, (b) annealed at 830℃, (c) annealed at 870℃. Patterns were observed along the beam incidence [100] in (a) and (b), and [110] in (c).

The film was then annealed at 830 and 870℃ in oxygen, and changes in crystallinitythe were examined. RHEED patterns of the film and rocking curves

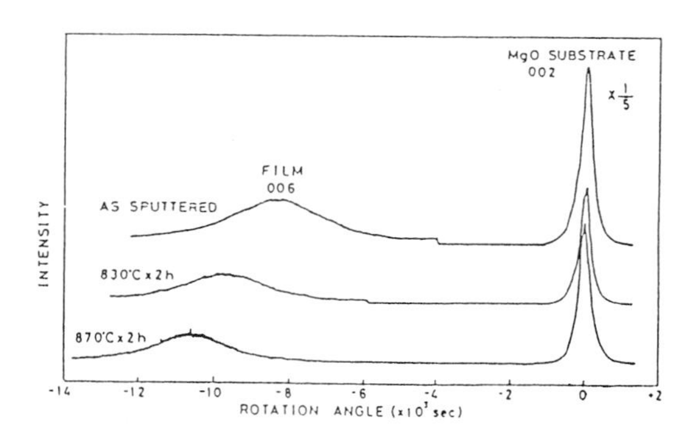

Fig.2 Rocking curves of the films and the substrates measured by the X-ray double crystal method. Cr K α_1 radiation was used.

Table I Changes in plane spacings and dipersion angles during annealing

	PLANE SPACING d_{001} (nm)			DISPERSION ANGLE $\Delta\phi$
	DIFFRACTOMETER METHOD	DOUBLE CRYSTAL METHOD	Δd_{001}	
AS SPUTTERED	1.192 nm	1.191 nm	±0.003	0.78°
830℃,2h	1.175	1.179	±0.003	0.74°
870℃,2h	1.172	1.172	±0.002	0.64°

MgO d_{200}=0.211 nm

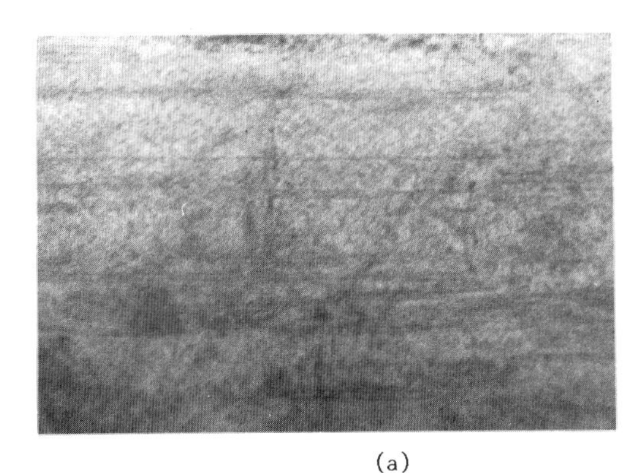

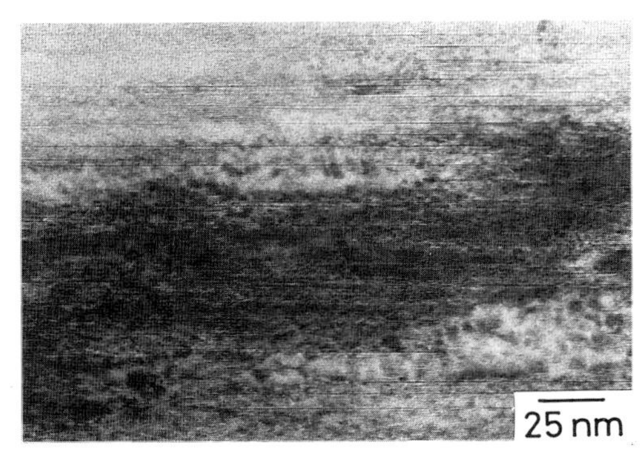

(a) (b)

Fig.3 Cross-sectional TEM photographs of the as-sputtered and annealed films
(a) as-sputtered, and (b) annealed at 830℃.

Table II Effects of film thickness on plane spacing

FILM THICKNESS (μm)	PLANE SPACING d_{001} (nm)	
	AS-SPUTTERED	AFTER ANNEALING
0.07	1.188	1.173
0.16	1.189	1.171
0.195	1.190	1.171
0.455	1.191	1.171

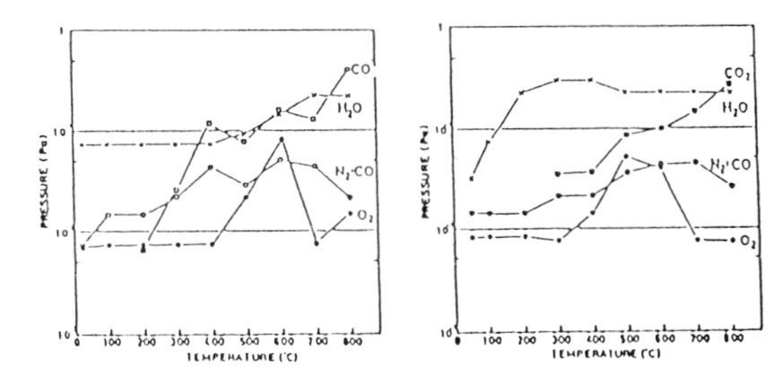

Fig.4 Gas analyses in the films.
(a) as-sputtered, and (b) annealed.
Annealing was done at 830℃ for 8 hours.

obtained by the double crystal method are shown in Figs. 1 and 2, respectively. The numerical results are summarized in Table I . Rotation angle differences between the film and the substrate increased because of annealing, as shown in Fig. 2. The lattice constants c_0 decreased to 1.175 and 1.172 nm, and plane spacing deviation Δd decreased. Improvement in crystallinity in the annealed films was also shown by TEM observation. Cross -sectional TEM photographs of as-grown and annealed films are shown in Fig. 3. The lattice image was sharpened by annealing. On the other hand, X-ray rocking curves of the film were still broad. There are faint Debye rings in the RHEED patterns of the annealed films. This means some subgrains were in the films. Orientation dispersion in the annealed films measured by the double crystal method decreased. These results indicate that the film was deposited epitaxially during sputtering, but was not of a single crystal type. Also, recrystallization of the amorphous region in the as-sputtered film occurred during annealing. The following factors are assumed to be causes of large plane spacing. They are thermal stress due to differences in thermal expansion coefficients between the film and the substrate, and internal stress due to sputtering. The thermal expansion coefficient of MgO is smaller than that of the film. Therefore, thermal stress generates tension in the films, and decreases c_0.

In order to examine the effects of the thermal expansion coefficient on plane spacing, a film was deposited on a CaF_2 substrate with a thermal expansion coefficient that was larger than that of the film. To prevent a reaction between the film and the substrate, the MgO thin interlayer was formed. The film became polycrystal, and the lattice constant c_0 was 1.173 nm, which is the same as that of the annealed film. The film thickness on a MgO substrate decreased. Lattice constants decreased slightly, as shown in Table II . The effects of substrate temperature on plane spacing were examined by varying the substrate holder temperature from 730 to 830℃. Lattice constants were almost the same, and there was no systematic change.

It was also assumed that plane spacing enlargement of as-sputtered films results from compression generated during sputtering. Included sputtering gas was a possible cause. Gas analysis was then carried out. The as-sputtered and the annealed films were heated to 800℃. Detected gas pressures at each temperature are shown in Fig. 4. Sputtering argon gas was not detected in either of the films. Increases in oxygen pressure at high temperature were due to oxygen release corresponding to the orthorombic to tetragonal phase transition . These results showed that the thermal expansion coefficient had little effect on the lattice constant, and that plane spacing enlargement relates to epitaxial growth. This is because there was little interaction between the polycrystal film and the substrate.

Effects of Heat Treatment on Impurity Concentration

Impurity distribution in films were examined by secondary ion mass spectrometry (SIMS). The SIMS profiles of the as-sputtered and annealed films are shown in Fig. 5. Er, Ba, and Cu signals are constant with depth, indicating good film uniformity. Si and Al are impurities from the substrate and enrichment of Mg at the surface are observed. This is due to the high substrate temperature. In the annealed film, interdiffusion and enrichment of Mg at the surface were significant. Al and Si are also enriched at the film surface. These impurity increases deteriorate the superconducting properties of the film.

Therefore, superconducting transition temperature Tc was affected by the film thickness. When the film thickness was 0.7 μm, Tc of the annealed film was 83 K. However, Tc of the 0.16 μm film was lowered to 30 K.

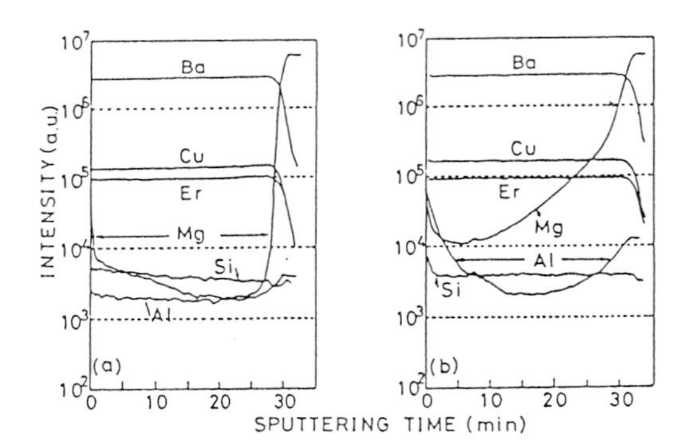

Fig.5 SIMS profiles of the as-sputtered and annealed films. (a) as-sputtered, and (b) annealed.

SUMMARY

$ErBa_2Cu_3O_7$ films were prepared by r.f. magnetron sputtering, and changes in crystallinity during annealing were examined by an X-ray diffraction and an RHEED method. The enlargement of plane spacing parallel to film plane (001) of the as-sputtered films was caused by a little thermal expansion difference between the film and the substrate, and by a stress relating to epitaxial growth.

ACKNOWLEDGEMENTS

The authors are grateful to Dr. T. Suganuma for his useful discussions. Thanks are also extended to Mr. T. Shimotu and M. Hiratani for TEM observation and resistivity measurement.

REFERENCES

[1] J. G. Bedonorz and K. A. Muller, Z. Phys. B64, 189 (1986).
[2] M. K. Wu, J. R. Ashburn, C. J. Torng, P. H. Hor, R. L. Meng, L. Gao, Z. J. Huang, Y. Q. Wang, and C. W. Chu, Phys. Rev. Lett. 58, 908 (1987).
[3] Y. Enomoto, T. Murakami, M. Suzuki, and K. Moriwaki, Jpn. J. Appl. Phys. 26, L1248 (1987)
[4] H. Adachi, K. Hirochi, K. Setsune, M. Kitabatake, and K. Wasa, Appl. Phys. Lett. 51 2263 (1987).
[5] P. Chaudhari, R. H. Koch, R. B. Laibowutz, T. R. McGuire, and R. J. Gambino, Phys. Rev. Lett. 58, 2684(1987).
[6] G. J. Clark, R. H. Koch, P. Chaudhari, and R. J. Gambino, Phys. Rev. B35, 8821 (1987).
[7] D. Dijkkamp, T. Venkatesan, X. D. Wu, S. A. Shaheen, N. Jisrawi, Y. H. Min-Lee, W. L. Mclean, and M.Croft, Appl. Phys. Lett. 51, 619 (1987).

[8] T. Terashima, K. Iijima, K. Yamamoto, Y. Bando, and H. Mazaki, Jpn. J. Appl. Phys. 27, L91 (1988).
[9] S. Tanaka, and H. Itozaki, Jpn. J. Appl. Phys. 27, L622 (1988).
[10] K. Nakamura, T. Hatano, A. Matsushita, T. Oguchi, T. Matsumoto, and K. Ogawa, Jpn. J. Appl. Phys. 26, L.791 (1987).

Microstructures and Superconductivities of $LnBa_2Cu_3O_{7-x}$ Thin Films

TOSHIYUKI AIDA, TOKUUMI FUKAZAWA, AKIRA TSUKAMOTO, KAZUMASA TAKAGI, TERUHO SHIMOTSU, and TSUNEO ICHIGUCHI

Central Research Laboratory, Hitachi, Ltd., Kokubunji, Tokyo, 185 Japan

ABSTRACT

$YBa_2Cu_3O_{7-x}$ and $ErBa_2Cu_3O_{7-x}$ thin films are deposited on $SrTiO_3$ and MgO single crystal substrates by rf-magnetron sputtering. $YBa_2Cu_3O_{7-x}$ film is grown epitaxially on $SrTiO_3$ substrate. $ErBa_2Cu_3O_{7-x}$ film is highly c-axis oriented perpendicular to MgO substrate. A lot of twins on a-b plane are observed by TEM. Jc seems to be affected by the subgrain defects rather than by the twin defects. The critical current dependency on the direction of applied magnetic field is also studied. Twin boundaries are thought not to be strong pinning centers of magnetic flux. Crystallographic anisotropy of oxygen-defect perovskite contributes to flux pinning.

INTRODUCTION

Since the first discovery of the high Tc ceramic superconductor with a modified perovskite structure, many extensive studies have been conducted on growing high quality thin films. Up to now, many film preparation techniques have been proposed; sputtering [1,2,3], MBE[4,5], reactive evaporation[6], pulsed laser evaporation[7], metalorganic deposition[8] and so on.

The purpose of this report is to show that single crystal films can be prepared by magnetron sputtering from a single target, keeping the substrate at a high temperature. Ceramic superconducting materials employed in this experiment are $YBa_2Cu_3O_{7-x}$ and $ErBa_2Cu_3O_{7-x}$. Substrates are $SrTiO_3$ with the same lattice parameters as the films, and MgO well known as slight reaction with the film. Prepared films are characterized by X-ray diffraction analysis and transmission electron microscopy observation.

The influence of microstructures and external magnetic fields on superconductive properties is investigated. In the results, Jc is affected by the subgrains rather than by the twin defects. Subboundaries act as weak couplings of superconductors and decrease Jc. A large effect of crystal orientation on Jc under an external magnetic field is also found. This is concerned with the characteristic properties of layered superconductors. Twin boundaries are thought not to be strong pinning centers of magnetic flux. Crystallographic anisotropy of oxygen-defect perovskites will contribute to flux pinning.

EXPERIMENTALS

Apparatus used in this experiment is a conventional rf-magnetron sputtering. The films are sputtered from $YBa_2Cu_{4.5}O_{7-x}$ and $ErBa_2Cu_{4.5}O_{7-x}$ targets, prepared by sintering a mixture of Y_2O_3, Er_2O_3, $BaCO_3$ and CuO. Typical sputtering conditions are listed in Table 1. After deposition, the films are annealed at 800-900 °C for 2-40h in an oxgen gas atmosphere. The crystal structure is determined by a standard X-ray diffractometer of Cu K_α and transmission electron microscopy.

Table 1. Sputtering conditions

Sputtering gas	Ar + 50%O_2
Gas pressure	4 Pa
rf input power	75W
Substrate temperature	750 °C
Film thickness	0.7µm
Growth rate	0.15µm/h
Postannealing	800 ~ 900 °C 2 ~ 40h O_2 atmosphere

Electrical resistivity is measured by a conventional four point probe technique. External magnetic field is applied to a maximum value of 1 Tesla.

RESULTS AND DISCUSSIONS

X-ray diffraction

The X-ray diffraction pattern of the samples as sputtered and annealed is shown in Fig.1. In $YBa_2Cu_3O_{7-x}$ film on $SrTiO_3$ (110), only (110) and (220) diffraction peaks were observed, because the film was grown on the $SrTiO_3$ substrate epitaxially. However, after heat treatment at 850℃, the diffraction peaks in the film shift to the higher angle.

$ErBa_2Cu_3O_{7-x}$ film on MgO (100) is highly c-axis oriented perpendicular to the substrate. Due to RHEED analysis, the a and b axes in the film were also oriented to the crystal orientation of the substrate. $ErBa_2Cu_3O_{7-x}$ film grew in a single crystalline state in spite of a large lattice mismatch between the film (0.39nm) and the substrate (0.42nm). This is the same result as $HoBa_2Cu_3O_{7-x}$ film on MgO substrate[2]. The peak shift in annealed sample was observed in the same way as $YBa_2Cu_3O_{7-x}$ film. This peak shift phenomenon is considered to correspond to the stress due to the epitaxial growth or oxygen gas absorption[9].

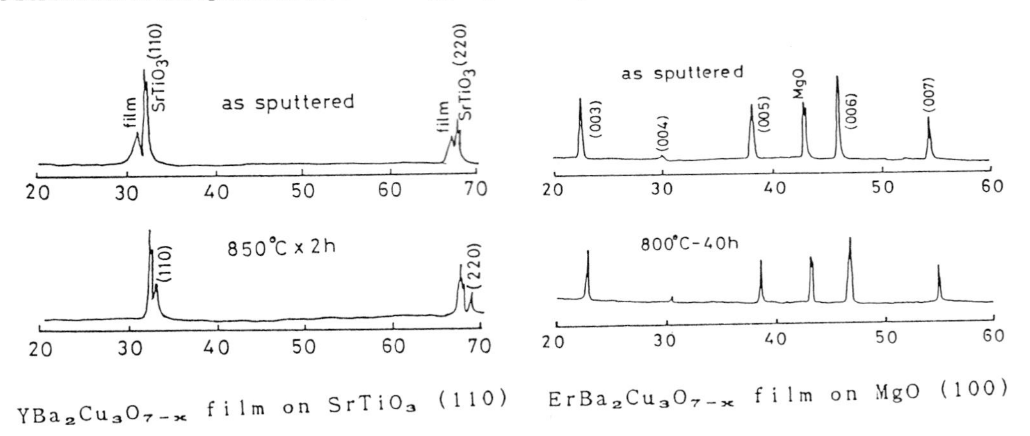

Fig. 1 X-ray diffraction patterns of $YBa_2Cu_3O_{7-x}$ and $ErBa_2Cu_3O_{7-x}$ films as sputtered and annealed.

TEM image

A cross sectional TEM image of (110) and (001) planes of $YBa_2Cu_3O_{7-x}$ thin film is shown in Fig.2. In the (110) plane image, good continuity of the lattice image between the substrate and the film was found. In the film, the three layers contrast is also seen, which corresponds to the c-axis unit cell length of the superconductive perovskite structure. The film grew on the substrate epitaxially. However, it is also seen in Fig.2 that the microcracks perpendicular to the c-axis were generated. These defects are considered to be caused by the thermal stress due to the large difference in the thermal expansion coefficients. In the (001) plane image, the directions in a and b axes tilt upward from the substrate surface by 45 degrees. The saw-toothed pattern reflects the microtwins between a and b axes. The periodical contrasts along a and b axes are also seen. This means some ordering lattice defects due to the diffusion of Sr and Ti impurities from the substrate[10].

Plan view and cross sectional TEM images of $ErBa_2Cu_3O_{7-x}$ thin film are shown in Fig.3. In the plan view image of a (001) plane, there are many stripe patterns which show the [110] reflection microtwins and some domains where microtwin stripes contact each other at 90 degrees. This is the same pattern as the powder sample[11]. In the cross sectional view of a (100) plane, the lattice image of a-b basal plane perpendicular to the c-axis is quite smooth and flat.

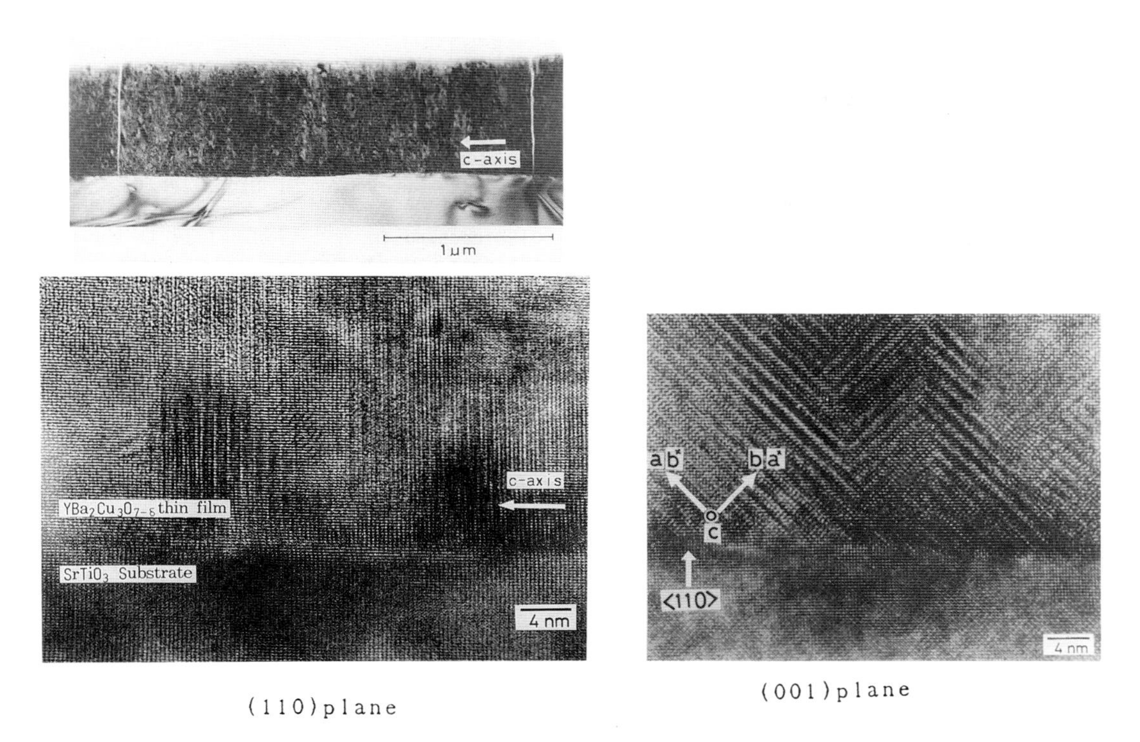

Fig. 2 Cross sectional TEM images of (110) and (001) planes of $YBa_2Cu_3O_{7-x}$ films on $SrTiO_3$ (110) substrate.

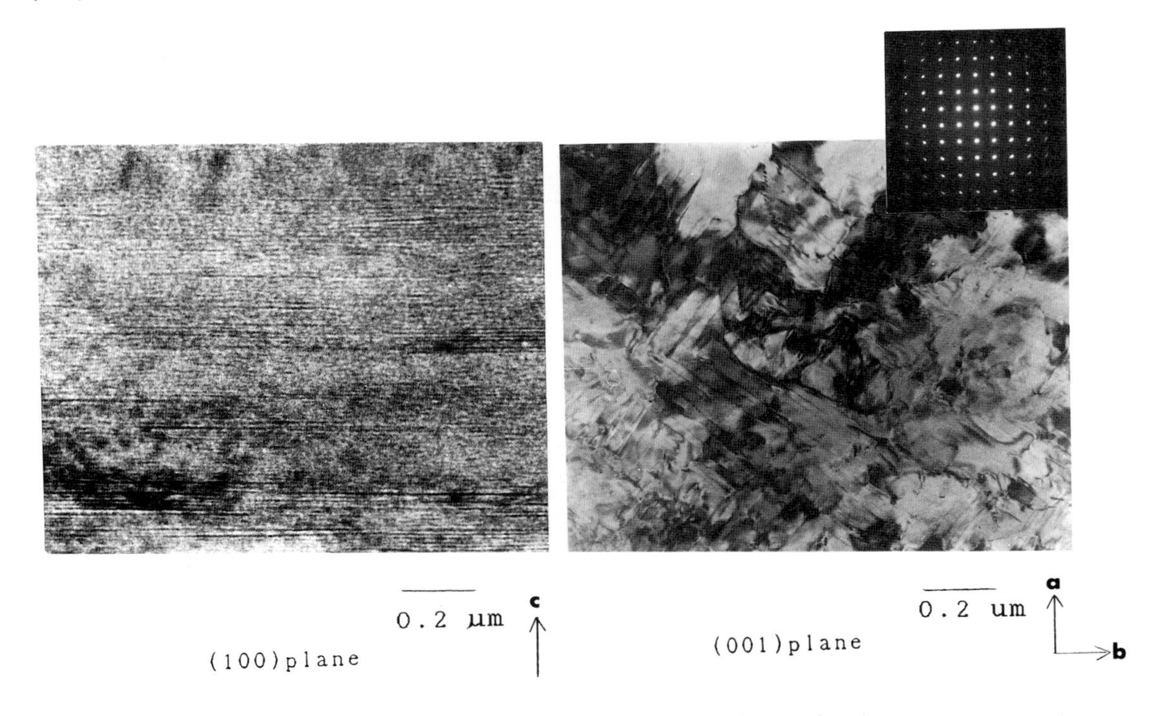

Fig. 3 Plan view and cross sectional TEM images of (100) and (001) planes on MgO (100)

Temperature and microstructure dependence on Jc

Temperature dependence of critical current density Jc in $YBa_2Cu_3O_{7-x}$ thin film is shown in Fig.4. Jc measured perpendicular to the c-axis is larger than that in parallel to the c-axis. This indicates the existence of the anisotropy in current flow in this ceramic superconductor or the interruption of the current path due to microcracks as shown in Fig.2. However, the anisotropy was not larger than that reported by Enomoto[1].

Temperature dependence of Jc along the [100] direction in $ErBa_2Cu_3O_{7-x}$film is shown in Fig.5. High Jc of 10^5-10^6 A/cm^2 was obtained. This high value is thought to originate in the smoothness and flatness in basal planes in which superconductive currents flow. However, the low Jc sample of 10^2 A/cm^2 was sometimes produced. A plan view TEM image of this low Jc sample is shown in Fig.6. Twin spacing is almost the same value of 50nm as that of high Jc sample, but subgrains with low angle misorientations were produced. So, Jc is considered to be affected by the subgrain defects rather than by the twin defects. In this ceramic superconductor, coherent length is quite short only few angstroms[12]. For this reason, the subboundary acts as a weak coupling and decreases the superconductive current.

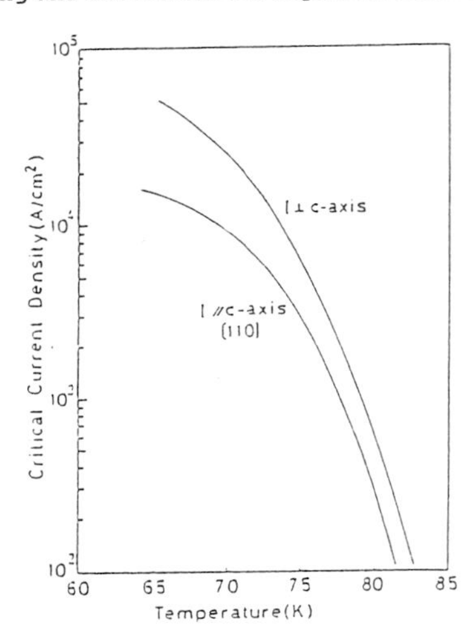

Fig. 4. Temperature dependence of Jc in $YBa_2Cu_3O_{7-x}$ thin films

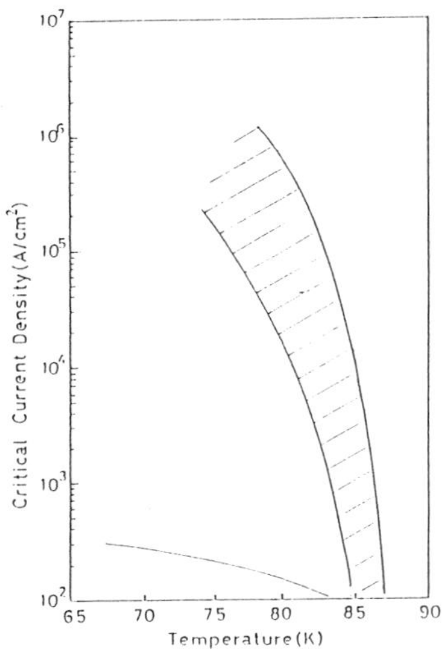

Fig. 5. Temperature dependence of Jc in $ErBa_2Cu_3O_{7-x}$ thin films.

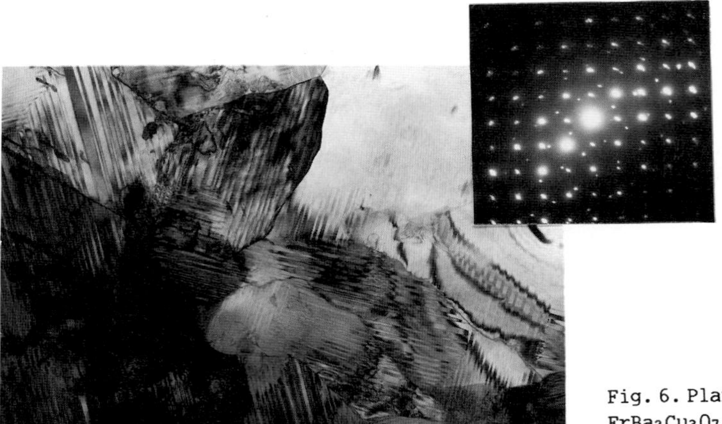

Fig. 6. Plan view TEM image of low Jc $ErBa_2Cu_3O_{7-x}$ thin film.

The external magnetic field on Jc

The external magnetic field B was applied on $YBa_2Cu_3O_{7-x}$ film on $SrTiO_3$ substrate and $ErBa_2Cu_3O_{7-x}$ on MgO substrate during Jc measurement. The result is shown in Fig.7. Applied current direction was perpendicular to the c-axis. In both cases, Jc was remarkably decreased when B was applied perpendicular to the basal plane. Basal plane in the $YBa_2Cu_3O_{7-x}$ film is normal to the substrate. On the contrary, basal plane in $ErBa_2Cu_3O_{7-x}$film is parallel to the substrate. So, the large anisotropy in Jc decrease originates the intrinsic properties of this oxide superconductor. This oxide is a second type superconductor in which H_{c1} is about 80 gauss[13].In higher B regions over H_{c1}, magnetic flux penetrates into the film as a vortex. When high B is applied to the film, some voltage will be generated according to the magnetic flux motion due to the Lorentz force J x B and electric power of V x J will be consumed in the film. As a result, the film is heated up and Jc decreases. Superconductive current in this oxide superconductor is known to flow in the inner two dimensional a-b basal plane. The results in Fig.7 can be interpreted clearly considering the magnetic flux moves in a and b axes directions more easily than in c axis direction. This is reasonable when we consider the following. When B is applied parallel to the basal plane, flux lines penetrate inter basal planes and are difficult to move because superconductive current is fixed in the inner basal plane, as pointed out by Suenaga[12]. Angular dependence of breakdown voltage when the direction of magnetic field changed continuously from parallel to normal to the substrate is shown in Fig 8. The sample was $ErBa_2Cu_3O_{7-x}$ film on MgO substrate. In this experiment, magnetic field was 0.5 T and current density was 10^5 A/cm². Superconductivity when magnetic field was parallel to a-b basal plane was broken more rapidly as the direction of magnetic field is closer to the c-axis direction. This phenomenon is quite similar to the result on a superconductor with a layered structure of Nb/Cu explained by the anisotropic Ginzburg-Landau model [14,15]. From this viewpoint, twin defects seem to have no great role in pinning the magnetic flux motion. Crystallographic anisotropy of this superconductor seems to have a strong intrinsic flux pinning center.

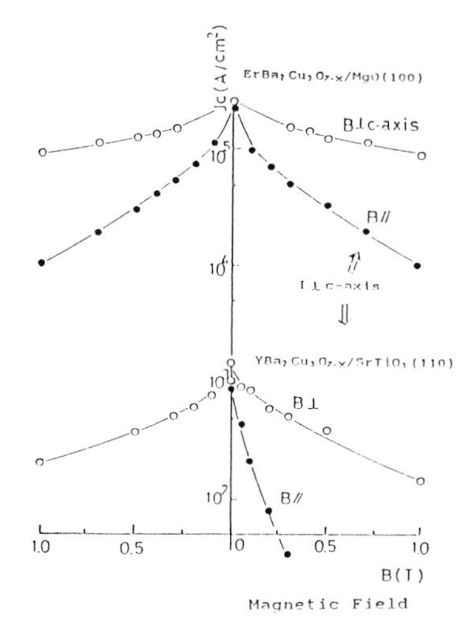

Fig. 7 Effect of external magnetic field on Jc of $YBa_2Cu_3O_{7-x}$ and $ErBa_2Cu_3O_{7-x}$ films.

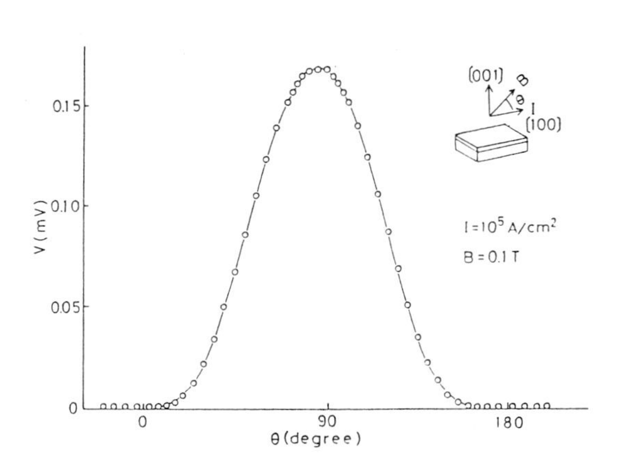

Fig. 8 Angular dependence of the beakdown voltage when the direction of magnetic field changes from parallel to normal to the substrate.

CONCLUSION

$YBa_2Cu_3O_{7-x}$ film on $SrTiO_3$ was grown epitaxially. Microcracks normal to the c-axis were generated. This seems to influence the anisotropy in current flow. $ErBa_2Cu_3O_{7-x}$ film on MgO substrate was grown in a single crystalline state despite the large mismatch of lattice parameters. In both films, a lot of microtwins were generated in the inner a-b basal planes and the lattice image of a basal plane normal to the c-axis was quite smooth and flat. Jc in $ErBa_2Cu_3O_{7-x}$ film ranged from 10^2 to 10^6 A/cm^2. Twin spacing in a low Jc sample was the same as that in a high Jc sample. However, subgrains were produced in a low Jc sample. Jc is considered to be affected by the subgrain defects rather than by the twin defects. Jc decreased rapidly when an external magnetic field was applied perpendicular to the basal plane of the film. This corresponds to a superconductor with a layered structure. Twin defects seem to have no great role in pinning the magnetic flux motion. Crystallographic anisotropy of this oxide superconductor contributes to flux pinning.

ACKNOWLEDGEMENTS

The authors would like to express their sincere appreciation to Prof.M. Suenaga of Brookhaven National Laboratory, USA, for his helpful discussions, to Dr.T. Suganuma, Dr.K. Miyauchi, Dr.Y. Tarutani, and to the members of Material Analysis Center in our laboratory.

REFFERENCES

[1] Y.Enomoto, T.Murakami, M.Suzuki and K.Moriwaki; Jpn. J. Appl. Phys. 26.7(1987)L1248
[2] S.Tanaka and H.Itozaki; Jpn. J. Appl. Phys. 27.4(19888)L622
[3] T.Aida, T.Fukazawa, K.Takagi and K.Miyauchi; Jpn. J. Appl. Phys. 26.9(1987)L1489
[4] R.B.Laibowitz, R.H.Koch, et al; Phys. Rev. B36(1987)8821
[5] J.Kwo, T.C.Hsieh, et al; Phys. Rev. B36(1987)4039
[6] T.Terashima, K.Iijima, et al; Jpn. J. Appl. Phys. 27.1(1988)L91
[7] X.D.Wu, D.Dijkkamp, et al; Appl. Phys. Lett. 51(1987)L861
[8] A.H.Hamdi, J.V.Mantese, et al; Appl. Phys. Lett. 51.21(1987)2152
[9] T.Takagi, T.Ishiba, et al; To be submitted in this proceedings.
[10] T.Aida, T.Fukazawa et al; Proceeding in 1988 ACeS annual meeting Cincinnati, USA , to be published.
[11] H.A Hoff, A.K.Singh and C.S.Pande; Appl. Phys. Lett. 52.22(1988)669
[12] M.Suenaga; Proceeding in 1988 MRS spring meeting, Tokyo, Japan, to be published
[13] Y.Ishikawa, K.Mori, K.Kobayashi and K.Sato; Jpn. J. Appl. Phys. 23.7(1988)L403
[14] Y.Iye, T.Tamegai, H.Takeya and H.Takei; Jpn. J. Appl. Phys. 26.11(1987)L1850
[15] C.S.L.Chun, G.G.Zheng, J.L.Vicent and I.K.Schuller; Phys. Rev. 29.9(1984)4915

Properties of High Jc BiSrCaCuO Thin Film

H. ITOZAKI, K. HIGAKI, K. HARADA, S. TANAKA, N. FUJIMORI, and S. YAZU

Sumitomo Electric Industries Ltd., 1-1-1, Koya-Kita, Itami, 664 Japan

ABSTRACT

BiSrCaCuO thin films were grown on MgO substrate by an RF magnetron sputtering. Its Jc was 1.9 million A/cm2 at 77.3K and 21 million A/cm2 at 40K. Anisotropic degradation of Jc was observed. In the most severe case, Jc decreased to 0.1 million A/cm2 as applied magnetic field increased to only 0.5 Tesla. X-ray analysis showed that this film consists of almost single high Tc 2223 phase with c=37.2A. RHEED, SEM and TEM observation showed that this high Jc thin film was c axis oriented polycrystalline. The grain boundaries and stacking faults were not barriers for supercurrent to flow but they might act as pinning places for magnetic flux.

INTRODUCTION

After the discovery of high Tc superconductive materials(1,2), many studies have been done in the field of thin films. Because high Tc superconducting thin film has big potential in electronics, such as sensors(3) or high speed devices(4). LnBaCuO has been well studied(5) to obtain high Jc (critical current density). The authors have studied epitaxial growth of HoBaCuO thin film and get a single crystal thin film(6-8). As it has no grain boundaries which work as a weak link, the Jc of this single crystal thin film was 3.5 million A/cm2. Recently Maeda(9,10) found rare earth free superconductor BiSrCaCuO which has higher Tc than LnBaCuO. Its Tc is about 110K(11,12). It is rather difficult to obtain single phase material of this compound. We have tried to obtain BiSrCaCuO thin films and have gotten high Jc single phase polycrystalline film. Here we report on the preparation conditions of BiSrCaCuO thin film and its electrical properties and structure.

EXPERIMENTAL

BiSrCaCuO thin films have been prepared by an RF magnetron sputtering with a single target. The target was prepared by sintering mixed powder of Bi2O3, SrCO3, CaO and CuO. The composition of the target was Bi:Sr:Ca:Cu=1.4:1:1:1.5. The target size was 4 inches in diameter. MgO single crystals were used for substrates. The substrates temperature was 600 to 800°C. the sputtering gas was 80% argon and 20% oxygen. The glow discharge was radio-frequency exited (13.56MHz). The film thickness was typically 200nm. The temperature dependence of the resistivities and the critical current densities were measured by a four-probe method in a cryostat. For these measurements, an annealed thin film was etched with a diluted HCl (0.5 vol.%) solution to form a narrow region with an area of about 20μm in length and 20 μm in width. The structure and morphology were analyzed by X-ray diffraction (XRD), reflection high-energy electron diffraction (RHEED), scanning electron microscopy (SEM) and transmission electron microscopy (TEM).

RESULTS AND DISCUSSION

Film Growth Temperature and Post Annealing Temperature

Crystallinity of thin film prepared by physical vapor deposition such as sputtering depends on the temperature of film growth and post annealing temperature(13). Firstly, the substrate temperature dependence of Jc was investigated. Samples were post annealed at 885°C for 60 min. Fig.1 shows that the Jc is very sensitive to the substrate temperature. Suitable temperature exists around 690°C to get high Jc film.

Secondly, post annealing temperature was investigated. All of the films were grown at the substrate temperature of 700°C. As the post annealing temperature increases from 885°C to 900°C, Jc increases, but Jc becomes zero suddenly at the post annealing temperature of 905°C shown in fig.2. The surface of the film becomes very rough by annealed at 905°C. The film might melt partially at the annealing temperature. Therefore it is important to control the process temperature such as growth temperature and post annealing temperature to obtain a high Jc BiSrCaCuO thin film.

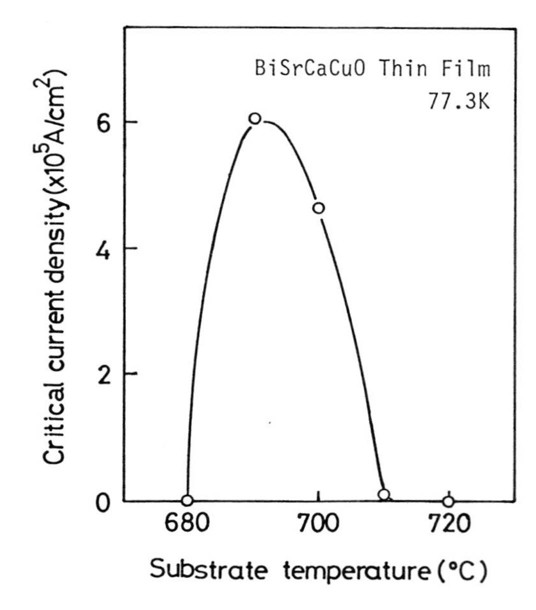

Fig.1 Substrate temperature dependence of critical current density for BiSrCaCuO thin films annealed at 885°C for 60 min.

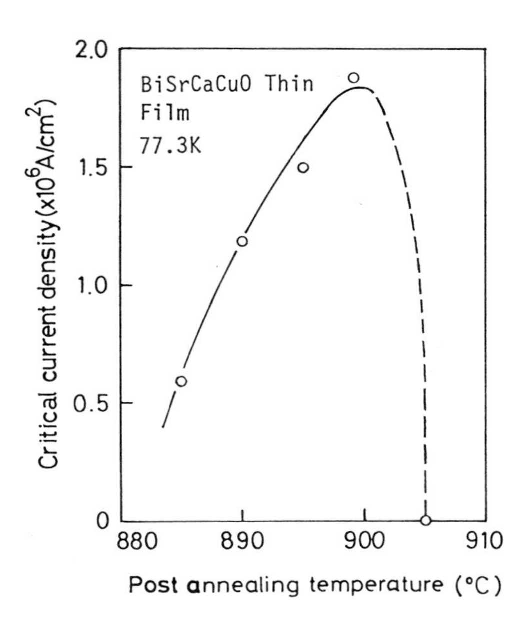

Fig.2 Post annealing temperature dependence of critical current density for BiSrCaCuO thin films annealed for 60 min which were grown at the substrate temperature of 700°C.

High Jc BiSrCaCuO Thin Film

A high Jc thin film was obtained by optimizing the preparation conditions, such as upper description. The following is its electrical and physical properties.

Electrical Properties

Temperature dependence of critical current density Jc was shown in Fig.3. Jc decreases monotonically as temperature increases. Jc at 40K is 21 million A/cm2 and Jc at the liquid nitrogen temperature of 77.3K is 1.9 million A/cm2. This is the highest Jc of BiSrCaCuO to be reported. Jc gradually decreases and becomes zero at 95K as temperature increases from 75K to 95K. This 95K is transition temperature of this film.

Fig.4 shows the temperature dependence of resistivity. Onset of Tc is more than 115K and resistivity drop has a tail to 95K. But there is no step which is usually shown in the resistivity-temperature curve of a bulk sample. This means that this film has only high Tc phase but includes some defects which make the tail of the resistivity-temperature curve.

Magnetic field dependence of critical current density at 77.3K was also investigated. As the applied magnetic field increases, the Jc decreases as shown in fig.5. The degradation of Jc depends on the direction of the applied field to the substrate and to the direction of the current flow. If the current flows parallel to the magnetic field, no Lorentz force comes out and Jc should not be affected by the magnetic field. But fig. 5 shows that the degradation in this case has been occurred. This indicates that the film has many obstacles for the current to flow and the actual current passes in the zigzag way. Therefore the magnetic flax gets force and the degradation of the Jc occurs. This film is polycrystalline described later and many grain boundaries might act as these obstacles. In the other cases where the current and the magnetic field are perpendicular to each other, the degradation is larger than above case. Comparing the two cases (b) and (c) in the fig. 5, degradation of (c) is larger than that of (b). The magnetic flux which are parallel to the surface are pinned stronger than those which are perpendicular to the surface. The magnetic flux might be pinned at grain boundaries or stacking faults, which are mainly located parallel to the surface. Jc becomes 0.1 million A/cm2 under the magnetic field of even 0.5 Tesla. This degradation is much larger than that of HoBaCuO thin film which can sustain more than 1.5million A/cm2 under the magnetic field of 1 Tesla.

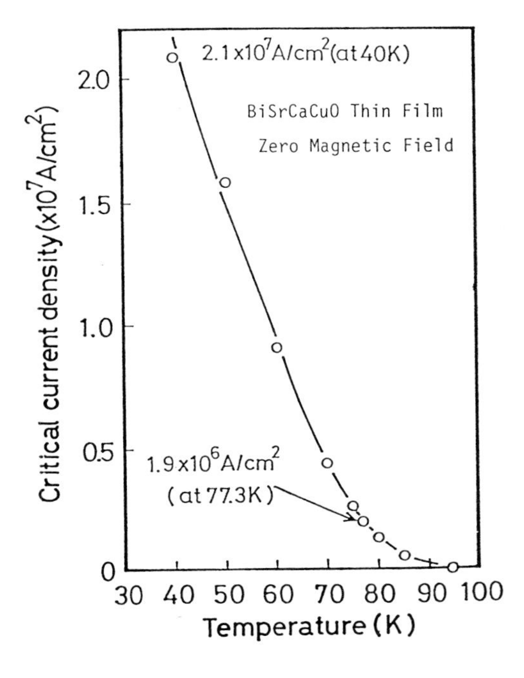

Fig.3 Temperature dependence of critical current density for an annealed BiSrCaCuO thin film.

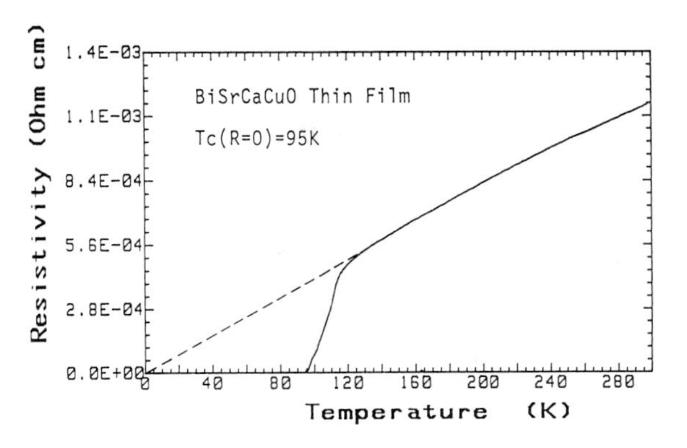

Fig.4 Temperature dependence of resistivity for an annealed BiSrCaCuO thin film.

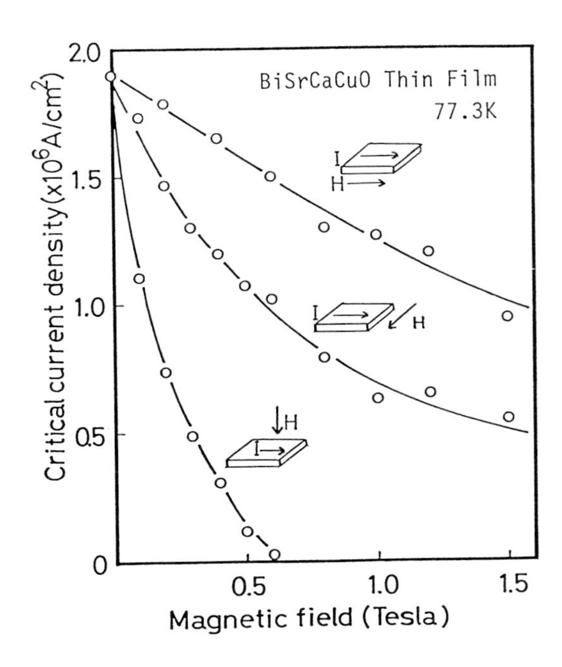

Fig.5 Magnetic field dependence of critical current density for an annealed BiSrCaCuO thin film. Directions of applied magnetic field are (a) parallel to the current flow, (b) perpendicular to the current flow and a c-axis, and (c) perpendicular to the current flow and parallel to a c-axis of the film.

Structure

Structure of the high Jc BiSrCaCuO thin film was analyzed by X-ray, RHEED, SEM and TEM. Fig. 6 shows the X-ray diffraction of BiSrCaCuO thin film. This pattern shows only (00n) peaks which comes from the high Tc phase of composition of 2223. The length of c-axis is calculated as c=37.2A. Therefore this high Jc film has c-axis oriented structure with single 2223 phase.

Fig. 7 shows RHEED patterns of the as grown thin film and the annealed thin film. The RHEED pattern of the as grown film is fine ring pattern. This means that the as grown film consists of very small grains. The RHEED pattern of the annealed film becomes many spots. This shows that very fine grains grow at the post annealing. Although the pattern becomes spots, spot rings are still exist in the RHEED pattern. Therefore film does not textured in the a or b axis direction, but it does only in c axis after post annealing.

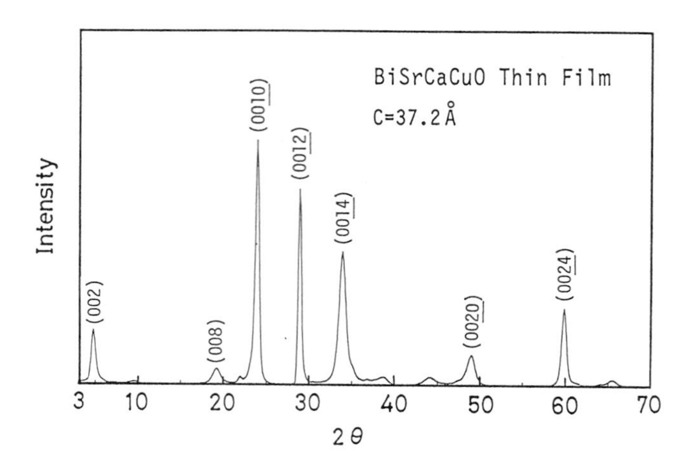

Fig.6 X-ray diffraction pattern for an annealed BiSrCaCuO thin film.

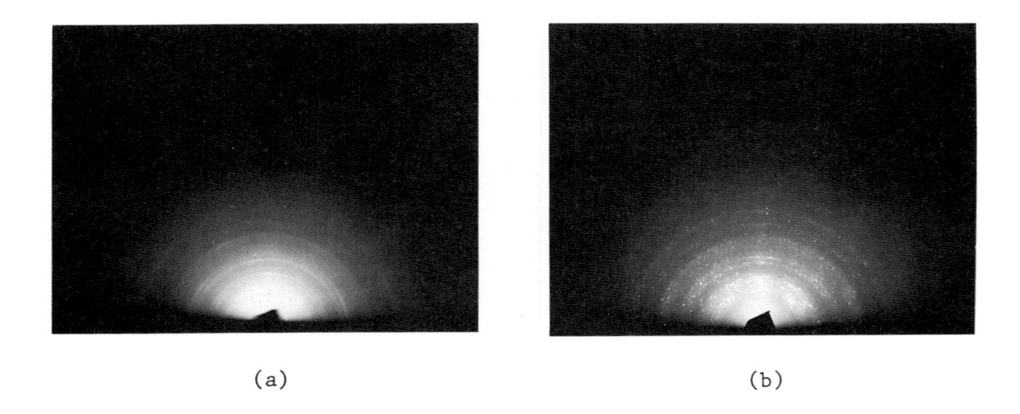

(a) (b)

Fig.7 Reflection high-energy electron diffraction patterns for (a) an as grown and (b) an annealed BiSrCaCuO thin film.

Morphology of the as grown sample and the annealed sample were observed by SEM, as shown in fig.8. Surface of the as grown film has many tiny particles. These composition was analyzed by EDX. These particles have almost the same composition as that of the film itself. The film does not grow uniformly, but nucleation growth occurs and particles grow on the surface. Grains grow at the post annealing as shown in fig.8(b). The size of the grains is about a half micron. The particles on the surface become larger and its number decreases.

Microstructure of the annealed film was also observed by TEM, as shown in fig.9. The TEM bright image shows some small grains and fringes in the grains. This fringes come from grain boundaries or stacking faults. A micro diffraction pattern of this region shows overlapped diffraction pattern of some grains. The c axis orients perpendicular to the film surface. Since Jc of more than a million A/cm2 can flow at 77.3K through many grain boundaries, they does not work as barriers or walls for supercurrent to flow, but work as pinning places of magnetic flux.

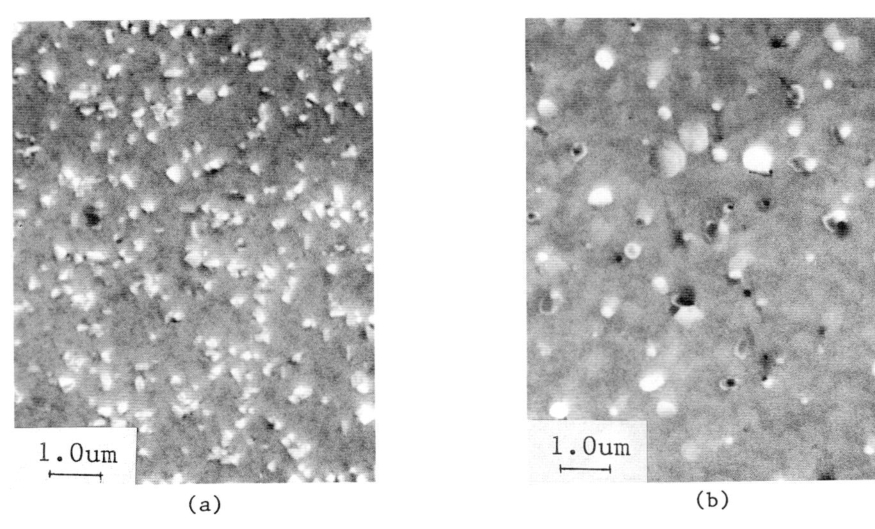

Fig.8 Scanning electron micrographs for (a) an as grown and (b) an annealed BiSrCaCuO thin film.

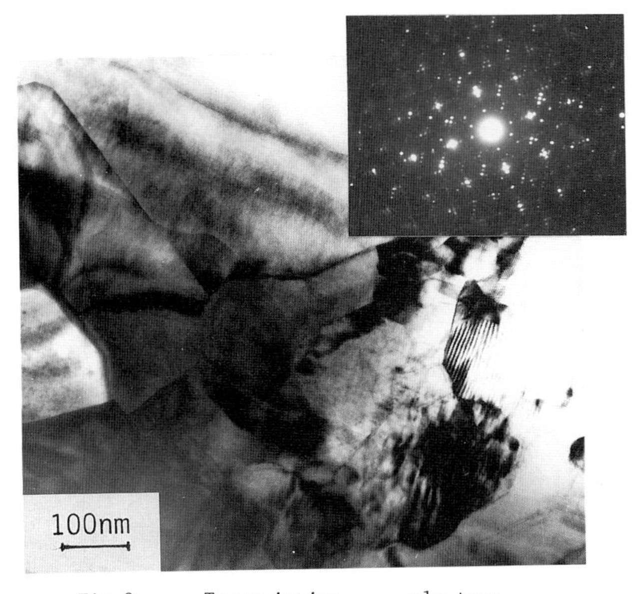

Fig.9 Transmission electron microscopy of BiSrCaCuO thin film. The inset is its micro electron diffraction pattern.

SUMMARY

We have successfully obtained a high Jc BiSrCaCuO thin film on a MgO substrate by an RF magnetron sputtering. Its critical current densities were 1.9 million A/cm2 at77.3K and 21 million A/cm2 at 40K. Anisotropic degradation of Jc was observed. In the most severe case, Jc decreased to 0.1 million A/cm2 as applied magnetic field increased to only 0.5 Tesla.

Structure of this high Jc film was analyzed. It was c-axis textured polycrystalline film and consisted of an almost single phase of high Tc 2223 phase with c=37.2A.

We investigated the process conditions to obtain a high Jc thin film and found that it is important to select and control process temperatures such as growth temperature and post annealing temperature.

ACKNOWLEDGMENT

We would like to thank T.Nishikawa J.Matsumoto and K.Yamaguchi for their help with the X-ray TEM and RHEED analyses. We greatly appreciate T. Yamaguchi for his advice and encouragements.

REFERENCES

1) J.G.Bednorz and Muller,Z.Phys.B64,189(1986).
2) M.K.Wu,J.R.Ashburn,C.W.Chu,Phys.Rev.Lett.58,908(1987).
3) Y.Enomoto,T.Murakami and M.Suzuki,in Proceedings of 5th International Workshop on Future Electron Devices -High-Temperature Superconducting Electron Devices-,Miyagi-Zao,June 1988,p325.
4) S.Takada,in Proceedings of 5th International Workshop on Future Electron Devices -High-Temperature Superconducting Electron Devices-,Miyagi-Zao,June 1988,p305.
5) S.Jin,T.H.Tiefel,R.C.Sherwood,M.E.Davis,R.B.vanDover,G.W.Kammlott,R.A.Fastnacht,and H.D.Keith,Appl.Phys.Lett.52,2074(1988).
6) S.Tanaka,H.Itozaki,Jpn.J.Appl.Phys.27,L622,(1988).
7) H.Itozaki,S.Tanaka,K,Higaki,andS.Yazu,Physica C153-155,1155(1988).
8) H.Itozaki,S.Tanaka,K.Harada,K.Higaki,N.Fugimori,andS.Yazu,in Proceedings of 5th International Workshop on Future Electron Devices -High-Temperature Superconducting Electron Devices-, Miyagi-Zao,June 1988,p149.
9) H.Maeda,Y.Tanaka,M.Fukutomi,andT.Asano,Jpn.J.Appl.Phys.27,L209(1988).
10) K.Togano,H.Kumakura,H.Maeda,K.Takahashi,andM.Nakao,Jpn.J.Appl.Phys.27,L323(1988).
11) Kitazawa,S.Yaegashi,K.Kishio,T.Hasegawa,N.Kanazawa,K.Park,andK.Fueki,in Proceedings of Amer.Ceram.Soc.90th Annual Meeting,Cincinnati,Ohio,May.1988.
12) K.Kuwahara,S.Yaegashi,K.Kishio,T.Hasegawa,andK.Kitazawa,in Proceedings of Latin-American Conf. on High Temperature Superconductivity,Rio de Janeiro,May,1988.
13) J.W.Matthews,Epitaxial Growth(Academic Press,New York,1975)p109.

Characteristics of Rare-Earth-Free Superconducting Thin Films

M. NEMOTO, H. KUWAHARA, K. KAWAGUCHI, Y. MATSUTA, and M. NAKAO

Sanyo Tsukuba Research Center, 2-1, Koyadai, Tsukuba, 305 Japan

ABSTRACT

Superconducting Tl-Ca-Ba-Cu-O thin films deposited by rf magnetron sputtering from $Tl_2Ca_2Ba_2Cu_3O_x$, $Tl_3Ca_2Ba_2Cu_3O_x$ and $Tl_2Ca_3Ba_1Cu_3O_x$ targets were investigated. After post-annealing, the composition of films prepared from the $Tl_2Ca_2Ba_2Cu_3O_x$ target was closest to that of the high-T_c phase in the targets used. A Tl-Ca-Ba-Cu-O film with zero resistivity at 116 K exhibited the critical current densities of ~5.5×10^6 A/cm^2 at 5.0 K and ~1.5×10^6 A/cm^2 at 50 K. The critical current densities of a randomly oriented Tl-Ca-Ba-Cu-O film, ~3.5×10^6 A/cm^2 at 5.0 K and ~8.0×10^5 A/cm^2 at 50 K, were larger than those of a highly oriented Bi-Ca-Sr-Cu-O film, ~1.1×10^6 A/cm^2 at 4.5 K and ~2.0×10^5 A/cm^2 at 40 K. Both the films were comprised of the low-T_c phase.

INTRODUCTION

Since the discoveries of high-T_c superconducting La-Ba-Cu-O [1] and Y-Ba-Cu-O [2] systems, many efforts have been made for preparation of superconducting films. Recent discoveries of superconducting Bi-Ca-Sr-Cu-O [3] and Tl-Ca-Ba-Cu-O [4] systems without any rare-earth elements have led to much new activity in preparation of thin films with T_c above 100 K. Deposition and characterization of thin films of these materials are very important for the study of superconducting mechanisms as well as a variety of applications.

The preliminary results of thin films of these systems have been already reported. [5] [6] [7] In this paper, we report on characteristics of Tl-Ca-Ba-Cu-O thin films made by rf magnetron sputtering from sintered targets with three different compositions. Characterization of the films was carried out using inductively coupled plasma-emission spectrometry (ICP), x-ray diffraction (XRD), magnetization and resistivity measurements.

EXPERIMENTAL

Sputtering was carried out using a conventional rf magnetron sputtering apparatus under conditions similar to previous works. [6] [7] Targets with compositions of $Tl_2Ca_2Ba_2Cu_3O_x$ (2223), $Tl_3Ca_2Ba_2Cu_3O_x$ (3223) and $Tl_2Ca_3Ba_1Cu_3O_x$ (2313) were used. Thin films ~3 μm thick were deposited onto (100) MgO, (100) $SrTiO_3$ and (100) yttria-stabilized ZrO_2 (YSZ) substrates at ambient temperature with a typical deposition rate of 400 Å/min. The as-deposited films were not conducting. The films were wrapped in gold foil. Annealing was carried out in an oxygen atmosphere at 900 - 950 ℃ for 10 min, followed by cooling to room temperature in a furnace.

The temperature dependence of the film resistivity was measured by the standard four-probe technique using pressed indium contacts. The samples were mounted on a cooling head of a closed-cycle helium refrigerator system. The temperature was measured by a Au + 0.07 % Fe - Chromel thermocouple. The applied constant current density ranged from 0.5 A/cm^2 to 5 A/cm^2

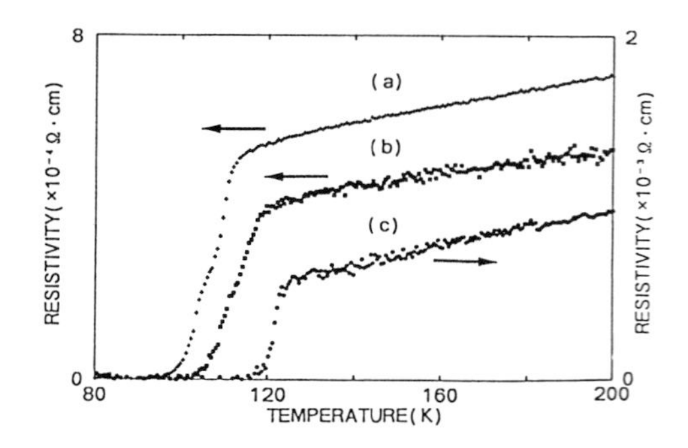

Fig. 1. Temperature dependence of resistivity for Tl-Ca-Ba-Cu-O films deposited on (100) MgO from targets with compositions (a) $Tl_2Ca_3Ba_1Cu_3O_x$, (b) $Tl_3Ca_2Ba_2Cu_3O_x$ and (c) $Tl_2Ca_2Ba_2Cu_3O_x$.

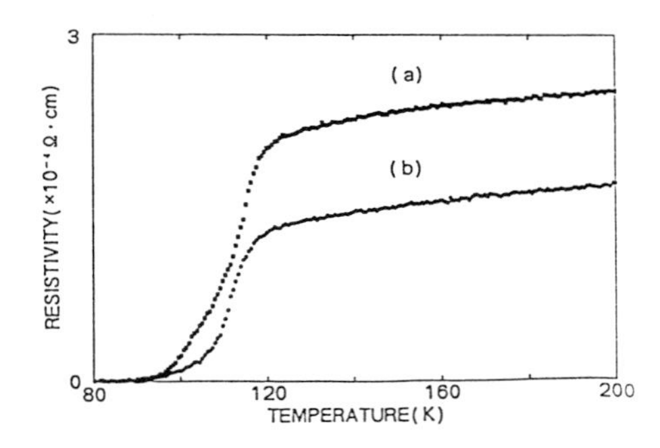

Fig. 2. Temperature dependence of resistivity for Tl-Ca-Ba-Cu-O films deposited on (a) (100) YSZ and (b) (100) $SrTiO_3$ substrates from a $Tl_2Ca_2Ba_2Cu_3O_x$ target.

depending on sample resistivity. X-ray diffraction measurements were carried out in θ-2θ geometry with Cu-K_α radiation (30 kV - 20 mA). The composition of the films was determined by ICP measurements. Magnetic susceptibility was determined by a superconducting quantum interference device (SQUID) magnetometer. The film thickness was determined by a Dektak profilometer.

RESULTS AND DISCUSSION

Figures 1 and 2 show the temperature dependence of resistivity for Tl-Ca-Ba-Cu-O films deposited from the three targets. The best electrical behavior was observed in the film on (100) MgO. The thin films were annealed at 930 ℃ for 10 min. The films deposited on (100) MgO from the 2313, 3223 and 2223 targets displayed zero resistivity at 95 K, 104 K and 116 K, respectively. The films deposited on (100) YSZ and (100) $SrTiO_3$ from the 2223 target displayed zero resistivity at 90 K and 92 K, respectively. In the previous work,[7] annealing of thin films deposited from the 3223 target was carried out in a covered alumina vessel containing a pellet with a composition of $Tl_2Ca_3Ba_1Cu_3O_x$. The highest zero-resistivity temperature was 98 K for a film on (100) MgO.

X-ray diffraction patterns of these films are shown in Figs. 3 and 4. Most of the peaks observed can be indexed as the low-T_c (2122) phase or the high-T_c (2223) phase using the crystal structure data by Hazen *et al.*[8] The film prepared from the 2313 target is comprised of only the low-T_c phase, while the films prepared from the 2223 and 3223 targets contain both the phases.

Table I shows the compositions of these films on (100) MgO. The film composition little changed through annealing process for films prepared from any target used. The composition of

Table I. Compositions of thin films on (100) MgO.

Target	before annealing	after annealing
$Tl_2Ca_3Ba_1Cu_3O_x$	1.7 : 3.7 : 1.2 : 3.0	1.9 : 3.6 : 1.2 : 3.0
$Tl_3Ca_2Ba_2Cu_3O_x$	2.0 : 2.4 : 2.4 : 3.0	2.1 : 2.4 : 2.1 : 3.0
$Tl_2Ca_2Ba_2Cu_3O_x$	2.3 : 2.3 : 2.1 : 3.0	2.0 : 2.2 : 1.9 : 3.0

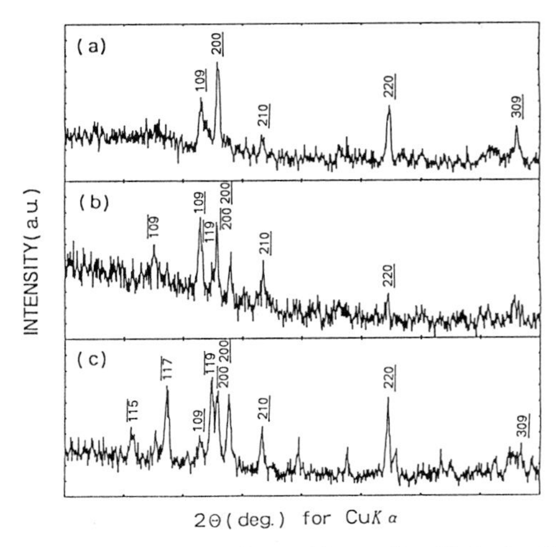

Fig. 3. X-ray diffraction patterns for Tl-Ca-Ba-Cu-O films deposited on (100) MgO from targets with compositions (a) $Tl_2Ca_3Ba_1Cu_3O_x$, (b) $Tl_3Ca_2Ba_2Cu_3O_x$ and (c) $Tl_2Ca_2Ba_2Cu_3O_x$.

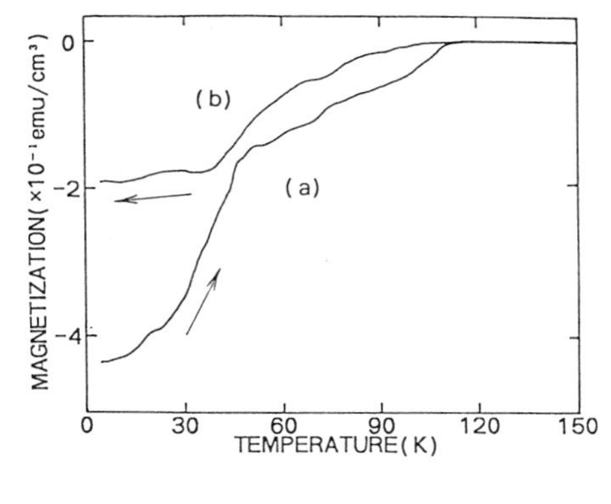

Fig. 5. Temperature dependence of magnetization for a Tl-Ca-Ba-Cu-O film. (a) Diamagnetic shielding data were obtained during warming in a field of 20 Oe after cooling in zero field. (b) Meissner data were obtained during cooling in the same field.

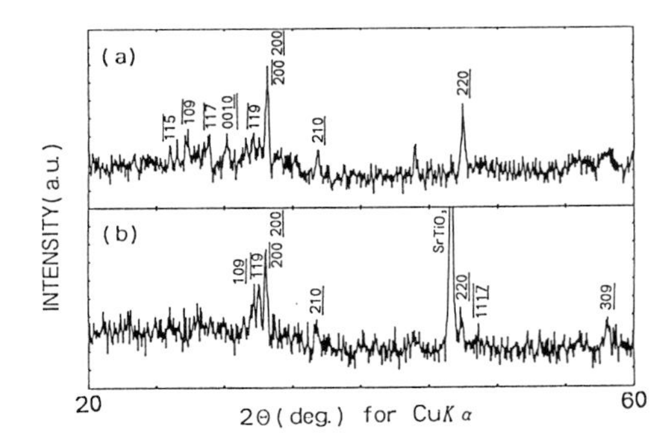

Fig. 4. X-ray diffraction patterns for Tl-Ca-Ba-Cu-O films deposited on (a) (100) YSZ and (b) (100) $SrTiO_3$ from a $Tl_2Ca_2Ba_2Cu_3O_x$ target.

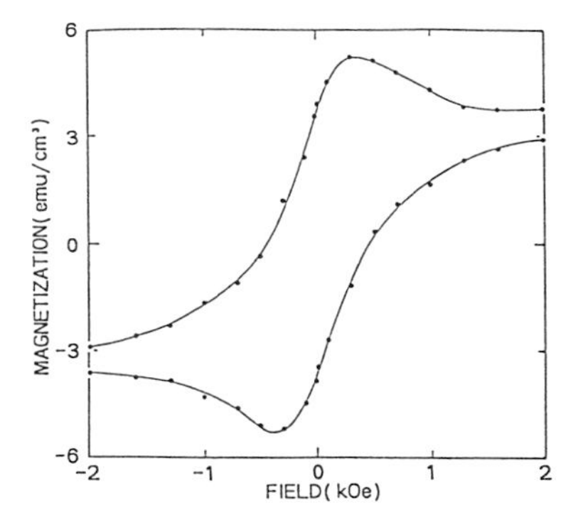

Fig. 6. Magnetization curve at 5.0 K of a Tl-Ca-Ba-Cu-O film for fields of up to ±2 kOe applied parallel to the film plane.

the film deposited from the 2223 target was closest to that of the high-T_c phase. Under the preparation condition in this study, the 2223 sputtering target was found to be most appropriate to form the high-T_c phase because of the good agreement between the composition of the annealed films and that of the high-T_c phase.

Magnetic properties of the films prepared from the 2223 target were studied using a SQUID magnetometer in a magnetic field parallel to the film plane. Figure 5 shows the temperature dependence of magnetization for the same film as in Fig. 1(c). The curve shown in Fig. 5(a) was obtained when the sample was cooled to 5.0 K in zero field and then warmed in a field of 20 Oe (diamagnetic shielding effect). The curve in Fig. 5(b) was obtained when the sample was cooled in the same field (Meissner effect). A drop in magnetization is observed at ~115 K, in good agreement with the result shown in Fig. 1 (c). The diamagnetic shielding and Meissner signals at 5.0 K are estimated to be ~11 % and ~24 % of those for a perfect diamagnet of the same shape and volume, respectively.

A magnetization curve at 5.0 K for a sample comprised of the low-T_c phase in fields of up to ±2 kOe is shown in Fig. 6. The critical current density of the film was determined from the magnetization hysteresis loop using the critical state model.[9] The film with the highest zero-resistivity temperature exhibited critical current densities of ~5.0×10^6 A/cm^2 at 5.0 K and ~1.2×10^6 A/cm^2 at 50 K in a field of 300 Oe, and extrapolates to ~5.5×10^6 A/cm^2 at ~5.0 K and ~1.5×10^6 A/cm^2 at 50 K in zero field. The critical current densities of a highly oriented Bi-Ca-Sr-Cu-O thin film were ~1.1×10^6 A/cm^2 at 4.5 K and ~2.0×10^5 A/cm^2 at 40 K,[7] while the corresponding values of the randomly oriented Tl-Ca-Ba-Cu-O thin film were ~3.5×10^6 A/cm^2 at 5.0 K and ~8.0×10^5 A/cm^2 at 50 K. Both the films were comprised of the low-T_c phase.

CONCLUSION

Superconducting Tl-Ca-Ba-Cu-O thin films prepared by rf magnetron sputtering from sintered targets with compositions of $Tl_2Ca_2Ba_2Cu_3O_x$, $Tl_3Ca_2Ba_2Cu_3O_x$ and $Tl_2Ca_3Ba_1Cu_3O_x$, exhibited zero resistivity at up to 116 K. The film deposited from the $Tl_2Ca_2Ba_2Cu_3O_x$ target displayed the highest zero-resistivity temperature, and the critical current densities of ~5.5×10^6 A/cm^2 at 5.0 K and ~1.5×10^6 A/cm^2 at 50 K. The critical current densities of a randomly oriented Tl-Ca-Ba-Cu-O film comprised of the low-T_c phase were larger than those of a highly oriented Bi-Ca-Sr-Cu-O film comprised of the corresponding phase. The critical current densities of the Bi-Ca-Sr-Cu-O and Tl-Ca-Ba-Cu-O films, as well as $Ln_1Ba_2Cu_3O_x$ films, exceed those of bulk samples by a factor of 100~1000. The high critical current densities of the films have been thought to be due to the highly preferred orientation. The result in this study indicates, however, that they are explained by the surface pinning effect. The optimization of annealing conditions to form single-phase films comprised of the high-T_c phase is in progress.

REFERENCES

1. J. G. Bednorz and K. A. Müller, Z. Phys. B 64, 189 (1986).
2. M. K. Wu, J. R. Ashburn, C. J. Torng, P. H. Hor, R. L. Meng, L. Gao, Z. J. Huang, Y. Q. Wang, and C. W. Chu, Phys. Rev. Lett. 58, 908 (1987).
3. H. Maeda, Y. Tanaka, M. Fukutomi and T. Asano, Jpn. J. Appl. Phys. 27, L209 (1988).
4. Z. Z. Sheng and A. M. Hermann, Nature 332, 138 (1988).
5. M. Nakao, H. Kuwahara, R. Yuasa, H. Mukaida and A. Mizukami, Jpn. J. Appl. Phys. 27, L378 (1988).
6. M. Nakao, R. Yuasa, M. Nemoto, H. Kuwahara, H. Mukaida and A. Mizukami, Jpn. J. Appl. Phys. 27, L849 (1988).
7. H. Kuwahara, M. Nakao and A. Mizukami, '88 MRS International Meeting on Advanced Materials Symposium D, Proceedings (to be published).
8. R. M. Hazen, L. W. Finger, R. J. Angel, C. T. Prewitt, N. L. Ross, C. G. Hadidiacos, P. J. Heaney, D. R. Veblen, Z. Z. Sheng, A. El Ali and A. M. Hermann, Phys. Rev. Lett. 60, 1657 (1988).
9. C. P. Bean, Phys. Rev. Lett. 8, 250 (1962).

Preparation and Electrical Properties of 107K-BiSrCaCuO Superconducting Thin Films

MASASHI MUKAIDA, KEN-ICHI KURODA, YASUO TAZOH, and SHINTARO MIYAZAWA

NTT LSI Laboratories, 3-1, Morinosato Wakamiya, Atsugi, 243-01 Japan

ABSTRACT

BiSrCaCuO superconducting oxide thin films with a zero-resistance temperature of 107K were prepared on MgO(100) substrates by the sequential deposition technique of Bi, SrF_2, CaF_2 and Cu, followed by annealing in O_2. The temperature dependence of the Hall coefficient and resistivity were measured. The Hall coefficient increased as the temperature decreased. The carrier in the film was hole-like. The temperature dependence of the inverse Hall coefficient was linear and the extrapolation to zero degrees passed through the origin. In addition, the extrapolation of the normal resistance to zero degrees always intersected the resistance axis on the negative side.

1. INTRODUCTION

Following the discovery of the high T_C superconducting BiSrCaCuO system by Maeda et al. [1], reserch activities on the preparation of this system with a zero-resistance temperature above 100K have expanded. The BiSrCaCuO system has three superconducting phases: a 20K-T_C phase $Bi_2Sr_2Cu_1O_x$ (abbreviated 2201 hereafter) [2,3], a 80K-T_C phase (2212) [4] and a 110K-T_C phase (2223) [5]. A great deal of effort has been focused on the preparation of films with a T_C of around 110K by such methods as sequential deposition, coevaporation, sputtering, chemical vapor deposition and laser deposition. Reports have appeared on the resistivity and critical magnetic field of the 110K phase. However so far, the electrical properties, such as the carrier concentration and/or Hall coefficient have only been measured for the 80K phase [6], not the 110K phase.

In this paper, the composition necessary to obtain a film with a T_C above 100K was investigated and the nominal composition ratio exhibiting a zero-resistance temperature above 100K was determined. To characterize the prepared films, the resistivity, magnetic susceptibility and Hall coefficient were measured as a function of the temperature.

2. EXPERIMENTAL

A 10-KW single electron gun with four hearths was used. The raw materials Bi, Cu, SrF_2 and CaF_2 were sequentially deposited to a total thickness of about 200nm on MgO(100) substrates at a vacuum of 10^{-4} Pa. During the deposition, the substrates were at the ambient temperature. SrF_2 and CaF_2 were used as the sources of Sr and Ca, because these fluorides are very stable even after deposition and they are also easily evaporated by an electron beam. On the other hand, films deposited by using Sr and Ca metal as sources rapidly corrod in air and change into oxides, hydroxides or carbonates. In this study, the deposited films were annealed after deposition at approximately 875°C for about an hour in O_2. More details about the annealing processes have been reported elsewhere [7,8].

The nominal film composition of cations ranged from 1112 to 1312 in the order Bi-Sr-Ca-Cu. The chemical composition of the as-deposited and annealed films was determined by X-ray fluorescence spectroscopy analysis (XFA), which was calibrated using inductive coupled plasma spectorscopy (ICP). No deviation in composition and no fluorine were detected in the films after annealing.

The electrical resistance was measured as a function of temperature with the standard four-probe pressure contact technique. The structure and crystallinity of the films were studied by X-ray diffraction (XRD) with $Cu_{K\alpha}$ radiation. The magnetic susceptibility of the films was measured with a SQUID magnetometer under an magnetic field of 100Oe. The Hall coefficient of the film was measured at various temperatures by the conventional van der Pauw method with Ag-contacts under an applied magnetic field of 5000Oe.

3. RESULTS AND DISCUSSION

3.1 CRYSTALLINITY OF THE FILMS

A scanning electron micrograph of an annealed film exhibiting multi-transition is shown in Fig. 1 (a). The annealed film is composed of various types of crystallites, namely very thin sheets (A) which occupy a large area of the film, platelets (B), needles (C) and bulky grains (D). The chemical composition of some of the crystals in the films was determined by electron probe microanalysis (EPMA) and micro-probe Auger electron spectroscopy (μAES) analysis. The EPMA and μAES analyses revealed that the needle-shaped crystals were composed of only Ca and Cu. On the other hand, the other crystals contained all 4 elements. When a film was deposited to a thickness of 500nm of more, many platelets, which a micro-probe X-ray deffraction (μXRD) pattern showed that of 80K phase, and bulky grains appeared in the annealed film. These films over 500nm thick exhibited no transition at 107K but only at 80K. The very thin sheets adhering to the substrate surface shown in Fig. 1 (b) were always found in films exhibiting a T_C of 107K. So, the very thin sheets are probably the high-T_C phase. The thickness of the very thin sheets was about 200-500Å as determined by scanning electron microscopy (SEM), while the size of the bulky grains was more than 1 μm. Therefore, it is likely that the fraction of high-T_C phase in the films is small.

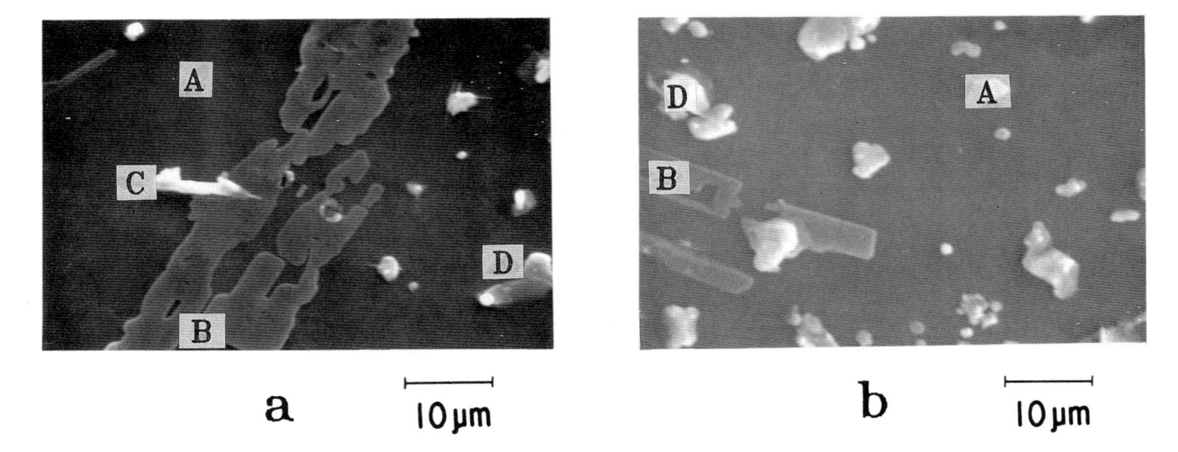

Fig. 1 (a) Scanning electron micrograph of the film. There are very thin sheets (A), platelets (B), needles (C) and bulky grains (D). (b) Micrograph of a film with a zero-resistance temperature at 107K. The bar in the figure indicates 10μm.

Also, there were no needle-shaped crystals in these films, as can be seen in Fig. 1 (b).

A typical X-ray diffraction pattern (XRD) of a film with a T_C of 107K is shown in Fig. 2. Each major peak can be assigned to either c-axis textured $Bi_2Sr_2Ca_1Cu_2O_x$ (low-T_C phase), $Bi_2Sr_2Ca_2Cu_3O_x$ (high-T_C phase) or $Bi_2Sr_2Cu_1O_x$, but the very weak peaks are unidentified. A broad peak around 17° is maybe from an amorphous like substance. From the (002) XRD peak at 2θ=4.8°, the lattice constant of the film was estimated to be about 37Å and odd reflections were missing. The lattice constant of the low-T_C phase was determined from the XRD peaks to be 30.61Å. The (00L) peaks showed that both the 110K phase and the 80K phase were highly oriented.

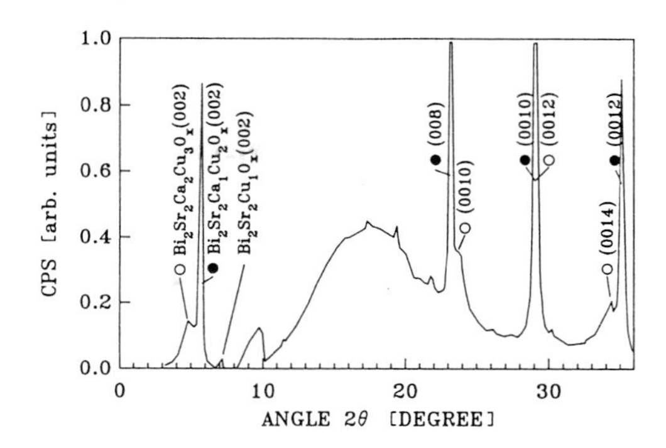

Fig. 2 X-ray diffraction pattern of a film with a T_C of 107K. The XRD peaks correspond to $Bi_2Sr_2Ca_2Cu_3O_x$ (○:high-T_C phase) and $Bi_2Sr_2Ca_1Cu_2O_x$ (●:low-T_C phase). Both phases are highly oriented along the c-axis perpendicular to the (001) substrate.

3.2 COMPOSITION RANGE FOR 110K

A compositional diagram for the transition temperature is shown in Fig. 3. The Bi to Cu ratio was fixed at roughly 1 to 2 in the diagram. In the figure, solid circles, open triangles, open squares and crosses represent a critical temperature above 100K (high-T_C phase), a critical temperature at around 80K (low-T_C phase), multi-transition at around 110K and around 80K (multi-phase) and insulator, respectively. It was found that the compositional ratio is as important in preparation of thin films with a T_C above 100K as the annealing temperature. It can clearly be seen that the compositional ratio Ca/(Cu+Bi) for films with a T_C above 100K was around 1/3, while Sr ranged to some extent. This means that Ca plays a significant role in high-T_C BiSrCaCuO thin films. In fact, the T_C of $Bi_2Sr_2Cu_1O_x$ film, which has no Ca, was reported to be as low as 20K [2,3].

3.3 RESISTIVITY MEASUREMENTS

Figure 4 shows the typical temperature dependence of the resistance for the films indicated by solid circles in Fig. 3. The nominal compositions were 1212, 1(2.5)12 and 1312 for the order Bi-Sr-Ca-Cu. All the films showed a zero-resistance temperature at 107K, and above this transition temperature, the resistance curve is quite linear. The linear characteristics of the new high-T_C oxide superconductors have been discussed in detail elsewhere [9]. It is worth noting that the extrapolation of the normal resistance to zero degrees was always negative, while the extrapolation of that of

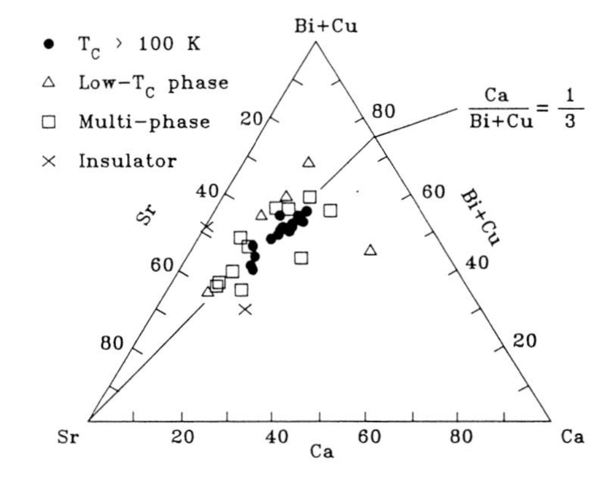

Fig. 3 (Cu+Bi)-Sr-Ca compositional diagram showing the compositions studied. The Bi to Cu ratio was fixed at 1 to 2. The line corresponds to Ca/(Bi+Cu)=1/3.

multi-phase superconducting films to zero degrees was positive [7]. In regard to other high-quality oxide superconducting single crystals and films, a zero or negative extrapolation seems to be characteristic of two-dimensional, highly oriented materials [10,11,12]. It is difficulte to calculate resistivity, for its roughness. Then, we will discuss resistance. The normalized R(300K)/R(150K) for 1212, 1(2.5)12 and 1312 films was 2.1, in spite of their compositional differences. This means that the resistance is probably related to the same material in these films.

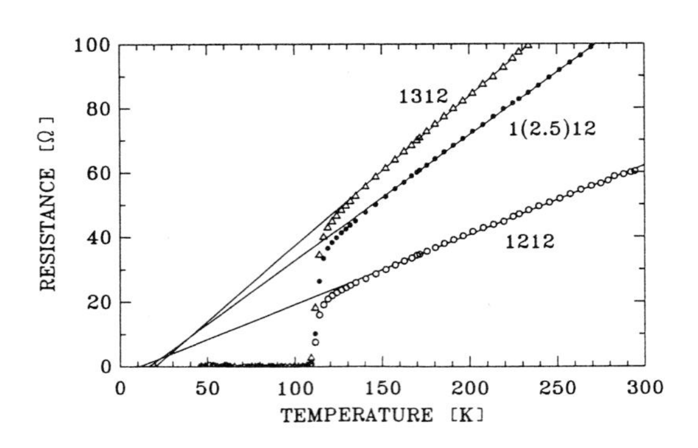

Fig. 4 Temperature dependence of resistance for BiSrCaCuO thin films with a nominal composition of 1212, 1(2.5)12 and 1312. Note that the extrapolation of the normal resistance to zero degrees is negative.

3.4 MAGNETIC SUSCEPTIBILITY

The magnetic susceptibility of a film exhibiting a T_C of 107K was measured from 20K to 150K, as shown in Fig. 5. As indicated in the insert, the magnetic susceptibility showed a superconducting transition at around 110K. This implies the existence of a superconductor with a T_C at around 110K. The slope of the curve also changed at around 80K showing the existence of the 80K phase.

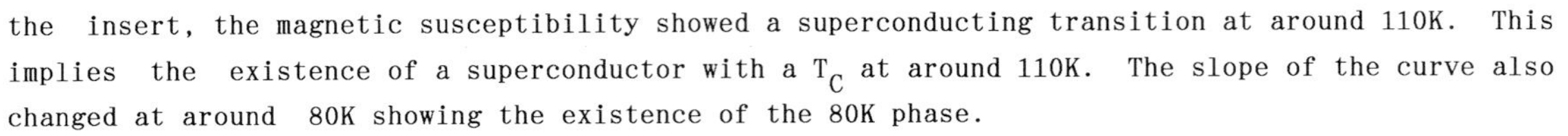

In comparison to the low-T_C phase, the volume fraction of 107K superconductor, as estimated from the magnetic susceptibility, was very small, probably only 10% or less. However, the 110K phase is most likely contained in the thin sheets, which are probably thinner than the penetration depth. In addition, the H_{C1} of the 110K phase was very small, only 60e [13], in comparison to the applied magnetic field (100Oe) used in the measurement; so the volume fraction of the 110K phase was probably underestimated.

3.5 HALL MEASUREMENTS

The film shows a hole-type Hall effect similar to the $La_{2-x}Sr_xCuO_4$ and $YBa_2Cu_3O_x$ systems [14,15]. The Hall coefficient of the film with a zero-resistance temperature at 107K showed a strong temperature dependence, in contrast to that of 80K phase single crystal [6], as shown in Fig. 6. From the temperature dependences, it is clear that the prepared film has characteristics which are quite different from the 80K phase.

The measured Hall coefficient is thought to depend mainly on the Hall coefficient of the very thin sheets. This can be explained as follows: if the film is composed of two phases, one of them is very thin but covers a large area and other is very thick, like the toll island shown in Fig. 1. The volume of

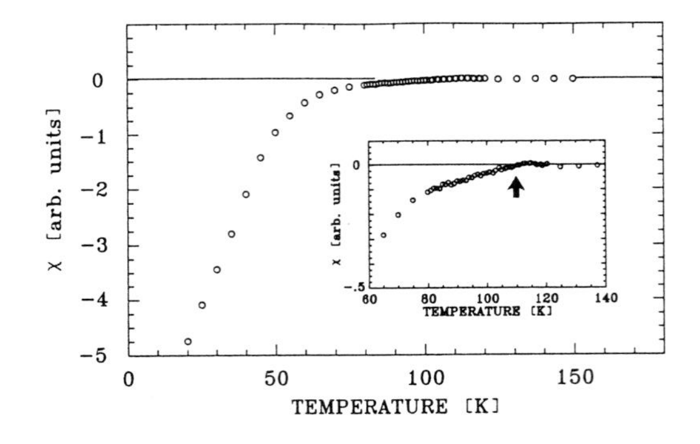

Fig. 5 Temperature dependence of magnetic susceptibility for a BiSrCaCuO thin film with a zero-resistance temperature at 107K. The insert shows a detail around 100K.

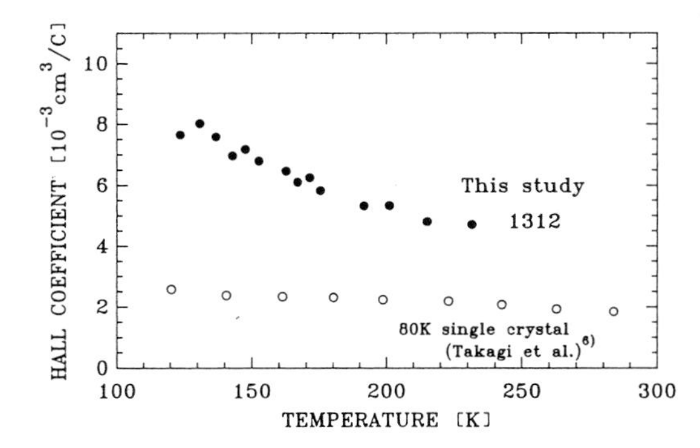

Fig. 6 Temperature dependences of the Hall coefficient for a film with a T_C of 107K and for 80K phase single crystal[6] for comparison (○).

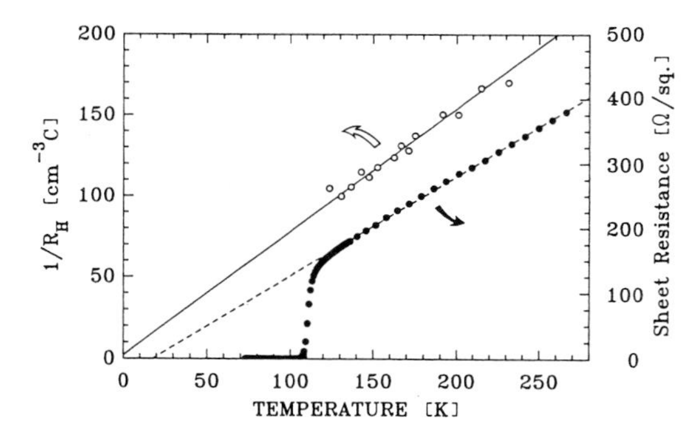

Fig. 7 Temperature dependence of the invese Hall coefficient ($1/R_H$) for a BiSrCaCuO thin film with a nominal composition of 1312. The extrapolation of the $1/R_H$ almost passes through the origin.

the thick part is greater than that of the thin part. The measured Hall voltage is the sum of the voltage from the thin parts and that from the thick part. The voltage depends on the applied magnetic field, Hall coefficient of the material, the current density in the material and the width of the material which the current passes through. Since the voltage is independent of the volume, but is dependent of the width, the measured Hall coefficient mainly depends on the part covering a large area. Therefore, the main contribution to the measured Hall coefficient must come from the very thin sheets.

Figure 7 shows the temperature dependence of the inverse Hall coefficient ($1/R_H$) of the film. It was very interesting to note that the temperature dependence of the inverse Hall coefficient was linear and the line to zero degrees almost passed through the origin. The same temperature dependence has also been obtained in $YBa_2Cu_3O_x$ ceramics [15,16]. This is probably one of the unique characteristics of two-dimensional high-T_C oxide superconductors, but the physical meaning is still unclear.

Based on the single-band model, for which the carrier concentration is determined from $1/(R_He)$, the carrier concentration derived from Fig. 7 decreases as the temperature decreases. However, this is inconsistent with the metallic temperature dependence of the resistance. So, to explain the temperature dependence of the carrier concentration, a multi-band model and/or another model is required.

SUMMARY

BiSrCaCuO superconducting thin films with a T_C of 107K have been successfully and reproducibly synthesized by the sequential deposition of Bi, SrF_2, CaF_2 and Cu. The film composition necessary to obtain a film with a T_C above 100K was in the region between 1112 and 1312 for the order Bi-Sr-Ca-Cu. The extrapolation of the normal resistance to zero degrees was always negative only for the films with zero resistance at 107K. The existence of a high T_C phase was confirmed from XRD peaks and magnetic susceptibility measurements. The Hall coefficient of these films was measured 80K phase single crystal reported so far. It was also found that the temperature dependence of the inverse Hall coefficient ($1/R_H$) was linear and, when extrapolated to zero degrees, almost passed through the origin.

ACKNOWLEDGEMENTS

The authors would like to thank M. Oda for the magnetic susceptibility measurements. They are also grateful to M. Yamamoto, K. Hohkawa, A. Matsuda, and A. Ishida for their useful discussions, comments and encouragement.

REFERENCES

[1] H. Maeda, Y. Tanaka, M. Fujitomi, and T. Asano, Jpn. J. Appl. Phys.**27**(1988)L209

[2] C. Michel H. Hervieu, M. M. Borel, A. Grandin, F. Deslandes, J. Provost and B. Raveau, Z. Phys. **B68**(1987)421

[3] J. Akimitsu A. Yamazaki, H. Sawa and H. Fujiki, Jpn. J. Appl. Phys. **26**(1987)L2080.

[4] E. Takayama-Muromachi, Y. Uchida, A. Ono, F. Izumi, M. Onoda, Y. Matsui, K. Kosuda, S. Takekawa and K. Kato, Jpn. J. Appl. Phys. **27**(1988)L556

[5] Y. Matsui, H. Maeda, Y. Tanaka and S. Horiuchi, Jpn. J. Appl. Phys. **27**(1988)L361

[6] H. Takagi, H. Eisaki, S. Uchida, A. Maeda, S. Tajima, K. Uchinokura and S. Tanaka, Nature. Vol. **332**. 24. March(1988)334

[7] K. Kuroda, M. Mukaida, M. Yamamoto, and S. Miyazawa, Jpn. J. Appl. Phys. **27**(1988)L625

[8] K. Kuroda, M. Mukaida and S. Miyazawa, Ext. Abst. 20th Conf. Solid State Devices and Materials, Tokyo(1988)431

[9] R. Micnas and J. Ranninger, Phys. Rev. Rapid Comm. Vol. **36**(1987)4051

[10] M. Naito, R. H. Hammond, B. Oh, M. R. Han, J. W. P. Hsu, P. Rosenthal, A. F. Marshall, M. R. Beasley, T. H. Geballe and A. Kapitulnik, J. Material Res. **2**(1987)713

[10] P. M. Mankiewich, R. E. Howard, W. J. Skocpol, A. H. Dayem, A. Ourmazd, M. G. Young and E. Good, Mat. Res. Soc. Symp. Proc. Vol **99**(1988)119

[12] S. W. Tozer, A. W. Kleinsasser, T. Penney, D. Kaiser and F. Holtzberg, Phys. Rev. Lett. **59**(1987)1768

[13] N. Murayama et al., Jpn. J. Appl. Phys. (to be published)

[14] M. Suzuki, and T. Murakami, Jpn. J. Appl. Phys. **27**(1987)L524

[15] S-W. Cheong, S. E. Brown, Z. Fisk, R. S. Kwok, J. D. Thompson,E. Zirngiebl, G. Gruner, D. E. Peterson, G. L. Wells, R. B. Schwarz, and J. R. Cooper, Phys. Rev. **B36**(1987)3913

[16] K. Char, Mark Lee, R. W. Barton, A. F. Marshall, I. Bozovic, R. H. Hammond, M. R. Beasley, T. H. Geballe, A. Kapitulnik and S. S. Landrman, Phys. Rev. Rapid Comm. Vol. **38**(1988)834

Characteristics in the Growth of the $YBa_2Cu_3O_x$ and $YBaSrCu_3O_x$ Crystalline Films

MASASHI YOSHIDA

Advanced Technology Research Laboratories, Sumitomo Metal Industries,
1-3, Nishinagasu Hondori, Amagasaki, Hyogo, 660 Japan

ABSTRACT

The change of the structure of the sputter-deposited amorphous films of $YBa_2Cu_3O_x$ and $YBaSrCu_3O_x$ by the thermal annealing was investigated using the SEM and the X-ray diffraction spectrometry. The $YBa_2Cu_3O_x$ films crystallized in the distorted perovskite structure when annealed above 920°C while the $YBaSrCu_3O_x$ above 980°C. The merging of (200), (020) and (006) lines was observed in the $YBa_2Cu_3O_x$ annealed at 800°C or in the $YBaSrCu_3O_x$ annealed between 800°C and 940°C. The existence of the antistructure defects between Y and Ba (Sr) was pointed out in these films. It was also found that by the annealing of the films of $YBaSrCu_3O_x$, the $YBa_2Cu_3O_x$ crystals grew preferentially below 900C while the $YBaSrCu_3O_x$ crystals grew above 900°C.

INTRODUCTION

The recent discovery of the superconductivity in $YBa_2Cu_3O_x$ with distorted perovskite structure [1] has generated tremendous interest in the compounds with the same crystal structure. Among the initial studies there are many involving substitution of elements on the Y and Ba sites. [2-5] It has been revealed that the change of superconducting transformation temperature (Tc) rarely occurs by the substitution of elements on the Y site. [2,3] On the other hand, when the Ba atom is substituted with Sr, the Tc decreases from 95K ($YBa_2Cu_3O_x$) to 83K ($YBaSrCu_3O_x$). [4,5]

It has been clarified that the structure transformation occurs in these compounds from the tetragonal to the orthorhombic phases with the transition temperature (Ts) around 600°C. [6] The low temperature orthorhombic phase is the superconductor with Tc around 90K, while the high temperature tetragonal phase is a semiconductor. It is also shown that the Tc of the orthorhombic phase depends critically on the annealing procedure below Ts. [7,8]

On the other hand, in the experiments of thin films of $YBa_2Cu_3O_x$, it has been revealed that the Tc of films depends on the maximum annealing temperature as well as the annealing procedure below Ts. For example, the films annealed at 940C have Tc of 84K while the films annealed at 920°C have a Tc as low as 36K. [9]

In the present experiments, we investigated the change of the structure and the electric transport property of films of $YBa_2Cu_3O_x$ and $YBaSrCu_3O_x$ annealed between 800°C and 980°C in order to clarify the origin of the dependence of the Tc on the maximum annealing temperature.

EXPERIMENTAL

Thin films of $YBa_2Cu_3O_x$ and $YBaSrCu_3O_x$ were prepared by the diode sputtering method. The targets were fabricated by mixing stoichiometric amounts of $BaCO_3$, $SrCO_3$, Y_2O_3 and CuO powders, and heating in air at 900C for 20 hours, and the substrates MgO (100) planes were used. Sputtering was carried out by argon ions. The substrates were kept at room temperature during the deposition. The deposited films were annealed in an O_2 environment at a fixed temperature for an hour and furnace cooled to 800°C and, then, cooled down to room temperature at the rate of 1°C/min. The structure of the films was investigated by X-ray diffraction spectrometry using Rigaku RU-500 with a Co radiation source. The electric resistance measurements of the films were done by a four-point probe technique with indium current and voltage contacts.

RESULTS AND DISCUSSION

In Fig. 1 are shown X-ray diffraction spectra of the as-deposited film, the films annealed at 800°C, 920°C and 980°C, respectively, of $YBa_2Cu_3O_x$. The as-deposited film was found to be amorphous. It was confirmed by EPMA analysis that the as-deposited film had the same composition of metals as the sputtering target. In the spectrum of the films annealed at 800°C, the diffraction lines from the $YBa_2Cu_3O_x$ crystals appear as well as the lines from CuO, $BaCO_3$ and Y_2BaCuO_5. The intensity of the lines from $YBa_2Cu_3O_x$ increases with increasing temperature in comparison with other lines. In the spectrum of the film annealed at 920°C, the population of the intensity of lines from the $YBa_2Cu_3O_x$ agrees with that of the bulk sintered samples [6], indicating a growth of polycrystalline with random orientations. On the other hand, in the spectrum of the film annealed at 980°C, the intensities of (00n) lines (n=3-6) increase, which indicates the oriented growth of the crystals with the c-axis perpendicular to the plane of the substrate. The oriented growth of $YBa_2Cu_3O_x$ on the MgO (100) plane above 980°C agrees with the results reported in ref. 9.

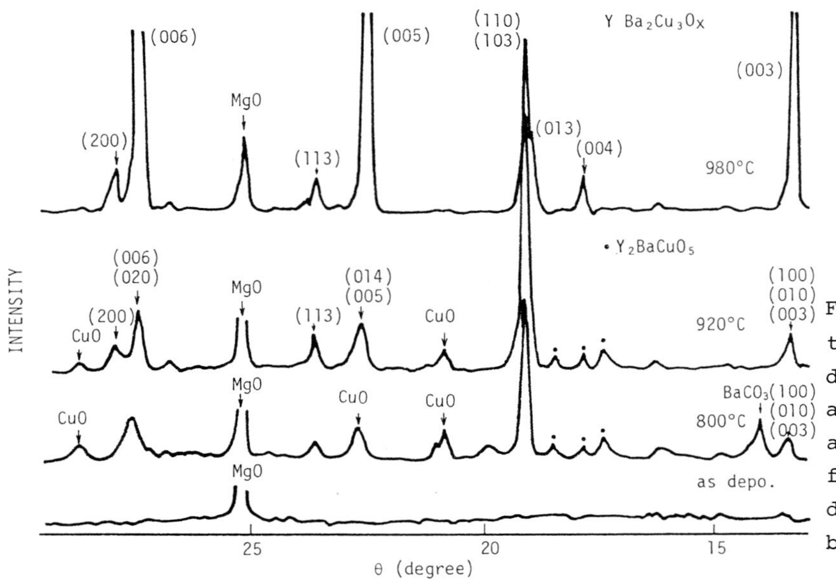

Fig. 1 X-ray diffraction spectra of the as-deposited film, films annealed at 800°C, 920°C and 980°C of $YBa_2Cu_3O_x$ for an hour and cooled down to room temperature by the rate of 1°C/min.

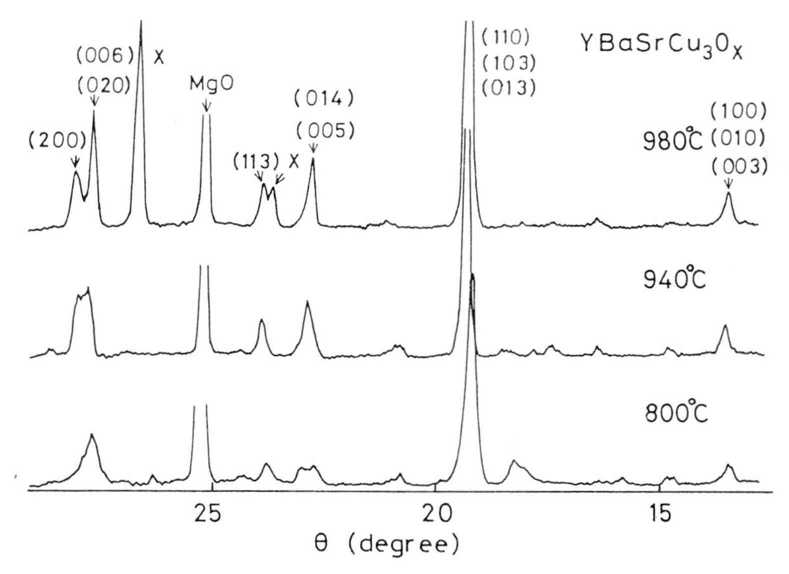

Fig. 2 X-ray diffraction spectra of the $YBaSrCu_3O_x$ films annealed at 800°C, 940°C and 980°C for an hour and cooled down to room temperature by the rate of 1°C/min.

In Fig. 2 are shown X-ray diffraction spectra of the $YBaSrCu_3O_x$ films annealed at 800°C, 940°C and 980°C, respectively. In the $YBaSrCu_3O_x$ films, a diffraction line appears at around 19° by the annealing above 800°C, which corresponds to the unresolved (103), (013) and (110) diffraction lines of the perovskite structure. It should be noticed that these lines shift to the higher angle side with increasing the annealing temperature. The origin of this peak shift will be discussed later. In the $YBaSrCu_3O_x$ film, the oriented growth does not occur by the annealing at 980°C being different from the case of $YBa_2Cu_3O_x$. On the other hand, unknown lines labeled by X appear in the $YBaSrCu_3O_x$ films by the annealing at 980°C.

In Figs. 1 and 2, it should be noticed that in the films annealed at 800°C, the splitting of (200), (020) and (006) does not occur but one peak appears at around 27°. We investigated the change of the three axes of $YBa_2Cu_3O_x$ and $YBaSrCu_3O_x$ crystals with different annealing temperature. In Figs. 3 (a) and (b) are shown the X-ray diffraction spectra of the $YBa_2Cu_3O_x$ and $YBaSrCu_3O_x$ films, respectively, with diffraction angle between 25 and 29 degrees. In the spectrum of the $YBa_2Cu_3O_x$ film annealed at 800°C, no splitting of three lines is seen but one peak appears at 27.6°. When annealed at 850C, peaks appear at 27.4° and 27.9° in addition to the peak at 27.6°. The peak at 27.6° disappears when annealed above 940°C. The lattice constants of the films annealed at 940°C are obtained as a=3.82A and b=c/3=3.88A, assuming that the peak at 27.4° is the unresolved (020) and (006) diffraction peaks. These values agree with those of the bulk sintered samples. [10]

A similar result was obtained also in $YBaSrCu_3O_x$. In the spectrum of the film of $YBaSrCu_3O_x$ annealed at 800C, the (200), (020) and (006) lines do not split but one peak appears at 27.7°. It should be noted that the splitting of lines does not occur even at 940°C in the $YBaSrCu_3O_x$ films, differing from the case of $YBa_2Cu_3O_x$. On the other hand, in the spectrum of the film annealed at 980°C, the (200) line resolves from the (020) or (006) lines. The lattice constants of the $YBaSrCu_3O_x$ annealed at 980°C are obtained as a=3.79 and b=c/3=3.84A, which agree with the results in ref. 4

The merging of the (200), (020) and (006) diffraction lines of the $YBa_2Cu_3O_x$ film annealed at 800℃ or the $YBaSrCu_3O_x$ films annealed between 800℃ and 940℃ is supposed to be caused by the existence of defects in these films because of the relatively low annealing temperature. It has been reported that the lattice constants of the $YBa_2Cu_3O_x$ depend on the oxygen content x, which depends critically on the cooling rate of the samples around the tetragonal-orthorhombic phase transition temperature. [6] However, the merging of the diffraction lines cannot be explained by the oxygen deficiency caused by the rapid cooling because the $YBa_2Cu_3O_x$ films annealed above 940C or the $YBaSrCu_3O_x$ annealed at 980℃ have the orthorhombic structure with the proper lattice constants by the cooling with the rate of 1℃/min.

It has been pointed out that there exist three types of defects in the compound crystals; vacancy, interstitial and antistructure. [11] The vacancy and the interstitial, however, cannot explain the results shown in Fig. 3, since, if there exist many vacancies or interstitials, the lattice distortion is expected, which causes the broadening of the X-ray diffraction lines rather than the merging of lines. On the other hand, the merging of the (200), (020) and (006) lines may be explained by the antistructure disorder as follows. It should be noticed that in the high temperature tetragonal phase of the $YBa_2Cu_3O_x$, the sequence of atoms -Y-Ba-Ba- exists along the c-axis, [10] which distinguish the c-axis from the other axes. By the phase transition from the tetragonal to the orthorhombic structure, a- and b-axes are distinguished from each other by the inclusion of oxygen along the b-axis but no change occurs along the c-axis except a small correction for the lattice constant. [10] If there exist many antistructure defects between Ba and Y atoms, the triple layerd structure cannot be obtained and the difference between the c-axis and the other two axes is obscured. In such cases the formation of a- and b-axes in the orthorhombic

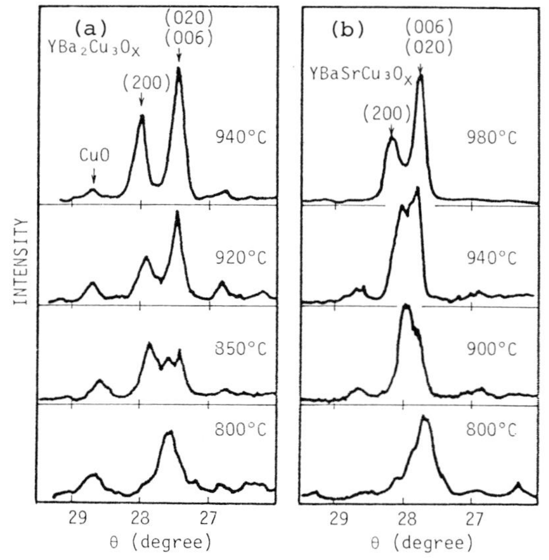

Fig. 3 X-ray diffraction spectra of $YBa_2Cu_3O_x$ (a) and $YBaSrCu_3O_x$ (b) with diffraction angle between 25 and 29.

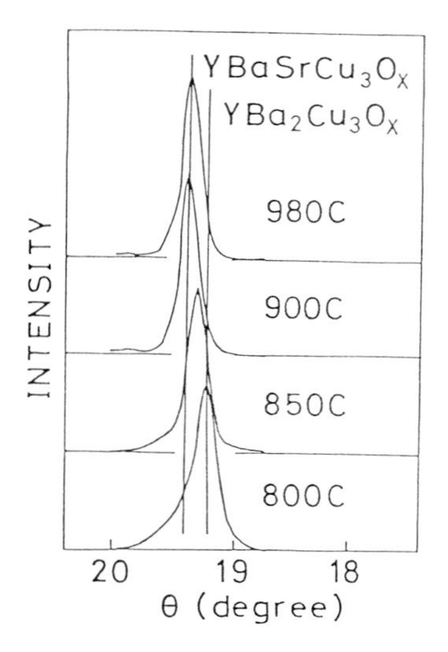

Fig. 4 X-ray diffraction spectra of $YBaSrCu_3O_x$ with diffraction angle around 19.

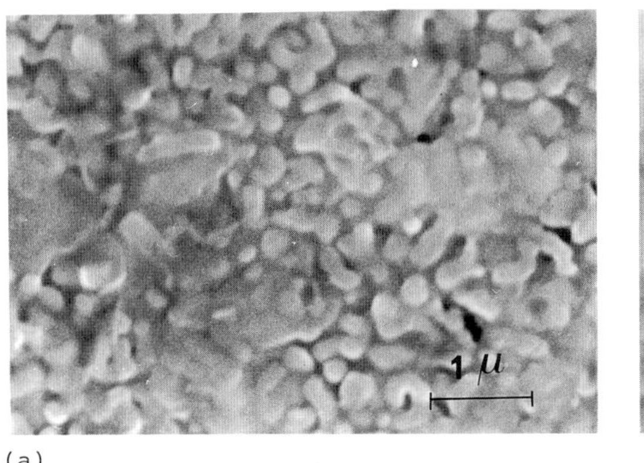

(a) (b)

Fig. 5 SEM images of the $YBaSrCu_3O_x$ films annealed at 850°C (a) and 900°C (b).

phase cannot be realized. Thus, the three axes become identical and the splitting of (200), (020) and (006) lines does not occur. The fact that the merging of three lines occurs in the $YBa_2Cu_3O_x$ annealed at 800°C or in the $YBaSrCu_3O_x$ annealed between 800°C and 940°C suggests that Y and/or Ba (Sr) atoms cannot migrate in the crystals at these temperatures so that the ordered arrangement of Y and Ba (Sr) along the c-axis cannot be achieved.

In Fig. 2 it was pointed out that the diffraction peak of the perovskite structure shift to the higher angle side with increasing the annealing temperature. In Fig. 4 are shown diffraction spectra of $YBaSrCu_3O_x$ with diffraction angle around 19°. In this figure, the unresolved three lines (110), (103) and (013) are seen. The peak at 19.3° in the film annealed at 800°C shifts to the higher angle side until 19.5° of the film annealed at 900°C. Above 900C, the position of lines does not change within the experimental uncertainty. The position of line of the $YBaSrCu_3O_x$ film annealed at 800°C nearly coincides with that of $YBa_2Cu_3O_x$. On the other hand, the position of line of the film annealed at 900°C or 980°C coincides with that of the sintered bulk $YBaSrCu_3O_x$. [4] This fact indicates that by annealing at 800°C, growth of $YBa_2Cu_3O_x$ rather than $YBaSrCu_3O_x$ occurs, which suggests that the potential barrier of the reaction of Sr with Y, Cu and O to form the compound is higher than that of Ba. In Fig. 5 (a) and (b) are shown SEM images of the $YBaSrCu_3O_x$ films annealed at 850°C and 900°C, respectively. In Fig. 5 (a) it is found that amorphous regions remain around the crystals with the dimensions of several hundreds nm. We suppose that the unreacted Sr atoms remain in these amorphous regions. On the other hand, in the film annealed at 900°C, no such amorphous regions are seen.

In Fig. 6 is shown the change of the electric resistivity in the temperature range between 10K and 150K of the $YBa_2Cu_3O_x$ film annealed at 920°C and $YBaSrCu_3O_x$ annealed at 980°C. In both films, resistivity decreases below 60K and zero resistivity is obtained at around 20K. In Fig. 7 is shown the temperature dependence of resistivity of the $YBa_2Cu_3O_x$ films annealed at 940°C, 960°C and 980°C. It should be noticed that, in the $YBa_2Cu_3O_x$ films annealed above 940°C, resistivity decreases below 90K (Tc-onset=90K). These results concerned with the $YBa_2Cu_3O_x$ films agree with those in ref. 9 though the Tc's of the films obtained in the present experiments are a little lower than those in ref. 9. In the $YBa_2Cu_3O_x$ film annealed at 920°C or in the $YBaSrCu_3O_x$ annealed at 980°C, the splitting of (200), (020) and (006) is not satisfactory in comparison with

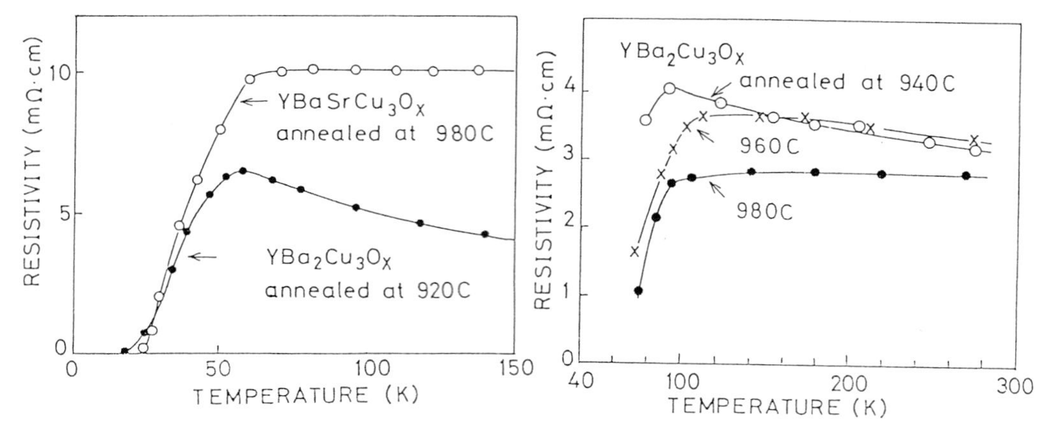

Fig.6 Temperature dependence of the resistivity of the $YBa_2Cu_3O_x$ annealed at 920°C and $YBaSrCu_3O_x$ annealed at 980°C.

Fig.7 Temperature dependence of the resistivity of the $YBa_2Cu_3O_x$ annealed at 940°C, 960°C and 980°C.

the $YBa_2Cu_3O_x$ film annealed above 940°C (see Fig. 3). Thus, it is considered that there remain still many defects in these films, which lower the Tc of these films. Since the temperature at which the splitting of the (200), (020) and (006) lines occurs is higher in the $YBaSrCu_3O_x$ than in the $YBa_2Cu_3O_x$, the higher temperature is needed to obtain the superconductor with Tc around 80K of $YBaSrCu_3O_x$ films.

References

[1] M. K. Wu, J. R. Ashburn, C. J. Torng, P. H. Hor, R. L. Meng, L. Gao, Z. J. Huang, Y. Q. Wang and C. W. Chu, Phys. Rev. Lett. 58, 908 (1987).

[2] H. Takagi, S. Uchida, H. Sato, H. Ishii, K. Kishio, K. Kitazawa, K. Fueki and S. Tanaka, Jpn. J. Appl. Phys. Lett. 26, L601 (1987).

[3] S. Kanbe, T. Hasegawa, M. Aoki, T. Nakamura, H. Koinuma, K. Kishio, K. Kitazawa, H. Takagi, S. Uchida, S. Tanaka and K. Fueki, Jpn. J. Appl. Phys. Lett. 26, L613 (1987).

[4] A. Ono, T. Tanaka, H. Nozaki and Y. Ishizawa, Jpn. J. Appl. Phys. Lett. 26 L1687 (1987).

[5] B. W. Veal, W. K. Kwok, A. Umezawa, G. W. Crabtree, J. D. Jorgensen, J. W. Downey, L. J. Nowicki, A. W. Mitchell, A. P. Paulikas and C. H. Sowers, Appl. Phys. Lett. 51, 279 (1987).

[6] E. T. Muromachi, Y. Uchida, K. Yukino, T. Tanaka and K. Kato, Jpn. J. Appl. Phys. Lett. 26, L665 (1987).

[7] E. T. Muromachi, Y. Uchida, M. Ishii, T. Tanaka and K. Kato, Jpn. J. Appl. Phys. Lett. 26, L1156 (1987).

[8] Y. Kubo, T. Yoshitake, J. Tabuchi, Y. Nakabayashi, A. Ochi, K. Utsumi, H. Igarashi and M. Yonezawa, Jpn. J. Appl. Phys. Lett. 26, L768 (1987).

[9] M. Komuro, Y. Kozono, Y. Yazawa, T. Ohno, M. Hanazono, S. Matsuda and Y. Sugita, Jpn. J. Appl. Phys. Lett. 26, L1907 (1987).

[10] E. T. Muromachi, Y. Uchida, Y. Matsui and K. Kato, Jpn. J. Appl. Phys. Lett. 26, L476 (1987).

[11] F. A. Kroger, The Chemistry of Imperfect Crystals, (Wiley, New York, 1984) p406.

Preparation and Characterization of Bi-Sr-Ca-Cu-O Thin Films by Ion Beam Sputtering

KAZUKI YOSHIMURA[1], HAJIME KUWAHARA[2], and SAKAE TANEMURA[1]

[1] Ceramic Science Department, Government Industrial Research Institute, Nagoya, 1, Hirate-cho, Kita-ku, Nagoya, 462 Japan
[2] Department of Development of Thin Film Application, Nisshin Electric Corp. Ltd., 47, Umezu, Takaku-cho, Ukyo-ku, Kyoto, 615 Japan

ABSTRACT

Thin films of the Bi-Pb-Sr-Ca-Cu-O system have been prepared by ion beam co-sputtering on MgO (100) with a sintered $Bi_2SrCaCu_2O_y$ and a metal Pb as targets. Post deposition *ex situ* annealing were required for obtaining Bi-Sr-Ca-Cu-O superconducting films. The characterization of the prepared films were carried out by measuring the temperature dependence of the resistivity, X-ray diffractometry, electron diffractometry, SEM observation and EPMA spectra. These results indicate that obtained film has highly [001] orientation with a lattice constant c=30.86 Å and consists of a low Tc likely single phase of Bi-Sr-Ca-Cu-O system. The optimum annealing temperature was 840°C in oxygen environment. The pre-sputtering of the substrate has not played any important role for the increasing of the volume fraction of the film with a preferential orientation.

INTRODUCTION

Following the recent discovery of a new type superconducting material, Bi-Sr-Ca-Cu-O system by Maeda *et al.* [1], extensive investigations have been done about this system. To apply this material to the superconducting devices, it is indispensable to establish the preparation technique of Bi-Sr-Ca-Cu-O thin films on any substrate. Some film preparation techniques such as rf magnetron sputtering, electron beam evaporation, screen printing and pyrolysis of organic acid salts have been known to be applicable [2-8].

As already reported [9] we successfully prepared the high quality Y-Ba-Cu-O thin films by ion beam sputtering (IBS). IBS has some superior points compared with conventional sputtering methods ; (1) IBS is available in high vacuum of 1×10^{-5} Torr which makes possible to prepare the films with less impurity, (2) the deposition rate of IBS is faster than conventional sputtering methods by several times, (3) IBS has possibility of fine control of oxygen composition by using a subsidiary oxygen ion source.

We have started series of investigations of Bi-Sr-Ca-Cu-O thin films by means of IBS to confirm the above general merits. In this report, results of the fundamental experiments by ion beam co-sputtering on MgO(100) with a sintered $Bi_2SrCaCu_2O_y$ and a metal Pb are presented as the first step for the further investigations by using a subsidiary oxygen ion source for the *in situ* preparation of superconducting films with or without low heating process. The deposited films were characterized by measuring the temperature dependence of the resistivity, X-ray diffractometry, electron diffractometry, SEM observation and EPMA analysis.

SAMPLE PREPARATION

Pb is well known as an assist material in synthesis of Bi compounds. Sunshine *et al.*[10] and Takano *et al.* [11] have recently reported that the addition of element Pb results in the preferential formation of the (2223)-phase with Tc of 110 K in a bulk case, and accordingly we have applied co-sputtering technique to deposition of Bi-Pb-Sr-Ca-Cu-O thin films. The targets used were sintered $Bi_2SrCaCu_2O_y$ and metal Pb.

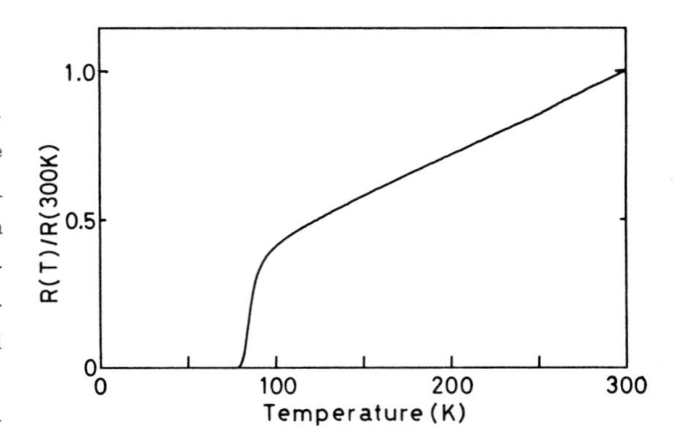

Fig. 1 Temperature dependence of resistivity of the sample annealed at 840°C

A MgO(100) single crystal was chosen as a substrate. As the morphology of sputtered films were strongly affected by the substrate surface condition, we used two different types of substrates : one is the substrate not surface-treated, the other is the substrate pre-sputtered by ion beam to etch strained surface layers (about 0.1 μm in thickness) of the substrate.

The applied conditions for co-sputtering were as follows: main ion source of 500 eV in accelerating voltage and 30 mA in current, vacuum of ion source about $5.5x10^{-5}$ Torr, vacuum around target and substrate being $1.1x10^{-5}$ Torr of O_2 gas, and substrate temperature of 300°C. The growth rate and the film thickness were about 50 Å/min and about 0.8 μm, respectively.

As the deposited film had a mirror-like surface colored yellow which indicated the little oxygen content, and insulating. Inductively coupled plasma-emission spectrometry (ICP) showed that the composition of as-sputtered film was Bi:Sr:Ca:Cu:Pb = 2.00:1.70:1.86:2.90:0.50. After deposition the samples were annealed in an *ex situ* manner at 800°C, 820°C, 840°C and 860°C, respectively, for 2 hours in a pure oxygen atmosphere. The increment and decrement of the temperature was about 100°C per an hour in every run. The film annealed at 860°C was visually transparent due to the evaporation of large amount of the film material during the annealing process.

TEMPERATURE DEPENDENCE OF THE RESISTIVITY

The resistivity measurement was done using a standard four prove method with Pt electrodes deposited on the film surface. The temperature dependence of resistivity was critically changed by the post-annealing temperature. Resistivity curve of the sample annealed at 800°C, 820°C and 860°C has two step structures, in which the resistivity drops around 80 K and 110 K and some resistivity remains even at 4 K for these samples. Two steps structures suggest the coexistence of the well known two superconducting phases in Bi-Sr-Ca-Cu-O system such as low Tc phase (Tc=80 K) and high Tc phase (Tc=110 K). As mentioned in the section of SEM observation later, some part of deposited film flied off through post-annealing treatment and the substrate MgO was exposed in some places. These surface morphology cause the breaking of superconducting path and hide their superconductivity.

On the other hand the sample annealed at 840° C designates superconductivity, as shown in Fig. 1. The on-set and end point in Tc was 91 K and 78 K, respectively. There is no structure around 110 K in resistivity curve, which implies that this film consist of a single phase Bi-Sr-Ca-Cu-O system.

X-RAY DIFFRACTOMETRY

Two-axis diffractometer collimated by a slit with 0.2 mm in width for the sample incidence and by a vertical Soller slit with 0.34° in slit angle for the scattered beam impinging on the graphite

(0002) monochrometer was used. The system is the parallel beam method with an incident angle of 86° in arc. Data was acquired with Cu Kα radiation at 0.02° in step width, one step for 2 seconds, over a 2θ angle from 2° to 65° with four repeats at room temperature (273 K).

The diffraction pattern of the as-deposited Bi-Pb-Sr-Ca-Cu-O films showed amorphous-like pattern where no characteristic peaks appeared. The diffraction pattern of the sample deposited on the not surface-treated substrate and annealed at 840°C which showed a superconducting behavior in dc-resistance is exemprified in Fig. 2. The principal index of the net planes and the interplaner d spacing for the observed respective peak as well as the intensity ratio I/I_0 are given in Table 1. For the assignment of the index, the results of the crystal structure analysis of the subcell of low temperature (2212)-phase of Bi-Sr-Ca-Cu-O system by other authors [12-16] were referred. Strong (00l) (l=2n, n=integer) reflections and very weak (115), (117) and (200) reflections clearly indicates that the most part of the present film being the preferentially oriented (00l) film. The averaged lattice constant c=30.86 Å is obtained from the d spacings of the major (00l) reflections with $I/I_0 > 0.18$ cited in Table 1. The value shows good agreement with the value determined by Sunshine *et al.* [10] for a single crystal of $Bi_{2.2}Sr_2Ca_{0.8}Cu_2O_8$. Intensity ratio from the reflection from (00 16), (00 18), (00 20) are different from in order comparing with the single crystal X-ray patterns [10,17]. This anomaly have not yet confirmed to be originated from either the small displacement of z location of Bi(1), Sr, Cu, O(1), O(2), and O(3) atoms in $Bi_2Sr_2CaCu_2O_8$ subcell proposed by Sunshine et al. [10],the oxygen occupancy, or the distortion associated with the super lattice.

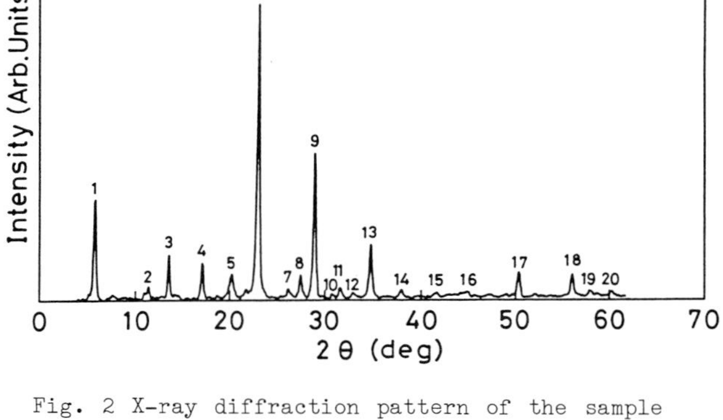

Fig. 2 X-ray diffraction pattern of the sample annealed at 840°C

Peak No.	(h	k	l)	2θ(°)	d(Å)	I/I_0
1	0	0	2	5.72	15.44	34
2	0	0	4	11.44	7.72	5
3 *				13.68	6.47	14
4	0	0	6	17.22	5.15	12
5 *				20.34	4.36	8
6	0	0	8	23.02	3.86	100
7 *				26.18	3.40	3
8	1	1	5	27.48	3.24	8
9	0	0	10	28.92	3.08	47
10	1	1	7	30.94	2.89	1
11 *				31.72	2.82	4
12	2	0	0	33.02	2.71	2
13	0	0	12	34.88	2.57	18
14 *				38.02	2.36	3
15	0	0	14	41.32	2,18	2
16	0	0	16	47.02	1.93	2
17 *				50.4	1.81	9
18	0	0	18	56.16	1.64	8
19 *				57.9	1.59	3
20	0	0	20	60.34	1.53	3

Notation I_0 means the maximum peak intensity for (0 0 8) after the background subtraction.

Asterisk stands for unidentified peaks.

Table 1 X-ray diffraction data of the sample annealed at 840°C.

The unidentified peaks are not assigned from the crystallines among (2223)-phase of Bi-Sr-Ca-Cu-O system [18-20], Bi_2Sr_2CuO [21-23], $PbSrCaCu_2O$ [24] and other possible impurities such as, $CaCO_3$, Bi_2O_3, BiO, CuO, Cu_2O, Ca_2CuO_3, Sr_2CuO_3, (CaSr)O and Ca_2PbO_4 [25]. Although Sunshine *et al.* [10] reported the weak peaks likely assigned as (00 2n+1) which violate pseudosymmetry Fmmm of the single crystal with the (2212)-phase, it is difficult for us to confirm that the present unidentified peaks listed in Table 1 which have a possibility to be assigned as (00 2n+1) have the similarity of their observation due to the limitation of the numbers of the observed peaks in the present experiments.

XRD of the samples deposited on the not surface-treated substrate and annealed at both 800°C and 820° C shows the preferentially oriented structure except the intensity ratio between (002) and (008) reflections being relatively small than in the previous case and that between (00 16) and (008) being slightly increasing comparative to the previous case. In the case of the XRD of the sample deposited on the non surface etched substrate and annealed at 860°C, both polycrystalline and c-axis orientation perpendicular to the substrate surface are observed. This implies that the *ex situ* heat treatment lower than 860° C is preferable for the epitaxial growth of the Bi-Sr-Ca-Cu-O system in the predent experiment.

The high Tc (2223)-phase was not clearly identified in XRD for those samples annealed at 800° C, 820°C and 860°C, respectively, although the temperature dependence of the resistivity and reflection electron diffraction pattern indicated the coexistence of high and low Tc phase of a Bi-Sr-Ca-Cu-O superconductor.

The diffraction pattern of the sample deposited on the surface etched substrate and annealed at the same temperature applied to the sample previously discussed shows the similarity to the pattern described above, although (200) reflection of MgO (Perciclase) is not obvious for both substrate with or without surface-dry-etching process. Accordingly, the surface-dry-etching process seems not play any important role for increasing the volume fraction of the film with c-axis orientation perpendicular to the substrate surface.

ELECTRON DIFFRACTOMETRY

Reflection diffraction pattern of the sample deposited on the non surface treated substrate and annealed at 840°C are exemplified in Fig. 3. The accelerated voltage applied was 200 KV. The fairly large area of the sample surface was illuminated by electron beams which was almost parallel to the sample surface because the sample was inserted after the condenser lens in the different manner from TEM mode. The streaky pattern parallel to crystal direction [001] and Kikuchi pattern indicates that the film is epitaxially grown on the substrate, a c-axis oriented crystallite is fairly large in size and the surface contributed to the present diffraction pattern is smooth on atomic scale. Electron diffraction pattern obtained by the other geometries of the setting of the above discussed sample shows the superimposing of the ring pattern and the pattern, as shown in Fig. 3. The ring pattern might be originated from the polycrystalline hills which are likely identified by the SEM image with small magnification.

Electron diffraction patterns obtained for the other samples which show the epitaxial growth evidence in XRD are essentially similar to those for the above discussed samples. Moreover, evidence of the coexistence of (2212) and (2223)phases of Bi-Sr-Ca-Cu-O system are observed.

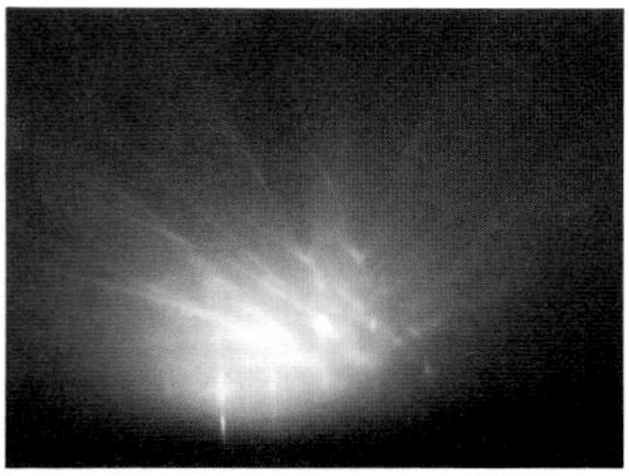

Fig. 3 Electron diffraction pattern of the sample annealed at 840°C

SEM OBSERVATION AND EPMA ANALYSIS

The surface of the sample annealed at 800°C, 820°C, and 860° C were rather rough and have less uniformity in composition.

Fig. 4 SEM image of the sample annealed at 800°C

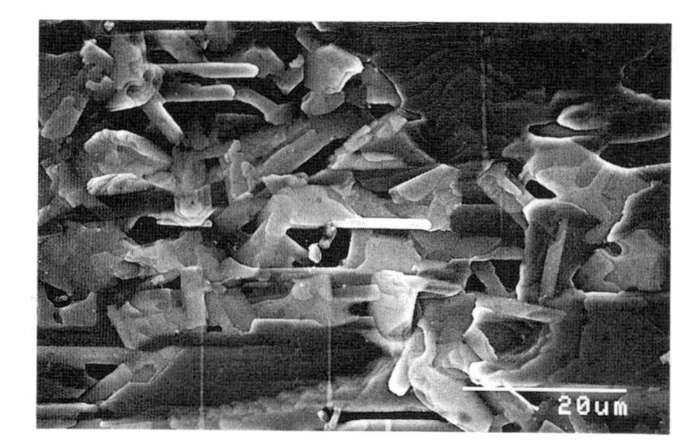

Fig. 5 SEM image of the sample annealed at 840°C

Fig. 4 is a typical SEM image of the sample annealed at 800°C. In this image black part is substrate MgO and two kinds of structures are observed. One has needle-like structure and the other has granular-type structure. Composition analysis was carried out for this sample by means of EPMA (Electron Prove Micro Analyser). As results, the composition ratio of needle-type structure is Bi:Sr:Ca:Cu = 2.0:2.1:1.0:2.5, that of granular type structure is Bi:Sr:Ca:Cu = 2.0:2.4:1.8:3.6. The former structure may corresponds to the low Tc phase and the latter structure may corresponds to the high Tc phase. Pb was not detected in each structure, which means Pb element being evaporated through post annealing process.

Fig. 5 shows the SEM image of the sample annealed at 840°C. The surface topography of this sample was rough as well as other samples in the present magnification level, although reflection diffractometry confirmed the existence of the small smooth area. This sample, however, consists of uniform composition and EPMA analysis indicates that the composition ratio of this sample is Bi:Sr:Ca:Cu = 2.0:1.9:1.0:1.9 and that Pb was not detected as well as other samples. This ratio shows good agreement with the composition ratio of the low Tc phase Bi-Sr-Ca-Cu-O system (Bi:Sr:Ca:Cu = 2:2:1:2). When we rewrite above composition as $Bi_2(Sr_{1-x}Ca_x)_3Cu_2O_y$, we can estimate the value x as 0.36. This result fairly agrees with the results reported by Yoshitake *et al.* [26] for the relation between the midpoint of Tc and value x and on the non-existence of high Tc phase under the condition $x < 0.5$.

CONCLUSIONS

The superconducting Bi-Sr-Ca-Cu-O thin films were successfully prepared on the MgO substrate by using ion beam co-sputtering technique with a target of $Bi_2SrCaCu_2O$ and a metal Pb, and post deposition *ex situ* annealing. As deposited film includes Pb element about 25% of Bi composition, element Pb evaporates entirely through post annealing process.

The samples after *ex situ* annealing process in oxygen flow at a temperature ranged from 800°C to 860°C shows the evidence of the epitaxial growth both on the surface treated and the not surface-treated substrate. The *ex situ* annealing lower than at 840°C are preferable for the formation of c-axis [001] oriented film perpendicular to the substrate. The sample annealed at 840°C shows superconducting behavior (on set was 91 K, end point was 78 K). The XRD analysis and composition analysis by EPMA shows this sample consists of single phase of low Tc phase Bi-Sr-Ca-Cu-O system with a lattice constant c=30.86 Å.

Although we added Pb element in sputtering process for the purpose of improving the synthesis of high Tc phase of Bi-Sr-Ca-Cu-O system, the optimum annealing condition to accomplish this purpose can not be discovered. It is considered to be necessary to apply other targets with different Bi-Sr-Ca-Cu compositions to obtain the superconducting film with the high Tc phase Bi-Sr-Ca-Cu-O system under the present sputtering and post-annealing conditions.

ACKNOWLEDGMENT

We are grateful to Professor U. Mizutani (Nagoya University) for his help with measuring the temperature dependence of the resistivity, Dr. N. Ishizuka for ICP spectra and Dr. Y. Murase for his assistance of performing the electron diffractometry, respectively.

REFERENCES

[1] H.Maeda, Y.Tanaka, M.Fukutomi and T.Asano, Jpn. J. Appl. Phys. 27, L209 (1988)
[2] T.Hashimoto, T.Kosaka, Y.Yoshida, K.Fueki and H.Koinuma, Jpn. J. Appl. Phys. 27, L384 (1988)
[3] H.Koinuma, M.Kawasaki, S.Nagata, K.Takeuchi and K.Fueki, Jpn. J. Appl. Phys. 27, L376 (1988)
[4] M.Nakao, H.Kuwahara, R.Yuasa, H.Mukaida and A.Mizukami, Jpn. J. Appl. Phys. 27, L378 (1988)
[5] M.Fukutomi, J.Machida, Y.Tanaka, T.Asano, H.Maeda and K.Koshino, Jpn. J. Appl. Phys. 27, L632 (1988)
[6] K.Kuroda, M.Mukaida, M.Yamamoto and S.Miyazawa, Jpn. J. Appl. Phys. 27, L625 (1988)
[7] H.Adachi, Y.Ichikawa, K.Setune, S.Hatta, K.Hirochi and K.Wasa, Jpn. J. Appl. Phys. 27, L623 (1988) / Y.Ichikawa, H.Adachi, K.Hirochi, K.Setune, S.Hatta and K.Wasa, Phys. Rev. B38, L765 (1988)
[8] H.Nasu, S.Makida, Y.Ibara, T.Kato, T.Imura and Y.Osaka, Jpn. J. Appl. Phys. 27, L536 (1988)
[9] K.Yoshimura, S.Nogawa and S.Tanemura, *High Tc Superconductivity: Thin Films & Devices*, Proc. SPIE 948, 99 (1988)
[10] S.A.Sunshine *et al.*, Phys. Rev. B38, 893 (1988)
[11] M.Takano *et al.*, Jpn. J. Appl. Phys. 27, L1041 (1988)
[12] Y.Syono *et al.*, Jpn. J. Appl. Phys, 27, L569 (1988)
[13] T.Kajitani, K.Kusaba, M.Kikuchi, N.Kobayashi, Y.Syono, T.B.Williams and M.Hirabayashi, Jpn. J. Appl. Phys. 27, L589 (1988)
[14] M.Onoda, A.Yamamoto, E.Takayama-Muromachi and S.Takenaka, Jpn. J. Appl. Phys., 27, L833 (1988)
[15] J.M.Tarascon *et al.*, Phys. Rev. B27, 9382 (1988)
[16] R.M.Hazen *et al*, Phys. Rev. Lett. 60, 1174 (1988)
[17] S.Sueno, R.Ishizaki, I.Nakai, K.Ohishi and A.Ono, Jpn. J. Appl. Phys. 27, L1463 (1988)
[18] E.Takayama-Muromachi *et al.*, Jpn. J. Appl. Phys. 27, L556 (1988)
[19] A.Ono, K.Kosuda, S.Sueno and Y.Ishizawa, Jpn. J. Appl. Phys. 27, L1007 (1988)
[20] Y.Matsui, H.Maeda, Y.Tanaka, E.Takayama-Muromachi, S.Takekawa and S.Horiuchi, Jpn. J. Appl. Phys. 27, L827 (1988)
[21] J.Akimitsu, A.Yamazaki, H.Sawa and H.Fujiki, Jpn. J. Appl. Phys. 26, L2080 (1987)
[22] C.Michel, M.Hervieu, M.Borol, A.Grandin, F.Deslandes, J.Provost and B.Raveau, Z. Phys. B68, 421 (1987)
[23] J.P.Franck, J.Jung, W.A.Miner and M.A.K.Mohamed, Phys. Rev. B38, 754 (1988)
[24] Y.Yamada and S.Murase, Jpn. J. Appl. Phys. 27, L996 (1988)
[25] for example, *Powder Diffraction File, Inorganic, Search Manual, Inter. Centr. for Diffraction Data*, Swarthmore, USA, (1987)
[26] T.Yoshitake, T.Satoh, Y.Kubo and H.Igarashi, Jpn. J. Appl. Phys. 27, L1261 (1988)

Observations of Y-Ba-Cu-O Films by Scanning Electron Microscopy and Auger Electron Spectroscopy

KENGO ISHIYAMA, YASUYUKI KAGEYAMA, and YASUNORI TAGA

Toyota Central Research and Development Laboratories, Inc., 41-1, Nagakute-cho, Aichi, 480-11 Japan

ABSTRACT

Structures of Y-Ba-Cu-O sputter-deposited films were investigated by Scanning Electron Microscopy (SEM) and Auger Electron Spectroscopy (AES). After heat treatment, the films showed different morphology for different substrates, i.e., Al_2O_3 and MgO, and for different thicknesses of 0.1 μm and 1.2 μm. Only for YBaCuO(0.1μm)/MgO, highly oriented crystal growth is seen, and for other systems, inhomogeneous crystal growths are seen. An interface roughning caused by preferential interdiffusion of Ba into an Al_2O_3 substrate was observed. A surface enrichment of Cu is found for the thick films. These differences in morphology are considered to be due to the differences in the diffusion processes.

INTRODUCTION

Formations of the films of Y-Ba-Cu-O high-Tc superconductor have been carried out by various techniques [1][2][3] and their structures were studied by X-ray and electron diffractometry, transmission electron microscopy (TEM) [4] and scanning electron microscopy (SEM) [5], and other techniques. From the studies of electrical property, noticeable dependence of superconductive properties on the substrate materials is already found out. The films deposited on an Al_2O_3 substrate shows a remarkable increase of resistivity at room temperature and lowering of transition temperature in comparison with other substrates, such as MgO, $SrTiO_3$ and YSZ [6]. It is supposed that the differences in these properties are due to differences in interactions between the films and the substrates.

In this paper we report the results of studies on sputter deposited Y-Ba-Cu-O films on Al_2O_3 and MgO by scanning electron microscopy (SEM) and Auger electron spectroscopy (AES), and also discuss the formation mechanisms of the morphology of the films.

EXPERIMENTAL

Y-Ba-Cu-O films were formed by the dc-magnetron sputtering techniques. The target used was a sintered disk of stoichiometric $YBa_2Cu_3O_X$. Ar of 99.999% purity was used as the sputtering gas. Two kinds of substrates, i.e., mirror finished Al_2O_3(0001) plane and MgO(100) plane cleaved out from a single crystal cube in air were used without heating. After evacuation down to 5×10^{-6} Torr, Ar gas was introduced to the production chamber up to 3×10^{-3} Torr. Substrates were cleaned by rf-sputtering before deposition of the film. Specimen films of 0.1 μm and 1.2 μm thick were deposited with a deposition rate of 70 Å/min. A heat treatment of the films at 900°C for 2 hours and cooling down to 300°C in 5 hours was carried out in oxygen flow of 960 sccm at 1.1 atm. Samples for SEM observations were cleaved just before introduction to the SEM column in order to avoid contaminations. A scanning Auger microprobe equipped with a sputter ion gun was used for depth analyses of the films. Total sputtering times of the depth analyses for the films of 0.1 μm and 1.2 μm are 50 minutes and 8 hours, respectively.

RESULTS

Scanning electron micrographs of fractured sections of YBaCuO films are shown in Fig.1. Figs.1(a) and 1(d) show the films of 0.1 μm as deposited on Al_2O_3(0001) and MgO(100), hereafter described as YBaCuO(0.1μm)/Al_2O_3 and YBaCuO(0.1μm)/MgO, respectively. The YBaCuO films as deposited have no remarkable morphology on their surfaces and sections. Fig.1(b) shows YBaCuO(0.1μm)/Al_2O_3 after heat treatment. The films are composed of grains and the surface is roughened. The grains show no remarkable orientation with relation to the substrate plane. A film/substrate interface is shown as a roughened boundary between the dark film and the bright substrate, as indicated by an arrow. The amplitude of the interface roughness is hundreds of angstroms. In the case of YBaCuO(1.2μm)/Al_2O_3, as shown in Fig.1(c), grains in the film grew larger compared with those in the film of 0.1 μm. Remarkable tendency of orientation is not appeared. The film/substrate interface is complexly roughened similar to that of the film shown in Fig.1(b). It is considered that the interface roughening indicate that remarkable dif-

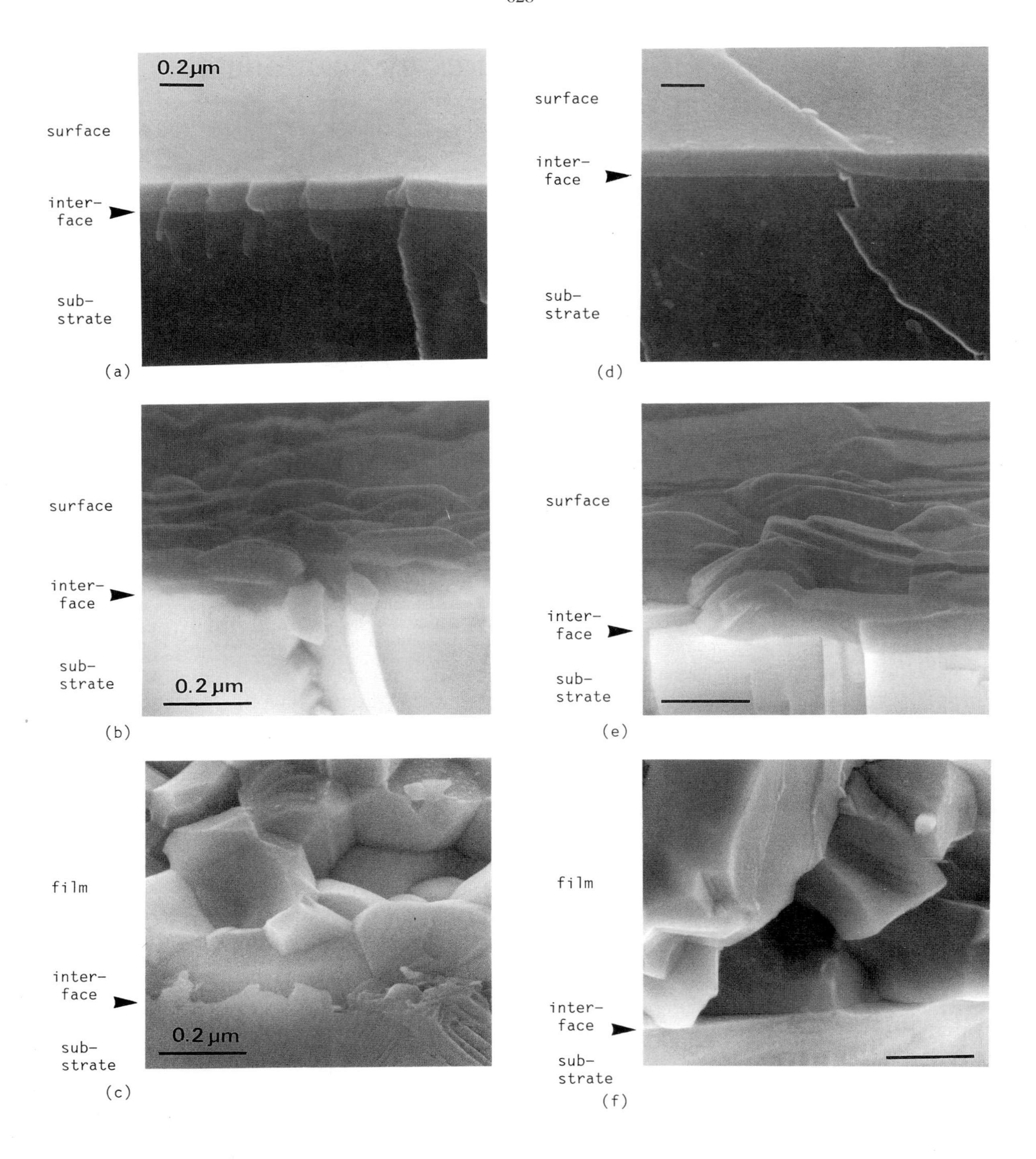

Fig.1. SEM graphs of YBaCuO films
(a),(b) YBaCuO(0.1μm)/Al_2O_3(0001), (c) YBaCuO(1.2μm)/Al_2O_3(0001)
(d),(e) YBaCuO(0.1μm)/MgO(100), (f) YBaCuO(1.2μm)/MgO(100)
(a),(d) as deposited
(b),(c),(e),(f) after heat treatment (900°C, 2 hours in O_2)

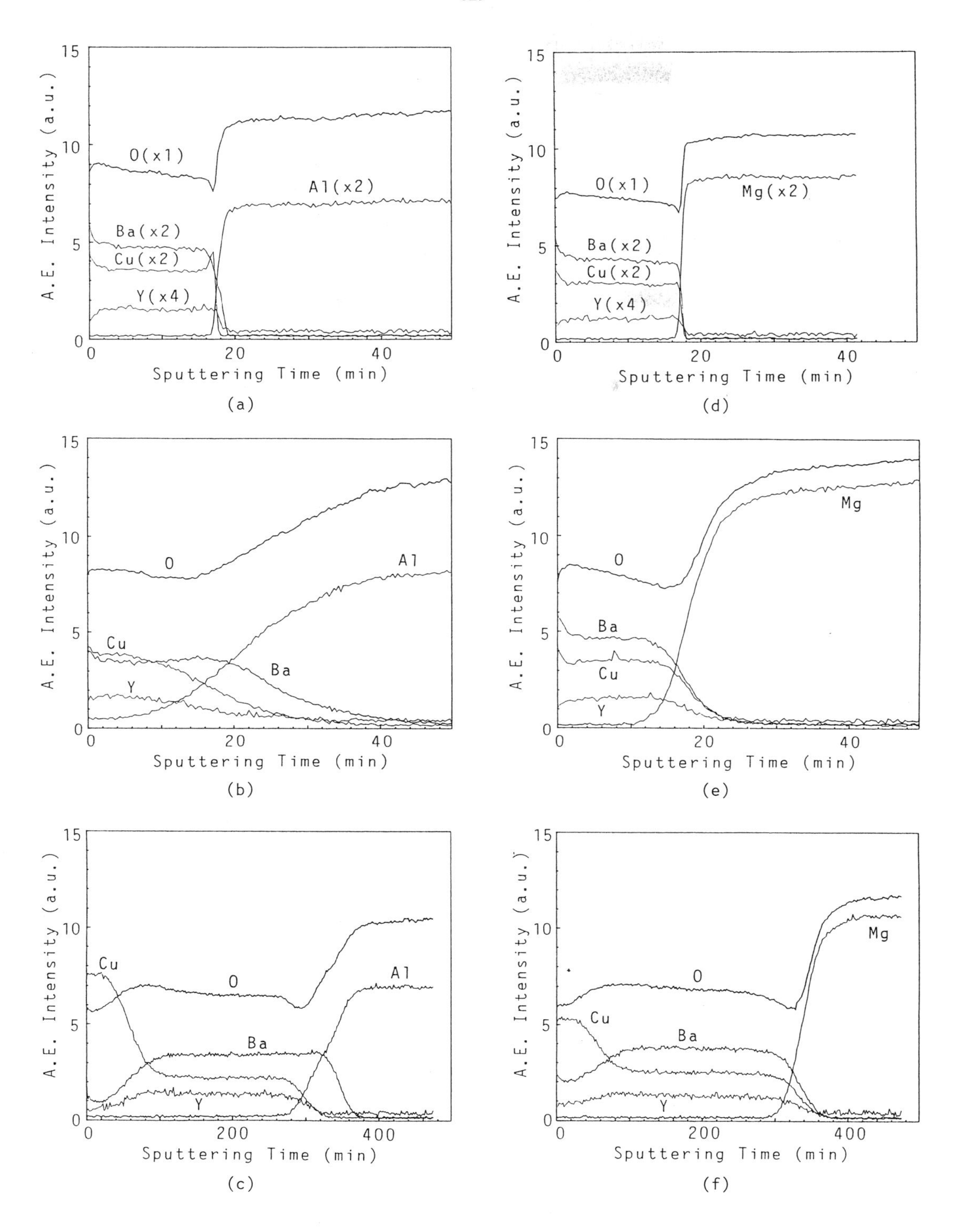

Fig.2. AES depth profiles of YBaCuO films
(a),(b) YBaCuO(0.1μm)/Al_2O_3(0001), (c) YBaCuO(1.2μm)/Al_2O_3(0001)
(d),(e) YBaCuO(0.1μm)/MgO(100), (f) YBaCuO(1.2μm)/MgO(100)
(a),(d) as deposited, (b),(c),(e),(f) after heat treatment

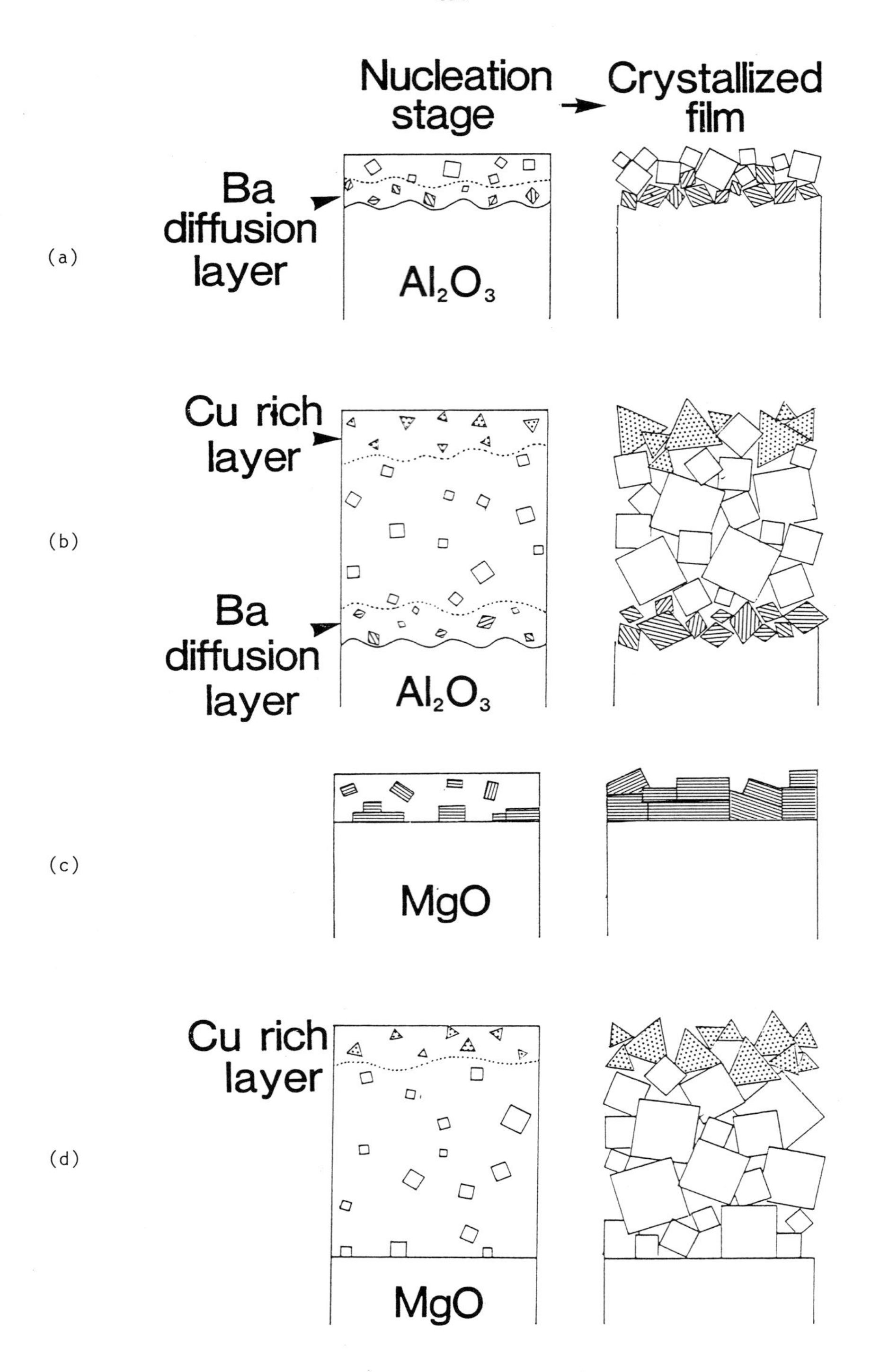

Fig.3. Schematic pictures of crystallization processes of YBaCuO films
(a) YBaCuO(0.1μm)/Al_2O_3, (b) YBaCuO(1.2μm)/Al_2O_3
(c) YBaCuO(0.1μm)/MgO, (d) YBaCuO(1.2μm)/MgO

fusion takes place across the interface. Fig.1(e) shows a sectional view of YBaCuO(0.1μm)/MgO after heat treatment. Most of grains are platelets with their planes nearly parallel to the substrate plane, so that the surface roughness of the film is smaller than that of YBaCuO(0.1μm)/Al_2O_3, as shown in Fig.1(b). The platelets have step structures. No features are seen on the dark section of the film, so that it is considered that the film has a homogeneous structure in its thickness. The film/substrate interface is seen as a clear and flat line, where its roughness is not larger than several tens of angstroms. In the film of 1.2 μm thick, as shown in Fig.1(f), grains in the film show no remarkable orientation. The interface is almost flat. The flatness of the YBaCuO/MgO interface indicates its low reactivity.

Figs.2(a)-(f) show Auger depth profiles for the samples corresponding to Figs.1(a)-(f), respectively. In both cases of YBaCuO(0.1μm)/Al_2O_3 and YBaCuO(0.1μm)/MgO as deposited, as shown in Figs.2(a) and 2(d), the Auger intensities of component elements of the films are approximately constant in the film. Quick falling down of them around the interface indicate that no remarkable interface reaction takes place. After heat treatment, as shown in Fig.2(b) for the Al_2O_3 substrate, Auger intensities vary during the measuring time. It is seen that Ba diffuse into the Al_2O_3 substrate deeper than the other elements. In Fig.2(c), in the case of YBaCuO(1.2μm)/Al_2O_3, a preferential diffusion of Ba is clearly seen, and a surface enrichment of Cu is also seen. For YBaCuO(0.1μm)/MgO shown in Fig.2(e), interdiffusion zone around the interface is smaller than that of YBaCuO/Al_2O_3 systems. A zone with nearly homogeneous composition is seen in the film. In thick film of 1.2 μm in thickness, as shown in Fig.2(f), a surface enrichment of Cu is seen. The Auger intensity ratio I(Y):I(Ba):I(Cu) of the as-deposited film is approximately 1:8:6. The ratios in the zones with nearly homogeneous composition in the annealed films of YBaCuO(1.2μm)/Al_2O_3, YBaCuO(0.1μm)/MgO and YBaCuO(1.2μm)/MgO are 1:6.5:4, 1:7:6 and 1:8:5, respectively. The compositional variations by heat treatment are large for YBaCuO/Al_2O_3 systems in comparison with YBaCuO/MgO systems.

DISCUSSION

From the results taken by AES and SEM, we derived the following mechanisms of the morphology formation of YBaCuO films, as schematically descrived in Figs.3(a)-(d). Two kinds of remarkable diffusion processes were observed by depth analyses of the YBaCuO films. Those are the surface enrichment of Cu commonly seen for two kinds of substrates, and the preferential diffusion of Ba into the Al_2O_3 substrate that causes the interface roughening. The former process is considered to be related to reactivity of Cu and oxygen and the latter is to reactivity of Ba and the Al_2O_3 substrate. It is also seen that the compositional redistribution takes place in the YBaCuO films. Differences in morphology of the films observed by SEM are considered to be due to these differences in diffusion processes. The differences in grain shapes are remarkably seen between those of YBaCuO(0.1μm)/MgO and those of other systems. It is considered that the platelet shown in Fig.1(e) is a characteristic shape corresponding to the composition nearly equal to that of the film as deposited. The grain shapes for other systems are supposed to be due to the considerably deviated compositions. The remarkable interface roughnesses of YBaCuO/Al_2O_3 systems indicate that the diffusion of Ba into Al_2O_3 is inhomogeneous on the interface. Besides, the flatness of the YBaCuO/MgO interfaces is to be explaned by a small and homogeneous interdiffusion. Additionally, in this system, at the nucleation stage, nucleation sites are limited within a small distance from the substrate plane. So that it is considered that the following crystal growth is strongly influenced by the substrate [5][7]. The homogeneity of the composition and nucleation close to the substrate are considered to be responsible for homogeneous crystal growth with high orientation. They are supposed to take place for the YBaCuO(0.1μm)/MgO system.

CONCLUSION

We studied on the microstructures of Y-Ba-Cu-O sputter-deposited films on Al_2O_3 and MgO substrates annealed in oxygen atmosphere. Two kinds of diffusion process in the films are seen. One is the surface enrichment of Cu for both substrates, and another is the preferential diffusion of Ba into Al_2O_3 that causes remarkable interface roughness. It is considered that these diffusion processes cause the composition redistribution in the films and their differences cause the differences in morphology of the films. For formation of homogeneous structures with high orientation, it is necessary to control the oxidation process of the film and to choose the substrate of low reactivity with the film.

REFERENCES

[1] C. X. Qiu and I. Shih: Appl. Phys. lett., 52(7) (1988) 587.
[2] L. Lynds, B. R. Weinberger, G. G. Peterson and H. A. Krasinski: Appl. Phys. Lett., 52(4) (1988) 320.
[3] O. Eryu, K. Murakami, K. Takita, K. Matsuda, H. Uwe, H. Kubo and T. Sakudo: Jpn. J. Appl. Phys., 27(4) (1988) L628.
[4] M. Tomita, T. Hayashi, H. Takaoka, Y. Ishii, Y. Enomoto and T. Murakami: Jpn. J. Appl. Phys., 27(4) (1988) L636.
[5] M. Futamoto and Y. Honda: Jpn. J. Appl. Phys., 27(4) (1988) L73.
[6] T. Aida, T. Fukazawa, K. Takagi and K. Miyauchi: ISPC-8, Tokyo (1987) 2345.
[7] K. L. Chopra: Thin Film Phenomina (McGraw-hill, New York, 1961)

4.11 Chemical Reactions and Superconductor/ Substrate Interaction

Comparison of The Stability of $Bi_2Sr_2CaCu_2O_{8+y}$ with $YBa_2Cu_3O_{6,5+y}$ in Various Solutions

H.K. LIU[1,2], S.X. DOU[1,2], A.J. BOURDILLON[1], and C.C. SORRELL[1]

[1] School of Materials Science and Engineering, University of New South Wales, P.O.Box 1, Kensington, NSW 2033, Australia
[2] Visiting Professor, Northeast University of Technology, Shenyang, People's Republic of China

ABSTRACT

The chemical behaviour of both $YBa_2Cu_3O_{6.5+y}$ {123} and $Bi_2Sr_2CaCu_2O_{8+y}$ {2212} in various solvents was studied in a comparative manner. In both compounds superconductivity apparently depends on a high oxidation state of Cu. Electrochemical corrosion rates were measured by a variety of techniques and are related to the concentration of labile Cu^{3+} ions in the specimens. Generally {2212} is the more stable compound, except in strongly basic solutions owing to the amphoteric nature of Bi. In the presence of acidic or basic solutions Bi^{3+} is oxidized to Bi^{4+} or Bi^{5+} by the high oxidation state of Cu in {2212} (similar to oxidation by Na_2O_2), demonstrating a common feature of cuprate superconductors. The concentration of labile ions was determined by the volumetric measurement of evolved oxygen from acid solution, and the technique was extended to measure the concentration of undecomposed or re-formed carbonates after sintering.

INTRODUCTION

The corrosion of superconducting oxide ceramics is important for both scientific and technological reasons: on the one hand corrosion products from specimens placed in various environments provide information about the chemical bonding and oxidation states of the ceramics (particularly with regard to the Cu^{3+} d^8 state), while on the other hand resistance to corrosion is a desirable feature particularly in the region surrounding electrical contacts where corrosion can be enhanced by effective electrochemical cells.

Superconducting $YBa_2Cu_3O_{6.5+y}$, {123}, whether in powdered or sintered form, has been observed to react vigorously with water at room temperature [1,2,3]. In reactions between {123} and acid solutions, oxygen is evolved, and the volume of gas evolved can be used to determine the concentration of labile (Cu^{3+}) ions, which generally correlates with superconducting properties [4]. These labile ions, which are apparently essential for high temperature superconductivity, also result in chemical instability. Applications may require protective coatings. Here we report comparative corrosion studies of the new superconductor $Bi_2Sr_2CaCu_2O_{8+y}$, {2212}, in various aqueous solutions.

EXPERIMENTAL PROCEDURE

Superconducting {2212} samples were prepared as previously described [5]. The samples were characterized in several ways: by four terminal resistance measurement to have a transition temperature (T_c, at midpoint) of 84 K; by analytical scanning electron microscopy (SEM) to have a phase composition of $Bi_2Sr_{3-x}Ca_xCu_2O_{8+y}$ with $0.4<x<0.8$ and y unmeasured; and by X-ray diffraction showing lattice parameters of 0.38 x 0.38 x 3.08 nm. A sample of {123} of similar density (~ 85 °/o) was used for comparison and had a T_c of 93 K and orthorhombic crystal structure [6]. Investigations were divided into three types of experiment as follows.

In a first set of experiments powders of {123} and of {2212} were immersed at room temperature into (a) 2M HCl, (b) 2M NaCl, (c) distilled water, (d) CH_3OH and (e) 2M NaOH. Morphological changes were observed.

In a second experiment the volumetric measurement technique [4] was used to determine leaching rates in 10°/o HCl of sintered pellets of {123} and {2212}.

In a third experiment electrochemical cells were constructed by ultrasonically bonding with indium pairs of tinned copper wires to the silver painted ends of rectangular bars of {123} and of {2212}. To reduce electrochemical effects at the contacts, these were coated with epoxy resin so as to protect them from corrosion failure during tests. The two leads were connected to terminals for room temperature resistance measurement while the samples and leads were immersed in distilled water. The resistance of the sample was recorded as a function of time.

Further corrosion protection is obtained if the whole {123} or {2212} specimen is covered with epoxy resin. To assess the effectiveness of such protection a piece of {2212}, connected with leads as before, was coated with a layer of epoxy 1 mm deep and the resistance measured as a function of time.

RESULTS AND DISCUSSION

Qualitative results for the corrosion of {123} and {2212} in the aqueous solutions are displayed in table I.

TABLE I. Reactions of $YBa_2Cu_3O_{6.5+y}$ and of $Bi_2Sr_{2.2}Ca_{0.8}Cu_2O_{8+y}$ with some solvents at room temperature.

	$YBa_2Cu_3O_{6.5+y}$	$Bi_2Sr_{2.2}Ca_{0.8}Cu_2O_{8+Y}$
2M HCl	evolution of gas giving blue solution	gas with brown precipitate dissolving after 2 days
2M NaCl	weak evolution of gas after 2 days	no reaction observed
H_2O	slow evolution of gas	very slow evolution of gas
CH_3OH	white precipitate on drying without gas evolution	no reaction observed
2M NaOH	weak evolution of gas	blue solution and brown precipitate

a) In 2M HCl, O_2 is more rapidly evolved from {123} than from {2212}. However the latter reaction occurs in two stages. At first a brown solid precipitates, but after 30 hours this dissolves. Most likely, the brown solid is Bi_2O_4 or Bi_2O_5 formed in the presence of a powerful oxidizing agent (Cu^{3+}) from Bi^{3+}. Bi_2O_5, besides being insoluble in water, is a very strong oxidizing agent. When the Cu^{3+} is exhausted the following reaction can take place:

$$Bi_2O_5+6HCl = 2BiCl_3+3H_2O+O_2 \qquad (1)$$

The $BiCl_3$ passes into solution and the brown precipitate disappears. As supporting evidence for this interpretation, it was found that Bi_2O_3 can be dissolved in HCl without precipitation.

The oxygen evolution is used in the volumetric technique for measuring the concentration of labile (Cu^{3+}) ions, which generally correlates with superconducting properties [4,5]. The technique can also be used to determine the concentration of carbonates in calcined or sintered superconductors by separating the CO_2 gas over NaOH. Results of such measurements from {123} sintered at various temperatures and for various times are shown in table II. Complete dissociation of $BaCO_3$ requires 50 h at 940 °C for samples of this size (20 mm diameter x 4 mm thick) and density (85°/₀ of theoretical). Samples sintered at lower temperatures for shorter times, e.g. at 880°C for 34 h, showed not only the presence of $BaCO_3$, but also a low concentration of Cu^{3+}. The presence of $BaCO_3$ on grain boundaries will have detrimental effects on critical current densities.

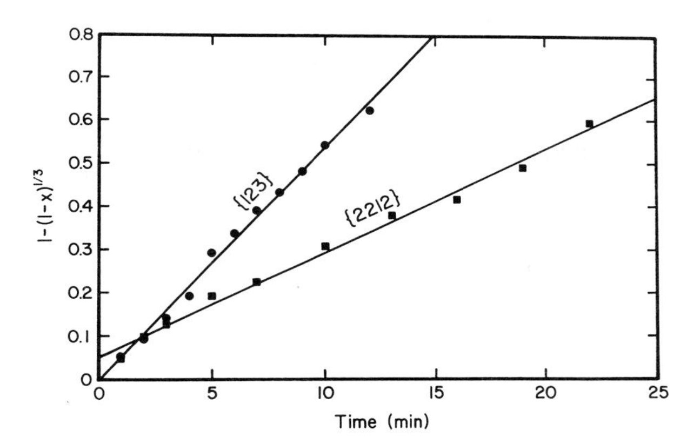

Figure 1. Leaching rates represented by $1-(1-x)^{1/3}$ plotted against time for {123} (●) and {2212} (■).

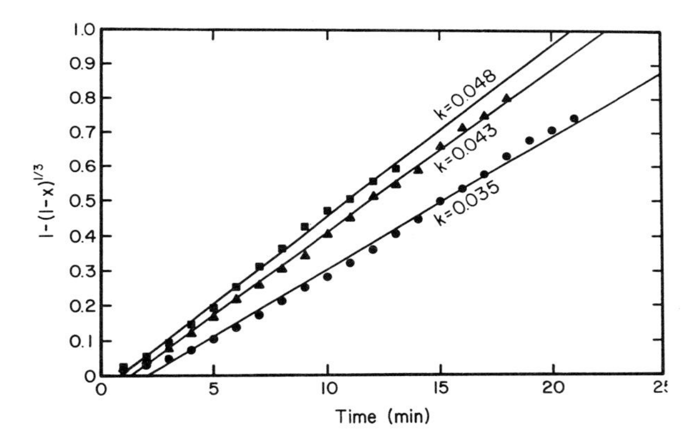

Figure 2. Leaching rates, k, compared for samples of {123} with varied density and Cu^{3+} concentrations with: ■, density = 5.17 gcm^{-3} and Cu^{3+} = 23.9 %; ▲, density = 4.66 gcm^{-3} and Cu^{3+} = 23.3 %; ●, density = 4.25 gcm^{-3} and Cu^{3+} = 20.8 %

In the third experiment the electrical resistance of the specimens was monitored during reaction. The resistance at time t (R_t) was normalized against initial resistance (R_o) as follows (see figure 3):

$$\Delta R = (R_t - R_o)/R_o \tag{3}$$

ΔR increased ten times faster in {123} than in {2212}. This number correlates with the ratio of Cu^{3+} ions present in these materials, which is expected if those ions are the driving force for the electrochemical cell reactions:

$$YBa_2Cu_3O_7 + e^- = YBa_2Cu_3O_{6.5} + {}^1/_4 O_2 \tag{4}$$

at the cathode; and

$$ {}^1/_2 Cu = {}^1/_2 Cu^{2+} + e^- \tag{5}$$

at the anode.

Table II. Concentration of $BaCO_3$ and Cu^{3+} ions in $YBa_2Cu_3O_{6.5+y}$ after (a) calcining and (b) various sintering procedures.

(a)

Calcining Temperature(°C)	880	900	920	940
Sintering Time (h)	34	34	34	34
$BaCO_3$ (wt.°/o)	0.55	0.42	0.43	0.39
6.5+y (O_2)	6.60	6.68	6.72	6.73
Cu^{3+} (wt.°/o)	6.48	12.29	14.65	15.3

(b)

Sintering Temperature (°C)	935	935	935	935
Sintering Times (h)	15	15	15	15
$BaCO_3$ (wt.°/o)	0.082	0.071	0.139	0.136
6.5+y (O_2)	6.72	6.80	6.81	6.83
Cu^{3+} (wt.°/o)	14.74	19.75	20.93	22.21

Leaching rates from sintered pellets are quantified below, but qualitative comparisons are first made for relative stabilities of {123} and {2212} in other aqueous environments.

b) In 2M NaCl, no reaction was observed from {2212}, while slow O_2 evolution from {123} resulted from reaction with water.

c) In distilled water, O_2 was evolved, but more slowly than in case (a), especially for {2212}.

d) Whereas white particles precipitated on {123} after immersion in methanol and drying, no reaction was observed on {2212}. The precipitation in the former case was ascribed to a reaction with water absorbed from air, forming $Ba(OH)_2$ and subsequently $BaCO_3$.

e) In contrast to the descriptions in (a) to (e), alkaline 2M NaOH reacted less strongly with {123} than with {2212}. In the latter case a dark blue solution with brown precipitates was observed. Again the precipitation suggests insoluble compounds with high oxidation states as follows.

It is known that Bi_2O_3, when treated with peroxide, e.g. Na_2O_2, in the presence of NaOH solution, forms $NaBiO_3$ [7]. It is likely that Cu^{3+} ions have similar oxidizing power to Na_2O_2, and in the presence of NaOH oxidize Bi^{3+} to form Bi_2O_4, which is insoluble in NaOH [8], and $NaBiO_3$ which is soluble in NaOH and may form a complex with Cu^{2+} as a blue solution. In support, no reaction was observed when Bi_2O_3 was immersed in 2M NaOH.

These results illustrate the effects of high oxidation state which results from pseudoperovskite structures in the oxide ceramics examined. In a second experiment leaching rates of sintered pellets of {123} and {2212} in 10°/o HCl allow more quantitative comparisons. The leaching rate, x , was determined by the volume fraction of oxygen evolved from specimens of similar density (examined further below) and sample size, in conditions of constant stirring rate and approximately constant solution concentration. Assuming the leaching process to be controlled by surface reaction of first order, the shrinking core model [9] may be used to describe the rate as follows:

$$1-(1-x)^{1/3} = k_l\ C\ t/r_o = k\ t \tag{2}$$

where C represents the solution concentration (in mole cm^{-3}), r_o the initial radius of the spherically shaped sample (in cm), k_l the linear rate constant (in cm^4 $mole^{-1}$ min^{-1}) and k the apparent rate constant (in min^{-1}). A plot of the left side of equation (2) against time results in a straight line whose slope measures the reaction rate (figure 1). The model fits well up to 95 °/o leaching. The reaction constant (1.0 min^{-1}) for {2212} was less than half of that for {123} (2.13 min^{-1}).

In {123} the leaching rate in fact depends critically on the concentration of Cu^{3+} but hardly at all on the specimen density (or porosity). This was demonstrated by measuring the leaching rate on three specimens of different density and Cu^{3+} concentration as determined by the volumetric technique (figure 2). The leaching rate constants correlate more strongly with the Cu^{3+} concentrations than with the density dependent surface areas.

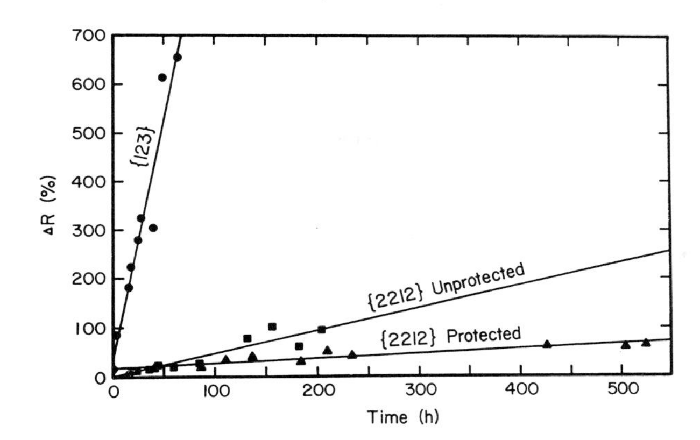

Figure 3. Change in resistance with time, ΔR, for {123} (●), {2212} (■),and protected (2212) (▲).

Support for these reactions arises from observations on the two leads. In the case of the {123} cell the 0.3 mm diameter leads were completely corroded in 55 h, while for the {2212} the time taken was four times greater, 220 h. This correlates again with relative Cu^{3+} concentrations.

Finally the resistance of the {2212} specimen was monitored after the specimen and contacts had all been coated with 1 mm thick epoxy resin and, after setting, immersed in distilled water. Though no obvious signs of corrosion of the leads were visible till the time of writing (40days), the specimen resistance increased after 40 days to a value only 30 per cent as high as that of an unprotected sample after 10 days. While the coating effectively slows down the corrosion rate, electrolyte leakage and contact effects seriously degrade the sample.

In summary, observations on {123} and {2212} cuprate superconductors in chemical and electrochemical reaction with various solvents, can be interpreted consistently through chemical instabilities caused by labile Cu^{3+} ions. In {123} these ions are important for superconductivity, so their observed presence in the non-rare earth superconductor {2212} indicates again the importance of chemical bonding in the mechanism for high temperature superconductivity. The electrochemical behaviour is of technical importance since corrosion around metal-superconductor contacts will be crucial, and optimized protective coatings can be used to diminish corrosion.

ACKNOWLEDGEMENTS. We are grateful to Metal Manufactures Ltd. for support (S.X.D.). H.K.L. is supported by a GIRD grant.

REFERENCES

1. K.G.Frase, E.G.Liniger and D.R.Clarke (1987) Adv.Ceram.Mat. 2 698-700
2. R.L.Barns and R.A.Laudise (1987) Appl.Phys.Lett. 51 1373-1375.
3. S.X.Dou, H.K.Liu, A.J.Bourdillon, N.X.Tan, J.P.Zhou, C.C.Sorrell and K.E.Easterling (1988) Mod.Phys.Lett.B 1 363-367.
4. S.X.Dou, H.K.Liu, A.J.Bourdillon, J.P.Zhou, N.Savvides and C.C.Sorrell (1988) submitted to Sol.Stat.Comm.
5. S.X.Dou, H.K.Liu, A.J.Bourdillon, N.Savvides and C.C.Sorrell (1988) Sup.Sci.& Tech., in press.
6. S.X.Dou, A.J.Bourdillon, C.C.Sorrell, K.E.Easterling, S.Ringer, N.Savvides, J.B.Dunlop and R.B.Roberts (1987) Appl.Phys.Lett 51 535-537.
7. B.H.Mahan (1979) University Chemistry, 3rd ed. Addison-Wesley
8. R.C.Weast (1977) CRC Handbook of Chemistry and Physics, 58th ed. p B-95.
9. H.Y.Sohn and M.E.Wadsworth (1979) Rate Processes of Extractive Metallurgy. Plenum. p.133.

Formation of Crystallized Buffer Layer for High-Tc Superconducting Thin Film

JŌJI SHINOHARA, YUJI IKEGAMI, and TERUAKI KAWAMOTO

Research Institute, Ishikawajima-Harima Heavy Industries Co., Ltd., 3-1-15, Toyosu, Koto-ku, Tokyo, 135 Japan

1.ABSTRACT

For the formation of high-Tc superconducting thin films, ceramic single crystal substrates which are made of MgO, $SrTiO_3$, YSZ(Yttrium Stabilized Zirconia),etc.,have been used. For the application of the high-Tc superconducting thin film for electric devices, superconducting tapes,electromagnetic shields,and etc.,however,it is necessary to develope the techniqus which can make the high-Tc superconducting thin films on Si wafer,metal,and/or glass substrates.

To get the YSZ and MgO films which have adequate crystallization and orientation as a bufer layer, coating conditions of RF magnetron Ion plating method and RF magnetron sputtering method were investigated.

Y-Ba-Cu-O thin films by RF magnetron sputtering were grown on the buffer layer and those properties were examined.

2.INTRODUCTION

Recently,it has been shown by varios works that high-Tc superconducting oxide films have been fabricated on single crystal substrates such as $SrTiO_3$, ZrO_2, and MgO [1-3]. However heat treatment process at temperature above 600-900°C has been necessary to prepare high-Tc superconducting films.

It has been difficult to form the superconducting films on metal and Si substrates because of causing interdiffusion between superconducting films and the substrates[4,5].

The preperations of high-Tc superconducting film on Si and metal substrates have large potential for application such as electronic devices, superconducting magnets, electromagnetic shields,etc.

In this paper we report on the preliminary results of the preparing buffer layers for high-Tc superconducting film on metal and quartz glass substrates. The buffer layers which were YSZ and MgO were fablicated on Nickel base superalloy IN-100 and quartz substrates. Y-Ba-Cu-O film were deposited on these buffer layers by RF magnetron sputtering and the properties were examined.

3.EXPERIMENTAL

3.1 Preperation of buffer layer

YSZ and MgO film were grown and evaluated for the candidates of buffer film. YSZ films were prepared by RF ion-plating method.Ion-plating was carried out at 350°C in Ar+40% O_2 atomosphere at pressure of 0.067 Pa using Zr+10wt% Y evaporant. The typical ion-plating conditions are shown in Table I. MgO films were prepared by magnetron sputtering using MgO compound target.

The sputtering conditions were as follows; substrate temperature 150 C,sputtering gass, Ar+ 20-40% O_2 at 0.067 Pa, film thickness, 0.2-1 µm.

The substrates of buffer layer were IN-100 plates (30x40x1 mm) and quartz glass (76x25x1 mm) The substrates of IN-100 plate were polished to Ra:0.5 µm.

Table I Ion-plating conditions.

Evaporant	Zr+10wt% Y
Ion-plating gas	Ar+40% O_2
Gas pressure	0.067 Pa
Substrate temperature	350°C
Substrate bias voltage	150-500 V
Growth rate	4 - 6 Å/sec

Fig.1 A cross-section of YSZ film coated by the ion plating process

3.2 Preperation of Y-Ba-Cu-O film

Y-Ba-Cu-O films were fablicated on IN-100 and quartz glass substrates with buffer layer by RF magnetron sputtering method. The sputtering target was a sintered Y-Ba-Cu-O compound which composition was $Y_1Ba_2Cu_{3.5}O_x$. Sputtering conditions were as follows; sputtering gass, Ar+20-50%O_2 at 0.1-0.67 Pa, RF input power, 200-300 watts. The temperature were measured by Pt-13%Rh thermocople located behind the substrate. The true temperature of the substrate is likely to be high due to irradiating by sputtering plasma.

The growth rate were 0.1 - 0.25 µm/hr and the film thickness were 0.3 -1 µm. The composition of Y-Ba-Cu-O films were analyzed by indused coupled plasma emission spectoscopy (ICP).

4. RESULTS AND DISCUSSION

4.1 Buffer layer

Fig.1 shows the cross section of YSZ buffer film which was deposited by ion-plating on quartz glass. Fig.2 shows X-ray diffraction patterns of YSZ film. These diffraction patterns show that the YSZ film was cubic structure which had a preferential oriented peak of (200). By alternating the ion-plating conditions, the preferential orientation of YSZ was obtained.

Table II. shows the peak ratios of (200)/(111) in the diffraction patterns compared with the bias voltage of substrate. As the bias voltage increased, the (200)/(111) ratio became large at the growth rate of 4 Å/sec.

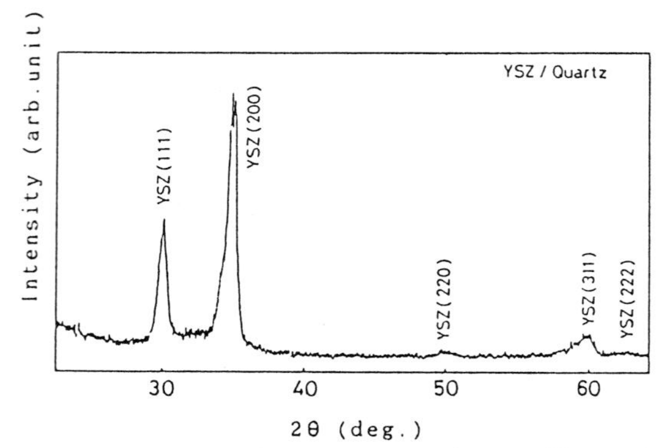

Fig.2 X-ray diffraction patterns of YSZ Film on Quartz

Table II Crystallization dependence of acceleration voltage in the YSZ film coating

Peak Ratio \ Vacc (V)	500	300	150
(200)/(111)	1.80	0.78	0.17
(220)/(111)	0.14	0.10	0.08
YSZ/Quartz	Rate 4Å/sec		

Table III Crystallization dependence of growth rate in the YSZ film coating

Peak Ratio \ Rate (Å/s)	4	8	16
(200)/(111)	1.80	1.34	1.10
YSZ/Quartz	Vacc=500V		

Table III. shows the peak ratios of (200)/(111) compared with the growth rate. In this table the low growth rate was likely to cause (200) preferential orientation of YSZ buffer layer. For substrates of IN-100 the similar results were obtained

Fig.3,4 show the X-ray diffraction patterns of MgO buffer layers which were fablicated by sputtering process. The peak corresponding to MgO (200) reflection was obsereved in both patterns. The profiles of diffraction peak of MgO (200) IN Fig.3,4 were broad slightly.

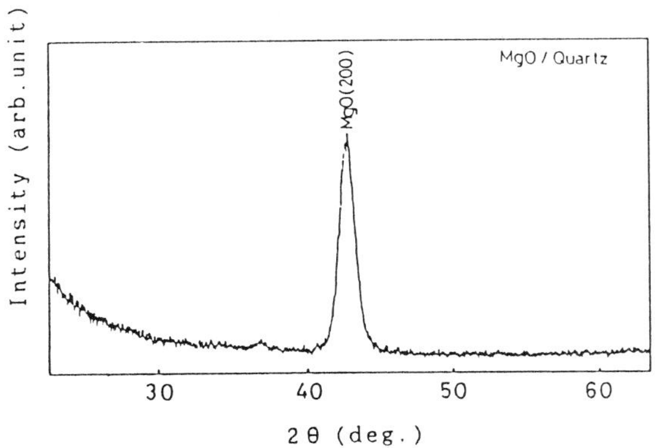

Fig.3 X-ray diffraction patterns of MgO film on quartz glass

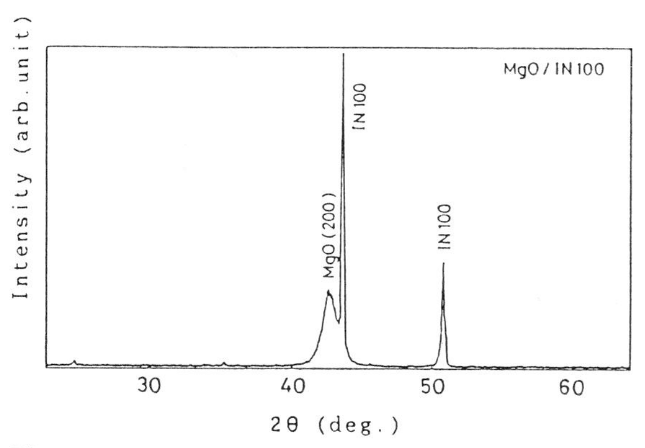

Fig.4 X-ray diffraction pattern of MgO film on IN-100

(a) as deposited 1 μm

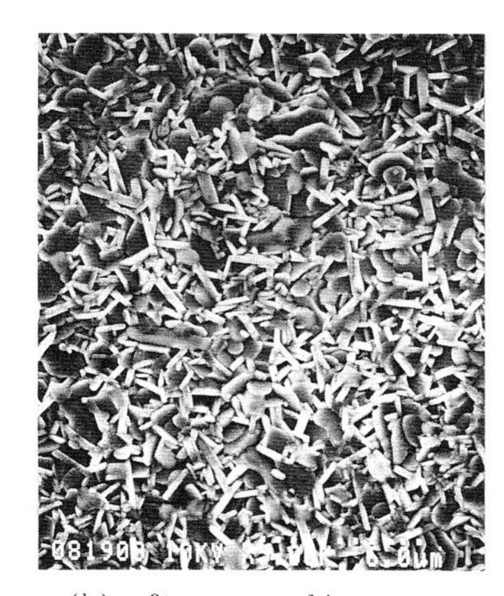

(b) after annealing (930°C x 10 min) 2 μm

Fig.5 SEM photographs of Y-Ba-Cu-O film on MgO / IN-100 substrate

4.2 Y-Ba-Cu-O film

Fig.5(a) shows SEM photographs of Y-Ba-Cu-O film on MgO/IN-100 substrate. The surface of as deposited film was smooth and the grain size of the film were below sub-microns. After annealing (930°C x 10 min), giant grains were grown on the surface. Fig.6 shows the X-ray diffraction patterns of Y-Ba-Cu-O film after annealing.The film showed the diffraction patterns corresponding to polycrystal stracture of orthorhombic Y-Ba-Cu-O film.

The film composition which was analyzed by ICP method was Y:Ba:Cu=1:1.94:3.65.

Fig.7 (a),(c) show the cross section image of Y-Ba-Cu-O film on YSZ/IN-100 substrate observed by EDX analysis and (b),(d) show SEM photographs respectively. These images show the mappings of Cu,Zr and Ni elements, which are corresponded to Y-Ba-Cu-O film,buffer layer of YSZ and IN-100 substrate. In Fig.7(a), the interfaces between each elements were clear, so there is no interdiffusion. In Fig.7(c) no interdiffusion was observed. Fig.8 shows AES (Auger Electron Spectroscopy) analysis of Y-Ba-Cu-O/MgO/IN-100 substrate (650°Cx2 hrs). In this AES depth profile, it is confirmed that buffer layer of MgO prevented inter diffusion between substrate and Y-Ba-Cu-O film.

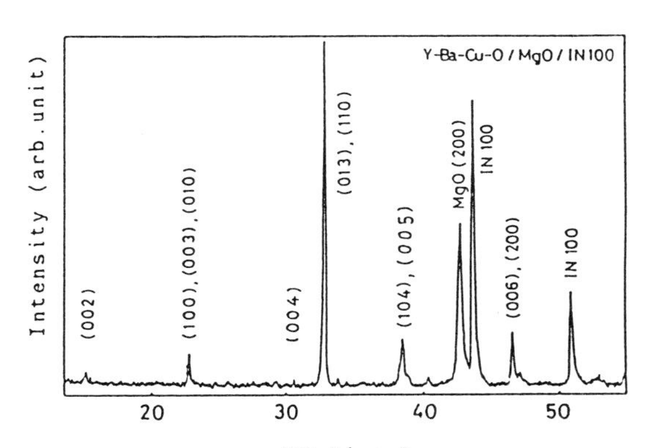

Fig.6 X-ray diffraction patterns of Y-Ba-Cu-O Film on MgO buffer layer / IN100

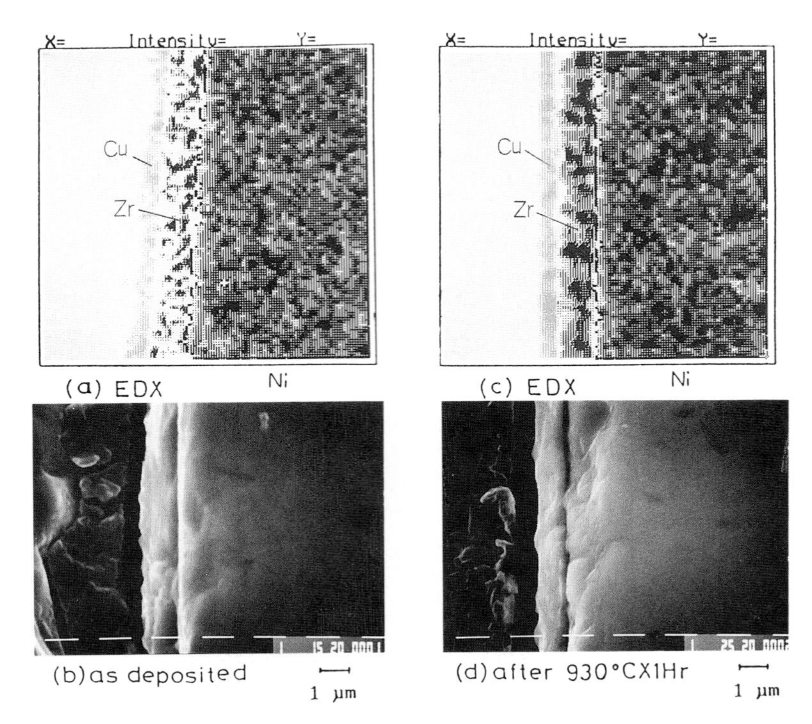

Fig.7 Cross section image of Y-Ba-Cu-O / YSZ films on IN100 substrate by EDX analysis

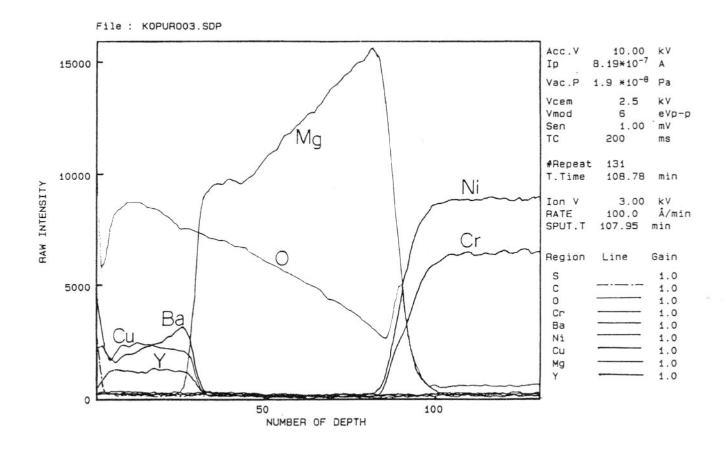

Fig.8 AES depth profile of Y-Ba-Cu-O film on MgO buffer layer / IN100 substrate (sputtering temperature of Y-Ba-Cu-O film : 650°C x 2 hrs)

5. CONCLUSION

We have prepared YSZ and MgO buffer layers and comfirmed its properties. An ion-plating method and a sputtering method were used to fablicated the films. Y-Ba-Cu-O film was fablicated on YSZ and MgO buffer layers. YSZ and MgO films were effective in protecting Y-Ba-Cu-O film from interdiffusion of elements of the substrate. YSZ and MgO films are promising for use in buffer layer of superconducting films.

ACKNOWLEGEMENT

The authors wish to thank M.Ayabe,K.Ishige,A.Yamanishi for usefull discussions.

REFERENCES

(1)M.Suzuki and T.Murakami;Jpn.J.Appl.Phys.26 (1987)L524.
(2)K.Char,A.D.Kent,A.Kapitulnik,M.R.Beasley and T.H.Geballe;Appl.Phys.Lett.51(1987)1370
(3)T.Terashima,K.Iijima,K.Yamamoto,Y.Bando and H.Mazaki;Jpn.J.Appl.Phys.27(1988)L91
(4)H.Nasu,H.Myoren,Y,Ibara,S.Makida,Y.Nishyama,T.Kato,T.Imura and Y.Osaka;Jpn.J.Appls.27(1988) L634
(5)T.Kitagawa,S.Shibata,H.Okazaki,T.Kimura,Y.Enomoto;Jpn.J.Appl.Phys.27(1988)L1113

High Tc Y-Ba-Cu-O Superconducting Thick Films Fabrication and Film/Substrate Interactions

YUNG-HAW HU and CHARLES L. BOOTH

E. I. Du Pont de Nemours & Company, Inc., Electronics Department, Experimental Station, Wilmington, DE 19898, USA

ABSTRACT

High Tc Y-Ba-Cu-O superconducting thick films having Tc at 87 K and R=0 at 79 K have been successfully made by using conventional thick film screen printing technology on MgO and ZrO_2 substrates. Film/substrate interactions were examined by electron microprobe analysis as a function of starting powders, heat treatments and substrates. At processing temperatures of 950°C, Ba and Y enrich at the substrate reaction boundary layer (~ 2 µm thick) with depletion of Cu in the superconductor for all the substrates we examined. On the other hand, Cu segregates more at the interface layer than Y and Ba in films fired at 900°C. The fired film microstructures were found to vary strongly with starting powders which demonstrated quite a difference in both densification and grain growth rates. Good superconducting films were never achieved on an alumina substrate. Limited film/substrate interactions were proved to preserve superconducting integrity and to have adhesion greater than 20 newtons, which is adequate for most of microelectronic applications.

Since the discovery of high Tc superconductivity in the class of copper-containing perovskite oxides ($YBa_2Cu_3O_{7-x}$) [1,2], a major focus of research effort has been the preparation of good quality films. The first practical application of these high Tc superconductor will likely occur in areas related to novel discrete electronic devices, such as SQUIDs, infrared devices, microwave circuits and other high-frequency devices. Many of these applications require the development of technology for producing and patterning thin layers of dense superconducting material on insulating substrates. A number of papers [3-6] have reported the successful preparation of Y-Ba-Cu-O films using different sputtering, spinning and dipping techniques. However, screen printing is considered to be the most economic way of patterning devices. Only a few papers [7,8] have demonstrated this conventional thick film technology. None of them addressed the adhesion problem, which is a vital property for the proper performance of the devices. Traditionally, the necessary film/substrate adhesion is formed by the addition of glasses and oxides to the paste formulations to initiate a glass, oxide or mixed bonded adhesion mechanism. We have found that Y-Ba-Cu-O superconducting phase is very reactive to most oxide and glass additives. Clearly, a study of film/substrate interactions will lead to adequate adhesion while retaining superconductivity.

The thick film pastes which we used to fabricate superconducting films were made by using Y-Ba-Cu-O powders or organometallic precursors with organic vehicles (surfactant, binder and solvent). Well-mixed pastes were achieved by three roll milling. Film thickness was controlled via viscosity control of the paste. Thick film patterns were screen printed on substrates, such as single crystal MgO, cubic ZrO_2 and Al_2O_3. Printed films were dried at 120°C for 20 minutes. Dry films were heat treated at 500°C for organic burnout, then sintered at either 900°C or 950°C under flowing oxygen. Fired films were furnance cooled to 450°C, annealed for 6 hours, then slowly cooled to room temperature. Resistivity was determined with a standard dc four-point probe method with a current I=0.5 mA applied to the sample. The best results to date have been obtained

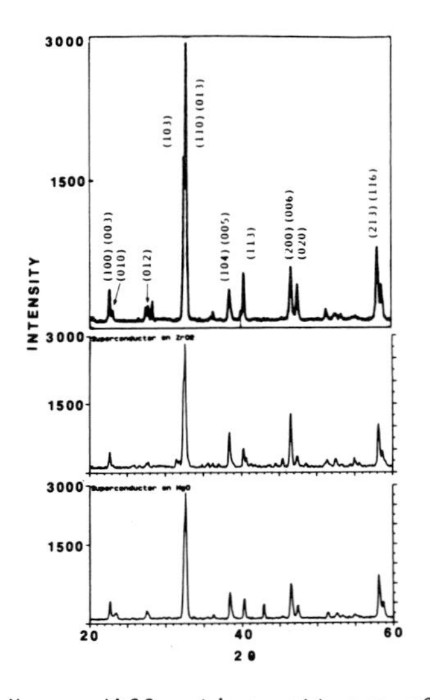

FIG. 1 - X-ray diffraction patterns of
(a) the standard Y-Ba-Cu-O powder,
(b) Y-Ba-Cu-O film on ZrO_2 substrate,
(c) Y-Ba-CU-O film on MgO substrate

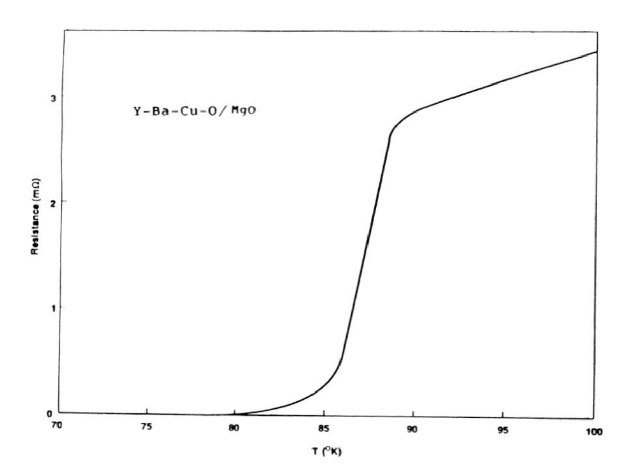

FIG. 2 - Typical resistance as a function of temperature of a film on MgO substrate

on MgO and ZrO_2 substrates. Fig. 1 portrays the temperature dependence of resistance of Y-Ba-Cu-O a thick film (~15 μm thick) on MgO substrate. The onset temperature (Tc) of superconductivity on this film is around 87 K, and the superconducting transition is complete (R=0) at 79 K. The X-ray diffraction patterns (Fig. 2) show that except for the characteristic peaks of the substrate, the other major peaks can be assigned to the orthorhombic superconducting phase. Films on alumina substrate do show a drop in resistance at about 80 K, but the resistance never reached a "zero" state. The reaction layer compositions were characterized by energy dispersive spectroscopy (EDS) with an electron microprobe.

The printed films made from Y-Ba-Cu-O powders do not densify up to a 950°C sintering temperature on all the substrates used in our studies, although high density clusters are observed in each films as shown in Fig. 3a. The density of the fired film is only around 85% of theoretical density. The textured nature of the superconducting grains shows up very clearly in the SEM secondary electron image of the polished cross-section, Fig. 3a. The grains have a rectangular shape (~ 5 μm in length) similar to that of the orthorhombic unit cell and are randonly arranged. Figs. 3b, 3c and 3d show the corresponding X-ray compositional maps for Ba, Y and Cu. The associated microprobe analysis of a superconducting phase $YBa_2Cu_3O_{7-x}$, a Y-rich phase and a small amount Cu rich phase have been identified. The cross-section X-ray maps indicate that the fired film has a homogeneous composition. This confirms the good quality of our starting powders.

We observed that films made on alumina at 950°C have the most pronounced interdiffusion. The interface boundary layer extends to 5 μm in thickness. The high degree of interdiffusion indicated by X-ray maps and the existance of Y and Cu rich second phases appear to deteriorate the superconducting properties of the films on alumina. As for films on Mgo and ZrO_2, a notable reaction boundary layer (1.5 μm) still occurs, although the degree of interaction is much less (Fig. 3a.) Ba diffuses much faster than Y and Cu and can be seen not only at the interface layer, but also in the substrate, Fig. 3b. Y, which is enriched at the reaction zone, does not diffuse into the substrate and shows an intermediate diffusion. In all cases, Cu concentration at the interface layer is very low since Ba and Y rich phases probably block the diffusion of Cu. The heat treatment at higher temperatures seems to make the film more homogeneous, but it also leads to more extensive interdiffusion between the film and the substrate. This interdiffusion appears to occur in both directions with Ba going into the substrate and Mg and Al into the interface layer.

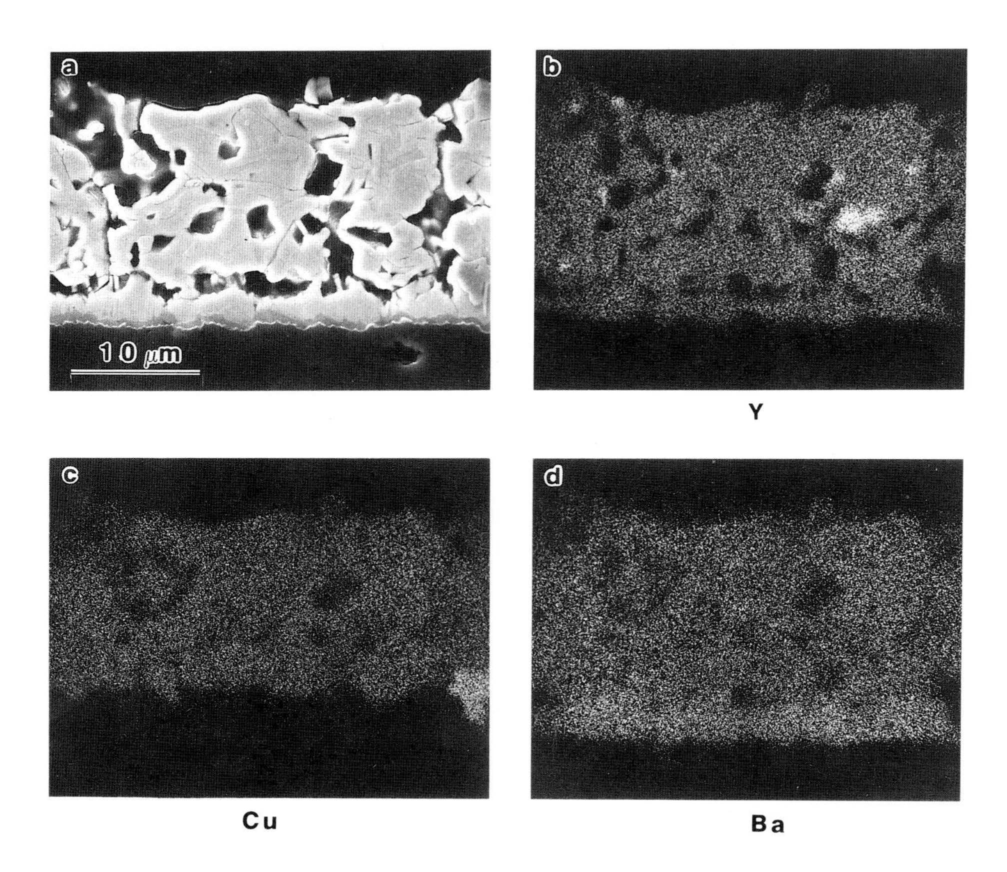

FIG. 3 - A cross-section scanning electron microscopy image and corresponding electron microprobe compositional maps, (b) yittrium, (c) copper, (d) barium for the film made by using 1-2-3 powder

In an attempt to minimize interdiffusion between Y-Ba-Cu-O films and substrates, the effectiveness of using precursor powders was explored. Films made from precursor pastes yielded good densification and high grain growth rates in all substrates when fired at 900°C. Hot stage optical microscopy shows that significant densification started as early as 840°C, followed by fast lateral grain growth. These results confirm the high degree of reactivity of the precursor powders. Fig. 4a shows the elongated grains having dimensions of 20-30 μm in length and 5 μm in width for a film fired on MgO. All the elongated grains orient parallel to the substrate surface and stack vertically and next to each other. This extraordinary stacking accounts for the smooth surface and high sintered density of the fired films. This unique microstructure not only improved the superconducting quality but also significantly improved the printability and solderability of the films. Isolated yittrium-rich and copper-rich phases can be seen in Fig. 4 with a copper-rich phase segrated along the reaction interface. The incomplete conversion of the precursors to the Ba-Y-Cu-O superconducting phase is one major drawback. Fig. 4 shows minor diffusion of all constituents between the films and MgO substrates and some vertical inhomogeneity in the film. The interdiffusion appears to occur in both directions, Cu richer and Y and Ba poorer toward the interface with Mg diffused into the Y-Ba-Cu-O film. The interdiffusion on most substrates was suppressed by using highly reactive precursor powders coupled with lower processing temperature (900°C), but this was not the case for films on alumina substrate. A 4 μm reaction layer extended into the substrate with Al diffused heavily into the superconducting film. The data demonstrate that the interdiffusion on Al_2O_3 can be reduced but will still occur for precursor powders, giving a long tail in the resistive drop.

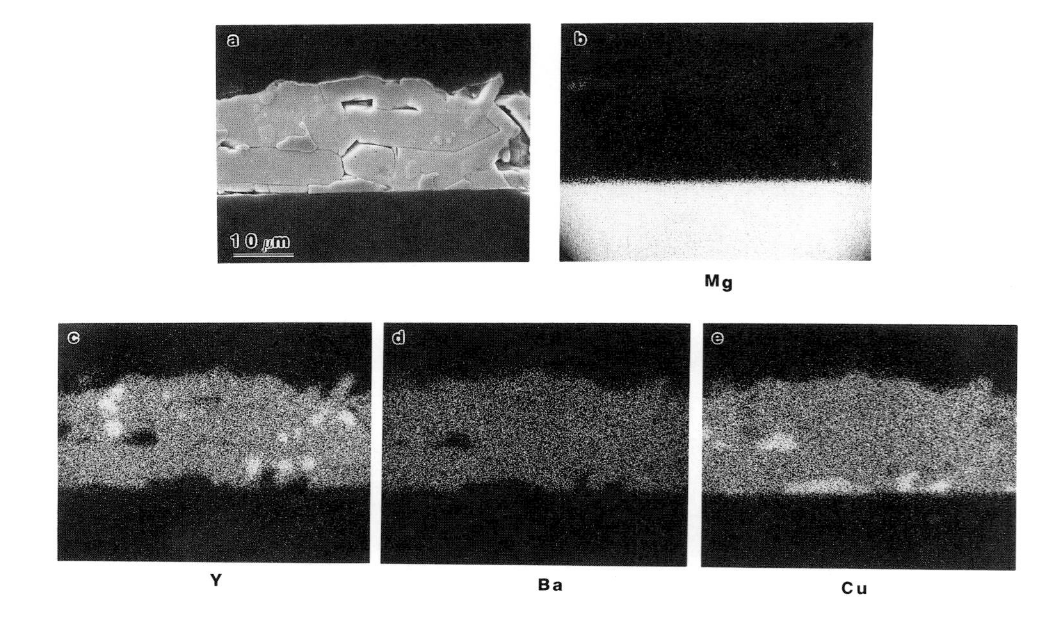

FIG. 4 - A cross-section scanning electron microscopy image and corresponding electron microprobe compositional maps, (b) magnesium, (c) yittrium, (d) barium, (e) copper for the film made by using precursor powder

We found that the general requirement for obtaining high adhesion, quality films is to avoid massive diffusion between the film and substrate. The adhesion of the film on MgO substrate has been tested at greater than 20 newtons. Obviously, the thin reaction boundary layer mainly enriched in Cu created suitable bonding.

In conclusion, we have successfully made Y-Ba-Cu-O superconducting thick films (15-20 μm) with R=0 at 79 K on MgO and ZrO_2 substrates. Interface boundary compositions vary with processing starting powders, processing temperatures and substrates. Ba and Y diffuse faster than Cu at the interface layer for films fired at 950°C. By contrast, the interdiffusion of the Y and Ba seems to be less than Cu at the film/substrate interface when fired at 900°C. Interface reaction can be suppressed with highly reactive precursor powders, which can be processed at lower temperatures to achieve high quality superconducting films. We demonstrated that the limited reactive layer not only gave adequate adhesion but also preserved the superconducting integrity of the film.

REFERENCES

1. M. K. WU, L. R. Ashburn, C. Y. Torng, P. H. Hor, R. L. Meng, L. Gao, Z. J. Huang, Y. Q. Wang and C.W. Chu, Phys. rev. lett., 58 908 (1987).

2. H Takagi, S. Uchida, K. Kilozawa and S Tanaka, Jpn. Appl. Phys. Lett., 26 L123 (1987).

3. M. Gurvitch and A Fiory, Appl. Phys. Lett., 51 1027 (1987).

4. A. Gupta, g. Koren, E. A. Giess, N. R. Moore, E. J. M. O'Sullivian and E. I. Cooper, Appl. Phys. Lett. 52 163 (1988).

5. C. E. Rice, R.B. Van Dover and G. J. Fisanick, Appl. Phys. Lett. 51 1842 (1987).

6. K. Koinuma, M. Kawasaki, T. Hashimoto, S. Nagata, K. Kitazawa, K. Fueki, K Masubuchi, and M. Kudo, Jpn. J. Appl. Phys., 26 L763 (1987).

7. R. C. Budhani, Sing-Mo H. Tzeng, H, J. Doerr and R. F. Bunshah, Appl. Phys. Lett., 51 1277 (1987).

8. I. Shih and C. X. Qiu, Appl. Phys. Lett. 52 748 (1988).

Substrate-HTcS Interaction Studies by (S)TEM

P.P.J. RAMAEKERS and D. KLEPPER

Centre for Technical Ceramics, P.O. Box 513, 5600 MB Eindhoven, The Netherlands

ABSTRACT

This paper concerns with compatibility aspects between HT_cS thin film and their substrates. The influence of substrate-thin film interaction and thin film microstructure on the superconducting properties is discussed. In this respect, data based on (S)TEM observations are presented. It is concluded, that it is of the utmost importance to exactly control the growth and crystallisation process of a HT_cS thin film. To be able to do this one should at least have regular access to a (scanning) transmission electron microscope.

INTRODUCTION

In the technology of thin film devices based on high-temperature superconductors (HTcS) the choice and preparation of the substrate is one of the key factors. Both chemical interaction, thin film orientation and thin film microstructure are critical for the superconducting properties (Tc, Jc, etc.) and these can be strongly influenced by the compatibility of substrate and superconducting layer.
Presently, we are studying chemical interactions and interdiffusion between $YBa_2Cu_3O_7$ thin layers and $SrTiO_3$ (100), Al_2O_3 (sapphire), ZrO_2(YSZ) and Si and between $BiCaSrCuO_x$ layers and $SrTiO_3$ (100) substrates. In addition we plan to examine the effects of nitride and noble metal intermediate layers as diffusion barriers. In this paper some results obtained by (S)TEM studies are presented.

Another key factor in the construction of hybrid HTcS devices is the choice of the thin film technology. In this paper we will summarize literature data on how the deposition method and the annealing procedure are related to the superconductor's microstructure and thus its superconducting behaviour.

EXPERIMENTAL

- Thin film preparation and characterisation

$YBa_2Cu_3O_{7-x}$ thin films were deposited using UHV triple E-gun deposition with BaF_2, Cu and Y as source materials. Oxygen was supplied close to the substrate, a

pressure of 10^{-7} mbar was maintained during deposition. The experimental results in this paper report on 0.25 m thick layers on $SrTiO_3$ (100).
Correct stoichiometry was obtained by firing in a 1 bar O_2/H_2O mixture at 850°C, as checked by Rutherford Backscattering. Overall thin film orientation was determined by X-ray diffraction.

- TEM specimen preparation

Small pieces of a wafer were cemented with Epox 812 with their thin film faces together. The two opposite sides of the "sandwich" were flatted along one of the major crystallographic axes with wet 600 grit silicon carbide abrasive paper.

One side was dimpled until about 25 microns with 6 micron diamond paste and ending with 0.05 micron alumina (1).
With this face the sandwich was cemented using wax on a dimpler pin. The specimen was thinned to 150 microns. A second dimple was made until about 20 microns of material remained between the two.

After dimpling and cleaning a copper ring (one hole grid) was fixed to the specimen with epoxy. The specimen was placed in a Gatan duo ion mill during about 8 hours at 6 kV and 0.5 mA. Specimens are observed in a JEOL 2000 FX at 200 kV (2).

RESULTS AND DISCUSSION

1. Literature survey on substrate-HT_cS compatibility.

The first phenomenon to be considered when studying substrate-thin film compatibility is the possibility of chemical reaction, interdiffusion or segregation. Although many researchers have described chemical interactions between substrate and HT_cS thin films, very few elaborate studies using a variety of substrates have been performed.

Gurvitch and Fiory (3) report on a number of substrates and possible buffer layers, giving a description of substrate-film reactions and interdiffusivity problems. Dam et al. (4) describe results obtained on thin films prepared by DC sputtering, substrates being Al_2O_3 (0001), MgO (100) and (110), Yttria stabilized ZrO_2 (111), Si (100), Gadolinium Gallium Garnet (100) and $SrTiO_3$ (100). Invariably the best results (minimal interaction and deterioration of superconducting properties) were found using monocrystalline $SrTiO_3$ substrates. In the case of $SrTiO_3$ Gurvitch and Fiory (3) observe sharp resistive transitions but mention poisoning of the superconductor by Ti, and loss of Cu from $YBa_2Cu_3O_7$. Poisoning by Ti and also by Sr was found by Gavaler et al. (5), whereas Agatsuma et al. (6) even found thin reaction layers of Bariumtitaniumoxide on $SrTiO_3$ (100) substrates. In this case, however, pastes containing non-reacted Y_2O_3, $BaCO_3$ and CuO powder were used, which might result in direct reaction with the $SrTiO_3$.

Also in the case of monocrystalline ZrO_2 substrates good results have been reported. Gurvitch and Fiory (3) give a ranking where ZrO_2 takes first place and $SrTiO_3$ second. Degradation of the HT_cS thin film, however, always takes place and progressively so with higher annealing temperatures and longer annealing times. That is why these authors recommend the use of a 300 A thin Ag or Nb buffer layer. ZrO_2 itself is also in the picture as a diffusion barrier in hybrid Si/or SiO_2/superconductor devices (7).

Important in this respect is the possibility that impurities are introduced into the HT_cS film during deposition or annealing, or coming from improper treatment of the substrate. Notorious is the inclusion of Ar^+ during sputtering, which, although in many cases partly removed by annealing can strongly decrease the critical current density. The quality of the polishing, cleaning, etching and drying procedure of the substrate to be covered will define the integrity of the interface and may also be a source of problems, not well realised by some research groups.

A second compatibility aspect of prime importance for the superconducting properties is the microstructure of the superconductor. The following microstructural features can be influenced by the substrate and by the deposition/annealing procedure: density, stoichiometry, crystallite orientation, presence of second phases, crystallite defects and lattice strains. In the context of this publication it is impossible to systematically treat all the relationships between these features and the thin film technology. Therefore we will refrain ourselves to a discussion of the most prominent effects described in the literature. The substrate's choice is not only important from the interaction point of view. Thin film orientation, lattice mismatch and strain, and macrostresses caused by thermal expansion differences should be also considered. For the example of $YBa_2Cu_3O_{7-x}$ deposition on monocrystalline $SrTiO_3$ many comments on the orientation dependence of the substrate and the deposition/annealing technology are known.

A (001) orientation of a $YBa_2Cu_3O_{7-x}$ thin film is one of the prerequisites for a high Jc. On $SrTiO_3$ this orientation is often preferentially obtained when starting with a (100) or (001) substrate. At least this is true for thinner films: the thicker the film, the more (h00) orientation is observed, although at the interface still (001) can be found (8).

Kentgens et al. (8) clearly indicate, for sputtered films, that a preferential c orientation is less advantageous for reasons of lattice mismatch. The fact that this orientation still appears for thinner films indicates that other factors have to be considered. Kentgens et al. postulate that a (001) $YBa_2Cu_3O_7$ orientation is more favourable for interfacial free energy reasons because of the different metal oxide surface structure, as compared to the (h00) orientation. Obviously, not only nucleation but also subsequent (or simultaneous) crystallization processes will determine the ultimate thin film orientation. Systematic experiments where growth temperature, deposition time, annealing characteristics (time, temperature, atmosphere) and perhaps also substrate preparation are varied under well defined conditions are needed to resolve this matter. Following the suggestion of Fujita et al. (9) that the lattice constants of $YBa_2Cu_3O_{7-x}$ have a temperature dependent and an oxygen deficiency dependent portion one can understand why it is so extremely important to be able to exactly control the growth process of a HT_cS thin film.

$YBa_2Cu_3O_{7-x}$ is an exciting material since its physical properties are strongly anisotropic. For instance the anisotropy in the thermal expansion coefficient (10) is one of the critical factors which determine the microstructural (and the superconducting) properties during thermal treatment. Differences in thermal expansion between substrate and HT_cS layer should also be considered for technological applications. In a recent paper Hashimoto et al. (11) presented some data (also temperature dependent) on this. Clearly the use of substrates with a large thermal expansion coefficient (MgO, $SrTiO_3$, YSZ) is favourable. A short annealing cycle at a sufficiently low temperature (preferably lower than the orthorhombic-tetragonal transition in $YBa_2Cu_3O_{7-x}$) is also recommended; this is also benificial for the reduction of substrate-film interactions.

2. (S)TEM Analysis Results

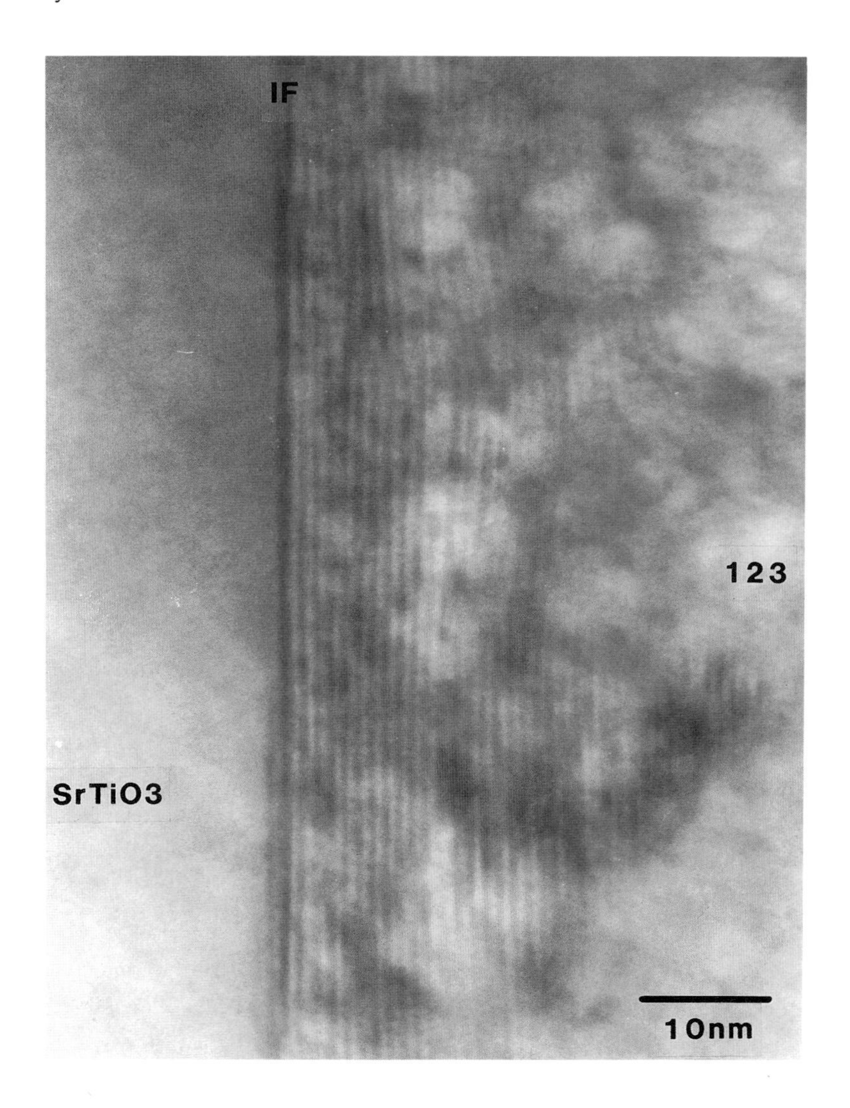

INTERFACE (IF). RIGHT YBa2Cu3O7 (123), LEFT SrTiO3

CONCLUSIONS

1. For the optimisation of superconducting properties of HT_cS thin films one should minimise substrate-thin film interaction. If on Si or SiO_2 substrates annealing above 900 K is needed the use of a buffer layer is recommended. Also with other substrates, even with $SrTiO_3$ or YSZ, a diffusion barrier may be needed.
2. For the optimisation of superconducting properties of HT_cS thin films it is of the utmost importance to control the growth and crystallisation process during and after deposition. In this way microstructural control is possible.
3. Regular access to a (scanning) transmission electron microscope and experience with HT_cS thin film specimen preparation and handling is one of the prerequisites for the development of device technology based on high-temperature superconductors.

REFERENCES

1. L.R. Nazar, Proceedings of 46th Annual meeting of EMSA (1988) p. 862.
2. M. Sarikaya, Proceedings of 46th Annual meeting of EMSA (1988) p. 858.
3. M. Gurvitch and A.T. Fiory, Mat. Res. Symp. Proc., Vol. 99, (1988) 297-301.
4. B. Dam, H.A.M. van Hal and C. Langereis, Europhys. Lett. 5 (1988) 455-460.
5. J.R. Gavaler, A.I. Braginski, J. Telvacchio, M.A. Janocko, M.G. Forrester and J. Greggi, Extended Abstracts High-Temperature Superconductors II, April 5-9, 1988, Reno (Nevada), 193-196.
6. K. Agatsuma, T. Ohara, H. Tateishi, K. Kaiho, K. Ohkubo and H. Karasawa, Physica C 153-155 (1988) 814-815.
7. A. Mogro-Campero and L.G. Turner, Appl. Phys. Lett. 52 (1988) 1185-1186.
8. A.P.M. Kentgens, A.K. Carim and B. Dam, J. Crystal Growth, accepted for publication.
9. J. Fujita, T. Yoshitake, A. Kamijo, T. Satoh and H. Igarashi, Extended Abstracts High-Temperature Superconductors II, April 5-9, 1988, Reno (Nevada), 109-112.
10. W. Wong-ng and L.P. Cook, Adv. Ceram. Mater. 2 (1987) 624; H.M. O'Bryan and P.K. Galagher, ibidem 2 (1987) 640.
11. T. Hashimoto, K. Fueki, A. Kishi, T. Azumi and H. Koinuma, Japanese Journal of Applied Physics 27 (1988) L214-L216.

Electrochemical Modification of Superconductive Oxides

W. Kinzy Jones

Department of Mechanical Engineering, Florida International University, The State University of Florida at Miami, Miami, FL 33199, USA

ABSTRACT

The electrochemical potential between highly electropositive cations and conductive oxides has been the basis for numerous electrochemical batteries (Li/MnO_2, Li/V_2O_5). A new cell, based on the superconductive oxide, $YBa_2Cu_3O_x$, as a cathode and lithium as an anode, has been developed. The cell exhibited high open-circuit voltage (2.8 volts) and long-term loaded voltage stability under low drain conditions at room temperature. The cell was discharged four months before destructive analysis. Room temperature operation was initially performed due to the use of a liquid electrolyte.

This initial work evaluated the effect of lithium interaction with the superconductive properties of the reacted cathode. No shift in transition temperature (93°K) has been observed in the lithium reaction $YBa_2Cu_3(Li_{0.87})O_x$ (x~6.7), although all superconductive properties were lost within two weeks after exposure to ambient conditions. No new phases were identified, indicative of lithium intercalation into the superconductor structure. This technique allows controlled introduction of cations species (Li^+, Ca^{++}, etc.) into a superconductive oxide structure for property modification studies.

INTRODUCTION

The use of conductive oxides as cathodes in electrochemical cells, such as the lithium anode-oxide cathode systems (Li/MnO_2), (Li/Ag_2CrO_4) and (Li/V_2O_5) have been widely investigated [1]. In this work, the superconductive oxide, $YBa_2Cu_3O_x$, has been used as a cathode in a lithium anode cell, providing a technique for structural and electronic property modification by the introduction of lithium cations into the oxide structure as

$$y\ Li + YBa_2Cu_3O_x \dashrightarrow YBa_2Cu_3(Li_y)O_x$$

with a corresponding change in valence states of the multivalent cations, such as copper, to maintain neutrality.

Electrochemical processes have been investigated to modify or control the oxygen concentration in $YBa_2Cu_3O_x$ formations [2]. Pre-fired yttrium-barium-copper oxide was utilized as an anode with stabilized zirconia as a cathode. Under the application of voltage, oxygen was transferred from the zirconia into the Y-Ba-Cu-O system, producing a stable superconductive oxide without high temperature sintering. Additionally, work has been reported [3] on the possibility of electrochemical synthesizing of superconductors, with films of yttrium and barium-captured copper oxide being fabricated by anodic film growth.

EXPERIMENTAL

Superconductive oxide cathodes were fabricated utilizing techniques described elsewhere [4]. The transition temperature of the finished cathode was observed at approximately 93°K, as shown in Fig. 1. Two types of cells were fabricated: all glass cell utilizing a pellet cathode to show proof of concept and a hermetic stainless steel cell for controlled loaded discharge evaluation. The cell chemistry and construction, except for the cathode material, was identical to a battery system utilized in a high-reliability implantable medical application [5].

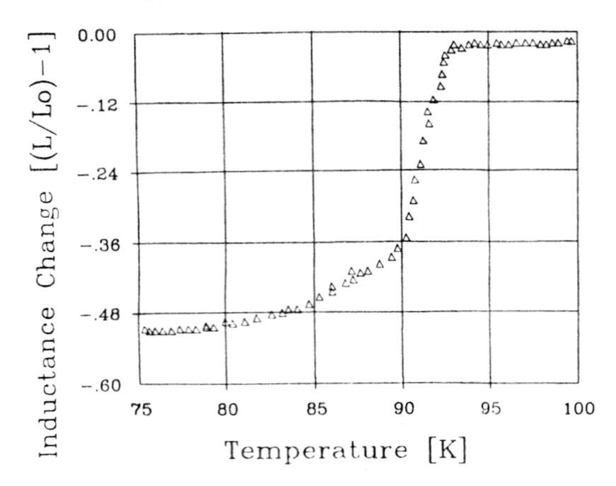

Figure 1. Transition temperature for starting material

Initially, a glass cell was fabricated to evaluate the electrochemical potential of the cell. A lithium disc was pressed into a 304 stainless steel lead wire and a (1/8 diameter x 1/4 length) pellet cathode was held by a stainless steel clip. A polypropylene mesh was used to provide physical separation. The glass cell was evacuated, back-filled with argon and 1, 3 dioxolane was introduced as an electrolyte. This cell demonstrated a loaded cell potential of 2.8, and voltage and current stability.

Hermetic cells were constructed by bonding the lithium anode to a stainless steel current collector attached to a hermetic insulating feed-through, which served as the negative terminal. A polypropylene separator was placed around the lithium and pressed cathodes were placed on each side of the protected lithium. The assembly was placed into a stainless steel can and hermetically sealed except for an electrolyte-filled hole. The cathodes were in physical contact with the can, making the can the positive terminal. The cells were evacuated, back-filled with the dioxolane electrolyte and the fill hole was hermetically sealed. (Fig. 2)

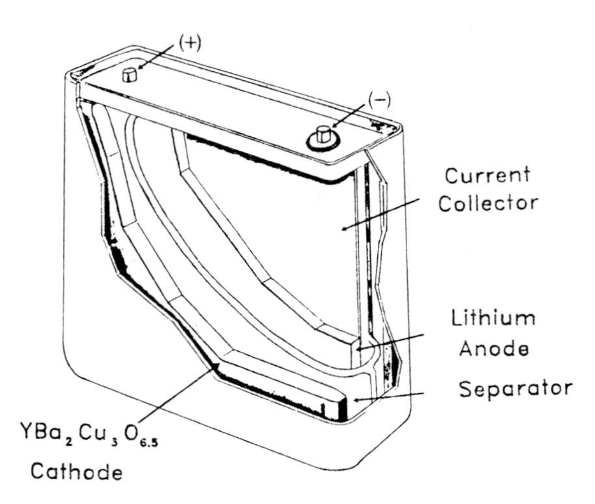

Figure 2. Fabricated hermetic cell

The cells were placed on loaded discharge (10KΩ, 20KΩ loads) and monitored until the 10KΩ cell had experienced a discharge of 150 mA hours (approximately 1,000 hours). The cells were subjected to destructive tear-down analysis and the cathodes were evaluated for changes in superconductivity properties. After tear down, the cathodes were stored in an argon back-filled n-heptane filled container until the first transition temperature measurement and X-ray diffraction analysis was made. Subsequently, the sample was restored in heptane without an argon back-fill, exposing the heptane to the atmosphere.

RESULTS

The discharge curve for the $Li/YBa_2Cu_3O_x$ system is given in Fig. 3. An apparently linear discharge curve was obtained under low drain rate conditions. The two marked data prints are the end conditions of the cells at tear down.

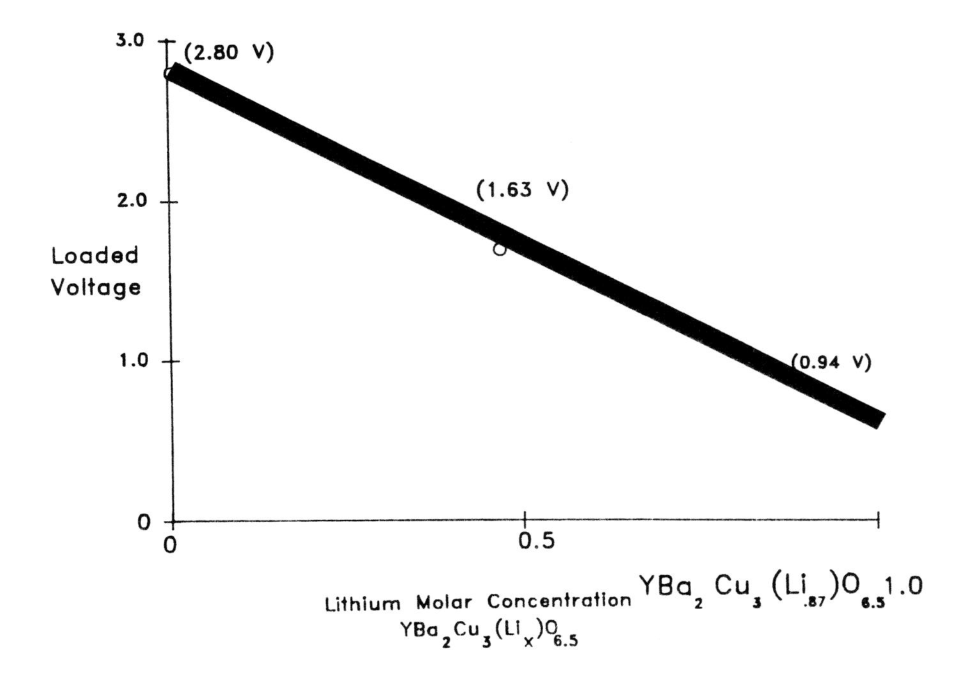

Figure 3. Discharge curve

The transition temperature of the reacted oxide after tear down, when measured by magnetic suseptibility, indicated no shift in transition temperature (93°K). For correlation, the transition temperature was remeasured two weeks later where the sample was found to be an insulator. During the two-week period, the samples had been maintained in heptane but were exposed to atmospheric conditions where oxygen or water absorption into the heptane reacted over time to quench the superconductive property. Analysis of the reaction product is on-going.

X-ray diffraction analysis between the starting and reacted superconductivity state cathode material indicated no new phase was present (Fig. 4), supporting lithium intercalation into the structure. Additional work to evaluate the lithium intercalation process is being performed.

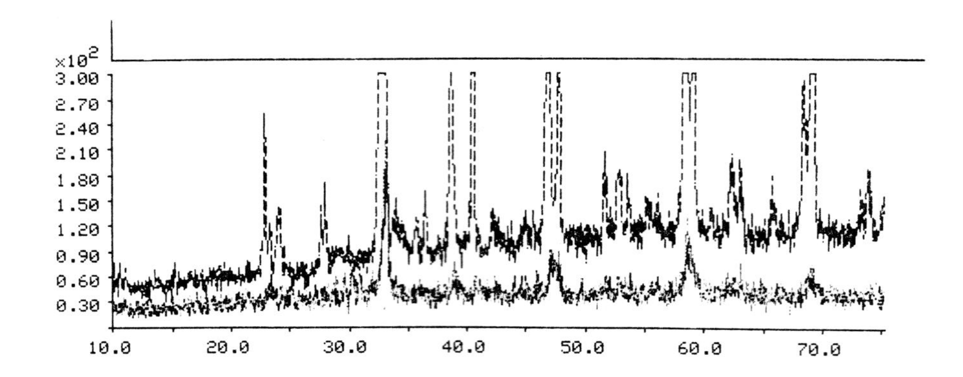

Figure 4. X-ray diffraction:
Upper pattern - initial $YBa_2Ca_3O_x$ (from powder pattern)
Lower pattern - reacted cathode (in pressed form)

CONCLUSION

Upon completion of this study, the following conclusions were reached:

1. Electrochemical methods can be utilized to introduce cations into superconductive oxides. These methods could be used to modify crystal structure or valence ratio of multivalent cations.

2. The introduction of lithium did not shift the superconductive transition temperature until environmental exposure, at which time all superconductive properties were lost.

3. Lithium reacted by intercalation into the oxide structure, which could provide a reversible mechanism for cation injection.

ACKNOWLEDGEMENT

The author wishes to thank researchers at MARTECH, Florida State University, Tallahassee, FL, especially Prof. Louis Testardi, for their technical support of this project and researchers at the University of Florida, Gainesville, FL, especially Profs. David Tanner and Greg Steward, for their support on magnetic suseptibility measurements.

REFERENCES

1. H.V. Ventatasaty, Lithium Battery Technology, The Electrochemical Society, J. Wiley, NY, pp. 63-74 (1984).

2. H. Kitazawa, et al., "Superconductor Electrochemically Produced; Osaka University," Nihon Keizai Simbun, Dec., p. 9, (1987).

3. D.J. Zurawski, and A. Wieckowski, Journal of the Electrochemical Society, June, (1988).

4. M.K. Wu, J.R. Ashburn, C.J. Toug, P.H. Hor, R.L. Meng, L. Gao, Z.J. Huang, Y.Q. Wang, and C.W. Chu, Phys. Rev. Lett. 58, 908 (1987).

5. Batteries for Implantable Biomedical Device, pp. 196-199, edited by B.B. Owens, Plenum Press, (1986).

4.12 Devices and Applications

Applications of Oxide-Superconductors to Electronic Devices; Optical Detector

HIROSHI IWASAKI

NTT Opto-Electronics Laboratories, Nippon Telegraph and Telephone Corporation,
Tokai, Ibaraki, 319-11 Japan

ABSTRACT

Many electronic applications for superconductors have been proposed. However, the many problems associated with using oxide superconductors for electronic devices limit their use. A potential application of oxide superconductors to optical detectors is reviewed.

INTRODUCTION

The development of superconducting materials is expected to bring forth revolutionary advanced electronic devices. This expectation of new superconducting electronic devices has increased remarkably since the discovery of the new high-T_c superconductors because these devices can operate at liquid nitrogen temperature. Many electronic applications for high T_c superconductors have been proposed. However, the many difficulties associated with using oxide superconductors for electronic devices limit their use. These include the problems associated with chemical and physical properties different from those of conventional superconducting materials which prevent their use of in conventional fabrication processes. Initial potential applications of high-T_c oxide superconductors to electronic devices are expected to start with the simulation of the already demonstrated low-T_c applications with overcoming the stated issues. This paper discusses an optical detector as a typical example of a simple Josephson device, (simple means this device can be fabricated with only one polycrystalline film using grain boundary junctions) comparing conventional low T_c material application and high T_c oxide material application.

OPTICAL DETECTOR

Testardi[1] discovered in his experiment, the radiation effect caused by the light absorption appearing in Pb thin films. Since this finding, this type of experiment has frequently been used as a means to study nonequilibrium phenomena in superconductors.[2] In a conventional metal superconductor, however, incident light is reflected from the surface of the superconductor as a result of the plasma oscillation of free electrons and light cannot penetrate into it. On the other hand, as oxide superconductors, such as $Ba(Bi,Pb)O_3$ (BPB) and recently discovered high T_c oxide materials, have small reflective coefficients, light can easily penetrate into the these materials and effectively excite quasi-particles. Thus BPB films and new high T_c materials can be used as optical detectors.[3] Recently, M. Leung et al., reported the result of optical detection in BYCO granular films.[4] Their experiments showed that the upper boundary of the minimum detectable power was 1 μW and the response time is of the order 20 ns. Table 1 shows the characteristics of superconducting optical detectors reported to date. Early experiments on BPB detectors are reviewed and preliminary results on high T_c material films are described.

a) Detection Principle

Generally, a superconductor has an energy gap between the Cooper pairs and the quasi-particles which is equal to the superconducting order parameter 2Δ. When photons penetrate into the superconductor, the Cooper pairs are broken into two quasi-particles and the number of excess quasi-particles thus produced is in proportion to the incident optical power. This increase in quasi-particles leads the superconductor to a nonequilibrium state and to a decrease in Δ. This nonequilibrium state in the superconductor is studied experimentally as well as theoretically. When the excess quasi-particle density n_{qp} is small, a decrease in the order parameter $\delta\Delta$ is described approximately by

$$\delta\Delta = n_{qp}/N(0)$$

Table 1 CHARACTERISTICS OF SUPERCONDUCTING OPTICAL DETECTORS

MATERIAL	BPB	LSCO	YBCO	NbN
RESPONSIVITY	10^4V/W	10^3V/W (same as BPB)	10V/W(at present) 10^6 V/W*	0.7V/W
NOISE	at 6K $\delta Vn \cong 10^{-9} V/Hz^{1/2}$	at 6K $\delta Vn \cong 10^{-9} V/Hz^{1/2}$	at 6K $\delta Vn \cong 10^{-9} V/Hz^{1/2}$ at 60K $\delta Vn \cong 10^{-8} V/Hz^{1/2}$	—
RESPONSE TIME	10^{-9}sec	below 10^{-6}sec	10^{-3}sec (NRL 10^{-7}sec)	10^{-9} sec
WAVE REGION	1~500 μm	1~100 μm	1~30 μm at High temp. λ max is larger	50~500 μm
INSTITUTION	NTT [3]	NTT [5]	NTT [6] NRL [4] HITACHI [7]	NRL [8]

* assuming junctions in the unit cell

Table 2
SUPERCONDUCTING PROPERTIES RELATED TO SENSING CHARACTERISTICS

- SENSITIVITY: S

$$S \propto \frac{N}{N(0) \cdot V} (1-R)$$

N(0): THE DENSITY OF STATES AT THE FERMI LEVEL
R : REFLECTION COFFICIENT
N : THE NUMBER OF GRAIN-BOUNDARY JUNCTION
V : VOLUME OF SENSING PART (RELATED TO FILM THICKNESS:t)

- RESPONSE : $1/\tau$

$$\tau \cong \tau_{qp} + \tau_{ph}$$

τ_{qp}: RELAXATION TIME OF QUASIPARTICLE ($\sim kTc/\Delta_0$)
τ_{ph}: THERMAL RELAXATION TIME OF PHONON ($\sim t$)

- WAVE REGION (MAXIMUM WAVELENGTH) : λ max

$$\lambda \max \propto \frac{1}{\Delta}$$

- WORKING TEMPERATURE : T

$$T < Tc$$

- NOISE : δ Vnoise

$$\delta Vnoise \propto kT^{1/2} \cdot e^{-\frac{\Delta}{kT}} \cdot \sqrt{N}$$

T: WORKING TEMPERATURE
Δ: GAP ENERGY (ORDER PARAMETER OF SUPERCONDUCTIVITY)

here N(0) is the density of states at the Fermi level. In oxide superconductors $\delta\Delta$ is expected to be rather large compared with that of metallic superconductors because of the small N(0). The decrease $\delta\Delta$ changes the I-V characteristic of the superconducting tunnel junction. Therefore, optical signals can be transformed to electrical ones through the order parameter change. Further, in a polycrystalline film, the changes are larger due to the accumulation effect of many series-connected grainboudary Josephson junctions within the irradiated area.

For example, the N(0) of BPB is about 0.15 states /unit cell eV spin for x = 0.25 ($Ba(Pb_{1-x}Bi_x)O_3$) which is one order of magnitude smaller than that for conventional superconductors. So, from above equation, $\delta\Delta$ in BPB becomes larger for the same value of n_{qp} than that in other superconductors.

Table 2 summarizes the other principle related to optical sensing abilities of the optical detector using the grain boundary Josephson junctions.

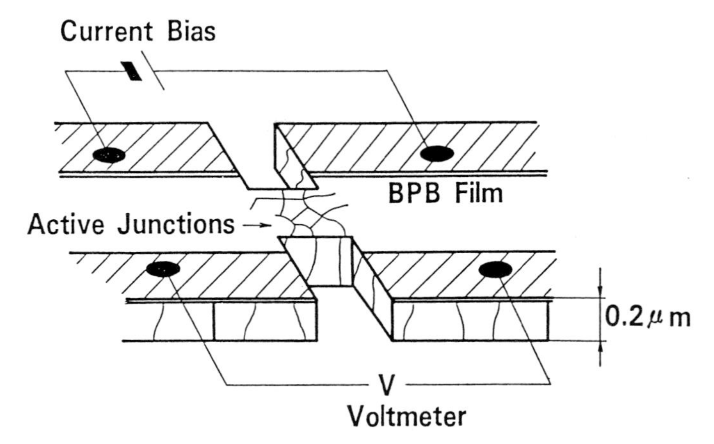

Fig. 1 Experimental setup for a BPB detector

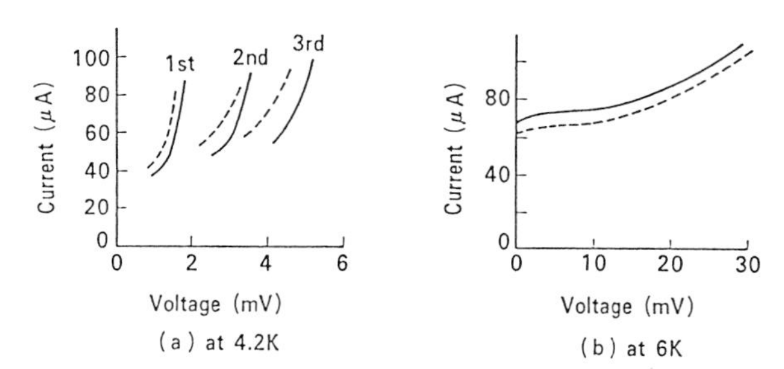

Fig. 2 Change in the I-V characteristics by Irradiation at two different temperatures.

b) BPB optical detector

The experimental setup for a BPB detector is shown in Fig. 1. Polycrystalline BPB films are deposited onto sapphire substrates by RF sputtering in a mixture of Ar and O_2.[5] The film is etched into the pattern shown in Fig. 1. Figure 2 shows the typical experimental results of the I-V curve change by irradiation.[2] The I-V curve changes from the solid lines to the dotted ones in Fig. 2. Each curve corresponds to a single particle tunneling of series-connected Josephson junctions having uniform areas and uniform critical current densities. The relationship between the incident optical power and the voltage shift for detectors having different film thicknesses is shown in Fig. 3. It is clear that the voltage shifts increase proportionally to the optical power. The responsivity i.e., the ratio between the voltage shift and optical power, is determined from this linear relationship. The responsivity increases with decreasing film thickness, rising to 10^3 V/W for a 0.15 μm thick film. The increase in responsivity with decreasing film thickness is probably due to increased density in the excess quasi-particles, n_{qp}, which increases according to a factor of 1/t (the reciprocal of the detector volume). The wave length dependence of the responsivity is shown in Fig. 4.[2] The responsivity increases slightly with wavelength in the 1 to 8 μm range. This is because the photon number per unit energy increases with the wavelength. In Fig. 5, frequency dependence of a BPB detector responsivity is shown for sinusoidal signals from a laser diode of 1.3 μm wavelength. The responsivity exhibits only a small decrease over the 50-700 MHz frequency region.[2] The abrupt decrease is caused by the unconformity of our amplifier used. Consequently, the response time is considered to be shorter than 1 ns.

The lowest detectable power level is one of the important characteristics. Though the noise level estimation is very difficult because of a lack of the knowledge related to the noise generation mechanism, calculating noise equivalent power (NEP) and D*(detectivity) is attempted using the noise predicted by the resistivity shunted junction model(RSJ model). Table 3 shows the calculation scheme of NEP of the BPB optical detector. Figure 6 shows the relationship between detectivity(D*) and wavelength for various kinds of semiconductive infrared detectors compared with the BPB optical detector.

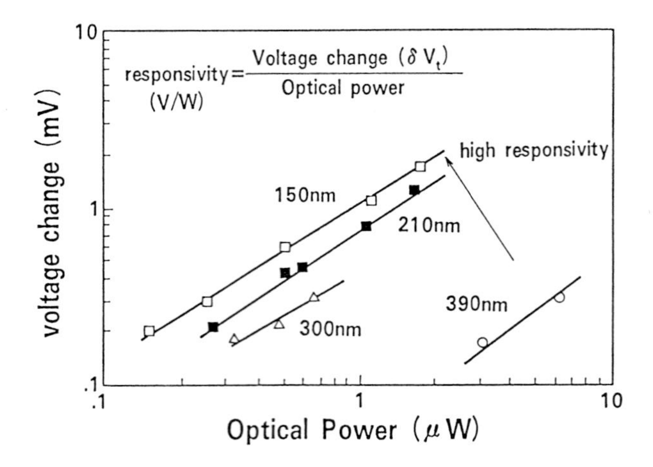

Fig. 3 Voltage change for BPB detectors with various thickness

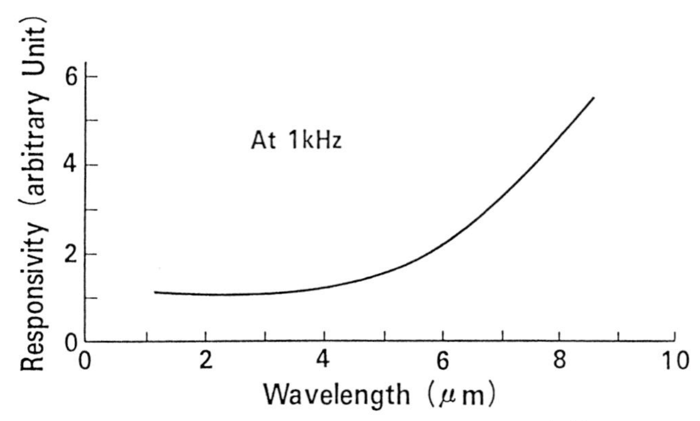

Fig. 4 Wavelength dependence of responsivity

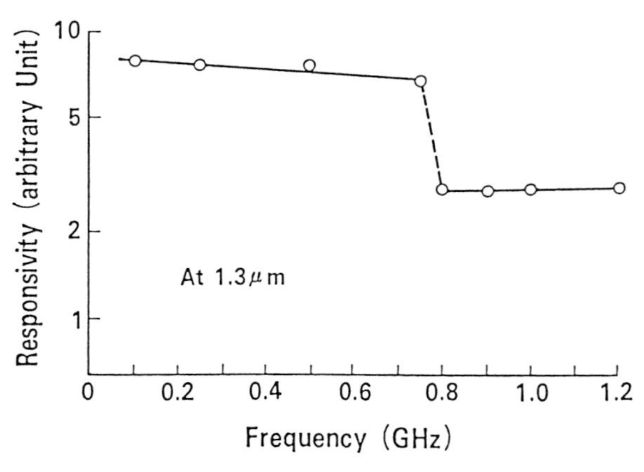

Fig. 5 Frequecy dependence of responsivity

Table 3

ESTIMATION OF NEP(NOISE EQUIVALENT POWER)

• DEVICE CONDITIONS

SENSITIVE AREA: $10\times10\mu m^2$	
$Rt=NRn: 2.5\times10^2\Omega$	(Rt: TOTAL RESISTANCE OF SERIES-CONNECTED JUNCTON ARRAY
	N : NUMBER OF JUNCTION
	Rn: NORMAL RESISTANCE OF JUNCTION)
Rd/Rt : 5	(Rd: DYNAMIC RESISTANCE AT BIAS POINT)
Ib/Ic : 1	(I_b : BIAS CURRENT, Ic: CRITICAL CURRENT)
T : 6K	WORKING TEMPERATURE

• ESTIMATED NEP AND D*

$\delta V_n \cong \delta V_{nj} = (R_d/R_n)\times(1+I_b^2/2I_c^2)^{1/2}(4kTR_nB)^{1/2}$

B: WAVE BAND WIDTH

USING ABOVE CONDITIONS

$\delta Vn = 1\times10^{-9}V/Hz^{1/2}$

$NEP = 7\times10^{-14}W/Hz^{1/2}$

$D^* = 3\times10^{11}cmHz^{1/2}/W$

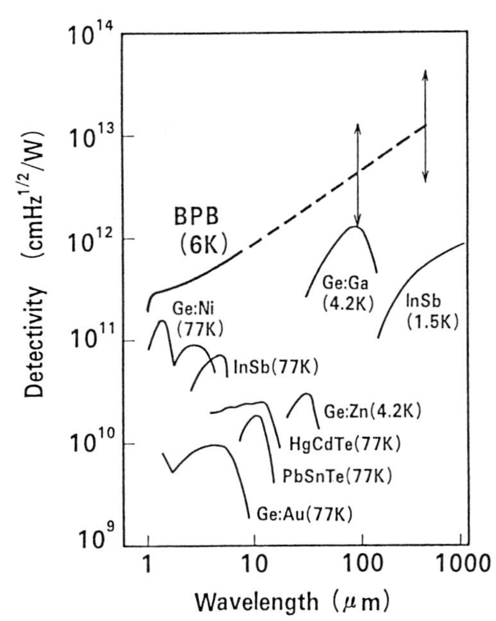

Fig. 6 Wavelength dependence of dectivity for various kinds of detectors

c) Preliminary results for high T_c oxide materials [6]

Optical reflectance spectra of a BYCO($Ba_2YCu_3O_\delta$) single crystalline thin film are shown in Fig. 7. The symbols denote "parallel and perpendicular to the *a-b* plane." As expected from the crystal structure, optical reflectance spectra of BYCO thin film show large anisotropic properties. When the polarization is parallel to the *a-b* plane, the reflectance is similar to the case of BPB. The other spectra exhibit characteristics of insulators, and light can penetrate deeply into films.

In the experiment, 1.3 μm laser light with various power levels is irradiated onto the 100 μm^2 area of a BYCO single crystalline film which has a T_c of 80 K and a film thickness of 700 nm. The change of the I-V curve along the *c*-axis is measured. The I-V curve along the *c*-axis is similar to that of a polycrystalline BPB film at 6 K which shows an I-V curve of series-connected Josephson junctions. Figure 8 shows voltage change due to optical irradiation power. The bias current is fixed to the value which gives maximum voltage change. The responsivity of this BYCO film is smaller by factor 4 than those of BPB films. This small responsivity can be due to the film thickness , the wide pattern and the reflectance anisotropy. Figure 9 shows frequency dependencies of light irradiation responsivity of a BYCO film. The responsivity of the BYCO film exhibits dispersions at around 100 KHz-1 MHz, and these frequencies are three orders of magnitude smaller than in the case of BPB. This slow response suggests that the BYCO film works as a bolometer.

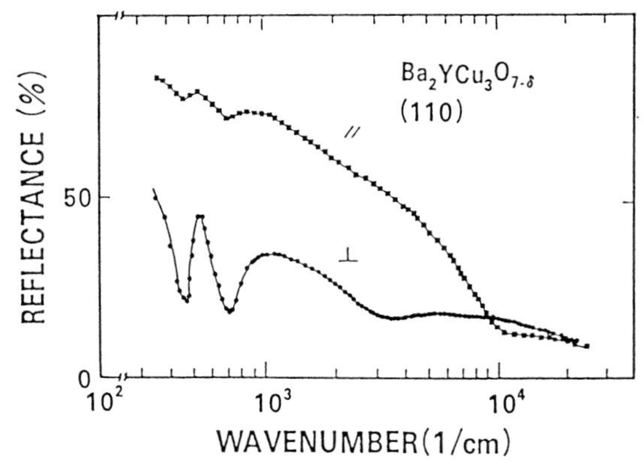

Fig. 7 Optical reflectance of a BYCO thin film

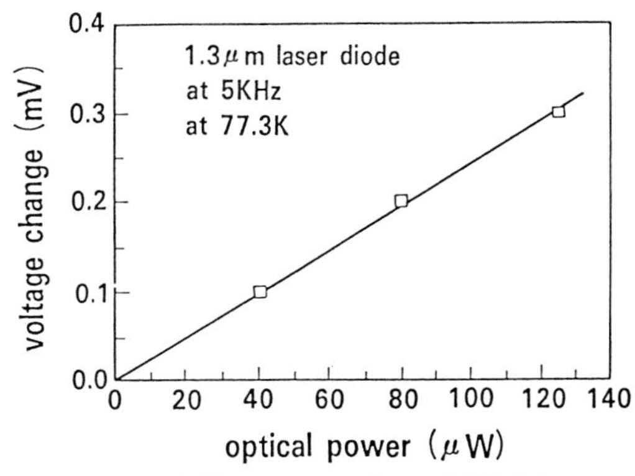

Fig. 8 Voltage change for a BYCO film (film thickness;700 nm irradiated areaa;100x100 μmm^2)

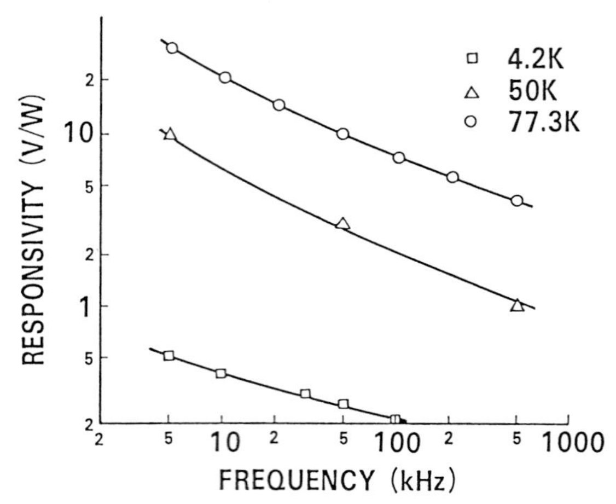

Fig. 9 Frequency dependencies of a BYCO thin film

CONCLUSION

As an example of application of oxide-superconductors to electronic devices, optical detectors are reviewed. A BPB optical detector can respond up to more than 1.3 GHz at the 1.3 μm wavelength. This fast responce results from the strong interaction among quasi-particles, phonons and Cooper pairs in BPB, which is a strong coupling superconductor. A BYCO thin film can also detect optical power, but it is inferior to a BPB film in the points of responsivity and frequency characteristic at present, but we can expect further research efforts for improvement of the film quality making it possible to realize highly responsive and wide band infrared detectors of high T_c oxide materials.

REFERENCES

[1] L.R. Testardi, Phys. Rev. 134 2189 (1971)
[2] D.N. Langenberg, In Low Temperature Physics-LT14 15 223
[3] Y. Enomoto and T. Murakami, J. Appl. Phys. 59 3807 (1986)
[4] M. Leung, P.R. Broussard, J.H. Claassen, M. Osofsky, S.A. Wolf and U. Strom, Appl. Phys. Lett. 51 2046 (1987)
[5] k. Moriwaki, Y. Enomoto and T. Murakami, Jap. J. Appl. Phys. 26 1147 (1897)
[6] Y. Enomoto, T. Murakami and M. Suzuki, Physica C 153-155 1592 (1988)
[7] U. Kawabe, Physica C 153-155 1586 (1988)
[8] J.H. Classen, J. Appl. Phys., 556 3367 (1984)

The Possibility of Superconducting Transistor and Squid with High-Tc Oxide Superconductors

USHIO KAWABE

Central Research Laboratory, Kokubunji, Tokyo, 185 Japan

I. Introduction

Application of superconducting oxides to analog and digital devices has been extensively given attention since Bednortz and Muller discovered high-critical-temperature superconductivity in the La-Ba-Cu-O system[1). Subsequently superconductors having a critical temperature above liquid-nitrogen temperature were discovered in the Y-Ba-Cu-O system by Wu et al.[2), in the Bi-Ca-Sr-Cu-O system by Maeda[3), and in the Tl-Ca-Ba-Cu-O system by Sheng et al[4). Liquid nitrogen is about 1/30 in the cost of liquid helium and more available in quantity of resource. If such materials can be applied to analog and digital devices, we can use the superconducting phenomena easily and profitably.

In this paper, it is worthwhile to discuss the possibility of typical analog and digital devices, taking two typical examples of a multi-layered dc-SQUID with superconducting $YBa_2Cu_3O_x$ films and a proximity-effect superconducting transistor, which are much expected as a candidate for the application of high-temperature superconductors. The present situation and the research further needed are duscussed.

II. High-temperature superconducting oxide films

The oxide superconductors discovered by Wu et al.[2), Maeda[3) and Sheng et al.[4) are the 90K-, 120K- and 125K-class superconductors, respectively. These critical temperatures are over liquid nitrogen temperature. Liquid nitrogen is about 1/30 the cost of liquid helium and more available in quantity of resource. If we can make good use of such materials, the application of superconductivity must become more prosperous in the field of a highly sensitive analog device and a high-speed, high-integration digital device.

It is two properties of superconducting films; the critical temperature and the critical current density at the zero magnetic field that is most important in the field of electronic-device use. The critical temperature of high-temperature-superconducting films is a function of the substrate used, the compound composition, the oxygen partial pressure, the substrate temperature and the temperature cooling program. The critical current density has a function of the Cu-O layer structure, the oxygen content, the film stress and the micro-structure containing twin, defect, impurity grain boundary.

To form a high-quality, high-temperature-superconducting oxide film for electronic-device use, however, there are still some problems that a specific substrate, such as MgO, $SrTiO_3$ or Y-stabilized Zr_2O_3, is needed, that the substrate temperature is high, and that the film morphology depends on the crystal orientation of the substrate and on the lattice mismatch between film and substrate. The high-temperature-superconducting oxide film thus obtained has some intrinsic properties that the anisotropy of current flow is strong between the ab-plane and the c-axis due to the orthorhombic crystal structure, and that there is a strong correlation among superconducting properties, oxygen content and film stability. At the present stage, it is still difficult to make a complex device with the high-temperature superconductors.

III. Posasible application of oxide superconductors to SQUIDs

The possible application of oxide superconductors to an analog device is discussed taking a typical example of SQUIDs. The SQUIDs are very useful as an ultra-sensitive magnetometer for biomagnetism and geomagnetism. Both rf and dc SQUIDs have recently been fabricated using $YBa_2Cu_3O_x$ superconductors, and investigated around liquid nirtogen temperature[5)6)]. However, there are still many problems for obtaining high-resolution SQUIDs with a superconducting $YBa_2Cu_3O_x$ film. The magnetic flux sensitivity of new oxide SQUIDs is one thousandth lower than that of a conventional metal SQUID, which is made from either Nb or Pb alloy superconductor.

The peculiar superconducting properties of oxide superconductors and an increase in noise with increasing the operation temperature are obstacles to SQUID design. The high formation temperature of a superconducting phase near 900℃ makes the device fabrication technologies very difficult. Layered structures comprising superconductors and insulators, and sandwich-type Josephson junctions, S-N-S and S-I-S, could not be realized due to the high annealing temperature of $YBa_2Cu_3O_x$ thin films. Oxygen concentration instability of the $YBa_2Cu_3O_x$ thin film breaks down the superconducting qualities during the device fabrication. Very short superconducting coherence length of near 1 nm demands superconductors of highly uniform condition. There is also an increase in thermal noise when the SQUIDs are operated at liquid nitrogen temperature[6)].

From our experimental results[7], the problem of a dc SQUID using the $YBa_2Cu_3O_x$ thin film was considerably clarified. Grain boundary Josephson junctions were used for the dc SQUID. The SQUID performance was correlated with the Josephson junction characteristics. The $YBa_2Cu_3O_x$ thin film SQUID operated at high temperatures was noisier than Nb or Pb alloy thin film SQUID operated at liquid helium temperature, as above-described. The noise problem related to SQUID quality and to thermal noise has to be further studied. A layered structure including insulating layer and input coil in addition to a SQUID inductor was also examined from a view-point of fabrication processing. The close structure between the input coil and the SQUID loop is necessary to increase the inductive coupling and to decrease the SQUID inductance. A dry-etching method for fine pattern formation was especially investigated for miniaturization of the SQUID inductor.The fine pattern was found to be thereby obtained up to 0.8 μm in line width. The protection layer of a MgO film has to be formed between $YBa_2Cu_3O_x$ films so as to prevent the mutual diffusion between layers and the critical temperature detrioration of $YBa_2Cu_3O_x$ films. A gold film was selected as the protection layer. A tri-layer of $YBa_2Cu_3O_x$-Au-MgO was tried. The critical temperature of the tri-layer was found to decrease on annealing with increasing the gold film thicknesss. The decrease in critical temperature was kept within 10K for the tri-layer film with a 350nm-thick gold film. The primitive tri-layer dc-SQUID composed of the SQUID inductor, the insulating layer and the input coil layers has already been fabricated and tested. Further improvement is needed.

IV. Possible application of oxide superconductors to superconducting transistor

THe possible application of high-critical-temperature oxide superconductors to a digital device is discussed, taking an example of typical field-effect superconducting transistors which are much expeted as a candidate for the high-temperature superconductors.

The possible field-effect superconducting transistor was first proposed by Clark[8]. He especially proposed the possibility of MOS (Metal Oxide Semiconductor)-type and MES (Metal Semiconductor)-type JOFET (Hybrid Josephson Field Transistor) obtained by a proximity combination of two long-coherent metal superconductors with either InAs or InSb semiconductor having high-electron mobility. However, the first realization was done in the proximity combination of two Pb-ally superconductors with a p-type Si semiconductor in 1984[9], and immedeately in the basic operation of a 0.5-μm-gate-length superconducting transistor with n-type InAs by NTT[10]. In 1982, Chalmer University of Technology did already in the basic experiment of 0.4-μm-gate-superconducting three-terminal device[11]. However, these devices had longer gate length because of high-electron-mobility compound semiconductors, into which the superconducting electron wavefunctions were easy to penetrate deeply. This simplified the basic experiments, but it had difficulties in higher speeds and higher levels of integration from the viewpoint of a future device.

It was, however, difficult to make a circuit with many transistors because the transisor used for our basic operation had the simple structure that the oxide-insulated gate was not on the same Si plane with source and drain electrodes. In order to resolve this problem, the fabrication and test were tried for a 0.1 μm-gate-length field-effect superconducting transistor having an oxide-insulated gate on the same Si plane with source and drain electrodes[5]. For fabricating the structure of this superconducting tansistor, a thin oxide-insulated gate electrode with the stencil structure of a special shape was formed on a Si substrate, from which top As ions were implanted and two thin layers of almost metallic conduction were formed. These two layers have to be made proximate in such a 0.1 μm-long range that two superconducting electron waves can be controlled by the electric field. Therefore, the finest insulated gate electrode of a gate length 0.1 μm was realized by electron beam lithography. After the surface cleaning of a Si semiconductor, from the top a Nb film was formed as the superconducting source and drain electrodes. From our experiments, the structure adequate for high level of integration was made by forming the shorter insulated gate electrode on than the usual gate length the same Si plane with source and drain electrodes. It is thereby possible to fabricate high speeds and higher integarations of the superconducting transistor. From the measured drain current-voltage characteristics featuring a superconducting transistor, the drain current increased abruptly near zero drain voltage with increasing the gate voltage, and it remained to be nearly zero at a gate voltage below 0.5 V. This value is still too high for a real transistor. This has to be solved by further research.

This transistor has a small internal impedance and source-drain electrodes made of a superconductor and so connections between transistors are possible by a lossless superconducting wiring, which is adequate for the transmission lines of high speed circuit networks. The transistor was also tested at 4.2 K because a superconducting Nb metal film with the critical temperature of 9.2 K was used for electrode materials. In a practical use, it is necessary that the superconducting transistor can be economically operated at liquid nitrogen temperature rather than at liquid helium temperature. It is, therefore, expected to apply the recently discovered high-temperature superconducting oxides to the electrode materials having the critical temperature above liquid nitrogen temperature. This superconducting oxides have a strong anisotropy in current flow due to the crystal-structural anisotropy, which is different from the isotropy in current flow for the superconducting Nb metal used in our experiments. Further research is needed for the realization of HTSC transistors. Many technologies have to be solved on 0.1 μm-level fine patterning, small-gate-electrode formation, epitaxial formation of superconducting source and drain electrodes, strain-free semiconductor and buffer materials, stable contact between semiconductor and superconductor, and lossless superconducting wire-bonding.

V. Further research needed for high-temperature-superconducting devices

Further material research is needed for digital and analog devices with high-temperature superconductors(HTSC).

(1) Possible low-temperature epitaxial formation of a clean HTSC thin film having a critical current density above 10^6 A/cm² and a critical temperature above 150K.

(2) Stable clean surface of oxygen-deficient HTSCs.

(3) Possibility of current noise reduction with increasing temperature.

REFERENCES

1)J.G.Bednorz and K.A.Muller: Z.Phys. B64, 189(1986)

2)M.K.Wu, J.R.Ashburn, C.J.Torng, P.H.Hor, P.L.Meng, L.Gao, Z.J.Huarg, Y.Q.Wang, and C.W.Chu: Phys.Rev.Lett. 58, 908(1987)

3)H.Maeda, Y.Tanaka, M.Fukutomi and T. Asano: JPN.J.Appl.Phys.Lett. 27, L209 (1988)

4)Z.Z.Sheng, A.M.Hermann, A.Elali, C.Almason, J.Estrada, T.Datta and R.J.Matson: Phys.Rev.Lett. 60, 937(1988).

5)R.H.Koch, C.P.Umbach, G.J.Clark, P.Chaudhari, and R.B.Laibowitz: Appl.Phys. Lett: 51, 200(1987).

6)H.Nakane, Y.Tarutani,T.Nishino, H.Yamada and U.Kawabe: Jpn.J.Appl.Phys. 26, L1925(1987).

7)Y.Tarutani, H.Nakane, T.Yamashita, T.Aida and U.Kawabe: Intl. Meeting on Advanced Materials, Tokyo(1988).

8)T.D.Clark: J.Appl.Phys., 51, 2736(1980)

9)NIKKEI ELECTRONICS(in Japanese) 12, 105(1984)/T.Nishino, M.Miyake, Y.Harada, U.Kawabe: IEEE Electron Devices Lett. 6, 297(1985).

10)H.Takayanagi and T.Kawakami: Phys.Rev.Lett. 54, 2449(1985)

11)Z.Ivanov, T.Claeson and T.Anderson: Jpn.J.Appl.Phys. Suppl.26, 1619(1987)

12)T.Nishino, M.Hatano, H.Hasegawa, U. Kawabe, F.Murai, T.Kue and K.Yagi: to be submitted to IEEE Electron Device Lett.

Magnetic Properties of High Temperature Superconductors: Implications for 1/f Noise in Squids

R.H. KOCH and A.P. MALOZEMOFF

IBM Research Division, Thomas J. Watson Research Center, Yorktown Heights, NY 10598-0218, USA

ABSTRACT

Relatively weak flux pinning in the new high temperature superconductors, coupled with a distribution in the pinning barriers strengths, is shown to imply large 1/f flux noise in Superconducting QUantum Interferences Devices (SQUIDs). A simple model is developed that is in order-of-magnitude agreement with experiments on thin-film SQUID devices.

PACS numbers : 74.60.Ge , 74.70.Vy

INTRODUCTION

Magnetic measurements of high temperature superconductors show unusually large time-logarithmic relaxation. This is true not only in ceramics,[1-4], where weak coupling between grains provides an easy path for flux motion, but also in crystals[5-6] and epitaxial films.[7] This relaxation has been interpreted in terms of a thermally activated flux creep model[5,8-11] and appears to be large because of the low pinning energies and relatively high temperatures. There is also a likelihood that a distribution of pinning barrier heights[11-12] may dominate the experimental behavior, particularly at temperatures approaching T_c .

In this article, we explore in a preliminary way, the implications of these phenomena for applications, in particular how the flux motion can give rise to 1/f noise which limits the performance of electronic devices like SQUIDs at low signal frequencies. A simple model allows us to estimate the magnitude of the 1/f noise in order-of-magnitude agreement with recent experiments[13-15] on thin film devices made of YBaCuO.

While magnetic relaxation arises from a non-equilibrium distribution of flux lines in a Type II superconductor, noise will arise even from an **equilibrium** distribution, because of random thermal activation over energy barriers of height U. We seek to calculate the noise energy per unit frequency, $\varepsilon(f)$, at a signal frequency f, for comparison with recent data[13-15] which has shown a strong 1/f spectrum, with amplitudes in the range of 10^{-27} to 10^{-26} J/Hz at 1 Hz and 77 K in good quality YBaCuO epitaxial films and SQUIDs.

MODEL

We suppose that the noise is flux noise arising from hopping of flux lines. The hopping of a single flux line with an average rate $1/\tau$ gives a contribution to ε of

$$\varepsilon_1(\omega) = \frac{(\Delta\phi)^2}{2L} \frac{\tau}{\pi(1 + \omega^2\tau^2)} , \tag{1}$$

where L is the inductance of the SQUID loop, ω is the angular frequency ($\omega = 2\pi f$), and $\Delta\phi$ is the change in flux in the loop due to each flux jump.

We presume that the relaxation times are exponentially dependent on a barrier height U according to

$$\tau(U) = \tau_0 \exp(U/kT) , \tag{2}$$

where τ_0 is an attempt time and T is the temperature. We furthermore presume the barrier heights are distributed over some broad range described by a distribution function P(U) normalized such that

$$\int_0^\infty P(U)\,dU = 1 \ , \tag{3}$$

and with an average $\overline{U}$ defined as

$$\overline{U} = \int_0^\infty U\,P(U)\,dU \ . \tag{4}$$

The total energy spectrum of the device is then Eq. 1, multiplied by the total number of flux lines N and integrated over U. For a sufficiently broad spectrum of U, this is a well-known problem[16] whose solution is

$$\varepsilon(\omega) = \frac{(\Delta\phi)^2}{2L}\frac{NkT}{\omega} P[kT \ln(1/\omega\tau_0)] \ . \tag{5}$$

This expression gives the 1/f noise spectrum and shows that the temperature dependence of the spectrum allows one in principle to determine the shape of the barrier distribution.

To establish a closer relation of this expression to experimentally measurable quantities, we express the number of flux lines N in terms of the DC bias current I_B in the SQUID using Ampere's law,

$$N \simeq \overline{\mu}\mu_0 I_B CD/4t\Phi_0 \ , \tag{6}$$

where $\overline{\mu}\mu_0$ is the effective permittivity of the body of the SQUID, C is the perimeter of the SQUID loop, D is the width of the loop, t is the film thickness, and Φ_0 is the flux quantum. We also estimate the change in flux in the loop due to a flux jump of distance d to be

$$\Delta\phi = \Phi_0 d/D \ . \tag{7}$$

Finally, we note that for a simple flat distribution out to a maximum value $2\overline{U}$, the amplitude of P(U) is just $1/2\overline{U}$. Combining the last three equations and the estimate for P(U), we have

$$\varepsilon(f) = \frac{\overline{\mu}\mu_0 I_B \Phi_0}{2L}\frac{d^2C}{4Dt}\frac{kT}{f2\overline{U}} \tag{8}$$

where f is the measurement frequency. Expressing this result using the usual SQUID parameter $\beta = 2I_0L/\Phi_0 \simeq 1$ and approximating I_B as I_0 and L as $\mu_0 C$ we find

$$\varepsilon(f) = \frac{\Phi_0^2}{2L}\frac{\beta d^2}{8Dt}\frac{\overline{\mu}kT}{2\overline{U}}\frac{1}{f} \ . \tag{9}$$

We next estimate the size of the 1/f flux noise at 1 Hz. Values of $\overline{U}$ around 0.02 eV have been deduced from magnetic relaxation and $J_c(T)$ measurements in crystals.[5] Similar values, 0.01 to 0.02 eV, have been measured for the mean activation energy for flux motion noise in epitaxial thin films.[17] We also use a typical SQUID inductance of about 200 pH, a film cross section of 1 μm by 10 μm, $\overline{\mu}$ as 1, and a flux hopping distance of 0.01 μm. This gives at 77 K a value of 3×10^{-27} J/Hz, in order-of-magnitude agreement with experiment. Of course this result depends strongly on the choice of the flux hopping distance, which is not really known at present. Our guess of about an order-of-magnitude times the in-plane coherence length seem reasonable, considering the likelihood of core pinning in these materials.

CONCLUSIONS

While large 1/f noise has been known for some time in electronic devices made from high temperature superconductors, our calculation appears to give the first insight into the origin of this noise, which sets a serious limitation on the operation of superconducting devices, at least in the low frequency regime. The calculation supports the idea that the noise is flux noise, and is closely related to the relatively low pinning energies and to the giant flux creep observed in these materials.

This connection makes it obvious that understanding flux pinning is required to minimize the noise. So far most measurements of flux creep have shown comparable behavior, and it is not yet clear to what extent the flux pinning can be increased. The present estimates have in one sense taken the most favorable case, namely values of $\overline{U}$ for bulk - rather than ceramic - material. In ceramic material, where flux can move more easily at grain boundaries, one could expect even greater flux motion, and indeed recent experiments[14] seem to show a correlation between the $1/f$ noise level and the epitaxial quality of the film. Clearly it is of the greatest importance to understand more precisely the nature of flux pinning in the high temperature superconductors so as to optimize the operation of SQUIDs and other devices.

The authors thank W. W. Webb for a valuable conversation which initiated this line of thought, J. Clarke for communication of experimental results before publication, and T. Worthington, R. Yandrofski and Y. Yeshurun for discussion of the barrier distribution problem.

REFERENCES

1. K. A. Müller, M. Takashige and J. G. Bednorz, Phys. Rev. Lett. **58**, 1143 (1987).
2. A. C. Mota, A. Pollini, P. Visani, K. A. Müller and J. G. Bednorz, Phys. Rev. B **36**, 401 (1987).
3. C. Giovannella, G. Collin, P. Rouault and I. A. Campbell, Europhys. Lett. **4**, 109 (1987).
4. M. Tuominen, A. M. Goldman and M. L. Mecartney, Phys. Rev. B **37**, 548 (1988).
5. Y. Yeshurun and A. P. Malozemoff, Phys. Rev. Lett. **60**, 2202 (1988); Y. Yeshurun, A. P. Malozemoff and F. Holtzberg, J. Appl. Phys., to be published (Proceedings of the Joint Intermag/3M Conference, Vancouver, Canada, July 12-15, 1988); Y. Yeshurun, A. P. Malozemoff, F. Holtzberg and T. Dinger, Phys. Rev. B, submitted.
6. M. Tuominen, A. M. Goldman and M. L. Mecartney, Physica, to be published;
7. C. Rossel and P. Chaudhari, Physica, to be published.
8. A. P. Malozemoff, T. K. Worthington, Y. Yeshurun, F. Holtzberg and P. H. Kes, Phys. Rev. B, to be published; T. K. Worthington, Y. Yeshurun, A. P. Malozemoff, W. J. Gallagher, R. M. Yandrofsky, F. Holtzberg, T. R. Dinger and D. L. Kaiser, submitted to the International Conference on Magnetism, Paris, France, July 25-29, 1988.
9. A. P. Malozemoff, T. K. Worthington, R. M. Yandrofski and Y. Yeshurun, submitted to Proceedings of ICTP Workshop on High Temperature Superconductivity, Trieste, July 25-28, 1988.
10. M. Tinkham, Helv. Phys. Acta, to be published.
11. D. Dew-Hughes, Cryogenics, to be published.
12. M. Foldeaki, M. E. McHenry and R. C. O'Handley, preprint.
13. R. H. Koch, C. P. Umbach, G. J. Clark, P. Chaudhari and R. B. Laibowitz, Appl. Phys. Lett. **51**, 200 (1987); R. H. Koch, C. P. Umbach, M. M. Oprysko, J. D. Mannhart, B. Bumble, G. J. Clark, W. J. Gallagher, A. Gupta, A. Kleinsasser, R. B. Laibowitz, R. L. Sandstrom and M. R. Scheuermann, Physica, to be published.
14. M. J. Ferrari, M. Johnson, F. C. Wellstood, J. Clarke, P. A. Rosenthal, R. H. Hammond and M. R. Beasley, Appl. Phys. Lett., to be published.
15. J. Clarke and R. H. Koch, Science, to be published.
16. P. Dutta and P. M. Horn, Rev. Mod. Phys., 53, 497 (1981).
17. R. H. Koch and W. J. Gallagher, to be published.

Liquid Nitrogen Temperature DC SQUID's Using Single-Target-Sputtered Films

W.J. GALLAGHER, R.H. KOCH, R.L. SANDSTROM, R.B. LAIBOWITZ, A.W. KLEINSASSER, B. BUMBLE, and M.F. CHISHOLM

IBM Thomas J. Watson Research Center, P.O. Box 218, Yorktown Heights, NY 10598, USA

ABSTRACT

For the reliable fabrication of thin-film devices with high temperature superconductors, we have developed a simple, single-target rf-magnetron sputtering process involving an unconventional sputtering geometry. The process allows reproducible formation of high transition temperature $YBa_2Cu_3O_{7-x}$ films from nearly stoichiometric targets and lends itself both to film growth with high temperature post anneals and to low temperature in situ film growth. The post-anneal process has been optimized to routinely yield epitaxial films on lattice matched substrates that are fully superconducting at 86~91 K with current densities at 77 K up to $8x10^5$ A/cm^2. Single-level dc SQUID's have been made by patterning loops in lower current density films with conventional photolithography and ion milling. Noise measurements at 77 K indicate a flux noise level of $3 \times 10^{-4}\Phi_0/\sqrt{Hz}$ at 20 Hz, dominated by low frequency (1/f) noise. The origin of the SQUID noise is elucidated by studying current-biased thin film stripes, which also show low-frequency noise. Both the current-voltage characteristics and the noise voltage spectrum of the stripes are consistent with expectations from flux-flow but not from Josephson coupling between grains. Further improvements in SQUID noise performance will probably require the use of higher current density films with localized Josephson regions.

INTRODUCTION

Modern high performance dc SQUID devices are sophisticated integrated circuits made of five to seven or more thin film layers. Considerable process control is required to meet the demands of producing nearly optimized devices. The fabrication of electronic devices with high temperature copper-oxide-based superconductors will undoubtedly require a similar degree of control, the first element of which is the the routine production of high quality films with uniform and predictable properties. Low processing temperatures will be helpful in order to minimize interdiffusion problems and thereby facilitate the use of substrates with desirable dielectric properties and the fabrication of device structures with abrupt interfaces and ultrathin layers. Many types of processes have now been used to fabricate high quality high temperature superconducting films [1-13] and the processing temperatures involved have in some cases been kept at least as low as 550-650°C[8-10,13]. For fabricating multi-level devices, simple, reliable processes are desirable, and single-target sputter deposition processes are likely to be preferred. This paper summarizes our use of single target rf magnetron sputtering for fabricating films and simple, single-level device structures out of $YBa_2Cu_3O_{7-x}$. Studies of the current-voltage characteristics and noise properties of simple SQUID's and isolated thin film stripes are also described.

SINGLE TARGET SPUTTERING PROCESS

The application of single-target sputter deposition techniques to the high temperature superconducting oxide materials is not straightforward. Difficulties stem from the fact that in sputtering compounds containing elements with a large electronegativity difference, negative ions are formed which are accelerated away from the target as an energetic particle flux that, in the usual sputtering geometry, is directly incident on the substrate. The film composition is modified by selective resputtering of high sputtering yield constituents. Heavily compensating the target composition for different sputtering yields at the substrate surface or sputtering in high pressures (80~100 mTorr), such that collisions slow down the energetic particles, are approaches often taken to circumvent this resputtering effect. Other approaches to circumventing the negative-ion problem involve the use of unconventional sputtering geometries in which either the direct substrate bombardment by the energetic particles is avoided or symmetry is used to make deposition and resputtering entirely isotropic.

We have used the geometric approach of sputtering in the off-axis geometry illustrated in Figure 1 to avoid the negative-ion resputtering effect. In the illustrated geometry, the substrates are placed outside the region of head-on negative ion flux from an rf magnetron source, but are still immersed in the outer edge of the plasma region. Substrate wafers are held on individually heated holders that are rotated to facilitate the achievement of uniform thickness and composition. The sputtering process is carried out in a small (~50 liter) chamber evacuated with a 360 liter/sec turbomolecular pump and having a base pressure of 10^{-7} Torr. Typical plasma parameters are 100 watts of rf power at 10 MHz, -165 volts dc self-bias, with a 98% Ar/2% O_2 gas flow mixture at 6 mTorr total pressure. Deposition rates are 7-10 nm/min, so that run durations of 40-60 minutes are required to yield the ~400 nm thick films we have studied.

Sputter targets used were fabricated by a simple one-step reactive sintering process. Nominally 99.5 +% pure powders of yttrium, barium, and copper oxides were homogenized either in a ball mill for 1-2 hours or in a small analytical mill for 30 seconds. The mixtures were then cold pressed at 827 bars to form 50 mm diameter discs. The discs were fired in flowing O_2 at 950°C for 12 hours and then cooled slowly in flowing O_2 for 4-6 hours. The targets, typically 80-85 % dense, were then ground flat and epoxied to titanium backing plates for better cooling. Sputtering from an initial target that was stoichiometric $YBa_2Cu_3O_{7-x}$ yielded reasonably stoichiometric $YBa_2Cu_3O_{7-x}$ films: 12% Y and 3% Ba deficient according to a chemical analysis using inductive plasma emission spectroscopy[14] that is estimated to have a 2-4% precision. These films, when post annealed properly, had T_C's (for zero resistance) of ~82-84 K. (Post anneal schedules that optimized the T_C of our sputtered films were similar to some conditions reported by others[2] and involved a gradual warming up to 850°C (2-12 hours from 650°C to 850°C) followed by a hold at 850°C for 1-3 hours and then by a slow furnace cool over ~2 hours, all in flowing oxygen.) Film composition and transition temperature were further improved by using targets that were made slightly off $YBa_2Cu_3O_{7-x}$ stoichiometry. A target that was slightly enriched in Y and deficient in Ba (+6% Y, -7% Ba) resulted in films closest to $YBa_2Cu_3O_{7-x}$ stoichiometry (~4% Y poor) and yielded the highest transition temperatures, zero resistance at 86-89 K for 400 nm thick films.

A given target yielded film results that were quite reproducible in composition (to the precision of the analytical technique) and resistive transition, although some unpredictability was associated with exactly what film composition was achieved with a given initial target composition. The reproducibility of resistive transitions for 400 nm thick films on $SrTiO_3$ substrates is illustrated in Figure 2. Shown are results from eleven consecutive runs with the same sputtering and post annealing conditions processed over a four month period. In all cases the films were fully superconducting at 86-89 K. These films had sheet resistivities at 100 K of 400-700 $\mu\Omega$-cm and resistivity ratios [R(300 K)/R(100 K)] between 1.9 and 2.3. Slight variations of the annealing procedure involving longer hold times at high temperature seemed to be associated with slightly lower resistivities and larger values of the resistivity ratio.

The largest value of critical current density we have measured in these films at 77 K (using a 2 μV criterion) is $8x10^5 A/cm^2$ for a 15 μm wide 20 μm long line patterned with conventional photolithography and ion milling. Typical critical current densities at 77 K are in excess of 10^4 A/cm^2. Films with the higher resistance ratios and sharper transitions have shown higher current densities. The narrowest lines (~5 μm wide) appeared to show some reduction of T_C and critical current density.

The microstructure of the high temperature post annealed films showed some roughness associated with occasional submicron boulder-like protrusions from an otherwise smooth underlying film. The size and area density of these protrusions was somewhat variable from sample to sample. Energy dispersive x-ray analysis did not indicate any compositional variation in these regions. Some microcracking was also evident in the high magnification secondary electron images. X-ray diffraction scans as well as texturing evident in high magnification images indicated a mixed epitaxy of c-axis normal and c-axis in-plane grains that are in register with the underlying lattice of the [100] $SrTiO_3$ substrate. High resolution transmission microscopy revealed that the c-axis normal portion of the film is near the substrate interface, while away from the interface the film is composed of two orthogonal variants with their c-axes parallel to the substrate surface.

We have also had good results with the above process using a new lattice-matched substrate material, $LaGaO_3$[15]. In addition, we have had encouraging initial results with <u>in-situ</u> processing in the off-axis geometry with substrate temperatures from 500 to 650°C[13].

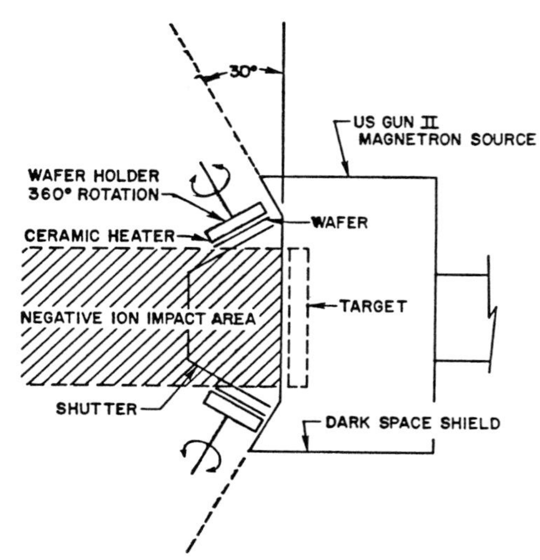

Figure 1: Sputter system geometry that avoids negative ion resputtering effect.

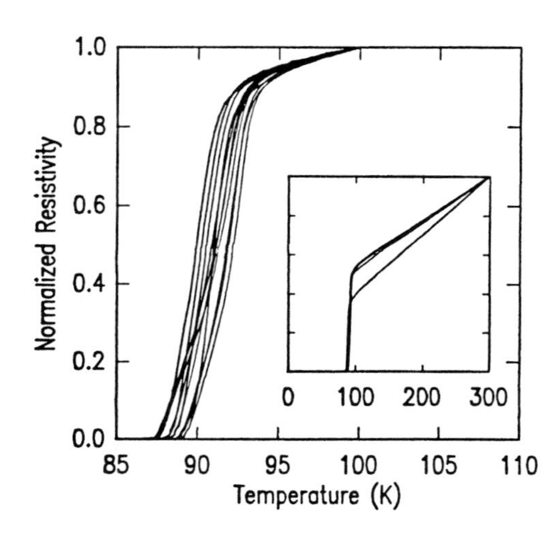

Figure 2: Resistance as a function of temperature for eleven consecutive depositions of 400 nm thick $YBa_2Cu_3O_{7-x}$ films onto $SrTiO_3$ substrates from a single target.

Figure 3: Optical micrograph of SQUID pattern ion milled in a $YBa_2Cu_3O_{7-x}$ film on $SrTiO_3$.

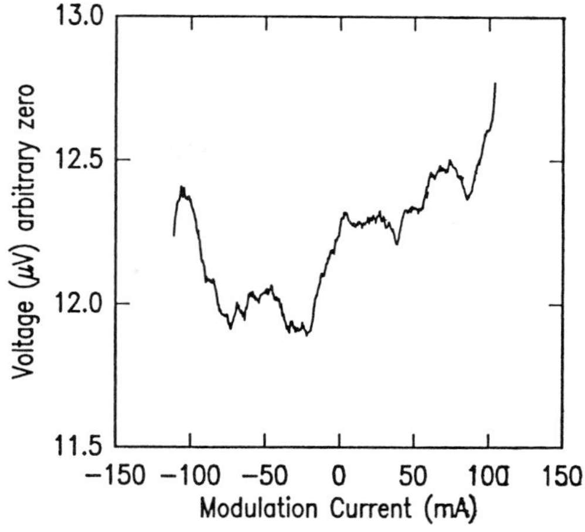

Figure 4: Flux-voltage transfer characteristics for a SQUID at 77 K.

SINGLE-LEVEL SQUID STUDIES

Using these high T_C films, we fabricated and studied a number of simple single-level dc-SQUID structures and study their operation. Figure 3 is an optical micrograph of a patterned $YBa_2Cu_3O_{7-x}$ film with Au contacts and wirebonds. SQUID loops with no intentional localized Josephson elements were formed by connecting two large superconducting pads with two bridging lines of widths 5-20 μ and lengths 5 to 100 μm. The pattern used contains 24 dc SQUID loops in addition to isolated thin film stripes. Contacts with resistance adequately low for current density and noise measurements ($\sim 10^{-6}\ \Omega\text{-cm}^2$) were made with a mild argon rf sputter preclean followed by a gold or silver deposition similar to that used by others[16].

Consistent with our earlier observations,[17-18], loops containing lines that could support high supercurrent densities did not display Josephson properties at 77 K. However loops in films with current densities of a few times 10^4 A/cm^2 displayed SQUID-like behavior at 77 K. The 77 K flux-voltage transfer characteristic for a SQUID made on a lower current density film with 5μm wide, 40 μm long links, can be seen in Figure 4, where the SQUID ouput voltage at fixed current bias is plotted as a function of current applied to an external magnet field coil. As with earlier high T_C polycrystalline grain boundary dc SQUIDs[17,18] the characteristics are not perfectly periodic and they were somewhat hysteretic. A fundamental period of about 45 mA is clearly observed, however, and this magnitude is consistent with estimates of the expected periodicity for a 40 μm sized loop in our field coil. The maximum slope of the transfer characteristic shown at 77 K is about 1 μV per flux quantum Φ_0, and noise measurements indicated a noise level of $3\times10^{-4}\Phi_0/\sqrt{Hz}$ at 20 Hz and a 1/f-like noise spectrum. For comparison, the expected white flux noise at 77 K for an optimized dc SQUID with a loop the size of this SQUID should be about $10^{-5}\Phi_0/\sqrt{Hz}$. Based on the results of studies of $YBa_2Cu_3O_{7-x}$ thin film stripes reported in the next section, we believe the low frequency noise in these structures is associated with flux motion in the superconducting films. Substantial further improvements in noise performance at 77 K will probably require the use of high current density films interrupted by localized Josephson regions.

NOISE PROPERTIES OF $YBa_2Cu_3O_{7-x}$ THIN FILM STRIPES

In order to elucidate the nature of the source of low frequency noise in thin films of $YBa_2Cu_3O_{7-x}$ we have undertaken careful measurements of the average voltage and the noise spectrum of current biased thin films stripes. Five stripes prepared on two separate deposition runs were measured in detail, though we only present data here from a single stripe. This stripe was 40 μm long and 10 μm by 0.4 μm in cross-section. Data from the other four experimental stripes is similar, except the critical current densities and noise voltage magnitudes differ.

Figure 5 shows the current-voltage characteristics for nine values of temperature. These I-V curves are of the form

$$V \simeq V_0(T)(I/I_0 - 1)^{\alpha} \tag{1}$$

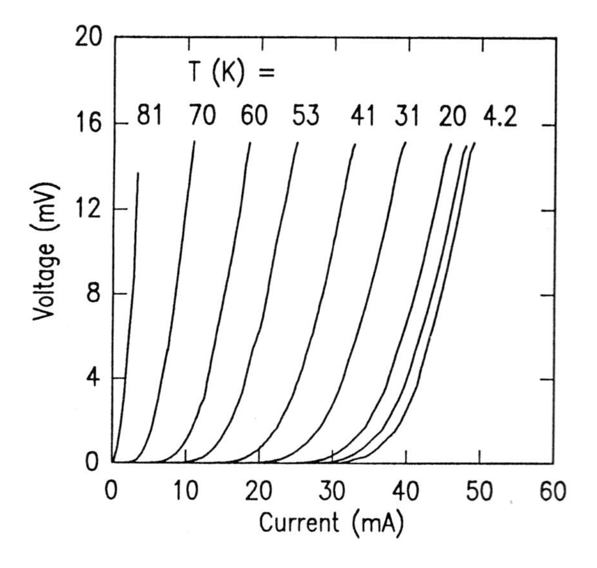

Figure 5: The voltage vs current characteristic for the stripe at 9 temperatures. The temperature for the unmarked characteristic is 10 K.

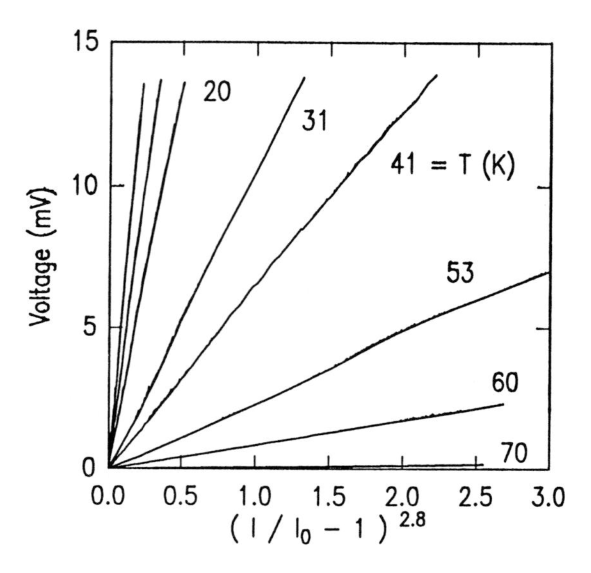

Figure 6: A plot of $(I/I_0 - 1)^{2.8}$ as a function of voltage for the same temperatures as in Figure 5.

where $V_0(T)$and I_0 are fitting parameters and $\alpha = 2.8 \pm 0.5$ as shown in Figure 6. The independence of the shape of the I-V curve with temperature is shown in Figure 7(a), where the log of the dynamic resistance, $R_D = dV/dI$, is plotted against log V. The dynamic resistance was measured at 43 Hz using a lock-in amplifier. For almost three orders of magnitude of voltage we find that $R_D \sim V^{(0.6-0.7)}$. From Equation 1 we expect the power law $R_D \sim V^{1-1/\alpha} = V^{0.64}$, as plotted in Figure 7(a). It is striking to notice that most good quality films that we have measured have the same power law behavior of R_D on V with the same exponent. The power spectrum of the voltage fluctuations $S_V(f)$ across the sample when current biased was found to be 1/f-like with no significant structure over the frequency range from 0.1 Hz to 100 KHz. Figure 7(b) plots S_V measured at 10 KHz vs V. There was no measurable noise in the sample below the transport-measured critical current.

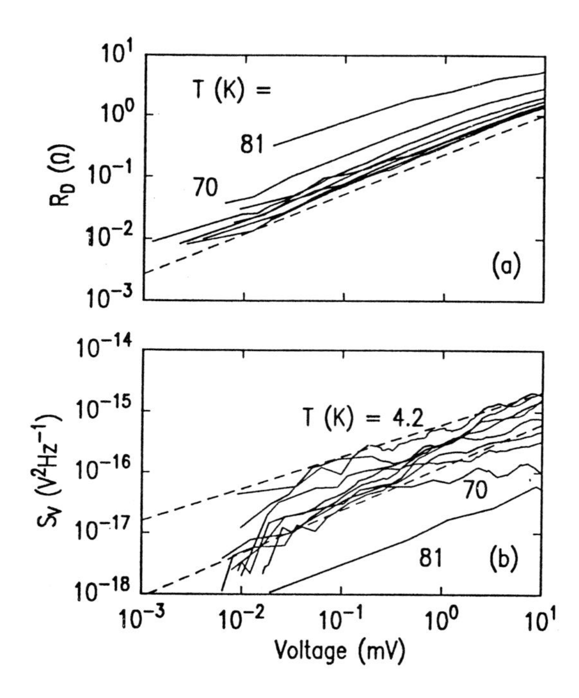

Figure 7 (a): The dynamic resistance, $R_D = \partial V/\partial I$ as a function of voltage. The dashed line has a slope of 0.64. (b): The voltage noise power at 10 KHz versus voltage for the same temperatures as Figure 5. The dashed lines have slopes of 0.71 and 0.53.

We first note that a Josephson junction model is clearly inconsistent with this data. The power law dependence of V on (I/I_0-1) is totally different than that predicted by models for nonhysteretic Josephson junctions or a distribution of junctions. Such models generally predict that, above the noise rounded critical current, the dynamic resistance R_D of a resistively shunted junction varies like I/V and the 1/f voltage noise power varies like $(I_0/V)^2$, which is exactly opposite to the power-law increase with voltage that measured values of these quantities exhibit. It is interesting to note that if the curvature in the I-V curves is interpreted as thermal noise rounding, the amount of rounding is much too large and would have to decrease with increasing temperature.

As we have recently argued[19] however, a flux flow model can account for both the observed shape of the I-V characteristics and for the power law relationship between voltage noise and average voltage. To derive the interdependence of average current and voltage, Maxwell's equations and the notions of flux flow need to be combined with some relationship between magnetic field strength H and magnetic induction B. Measured magnetization data[20] is consistent with there being a power law relationship between B and H above the lower critical field H_{c1}. In the case of strong pinning, assuming that $B(H) = B_1(H/H_{c1} - 1)^\beta$ it can be shown that $V = V_0(I/I_0 - 1)^\beta$. in the limit $V \to 0$. (In the case of weak pinning, a power law current-voltage relation is again obtained, this time with an exponent of $\beta + 1$.) The noise is assumed to be due to fluctuations in the number of pinned vortices, which intern causes fluctuations in the local electric fields. The total voltage noise is found[19] to scale as $S_V \sim V^{(\frac{2}{\beta+1})}$. The spectrum of fluctuations is 1/f-like due to a 1/f spectrum for the pinned occupancy power spectral density. Thus both the power law dependence of average voltage on $(I/I_0 - 1)$ shown in Figure 6 and the power law relationship between the voltage power spectral density and the voltage are consistent with the predictions of the flux flow theory. Further information about pinning strengths and times can be derived from more detailed considerations[19].

CONCLUSIONS

In summary, we have described a simple, reliable, single-target sputtering process for high temperature superconducting oxide films. Lower current density $YBa_2Cu_3O_{7-x}$ films made with this process are suitable for simple dc SQUID structures (with no intentional Josephson elements) that function at 77 K with fairly low noise. The average and fluctuating voltages that develop in thin film stripes like those that comprise legs of these SQUIDs were shown to be consistent with the expectations of a flux flow analysis. Further significant progress in lowering SQUID device noise at 77 K will probably require the introduction of localized weak-link interruptions in otherwise high current density loops.

The authors wish to thank T.R. Dinger, E.A. Giess, M. Plechaty, T.R. McGuire, W.H. Price, A.P. Segmuller, S. Shinde, D. Goland, T. Jackson, D. Chance, H. Harris, D. DeLaura, A. Moldovan, J. Pucyloski, and E. Kummer for contributions to this work. Partial support for this work was provided by U.S. Office of Naval Research.

REFERENCES

1. R.B. Laibowitz, R.H. Koch, P. Chaudhari, and R.J. Gambino, Phys. Rev. B35, 8821 (1987); P. Chaudhari, R.H. Koch, R.B. Laibowitz, T.R. McGuire and R. J. Gambino, Phys. Rev. Lett. **58**, 2684 (1987).
2. B. Oh, M. Naito, S. Arnason, P. Rosenthal, R. Barton, M. R. Beasley, T. H. Geballe, R. H. Hammond and A. Kapitulnik, Appl. Phys. Lett. **51**, 852 (1987); M. Naito, R.H. Hammond, B. Oh, M.R. Hahn, J.W.P. Hsu, P. Rosenthal, A.F. Marshall, M.R. Beasley, T.H. Geballe, and A. Kapitulnik, J. Mat. Res. **2**, 713 (1987).
3. R.M. Silver, J. Talvacchio, and A.L. de Lozanne, Appl. Phys. Lett., **51**, 2149 (1987).
4. M.R. Scheuermann, C.C. Chi, C.C. Tsuei, D.S. Yee, J.J. Cuomo, R.B. Laibowitz, R.H. Koch, B. Braren, R. Srinivasen, and M.M. Plechaty, Appl. Phys. Lett., **51**, 1951 (1987).
5. D. Dijikkamp, T. Venkatesan, X. D. WU, S. A. Shaheen, N. Jisrawi, Y. H. Min-lee, W. L. McLean and M. Croft, Appl. Phys. Lett. **51**, 619 (1987).
6. Y. Enomoto, T. Murakami, M. Suzuki, and K. Morwaki, Jap. J. Appl. Phys. **26**, L1266 (1987).
7. P. K. Mankiewich, J. H. Scofield, W. J. Skocpol, R. E. Howard, A. H. Dayem, and E. Good, Appl. Phys. Lett., **51**, 1753 (1987).
8. D. K. Lathrup, S. E. Russek and R. A. Buhrman, Appl. Phys. Lett., **51**, 1554 (1987).
9. H. Adachi, K. Hirochi. K. Setsune, M. Kitabatake, and K. Wasa, Appl. Phys. Lett., **51**, 2263 (1987).
10. T. Terashima, K. Iijima, K. Yamamoto, Y. Bando, and H. Mazaki, Jpn. J. Appl. Phys. **27**, L91 (1988).
11. G.K. Wehner, Y.H. Kim, D.H. Kim, and A.M. Goldman, Appl. Phys. Lett., **52**, 1187 (1988).
12. D.C. Bullock, C.T. Rettner, V.Y. Lee, G. Lim, R.J. Savoy, and D.J. Auerbach, in Thin Film Processing and Characterization of High-Temperature Superconductors, AIP Conference Proceedings No. 165, J.M.E. Harper, R.J. Colton, and L.C. Feldman, eds, American Institute of Physics (New York, 1988) pp. 71-79; W.Y. Lee, J. Salem, V. Lee, C.T. Rettner, G. Lim, R. Savoy, and V. Deline, ibid. pp. 95-105.
13. R.L. Sandstrom, W.J. Gallagher, T.R. Dinger, R.H. Koch, R.B. Laibowitz, A.W. Kleinsasser, R.J. Gambino, B. Bumble, and M.F. Chisholm, Applied Phys. Lett. **53**, 444 (1988).
14. M.M. Plechaty, B.L. Olson and G.J. Scilla, (unpublished).
15. R.L. Sandstrom, E.A. Giess, W.J. Gallagher, A. Segmüller, E.I. Cooper, M. Chisholm, A.Gupta, S.Shinde, and R.B. Laibowitz Appl. Phys. Lett. (submitted).

16. J.W. Ekin, A.J. Panson, and B.A. Blankenship, Appl. Phys. Lett., **52**, 331 (1988).
17. R. H. Koch, C. P. Umbach, G. J. Clark, P. Chaudhari, and R. B. Laibowitz, Appl. Phys. Lett. **51**, 200 (1987).
18. R.H. Koch, C.P. Umbach, M.M. Oprysko, J.D. Mannhart, B. Bumble, G.J. Clarke, W.J. Gallagher, A. Gupta, A. Kleinsasser, R.B. Laibowitz, R.L. Sandstrom, and M.R. Scheuermann, **Physica C 153-155**, 1685 (1988).
19. R.H. Koch and W.J. Gallagher (to be published).
20. W.J. Gallagher, T.K. Worthington, T.R. Dinger, F. Holtzberg, D.L. Kaiser, and R.L. Sandstrom, Physica 148B, 228 (1987).

Prospects for Electronic Applications of Rare-Earth-Free Superconducting Thin Films

MASAO NAKAO

Sanyo Tsukuba Research Center, 2-1, Koyadai, Tsukuba, 305 Japan

ABSTRACT

Thin films of the high-T_c superconductors Bi-Ca-Sr-Cu-O (BCSCO) and Tl-Ca-Ba-Cu-O (TCBCO) were prepared and characterized. A zero resistance state was achieved at 116 K in the TCBCO films. The crystal structures of these films are pseudo-tetragonal, so that the surface morphologies are essentially twin free. Using a SQUID magnetization measurement in a parallel field, a randomly oriented polycrystalline TCBCO thin film showed a critical current density (J_c) in excess of 1.2×10^6 A/cm^2 at 50 K. This large J_c indicates the presence of surface pinning, which is much stronger than any bulk pinning observed by a measurement in a perpendicular field. Anisotropy of J_c is not only dependent on the degree of preferred orientation of the films, but also the presence of surface effects caused by a relatively large penetration depth. Microbridge dc SQUID's have been fabricated from the BCSCO and TCBCO thin films patterned by maskless etching using focused ion beam (FIB) processes. The SQUID operation observed up to 95 K in TCBCO shows promise for early applications of the films provided that 1/f noise can be reduced.

INTRODUCTION

Following the discoveries of superconductivity above 30 K in the La-Ba-Cu-O system [1] and above 90 K in the Y-Ba-Cu-O system [2], there have been many reports of superconductivity at 90-95 K in Ln-Ba-Cu-O systems with Ln = Nd, Sm, Eu, Gd, Dy, Ho, Er, Tm, Yb and Lu. The commonality of including rare-earth elements suggested strongly that rare-earth elements are tied to the attainment of high transition temperatures.

However, a new breakthrough discovery of superconductivity at 105 K has been reported by Maeda *et al.* [3] in the Bi-Ca-Sr-Cu-O (BCSCO) system without including any rare-earth elements. Structural studies [4] indicated that the superconducting phase differs from the K_2NiF_4 and $LnBa_2Cu_3O_{7-\delta}$ structures. Shortly after that, a zero resistance state above 120 K was achieved in the Tl-Ca-Ba-Cu-O (TCBCO) system [5].

Fabrication of high-quality thin films of these rare-earth-free materials is of crucial importance for accurate determination of their fundamental superconducting and structural properties, as well as for a variety of electronic applications. Almost all cryoelectronic devices are based on the use of superconducting thin films and, moreover, layered film structures combined with insulating, normal-metallic and/or semiconducting films. The purpose of this paper is to evaluate the performance of BCSCO and TCBCO thin films from the materials point of view: critical currents, energy gaps and grain boundary junctions.

THIN FILM FABRICATION

The thin films of TCBCO and BCSCO were deposited by rf magnetron sputtering. The procedures have been described previously [6,7]. Briefly, typical sputtering conditions are listed in Table I. As deposited, the films are highly disordered or amorphous and must be annealed carefully in an oxygen atmosphere in order to obtain the superconducting phase. In particular, since thallium is volatile, the TCBCO films are wrapped tightly in gold foil to prevent thallium from evaporating during the anneal and short heat treatments are effective. The alternative is to anneal the films in a sealed alumina crucible containing pellets of the same composition as the target [7].

The films are multiphasic with two significant superconducting phases referred to as the 2122 phase (e.g., $Tl_2Ca_1Ba_2Cu_2O_8$) and the 2223 phase (e.g., $Tl_2Ca_2Ba_2Cu_3O_{10}$) [8]. The 2122 phase with the lower T_c, around 80 K in BCSCO and 108 K in TCBCO, is dominant in most samples synthesized so far. The best superconducting behavior was observed in the TCBCO films grown on (100) MgO substrates.

In this case, the T_c onset was around 125 K and the full transition completed by 116 K. The crystal structures are pseudo-tetragonal except for an incommensurately modulated superstructure observed along the b axis in BCSCO, so that the surface morphologies are essentially twin free.

In spite of many similarities, the TCBCO films and the BCSCO films differ in X-ray diffraction analysis. That is, the BCSCO films tend to have a prefered orientation, where the c-axis points normal to the plane of the substrate [6]. On the other hand, the TCBCO films in general are polycrystalline materials with randomly oriented grains [7]. Comparing the surface morphologies, we consider that there is a layer growth in the BCSCO films, while there is a random growth with different directions in the TCBCO films. It seems that this difference is closely related to that in the intersheet bonding of the Bi-O and Tl-O double layers.

Table I. Sputtering and post-annealing conditions.

	Bi-Ca-Sr-Cu-O	Tl-Ca-Ba-Cu-O
Target	$Bi_1Ca_1Sr_1Cu_2O_x$ sintered disk	$Tl_2Ca_2Ba_2Cu_3O_x$ sintered disk
Sputtering gas and pressure	pure Ar (30 mTorr)	pure Ar (30 mTorr)
Deposition rate	270 Å/min	400 Å/min
Film thickness	0.5 - 2 μm	0.5 - 3 μm
Annealing condition	880 ℃ - 1 h in oxygen	930 ℃ - 10 min in oxygen
Substrate	(100) MgO, (100) YSZ, (100) $SrTiO_3$ and (110) $SrTiO_3$	

CRITICAL CURRENTS

The critical current data were deduced from magnetization hysteresis loops obtained using a SQUID magnetometer. This technique has the advantage that it eliminates contact resistance problem, i.e., heating and consequent quenching at the contacts. In order to calculate the critical current density (J_c) from the magnetization loops, we employed the Bean formula [9,10] given by

$$J_c(H) = 30M(H)/R \quad \text{(in a perpendicular field)}, \tag{1}$$

and

$$J_c(H) = 40M(H)/d \quad \text{(in a parallel field)}, \tag{2}$$

with

$$M(H) = \Delta M(H)/2. \tag{3}$$

where $J_c(H)$ is given in A/cm^2, $M(H)$ is half the difference in magnetization density in emu/cm^3 for increasing and decreasing fields, R is the average radius of the film in cm and d is the film thickness in cm. The Bean formula is valid only for the critical state and assumes a uniform current density throughout the film. If the film is anisotropic or inhomogeneous, the measured J_c would depend on whether the applied field is parallel or perpendicular to the film plane. In fact, strong anisotropy was found in J_c of a single crystal, being much larger when the applied field is parallel to the c axis than when it is perpendicular to this axis [11]. Conversely, J_c of a polycrystalline material with randomly oriented grains would be independent of the orientation of the applied field.

However, the parallel field measurements revealed that randomly oriented polycrystalline TCBCO thin films give parallel J_c's as high as 5.0×10^6 A/cm^2 at 5.0 K and 1.2×10^6 A/cm^2 at 50 K [12]. These values are a factor of over 1000 larger than J_c's reported in bulk samples [13]. This surprising result indicates the presence of surface pinning, which is much stronger than any bulk pinning observed by the measurements in a field perpendicular to the film plane. Though various growth morphologies have been observed in these films, the surface can be regarded as substantially flat within a relatively large penetration depth. That is, anisotropy of J_c is not only dependent on the degree of preferred orientation of the films, but also the presence of surface effects.

It is worth listing possible pinning mechanisms as shown in Table II. In contrast to the Y-Ba-Cu-O system, it is hardly conceivable that twins could provide barriers to vortex motion in these materials because of the pseudo-tetragonal symmetry. The critical current data, however, demonstrated convincingly that the rare-earth-free systems also have high thermodynamic critical currents and that vortices can be pinned.

Table II. Possible pinning mechanisms in oxide superconducting thin films.

Pinning center		Remarks
Planar defect	Grain boundary	Josephson-like weak links
	Twin boundary	Anisotropy of twin lamellas in the *a*-*b* planes
	Stacking fault	Intergrowth of multiphases along the *c* axis
	Precipitate	No evidence
Linear defect	Dislocation	Strain inhomogeneity
Point defect	Oxygen vacancy	Disorder at Cu-O chains or planes
	Substitutional impurity	Fluctuation in the number of impurities
Others	Surface	Surface barrier caused by a large penetration depth
	Anisotropy	Random crystallographic grain orientation

ENERGY GAPS

The energy gaps of oxide superconductors are among the most important parameters characterizing the behavior and performance of superconducting devices. In order to estimate the superconducting energy gap (2Δ) of TCBCO thin films, electron tunneling measurements were performed.

The junctions used in this measurements were formed in a superconductor/insulator/normal-metal (SIN) configuration. To make tunnel barriers, 100-Å-thick Al_2O_3 insulating films were deposited by rf magnetron sputtering on the post-annealed TCBCO films. No special cleaning or other surface treatments were employed prior to the deposition. The Ag counterelectrodes were then deposited.

Figure 1 shows the differential tunneling conductance dI/dV as a function of junction voltage at 77 K. Two distinct gap structures can be seen in the spectrum. Estimating the energy gaps by the peak separation, we obtain $2\Delta_1$=0.11 eV and $2\Delta_2$=0.14 eV. These extremely large gap values are comparable to the band gaps of semiconductors such as InSb. It is likely that the two energy gaps observed are attributable to gap anisotropy and/or layered intergrowth of 2122 and 2223 phases along the *c* axis. Further research is anticipated to better understand the detailed gap structures.

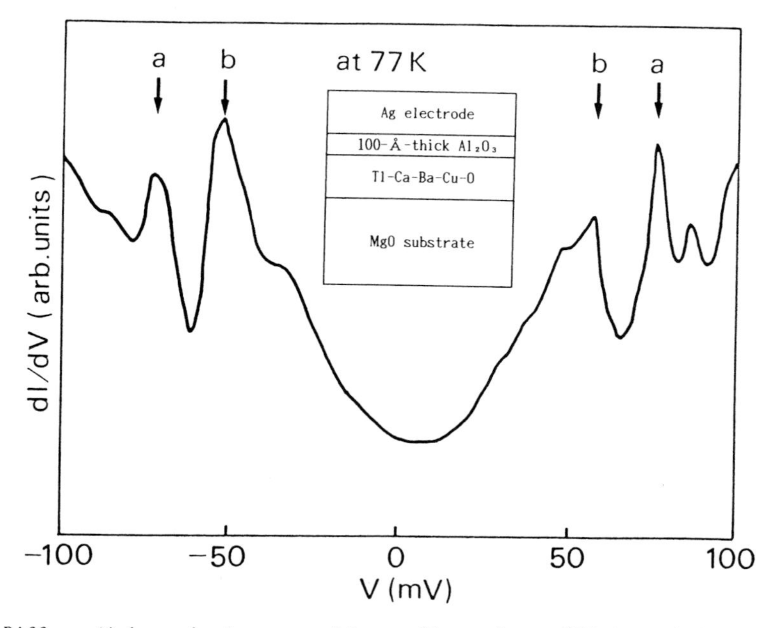

Fig. 1. Differential conductance vs bias voltage for a SIN tunnel junction at 77 K.

SQUID FABRICATION

One of the best ways to demonstrate the feasibility of the TCBCO and BCSCO thin films for electronic applications is to fabricate some device actually operating at the liquid nitrogen temperature (77 K). Among other active devices, microbridge SQUID's show promise for near-term electronic applications.

Fabrication of the microbridge SQUID requires high resolution lithography to write a microbridge. Maskless etching using focused ion beam (FIB) lithography was employed for this purpose [14]. The SQUID patterns were delineated by repetitive raster scanning of a 80 keV Au^+ FIB from a liquid metal ion source of a Au-Si alloy, where Au^+ ions were selected by using an $E\times B$ massfilter. A typical probe current was 1-2 nA and a beam diameter was about 1 μm. Total dose of Au^+ ions was 1.0×10^{18} ions/cm^2. The SQUID's examined have a 50 μm by 50 μm square hole and 20-μm-wide microbridges.

The temperature dependence of the Josephson critical current (I_c) behaves like a granular microbridge form: $I_c(T) \sim (1-T/T_c)^{1.5}$. A typical result in TCBCO is shown in Fig. 2. Thus, we find that the Josephson junction is formed in grain boundaries at the microbridge. The quantum interference effects were observed up to 95 K in the TCBCO-film dc SQUID. There was no remarkable degradation of superconductivity in the FIB etching processes. However, the 1/f noise probably due to magnetic flux trapping was high and increased rapidly with temperature. The BCSCO-film dc SQUID, on the other hand, did not operate above 12 K. The high degree of preferred orientation of the BCSCO films is presumably the prime reason for the disappearance of grain boundary junctions.

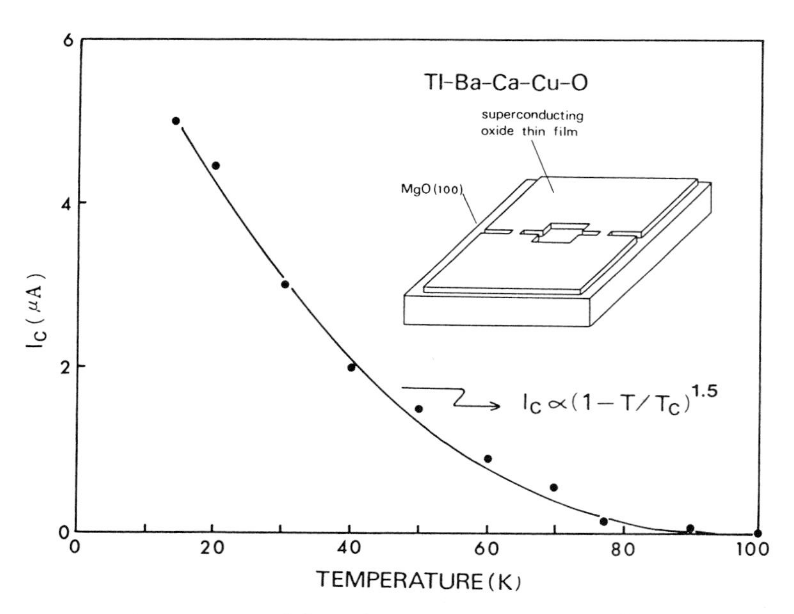

Fig. 2. Temperature dependence of Josephson critical current for a TCBCO-film dc SQUID.

CONCLUSIONS

1. The highest zero-resistance temperature was 116 K in the TCBCO thin films. There seem to be intergrowths of the high-T_c 2223 phase and the low-T_c 2122 phase.
2. The critical current densities of randomly oriented polycrystalline TCBCO thin films can be remarkably high and reflect strong surface pinning in the parallel direction.
3. The energy gaps of TCBCO are comparable to the band gaps of narrow-gap semiconductors. A new heterojunction between a superconducting film and a semiconducting film would be possible, where the energy gap at the Fermi surface and the band gap at the Brillouin zone boundary are effectively connected.
4. The TCBCO-film dc SQUID actually operated above 77 K except that the 1/f noise increased rapidly with temperature. Noise reduction will depend entirely upon the improvement of film fabrication technology.

ACKNOWLEDGEMENTS

This work was greatly assisted by the continued efforts of R. Yuasa, M. Nemoto, H. Mukaida, K. Shikichi, H. Kuwahara, H. Furukawa, K. Kawaguchi, Y. Matsuta and I. Yoshida. I would like to thank A. Mizukami for encouraging this work.

REFERENCES

1. J. G. Bednorz and K. A. Müller, Z. Phys. **B 64**, 189 (1986).
2. M. K. Wu, J. R. Ashburn, C. J. Torng, P. H. Hor, R. L. Meng, L. Gao, Z. J. Huang, Y. Q. Wang, and C. W. Chu, Phys. Rev. Lett. **58**, 908 (1987).
3. H. Maeda, Y. Tanaka, M. Fukutomi and T. Asano, Jpn. J. Appl. Phys. **27**, L209 (1988).
4. A number of papers appeared in Jpn. J. Appl. Phys. **27**.
5. Z. Z. Sheng and A. M. Hermann, Nature (London) **332**, 138 (1988)
6. M. Nakao, H. Kuwahara, R. Yuasa, H. Mukaida and A. Mizukami, Jpn. J. Appl. Phys. **27**, L378 (1988).
7. M. Nakao, R. Yuasa, M. Nemoto, H. Kuwahara H. Mukaida and A. Mizukami, Jpn. J. Appl. Phys. **27**, L849 (1988).
8. R. M. Hazen, L. W. Finger, R. J. Angel, C. T. Prewitt, N. L. Ross, C. G. Hadidiacos, P. J. Heaney, D. R. Veblen, Z. Z. Sheng, A. El Ali and A. M. Hermann, Phys. Rev. Lett. **60**, 1657 (1988).
9. C. P. Bean, Phys. Rev. Lett. **8**, 250 (1962).
10. W. A. Fietz, M. R. Beasley, J. Silcox and W. W. Webb, Phys. Rev. **136**, A335 (1964).
11. K. Kawaguchi, S. Sasaki, H. Mukaida and M. Nakao, Jpn. J. Appl. Phys. **27**, L1015 (1988).
12. M. Nemoto, H. Kuwahara, K. Kawaguchi, Y. Matsuta and M. Nakao, this proceedings.
13. K. Takahashi, M. Nakao, D. R. Dietderich, H. Kumakura and K. Togano, Jpn. J. Appl. Phys. **27**, L1457 (1988).
14. R. Yuasa, M. Nakao, S. Fujiwara, K. Kaneda, S. Suzuki and A. Mizukami, Ext. Abs. FED HiTcSc-ED WORKSHOP, 245 (1988).

Current Limiting Function of Y-Ba-Cu-O Superconducting Ceramics

HIDEO OKUMA[1], MASARU OKAMOTO[1], KUMI OKUWADA[2], TSUTOMU FUJIOKA[3], KOJI HIGASHIBATA[1], and HIRONORI SUZUKI[1]

Heavy Apparatus Engineering Laboratory[1], Research and Development Center[2], and Power and Fusion Technology Development Department[3], Toshiba Corporation, 2-1, Ukishima-cho, Kawasaki-ku, Kawasaki, 210 Japan

ABSTRACT

Meander-shaped superconductive ceramics, $YBa_2Cu_3O_{7-x}$, with 30 cm in length were fabricated from green sheets by the doctor blade method. These ceramics were sintered and provided with electrode leads to make superconductive current limiting device (CLD).
The resistance of CLD increases in proportion to the applied current above the critical current, Ic, and reaches to the constant up to the currents about six times the Ic.
Tests were made to verify that CLD could reduce currents as they became resistor at currents exceeding the Ic.

INTRODUCTION

Attempts have been made to obtain Y-Ba-Cu-O based superconducting ceramic wires in silver-sheathed form [1]. Non-sheathed wires of such ceramics can be fabricated by the doctor blade method [2], an alternate to the conventional extrusion process [3].
In a current limiting test, samples used must be long and have a small cross-section enough to obtain high resistances in normal state.
This paper reports current characteristics of meander-shaped ceramics and their testing results as current limiting devices.

EXPERIMENTS

Sample Preparation

Y_2O_3, Ba_2CO_3 and CuO in powder form were mixed with each other, calcined, and ground in a ball mill. The calcined powder then was mixed with binder, and formed into sheets by the doctor blade method. The sheets were laminated, cut into a meander shape, and sintered. As shown in Fig. 1, a sample unit thus prepared is constructed of one superconducting ceramic rod (having a cross-sectional area of 0.9 mm x 1.2 mm and an effective length of about 300 mm) which is folded 11 times. The electrodes consisted of Ag-Pd paste. For easy handling, the electrode portions were fixed by soldering on alumina substrates. Three such units were connected in series to make one current limiting device (CLD) for testing its function.

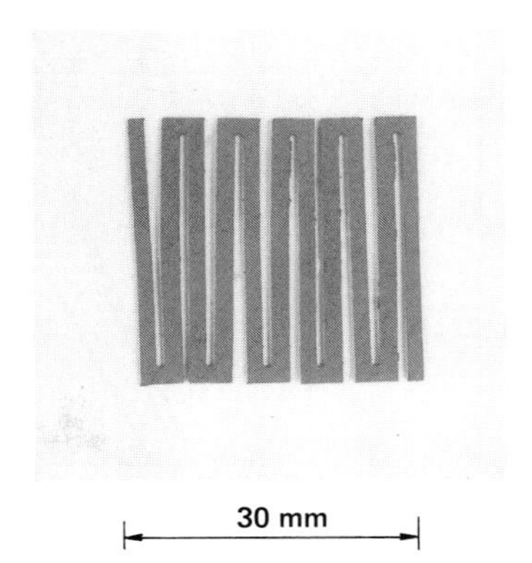

30 mm

Fig. 1 Sintered meander shape Y-Ba-Cu-O ceramics

Measurement

A DC current-resistance (I-R) characteristics was measured in liquid nitrogen by four-probe method. In the largecurrent region, a pulsed circuit was used as shown in Fig. 2. Tests of the current limiting function was conducted using the circuit, as shown in Fig. 3.

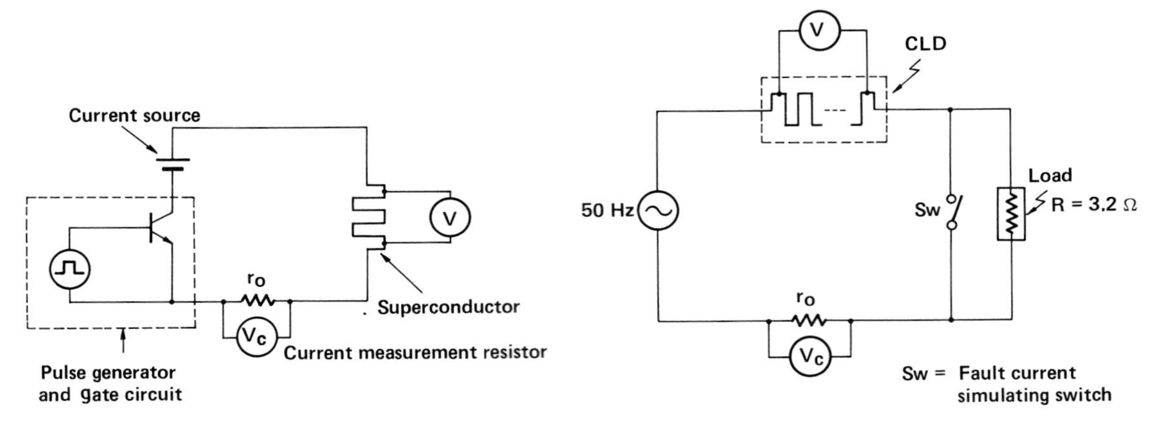

Fig. 2 Pulse measurement circuit

Fig. 3 Current limiting test circuit

RESULTS AND DISCUSSIONS

Current-Resistance Characteristics

The I-R characteristics of three samples are shown in Fig. 4. Though some extent of inconsistency is found in characteristics among the three samples, they have all the same critical current of about 1 A. Their resistance increase with applied currents above the critical current. Fig. 5 shows the current-resistance curve of the CLD consisting of the three untis connected in series.

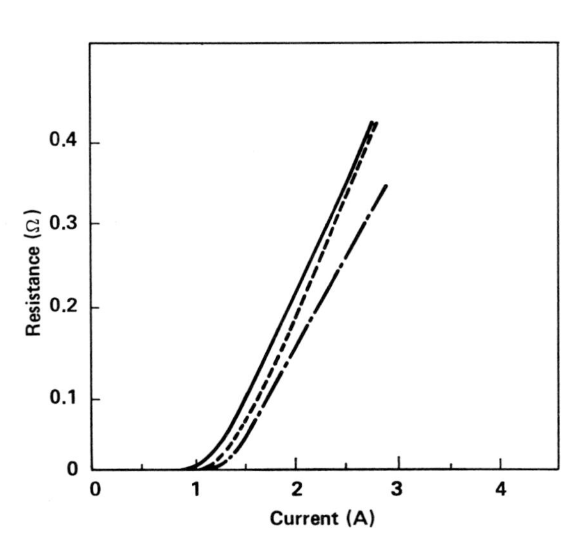

Fig. 4 Current-resistance characteristics of three samples

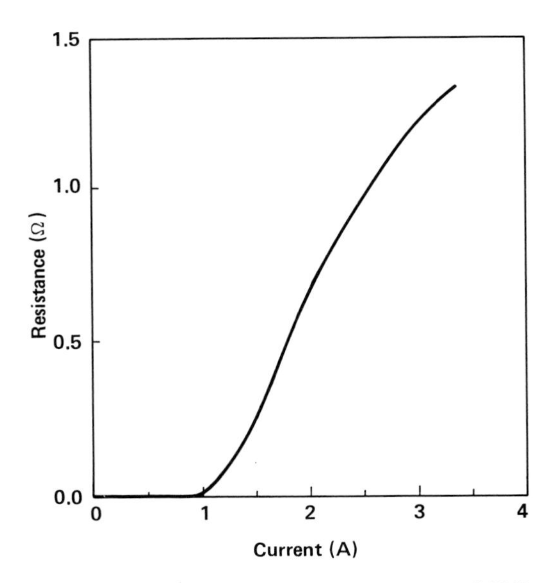

Fig. 5 Current-resistance characteristics of CLD

A short bar of the sample was subjected to resistance measurement by the pulse method in the largecurrent region. As shown in Fig. 6, the resistance no longer depends on applied current but becomes the constant up to the currents six times the Ic. Fig. 7 shows the response curve for the pulse waveforms observed. Within experimental error, that was 20 usec, no delay of quenching was observed. The resistance in the largecurrent region is about 1/10 of the value at room temperature (i.e., 2.3 ohms).

Resistances of the YBCO at currents exceeding the critical current (or after current quenching) may be considered as follows:

The actual resistance of a YBCO ceramics is the complicated sum of the resistances due to grains and grain boundaries, and the critical current in the grains differs from that in the grain boundaries. According to the magnetization measurement [4], the critical current in the grains is about 7×10^4 A/cm^2. Our pulsed current measurements show that the current did not exceed this grain's critical current density. Thus it may be considered that the resistance measured after current quenching represents the resistance in the grain boundaries.

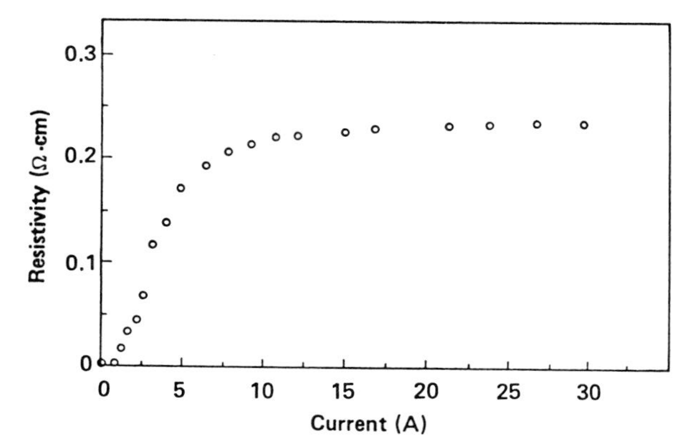

Fig. 6 Pulse current-resistance characteristics of the bar sample

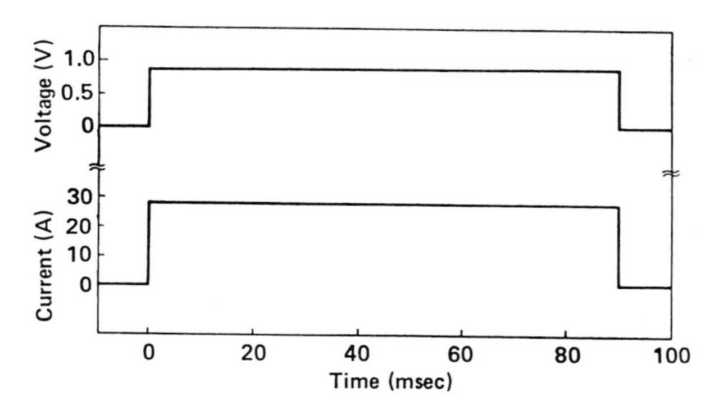

Fig. 7 Pulse waveform response of the bar sample

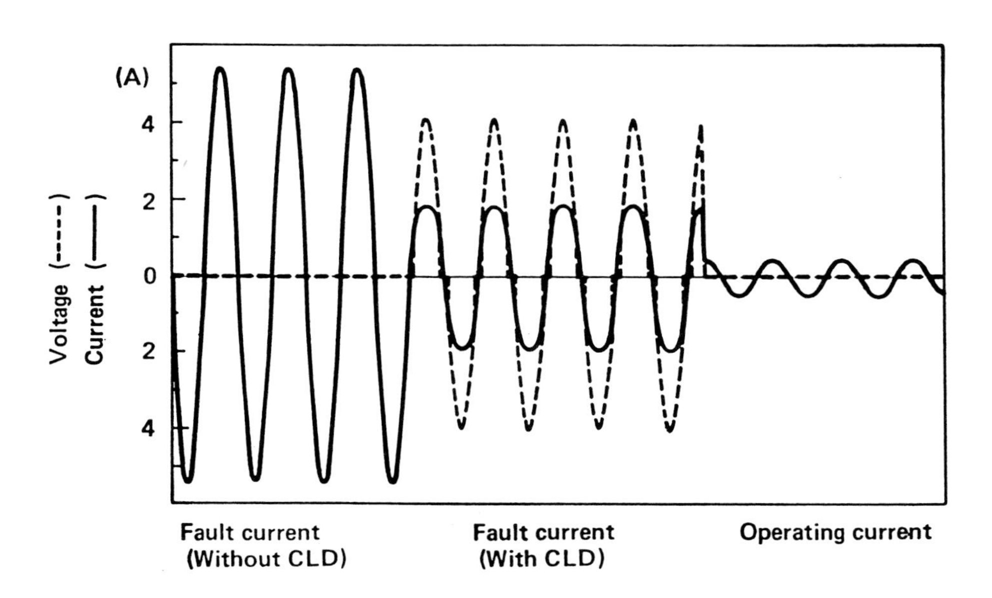

Fig. 8 Waveform of current limiting test

Current Limiting Test

In the test circuit shown in Fig. 3, the normal (operating) current was 0.5 A in peak. When both the CLD and a load were replaced by the conductors, currents through the curcuit were 5.4 A in peak. Then, the CLD on, currnts were 1.8 A in peak. This tells that the short-circuit currents were reduced by the CLD. Operating current flowed immediately after the load was freed from short-circuiting as shown in Fig. 8. In Fig. 8, the broken line gives the voltage response of the CLD. A voltage appear when the current is exceeded the Ic in each cycle. When the current decreases to below the Ic, the voltage goes to zero.

Under such cooling condition as above, the CLD's temperature staged under the critical temperature and no significant evaporation of liquid nitrogen was observed during the CLD operation.

CONCLUSION

Meander-shaped current limiting devices (CLDs) of YBCO were fabricated by the doctor blade method. Their resistance increases in proportion to the applied currents where it exceeds the critical current, and becomes constant where the current reaches about six times the Ic. This resistance is considered to have been generated in the grain boundary portion of the CLD. Experiments have shown that the CLD can automatically reduce currents exceeding the Ic as it becomes a resistor in such current regions.

REFERENCES

[1] Y. Yamada et. al.; Jpn. J. Appl. Phys. 26 (1987) Suppl. 26-3 1205.
[2] K. Okuwada et al.; Submitted to Proc. International Meeting on Advanced Materials (MRS) (Tokyo, 1988)
[3] Y. Tanaka; Ohm 88 (1988) (5) 25
[4] D. Ito et al.; Proc. Japan-US Workshop on High-field Superconducting Materials for Fusion (Fukuoka, 1987)

Fundamental Properties of a New Superconducting Motor

AKIO TAKEOKA, AKIRA ISHIKAWA, MASAHARU SUZUKI, YASUO KISHI, and YUKINORI KUWANO

Sanyo Electric Co., Ltd., Functional Materials Development Center, 100 Dainichi Higashimachi, Moriguchi, Osaka, 570 Japan

ABSTRACT

We developed a brand new superconducting motor using high-Tc ceramic superconductors for the first time. This motor, named the "Meissner Motor", utilizes the repulsive force caused by the Meissner effect, which appears below Tc and disappears above that. The motor rotated at a maximum speed of 40 rpm and generated a torque below 0.66 gf-cm. Though the repulsive force to drive the motor increased with the decrease of temperature or the increase of the gradient magnetic field, it was only about 1.1 gf/g at 77 K in 3500 G/cm and differs a little with the processes of magnetization. The rotating speed of the motor is limited by heating ability and the torque is limited by cooling ability.

INTRODUCTION

Since the discovery of high-Tc ceramic superconductors in 1986, they have been investigated rapidly and widely. We developed a new superconducting motor using ceramic superconductors in a Y-Ba-Cu-O system. This motor utilizes the repulsive force caused by the Meissner effect and transitions between a normal state and a superconducting state. We call it the "Meissner Motor". We have not seen any prior investigation on a motor based on this idea except the precedent investigation of thermomagnetic motors by Murakami et al. of Tohoku University.[1] We developed it using high-Tc ceramic superconductors at the temperature of liquid nitrogen for the first time. We expect this motor to be energized only by natural cooling and heating sources without any consumptive energy. This requires superconductors with a critical temperature Tc above room temperature. Inevitably we used ceramic superconductors with a Tc near 100 K, liquid nitrogen for cooling and a heater. In this paper, we will discuss the torque and the rotating speed of the Meissner Motor experimentally and theoretically, and clarify the future of the Meissner Motor. The torque was determined by measuring repulsion dependence on temperature or on magnetic field, and was examined theoretically.

STRUCTURE OF THE MEISSNER MOTOR

The structure of the Meissner Motor is shown in Fig.*1*, and a photograph of the motor is shown in Fig.*2*. One side of a Delrin wheel was notched at 30° angles to the radius direction and 16 disk-shaped superconducting plates were inserted to provide a rotor. A 190 W heater and liquid nitrogen were prepared above and below the rotor respectively. A rare-earth magnet of neodymium was set near the place where the superconducting plates enter the liquid nitrogen. Other components are listed in Table I. The rotor was suspended by a pivot which has low torque at a low temperature. The motor was

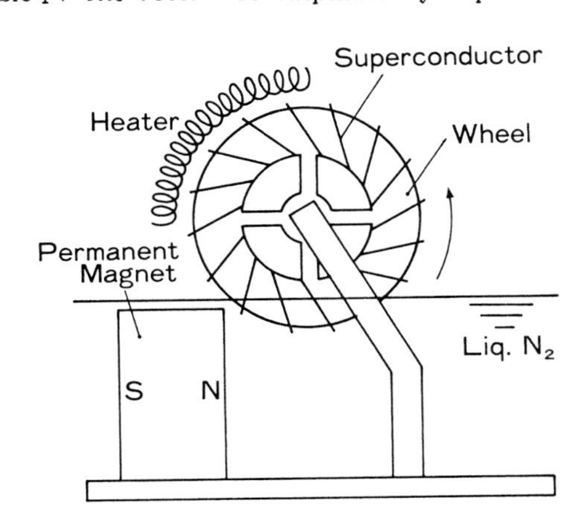

Fig.*1* Schematic diagram of the Meissner Motor

Fig.*2* Photograph of the Meissner Motor

set in an adiabatic dewar made of hard glass. The level of liquid nitrogen was automatically controlled with a level sensor. The superconductors were prepared by pressing and pre-sintering a coprecipitated compound of $YBa_2Cu_3O_{7-x}$ at 960 °C for 3 hours and sintering at 980 °C for 1 hour. The critical temperature Tc was 93 K as determined by their resistivity measurements.

Cooled down in liquid nitrogen, the superconducting plates near the magnet were repelled by the magnetic field due to the Meissner effect and this provides the motor with a torque. The rotating superconducting plates are heated in turn at the top of the wheel and return to the normal state. In this state, they can easily move into the magnetic field. The motor utilizes the repulsive force caused by the Meissner effect as a torque and rotates due to the state transition of the superconductors.

Table I . Specifications of the Meissner Motor.

Component	Material	Size (mm)
Superconductor	Y-Ba-Cu-O System	ϕ 15 × 2
Wheel	Delrin	ϕ 54 × 3
Heater	Nichrome Wire	ϕ 4 × 60
Magnet	Neodymium Magnet	30×20×50

EXPERIMENTAL

Rotating speed and torque are important properties to show the capability of the motor. As the torque is very small and difficult to measure, the starting torque is used to approximate the value of the torque. The rotating speed dependence on levels of liquid nitrogen was measured with the heater outputs as parameters. The torque was obtained as follows. A string with a small weight at one end was hung over the motor axle and the other end of the string was connected to a strain gauge. The minimum frictional force was measured when the motor was stopped and the weight was subtracted and multiplied by the radius of the motor axle to give the torque.

The repulsive force is proportional to the dia-magnetization of the superconductor, and it depends on the temperature, magnetic field intensity, and gradient of the magnetic field. The temperature and the external magnetic field dependence of the repulsive force was measured to investigate the torque of the motor. The repulsive force was measured with an apparatus as shown in Fig.3. A superconductor of 18 mm×4.7 mm×2 mm was set into a block of copper with a thermocouple and laid on the bottom of a Fiber Reinforced Plastic dewar, under which a neodymium permanent magnet of ϕ 20 mm×90 mm was set to supply a magnetic field. The repulsive force was measured by reading an electric load scale under the magnet. The temperature gradually increased in accordance with the evaporation of liquid nitrogen in the dewar. The gradient field dependence of the repulsive force was measured by moving the dewar, with sufficient liquid nitrogen, up and down in the gradient magnetic field. The dia-magnetization of the superconductor was measured with the Vibrating Sample Magnetometer at 77 K.

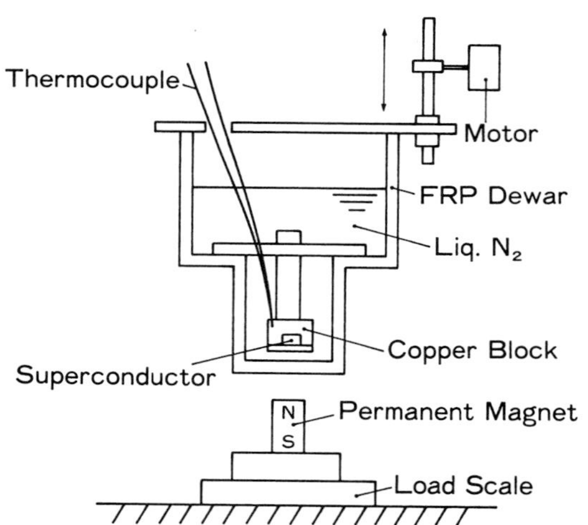

Fig. 3 Apparatus to measure the repulsive force caused by the Meissner effect.

RESULTS AND DISCUSSION

Fundamental operations of the Meissner Motor

Figure 4 shows the rotating speed dependence on the levels of liquid nitrogen with the heater outputs as parameters. The level of liquid nitrogen was measured from the lower end of the wheel. The rotating speed had a maximum of 40 rpm when the level of liquid nitrogen was 7 mm and the heater output was 190 W. The rotating speed ought to increase as the level of liquid nitrogen increases. However the rotating speed decreases in most cases. This is considered as follows. An increase in the cooled volume of superconducting plates and the fixed heater power keep the plates in a superconducting state all through the process and the plates will be repelled by the magnetic field and move to the opposite direction when they approach the magnet. Thus the balance of cooling and heating power was found to be very important to rotation of the Meissner Motor. In this particular motor, the rotating speed is limited by heating ability.

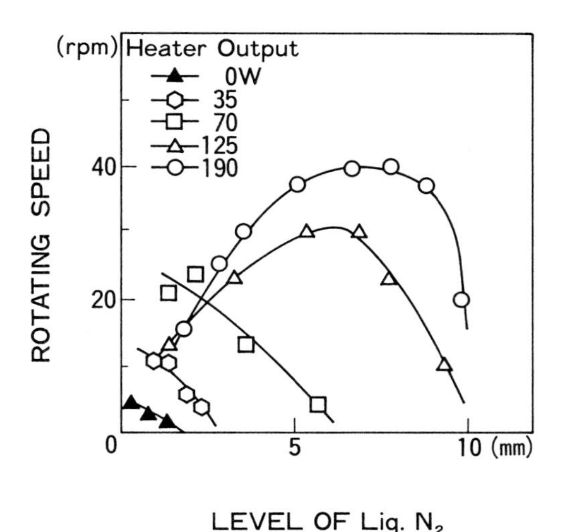

Fig.4 Rotating speed versus level of liquid nitrogen at various heater powers.

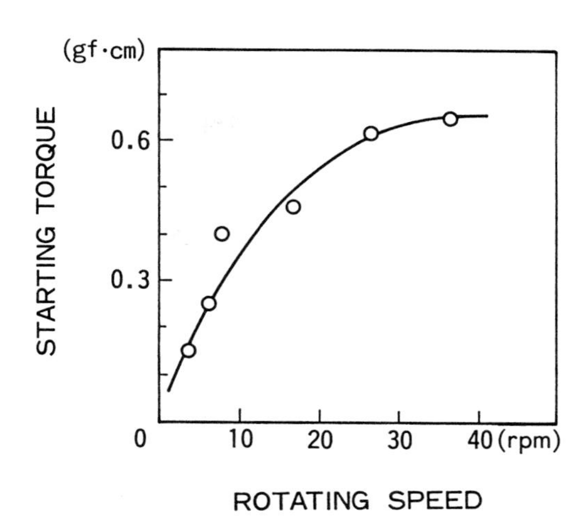

Fig.5 Starting torque versus rotating speed of the Meissner Motor.

The starting torque of the Meissner Motor is related to the no-load rotating speed as shown in Fig.5. The rotating speeds were read at the maximum speeds of various heater outputs in Fig.4. The torque increased in proportion to the rotating speed. However, the torque was very small and only 0.66 gf-cm at the maximum rotating speed of 40 rpm. The torque is considered to be subject to the viscous drag of liquid nitrogen.

Repulsive force caused by the Meissner effect

The temperature dependence of the repulsive force caused by the Meissner effect is shown in Fig.6. The gradient magnetic field operates as a parameter. The repulsive force appears below 90 K,

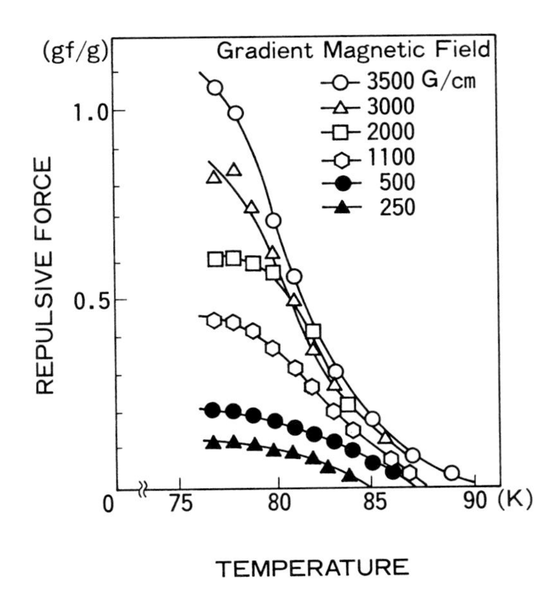

Fig.6 Temperature dependence of the repulsive force at various gradient magnetic fields.

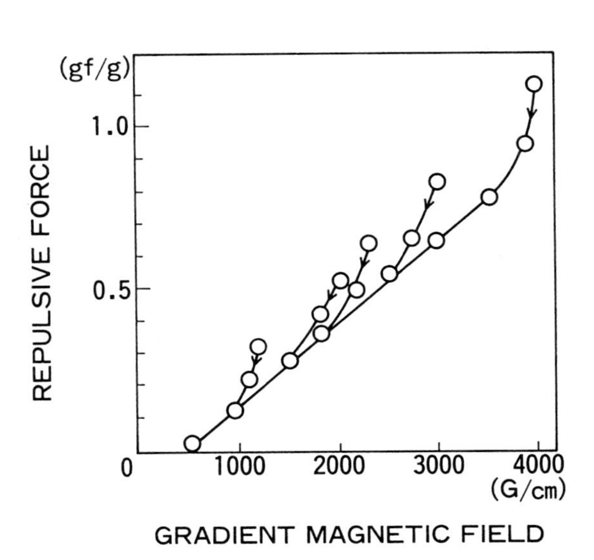

Fig.7 Gradient magnetic field dependence of repulsive force.

and increases when temperature decreases or the gradient magnetic field increases. The increase of the repulsive force at lower temperatures is considered to correspond to the increase of the dia-magnetization of the superconductor. The repulsive force is only 1.1 gf/g at 77 K in 3500 G/cm, which shows that reducing the operating temperature and increasing the gradient of the magnetic field are required to increase motor torque.

The repulsive force in gradient magnetic fields is shown in Fig.7. Here, the superconductor was moved in a magnetic field and the initial gradient magnetic field was changed. Though the repulsive force was maximum at any initial gradient magnetic field just after cooling in liquid nitrogen, it abruptly decreased when the gradient magnetic field was changed, and decreased in proportion to the gradient magnetic field. Those repulsive forces differ about 0.15 gf/g in the same magnetic field. This change in the repulsive force is considered as follows. Figure 8 shows the σ-H characteristics of a superconductor measured with the V.S.M. at the temperature of liquid nitrogen. The solid line with dots represents dia-magnetization when the temperature of the superconductor is decreased to the temperature of liquid nitrogen after applying the magnetic field. Dia-magnetization σ_1 on the solid line with dots is larger than σ_2 on the solid line, and σ_1 decreases down to σ_2 when the external magnetic field decreases in liquid nitrogen. The difference between σ_1 and σ_2 is considered to result from the trapped magnetic flux in a superconductor. It is considered that the abrupt decrease of the repulsive force in Fig.7 corresponds to the dia-magnetization shift from σ_1 to σ_2. We are investigating the trap phenomena of magnetic flux in different processes in the dia-magnetization state.

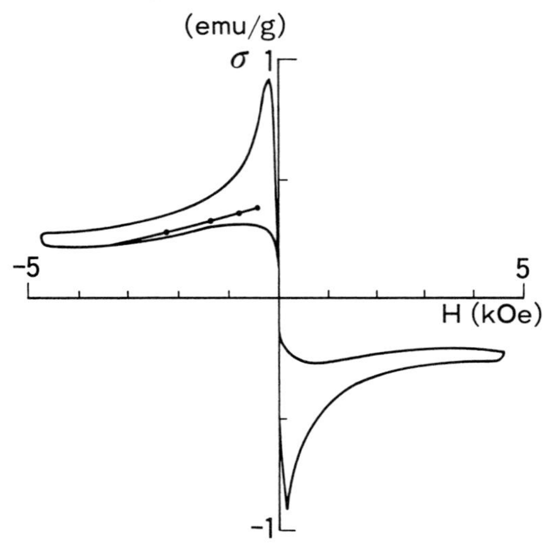

Fig.8 Magnetization curve for a superconductor measured at 77 K.

The lower critical field Hc_1 of the superconductor is about 80 G and much smaller than the magnetic field of the permanent magnet. Then a superconductor with Hc_1 which exceeds the magnetic field will improve the torque of the motor remarkably. The repulsive force of the Meissner Motor is considered to be at maximum just as the superconductor approaches the magnet in liquid nitrogen and it decreases abruptly as discussed above. However the repulsive force variation is complicated because it actually takes some time to be cooled and the external magnetic field changes during that time. Thus the Meissner Motor will require superconductors with larger dia-magnetization and higher Hc_1.

Torque fluctuation

In the Meissner Motor, the torque fluctuation may appear as a result of variation in the repulsive force as discussed above. The torque fluctuation was calculated and estimated when the level of liquid nitrogen was 7 mm, the heater power was 190 W, and the rotating speed was 40 rpm. The weight and the surface area of a superconductor cooled down in liquid nitrogen varies with its rotational position. Though the actual weight and surface area of the superconductor are 1.5 g and 4.5 cm^2 respectively, the average weight W in liquid nitrogen is defined as 0.5 g and the average surface area S is 2.0 cm^2. Figure 9 shows a calculating model, where r is 2.7 cm long and is the average orbital radius of the superconductor, and θ is $\pi/3$ radian. The superconductor begins to cool in liquid nitrogen at point A and is completely cooled at point B. Assuming that the temperature of the superconductor reaches T_1 K at A and T_2 K at B, the temperature changes of the superconductor are obtained from the following heat transfer equation

$$Q\,S\,t = \int_{T_1}^{T_2} m\,C\,p\,dT,$$

where Q is the heat flow density, t the cooling time, m the mole number of the superconductor and Cp the specific heat. Though the heat flow density Q varies according to temperatures of the superconductor between T_1 and T_2, it is assumed as a fixed value here for easy estimation. Though the temperature T_1 is above Tc theoretically, it is

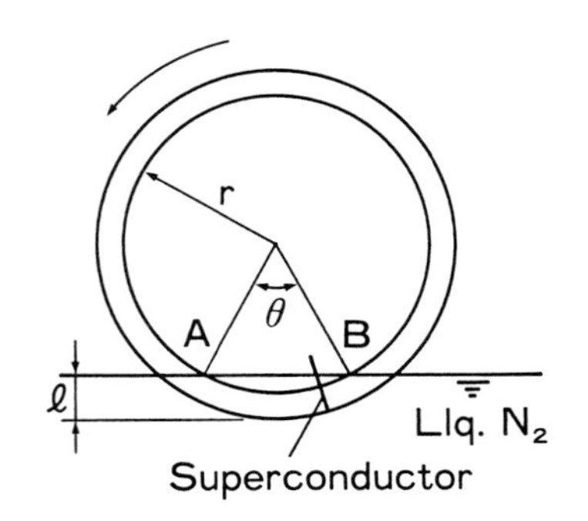

Fig.9 Configuration of a calculation model for the temperature changes of the Meissner Motor.

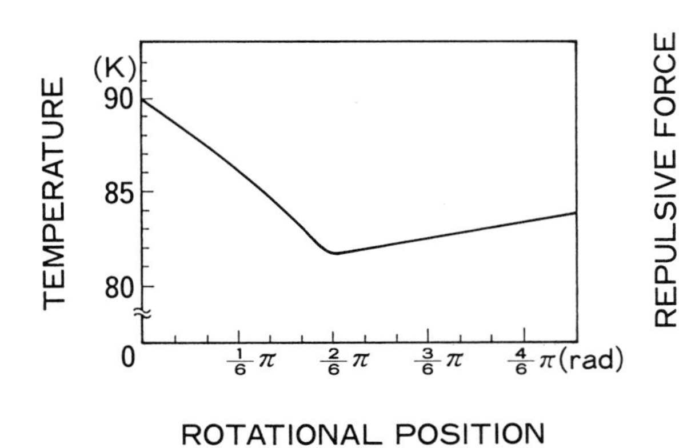

Fig.*10*. Temperature versus rotational position of the Meissner Motor.

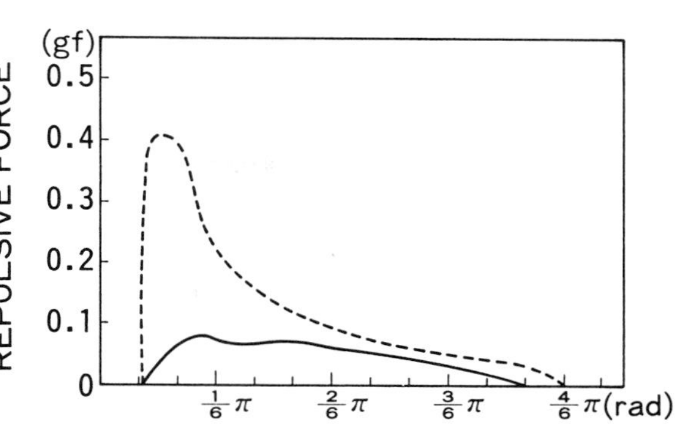

Fig.*11*. Repulsive force versus rotational position of a superconductor.

assumed to be equal to the temperature 90 K at which the repulsive force is zero as shown in Fig.*6*. The cooling time is calculated as 0.25 seconds using a rotating speed of 40 rpm and the degree in liquid nitrogen of $\pi/3$ radian. The mole number is obtained as 7.6×10^{-4} mole from the composition of the superconductor $Y_1Ba_2Cu_3O_{7-x}$. As the specific heat Cp is 112.5 J/mol • K at 83 K and 124 J/mol • K at 90 K [2], assuming that it varies linearly between T_1 and T_2, it is given as $Cp=1.6T-20$. The heat flow density Q is calculated as 1.5 w/cm^2 from the time required to vary the temperature of a superconductor from 90 K to 77 K in another experiment. The temperature T_2 at B is calculated as 81 .6 K under the above conditions. The temperature variation in one rotational cycle is obtained similarly as above and is as shown in Fig.*10*. The origin of the X-axis corresponds to point A. Figure *10* shows that the temperature of the superconductor decreases very slowly in liquid nitrogen and the superconductor cannot obtain sufficient repulsive force.

The position variation of the repulsive force is obtained from Fig.*6* using the above temperature variation. The calculated results are shown in Fig.11. The dashed line shows an imaginary repulsive force when the temperature of the superconductor is assumed to be 77 K all through the process. The repulsive force of a superconductor acts as a torque within $\pi/2$ radian and is much smaller than the imaginary force, which shows that the repulsive force is limited by the cooling ability. The Meissner Motor has 16 superconducting plates at a distance of $\pi/8$ radian, and their repulsive forces generate a torque with a lag of $\pi/8$ radian.

Figure *12* shows the total repulsive force characteristics obtained from the repulsive forces shown as the dashed line in the same figure. The total repulsive force varies cyclically between 0.20 gf and 0.23 gf. The calculated torque from the total repulsive force changes between 0.54 gf-cm and 0.62 gf-cm with a period of about 0.1 second. As the torque is assumed to be smaller than the starting torque of 0.66 gf-cm as discussed above, the calculated torque is in good agreement with the experiment. It is evident from Fig .*12* that a narrower distance between the superconductors will provide more torque and less torque fluctuation.

The maximum torque of the Meissner Motor is theoretically calculated assuming that the temperatures of the 16 superconductors are 77 K all through its rotation. The weight of the superconductor was 1.5 g and the length from the center of the superconductor to the center of the motor axle was 2.4 cm. The maximum total torque, easily calculated from the repulsive force shown as the dashed line in Fig.*11*, is about 5.0 gf-cm.

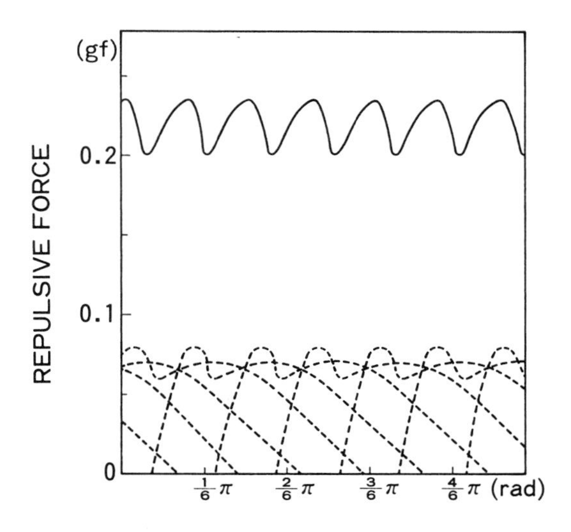

Fig.*12* Repulsive force versus rotational position of the Meissner Motor. The solid line shows the total repulsive force and the broken line shows the repulsive force of each superconductor.

CONCLUSION

We have developed a Meissner Motor using high-Tc superconductor and investigated its fundamental characteristics. It rotated at a maximum speed of 40 rpm and generated a starting torque of about 0.66 gf-cm. The rotating speed of this motor is limited by heating ability and the torque is limited by cooling ability. The repulsive force caused by the Meissner effect depends on the temperature of the superconductor and the gradient of the magnetic field, and has a value of about 1.1 gf/g at 77 K in 3500 G/cm. The repulsive force differs with the process of magnetization: cooled down in a magnetic field or magnetized after cooling. The theoretical torque of this motor is about 5.0 gf-cm, which will be realized by improving the cooling method and the design of the magnetic field. Superconductors with higher Tc and larger Hc_1 are also required to improve the actual torque.

REFERENCES

[1] K.Murakami and M.Nemoto, IEEE Trans. on Magn. MAG-8, 378 (1972).
[2] K.Kitazawa, M.Sakai, S.Uchida, H.Takagi, K.Kishino, S.Kanbe, S.Tanaka and K.Fueki, Jpn. J. Appl. Phys. 26, L748 (1987).

High-Tc Superconductor Lenses and Guides for Intense Charged Particle Beams

HIDENORI MATSUZAWA, OSAMU OHMORI, HIDEYUKI YAMAZAKI, JUN UENO, AKIHITO FURUMIZU, AKIHIRO SAITO, TOMOTA TAKAHASHI, and TETSUYA AKITSU

Faculty of Engineering, Yamanashi University, 4-3-11, Takeda, Kofu, 400 Japan

ABSTRACT

A new type of lenses and guides for charged particle beams, especially for relativistic electron beams (REBs) was proposed. When REBs go through small apertures of rings or tubes of high-Tc superconductor, self-induced magnetic fields of the REBs cannot penetrate into the rings and tubes owing to the Meissner effect or the skin effect. The magnetic fields are then compressed and focus the REBs. Copper rings, also, prevent pulsed magnetic fields from penetrating into the rings because of finite time constants for penetration of magnetic field. Copper electrodes have, however, electrical resistivity, and currents induced are reduced in amplitude in these electrodes. Superconductor electrodes are accordingly superior in focusing ability to copper electrodes. These ideas were experimentally confirmed for REBs using electrodes made of Y-Ba-Cu-O compound.

INTRODUCTION

Focusing and transport of intense charged particle beams [1-4] have been one of important subjects in beam technology for high-power lasers, X-ray generation, thermo-nuclear fusion research, and high-power microwave generation. In the present paper, we propose a new type of lenses and guides for intense charged particle beams, especially for relativistic electron beams (REBs). We further demonstrate experimentally the utility of these novel devices [5, 6].

REBs are accompanied by self-induced magnetic field $B\theta$. When the REBs are injected into superconducting rings or tubes having small apertures, the magnetic field cannot penetrate into the superconductors owing to the Meissner effect. The magnetic field is then compressed by the rings or tubes of superconductor. The REBs are therefore focused and transported with little loss.

When REBs are injected into the apertures of tubes of electrical conductor, electromotive forces are induced in the inner surface of the tubes. Such voltage induced in cooled anodes is equal in amplitude to that at room temperature. Induced currents will hence be greater in amplitude at the boiling point of N_2 than those at room temperature. This is because electrical resistivity of, for example, copper is lower at liquid nitrogen temperature than that at room temperature, by a factor of about five. Focusing is, therefore, to be more effective with cooled anodes. Superconducting anodes are consequently expected to be superior in focusing force to such normal metal anodes as copper ones.

Focusing of REBs with metal tubes was tried and discussed by other authors [7-10]. In their papers, metal electrodes were treated to be ideal conductors of electricity. Normal metals have, however, electrical resistivity of finite values. Superconductors are, on the contrary, of no resistivity. Therefore, experimental studies using both copper and superconducting anodes will give us some profitable results.

Recently superconductors of Tc higher than 100 K have been reported [11, 12]. Those high-temperature characteristics are certainly very attractive, but procedures are not easy for sintering and obtaining pure superconducting phases as compared with Y-Ba-Cu-O compound. We adopted then the latter compound. We prefer thick or bulk superconductors to thin-film ones in the present work, because thin-film anodes are sputtered and damaged in a few shots by operation of high-power beam diodes. The Y-Ba-Cu-O compound is degraded by moisture or water. The compound should not be in direct contact with liquid nitrogen, because the compound becomes wet just after liquid nitrogen is exhausted. We used copper-made heat sinks on which superconductors were soldered with electrically conducting epoxy resin.

The beam diode in the present experiment has been operated at high pressures of the order of 10^{-1} -10^{-2} Torr of Ne gas. These high-pressure operation has two main features [13]: One of them is that REBs generated are higher in amplitude than those for normal in-vacuum operation. These result from electron multiplication near cathodes, via ionization collisions between primary electrons of kinetic energies of about 150 eV and neutral gas molecules introduced into the diode chamber. The other is that charge neutralization [14] is achieved with ions produced via ionization collisions, and that electron beams are influenced by weaker forces than those for in-vacuum operation. Neon was chosen among the gases of He, H_2, and Ne which do not liquify at the boiling point of N_2. With Ne of the lowest thermal conductivity among them [15], the anode can be kept at lower temperatures during experiments.

EXPERIMENTAL APPARATUS

Figure 1 shows the beam diodes used. We prepared five different anodes. Three of them had an inner diameter of 20 mm. Two of the three had 25-mm-long tubes of copper and of superconductor, respectively (Fig. 1a). The third had a 145-mm-long superconductor tube (Fig. 1b). Another two anodes were 8.5 mm in inner diameter and were 25 mm in axial length: One of them was made of copper only and the other had superconductors similar to that in Fig. 1a. Experimental results using these two anodes will be reported elsewhere.

High-voltage pulses (450 kV, duration time of 40 ns, 82 J/pulse) were generated from a coaxial Marx-type generator [16]. The cathode was made of graphite, covered with silicon resin on the truncated conical surface. The top of the cathode was spaced at zero millimeter from that of the anodes. The anodes were suspended with 1-mm-thick perforated stainless steel plates. Temporal waveforms of REBs were observed with a high-speed storage oscilloscope (Tektronix 7934). REBs generated were detected with a Faraday cup which was isolated from the diode chamber with a 20-μm-thick titanium foil. The Faraday cup was kept at a pressure of 10^{-4} Torr. REBs thus collected have hence kinetic energies higher than 60 keV [17].

EXPERIMENTAL RESULTS

In the previous paper[13], we reported that there exists an optimum gas pressure p_m for each of the gases introduced into the diode chamber. The optimum pressures p_m were inversely proportional to relative total ionization cross-sections at some hundreds keV of electrons, and the p_m for Ne was 0.15 Torr for the configuration of titanium-foil anode. In the present paper, we tried first to confirm that such relationships hold even for foil-less superconducting anodes.

(a)

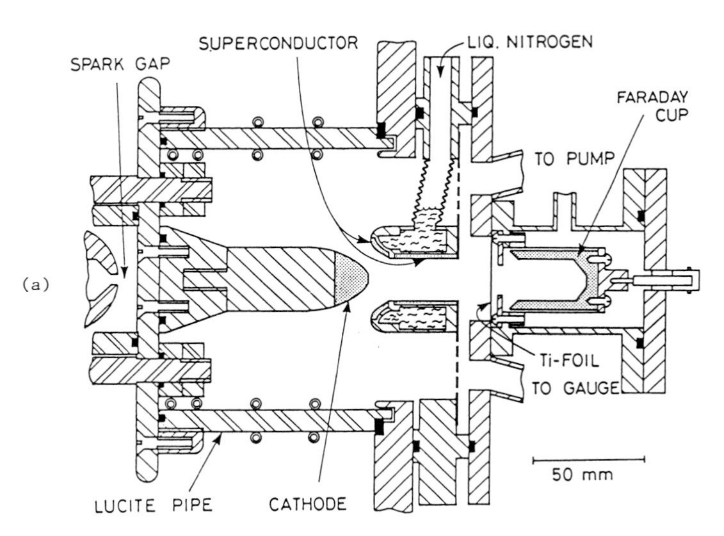

(b)

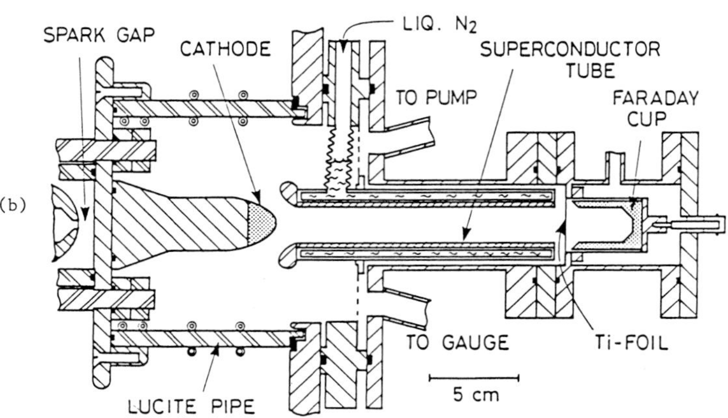

Fig. 1. REBs diodes used.
(a) short anode, (b) long anode.

Short anodes

Figure 2 shows the pressure dependences of REBs generated for the superconductor anode (Fig. 2a) and for the copper anodes at -182.5 °C (Fig. 2b) and at room temperature (Fig. 2c). In either case, the diodes had the same optimum pressure (0.15-Torr-Ne) as that for conventional foil-anode-diodes[13]. REBs for cooled anodes (Figs. 2a and 2b) are higher in amplitude than those for room temperature operation (Fig. 2c) by a factor of two at 0.15 Torr. For room temperature operation of the superconductor anode, pressure dependences were obtained which were almost the same as Fig. 2c. When Fig. 2a is compared with Fig. 2b, there are seemingly few differences in the ability of focusing between superconducting anodes and normal conducting ones. From Figs. 2b and 2c, copper electrodes have higher ability of focusing when they are cooled with liquid nitrogen. The aperture in front of the Faraday cup was 26 mm in diameter.

In order to know cross-sectional distribution of REBs, plastic films were set just downstream the 20-μm-thick titanium foil in front of the aperture. Figure 3 shows the damage patterns for three shots of REBs for the superconducting anode (Fig. 3a) and for the copper anode cooled with liquid nitrogen (Fig. 3b). The plastic film in Fig. 3a had a hole damage caused by irradiation of REBs, keeping the titanium foil with no hole. The plastic film in Fig. 3b had, on the other hand, no hole but only a faint and dispersed pattern. From these results the superconducting anode is surely superior in the ability of focusing of REBs to the copper anode. Little differences between Figs. 2a and 2b are due to the aperture which was too big to distinguish the cross-sectional distributions of REBs.

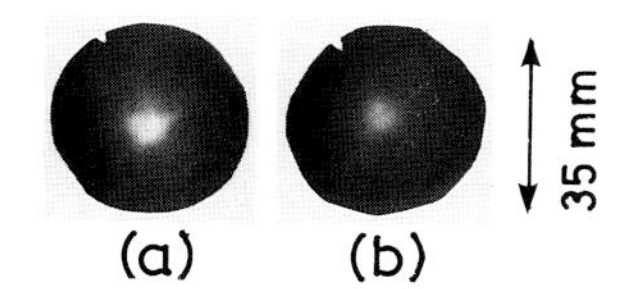

Fig. 3. Damage patterns of plastic films caused by REBs irradiation. (a) superconducting anode and (b) copper anode cooled with liquid nitrogen.

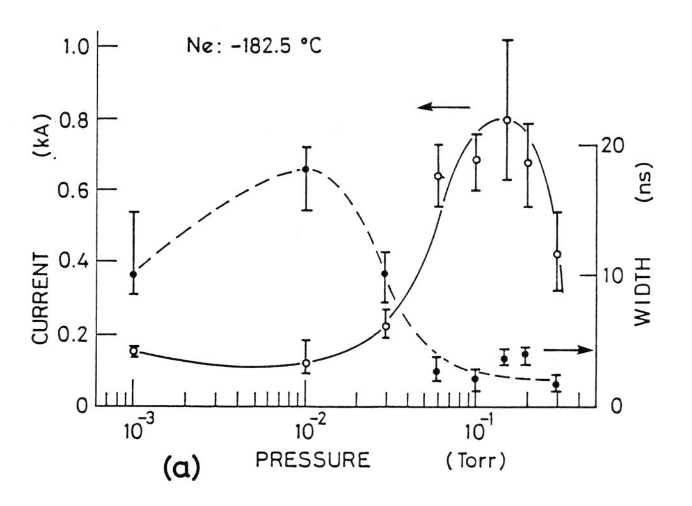

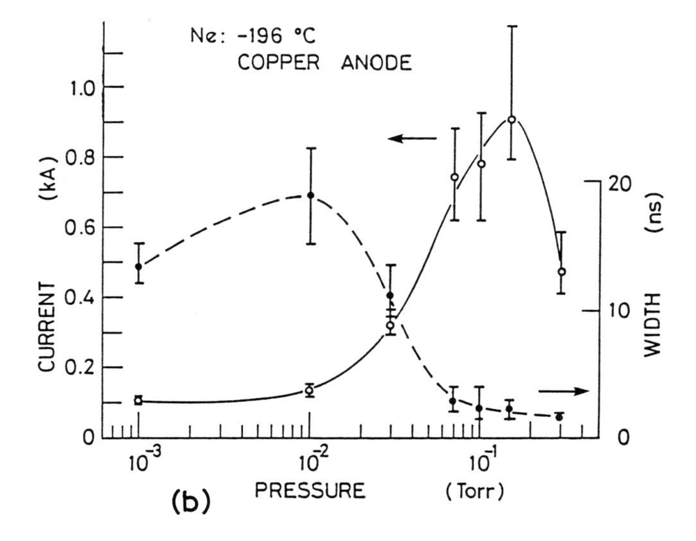

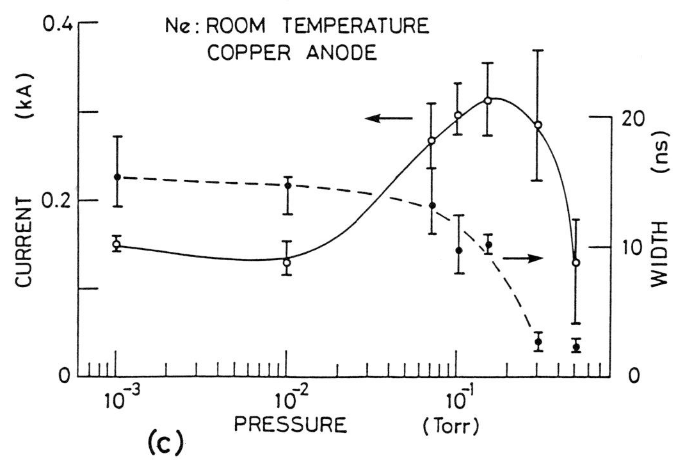

Fig. 2. Pressure dependences of REBs generated. (a) superconducting anode, (b) copper anode cooled with liquid nitrogen, and (c) copper anode at room temperature.

Figure 4 shows examples of waveforms of the REBs, where the superconducting anode produces REBs of short duration time (Fig. 4b') and of rippled envelope. There may occur instabilities, but no discussion is given here because of insufficient experimental data.

Figure 5 shows the duration times (full width at half maximum) of and peak values of the REBs as a function of the temperature of anodes. These results were obtained while the temperatures of the anodes were rising after liquid nitrogen delivered into the heat sink was exhausted. The peak values of the REBs decreased with increasing temperature, whereas the duration times changed in reversed behavior. These crossing characteristics suggest presence of mechanisms in which REBs are bunched while they are moving along the axes of focusing tubes.

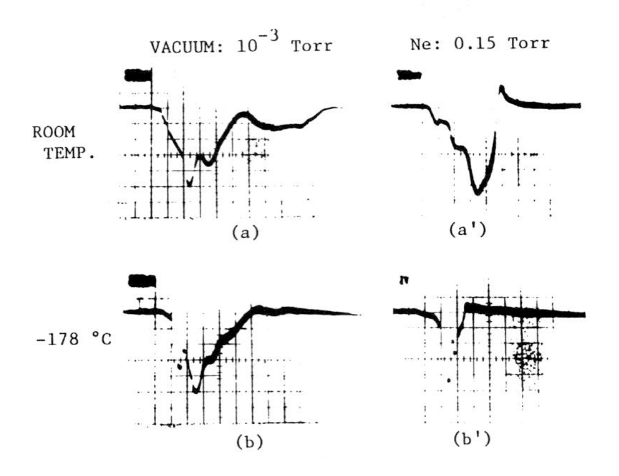

Fig. 4. Waveforms of REBs.

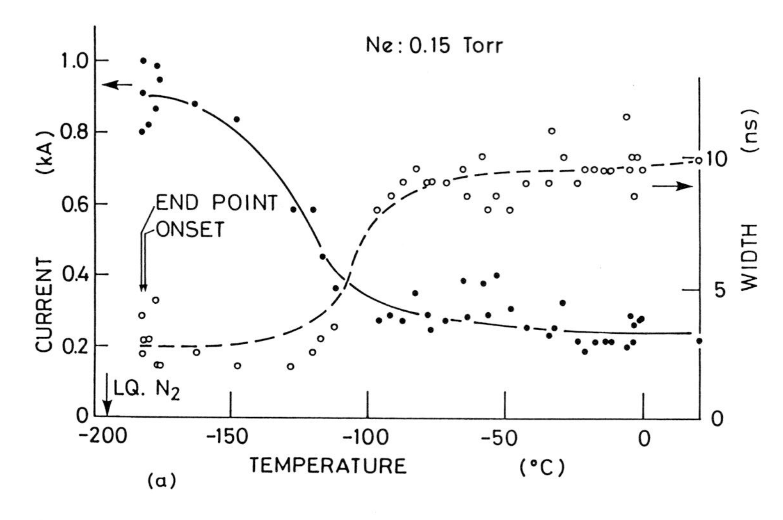

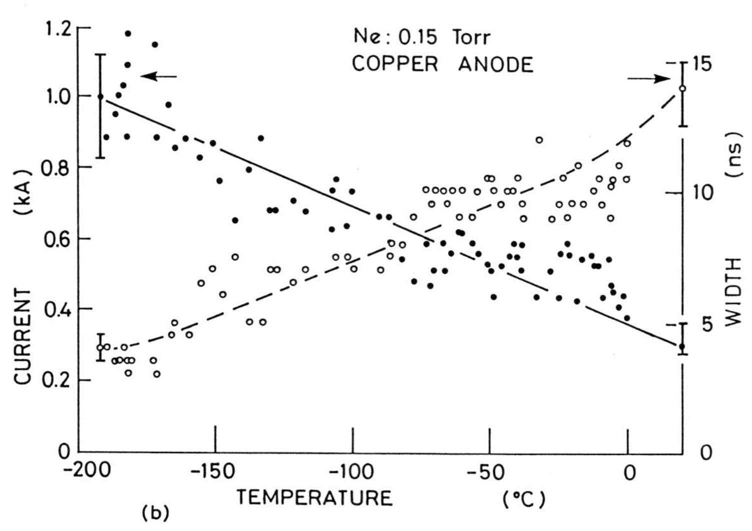

Fig. 5. Peak values of current (solid circles) and pulse width (full width at half maximum, open circles) as a function of temperatures of anodes.
(a) superconducting anode and
(b) copper anode, at 0.15-Torr-Ne.

A long anode

In developing lenses and guides for REBs, experimental investigation is necessary concerning other type of superconductors, in addition to short anodes. Figure 6 shows characteristics for the long anode (Fig. 1b). For this long anode, also, the optimum gas pressure of Ne (0.15 Torr) was nearly equal to that for the short anodes (Fig. 2). The long anode had a transport efficiency comparable to those for the short anodes. Pressure ranges over which REBs were transported effectively became narrower than those for short anodes. This narrowing is attributable to charge neutralization which should be satisfied more strictly over the longer paths of REBs.

When Fig. 6c is compared with Fig. 5a, peak currents at low temperatures did not saturate, in contrast to the temperature dependences for the short anode. These results are probably because the superconductor anode was not cooled enough below Tc along the long axis. Experimental results are still preliminary. Nevertheless, Fig. 6 shows the possibility of long guides and lenses for REBs.

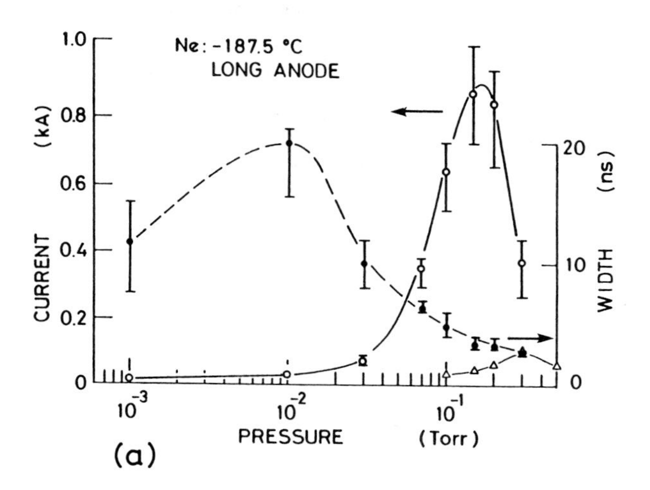

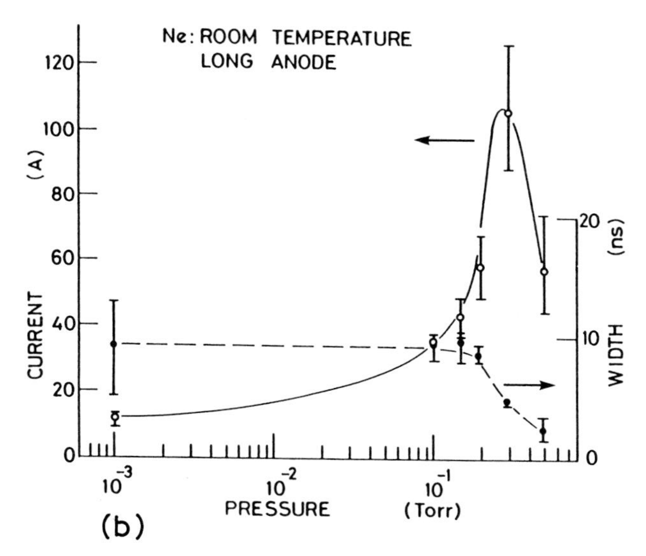

Fig. 6. Pressure dependences of REBs generated for long anode at -187.5 °C (a) and at room temperature (b). Temperature dependences of REBs are shown in (c) where solid cirlces are for peak currents and open circles are for pulse width (full width at half maximum).

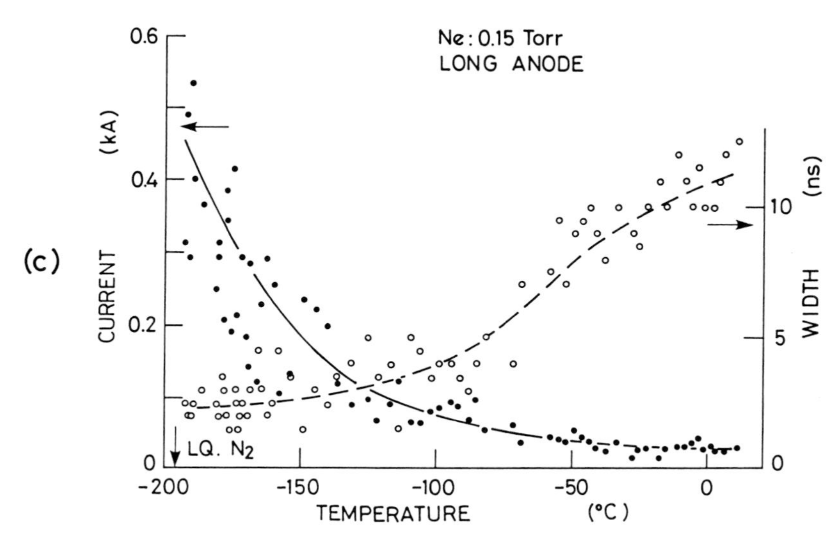

SUMMARY

We proposed a new type of superconductor lenses and guides for intense charged particle beams, and demonstrated their utility for REBs under the condition of high-pressure operation. We also showed that even copper electrodes have higher performance at lower temperatures, although metal guide-tubes were tried by other authors at room temperature only. From these facts anode tubes should be as low in electrical resistivity as possible as in superconductor anodes.

In the present paper, the superconductor lenses and guides were demonstrated for only pulsed electron beams. Those ideas are, however, applicable to continuous electron currents: The skin effect is effective for only pulsed currents, whereas the Meissner effect holds for both pulsed and continuous magnetic fields.

ACKNOWLEDGMENTS

We would like to thank Associate Professor Y. Saito, Yamanashi University and his students for teaching us the methode to prepare the high-Tc superconductor, and Y. Matsuno and Y. Chino for their cooperative technical support in preparing the experimental apparatus. The present work has been partially supported by Grant-in-Aids for Scientific Research (Nos. 63055016 and 63580004) from the Ministry of Education, Science and Culture.

REFERENCES

1. S. Humphries, Jr., J. Appl. Phys. 63, 583 (1988).
2. W. W. Destler, P. G. O'Shea, and Z. Segalov, J. Appl. Phys. 61, 2458 (1987).
3. J. D. Miller and R. M. Gilgenbach, Phys. Fluids 30, 3165 (1987).
4. H. Matsuzawa, T. Akitsu, Y. Suzuki, and S. Fuki, J. Appl. Phys. 61, 45 (1987).
5. H. Matsuzawa, O. Ohmori, H. Yamazaki, J. Ueno, A. Furumizu, A. Saito, T. Takahashi, and T. Akitsu, the paper of Technical Meeting on Optical and Quantum Devices, OQD-88-15, IEE Japan (1988) (in Japanese).
6. H. Matsuzawa, O. Ohmori, H. Yamazaki, J. Ueno, A. Furumizu, A. Saito, T. Takahashi, and T. Akitsu, in *Proc. 7th International Conference on High-Power Particle Beams*, Karlsruhe, July 4-8, 1988 (to be published).
7. L. P. Bradley, T. H. Martin, K. R. Prestwich, J. E. Boers, and D. L. Johnson, in *Record of 11th Symposium on Electron, Ion, and Laser Beam Technology*, edited by R. F. M. Thornley (San Francisco Press, San Francisco, 1971), p. 553.
8. C. L. Olson, Phys. Fluids 16, 529 (1973).
9. C. L. Olson, Phys. Fluids 16, 539 (1973).
10. A. C. Greenwald, J. Appl. Phys. 50, 6129 (1979).
11. H. Maeda, Y. Tanaka, M. Fukutomi, and T. Asano, Jpn. J. Appl. Phys. 27, L209 (1988).
12. Z. Z. Sheng and A. M. Hermann, Nature 332, 138 (1988).
13. H. Matsuzawa and T. Akitsu, J. Appl. Phys. 63, 4388 (1988).
14. R. B. Miller, *An Introduction to the Physics of Intense Charged Particle Beams* (Plenum, New York, 1982), Chaps. 1, 3, and 4.
15. For example, Rika-nenpyo (Maruzen, Tokyo, 1987), p. 480 (in Japanese).
16. H. Matsuzawa and T. Akitsu, Rev. Sci. Instrum. 56, 2287 (1985).
17. L. Pages, E. Bertel, H. Joffre, and L. Sklavenitis, Atomic Data 4, 1 (1972).

Anisotropy and Hysteresis Behavior of Galvanomagnetic Effect in Super Magneto-Resistor

HIDEO NOJIMA, SHOEI KATAOKA, MASAYA NAGATA, SHUHEI TSUCHIMOTO, and NOBUO HASHIZUME

Central Research Laboratories, Sharp Corporation, 2613-1, Ichinomoto-cho, Tenri, 632 Japan

ABSTRACT

The hysteresis behavior, fluctuation and their anisotropy of galvanomagnetic effect in novel magnetic sensor using bulk and film samples of ceramic superconductor Y-Ba-Cu-O have been measured. For the bulk element, the hysteresis was observed and its property was independent of the direction of magnetic field. For the film element, when the magnetic field direction was perpendicular to the element surface, the hysteresis was observed in a way similar to the bulk element, but when the magnetic field direction was parallel to the element surface, the hysteresis was eliminated to zero and fluctuation of output voltage was greatly reduced. These properties can be explained by considering the dynamics of magnetic flux trapping in the element.

INTRODUCTION

We have studied a fundamental galvanomagnetic effect of an Y-Ba-Cu-O ceramic superconductor and proposed a some applications to magnetic sensor based on this effect[1,2]. This sensor has following features: 1) it undergoes a very abrupt change in resistance at low magnetic fields, leading to a very high sensitivity; 2) it can be used for analogue as well as digital operations; 3) it has a very simple structure; 4) it can be very easily operated. We named this sensor "Super Magneto-Resistor". Furthermore[3,4], we have also shown that the sensor fabricated from a $Y_1Ba_2Cu_3O_{7-x}$ film patterned to a meander shape had a extremely improved sensitivity.

In those works, it was concluded that the magnetoresistance at an applied magnetic field lower than Hc1 is attributed to the weak couplings between superconductive grains in a ceramic superconductor, where Hc1 is a lower critical magnetic field. Thus it was considered that ceramic superconductor is a multiconnected Josephson network. It is expected that magnetic flux is trapped in a Josephson junction and/or in a Josephson network and the trapped flux induces unfavorable properties for sensing devices such as hysteresis and fluctuation. Therefore in order to realize a highly sensitive and stable magnetic sensor, investigation of dynamics of flux trapping is important.

In this paper, we report on hysteretic behavior, fluctuation of magnetoresistive effect and their anisotropy of this sensor using bulk and film samples of ceramic superconductor $Y_1Ba_2Cu_3O_{7-x}$. We discuss an origin of these properties, and present a method for eliminating the hysteresis and the fluctuation.

EXPERIMENTAL

In this experiment, $Y_1Ba_2Cu_3O_{7-x}$ bulk and film samples were prepared by standard solid-state reaction method and spray pyrolysis method, respectively. The preparation procedure of bulk element was described in detail in the previous paper[2]. Spray pyrolysis method is a convenient technique to obtain superconductive films. An aqueous solution containing Y, Ba, Cu nitrates was prepared by dissolving $Y(NO_3)_3 6H_2O$, $Ba(NO_3)_2$ and $Cu(NO_3)_2 3H_2O$ into water. The aqueous solution was sprayed over yttrium-stabilized zirconia(YSZ) substrates which were kept at about 350°C. After heating at 800°C for 5 min in air, the aqueous solution was sprayed again over the film under the same condition. Finally, the film was heated at 950°C for 5 min in air. Final thickness of the film was about 5 μm.

The sample characterization was performed by EPMA, X-ray diffraction spectroscopy and SEM observation. The bulk sample investigated in this paper has a composition of $Y_1Ba_2Cu_3O_{7-x}$ and an orthorhombic crystal system with a=3.83A, b=3.90A and c=11.7A. The film sample also has a composition of $Y_1Ba_2Cu_3O_{7-x}$ with c-axis oriented perpendicular to the substrate surface. The SEM observation showed that both elements consisted of many grains ranging from 2μm to 5μm in diameter.

The dimensions of the bulk element and the film element were $1x7x0.7\ mm^3$ and $5x10x(5x10^{-3})\ mm^3$, respectively. The schematic structure of the both elements and magnetic field direction are shown in Fig.1. Ohmic electrodes were formed by Ti evaporation onto the element. The contacts between the Ti electrodes and the electrical lead wires were made by silver paste. The zero resistance temperatures $Tc_{(zero)}$ were 83 K and 81 K for the bulk element and the film element, respectively.

The galvanomagnetic effects were measured in a uniform magnetic field generated by the Helmholtz' coil. The anisotropy measurements were performed with the magnetic field direction in a x-y plane shown in Fig.1. The instrumental noise level was about 10^{-7}V.

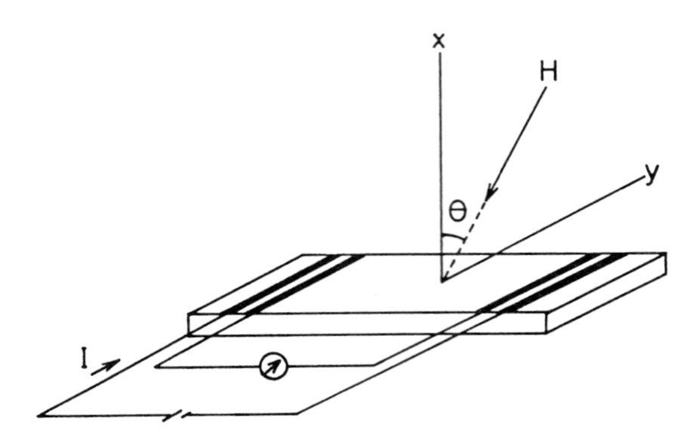

Fig.1. Schematic structure of the element and the magnetic field direction.

RESULTS AND DISCUSSION

a) BULK ELEMENT

Figure 2 shows the magnetic field dependence of critical current Ic of the bulk element at 77 K. Critical currents Ic were determined from the measurements of current-voltage characteristics at each applied magnetic field. The time for sweeping the current on each I-V measurement was about 100 sec. It is found that the critical currents for increasing field and decreasing field are different. As shown, when the magnetic field was increased, the critical current decreased steeply, and when the magnetic field was decreased from the maximum applied field Hmax of 150 Oe, the critical current increased with decreasing field up to 15 Oe. However, below 15 Oe the critical current decreased with decreasing field. The critical current for decreasing magnetic field was higher than that for increasing field for most of the field range observed. This relation did not hold for small magnetic fields (less than about 9.3 Oe).

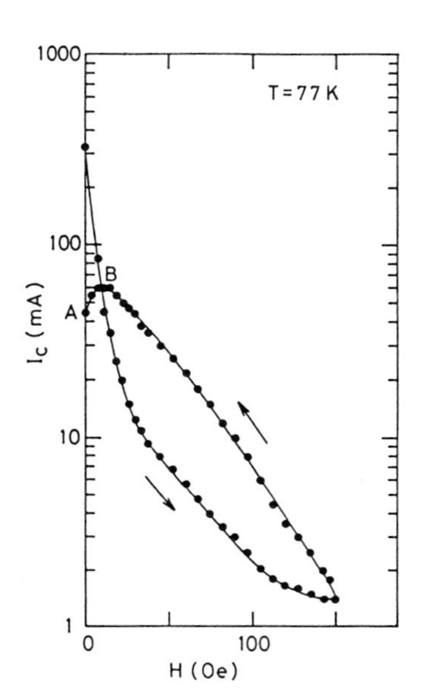

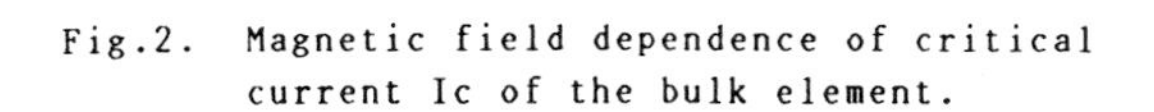

Fig.2. Magnetic field dependence of critical current Ic of the bulk element.

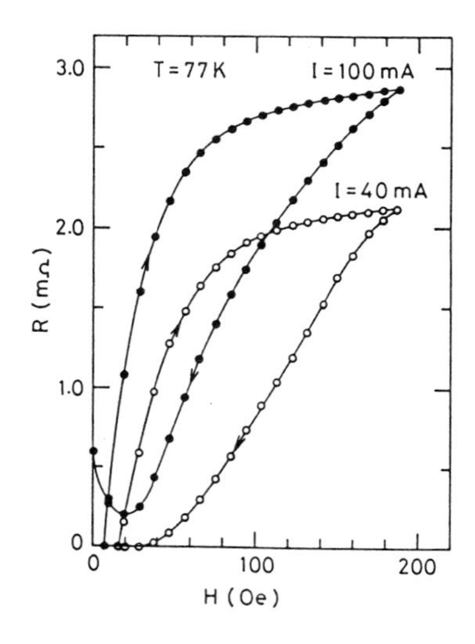

Fig.3. Hysteretic behavior of magneto-resistive characteristics of the bulk element. Currents through the element are 100mA (●) and 40mA (○).

The physical mechanism of these complicated characteristics is not clear yet, but it is considered as follows. When a magnetic field lower than Hcl is applied to the element, the magnetic flux penetrates into the element along the grain boundaries or along the regions unfilled by the superconductive grains. When the field is subsequently decreased, the penetrating magnetic flux is retained by electromagnetic induction and trapped at the grain boundaries or at the unfilled regions. The trapped flux induces an internal magnetic field. The field as observed at the untrapped regions has the opposite polarity to the external magnetic field. The effective internal field at the untrapped regions was 15 Oe and the absolute value of the total effective magnetic field became lowest at the external magnetic field of 15 Oe. Since the total effective magnetic field is smaller than the external magnetic field at a field from Hmax to a certain value, 9.3 Oe in this case, the critical current for decreasing field is higher than that for increasing field in this range. At an external magnetic field lower than 9.3 Oe, the total effective magnetic field becomes larger than external field, and thus the critical current for decreasing field becomes lower than that for increasing field. As a result, the magnetic field dependence of the critical current Ic has the hysteresis as shown in Fig.2.

The magnetoresistive effect of the bulk element at 77 K is shown in Fig.3. This shows the resistance, R, of the bulk element, with a current I through the element, for increasing and decreasing magnetic fields. It is found that the magnetoresistive effect has hysteresis for both I=40 mA and 100 mA. Figure 3 shows typical two cases. When I=40 mA, the resistance is equal to zero up to a magnetic field of about 14 Oe, and then increases rapidly, and with the decrease of field, the resistance decreases monotinically and becomes zero at about 28 Oe. It is shown that the resistance for decreasing field are smaller than that for increasing field in range measured in this experiment. When I=100 mA, with the increase of field, the resistance is equal to zero up to a magnetic field of about 7 Oe and then increases. With the decrease of field, the resistance decreases up to a certain magnetic field, 20 Oe, and then increases. At zero magnetic field, the resistance retains a finite value. The resistance for decreasing field is lower than that for increasing field for most of the magnetic field range measured in this experiment. However, for small magnetic field the resistance for decreasing field becomes higher than that for increasing field.

These properties can be explained by the magnetic field dependence of critical current Ic shown in Fig.2. When the current I is lower than Ic, naturally the resistance is equal to zero, and when the current I becomes higher than Ic, the resistance becomes finite value. The larger the difference between I and Ic is, the larger the resistance becomes. When I=40 mA, the current I is lower than a point A in Fig.2 (Ic at zero magnetic field after experiencing the applied field). Then with the increase of field, the critical current which is initially higher than I decreases and becomes lower than I, and then with the decrease of field from Hmax the critical current increases and becomes higher than I. Thus the resistance suddenly appears and increases steeply with the increase of field and decreases monotonically to zero with the decrease of field. When I=100 mA, the current I is higher than point B in Fig.2 (Ic of peak value for decreasing field). Then with the increase of field, the critical current which is initially higher than I decreases and becomes lower than I, and then with the decrease of field from Hmax the critical current increases and approaches to I up to about 15 Oe and then decreases. However the critical current Ic never reaches the current I. Therefore in this case the resistance increases in a way similar to the case of I=40 mA for increasing field, but for decreasing field the resistance decreases up to 20 Oe and then increases as shown in Fig.3. The magnetic field at which the critical current Ic becomes highest and the resistance R becomes lowest are different, since the maximum applied field are different in Fig.2 and in Fig.3.

We have measured the dependence of these properties on the magnetic field direction. They were independent of the field direction in x-y plane in Fig.1.

b) FILM ELEMENT

The hysteretic behavior of magnetoresistive effect of the film element is shown in Fig.4 when the direction of magnetic field is perpendicular (a) and parallel (b) to the element surface. For the perpendicular field, the hysteretic property is similar to that of the bulk element. However, for the parallel field, the hysteresis is reduced to zero (to an accuracy of our instrumental noise level) without large degradation of the magnetoresistive sensitivity.

The fluctuation of output voltage of the film element with a magnetic field perpendicular (a) and parallel (b) to the element surface at 77 K is shown in Fig.5. As shown, the fluctuation for perpendicular field is much larger than that for parallel field. The fluctuation for parallel field was very small and comparable to our instrumental noise level.

Since the hysteresis is considered due to the trapped flux and motion of trapped flux induces fluctuation of output voltage, these experimental results on film element imply that when the magnetic field direction is parallel to the element surface, magnetic flux is not trapped in the element. As for the reasons for this effect, we consider as follows. The penetrating flux is trapped at the grain boundaries or at the unfilled regions as discussed above. But for the parallel field, the flux cannot be trapped at the unfilled regions. The film thickness is about 5 μm and the diameter of the superconductive grain ranges from 2 μm to 5 μm. Thus the film is consisted of almost one layer of grain on the substrate. As a result, when the magnetic field direction is parallel to the element surface, the superconducting current for trapping flux at an unfilled region between grains cannot form a loop. Thus the flux cannot be trapped at an unfilled region. On the other hand, the current for trapping flux at a grain boundary can form a loop. However, in this case the direction of the trapped flux is perpendicular to the current through the element and thus the Lorentz's force is perpendicular to the element surface. Therefore the magnetic flux is expelled out of the element along the grain boundary. The pinning force along the grain boundary is considered to be weak. For the reasons discussed above, the flux is not trapped or even if it is trapped it will be expelled rapidly out of the element when the magnetic field direction is parallel to the film element surface. As a result, we can obtain non-hysteretic property and excellent stability.

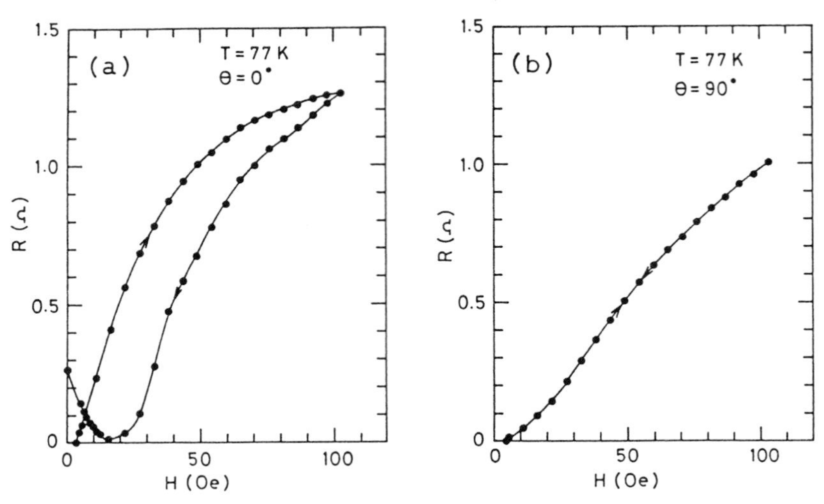

Fig.4. Hysteretic behavior of magneto-resistive characteristics of the film element. The magnetic field direction is perpendicular (a) and parallel (b) to the element surface.

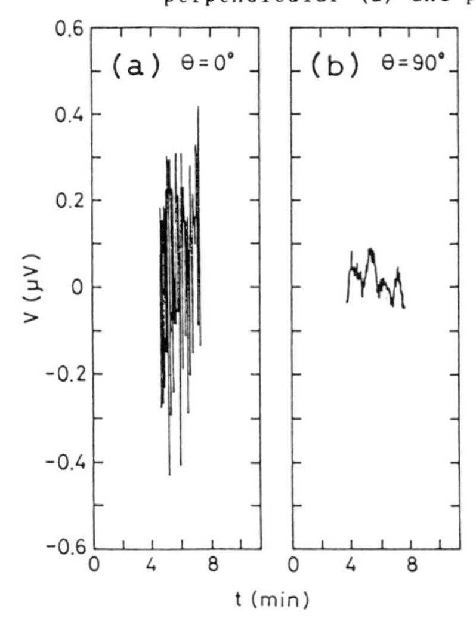

Fig.5. Fluctuation of output voltage with applied magnetic field perpendicular (a) and parallel (b) to the element surface. For the parallel field, the fluctuation is comparable to our instrumental noise level.

CONCLUSION

We have measured the hysteretic behavior, fluctuation and their anisotropy of galvanomagnetic effect in novel magnetic sensor using ceramic superconductor $Y_1Ba_2Cu_3O_{7-x}$. For the bulk element, the hysteresis of magnetoresistive effect was observed and their property was independent of the directions of magnetic field. However for the film element, we are successful in obtaining non-hysteretic property and excellent stability without degrading the magnetoresistive sensitivity when the magnetic field direction is parallel to the element surface. These properties can be explained by considering the dynamics of magnetic flux trapping in the element.

ACKNOWLEDGEMENTS

We would like to thank H.Shintaku for technical assistance on electrical measurements and also R.Kita and E.Ohno for valuable discussions.

REFERENCES

1) S.Tsuchimoto, S.Kataoka, H.Nojima, R.Kita, M.Nagata and H.Shintaku: IEDM Tech.Digest 12.7 (1987), 867.
2) H.Nojima, S.Tsuchimoto and S.Kataoka: Jpn.J.Appl.Phys.27 (1988) 746.
3) R.Kita, M.Nagata, H.Shintaku, H.Nojima, S.Tsuchimoto, N.Hashizume and S.Kataoka: 5th international workshop on Future Electron Devices (1988) 231.
4) S.Kataoka, S.Tsuchimoto, H.Nojima, R.Kita, M.Nagata and H.Shintaku: Sensors and Materials 1 (1988) 7.

Some Magnetic Applications of High-Tc Superconductors: Foils, Ductile Materials, Filaments and Magnets

J. GYÖRGY[1], J.MATRAI[1], I. KIRSCHNER[2], G. SZENTGYÖRGYI[3], T. PORJESZ[2], M. LAMM[1], I. MOLNAR[3], R. LAIHO[4], GY. KOVACS[2], T. TRAGER[1], P. LUKACS[3], T. KARMAN[2], I. P-GYÖRGY[1], M. TAKACS[3], and G. ZSOLT[2]

[1] Central Research and Design Institute of Silicate Industry, Budapest, 1032 Bécsi út 126/128, Hungary
[2] Department for Low Temperature Physics, Roland Eötvös University, Budapest, 1088 Puskin u. 5-7, Hungary
[3] Hungalu, Engineering and Research Centre, Budapest, 1133 Pozsonyi út 56, Hungary
[4] Wihuri Physical Laboratory, University of Turku, 20500 Turku, Finland

ABSTRACT

Foils of different thickness have been elaborated by mixing with organic filling components based on Y-Ba-Cu-O compounds. Pieces of them are capable to levitate Co-Sm magnets and have good properties for magnetic screening. These foils can be produced on a semi-plant level.

Ductile rods of different length and thickness have also been produced for magnetic purposes.

First steps of producing thin filaments were successful for making different high-Tc superconducting tools.

After the first experiments, complicated magnet systems have been built based on superconducting ceramic spirals and rings of different size. They are energized partly by electric and partly by magnetic ways.

SUPERCONDUCTING SAMPLES

Most of the samples made for application consists of Y1Ba2Cu3O7 material. The preparation process of the samples was started from chemicals of 99.99 % purity which were mixed in the ratio of 1(Y2O3)+4(BaCO3) and ground in an acetonic medium by a ball mill during 8-10 hours. This process ensures the homogeneity of the mixture as well and the average diameter of the grains fell between 10 and 20 µm.

For inducing a solid-state reaction the primary heat treatment was performed at 950 °C for 3 hours in air at which the rate of temperature change was 100 °C/h, both at heating up and cooling down. The reaction products were ground again in a ball mill for 24 hours resulting in an average grain diameter less than 10 µm. After drying samples (rods, prisms, rings etc.) of different forms and sizes (see Fig.1) were pressed of the perfectly mixed powder under 100 MP pressure. The samples were sintered at 980 °C in air for 4 hours, using a heating up rate of 100 °C/h and cooling slowly down for 24 hours. According to the X-ray diffractogram (Jeol JDX-85) this procedure brought about a single phase orthorhombic Y1Ba2Cu3O7-x material [1]. The ceramic samples obtained in this way are mechanically stable compact, have a density of 4.85 gr/cm3 and become superconducting at the critical temperature Tc>90 K. Very dense, compact and hard samples were made by isostatic pressing at 500-600 MPa pressure in different forms and sizes too.

CRITICAL MAGNETIC FIELDS

Connecting with the prospective magnetic application, it is very important to know the values of the critical magnetic fields. Measurements of the upper and lower field depending on the temperature were carried out by a SQuID-technique.

Experiments concerning the lower critical magnetic field led to the conclusion that its temperature dependence departs from the conventional square function (Fig.2). In this figure the normalized values h=H(T)/H(Tc) and t=T/Tc are demonstrated. By these experiments the function is nearly linear.

The upper critical field provides very high values (Fig.3)

FOLIES

The prepared Y-Ba-Cu-O powder mixture was ground down to the average particle size of 5 microns. The ground powder was mixed with the PVC paste of type SOLVIC 376 NB 75 wt % quantity as a filling material. The homogeneized mixture was poured onto a glass plate. The poured folie with glass plate was heated at 140 °C for 40 minutes. The product is flexible (elastic) at room temperature (Fig.4).

1

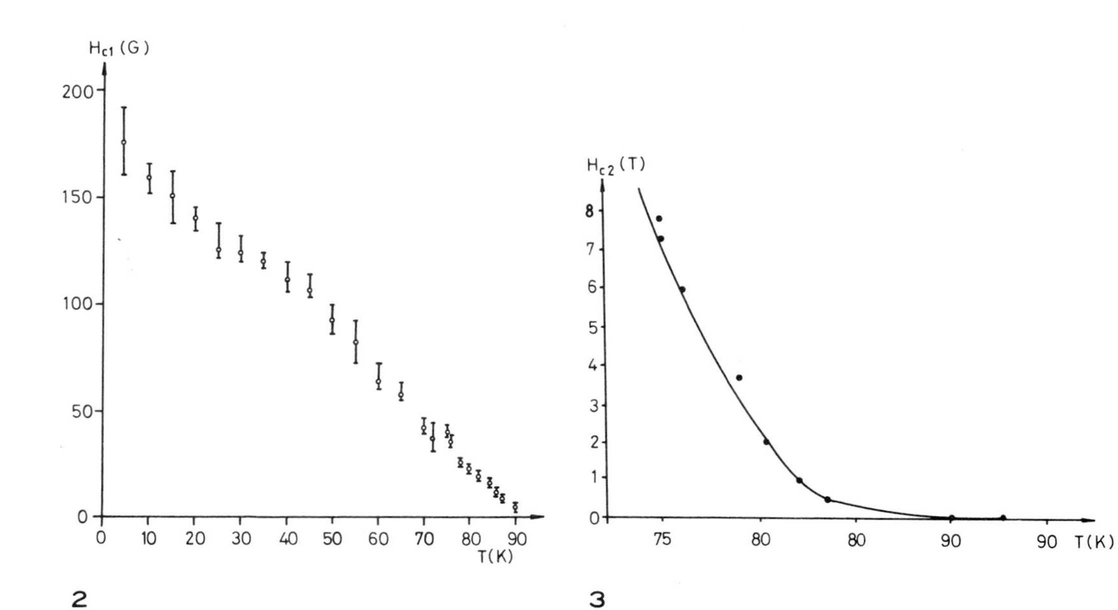

2 3

ELASTIC RODS

From Y-Ba-Cu-O mixture ground below 5 microns using SOLVIC 376 NB PVC suspension in 15 wt % quantity a homogeneous elastic paste was produced. The paste was formed by an extruder to rod with 10 mm diameter then it was heated at 140 °C for 40 minutes. The product is flexible at room temperature (Fig.5).

FILAMENTS

Initial experiments for making high-Tc superconducting filaments were performed. On the basis of the first experiences a possibility exists to accomplish this task. For obtaining good quality filaments this research must be continued in different directions. The first results are hopeful form the point of view of establishing useful filaments (Fig.6). The extrusion of slurry described previous publications resulted coars grain-size and consequently knotty surfaces. To avoid this drewbacks, we have used a precursor ($YBa_2Cu_3O_{7-\delta}$) with grainsize diameter smaller than 1 μ. Before the extrusion more than 80 wt % precursor was homogeneized in benzene-solution of nitro-cellulose. The filaments was dried and fired in an oxygene atmosphere at 840 °C. The average diameter of filaments was 14 μm.

4

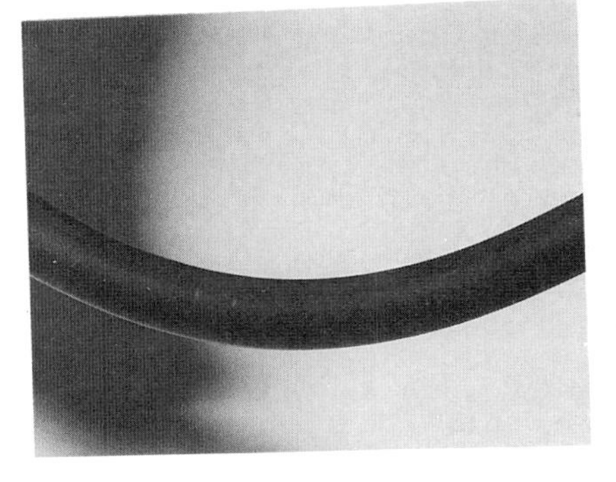

5

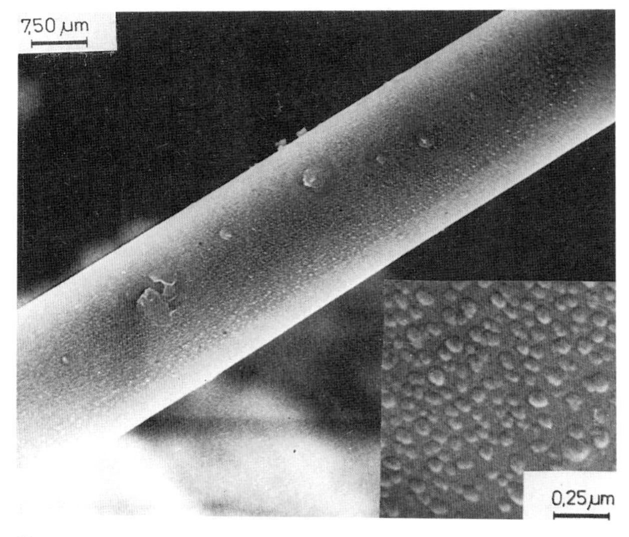

6

MAGNETS

The experimental results show that the value of the magnetic field intensity inside of the rings depends on their composition and material quality determined by preparation technology. Increasing the external magnetic field intensity does no imply an increase of the magnetic field produced by the superconducting current in the rings [2].

According to the measurements, the rings contain a magnetic field intensity of 1.5-1.8 mT, which corresponds to a superconducting current of 54-71 A, and a bulk current density of 86-182 A/cm2.

Figure 7 represents a superconducting magnet constructed of ceramic rings. This consists of 15 rings having an outer diameter of 4.8 cm, inner diameter of 2.4 cm and thickness of 0.45 cm, and after mounting them together the magnet has a net height of 7.0 cm. The individual rings have an average magnetic field. of 1.62 mT and a superconducting current of 62 A respectively. These values provide a 16.6 mT calculated magnetic field for the whole system, which is in a good agreement with the measured field intensity of 16.1 mT. An other magnet was put together of 10 rings having 0.4 cm and of 10 other having 0.3 cm thickness. Their outer diameter was in general 4.8 cm and their inner diameter 2.5 cm. Corresponding to the magnetic field of 17 Oe and the superconducting current of 65 A of the individual rings and to the net height of the magnet which is 7.7 cm, the calculated value of the magnetic field is 21.2 mT, while the experimentally measured one is 20.7 mT.

The superconducting magnets of this type may perhaps be efficiently developed further if a more complicated system of rings is formed instead of the simple one outlined above. This latter would consists of layers formed by concentrically joined rings of different size and then these layers are put on top of the others. In this method the number of the amper-turns and thus both the magnetic field intensity and the energy storing capacity can be increased significantly.

In the Figure 8 a spiral-based magnet can be seen. Its fundamental structural elements is a ceramic spiral where good electrical contacts are provided by packings made of soft indium. The smaller magnet of this type has 5.5 windings in 4.8 cm length and produces a magnetic field of 0.15 mT with 1 A supply current. The bigger one has also 5.5 effective turns in 5.8 cm length and in a case of 25 A current a magnetic field of 3.2 mT belongs to it [3].

This magnet building system can be developed further, if a few spirals of different diameter are put into one another, forming "solenoids" by which more intense magnetic field can be produced. Using spirals of different length as correction coils even some minimal requirements of homogeneity can be fulfilled too.

7

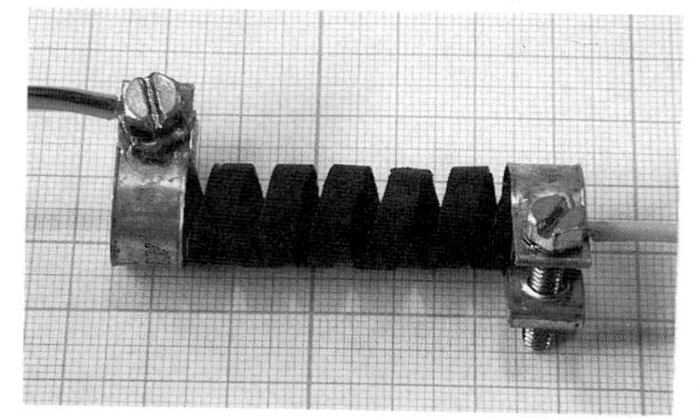

8

REFERENCES

1. Kirschner I., Träger T., Mátrai J., Porjesz T. and György J.
   Physica C153 1419 (1988)
2. Kirschner I., Lamm M., Porjesz T., Mátrai J., Kovács Gy., Träger T.,
   Kármán T., György Z. and Zsolt G.
   Physica C153 1417 (1988)
3. Kirschner I., Mátrai J., Szentgyörgyi G., Porjesz T., Lamm M., Molnár I.,
   Kovács Gy., Träger T., Lukács P., Kármán T., György J., Takács M. and
   Zsolt G.
   Cryogenics 28 No.9 (1988)

The Potential Utilization of High Temperature Superconductivity

GEORGE O. ZIMMERMAN

Physics Department, Boston University, Boston, MA 02215, USA

ABSTRACT

The large scale practical application of high T_c materials is discussed and two possible examples are given. One of the examples points to a new way of installation, while the other points to a way of using the properties of superconductivity to overcome some of the mechanical defficiencies of the materials.

TEXT

I have been asked to talk about the potential utilization of the recently discovered high temperature superconductors. When they were first discovered, their practical utilization seemed to be just around the corner; however, once their properties were looked at more closely, people have realized that it will take some time to put the new substances to use. There is no doubt that there is superconducting material which will, in the future, have great utility, that the material is relatively easy to prepare, and that the energy cost for keeping the material superconducting is possibly smaller than the cost of the losses due to power dissipation in normal conductors. The great advance came when, after the discovery of the 30K to 40K superconductor by J.G. Bednorz and K.A. Müller, C.W. Chu and his coworkers at the University of Houston pushed the transition temperature above 77K, the boiling point of nitrogen. However, the properties of the high T_c superconductors such as their brittleness, and the low current carrying capacity of the easily prepared samples, form a formidable barrier against their immediate use.

There are two general types of usage the new materials can be put to. They are large scale and small scale. Large scale use includes, for example, electric power transmission, motors, generators, etc. Small scale usage would involve electronics, microcircuits, Josephson-junction computers, and other similar uses. Except for very large computers, and the amplification of signals in the millimeter and submillimeter wavelangth range for radioastronomy, where a large energy gap is of advantage, (see Paul L. Richards, Physics Today p.54, March 1986), I can see relatively little advantage to the use of high T_c materials over conventional superconductors with transition temperatures below 20K. The reason is that cryogenic techniques have cut the cost and maintenance time of helium cooled systems to a negligible level. Moreover, because of the higher temperature at which the systems could be operated, the devices could become much noisier than their conventional counterparts, and thus the advantage of higher temperature would result in a poorer performance of the device. In this I share the views expressed by John Clarke, (see Nature, 333, 29, (1988)). I will thus concentrate on the large scale application possibilities of the new materials.

Historically, new technologies, when first introduced, coexist with the old ones for a long time after they have been introduced. They also bring with them new possibilities which are gradually realized. Also, it is important during the introductory stage to consider the compatibility of the new systems with those they are designed to replace, as well as their cost and availability. Cost and availability are connected.

Let us first look at the availability of the materials. Currently there are three contenders for practical high T_c materials with more to come. Those are the YBaCuO 123 compound, with Y replaced with various other rare earth elements, the BiAlCaSrCuO compound, and the TlBaCaCuO compound, the last two having been reported in literature in March 1988, (see Chu et al, Phys. Rev. Letters 60, 941 and Submaranian et al Nature 333, 420). If one looks at the composition of the earth's crust, (Principles of Geochemistry, Brian Mason, John Wiley and Sons, (1952)), one sees that the rarest element in the first compound is Y, which constitutes 28ppm of the earth's crust. The abundance of Y is only half that of Cu. In contrast, the second compound has Bi, which constitutes only 0.2ppm and the third has Tl which constitutes only 0.6ppm of the earth's crust. I believe that the 123 compound might be the most economical of the currently existing high T_c superconductors, barring the revelations of some new properties.

I believe that the low current carrying capacity of the new superconductors will be remedied by better synthesizing techniques, since films have been obtained with current carrying capacities of over a million amps per square centimeter, and it is generally believed that the current limit is set by the barriers between the superconducting grains. Leaving this problem, (to be solved in the near future,) aside, several problems concerned practical utilization still remain. The main ones I see are the brittleness of the materials, their electrical contact to conventional conductors, and impedance matching between superconductors and conventional conductors, since we assume that any superconducting system would have to eventually connect to a conventional electrical system.

On the one hand we have difficulties to overcome, on the other, with the new technology we have opportunities to improve and augment the old systems and put the new ones to better use. Old technologies might be the use of Cu or Al in electrical power transmission, electrical machinery and communication. New technology might be the use of superconductivity in new ways or in new combinations. One natural combination would be the generation of electrical power by photovoltaic means and superconducting transmission lines. Photovoltaic generation takes place at a low voltage, and a superconducting power transmission line would be an ideal means of propagation for this power. The problem would come at the load end, where the electrical potential would have to be increased. There are standard techniques for doing that, and the most suitable ones could be employed.

Since I have been asked to "dream" of some possible hight T_c applications, let me concentrate on two examples. The first illustrates a possible enhanced use of conductors, the second illustrates a way of coping with the shortcomings of the present material.

1) A Superconducting Line Along a Road Right of Way.
When a new road is built, one could incorporate a strip of HTSC along the side of the road. Such a strip could be used for power transmission, sensing, as well as for communications, (cellular phones, for instance). If automobile technology progresses sufficiently, the strip could also be used for the guidance of vehicles, or possibly the levitation of vehicles. Because of the small power loss of the superconductors, even for rf signals, the number of amplification stations could be held to a minimum. Dc would be used for power transmission while ac of various frequencies could be used for sensing, control and communication. The power transmission could take place at a very low potential because there is no resistance in a superconductor. Electrical insulation would be provided by any substance which is not superconducting.

The paving and the surrounding soil should provide sufficient insulation for the line to remain at superconducting temperatures. Typically there would be a a heat leak of about 100 watts for each meter of line due to heat from the surroundings. The refrigerators providing the cooling could thus be solar powered if we assume a collection area 1 meter wide and an efficiency of 10%, since the average power due to solar radiation is approximately 1Kw per square meter. That collection area could be built along the emergency lane.

The installation of such a system could be performed either while installing a new roadbed or in old roadbeds. While paving the road, one would leave a groove near the side of the road about $\frac{1}{2}$" wide and 1" deep. One will deposit into this groove some $YBa_2Cu_3O_x$, for instance, and fire it in place so that it becomes a superconductor when cooled down to 90K. The groove would then be filled in so that a tube of about $\frac{1}{2}$ inch remains above the superconductor. The refrigeration will be provided by solar powered refrigerators which would get their power from a solar strip, either photovoltaic or radiation absorbtion by means of a circulating fluid.

Figure a shows such a system installed along a road with subsequent enlargment of the details. Figure b shows an authors concept of how an installation unit would be designed and how it would operate.

Although the installation and operation operation of the system is concieved to be relatively "low-tech", specifically designed to fit with the equipment and sophistication of a roadpaving crew, several components of the system would have to be developed.

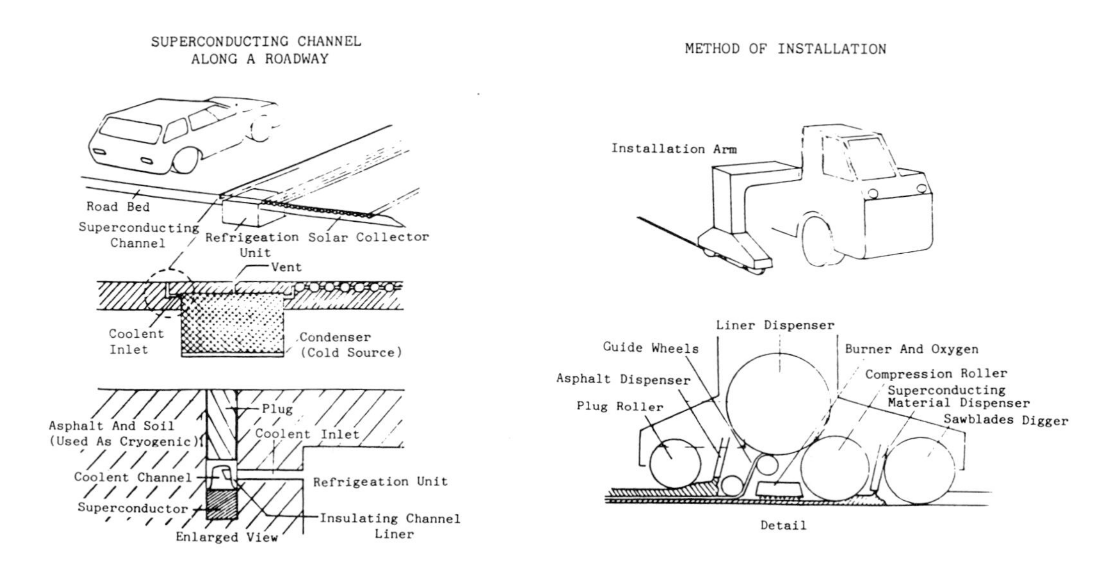

Figure a

Figure b

a) One would need to develop a reliable sequence of scintering and cooling of the component powder so it becomes a superconductor. The initial material would be compounds of Y, Ba and Cu, for $YBa_2Cu_3O_7$, or other materials which will provide the initial starting point for new high T_c materials to be discovered. The recently discovsred Bi-Al-Ca-Sr-Cu-O compounds or the Tl compounds would be equally good candidates. Usually superconductivity comes about after several steps of scintering, grinding, pressing, scintering etc. This process would have to be reduced to one continuous process. Presently this could be done by prepreparing the material so that only the last step would have to be performed while laying the road. It would be preferable, however, if the initial chemicals could be poured into a hopper and the whole process took place during the installation.

b) The refrigeration system will have to be developed. Solar powered refrigerators have been developed in connection with low grade refrigeration, -10°C. two stage refrigerators might have to be developed in order to cool to -150°C or -200°C. Those refrigerators would have to be low cost, reliable, and would have to operate with minimal maintenance.

2)Electric Current Transmission by dc Magnetic Coupling in Hight T_c Superconductors

Introduction: The goal of this project is to design a prototype of a superconducting, flexible, transmission line using the recently discovered High T_c Superconductors (HTSC) which become superconducting at temperatures above that of the boiling point of liquid nitrogen, about 77K. All such compounds discovered to date are inflexible and brittle, and attempts to alter those mechanical properties have met with little success. Our approach is to accept those limitations and to make a flexible conductor out of this material by constructing a chain out of a set of links made out of HTSC. Because there is considerable contact resistance between the individual links, the links are coupled by means of a magnetic field. This coupling takes advantage of the fact that superconductors can sustain a current without resistive dissipation and that they exhibit the Meissner effect which tends to exclude a magnetic field from the interior of a superconductor.

In this project, work on which started in December of 1987, we have proceeded along several parallel paths. One is to build a model circuit out of conventional superconducting material, obtain its characteristics, improve and optimize the design for power transmission, and build a prototype out of conventional superconducting materials. The second path is to explore the various ways one can manufacture and assemble the links into a chain made out of the (HTSC) material. We have made considerable progress on both.

Apparatus: The apparatus is shown in Fig.1 and Fig.2. The two configurations of the measuring instruments were necessitated by the fact that we wanted to characterize the circuit by performing both dc and ac measurements. The central core in both configuraitons is a primary input, with part of the primary being superconducting. The secondary loop is completely superconducting with a mutual inductance between it and the primary as well as between it and the tertiary circuit. There is an extra loop in the secondary circuit for the insertion of a magnetometer probe (Hewlett Packard 428 Clip-on dc Milliammeter) to measure the dc current in that circuit. That circuit is in turn coupled to a tertiary by means of a mutual inductance. The tertiary consists of a superconducting coil and a load resistance across which a voltage is measured. This circuit will thus simmulate one link of the eventual chain, into which both the source and the load are coupled. The design tested is not optimal and work is being done on improvements.

We have investigated the characteristics of such a system with the inductors wound on top of each other on a dielectric. The superconducting components of the apparatus were made out of lead wire and the circuit characteristics were measured at helium temperature where lead is superconducting. The dashed envelope on the circuits denotes the parts at low temperatures. The charactristics were also measured at 77K and 300K to eliminate the behavior not due to superconductivity. Here we only report the superconducting behavior. Because we want to transmit power from a source which will have a characteristic impedance, to a superconducting link which will have only a reactive impedance, to a load which again has a resistive component, impedance matching becomes important. The impedances are reflected in the circuits as N^2 where N is the ratio of turns in the transformer or mutual inductance. The turn ratio in the measurements which we present here was 80:10:10:10 and 80:10:10:80, with Primary:First Secondary:Second Secondary: Tertiary. The inductances had values of 5μ-henries for the 80 turn coil, 0.08μ-henries for the 10 turn coil and the mutual inductance was about 10% of ideal. We believe we can improve that to close to 100%.

dc Characteristics: The purpose of this measurement was to determine the basic idea of whether dc currents can be induced in the secondary. There was also a hope that we could see acurrent in the tertiary circuit. The latter results are ambiguous. The measurements were made with the circuit shown in Fig.1. In order to calibrate the magnetometer a one-turn coil was wound around the magnetometer probe powered by the "DC2" power supply. "DC1" is the primary power supply with an ammeter, and "V" measures the voltage across the superconducting coil. Obviously the reading of "V" was zero thus proving that we were in a superconducting state. We used currents up to 8 amperes without the system becoming normal.

Fig.3 shows the output of the magnetometer with the primary current indicated at the steps. This measurement indicated that there is a current in the secondary which has a value of 0.07 amps for one amp applied to the primary. This is consistent with the approximate 10% efficiency of the coupling. We also determined that the current in the secondary was linear with the primary current. The secondary current was measured with the tertiary circuit both open and closed. Some trends were detected which indicate the coupling of the tertiary circuit to the secondary, pointing to possible dc power transmission.

ac Characteristics: The ac behavior was measured with the circuit shown in Fig.2. The ac signal from a function generator was put through an amplifier and applied to the primary circuit. The voltage across the load resistor "R_L" in the tertiary circuit was the output measured. We found that the output was greater for a square wave than for a sine wave. This difference, especially at low frequencies, could not be accounted for by the harmonic content of the Fourier components in the square wave.

Fig.4 shows the output as a function of frequency. "1" denotes the square wave while "2" is the sine wave. The outputs are approximately linear with frequency between 1KHz and 10KHz. The low frequency portion of the graph is magnified in Fig.5. Here we see a rapid drop-off of the sine signal between 600Hz and 1000Hz while the square wave signal levels off. There is even an indication of an increase in the square wave signal near 10Hz. Below this frequency the reliability of our instruments is questionable. We are currently setting up apparatus to measure the low frequency region. This is of importance in this investigation because it will point towards a true dc tranmission line.

Fig.6 shows the power measured at the load resistor "R_L" as a function of the resistance at various frequencies. One observes that there is a maximum in the power as a function of the load resistance and that maximum tends to lower resistance as the frequency increases.

Manufacturing: Because our eventual aim is to build a transmission line out of HTSC material, we developed a system for the machining of the individual components from bulk material. Fig.7 and Fig.8 show the various stages of this process. The material is cut into rectangular parallelopipeds and holes drilled and tapped at either end. The material is then cut away so that there are only two strips left connecting the tapped holes, one on each end. Finally the material around the holes is machined away so that only the threads are left, thus forming a spiral. One is thus left with a link, shown at the bottom of Fig.7. Those links are then joined at the spiral with a pin forming the pivot. This demonstrates the feasibility of making the transmission line out of bulk material. The nickel in the two illustrations is shown for the sake of size comparison.

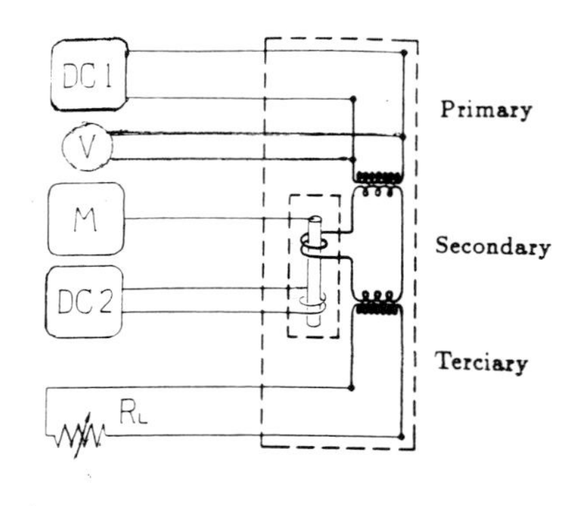

Figure 1
Set up for DC Measurements

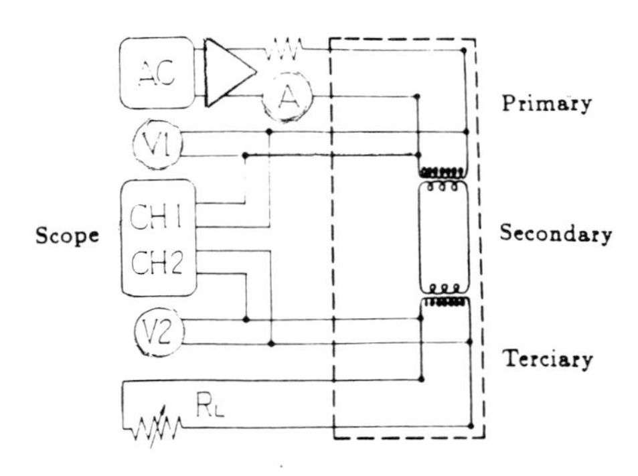

Figure 2
Set up for AC Measurements

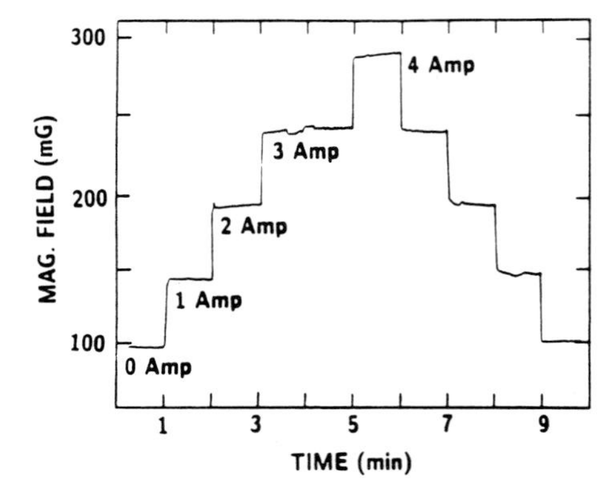

Figure 3
Magnetometer Secondary Output as a Function of the Applied Primary DC Current

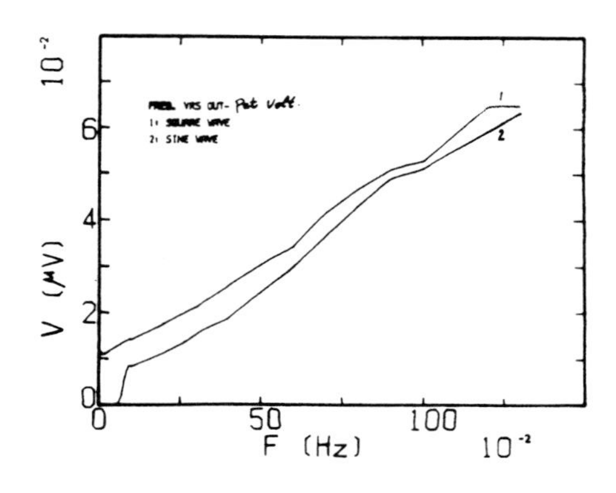

Figure 4
Voltage Output as a Function of Frequency

Conclusions: The measurements presented here show the feasibility of a superconducting power transmission line. We have shown that at low frequencies the square wave is more efficient than the sine wave. We are in the process of improving the efficiency of the line and investigating the low frequency characteristics, below 10 Hz.

The application of hight T_c superconductivity for large scale usage will reqiure imagination, ingenuity, and a thorough knowledge of the properties and potential of the material, in addition to hard work.

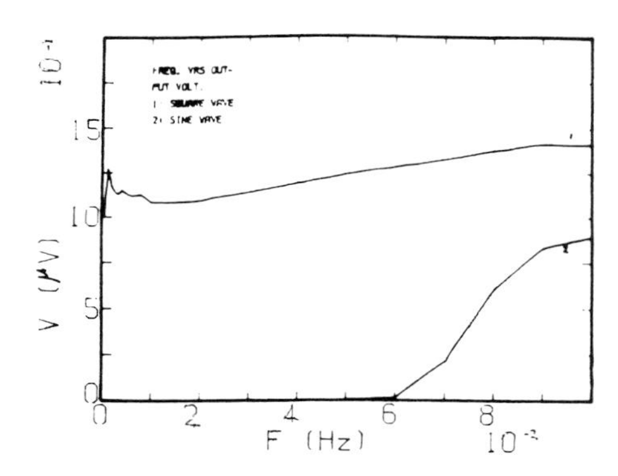

Figure 5
Low Frequency Magnification
of Figure 4

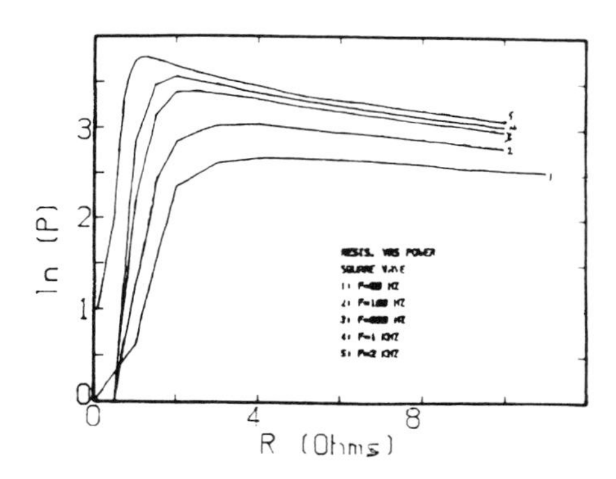

Figure 6
Power Output as a Function of
Load Resistance

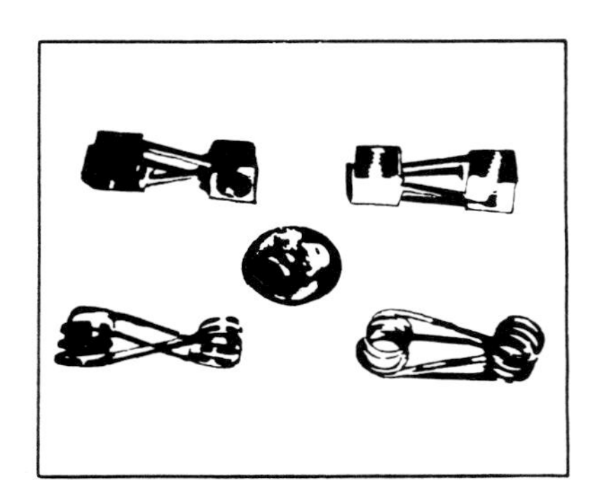

Figure 7
Stages of Fabrication of Links

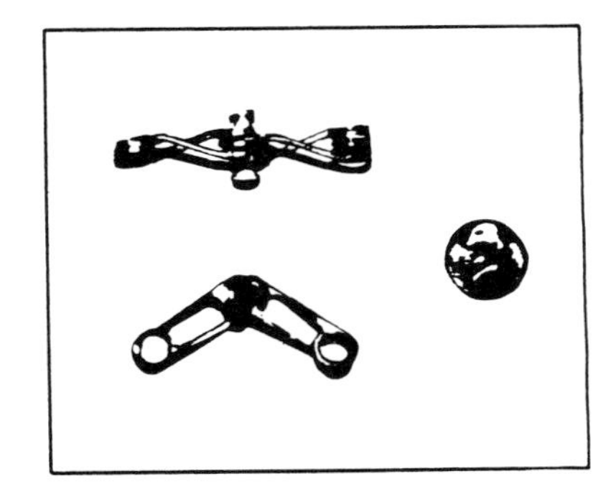

Figure 8
Model of Joint Links

Development of Superconducting Materials and Impacts on the Social Economy

TSUNEO NAKAHARA

Executive Vice President, Sumitomo Electric Industries, Ltd., 1-1-3, Shimaya, Konohana-ku, Osaka, 554 Japan

ABSTRACT

This report deals with the recent development of high Tc superconductors and the problems that remain. Thin film and wiring technologies are reviewed as the key technologies for practical application. In addition, impacts on the social economy are discussed.

1. INTRODUCTION

Since the discovery of superconductivity in 1911, various kinds of metallic superconductors have been developed, and a few have been put to practical use as magnets for research in liq. He. The market for metallic superconductors had been limited with the exception of the MRI magnet.

This situation changed with the discovery of abrupt reduction of resistivity at 30K (1) and confirmation of superconductivity (2) in 1986. Y, Bi and Tl based compounds have been reported(3), (4), (5) and the critical temperature has been rising year by year. Market expansion has become a greater possibility.

In this presentation thin film and wiring technologies, which are thought to be the key technologies for application, will be reviewed, and the impacts of the high critical temperature superconductor (HTSC) on the social economy will be mentioned.

2. KEY TECHNOLOGY FOR APPLICATION

After studying various applications of HTSC, such as electronic devices, sensors, cables and magnets, we have come to the conclusion that film and wiring technologies are the key technologies for these applications, as shown in Fig.1. In order to apply new HTSC materials for practical use, we must have precise process control, atomic order chacterization, and material design, such as computer graphics.

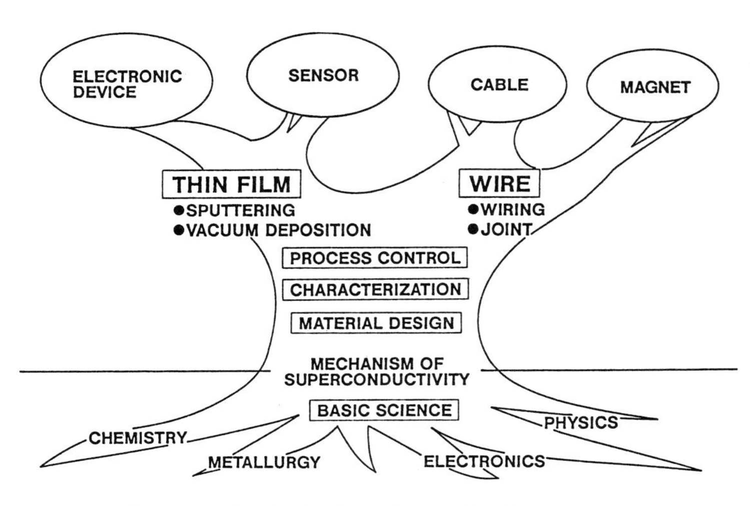

Fig.1 Key technology for applications

The reported and confirmed top data are shown in Table 1 (6), (7). The critical current density (Jc) in film superconductors shows more than 10^6 A/cm², and this characteristic is thought to render the superconductors suitable material as the electric devices or sensors.

As for wire superconductors, the reproducible and stable Jc data are still in the order of a few thousand A/cm² and there are many problems to be solved in order to produce a good superconducting wire.

Table 1 Top data in superconductors

(Tc, Jc AT O TESLA, REPORTED AND CONFIRMED DATA)

	CHARACTERISTICS	Y-SYSTEM	Bi-SYSTEM	TI-SYSTEM	ORGANIC
FILM	Tc(K)	TYPICAL DATA 91	NEC 107	IBM 120	——
	Jc(A/cm^2) (77.3K)	SUMITOMO $^*4\times10^6$	SUMITOMO 1.9×10^6	SUMITOMO $^*3.2\times10^6$	
BULK	Tc(K)	TYPICAL DATA 93	NRIM 110	IBM, SUMITOMO 125	UNIV. OF TOKYO 13
	Jc(A/cm^2) (77.3K)	AT&T 1.7×10^4		SANDIA NATIONAL LAB. 2.2×10^4	
WIRE	Tc(K)	SUMITOMO 91	SUMITOMO 104	HITACHI 107	——
	Jc(A/cm^2) (77.3K)	*4140	*4400	3500	

★ FIRST ANNOUNCED AT THIS CONFERENCE

3.THIN FILM TECHNOLOGY

Various processes have been proposed and attempted to make a good film superconductor that is highly pure, and whose composition and crystal direction are well-controlled.

Various problems remain unsolved, however, as shown in Fig. 2.

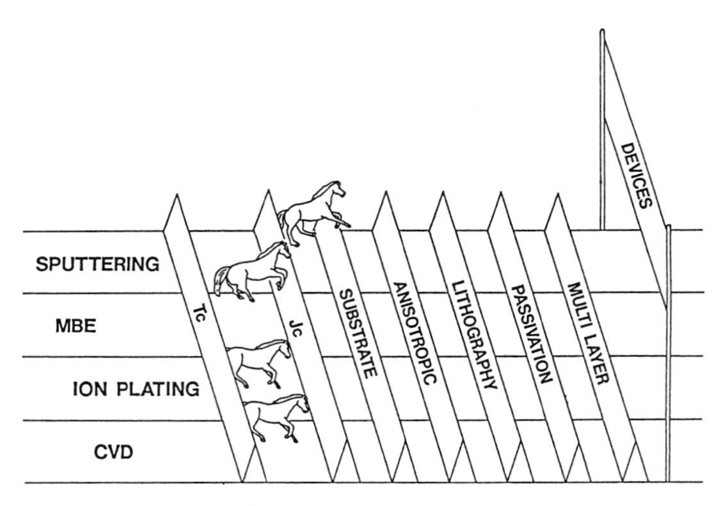

Fig.2 Race for High Tc Superconductor Film to be used in devices

By use of the sputtering method, Tc and Jc hurdles have been overcome, but there are other hurdles to get over, such as substrate limitation, processing and multilayers in the SIS (Superconductor-Insulator-Superconductor) structure, in order to make a device. The temperature and magnetic field dependencies of Jc in our YBaCuO film are shown in Fig.3 and Fig.4, respectively. You can see 4×10^6 A/cm² at 77.3K without a magnetic field, and also 0.71×10^6 A/cm² at 77.3K under a vertical magnetic field of 8T. However, when Jc remains low without a magnetic field, it decays very rapidly as the magnetic field increases. The Jc of the YBaCuO film superconductor shows the possibility of having a higher numerical Jc value in magnetic fields over 10T, than that of Nb_3Sn, and is thought to be sufficient for general electronic applications.

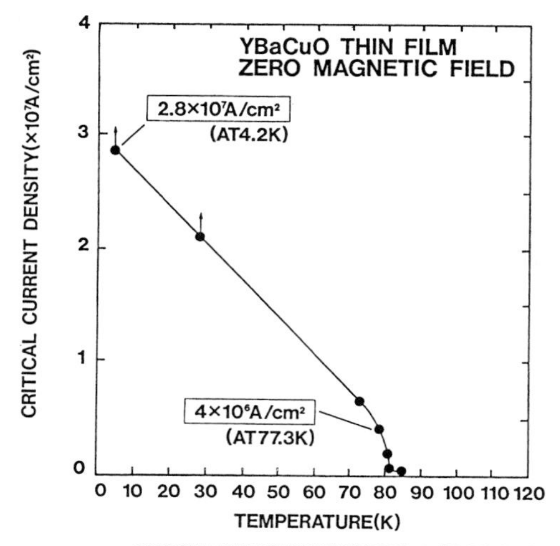

CO-WORKED WITH HOKURIKU ELECTRIC POWER CO.,INC., KANSAI ELECTRIC POWER COMPANY CO., INC. AND SHIKOKU ELECTRIC POWER CO., INC.

Fig.3 Temperature dependency of critical current density in YBaCuO thin film

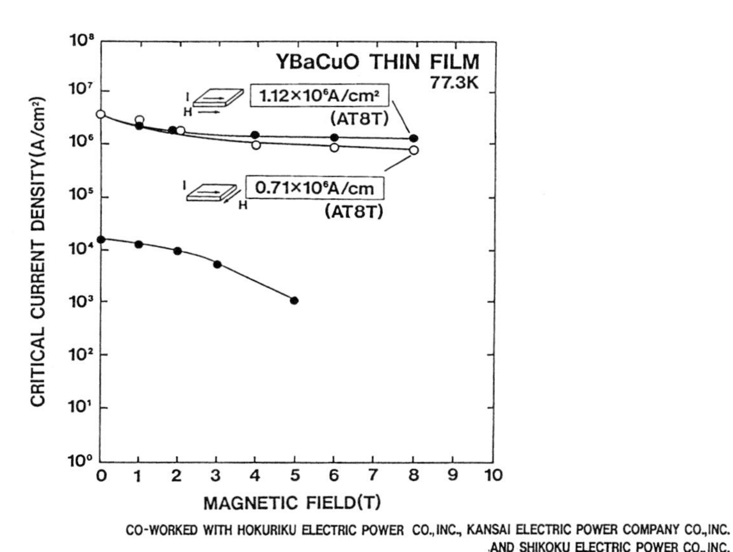

CO-WORKED WITH HOKURIKU ELECTRIC POWER CO.,INC., KANSAI ELECTRIC POWER COMPANY CO.,INC. AND SHIKOKU ELECTRIC POWER CO.,INC.

Fig.4 Magnetic field dependency of critical current density in YBaCuO thin film

The Jc's dependence on temperature in our Bi and Tl based compound films is shown in Fig.5 and Fig.6, respectively. Both of them show more than a million A/cm² of Jc at 77.3K without a magnetic field and, are thought able to solve one of the problems in practical usage. We must get over the next hurdle to produce devices.

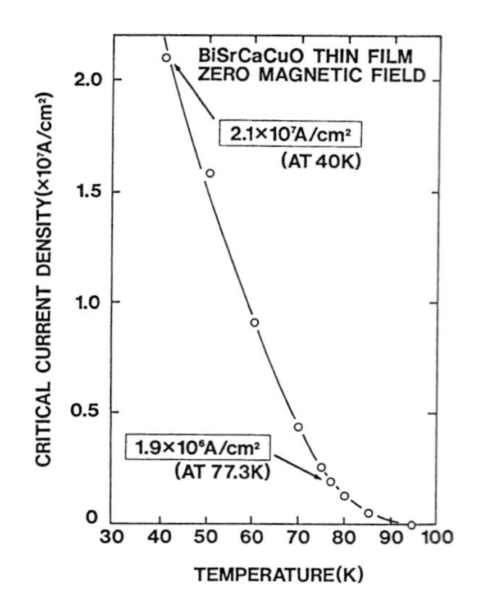

Fig.5 Temperature dependency of critical current density in BiSrCaCuO thin film

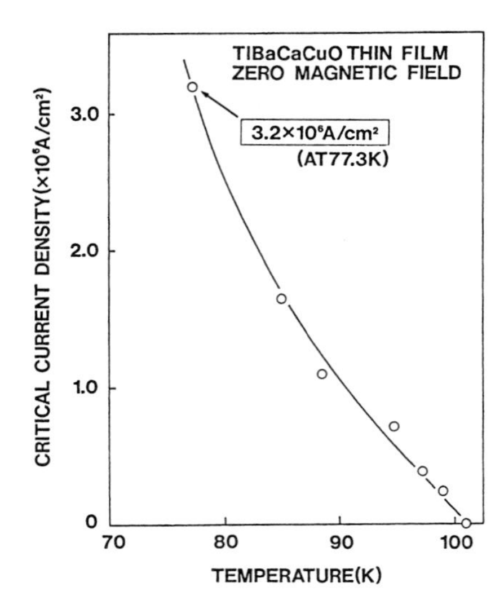

Fig.6 Temperature dependence of critical current density in TlBaCaCuO thin film

4.WIRING TECHNOLOGY

There are many barriers, which are shown in Fig.7, to making a superconductor wire. These are Jc in a high magnetic field, total current(Ic), anisotropy, strain, stability and wire length.

There are three major methods of making a superconducting wire, and each applies a different technology; film, melt, and powder.

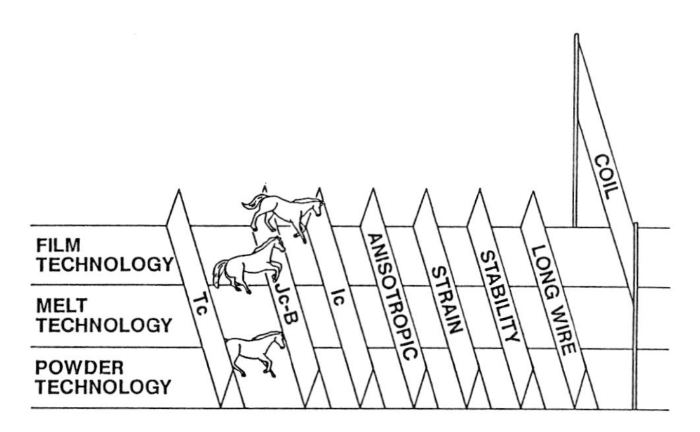

Fig.7 Race for the high-Tc superconductor wire

With powder technology, Jc becomes higher as the reduction ratio increases (Fig.8). However, the stable and reproducible data are in the order of a few thousand A/cm² at 77.3K without a magnetic field.

The longitudinal distribution of Ic is also important. Our Ic data are within 3% of distribution in wire about 10m long. Using this wire, we made a coil of 50mm bore diameter. The coil can generate 64mT of magnetic field on a conductor carrying 35A of transport current (Fig.9). We must clear the high Jc hurdle in high magnetic fields for practical applications, and solve the various remaining problems, such as grain boundary, weak link and so on.

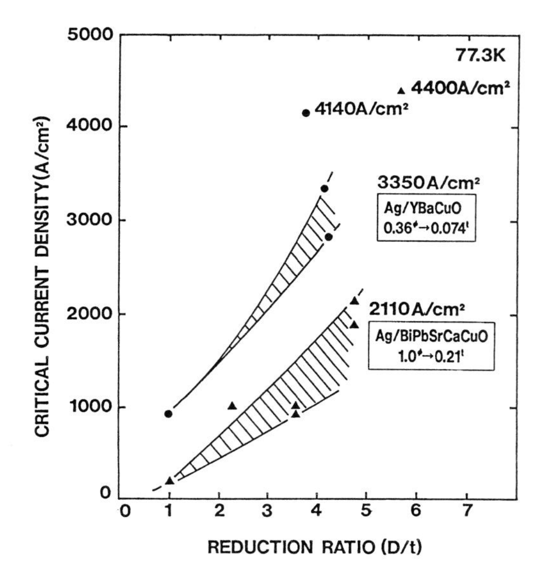

Fig.8 Reduction ratio dependency of critical current density in Ag sheathed YBaCuO and BiPbSrCaCuO tapes.

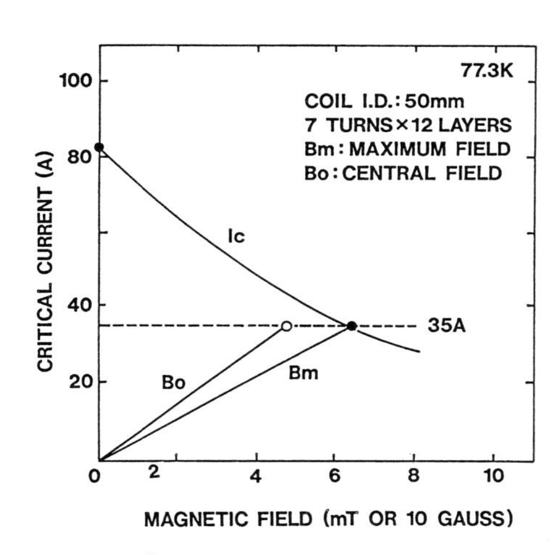

Fig.9 Test coil made by Ag sheathed YBaCuO tape superconductor

5. IMPACTS ON THE SOCIAL ECONOMY AND CONCLUSION

The impacts of the HTSC used in liq. N_2 are shown in Table 2. HTSC is thought to be superior in the electronics field in the form of devices or sensors, and in areas which require high magnetic fields.

The superconductor market will grow to be ten times that which it is today, if it can be used in liq. N_2 (Table 3).

We think the world market will increased by about 36 billion dollars a year if a superconductor that does not require refrigeration is developed. If we can develop a higher critical temperature superconductor, the marketwill grow very rapidly and will be one of the leading industries.

We think the development of material should be given more attention in the near future, since the market will grow as the critical temperature grows.

Table 2 Impacts expected by high Tc superconductor
○ SIGNIFICANT △ IMPROVEMENT × NEGLIGIBLE

APPLICATIONS	KARLSRUHE NUCLEAR RESEARCH CENTER	SIEMENS AG	NIKKEI INDUSTRY RESEARCH INSTITUTE	SUMITOMO	
				MARKETING GROUP	RESEARCH GROUP
POWER ELECTRONICS					
·TURBO-GENERATOR	△	○	△	△	△
·ENERGY-CABLES	△	△	×~△	×	△
·TRANSFORMERS	○	△	×~△	△	△~○
·POWER SWITCHES	○	○		○	△
·S M E S	△	○	×	×	○
·MHD-GENERATOR	×	×	×	×	×
ENERGY					
·FUSION REACTORS	×	×		×	○
·PARTICLE ACCELERATORS	△	△	△	△	×~△
TRANSPORTATION					
·LEVITATED TRAIN	×	×	○	×	△
·ELECTRO MAGNETIC THRUSTER SHIP			×	×	△
MEDICAL					
·M R I	△	△	○	△	×~△
·SQUID				○	○
MICRO ELECTRONICS					
·HF-TECHNIQUE	○	○	○	○	○
·LEAD WIRE			○	○	○
OTHERS					
·S R			○	△	△
·HIGH FIELD MAGNETS	○	○		○	○
·MOTOR			×~△	△	△

ASSUMPTIONS
·OPERATIONS TEMPERATURE ~77K
·CRITICAL CURRENT DENSITIES AT 77K 10^4 ~ 10^6 /cm² IN MAGNETIC FIELDS UP TO 20T
·STABILIZED CONDUCTORS AVAILABLE
·AC LOSSES COMPARABLE WITH THOSE OF METALLIC SUPERCONDUCTORS
·WORKABLE CONDUCTORS

Table 3 World market forecast

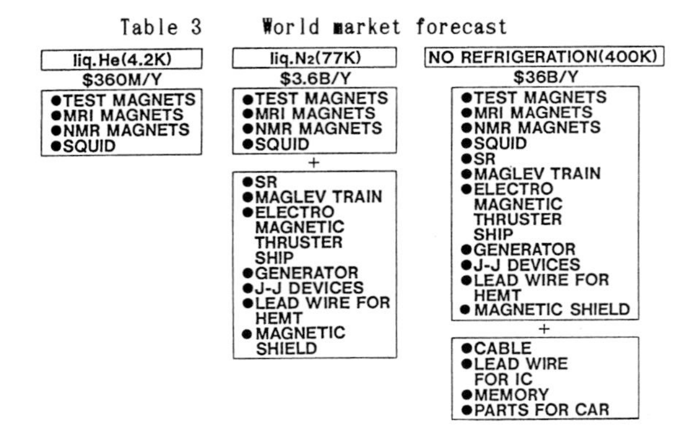

REFERENCES

(1) J.G.Bednorz and K.A.Muller, Z.Phys., B64, 189 (1986)
(2) H.Takagi et.al, Jpn.J.Appl.Phys.Part2, 26 L123 (1987)
(3) M.K.Wu, C.W.Chu et. al, Phys. Rev. Lett. 58, 908 (1987)
(4) H.Maeda et.al, Jpn.J.Appl. Phys., 27 L209 (1988)
(5) Z.Z.Sheng and A.M.Hermann, Nature 332, 138 (1988)
(6) S.Jin et.al, Appl. Phys. Lett 52, 2074 (1988)
(7) S.S.P.Parkin et.al, to be pablished in Phys. Rev. Lett. (1988)

4.13 Tunneling and Tunneling Junction

Electron Tunneling into Superconducting Bismuth-Copper Oxide Systems

TOSHIKAZU EKINO and JUN AKIMITSU

Department of Physics, Aoyama-Gakuin University, 6-16-1, Chitosedai, Setagaya-ku, Tokyo, 157 Japan

ABSTRACT

The energy gaps of the polycrystalline Bi-Sr-Ca-Cu-O and Bi-Sr-Cu-O superconductors have been measured by means of the point-contact tunneling technique. We find that the energy gaps at T=0K are 2Δ=76-81meV in $Bi_2Sr_2Ca_1Cu_2O_y$(Tc=76K), 91-100meV in $Bi_2Sr_2Ca_2Cu_3O_y$(Tc=105K) and 3.0-3.5meV in $Bi_2Sr_2Cu_1O_y$ (Tc=6.5K). Temperature variations of these energy gaps are also extracted. The ratios $2\Delta/kTc$ are 11.6-12.4 for $Bi_2Sr_2Ca_1Cu_2O_y$, 10.0-11.0 for $Bi_2Sr_2Ca_2Cu_3O_y$, extremely larger than BCS value. $2\Delta/kTc$ of $Bi_2Sr_2Cu_1O_y$ is 5.4-6.2, larger than those of the usual strong coupling superconductors. These results give the information about the difference of the pairing nature on the superconducting state between these materials which have multiple CuO_2 layers or a single CuO_2 layer between the SrO layers.

INTRODUCTION

It is believed that the two dimensional CuO_2 planes are essential for the occurrence of the high temperature superconductivity. A family of these superconducting oxides is the Bi-Sr-Ca-Cu-O system[1], $Bi_2Sr_2Ca_1Cu_2O_y$(Tc=76K) and $Bi_2Sr_2Ca_2Cu_3O_y$(Tc=105K) and Ca-absent $Bi_2Sr_2Cu_1O_y$(Tc=6.5K)[2,3]. Electronic properties in the superconducting state are mainly dominated by the energy gap at the Fermi level. The measurement of the energy gap can elucidate that the superconductivity in these materials is attributed to the BCS theory or not. Indeed, the possibility is suggested that the spin or charge mediated pairing mechanism works on the cupro-oxide systems[4,5]. One of the most direct measures of the energy gaps is the electron tunneling. Many workers have investigated the energy gaps of $La_{2-x}Sr_xCuO_y$ and $Y_1Ba_2Cu_3O_y$ by means of electron tunneling[6]. Some data show the asymmetric current-voltage curves, the multiple peaks in tunnel conductances[7,8] and the larger values of energy gaps than that of the BCS theory. It is worthy of survey that these noticeable characteristics are intrinsic or not for the another newly discovered oxides.

In this paper, we report the electron tunneling measurements of the superconductivity of polycrystalline Bi-Sr-Ca-Cu-O and Ca-absent Bi-Sr-Cu-O systems. The comparison of the energy gaps can draw the distinction between Bi-Sr-Ca-Cu-O and Bi-Sr-Cu-O systems.

EXPERIMENTAL

Measurements were made by the point-contact configuration with Al electrode[9]. Samples were prepared by solid state reaction from mixtures of Bi_2O_3, $SrCO_3$, $CaCO_3$ and CuO. Nominal compositions of Bi:Sr:Ca:Cu=1:1:1:2 were mixed and pressed into pellets and fired at 880-890°C for 12-24 hours. Pellets of Bi-Sr-Cu-O, Bi:Sr:Cu=2:2:1.3, were fired at 900°C for 12 hours. All samples were furnace cooled to room temperature. Tc's of these samples were measured by the standard DC four probe method. $Bi_1Sr_1Ca_1Cu_2O_y$ shows the double transitions. Their onset temperatures are 110K and 85K, the end points are 105K and 76K at each transition. $Bi_2Sr_2Cu_{1.3}O_y$ shows the onset temperature is 8.0K and the end point is 6.5K.

RESULTS AND DISCUSSION

Figs.1 and 2 show typical examples of the I(V) and dI/dV(V) curves for Bi-Sr-Ca-Cu-O, where I and V are the tunnel current and the DC bias voltage, respectively. The dI/dV(V) curves in Figs.1 and 2 correspond to lower Tc phase $Bi_2Sr_2Ca_1Cu_2O_y$ and higher Tc phase $Bi_2Sr_2Ca_2Cu_3O_y$ determined from the temperature variations of dI/dV(0mV). Tunnel resistances in Figs.1 and 2 at high bias voltages are

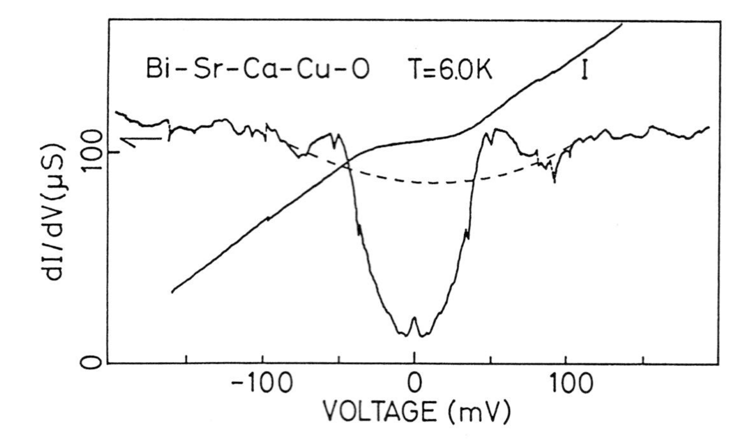

Fig. 1. The I(V) and dI/dV(V) curves for $Bi_2Sr_2Ca_1Cu_2O_y$. Broken line shows the background conductance curve interpolated from the higher bias regions.

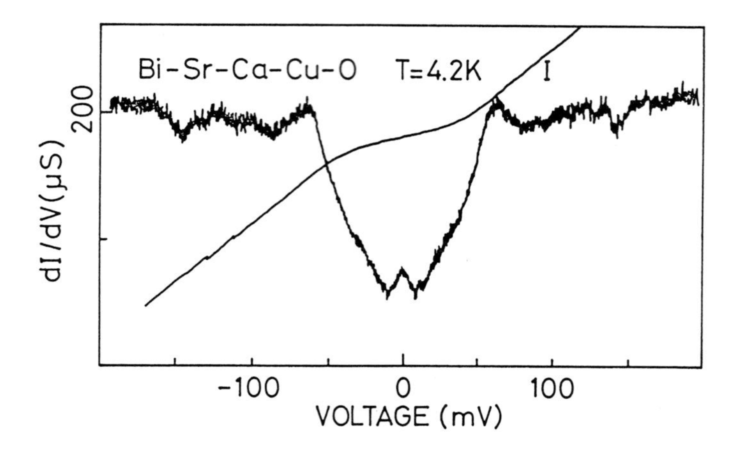

Fig. 2. The I(V) and dI/dV(V) curves for $Bi_2Sr_2Ca_2Cu_3O_y$.

about 10k ohms and 5k ohms, respectively. As indicated in Figs.1 and 2, flat background conductances against the bias voltages were sometimes observed, which were rarely observed in the cases of $La_{2-x}Sr_xCuO_y$ and $Y_1Ba_2Cu_3O_y$. Moreover, because the present materials are hardly damaged at the surface and insensitive to the oxygen stoichiometry, the large reduction of dI/dV(0mV), about 90% of the higher bias regions, can be easily attained.

In order to obtain the correct energy gap 2Δ, we tried to fit the broadened BCS density of states $D(E,\Gamma)$ proposed by Dynes et al[10] to the experimental data. $D(E,\Gamma)$ is expressed as

$$D(E,\Gamma)=\mathrm{Re}[(E-i\Gamma)/((E-i\Gamma)^2-\Delta^2)^{0.5}] \tag{1}$$

where Γ and Δ are the broadening parameter and energy gap, respectively. The temperature factor was also considered for the fitting calculations.

Fig.3 shows the fitting results of $D(E,\Gamma)$ to the dI/dV(V) curve normalized by the interpolated background conductance curve as shown by the broken line in Fig.1. The final results are 2Δ=76.0meV, Γ=13.3meV. Employing the values of Tc=76.0K and taking the spreads of data into account, the ratio $2\Delta/kTc$=11.6-12.4 is obtained. It is extremely larger than the BCS value and suggests the exotic nature of the pairing mechanism in this superconductor. The gap value obtained from the above procedure is almost consistent with the interval between inner points where the superconducting and the interpolated background conductance curves cross each other.

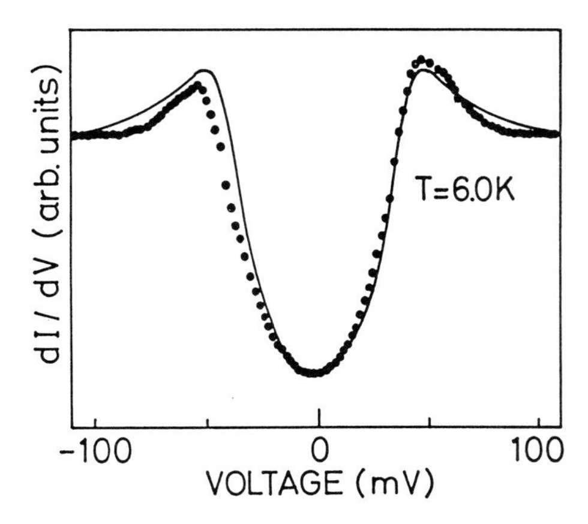

Fig. 3. Fitting result to the normalized dI/dV(V) curve in Fig.1. Filled circles are experimental data from the dI/dV(V) curve normalized by the interpolated background conductance curve. Solid line shows Eq.(1) with thermal broadening. Optimum values in this case are Δ=38.0meV and Γ=13.3meV.

Applying this manner to Fig.2, since the fitting procedure can not be successful in this case , the energy gap is 2Δ=91.0-100.0meV. The ratio $2\Delta/kTc$ is found as 10.0-11.0, close to that of the lower Tc phase.

Temperature dependence of 2Δ using the fitting procedures in lower Tc phase $Bi_2Sr_2Ca_1Cu_2O_y$ is shown in Fig.4. Data points coincide with the BCS curve. The similar temperature dependence is also observed in the higher Tc phase. It should be noted that 2Δ's at low temperatures are well above the BCS values.

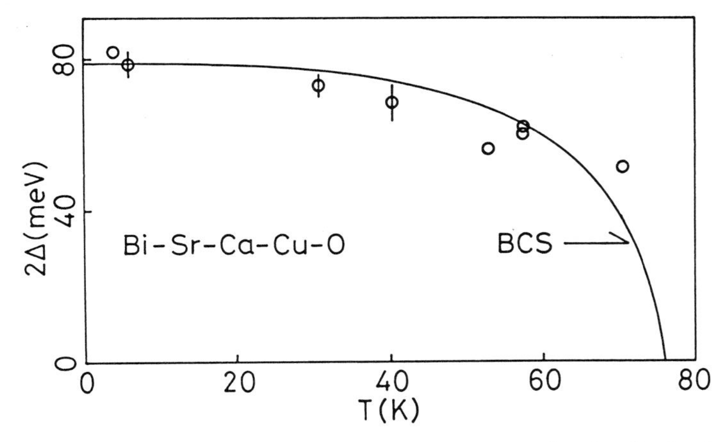

Fig. 4. Temperature dependence of the energy gap 2Δ for $Bi_2Sr_2Ca_1Cu_2O_y$.

Fig.5 shows the dI/dV(V) curve for $Bi_2Sr_2Cu_{1.3}O_y$. The dI/dV(0mV) does not drop so drastically compared with the data in Figs.1 and 2. This may be due to the current leak through the junction. The superconducting energy gap , however, can be determined by the similar manner used in Fig.2 . The value obtained is 2Δ=3.0-3.5meV. The ratio $2\Delta/kTc$ is 5.4-6.2 for Tc=6.5K, again larger than that expected from the BCS theory.

We finally mention about the energy gaps of these materials. In these experiments the polycrystalline samples are used, therefore the anisotropies of gaps should be smeared out. The ratio $2\Delta/kTc$ in $Bi_2Sr_2Cu_{1.3}O_y$, although the Tc is low, is almost equal to or larger than that of $La_{2-x}Sr_xCuO_y$ and $Y_1Ba_2Cu_3O_y$ [11,12]. This strong pairing interaction in Ca-absent $Bi_2Sr_2Cu_{1.3}O_y$ potentially leads to the high Tc of about 100K as a result of the insertion of more CuO_2 layers in the unit cell in Ca-contained Bi-Sr-Ca-Cu-O systems. The group of materials having multiple CuO_2 layers between SrO layers have more strong pairing interaction than that of single CuO_2 layer. One of the reasons may be the increase of the carrier density in this system.

In conclusion, we observed the superconducting energy gaps in bismuth-copper oxide systems. The ratios $2\Delta/kTc$ are 10.0-11.0 for $Bi_2Sr_2Ca_2Cu_3O_y$, 11.6-12.4 for $Bi_2Sr_2Ca_1Cu_2O_y$ and 5.4-6.2 for $Bi_2Sr_2Cu_1O_y$. These are enough larger than the BCS value. Further studies to explain the large

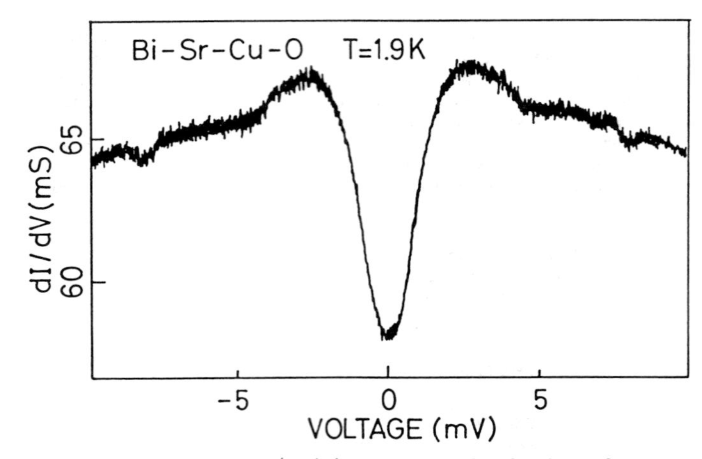

Fig. 5. The dI/dV(V) curve for $Bi_2Sr_2Cu_{1.3}O_y$.

energy gaps and to clarify the anisotropies by using the single crystals are needed.

This work was partially supported by the research project for high-Tc superconducting oxides of the Ministry of Education, Science and Culture, Japan.

REFERENCES

1. H. Maeda, Y. Tanaka, M. Fukutomi and T. Asano, Jpn. J. Appl. Phys. 27, L209(1988).
2. C. Michel, M. Herview, M.M. Borel, A. Gradin, F. Deslandes, J. Provost and B. Raveau, Z. Phys. B68, 421(1987).
3. J. Akimitsu, A. Yamazaki, H. Sawa and H. Fujiki, Jpn. J. Appl. Phys. 26, L2080(1987).
4. P.W. Anderson, Science 235, 1196(1987).
5. C.M. Varma, S. Schmidt-Rink and E. Abrahams, Solid State Commun. 62, 681(1987).
6. See for example, Novel Superconductivity, edited by S. A. Wolf and V. Z. Kresin (Plenum, New York, 1987).
7. M. Naito, D.P.E. Smith, M.D. Kirk, B. Oh, M.R. Hahn, K. Char, D.B. Mitzi, J.Z. Sun, D.J. Webb, M.R. Beasley, O. Fischer, T.H. Geballe, R.H. Hammomd, A. Kapitulnik and C.F. Quate, Phys. Rev. B35, 7228(1987).
8. I. Iguchi, H. Watanabe, Y. Kasai, T. Mochiku, A. Sugishita and E. Yamaka, Jpn. J. Appl. Phys. 26, L645(1987).
9. T. Ekino, J. Akimitsu, Y. Mastuda and M. Sato, Solid State Commun. 63, 41(1987).
10. R.C. Dynes, V. Narayanamurti and J.P. Garno, Phys. Rev. Lett. 41, 1509(1978).
11. T. Ekino, J. Akimitsu, M. Sato and S. Hosoya, Solid State Commun. 62, 535(1987).
12. T. Ekino and J. Akimitsu, Jpn. J. Appl. Phys. 26, L452(1987).

Energy Gap Measurement Made on "Clean" and Oriented $YBa_2Cu_3O_{7-\delta}$ Surfaces

JAW-SHEN TSAI[1], ICHIRO TAKEUCHI[1], JUNICHI FUJITA[2], TSUTOMU YOSHITAKE[2], TETSURO SATO[3], SHINICHI TANAKA[1], TAKAHITO TERASHIMA[4], YOSHICHIKA BANDO[4], KENJI IIJIMA[5], and KAZUNUKI YAMAMOTO[5]

[1] Microelectronics Research Laboratories, NEC Corporation, Kawasaki, 213 Japan
[2] Fundamental Research Laboratories, NEC Corporation, Kawasaki, 213 Japan
[3] Resources and Environment Protection Research Laboratories, NEC Corporation, Kawasaki, 213 Japan
[4] Institute for Chemical Research, Kyoto University, Uji, 611 Japan
[5] Research Institute for Production Development, Kyoto, 606 Japan

ABSTRACT

Superconductive energy gap of $YBa_2Cu_3O_{7-\delta}$ is measured in a novel broken film edge junction. (001), (103) and (110) oriented films are broken in a cryogenic environment along the appropriate directions together with the $SrTiO_3$ substrate. Pb electrode is brought close in situ to the clean broken film edge. The normalized energy gap $2\Delta(0)/k_BT_c$ measured in the direction along and perpendicular to the Cu-O plane are found to be 5.9±0.2 and 3.6±0.2 respectively. These values are independent of the variation in the values of T_c within the examined range of 40K~90K. The gap difference structure at $\Delta_{YBCO}-\Delta_{Pb}$ is observed which help identifying the value of energy gap of the oxide superconductor unambiguously. Identification of the gap energies in BiSrCaCuO system by the same technique was attempted.

INTRODUCTION

The high-T_c oxide superconductors such as $YBa_2Cu_3O_{7-\delta}$ have a marked anisotropic crystal structure. The more recently found Bi compounds and Tl compounds have even stronger anisotropy. The existence of the layered Cu-O planes in those material might suggest the presence of notable anisotropic electrical properties in these crystals. From the magnetic measurements, anisotropy in critical field, coherence length and penetration depth were discovered [1]. Anisotropy in critical current was also found in a thin film samples [2]. Ebisawa *et al.* [3] have predicted an anisotropic energy gap, based on the discrepancy in the observed gap data obtained from tunneling experiments and infrared experiments. Several attempts were made [4] but no evidence of gap anisotropy was discovered. In the previous tunneling studies, varieties of the gap values $2\Delta(0)/k_BT_c$ were reported, ranging typically from 4 to 6 [5]. Some of the reported ambiguity may be attributed to inhomogeneity and anisotropy of the samples. The multiple gap like structures observed by many authors might have complicated the issue further. We have developed a novel broken film edge junction technique, with which a clean and preferentially oriented tunneling interface was realized. With such junction, $\Delta(T)$ was monitored so that T_c where $\Delta(T)=0$ could be extrapolated. The reduced energy gap $2\Delta(0)/k_BT_c$ obtained in this way showed systematic and reproducible anisotropy in the two chosen crystal orientations, namely, along c-axis and c-plane.

EXPERIMENTAL

Superconductive $YBa_2Cu_3O_{7-\delta}$ films (thickness 100~300nm) were prepared by ion beam sputtering [6] and activated reactive evaporation [7]. They were epitaxially grown on $SrTiO_3$ substrates and X-ray studies and RHEED analysis proved these films to be single-phase having single orientation. In order to obtain the clean surfaces for tunneling, films were broken at liquid He temperature. The Pb electrodes were then brought to contact with the exposed film edges to form tunnel junctions. Use of pre-cut grooves in substrates made tearing of the films along desired directions in straight lines possible. SEM photomicrographs of severed film edges revealed that the breaks actually took place in the middle of the grains rather than at the grain boundaries (Fig. 6). (001) films were used to prepare tunneling surfaces parallel to the Cu-O plane (ab-plane), and breaking (110) films in appropriate directions provided surfaces either parallel or perpendicular to the Cu-O plane. The adhesion of the film to the substrate was very good. The direction of the breakes were along the crystal plane. So breaking (110) film along c-plane would obtain smooth surface with occasional atomic-size steps. For the film edge made in along c-axis, the contribution of (001) plane should be quite limited for the same reason. The direction of the breake made in (001) films was randomly chosen, but such film edges should not contain much of the (001) plane either. The resistance of the tunnel junctions was controlled manually. It was demonstrated from the contact resistance measurement on top surfaces of the films that virtually all samples were covered with non-superconductive layer. This indicates that the direction-controlled tunnelings at low-impedance broken film edges are quite legitimate. dI/dV-V measurements were conducted to obtain the gap voltage. Temperature dependence of the energy gaps were traced and they were extrapolated to provide the values of T_c which represent the dominant phase along the junction. By doing so, one can obtain the T_c of the tunneling surface itself.

RESULTS AND DISCUSSIONS

Figure 1 (a) (b) show dI/dV-V curves of a junction made with a (110) film broken along (001) plane. The curve (a) taken at 4.2K reveals small peaks near 5mV and another structures near 8mV which correspond to the gap difference peaks $(\Delta_{YBCO}-\Delta_{Pb})/e$ and the gap sum peaks $(\Delta_{YBCO}+\Delta_{Pb})/e$ respectively. At temperature above the T_c of Pb(T=10K), the small peak in dI/dV vanished, and the position of the main gap structure shifted by the amount of Δ_{Pb}/e (Fig. 1(b)). The observation of such gap difference and sum structures strongly indicate that one is observing no other than the energy gap of $YBa_2Cu_3O_{7-\delta}$. Since the energy gap of Pb is well established, the energy gap of $YBa_2Cu_3O_{7-\delta}$ can be determined conclusively from these curves. Energy gap in the Cu-O plane was obtained using (001) films and (110) films broken along the c-axis. Fig. 1(c) shows dI(V)/dV curve of a junction formed by breaking the same film used in Fig. 1(a) and (b) along (110) plane. A larger gap value was observed compared to the Fig. 1(b).

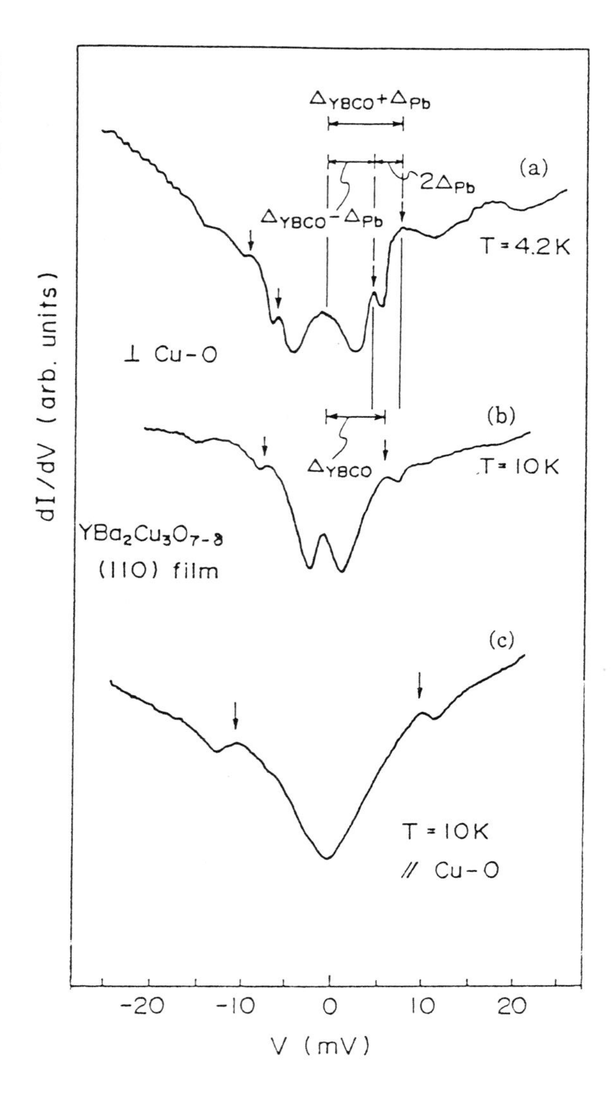

Fig. 1 dI(V)/dV of junctions made with (110) film broken along c-plane (a), (b); and broken along c-axis (c). (a) and (b) are dI(V)/dV curves taken below and above the T_c of Pb. Gap difference and sum structures can be seen in (a).

To obtain the normalized value of the gap $2\Delta(0)/K_BT_c$, a valid determination of T_c is required. Temperature dependence of the gap $\Delta(T)$ was traced in order to find $T_c=T(\Delta=0)$, for T_c obtained this manner would most likely represent the T_c of the dominant phase at the junction surface. Fig. 2 shows such $\Delta(T)$. Gap structure near T_c was quite difficult to make out thus, T_c value was extrapolated from $\Delta(T)$ with a appropriate error as shown in Fig. 3. $2\Delta(0)/k_BT_c$ values for tunneling parallel to the Cu-O plane and perpendicular to the Cu-O plane were scattered about 6 and 3.6 respectively signifying the anisotropy. This is further illustrated in Figure 3 where $\Delta(0)$ were plotted against the T_c. In Fig. 3, both of the parameters were extrapolated from $\Delta(T)$ curves like Fig. 2 for each point. Even with large error values in critical temperatures, they fit quite accurately on two straight lines (top for parallel and bottom for perpendicular to the Cu-O plane) passing through the origin. Two distinct values of gap energy were observed along two different crystal orientations at the surfaces that were carefully prepared. It seems that the observed gap voltage systematically correlates with the direction of the tunneling. The smaller normalized gap value along c axis was also reported recently in a sandwich type tunnel junction [8]. Perhaps such data

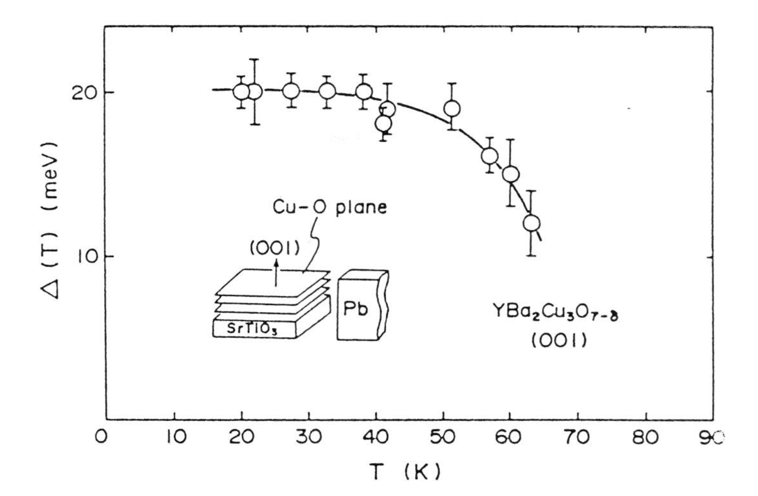

Fig. 2 Temperature dependence of the gap voltage. This $\Delta(T)$ curve was for a junction made with (001) film.

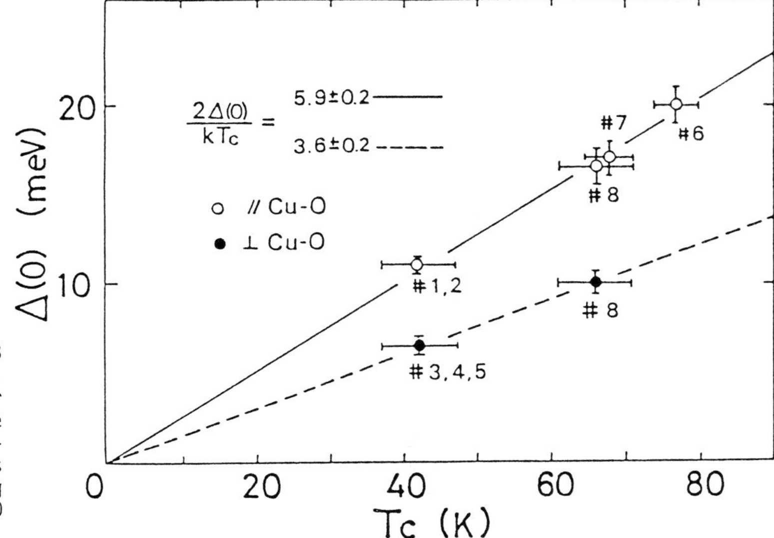

Fig. 3 $\Delta(0)$ vs T_C for 8 different samples. Closed circles are for the tunneling done perpendicular to the Cu-O plane; open circles are for the tunneling parallel to the Cu-O plane.

indicate that in our experiment, the electrons were truly tunneling through the surface barrier directionally along the chosen crystal axis. As for the physical meaning of the apparent anisotropy observed in our experiment, one can only speculate some of the possibilities. We might be observing the true energy gap anisotropy of the Y-Ba-Cu-O system, as previously observed in a SN sample by Binning and Hoenig [9]. Anisotropic gap suppression near S/I interface due to short and anisotropic coherence length [10] might also be responsible for the observed effect.

Impedance of the tunnel junction could be controlled externally, but the observed gap voltage was independent of the impedance values ranging between $10\text{~}10^6\Omega$. Fig. 4 shows the impedance dependence of observed gap voltage for two different junctions. Further reduction in the impedance of the junction would lead to an observation of Josephson current. In some samples, the Josephson current between Pb/Y-Ba-Cu-O junction remained finite even above the T_C of the Pb electrode. Fig. 5 shows the observed $I_C(T)$. AC Josephson effect was also observed up to about 9.2K. The Josephson frequency we found in such junction was observed up to about 9.2K. The Josephson frequency we found in such junction was usual 2eV/h. One interpretation of such phenomenon is the proximity-induced Josephson effect discussed by Han *et al.* [11].

Electron-gun evaporated Bi-Sr-Ca-Cu-O films [12] were also used to form the broken-film-edge junction. These films were coevaporated on the MgO substrate with T_C end point of 103K and the (001) preferred orientation. However, they were not the epitaxial films as in the case of films

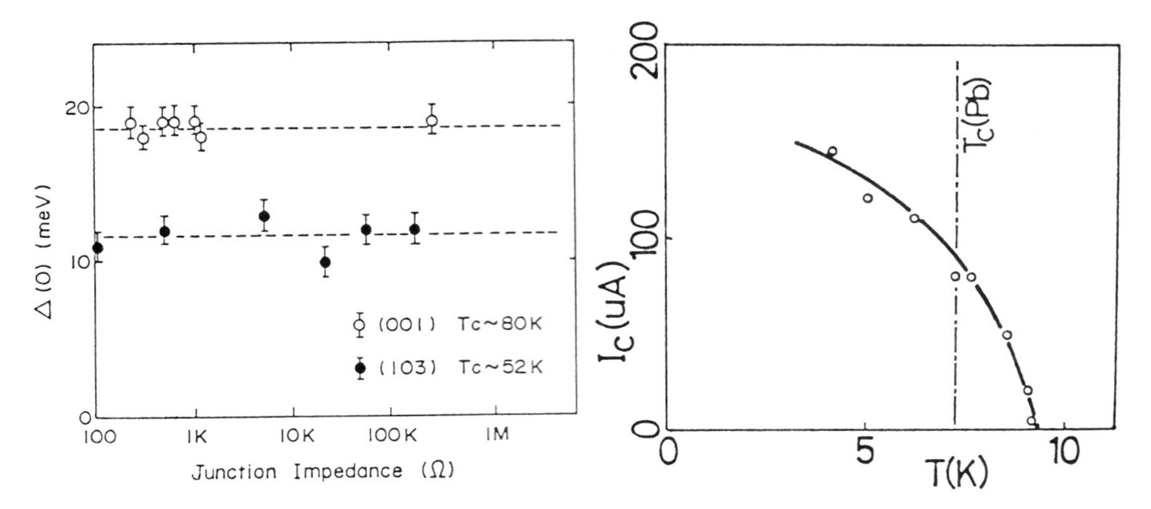

Fig. 4 Impedance dependence of the gap voltages for two samples.

Fig. 5 Temperature dependence of the Josephson current I_C of the junction. T_C enhancement was observed.

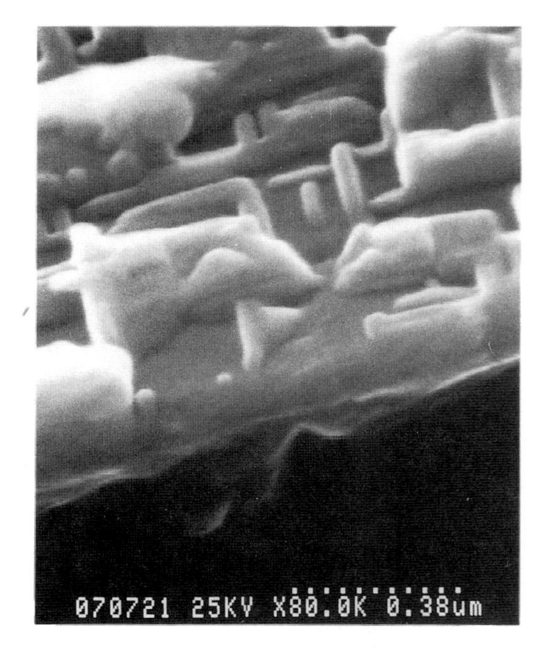

Fig. 6 SEM micrograph of the broken edge of a (001) Y-Ba-Cu-O film.

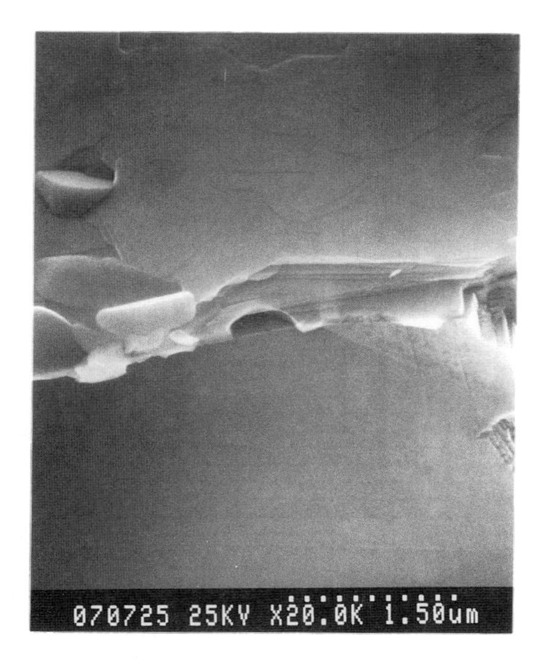

Fig. 7 SEM micrograph the broken edge of a (001) Bi-Sr-Ca-Cu-O film.

used in the Y-Ba-Cu-O junctions. Figure 6 and 7 show the SEM micrographs of the broken-edges of the Y-Ba-Cu-O/$SrTiO_3$ and the Bi-Sr-Ca-Cu-O/MgO respectively. Both films were (001) films. As one can see from Fig. 6, the Y-Ba-Cu-O film were cleaved cleanly along the (100)/(010) plane, at least at the present magnification. As for the Bi-Sr-Ca-Cu-O film, Fig. 7 shows the broken film edge contained many new cleaved surfaces with many different orientations. Electrically, the Bi-Sr-Ca-Cu-O film edge junction did not provide us a consistent gap energy value. Figure 8 shows one of a typical (dI/dV)/V characteristics of a Bi-Sr-Ca-Cu-O broken-film-edge junction. It shows two gap like structures with a large sub-gap conductance peak. However, these gap values were not quite reproducible as in the Y-Ba-Cu-O junctions. Perhaps, the difference in the way the films cleaved, one with a directional cleavage orientation, the other with a random orientation, was responsible for the difference in the consistency of the tunneling data. The inherent multi-phase nature of the Bi-Sr-Ca-Cu-O might also contribute to the difficulties in the tunneling experiment.

CONCLUSION

In conclusion we have measured energy gap in $YBa_2Cu_3O_{7-\delta}$ by using a unique thin film breaking technique to prepare clean and preferentially oriented superconducting surface. In situ tunneling measurements were carried out on such surfaces with Pb counterelectrode. The normalized energy gap $2\Delta(0)/k_BT_c$ of 5.9±0.2 and 3.6±0.2 was observed consistently and reproducibly in the direction parallel and perpendicular to the Cu-O plane, respectively. These values are independent of T_c, even when T_c was as low as 40K in some samples where the one dimensional oxygen chain cease to exist. The anisotropy ratio of the gaps in two crystal orientations was 1.6±0.1 by this tunneling studies.

ACKNOWLEDGEMENTS

We are in debt to H. Igarashi for the most helpful discussions. We are also grateful for the supports from H. Abe and M. Yonezawa, throughout the course of this research.

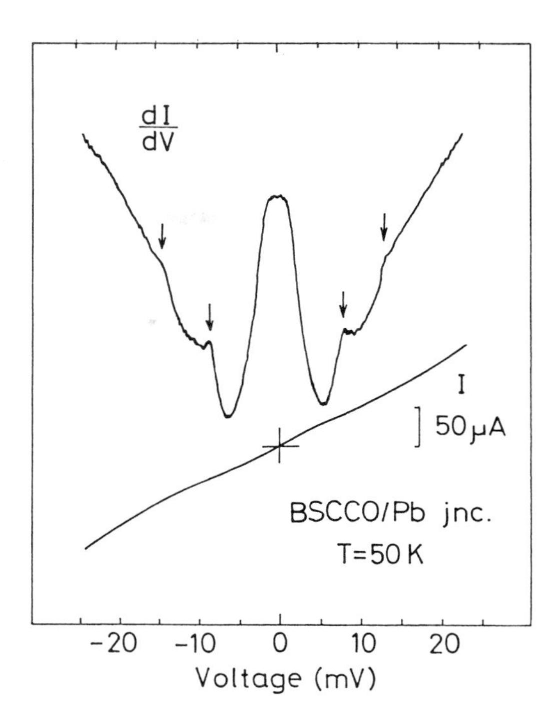

Fig. 8 (dV/dI)V and I/V characteristics of a Bi-Sr-Ca-Cu-O broken-film-edge junction.

REFERENCES

[1] T.K. Worthington, W.J. Gallagher and T.R. Dinger Phys. Rev. Lett. 59, 1160 (1987)
[2] Y. Enomoto, T. Murakami, M. Suzuki and K. Moriwaki, Jpn. J. Appl. Phys. 26, L1248 (1987)
[3] H. Ebisawa, Y. Isawa and S. Maekawa, Jpn. J. Appl. Phys. 26, L992, (1987)
[4] J.R. Kirtley, W.J. Gallagher, Z. Schlesinger, R.L. Sandstrom, T.R. Dinger and D.A. Chance, 1987, preprint. Phys. Rev. B35, 8846 (1987)
Z. Schlesinger, R.T. Collins, D.L. Kaiser, and F. Holtzberg, Phys. Rev. Lett. 59, 1958 (1987)
[5] A. Barone, Physica C, 153-155, 1712 (1988)
[6] J. Fujita, T. Yoshitake, A. Kamijyo, H. Igarashi and T. Satoh, to be published, Extended Abstracts MRS Spring Meeting (1988)
[7] T. Terashima, K. Iijima, K. Yamamoto, Y. Bando and H. Mazaki, Jpn. J. Appl. Phys. 27, L91, (1988)
[8] J. Takada, H. Mazaki, T. Terashima, K. Iijima, K. Yamamoto, K. Hirata and Y. Bando, Extended Abstracts of the 20th Conference on Solid State Devices and Materials, Tokyo, 455 (1988)
[9] G. Binnig and H.E. Hoenig, Z. Physik. B32, 23 (1978)
[10] G. Deutscher and K.A. Müller, Phys. Rev. Lett. 59, 1745 (1987)
[11] S. Han, K.W. Ng, E.L. Wolf, A. Millis, J.L. Smith and Z. Fisk, Phys. Rev. Lett. 57, 238 (1986)
[12] T. Yoshitake, T. Satoh, Y. Kubo and A. Igarashi, Jpn. J. Appl. Phys . 27, L1089 and L1262 (1988)

AC-DC Conversion Effect in Ceramic Superconductor

S. Ikegawa[1], T. Honda[2], H. Ikeda[3], A. Maeda[4], H. Takagi[5], S. Uchida[5], K. Uchinokura[4], and S. Tanaka[5,6]

[1] Research and Development Center, Toshiba Corporation, Kawasaki, 210 Japan
[2] Materials & Electronic Devices Laboratory, Mitsubishi Electric Corporation, Sagamihara, 229 Japan
[3] Engineering Research Center, Tokyo Electric Power Company, Chofu, 182 Japan
[4] Department of Applied Physics, The University of Tokyo, Hongo, Bunkyo-ku, Tokyo, 113 Japan
[5] Engineering Research Institute, The University of Tokyo, Hongo, Bunkyo-ku, Tokyo, 113 Japan
[6] Present address: Tokai University and also ISTEC

ABSTRACT

The dc voltages(V_{dc}) induced by an rf current(2-20MHz), which had been previously interpreted as the reverse ac Josephson effect, were investigated in $BaPb_{1-x}Bi_xO_3$ and $YBa_2Cu_3O_{7-y}$, by using second harmonic superposition on the rf current and differential resistance measurement. In both cases, the V_{dc} is not due to the reverse ac Josephson effect but due to nonlinear current-voltage characteristics. In $BaPb_{1-x}Bi_xO_3$, the nonlinear but symmetric I-V curve is induced by the superconducting transition($\simeq$10 K) and thus the V_{dc} is observed only at around 10 K. In $YBa_2Cu_3O_{7-y}$, local nonlinear resistance near two electrical contacts for potential measurement is responsible for this effect and thus the V_{dc} is observed over wide temperature range up to 300 K.

INTRODUCTION

There have been several reports on possible superconductivity at temperatures above 100 K in RE-Ba-Cu-O (RE=Y, rare-earth elements) system. Among them, Chen et al. observed that a dc voltage (V_{dc}) induced by an rf current across $Y_{1.8}Ba_{0.2}CuO_{4-y}$ ceramics up to 240 K[1]. They claimed that the effect was due to so-called "reverse ac Josephson effect"[2] and superconductivity in granular form exists at 240 K[1]. The evidence for this interpretation was a polarity change of the V_{dc} as functions of rf amplitude, frequency and temperature in a random fashion[3,4]. In favor of reverse ac Josephson effect in granular superconductor, it was proposed that a quantized dc voltage $V_J(f)=Nhf/2e$, where N is an integer, was induced by radiation with frequency f at every part of weak coupling[2] and up to one million of series-connected junctions gave observable dc voltage[1,3,4]. It is known, however, that the single junction system does not stay at a fixed value of N and undergoes a transition to another state with opposite sign of N in a short period of time[2]. The V_J(f=10 MHz, N=1) is 20.7 nV and about 10^{-6} times smaller than the thermal energy k_BT at 240 K. Thus, the total dc voltage induced by the reverse ac Josephson effect in megahertz region at such a high temperature as 240 K is hard to exist stably and unlikely to be observable.

There is an alternative explanation of this rf-to-dc conversion effect: a nonlinear current-voltage characteristic including rectification[5]. However, there has been few systematical study on this effect. In this paper, we investigated the rf-to-dc conversion effect for well-characterized oxide superconductor $BaPb_{1-x}Bi_xO_3$ and high-temperature superconductor RE-Ba-Cu-O by ordinary resistance measurement, differential resistance measurement, and also V_{dc} measurement. In a particular case of the V_{dc} measurements, the second harmonic wave was superposed on the rf current. Based on these results, we compare the two cases of $BaPb_{1-x}Bi_xO_3$ ($T_c\simeq$10 K) and RE-Ba-Cu-O ($T_c\simeq$90 K) and discuss the origin of the V_{dc}.

In general, an alternating current including a second harmonic in the form

$$I = I_{rf1}\sin(\omega t) + I_{rf2}\sin(2\omega t+\phi), \tag{1}$$

generates a dc voltage across a nonlinear resistance element, where I_{rf1} and I_{rf2} are amplitudes of fundamental wave current and second harmonic wave current, respectively. A phase difference of the fundamental and the second harmonic wave is represented by ϕ. If the nonlinear current-voltage relation V(I) is assumed to be Maclaurin expansion up to the third order, dc component of voltage is given by

$$V_{dc} = \left.\frac{d^2V}{dI^2}\right|_0 \left(\frac{I_{rf1}^2 + I_{rf2}^2}{4}\right) + \left.\frac{d^3V}{dI^3}\right|_0 \left(-\frac{I_{rf1}^2\times I_{rf2}}{8}\sin\phi\right). \tag{2}$$

The first term in Eq. (2) means rectification. The second term, which depends on ϕ with $-\sin\phi$ behavior, comes from the interplay between the fundamental and the second harmonic wave at nonlinear element and we hereafter call this effect "nonlinear interference effect". On the other hand, in the case of reverse ac Josephson effect, the induced dc voltage does not depend on ϕ. Therefore, we have investigated the dependence of V_{dc} on ϕ and I_{rf2}.

EXPERIMENTAL

In the present experiment, we used a $BaPb_{0.75}Bi_{0.25}O_3$ ceramic sample and single-phase $YBa_2Cu_3O_{7-y}$ ceramic samples prepared by solid state reaction. Other RE-Ba-Cu-O ceramics made by solid state reaction were also used. The resistance of the sample was measured by a standard four-probe technique using a dc current. The contact resistance of each potential contact was evaluated from the difference between four-probe resistance measurement and three(or two)-probe measurement[6,7]. The electrical contacts to $BaPb_{0.75}Bi_{0.25}O_3$ were made by silver paste (du Pont 7713) with post heat treatment at 550 °C. Four kinds of electrical contacts were used for RE-Ba-Cu-O, whose preparation conditions including post heat treatment are listed in Table I.

The V_{dc} was defined as the difference of the dc voltage between on- and off-states of the oscillator, in order to subtract offset dc voltage due to thermoelectromotive force etc. A digital voltmeter (YEW-2501A) which has 10 nV resolution was used for voltage measurement. When a second harmonic wave was superposed on the rf current, two function generators(HP-3325A), which were locked to the same reference signal(1 MHz), were connected through a power divider. The rf-to-dc conversion effect for the frequency range 2-20MHz was measured by using two different circuits. Figure 1(a) shows the "type-A" circuit, which is almost the same as that used by Chen et al[1]. The "type-B" circuit(Fig. 1(b)) incorporates a 50 Ω coaxial transmission line, which is terminated near the sample, and does not include any transformer in order to suppress the internal generation of harmonics in the circuit. A pair of inductance coils is inserted in potential leads of the four-probe configuration in order to suppress an rf current through the circuit composed of potential leads and a digital voltmeter.

Differential resistance measurements were performed by a conventional method using a modulation current with frequency of about 1 kHz and a lock-in detector (EG&G-5205). In all kinds of experiments, the temperature was measured with a Au+0.07%Fe-Chromel thermocouple.

TABLE I. Electrical contacts for RE-Ba-Cu-O

No.	Contact material	Heat treatment	R(Ω)
1	Silver paste(duPont 4922)	50°C in air	0.34
2	Silver paste(duPont 6838)	300°C in O_2	0.33
3	Gold paste	900°C in O_2	0.040
4	Ultrasonic soldering	No treatment	2.3

R: Contact resistance for $YBa_2Cu_3O_{7-y}$ at 100 K.
Typical contact area is 0.7 × 2.4 mm^2.

FIG. 1. Circuit diagrams for rf-to-dc conversion effect.

RESULTS ON $BaPb_{1-x}Bi_xO_3$

Figure 2(a) shows resistivity as a function of temperature in various magnetic fields. In a zero field, T_c^{onset} is about 11 K and zero-resistance state is achieved at about 10 K. The normal-state resistivity of the sample is ten times as large as that of the ordinary polycrystalline sample of the same composition[8] and the resistive transition is significantly broadened by a weak magnetic field up to 0.5 T. Therefore, the sample can be regarded as composed of weakly-linked superconducting particles.

Figure 2(b) shows the V_{dc} due to rf-to-dc conversion effect when the type-A circuit with only one oscillator for a sinusoidal current of 10 MHz were used. In a zero field, with decreasing temperature, the V_{dc} appears at T_c^{onset} and attains its maximum at around 10 K. When the magnetic field is applied, V_{dc} is strongly suppressed. This suggests that the observed V_{dc} is correlated with the superconducting transition.

When a second harmonic wave was superposed on the rf current, it was found that the V_{dc} depends periodically on the phase ϕ. Since the impedance was not matched in the type-A circuit, the phase and the amplitude of the rf current, monitored across a 200 Ω resistance(see Fig. 1(a)), may not be the exact values at the position of the sample. Hence, we constructed another circuit(type-B) incorporating an impedance-matched transmission line. The phase and the amplitude were calibrated by using a dummy circuit composed of a coaxial line and a termination load. Figure 3 shows the V_{dc} vs. ϕ characteristics, when a second harmonic was

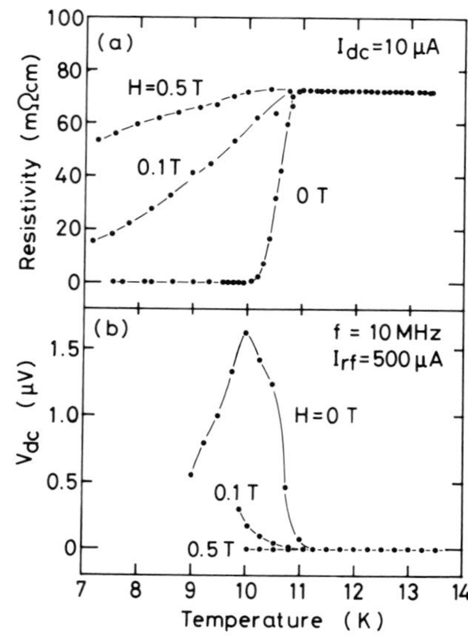

FIG. 2. Temperature-dependences of (a) dc resistivity and (b) V_{dc} for $BaPb_{0.75}Bi_{0.25}O_3$.

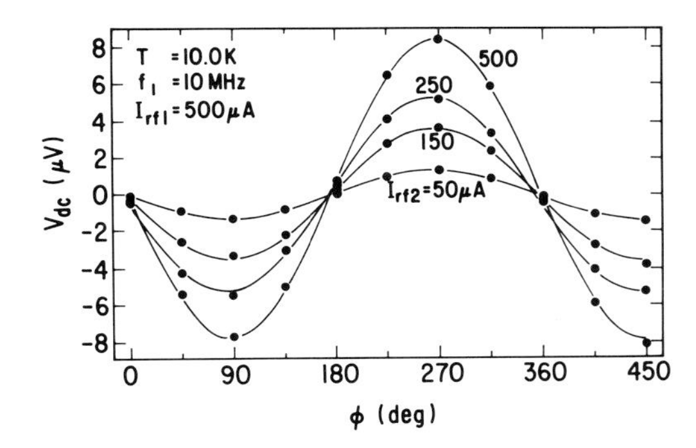

FIG. 3. V_{dc} vs. ϕ using type-B circuit for $BaPb_{0.75}Bi_{0.25}O_3$.

superposed on the rf current in the form of Eq. (1) using the type-B circuit with an additional oscillator. The solid curves in Fig. 3 show fits of experimental data (closed circles) to the form

$$V_{dc} = V_{const} - V_\phi \sin(\phi+\Delta\phi), \tag{3}$$

using least-squares method, where $\Delta\phi$ is a fitting parameter for the actual phase shift. The values of V_ϕ, which is the amplitude of the ϕ-dependent part of V_{dc}, are plotted with closed symbols in Fig. 4 as a function of I_{rf2}. In the temperature range of resistive transition, V_ϕ exhibits various types of I_{rf2}-dependence. In all of the V_{dc} vs. ϕ characteristics, $|\Delta\phi|$'s were less than 10 deg and $|V_{const}|$'s were less than 0.3 μV. Thus the V_{dc} was found to depend on the phase ϕ with $-\sin\phi$ behavior, which is consistent with the prediction of nonlinear interference effect.

Therefore, we shall perform the comparison between the "nonlinear interference effect" model and experimental results when the second harmonic was superposed. Figure 5 shows the differential resistances of the same bulk sample as a function of dc bias current. For temperatures less than 11.0 K, various nonlinear behaviors appear owing to superconducting transition. In various magnetic fields at 10.0 K, nonlinearity is strongly suppressed with increasing magnetic field. It should be noted that all the curves are symmetric within the experimental resolution. Figure 5 shows that d^3V/dI^3 changes significantly even within the range of the applied rf current, which means that the Maclaurin expansion is not appropriate in the present case. Therefore, we estimated the dc voltages numerically as follows. First, the differential resistance is integrated to yield an I-V curve. Next, a waveform of an alternating voltage is computed on the assumption that an alternating current in the form of Eq. (1) is fed through the element with the above I-V curve. Finally, the dc component of voltage is obtained from time average of the above alternating voltage over one period. Thus obtained V_{dc} is found to be described as $V_{dc}=-V_\phi \sin\phi$. In Fig. 4, the V_ϕ derived from experimental I-V curve are plotted with open symbols as a function of I_{rf2}. Apart from the quantitative discrepancy with a factor of about two, the I_{rf2}-dependence of the estimated V_ϕ agrees with experimental values qualitatively: at 9.5 K both the experimental and estimated V_ϕ-I_{rf2} curves have positive curvature, whereas at 10.0 K both curves have negative curvature, and at 10.5 K both curves have a peak. These changes arise from the different nonlinear characteristics at different

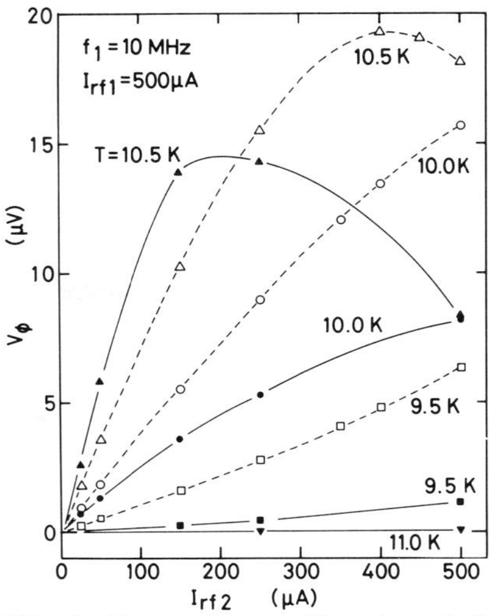

FIG. 4. V_ϕ vs. I_{rf2}. Closed symbols indicate experimental values. Open symbols indicate estimated values from I-V curves of $BaPb_{0.75}Bi_{0.25}O_3$.

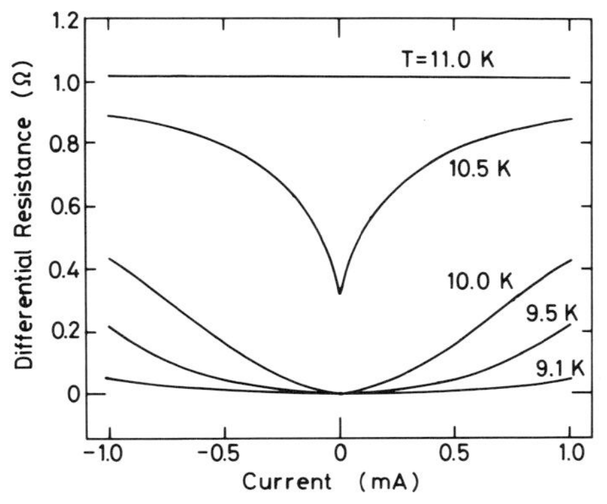

FIG. 5. Differential resistances as a function of dc bias current for $BaPb_{0.75}Bi_{0.25}O_3$.

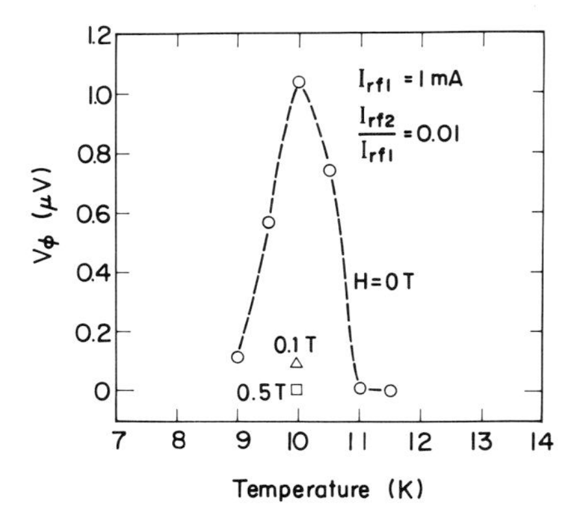

FIG. 6. Temperature dependence of V_ϕ derived from I-V curves of $BaPb_{0.75}Bi_{0.25}O_3$ under the condition that I_{rf2} is 1% of I_{rf1}.

temperatures. For the data taken under applied magnetic fields, we also performed the same kind of numerical estimation and the results also agree qualitatively well with the experimental V_ϕ.

Next, we compare the V_{dc} when only one oscillator was used(Fig. 2(b)) and the nonlinear interference effect model. Figure 6 shows temperature dependence of the V_ϕ which is derived from the experimental I-V curves. The observed V_{dc} in Fig. 2(b) depends on temperature and magnetic field in the same way as the estimated V_ϕ in Fig. 6. This suggests that the V_{dc} in Fig. 2(b) also originates from the nonlinear interference effect, in spite of the fact that harmonic waves are not superposed intentionally. The second harmonic was considered to be generated in the type-A circuit itself, which is composed of a transformer, a resistance, a capacitance, a pair of transmission lines, electrical contacts and the sample. From the observed V_{dc}, the magnitude of the second harmonic generated in the type-A circuit was estimated to be 1 % of the fundamental wave, which is a possible value for an rf circuit including a transformer.

RESULTS ON RE-Ba-Cu-O

Figure 7(b) shows V_{dc} for $YBa_2Cu_3O_{7-y}$ with the electrical contacts No. 2. When the type-A circuit and only one oscillator are used(closed circles in Fig. 7(b)), the V_{dc} is observed at room temperature and increases gradually with decreasing temperature across the resistive transition(Fig. 7(a)). As will be shown later(Fig. 9), the V_{dc} depends on frequency with a random manner, which had been pointed out by Chen et al[1]. On the other hand, if we use the type-B circuit, V_{dc} does not appear within the experimental resolution (<0.04 μV) over the whole temperature range (open circles in Fig. 7(b)) for any frequency. Therefore, we hereafter concentrate on the case when the type-A circuit is used.

Similar V_{dc}'s are observed in other RE-Ba-Cu-O ceramic samples with contact No. 1 or No. 2. They are shown in Fig. 8, where T_c is the midpoint of resistive transition. It should be stressed that there are no clear structures in the temperature dependence nor a relation between resistive T_c and V_{dc} for any sample. Next, the effect of applied magnetic fields was investigated for $La_{1+x}Ba_{2-x}Cu_3O_y$, which shows the largest V_{dc}(Fig. 8(a)) among the measured samples. The V_{dc}'s are found to be independent of magnetic field up to 6 T above and below T_c. The V_{dc} in RE-Ba-Cu-O depends on the type of electrical contacts rather than sample composition. Figure 9 shows the dependence on electrical contacts using $YBa_2Cu_3O_{7-y}$ samples from the same batch. Their contact resistances are listed in Table I. The V_{dc} is observed only in the samples with contact Nos. 1 and 2.

In order to investigate behavior of the V_{dc} in more detail, we use $YBa_2Cu_3O_{7-y}$ with contact No. 2, which have rather stable contact resistance against temperature cycles. The V_{dc} is found

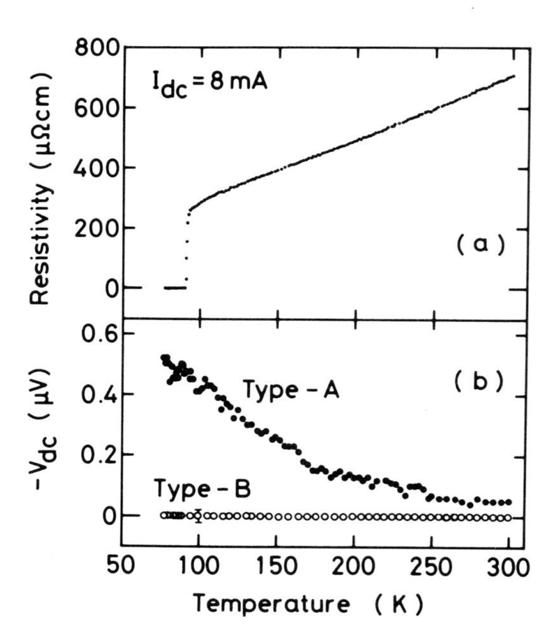

FIG. 7. (a)dc resistivity vs. temperature. (b)V_{dc} vs. temperature using type-A circuit (●: f=11 MHz, I_{rf}=200 μA), type-B circuit (○: f=14 MHz, I_{rf}=500 μA), for $YBa_2Cu_3O_{7-y}$.

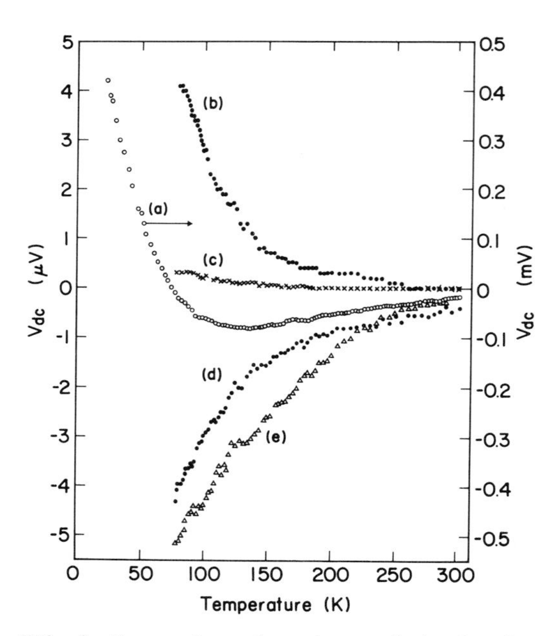

FIG. 8. Temperature dependence of the V_{dc} for (a)$La_{1+x}Ba_{2-x}Cu_3O_y$: x≃0.2 (T_c=35 K), (b)$Y_{1.8}Ba_{0.2}CuO_y$ (not superconducting), (c)$GdBa_2Cu_3O_{7-y}$ (T_c=93 K), (d)$La_{1.85}Ba_{0.15}CuO_4$ (T_c=30 K), (e)$YBa_2Cu_3O_y$ (different from the sample in Fig. 7, T_c=92 K). I_{rf}=200-500 μA and f=6-18 MHz.

to be proportional to I_{rf} squared at fixed temperature. When a second harmonic wave is superposed on the rf current in the form of Eq. (1) using type-A circuit with an additional oscillator, the V_{dc} is found to include a component which depends periodically on ϕ (Fig. 10). The solid curves in Fig. 10 are fitted to Eq. (3). The ϕ-dependent part (V_{ϕ}) is proportional to I_{rf2}, while V_{const} increases only very slightly with increasing I_{rf2}. These features are consistent with Eq. (2), which expresses rectification and nonlinear interference effect, except for the existence of $\Delta\phi$. Finally, we show the dependence on measuring circuit again. Figure 11 shows the V_{dc} vs. ϕ characteristics when a second harmonic wave is superposed using the type-B circuit with an additional oscillator. We could not observe clear ϕ-dependence for $I_{rf1} \leq 1000$ μA (Fig. 11(b)). However, when the two coils are eliminated from the potential leads of the type-B circuit, the V_{dc} increases and periodical dependence of V_{dc} on ϕ is observed, as is shown in Fig. 11(a). Indeed, if the coils (2 mH) are inserted in the potential leads of the type-A circuit in the same way as in Fig. 1(b), V_{const} decreased down to 1 % and periodical dependence of the V_{dc} is not observed.

DISCUSSION

In $BaPb_{0.75}Bi_{0.25}O_3$, which is denoted by "BPBO" hereafter, the numerical estimation based on nonlinear I-V curves of the sample agrees well with the observed V_{dc}. We conclude, therefore, that the observed V_{dc} originates not from the reverse ac Josephson effect but from the nonlinear interference effect in the bulk sample with a nonlinear I-V characteristic induced by the superconducting transition. Even if the second harmonic wave is not superposed intentionally, the V_{dc} is generated by the interplay between the fundamental and the internally generated second harmonic wave.

In RE-Ba-Cu-O, the results in Fig. 10 suggest that the V_{dc} comes from the nonlinear I-V characteristics, as was in the case of BPBO. However, the properties of the V_{dc} in RE-Ba-Cu-O are different in several aspects from those in BPBO. First, the V_{dc} is observed over wide temperature range irrespective of the bulk superconducting transition of the material in RE-Ba-Cu-O(Fig. 8), while the V_{dc} was observed only near T_c in BPBO(Fig. 2(b)) because a I-V characteristic of the sample becomes nonlinear only near T_c owing to superconducting transition. In $YBa_2Cu_3O_{7-y}$, nonlinearity of the bulk resistance was observed only in the very narrow temperature range (<2 K) of resistive transition (≃92 K) within the experimental bias current of 10 mA. Secondly, the V_{dc} in RE-Ba-Cu-O is independent of applied magnetic field, while in BPBO the suppression of the V_{dc} by the application of magnetic field (Fig. 2(b)) is understood as the suppression of the nonlinearity in I-V characteristics.

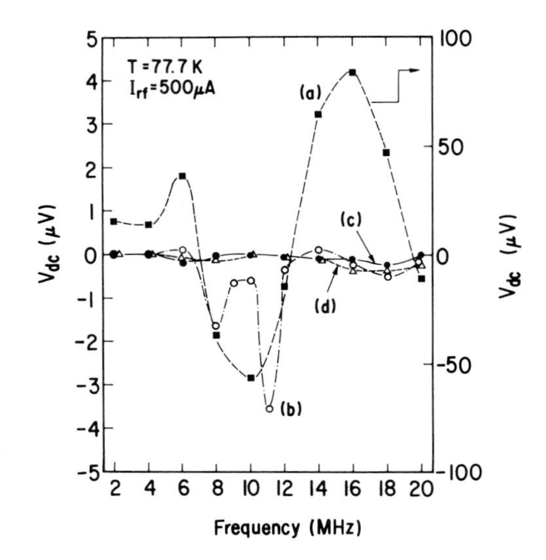

FIG. 9. V_{dc} vs. frequency of rf current for four types of electrical contacts with $YBa_2Cu_3O_{7-y}$ samples from the same batch; (a)No. 1, (b)No. 2, (c)No. 3, and (d)No. 4 electrical contacts.

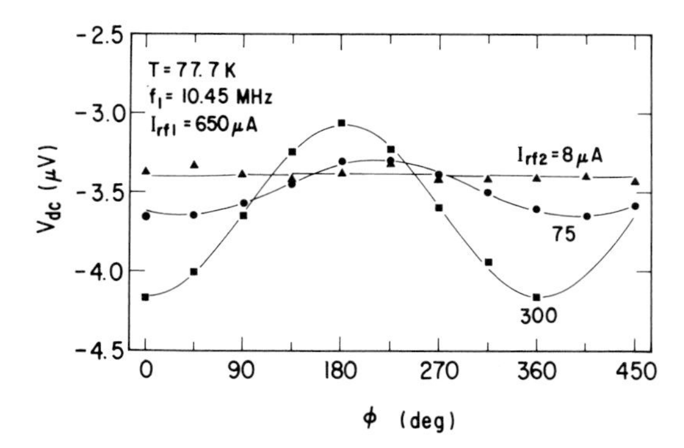

FIG. 10. V_{dc} vs. ϕ using type-A circuit. The sample and the electrical contacts are the same as in Fig. 7.

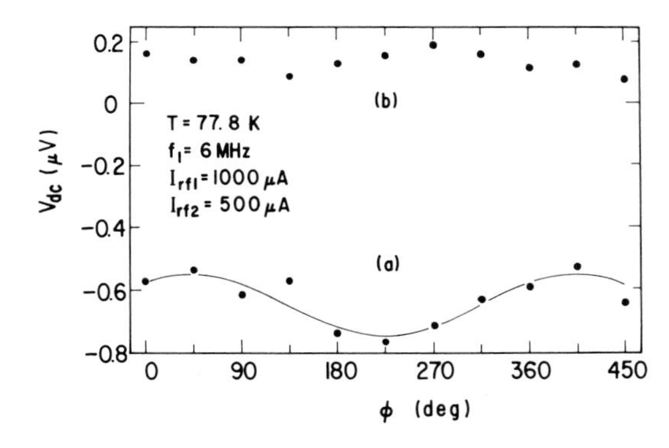

FIG. 11. V_{dc} vs. ϕ using type-B circuit: (a)eliminating the coils from potential leads, (b)two coils are inserted in potential leads. The sample and the electrical contacts are the same as in Fig. 7.

Above two points suggest that the V_{dc} in RE-Ba-Cu-O is not related to the superconductivity which can be detected by ordinary resistance measurement. Thirdly, when the second harmonic wave is superposed using type-B circuit, V_{const} is a dominant part of the V_{dc} and definite phase shift $\Delta\phi$(≃106 deg) is observed in $YBa_2Cu_3O_{7-y}$(Fig. 11(a)), while they are negligibly small in BPBO (Fig. 3). Finally, from the resistance measurements, a contact resistance of each potential contact for BPBO is much smaller than the resistance of the sample itself(≃1 Ω), while the contact resistances in Table I are larger than the $YBa_2Cu_3O_{7-y}$ sample(≃4 mΩ).

The V_{dc} in $YBa_2Cu_3O_{7-y}$ is strongly suppressed by the two coils inserted in potential leads in both cases of type-A and type-B circuits. The digital voltmeter used in the present work has an equivalent input capacitance of about 100 pF. Thus, input impedance for 10 MHz is reduced to 160 Ω. Therefore, an rf current can flow through the circuit composed of the digital voltmeter and potential leads. The coils suppress this rf current. Our experimental results show that the V_{dc} is observable with particular types of electrical contacts(Fig. 9) and only when the rf current can flow through the potential contacts. This suggests that the V_{dc} is generated from the rf current flowing through two potential contacts and not from the rf current of regular path.

We conclude, therefore, that the V_{dc} in RE-Ba-Cu-O comes from rectification and nonlinear interference effect near the two potential electrical contacts. The above mentioned differences between RE-Ba-Cu-O and BPBO can be explained by this interpretation. It is well known that low-resistance ohmic contact to RE-Ba-Cu-O is difficult to be made without special technique[6,7], because semiconducting or insulating layer exists on the sample surface. Some reports remarked that Schottky barriers are formed at the interface between metal electrodes and La-Sr-Cu-O[9] or Y-Ba-Cu-O[10]. In the case of Schottky barrier, nonlinearity of a current-voltage characteristic increases with decreasing temperature. Therefore, the observed temperature dependence of V_{dc} can be explained by the presence of Schottky barrier at the potential contacts.

The strange behavior of sign and magnitude of V_{dc}, such as random dependence of V_{dc} on frequency(Fig. 9), can be explained as follows. The V_{dc} from rectification corresponds to the difference of rectification at two potential contacts. Therefore, if the dependences of rectification on rf amplitude, frequency and temperature at the two contacts is different from each other, the sign and magnitude of the V_{dc} change in a random fashion. Nonlinear interference effect between the fundamental and the internally generated second harmonic wave is the source of the V_{dc} in BPBO and also contributes to the V_{dc} in RE-Ba-Cu-O(but is small in our experiments). The sign and magnitude of the V_{dc} from nonlinear interference effect strongly depend on the phase difference of fundamental wave and internally generated second harmonic wave. Thus they depend on rf amplitude, frequency and temperature through the nonlinear characteristic of the circuit impedance.

CONCLUSION

The rf-to-dc conversion effect was investigated in the frequency range from 2 to 20 MHz for $BaPb_{1-x}Bi_xO_3$ and RE-Ba-Cu-O ceramic samples. We conclude that the observed V_{dc} is not due to reverse ac Josephson effect, which was proposed previously, but due to nonlinear current-voltage characteristics. In $BaPb_{1-x}Bi_xO_3$, it is due to the nonlinear interference effect, which is an interplay between an alternating current and its second harmonic in the sample with the nonlinear I-V characteristic induced by the superconducting transition. In RE-Ba-Cu-O, the V_{dc} is due to rectification and nonlinear interference effect at local nonlinear resistance near the potential electrical contacts, for example Schottky barrier. The polarity change of the V_{dc} in a random fashion can be explained by unbalance of rectification at two potential electrical contacts and phase sensitivity of nonlinear interference effect. Therefore, one can not insist on the validity of the reverse ac Josephson effect instead of the nonlinear current-voltage characteristic only on the grounds of a polarity change of the V_{dc} as functions of rf amplitude, frequency and temperature. More details will be published elsewhere([11] for $BaPb_{1-x}Bi_xO_3$ and [12] for RE-Ba-Cu-O).

REFERENCES

[1] J. T. Chen, L. E. Wenger, C. J. McEwan and E. M. Logothetis, Phys. Rev. Lett. 58, 1972 (1987).
[2] J. T. Chen, R. J. Todd, and Y. W. Kim, Phys. Rev. B 5, 1843 (1972).
[3] H. Sadate-Akhavi, J. T. Chen, A. M. Kadin, J. E. Keem, and S. R. Ovshinsky, Solid State Commun. 50, 975 (1984).
[4] A. M. Saxena, J. E. Crow, and Myron Strongin, Solid State Commun. 14, 799 (1974).
[5] A. Khurana, Phys. Today 40 (7), 17 (1987).
[6] J. van der Mass, V. A. Gasparov, and D. Pavuna, Nature 328, 603 (1987).
[7] J. W. Ekin, A. J. Panson, and B. A. Blankenship, Appl. Phys. Lett. 52, 331 (1988).
[8] Truong D. Thanh, A. Koma, and S. Tanaka, Appl. Phys. 22, 205 (1980).
[9] K. Takeuchi, Y. Okabe, M. Kawasaki, and H. Koinuma, Jpn. J. Appl. Phys. 26, L1017 (1987).
[10] M. D. Kirk, D. P. E. Smith, D. B. Mitzi, J. Z. Sun, D. J. Webb, K. Char, M. R. Hahn, M. Naito, B. Oh, M. R. Beasley, T. H. Geballe, R. H. Hammond, A. Kapitulnik, and C. F. Quate, Phys. Rev. B 35, 8850 (1987).
[11] S. Ikegawa, T. Honda, H. Ikeda, A. Maeda, H. Takagi, S. Uchida, K. Uchinokura, and S. Tanaka, J. Appl. Phys. 64, Nov. 15, (1988) (in press).
[12] S. Ikegawa, T. Honda, H. Ikeda, A. Maeda, S. Uchida, K. Uchinokura, and S. Tanaka, submitted to J. Appl. Phys.

Structural Parameter Dependence of the Quantum Behavior of Small Superconductor Junctions

MASANORI SUGAHARA and NOBUYUKI YOSHIKAWA

Faculty of Engineering, Yokohama National University, Tokiwadai 156, Hodogaya-ku, Yokohama, 240 Japan

ABSTRACT

Quantum-mechanical study is made on the junction parameters of Josephson junction and its dual device, Phase-Quantum-Tunneling junction concerning both tunnel-type and bridge-type devices. The theoretical results agree with experiments fairly well.

I. INTRODUCTION

The discovery of high T_c oxide superconductor[1] brought about a new possibility of high-temperature superconductor electronics. One of the most fascinating dream of this field is the superconductor VLSI technology[2]. In the refined superconductor circuits is used the flux-quantizing characteristic of Josephson junction[3] in the unit of flux quantum $\Phi^*=2.07\times10^{-15}$Wb. Although the value of Φ^* is felt small, it appears fairly large information carrier in an advanced VLSI. For example, researchers would have difficulty in designing a highly packed superconductor memory VLSI using gate of μm size. Therefore a superconductor circuit with information carrier effectively much smaller than Φ^* would be welcomed in developing superconductor VLSI technology. In this report are discussed the parameter conditions of Josephson junction and its dual device, "Phase-Quantum-Tunneling (PQT)" junction[4,5] which is one candidate of the smaller information-carrier devices. The characteristic of a Josephson junction is given with the relation $I=I_c\sin\theta=I_c\sin(2\pi\phi/\Phi^*)$ between supercurrent I and the phase difference θ (or the flux ϕ passed across the junction). On the other hand a PQT junction has a relation $V=V_c\sin(2\pi q/e^*)$ between the voltage V and the charge q passed through the junction. In superconductors e^* is the pair charge 2e. One may use e^* as the information carrier in PQT circuits[6]. The feasibility of Josephson and PQT VLSI depends on the limit of the parameter range of each elementary junction. In order to clarify the range, parameter limit for Josephson and PQT junctions are studied based on the quantum-mechanical treatment of the quantum fluctuation of (q, ϕ) concerning the tunnel-type and bridge-type junctions.

II. JOSEPHSON EFFECT AND COHERENCE OF QUANTUM FLUCTUATION

In the derivation of Josephson effect[7], Josephson presupposed the following conditions for the superconductor tunneling junction with two electron-pair systems 1 and 2.

(i) The variation of θ at pair tunneling is negligible.

(ii) As the result, θ is considered to be strongly "degenerate" with many junction states represented with different pair-tunneling number n.

(iii) The junction state $|\Psi\rangle$ is given with the superposition

$$|\Psi\rangle=\sum_{n=-\infty}^{\infty} w_n{}^{1/2}\exp(in\theta)|\Psi_1(N_1+n)\Psi_2(N_2-n)\rangle, \tag{1}$$

where w_n is the probability of the appearance of the state of n tunneling pairs where the pair number in system 1 is N_1+n, and in 2, N_2-n. The strong degeneracy means that w_n is nearly constant in wide range of n.

(iv) Getting the expectation value of the pair-tunneling Hamiltonian with respect to $|\Psi\rangle$, one find the coupling energy $E_j\cos\theta$ which yields the current-phase relation $I=I_c\sin\theta$.

It is easy to show that these conditions are not always satisfied unconditionally. For example, if one diminishes the size of the tunneling junction, the decreasing junction capacitance C can elevate the electrostatic energy $e^{*2}/2C$ at the pair tunneling so high that the conditions (i)-(iii) are unjustifiable.

The quantum state $|\Psi\rangle$ of Eq.(1) in which many n states are superposed with non-negligible $w_n{}^{1/2}$ just corresponds to a charge fluctuation where one observes the tunneling of ne^* at a probability w_n. In order that a macroscopic quantum effect appears from the quantum fluctuation, the quantum state should resemble a classical state as closely as quantum mechanics allow. Therefore the quantum fluctuation of charge q and flux ϕ (or θ) in Josephson junction should be one kind of the coherent states[8] where the least uncertainty relation

$$\delta q \delta \phi = \hbar/2 \quad (2)$$

is always satisfied. The conditions (i)-(iii) are replaced with the restriction for the (q, ϕ) fluctuation

$$\delta q >> e^{*}, \quad \delta\phi << \Phi^{*}. \quad (3)$$

On the other hand, in case of another type of coherent fluctuation with a different condition

$$\delta q << e^{*}, \quad \delta\phi >> \Phi^{*}, \quad (4)$$

one may expect a new macroscopic quantum effect (PQT effect) dual to the Josephson effect.

III. MACROSCOPIC QUANTUM EFFECTS IN TUNNEL JUNCTION

In Fig.1 is seen a model of a superconductor tunnel junction with capacitance C whose electrode separation is d. The energy loss mechanism interacting with the tunneling charge is expressible with an ensemble of many harmonic oscillators (loss oscillators)[9]. The conductor loop connected to the capacitance electrodes has a large inductance L. A current source I_s is connected to the loop. The energy dissipation is supposed to be neglected except in the tunnel region. The Hamiltonians of the LC electromagnetic system (F), of the loss mechanism (L), of the interaction between F and L, and of the interaction between F and the source (S) are given

$$H_F = (q^2/C + \phi^2/L)/2$$

$$H_L = \sum_i (p_i^2/M_i + K_i x_i^2)/2$$

$$H_{FL} = -\sum_i (q_i/M_i)\vec{p}_i \cdot \vec{A} = -\sum_i (q_i/M_i d) p_i \phi$$

$$H_{FS} = -I_s \phi$$

$$[q, \phi] = i\hbar, \quad [x_i, p_i] = i\hbar,$$

where $(p_i^2/M_i + K_i x_i^2)/2$ is the energy of i'th loss oscillator, and $A=\phi/d$ is the vector potential. Using the total Hamiltonian $H=H_F+H_L+H_{FL}+H_{FS}$, the equations of motion for q, ϕ, x_i, p_i are found:

$$\dot{q} = \phi/L - \sum_i (q_i/M_i d) p_i - I_s$$

$$\dot{\phi} = -q/C$$

$$\dot{x}_i = p_i/M_i - (q_i/M_i d)\phi$$

$$\dot{p}_i = -K_i x_i,$$

where q_i is the charge of i'th oscillator. Obtaining solutions for $(x_i(t), p_i(t))$ and substituting them in the equations for $q(t)$, $\phi(t)$, we find after partial integration

$$\ddot{\phi} + 2\gamma\dot{\phi} + \omega^2\phi = f(t) \quad (5)$$

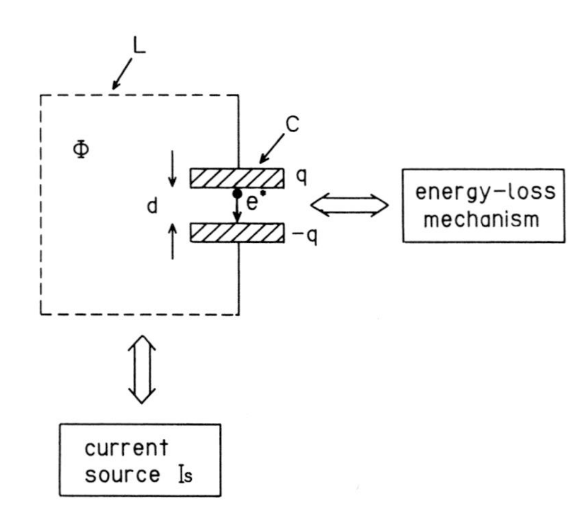

Fig.1 A model of a tunnel junction with energy-loss mechanism.

with

$$\omega^2=(LC)^{-1}-\kappa\Omega_{max}/\pi C$$

$$2\gamma=\kappa/2C$$

$$\kappa=\pi q^{*2}\sigma/M^*d^2$$

$$f(t)=I_s+F(t)-\kappa\phi(0)\delta(t)$$

$$F(t)=\int_0^{\Omega_{max}}(q_i/M_i d)p_i(0)\sigma d\Omega,$$

where we assumed $q_i{}^2/M_i=q^{*2}/M^*$, and the frequency(Ω) distribution of the loss oscillators

$$\sigma(\Omega)=\begin{cases}\sigma & 0\leqq\Omega\leqq\Omega_{max}\\ 0 & \Omega>\Omega_{max}.\end{cases}$$

$\sigma(\Omega)$ corresponds to the "resistance" case[9]. Using ϕ found from Eq.(5), we get in the limit of $L\to\infty$ the squared average of the flux fluctuation $\langle\delta\phi^2\rangle_n$ for the mode with energy $E_n=\hbar\Omega(n+1/2)$

$$\langle\delta\phi^2\rangle_n=\int_0^{\Omega_{max}}(\Phi^{*2}G_c/2\pi^2\kappa)(2n+1)\rho_{nn}(\Omega)[\gamma^2/\Omega(\Omega^2+4\gamma^2)]d\Omega, \tag{6}$$

where $G_c=e^{*2}/h$, $\rho_{nn}(\Omega)\equiv[1-\exp(-\hbar\Omega/k_BT)]\exp(-\hbar\Omega n/k_BT)$ is the density matrix element of the oscillator on E_n level. In the similar manner one can obtain the squared average of the charge fluctuation ($L\to\infty$)

$$\langle\delta q^2\rangle_n=\int_0^{\Omega_{max}}(e^{*2}\kappa/8\pi^2G_c)(2n+1)\rho_{nn}(\Omega)[\Omega/(\Omega^2+4\gamma^2)]d\Omega. \tag{7}$$

The zero-point oscillation mode satisfies the least uncertainty relation Eq.(2)[10], and therefore is one of the coherent mode. We must note that Glauber's coherent state[8] is one of the shifted mode of the ground state mode[11]. From Eqs.(6) and (7) is found the fluctuations of the ground-state mode

$$\langle\delta\phi^2\rangle_c=\int_0^{\Omega_{max}}(\Phi^{*2}G_c/2\pi^2\kappa)[\gamma^2/\Omega(\Omega^2+4\gamma^2)]d\Omega \tag{8}$$

$$\langle\delta q^2\rangle_c=\int_0^{\Omega_{max}}(e^{*2}\kappa/8\pi^2G_c)[\Omega/(\Omega^2+4\gamma^2)]d\Omega. \tag{9}$$

The necessary condition (3) of the Josephson effect is rewritten

$$\langle q^2\rangle_c > e^{*2} \tag{10}$$

With this condition the least uncertainty relation gives $\langle\delta\phi^2\rangle_c<(\Phi^*/4\pi)^2$ which denotes that the phase fluctuation in the coherent mode is negligible. Ω_{max} appearing in Eqs.(6)-(9) is the maximum frequency of the loss oscillators interacting with tunneling charges. This may correspond to the reciprocal of the minimum pair-tunneling time τ_p. Since the minimum pair size is the coherence length ξ, we have $\tau_p\cong\xi/v_F$ (v_F Fermi velocity), and find $\Omega_{max}\cong\pi\Delta/\hbar\cong$ gap frequency. With pair tunneling probability K ($\leqq\tau_p^{-1}$), the pair tunneling Hamiltonian is $H_t=hK(b_1^\dagger b_2+b_2^\dagger b_1)$. The coupling energy E_J is the expectation value of H_t with respect to Eq.(1). $|E_J|_{max}$ is found to be

$$|E_J|_{max}=h\tau_p^{-1}\langle q^2\rangle_c/e^{*2}=\int_0^{\Omega_{max}}(\hbar\pi\Delta\kappa/2e^{*2})[\Omega/(\Omega^2+4\gamma^2)]d\Omega,$$

which is near to Ambegaokar-Baratoff result[12]. The stability of the coupling is maintained when

$$|E_J|_{max}>k_BT. \tag{11}$$

The condition of the negligibility of thermal fluctuation of θ is found from Eq.(6)

$$\langle\delta\phi^2\rangle_T=\sum_{n=1}^{\infty}\int_0^{\Omega_{max}}(\Phi^{*2}G_c/2\pi^2\kappa)2n\rho_{nn}(\Omega)[\gamma^2/\Omega(\Omega^2+4\gamma^2)]d\Omega\ll\Phi^* \tag{12}$$

The requirements Eqs.(10)-(12) determine the limit of the junction (CG Josephson junction), which is displayed in the CG plane of Fig.2. Solid circles denote the parameters of experimentally confirmed Josephson junction. All those data points are subsumed within the theoretical limit.

Next we consider the case

$$\langle\delta\phi^2\rangle_c > \Phi^*, \tag{13}$$

where coherent charge fluctuation is negligible ($\langle\delta q^2\rangle_c < (e^*/4\pi)^2$). The wave function of the zero-point oscillation of the flux ϕ is $\Psi(\phi)\propto \exp[i\phi Q/\hbar]$ [4,5], where Q is the total charge passed through the junction ($Q=\int^t q(t')dt'$). Q has some arbitrariness depending on the selection of the initial state. $\Psi(\phi)$ should be substantially unchanged by the replacement $Q\to Q+me^*$ (m, integer). ϕ is forced to take $\phi=m'\Phi^*$ ($m'=0, 1, 2,\ldots$). Therefore the large coherent flux fluctuation of Eq.(13) appears to be caused by the "tunneling" of many flux quanta. The expectation value of the tunneling Hamiltonian of the flux quanta in the coherent state gives the coupling energy $E_p(Q)$ which leads to a V-Q relation of PQT junction of tunnel type (CG-PQT) $V=V_c\sin(2\pi Q/e^*)$. $|E_p|_{max}$ and $|V_c|_{max}$ is respectively $h\tau_p^{-1}\langle\delta\phi^2\rangle_c/\Phi^{*2}$ and $(4\pi^3\Delta/e^*)\langle\delta\phi^2\rangle_c/\Phi^{*2}$. Beside Eq.(13) the necessary conditions of CG-PQT are

$$|E_P|_{max} > k_BT \tag{14}$$

$$\langle\delta q^2\rangle_T = \sum_{n=1}^{\infty}\int_0^{\Omega_{max}} (e^{*2}k/8\pi^2G_c)2n\rho_{nn}(\Omega)[\Omega/(\Omega^2+4\gamma^2)]d\Omega < e^{*2} \tag{15}$$

The limit of the junction parameters for CG-PQT is displayed in Fig.2. Open circles denotes the experimental data of small tunnel junctions in the experiment of Tinkham's group[13]. These junctions revealed unusual behavior different from Josephson junction. Triangles are estimated value of granular PQT junction obtained in our experiment[14].

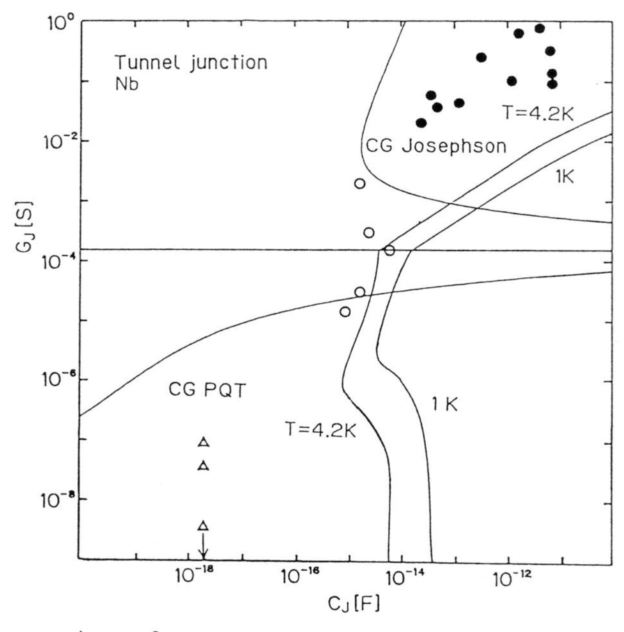

Fig.2 The parameter regions of superconductor tunnel junction where macroscopic quantum effects are expected. Josephson effect should appear in the region which is surrounded by a concave curve and curves with temperature. PQT effect should appear in the region which is surrounded by a convex curve and curves with temperatures.

IV. MACROSCOPIC QUANTUM EFFECT IN BRIDGE JUNCTION

In Fig.3 is given a model of superconductor bridge junction with the energy loss mechanism in the bridge part which is represented with a small inductance L. The magnetic inductance L_m with respect to the bridge part is very large ($L_m\to\infty$) because of the largeness of the "flux-interlinkage space". Therefore L_m can be ignored in comparison with the small "kinetic" inductance $L_k=\mu_0\ell\lambda^2/S$ which shunts L_m, where μ_0 is the vacuum permeability, ℓ is bridge length, λ is penetration depth, and S is bridge cross section. Representing the loss mechanism with an ensemble of small oscillators, and considering the Hamiltonian of the total system, we can obtain the coherent fluctuation ($\langle\delta q^2\rangle_c$, $\langle\delta\phi^2\rangle_c$) and thermal fluctuation ($\langle\delta q^2\rangle_T$, $\langle\delta\phi^2\rangle_T$) of charge and flux in the bridge junction after the calculations in §III. It must be noted that the role of pair tunneling in the tunnel junction is substituted in the bridge junction for the frequent transition of phase difference along the bridge in the unit of 2π. In the following discussion, the transition of the phase difference is expressed with "flux-quantum" tunneling in the bridge junction. Since a

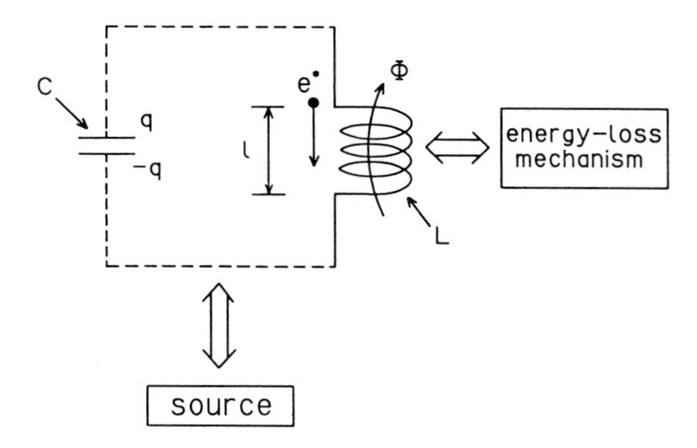

Fig.3 A model of bridge junction with energy-loss mechanism.

tunneling flux quantum has a size which spreads over area λ around the bridge and electrodes, the minimum tunneling time τ_f may be λ/c (c light velocity). The maximum frequency of the loss oscillators interacting with the tunneling flux may be $\tau_{max}^{-1} \cong \tau_f^{-1} \cong c/\lambda$.

The PQT effect in the bridge (LR-PQT) should appear when

$$\text{coherent fluctuation of flux} > \Phi^* \tag{16}$$

$$|\text{junction energy}|_{max} > k_B T \tag{17}$$

$$\text{thermal fluctuation of charge} < e^* \tag{18}$$

In Fig.4 the limit of these conditions are displayed in L-R plane.

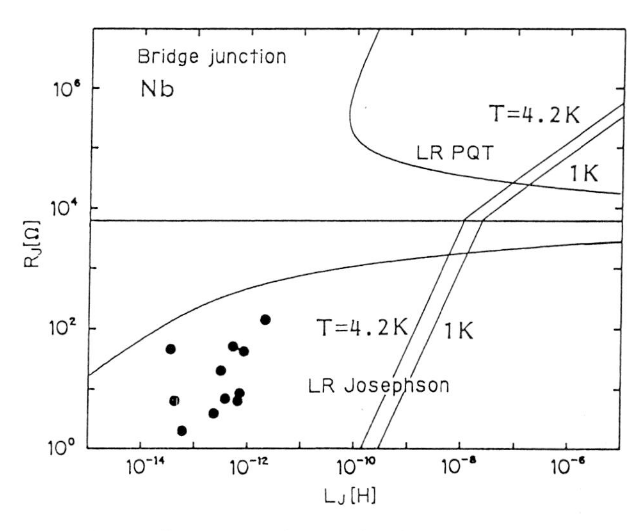

Fig.4 The parameter regions of superconductor bridge junction where macroscopic quantum effects are expected. PQT effect should appear in the region surrounded by a concave curve and curves with temperatures. Josephson effect should appear in the region which is surrounded by a convex curve and curves with temperatures.

On the other hand Josephson effect in bridge (LR-Josephson) is expected when

$$\text{coherent fluctuation of charge} > e^* \tag{19}$$

$$|\text{junction energy}|_{max} > k_B T \tag{20}$$

$$\text{thermal fluctuation of flux} < \Phi^* \tag{21}$$

The limits given with these requirements are displayed in Fig.4. The solid circles represent the parameters of Josephson bridges obtained from many published experiments. It is found that all the data points are subsumed within the limits.

V. CONCLUSION

The limit of the junction parameters necessary to expect macroscopic quantum effects (Josephson effect and its dual, PQT effect) in superconductor tunnel and bridge junction is discussed based on quantum mechanical treatment of electromagnetic quantities in the junctions with energy loss mechanism. It is found that the parameters of experimentally observed Josephson junction of tunnel and bridge type are always subsumed within the theoretical limit.

Our theory indicates that Josephson effect necessitates the following conditions. 1) The existence of two systems of charged bosons in wave-like quantum-condensation (coherent) states. 2) The existence of charge transition between the two systems through the barrier region, which is repellent against the intrusion of the wave-like charge, and in which tunneling charges lose wave-like character being quantized by number. 3) The existence of large coherent fluctuation of charge inside the junction where the interaction with flux (or phase) takes place. 4) Low thermal disturbance. It seems impossible to satisfy above requirements without using superconductor tunnel or bridge junctions in any temperature circumstances. On the other hand, the requirements for PQT junctions are as follows. 1) Existence of two separated wave-like flux (or phase) systems where ground state flux forms coherent state. 2) The existence of flux transition (flux-tunneling) across flux repellent region where a quantization of flux takes place. 3) Existence of large coherent fluctuation of flux inside the junction where the flux interacts with charge. 4) Low thermal disturbance. These requirements can be satisfied in some superconductor tunnel and bridge junctions. It is known, however, that the quantizations of flux takes place in small specimens of normal metals at very low temperature[15]. Therefore there is some possibility that PQT effect in observable at very low temperature even in normal metal junction[16]. At higher temperature where the flux quantization is absent, however, PQT junction using normal metal may be hopeless.

REFERENCES

1) J. G. Bednorz and K. A. Muller, Z. Phys. 1364 (1986) 189.
2) see for example IBM J. Res. Develop. 24 (1980, Special issue on Josephson Computer Technology).
3) B. D. Josephson, Phys. Lett. 1 (1962) 251.
4) M. Sugahara and N. Yoshikawa, Jpn. J. Appl. Phys. 26, Supplement 26-3 1987) 1411.
5) M. Sugahara and N. Yoshikawa, Pros. World Congress on Superconductivity, Houston, 1988.
6) N. Yoshikawa, T. Murakami and M. Sugahara, to be published in proc. ASC'88, San Francisco, 1988.
7) B. D. Josephson, Thesis in Superconductivity, vol.1, p.423 (R. D. Parks ed. Marcel Dekker, Inc., New York, 1969)
8) R. J. Glauber, Phys. Rev. 130 (1963) 2529; 131 (1963) 2766.
9) A. J. Leggett, S. Chakravarty, A. T. Dorsey, Matthew P. A. Fisher, Anupam Garg and W. Zwerger, Rev. Mod. Phys. 59 (1987) 1.
10) L. I. Schiff, Quantum Mechanics (McGraw-Hill Book Co., New York, 1968).
11) M. Sargent III, M. O. Scully and W. E. Lamb, Jr., Laser Physics (Addison-Wesley, Reading, Massachusetts, 1974).
12) V. Ambegaokar and A. Baratoff, Phys. Rev. Lett. 10 (1963) 486.
13) C. Iansti, A. T. Johnson, W. F. Smith, H. Rogalla, C. J. Lobb and M. Tinkham, Phys. Rev. Lett. 59 (1987) 489.
14) N. Yoshikawa, T. Murakami and M. Sugahara, Jpn. J. Appl. Phys. lett. 26 (1987) L1701.
15) R. A. Webb, S. Washburn, C. P. Umbach and R. B. Laibowitz, Phys. Rev. Lett. 54 (1985) L1701.
16) T. A. Fulton and G. T. Doran, Phys. Rev. Lett. 59 (1987) 109.

An Observation of Quasi-Particle Tunneling Characteristics in All Y-Ba-Cu-O Thin Film Tunnel Junction

T. SHIOTA, K. TAKECHI, Y. TAKAI, and H. HAYAKAWA

Department of Electronics Engineering, Nagoya University, Furo-cho, Chikusa-ku, Nagoya, 464-01 Japan

ABSTRACT

The first results are shown on the tunnel junction using high Tc superconducting oxide (Y-Ba-Cu-O) thin films for both base and counter electrodes. The tunnel junction with electrodes made of two crossing stripes of Y-Ba-Cu-O thin films was fabricated by an RF magnetron sputtering on an MgO substrate. A highly resistive thin barrier layer was prepared by plasma-fluorination of the surface of the Y-Ba-Cu-O base electrode prior to the deposition of the counter electrode.

A clear tunnel characteristic in an I-V curve with a well defined gap voltage and a small sub-gap leakage current was observed at 4.2K, although Josephson currents were not observed. The gap voltage was about 18mV at 4.2K, decreasing as the temperature was increased. The gap structure was observed up to 77K which was about the superconducting on-set temperature of the electrode films.

INTRODUCTION

Recent discoveries of high temperature oxide superconductors have motivated many investigators to develop various applications of high temperature operating superconductors. There are many attempts to make Josephson junctions for electronics applications. Immediately after the discovery of Y-Ba-Cu-O superconductors, a crack junction was formed with an aim of an all ceramics Josephson junction [1]. Lately, there have been various reports on the all Y-Ba-Cu-O junctions, such as, grain boundary, or weak link junctions using superconducting oxide thin films [2,3], point contact junctions and also metal-insulator-oxide superconductor thin film tunnel junction [4] or metal-oxide superconductor point contact junctions [5]. In spite of these extensive work, there have been no experimental results on the all oxide superconductor thin film Josephson junctions or even, quasi-particle tunneling junctions. This is partly because of the difficulties in making superconducting films of a good quality and obtaining a suitable insulating barrier layer as well as the inherent short coherence length of oxide superconductors.

This paper reports the first result of a tunnel junction made by using high Tc superconducting oxide (Y-Ba-Cu-O) thin films for both base and counter electrodes with a unique highly resistive interfacial barrier layer.

EXPERIMENTAL

Both upper and lower Y-Ba-Cu-O thin films (about 1 μm thick) constructing the tunnel junction were fabricated on a (100) MgO substrate by an RF magnetron sputtering under the following condition, i.e., total working pressure 50mTorr ($Ar:O_2$=4:1), an RF input power 150W, a target composition $Y_1Ba_3Cu_5O_x$, a heater block temperature 610℃ and a target-substrate distance about 2.5cm. A metal mask was used to get about 1mm wide stripes of oxide superconductors. After sputtering, the film was cooled down to room temperature in oxygen at a pressure of 1 Torr. Post-annealing was not carried out on the deposited oxide films. The as-deposited film has a crystal structure, partially oriented along the c-axis. The composition of the films was nearly the same as that of the target.

The tunnel junction was fabricated in the following sequence. Firstly, a lower Y-Ba-Cu-O film was deposited in the way mentioned above and the surface of the film was plasma-fluorinated to make a highly resistive thin barrier layer. Finally, an upper Y-Ba-Cu-O film was deposited to form a sandwich type junction as shown in Fig. 1. The deposition of the upper Y-Ba-Cu-O film was done in the same condition as in the case of lower films.

The surface fluorination was carried out in a 8cm wide cylindrical glass reactor by using a CF_4 gas plasma at a pressure of 0.5 Torr. The RF power was 100W and a 20 minutes-treatment was enough to observe the tunneling characteristics.

Current-voltage characteristics were measured by the 4 terminal method as shown in Fig. 1 at various temperatures from 4.2K to about 80k.

RESULTS AND DISCUSSIONS

A superconducting property of the lower Y-Ba-Cu-O film is shown in Fig. 2. Since the composition of the film is not precisely $Y_1Ba_2Cu_3O_x$, a considerably wide superconducting transition region is observed, but a relatively high superconducting onset temperature (90K) and a zero resistance temperature (50K) are confirmed.

The fluorination was confirmed to increase the surface resistance of Y-Ba-Cu-O films and fluorine atoms were detected on the surface of the treated Y-Ba-Cu-O films by a measurement of ESCA.

The tunnel junction composed of these Y-Ba-Cu-O films shows a clear quasi-particle tunneling characteristic in an I-V curve with a well defined gap voltage and a small sub-gap leakage current as shown in Fig. 3a), which was observed at 4.2K. Josephson currents ,however, are not observed. The quasi-particle tunneling characteristics smear out as the temperature is increased as shown in Fig. 3b). The gap voltage is about 18mV at 4.2K, decreasing as the temperature is increased. The gap structure was observed up to 77K which was about the superconducting on-set temperature of the electrode films. Figure 4 shows a temperature dependence of the gap voltage. Similar characteristics were observed also in the junction fabricated on a $SrTiO_3$ substrate.

The gap voltage evaluated in this experiment, however, seems smaller than the reported values[6]. There are some possibilities for this point. The present junction might be an NIS junction but not an SIS junction. This is, however, not likely because the lower Y-Ba-Cu-O films show a superconducting property and the upper ones show a superconducting onset. Another possibility is concerned with a relatively poor quality of our Y-Ba-Cu-O films. Although they show a superconducting transition, the transition temperature is not so high as 90K. The crystal structure also is not of highly c-axis oriented. Tsai et al.[7] showed results on the Pb-YBCO tunneling junctions, having different gap voltages depending on the quality of the superconducting films. The superconducting but not c-axis oriented Y-Ba-Cu-O films having a low Tc 52K show a gap voltage about 22mV similar to the present gap voltage, while those with a high Tc 80K show about 38mV for the gap voltage.

In this preliminary experiment, it may be difficult to get a definite conclusion on the details of the chemical and physical properties of the junctions. Further experiments are now in progress to fabricate high quality tunnel junctions by the improvements of experimental techniques for preparing superconducting films with high qualities, a precise control of the plasma-fluorination and micro-fabrications.

CONCLUSIONS

An all Y-Ba-Cu-O thin film tunnel junction was fabricated, by using an RF magnetron sputtering and a plasma-surface-fluorination technique. A clear gap voltage about 18mv at 4.2k and a small sub-gap leakage current were observed. The gap voltage smeared out at 77k. The present results strongly indicate a possibility to realize all high temperature oxide superconductor (Josephson) tunnel junctions which are quite useful for electronics applications.

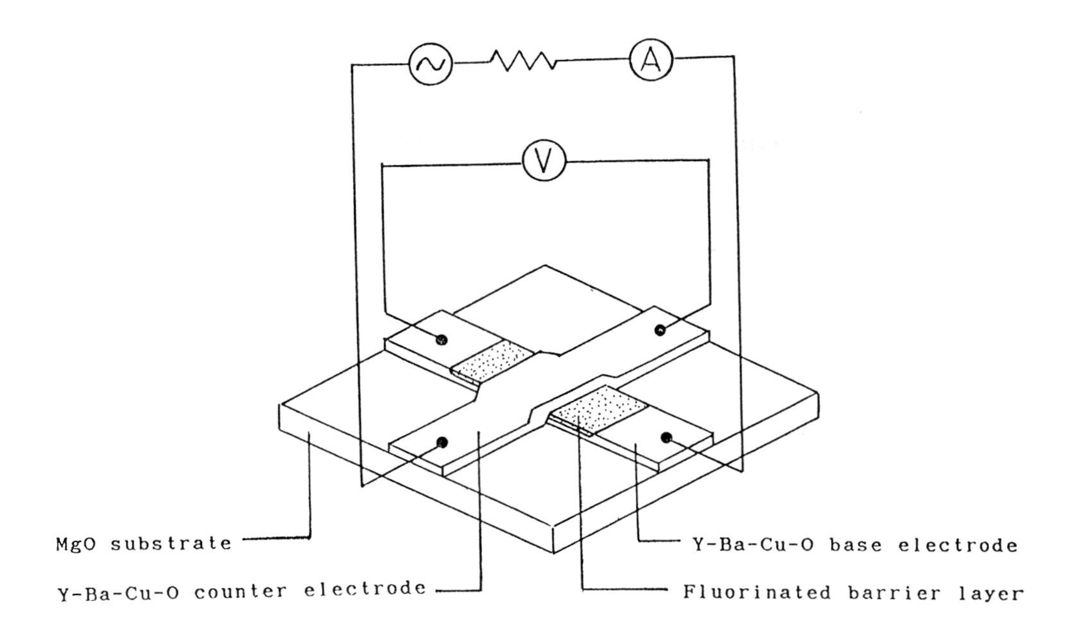

Fig.1 Configuration of the tunnel junction and the electrode for the measurement of the I-V characteristics.

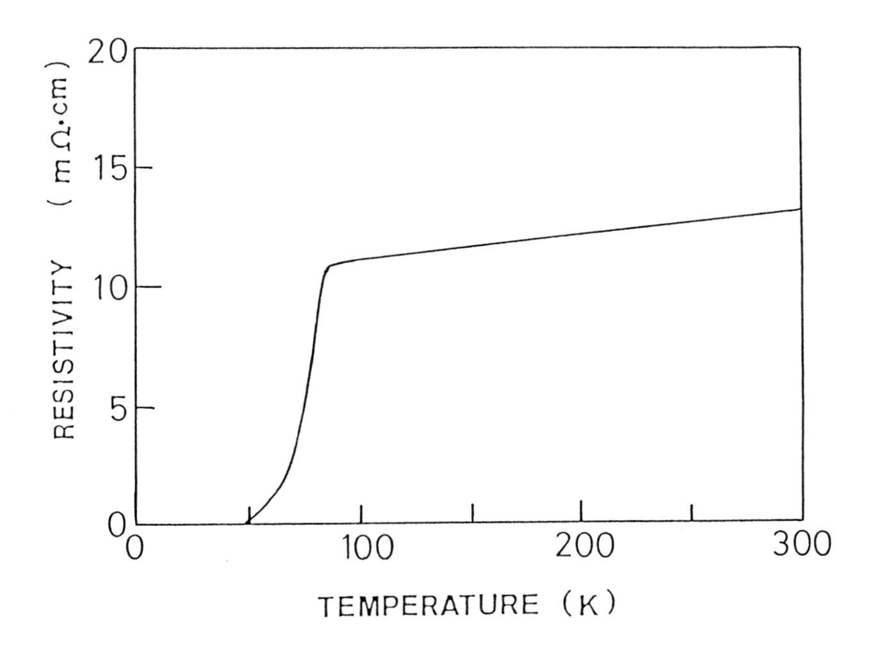

Fig.2 Temperature dependence of the resistivity of the lower Y-Ba-Cu-O film.

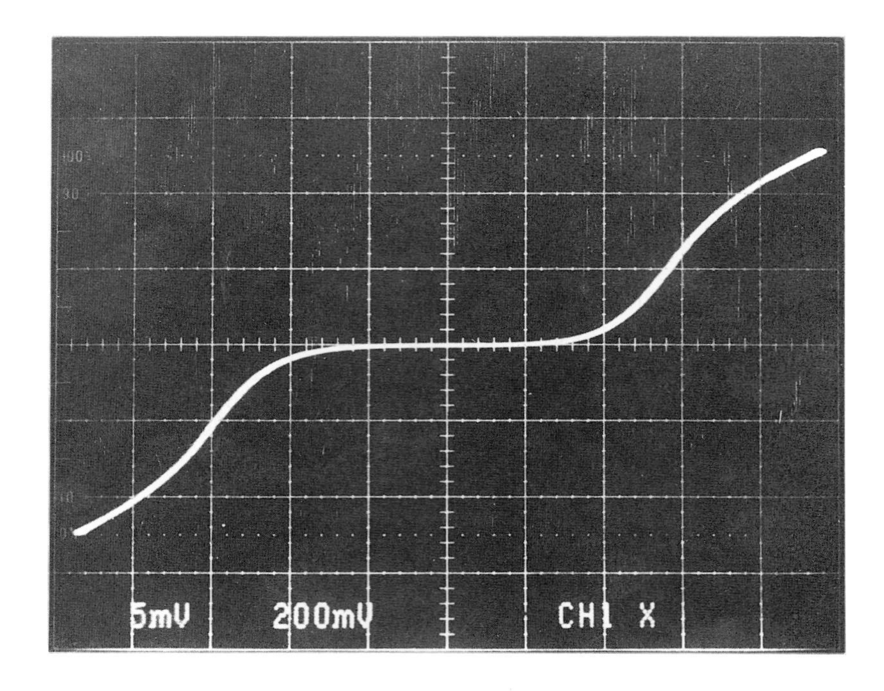

(a) T=4.2K horizontal 5mV/div vertical 2mA/div

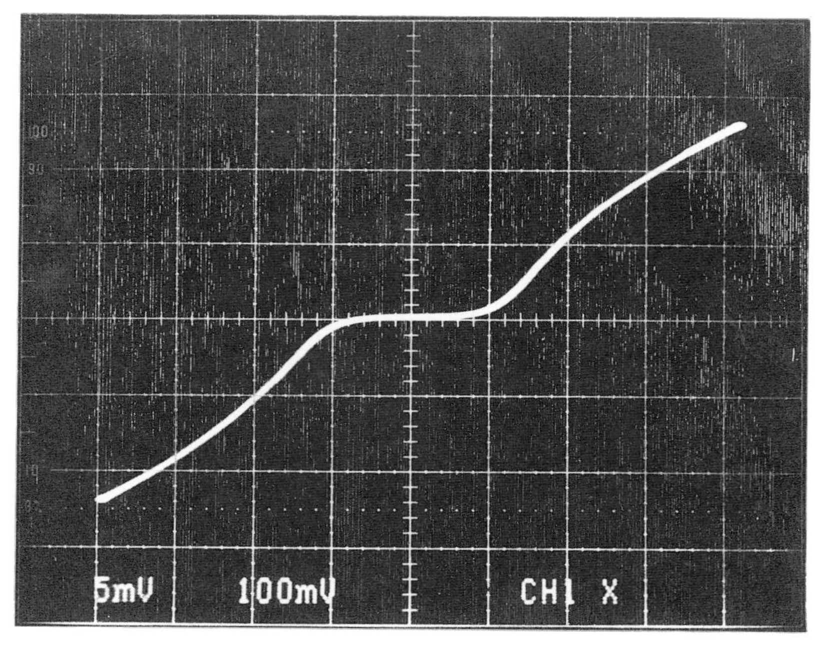

(b) T=30K horizontal 5mV/div vertical 1mA/div

Fig.3 I-V characteristics of the tunnel junction.

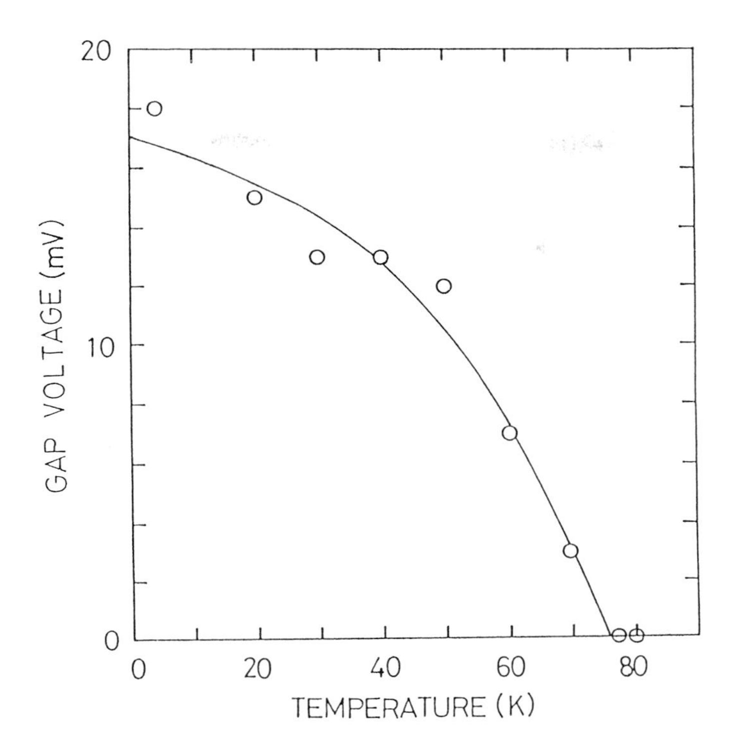

Fig.4 Temperature dependence of the gap voltage

ACKNOWLEDGMENT

This work was partly supported by a Grant-in-Aid for Special Research from the Ministry of Education, Science and Culture.

REFERENCES

[1] J.S.Tsai, Y.Kubo and J.Tabuchi "All-Ceramic Josephson Junctions Operative up to 90K" Jpn.J.Appl.Phys. 26 (1987) L701

[2] H.Tanabe, S.Kita, Y.Yoshizako, M.Tonouchi and T.Kobayashi "Grain Boundary Josephson Junctions Using Y-Ba-Cu-O Films Operative at 77K" Jpn. J. Appl. Phys. 26 (1987) L1961

[3] T.Yamashita, A.Kawakami, S.Noge, M.Takata, T.Komatsu and K. Matusita "High-Tc Superconducting Films and Devices at 77K" 5th Int. Workshop on Future Electron Devices 1988 p161

[4] A. Nakayama, A. Inoue, K. Takeuchi and Y. Okabe "Y-Ba-Cu-O/AlOx/Nb Josephson Tunnel Junctions" Jpn. J. Appl. Phys. 26 (1987) L2055

[5] T.Ekino, J.Akimitsu "Superconducting Tunneling in Y-Ba-Cu-O System" Jpn. J. Appl. Phys. 26 (1987) L452

[6] I.Iguchi, H.Watanabe, Y.Kasai, T.Mochiku, A.Sugishita and E.Yamaka "Tunneling Spectroscopy of Y-Ba-Cu-O Compound" Jpn. J. Appl. Phys. 26 (1987) L645

[7] J.S.Tsai, I.Takeuchi, J.Fujita, Y.Yoshitake, S.Miura, S.Tanaka,T.Terashima, Y.Bando, K.Iijima and K.Yamamoto "Energy of Gap Anisotropy in $YBa_2Cu_3O_{7-\delta}$" 5th Int. Workshop on Future Electron Devices (1988) p219

4.14 Bi- and Tl-Based Cuprate Superconductors

Tl-Based Cu-O High Temperature Superconductors

A.M. Hermann and Z.Z. Sheng

Department of Physics, University of Arkansas, Fayetteville, AR 72701, USA

Abstract We have discovered reproducible and stable bulk superconductivity in the Tl-Ca-Ba-Cu-O system with zero resistance above 120 K. Magnetic and electronic transport properties including thermoelectric power and critical current of the new superconductors are presented. The Tl-Ca-Ba-Cu-O system can form a number of superconducting compounds, including $Tl_mCa_{n-1}Ba_2Cu_nO_{1.5m+2n+1}$ with m = 1 and 2, and n = 1, 2, 3, and 4. The Tc increases with m and n. The 2223 (Tl:Ca:Ba:Cu) phase has the highest Tc (125 K) to date. We present Josephson junction studies on the 2223 phase showing the existence of electron pair supercurrents and demonstrating weak link behavior at 77 K. We discuss an unusual levitation phenomenon in which the 2223 phase can be suspended, above, below, or to the side of a ring shaped magnet. We also discuss a new Tl-based superconducting system, in which Sr replaces the Ba, the Tl-Ca-Sr-Cu-O system.

INTRODUCTION

Discoveries of 30-K La-Ba-Cu-O superconductor [1] and 90-K Y-Ba-Cu-O superconductor [2] have stimulated a worldwide race for higher temperature superconductors. Breakthroughs were recently made by the discoveries of the 90-K Tl-Ba-Cu-O system [3,4], 110-K Bi-Ca-Sr-Cu-O system [5,6] and 120-K Tl-Ca-Ba-Cu-O system [7-9]. The Tl-Ba-Cu-O system is the first rare earth-free system which reaches zero resistance above liquid nitrogen boiling point, while the Tl-Ca-Ba-Cu-O system is the first system which reaches zero resistance above 100 K and has the highest zero-resistance temperature (125 K). In this paper, we present the preparation procedures and some properties of the 120-K Tl-Ca-Ba-Cu-O superconductors, we discuss unusual levitation phenomena due to flux pinning, and we present Josephson junction studies on Tl-Ca-Ba-Cu-O which show the existence of electron pair supercurrents and weak link behavior at 77 K. We also present preliminary data on a new Tl-based superconducting system, Tl-Ca-Sr-Cu-O.

PREPARATION

Tl-Ca-Ba-Cu-O superconductive compounds form easily; there are many ways to make good-quality superconducting samples. One of typical procedures in preparing the Tl-Ca-Ba-Cu-O samples which we use is the following. Appropriate amounts of Tl_2O_3, CaO and Ba-Cu oxide (depending on the desired stoichiometry) were completely mixed and ground, and pressed into a pellet with a diameter of 7 mm and a thickness of 1-2 mm. The pellet was then put into a tube furnace which had been heated to 880-910 °C, and was heated for 2-5 minutes in flowing oxygen, followed by furnace cooling to below 200 °C. Quenching in air from 900 °C to room temperature depresses Tc only slightly. A number of samples with different stoichiometry, including a series of samples with nominal compositions of $Tl_2Ca_yBaCu_3O_{7+y+x}$ with y = 1, 1.5, 2, 3 and 4, were prepared. Samples with nominal compositions of $Tl_{1.86}CaBaCu_3O_{7.3+x}$, $Tl_2CaBa_2Cu_2O_{8+x}$ and $Tl_2Ca_2Ba_2Cu_3O_{10+x}$ were also synthesized. All samples readily levitate over a magnetic field. The strong Meissner effect observed is due to both a large volume fraction of superconductive phase and a higher transition temperature. Recently, most samples were prepared using precursor Ba-Ca-Cu-oxides.

STRUCTURE

The Tl-Ca-Ba-Cu-O system can form a number of superconducting phases. Two phases, $Tl_2Ca_2Ba_2Cu_3O_{10+x}$ (2223) and $Tl_2Ca_1Ba_2Cu_2O_{8+x}$ (2122), were first identified [10]. The 2223 superconductor has a 3.85 x 3.85 x 36.25 A tetragonal unit cell. The 2122 superconductor, which appears to be structurally related to $Bi_2Ca_1Sr_2Cu_2O_{8+x}$, has a 3.85 x 3.85 x 29.55 A tetragonal unit cell [10,11]. The 2223 phase is related to 2122 by addition of extra calcium and copper layers. In addition, the superconducting phase in the Ca-free Tl-Ba-Cu-O system is $Tl_2Ba_2CuO_{6+x}$ (2021) [10,12]. Fig. 1 shows schematically the arrangements of metallic atoms in these three Tl-based superconducting phases. The 2021 phase has a zero-resistance temperature of about 80 K, whereas the 2122 and 2223 phases have zero-resistance temperatures 108 K and 125 K, respectively [10-15]. It appears that the addition of a Ca and Cu layer increases the transition temperature about 20 K. If this trend continues linearly, it might be expected that 2324 phase will have a transition temperature at 140-150 K, and, based on density of states arguments within the BCS framework, a Tl-Ca-Ba-Cu-O phase with 9-10 Cu and Ca layers would have a superconducting transition above 200 K [16].

A new series of superconducting compounds with a single Tl-O layer, which we denote by $TlCa_{n-1}Ba_2Cu_nO_{2n+2.5}$, were recently also reported [17,18]. The Tl-Ca-Ba-Cu-O superconducting series should be represented using a general formula of $Tl_mCa_{n-1}Ba_2Cu_nO_{1.5m+2n+1}$ with m = 1 and 2, and n = 1, 2, 3, and 4. Tc of the single Tl-O layer compounds also increases with the number of Cu-Ca layers, and is slightly lower than that of the corresponding double Tl-O layer compounds. Therefore, an increase of Tc might be achieved not only by increasing Cu-Ca layers, but also by increasing the number of Tl-O layers. The role of the interlayer spacings has not been clarified yet.

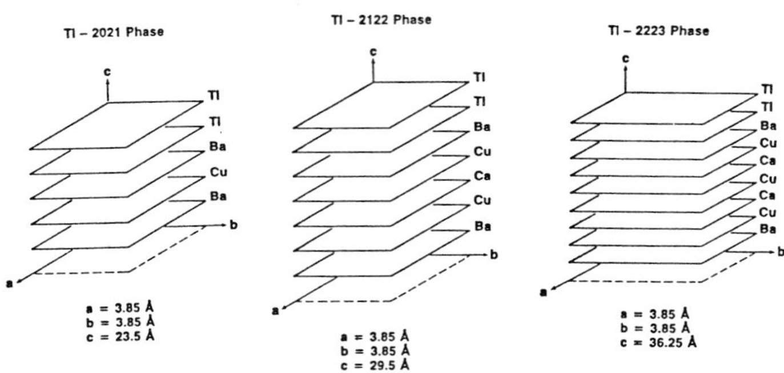

Figure 1 Schematic arrangements of the Tl-based superconducting phases 2021, 2122 and 2223.

RESISTANCE

Fig. 2 shows resistance-temperature variation for a nominal $Tl_{2.2}Ca_2Ba_2Cu_3O_{10.3+X}$ sample. This sample has an onset temperature near 140 K, midpoint of 127 K, and zero resistance temperature at 122 K. The highest zero resistance reported so far for the Tl-Ca-Ba-Cu-O system is 125 K [14,15].

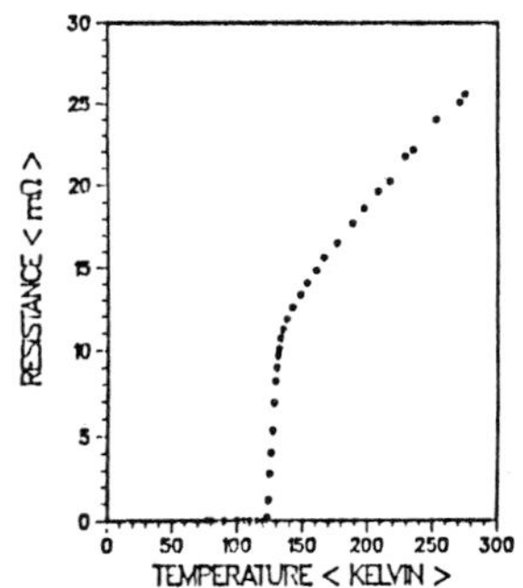

Figure 2 Resistance-temperature dependence of a nominal $Tl_{2.2}Ca_2Ba_2Cu_3O_{10.3+X}$ sample.

MAGNETIZATION

Magnetization measurements on the Tl-Ca-Ba-Cu-O samples were performed utilizing a SQUID magnetometer manufactured by BTI Corp., San Diego, CA [9]. Fig. 3 shows DC magnetization (field cooled and zero field cooled) as a function of temperature for an applied field of 1 mT for a nominal $Tl_2Ca_4BaCu_3O_{11+X}$ sample. The insert of Fig. 3 shows data traces for the same sample, and also for a well-prepared $EuBa_2Cu_3O_{7-X}$ sample, where the vertical axis represents the dX"/dH signal of an EPR spectrometer. As is seen from the insert, the onset temperature for the sample A ($Tl_2Ca_4BaCu_3O_{11+X}$) is 118.3 K, 23.9 K higher than that of the sample B ($EuBa_2Cu_3O_{7-X}$, whose onset temperature is 94.4 K). This onset temperature is consistent with those measured by resistance-temperature variations.

Fig. 4 shows a similar transition temperature for the 2223 phase by 5 kHz AC susceptibility measurements. The onset of the transition is 123 K.

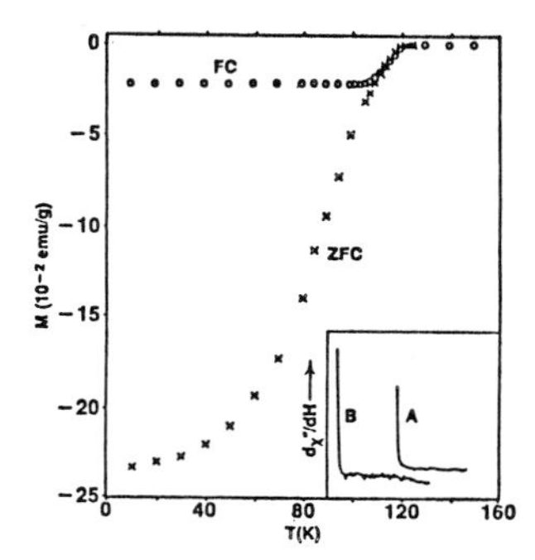

Figure 3 Field cooled (FC) and zero field cooled (ZFC) magnetization as a function of temperature for a DC field of 1 mT for a nominal $Tl_2Ca_4BaCu_3O_{11+x}$ sample. The two data traces in the insert illustrate the sharp onset of superconductivity observed by the microwave technique as described in the text. Sample A is $Tl_2Ca_4BaCu_3O_{11+x}$ with onset temperature of 118.3 K, and for comparison, data for a sample of $EuBa_2Cu_3O_{7-x}$ (sample B) with onset temperature of 94.4 K is plotted. Note that the difference in onset temperatures is 23.9 K.

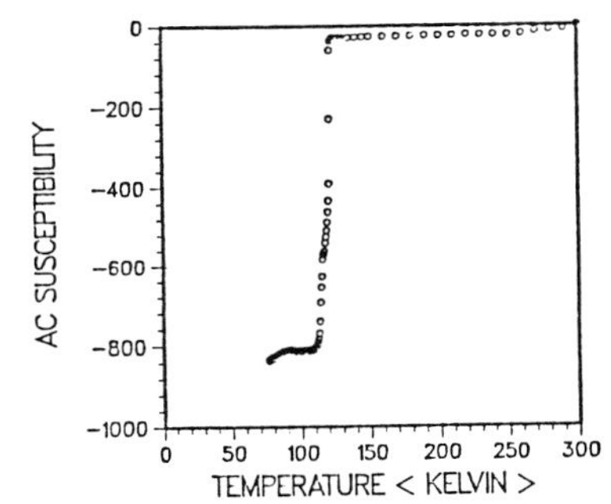

Figure 4 Temperature dependence of AC susceptibility for a nominal $Tl_2Ca_2Ba_2Cu_3O_{10+x}$ sample.

THERMOELECTRIC POWER

Fig. 5 shows thermoelectric power as a function of temperature for a nominal $Tl_2Ca_2Ba_2Cu_3O_{10+x}$ sample [19]. The normal-state thermoelectric power is positive, indicating dominant hole conduction. At least three separate ranges of temperature-dependent behavior are apparent. Below the transition (the midpoint of the transition was determined to lie at 118 K), the thermoelectric power is zero. From the transition temperature to about 175 K, the thermoelectric power is a increasing function of temperature. Finally, from 175 K to room temperature, the thermoelectric power decreases linearly with increasing temperature. The temperature dependence of the Tl-Ca-Ba-Cu-O superconductor is qualitatively similar to that of Y-Ba-Cu-O samples [20].

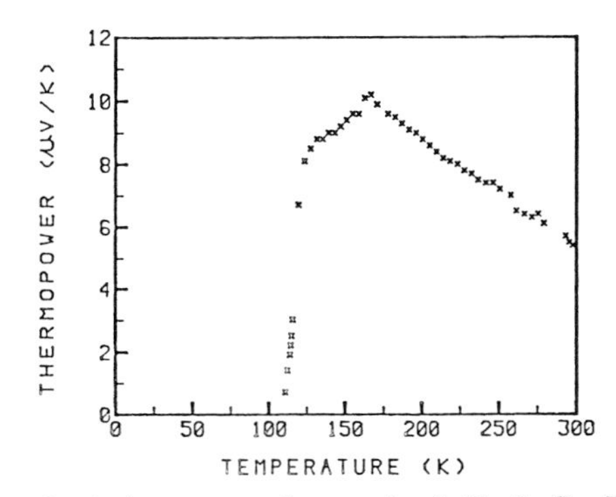

Figure 5 Thermoelectric power of a nominal $Tl_2Ca_2Ba_2Cu_3O_{10+x}$ sample as a function of temperature.

CRITICAL CURRENT

Fig. 6 shows the magnetic field dependence of the critical current density of the 2223 phase; also shown is the dependence for a high quality $YBa_2Cu_3O_{7-x}$ (Y-123) sample for comparison. While the 2223 phase critical current densities shown are relatively low, the weak sensitivity to magnetic field as compared to those of the Y-123 are very promising. Extremely high critical current densities have recently been found in 2122 polycrystalline thin films [21]

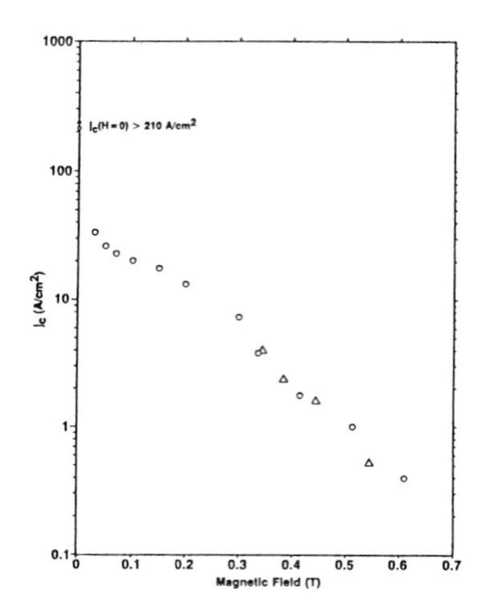

Figure 6 Magnetic field dependence of critical current density of a nominal $Tl_2Ca_2Ba_2Cu_3O_{10+x}$ sample.

LEVITATION

Properly prepared Tl-Ca-Ba-Cu-O samples can easily levitate over a magnet, as mentioned previously. There are two models that can explain the levitation above the magnet. We treat for simplicity the case of levitation of a spherical magnet over a disk superconductor.

Flux Exclusion Model. Consider the force on the magnet due to gravity and to its image dipole (below the superconductor). At equilibrium,

$$F=[-4\pi(-M4\pi R^3/3)MR^3(2h_o)^{-4}-mg]e_r=0$$

This gives an equilibrium h_o of

$$h_o=[\pi^2M^2R^6/(3mg)]^{1/4}$$

For M=400G, R=0.15cm, m=0.17gm, h_o is 0.44cm.

Flux Penetration Model. Assume flux enters the superconductor as it would empty space, except that the vortex energy is given [22] as

$$E_1=H_{c1}\Phi_o/4\pi$$

For a thin superconducting disk of thickness T,

$$E_1=4R^3MH_{c1}T/3h, \text{ or}$$

at equilibrium,

$$h_o=[4MH_{c1}LR^3/(3mg)]^{1/2}$$

With the same values as before and H_{c1}=100G, h_o=0.33cm. Both of these values are within the range determined experimentally.

Careful observations have shown that for some Tl-based samples, the force between the sample and the magnet is complicated [23]. Fig. 7a shows a small magnet levitating over a small Tl-based sample. The lateral stability of the magnet cannot explained only by the Meissner effect. Fig. 7b shows two Tl-based superconductors suspended horizontally above and below a ring-shaped magnet. A similar magnetic effect showing the coexistence of repulsive and attractive forces was reported for some Y-Ba-Cu-O/AgO samples [24]. Fig. 7c shows two Tl-based superconductors suspended vertically near a ring-shaped magnet. The downward flowing nitrogen

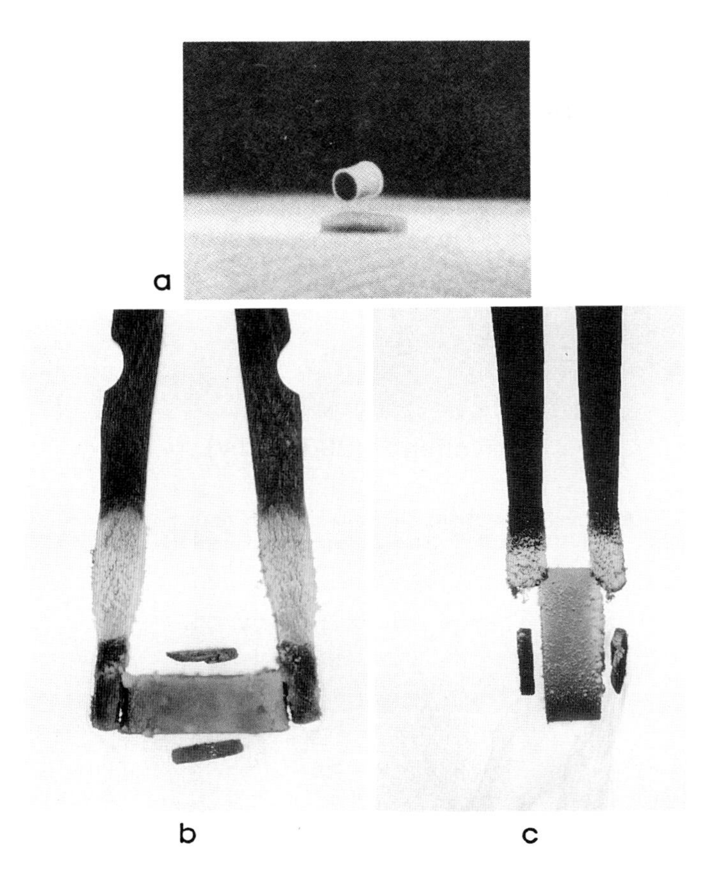

Figure 7 (a) a magnet levitating over an unusual Tl-based superconductor, (b) two Tl-based superconductors suspended horizontally above and below a ring shaped magnet, and (c) two Tl-based superconductors suspended vertically near a ring shaped magnet.

vapor is evident in figures 7b and 7c. The levitation beneath or at the side of the magnet clearly involves flux penetration into the non-superconducting regions and corresponding attractive supercurrents which are pinned. Corresponding large residual positive magnetic susceptibility following application of a magnetic field has been observed experimentally in the 2223 phase [25].

JOSEPHSON JUNCTION STUDIES

We have made [26] point contacts between a classical BCS superconductor (Pb) and $Tl_2Ca_2Ba_2Cu_3O_{10+x}$. A needle shaped piece of Pb was mounted in an apparatus designed to form bulk tunnel junctions [27]. A small coil was mounted on top of the point contact to expose the junction to a dc magnetic field (up to $2x10^{-6}$ T).

At 4.2 K, we formed several stable Josephson junctions. Fig. 8 shows the I-V characteristics of a typical junction at 4.2 K. A critical current of 1.455 mA was found with no magnetic field.
Application of small external magnetic field gave the periodic behavior predicted by pair tunneling between two superconductors [28], as shown in Fig. 9.

Above 10 K, the current-voltage behavior always became linear. Hence we conclude that, below the critical temperature of Pb the supercurrent consisted of electron pairs which tunnel from Pb into $Tl_2Ca_2Ba_2Cu_3O_{10+x}$ and that the transport in the $Tl_2Ca_2Ba_2Cu_3O_{10+x}$ is accomplished by electron pairs.

We have also found tunnel junctions with $Tl_2Ca_2Ba_2Cu_3O_{10+x}$ only. In this case, the Pb was replaced by a needle-shaped $Tl_2Ca_2Ba_2Cu_3O_{10+x}$ sample formed using sandpaper. The 4.2 K I-V characteristics of the one Josephson junction we were able to form using this geometry are shown in Fig. 10. Fig. 11 shows the I-V characteristics for such a junction at 77 K. Here weak-link behavior is evident.

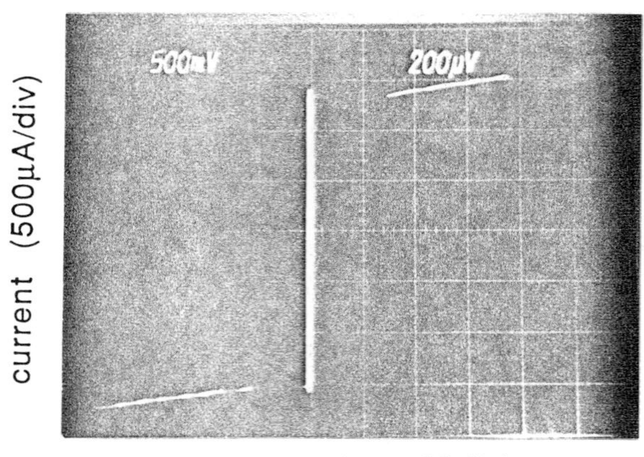

Figure 8 Current-voltage characteristics of a bulk $Pb/Tl_2Ca_2Ba_2Cu_3O_{10+x}$ Josephson junction at 4.2 K.

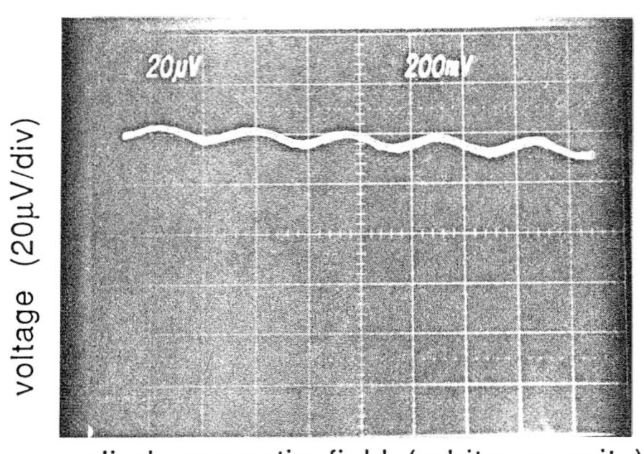

Figure 9 Voltage versus external dc magnetic field for the junction of Fig. 8 at a bias current of 1.458 mA.

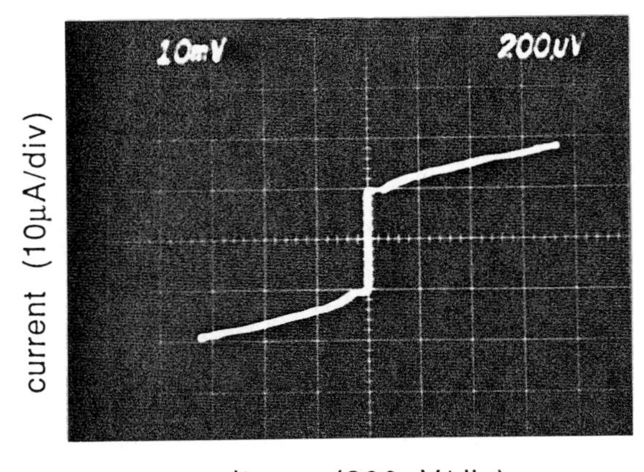

Figure 10 Current-voltage characteristics for a Josephson junction formed by a point contact of two pieces of bulk $Tl_2Ca_2Ba_2Cu_3O_{10+x}$ at 4.2 K.

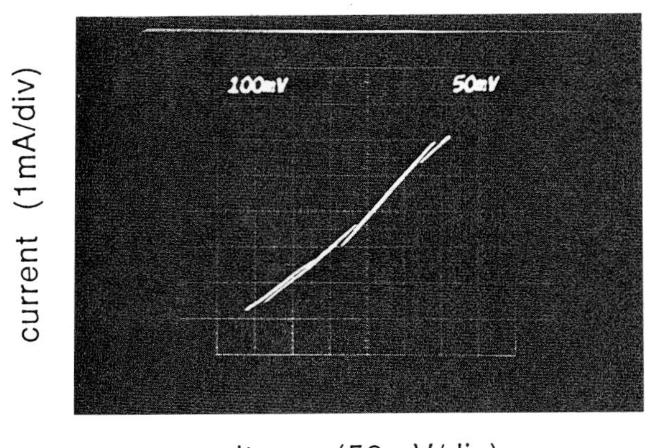

Figure 11 Current-voltage characteristics for a point contact of two pieces of $Tl_2Ca_2Ba_2Cu_3O_{10+x}$ at 77 K displaying weak link behavior.

TL-CA-SR-CU-O SUPERCONDUCTING SYSTEM

Using similar sample preparation procedures as described above we have observed unambiguous superconductivity at 20 K in the Tl-Sr-Ca-Cu-O system [29]. A superconducting transition at 70 K was also observed in the same system.

In Fig. 12a we present resistance versus temperature for sample #8165 (with a nominal composition of $Tl_2Sr_2Ca_2Cu_3O_{10+x}$). The onset temperature is about 22 K and zero resistance is reached at 10 K. In Fig. 12b we present the field cooled and zero field cooled dc susceptibility versus temperature. As can be seen there is an onset of diamagnetism at about 25 K. The diamagnetic shielding at the lowest temperatures is weak and corresponds to a superconducting volume fraction of less than 1%. In Fig. 12c we present the microwave spectrometer signal versus temperature. We find that there is a weak but abrupt change in dX"/dH (expended scale), indicating the onset of superconductivity at about 70 K. This is followed by a large increase in signal at lower temperatures, indicating an increasing fraction of the sample becomes superconducting at temperatures below 20 K. This is consistent with both the susceptibility data (Fig. 12b) and the observed drop in the resistance (Fig. 12a).

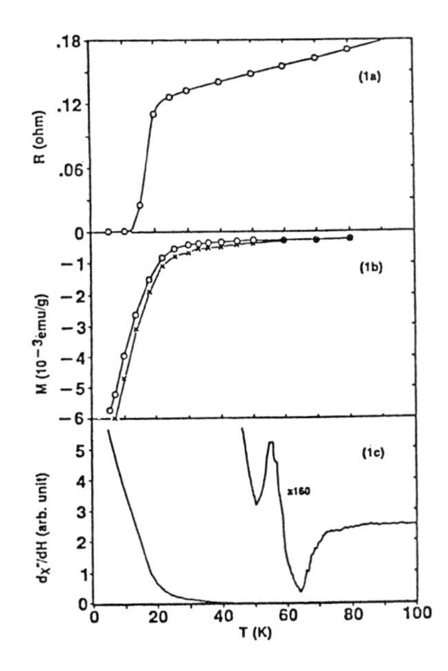

Figure 12
(a) Resistance versus temperature for sample #8165 (with a nominal composition of $Tl_2Sr_2Ca_2Cu_3O_{10+x}$). Zero resistance, which corresponds to resistivity $<10^{-8}$ ohm-cm, is reached at 10 K. The linear region extends up to 300 K.
(b) Zero field cooled (cross points) and field cooled (circle points) dc susceptibility versus temperature for an applied dc magnetic field of 100 G for sample #8165.
(c) Microwave spectrometer output (the field derivative of the absorption signal) versus temperature for an applied dc magnetic field of 40 G for sample #8165.

Figs. 13a, b and c show resistance, magnetization, and microwave absorption data, respectively, versus temperature for sample #8107 (with a nominal composition of $Tl_2SrCa_2Cu_3O_{9+x}$). The temperature dependence of resistance (Fig. 13a) exhibits a metallic-like behavior down to 4.2 K, but with a step centered at 60 K (with a 20 K width). A change in slope near 75 K was also observed. The dc susceptibility shown in Fig. 13b indicates an onset of

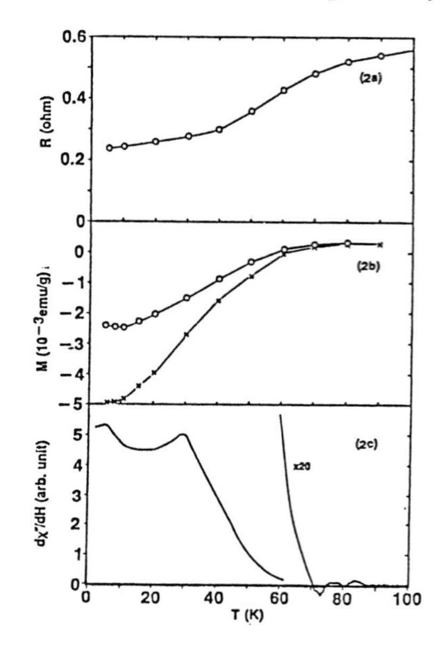

Figure 13
(a) Resistance versus temperature for sample #8107 (with a nominal composition of $Tl_2SrCa_2Cu_3O_{9+x}$). The linear region above about 80 K extends up to 300 K.
(b) Zero field cooled (cross points) and field cooled (circle points) dc susceptibility versus temperature for an applied dc magnetic field of 100 G for sample #8107.
(c) Microwave spectrometer output (the field derivative of the absorption signal) versus temperature for an applied dc magnetic field of 40 G for sample #8107.

diamagnetism at about 70 K. Below 10 K the fraction of the sample which participates in diamagnetic shielding is 0.3% with a Meissner fraction of about half of that. The microwave absorption data of Fig. 13c indicate an onset temperature (expended scale) of about 70 K. A additional sample #8177 (with a nominal composition of $Tl_2Sr_2Ca_2Cu_3O_{10+x}$) also had similar characteristic indications of the onset of superconductivity (data not shown): i.e., a step in the resistance at 77 K, an onset of diamagnetism at about 70 K, and an onset of the microwave signal at about 80 K.

We interpret the above data as indicating that there are at least two superconducting phases, one with onset temperature at about 70 K and the other about 20 K, in the Tl-Sr-Ca-Cu-O system. We believe that the superconducting phases are Tl-Sr-Ca-Cu-O, because (1) Tl-Sr-Cu-O samples are not superconducting down to 4.2 K, although under similar preparation conditions they have a metallic-like resistance-temperature behavior above liquid nitrogen temperature, and although the Bi-Sr-Cu-O system is superconducting at about 20 K; and (2) there is no contamination source which can cause the observed superconductivity. However, in all samples investigated down to 4.2 K, the Meissner and flux shielding signals are small, typically corresponding to less than 1% superconducting volume fraction. For this reason, we have not yet carried out chemical and structural identification of the superconducting phases, rather we are attempting to increase the portion of the superconducting phases, particularly that at about 70 K. Nevertheless, it may be surmised that if the Tl-Sr-Ca-Cu-O superconductors have structures similar to those of the corresponding Tl-Ba-Ca-Cu-O and Bi-Sr-Ca-Cu-O superconductors, their Cu-O sheets would be more buckled reflecting their lower Tc [12-14], or alternatively, they may simply have their own separate structures.

REFERENCES

1. J.G.Bednorz and K.A.Muller, Z. Phys. B 64, 189 (1986).
2. M.K.Wu, J.R.Ashburn, C.T.Torng, P.H.Hor, R.L.Meng, L.Gao, Z.J.Huang, Y.Q.Wang, and C.W.Chu, Phys. Rev. Lett. 58, 908 (1987).
3. Z.Z.Sheng and A.M.Hermann, Nature 332, 55 (1988).
4. Z.Z.Sheng, A.M.Hermann, A.El Ali, C.Almason, J.Estrada, T.Datta, and R.J.Matson, Phys. Rev. Lett. 60, 937 (1988).
5. H.Maeda, Y.Tanaka, M.Fukutomi, and T.Asano, Jpn. J. Appl. Phys. Lett. 27, L207 (1988).
6. C.W.Chu, J.Bechtold, L.Gao, P.H.Hor, Z.J.Huang, R.L.Meng, Y.Y.Sun, Y.Q.Wang, and Y.Y.Xue, Phys. Rev. Lett. 60, 941 (1988).
7. Z.Z.Sheng and A.M.Hermann, Nature 332, 138 (1988).
8. Z.Z.Sheng, W.Kiehl, J.Bennett, A.El Ali, D.Marsh, G.D.Mooney, F.Arammash, J.Smith, D.Viar, and A.M.Hermann, Appl.Phys.Lett. 52, 1738 (1988).
9. A.M.Hermann, Z.Z.Sheng, D.C.Vier, S.Schultz, and S.B.Oseroff, Phys.Rev.B 37, 9742 (1988).
10. R.M.Hazen, L.W.Finger, R.J.Angel, C.T.Prewitt, N.L.Ross, C.G.Hadidiacos, P.J.Heaney,

D.R.Veblen, Z.Z.Sheng, A.El Ali, and A.M.Hermann, Phys. Rev. Lett. 60, 1657 (1988).
11. L.Gao, Z.J.Huang, R.L.Meng, P.H.Hor, J.Bechtold, Y.Y.Sun, C.W.Chu, Z.Z.Sheng, and A.M.Hermann, Nature 332, 623 (1988).
12. C.C.Torardi, M.A.Subramanian, J.C.Calabrese, J.Gopalakrishnan, E,M.McCarron, K.J.Morrissey, T.R.Askew, R.B.Flippen, U.Chowdhry, and A.M.Sleight, Phys. Rev. B. (in press, 1988).
13. M.A.Subramanian, J.C.Calabrese, C.C.Torardi, J.Gopalakrishnan, T.R.Askew, R.B.Flippen, K.J.Morrissey, U.Chowdhry, and A.M.Sleight, Nature 332, 420 (1988).
14. C.C.Torardi, M.A.Subramanian, J.C.Calabrese, J.Gopalakrishnan, K.J.Morrissey, T.R.Askew, R.B.Flippen, U.Chowdhry, and A.M.Sleight, Science 240, 631 (1988).
15. S.S.P.Parkin, V.Y.Lee, E.M.Engler, A.I.Nazzal, T.C.Huang, G.Gorman, R.Savoy, and R.Beyers, Phys.Rev.Lett. 60, 2539 (1988).
16. P.Grant (unpublished).
17. Y.Luo, Y.L.Zhang, J.K.Liang, and K.K.Fung (submitted, 1988).
18. R.Beyers, S.S.P.Parkin, V.Y.Lee, A.I.Nazzal, R.Savoy, G.Gorman, T.C.Huang, and S.La Placa (submitted, 1988).
19. N.Mitra, J.Trefny, B.Yarar, G.Pine, Z.Z.Sheng, and A.M.Hermann, Phys.Rev.B. (to be published, 1988).
20. N.Mitra, J.Trefny, M.Young, and B.Yarar, Phys. Rev. B 36, 5581 (1987).
21. See Superconductor Week 2, No.18 (May 9, 1988), p.1.
22. E.Hellman, E.M.Gyorgy, D.W.Johnson,Jr., H.M.O´Bryan, and R.C.Sherwood, J.Appl.Phys. 63, 447 (1988).
23. W.G.Harter, A.M.Hermann, and Z.Z.Sheng, Appl.Phys.lett. (te be published, 1988).
24. P.N.Peters, R.C.Sisk, E.W.Urban, C.Y.Huang, and M.K.Wu, Appl.Phys.lett. 52, 2066 (1988).
25. S.Schultz et al. (unpublished)
26. W.Eideloth, F.S.Barnes, Z.Z.Sheng, and A.M.Hermann (submitted, 1988).
27. W.Eideloth, F.S.Barnes, S.Geller, L.C.Xu, and E.A.Kraut (submitted, 1988).
28. A.Barone, G.Paterno, Physica and Applications of the Josephson Effect, (Wiley, New York, 1982).
29. Z.Z.Sheng, A.M.Hermann, D.C.Vier, S.S.Schultz, and S.B.Oseroff, D.J.George, and R.M.Hazen Phys.Rev. B (to be published, 1988).

Synthesis and Structure-Property Relationships of Tl and Bi Containing Copper-Oxide Superconductors

M.A. SUBRAMANIAN, J. GOPALAKRISHNAN, C.C. TORARDI, J.C. CALABRESE, K.J. MORRISSEY, J. PARISE, P.L. GAI, and A.W. SLEIGHT

Central Research and Development Department, E.I. du Pont de Nemours and Company, Experimental Station, Wilmington, Delaware 19898, USA

ABSTRACT

Synthesis, structure and superconducting properties of two series of oxides, $T\ell_2Ba_2Ca_{n-1}Cu_nO_{2n+4}$ and $T\ell Ba_2Ca_{n-1}Cu_nO_{2n+3}$, are described. The structures consist of perovskite-like $Ba_2Ca_{n-1}Cu_nO_{2n+2}$ units which alternate with either bilayer ($T\ell_2O_2$) or monolayer ($T\ell O$) units. A new superconductor ($T_c \sim 120K$) with approximate formula $(T\ell,Pb)Sr_2Ca_2Cu_3O_x$ has been characterized for the first time. Compositions in the $Bi_2Sr_{3-x}Y_xCu_2O_8$ system which traverse the insulator-superconductor boundary have been investigated.

INTRODUCTION

The discoveries of superconductivity in Bi-Sr-Ca-Cu-O and Tℓ-Ba-Ca-Cu-O systems early this year (1,2) have not only provided compounds with the highest transition temperatures ($T_c \sim 125$ K) but also revealed a structural pattern that is common to all the superconducting copper oxides. Based on structural investigations of these systems (3-11), two homologous series of superconducting copper oxides have been identified. The first series, which is common to both bismuth- and thallium-containing oxides, is $A_2^{III}B_2^{II}Ca_{n-1}Cu_nO_{2n+4}$ where double Tℓ-O (Bi-O) layers separate perovskite-like $B_2Ca_{n-1}Cu_nO_{2n+2}$ slabs (B = Ba when A = Tℓ and B = Sr when A = Bi). The second series, $T\ell Ba_2Ca_{n-1}Cu_nO_{2n+3}$, consists of monolayers of Tℓ-O separating $Ba_2Ca_{n-1}Cu_nO_{2n+2}$ slabs. In both the series, the general trend appears to be that the T_c increases with increasing number of consecutively stacked CuO_2 layers (n), the highest T_c (125 K) being exhibited by $T\ell_2Ba_2Ca_2Cu_3O_{10}$ (4).

Although the gross structural features of these compounds have been revealed by a combination of x-ray and neutron diffraction methods, the structure-property relationships are not completely understood at present. Thus, mechanism giving rise to Cu(III) (or holes), role of Tℓ-O (Bi-O) layers in producing high T_c, and the relation between magnetism and high T_c in these cuprates are some of the issues that are not fully resolved. In this paper, we present some of the recent results of our investigations on these systems. First, we provide a summary of our work on the thallium- and bismuth-containing systems. This is followed by details of synthesis and structures of new Tℓ and (Tℓ,Pb) containing cuprates exhibiting T_c's upto 122K. The last section presents an investigation of $Bi_2Sr_{3-x}Y_xCu_2O_8$ where the compositions traverse an antiferromagnetic insulator-superconductor boundary as a function of composition.

RESULTS AND DISCUSSION

Tℓ-Containing Copper Oxide Superconductors

Tℓ-Ba-Ca-Cu-O system

Powder samples in the Tℓ-Ba-Ca-Cu-O (4,5,7) systems were prepared by heating at 850-925°C for 1 to 12 hrs stoichiometric mixtures of high purity oxides in sealed gold tubes. Crystals were grown from off-stoichiometric (copper-rich) mixtures. Superconducting transition temperatures were determined by flux exclusion and four-probe electrical resistivity measurements. Crystals were examined by electron microscopy as well as x-ray diffraction using both precession photographs and axial oscillation photographs. Single crystal x-ray diffraction data were obtained as previously described for $T\ell_2Ba_2CuO_6$(7), $T\ell_2Ba_2CaCu_2O_8$ (5), $T\ell_2Ba_2Ca_2Cu_3O_{10}$(4)

TABLE I Crystallographic information and superconducting transition temperatures of Tℓ-containing copper oxide superconductors.

Compound[a]	a(Å)	b(Å)	c(Å)	Space group	T_c (K)
$Tℓ_2Ba_2CuO_6$	3.866(1)		23.239(6)	I4/mmm	84-90
	5.4967(3)	5.4651(3)	23.246(1)[a]	Abma	
$Tℓ_2Ba_2CaCu_2O_8$	3.8550(6)		29.318(4)	I4/mmm	98-110
	3.8559(1)		29.420(1)[a]		
$Tℓ_2Ba_2Ca_2Cu_3O_{10}$	3.8503(6)		35.88(3)	I4/mmm	105-125
	3.8487(1)		35.662(2)[b]		
$TℓBa_2CaCu_2O_7$	3.85		12.73	P4/mmm	70-80
$TℓBa_2Ca_2Cu_3O_9$	3.853(1)	-	15.913(4)	P4/mmm	110
$(Tℓ,Pb)Sr_2CaCu_2O_7$	3.80(1)		15.10(5)	P4/mm	80-90
$(Tℓ,Pb)Sr_2Ca_2Cu_3O_9$	3.808	-	15.232	P4/mmm	115-122

[a]From neutron powder diffraction data, all others from single crystal x-ray data.

and $TℓBa_2Ca_2Cu_3O_9$ (11). Unit cell parameters for these phases are given in Table I along with their superconducting transition temperatures. For a same value of n, $TℓBa_2Ca_{n-1}Cu_nO_{2n+3}$ compounds have lower T_c's than $Tℓ_2Ba_2Ca_{n-1}Cu_nO_{2n+4}$ compounds. However, within each series T_c increases with increasing n.

In contrast to the Bi containing superconductors, crystals of $Tℓ_2Ba_2Ca_{n-1}Cu_nO_{2n+4}$ (n = 1, 2, and 3) and $TℓBa_2Ca_{n-1}Cu_nO_{2n+3}$ (n = 3) are plate-shaped but not micaceous. No obvious superstructure is observed and no twinning occurs because of the tetragonal symmetry. Lattice imaging of the n = 3 phase in the $Tℓ_2Ba_2Ca_{n-1}Cu_nO_{2n+4}$ series clearly shows the Tℓ-O double layers and the Cu-O triple layers (4). However, a prominent defect in some of the crystals is the presence of five consecutive Cu-O layers. Electron diffraction from regions of intergrowth showsa c-axis spacing of 48 Å, as expected for a(n = 5)phase.

Crystal structures of the $Tℓ_2Ba_2Ca_{n-1}Cu_nO_{2n+4}$ phases with n = 1, 2, and 3 are shown in Figure 1. The structures differ from one another by the number of consecutive Cu-O sheets. For example in $Tℓ_2Ba_2Ca_2Cu_3O_{10}$ (4), triple sheets of corner-sharing square-planar CuO_4 groups are oriented parallel to the (001) plane. Additional oxygen atoms are located above and below the triple Cu-O sheets. There are no oxygen atoms between the these Cu-O sheets. A small amount of thallium (~ 12%) substitutes for the calcium found between the Cu-O layers of the n = 2 and 3 phases. It is not yet clear if the thallium sites are partially vacant or if there is some calcium substitution (~ 15%) for Tℓ. Barium ions are found above and below the Cu-O single, double, or triple sheets in nine-coordination with oxygen. These slabs alternately stack with a double thallium-oxygen layer along the c axis. Thallium bonds to six oxygen atoms in a distorted octahedral arrangement where the octahedra share edges within and between each sheet of the double layer. A subtle ordering is observed in the Tℓ-O sheets where the atoms are displaced from thier ideal positions to create a more favorable bonding environment (4, 5, 7). Recent neutron scattering studies (12) on $Tℓ_2Ba_2CaCu_2O_8$ showed strongly correlated local displacements of both the Tℓ and oxygen atom in the Tℓ-O plane from their high symmetry crystallographic sites to form Tℓ-O chains. This indicates that local symmetry is no longer tetragonal.

Neutron powder diffraction data for $Tℓ_2Ba_2Ca_{n-1}Cu_nO_{2n+4}$ (n = 1, 2 and 3) phases were obtained at Brookhaven National Laboratory. The results (6,13) are essentially in agreement with the structural refinements based on single crystal x-ray diffraction data and confirm the positional disorder of the oxygen atoms in the Tℓ-O planes and the absence of oxygen between consecutive copper-oxygen sheets. The low temperature refinements for $Tℓ_2Ba_2Ca_2Cu_3O_{10}$ indicate that the symmetry remains tetragonal down to 13 K and there is no significant structural change or discontinuity in the cell parameters through the superconducting critical temperature near 125 K. In contrast to this, $Tℓ_2Ba_2CuO_6$ is orthorhombic from 298 to 12 K. Presumably, this polycrystalline sample possess a slightly different composition from the crystal used for our single crystal x-ray studies.

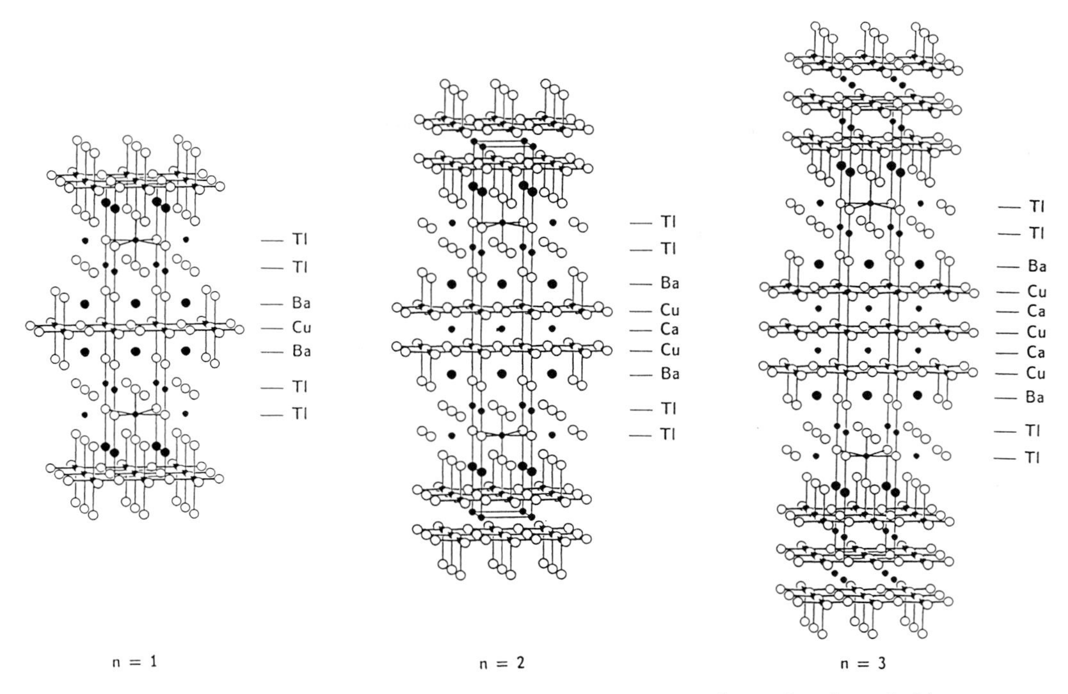

Figure 1 Structures of $T\ell_2Ba_2Ca_{n-1}Cu_nO_{2n+4}$ (n = 1, 2 and 3). Metal atoms are shaded and Cu-O bonds are shown.

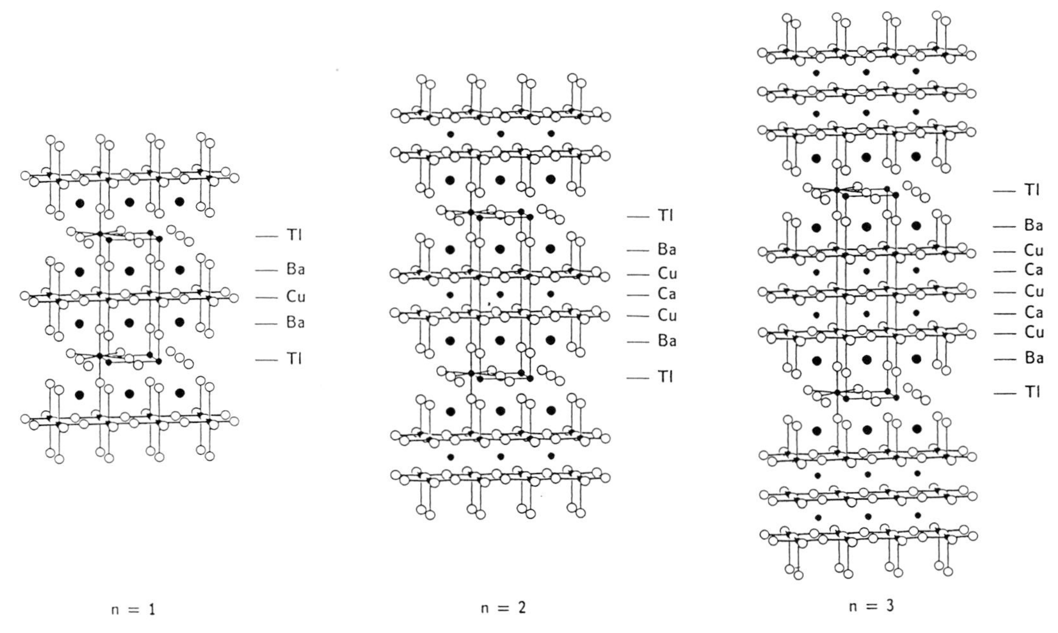

Figure 2 Structures of $T\ell Ba_2Ca_{n-1}Cu_nO_{2n+3}$ (n = 1, 2, and 3). Metal atoms are shaded and Cu-O bonds are shown.

Crystal structures of $TlBa_2Ca_{n-1}Cu_nO_{2n+3}$ phases with n = 1, 2, and 3 have been proposed on the basis of powder x-ray diffraction data (9). The perovskite-like $Ba_2Ca_{n-1}Cu_nO_{2n+2}$ slabs in $TlBa_2Ca_{n-1}Cu_nO_{2n+3}$ alternate with a single Tl-O layer whereas in $Tl_2Ba_2Ca_{n-1}Cu_nO_{2n+4}$, these slabs alternate with double Tl-O sheets (Figure 2). We have grown single crystals of $TlBa_2Ca_2Cu_3O_9$ (n = 3) and refined the structure from single crystal x-ray diffraction data (11). The structure contains triple copper-oxygen sheets separated by Ca ions which show ~5% Tl substitution. These units alternate with single Tl-O layers. Similar to $Tl_2Ba_2Ca_{n-1}Cu_nO_{2n+4}$, the existence of correlated atomic displacements within the thallium-oxygen sheets is suggested by a disordered Tl-O arrangement in the average structure. The Cu-O sheet units in $TlBa_2Ca_{n-1}Cu_nO_{2n+3}$ are stacked directly above each other, whereas in $Tl_2Ba_2Ca_{n-1}Cu_nO_{2n+4}$, which contains double Tl-O sheets, alternate units are shifted by $(a_1+a_2)/2$ which causes the approximate doubling of the c lattice dimension for the latter compounds (Table I).

Tl-Pb-Sr-Ca-Cu-O Superconductors

Recently Sheng *et al.* (14) reported superconductivity at 20 K (some with onset at 70 K) in the Tl-Sr-Ca-Cu-O system with a very weak Meissner effect (<1%). Our attempts to synthesize $Tl_2Sr_2Ca_{n-1}Cu_nO_{2n+4}$ (n = 1, 2 and 3) and $TlSr_2Ca_{n-1}Cu_nO_{2n+3}$ (n = 2 and 3) were not successful. In addition to the above compositions, we also tried numerous combinations by varing Tl:Sr:Ca:Cu ratios. A few samples showed weak Meissner effect around 20K and we did not observe superconductivity above 20 K in any of the samples. However, we did observe bulk superconductivity up to 122 K in the system Tl-Pb-Sr-Ca-Cu-O. We have identified at least two superconducting phases by X-ray and electron diffraction studies. We have also grown single crystals of one of the phases whose composition can approximated as $(Tl,Pb)Sr_2Ca_2Cu_3O_x$ and determined its crystal structure.

Compositions in the Tl-Pb-Sr-Ca-Cu-O system were prepared by reacting Tl_2O_3, CaO_2, SrO_2 PbO_2, and CuO in various proportions at 850-915°C in sealed gold tubes. Powder x-ray diffraction revealed that most of the products were mixtures containing mainly phases with characteristic reflections at 12.1 Å and 15.2 Å. Flux exclusion measurements showed superconductivity in all our samples prepared in the system Tl-Pb-Sr-Ca-Cu-O. Meissner fractions were large (>30%) and transition temperatures were 80 to 90 K in some samples and 115 to 122 K in others. However, when we excluded one of the metal components, no superconductivity was observed. By combining x-ray and Meissner data we conclude that the 12.1 Å phase has a T_c of 80 to 90K whereas the phase with 15.2 Å has a T_c around 120K.

Powder x-ray diffraction data of both the 12.1 Å and 15.2 Å phases could be indexed on the basis of primitive tetragonal unit cells with a = 3.80, c = 12.1 Å and a = 3.80Å, c = 15.1 Å respectively. This suggests that these phases are probably similar to $TlBa_2CaCu_2O_7$ and $TlBa_2Ca_2Cu_3O_9$. Partial substitution of Pb for Tl (or Ca) in the lattice appears essential to stabilize $(Tl,Pb)Sr_2Ca_{n-1}Cu_nO_x$ phases. Single crystals of the 15.2Å phase were grown from an oxide mixture in the molar ratio 1:1:2:3:4 (Tl:Pb:Sr:Ca:Cu) in a sealed gold tube. The mixture was heated to 910°C for 6 hours and cooled at 2°C/min. The structure of this compound determined from single crystal x-ray data is essentially the same as that of $TlBa_2Ca_2Cu_3O_9$ (Figure 2, n = 3) consisting of single layers of (Tl,Pb,O) which separate units of three CuO_2 layers in the c direction.

Superconductor-To-Insulator Transition in $Bi_2Sr_{3-x}Y_xCu_2O_{8+y}$

The compositions $Bi_2Sr_{3-x}Y_xCu_2O_8$ have been prepared with the structure of superconducting $Bi_2Sr_{3-x}Ca_xCu_2O_8$. The range of x in the $Bi_2Sr_{3-x}Y_xCu_2O_8$ system was determined from x-ray diffraction data. The a, b and c lattice parameters varied smoothly from x = ~0.2 to 1.0 (Fig. 3a and b). Outside this range, impurity phases were detected in addition to the n = 2 phase. The c axis decreases with increasing x as might be expected since Y^{3+} is smaller than Sr^{2+}. However, there is an increase in the a axis with increasing x which can be attributed to the decrease in copper oxidation state which lead to longer Cu-O distances within the copper oxygen sheets. Compositions are superconducting for x 0.2 to 0.4. ($T_c \sim 65 - 72$ K). Chemical analysis showed a substantial Cu^{III} concentration for superconducting compositions such as $Bi_2Sr_{1.7}Cu_{0.3}Cu_2O_8$. However, the Cu^{III} concentration decreases sharply with increasing concentration of Y (Fig. 4). The decrease in Cu^{III} concentration also results in the loss of superconductivity and a changeover to insulating behavior.

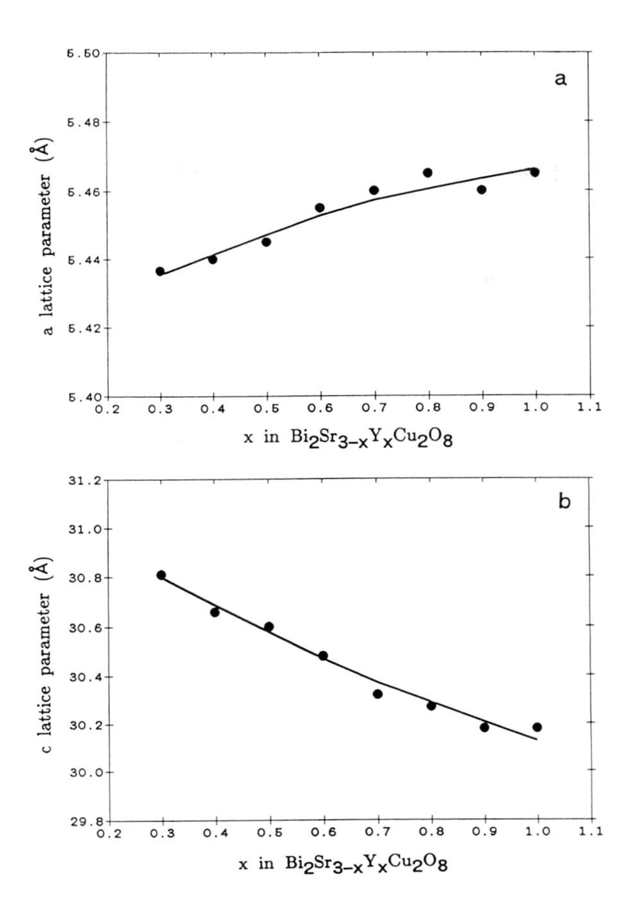

Figure 3 a) Variation of 'a' (average a and b) lattice parameter as a function of x for $Bi_2Sr_{3-x}Y_xCu_2O_{8+y}$.

b) Variation of c lattice parameter as a function of x for $Bi_2Sr_{3-x}Y_xCu_2O_{8+y}$.

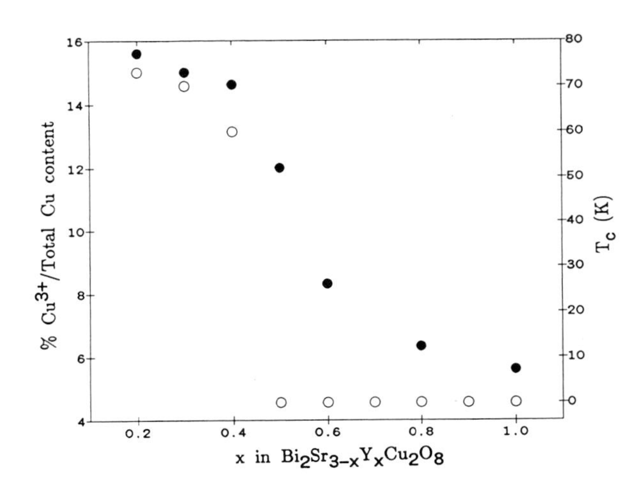

Figure 4 Variation of Cu^{3+} content (filled circles) and T_c (open circles) as a function of x for $Bi_2Sr_{3-x}Y_xCu_2O_{8+y}$.

$Bi_2Sr_{3-x}Ca_xCu_2O_8$ phases are known over a range of x that does not include x =0.0 (3). The nonexistence of $Bi_2Sr_3Cu_2O_8$ is because the cation site between the adjacent CuO_2 layers requires a relatively small ion such as Ca^{2+}. Although some substitution of Ca^{2+} by Sr^{2+} occurs, complete substitution is unknown in any of these layered copper oxides including n = 'infinity' member, $(Sr,Ca)CuO_2$ (15). However, Y^{3+} is smaller than Sr^{2+} and thus Y^{3+} can serve to decrease the average size of the cation between the adjacent CuO_2 layers.

An understanding of the $Bi_2Sr_{3-x}Y_xCu_2O_8$ systems is complicated by our lack of understanding of the $Bi_2Sr_{3-x}Ca_xCu_2O_8$ system where there are interrelated unresolved compositional and structural issues. Written as $Bi_2Sr_{3-x}Ca_xCu_2O_8$, there would be no Cu^{III}. This would seem inconsistent with the observed superconductivity; furthermore, chemical analysis shows a significant Cu^{III} content. The defect giving Cu^{III} might be oxygen interstials, but current evidence suggests that Cu^{III} is probably present even without such interstials. Structural refinements show that there is no significant oxygen content between the adjacent CuO_2 layers. There is evidence for oxygen between the adjacent Bi-O layers in certain preparations (16). However, this interstitial oxygen causes a decrease in T_c and presumably has resulted in oxidation of Bi^{III} to Bi^{V} rather than of Cu^{II} to Cu^{III}. Another defect mechanism for producing Cu^{III} would be cation vacancies on the Sr^{2+} site. Some microprobe data appear to support the possibility (17), but recent crystallographic results (18) indicate that a deficiency of Bi^{3+} is more likely than a deficiency of A^{2+} cations. The suggestion that Bi^{3+} might substitute for Ca^{2+} between adjacent CuO_2 layers adds a further complication since this would push the average oxidation state of copper below two.

Given the evidence for Cu^{III}, the $Bi_2Sr_{3-x}Ca_xCu_2O_8$ and $Bi_2Sr_{3-x}Y_xCu_2O_8$ formulations must be regarded as idealized. Whatever be the cause for Cu^{III}, the substitution of Y^{3+} for A^{2+} should result in a decreased average oxidation state for copper. In fact a decreased Cu^{III} content is observed both for the $Bi_2Sr_{3-x}Y_xCu_2O_8$ and $Bi_2(Sr,Ca)_{3-x}Y_xCu_2O_8$ (19) systems. For the $Bi_2Sr_2YCu_2O_8$ end member one might expect some Cu^{I} content. We have XANES experiments underway on $Bi_2Sr_2YCu_2O_8$ to search for Cu^{I}.

Another suggestion based on band structure calculations (20) is that the Cu^{III} content in $Bi_2Sr_2CaCu_2O_8$ results from an overlap of the Bi 6p band with the Cu $d_{x^2-y^2}$ - O $2p\sigma$ band at the Fermi level. This proposal makes little sense on chemical grounds since it would mean oxidation of Cu^{II} by Bi^{III}. The band structure calculations are misleading because they are based on an idealized structure which has Bi-O distances too different from the real structure. In an oxidized system, the Bi 6p band is expected to be well above the Fermi level. The insulating regions in the $Bi_2(Sr,Ca)_{3-x}Y_xCu_2O_8$ systems is further proof that the Bi 6p band does not overlap the Fermi level.

The metal-insulator boundary has now been found in the systems $La_{2-x}A_xCuO_4$, $YBa_2Cu_3O_{6+x}$ and $Bi_2Sr_{3-x}Y_xCu_2O_8$. In the former two systems, it has been shown that the insulating state is associated with antiferromagnetism (21,22). Recent muon spin resonance studies have shown that $Bi_2Sr_2YCu_2O_8$ is also antiferromagnetic with a Neel temperature of about 210K (23). Thus in these three systems long range magnetic order must be destroyed through Cu^{III} doping before superconductivity can arise. We find insulator-superconductor boundary exists in the $Bi_{2-x}Pb_xSr_{3-x}Y_xCu_2O_8$ and $Bi_{2-x}Cd_xSr_{3-x}Y_xCu_2O_8$ systems (24). Insulating analogues in the Tℓ/Ba/Ca/Cu/O system have not yet been found. However, it is likely in these systems that the Tℓ 6s band overlaps the Fermi level in which case the insulating region may elude us.

REFERENCES

1. H. Maeda, Y. Tanaka, M. Fukutomi, and T. Asano, Jap. J. Appl. Phys., 27, L209 (1988).

2. Z. Z. Sheng and A. M. Hermann, Nature (London), 332, 138 (1988).

3. M. A. Subramanian, C. C. Torardi, J. C. Calabrese, J. Gopalakrishnan, K. J. Morrissey, T. R. Askew, R. B. Flippen, U. Chowdhry and A. W. Sleight, Science, 239, 1015 (1988).

4. C. C. Torardi, M. A. Subramanian, J. C. Calabrese, J. Gopalakrishnan, K. J. Morrissey, T. R. Askew, R. B. Flippen, U. Chowdhry and A. W. Sleight, Science 240, 631 (1988).

5. M. A. Subramanian, J. C. Calabrese, C. C. Torardi, J. Gopalakrishnan, T. R. Askew, R. B. Flippen, K. J. Morrissey, U. Chowdhry and A. W. Sleight, Nature, 332, 420 (1988).

6. D. E. Cox, C. C. Torardi, M. A. Subramanian, J. Gopalakrishnan and A. W. Sleight, Phys. Rev. B., in press.

7. C. C. Torardi, M. A. Subramanian, J. C. Calabrese, J. Gopalakrishnan, E. M. McCarron, K. J. Morrissey, T. R. Askew, R. B. Flippen, U. Chowdhry and A. W. Sleight. Phys. Rev. B 38, 225 (1988).

8. S. S. P. Parkin, V. Y. Lee, E. M. Englar, A. I. Nazzal, T. C. Huang, G. Gorman, R. Savoy and R. Beyers Phys. Rev. Lett., 61, 750 (1988).

9. S. S. P. Parkin, V. Y. Lee, A. I. Nazzal, R. Savoy, R. Beyers and S. J. La Placa, Phys. Rev. Lett. 61, 750 (1988).

10. M. Hervieu, A. Maignan, C. Martin, C. Michel, J. Porvost and B. Raveau, J. Solid State Chem. 75, 212 (1988).

11. M. A. Subramanian, J. B. Parise, J. C. Calabrese, C. C. Torardi, J. Gopalakrishnan and A. W. Sleight, J. Solid State Chem., in press.

12. W. Dmowski, B. H. Toby, T. Egami, M. A. Subramanian, J. Gopalakrishnan and A. W. Sleight, Phys. Rev. Lett., submitted.

13. J. B. Parise, J. Gopalakrishnan, M. A. Subramanian and A. W. Sleight, J. Solid State Chem., in press.

14. Z. Z. Sheng, A. M. Hermann, D. C. Vier, S. Schultz, S. B. Oseroff, D. J. George and R. M. Hazen, preprint.

15. T. Siegrist, S. M. Zahurak, D. W. Murphy and R. S. Roth, Nature, 334, 231 (1988).

16. P. Coppens *et al.*, submitted.

17. A. K. Cheetham, A. M. Chippendale, and S. J. Hibble, Nature, 333, 21 (1988).

18. Y. Gao, P. Lee, P. Coppens, M. A. Subramanian and A. W. Sleight, Science, in press.

19. R. Yoshizaki, Y. Saito, Y. Abe and H. Ikeda, Physica C 152, 408 (1988); A. Manthiram and J. B. Goodenough, Applied Phys. Lett. 55, 420 (1988).

20. M. S. Hybertsen and L. F. Mattheiss, Phys. Rev. Lett., 60, 1661 (1988); F. Herman, R. V. Kasowski, and W. Y. Hsu, Phys. Rev. B, 38, 204 (1988).

21. Y. J. Uemura, *et al.*, Phys. Rev. Lett. 59, 1045 (1987).

22. J. H. Brewer, *et al.*, Phys. Rev. Lett., 60, 1073 (1988).

23. Y. J. Uemura, *et al.*, J. Phys. (Paris), in press.

24. J. Gopalakrishnan *et al.*, to be published.

Structural Study of $Tl_2Ba_2Ca_{n-1}Cu_nO_{4+2n}$ (n=1, 2 and 3) by Powder X-Ray Diffraction

Y. SHIMAKAWA, Y. KUBO, T. MANAKO, Y. NAKABAYASHI, and H. IGARASHI

Fundamental Research Laboratories, NEC Corporation, 4-1-1, Miyazaki, Miyamae-ku, Kawasaki, 213 Japan

ABSTRACT

Crystal structures of $Tl_2Ba_2Ca_{n-1}Cu_nO_{4+2n}$ (n=1, 2 and 3) determined by powder X-ray diffraction were refined using Rietveld analysis. The results of the Rietveld analysis revealed that as n increased the lattice parameter, a, and the interatomic distance Cu(1)-O(2) became shorter, and the Tl-O and Ba-O layers showed a tendency to become flatter. Possible substitution of about 10% of the Ca-sites with Tl ions in the 2212 and 2223 phases was also suggested. In our experiments, some samples with similar diffraction patterns showed different T_c's. In particular, in the Tl-Ba-Cu-O system, a wide range of behaviors varying between non-superconductivity and a T_c of 80K was observed. These difference in T_c's may be related to substitution effects, modulated structures, and intergrowth structures.

INTRODUCTION

Since the recent discovery of two new systems of superconducting materials, Bi-Sr-Ca-Cu-O [1] and Tl-Ba-Ca-Cu-O [2], many studies have been made on the phases present in these systems and the correlation between their crystal structures and T_c's [3-12]. In the Tl-Ba-Ca-Cu-O system, six oxides with nominal compositions $TlBa_2CuO_5$ (1201 phase), $Tl_2Ba_2CuO_6$ (2201 phase), $TlBa_2CaCu_2O_7$ (1212 phase), $Tl_2Ba_2CaCu_2O_8$ (2212 phase),$TlBa_2Ca_2Cu_3O_9$ (1223 phase) and $Tl_2Ba_2Ca_2Cu_3O_{10}$ (2223 phase) have been reported thus far. The 1201, 1212 and 1223 phases are made up of respectively one, two, and three Cu perovskite-like units separated by Tl-O monolayers, whereas the 2201, 2212 and 2223 phases are made up of respectively one, two and three Cu perovskite-like units separated by Tl-O bilayers. In contrast, in the Bi-Sr-Ca-Cu-O system, Bi-O bilayer phases are isostructual with the phases present in the Tl-Ba-Ca-Cu-O system, but Bi-O monolayer phases have not been reported thus far.

These phases showed a tendency that as the number of Cu perovskite-like unit increased the superconducting transition temperatures (T_c's) became higher. However, the correlation between crystal structures and T_c's in detail is still uncertain. In the present work, we synthesized the three phases: $Tl_2Ba_2CuO_6$ (2201 phase), $Tl_2Ba_2CaCu_2O_8$ (2212 phase) and $Tl_2Ba_2Ca_2Cu_3O_{10}$ (2223 phase), and examined their crystal structures and the correlation between crystal structures and T_c's. Crystal structures determined by powder X-ray diffraction were refined using Rietveld analysis.

Moreover, different T_c's in the same phase were observed. In particular in the Tl-Ba-Cu-O system, behavior varying between non-superconductivity and a T_c of 80K have been reported [10-14], and have been ovserved in this laboratory. Similarly, T_c's of 8K [15] or 20K [16] have been reported in the Bi-Sr-Cu-O system. These difference in T_c's will be discussed later.

EXPERIMENTAL

Samples were prepared by solid state reaction. Mixtures of Tl_2O_3, BaO, CaO and CuO powders with appropriate nominal compositions were pressed into pellets and wrapped in Pt or Au foil, then heated at temperatures between 845°C and 920°C.

Samples used for the crystal structures refinement with the general formula $Tl_2Ba_2Ca_{n-1}Cu_nO_{4+2n}$ were prepared by mixing raw materials in the stoichiometric molar ratios Tl:Ba:Ca:Cu=2:2:0:1 (n=1), 2:2:1:2 (n=2) and 2:2:2:3 (n=3), then heating at 900°C for 5 min (2201 phase), or at 890°C for 10 hrs (2212 and 2223 phases). Samples for the experiments in the Tl-Ba-Cu-O system were prepared by mixing raw materials in the molar ratios Tl:Ba:Cu=2:2:1, 2:2:2, and 2:2:3, then heating at 845°C, 890°C or 910°C for 1hr.

Resistivity measurements were carried out using a DC four-probe method, and AC susceptibility measurements were performed using a self-inductance method at 1kHz with Hac=0.1, 1 and 10Oe.

X-ray diffraction data for these samples were collected using Cu Kα radiation with a curved graphite monochromator. The diffractometer stepped 0.02° in 2θ every 4sec over a 2θ range from 10° to 100°.The results were refined using Rietveld analysis computer program RIETAN [17].

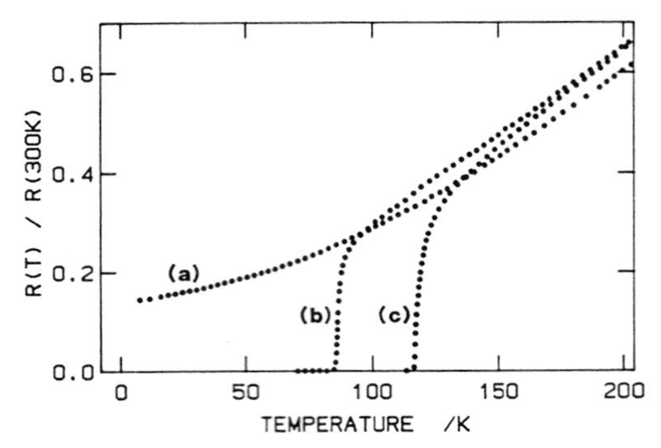

Fig.1 Temperature dependences of the DC resistivity for $Tl_2Ba_2Ca_{n-1}Cu_nO_{4+2n}$. (a) n=1; $Tl_2Ba_2CuO_6$, (b) n=2; $Tl_2Ba_2CaCu_2O_8$, (c) n=3; $Tl_2Ba_2Ca_2Cu_3O_{10}$.

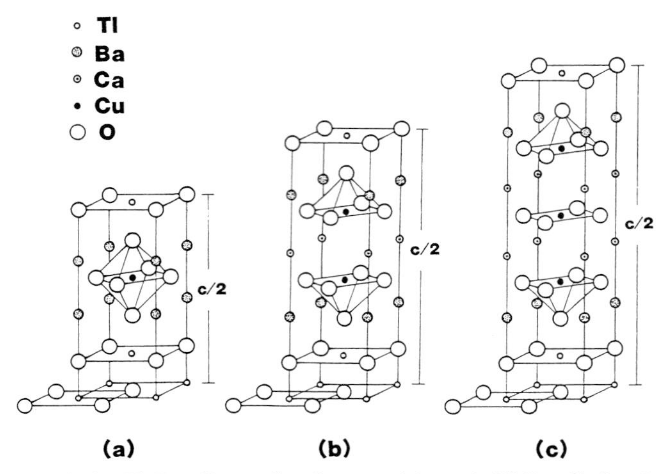

Fig.2 The structure models for $Tl_2Ba_2Ca_{n-1}Cu_nO_{4+2n}$. (a) n=1; $Tl_2Ba_2CuO_6$, (b) n=2; $Tl_2Ba_2CaCu_2O_8$, (c) n=3; $Tl_2Ba_2Ca_2Cu_3O_{10}$.

RESULTS AND DISCUSSION

1.CRYSTAL STRUCTURES REFINEMENT OF $Tl_2Ba_2Ca_{n-1}Cu_nO_{4+2n}$ (n=1, 2 AND 3)

The results of the resistivity measurements are shown in Fig.1. The 2223 phase had a T_c (R=0) of 116K and the 2212 phase had a T_c (R=0) of 85K. The 2201 phase was a metallic conductor, and did not exhibit superconductivity down to the liquid helium temperature.

To refine these crystal structures, the structure models were assumed to be those shown in Fig.2 on the basis of the results of TEM observations [18-20]. Thermal parameters (B) were assumed to be isotropic. Because these phases include heavy atoms like Tl and Ba, it is hard to refine the occupation factors (g) for the oxygen ions. Thus, only the occupation factors of the metal ions were considered, and the occupation factors of oxygen ions were taken to be 1.0. Final crystal structure parameters and R-factors for each samples are listed in Tables 1, 2 and 3. The observed and calculated diffraction patterns for each sample are shown in Fig.3.

From the calculated occupation factors for the metal ions, it should be noted that the occupation factors of Ca ions in the 2212 and 2223 phases are considerably larger than 1.0, which suggests that some Ca-sites are occupied by other metal ions in those phases. Judging from the ionic radii, Tl(III) ions are likely candidates for substitution in Ca(II) ion sites. Moreover occupation of Tl-sites by Ca ions is also suggested by the observation that the occupation factors of the Tl ions in the 2212 and 2223 phases are smaller than 1.0, and are lower than those of other metal ions. The results of the refinement allowing for substitution between Tl and Ca ions in the 2212 and 2223 phases are as follows. In the 2212 phase, 12% of the Ca-sites were occupied substitutionally by Tl ions and 3% of the Tl-sites by Ca ions. The composition formula is thus $Tl_{2.06}Ba_2Ca_{0.94}Cu_2O_x$. In the 2223 phase, 8% of the Ca-sites were occupied substitutionally by Tl ions and 3% of the Tl-sites by Ca ions. The composition formula is thus $Tl_{2.10}Ba_2Ca_{1.94}Cu_3O_x$. Recent study of the 2212 phase by neutron diffraction reported occupation of about 9% of the Ca-sites by Tl ions [21], which agrees well with the present results. On the other hand, in the 2201 phase, the occupation factors of all metal ions are close to 1.0. Thus no substitution seems to have occurred in this phase.

Table 1 Crystallographic data for $Tl_2Ba_2CuO_6$ (2201 phase). Space group I4/mmm; a = 3.85944(6), c = 23.1272(4)Å. Numbers in parentheses are uncertainties in the last decimal place.

Atom	Site	x	y	z	B/Å^2	g
Tl	4e	0	0	0.2031(2)	1.18(11)	0.995(6)
Ba	4e	0.5	0.5	0.0841(2)	0.59(5)	1.009(9)
Cu	2a	0	0	0	0.29(3)	0.994(2)
O(1)	4e	0.5	0.5	0.212(2)	6.9(18)	1
O(2)	4e	0	0	0.114(2)	3.4(9)	1
O(3)	4c	0	0.5	0	1.7(4)	1

$R_{wp} = 8.35$ $R_p = 6.19$ $R_e = 5.74$ $R_I = 3.25$ $R_F = 3.01$

Table 2 Crystallographic data for $Tl_2Ba_2CaCu_2O_8$ (2212 phase). Space group I4/mmm; a = 3.85032(7), c = 29.2997(6)Å.

Atom	Site	x	y	z	B/Å^2	g
Tl	4e	0.5	0.5	0.2139(2)	0.91(17)	0.96(1)
Ba	4e	0	0	0.1220(3)	0.45(8)	1.05(1)
Cu	4e	0.5	0.5	0.0545(7)	0.22(4)	1.05(3)
Ca	2a	0	0	0	0.45(8)	1.40(5)
O(1)	4e	0	0	0.216(4)	4.5(21)	1
O(2)	4e	0.5	0.5	0.143(3)	2.2(10)	1
O(3)	8g	0	0.5	0.052(2)	1.1(5)	1

$R_{wp} = 8.46$ $R_p = 6.14$ $R_e = 5.07$ $R_I = 5.25$ $R_F = 3.82$

Table 3 Crystallographic data for $Tl_2Ba_2Ca_2Cu_3O_{10}$ (2223 phase). Space group I4/mmm; a = 3.84775(8), c = 35.5859(7)Å

Atom	Site	x	y	z	B/Å^2	g
Tl	4e	0	0	0.2201(2)	2.15(21)	0.96(1)
Ba	4e	0.5	0.5	0.1452(2)	1.07(11)	1.04(1)
Cu(1)	4e	0	0	0.0903(5)	0.53(5)	1.03(3)
Ca	4e	0.5	0.5	0.0456(6)	1.07(11)	1.23(4)
Cu(2)	2a	0	0	0	0.26(3)	1.03(4)
O(1)	4e	0.5	0.5	0.220(2)	3.5(19)	1
O(2)	4e	0	0	0.160(2)	1.7(10)	1
O(3)	8g	0	0.5	0.088(1)	0.8(5)	1
O(4)	4c	0	0.5	0	0.8(5)	1

$R_{wp} = 8.93$ $R_p = 6.30$ $R_e = 4.41$ $R_I = 7.24$ $R_F = 5.22$

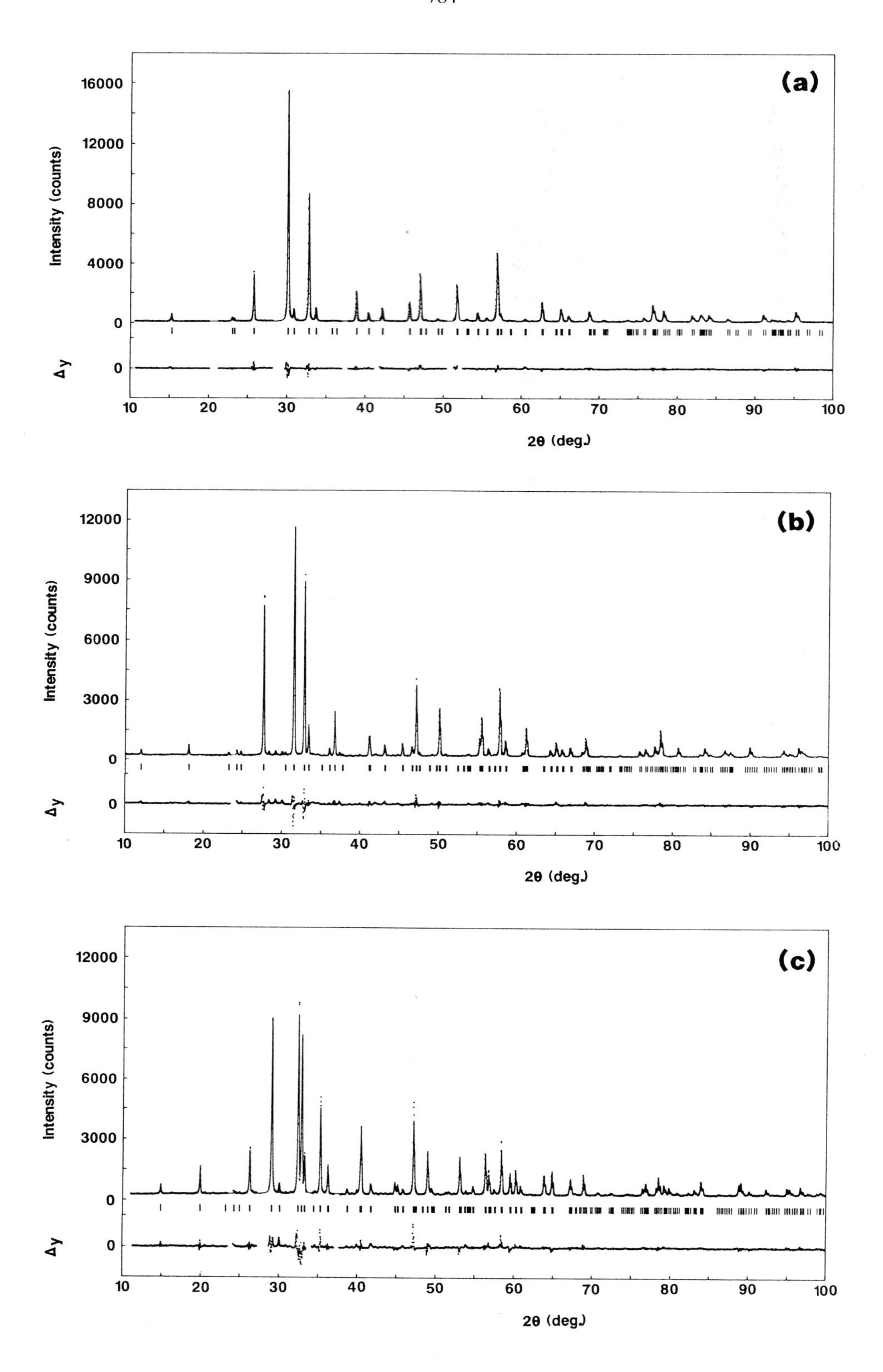

Fig.3 Rietveld refinement patterns for $Tl_2Ba_2Ca_{n-1}Cu_nO_{4+2n}$. The dots are observed intensities, and the solid lines calculated intensities. (a) n=1; $Tl_2Ba_2CuO_6$, (b) n=2; $Tl_2Ba_2CaCu_2O_8$, (c) n=3; $Tl_2Ba_2Ca_2Cu_3O_{10}$.

Table 4. Crystallographic information for $Tl_2Ba_2Ca_{n-1}Cu_nO_{4+2n}$ including lattice parameters, a and c, metal-sheets separations, Tl-Tl, Tl-Ba, Ba-Cu(1), Cu(1)-Ca, Ca-Cu(2), roughnesses of layers, Tl, Ba, Cu(1), Cu(2), and Cu(1)-O(2) interatomic distances (Å). The roughness of layer is defined as the difference of z-coordinates between the metal and oxygen ions of the layer.

n	a	c	Tl-Tl	Tl-Ba	Ba-Cu(1)	Cu(1)-Ca	Ca-Cu(2)	Tl	Ba	Cu(1)	Cu(2)	Cu(1)-O(2)
1	3.8594	23.127	2.169	2.752	1.944	—	—	0.21	0.69	0	—	2.64
2	3.8503	29.300	2.115	2.692	1.977	1.596	—	0.06	0.62	0.07	—	2.59
3	3.8478	35.586	2.128	2.665	1.953	1.590	1.622	~0	0.53	0.08	0	2.48

Crystallographic information obtained from the refinements are listed in Table 4. Some specific results of the Rietveld refinement should be noted. First, the separation between Tl ion sheets, about 2.1Å, is shorter than that between Bi ion sheets in the Bi-Sr-Ca-Cu-O system,about 3.2Å [3]. This short separation suggests the Tl-O layer makes up a simple rock-salt type layer. In contrast, the long separation in the Bi-O sheets suggests a more complicated structure of the Bi-O layer, for example, present in the $Bi_4Ti_3O_{12}$-type structure, where two Bi-O sheets are needed. This may be the reason that a Bi-O monolayer phase does not exist. Second, the Ba-O layer is fairly rough in comparison with other layers, and becomes flatter as n increases. This tendency is also observed in the roughness of the Tl-O layer, though the level of roughness is less. Third, as n increases, the lattice parameter, a, and Cu(1)-O(2) interatomic distance become shorter. However, the lattice parameter (3.82Å) and the Cu(1)-O(2) distance (2.2Å) in the Bi-Sr-Ca-Cu-O system [3-4] are still shorter than those in $Tl_2Ba_2Ca_2Cu_3O_{10}$. Thus the correlation between the interatomic distances and T_c's is still uncertain.

An incommensurate superstructure modulation was observed by electron microscopy [20], although this modulation was weaker than that observed in the Bi-Sr-Ca-Cu-O system [22]. In order to examine the structural distortion, the structural symmetry was assumed to be orthorhombic for these phases. But the refined lattice parameters, a and b, were equal within the uncertainty, so the structural distortion, if any, is considered to be very small.

2.VARIATION IN T_c

In our experiments, some samples with similar diffraction patterns and identical compositions had different T_c's. In addition to the samples discussed above, we found that, for example, the 2201 phase exhibited superconductivity, and the 2212 phase had a T_c of about 100K.

Particularly in the Tl-Ba-Cu-O system, a wide range of T_c values was observed. Samples with the starting composition Tl:Ba:Cu=2:2:1 showed a T_c of about 10K if heated at 890°C for 1 hr, and a T_c of about 20K if heated at 900°C for 1 hr. On the other hand, samples with Cu-rich starting compositions showed higher transition temperatures. Figure 4 shows the results of the DC resistivity and the AC susceptibility measurements of sample with the starting composition Tl:Ba:Cu=2:2:3 made by heating at 910°C. The $T_{c\ onset}$ was about 80K, and the large Meissner signal at low temperature suggested that the superconductivity had a bulk nature. However, the diamagnetic response increased gradually, and its dependence on the external magnetic field was very large, which suggested that the superconductivity was rather percolative, and the 80K phase regions were distributed in the sample and weakly connected. This sample had a diffraction pattern similar to that of the 2201 phase, although CuO peaks also appeared to be present. Thus the superconductivity at 80K should be attributed to the 2201 phase, and the difference in T_c's may be related to substitution effects, modulated structures, and intergrowth structures. The samples which exhibited relatively high T_c's were partially melted during processing; liquid phase reactions may have promoted these substitution, modulation, and intergrowth effects. The Cu-rich starting composition should play a role in lowering the melting-point. Detailed study of the correlation between the different Tc's and crystal structures of these materials is now in progress.

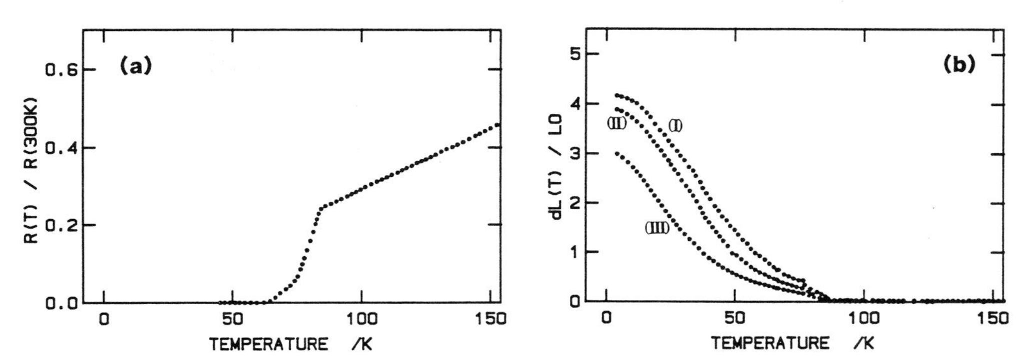

Fig.4 Temperature dependences of (a) the DC resistivity and (b) the AC susceptibility for the sample with starting composition Tl:Ba:Cu=2:2:3. The external AC magnetic field was (I) 0.1, (II) 1 or (III) 10 Oe.

CONCLUSION

We synthesized the following three phases in the Tl-Ba-Ca-Cu-O system: $Tl_2Ba_2Ca_2Cu_3O_{10}$ (2223 phase) with a T_c of 116K, $Tl_2Ba_2CaCu_2O_8$ (2212 phase) with a T_c of 85K, and $Tl_2Ba_2CuO_6$ (2201 phase) which did not exhibit superconductivity. The results of the Rietveld analysis revealed that as the number of Cu perovskite-like unit increased the lattice parameter, a, and the interatomic distance Cu(1)-O(2) became shorter, and the Tl-O and Ba-O layers showed a tendency to become flatter. Possible substitution of about 10% of the Ca-sites with Tl ions in the 2212 and 2223 phases was also suggested. Moreover in the 2201 phase, a wide range of behaviors varying between non-superconductivity and a T_c of 80K was observed.

ACKNOWREDGMENTS

We thank F. Izumi of NIRIM for giving us the Rietveld analysis program. We gratefully acknowledge useful discussions with S. Iijima. We also thank M. Yonezawa and K. Utsumi for their support of this work.

REFERENCES

[1] H. Maeda, Y. Tanaka, M. Fukutomi and T. Asano: Jpn.J. Appl. Phys. 27 (1988) L209.
[2] Z.Z. Sheng and A.M. Hermann: Nature 332 (1988) 138.
[3] M.A. Subramanian, C.C. Torardi, J.C. Calabrese, J. Gopalakrishnan, K.J. Morrissey, T.R. Askew, R.B. Flippen, U. Chowdhry and A.W. Sleight: Science 239 (1988) 1015.
[4] J.M. Tarascon, Y. Le Page, P. Barboux, B.G. Bagley, L.H. Greene, W.R. McKinnon, G.W. Hull, M. Giroud and D.M. Hwang: Phys.Rev. B37 (1988) 9382.
[5] M.A. Subramanian, J.C. Calabrese, C.C. Torardi, J. Gopalakrishnan, T.R. Askew, R.B. Flippen, K.J. Morrissey, U. Chowdhry and A.W. Sleight: Nature 332 (1988) 420.
[6] C.C. Torardi, M.A. Subramanian, J.C. Calabrese, J. Gopalakrishnan, K.J. Morrissey, T.R. Askew, R.B. Flippen, U. Chowdhry and A.W Sleight: Science 240 (1988) 631.
[7] Y. Kubo, Y.Shimakawa, T. Manako, T. Satoh and H. Igarashi: Jpn. J. Appl. Phys. 27 (1988) L591.
[8] B. Morosin, D.S. Ginley, P.F. Hlava, M.J. Carr, R.J. Baughman, J.E. Schirber, E.L. Venturini and J.F. Kwak: Physica C152 (1988) 413.
[9] S.S.P. Parkin, V.Y. Lee, A.I. Nazzal, R. Savoy, R. Beyers and S.J. La Placa: Phys. Rev. Lett. 61 (1988) 750.
[10] Y. Shimakawa, Y. Kubo, T. Manako, Y. Nakabayashi and H. Igarashi: Physica C156 (1988) 97.
[11] C.C. Torardi, M.A. Subramanian, J.C. Calabrese, J. Gopalakrishnan, E.M. McCarron, K.J. Morrissey, T.R. Askew, R.B. Flippen, U. Chowdhry and A.W. Sleight: Phys. Rev. B38 (1988) 225
[12] R. Bayers, S.S.P. Parkin, V.Y. Lee, A.I. Nazzal, R. Savoy, G. Gorman, T.C. Huang and S. La Placa: Appl.Phys.Lett. 53 (1988) 432.
[13] S. Kondoh, Y. Ando, M. Onoda, M. Sato and J.Akimitsu: Solid State Comm. 65 (1988) 1329.
[14] Z.Z. Sheng and A.W. Hermann: Nature 332 (1988) 55.
[15] J. Akimitsu, A. Yamazaki, H. Sawa and H. Fujiki: Jpn. J. Appl. Phys. 26 (1987) L2080.
[16] C. Michel, M. Hervieu, M.M. Borel, A. Grandin, F. Deslandes, J. Provost and B. Raveau: Z. Phys. B68 (1988) 421.
[17] F. Izumi: J. Crystallogr. Soc. Jpn. 27 (1985) 23 [in Japanses].
[18] S. Iijima, T. Ichihashi and Y. Kubo: Jpn. J. Appl. Phys. 27 (1988) L817.
[19] S. Iijima, T. Ichihashi, Y. Shimakawa, T. Manako and Y. Kubo: Jpn. J. Appl. Phys. 27 (1988) L837.
[20] S. Iijima, T. Ichihashi, Y. Shimakawa, T. Manako and Y. Kubo: Jpn. J. Appl. Phys. 27 (1988) L1061.
[21] A.M.Hewat, E.A. Hewat, J. Brynestad, H. Mook and E.D. Specht: Physica C152 (1988) 438.
[22] Y. Matsui, H. Maeda, Y. Tanaka and S. Horiuchi: Jpn. J. Appl.Phys. 27 (1988) L361.

Crystal Structure of Sr-Ca-Cu-O: A Comparison Between That of Sr-Ca-Cu-O and of Bi-Sr-Ca-Cu-O

ATSUHIRO KUNISHIGE[1]*, HIROSHI YOSHIKAWA[1], TOSHIHIKO ANNO[1], ITSUHIRO FUJII[1], HIROSHI DAIMON[1], SHIZUKA YOSHII[2]

[1] Ube Research Laboratory, Ube Industries, Ltd., 1978-5 Kogushi, Ube, 755 Japan
[2] Corporate Research & Development, Ube Industries, Ltd., 1-12-32 Akasaka, Minato-ku, Tokyo, 107 Japan

ABSTRACT

A new oxide-superconductor without rare-earth elements, Bi-Sr-Ca-Cu-O, has layered structure with alternative stacking of perovskite-like layers, consisting of Sr, Ca, Cu and O, and Bi_2O_2 layers. We are interested in the contribution of the Bi_2O_2 layer in this compound to the crystal structure that is closely related to superconductivity. In order to understand the role of the Bi_2O_2 layer in this type of superconductors, we synthesized Sr-Ca-Cu-O compounds and made a comparison between the structures of Sr-Ca-Cu-O and Bi-Sr-Ca-Cu-O.

The crystal structure of the samples was refined by the Rietveld analysis of their X-ray powder diffraction patterns. The crystal structure of Sr-Ca-Cu-O is much different from that of Bi-Sr-Ca-Cu-O. We have found zigzag band like structure consisting of Cu-O in Sr-Ca-Cu-O instead of $Cu\text{-}O_4$ sheets in Bi-Sr-Ca-Cu-O. Bi-Sr-Ca-Cu-O synthesized from Bi_2O_3 and Sr-Ca-Cu-O powder by the solid state reaction showed superconductivity. This result suggests that the Bi_2O_2 layer plays an important role in forming the $Cu\text{-}O_4$ sheets that is essential for the superconductivity of the high-Tc oxide-superconductors.

INTRODUCTION

Since Bednortz and Muller first reported the Ba-La-Cu-O system as being high-Tc superconducting oxides [1], many investigations have been carried out in order to find new high-Tc superconducting oxides . Y-Ba-Cu-O was one of the most fruitful results of these investigations [2]. Discovery of the new type of superconducting Bi-Sr-Cu-O system [3,4] seems much significant in spite of its rather low critical temperature because of leading to the subsequent discovery of the Bi-Sr-Ca-Cu-O system with no rare-earth element [5] that has higher-Tc than that of Y-Ba-Cu-O system.

Crystal structure of Bi-Sr-Ca-Cu-O system was supposed to be the $Bi_4Ti_3O_{12}$-type and has layerd structure with alternative stacking of perovskite-like layers, consisting of Sr,Ca,Cu and O, and Bi_2O_2 layers. We are interested in the contribution of the Bi_2O_2 layer in this compound to the crystal structure that is closely related to superconductivity. We synthesized Sr-Ca-Cu-O compound and made a comparison between the structure of Sr-Ca-Cu-O and that of Bi-Sr-Ca-Cu-O.

EXPERIMENTAL

The samples of Sr-Ca-Cu-O were prepared from powders of $SrCO_3$, $CaCO_3$ and CuO, which were mixed with the nominal composition of Sr:Ca:Cu=1:1:2. The mixed powder was calcined at 980°C for 12 hours in air.

X-ray powder diffraction measurement was performed with these samples by using CuKα radiation. The X-ray powder diffraction pattern seemed to be single phase with no signal from the raw materials. Electorical resistivity of the Sr-Ca-Cu-O pellets sintered at 980°C was measured by using a conventional four-probe method.

---

*Present address, UBE SCIENTIFIC ANALYSIS LABORATORY, 1978-5 Kogushi, Ube City, Yamaguchi Prefectureture, 755, Japan.

Finally we synthesized Bi-Sr-Ca-Cu-O from the Sr-Ca-Cu-O powder and Bi_2O_3 by solid state reaction. The powders were mixed with the nominal composition of Bi:Sr:Ca:Cu=1:1:1:2, which was originally reported by Maeda et al., and calcined at temperatures in a range from 800°C to 870°C in air for 6 hours. The calcined powder was pulverized and pressed into pellets. The pellets were sintered at temperatures in a range from 820°C to 890°C for 12 hours in air.

X-ray powder diffraction, electric resistance and magnetic susceptibility measurements were also carried out with the samples of Bi-Sr-Ca-Cu-O obtained.

RESULTS AND DISCUSSION

For determination of the structure of Sr-Ca-Cu-O obtained, its X-ray powder diffraction pattern was compared with those of the oxides consisting of at least two metal elements out of Sr, Ca and Cu. The X-ray powder diffraction pattern of $SrCuO_2$ [6] found resemble to that of Sr-Ca-Cu-O obtained. The Rietveld analysis of the X-ray diffraction pattern was performed on the basis of the cystal structure of $SrCuO_2$ using computer program named "RIETAN" [7].

A fairly good agreement between observed and calculated intensities was attained. Figure 1 shows the X-ray powder diffraction pattern and its fitting curve. Table I gives the refined structural parameters of Sr-Ca-Cu-O.

Table I. Refined structural parameters for Sr-Ca-Cu-O. a=3.446(1)A. b=16.10(7)A. c=3.868(1)A. Space group Cmcm.

Atom	site	X	Y	Z	B(A^2)	Occupancy
Sr/Ca	4c	0	0.92(9)	1/4	0.5	0.4(8)/0.5(2)
Cu	4c	0	0.06(1)	1/4	0.5	1.0
O(1)	4c	0	0.94(0)	1/4	1.0	1.0
O(2)	4c	0	0.18(0)	1/4	1.0	1.0

Reliability factors

R_{WP}=15.39 R_P=12.11 R_E=7.84 R_I=9.25 R_F=6.38

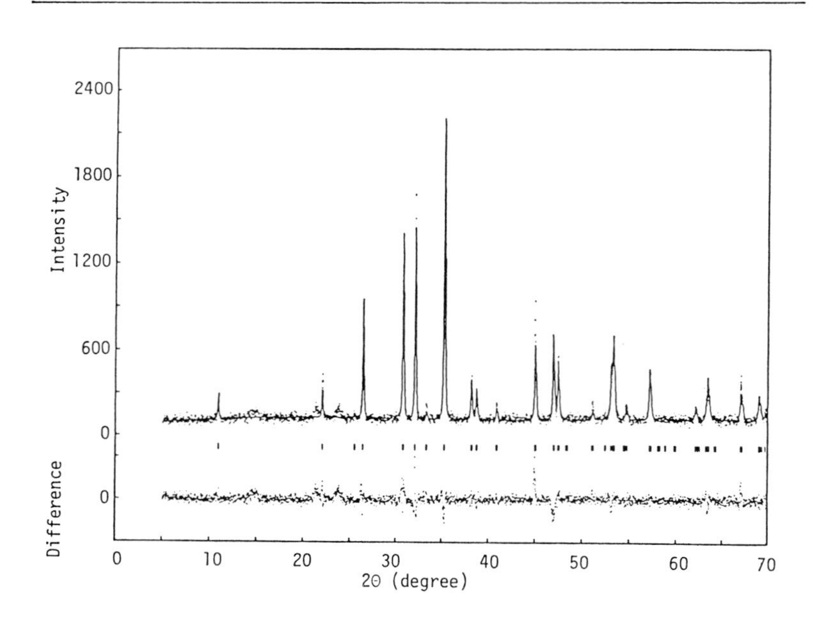

Fig. 1 X-ray powder diffraction pattern and fitting curve of Sr-Ca-Cu-O.

In the figure the solid line represents calculated intensities, dots overlying them are observed intensities. The difference between the observed and the calculated results represented below them as dotted pattern. In the Rietveld analysis we took the probability of occupancy of Sr and Ca at the lattice site of Sr in $SrCuO_2$ as one of the fitting parameters. From the results of the analysis, the probability of occupancy of Sr and that of Ca were 0.4(8) and 0.5(2) respectively. The electrical resistivity of Sr-Ca-Cu-O increases as lowering temperature and never shows the superconducting transition down to 10K.

The samples of Bi-Sr-Ca-Cu-O obtained were apparently almost the mixtures of the high-Tc and low-Tc phases. Their typical temperature dependence of the electrical resistivity and the magnetic susceptibility are represented in Figs. 2 and 3 respectively. It is difficult in general to determine the crystal structure of such mixtures. So the Rietveld analysis was performed with the X-ray powder diffraction pattern of Bi-Sr-Ca-Cu-O consisting only of the low-Tc phase, whose temperature dependence of the electrical resistivity and the magnetic susceptibility are shown in Figs. 4 and 5

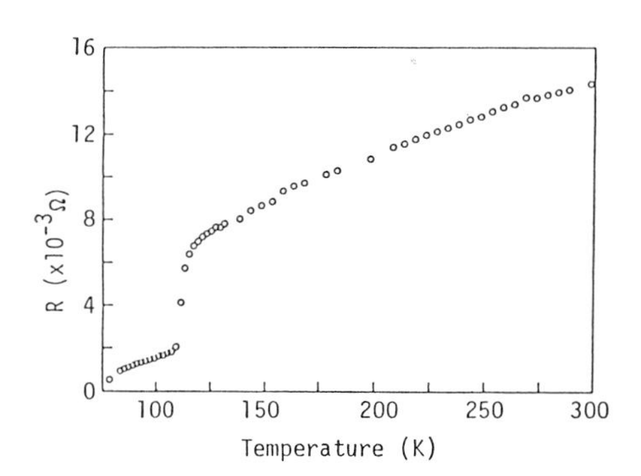

Fig. 2 Typical temperature dependence of the electrical resistivity for Bi-Sr-Ca-Cu-O containing the low-Tc and high-Tc phases.

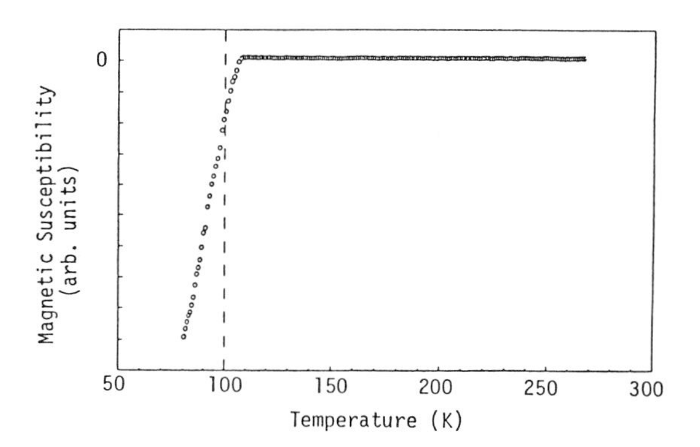

Fig. 3 Typical temperature dependence of the magnetic susceptibility for Bi-Sr-Ca-Cu-O containing the low-Tc and high-Tc phases measured by using a vibrating sample magnetometer under the constant magnetic field of 100 Oe.

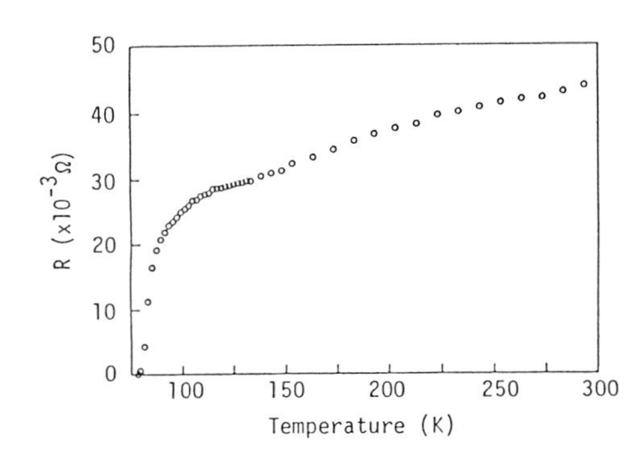

Fig. 4 Temperature dependence of the electrical resistivity for Bi-Sr-Ca-Cu-O containing only the low-Tc phase.

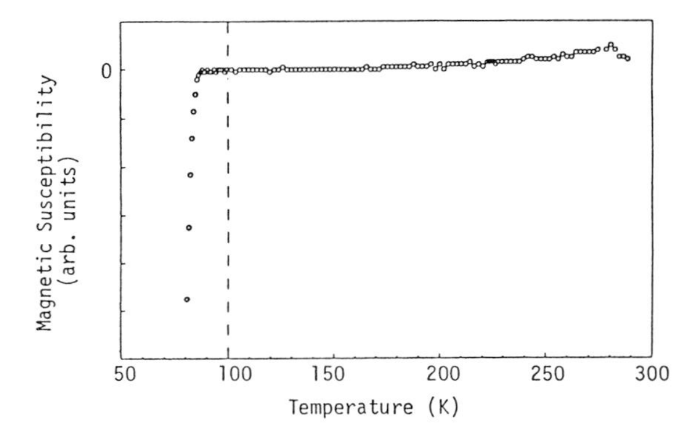

Fig. 5 Temperature dependence of the magnetic susceptibility for Bi-Sr-Ca-Cu-O containing only the low-Tc phase measured by using a vibrating sample magnetometer under the constant magnetic field of 100 Oe.

respectively. The analysis was carried out on the basis of the results of T. Kajitani et al. [8]. A crystal lattice modulation along the b-axis was found with a period of about 27A in Bi-Sr-Ca-Cu-O. But this modulation was averaged and the analysis was carried out with a sub-sell, 1/5 of the real unit cell, for simplification. The results are shown in Table II and Fig. 6.

Table II. Refined structural parameters for the low-Tc phase of Bi-Sr-Ca-Cu-O.
a=5.39(8)A. b=5.39(7)A. c=30.66(1)A. Space group Fmmm.

Atom	site	X	Y	Z	B(A^2)	Occupancy
Bi	8i	0	0	0.19(7)	0.6	1.0
Cu	8i	0	0	0.05(6)	0.45	1.0
Sr/Ca(1)	8i	0	0	0.39(2)	0.6	0.6/0.4
Sr/Ca(2)	4b	0	0	1/2	0.6	0.6/0.4
O(1)	8f	1/4	1/4	1/4	1.0	0.(0)
O(2)	16j	1/4	1/4	0.04(7)	1.0	1.0
O(3)	8i	0	0	0.1(4)	1.0	1.0
O(4)	8i	0	0	0.3(2)	1.0	1.0

Reliability factors
R_{WP}=21.16 R_P=16.09 R_E=4.96 R_I=21.42 R_F=11.29

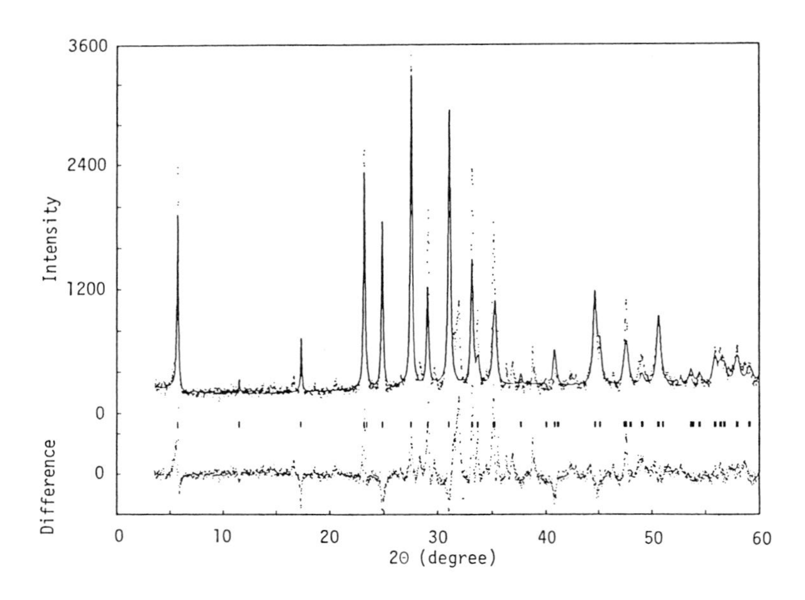

Fig. 6 X-ray powder diffraction pattern and fitting curve of the low-Tc phase of Bi-Sr-Ca-Cu-O

The crystal structures refined with the Rietveld analysis are shown in three dimensions in the Figs. 7 and 8. Figures were drawn with a computer using the computer program of "COSMIC", the supporting system for the development of new materials with designing molecules, developed by UBE Industries, Ltd.. In these figures every element was drawn as the individual sphere proportional to its ion radius.

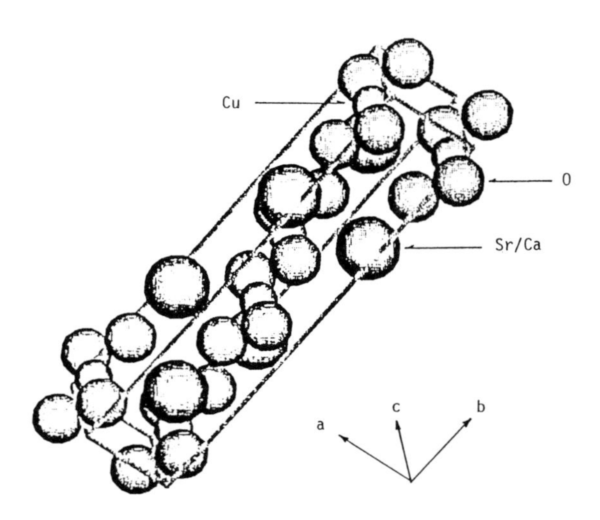

Fig. 7 Schematic picture of the structural model for Sr-Ca-Cu-O using the refined structural parameters shown in Table I.

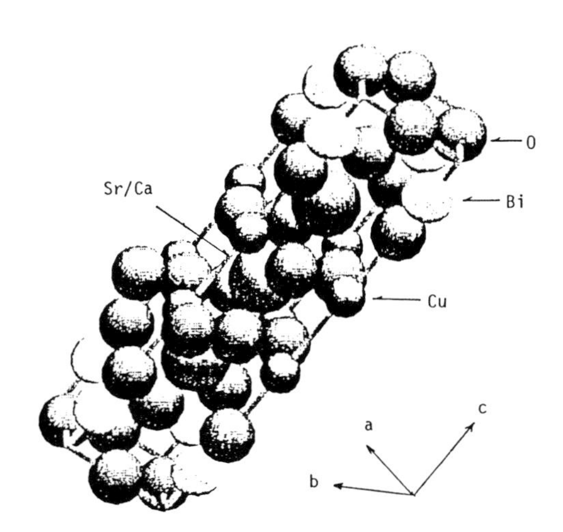

Fig. 8 Schematic picture of the structural model for the low-Tc phase of Bi-Sr-Ca-Cu-O using the refined structural parameters shown in Table II.

It is found in these figures that Sr-Ca-Cu-O dose not have the perovskite-like structure, which contains $Cu\text{-}O_6$ octahedron or $Cu\text{-}O_5$ pyramidal structure found in the superconducting oxides like as Ba-La-Cu-O and Y-Ba-Cu-O. $Cu\text{-}O_4$ planer structure exists in Sr-Ca-Cu-O. But every $Cu\text{-}O_4$ holds sides of neighboring $Cu\text{-}O_4$'s in common and expands like as the zigzag band. The superconducting oxides have also planer $Cu\text{-}O_4$'s, which are in most cases as the parts of $Cu\text{-}O_6$ octahedron or $Cu\text{-}O_5$ pyramidal structure. The $Cu\text{-}O_4$ in the superconducting oxides, in contrast to that in Sr-Ca-Cu-O, holds a oxygen at the vertex of the neighboring planer $Cu\text{-}O_4$'s in common and expands like as a sheet.

In conclusion the Bi_2O_2 layer plays an important role in forming the $Cu\text{-}O_4$ sheets that is essential for the superconductivity of the high-Tc oxide-superconductors. More detailed experiments should be needed to clarify the mechanism of forming the $Cu\text{-}O_4$ sheets from the $Cu\text{-}O_4$ zigzag band like structure in the reaction of Sr-Ca-Cu-O and Bi_2O_3. Clarification of the reaction process may suggest a key point for fabricating the single high-Tc phase of Bi-Sr-Ca-Cu-O and searching for new higher-Tc superconducting materials.

ACKNOWLIGEMENT

We thank Dr. K. Uchida for his helpful advice on this work.

REFERENCES

1) J. G. Bednortz and K. A. Muller: Z. Phys. B64 (1987) 189.
2) M. K. Wu, J. R. Ashburn, C. J. Torng, P. H. Hor, R. L. Meng, L. Gao, J. Huang, Y. Q. Wang and C. W. Chu: Phys. Rev. Lett. 58 (1987) 908.
3) J. Akimitsu, A. Yamazaki, H. Sawa and H. Fujiki: Jpn. J. Appl. Phys. 26 (1987) L2080.
4) C. Michel, M. Hervieu, M. M. Borel, A. Grandin, F. Deslandes, J. Provost and B. Raveau: Z. Phys. B68 (1988) 421.
5) H. Maeda, Y. Tanaka, M. Fukutomi and T. Asano: Jpn. J. Appl. Phys. 27 (1988) L209.
6) C. L. Teske and Hk. Muller-Buschbaum: Z. Anorg. Allg. Chem. 379 (1970) 234.
7) F. Izumi: J. Cryst. Soc. Japan 27 (1985) 23 [in Japanese].
8) T. Kajitani, K. Kusaba, M. Kikuchi, N. Kobayashi, Y. Syono, T. B. Williams and M. Hirabayashi: Jpn. J. Appl. Phys. 27 (1988) L587.

New Tl-Ba-Ca-Cu-O (1234, 1245 and 2234) Superconductors with Very High Tc

H. IHARA[1], R. SUGISE[2], T. SHIMOMURA[3], M. HIRABAYASHI[1], N. TERADA[1], M. JO[1], K. HAYASHI[1], M. TOKUMOTO[1], K. MURATA[1], and S. OHASHI[4]

[1] Electrotechnical Laboratory, Umezono, Tsukuba, 305 Japan
[2] Ube Industries Ltd., Ube, 755 Japan
[3] Unitika R & D Center, Uji, 611 Japan
[4] R & D Division, Asahi Glass Co., Ltd., Hazawa, Kanagawa-ku, Yokohama, 221 Japan

Abstract

New $TlBa_2Ca_3Cu_4O_{11}$ (1234), $TlBa_2Ca_4Cu_5O_{13}$ (1245) and $Tl_2Ba_2Ca_3Cu_4O_{12}$ (2234) superconductors with Cu-O multilayers above four layers have been synthesized. These compounds have Tc values of 117, 122 and 115 K, respectively. The 1234 phase showed the highest Tc value in the $TlBa_2Ca_{n-1}Cu_nO_{2n+3}$ superconductor family. The Tc value of 2234 phase was lower than that of 2223 phase.

The single Tl-O layered compounds have a simple tetragonal phase of the space group of P4/mmm with the lattice constants of a = b = 3.85 Å and c = 19.1 Å for the 1234 phase and a = b = 3.85 Å and c = 22. 3 Å for the 1245 phase. The double Tl-O layered 2234 phase has a body-centered-tetragonal structure with the lattice constants of a = b = 3.85 Å and c = 42.0 Å. Electron diffraction patterns and lattice images from a high resolution TEM have confirmed that these compounds have an oxygen-deficient layered-perovskite structure. Their c-lattice constants follow the c-axis rule of linear relations of c = 6.3 +3.2 n Å (n = the multiplicity of CuO layers in aunit) in the single Tl-O layer compound family and c=2 (8.2 + 3.2 n) Å in the double Tl-O layer compound family.

Introduction

Search for new high-Tc superconductors is of great importance from both scientific and technological standpoints. Increasing the Tc value is a principal objective in materials design of oxide superconductors. For this purpose we have many empirical rules. One of the most promising rules is the relationship between Tc value and the number of Cu-O layers derived from the layered perovskite compounds of $Tl_2Ba_2Ca_{n-1}Cu_nO_{2n+4}$ and $Bi_2Sr_2Ca_{n-1}Cu_nO_{2n+4}$.[1-3] If we can prepare a superconducting compound with four or five Cu-O layers for example, the Tc value may become higher than that of the $Tl_2Ba_2Ca_2Cu_3O_{10}$ (2223) phase.

Another purpose of materials design for Tl compounds is the reduction of Tl content, because of its toxicity and shortage in natural resources. The reduction seems to be possible since Tl is thought to be merely a structural stabilizer and has no important direct role in superconductivity. Therefore the double Tl-O layers can be reduced to the single Tl-O layer. The increase in the number of Cu-O layers leads to a decrease in Tl content. Thus we intended to prepare a new superconductor with four or five Cu-O layers in single Tl-O layer compounds and also in double Tl-O layers compounds.[5]

In this work, we have synthesized new $TlBa_2Ca_3Cu_4O_{11}$ (1234), $TlBa_2Ca_4Cu_5O_{13}$ (1245) and $Tl_2Ba_2Ca_3Cu_4O_{12}$ (2234) superconductors with T_c of 122.1, 117 and 115 K, respectively; we have determined the lattice structures of these compounds from x-ray and electron diffraction and high resolution TEM(transmission) electron microscope measurements .

The idea of synthesis of single Tl-O layer compound is based on a condensation and polymerization reaction. At first, a double Tl-O layered compound with a small number of Cu-O layers is synthesized. Then the number of CuO layers is increased by evaporating the Tl component . Furthermore one of the two Tl-O layers of the double unit is evaporated to obtain the single Tl-O layer compound. A part of the single Tl-O layers is eliminated to further increase the multiplicity of Cu-O layers.

Experimental

The samples were prepared by firing the mixed powders of Tl_2O_3, CaO_2, BaO_2 and CuO with nominal compositions of $Tl_4Ba_2Ca_3Cu_4O_y$, $Tl_2Ba_2Ca_5Cu_6O_y$ and $Tl_{2.5}Ba_2Ca_4Cu_5O_y$ for 1234, 1245 and 2234 phases, respectively. Preparation of the samples has to be carried out with great care by using a glove-box and a furnace in a draft chamber with flowing oxygen, because of the toxicity of thallium compounds and the high reactivity of Ca and Ba oxides with CO_2. The mixed powders of starting materials were pressed into pellets of diameter 10 mm and thickness 1mm under a pressure of 2000 kg/cm^2. The pellets were fired at 890 °C for about 30 min in flowing oxygen gas. The firing time was determined by measuring the weight of the sample to know the quantity of evaporated Tl_2O_3. Then the samples were cooled in air to room temperature by taking them out of the furnace. This process is different from the previous one which includes furnace cooling,[5] but the present method is more controllable than the previous one.

The T_c of the sample was measured resistively and magnetically by using a conventional four-probe technique and a SQUID magnetometer. The onset of T_c (T_{co}), midpoint of T_c (T_c) and end point of T_c (T_{ce}) were defined as 10%, 50% and 90% drop-points of the resistive transition. The temperature was measured with a calibrated platinum resistance thermometer. The experimental error in T_c measurement was $\pm$ 0.2 K. The sample composition was determined by an electron probe micro-analyzer and inductively coupled plasma spectrometer. The oxygen content was measured by an inert gas fusion non-dispersive IR method. The calibrations of the composition analysis were performed by using 2212 and 2223 samples and $YBa_2Cu_3O_{6.95}$ samples with a single phase as standard samples. The error in composition analysis was estimated to be less than 3 %. The structure was determined by x-ray diffraction (XRD) with a Cu target, electron diffraction and high resolution

TEM techniques. Electron microscope observations were performed on a JEM 2000 EX at 200 kV and JEM 4000 EX at 400 kV. These high resolution TEM instruments have point resolutions of 2.1 and 1.9 Å, respectively.

Results and Discussion

$TlBa_2Ca_3Cu_4O_{11}$ (1234)

Superconductivity. Figure 1 shows the temperature dependence of the resistance for the present low-Tl sample with 1234 phase. The present result (a) in solid line is compared with the previous one (b)[5] in dashed line. The T_{co}, T_c and T_{ce} values are 123.6, 121.9 and 120.6 K, respectively. The Tc value is a little lower than the previous value of 122.1 K, and the Tce value is a little higher than the previous value of 120.4 K. The differences of these values, however, are within experimental error of 0.2 K. The resistance of the 1234 phase is much higher than that of the 2223 phase. The high resistance may be caused partly by impurity phases and partly by grain boundaries.

Figure 2 shows magnetization versus temperature at the field of 100 Gauss for the 1234 phase sample. The onset temperature of superconducting transition is 122.0 K, which corresponds to the resistively measured Tc value. The amplitude of the Meissner signal in the 1234 phase sample is comparable to that in a well prepared $YBa_2Cu_3O_{7-y}$ superconductor,[6] which indicates the bulk nature of superconductivity over 50 volume % of the 1234 phase because 2223 and 2212 phases included are a few volume %.

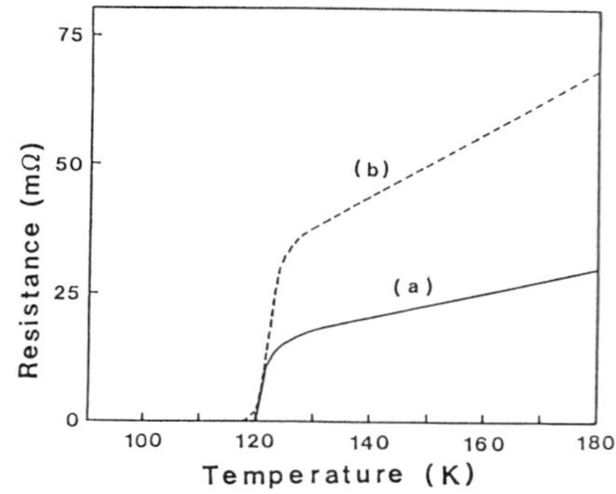

Fig. 1 Resistance versus temperature for the $TlBa_2Ca_3Cu_4O_{11}$ sample.

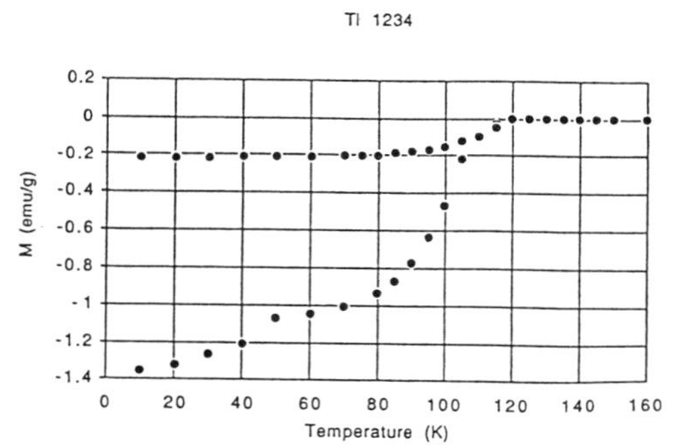

Fig. 2 Magnetization versus temperature of the $TlBa_2Ca_3Cu_4O_{11}$ (1234) sample under 100 Gauss.

Structure. Figure 3 shows the x-ray diffraction pattern for the sample with the 1234 phase. The diffraction pattern shows a tetragonal structure with the lattice constants of a = b = 3.85 Å and c = 19.1 Å. The phase is completely different from the previously reported 2212 and 2223 phases.[2, 6] The samples with the starting composition of $Tl_4Ba_2Ca_3Cu_4O_y$ have formed almost the single 1234 phase except for a small amount 2223, 2212 and amorphous phases as shown in Fig. 3. A trace of $BaCuO_2$ and amorphous phase was detected by the electron probe micro-analysis. Most of the peaks in Fig. 3 were identified only by the Miller indices of the 1234 phase.[5]

The composition of the tetragonal phase was determined as $TlBa_2Ca_3Cu_4O_{11}$ (1234) within an experimental error of 10 %. The result leads to the conclusion that the phase has single Tl-O layer and four Cu-O layers.

Figure 4 shows the electron diffraction pattern and the lattice image viewed along [010]. The electron diffraction pattern of [010] zone also gives the simple tetragonal structure with the lattice constant of a=3.9 Å and c=19 Å, because every (h0*l*) spot is observed.[7] A modulation with a period of (006) along the (00*l*) direction is obviously observed, which corresponds to the lattice parameter of 3.2 Å. The lattice image of Fig. 4 shows the periodic light and dark bands along the c-axis. The image

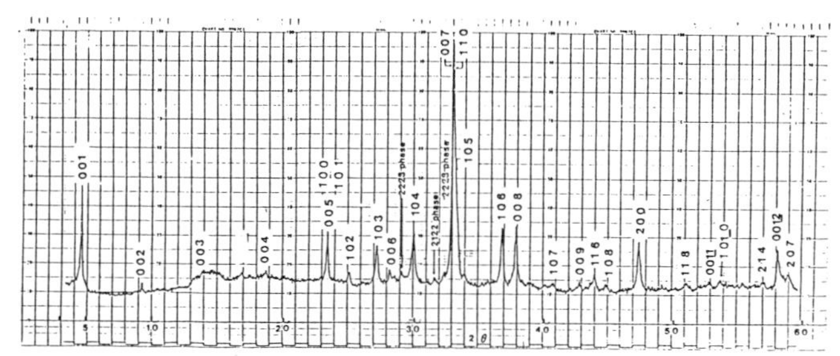

Fig. 3 X-ray diffractogram of the 1234 sample with the tetragonal-phase of a = b = 3.85 Å and c = 19.1 Å.

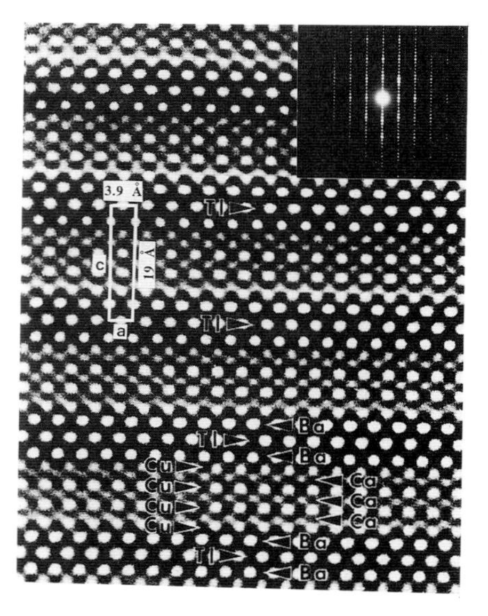

Fig. 4 The electron diffraction pattern and the lattice image of [010] zone of the 1234 crystal with the lattice constants of a = 3.9 Å and c = 19 Å.

can be interpreted as a projected charge density of the crystal atoms. The period of a colinear dark spot lines separated by a wide light band is 19 Å. The light band corresponds to the light atoms of Cu and Ca, and the dark band to heavy atoms of Tl and Ba. Three colinear thick dark lines in the dark band along the a-axis are attributed to the single Tl layer and double Ba layers on both sides of the Tl layer. Seven colinear dark lines in the light band are attributed to the four Cu-layers and three Ca-layers stacking alternatively. The image of the Ca atom is smaller than that of the Cu atom because of the lack of d-electrons in the Ca atom. Each dark colinear line along the a-axis is assigned to each component atom as shown in Fig. 4.

The electron diffraction pattern viewed along [001].gives a simple tetragonal structure with the lattice constant of a=b=3.9 Å, because every (hk0) spot is observed. The lattice image viewed along [001]. consists of a common face-centered square net of 3.9 Å x 3.9 Å. The dark spots corresponds to the A site atoms of Ba and Ca or the B-site atoms of Tl and Cu.[7] It is difficult to distinguish which spot is the A or B site. The light spots alternating with the dark spots correspond to the oxygen atom sites.

The distance between Tl layers is 19 Å, which is consistent with the result of c=19.1 Å from the XRD measurement. The average space between the Cu layers in the lattice image corresponds to 3.2 Å, which is equal to the distance evaluated from the (006) periodic modulation in the ($h0l$) diffraction pattern. Then the space between the Tl and Cu layers is estimated as 4.75 Å. Each of the component atoms has the common distance of 3.9 Å along a-axis, which is consistent with the lattice constant of a=3.85 Å. The rectangular square enclosed with a white line in Fig. 5 corresponds to the unit cell.

There are two kinds of light spots in the lattice image of Fig. 5 One is bright spots between each component atom except Ca; another is dim spots between Ca atoms. The bright spots correspond to the oxygen atoms which are located around each atom. The dim spots would be attributed to the oxygen vacancies in the Ca layer. Since the space of the oxygen vacancy is much smaller than that of the oxygen atom and oxygen is a light element with large ion radius, the image of the vacancy becomes smaller and darker than that of the oxygen atom.

From these results, we can derive the simple tetragonal lattice structure of the 1234 phase as shown in Fig. 5. The unit cell consists of five-fold simple tetragonal sub-cells. Three-fold sub-cells in the middle part are composed of Cu and Ca atoms, of which the sub-lattice parameter along the c-axis is 3.2 Å. The sub-lattice parameter c (marked with c_1) is consistent with the length of 3.2 Å which corresponds to the period for periodic (006) modulation in the ($h0l$) diffraction pattern. The

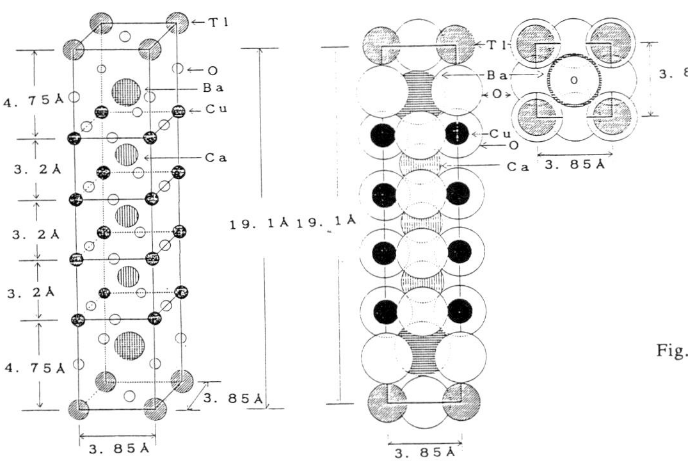

Fig. 5 (a) : The lattice structure of the 1234 crystal with the space group of P4/mmm and the lattice constants of a=b=3.85 Å and c=19.1 Å. (b) : Packing model of component ions of the 1234 crystal using realistic ion radius.

room between Cu-O layers is too small to be occupied by oxygen ions. Then oxygen vacancies are formed in the Ca layers of this 1234 phase as shown in Fig. 5 (b). Thus the 1234 phase is an oxygen deficient layered perovskite.[5, 8] On the other hand, both top and bottom tetragonal sub-cells are composed of Tl, Ba and Cu atoms, of which the sub-lattice parameter along the c-axis (marked with c_2) is estimated as 4.75 Å. From the lattice image of Fig. 5 we can evaluate the distances between the Tl and Ba layers and between the Ba and Cu layers as 2.75 Å and 2.0 Å, respectively. The oxygen content of the 1234 phase is evaluated as 2n+3=11 from the lattice structure. The oxygen content is near to the observed value of 11.5 ± 1 by an inert gas fusion and non-dispersive IR method.

The primitive tetragonal lattice of Fig. 5 have the four-fold symmetry around c-axis and two-fold symmetry around a- and b-axes and mirror symmetry for each (001), (010) and (001) plane. Then the space group of the 1234 phase is P4/mmm.[8,9] The space group is the same as that of tetragonal $YBa_2Cu_3O_6$.[7] The observed lattice parameter c is consistent with the c-axis rule of c = 6.3+3.2n =19.1 Å (n=4) for the single Tl-O layer compound $TlBa_2Ca_{n-1}Cu_nO_{2n+3}$.[5] This c-axis rule can be derived from the c_1 and c_2 lattice parameters as $c = (n-1)\ c_1 + 2\ c_2$

$TlBa_2Ca_4Cu_5O_{13}$ (1245)

Superconductivity. Figure 6 shows the temperature dependence of the resistance for the 1245 phase sample. The resistive curve shows two-step drops as shown in earlier mixed phase Bi-Sr-Ca-Cu-O sample.[3] The high temperature drop shows a T_c value of 117 K, and the low temperature drop a T_c value of 108 K. The T_c at 117 K is attributed to the 1245 phase and the T_c at 108 K to the 1256 phase. The 1256 phase is sometimes observed in the lattice images from the high resolution TEM for the 1245 sample as an intergrowth phase. Another possibility, however, is not necessarily excluded. The high T_c might be attributed to the 1234 phase and the low T_c to the 1245 phase, because the 1234 phase usually has a wide range of Tc value (117 ~ 122 K). The 1234 phase was sometimes observed in the lattice images of the 1245 sample, though the 1234 phase was not detected in the x-ray diffraction pattern. At any rate the results lead to the fact that the 1245 phase has not shown a higher Tc value than the 1234 phase.

Figure 7 shows magnetization versus temperature for the 1245 phase sample measured at a field of 10 Gauss. The magnetization had a tendency to saturate above 10 Gauss. The magnetization curve shows the starting drop point of 117 K. The point corresponds to the resistively measured Tc value. Even though the amplitude of the Meissner signal is small in comparison with the 1234 phase by about a factor of 20, the Tc of 117 K is attributed to a bulk superconductor from the major 1245 phase and not from other minor 1256 nor 1234 phases in consideration of a smaller applied field by a factor 10 than in Fig. 2.

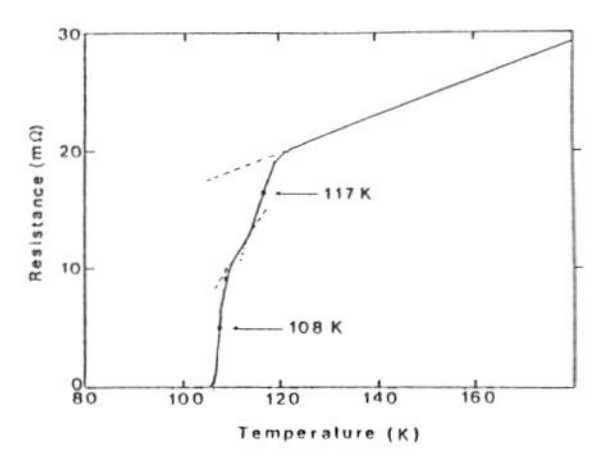

Fig. 6 Resistance versus temperature of the samples of $TlBa_2Ca_4Cu_5O_{13}$ (1245) phase.

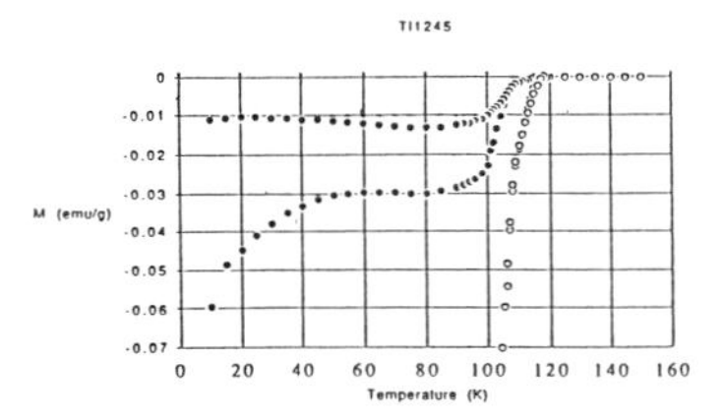

Fig. 7 Magnetization versus temperature of the 1245 sample under 10 Gauss.

Structure. Figure 8 shows the x-ray diffraction pattern of the sample with the 1245 phase. The diffraction pattern shows a tetragonal structure with the lattice constants of a = b = 3.85 Å and c = 22.3 Å. The phase is completely different from the previously reported 1234,[5] 2212 and 2223 phases[2,6] even from the 2234 phase. The samples have a large amount of 1245 phase and a small amount of such impurity phases as $BaCuO_2$ and amorphous phases. Almost all the strong peaks in Fig. 8 were identified by the Miller indices of the 1245 phase.

The composition of the new tetragonal phase was determined as $TlBa_2Ca_4Cu_5O_{13}$ (1245) within an experimental error of 10 %. The result leads to the fact that the phase has single Tl-O layer and five Cu-O layers.

Figure 9 shows the electron diffraction pattern and the lattice image viewed along [010] for the 1245 phase. The electron diffraction pattern of [010] zone gives the simple tetragonal structure with the lattice constant of a=3.9 Å and c=22 Å, because every (h0*l*) spot is observed as in 1234 phase. A modulation with a period of (007) along the (00*l*) direction is obviously observed, which corresponds to the sub-lattice parameter of 3.2 Å. The lattice image of Fig. 9 shows periodic light and dark bands along the c-axis. The period of the dark bands separated by a wide light band is 22 Å. The light band consists of the light atoms of Cu and Ca, and the dark band of heavy atoms of Tl and Ba. Each dark colinear line along the a-axis is assigned to the component atoms as shown in Fig. 9.

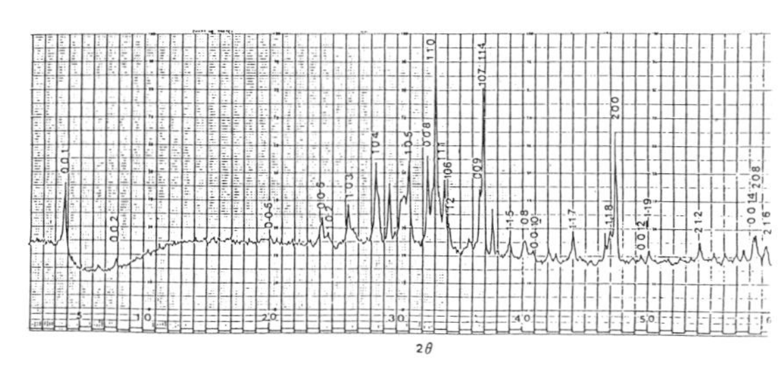

Fig. 8 X-ray diffractogram of the 1245 sample with the tetragonal-phase of a = b = 3.85 Å and c = 22.3 Å.

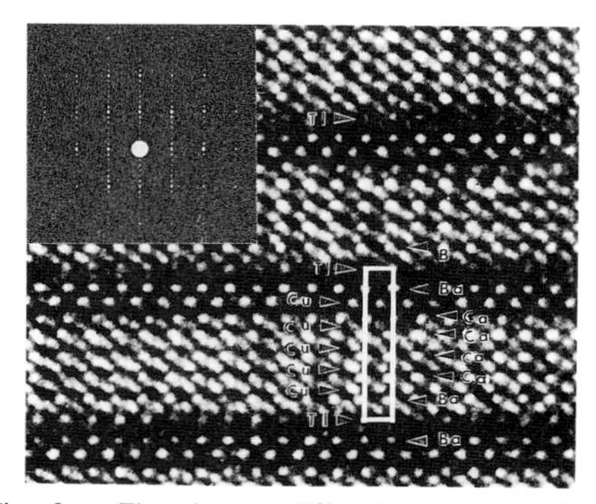

Fig. 9 The electron diffraction pattern and the lattice image of [010] zone of the 1245 crystal.

The distance between Tl layers is 22 Å, which is consistent with the result of c=22.3 Å from the XRD. The average space between the Cu layers in the lattice image corresponds to 3.2 Å, which is equal to the distance evaluated from the (007) periodic modulation in the (h0*l*) diffraction pattern. Then the space between the Tl and Cu layers is estimated as 4.75 Å as in the 1234 phase. Each of the component atoms has the common distance of 3.9 Å along a-axis, which distance is consistent with the lattice constant of a=3.85 Å. The rectangular square 3.9 x 22 $Å^2$ enclosed with a white line in Fig. 9 corresponds to the unit cell.

From these results, we can derive the simple tetragonal lattice structure of the 1245 phase as shown in Fig 10. The unit cell consists of six-fold simple tetragonal sub-cells. Four-fold sub-cells in the middle part consist of Cu and Ca atoms, of which the sub-lattice parameter along c-axis is 3.2 Å. The sub-lattice parameter c (marked with c_1) is consistent with the length of 3.2 Å which corresponds to the periodic (007) modulation in the (h0*l*) diffraction pattern . The room between Cu-O layers is too small to be occupied by oxygen ions. Then oxygen vacancies are formed in the Ca layers of this 1245 phase too, as in the 1234 phase. Thus the 1245 phase is also an oxygen deficient layered perovskite. On the other hand, both top and bottom tetragonal sub-cells consist of Tl, Ba and Cu atoms, of which the sub-lattice parameter along the c-axis (marked with c_2) is estimated to be 4.75 Å. The distances between the Tl and Ba layers and between the Ba and Cu layers are evaluated as 2.75 Å and 2.0 Å, respectively, as in the 1234 phase.

The oxygen content of the 1245 phase is evaluated as 2n+3=13 from the lattice structure. The oxygen content is near to the observed value of 13 $\pm$ 1 by an inert gas fusion and non-dispersive IR method. This and other impurity phases such as amorphous and $BaCuO_2$ phases will make the error in oxygen content determination large. The lattice images from the electron microscopic measurement sometimes show the intergrowth of 1234 and 1256 phases in the matrix of the 1245 phase.

The primitive tetragonal lattice of Fig. 10 has the four-fold symmetry around c-axis and two-fold symmetry around a and b-axes and mirror symmetry for each (001), (010) and (001) plane. Then the space group of the 1245 phase is P4/mmm as in

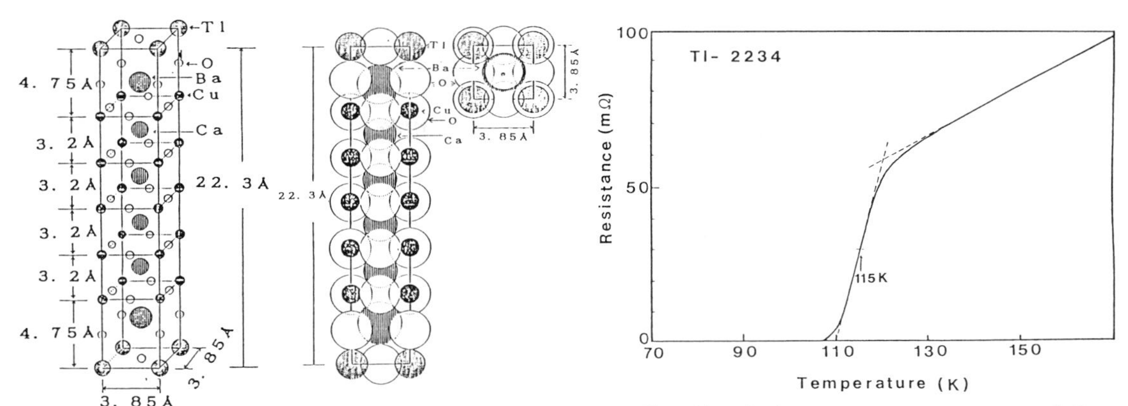

Fig. 10 (a) : The lattice structure of the 1245 crystal with the space group of P4/mmm and with the lattice constants of a=b=3.85 Å and c=22.3 Å. (b) : Packing model of component ions of the 1245 crystal using realistic ion radius.

Fig. 11 Resistance versus temperature of the sample of $Tl_2Ba_2Ca_3Cu_4O_{12}$ (2234) phase.

the 1234 phase.[2] The observed c-lattice parameter is consistent with the c-axis rule of c = 6.3+3.2n = 22.3 Å (n=5) for the single Tl-O layer compound $TlBa_2Ca_{n-1}Cu_nO_{2n+3}$.[7]

$Tl_2Ba_2Ca_3Cu_4O_{12}$ (2234)

Superconductivity. Figure 11 shows the temperature dependence of resistance for a 2234 phase sample. The T_{co}, T_c and T_{ce} values are 121, 115 and 111 K. The 2234 phase has not shown a higher Tc value than the 2223 phase does. The resistance of the 2234 phase was always a little higher than that of the 2223 phase. The high resistance would be caused partly by impurity phases and partly by grain boundaries. The magnetic susceptibility measurement showed an onset temperature of superconducting transition at 116 K, which corresponded to the resistively measured Tc value.

Structure Figure 12 shows the x-ray diffraction pattern (XRD) of the sample with the 2234 phase. The diffraction pattern shows a tetragonal structure with the lattice constants of a = b = 3.85 Å and c = 42.0 Å. The phase is clearly distinguished from the previously reported 2212, 2223, 1234 and 1245 phases[1-2]. The main peaks in Fig. 12 were identified by the Miller indices of the 2234 phase. The electron diffraction pattern of the sample showed a body-centered tetragonal patterns with lattice constants of a=3.9 Å and c=42 Å, since the extinction rule followed a body-centered-tetragonal type. A modulation with a period of (0013) along the (00*l*) direction is observed, which corresponds to the lattice parameter of 3.2 Å.

From the results, we can derive the body-centered-tetragonal lattice structure of the 2234 phase as shown in Fig. 13. The space group is I4/mmm which is the same as that of the 2223 phase. Three-fold sub-cells in the middle part are composed of Cu and Ca atoms, of which the sub-lattice c-parameter is 3.2 Å. The c-parameter (marked with c_1) is consistent with the distance periodic (0013) modulation in the (h0*l*) diffraction pattern . The value is nearly equal to the 3.17 Å distance between Cu and Cu layers in $Tl_2Ba_2CaCu_2O_8$ compound.[6] Then the oxygen atoms are thought to be eliminated in the Ca layers of this 2234 phase too. Thus the 2234 phase is also an oxygen deficient layered perovskite. Both top and bottom tetragonal sub-cells are composed of adjacent double Cu layers. Then the four adjacent Cu-O layer alternate with the double Tl-O layers. The distance (c_2) between Tl-O and Cu-O layers is 4.75 Å., which is almost the same as the 4.72 Å distance between Tl and Cu layers in $Tl_2Ba_2CaCu_2O_8$ compound.[6] The distance (c_3) between two Tl-O layers is 2.0 Å. From the lattice image of the 2234 phase we can evaluate the distances between the Tl and Ba layers and between the Ba and Cu layers as 2.75 Å and 2.0 Å, respectively.

The body centered tetragonal lattice of Fig. 13 has the four-fold symmetry around c-axis and two-fold symmetry around a and b-axes and mirror symmetry for each (001), (010) and (001) plane. Then the space group of the 2234 phase is I4/mmm, the same as the 2223 phase.[2] The space group is the same as that of tetragonal La_2CuO_4.[8] The observed c-lattice parameter is consistent with the c-axis rule of c = 2(8.2+3.2n)= 42.0 Å (n=5) for the double Tl-O layer compound $Tl_2Ba_2Ca_{n-1}Cu_nO_{2n+4}$.

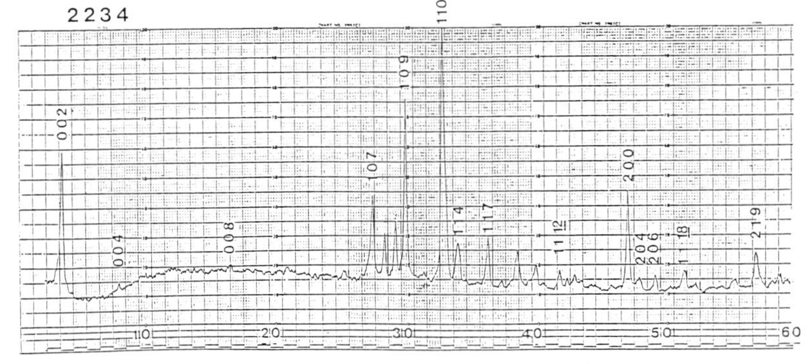

Fig. 12 X-ray diffractogram of the 2234 sample with the body-centered-tetragonal phase of a=b=3.85 and c=42.0 Å.

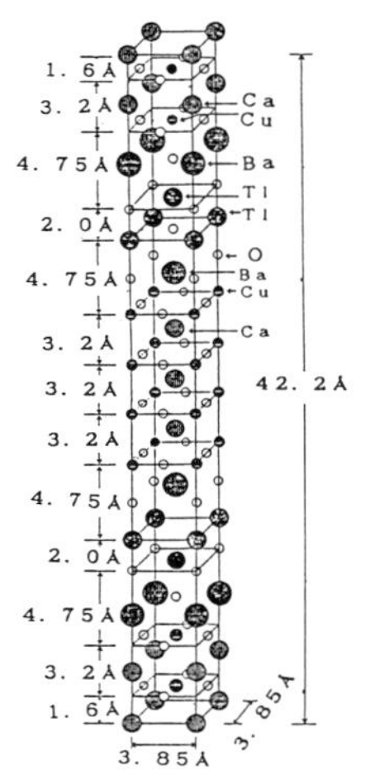

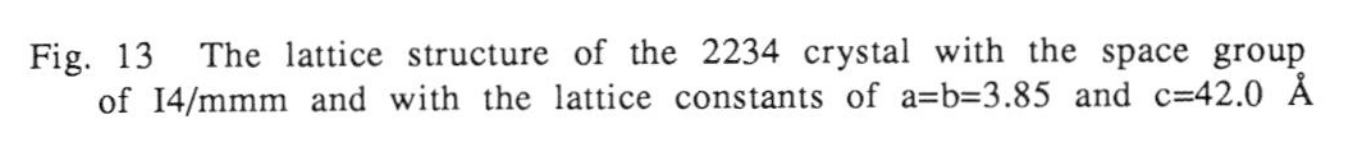
Fig. 13 The lattice structure of the 2234 crystal with the space group of I4/mmm and with the lattice constants of a=b=3.85 and c=42.0 Å

This c-axis rule was derived from the c_1, c_2 and c_3 lattice parameters by the equation of $c = 2((n-1)c_1 + 2c_2 + c_3)$. The oxygen content of the 2234 phase is also evaluated as 2n+4=12 from the lattice structure.

Discussion

The new superconductor family of $TlBa_2Ca_{n-1}Cu_nO_{2n+3}$ (n=1 ~ 5) have shown c lattice constants of 9.4, 12.6, 15.9, 19.1 and 22.3 Å and counter to Tc values of 13, 91, 116, 122 and 117 K for n=1, 2, 3, 4 and 5, respectively (as shown in Figs. 14 and 15). Their lattice constants followed well the c-axis rule of c = 6.3+3.2n for the single Tl-O layer compound. The calculated and experimental lattice constants c are in good agreement within 0.1 Å. The highest Tc value was obtained for the n=4 phase. This result is against the intuitive empirical rule. The average valence (z) of Cu ion in this $TlBa_2Ca_{n-1}Cu_nO_{2n+3}$ family is expressed as z = 2 + 1/n.[5] The structure with four Cu-O layers in a unit cell is responsible for the highest T_c value. This result would be explained by an average valence of the Cu ion of +2.25, or a Cu^{3+} ion concentration of 25 % .

The new high-Tc superconductors of single Tl-O layer compounds have less than half the Tl content of the previous 2223 phase. They are very important in science and technology, because of their new composition and structure, very high Tc and reduced thallium content.

The family of $Tl_2Ba_2Ca_{n-1}Cu_nO_{2n+4}$ (n=1 ~ 4) has been prepared and their lattice constants followed well the c-axis rule of c = 2(8.2+3.2n) for the double Tl-O layer compound. They have shown the c lattice constants of 23.2 , 29.3 , 35.6 and 42.0 Å and the Tc values of 80 K, 110 K, 127 K and 115 K for n=1, 2, 3 and 4, respectively as shown in Fig. 15. The calculated and experimental lattice constants c are in good agreement. The highest Tc value was obtained for the n=3 sample. This result would be explained by a proper carrier concentration. The stoichiometric double Tl-O layer compound $Tl_2Ba_2Ca_{n-1}Cu_nO_{2n+4}$ has always divalent Cu^{2+} ions in contrast to single Tl-O compound $Tl_1Ba_2Ca_{n-1}Cu_nO_{2n+3}$ in which the averge valence (z) of Cu ion is expressed as z = 2 + 1/n.. Tl vacancies or their occupation by Cu or Ca ions are thought to contribute to the formation of higher valence Cu^{3+} ions. This is consistent with Tl deficient concentration in the high-Tc samples of double Tl-O layered compounds.

In summary, the new 1234 and 1245 phase superconductors with single Tl-O layer and four or five Cu-O layers have been designed and prepared for the first time. They have T_c values of 122.1 and 117.0 K. Both phases have a simple tetragonal structure with lattice constants of c=19.1 or 22.3 Å with a common lattice constant of a=b=3.85 Å. The high-resolution electron microscopic measurement gave the simple tetragonal diffraction patterns, and the lattice images of the oxygen-deficient

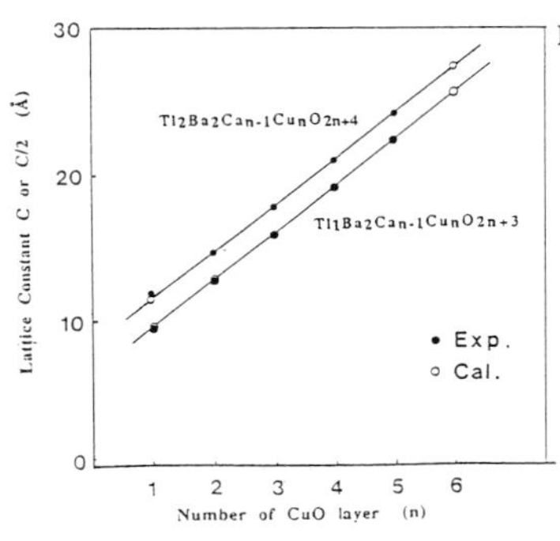

Fig. 14 The lattice constants c or c/2 versus the number of Cu-O layer for the $TlBa_2Ca_{n-1}Cu_nO_{2n+3}$ or $Tl_2Ba_2Ca_{n-1}Cu_nO_{2n+4}$ compounds, respectively. The lines are c=6.3+3.2n and c/2=8.2+3.2n. The filled circles are experimental ones and the open circles are calculated ones.

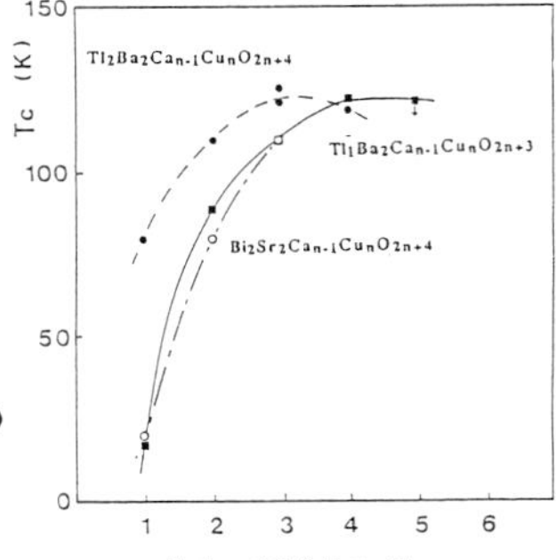

Fig. 15 Tc values versus the number (n) of Cu-O layers of $TlBa_2Ca_{n-1}Cu_nO_{2n+3}$ compounds compared with $Tl_2Ba_2Ca_{n-1}Cu_nO_{2n+4}$ and $Bi_2Sr_2Ca_{n-1}Cu_nO_{2n+4}$ compounds.

layered-perovskite structure with a single Tl-O layer alternating with four or five Cu-O layers follow the c-axis rule of c=6.3+3.2n [Å]. Their space group is P4/mmm. The phases have a large superconductor family of $TlBa_2Ca_{n-1}Cu_nO_{2n+3}$ (n=1, 2, 3, 4, 5, ----). The Tc value is strongly related to the valence of Cu ion.

New high Tc oxide superconductor $Tl_2Ba_2Ca_3Cu_4O_{12}$ (2234) with double Tl-O layer and four Cu-O layers has a body-centered tetragonal structure with the lattice constant of a=b=3.85 Å and c=42.0 Å and its space group is I4/mmm. The lattice constant c follows well the c-axis rule of c = 2(8.2+3.2n) for the double Tl-O layer compound. The Tc value of its superconductor was 117 K, unexpectedly lower than the 127 K of the three Cu-O layer 2223 compound.

The authors gratefully acknowledge Messrs. T. Oosuna and K. Ibe for electron microscope measurement at JEOL and Drs. M. Onoda, Y. Matui, H. Unoki and Y. Kimura for valuable discussions.

References

[1] S. kondoh, Y. Ando, M. Onoda and M. Sato, Solid State Comm. , 65 1329 (1988).

[2] Z. Z. Sheng & A. M. Hermann, Nature **332**, 55 (1988); R. M. Hazen, L. W. Finger, R. J. Angel, C. T. Prewitt, N. L. Ross, C. G. Hadidiacos, P. J. Heaney, D. R. Veblen, Z. Z. Sheng, A. Elali, A. M. Hermann, Phys. Rev. Letter. 60 1657 (1988).

[3] H. Maeda, Y. Tanaka, M. Fukutomi, T. Asano, Jpn. J. Appl. Phys. 27 209 1988.

[4] J. M. Wheatley, T. C. Hsu, P. W. Anderson, Nature 333 121 (1988).

[5] H. Ihara, R. Sugise, M. Hirabayashi, N. Terada, M. Jo, A. Negishi, M. Tokumoto, Y. Kimura, T. Shimomura, Nature 334 510 (1988). : H. Ihara, M. Hirabayashi, N. Terada, M. Jo, K. Hayasi, A. Negishi, M. Tokumoto, H. Oyanagi, R. Sugise, T. Shimomura and S. Ohasi, Proc. MRS Intern. Meeting on Advanced Materials (May 1988, Tokyo); IEEE Magn. MAG25 No.3 (1989).

[6] M. A. Subramanian, J. C. Calabrese, C. C. Torardi, J. Gopalakrishnan, T. R. Askew, R. B. Flippen, K. J. Morryssey, U. Chowdhry, A. W. Leight, Nature 332 420 (1988).

[7] H. Ihara, M. Hirabayashi, R. Sugise, N. Terada, M. Jo, K. Hayashi, M. Tokumoto, T. Shimomura and S. Ohashi, (in print Phys. Rev. B 38 No16 (1988)).

[8] J. D. Jorgensen, M. A. Beno, D. G. Hinks, L. Soderholm, K. J. Volin, R. L. Hitterman, J. D. Grace, Ivan K. Schuller, C. U. Segre, K. Zhang, M. S. Kleefisch, Phys. Rev. B 36 3608 (1987).

[9] M. Tokumoto, H. Ihara, T. Matsubara, M. Hirabayashi, N. Terada, H. Oyanagi, K. Murata, Y. Kimura, Jpn. J. Appl. Phys. 26 1565 (1987).

The Formation of High-Tc Superconducting Phases with Four Cu-O Layers in Tl-Ca-Ba-Cu-O Systems

P.T. WU[1], R.S. LIU[1], J.M. LIANG[1], and L.J. CHEN[2]

[1] Materials Research Laboratories, Industrial Technology Research Institute, 195 Chung-hsing Rd., Sec. 4, Chutung, Hsinchu 31015, Taiwan
[2] Department of Materials Science and Engineering, National Tsing Hua University, Hsinchu, Taiwan

ABSTRACT

Zero-resistance at temperatures up to 162 K was recorded for multiphase samples with nominal composition $TlCa_4Ba_3Cu_6O_x$. The compound with Tc-zero higher than 140 K was found to be highly repeatable. The compound was tentatively identified to be $TlCa_2Ba_3Cu_4O_x$, tetragonal in structure with a=b=0.394 nm and c=3.95 nm and of P4/mcc space group by TEM and EDS analysis. High resolution TEM images showed that the phase is primarily of a ten-subcell structure which is consistent with a structure model with four Cu-O layers interposed between Tl-O layers. Polytype structures, consisting of four- and five-subcell structures were also observed. The five-subcell structure was identified to be of P4/mmm space group with a=b=0.394 nm and c=1.97 nm.

INTRODUCTION

High temperature superconductivity has recently been discovered in Tl-Ca-Ba-Cu-O (TCBCO) system.[1-5] The compounds bear the distinction of possessing the highest Tc-zero to date (125 K). Because of the high toxicity that plagues Tl, the compounds are generally considered to be questionable candidate materials for commercial applications. Probably as a result, relatively small number of research groups are engaged in the research of Tl-containing superconducting oxides compared to those of vastly more popular Y-Ba-Cu-O (YBCO) and Bi-Sr-Ca-Cu-O (BSCCO) systems. Nevertheless, the system represents a novel high Tc superconducting system, similar to BSCCO system, which does not require simultaneous presence of rare earth (such as Y), and alkaline earth elements (such as Ca, Sr, Ba).[1] Furthermore, there are reports that the crystal structures of TCBCO compounds are generically related to YBCO as well as BSCCO compounds.[6-9] Initial study of Tl compounds indicated that the compounds are of simpler structures compared to those of much more complex BSCCO compounds. Study of thallium compounds may provide valuable information for the understanding of more difficult BSCCO compounds.

EXPERIMENTAL

Appropriate amounts of high purity $CaCO_3$, $BaCO_3$ and CuO powders were weighed stoichiometrically, ball-milled, filtered, dried and ground in an agate mortar. The well-mixed oxides were calcined at 925°C in air for 12 h with several intermittent grindings to obtain more homogeneous mixture of Ba-Ca-Cu-O powders. The presence of $BaCuO_2$ and CaO was found in calcined samples by x-ray diffraction. Differential thermal analysis (DTA) data showed that reactions of Ca-Ba-Cu-O powders occurred in the temperature range 920-930°C. The results provided guidance for selecting calcining temperatures. The Ba-Ca-Cu-O powders were then mixed with appropriate amount of Tl_2O_3 to yield mixtures with nominal cation stoichiometries of $TlCa_4Ba_3Cu_6$. The mixtures were ground and pressed into a cylindrical pellet, 2 mm in thickness and 8 mm in diameter, under a pressure of about 2 ton/cm^2. the pellets were then sintered in an enclosed gold crucible in flowing O_2 to alleviate possible decomposition of Tl_2O_3 to Tl_2O and O_2 for different periods of time. The annealing temperatures ranged from 880 to 970°C. The samples were then cooled by furnace cooling(3-5°C/min), controlled cooling (18°C/h) or cooling in air. DTA was conducted with a Du Pont Instruments 1090B thermal analyzer. A standard four point probe was used for electrical resistance measurements. Electrical contacts to the samples were made by fine copper wires attached to the samples with a conductive silver paint. The temperature was measured with a calibrated platinum resistor close to the sample. The detection limit for zero resistance was $10^{-6}\Omega$. Magnetization data were taken in an automatic superconducting quantum interference device (SQUID). X-ray diffraction (XRD) analysis was performed with a Philips x-ray diffractometer. JEOL-2000EX and JEOL-2000FX transmission electron microscopes operating at 200 kV were used. An EDAX 9100/70 energy dispersive spectrometer (EDS) was used to measure the atomic concentrations of the samples. Quantitative analysis was performed using standardless techniques with ZAF correction factors from EDAX 9100 program developed by EDAX laboratories.[10]

RESULTS AND DISCUSSION

Zero-resistance was recorded in the temperature range 130-162 K for samples annealed at 970°C for 4.5-10 min followed by furnace cooling. In samples annealed at 990°C for 4.5 min followed by furnace cooling, zero resistance was achieved at 152 K. Electrical resistance versus temperature curves for selected samples are plotted in Fig. 1. The zero-resistance at temperatures higher than 130 K was repeatedly observed for many samples in a time span of more than two months. However, efforts to obtain evidence of magnetic susceptibility change were not successful. The exact cause for the absence of Meissner effect is not clear at this time. $BaCuO_2$ was detected by both TEM and XRD in all sintered samples.

Two mirror planes and one four-fold axis were found in the CBED pattern along [001] direction of the grain with composition ratio of 1:2:3:4. The phase is herein referred to as 1:2:3:4 phase. The CBED pattern reveals a 4 mm symmetry in the whole pattern and the corresponding point group is either 4 mm or 4/mmm. Examinations of CBED patterns along [100] and [210] directions revealed that they are of 2 mm symmetry. It is therefore concluded that the symmetry of the crystal belongs to 4/mmm point group. Analysis of electron diffraction patterns along [100], [110], [001] and [210] directions indicated that the space group is P4/mcc. The conclusion was based on the observations that the point group of the lattice is of 4/mmm symmetry and the selection rules for the diffractions are 0kl:l=2n, hhl:l=2n, 00l:l=2n.[11] It is to be noted that strong (010) diffraction spot is evident in the [100] pattern for the 1:2:3:4 phase, whereas it is forbidden for the so-called 2:2:2:3 phase with a=b=0.385 nm and c=3.623 nm.[12] The lattice parameters of the 1:2:3:4 phase were determined to be a=b=0.394 nm, and c=3.95 nm.

High resolution lattice images of the 1:2:3:4 phase along [110] or [100] direction showed that the phase is of a ten-subcell structure. A subcell is herein defined as a tetragonal cell with metal atoms occupying the corners of the cell and a=b=0.394 nm and c about one-tenth of the length of the tetragonal unit cell of the phase. An example is shown in Fig. 2. However, polytypes of the phase with four- as well as five-subcell structures were also observed. The composition ratios of the regions with the presence of four- and five-subcell structures were measured to vary from 1:2:2:3 to 1:2:3:4 by EDS. Five-subcell structure was observed to be predominant in some of the grains. An example is shown in Fig. 3. The five-subcell structure was identified to be of P4/mmm space group with a=b=0.394 nm and c=1.97 nm.

The structure of 1:2:3:4 phase is distinctly different from the so-called 2:2:2:3 phase observed in samples exhibiting Tc-zero at about 110 K.[4-8,13] The 2:2:2:3 phase was detected by both electron diffraction and EDS in samples prepared under different conditions which will be described in details elsewhere.[14] An electron DP and a high resolution image of the structure of the 2:2:2:3 phase along [100] direction are shown in Fig. 4. In contrast to the [100] DP of the 1:2:3:4 phase as shown in the inset of Fig. 3, (010) spot is absent in the [100] DP of the 2:2:2:3 phase. The 2:2:2:3 phase was found to be of a triple perovskite-like layers interposed between Tl_2O_2 layers.[8]

The composition of the regions with the ten-subcell structure being predominant was measured by EDS to be close to $TlCa_2Ba_3Cu_4O_x$. Based on the P4/mcc space group and the observed lattice images along [100], [110] and [001] directions of the phase as well as the composition ratio being

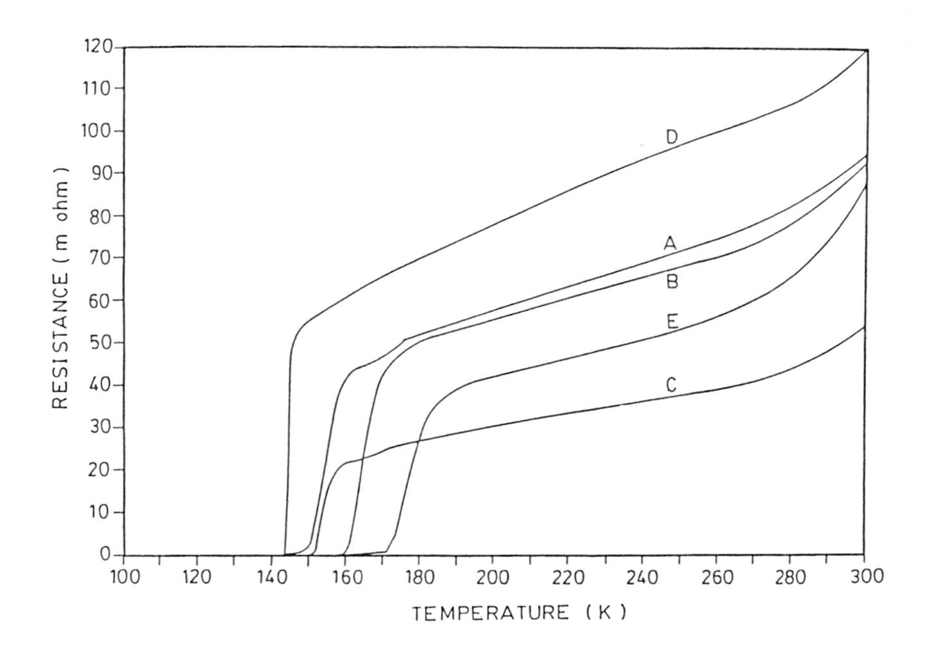

Fig. 1 Electrical resistance versus temperature curves for selected samples.

1:2:3:4, a probable structure model involving four layers of Cu-O is proposed. The sequence of cations along c-axis is suggested to be Tl-Ba-Cu-Ca-Cu-Ba-Cu-Ca-Cu-Ba-Tl. The model can account for the observed four-subcell structure by simply taking away one layer each of Ba and Cu near the central layers of the stacking sequence. The new sequence is Tl-Ba-Cu-Ca-Cu-Ca-Cu-Ba-Tl with an atomic ratio of 1:2:2:3. The schematical diagrams of the probable structures of 1:2:3:4 phase with ten- and five-subcell structures are highlighted in Figs. 2 and 3, respectively. The diagrams were drawn with the assumption that dark dots correspond to metal atoms. We note that identification of exact positions of metal atoms by lattice imagings requires extensive image simulation with known electron-optical and specimen parameters. However, by a comparison of the lattice images of 1:2:3:4 and 2:2:2:3 compounds, layers with the heavy Tl atoms could be identified since the 2:2:2:3 compound was well characterized previously.[8]

It was pointed out that a direct correlation between the copper-oxygen layers and the materials' superconductivity was apparent. The more copper-oxygen layers, the higher the Tc. For Tl-containing compounds, the Tc's were found to be 80, 110 and 125 K for compounds with one, two and three layer of Cu-O layers, respectively. It was predicted that a four-layer thallium compound would be superconducting at 150 to 160 K.[15] In this context, the 1:2:3:4 phase, which

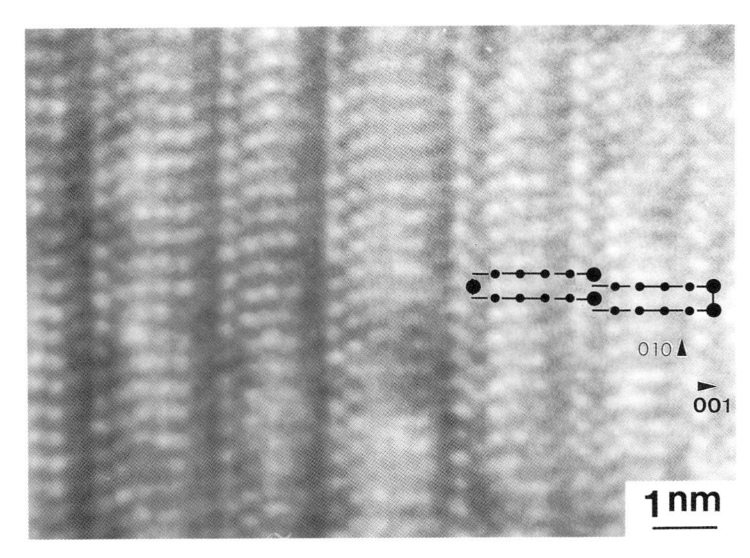

Fig. 2 High resolution (HR) lattice image showing the ten-subcell structure of the 1:2:3:4 phase along [100] direction. Inset is the corresponding diffraction pattern (DP).

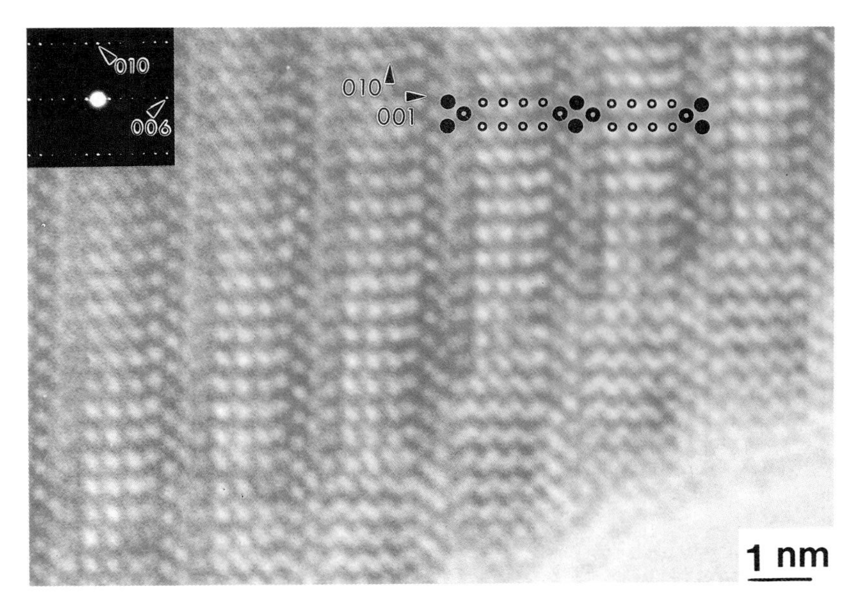

Fig. 3 HR image showing the five-subcell structure along [100] direction. Inset is the corresponding DP.

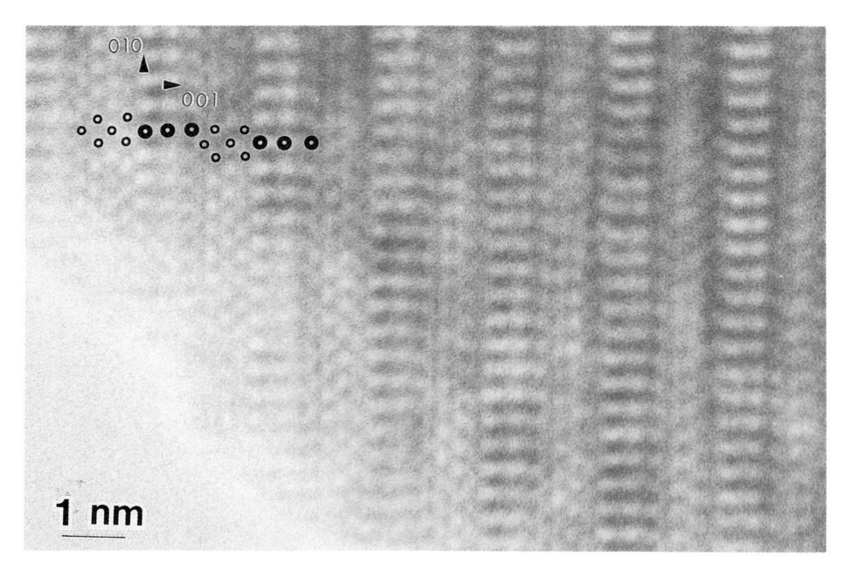

Fig. 4 HR image of a 2:2:2:3 phase along [100] direction. Inset is the corresponding DP.

was proposed to be of a four-layer structure, was found in all samples exhibiting Tc-zero above 140 K is likely to be the phase responsible for zero resistance at such high temperatures.

In summary, zero-resistance at temperatures up to 162 K was recorded for multiphase samples with nominal composition $TlCa_4Ba_3Cu_6O_x$. The compound with Tc-zero higher than 140 K was found to be highly repeatable. The compound was tentatively identified to be $TlCa_2Ba_3Cu_4O_x$, tetragonal in structure with a=b=0.394 nm and c=3.95 nm and of P4/mcc space group by TEM and EDS analysis. High resolution TEM images showed that the phase is primarily of a ten-subcell structure which is consistent with a structure model with four Cu-O layers interposed between Tl-O layers. Polytype structures, consisting of four- and five-subcell structures were also observed. The five-subcell structure was identified to be of P4/mmm space group with a=b=0.394 nm and c=1.97 nm.

REFERENCES

[1] Z.Z. Sheng and A.M. Hermann, Nature 332, 55 (1988).
[2] Z.Z. Sheng and A.M. Hermann, Nature 332, 138 (1988).
[3] R.S. Liu, W.H. Lee, P.T. Wu, Y.C. Chen and C.T. Chang, Jpn. J. Appl. Phys. 27, L1206 (1988).
[4] Z.Z. Sheng, W. Kiehl, J. Bennett, A. El Ali, D. Marsh, G.D. Money, F. Arammash, J. Smith, D. Viar, and A.M. Hermann, Appl. Phys. Lett. 52, 1738 (1988).
[5] P.T. Wu, R.S. Liu, J.M. Liang, W.H. Lee, L. Chang, L.J. Chen and C.T. Chang, Physica C 156, 109 (1988).
[6] L. Gao, Z.J. Huang, R.L. Meng, P.H. Hor, J. Bechtold, Y.Y. Sun, C.W. Chu, Z.Z. Sheng, and A.M. Hermann, preprint.
[7] A.M. Hermann, Z.Z. Sheng, D.C. Vier, S. Schultz, and S.B. Oseroff, preprint.
[8] M. Kikuchi, N. Kobayashi, H. Iwasaki, D. Shindo, T. Oku, A. Tokiwa, T. Kajitani, K. Hiraga, Y. Syono, and Y. Muto, Jpn. J. Appl. Phys. 27, L1050 (1988).
[9] S. Iijima, T. Ichihashi, and Y. Kubo, Jpn. J. Appl. Phys. 27, L817 (1988).
[10] A.O. Sandborg, R.B. Shen, and S.G. Maegdlin, The EDAX EDITOR, 10-3, 11 (1980).
[11] International Tables for Crystallography, vol. A, edited by T. Hahn (D. Reidel, Dordrecht, Holland, 1983).
[12] S.S. Parkin, V.Y. Lee, E.M. Engler, A.I. Nazzal, T.C. Huang, G. Gorman, R. Savoy, and R. Beyers, Phys. Rev. Lett. 60, 2539 (1988).
[13] R.M. Hazen, L.W. Finger, R.J. Angel, C.T. Prewitt, N.L. Ross, C.G. Hadidiacos, P.J. Heaney, D.R. Veblen, Z.Z. Sheng, A. El Ali, and A.M. Hermann, Phys. Rev. Lett. 60, 1657 (1988).
[14] J.M. Liang, unpublished work.
[15] R. Poole, Science 240, 146 (1988).

Superconductivity of Tl-Sr-Ca-Cu-O System in Relation to Tl-Ba-Ca-Cu-O and Bi-Sr-Ca-Cu-O Systems

SHIMPEI MATSUDA, SEIJI TAKEUCHI, ATUKO SOETA, TAKAAKI SUZUKI, KATSUZO AIHARA, and TOMOICHIKAMO

Hitachi Research Laboratory, Hitachi Ltd., Hitachi, Ibaraki, 319-12 Japan

ABSTRACT

It was found that Tl-Sr-Ca-Cu-O system gives rise to at least two superconducting phases, a high Tc phase at 100K and a low Tc phase at about 75K. The two phases are isostructural to those of Bi-Sr-Ca-Cu-O System. Comparing the Tc's and crystal structures of the Tl and Bi systems, it was suggested that Tc is mainly determined by the combination of two alkaline earth elements, i.e., Sr/Ca vs Ba/Ca, when the repetition unit in the layered perovskites is equal.

INTRODUCTION

A considerable number of superconductors based on copper oxide have been reported since the first discovery of a high Tc superconductor, La-Ba-Cu-O, by Bednorz and Müller[1]. Superconductors with higher Tc's were subsequently synthesized by replacing the La and/or Ba sites by other lanthanide and alkaline earth elements, for example, La-Sr-Cu-O (Tc=40K)[2], Y-Ba-Cu-O (90K)[3], and Y-Sr-Cu-O (80K)[4]. Recently superconducting transition exceeding 100K has been reported in Bi-Sr-Ca-Cu-O (105K, denoted as BSCC)[5] and Tl-Ba-Ca-Cu-O (120K, TBCC)[6, 7] systems. In both Bi and Tl containing systems it has been found that there are at least two crystal phases which give rise to superconducting state, i.e., a high Tc phase with 4-layered perovskite structure and a low Tc phase with 3-layered perovskite structure. It has been pointed out that Tc becomes higher as the repetition unit in the layered perovskite structure increases.

Comparing TBCC with BSCC system, the combination of Tl-Ba and Bi-Sr seems to be essential to determine the Tc's, since the remaining portion Ca-Cu-O is the same in both systems.[8] For establishing a mechanism of the high Tc superconductors and for designing a new higher Tc superconductor, it is significant to know whether Tl-Sr-Ca-Cu-O (TSCC) and/or Bi-Ba-Ca-Cu-O (BBCC) gives superconductivity, and if it does, what the Tc is.

The TSCC system was studied by Nagashima et al[9]. They reported that zero resistance was attained at 40K and a resistance drop was observed at about 100K. We have synthesized TSCC samples under various conditions, and found that there are at least two superconducting phases, i.e., a high Tc phase at 100K and a low Tc phase at about 75K. It is suggested that the combination of two alkaline earth elements (Ba/Ca or Sr/Ca) plays an important role in determining the Tc's in the Tl and Bi containing systems.

EXPERIMENTAL

The TSCC samples used in the present experiments were prepared as follows; (a) Appropriate amounts of $SrCO_3$, $CaCO_3$, and CuO were thoroughly mixed and calcined at 850°C for 5 hrs, (b) The resultant Sr-Ca-Cu-O was then mixed with Tl_2O_3 powder and pressed into a pellet (1 mm thickness and 30 mm diameter), (c) The pellet was placed in an alumina crucible with a cover and calcined in a furnace at 800 - 900°C for 1 - 20 hrs in a flowing air atmosphere. In some experiments Tl_2O_3 powder was placed in the crucible to make up a loss of Tl_2O_3 through vaporization. Typical nominal compositions were Tl/Sr/Ca/Cu (2/2/2/3) and (2/2/1/2). For the purpose of comparisons, various samples of TBCC and BSCC were also prepared in a similar fashion.

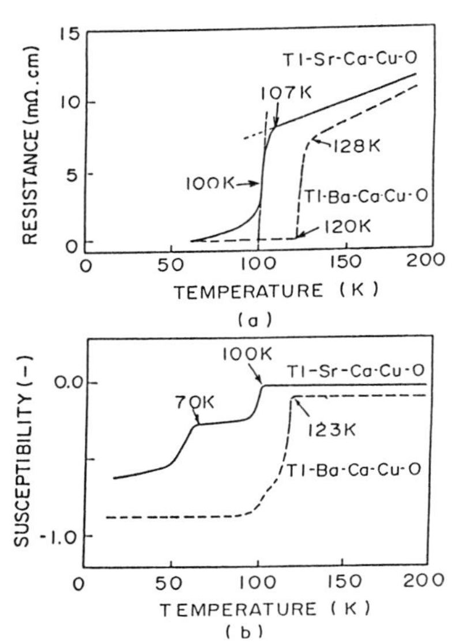

Figure 1. Temperature dependence of resistance (a) and susceptibility (b) for a Tl-Sr-Ca-Cu-O sample with a nominal composition (2/2/2/3). For the actual composition see text. Dashed lines for a Tl-Ba-Ca-Cu-O sample are shown for comparisons.

Eletrical resistance was measured by the standard four-probe method with indium soldering contact, and susceptibility change by the standard AC method as described previously[10]. The X-ray diffraction patterns were obtained using RIGAKU - RU 200 with Cu-Kα emission and SEM photographs using HITACHI S-800 electron micrograph.

RESULTS AND DISCUSSIONS

A number of samples consisting of Tl/Sr/Ca/Cu with various nominal compositions were prepared and their resistivity and susceptibility were measured between room temperature and liquid N_2 or He temperature. Typical resistivity and susceptibility changes for a TSCC (2/2/2/3) sample calcined at 870°C for 5 hrs are given in Figure 1. The composition analyzed by the ICP method was TSCC (1.2/1.8/2.0/3.0), showing a considerable loss of Tl through vaporization during the calcination. For a purpose of comparison the changes for a TBCC sample with the high Tc phase are also shown. The resistance of TSCC starts to deviate from a linear decrease at 107K, and decreases sharply at about 100K. A residual resistance which is often observed in a BSCC sample is also seen in the present sample. In Figure 1(b) it is seen that the susceptibility decreases in two steps, one at 100 K and the other at 70K. Superconducting volume between 100K and 4.2K is estimated to be about 60% of the total volume of the sample. The sample in disc form levitated on a magnet when cooled in liquid N_2. Taking into consideration the resistivity and susceptibility, we suppose that TSCC has two superconducting phases, i.e., a high Tc phase at 100K and a low Tc phase at about 70 K. Thus TSCC has a lower Tc than TBCC by about 20K in each phase, which is apparently caused by the difference of the combination of alkaline earth elements (Sr/Ca vs Ba/Ca).

Figure 2 shows a resistivity change of another TSCC sample calcined at 850°C for 3 hrs. A sharp decrease in resistivity is observed at 80 K, though zero resistance is not attained. The phase at 80 K is supposed to correspond to the low Tc phase of TSCC as indicated by the susceptibility change in Figure 1 (a). A typical resistivity curve of BSCC shown in Figure 2 clearly indicates a high Tc phase at 100K and low Tc phase at 80 K[5]. Combining the results shown in Figure 1 and 2 it might be said that both TSCC and BSCC have approximately the same superconducting transition temperature in the high and low Tc phases.

A number of SEM observation with EDX analysis were performed on the TSCC samples. Crystal grains consisting of Tl, Sr, Ca, and Cu were mainly observed, while minor grains consisting of Ca/Tl and Ca/Cu were also identified. Though the present TSCC samples apparently contained at least two superconducting phases, the quantitative analysis by EDX to distinguish the two phases was not possible.

An X-ray diffraction pattern of the TSCC sample shown in Figure 1 is given in Figure 3. The pattern is rather complicated, indicating the existence of several crystal phases. A tentative analysis was made based on the assumption that the sample was a mixture of two crystal phases shown in Figure 4; (a) a high Tc phase with 4-layered perovskite structure (c=15.6 A, a=b=3.79 A), and (b) a low Tc phase with 3-layered perovskite structure (c=12.1 A, a=b=3.79 A). Most of the peaks could be assigned to those crystal phases, though there were several unknown peaks remaining. The peak fitting was better when a single plane of TlO instead of $(TlO)_2$ was assumed.

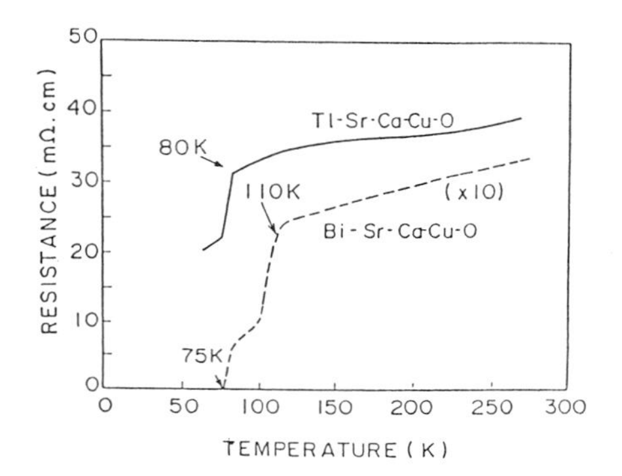

Figure 2. Temperature dependence of resistance for a Tl-Sr-Ca-Cu-O sample exhibiting the low Tc phase. Dashed line for a Bi-Sr-Ca-Cu-O sample is shown for comparisons. Nominal compositions are (2/2/2/3) for both samples.

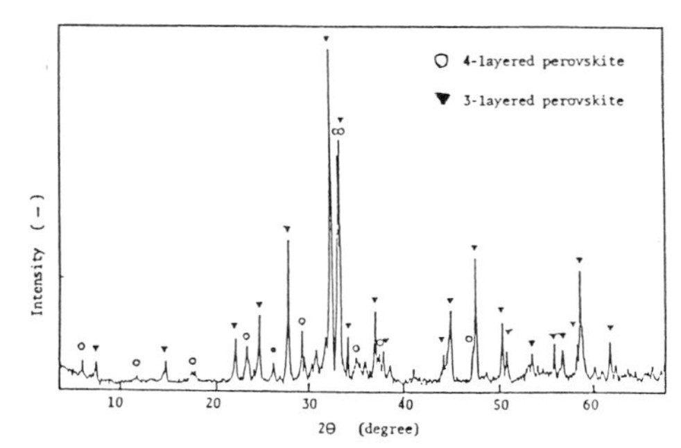

Figure 3. X-ray diffraction pattern of Tl-Sr-Ca-Cu-O for the sample shown in Figure 1. Peaks were assigned assuming 3- and 4-layered perovskite structures shown in Figure 4.

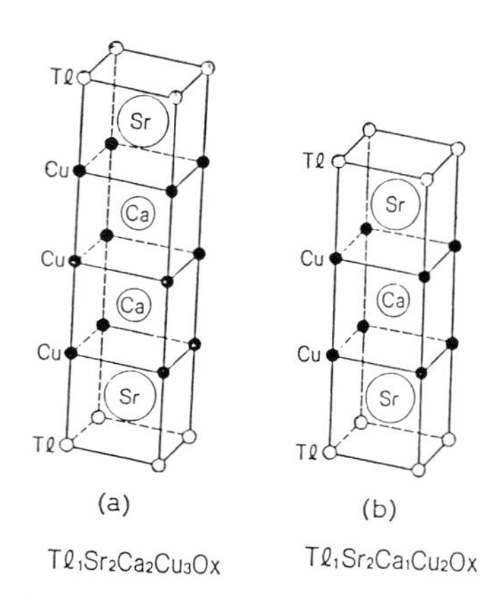

Figure 4. Crystal structures of Tl-Sr-Ca-Cu-O system assumed in analysis of X-ray diffraction pattern; (a) high Tc phase with 4-layered perovskite, (b) low Tc phase with 3-layered perovskite.

Table 1. Critical Temperature of (Tl,Bi)-(Ba/Ca,Sr/Ca)-Cu-O Systems

High Tc Phase

(4-Layered Perovskite)

	Ba/Ca	Sr/Ca
Tl/Cu	120 K	100 K*
Bi/Cu	—	105 K

Low Tc Phase

(3-Layered Perovskite)

	Ba/Ca	Sr/Ca
Tl/Cu	105 K	75 K*
Bi/Cu	—	80 K

* Midpoint temperature of transition.

The present findings that TSCC has two superconducting phases closely related to BSCC would suggest a few characteristic properties of the copper oxide based superconductors containing Tl or Bi. Combination of (Tl, Bi) and (Ba, Sr) makes four kinds of compounds, namely TBCC, TSCC, BSCC, and BBCC. Table 1 summarizes the critical temperature of the four compounds in the matrix form. First of all it is noted that BSCC and TSCC have nearly the same Tc, suggesting that Tc is determined by the portion, Sr-Ca-Cu-O. Secondly the Tc of TBCC is higher than that of TSCC by about 20 K, suggesting that the combination of Ba/Ca gives higher Tc than Sr/Ca. In Figure 5 the critical temperature of various layered perovskites as a function of the number of layeres is given. It is clearly shown that Tl-Sr-Ca-Cu-O system has a property parallel to Bi-Sr-Ca-Cu-O system in the 3-layered and 4-layered perovskites. We propose that the variation of Tc in the same number of layers is caused by the combination of A-site ions, i.e., Ba/Ca vs Sr/Ca. Considering the ionic radius of alkaline earth elements, i.e., Ba^{2+} 1.47 A, Sr^{2+} 1.25 A, and Ca^{2+} 1.07 A (12 coordination), it could be supposed that a larger difference in ionic radii between Ba^{2+} and Ca^{2+} causes a higher strain in crystal lattice and consequently a higher Tc. The fact that $YBa_2Cu_3O_x$ gives higher Tc than $YSr_2Cu_3O_x$ could be understood in the same way (Ba^{2+}/Y^{3+} vs Sr^{2+}/Y^{3+}, where Y^{3+} 0.99 A). In Figure 6 the dependence of critical temperature on the ratio of ionic radii of two A-site ions in the layered perovskites is shown. In the calculation of Ba/Y and Sr/Y ratio, the radius of Y is assumed to be 1.13A instead of normal trivalent radius 0.99A, taking into account the charge density difference (trivalent vs divalent ion). When a trivalent ion is compared with a divalent ion with the same ionic radius, the former is expected to exert a larger electrostatic force on the surrounding negative ions. A conversion factor 1.14 (= $(3/2)^{1/3}$) is multiplied to the normal trivalent radius. It is seen that Tc increases as the ratio of ionic radii increases in the 3-layered and 4-layered perovskite superconductors. A compound consisting of Bi-(Ba/Ca)-Cu-O has not been synthesized yet, but if it were realized, it might have similar superconducting properties to Tl-(Ba/Ca)-Cu-O.

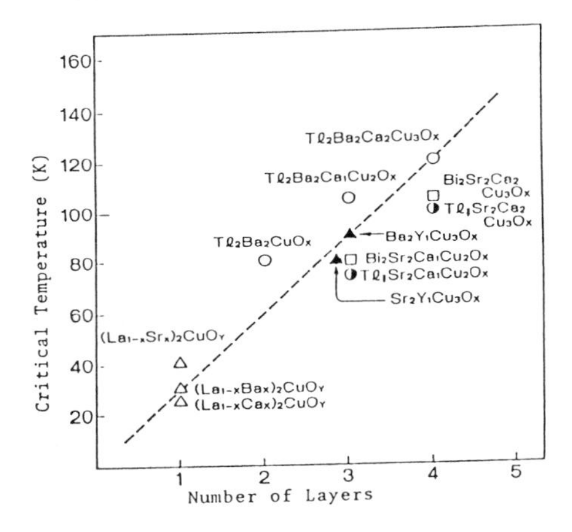

Figure 5. Critical Temperature of Layered Perovskite Superconductors

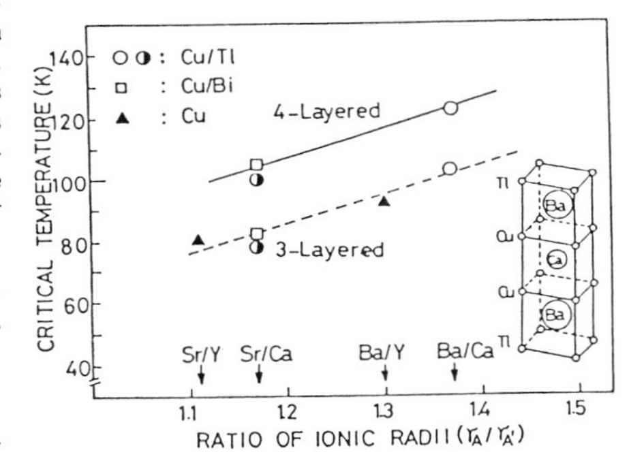

Figure 6. Dependence of critical temperature on ratio of ionic radii of two A-site ions in layered perovskites.

Acknowledgement: We gratefully acknowledge the following Hitachi colleagues for their assistance with this publication: Tsuneyuki Kanai, Teruo Kumagai, and Kazutoshi Higashiyama.

REFERENCES

(1) J.G. Bednorz and K.A. Müller : Z. Phys. B64 (1986) 189.
(2) S. Uchida, H. Takagi, K. Kitazawa and S. Tanaka: Jpn. J. Appl. Phys. 26 (1987) L1, ibid L151 and L123.
(3) M.K. Wu, J.R. Ashburn, C.J. Torng, P.H. Hor, R.L. Meng, L. Gao, Z. J. Huang, Y.Z. Wang and C.W. Chu: Phys. Rev. Lett. 58 (1987) 908.
(4) M. Oda, T. Murakami, Y. Enomoto, and M. Suzuki: Jpn. J. Appl. Phys. 26 (1987) L804.
(5) H. Maeda, T. Tanaka, M. Fukutomi and T. Asano: Jpn. J. Appl. Phys. 27 (1988) L209.
(6) Z.Z. Sheng and A.M. Hermann: Nature, 332 (1988) 138.
(7) Z.Z. Sheng, W. Kiehl, J. Bennett, A. Fl Ali, D. Marsh, G.D. Mooney, F. Arammash, J. Smith, D, Viar and A.M. Hermann: Appl. Phys, Lett. 52 (1988) 1738).
(8) M. Ronay and D.M. Newns: "A possible role of the Alkaline earth ions in high temperature superconductivity," submitted to Phys. Rev. B, Rapid communications.
(9) T. Nagashima, K. Watanabe, H. Saito and Y. Fukai: Jpn. J. Appl. Phys. 27 (1988) L1077.
(10) S. Matsuda, M. Okada, T. Morimoto, T. Matsumoto and K. Aihara : Mat. Res. Soc. Symp. Proc. Vol. 99, p. 695 (1988) Material Research Society.

Effect of Starting Materials and Heat Treatments on T1-Based High-Tc Superconductors

I. KIRSCHNER[1], I. HALASZ[2], GY. KOVACS[1], T. TRAGER[3], T. PORJESZ[1], G. ZSOLT[1], J. MATRAI[3], and T. KARMAN[1]

[1] Dept for Low Temperature Physics, Roland Eötvös University, 1088 Budapest, Puskin u. 5-7., Hungary
[2] Central Research Institute for Chemistry, 1025 Budapest, Pusztaszeri út 59/67., Hungary
[3] Central Research and Design Institute for Silicate Industry, 1034 Budapest, Bécsi út 126/128., Hungary

ABSTRACT

Superconducting samples of the same nominal composition of $TlCaBaCuO_{4.5\pm x}$ have been prepared by the aid of different starting materials and heat treatments. They have the critical temperatures of 0 K (no zero resistivity), 106 K, 111 K, 121 K, 92 K and 105 K respectively. It has been stated the highest Tc belongs to the most crystallized and essentially single-phase structure. In the most cases the phase of $Tl_2CaBa_2Cu_2O_8$ is responsible for the superconductivity. Certain content of the $Tl_2Ca_2Ba_2Cu_3O_{10}$ phase and some unidentified phases can also be found. A time-dependent character of some samples has been observed too.

INTRODUCTION

As it was observed in earlier papers [1,2], the superconducting properties of Tl-Ca-Ba-Cu-O compounds are influenced by the starting materials and heat treatment procedures. In the present paper the effect of these factors is investigated by a systematic change of the beginning chemicals and annealing processes at the same nominal composition. In order to realize this task, two different types of starting materials and chemical reactions, moreover four different types of heat treatment procedure were applied for changing preparation conditions.

SAMPLE PREPARATION

All samples have the nominal composition of $TlCaBaCuO_{4.5\pm x}$.

Among them the samples A, B, E and F were prepared by solid state reaction from starting materials of $TlNO_3$, CaO, $Ba(NO_3)_2 \times H_2O$ and CuO with the molar ratio of 1:1:1:1, while the samples C and D from $TlNO_3$, CaO and $BaCuO_2$ with the same molar ratio.

During the sample preparing process A, the starting materials were ground, mixed and the first heat-treatment was carried out at 165 °C, in air for 8.5 hours. This was followed by a quick cooling with a rate of 300 °C/min. Then it was heat-treated at 600 °C in air for 0.5 hour, during which it became sintered. After grinding, pressing under 5 MPa and a newer heat-treatment at 850 °C in air for 0.5 hour and cooling slowly for 16 hours we got back pellets of 1.5 cm diameter and 0.3 cm thickness.

The sample preparing process B, differed from the previous one in heat treatment method. During this preparation way, the first heat treatment happened at 350 °C in air for 1.5 hours, followed by a very rapid cooling with a rate of 500 °C/min. The second heat-treatment was carried out at 205 °C in air for 6.5 hours, then it was ground, pressed at 5 MPa pressure into pellets of 1.5 cm diameter and 0.3 cm thickness. The pellets were submitted to a heat-treatment at 205 °C, in air, for 8 hours, which was followed by a new one at 700 °C, in air for 1 hour and a slow cooling down. The final heat treatment was carried out 800 °C for 1.5 hours.

Based on the experiences of the first investigations simplified heat treatment processes were elaborated for the further sample preparation.

In the case of samples C and E starting mixtures were carefully ground, homogenized and preheated at 400 °C in air for 30 minutes. The next step was a slow cooling during 1 hour to room temperature and then the products were reground and heated again up to 900 °C in air for 10 minutes with a rate of 180 °C/min. This was followed by a quenching of 1000 °C/min down to liquid nitrogen temperature. After grinding, prisms of 2.5x0.5x0.2 cm sizes were pressed by 10 MPa, which were sintered at 900 °C in air for 3 minutes, followed by a very quick cooling (1000 °C/min) to liquid nitrogen.

The preparation process of the samples D and F differs from the earlier only in the last stage of the heat treatment. This latter consists of a gradual cooling to room temperature with a rate of 150 °C/hour.

SCANNING ELECTRON MICROSCOPY

The effect of different heat treatment techniques is reflected in the photographs taken on the different kinds of samples by a scanning electron microscope Jeol G-35. The pictures A, B, C, D, E and F of Fig.1 show the unlike character of microcrystalline structure of samples depending on the preparation process.

As it can be seen, the heat treatment of kind A resulted in forming microscrystals of different types, characterized by sharp contours, while the heat treatment of kind B led to rounded microcrystals and to a more homogeneous crystalline structure. This shows that a less homogeneous microcrystalline structure is given by the first heat treatment process, than by the second. Similarly to this, procedure C results in forming microcrystals of different shapes with more sharp edges. Samples D are characterized by a very homogeneous structure of strongly rounded microcrystals. A difference exists between the samples E and F too. Preparation process E provides a mixed structure of mainly two different microcrystals in a layered system. Samples F have a well-sintered picture without microcrystals of sharp contours. Collating the specimens E and F to each other, it seems to be clear that the latter represent a more homogeneous microcrystalline structure.

X-RAY DIFFRACTION

Analysis of the powder X-ray diffraction patterns (Philips PW-1360) provides useful informations on the possible phases and phase compositions established due to the different starting materials and heat treatment processes and related to superconductivity (see Fig.2 A, B, C, D, E and F). Comparision of the present diffraction data with those of Subramanian et al [3], Hazen et al [4] and ours [2] leads to some conclusions on the material content and crystallization in the specimens.

The characterizing three peaks in the diffractogram of samples A are at 2Θ values of 27.65, 31.45 and 32.81. There are some peaks which represent certain content of unidentified phases as well.

43 percents of the peaks of samples B belongs to $Ba(NO_3)_2$, while the other part of them does not agree with the known superconducting $Tl_2CaBa_2Cu_2O_{8\pm x}$ or $Tl_2Ca_2Ba_2Cu_3O_{10\pm x}$ phases.

The biggest part of peaks in samples C correspond to the Subramanian's type $Tl_2CaBa_2Cu_2O_8$ superconductor. Beside this less content of $Ba_2Cu_3O_5$ or $BaCuO_2$ and Hazen's type superconducting $Tl_2Ca_2Ba_2Cu_3O_{10}$ is supported by peaks. Samples D are the best crystallized ones. They can be regarded as almost pure singie-phase samples, in which beside the main superconducting material content of $Tl_2CaBa_2Cu_2O_8$ only an insignificant quantity of $BaCuO_2$ or $Ba_2Cu_3O_5$ and $Tl_2Ca_2Ba_2Cu_3O_{10}$ may be found.

In the samples E the majority phase consists of $Tl_2CaBa_2Cu_2O_8$ again and the other remnant phases, mentioned above, also occur.

Contrary to this, samples F do not have $Tl_2Ca_2Ba_2Cu_3O_{10}$ phase and only a very small amount of Ba-cuprate can be found in them.

Both of these superconducting phases have a tetragonal unit cell with lattice parameters a=3.86 Å and 5.41 Å, c=5.44 Å and 36.25 Å respectively.

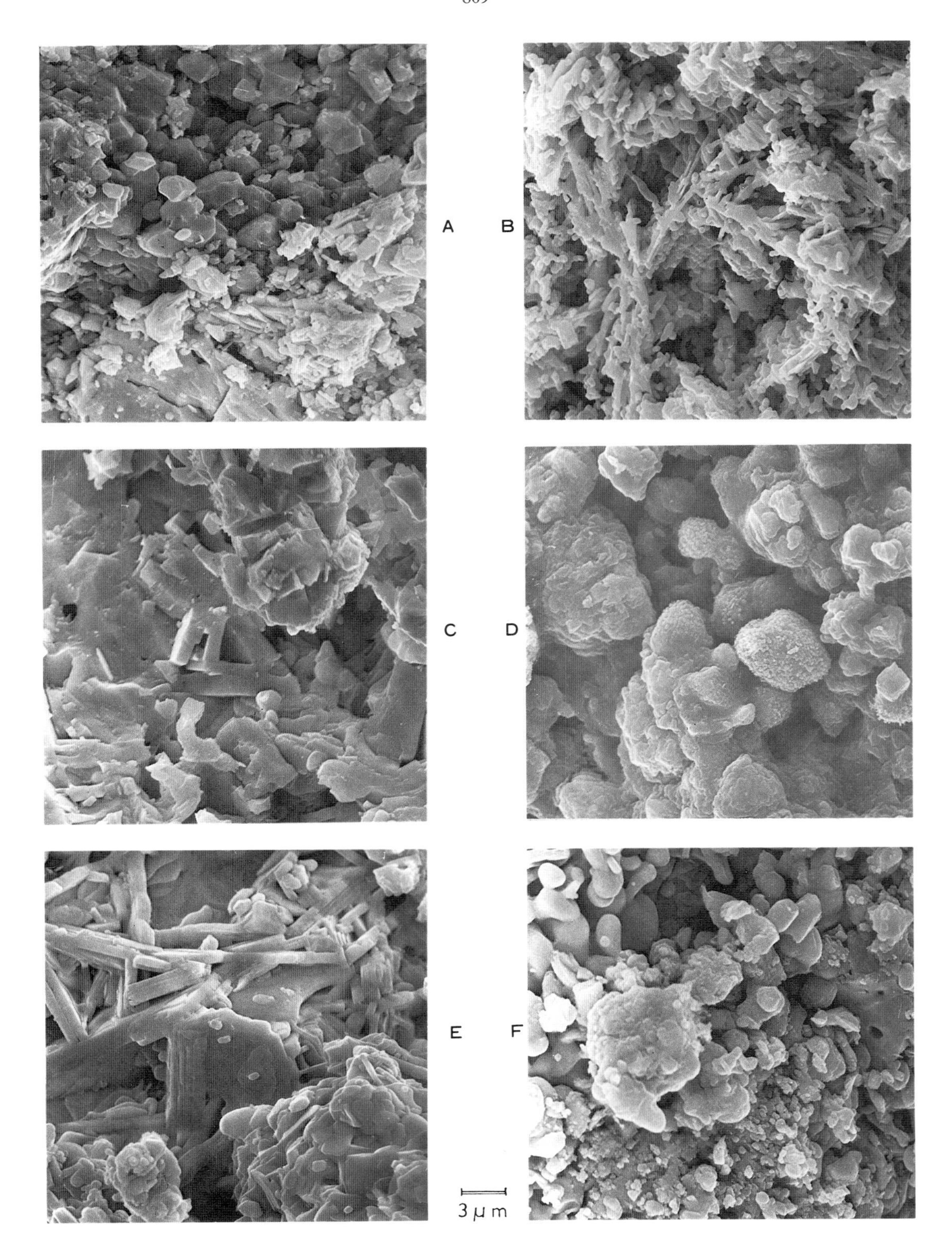

Fig.1 SEM photograph of samples kind A, B, C, D, E and F.

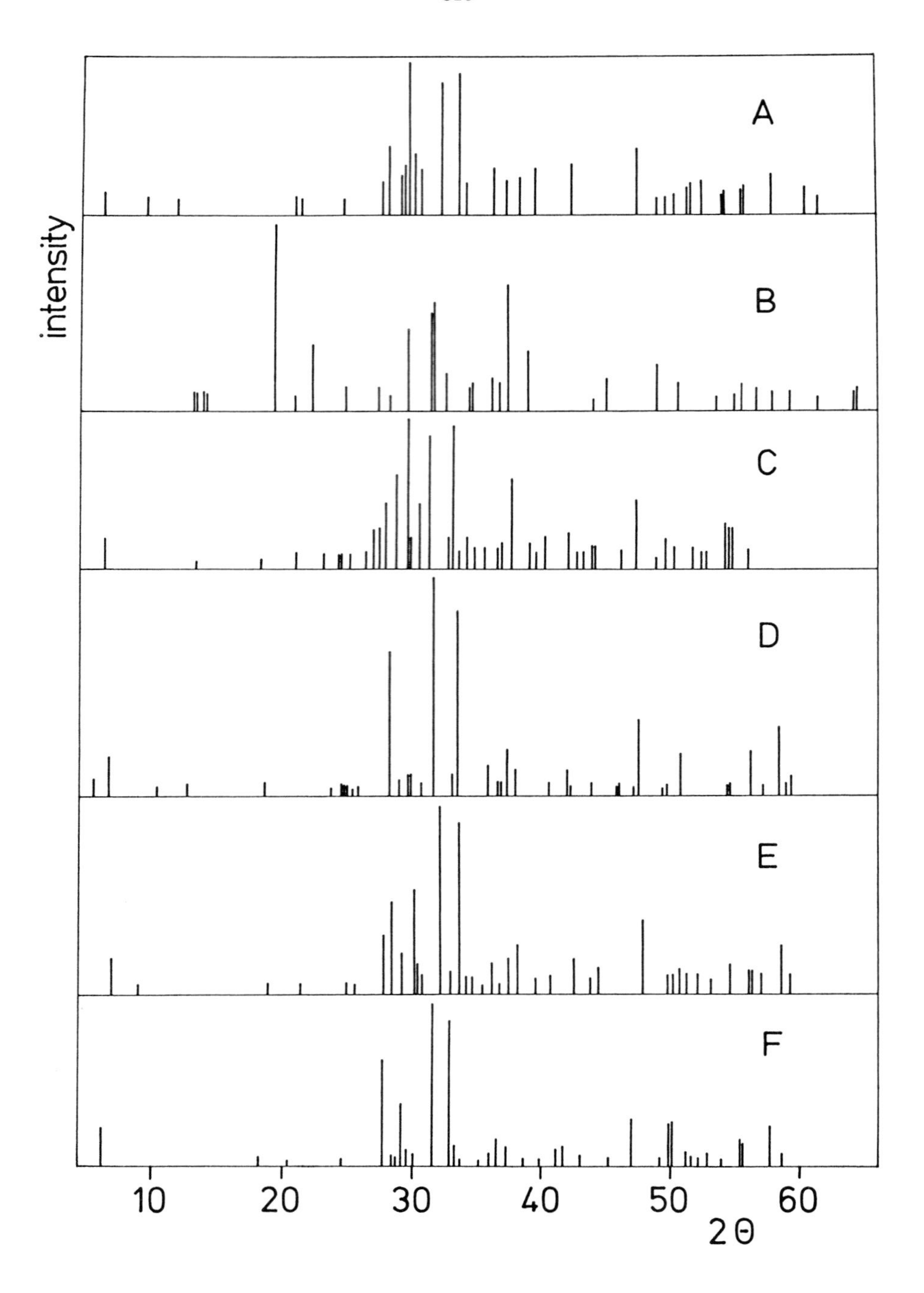

Fig.2 X-ray diffractogram of the samples.

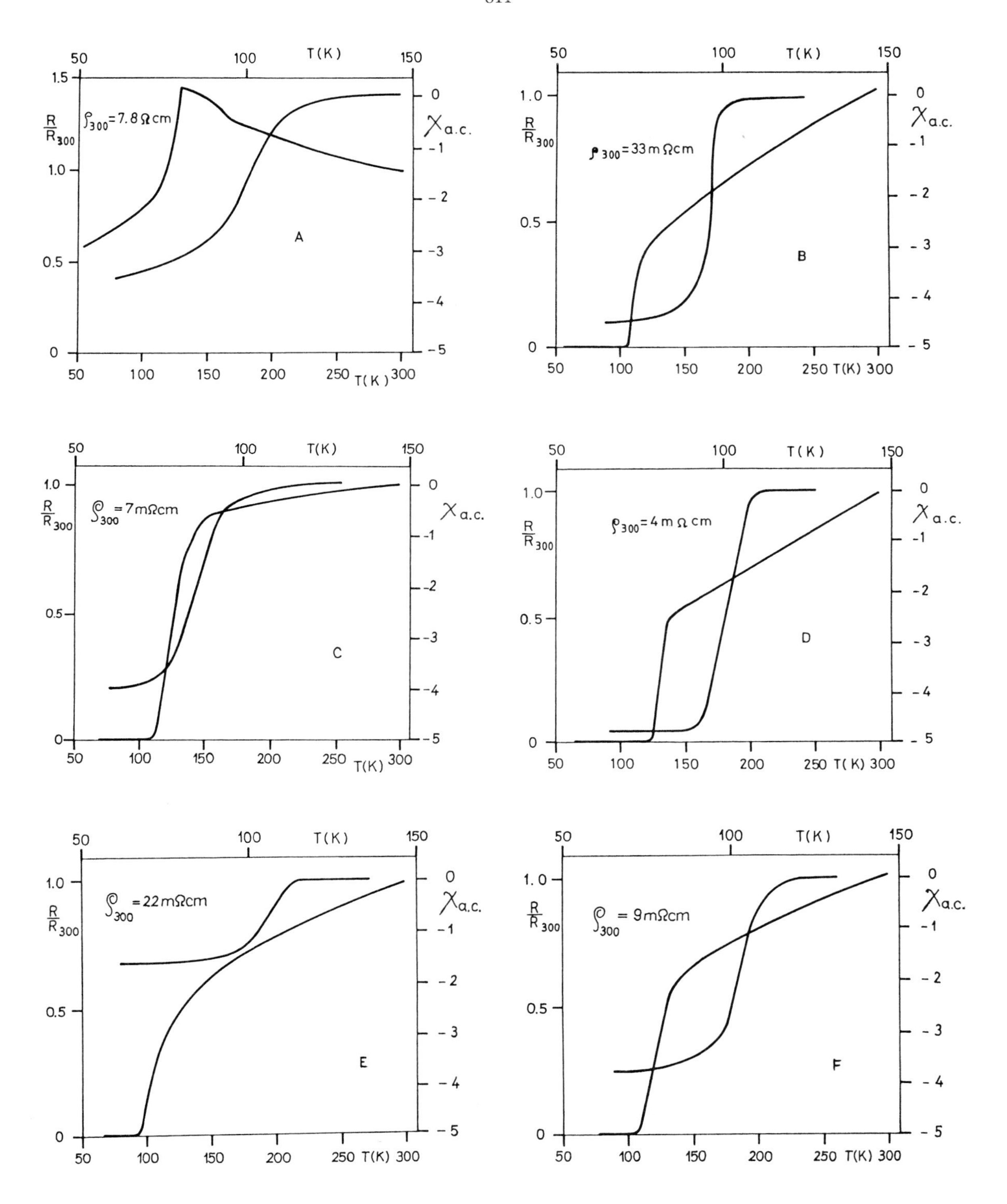

Fig.3 Resistance and a.c. susceptibility of the samples.

According to the results of earlier experiments on Y-Ba-Cu-O compounds it was expected that more homogeneous microcrystalline structure and better crystallized samples are more favourable from the point of view of high-Tc superconductivity in Tl-Ca-Ba-Cu-O systems too. This supposition can be controlled by electrical and magnetic measurements.

ELECTRICAL AND MAGNETIC PROPERTIES

The resistance-temperature (R-T) curves and the values of room temperature specific resistivity were determined by a conventional four-probe method using metallic press contacts to reach smaller contact resistances. During the electrical experiments 1-10 mA measuring currents were used. Fig.3 A, B, C, D, E and F demonstrate the typical R-T characteristics for the samples.

The magnetic measurements were performed by a low frequency a.c. method in a double-coil system. This method can determine directly the repulsion of the magnetic field from the samples and provides a.c. susceptibility signals which are also shown in Fig.3.

These investigations represent a direct correlation between the preparation technique and the superconductivity, inasmuch the different starting materials and different heat treatment processes bring forward different superconducting phases even at the same nominal concentration which play a fundamental role in the establishing actual value of the critical temperature.

TIME DEPENDENCE

A time-dependent behaviour was observed in one part of the samples, which manifests itself in the change of the characteristic parameters (namely the onset- and the critical temperature). It can happen either by a saturation character or by a transformation into a semiconductor.

REFERENCES

1. Sheng Z.Z. and Hermann A.M., Nature 332 138 (1988)
2. Kirschner I., Halász L., Kármán T., S.-Rozlosnik N., Zsolt G. and Trager T., Phys. Lett. A 130 39 (1988)
3. Subramanian M.A., Calabrese J.C., Torardi C.C., Gopalakrishnan J., Askew T.R., Flippen R.B., Morrissey K.J., Chowdhry U. and Sleight A.W., Nature 332 420 (1988)
4. Hazen R.M., Finger L.W.,, Anger R.J., Prewitt C.T., Ross N.L., Hadidiacos C.G., Heaney P.J., Veblen D.R., Sheng Z.Z., El-Ali A. and Hermann A.M., Phys. Rev. Lett. 60 1657 (1988)

Processing, Characterisation and Properties of the Superconducting Tl-Ba-Ca-Cu-O System

S.X. Dou[1,3], H.K. Liu[1,3], A.J. Bourdillon[1], N.X. Tan[1], N. Savvides[2], C. Andrikidis[2], R.B. Roberts[2], and C.C. Sorrell[1]

[1] School of Materials Science and Engineering, University of New South Wales, P.O. Box 1, Kensington, N.S.W. 2033, Australia
[2] CSIRO Division of Applied Physics, Lindfield, N.S.W. 2070, Australia
[3] Visiting Professor from Northeast University of Technology, Shenyang, People's Republic of China

ABSTRACT

The Tl-Ba-Ca-Cu-O system, when prepared by standard sintering procedures, shows mass loss by volatilisation of Tl even at sintering temperatures lower than the decomposition temperature of Tl_2O_3 (850°C). The disappearance of Tl results in difficulties in preparation, especially since Tl_2O_3 emission is severely toxic. The use of a two-stage sintering technique involves first the preparation of Ba-Ca-Cu-O by standard ceramic processing, so that Tl_2O_3 can be added to the sintered Ba-Ca-Cu-O, enclosed in a silver tube, and sintered. This technique ensures reproducible results and safety. Two superconducting phases were confirmed by T_c measurements, x-ray diffraction patterns, and electron microanalysis. The materials were characterised by DTA, electrical and magnetic measurement, and TEM. Prolonged sintering degraded the superconducting properties, probably due to thallium evolution. Transition broadening in applied magnetic fields is ascribed to anisotropy.

Superconducting transition temperatures (T_c) in the vicinity of 120K have been reported for the Tl-Ba-Ca-Cu-O system [1]. At least two superconducting phases with T_c > 100K have been identified [2] in this system ($Tl_2Ba_2Ca_2Cu_3O_{10+y}$, $Tl_2Ba_2Ca_1Cu_2O_{8+y}$) and layered structure similar to those of Bi-Sr-Ca-Cu-O system [3,4,5] have been proposed. In this paper we report on the processing techniques, electrical transport properties, and microstructures of thallium-based compounds with T_c (midpoint) up to 123K and zero resistance up to 120K. These compounds have features that are analogous to those observed in other high-temperature superconductors: $YBa_2Cu_3O_7$ and $Bi_2Sr_{3-x}Ca_xCu_2O_8$.

Thallic oxide (Tl_2O_3) is highly toxic and it has a high vapour pressure above 700°C. It decomposes above 875°C into toxic vapours, e.g., TlO, Tl. Thus, major difficulties are encountered in processing the thallium-based materials by standard ceramic procedures. We find, for example, that thallium is easily lost if calcining and sintering at temperatures above 850°C are carried out in an open crucible. We have overcome these difficulties by carrying out liquid phase sintering in closed silver crucibles.

Our technique consists of first preparing starting materials of Ba-Ca-Cu-O in appropriate ratios by calcining and sintering $BaCO_3$, $CaCO_3$, and CuO powders at ~890°C. Thallium is then added to these systems in two ways: (a) Tl_2O_3 is placed together with a block of the starting material in a thin-walled silver tube and sealed, or (b) the starting material is crushed to a powder, mixed with Tl_2O_3, and sealed in the silver tube. The final sintering is carried out at 800°-860°C for 20-180 minutes followed by slow cooling to 200°C at a rate of 1°-5°C/min. Since Tl_2O_3 melts at 717°C [6] the liquid oxide promotes liquid-phase sintering.

Figure 1 shows differential thermal analysis (DTA) data (obtained using a DuPont 900 differential thermal analyzer at 20°C/min) of a superconducting specimen with starting composition $Tl_2Ba_2Ca_2Cu_3O_{10+y}$. The data clearly show endothermic dissociation at ~815°C followed by decomposition of the specimen at ~925°C and explain the loss of thallium during sintering in an open environment.

Resistivity, a.c. susceptibility, and Meissner effect were measured on rectangular blocks of material of approximate dimensions 15 x 3 x 3 mm^3. The resistivity measurements were made by the standard four-probe technique with

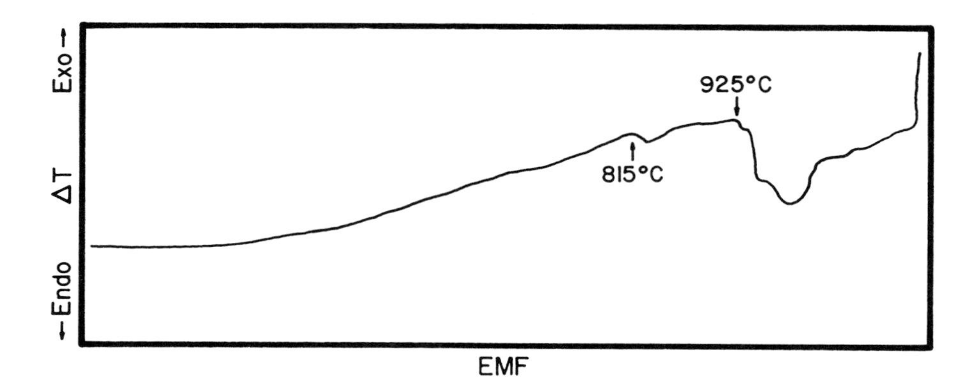

Figure 1. Differential thermal analysis of $Tl_2Ba_2Ca_2Cu_3O_{10+y}$.

probes ultrasonically bonded onto the specimens using indium. The specimens were connected to a copper block in which were embedded calibrated temperature sensors. Resistivity as a function of magnetic field was studied in the region of the resistive transition. Applied magnetic fields of 0-7T were provided by a superconducting coil. The a.c. susceptibility was measured by a mutual inductance technique at a frequency of 70 Hz. The volume fraction of superconducting phase was determined by comparison of the measured Meissner effect at 4.2K with that of a lead specimen of the same size [7].

Figure 2 shows the temperature dependence of resistivity for a specimen of nominal composition $Tl_1Ba_1Ca_2Cu_3O_x$ prepared under optimum conditions-880°C in flowing O_2 for 3 hours. The superconducting phase was identified as $Tl_2Ba_2Ca_2Cu_3O_{10+y}$. Prolonged sintering at 880°C or at higher temperatures tends to cause this phase to decompose, with accompanying degradation of electrical

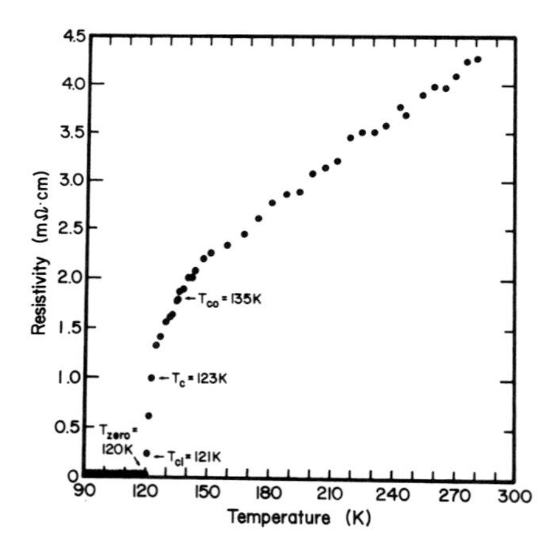

Figure 2. Temperature-dependent resistivity of $Tl_2Ba_2Ca_2Cu_3O_{10+y}$ showing transition temperature T_c = 120K.

properties. Figure 3(a,b) shows the effect of prolonged sintering on the resistivity and a.c. susceptibility of a typical specimen. Initially the specimen (sintered at 880°C for 3hours) has a low resistivity and T_c = 120K with a transition width (90-10%) ΔT_c = 8K. When the specimen is heat treated for an additional 3 hours both its normal state and superconducting properties degrade. The relative volume of superconducting phase (measured by Meissner effect at 4.2K) is found to change from ~20% to 15%. The ρ-T data indicate that the higher-T_c phase ($Tl_2Ba_2Ca_2Cu_3O_{10+y}$, {2223}, dissociates, giving rise to a second phase of composition $Tl_2Ba_2Ca_1Cu_2O_{8+y}$, {2212}, with T_c = 100K). X-ray diffraction data described below and microstructural examination confirm this finding.

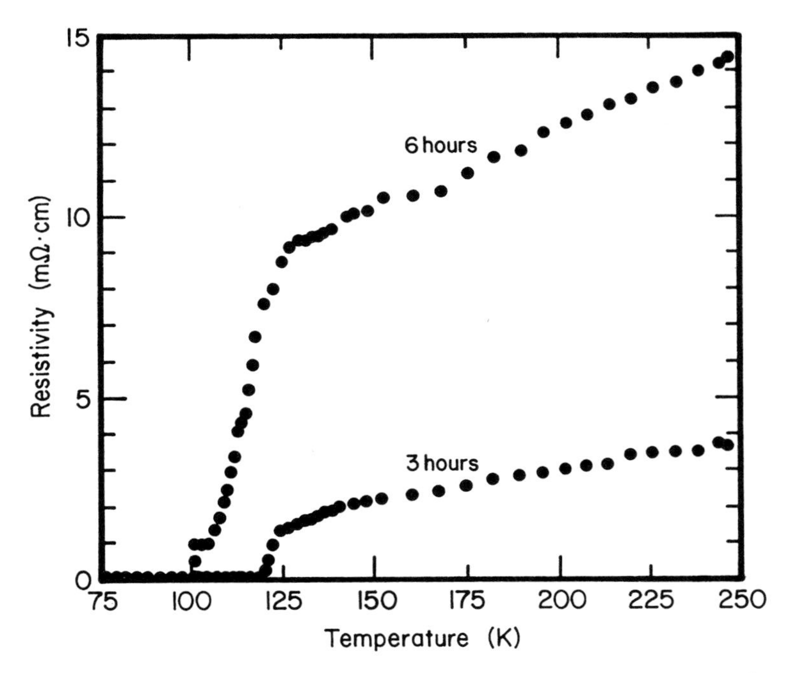

Figure 3a. Temperature-dependent resistivity of $Tl_2Ba_2Ca_2Cu_3O_{10+y}$ showing degradation in Tc after increasing sintering time from 3h to 6h.

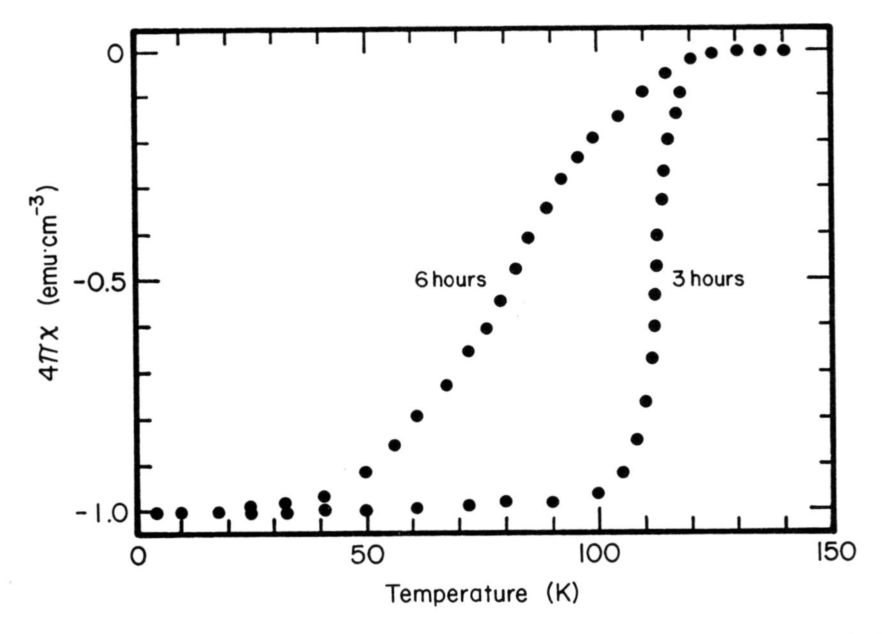

Figure 3b. Temperature-dependent a.c. susceptibility after sintering for 3h and 6h.

Figure 4 shows typical resistive transitions of a specimen in various magnetic fields between zero and 7T. At zero applied field the superconducting transition is reasonably sharp, with Tc = 8K, and Tc = 110K. With increasing magnetic field the transition broadens but the 90% onset shifts only slightly to lower temperatures. This behaviour is qualitatively similar to that observed in $Y_1Ba_2Cu_3O_{7-y}$ and its derivative compounds and suggests strong anisotropy in the critical field and critical current density.

Microstructures were studied with a JEOL JSM 840 scanning electron microscope (SEM) and JEOL 2000 FX transmission electron microscope (TEM), both equipped

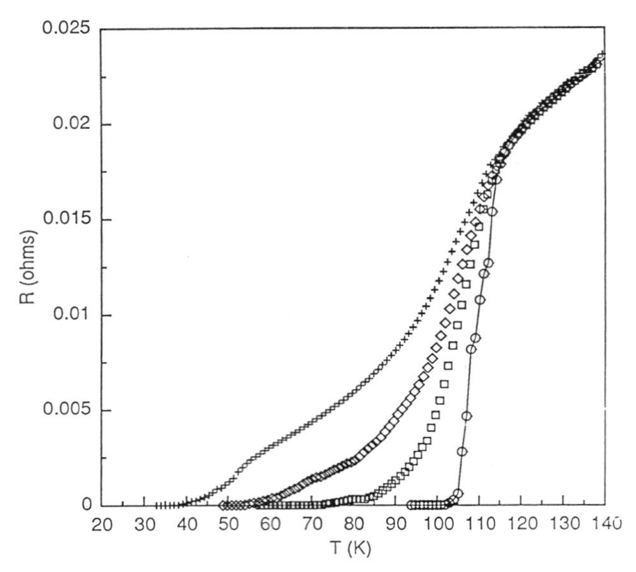

Figure 4. Temperature-dependent resistance of $Tl_2Ba_2Ca_2Cu_3O_{10+}$ with applied magnetic field at 0T (o), 0.1T (□), 1T (◇) and 7T (+).

with Link Systems energy dispersive spectrometers (EDS). The first sintered product, obtained prior to liquid-phase sintering with Tl_2O_3, contained a multiphase assemblage of binary compounds such as CaO, ternaries such as Ca_2CuO_3 and $BaCuO_2$, and quaternaries such as $Ba_3CaCu_3O_7$. When this product was reacted with Tl_2O_3 a multiphase superconducting material was formed. The sample was considerably denser and tougher than was achieved with the plate-like morphology of an otherwise similar material, $Bi_2Sr_{3-x}Ca_xCu_2O_{8+y}$ [3,4,5]. The dominant phase (~90%) showed a mean composition corresponding to $Tl_2Ba_2Ca_{0.8}Cu_2O_{8+y}$ or $Tl_2Ba_2Ca_2Cu_3O_{10+y}$. The sample also contained minority phases such as $BaCuO_2$, and Tl-Cu-O.

Confirmation of these structures was obtained in TEM from grains of material picked from suspension in methanol onto holey carbon. These grains are like $YBa_2Cu_3O_{7+y}$ in not showing superlattice reflections (which generally might be expected in oxide superconductors due to soft phonons), but they are unlike $Bi_2Sr_{3-x}Ca_xCu_2O_{8+y}$, which can show a superlattice, periodic in a [110] direction [4,8,9]. In this structure the incommensurate superlattice has periodicity about 5 x a, where a = 0.3839 nm, c = 3.0863 nm. Sometimes the superlattice can be observed in <110> directions in the tetragonal structural. In $Tl_2Ba_2CaCu_2O_{8+y}$, however, bright field images (figure 5) show rows of lattice planes due to the absorption of the layered heavy elements. The typical separation of these rows (1.67 nm, two rows per unit cell) matches the electron diffraction pattern. This pattern is consistent with I4/mmm symmetry, with lattice parameters of a = 0.389 + 0.002 nm and c = 3.378 + 0.022 nm (see also ref. 1).

Also seen in figure 5 are stacking faults, which showed conventional fringes when the specimen was tilted off-axis into a two-beam condition. Stacking faults can have a fundamental effect on the properties. Variations in lattice spacing observable in figure 5 imply that layers of different phases (and different T_c) can co-exist in the same crystal, one layer having more or less $CaCuO_2$ complexes within the unit cell. Such material is not stoichiometric, with a single crystal containing a multiphase layered assemblage, e.g., of {2212}, {2201}, {2223}, or even {2234}. The last of these may be expected to have the highest T_c.

X-ray powder diffraction patterns were obtained from a Philips type PW 1140/00 powder diffractometer with Cu Kα radiation. Indexed diffraction patterns for the tetragonal I4/mmm structure are given in tables I and II. Unit cell parameters were determined to be a = 0.3914 nm, c = 3.3644 nm for $Tl_2Ba_2Ca_2Cu_3O_{10+y}$ and a = 0.3864 nm, c = 2.9471 nm for $Tl_2Ba_2Ca_1Cu_2O_{8+y}$, respectively.

Figure 5. Bright field lattice image of (100) plane of $Tl_2Ba_2Ca_2Cu_3O_{10+y}$ showing stacking faults apparently due to multiphase layer of assemblage.

Finally, unstable ions have been observed in {2223} as in $YBa_2Cu_3O_{7-y}$ [10] and $Bi_2Sr_{3-x}Ca_xCu_2O_{8+y}$ [5]. Labile ions can be measured by the evolution of O_2 gas after immersion of a sample in dilute hydrochloric acid. In {2223} 0.91 ml of O_2 were evolved per gram of sample, which corresponds to 8.7% of the Cu ions being in the state Cu^{3+}, or to an average Cu valence of 2.09. However, the material was not single-phase and some variation is to be expected in material prepared by different means.

In conclusion, two phases of the Tl-Ba-Ca-Cu-O system have been formed reproducibly by sintering in closed silver crucibles. The product was not single-phase, but the dominant phase was identified for material with T_c = 123K, and Meissner effect measurements showed that the superconducting fraction was 20%. Superconducting properties were degraded by prolonged sintering, probably owing to loss of thallium. In magnetic fields the superconducting transition broadened, which is consistent with anisotropic conduction. It is to be expected that, with each new high-temperature superconductor discovered, the number of applicable theoretical models will be reduced.

Table I. Diffraction pattern for the $Tl_2Ba_2Ca_2Cu_3O_{10}$ superconductor based on a tetragonal unit cell with a = 0.39138 nm, c = 3.36439 nm.

2θ	I/I_o	$\sin^2\theta$	d_{obs}	d_{cal}	h	k	l
5.4	24.4	0.002219	16.3638	16.8220	0	0	2
29.46	85.2	0.06466	3.0315	3.0349	1	0	7
31.99	25.8	0.07592	2.7982	2.8037	0	0	12
33.26	100	0.08192	2.6936	2.7033	1	0	9
34.11	31.5	0.08602	2.6284	2.6288	1	1	4
36.15	25.4	0.09626	2.4844	2.4817	1	1	6
39.31	23.6	0.1131	2.2916	2.3118	1	1	8
42.04	21.3	0.1286	2.1491	2.1373	1	1	10
45.95	21.4	0.1524	1.9751	1.9695	1	1	12
49.28	48.0	0.1738	1.8491	1.8790	2	0	5
52.35	27.9	0.1946	1.7477	1.7503	2	1	0
53.39	41.7	0.2018	1.7162	1.7136	2	1	4
54.28	22.2	0.2081	1.6902	1.6939	2	1	5
55.1	27.6	0.2139	1.6668	1.6708	2	1	6
55.66	24.4	0.2180	1.6511	1.6446	2	1	7
56.69	29.5	0.2254	1.6236	1.6159	2	1	8
58.75	65.4	0.2406	1.5717	1.5852	2	1	9
59.73	36.2	0.2479	1.5483	1.5527	2	1	10
67.43	33.1	0.3081	1.3890	1.3790	2	2	2
69.25	39.4	0.3229	1.3567	1.3654	2	2	4

ACKNOWLEDGEMENTS: We are grateful to Metal Manufactures Ltd. for support (S.X.D), and to the Department of Industry, Techn ology and Commerce for support (H.K.L) under a general technology grant.

Table II. Diffraction pattern for The $Tl_2Ba_2CaCu_2O_8$ superconductor based on a tetragonal unit cell with a = 0.38637 nm, c = 2.94713 nm.

2θ	I/I_o	$\sin^2\theta$	d_{obs}	d_{cal}	h	k	l
29.39	88.1	0.06434	3.0386	3.0368	1	0	6
32.78	100	0.07962	2.7318	2.7320	1	1	0
33.61	91.4	0.08360	2.6666	2.6662	1	0	8
34.11	33.7	0.08602	2.6284	2.6322	1	1	3
34.83	20.4	0.08955	2.5765	2.5616	1	1	4
36.15	27.2	0.09626	2.4844	2.4787	1	1	5
39.31	25.2	0.1131	2.2916	2.2918	1	1	7
40.73	47.4	0.1211	2.2152	2.1944	1	1	8
43.84	39.9	0.1393	2.0651	2.0978	1	1	9
47.56	81.8	0.1626	1.9120	1.9128	1	1	11
49.28	51.3	0.1738	1.8491	1.8265	1	1	12
52.35	29.8	0.1946	1.7477	1.7446	1	1	13
53.39	44.6	0.2018	1.7162	1.7249	2	1	1
54.28	23.7	0.2081	1.6902	1.7018	2	1	3
55.1	29.5	0.2139	1.6668	1.6823	2	1	4
55.66	26.1	0.2180	1.6511	1.6581	2	1	5
56.69	63.6	0.2254	1.6236	1.6299	2	1	6
58.75	69.9	0.2406	1.5717	1.5644	2	1	8
69.98	27.9	0.3288	1.3444	1.3308	2	2	3

REFERENCES

1) Z.Z.Sheng and A.M. Hermann, Nature (1988) 332, 55-58, 138-39.
2) M.A.Sabramanian, J.C.Calabrese, C.C.Torardi, J.Gopalakrishnan, T.R.Askew, R.B.Flippen, K.J.Morrissey, U.Chowdry and A.W.Sleight, Nature (1988) 332, 420-22.
3) D.R.Veblen, P.J.Heaney, R.J.Angel, L.W.Finger, R.M.Hazen, C.T.Prewitt, N.L.Ross, C.W.Chu, P.H.Hor and R.L.Meng, Nature (1988) 332, 374-37.
4) J.M.Tarascon, Y.Le Page, P.Barbou, B.G.Bagley, L.H.Greene, W.R.McKinnon, G.W.Hull, M.Giroud and D.M.Hwang (1988), Crystal Substructure and Physical Properties of the Superconducting Phase $Bi_4(Sr,Ca)_6Cu_4O_{16+x}$. Preprint.
5) S.X.Dou, H.K.Liu, A.J.Bourdillon, N.X.Tan, J.P.Zhou, N.Savvides and C.C.Sorrell, Labile Ions in Superconducting Bi-Sr-Ca-Cu-O and its Composition and Properties. Superconductor Science and Technology, IOP (1988), in press.
6) CRC Handbook of Chemistry and Physics, 58th edition. Ed. R.C.Weast (1987) CRC, Cleveland.
7) S.X.Dou, A.J.Bourdillon, C.C.Sorrell, K.E.Easterling, S.Ringer, N.Savvides, J.B.Dunlop, and R.B.Roberts, Appl. Phys. Lett. (1987) 51, 535-37.
8) E.A.Hewat, J.J.Capponi, C.Chaillout, J.L.Hodeau, M.Marezio. Preprint.
9) J.M.Tarascon, P.Barboux, L.H.Greene, G.B.Bagley, G.W.Hull, Proc. Int. Conf. High-Temp. Supercond. Mater. Mech. Supercond. Feb. 29 - March 4, Interlaken, Switzerland, in press
10) S.X.Dou, H.K.Liu, A.J.Bourdillon, N.X.Tan, J.P.Zhou, C.C.Sorrell and K.E.Easterling, Mod. Phys. Lett. B (1988) 1, 363-67.

Substitution Effect of Y for Ca in Tl-Ba-Ca-Cu-O System

T. MANAKO [1], Y. SHIMAKAWA [1], Y. KUBO [1], T. SATOH [2], and H. IGARASHI [1]

Fundamental Research Laboratories [1], Resources and Environment Protection Laboratories [2], NEC Corporation, 4-1-1, Miyazaki, Miyamae-ku, Kawasaki, 213 Japan

ABSTRACT

The substitution effect of Y for Ca in $Tl_2Ba_2CaCu_2O_8$ was studied. With the substituion, T_c decreased and lattice parameters changed systematically. Another phase, however, was found in highly substituted samples. X-ray diffraction and EPMA studies revealed that this new phase has a chemical composition of $TlBa_2Y_{1-x}Ca_xCu_2O_7$ (1212 phase) and its structure is similar to both $Tl_2Ba_2CaCu_2O_8$ and $YBa_2Cu_3O_7$. This structure contains two Cu perovskite-like pyramids sandwiched by monolayer Tl-O sheets, in contrast to that there are bilayer Tl-O sheets in the $Tl_2Ba_2CaCu_2O_x$ and Cu-O chain layer in the $YBa_2Cu_3O_7$.

For x=0, electrical property of this phase is semiconductor-like and non-superconducting. With increasing x, it becomes metallic, and for $x>0.3$, it exhibits superconductivity.

INTRODUCTION

Since the discovery of superconductivity in the Tl-Ba-Ca-Cu-O system [1], many studies have been made on these oxides. Three types of crystal strucrures isostructural with Bi-Sr-Ca-Cu-O system have already been confirmed [2,3]; $Tl_2Ba_2CuO_6$ (2201phase), $Tl_2Ba_2CaCu_2O_8$ (2212 phase) and $Tl_2Ba_2Ca_2Cu_3O_{10}$ (2223 phase). These structures include some Cu-O pyramids or octahedrons sandwiched by Tl-O double layer sheets.

It is very interesting to substitute one element with other elements having different valences in oxide superconductor, because it will vary the average Cu valence and influence electrical properties. The Y substitution study for Ca in Bi 2212 phase [4] revealed that all of Ca can be substituted by Y and that T_c decreased and the lattice parameters changed successively with the substitution. This T_c decrease is considered to be caused by the decrease of the average Cu valence. Since the Ca site of the 2212 phase is quite similar to the Y site of the $YBa_2Cu_3O_7$ (1-2-3 structure) and especially Tl system have the same alkaline earth, so Ca site seems to be easily substituted by the Y ion.

We made substitution study of Y for Ca in the Tl 2212 phase, and prepared samples with the starting compositions of $Tl_2Ba_2Ca_{1-x}Y_xCu_2O_8$. In the small x region, single phase samples with 2212 structure were obtained. With increasing x, the transition temperatures decreased and the lattice parameters changed systematically with the substitution, as was observed in the Bi system. But the second phase appeared in the range of $x>0.5$, and for $x>0.8$ the samples became single phase with an unknown structure. Electrical property of this phase is semiconductor-like and exhibit no superconductivity. In this work, we determined its structure to be $TlBa_2YCu_2O_7$ (1212 structure) by a powder x-ray diffraction using the Rietveld analysis [5]. Also, we found that by substituting Ca for Y reversely, it becomes to exhibit superconductivity without changing its structure.

EXPERIMENTAL

Samples were prepared by solid state reaction. Tl_2O_3, BaO, Y_2O_3, CaO and CuO powders were mixed with the designed compositions. After mixing, the mixtures were pressed into pellets and wrapped in a gold foil and fired at 890°C for 3 hours in O_2 atmosphere. Some pellets were reground and fired once again. DC resistivity was measured by a conventional four probe method. Powder X-ray diffraction measurements were performed using Cu Ka radiation. The sampling interval was 0.02° in 2θ and fixed time was 4 sec for every step. The 2θ range was from 10° to 100°. Rietveld analysis was performed using the computer program RIETAN[6]. The chemical compositions of samples were examined by EPMA technique.

RESULTS AND DISCUSSION

1. CRYSTAL STRUCTURE OF THE $TlBa_2YCu_2O_7$

The X-ray diffraction pattern for the sample with the starting composition of $Tl_2Ba_2YCu_2O_x$ was quite differnt from that of the 2212 phase and successfully indexed as a tetragonal unit cell with the lattice parameters of $a_0=3.86$Å and $c_0=12.5$Å. This c_0 value is smaller than the half value of that of the 2212 phase by 2.1Å which is very near to the separation between Tl-O sheets in the $Tl_2Ba_2Ca_{n-1}Cu_nO_{4+2n}$ (n=1,2 and 3) [3]. It suggests that one of the Tl-O double layers in the 2212 phase may be missing in this system, that is, $TlBa_2YCu_2O_7$ (1212 phase) should have been obtained. Indeed, this XRD pattern was the same one as that of

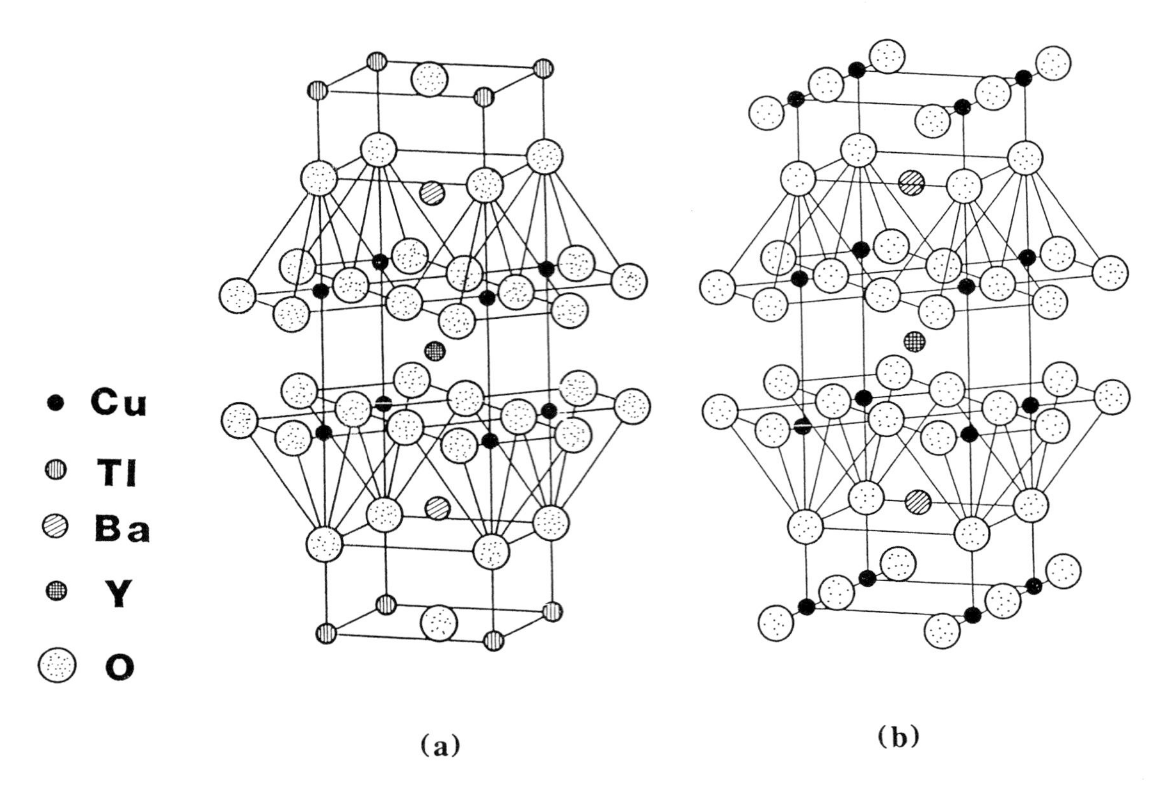

Fig.1 The structure models for (a) $TlBa_2YCu_2O_7$ and (b) $YBa_2Cu_3O_7$.

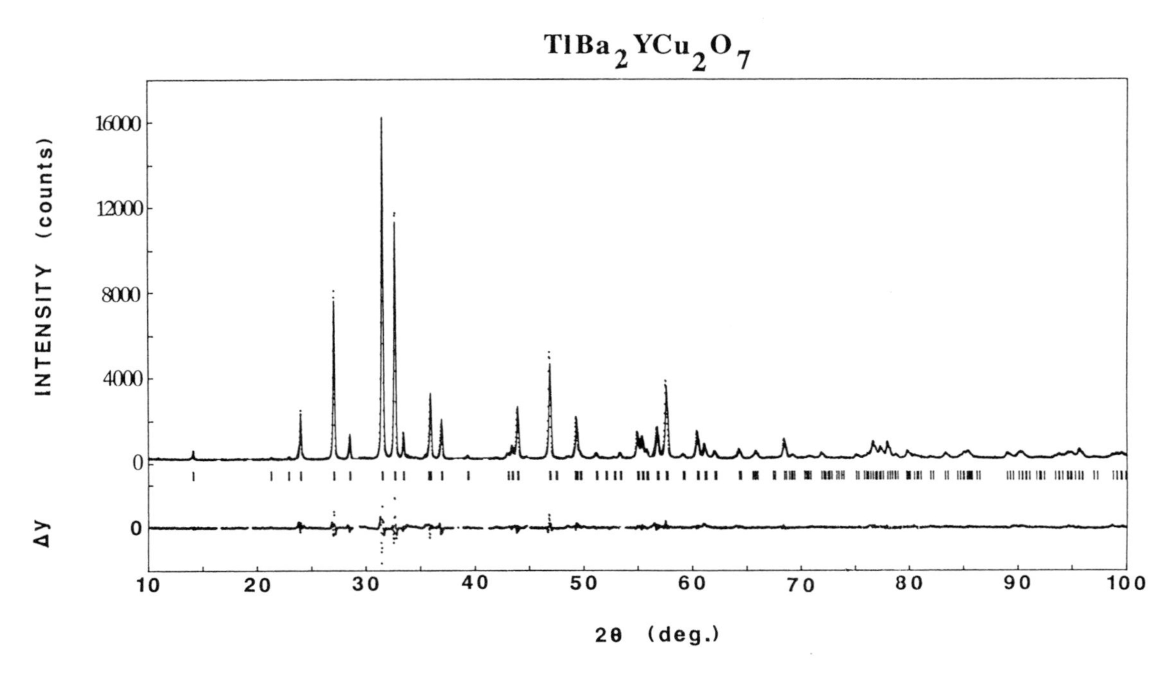

Fig.2 Rietveld refinement patterns for $TlBa_2YCu_2O_7$. The dots are observed intensities, and the solid lines calculated intensities.

Table 1 Crystallographic data for $TlBa_2YCu_2O_7$. Space group P4/mmm; a=3.86873(8), c=12.4732(3)Å. Numbers in parenteses are uncertainties in the last decimal place.

Atom	Site	x	y	z	B/Å^2	g
Tl	1d	0.5	0.5	0.5	2.13 (15)	0.96 (3)
Ba	2g	0	0	0.2902 (4)	0.85 (6)	0.98 (4)
Cu	2h	0.5	0.5	0.1305 (10)	0.42 (3)	0.97 (4)
Y	1a	0	0	0	0.64 (5)	1.02 (4)
O (1)	1b	0	0	0.5	2.55 (15)	1
O (2)	2h	0.5	0.5	0.3368 (37)	1.02 (6)	1
O (3)	4i	0	0.5	0.1145 (27)	0.51 (3)	1

R_{wp}=8.29 R_p=6.20 R_e=4.52
R_I=3.73 R_F=2.67

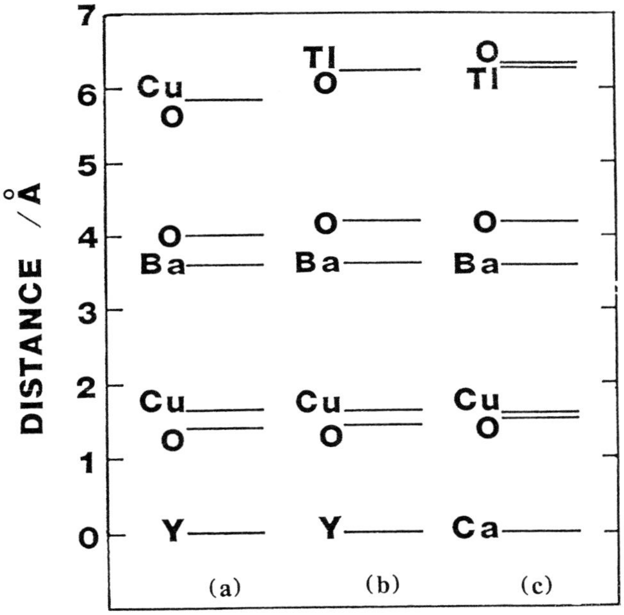

Fig.3 Interlayer separations of (a) $YBa_2Cu_3O_7$, (b) $TlBa_2YCu_2O_8$ and (c) $Tl_2Ba_2CaCu_2O_8$

the sample mode from the starting composition of 1212. This model has also been supported by the electron probe micro analysis which determined the chemical composition as $Tl_{0.9}Ba_2Y_{1.2}Cu_2O_{6.2}$.

The crystal structure model of this phase is shown in Fig.1 (a). It is very similar to that of $YBa_2Cu_3O_7$ as shown in Fig.1 (b). The Reitveld refinement was performed using a sample of starting composition of $TlBa_2YCu_2O_x$. In the refinement, the occupation factors of oxygen atoms were fixed to 1. The observed and calculated x-ray diffraction patterns are shown in Fig.2, and refined crystal structure parameters and R factors are listed in Table 1. A considerably good fitting is obtained. The occupation factors of metal ions are very close to 1. The calculated interlayer distances for the $TlBa_2YCu_2O_7$ are shown in Fig.3 (b), together with those for the 2212 and the 1-2-3 phases. It is clear that the crystal structure of the 1212 phase is similar to both 2212 and 1-2-3 phases. The roughness of Cu-O sheet in the 1212 phase is similar to that in the 1-2-3 phase but is larger than that in the 2212 phase. This should be related to the difference of ionic valence between Y and Ca ions. The roughness of Ba-O sheet resembles to the 2212 phase.

It is well known that the Tl-Ba-Ca-Cu-O system has many crystal structure variations. In our previous works, we discovered the exisistense of Tl-O monolayer sheets in this system [7,8,9]. But they were considered to be a kind of defect structure, because most of Tl-O sheets were found in a bilayer form. Recently, some studies also reported the existence of the Tl-O monolayer structure in the Tl-Ba(-Ca)-Cu-O system [10,11,12], although it was difficult to get a single phase sample. Therefore, it seems that Tl-O bilayer is more stable than monolayer in normal $Tl_2Ba_2Ca_{n-1}Cu_nO_{4+2n}$ system (n=1,2 and 3). Because the Tl-O sheet supplies the charge of +1 per layer, removement of one layer from the Tl-O bilayer in the 2201, 2212 and 2223 phases requires the increase of average Cu valence up to 3, 2.5 and 2.25, respectively. The Y substitution for Ca makes this charge compensation unnecessary and the Tl-O monolayer structure becomes stable.

2. Ca SUBSTITUTION EFFECT

The electrical property of $TlBa_2YCu_2O_7$ was semiconductor-like and showed no superconductivity although the crystal structure is very similar to that of the superconducting 1-2-3 phase. This should be related to the fact that the average Cu valence estimated from the chemical composition of $TlBa_2YCu_3O_7$ is +2. It seems possible to raise average Cu valence by substituting Ca for Y and to make the specimen exhibit superconductivity.

The Ca substituted samples with starting composition of $TlBa_2Y_{1-x}Ca_xCu_2O_7$ were prepared. Single phase samples were obtained in the range of $x<0.4$. For the larger x value, $BaCuO_2$ and trace of 2212 phase were recognized in XRD measurement. The lattice parameters a_0 and c_0 determined from XRD measurement are plotted against x in Fig.4. With increasing x, lattice parameters changed systematically, that is, a_0 became shorter and c_0 became longer. It is the same tendency as is observed in the 2212 phase. The results of DC resistivity measurement are shown in Fig.5. Normal resistivities at 300K and zero

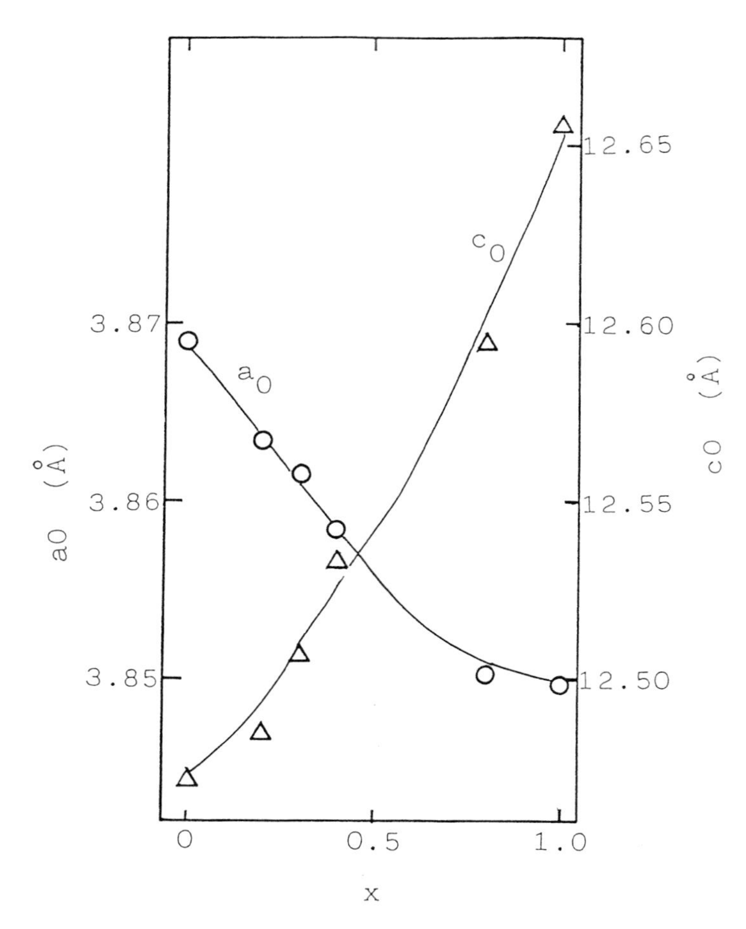

Fig.4 Lattice parameters plotted against the composition x in $TlBa_2Y_{1-x}Ca_xCu_2O_7$

Table 2 Resistivities at 300K and transition temperatures ($\rho=0$) of $TlBa_2Y_{1-x}Ca_xCu_2O_7$.

x	0	0.2	0.3	0.4	0.5	0.8	1.0
ρ(Ω·cm)	1060	0.0569	0.0219	0.0235	0.0111	0.0070	0.0069
T_c(K)	—	—	35	65	86	92	70

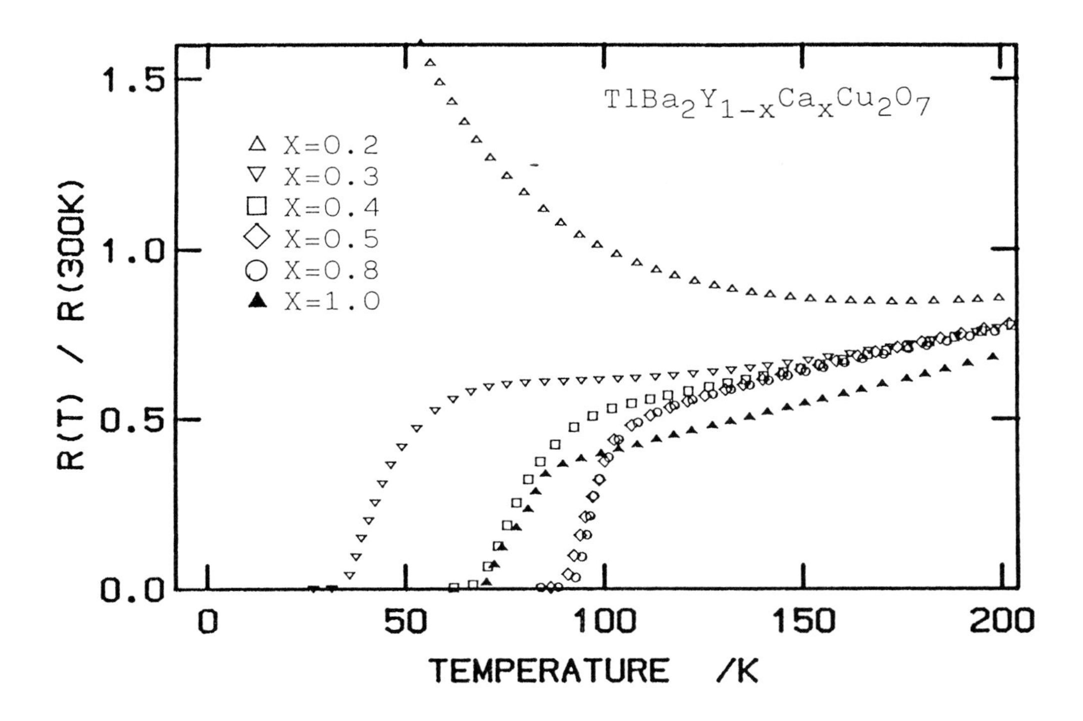

Fig.5 Temperature dependences of the dc resistivity for $TlBa_2Y_{1-x}Ca_xCu_2O_7$

resistivity temperatures are listed in Table 2. Normal resistivity decreased by the substitution. It should be due to the increase of carrier density. Samples with x larger than 0.3 exhibited superconductivity. T_c increased with increasing x, and when x=0.8, T_c reached the maximum value of 92K. However these T_c values were very sensitive to the preparation conditions [3], probably because of partial substitution effects, modulated structures and intergrowth structures. Further studies are required to establish the exact influence of substituion.

CONCLUSION

A single phase sample with a composition of $TlBa_2YCu_2O_7$ was obtained. The Rietveld analysis of x-ray diffractiion pattern revealed that the crystal structure of this phase is similar to both 2212 and 1-2-3 phases. Electrical property was semiconductor-like, but Ca substitution for Y caused a 90K superconductivity.

ACKNOWLEDGEMENT

We thank F. Izumi of NIRIM for giving us the Rietveld analysis program. We gratefully acknowledge useful discussions with S. Iijima. We also thank M. Yonezawa for the support of this work.

REFERENCES

[1] Z.Z. Sheng and A.M. Hermann, Nature 332(1988)138.
[2] Y. Kubo,Y. Shimakawa,T. Manako,T. Satoh and H. Igarashi, Jpn. J. Appl. Phys. 27(1988)L591.
[3] Y. Shimakawa, Y. Kubo, T. Manako, Y. Nakabayashi and H. Igarashi, Physica C, 156(1988)97.
[4] N. Fukushima, H. Niu and K. Ando, Jpn. J. Appl. Phys (submitted).
[5] T. Manako, Y. Shimakawa, Y. Kubo, T. Satoh and H. Igarashi, Physica C (in press).
[6] F. Izumi, J. Crystallogra. Soc. Jpn. 27(1985)23 [in Japanese].
[7] S. Iijima, T. Ichihashi and Y. Kubo, Jpn. J. Appl. Phys. 27(1988) L817.
[8] S. Iijima, T. Ichihashi, Y. Shimakawa, T. Manako and Y. Kubo, Jpn. J. Appl. Phys. 27(1988)L837.
[9] S. Iijima, T. Ichihashi, Y. Shimakawa, T. Manako and Y. Kubo, Jpn. J. Appl. Phys. 27(1988)L1054.
[10] R. Bayers, S.S.P. Parkin, V.Y. Lee, A.I. Nazzal, R. Savoy, G. Gorman and T.C. Huang, Phys. Rev. Lett. 61(1988)750.
[11] B. Morosin, D.S. Ginley, P.F. Hlava, M.J. Carr, R.J. Baughman, J.E. Schirber, E.L. Venturini and J.F. Kwak, Physica C 152(1988)413.
[12] M. Hervieu, A. Maignan, C. Martin, C. Michel, J. Provost and B. Raveau, J. Solid State Chem. 75(1988)212.

4.15 110K Phase of Bi-Sr-Ca-Cu-O Fabrication and Microstructure

Preparation and Properties of Pb-Doped Bi-Sr-Ca-Cu-O Superconductors

HIROSHI MAEDA, HIROAKI KUMAKURA, TOSHIHISA ASANO, HISASHI SEKINE, DANIEL R. DIETDERICH, SYOZO IKEDA, KAZUMASA TOGANO, and YOSHIAKI TANAKA

National Research Institute for Metals, Tsukuba Laboratories, 1-2-1, Sengen, Tsukuba, 305 Japan

ABSTRACT

Bi-Sr-Ca-Cu-O superconductors with a Pb addition were prepared by a conventional solid state reaction. Early in the sample heat treatment a large volume fraction of low-T_c phase ($Bi_2Sr_2Ca_1Cu_2O_x$) exists. With further sintering the volume fraction of the high-T_c phase ($Bi_2Sr_2Ca_2Cu_3O_x$) gradually increases. A zero resistance transition temperature of 110K was achieved for samples with a nominal composition of $Bi_{0.7}Pb_{0.3}Sr_1Ca_1Cu_{1.8}O_x$ and a sintering treatment of 845 °C for 300 h in air. The critical current density, J_c, of the sample at 77K in zero magnetic field is 210 A/cm^2, which is much larger than that of Pb-free sample. A pellet compaction process to densify and align the grains of high-T_c phase during an interrupted sintering treatment has been shown to be an effective method to increase J_c. A J_c greater than 1100 A/cm^2 can be easily achieved by this process. A new mode of modulation with a wave length of about 50 Å is seen in the Pb modified material in addition to the conventional modulation observed in the Pb-free system. Periodicities of the new, as well as, conventional modulation modes become less regular in the Pb-doped system.

INTRODUCTION

The discovery of superconductivity in the Bi-Sr-Ca-Cu-O (BSCCO) system with transition temperature, T_c above 100K has brought about a new phase in the research of high-T_c oxide superconductors.(1) The BSCCO system has at least two superconducting phases. i.e., one with a T_c of about 80K(low T_c phase) and the other of about 110K(high-T_c phase). These two oxides can be expressed by the general molecular formula $Bi_2Sr_2Ca_{n-1}Cu_nO_x$ (x=2n+4) with n=2 and 3, namely, the ideal metal atom compositions for the low-T_c and high-T_c phases are 2212 and 2223, respectively. However, the Sr to Ca ratio in both phases has some range about the ideal value and the T_c of both phases tends to decrease with increasing Ca content.(2) Both crystal structures are similar , differing only in the number of CuO_2-Ca-CuO_2 slabs (i.e. perovskite-like lamella) inserted between double Bi-O layers. T_c is observed to increase as the number of CuO_2 slabs inserted increases. In additon, these superconductors have a characteristic modulated structure with highly strained a-c lattice planes.(3).

The Bi oxides superconductors have some interesting features with regard to practical applications. For example, a phase transformation does not occur below the sample sintering temperature as in $YBa_2Cu_3O_7$(YBCO) system. This will permit better control of the materials properties. The grains have a very thin planer shape with their surfaces parallel to the c-plane which will facilitate grain alignment for J_c improvement. Since cleavage of these phases occurs between the Bi-O bi-layers along the c-plane it is possible to deform the pellet samples by compaction without interrupting the high J_c current axis of the material. This suggests the possibility of the development of a prefered grain orientation with the alignment of the c-planes of each grain in the longitudinal direction of a wire or a tape fabrication. Of course, most of the BSCCO films prepared by various techniques such as sputtering, evaporation and printing have a strong preferred orientation with the c-plane parallel to the substrate.

To make use of these favorable features of BSCCO, it is necessary to form the high-T_c phase. However, a sample with only high-T_c phase could not be prepared. A tail in the resistive transition curve usually appears due to a mixture of the low-T_c phase and high-T_c phase. The maximum volume fraction of high-T_c phase obtained in a Pb-free sample is about 40%. Therefore, methods to increase the volume fraction of the high-T_c phase were sought. Very recently, Takano et al. succeeded in preparing samples that were almost completely high-T_c phase by the addition of Pb.(4) In addition, the temperature range for optimum sintering was found to expand with reduced oxygen partial pressure.(5) Therefore, it is very important to perform research on Pb-modified BSCCO, especially on the the critical current density J_c. J_c in BSCCO bulk sample is still lower than that of YBCO samples.

In this report, we describe the preparation, the superconducting properties and the crystal structure of Pb-modified BSCCO oxides and one useful techniques to improve the J_c of the material.

EXPERIMENTAL PROCEDURE

Samples with the nominal composition of $Bi_{(0.7-1)}Pb_{(0.2-0.3)}Sr_{(0.8-1)}Ca_{(1-1.5)}Cu_{(1.4-2)}O_x$ were prepared by a conventional solid state reaction using high-purity powders of Bi_2O_3, PbO (or Pb_3O_4), $SrCO_3$, $CaCO_3$ and CuO. Disk-shaped pellets (20 mm in diameter and 1-2 mm in thickness) were calcined at 800 °C for about 10h and sintered in the temperature range 840 to 850 °C. Some pellets were re-pressed (i.e. compacted) during an interrupted sintering treatment. A sample is sintered at 845 °C for 100hr, cooled to room temperature, compacted about one-half in the thickness, and finally sintered again at 845 °C for 100 h in air. This sample will be called the pressed sample. Measurements of resistivity vs temperature curves and J_c were performed in magnetic fields using a standard four-probe resistive method. AC susceptibility was also measured at a frequency of 700 H_z. X-ray diffraction, scanning electron microscopy(SEM) and high resolution transmission electron microscopy(HRTEM) were carried out at room temperature.

RESULTS AND DISCUSSION

1. Superconducting properties

A sample's T_c and volume fraction of high-T_c phase strongly depend on its sintering treatment. Figure 1 shows the resistivity variation with temperature of samples with a nominal composition of $Bi_{0.7}Pb_{0.3}Sr_1Ca_1Cu_{1.8}O_x$ that received different sintering times at 845 °C. The transition temperature T_c (zero resistance criteria) increases with increasing sintering time and a T_c of 110K is attained after 300h. AC susceptibility and x-ray diffraction data also show that a large volume fraction of low-T_c phase exists in samples with a short sintering time. The volume fraction of the high-T_c phase increases as the sintering time is increased to finally account for about 90% of the sample volume. Figure 2 shows resistivity transition curves for sample D in Fig.1 at various magnetic fields up to 8T. The head of the transition curve (i.e. high temperature portion) is only slightly influenced by the magnetic field, while the tail of the curve shifts to lower temperature as the magnetic field is increased. The Pb addition reduces the magnitude of the low temperature, however, it could not be completely removed. The slope of the upper critical fields H_{c2} at T_c, $(dH_{c2}/dT)_{Tc}$ (H_{c2} is defined at the midpoint of the transition in Fig.2) is 1.8T/K. H_{c2} at 77K, estimated by a linear

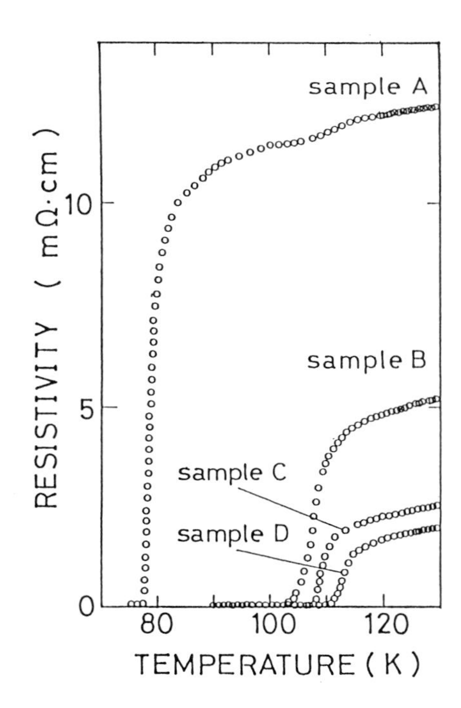

Fig.1 Resistivity vs. temperature curves for samples with the nominal composition of $Bi_{0.7}Pb_{0.3}Sr_1Ca_1Cu_{1.8}O_x$ with the following sintering conditions: 845 °C for 10h (sample A), 85h (sample B), 220h (sample C) and 300h (sample D)

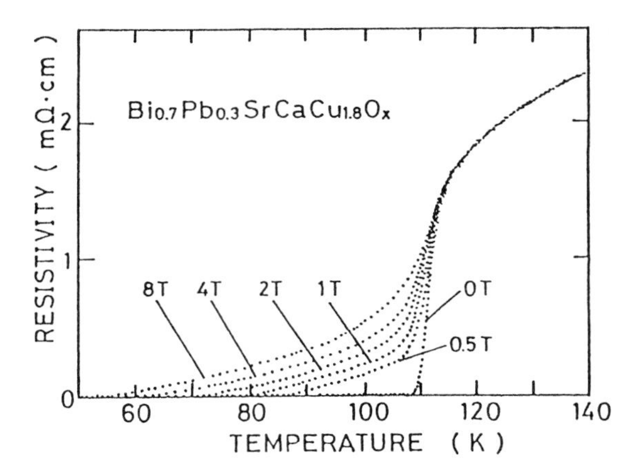

Fig.2 Resistivity vs. temperature curves at various magnetic fields for sample D of Fig.1.

extraporation is 60T and H_{c2} at 0K, using WHH theory, $H_{c2}(0)= -0.69T_c(dH_{c2}/dT)_{Tc}$, is 140T.

J_c of the sample D at 77K and zero magnetic field is about 210 A/cm^2, which is less than the best J_c obtained for YBCO samples. A low density, about 60% of the theoretical density, and preferred orientation, as seen in the SEM micrograph of Fig.3(b) is partially responsible for the low J_c value. In contrast, the pressed sample is dense and highly textured with layered grains laminated parallel to the pellet surface as shown in Fig.3(a). X-ray diffraction analysis of the pressed sample confirmed the sample is highly textured with strong (001) reflections and all others reflections weak. Figure 4 shows the J_c(77K) variation in magnetic fields up to 650 Oe for the pressed sample with the nominal composition of $Bi_{0.7}Pb_{0.3}Sr_1Ca_{1.5}Cu_2O_x$. The magnetic field was applied normal to the direction of current flow and parallel to the surface of the pellet(see Fig.4). A remarkable improvement in J_c, about 700 A/cm^2 at 77K and zero magnetic field is achieved which is about one order of magnitude larger than that of an un-pressed sample. Recently, we have obtained a J_c of 1100 A/cm^2 for small pressed samples with the nominal composition of $Bi_{0.8}Pb_{0.2}Sr_{0.8}Ca_1Cu_{1.4}O_x$. However, a drastic decrease in J_c is seen at low magnetic fields up to 100 Oe. This suggests that the superconducting coupling between grains is still very weak. This weak coupling between grains in the BSCCO system causes a rapid decrease in J_c in a similar manner to that of YBCO.

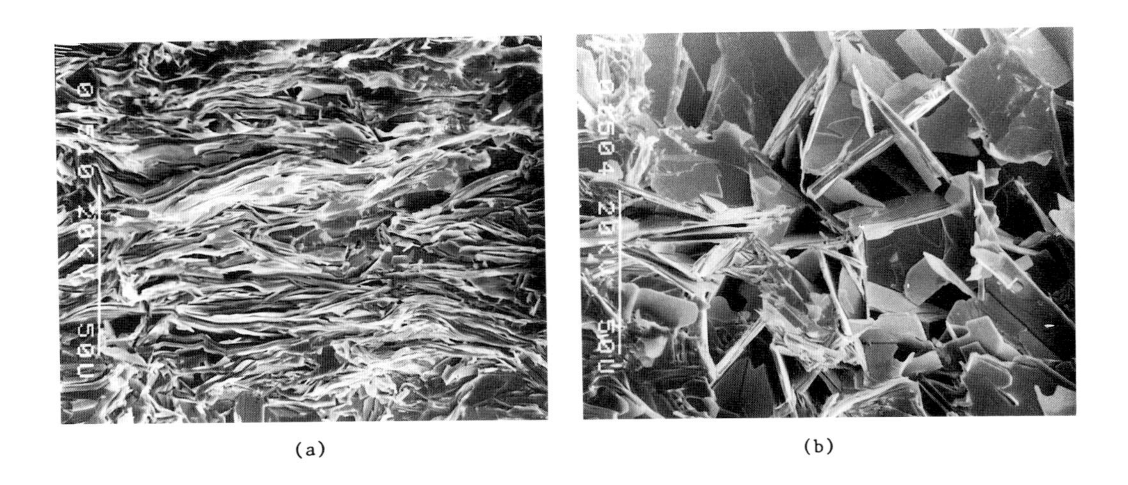

(a) (b)

Fig.3 SEM micrographs of the fractured surface of two samples with the nominal composition of $Bi_{0.7}Pb_{0.3}Sr_1Ca_1Cu_2O_x$.
(a) pressed 1 sample: sintered at 845 °C for 100h, pressed and subsequently sintered for another 100h at 845 °C.
(b) un-pressed sample: sintered at 845 °C for 200h.

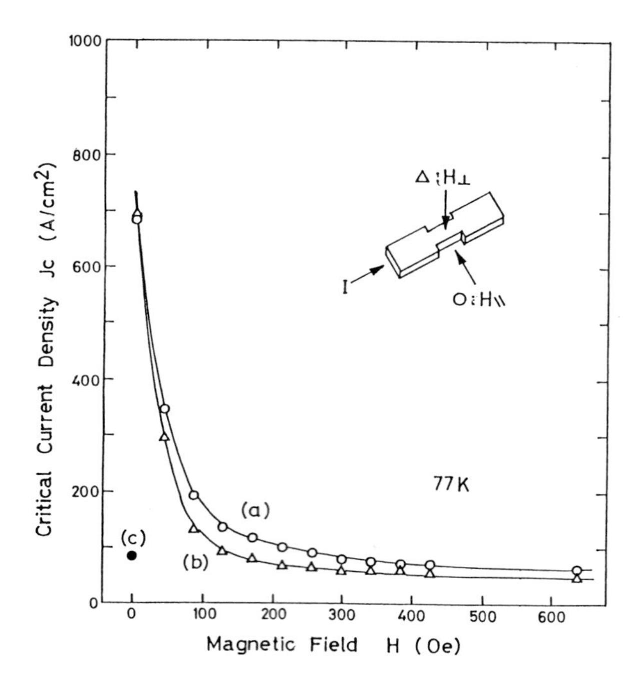

Fig.4 J_c(77K) vs. applied magnetic field for the pressed sample of Fig.3. Two data sets are plotted (a) magnetic field parallel to the sample surface, and (c) un-pressed sample.

Since grains of BSCCO can be easily deformed, not due to plastic deformation but perhaps due to cleavage occured between double Bi-O layers, it has been possible to fabricate an Ag-sheathed wire. Figure 5 shows the cross section of the 1330 filament wire made using powders with the nominal composition of $Bi_{0.8}Pb_{0.2}Sr_{0.8}Ca_1Cu_{1.4}O_x$ after the final deformation and heat treatment. The appropriate heat treatment temperature for the wire seems to be below 830 °C since a reaction between the oxide and Ag-sheath occurs above this temperature. This temperature is about 20 °C less than the optimum temperature obtained for bulk samples heat treated in air with no cladding. In this case, it is advisable to use semi-sintered powders (i.e. sintered at 845°C for 100 h in air) for wire fabrication.

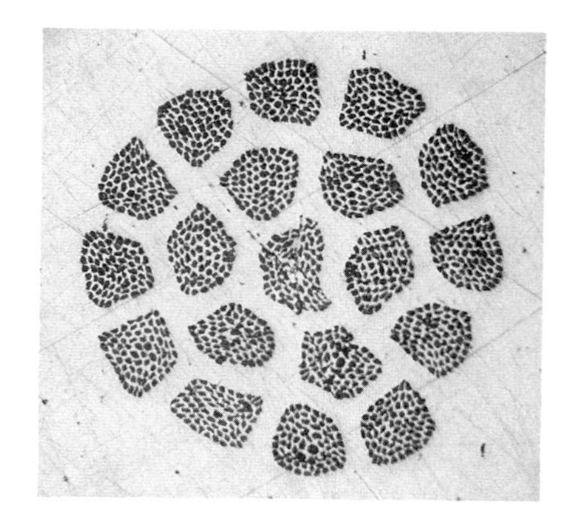

Fig.5 Cross-section of a wire with 1330 BSCCO filaments in a Ag sheath.

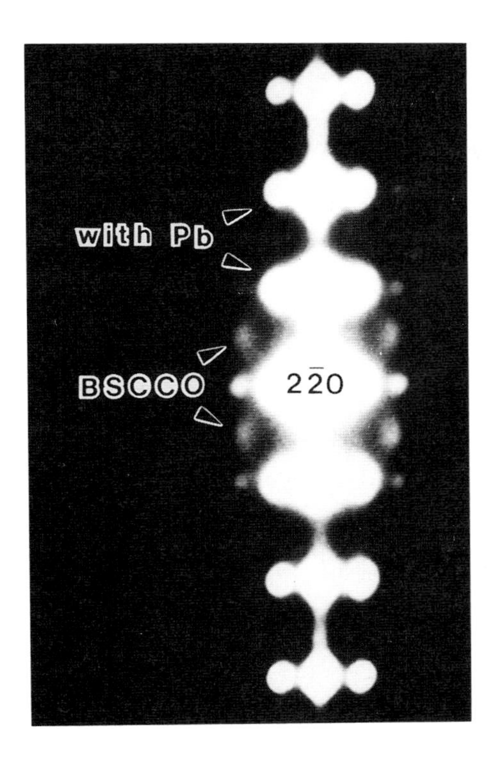

Fig.6 Electron diffraction pattern taken with the beam parallel to the perovskite [110] direction. The sample has the nominal composition of $Bi_{0.7}Pb_{0.3}Sr_1Ca_1Cu_{1.8}O_x$ and a sintering treatment of 845 °C for 85h.

2. Structure

Figure 6 shows an electron diffraction pattern taken with the incident beam normal to perovskite (110) planes. Satellite spots (+Pb) due to a modulation are seen along the [1$\bar{1}$0] direction (b^* direction in orthorhombic notation). From the satellite splitting the modulation wave length is estimated to be about 50 Å. This is different from that of the Pb-free system(BSCCO), i.e. 27Å.(6) Figure 7 shows a high resolution electron micrograph of the newly found modulation. As expected from the satellite splitting the dark contrast with a wave

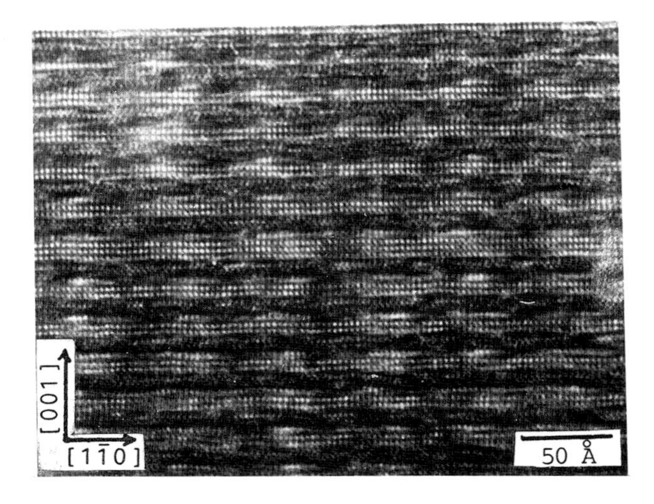

Fig.7 Newly found modulated structure in the BSCCO material with Pb.

length of about 50 Å is seen along [110] direction. However, the periodicity of the modulation is less defined than that of the Pb-free. Besides the new modulation the conventional modulation with a wave length of 27 Å is also observed. The periodicity of modulation is strongly disturbed by the Pb addition. When Pb is added the conventional modulation seems to become less stable and is replaced by the new modulation. In the transient state, both modulations appear to coexist. The distributed modulation must be related to strain energy reduction due to the Pb-addition. EDX analysis shows, of course, that Pb is completely incorporated into the high-T_c grains. However, it is not known which lattice site it occupies. EDX analysis suggests Pb substitutes for Bi. The Pb content seem to decrease considerably during long time sintering.

CONCLUSION

Since the discovery of high-T_c BSCCO superconductors many researchers have been investigating the oxides. However, many problems still remain to be solved. The main problems are the following. (1) T_c increases with increasing n (i.e. as the number of CuO_2 layers increase) up to 3. Will this trend continue with a further increase of n ?
(2) The formation of the high-T_c phase is enhanced with a Pb addition. This is perhaps related to the reduction of strain energy. The question remains as to which lattice site Pb occupies and if it orders on this site.
(3) It has been demonstrated that a dense highly textured material can easily be produced resulting in a J_c improvement. However, J_c decrease drastically with increasing magnetic field, showing the existence of superconducting weak links between grains. What is the origin of the weak links in BSCCO ?

REFERENCES

(1) H.Maeda, Y.Tanaka, M.Fukutomi and T.Asano: Jpn. J. Appl. Phys. 27(1988) L201.
(2) T.Satoh, T.Yoshitake, Y.Kubo and H.Igarashi: submitted to Appl. Phys. Lett.
(3) Y.Matsui, H.Maeda, Y.Tanaka and S.Horiuchi: Jpn. J. Appl. Phys. 27(1988) L372.
(4) M.Takano, J.Takada, K.Oda, H.Kitaguchi, Y.Miura, Y.Tomii and H.Mazaki: Jpn. J. Appl. Phys. 27(1988) L1041.
(5) U.Endo, S.Koyama and T.Kawai: submitted to Jpn. J. Appl. Phys.

Formation of the High Tc Phase of Bi-Pb-Sr-Ca-Cu-O Superconductors

SATOSHI KOYAMA[1], UTAKO ENDO[1], and TOMOJI KAWAI[2]

[1] Research and Development Department, Chemical Division, Daikin Industries, Ltd., 1-1, Nishi-hitotsuya, Settsu, 566 Japan
[2] The Institute of Scientific and Industrial Research, Osaka University, 8-1, Mihogaoka, Ibaraki, 567 Japan

ABSTRACT

Codecomposition of mixed nitrates of Bi, Pb, Sr, Ca and Cu around 830 °C under low oxygen pressure led to the formation of a high Tc superconducting phase of Bi-Pb-Sr-Ca-Cu-O with Tc(zero) at 107.0K. The best starting compositions were in the region close to $Bi_{1.84}Pb_{0.34}Sr_2Ca_2Cu_3O_y$ with a little excess of Ca and Cu. In this region, the samples showed the absence of the 80K and semiconducting phase, and there was no indication of impurities at all. The reaction under low oxygen pressure has an effect to lower the temperature with broad ranges to render the high Tc phase.

INTRODUCTION

Since the discovery of superconductivity at onset temperature above 100K in the Bi-Sr-Ca-Cu-O system[1], much effort has been made to single out the high Tc phase. Coexistence of the high Tc phase with the 80K phase or the semiconducting phase, however, has always been observed in this system. In order to increase the high Tc phase, an addition of a large excess of Ca and Cu[2,3], high pressure oxygen treatment[4,5] and long time sintering[6] have been attempted. Among these efforts, partial substitution of Pb for Bi has been reported to be effective in increasing the ratio of the high Tc phase[7].

We have applied a codecomposition method using metal nitrates as starting materials and also applied a low oxygen pressure treatment during the solid reaction[8]. These procedures on Bi-Pb-Sr-Ca-Cu-O system evidently affected the appearance of the high Tc phase, which appeared in the wide range of heating temperature and reaction time. We found that the pure 110K phase is obtained around the ideal composition, $(Bi,Pb)_2Sr_2Ca_2Cu_3O_y$, but with a little excess of Ca and Cu[9]. With this composition, the solid reaction under low oxygen pressure using the codecomposed powders led to the formation of the Bi-Pb-Sr-Ca-Cu-O superconductor which showed the X-ray diffraction pattern of a pure 110K phase with no indication of impurities at all.

EXPERIMENTAL

The effect of oxygen pressure on the formation of the high Tc phase was studied using $Bi_1Sr_1Ca_1Cu_2O_y$ compound[1] made by conventional solid reaction method. These standard $Bi_1Sr_1Ca_1Cu_2O_y$ powders were prepared by mixing Bi_2O_3, $SrCO_3$, $CaCO_3$ and CuO in an agate mortar, followed by heating at 810 °C for 16 hours in air. They were made into pellets and heated for 12 hours at various temperatures in the stream of $Ar\text{-}O_2$ mixture, with the oxygen partial pressure of 1 to 0 atm (1,1/2,1/5,1/13,1/30,1/100 and 0). After the heating, they were slowly cooled to 500 °C.

The Bi-Pb-Sr-Ca-Cu-O samples were prepared by codecomposition of metal nitrates. The powders of Bi_2O_3, PbO, $Sr(NO_3)_2$, $Ca(NO_3)_2 \cdot 4H_2O$ and CuO were dissolved in nitric acid with the desired cation ratio. The solution was stirred and heated until it became dry to form a solid with a light blue color. The nitrates were then co-decomposed at 800 °C for 30 min. The powder thus prepared was ground and pressed into pellets. They were heated at 828-843 °C for 36-130 hours under oxygen pressure of 1/13 atm and then were slowly cooled to room temperature.

The samples with the composition $Bi_xPb_ySr_2Ca_2Cu_3O_z$ were prepared varying x and y values and then with x and y fixed at 1.84 and 0.34, Sr, Ca and Cu ratios were varied as $Bi_{1.84}Pb_{0.34}Sr_xCa_yCu_zO_w$.

An effect of oxygen treatment during the cooling process was examined on several samples. After the solid reaction under the oxygen pressure of 1/13 atm, the samples were slowly cooled to 750 °C in the same atmosphere. Then the flowing gas was switched to pure oxygen and the samples were cooled to room temperature.

The structure and the superconducting properties of the samples thus prepared were evaluated by powder X-ray diffraction using CuKα radiation, resistivity measurement using a standard four-probe technique and ac susceptibility measurement at 200Hz. The dimension of the samples was 13mm in diameter and 1-2mm in thickness.

RESULTS AND DISCUSSION

Bi-Sr-Ca-Cu-O superconductors have been mainly prepared in air or under high oxygen pressure[1-7]. In order to know the effect of oxygen pressure, we have measured the diamagnetic signal due to the 110K phase of $Bi_1Sr_1Ca_1Cu_2O_y$ under various oxygen pressures(Fig.1). The closed circles in Fig.1 indicate the samples melted, the triangles partially melted, the open circles not melted though transformed and the double circles not melted at all. As is seen in the figure, the melting point becomes lower as the oxygen pressure is lowered. It becomes 810 °C in the absence of oxygen.

The resistivity and magnetic susceptibility measurements on these samples showed that the high Tc phase is formed in the samples indicated by the dotted area of Figure 1. It is formed at lower temperatures with broader ranges as the oxygen pressure decreases. For example, high Tc phase appears at the temperatures in the range of 15 to 20 °C at the lower oxygen pressure, whereas it appears in a narrow region of 5 °C in air. The high Tc phase appears even at 800 °C by decreasing oxygen pressure. The low oxygen pressure is effective in lowering the reaction temperature. As shown in the contour lines, the high Tc phase formation was most remarkable at the temperature just below the melting point under oxygen pressure of 1/13 atm. Thus we employed this oxygen pressure in the following reactions.

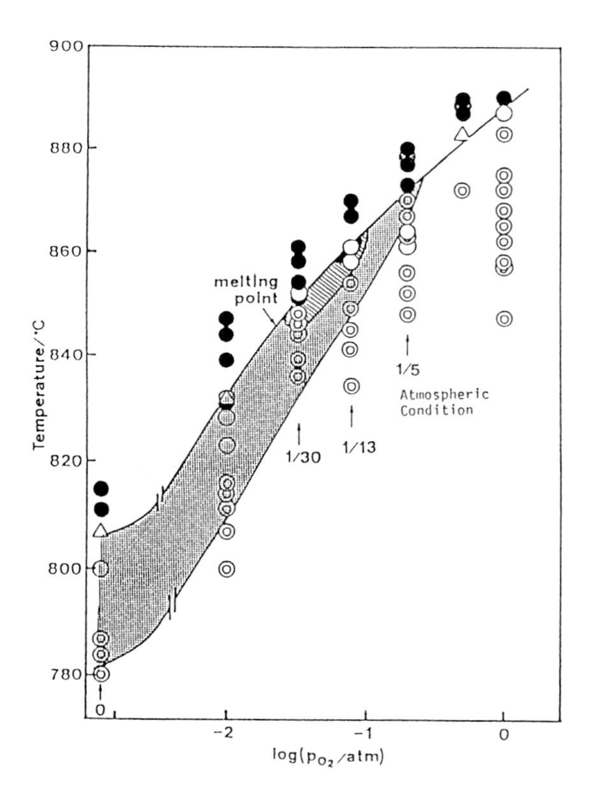

Fig.1 Effect of oxygen pressure and reaction temperature on the formation of $Bi_1Sr_1Ca_1Cu_2O_y$ superconductors. The closed circles indicate that the samples melted, the triangles partially melted, the open circles not melted though transformed and the double circles not melted at all. The high Tc phase appears in the dotted area and more preferably in the dashed area. The high Tc phase formation is most clear at the temperature just below the melting point under oxygen pressure of 1/13 atm.

Under this oxygen pressure, we have prepared Bi-Pb-Sr-Ca-Cu-O samples by codecomposition of metal nitrates. The pure 110K phase was formed in the nominal composition of $Bi_{1.84}Pb_{0.34}Sr_{1.91}Ca_{2.03}Cu_{3.06}O_y$, as was revealed by X-ray diffraction and magnetic susceptibility measurements. In Fig.2 are shown the X-ray patterns of the sample prepared by the codecomposition method. The series of peaks of the high Tc phase[6] are seen with the characteristic (002) peak at $2\theta=4.7°$. The low Tc and semiconducting phases are not observed whose characteristic peaks are at $2\theta=5.7°$ and $7.2°$, respectively[6]. The peaks at higher angles are indexed as the high Tc phase by the calculation based on the tetragonal unit cell with a=b=5.396Å and c=37.180Å. The volume fraction of the high Tc phase is estimated to be more than 99% taking the detection limit of X-ray diffraction into account.

The magnetic susceptibility of the sample drops sharply at 110K and it does not change at all around 80K(Fig.3). This gives another evidence that a high Tc phase exists without any 80K phase. Resistivity measurement on the samples showed zero resistivity at 107.0K (Fig.3).

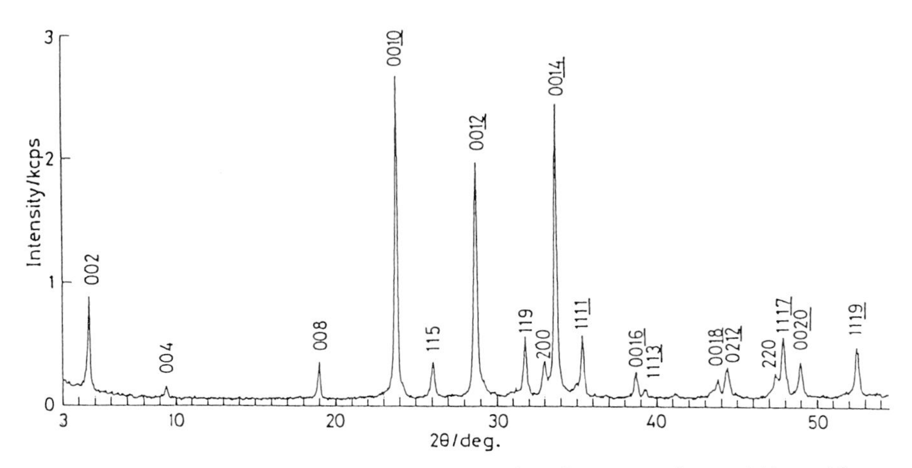

Fig.2 The X-ray diffraction pattern of the sample with the nominal composition $Bi_{1.84}Pb_{0.34}Sr_{1.91}Ca_{2.03}Cu_{3.06}O_y$ prepared at 830 °C for 84 hours (O_2=1/13 atm).

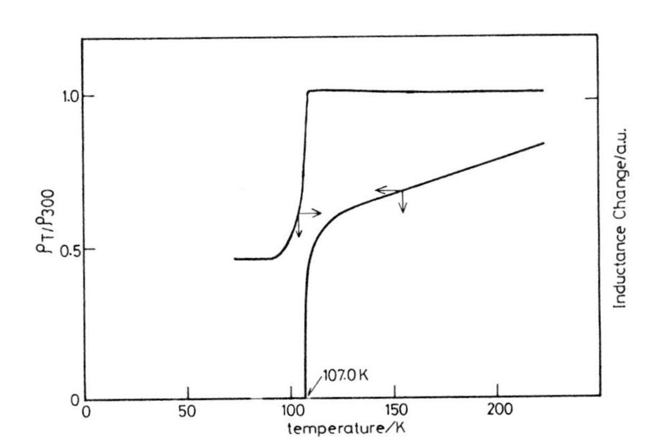

Fig.3 Resistivity-temperature and ac susceptibility-temperature curves of the sample.

Figure 4 shows the relation between the starting compositions and the X-ray reflection intensities of the 80K phase and Ca_2PbO_4 in Bi_xPb_y-$(Sr_2Ca_2Cu_3O_z)$. The circles in the figure represent the degree of the 80K phase formation and the curves are its "contour lines". The 80K phase was small inside these curves (point B,D). However, the Ca_2PbO_4 formation became clear as the Pb content was increased (point C,D). The optimum content for Bi and Pb, where no Ca_2PbO_4 and the smallest 80K phase is formed, is considered to lie in the shaded region in Fig.4. We chose the value 1.84 and 0.34 as the Bi and Pb content (point B in the figure) and varied the Sr, Ca and Cu ratio.

Fig.5 shows the relation between Sr-Ca-Cu compositions and X-ray reflection intensities at $2\theta=27.5°$ (80K phase), 17.8° (Ca_2PbO_4) and 21.9° (semiconducting phase) in the samples $Bi_{1.84}Pb_{0.34}Sr_xCa_yCu_zO_w$. The 80K phase was not present in the region below the solid curve (point C,D). The Ca_2PbO_4 formation was not observed above the dotted curve (point A,B,C). There appeared a reflection peak at $2\theta=21.9°$ below the broken curve (point D,E). This peak and one at $2\theta=7.2°$ are coupled and are assigned as that of the semiconducting phase[10] with a single Cu-O layer sandwiched by Bi_2O_2. In the narrow shaded region, the 80K phase, Ca_2PbO_4 or any other compound than the 110K phase was not observed in the X-ray pattern. That is, the compositions in this region give rise to the single 110K phase. When Ca and Cu are decreased, the 80K phase tends to appear, while improper amount of them leads to the formation of Ca_2PbO_4 and too much of them causes the formation of the semiconducting phase. It is necessary to choose proper starting compositions within such a narrow region to get the pure 110K phase.

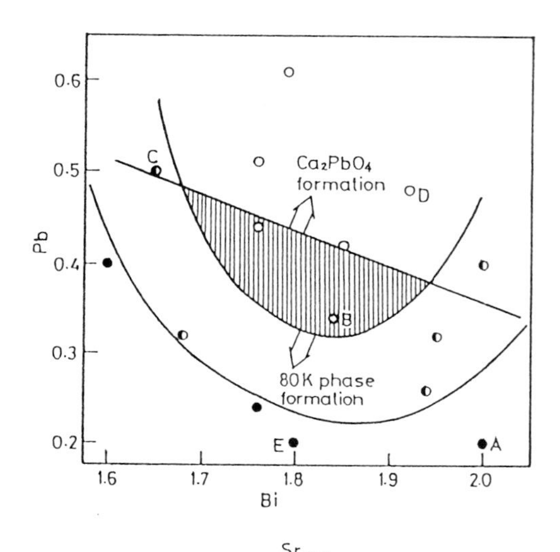

Fig.4 Relation between the amount of Bi and Pb (Sr:Ca:Cu=2:2:3) and the X-ray reflection intensities of the 80K phase in $Bi_xPb_ySr_2Ca_2Cu_3O_z$. Circles denote: ● ;the peak intensity >300(cps), ◐ ; >150, ○ ; <150.

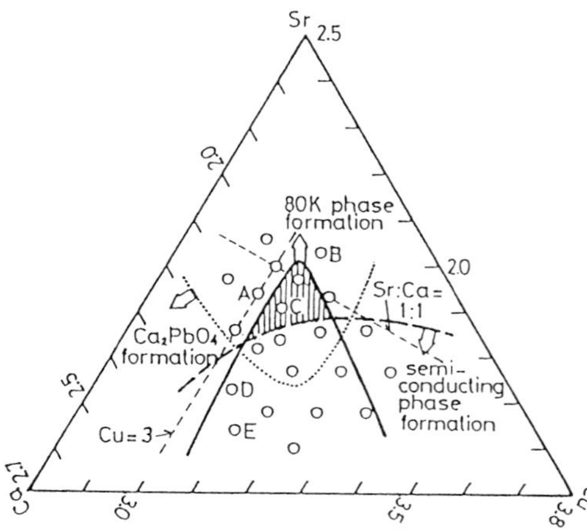

Fig.5 Relation between the ratios of Sr, Ca and Cu(Bi:Pb=1.84:0.34) and the intensities of the X-ray reflection peak at $2\theta=27.5°$ (80K phase), 17.8° (Ca_2PbO_4) and 21.9° (semiconducting phase) in the samples $Bi_{1.84}Pb_{0.34}Sr_xCa_yCu_zO_w$.

It should be noted that the high Tc phase with no low Tc phase was formed even at 828°C for 36 hours under low oxygen pressure by the reaction of the starting materials prepared by the co-decomposition of the nitrates with the starting composition $Bi_{0.8}Pb_{0.2}Sr_{0.8}Ca_{1.0}Cu_{1.4}O_y$. It means that the reaction for a short time at relatively low temperature is enough to give the high Tc phase.

These phenomena might be attributable to two reasons. One is the codecomposition method and the other is the low oxygen pressure during the reaction. By using the co-decomposition method, the powders can be mixed uniformly, retaining the original cation ratio. Because of this uniformity the reaction may proceed in a rather short time. As seen in the Bi-Sr-Ca-Cu-O system, the temperature of the high Tc phase formation becomes lower, the range being widened, as the oxygen pressure is lowered. It is stated that the high Tc phase is stable at lower temperature[4,5]. This may be one of the reasons why the lowering the reaction temperature under low oxygen pressure is effective in enhancing the high Tc phase formation.

The pure 110K phase was not obtained with the starting composition strictly kept as (Bi,Pb):Sr:Ca:Cu=2.0:2.0:2.0:3.0. Without a little excess of Ca and Cu, the 80K phase tends to appear, while too much Ca and/or Cu leads to the formation of Ca_2PbO_4. It is necessary to balance these reactions in forming the pure 110K phase.

The high Tc phase without the low Tc one was also obtained by a conventional solid state reaction at 842°C for 84 hours (O_2=1/13 atm) using the starting powders mixed in an agate mortar. So far, the mixing of the powders in an agate mortar has been recognized not to be an effective way for obtaining the high Tc phase. The point of our process again resorts to the low oxygen pressure and the choice of proper initial composition.

There observed some samples which have low temperature tailing in the R-T curve. Since we thought that some oxygen deficiencies in the 110K phase prevent a current flow, we have adopted the oxygen treatment (1 atmosphere of O_2) to the cooling process on these samples after the complete formation of the triple Cu-O layered structure. Then the tailing disappeared. The tailing seems to have a correlation with the intensity of a reflection at $2\theta=21.9°$ (semiconducting phase). The smaller this peak becomes, the smaller the tailing is.

Interestingly, We could find samples naturally oriented with the c-axis perpendicular to the surface or easily oriented when pressed. These samples were formed after heating just below the melting point. The grains had grown to form flakes. The peaks with the index (0 0 l) are sharp and strong compared to others. From the practical point of view, this phenomenon might be attractive, since the oriented superconducting ceramics uniquely having high Tc single phase are desirable for a wire fabrication.

REFERENCES

[1] H.Maeda, Y.Tanaka, M.Fukutomi and T.Asano: Jpn.J.Appl.Phys., 27 (1988) L209.
[2] A.Sumiyama, T.Yoshitoshi, H.Endo, J.Tsuchiya, N.Kijima, M.Mizuno and Y.Oguri: Jpn.J.Appl.Phys., 27 (1988) L542.
[3] N.Kijima, H.Endo, J.Tsuchiya, A.Sumiyama, M.Mizuno and Y.Oguri: Jpn.J.Appl.Phys., 27 (1988) L821.
[4] K.Kitazawa, S.Yaegashi, K.Kishio, T.Hasegawa, N.Kanazawa, K.Park and K.Fueki: Adv.Ceram.Mater. in press ,
[5] K.Kuwahara, S.Yaegashi, K.Kishio, T.Hasegawa and K.Kitazawa: Proc.Latin-American Conf. on High Temperature Superconductivity, in press
[6] H.Nobumasa, K.Shimizu, Y.Kitano and T.Kawai: Jpn.J.Appl.Phys., 27 (1988) L846.
[7] S.A.Sunshine, T.Siegrist, L.F.Schneemeyer, D.W.Murphy, R.J.Cava, B.Batlogg, R.B.van Dover, R.M.Fleming, S.H.Glarum, S.Nakahara, R.Farrow, J.J.Krajewski, S.M.Zahurak, J.V.Waszczak, J.H.Marshall, P.Marsh, L.W.Rupp, Jr. and W.F.Feck: in press.
[8] U.Endo, S.Koyama and T.Kawai: Jpn.J.Appl.Phys., 27 (1988) No.8 in press.
[9] S.Koyama, U.Endo and T.Kawai: submitted to Jpn.J.Appl.Phys.
[10] J.M.Tarascon, W.R.McKinnon, P.Barboux, D.M.Hwang, B.G.Bagley, L.H.Greene, G.Hull, Y.LePage, N.Stoffel and M.Giroud: submitted to Phys.Rev.B.

Synthetic Conditions and Structural Properties of the High-T_c Phase in the Superconducting Bi-Sr-Ca-Cu-(Pb)-O System

MASAAKI MIZUNO, HOZUMI ENDO, JUN TSUCHIYA, NAOTO KIJIMA, AKIHIKO SUMIYAMA, and YASUO OGURI

Mitsubishi Kasei Corporation (Formerly, Mitsubishi Chemical Industries Ltd.), Research Center, 1000 Kamoshida-cho, Midori-ku, Yokohama, 227 Japan

ABSTRACT

Synthetic conditions, structural and thermal properties of the high-T_c phase (T_c > 100 K) in the Bi-Sr-Ca-Cu-O system have been investigated. Three methods have been found to be effective in increasing the volume fraction of the high-T_c phase in the Bi-Sr-Ca-Cu-O system: (1)starting from a nominal composition with more Ca and Cu than $Bi_2Sr_2Ca_2Cu_3O_x$ (2) the addition of Pb to Bi-Sr-Ca-Cu-O system (3) annealing at 870°C under higher oxygen partial pressures than 0.2 atm. Sample with the nominal composition of $Bi_2Sr_2Ca_6Cu_8O_x$ fired at 870°C for 120 hours was found that it consisted of mainly the high-T_c phase. Unit cell dimensions of the high-T_c phase determined by X-ray powder diffraction and transmission electron diffraction measurement are a = 5.40 Å, b = 27.0 Å and c = 36.8 Å, indicating a pseudotetragonal symmetry. Transmission electron microscopy showed that there were stacking faults in the high-T_c phase. The high-T_c phase and the low-T_c phase (T_c ~ 80 K) decompose into $Bi_2Sr_2CuO_x$ by annealing under oxygen partial pressure of 0.02 and 0.1 atm at 870°C. Under higher oxygen partial pressures than 0.2 atm at 870°C, the high-T_c phase was remarkably formed from the possible disproportionation reaction of the low-T_c phase.

INTRODUCTION

Since the discovery of a slight signal of a high-T_c superconducting phase (T_c > 100 K) in the Bi-Sr-Ca-Cu-O system[1], much effort has been attempted to synthesize the single phase. It is known that there are three ways to increase the volume fraction of the high-T_c phase. Firstly, we reported that the volume fraction of the high-T_c phase is increased by starting from compositions with surplus Ca and Cu than the possible ideal composition of the high-T_c phase ($Bi_2Sr_2Ca_2Cu_3O_x$) and by a prolonged firing[2,3]. Secondly, Kitazawa et al.[4] and the authors[5] have reported that the high-T_c phase can be obtained by annealing in an atmosphere with a high oxygen partial pressure. Thirdly, Takano et al.[6] and the authors[7] have reported that the substitution of Pb for Bi or the addition of Pb to the $Bi_2Sr_2Ca_2Cu_3O_x$ also increases the high-T_c phase.

In this paper, we summarize those results of our studies on the synthesis of the high-T_c phase. We also report structural and thermal properties of the high-T_c phase.

EXPERIMENTAL

Samples were prepared by a solid state reaction. Starting materials were high purity powders of Bi_2O_3, $SrCO_3$, $CaCO_3$, CuO and PbO. These powders were mixed, pressed into pellets, preheated at 800°C for 24 hours and then fired at 870°C for 24-120 hours or fired at 855°C when the Pb was doped or added. They were slowly cooled down to room temperature (about 50°C/h) in the furnace. The ac susceptibility was measured using an ac Hartshorn type bridge. The absolute vale of susceptibility was determined using a piece of superconducting Sn-Pb alloy cut to the same dimensions with the specimen. The temperature was measured by the Pt-Co resistance thermometer. X-ray powder diffraction measurements were carried out with Philips 1700 diffractometer using CuKα radiation. The specimen used for transmission observation was prepared as follows: The sintered pellet, fired at 870°C for 120 hours, was crushed into powder in an agate mortar. The powder thus obtained was dispersed onto the holey carbon film. High-resolution electron microscopic observation were made by a HITACHI H-9000 transmission electron microscope operated at 300kV. Differential thermal analyses (DTA) and thermogravimetric measurements (TG) were carried out using ULVAC TGD 7000 under the various oxygen partial pressures. The atmospheres of the desired oxygen partial pressures were obtained by mixing O_2 and Ar. TG and DTA measurements were carried out as follows: the starting materials were heated to desired temperatures at the constant heating rate of 20°C/min, then the temperature was held for 5 hours and were brought down to the room temperature at the constant cooling rate of 20°C/min.

RESULTS AND DISCUSSION

Starting compositions

Figure 1 shows the relationship between starting compositions and the high-T_c phase synthesized by firing at 870°C for 24 hours. The open circles indicate the starting compositions at which the high-T_c phase was formed dominantly. The closed circle indicates the starting composition at which the high-T_c phase was not obtained. We obtained a trace of the high-T_c phase by starting from composition of $Bi_2Sr_2Ca_2Cu_3O_x$. On the other hand, the volume fraction of the high-T_c phase was increased by starting from nominal compositions with more Ca and Cu than $Bi_2Sr_2Ca_2Cu_3O_x$. Fig. 2 shows the X-ray diffraction pattern of samples with the nominal composition of $Bi_2Sr_2Ca_6Cu_8O_x$ fired at 870°C for 24 hours (sample A) and 120 hours (sample B). X-ray diffraction revealed that sample A consists of the high-T_c phase, the low-T_c phase, Ca_2CuO_3 and unreacted CuO. The high-T_c phase and the low-T_c phase are characterized by the broad peaks at $2\theta = 4.8°$ and the sharp peak $2\theta = 5.7°$, respectively. In the diffraction pattern of sample B, however, reflections of the low-T_c phase disappeared and those of the high-T_c phase became dominant. Fig. 3 shows the temperature dependence of the magnetic susceptibility of the sample A and B. The ac susceptibility of sample A shows a diamagnetic behavior below 107 K and has a shoulder around 80 K. On the other hand, the ac susceptibility of sample B decreases monotonically below 107 K. These results were consistent with those of the X-ray diffraction. This means that the volume fraction of the high-T_c phase is increased with increasing firing period.

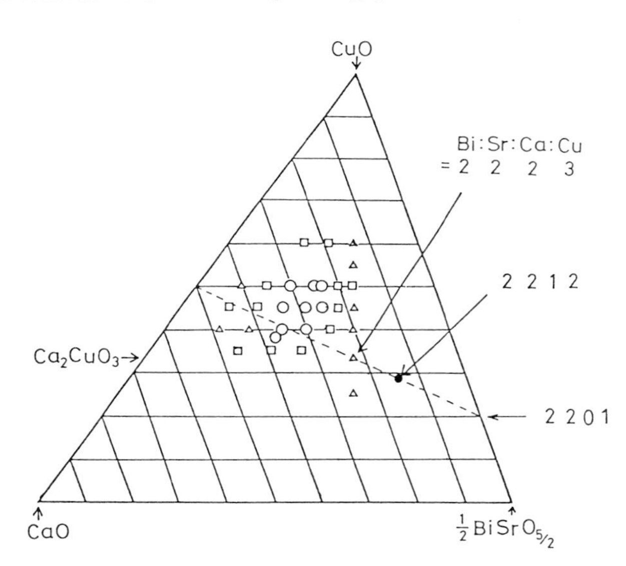

Fig. 1. Relationship between starting compositions and the high-T_c phase synthesized by firing at 870°C for 24 hours.
○: the high-T_c phase was dominant.
□: small amount of the high-T_c phase was detected.
△: a trace of the high-T_c phase was formed.
●: the high-T_c phase was not obtained.

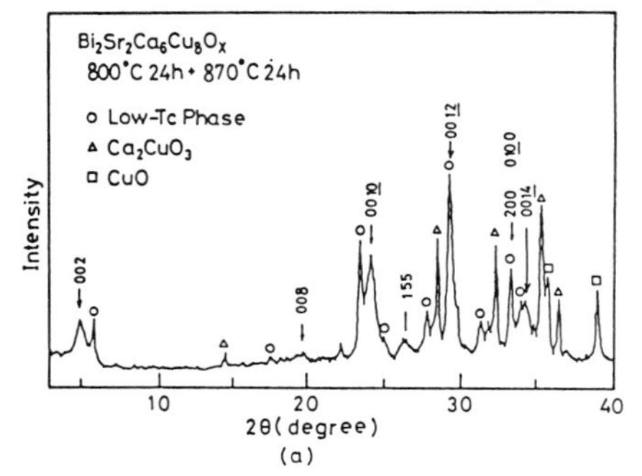

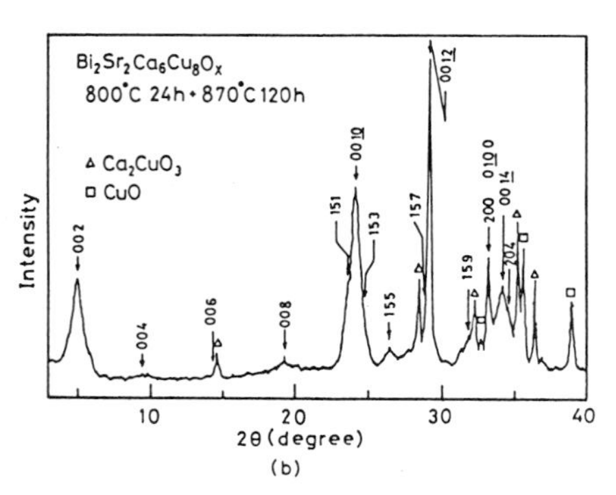

Fig. 2. X-ray powder diffraction patterns over the 2θ range from 3° to 40° obtained using CuKα radiation for sample A (a) and sample B (b). Reflections with the Miller indices are for the high-T_c phase.

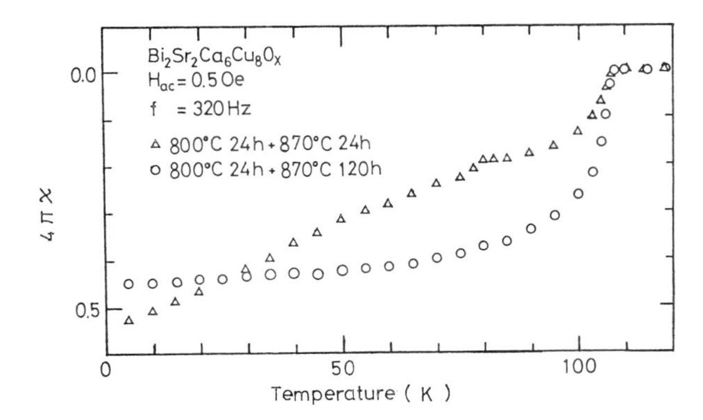

Fig. 3. Temperature dependence of the magnetic susceptibility for sample A(△) and sample B (○).

Structural properties

Figure 4 shows electron diffraction patterns obtained for the high-T_c phase in sample B. The primary reflections in the electron diffraction patterns are indexed using a body-centered unit cell of a = b = 3.82 Å, c = 36.8 Å. The presence of weaker satellite spots indicates a large unit cell with lattice parameters of a = 5.4 Å, b = 5a = 27.0 Å, c = 36.8 Å that is rotated by 45 degrees from the body-centered unit cell. However, the average lattice length of the b-axis is about 4.6a, which suggests the formation of an incommensurate superstructure. The observed X-ray powder diffraction pattern of sample B was completely indexed by the orthorhombic unit cell, as shown in Fig . 2(b), with the aid of the intensity information obtained from the electron diffraction. We observed the

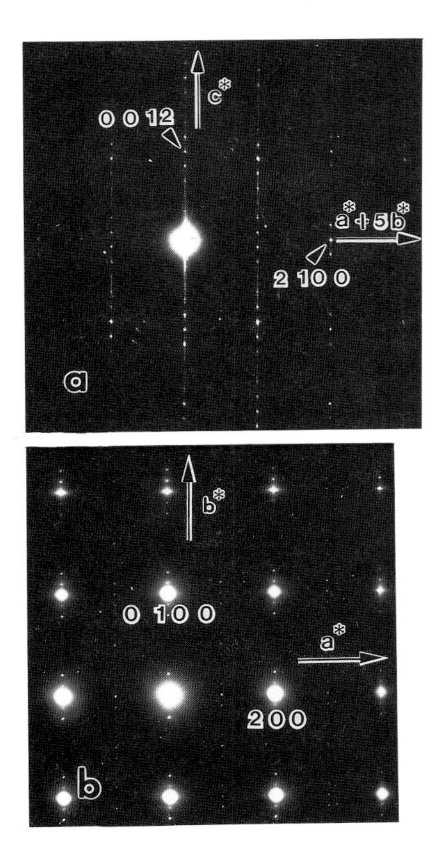

Fig. 4. Electron diffraction patterns of the high-T_c phase in sample B. Incident beams are parallel to the [510] direction (a) and [001] direction (b). The diffraction pattern from the [001] direction has satellite spots along the b^*-axis.

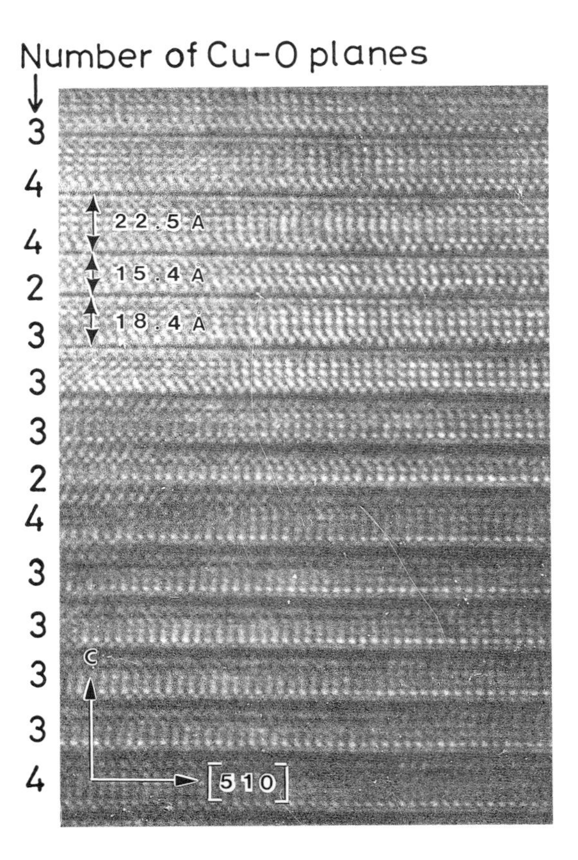

Fig. 5. High-resolution transmission electron microscope image of the high-T_c phase in sample B, taken with an incident beam along the [510] direction.

high resolution structure image with an electron beam along the [510] direction. The diffraction pattern from this direction has no satellite spots, as shown in Fig. 4(a). The resultant high-resolution structure image is given in Fig. 5. In the image, there are three types of layer structural units with different thicknesses. The 30.8/2 Å or 45/2 Å layer is observed between the main layers with a thickness of 36.8/2 Å. It seems that the broadening of the 002 reflection of the high-T_c phase in the X-ray powder diffraction pattern is due to stacking faults in the 36.8/2 Å layers, as shown in Fig. 5.

Bi-Sr-Ca-Cu-Pb-O system

Figure 6 shows the X-ray powder diffraction pattern of the samples with the nominal composition of $Bi_2Sr_2Ca_2Cu_3Pb_xO_y$ (x = 0.2, 0.4, 0.6) fired at 855°C for 120 hours. The low-T_c phase is dominant in the sample of x = 0.2. In the diffraction pattern of the sample of x =0.6, intensities of the peaks due to the high-T_c phase increase and those due to the low-T_c phase decrease. It should be noted that the diffraction peaks due to the high-T_c phase in the X-ray diffraction pattern of the samples with Pb are sharper than those of samples without Pb (e.g. sample B). It suggests that the addition of Pb promotes the crystallization of the high-T_c phase, that is to say, the 36.8/2 Å layered structures are formed with long-range coherency. Fig. 7 shows the results of DTA of the mixed powder of raw materials with the nominal composition of $Bi_2Sr_2Ca_2Cu_3Pb_xO_y$(x = 0, 0.2, 0.4, 0.6). Although these endothermic peaks have not been assigned yet, it is clear that all these peaks shift to a lower temperature with an increase in the Pb concentration. This suggests that Pb acts at least as a flux. The addition of Pb lowers the optimum firing temperature to form the high-T_c phase to about 855°C, which is close to the temperature of the endothermic peak denoted by arrows in Fig. 7.

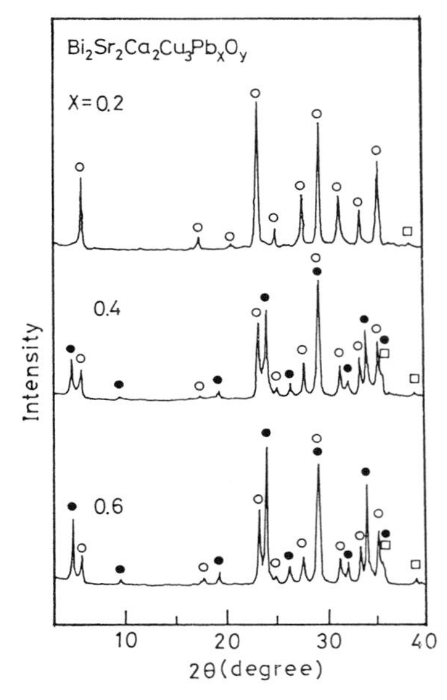

Fig. 6. X-ray powder diffraction patterns over the 2θ range from 3° to 40° obtained using a CuKα radiation for sample of X = 0.2, 0.4, 0.6. The peaks denoted by ●, ○ and □ correspond to the high-T_c phase, the low-T_c phase and CuO, respectively.

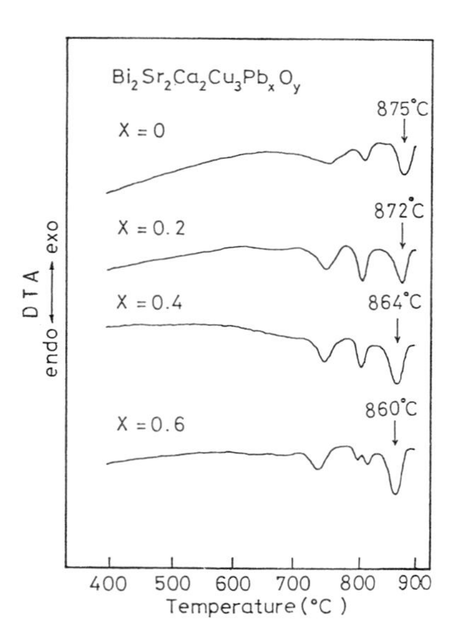

Fig. 7. DTA curves of the mixed powder of raw materials with the nominal composition of $Bi_2Sr_2Ca_2Cu_3Pb_xO_y$ (x = 0, 0.2, 0.4, 0.6). The heating rate is 20°C/min.

Thermal properties

Samples with the nominal composition of $Bi_2Sr_2Ca_6Cu_8O_x$ fired at 870°C for 24 hours were used for TG-DTA measurements, which consist of the high-T_c phase, the low-T_c phase, Ca_2CuO_3 and CuO, as shown in Fig. 8(a). Figure 8(b)-(e) show the X-ray diffraction patterns of samples after the TG-DTA measurements under various oxygen partial pressures at 870°C. Under oxygen partial pressure of 0.02 atm, both the high-T_c phase and the low-T_c phase disappeared, and eventually, $Bi_2Sr_2CuO_x$

appeared and the amount of Ca_2CuO_3 increased. Under 0.1 atm, although the high-T_c phase disappeared, the low-T_c phase still remained. Under 0.2 atm, there observed no change in the X-ray diffraction patterns before and after the TG-DTA measurements. Under 0.4 and 0.96 atm, the amount of the low-T_c phase, Ca_2CuO_3 and CuO decreased, and the high-T_c phase increased. In addition, a new peak appeared at $2\theta = 7.2°$ in the X-ray diffraction pattern, corresponding to $Bi_2Sr_2CuO_x$. These results suggest the occurrences of the disproportionation reaction of the low-T_c phase. That is, the low-T_c phase is not thermodynamically stable and causes the disproportionation reaction:

$$2Bi_2Sr_2CaCu_2O_x \rightarrow Bi_2Sr_2Ca_2Cu_3O_{x'} + Bi_2Sr_2CuO_{x''} \quad (1)$$

at 870°C. The decrease in the amount of Ca_2CuO_3 and CuO suggests that the high-T_c phase and the low-T_c phase may also be produced by the solid state reaction of the low-T_c phase, $Bi_2Sr_2CuO_x$, Ca_2CuO_3 and CuO in addition to the disproportionation reaction.

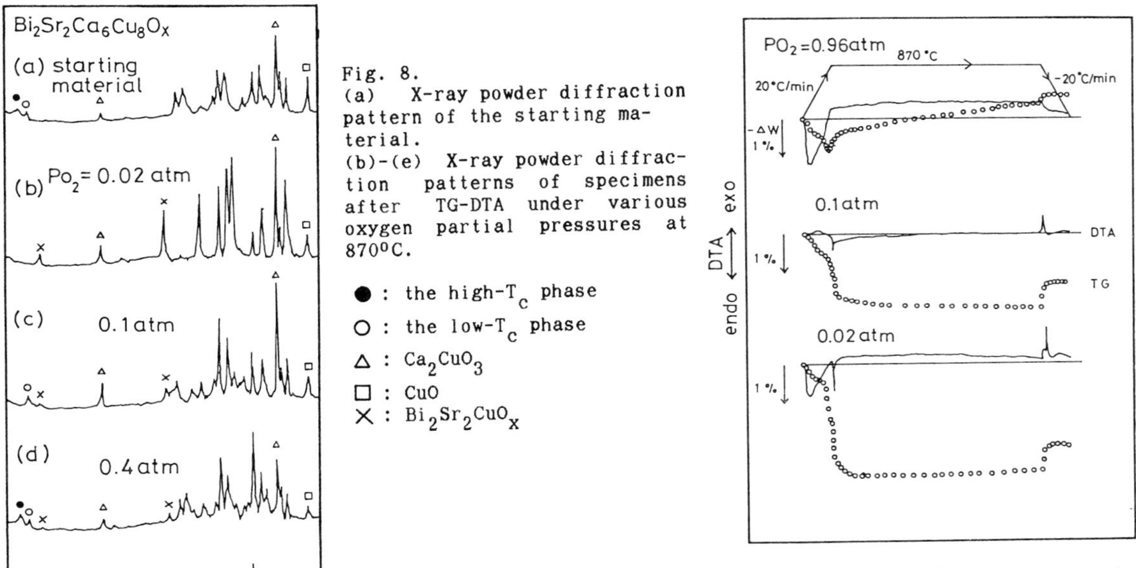

Fig. 8.
(a) X-ray powder diffraction pattern of the starting material.
(b)-(e) X-ray powder diffraction patterns of specimens after TG-DTA under various oxygen partial pressures at 870°C.

● : the high-T_c phase
○ : the low-T_c phase
△ : Ca_2CuO_3
□ : CuO
× : $Bi_2Sr_2CuO_x$

Fig. 9. The TG and DTA curves under the oxygen partial pressures of 0.96, 0.1 and 0.02 atm at 870°C. The open circles indicate the TG curves and the solid line with sharp peaks indicates the DTA curves.

Figure 9 shows the TG and DTA curves under the various oxygen partial pressures. All TG curves show a small amount of weight loss in the initial stage of heating-up. Under the oxygen partial pressure of 0.96 atm, the TG curve shows the gradual weight gain throughout holding temperature at 870°C and cooling process to room temperature. Under the oxygen partial pressures of 0.02 and 0.1 atm, the drastic weight loss and gain occurred at 870°C. This may be due to the desorption and absorption of oxygen throughout heating and cooling, respectively. These weight changes correspond to very strong endothermic and exothermic reactions. These results may suggest the melting of specimens occurred. The weight losses of specimens were not recovered in the cooling stage. For the purpose of clarifying the contribution of impurities(Ca_2CuO_3 and CuO) to the weight changes above described, TG-DTA measurement for the single phase of Ca_2CuO_3 and CuO were carried out under oxygen partial pressures of 0.02 and 0.96 atm at 870°C. These results showed that the contributions of them were negligible. In order to elucidate the weight loss and gain in the TG curves of $Bi_2Sr_2Ca_6Cu_8O_x$, the mean valence of copper ions was analysed with iodometry, assuming that charge state of Bi, Sr, Ca and O are 3+, 2+, 2+ and 2-, respectively. These results are shown in Table I. The mean valence of copper ions in the high-T_c phase is higher than that in the low-T_c phase, and the mean valence of copper ions in the low-T_c phase is higher than that in the lower-T_c phase

Table I. Mean valence of copper ions in the $Bi_2Sr_2Ca_6Cu_8O_x$. Sample C, D, and E indicates the as-sintered sample, the sample after TG and DTA under 0.96atm at 870°C and the sample under 0.02atm at 870°C, respectively.

	Mean valence of copper ions	
Sample C	2.07	(the high-T_c phase ≈ the low-T_c phase)
Sample D	2.29	(the high-T_c phase >> the low-T_c phase, $Bi_2Sr_2CuO_x$)
Sample E	1.86	($Bi_2Sr_2CuO_x$)

($Bi_2Sr_2CuO_x$). These results can be explained by the following reaction scheme:

$$Bi_2Sr_2CaCu_2O_{8+\delta} + [CaCuO_2] + 1/2(\delta' - \delta)[O_2] \rightarrow Bi_2Sr_2Ca_2Cu_3O_{10+\delta'} \quad (2)$$

$$Bi_2Sr_2CuO_{6+\delta''} + [CaCuO_2] + 1/2(\delta - \delta'')[O_2] \rightarrow Bi_2Sr_2CaCu_2O_{8+\delta} \quad (3)$$

Oxygen may participate in these reactions, because the increase in the mean valence of copper ions may be attributed to the additional oxygen described as (δ - δ') and (δ - δ'') in the eqs. (2) and (3). The relationship between the observed phase and oxygen partial pressures is summarized in Fig. 10. In the figure the closed circles show the specimens consisting of $Bi_2Sr_2CuO_x$, Ca_2CuO_3 and CuO. The half closed circles show the mixture of the low-T_c phase, $Bi_2Sr_2CuO_x$, Ca_2CuO_3 and CuO. The open circles show the specimens which did not have apparent differences from the starting material in the phases. The double circles indicate the specimens which show the growth of the high-T_c phase. The solidus line is also shown in Fig. 10. In the region of the double circles, it is speculated that $Bi_2Sr_2CuO_x$, which was deduced from the disproportionation of the low-T_c phase, reacted with Ca_2CuO_3 and CuO to form the low-T_c phase. Then the low-T_c phase thus formed was used to the subsequent disproportionation reaction, and accordingly the high-T_c phase grew and the amounts of Ca_2CuO_3 and CuO decreased during these reactions. However, we cannot exclude the possibility of the simultaneous solid state reaction of Eq. (2) to form the high-T_c phase.

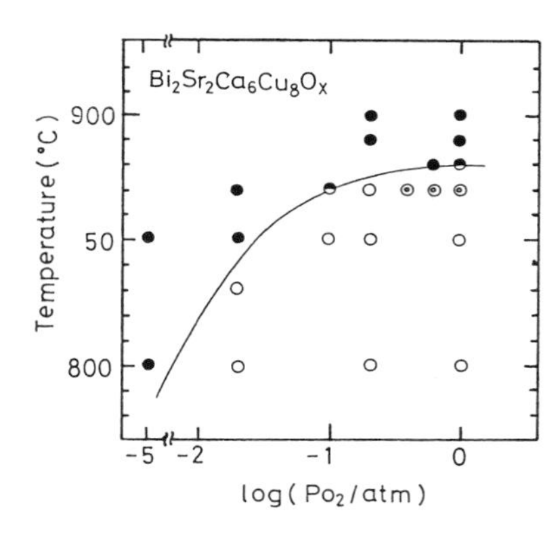

Fig. 10. Effect of oxygen partial pressure for the formation and the disappearance of the high-T_c phase in the $Bi_2Sr_2Ca_6Cu_8O_x$ system. The closed circles indicate that the sample melted, the half closed circles partially melted, the double circles not melted though sintered and the open circles not melted at all.

REFERENCES

[1] H. Maeda, Y. Tanaka, M. Fukutomi and T. Asano: Jpn. J. Appl. Phys. 27 (1988) L209.
[2] A. Sumiyama, T. Yoshitomi, H. Endo, J. Tsuchiya, N. Kijima, M. Mizuno and Y. Oguri: Jpn. J. Appl. Phys. 27 (1988) L541.
[3] N. Kijima, H. Endo, J. Tsuchiya, A. Sumiyama, M. Mizuno and Y. Oguri: Jpn. J. Appl. Phys. 27 (1988) L821.
[4] K.Kitazawa, S. Yaegashi, K. Kishio, T. Hasegawa, N. Kanazawa, K. Park and K. Fueki: Adv. Ceram. Mater. (in press)
[5] H. Endo, J. Tsuchiya, N. Kijima, A. Sumiyama, M. Mizuno and Y. Oguri: submitted to Jpn. J. Appl. Phys.
[6] M. Takano, J. Takada, K. Oda, H. Kitaguchi, Y. Miura, Y. Ikeda, Y. Tomii and H. Mazaki: Jpn. J. Appl. Phys. 27 (1988) L1041.
[7] M. Mizuno, H. Endo, J. Tsuchiya, N. Kijima, A. Sumiyama and Y. Oguri: Jpn. J. Appl. Phys. 27 (1988) L1225.

Pb Substituted Bi-Sr-Ca-Cu-O Superconductor

YUTAKA YAMADA, SATORU MURASE, MISAO KOIZUMI, MINORU TANAKA, DAISUKE ITO, SHIROU TAKENO, ISAO SUZUKI, and SHIN-ICHI NAKAMURA

Toshiba R & D Center, 4-1, Ukishima-cho, Kawasaki-ku, Kawasaki, 210 Japan

ABSTRACT

Pb substituted BiSrCaCuO with Tc above 100 K has been studied in terms of its microstructure and superconducting properties. Transmission electron microscopic observations revealed that the c-axis intergrowth structure, observed frequently in non-Pb-added Bi-Sr-Ca-Cu-O, was suppressed in the Pb introduced samples. However, lattice parameters were not obviously changed by Pb introducton and the modulation structure along the b-axis still existed. The effect of the magnetic field and temperature on transport critical current density Jc exhibited Josephson junction type behavior. An example of this junction, an anisotropic crystal distribution, is shown: a highly orientated specimen showed a strong anisotropic Jc for magnetic field directions parallel and perpendicular to the pellet surface (c-plane).

INTRODUCTION

It is well known that a high Tc (~110 K) phase of Bi-Sr-Ca-Cu-O co-exists with a low Tc (80 K) phase [1]. It is difficult to obtain only a high Tc phase. However, Pb addition to Bi-Sr-Ca-Cu-O was recently found to be effective in enhancing Tc [2]. Takano, et al.[3], reported that a Pb substituted sample showed a high Tc of 107 K and the volume fraction of the high Tc phase was about 85%. This paper reports the microstructure and superconducting properties for Pb introduced Bi-Sr-Ca-Cu-O superconductors.

EXPERIMENTAL

$(Bi_{1-x}Pb_x)_2Sr_2Ca_2Cu_3O_y$ samples with x=0, 0.3, 0.5 were prepared from a mixture of Bi_2O_3, PbO, CuO, $SrCO_3$, $CaCO_3$ powders. The samples were heated in air for 10 hrs at 850°C. The calcinated powders were reground into powder and pressed at 1.5 t/cm^2 into a 2 cm diameter pellet. The pellets were sintered at 850°C and subsequently slowly cooled at the rate of 3 °C/min to room temperature. The specimens, heat-treated for 165, 200 and 300 hrs, showed Tc above 100 K. X-ray diffraction analysis was performed on the ground powders. Transmission electron microscopy observation was made by a 400 kV transmission electron microscope (JEOL 400FX) on the samples obtained by grinding and ion milling. SEM observation and EDX analysis were also carried out on the fracture surface. For standard four probe resistivity measurements, a part of each sample was cut into rectangular bars, a typical dimension of 1x2x20 mm. Voltage and current leads were ultrasonically soldered to the specimen. The temperature was monitored by a Pt-Co resistive thermometer. The zero resistivity transition temperature Tc(end) was defined as the temperature at which the resistivity became 10^{-3} of Tc(onset). The critical current density Jc was obtained by both the resistive transport method and the magnetization method, using a SQUID magnetometer (Quantum Design, model MPMS). Magnetization measurements were made on the sintered pellets and powders obtained from the pellets.

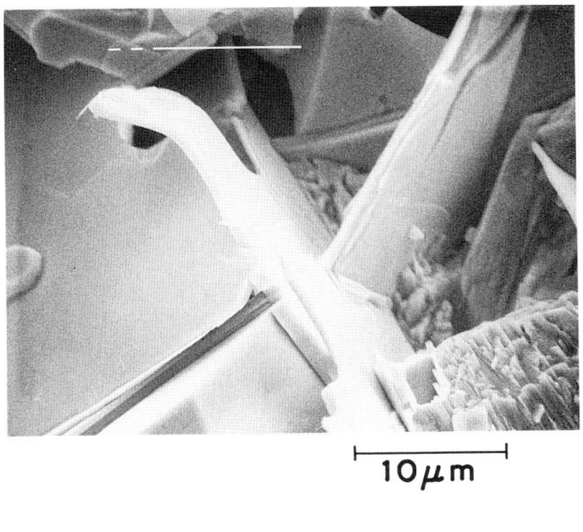

Fig.1 SEM micrograph showing fracture surface of $(Bi_{0.7}Pb_{0.3})_2Sr_2Ca_2Cu_3O_y$ heat-treated at 850°C for 165 hrs.

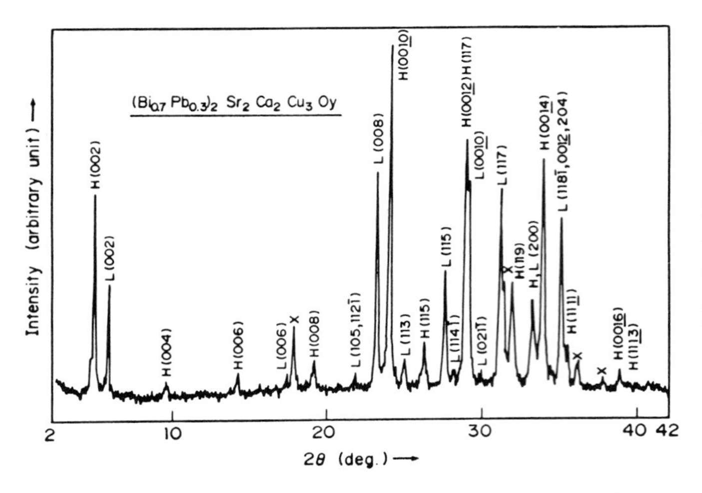

Fig.2 X-ray powder diffraction pattern for $(Bi_{0.7}Pb_{0.3})_2$ $Sr_2Ca_2Cu_3O_y$ heat-treated at 850°C for 200 hrs: H, high Tc phase; L, low Tc phase; x, unknown phase.

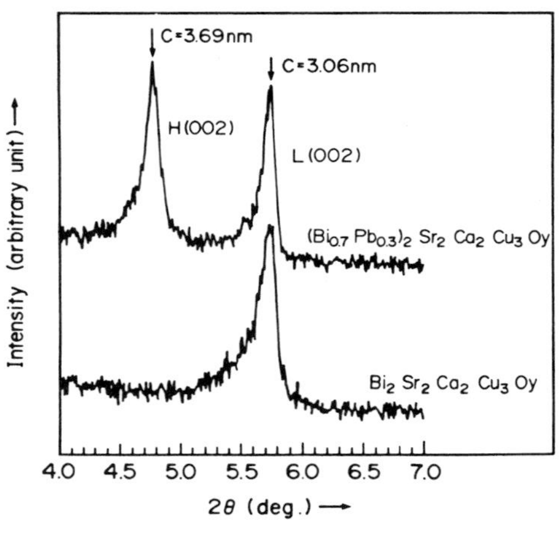

Fig.3 Comparison of (002) peaks for samples with and without Pb. Both were heat-treated at 850°C for 165 hrs.

RESULTS AND DISCUSSION

Microstructure

An SEM micrograph of the fracture surface for the sample with Tc of 108 K (x=0.3, heat treated at 850°C for 165 hrs) is shown in Fig.1. As had been observed also in non-Pb-additive samples, the microstructure consisted of micaceous flakes. Because of low heat treatment temperature, 850°C, the density was low and the sample was porous. All samples are denoted by the nominal composition in figure captions. However, in the final composition analyzed by EDX, the Pb and Ca content decreased from the starting nominal composition.

Figure 2 shows the powder X-ray diffraction pattern for the sample with Tc above 100 K (x=0.3, heat treated at 850°C for 200 hrs). Peaks of a high Tc (2223) phase and a low Tc (2212) phase were clearly observed. Moreover, an unknown phase denoted by X in the figure also appeared. Considering these peaks together the results in Ref.2, the X-ray diffraction pattern for the sample $PbSrCaCu_2O_y$, we think that this unknown phase possibly corresponds to Ca_2PbO_4.

Figure 3 illustrates a comparison of X-ray diffraction patterns for samples with and without Pb heat-treated at 850°C for 165 hrs. Although a Pb added specimen contains both a high Tc phase and a low Tc phase, a non-Pb-added one includes only the low Tc phase for this heat-treatment condition. The c-axis lengths were 3.69 nm and 3.06 nm for the high Tc phase and the low Tc phase, respectively, in the Pb added sample. The low Tc phase in $Bi_2Sr_2Ca_2Cu_3O_y$ had the same c-axis length, 3.06 nm. Therefore, Pb introduction has little effect on the c-axis length. Similarly, the a and b-axis lengths seemed to be unchanged with the TEM observation in the available resolution limit. However, the (002) peak width for Pb added sample was narrower than that for non-Pb-added sample. Therefore, the c-axis length fluctuation seems to be small. This was also confirmed by TEM, as shown in Fig.4. Figure 4 (a) shows the high Tc phase for the Pb added sample (x=0.3, heat treated at 850°C for 165 hrs). Little c-axis fluctuation (intergrowth) was observed, compared with the non-Pb-added low Tc phase shown in Fig.4 (b). This low Tc phase contained a longer Bi_2O_2 layers distance, 1.8 nm, than the usual Bi_2O_2 layers distance, 1.5 nm. This c-axis intergrowth structure was also observed by other researchers [4] for the non-Pb-added samples. Consequently, Pb introduction is considered to suppress microstructural fluctuation, especially for the c-axis. However, modulation structure along the b-axis, the black contrast at Bi_2O_2 layers, was also observed, as shown in Fig.4 (a). Moreover, in this specimen, some crystal defects such as amorphous phase and grain boundary, which affect critical current density, were observed, as shown in Figs.5 (a) and (b).

Superconducting Properties

Figure 6 shows the temperature dependence of resistivity and magnetic moment for the specimen with x=0.3 heat-treated at 850°C for 165 hrs. A clear Meissner effect appeared at 108 K, where the resistivity reached zero. Since this specimen contained a low Tc phase as well as a high Tc phase, as shown in Fig.2, a large diamagnetic signal appeared again at around 70 K.

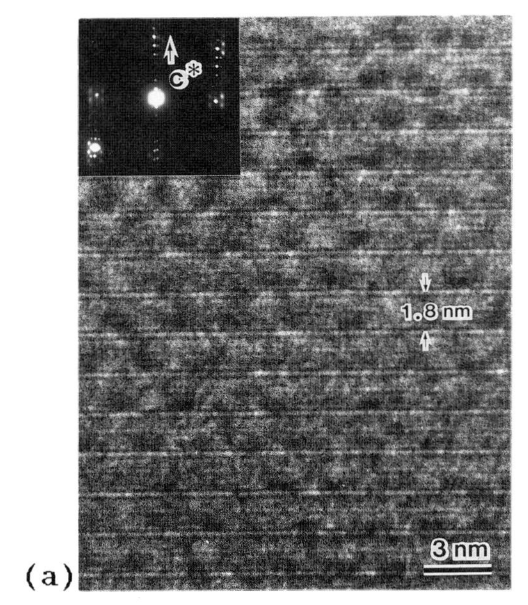

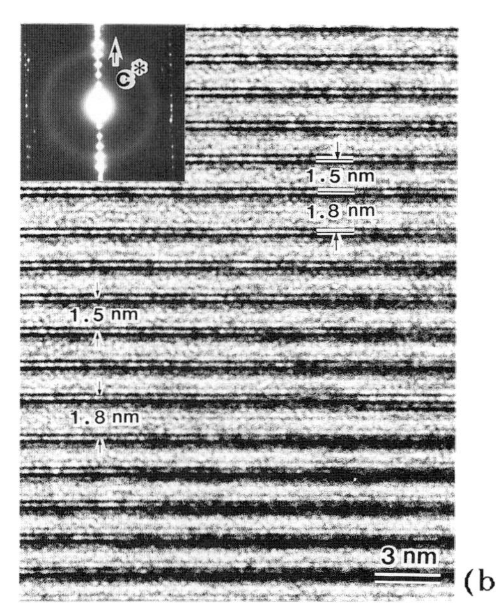

Fig.4 High resolution images of (a) highTc phase in $(Bi_{0.7}Pb_{0.3})_2Sr_2Ca_2Cu_3O_y$ and(b) low Tc phase in $Bi_2Sr_2Ca_2Cu_3O_y$. Both were heat-treated at 850°C for 165 hrs. Intergrowth structure along the c-axis is observed in (b) but not in (a). Modulation structure along the b-axis, the black contrast at Bi_2O_2 layers, also exists in Pb substituted sample (a).

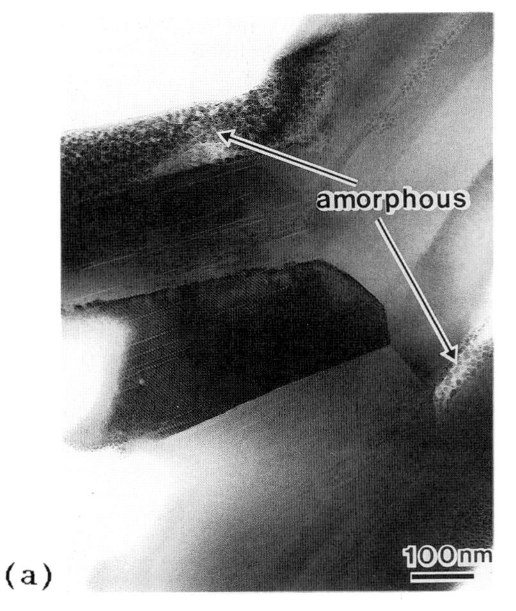

Fig.5 (a) amorphous phase and grain boundary and (b) high resolution image of amorphous phase in $(Bi_{0.7}Pb_{0.3})_2Sr_2Ca_2Cu_3O_y$ heat-treated at 850°C for 165 hrs.

The magnetic field dependences for transport Jc at 4.2 K and 77 K are shown in Fig.7. At 4.2 K, Jc was unchanged up to 12 T while the 0.1 T magnetic field decreased Jc by a factor of ten. For this specimen, Jc increased with increasing the magnetic field. This so-called peak effect was observed for other BiSrCaCuO specimens. For conventional superconductors, such as Nb_3Al and $Nb_3(Al,Ge)$, the peak effect has been observed in the case of a weak pinning system [5]. A Bi system may have weak pins, even at 4.2 K, compared with the Y system. On the other hand, Jc at 77 K was continuously and drastically decreased to $4x10^{-2}$ A/cm^2 at 0.5 T. This is the same case as in the Y system [6]. Therefore, for these oxide system, weak link coupling by Josephson junction, very sensitive to the magnetic field, is dominant especially at 77K.

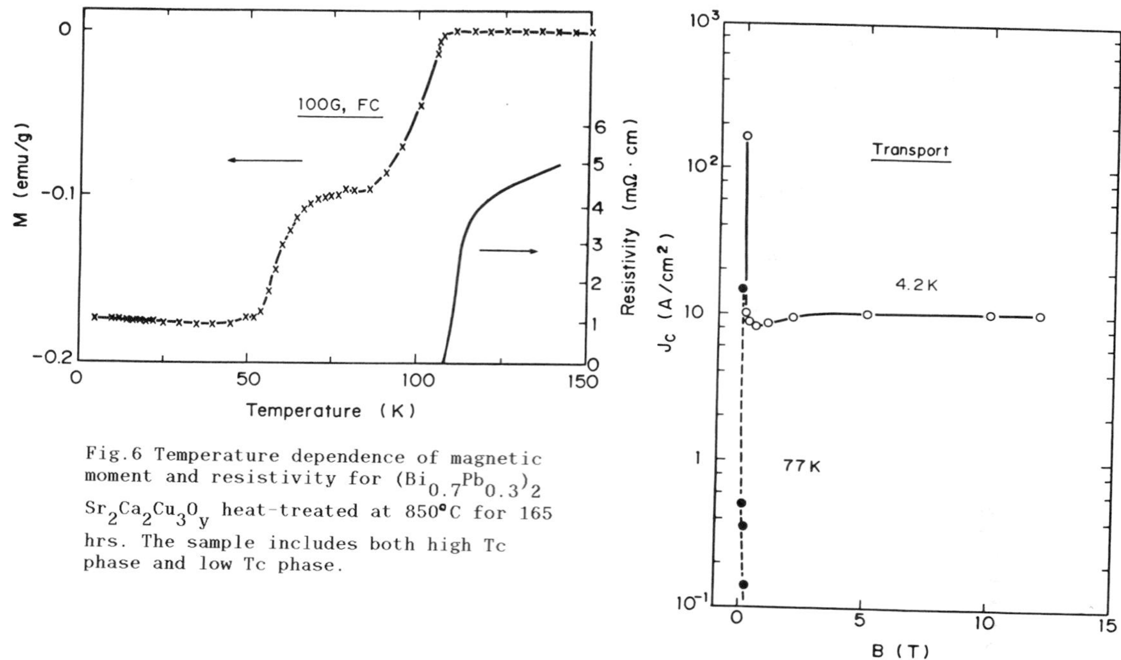

Fig.6 Temperature dependence of magnetic moment and resistivity for $(Bi_{0.7}Pb_{0.3})_2Sr_2Ca_2Cu_3O_y$ heat-treated at 850°C for 165 hrs. The sample includes both high Tc phase and low Tc phase.

Fig.7 Magnetic field dependence of transport critical current density at 4.2 K and 77 K for $(Bi_{0.7}Pb_{0.3})_2Sr_2Ca_2Cu_3O_y$ heat-treated at 850°C for 165 hrs.

Figure 8 shows the results of the magnetically obtained critical current density, Jcm^{bulk}, for the same sintered sample as in Fig.7. As in the transport critical current density, Jct, in Fig.7, the magnetic field dependence of Jcm^{bulk} was larger at 77 K than at 5 K. The magnetization value was unchanged after grinding the sample to powders of about 50 μm diameter. In this case, the Jcm^{powder} increased by a factor of 30, as indicated in the vertical axis of the figure. The superconducting current passes only inside the crystal even in the form of a sintered pellet. Therefore, the present Jcm-B dependence is considered to be the same as for a single crystal and to be intrinsic. On the contrary, Jcm for the Y system is less sensitive to the magnetic field even at 60 K and 80 K [7]. Bi system with the drastic magnetic field dependence of Jcm as well as Jct may have weak pins.

Figure 9 shows the temperature dependence for Jct and Jcm at 0 T for the same sample as in Figs.7 and 8. According to Deutcher and Muller [8], for a high Tc oxide superconductor with a short coherence length, the Josephson junction dominates the supercurrent and, thus, Jc shows the following temperature dependence; $Jc \propto (1-T/Tc)^2$. Their prediction is in good agreement with our transport Jc data in Fig.9, while the magnetization Jc deviates from the slope $(1-T/Tc)^2$ especially below 20 K. The origin of the Josephson junction current is generally considered to be the grain boundary and non-superconducting phases, as shown in Fig.5. Another case is anisotropy. Koike, et al.[9], reported a Bc_2 anisotropy of 20 T (for the magnetic field perpendicular to the c-plane) and 400 T (for the magnetic field parallel to the c-plane). For the crystal morphology in Fig.1, the crystals' c-planes cross each other. Then transport Jc will be affected by the magnetic field. Figure 10 demonstates an example of this effect: transport Jc in different magnetic field directions for a highly orientated specimen (but not containing Pb and mainly consisting of a low Tc phase). In the X-ray diffraction pattern, the ratio of peaks for the c-plane to that for other planes was about 3:1 and the c-plane was parallel to the pellet surface. The field dependence for Jc was smaller for the parallel magnetic field than for the perpendicular field. Furthermore, as the same case for a single crystal data, Jc for the magnetic field parallel to the c plane was larger than that for the perpendicular field. For these strong anisotropic materials, the random crystal distribution in the specimen increases the Josephson junction weak links. Thus, an orientated or textured oxide conductor is promising for Jc enhancement and industrial applications.

CONCLUSION

Pb substituted Bi-Sr-Ca-Cu-O specimens with high Tc above 100 K have been investigated in terms of microstructure and superconducting properties. Pb addition was considered to suppress the intergrowth structure in the c-axis while the modulation structure along the b-axis was still observed. The magnetic field and temperature dependence of Jc indicated that

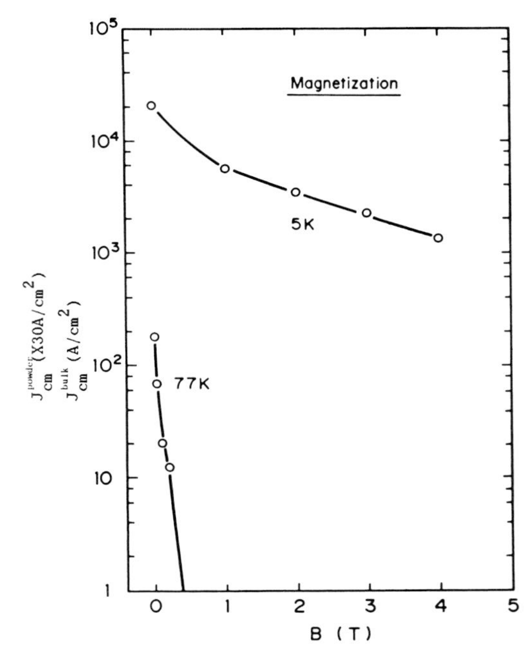

Fig.8 Magnetic field dependence of critical current density deduced from magnetization at 5 K and 77 K for the same sample as in Fig.7.

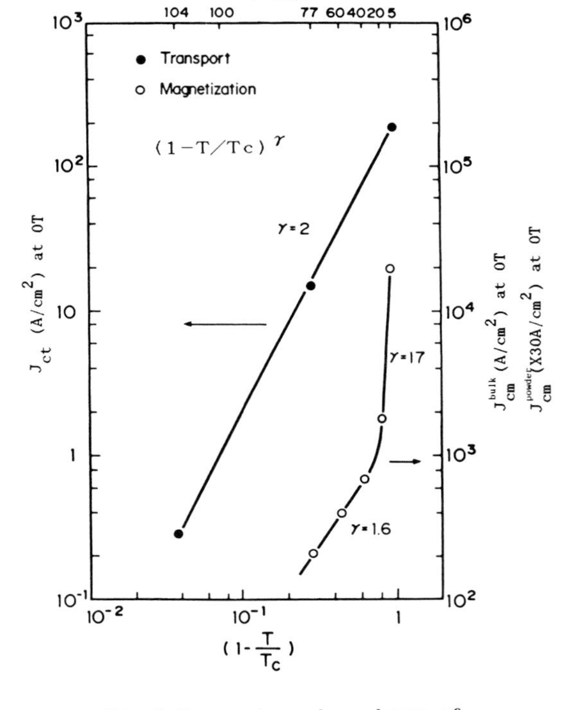

Fig.9 Temperature dependence of transport and magnetization critical currrent density at 0 Tesla for the same sample as in Fig.7.

transport Jc was restrained by the Josephson junction. An example of this weak link, caused by anisotropy, was demonstrated.

ACKNOWLEDGMENT

The authors are grateful to Dr.O.Horigami and Mrs.M.Tada, M.Hayashi, T.Nomaki, S.Shikanai and K.Yamamoto for their helpful discussions. They also would like to thank Mr.F.Umibe for his review of the English manuscript.

REFERENCES

1) H.Maeda, Y.Tanaka, M.Fukutomi and T.Asano: Jpn.J.Appl.Phys. 27, L209(1988).
2) Y.Yamada and S.Murase: Jpn.J.Appl.Phys. 27, L996(1988).
3) M.Takano, J.Takada, K.Oda, H.Kitaguchi, Y.Miura, Y.Ikeda, Y.Tomii and H.Mazaki: Jpn.J.Appl.Phys. 27, L1041(1988).
4) N.Kijima, H.Tsuchiya, A.Sumiyama, M.Mizuno and Y.Oguri: Jpn.J.Appl.Phys. 27, L821(1988).
5) Y.Yamada, S.Murase, M.Sasaki, E.Nakamura, H.Kumakura, K.Togano, K.Tachikawa: Proc.18th Int.Conf.on Low Temperature Physics, Kyoto, 1987, Jpn.J.Appl.Phys.26 Supplement 26-3, 1493(1987).
6) Y.Yamada, N.Fukushima, S.Nakayama, H.Yoshino and S.Murase: Jpn.J.Appl.Phys.26, L865(1987).

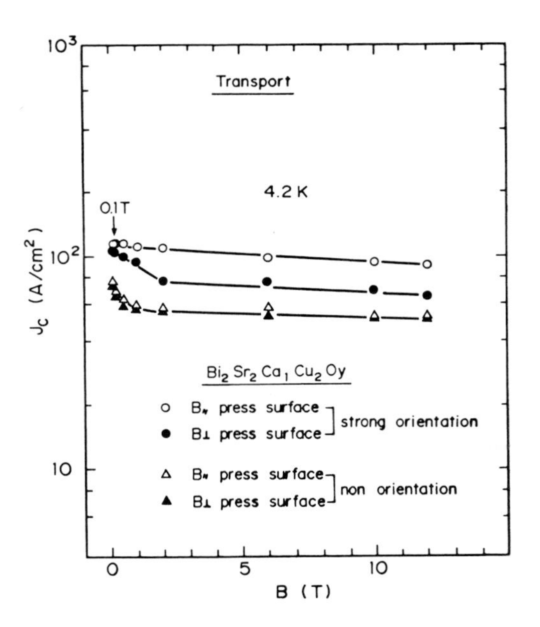

Fig.10 Magnetic field dependence of transport critical current density at 4.2 K for highly orientated and non-orientated $Bi_2Sr_2Ca_1Cu_2O_y$. C-plane is parallel to the press surface for the highly orientated sample.

7) I.Apfelstedt, R.Flükiger, H.Küper, R.Meier-Hirmer, B.Obst, C.Politis, W.Schauer, F.Weiss and H.Wühl: Proc.18th Int.Conf.on Low Temperature Physics, Kyoto, 1987, Jpn.J.Appl.Phys. 26 Supplement 26-3, 1181(1987); H.Küper, I.Apfelstedt, W.Schauer, R.Flükiger, R.Meier-Hirmer and H.Wühl: preprint submitted to Z.Phys.B.
8) G.Deutscher and K.A.Müller: Phys.Rev.Lett. 59, 1745(1987).
9) Y.Koike, T.Nakanomyo and T.Fukase: Jpn.J.Appl.Phys. 27, L841(1988).

Microstructure and Superconducting Properties of BiSrCaCuO Compound

A. Hayashi, M. Murakami, M. Morita, H. Teshima, K. Doi, K. Sawano, M. Sugiyama, H. Hamada, and S. Matsuda

R & D Laboratories-I, Nippon Steel Corporation, 1618 Ida, Nakahara-ku, Kawasaki, 211 Japan

ABSTRACT

Zero resistance temperature of 106 K was obtained for $BiSrCaCu_2O$ compound with a sharp signal of diamagnetism. High Tc phase appeared 880 °C, but subsequent annealing at 850 °C was effective to raise Tc(end). The samples were multi-phase, and the volume fraction of high Tc phase was small. The low Jc may be attributed to the microstructure, as well as to the small volume fraction of superconducting phase.

INTRODUCTION

The Bi-Sr-Ca-Cu-O system has attracted much attention since the discovery by Maeda et al. of superconducting phase with Tc = 105 K[1]. However, zero resistance temperature can be obtained at 80 K because of coexisting low Tc phase. Therefore, substantial work has been done to raise Tc(end) and to increase the volume fraction of high Tc phase. On the other hand, few study has been reported concerning the microstructure and its relation to superconducting properties of the Bi based compounds.

We report our results of measurement of superconducting properties for Bi compounds with respect to the microstructure observation. Results of magnetic properties and specific heat measurements are also presented.

EXPERIMENTAL

The samples were prepared by the conventional solid state reaction using Bi_2O_3, $SrCO_3$, $CaCO_3$, and CuO. The nominal cation ratio of the samples was fixed to Bi:Sr:Ca:Cu = 1:1:1:2 throughout this work. The powders were mixed in a ball mill with zirconia balls for 12 h with ethanol, then dried in vacuum. The mixtures were calcined at 850 °C for 8 h. The products were pulverized and pressed into disks 20 mm in diameter, followed by sintering at 800 to 900 °C in air.

The crystalline phases of the samples were identified by X-ray diffractometry.

Microstructure was observed with optical microscope and scanning electron microscope.

Electrical resistivity and critical current density(Jc) were measured by a standard four probe method using ultrasonic solder for the contact to electrical lead. The current value was 10 mA in resistivity measurement unless otherwise specified.

Magnetic measurement was carried out with a vibrating sample magnetometer at 77 K. AC magnetic susceptibility measurement was also performed using inductive method.

The specific heat was measured in the temperature range from 60 K to 120 K using a combination of AC calorimetric method and standard adiabatic method[2].

RESULTS AND DISCUSSION

Transport Measurements

Fig. 1 shows the temperature dependence of the resistivity for several samples sintered at different temperatures for 8 h and furnace cooled. The drop in resistivity at around 110 K was observed for the samples sintered at 850 - 880 °C. This indicates that high Tc phase is formed by the heat treatment. However, these curves have long tails which extend to lower temperatures and zero resistance was achieved at around 70 K.

The magnitude of the drop in resistivity increases with increasing sintering temperature and at 880 °C about 75 % drop was observed. However, the resistivity drop became hardly visible and the resistivity in normal state also increased when sintering temperature was raised to 890 °C, then at 900 °C the sample showed semiconductor-like behavior and no superconducting transition was observed down to 4.2 K. On sintering above 880 °C, pellets were bent or twisted indicating that the samples melted partially. Therefore, sintering temperature is considered to be critical for the growth of high Tc phase and the optimum temperature seems to be close to the partial melting point.

So, we fixed the initial sintering temperature at 880°C and studied the heat patterns. Fig. 2 shows the results for the samples sintered at 880°C followed by intermediate annealing at 850 C. As can be seen from the figure, sintering at 880°C for long period obviously enhanced the magnitude of the drop in resistivity at 110 K, while subsequent annealing at 850°C was much more effective to enhance the drop and also raised zero resistance temperature. Tc(end) increased with the annealing period and the optimum condition was a combination of sintering at 880°C for 98 h and annealing at 850°C for 198 h(Sample A), which resulted in the largest drop at 110 K(95 %) and highest Tc(end) = 96 K. The result implies that the 110 K material preferentially grows at around 850°C.

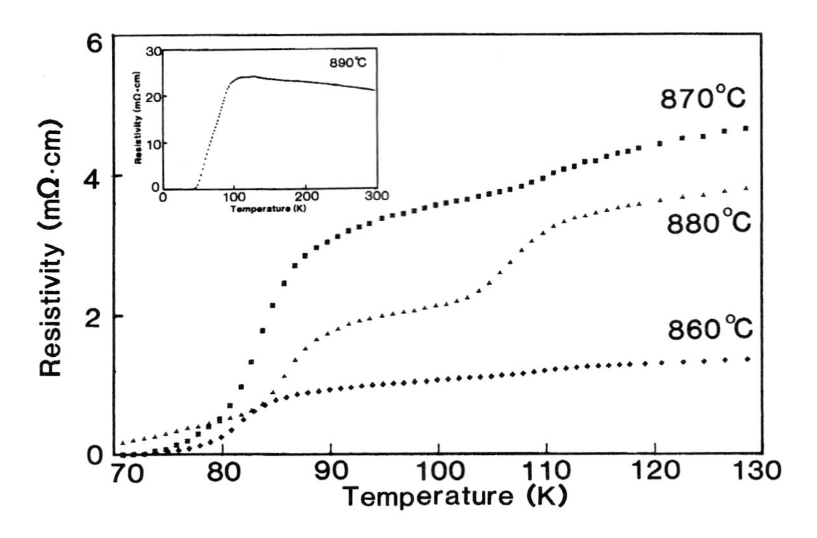

Fig. 1. Temperature dependence of electrical resistivity of $BiSrCaCu_2O_x$ sintered for 8 h.

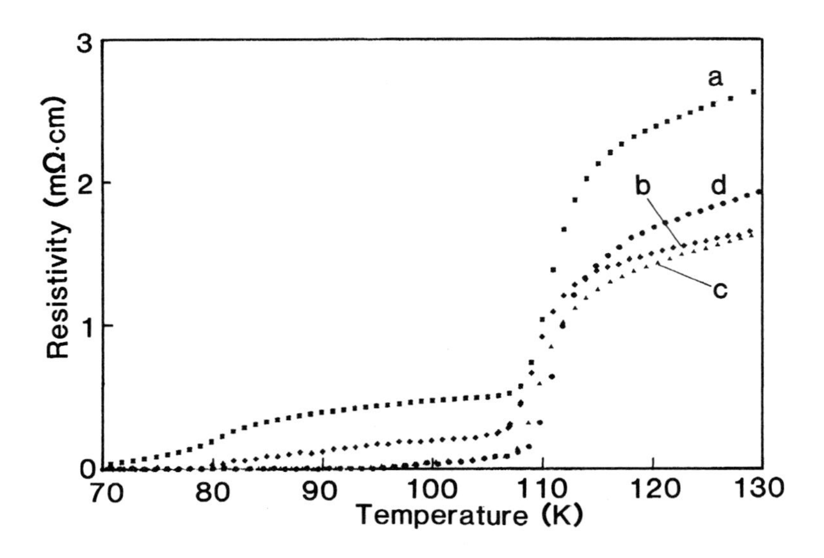

Fig. 2. Temperature dependence of electrical resistivity of $BiSrCaCu_2O_x$ sintered at: a)880°C for 98 h ; b)880°C for 8 h and 850°C for 8 h ; c)880°C for 98 h and 850°C for 98 h ; d) 880°C for 98 h and 850°C for 198 h.

It should be noted that results of resistivity measurements strongly depend on the measuring current. Fig. 3 shows the resistivity-temperature curve for the same sample using three different currents : 1, 10 and 100 mA. Tc decreases from 92 K to 70 K with slightly increasing measuring current. Similarly, for the sample A, Tc(end) reached 106 K with current value of 1 mA. Fig. 4 shows the temperature dependence of electric resistivity together with AC susceptibility data which demonstrate a large diamagnetic signal at 105 K.

The decrease in Tc with increasing the measuring current means very low Jc of the sample. In practice, transport Jc of the sample A was less than 10 A/cm^2 at 77 K in zero magnetic field.

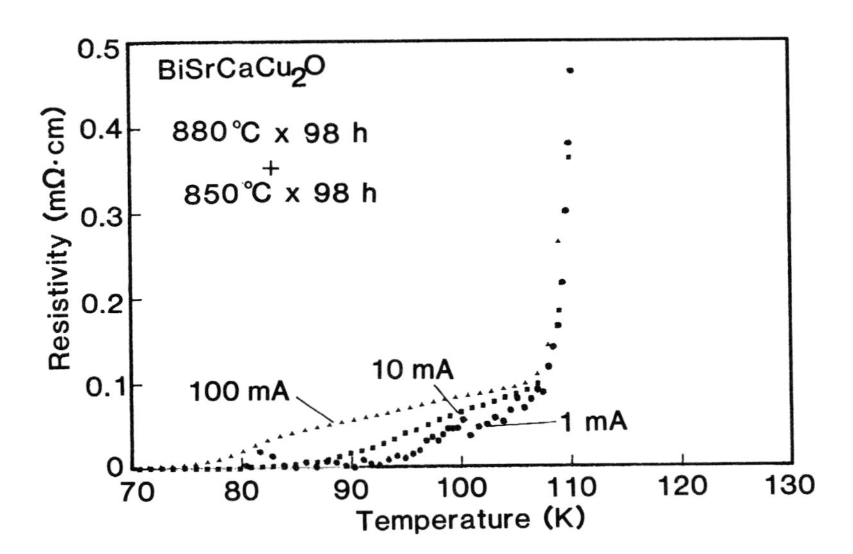

Fig. 3. Temperature dependence of electrical resistivity using different measuring currents : 1. 10 and 100 mA.

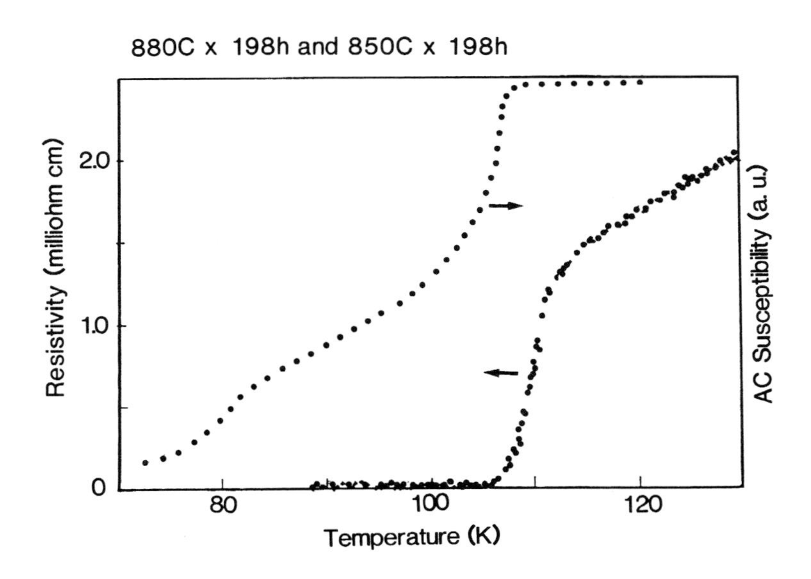

Fig. 4. Temperature dependence of electrical resistivity and AC susceptibility of the sample A(sintered at 880°C for 98 h and 850°C for 198 h).

Phase Identification

X-ray diffraction showed that all the samples consisted of multi phases. Even our best sample(sample A) has been found to be composed of at least three phases(Fig. 5). The peaks marked with open circle can be indexed on the structure of low Tc phase proposed by Tarascon et al.[3]($Bi_2Sr_2CaCu_2O$) with lattice parameters a = 3.8 Å and c = 30.6 Å. The peaks marked with closed circles correspond to the high Tc phase ($Bi_2Sr_2Ca_2Cu_3O$)[4] with c = 36.8 Å. CuO and another unidentified phase were also present. Judging from the X-ray diffraction pattern, the volume fraction of high Tc phase is fairly small even in the sample with Tc(end) = 106 K.

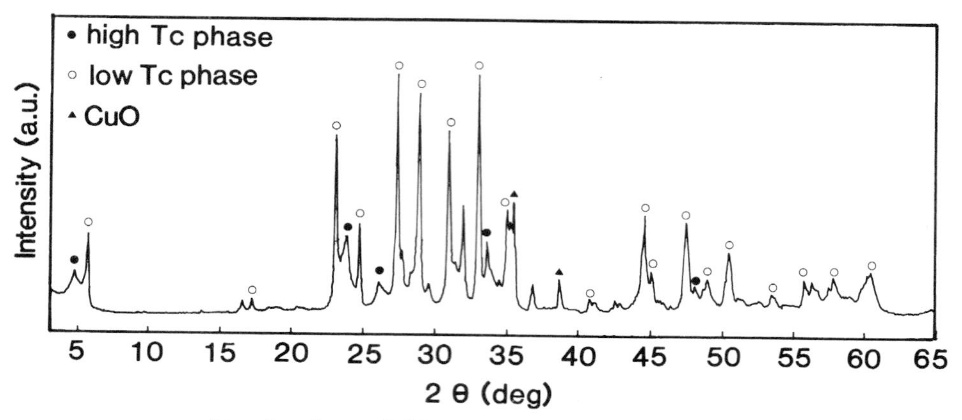

Fig. 5. X-ray diffraction pattern of the sample A.

Microstructural Observation

Fig. 6 shows the optical micrographs for the polished surfaces of the samples, which also reveals clearly multi-phased feature. Fig.6(a) shows microstructure of the sample sintered at 890 °C for 1 h, followed by annealing at 850°C for 4 h. The resistivity curve showed a large drop at 110 K in the sample. The microstructure is characterized by thin elongated grains about 20 - 50 μm, which has a tendency to align parallel to each other with some inclusion between them. Fig. 6(b) corresponds to the sample sintered at 890°C for 8 h showing no appreciable drop at 110 K and Tc(end) was 40 K. Thin grain seems to have decomposed into grain about 10 μm with smaller aspect ratio. From the above results, it can be deduced that thin elongated grain is responsible for the superconductivity, possibly both high- and low-Tc phase. In fact, further annealing at 850°C did not affect the microstructure, while it improved Tc(end). The observation of the fracture surface (Fig.7) shows similar result. Thin plate-like phase well developed at 880 °C disappeared during sintering at 890°C and the sample showed zero resistance below 70 K.

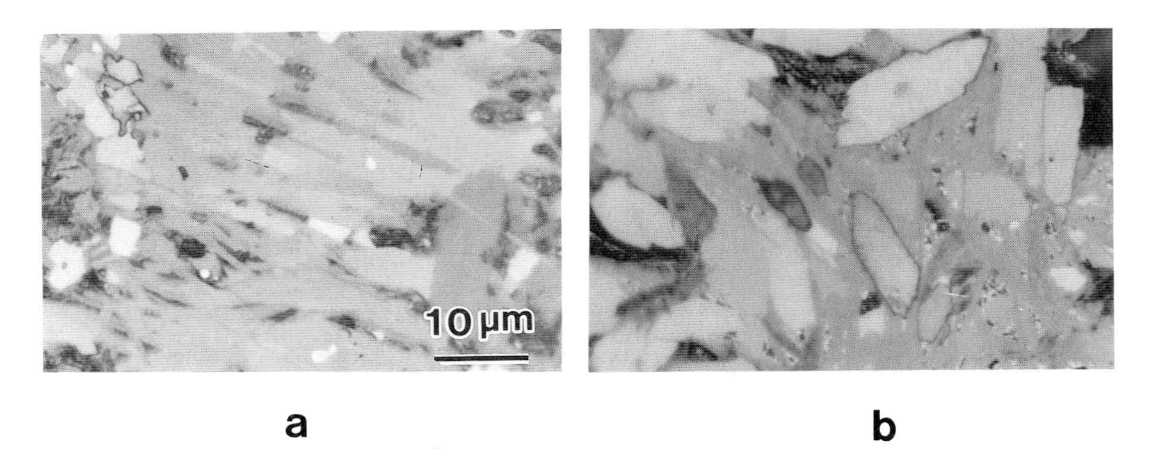

Fig. 6. Optical micrograph for the polished surface.
(a)890°C for 1 h and 850°C for 4 h ; (b)890°C for 8 h.

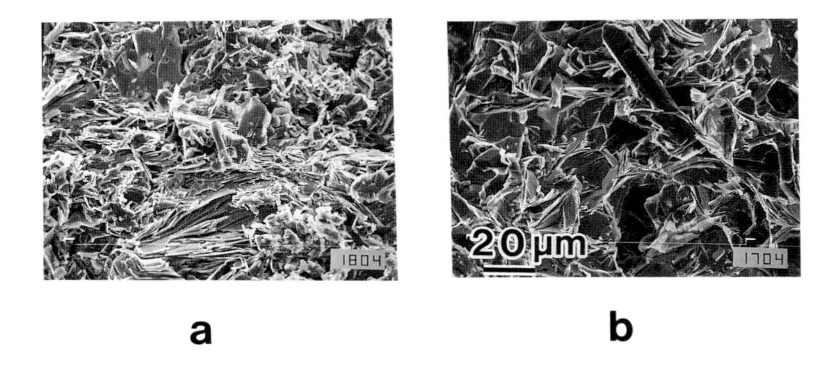

Fig. 7. SEM micrograph for the fracture surface.
(a)880°C for 8 h and 850°C for 4 h.
(b)890°C for 8 h.

Magnetization and Specific Heat Measurement

Fig.8 shows the magnetization curves of BiSrCaCuO compound at 77 K together with that of bulk sintered $YBa_2Cu_3O_x$ which is presented for comparison. Bi compound has significantly a small magnetization signal compared with $YBa_2Cu_3O_x$ and magnetic hysteresis is fairly small. It suggests that Jc is extremely low, which is consistent with the low value of measured transport Jc.

Fig. 9 shows the results of specific heat scan for the sample A. As shown in Fig.4, this sample shows a large drop in resistivity at 110 K and has Tc(end) of 106 K and also shows a large diamagnetic signal at 105 K. However, clear specific heat jump at Tc is not observed showing only a broad anomaly around 110 K. This result may also suggest that the volume fraction of 110 K phase is small.

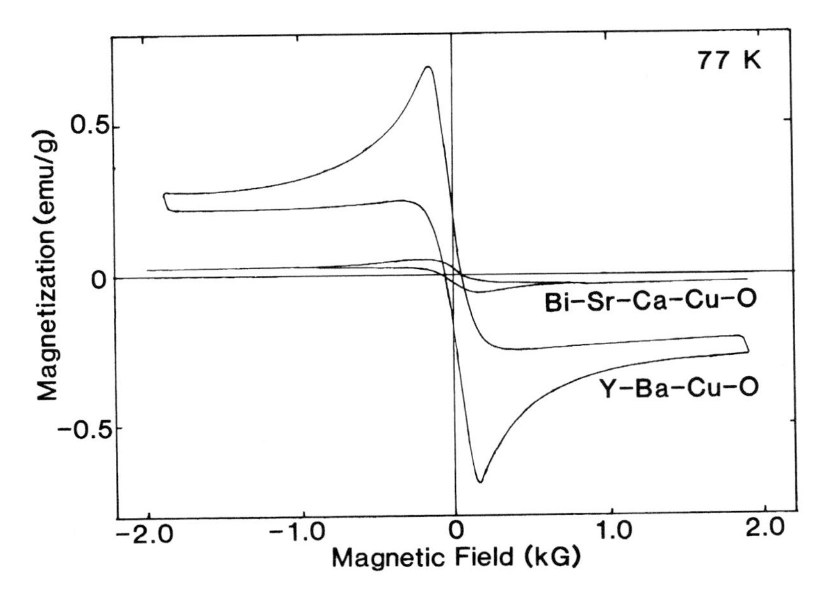

Fig. 8. The magnetization curve of $BiSrCaCu_2O_x$ at 77 K. Bulk sintered $YBa_2Cu_3O_x$ is shown for comparison.

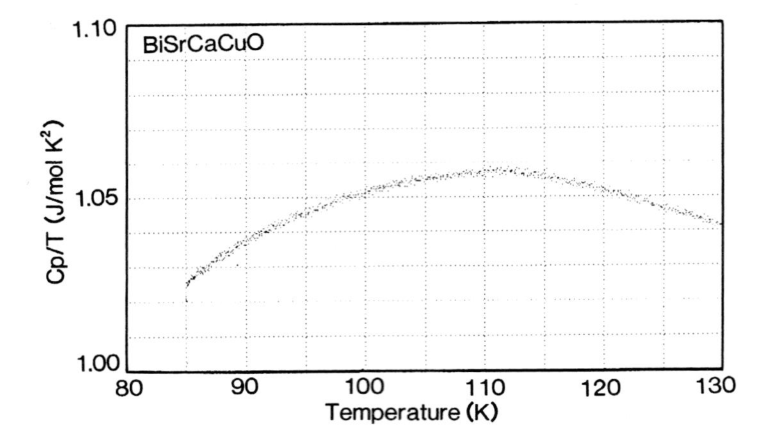

Fig. 9. The temperature dependence of specific heat over temperature of the sample A.

CONCLUSIONS

In Bi system, high Tc phase is considered to form at 880°C preferentially, perhaps under the partial melted condition. But once it is formed it can grow at lower temperature of around 850 °C and long period annealing at 850°C is effective to raise Tc(end). Zero resistance temperature is lowered with slightly increasing the measuring current suggesting a very low Jc of the material. It is also found that at least three phases are present in the Bi compound. The phases are high- and low-Tc phase and CuO. The volume fraction of the high Tc phase is fairly small judging from the results of X-ray diffraction and magnetization and specific heat measurements. Low Jc confirms this result. Microstructural observation showed that the superconducting phase grows in a form of thin grain and the connection between them was poor, which is also responsible for the low Jc, as well as the small volume fraction of high Tc phase. Therefore, for higher Jc, the control of microstructure to improve the connection between superconducting grains is needed, parallel with the effort to promote the formation of the high Tc phase. Growth from the melt or using some flux to lower melting points may be promising method.

REFERENCES

[1] H. Maeda, Y. Tanaka, M. Fukutomi and T. Asano, Jpn. J. Appl. Phys. 27, L209 (1988).
[2] M. V. Nevitt, G. W. Crabtree and T. E. Klippert, Phys. Rev. B36, 2398 (1987).
[3] J. M. Tarascon, Y. Le Page, P. Barboux, B. G. Bagley, L. H. Greene, W. R. Mckinnon, G. W. Hill, M. Giroud and D. M. Hwang, Phys. Rev. in press.
[4] e.g. H. Maeda, Bull. Jpn. Inst. Metal 27, 566 (1988).

Effect of Ambient Gas on Preparing Bi-Sr-Ca-Cu-O Superconductor

AKIRA FUKIZAWA, YOSHINOBU SAKURAI, and AKIRA OHTOMO

Central Research Laboratory, Showa Denko K.K., Ohta-ku, Tokyo, 146 Japan

ABSTRACT

Systematic decrease of the formation temperature of high Tc superconducting phase with zero resistance at 107K was observed in the Bi-Pb-Sr-Ca-Cu-O system by reducing the oxygen content in the N_2/O_2 mixing atmosphere following a reduction of melting temperature. Volume fraction of high Tc phase was exponentially increased with increasing heat treating time. It increased rapidly near the melting temperature but the critical current density decreased by the possible formation of weak links at the grain boundaries.

INTRODUCTION

Bi-Sr-Ca-Cu-O superconductor has given us a potentiality of application in various industrial fields because of its high Tc beyond 100K and chemical stability(1,2). Many efforts have been attempted to synthesize the high Tc(110K) mono-phase by elongating sintering time(3), changing basic composition(4) and adding other elements(5). Among them, addition of Pb was most effective to proceed the formation of 110K phase and reduce firing time(6). We have examined the influence of ambient gas on synthesis of the 110K phase. It was observed that about 20℃ higher temperature is necessary to form the 110K phase in O_2 atmosphere compared with sintering in air. In other gases, N_2, Ar and He, only a small fraction of 110K phase was obtained just below their melting points near 820℃(7). Recently, the relation between formation temperature of 110K phase and ambient oxygen pertial pressure was reported in the Ar/O_2 mixing atmosphere using Bi-Pb-Sr-Ca-Cu-O system(8). In this paper, we report the formation of 110K phase in the N_2/O_2 mixing atmosphere focusing on the transformation kinetics.

EXPERIMENTAL

Three different precursors having nominal composion Bi:Pb:Sr:Ca:Cu = 0.8:0.2:1:1:1.5, 0.9:0.1:1:1:1.5 and 0.8:0.2:1:1:2 were prepared by the tartaric acid gels which are product of metal nitrates and tartaric acid mixtures. Details of sample preparation were explained in elsewhere(7). Obtained black powders were heat treated at 800℃ in air for 12h, then, pulverized and pressed into pellet shape with 10 mm in diameter and 1 mm in thickness. These specimens, composing low Tc(80K) phase, were sintered at various temperatures ranging 790 to 860℃ more than 10h in the different atmosphere, N_2, N_2/O_2(including 8% oxygen) mixing gas and air. In the air atmosphere, dried and humidious conditions which were prepared by flowing air through silica-gel or water were compared. After heat treatment, all of specimens were cooled to room temperature at a rate of 30℃/min in the atmospheres. Superconducting transition temperature Tc was detected by the conventional four probe resistance measurements. Transport current measurements were made using a pulse method to minimize sample heating. Structure and composition analyses of samples were carried out by X-ray diffraction, SEM and ICP method. Thermal analysis(DTA) was also carried out.

RESULTS AND DISCUSSION

Melting behavior of the pre-sintered samples measured by DTA in air and N_2 atmosphere are shown in Fig.1. By substituting Bi by Pb with an atomic ratio of 0.1 and 0.2, no apparent change in melting points was obtained for the ideal 2223 composition. In the N_2 atmosphere, they shifted about 30℃ toward lower temperature. The same result was obtained in the $Bi_{0.8}Pb_{0.2}SrCaCu_2O_x$ system. It has been believed that suitable treating temperature to form the 110K phase is very close to melting point.

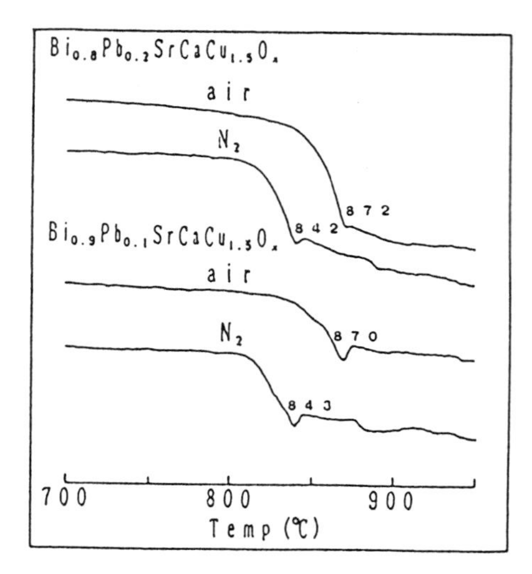

Fig.1 DTA curves of $Bi_{0.8}Pb_{0.2}SrCaCu_{1.5}O_x$ and $Bi_{0.9}Pb_{0.1}SrCaCu_{1.5}O_x$ system under air or N_2 atmosphere. Heating rate is 5℃/min.

That is why, the heat treating temperature within 50℃ below the melting point in each atmosphere was chosen for the present experiment. In this paper, we mainly report the transformation behavior on the $Bi_{0.8}Pb_{0.2}SrCaCu_{1.5}O_x$ system under various heat treating conditions. Other compounds also treated at the same conditions to compare the difference in superconducting properties among these compositions.

In the course of heat treatment, the weight loss of each specimen was increased with increasing heating time. About 2.5% of weight loss was

Table 1.Compositions of samples after sintering are analized by ICP. Asterisk denotes the starting material used.

Heat treated temp.(℃)	Heat treated time (h)	Analized compositional ratio Bi	Pb	Sr	Ca	Cu
* 800	12	0.87	0.21	0.94	1	1.5
855	13	0.87	0.21	0.95	1	1.5
855	37	0.87	0.19	0.93	1	1.5
855	96	0.88	0.18	0.95	1	1.5
855	133	0.86	0.17	0.93	1	1.5

observed after 100h sintering notwithstanding difference in treated temperature or atmosphere. This weight loss, less than 1 % for the Pb free specimen, was expected mainly causing an evaporation of Pb element during sintering. However, as shown in table 1, no remarkable depletion of Pb content was observed by ICP analysis for the specimens treated 13h to 133h at 855℃ in air. Another possible volatile element such as residual carbon should be analized(9).

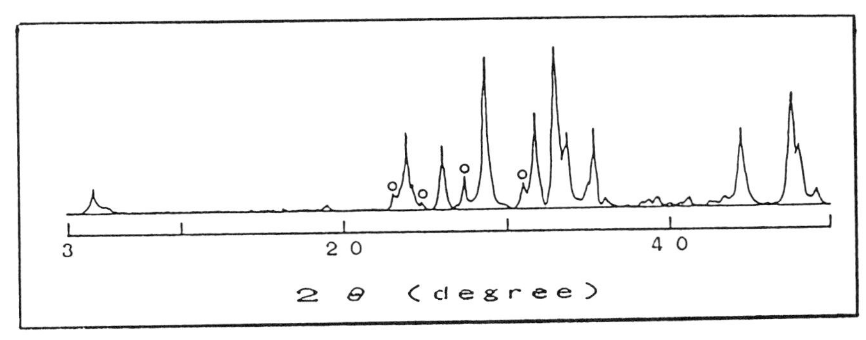

Fig.2 X-ray diffraction pattern of $Bi_{0.8}Pb_{0.2}SrCaCu_{1.5}O_x$ spesimen heat treated at 850℃ for 30h in N_2/O_2(8%) atmosphere. Open circle indicates residual low Tc(80K) phase.

The formation of 110K phase has generally required longer heating time sometimes exceeded a week. In our experiments, nearly single 110K phase was formed within 30h at 850℃ in N_2/O_2 atmosphere. Fig 2 shows the X-ray diffraction pattern of this specimen which includes small fraction of the 80K phase. No lowest Tc(20K) phase was observed during heat treatment except the melted samples. We found that the transformation rate from 80K phase to 110K phase was strongly dependent on treating temperature.

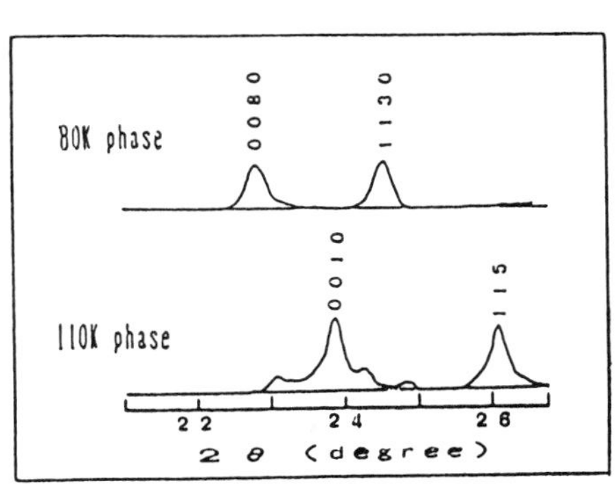

Fig.3 X-ray diffraction patterns of the 80K and 110K phases. Peak intensity ratio represented by I(110)/(I(110)+I(80)) is caluculated by using the peaks (0080),(1130) of 80K phase and (0010),(115) of 110K phase.

For comparing the volume change among both phases during transformation, as shown in Fig.3, integrated X-ray diffraction peak intensities of (0080),(1130) for 80K phase and (0010),(115) for 110K phase were calculated by using the gaussian fitting of each peaks. The peak intensity ratio of 110K phase vs 80K phase represented by I(110)/(I(110)+I(80)) is plotted on Fig.4 as a function of heat treating time. In the air atmosphere, for the data obtained at 855℃ and 850℃, volume content of the 110K phase is exponentially increased with increasing time. No influence in productivity of 110K phase between dried and humidious air are observed. Shortage of experimental data has a difficulty to fitting on the exponential formura. However, for better understanding, we fitted all of data on the semi-empirical exponential curves. In the N_2/O_2 atmosphere, the data obtained at 830℃ and 840℃ is closely related to the fitting curves of the data at 850℃ or 855℃ in air atmosphere. Above these temperatures,

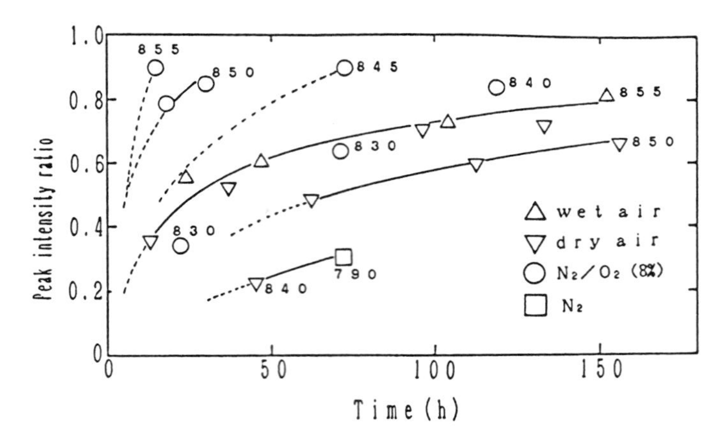

Fig.4 Peak intensity ratio vs heat treating time relation. Numbers in the figure denote treated temperature in each atmosphere.

nearly single 110K phase is obtained within a few tens of hours. In the case of N_2 atmosphere, the 110K phase was formed even at 790℃. At higher temperature above 800℃, no 110K phase was formed and the starting 80K phase was decomposed to the 20K phase above 810℃ following partial melt. It is noteworthy that the data at 790℃ is closely related to the fitting curve of the data obtained at 840℃ in air. As already pointed out in the Fig.1, the melting point of pre-sintered specimen was reduced about 30℃ by changing atmosphere from air to N_2. Downward shift on temperature of the transformation curves at reduced oxgen atmosphere can be explained by the influence of reduced melting points. Comparing these results, $Bi_{0.8}Pb_{0.2}SrCaCu_2O_x$ and $Bi_{0.9}Pb_{0.1}SrCaCu_{1.5}O_x$ samples represented almost same tendency on the transformation curves, however, the latter showed about 15% lower content of the 110K phase after the same heat treatment.

As shown in Fig.5(a), the specimen sintered at 840℃ for 119h in N_2/O_2 atmosphere shows typical resistivity vs temperature curve which is decreasing linearly with decreasing temperature dropping to zero at 107K when the current density is 20~200mA/cm^2. Even at higher current density of 0.9A/cm^2, zero resistance was obtained at 105K in the same specimen.

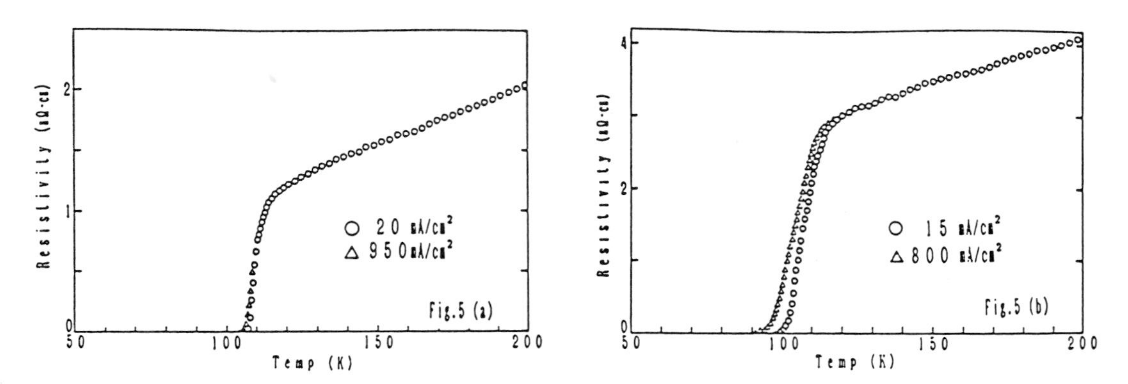

Fig.5 Resistivity vs temperature relation at different current densities. (a): sintered at 840℃ for 119h in N2/02(8%) atmosphere. (b):sintered at 840℃ for 45h in air.

Apparent zero resistance change at different measured current density was observed in the specimen including low volume fraction of 110K phase about 20%. As shown in Fig.5(b), the zero resistance temperature is decreased from 102K to 91K when the current density is increased 15 to 800mA/cm^2. When the specimen was sintered at 850℃ and 855℃ in N_2/O_2 atmosphere, the zero resistance temperature was steeply decreased to 96K and 76K even at current density of 20mA/cm^2, neverthless those includeing large volume fraction of 110K phase above 80%. These depression in resistivity may be explained by the formation of partial melted zone at the restricted area such as grain boundaries during sintering close to melting temperature.

Measured critical current defined as the current where 0.1μV is induced in liquid nitrogen(77K) is shown in Fig.6 as the I-V relation.

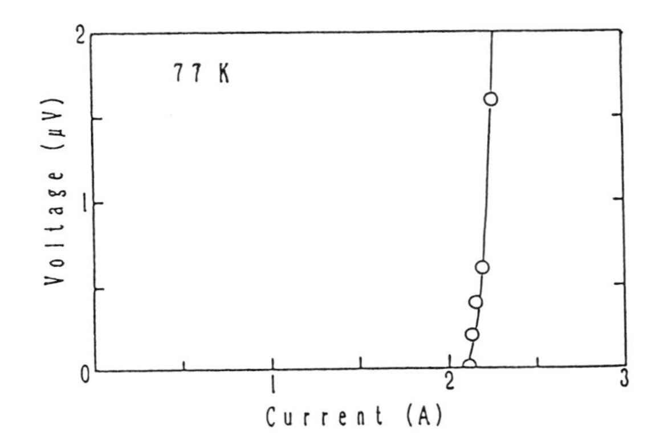

Fig.6 Transport current vs voltage relation at 77K.

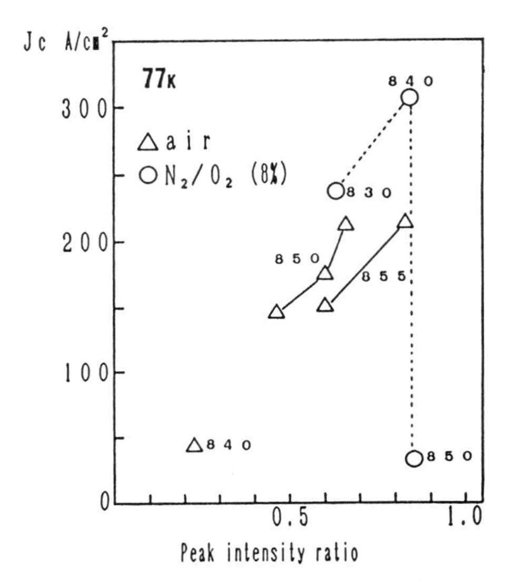

Fig.7 Jc vs peak intensity relation. Numbers in the figure denote treated temperature.

Critical current density(Jc) was calculated by deviding the current value with the cross section of the specimen about 0.7mm^2. Fig.7 shows the relation between Jc and I(110)/(I(110)+I(80)) where the data obtained from the specimen prepared at various conditions are plotted. There exist the tendency that Jc increases with increasing volume fraction of 110K phase. Highest Jc value of 310A/cm^2 was obtained for the specimen sintered at 840℃ for 119h in N_2/O_2 atmosphere. For the specimen heat treated at 850℃, however, Jc droped to 32A/cm^2 although which contains large volume fraction of 110K phase. As mentioned above, resistivity near Tc was apparently degraded above 850℃ in N_2/O_2 atmosphere. We think that the weak links such as partial melted zone or lower Tc phase formed at intra-grain or grain boundary in the specimen sintered close to the melting temperature resulted in depression of Jc, even though bulk composition was nearly single 110K phase. Same results were obtained both in $Bi_{0.9}Pb_{0.1}SrCaCu_{1.5}O_x$ and $Bi_{0.8}Pb_{0.2}SrCaCu_2O_x$ samples. Precise identification for the microstructure in these specimen should be necessary.

In conclusion, we observed systematic reduction of formation temperature of 110K phase by reducing oxegen content in the heat treating atmosphere. By addition of Pb to the Bi-Sr-Ca-Cu-O system, the 110K phase is thought to be stabilized, resulting wider synthsic region. The transformation from 80K to 110K phase was proceeded on the exponential curve with time. It proceeded more rapidly near the melting temperature, but causing the formation of weak links at grain boundaries for degrading Jc.

We appreciate to M.Shioya for ICP anlysis and H.Sakamoto for analysis supporting. We are also grateful H.Soga,K.Ichikawa and J.Satoh for their encouragement.

REFERENCES

(1) H.Maeda, Y.Tanaka, M.Fukutomi and T.Asano:Jpn.J.Appl.Phys.27(1988) L209.
(2) J.M.Tarascon,Y.Le-Page,P.Barboux,B.G.Bagley,L.H.Greene,W.R.Mckinnon, G.W.Hull, M.Giroud and D.M.Hwang: Phys.Rev.B 1988 in press.
(3) H.Nobumasa,K.Shimizu,Y.Kitano and T.Kawai:Jpn.J.Appl.Phys.,27(1988)L846.
(4) A.Sumiyama, T.Yoshitomi, H.Endo, J.Tsuchiya, N.Kijima, M.Mizuno and Y.Oguri:Jpn.J.Appl.Phys., 27(1988)L542.
(5) S.Jin, R.S.Sherwood, T.H.Tiefel, G.W.Kammoloti, R.A.Fastnacht, M.E.Davis and S.M.Zahurak:Apple.Phys.Lett., 52(19)1988,1628.
(6) M.Takano, J.Takada, K.Oda, H.Kitaguchi, Y.Miura, Y.Ikeda, Y.Tomii and H.Mazaki:Jpn.J.Appl.Phys., 27(1988)L1041.
(7) A.Fukizawa, A.Ohtomo and Y.Sakurai:Proceedings of Mater.Res.Soc.Meetings on Advanced Mater.(Tokyo.1988),in press. and Abstract of ISS'88.
(8) U.Endo, S.Koyama and T.Kawai:Jpn.J.Appl.Phys.,27(1988) in press.
(9) N.Uno, N.Enomoto, Y.Tanaka and H.Takami:Jpn.J.Appl.Phys.,27(1988)L1013.

Crystalline Structures and Superconducting Properties of Rapidly Quenched $BiSrCaCu_2O_x$ Ceramics

TSUNEYUKI KANAI, TERUO KUMAGAI, ATSUKO SOETA, TAKAAKI SUZUKI, KATSUZOU AIHARA, TOMOICHI KAMO, and SHINPEI MATSUDA

Hitachi Research Laboratory, Hitachi Ltd., Hitachi, Ibaraki, 317 Japan

ABSTRACT

Phase changes and superconducting properties of rapidly quenched $BiSrCaCu_2O_x$ ceramics have been studied. The rapidly quenched sample consists of an amorphous state which begins to crystallize around 510°C. The crystal system was found to be tetragonal with a=5.39Å and c=24.5Å. By annealing above 700°C, a tetragonal system with a=5.41Å and c=30.8Å was produced. The former crystal was a semiconductor and the latter was a superconductor with Tc=75K. A superconducting phase with Tc=105K was observed in a sample annealed at 840°C. The superconducting volume ratio in the samples had a maximum at 803-840°C. The decrease of the volume ratio at the higher temperatures is considered to result from a decomposition of the superconducting phase (75K) into a nonsuperconducting phase.

INTRODUCTION

Many studies have been reported on high-Tc superconducting oxides in the last two years. A pioneering discovery of the 40K class La-Ba(Sr)-Cu-O system [1,2] was followed by the attainment of superconductivity above 90K in the Y-Ba-Cu-O system [3]. Recently, Maeda et al. have discovered a new superconducting system, Bi-Sr-Ca-Cu-O with a Tc above 100K [4]. It is reported that the Bi-Sr-Ca-Cu-O system gives rise to at least four different crystal phases having a tetragonal or pseudotetragonal layered structure in samples prepared by a conventional powder mixture method [5,6,7]. Their a-cell dimensions are about 5.4Å, while their c-cell dimensions are around 18Å, 24Å, 30Å and 36Å.

In this paper, rapidly quenched samples with the composition of $BiSrCaCu_2O_x$ were annealed at various temperatures to clarify the relationship between crystalline structures and superconducting properties.

EXPERIMENTAL PTOCEDURE

Mixed powders of Bi_2O_3, $SrCO_3$, $CaCO_3$ and CuO were melted at 1050°C for 1h in air. The nominal composition was $BiSrCaCu_2O_x$. The completely molten material was quenched into liquid nitrogen to obtain amorphous samples. These were considered to have more compositional homogeneity than that from a conventional powder mixture method [8]. The sample was annealed at temperatures between 605°C and 874°C for 5h in air, where heating and cooling rates were around 300°C/h.

Phase changes in the as-quenched samples were examined by differential thermal and thermogravimetric analysis (DTA and TGA). Crystal phases of the sample were examined by X-ray diffraction analysis (XRD) using a Cu-Kα radiation. Magnetic and electrical properties were measured by AC inductive method (100Hz, 0.04mT) and a standard four-probe method (10mA), respectively. An absolute volume of a superconducting phase in the sample was calculated from eq. (1).

$$V = Vsc/Vsample = \frac{E/C}{W/\sigma} \times 100 \quad (\%) \qquad (1)$$

where E is an inductance change (dB), C is an equipment constant (21dB/cm^3), W is a sample weight (g) and σ is a theoretical density of the sample (g/cm^3). We assumed σ to be 6.55g/cm^3 [5].

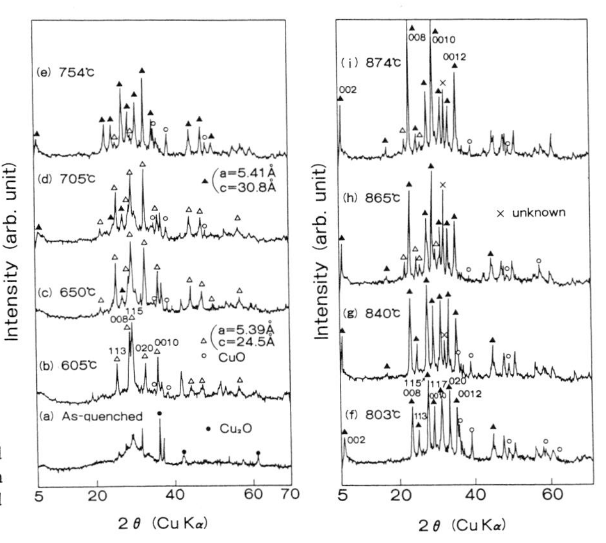

Fig.1. X-ray diffraction patterns of as-quenched samples annealed at various temperatures for 5h in air. Δ:crystalline phase with a=5.39Å and c=24.5Å, ▲: a=5.41Å and c=30.8Å, x:unknown.

RESULTS AND DISCUSSION

Figure 1 shows X-ray diffraction patterns of $BiSrCa_2Cu_x$ in an as-quenched state (a) and annealed states at various temperatures, (b)- (i). The pattern (a) shows that the as-quenched sample consists of an almost amorphous phase, as indicated by broad peak around 30°. However, Cu_2O crystalline peaks and other weak peaks were also seen in this X-ray pattern in spite of the rapidly quenched condition. The Cu_2O crystals changed into CuO by annealing in air.

The formation of two crystalline phases other than CuO was confirmed in the samples annealed at various temperatures. One is a tetragonal crystalline phase with a=5.39Å and c=24.5Å (hereafter referred to as phase A) which was typically observed in the sample annealed at 605°C, Fig. 1(b). The other is a=5.41Å and c=30.8Å(hereafter referred to as phase B), shown in Fig. 1(f). As the annealing temperature increased from 650°C to 754°C, the peaks from phase B emerged and increased, while the peaks from phase A decreased. At 803°C, only the peaks from phase B were observed. Above 865°C, phase A was formed again together with phase B and a strong peak marked X ($2\theta \sim 31.9°$) which was attributed to neither phase A, phase B nor other compounds with the combination of Bi, Sr, Ca, Cu, O element listed in JCPDS cards.

Figure 1 also shows that specific reflection indices of phase B are emphasized: 001 peaks are enhanced by increasing the annealing temperature, as shown in Fig. 1 (e)-(i). Since these X-ray patterns were obtained by a standard powder method, this intensity enhancement suggests that c-planes of phase B are easy to grow.

Figure 2 shows DTA and TGA curves of an as-quenched $BiSrCaCu_2O_x$ sample at a scanning rate of 5°C/min up to 1000°C in air. On the basis of the X-ray analysis, the broad exothermic peak around 510°C may be due to the structural change from an amorphous state to a crystalline structure of phase A. On the other hand, the endothermic reactions around 710°C and 780°C are considered to be closely related to the structural change from phase A to phase B. The large endothermic reaction at 875°C is probably due to the formation of a liquid phase. Furthermore, the peaks at 920°C and 970°C are due to melting reactions of residual materials. In the TGA curve, the weight of the sample gradually increases from 500°C, and has a maximum value around 750°C-850°C. Since the weight change is considered to be caused by the absorption and desorption of oxygen, this result means that the crystal phases A and B need more oxygen atoms than the amorphous or liquid phase.

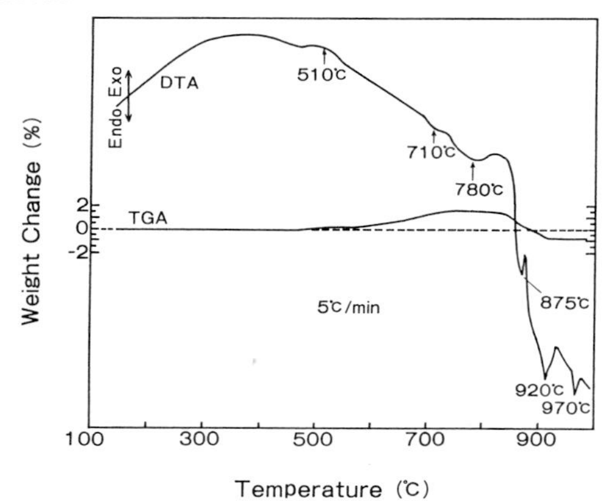

Fig.2. Differential thermal analysis (DTA) and thermogravimetric (TGA) curves of an as-quenched $BiSrCaCu_2O_x$ sample at a scanning rate of 5°C/min in air.

Figure 3 shows SEM images of the fracture surface of the annealed samples. The as-quenched sample has a flat and smooth fracture surface, which is expected for amorphous materials. On the other hand, the surface of the sample annealed at 650°C consists of grains of a few microns. By annealing at 754°C, laminated structures were grown with fine grains. Combining the results of the X-ray analysis, it is supposed that the laminated and the fine-grain phases correspond to phase B and phase A, respectively. As the annealing temperature increases from 803°C to 874°C, the laminated structure further developed (Figs. 3(d) and 3(e)). This tendency corresponds to the intensity enhancement of the 001 peaks in the X-ray analysis. It is also revealed that the samples annealed above 803°C had a lower density than that of the samples annealed

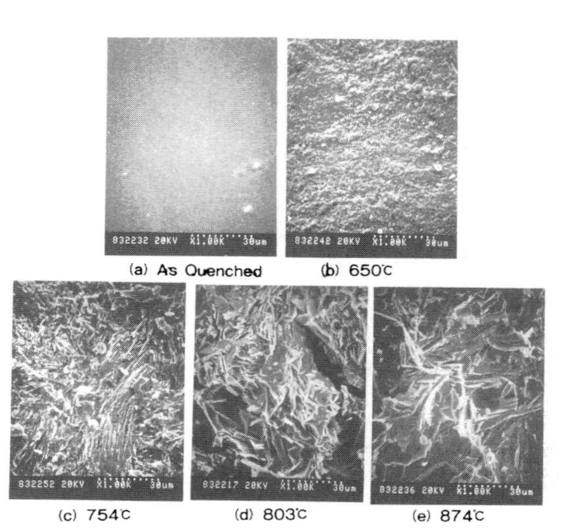

Fig.3. SEM images of fracture surfaces at various annealing temperatures.

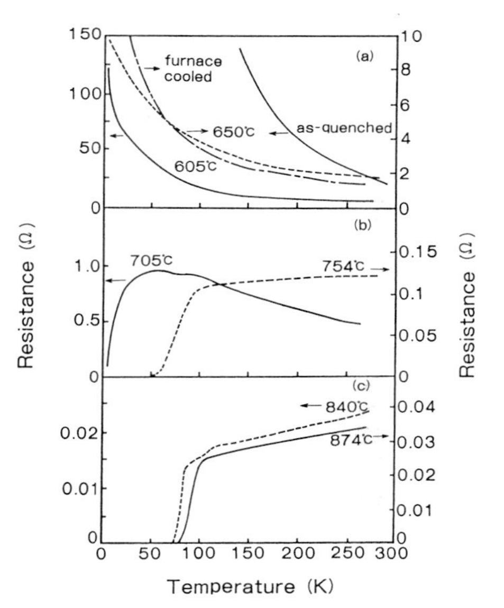

Fig.4. Temperature dependence of the resistance for the as-quenched samples annealed at various temperatures.

below 754°C. This may be related to the anisotropic growth of phase B. Since each grain grows in random directions, voids may tend to be produced among the plate-like crystal grains.

Figure 4 shows the temperature dependence of the resistance for the samples annealed at various temperatures and the furnace-cooled sample. The resistances of the as-quenched, the 605°C and 650°C annealed samples increase as the temperature decreases. While the sample annealed at 705°C showed somewhat complicated behavior, the resistance continued to increase at 100K and reached its maximum at 60K, zero resistance was not obtained even at 4.2K. By annealing above this temperature, all samples showed metallic behavior with zero resistance between 50K and 80K. The sample annealed at 840°C showed a small resistance drop at 120K which corresponds to the high-Tc phase of Bi-Sr-Ca-Cu system, and a sharp drop around 80K from the low-Tc phase. It was this annealing temperature that showed the high-Tc phase under our annealing conditions. It has been reported that the high-Tc phase has c-cell dimensions of 36Å [7] (hereafter referred to as phase C).

Figure 5 shows the temperature dependence of the susceptibility for the samples annealed at various temperatures. All the samples annealed above 705°C show a susceptibility decrease around 80K. The presence of the high-Tc phase was not observed in the magnetic method, probably due to the minute amounts of this phase.

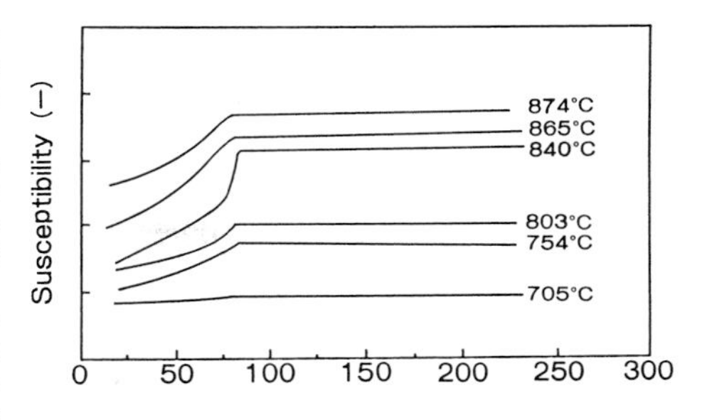

Fig.5. Temperature dependence of the susceptibility for the as-quenched samples annealed at various temperatures.

Figure 6 shows the volume ratio of the superconducting phase in the samples annealed at various temperatures. This volume ratio was calculated from the susceptibility change between 20K and the onset temperature using eq. (1). The superconducting volume ratio of the sample increases above 705°C and shows a maximum value of 70-75% at 803-840°C. The volume ratio decreases by further increasing the temperature. This suggests that the superconducting phase B and/or phase C may decompose into a nonsuperconductiong phase above around 840°C. In fact the X-ray diffraction analysis showed that phase A, which was a nonsuperconductiong phase, and an unknown peak marked X emerged above 840°C.

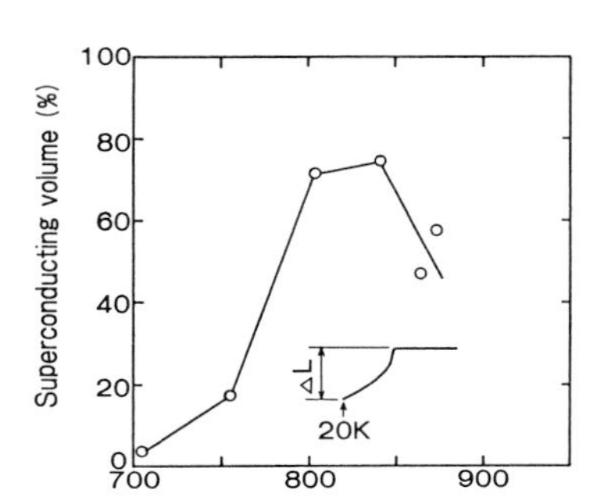

Fig.6. Superconducting volume change of $BiSrCaCu_2O_x$ ceramics annealed at various temperatures.

Figure 7 summarizes the results of the X-ray, the DTA and the electrical measurement. A crystal structure for phase B has been proposed by several groups [5,6,7] : the compound has two CuO_2 layers, two (Sr,Ca)O layers and a (Ca,Sr) layer stacked between Bi_2O_2 layers, $Bi_2(Ca,Sr)_3Cu_2O_x$. For phase A, the composition is $Bi_2(Ca,Sr)_2CuO_x$ and the structure is derived by removing a CuO_2 and a (Ca,Sr) layer from phase B [5,7]. On the other hand, phase C which is obtained by inserting a CuO_2 and a (Ca,Sr) layer into phase B is given by $Bi_2(Ca,Sr)_4Cu_3O_x$ [7].

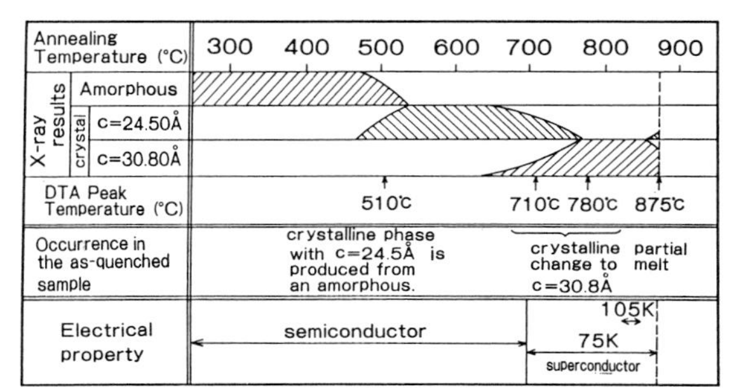

Fig.7. Summary of the X-ray, the DTA and the electrical properties of the rapidly quenched $BiSrCaCu_2O_x$ ceramics.

According to our results, the first crystalline structure formed from an amorphous state is considered to be phase A. By annealing above 700°C, phase A changed into phase B. Around 803-840°C, the total volume of the superconductiong phase reached a maximum, where phase C was observed. Above this temperature, the superconducting absolute volume ratio decreased, and the nonsuperconducting phase A and the other unknown phases were produced.

The reason why the total superconducting volume ratio decreases above 840°C may be due to the decomposition of the low-Tc superconductiong phase into two compounds:

$$Bi_4(Ca,Sr)_6Cu_4O_x \rightarrow Bi_2(Ca,Sr)_2CuO_x + Bi_2(Ca,Sr)_4Cu_3O_x \quad (2)$$

where the electrical property of $Bi_2(Ca,Sr)_2CuO_x$ is a semiconductor. The $Bi_2(Ca,Sr)_4Cu_3O_x$ phase is stable only in a limited temperture range, being unstable above 840°C, thus resulting in the further decomposition into nonsuperconducting materials.

REFERENCES

[1]J. G. Bednorz and K. A. Muller: Z. Phys. B64 (1986) 189.
[2]S. Uchida, H. Takagi, S. Tanaka and K. Kitazawa: Jpn. J. Appl. Phys. 26 (1987) L1.
[3]M. K. Wu, J. R. Ashburn, C. J. Torng, P. H. Hor, R. L. Meng, L. Gao, Z. J. Haung, Y. Q. Wang and C. W. Chu: Phys. Rev. Lett. 58 (1987) 909.
[4]H. Maeda, Y. Tanaka, M. Fukutomi and T. Asano: Jpn. J. Appl. Phys. 27 (1988) L209.
[5]E. Takayama-Muromachi, Y. Uchida, A. Ono, F. Izumi, M. Onoda, Y. Matsui, K. Kosuda, S. Takekawa and K. Kato: Jpn. J. Appl. Phys. 27 (1988) L365.
[6]T. Kajitani, K. Kusada, M. Kikuchi, N. Kobayashi, Y. Syono, T. B. Williams and M. Hirabayashi: Jpn. J. Appl. Phys. 27 (1987) L587.
[7]E. Takayama-Muromachi, Y. Uchida, Y. Matsui, M. Onoda and K. Kato: Jpn. J. Appl. Phys. 27 (1988) L556.
[8]T. Komatsu, R. Sato, K. Imai, K. Matushita, T. Yamashita: Jpn. J. Appl. Phys. 27 (1988) L550.

Properties of Pb-Stabilized Bi-Oxide Superconductors

N. MURAYAMA[1], E. SUDO[2], and Y. TORII[1]

[1] Government Industrial Research Institute, Nagoya, 1-1, Hirate-cho, Kita-ku, Nagoya, 462 Japan
[2] Tokyo Kokyu Rozai Co. Ltd., Asamuta, Ohmuta, 836 Japan

ABSTRACT

The initial magnetization curve of $Bi_{1.5}Pb_{0.5}Sr_2Ca_2Cu_3O_x$ superconductor with the T_c of 105 K was measured. The lower critical field, H_{c1}, was 6 Oe at 77 K, which was much smaller than that of $YBa_2Cu_3O_{7-y}$ (52 Oe). The volume fraction of the 105 K phase was estimated to be about 80 %.

Grinding the powder in an attrition mill resulted in a slight increase in Tc, although the densification of the sintered body did not ocurr. By hot-pressing at a pressure of 39 MPa at 800 °C in vacuum of 5 x 10^{-5} Torr for 2 h, the bulk density of the hot-pressed sample reached 6.21 g/cm^3, which was higher than 95 % of the theoretical density, and the T_c was 72 K. The grains were oriented with the c-axis along the pressing axis. After annealing at 835 °C in air for 40 h, the T_c was improved to be 108 K and the critical current density at 77 K was found to be 731 A/cm^2.

INTRODUCTION

Recently, Maeda et al. discovered a new high-T_c oxide superconductor in the Bi-Sr-Ca-Cu-O system [1]. It has been clarified that there are two superconducting phases with transition temperatures of 105 K (high-T_c phase) [2] and 75 K (low-T_c phase) [3]. The former phase has a longer c-axis than the latter phase. It is difficult to prepare the high-T_c phase in a single phase [4]. Moreover, this material is hard sinterable, because the powders are in the form of thin-plate.

From the measurement of an initial magnetization curve, a volume fraction of superconducting phases and a lower critical field, H_{c1}, can be estimated. Uehara et al. measured the H_{c1} of $BiSrCaCu_2O_x$ at 77 K. This compound consists of the high-T_c phase and the low-T_c phase [5]. Because the temperature, at which H_{c1} was measured, was close to the T_c of the low-T_c phase (75 K), it seems that the measured H_{c1} is affected by the low-T_c phase.

It has been found that the partial Pb-substitution in Bi-Sr-Ca-Cu-O superconductors resulted in the stabilization of the high-T_c phase [6]. In the present work, we report electrical and magnetic properties of Bi-Pb-Sr-Ca-Cu-O superconductor, which scarcely contains the low-T_c phase. In addition, we attemped to prepare the superconducting sintered bodies using an attrition mill and hot-pressing, and in the latter samples, the grain-orientation and critical current density were evaluated.

EXPERIMENTAL

Appropriate amounts of Bi_2O_3, PbO, $SrCO_3$, $CaCO_3$ and CuO of high purity (99.9 %) were wet-mixed in ethanol and calcined at 800 °C in air. Staring compositions of the samples for the measurement of magnetic properties and the experiment of sintering preperties were Bi:Pb:Sr:Ca:Cu = 1.5:0.5:2:2:3 and 1.6:0.4:1.6:2:2.8, respectively. The calcined powders were pressed, then heated at 830 °C in air and ground. This process (from pressing to grinding) was repeated several times to ensure a complete solid-state reaction, and the total firing time was about 70-100 h. Fine powder was prepared by grinding the as-received powder in an attrition mill using 3.0 mm diameter zirconia balls as media. The milling time was 12 h. The as-received and ground powders were prepared by pressing the powders at a pressure of 650 MPa and heating at 830-840 °C in air. In addition, the as-received powder was also uniaxially hot-pressed at a pressure of 39 MPa at 800 °C in vacuum of 5 x 10^{-5} Torr for 2 h using cylindrical dies made of carbon and cooled to room temperature in a furnace. Then the resulting samples were annealed at 835 °C in air for 40 h.

The lattice parameters were determined from X-ray powder diffraction data obtained at room temperature by graphite-monochromatized Cu radiation. The temperature dependence of the electrical resistivity was measured by a four-probe dc method at temperatures from 25 to 200 K. The measurement current was 10 mA. The magnetization measurements were performed on a normal sintered and cylindrical sample with dimensions of 4 mm diameter, 3 mm height and weight of 0.148g using a vibrating sample magnetometer (TOEI KOGYO CO. ,LTD., MODEL VSM-5-15 AUTO). The bulk densities were measured and grain orientation was evaluated by X-ray diffractometry. The critical current density of the hot-pressed sample annealed in air was measured at 77 K. The sample size was 0.99 mm x 1.7 mm x 39 mm and the spans between the electrodes to pick up voltage and between the electrodes to flow current were 1.1 mm and 35 mm, respectively.

RESULTS AND DISCUSSION

Figure 1 shows the X-ray diffraction pattern for $Bi_{1.5}Pb_{0.5}Sr_2Ca_2Cu_3O_x$. Most of the observed reflections were indexed on the basis of a pseudotetragonal unit cell with a = 5.410 Å and c = 37.18 Å. The low-T_c phase was not appreciably observed in the X-ray pattern. However, the coexistence of an unknown phase was confirmed. The temperature dependence of the resistivity for $Bi_{1.5}Pb_{0.5}Sr_2Ca_2Cu_3O_x$ is shown in Fig.2. The onset T_c and the offset T_c were 119 K and 105 K, respectively. Figure 3 shows the temperature dependence of the magnetization for $Bi_{1.5}Pb_{0.5}Sr_2Ca_2Cu_3O_x$ in the magnetic field of 100 Oe. The diamagnetism due to the Meissner effect was clearly observed below about 115 K. This transition temperature is in agreement with the onset T_c obtained in the resistivity measurement. Figure 4 shows a typical magnetization curve measured at 77 K and the initial magnetization curve is shown in Fig.5. The diamagnetism was proportional to the magnetic field below 6 Oe. This result indicates that the lower critical field, H_{c1}, of $Bi_{1.5}Pb_{0.5}Sr_2Ca_2Cu_3O_x$ is 6 Oe. Uehara et al. reported that the H_{c1} was 16.2 Oe at 77 K for $BiSrCaCu_2O_x$ which contains the high-T_c phase and the low-T_c phase [5]. This decrease of H_{c1} may be due to an increase in the high-T_c phase content or the effect of a partial Pb-substitution. As shown in Fig.6, the H_{c1} value of $YBa_2Cu_3O_{7-y}$ was 52 Oe, which was an order of magnitude larger than that of $Bi_{1.5}Pb_{0.5}Sr_2Ca_2Cu_3O_x$. It was estimated from the value of $-4\pi M/H$ below the H_{c1} (6 Oe) that the volume fraction of the 105 K phase was about 80 %. In the calculation, the theoretical density of the sample was 6.38 g/cm^3 and the demagnetizing factor was assumed to be 1/3 which value is derived theoretically for a spherical sample.

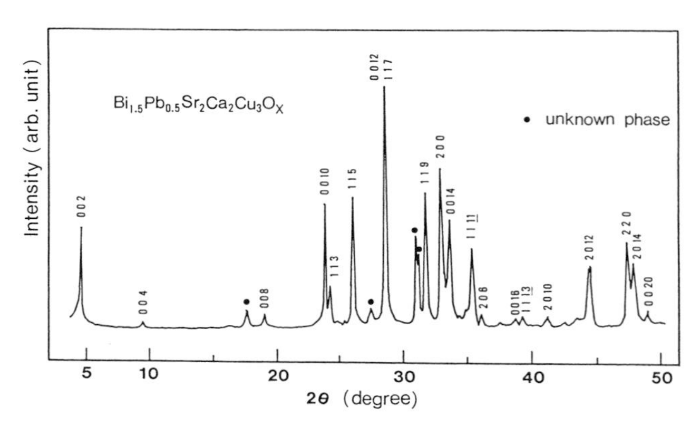

Fig.1. X-ray diffraction pattern for $Bi_{1.5}Pb_{0.5}Sr_2Ca_2Cu_3O_x$.

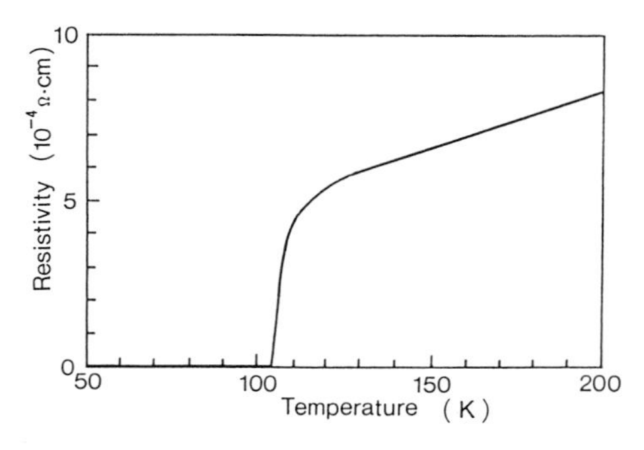

Fig.2. Temperature dependence of the resistivity for $Bi_{1.5}Pb_{0.5}Sr_2Ca_2Cu_3O_x$.

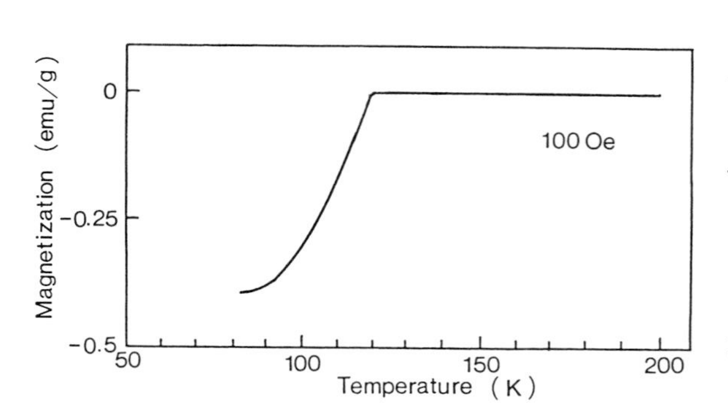

Fig.3. Temperature dependence of the magnetization for $Bi_{1.5}Pb_{0.5}Sr_2Ca_2Cu_3O_x$.

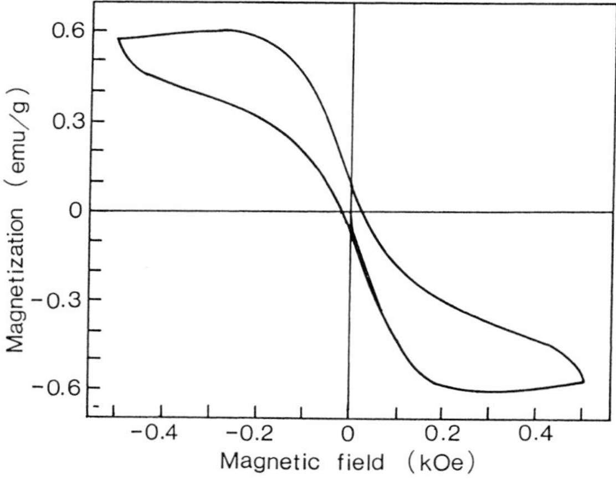

Fig.4. Magnetization curve at 77 K for $Bi_{1.5}Pb_{0.5}Sr_2Ca_2Cu_3O_x$.

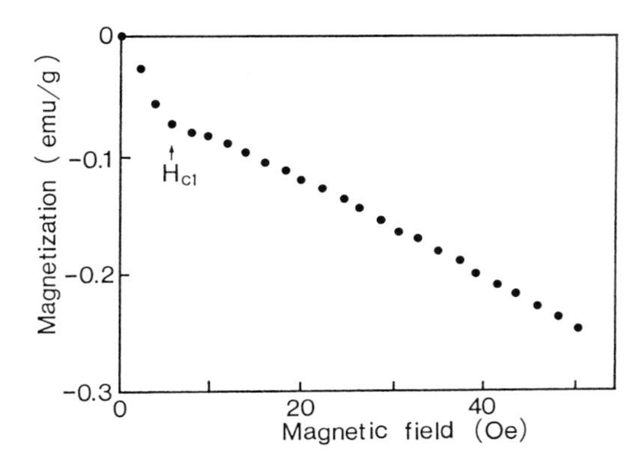

Fig.5. Initial magnetization curve at 77 K for $Bi_{1.5}Pb_{0.5}Sr_2Ca_2Cu_3O_x$.

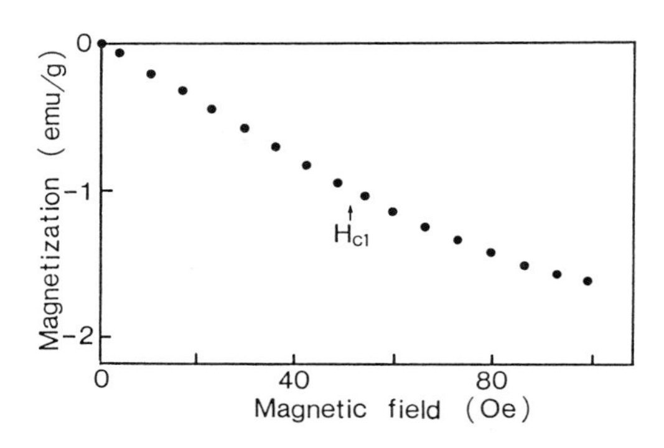

Fig.6. Initial magnetization curve at 77 K for $YBa_2Cu_3O_{7-y}$.

The as-received and attrition-milled powders were 8 μm and 2 μm in average size, respectively. Table I shows the bulk densities of the compacts, normal sintered and hot-pressed samples. The bulk density of the compact of attrition-milled powder was lower than that of the as-received powder. The sinterability was not improved by grinding the as-received powder, although the powder size became smaller. The bulk density of the hot-pressed sample was higher than 95% of the theoretical density. Despite hot-pressing in vacuum, the high-T_c phase were not decomposed and no appreciable change in the lattice parameters was observed.

Figure 7 and 8 show the temperature dependence of the resistivity of the compacts and the normal sintered samples. The compact of the as-received powder exhibited semiconductivity above 115 K and metallic conductivity below 115 K, although the resistivity was not zero above 25 K. The normal sintered sample of the as-received powder had the Tc of 103 K and exhibited metallic conductivity above the Tc. The resistivity of the compact of the attrition-milled powder was 1.2 x 10^3 Ω·cm at room temperature. The Tc of the normal sintered sample of the attrition-milled powder was observed at 108 K. It is concluded that the weak link between the grains occurs in the normal sintering and that grinding the powder results in the slight increase in T_c.

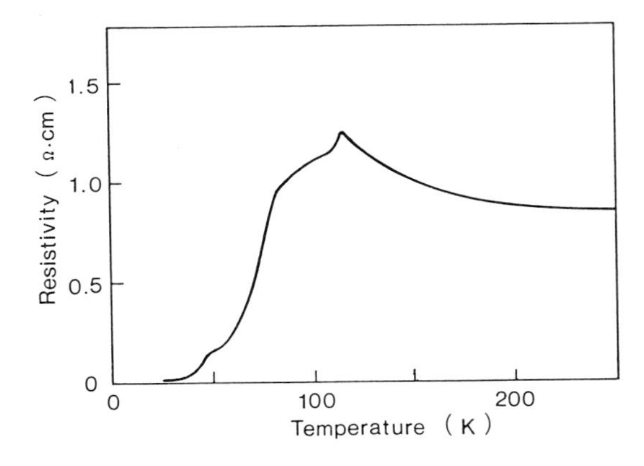

Fig.7. Temperature dependence of the resistivity for the compact of $Bi_{1.6}Pb_{0.4}Sr_{1.6}Ca_2Cu_{2.8}O_y$.

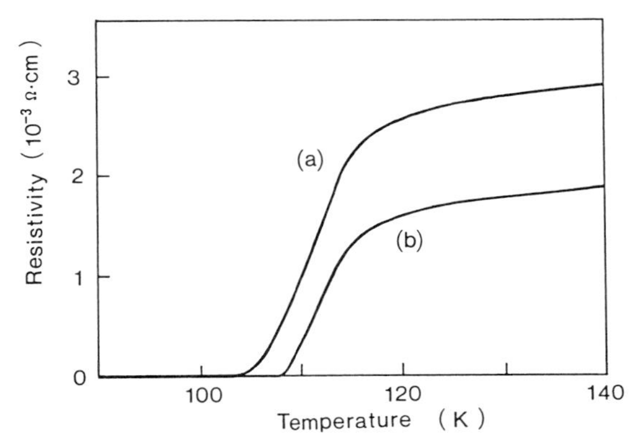

Fig.8. Temperature dependence of the resistivity for the normal sintered samples of $Bi_{1.6}Pb_{0.4}Sr_{1.6}Ca_2Cu_{2.8}O_y$. (a): as-received powder, (b): ground powder.

Table I. Bulk densities (g/cm^3) of the compact, the normal sintered and the hot-pressed samples.

	compact	normal sintered sample	hot-pressed sample
As-received powder	4.98	4.81	6.21
Attrition-milled powder	4.03	3.91	—

Figure 9 shows X-ray diffraction patterns for the perpendicular plane (a) and the parallel plane (b) to the pressing axis of the hot-pressed sample not annealed in air. The grains were found to be oriented with the c-axis along the pressing axis. Figure 10 shows the temperature dependence of the resistivity for the hot-pressed samples. In the sample before annealed in air, the T_c was observed at 72 K. When annealed in air, the T_c reached 108 K, which was higher by 5 K than that of the normal sintered sample, and the resistivity at room temperature was lower than that of the normal sintered sample. It seems hat the slight increase in T_c arised from the stronger link between the grains. Figure 11 shows the measurement current dependence of the resistivity at 77 K for the hot-pressed sample annealed in air. The superconductivity of the sample was broken above 12.3 A. It is estimated from this result that the critical current density is 731 A/cm^2. During flowing dc current of 11 A for 1 h, the sample kept superconductivity. The longer time test is now in progress. It is expected that J_c will increase by optimizing the hot-pressing condition.

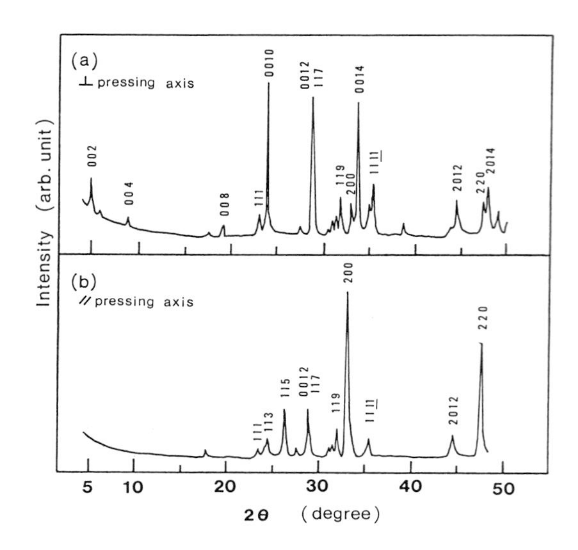

Fig.9. X-ray diffraction patterns for the perpendicular (a) and the parallel (b) planes to the pressing axis of the hot-pressed sample of $Bi_{1.6}Pb_{0.4}Sr_{1.6}Ca_2Cu_{2.8}O_y$.

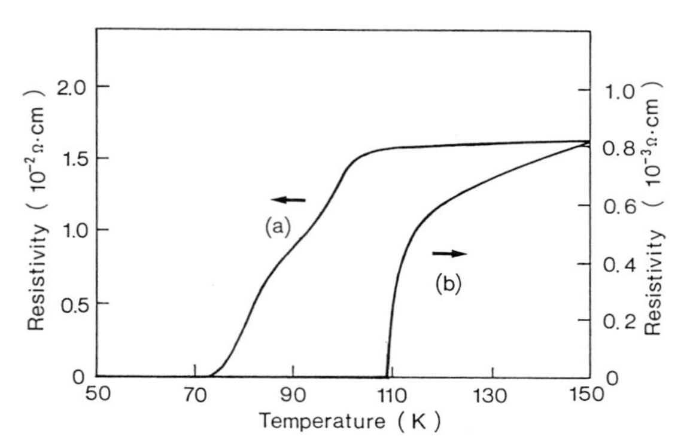

Fig.10. Temperature dependence of the resistivity for the hot-pressed sample of $Bi_{1.6}Pb_{0.4}Sr_{1.6}Ca_2Cu_{2.8}O_y$.
(a): before annealed in air,
(b): after annealed in air.

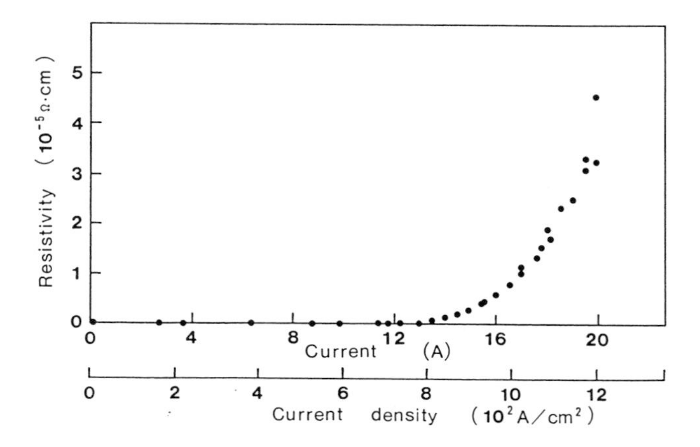

Fig.11. Measurement current dependence of the resistivity at 77 K for the hot-pressed sample of $Bi_{1.6}Pb_{0.4}Sr_{1.6}Ca_2Cu_{2.8}O_y$ annealed in air.

CONCLUSIONS

The initial magnetization curve of $Bi_{1.5}Pb_{0.5}Sr_2Ca_2Cu_3O_x$ superconductor with the T_c of 105 K was measured. The lower critical field, H_{c1}, was 6 Oe at 77 K, which was much smaller than that of $YBa_2Cu_3O_{7-y}$ (52 Oe). The volume fraction of the 105 K phase was estimated to be about 80 %.

Grinding the powder in the attrition mill resulted in a slight increase in Tc, although the densification of the sintered body did not ocurr. By hot-pressing at a pressure of 39 MPa at 800 °C in vacuum of 5 x 10^{-5} Torr for 2 h, the bulk density of the hot-pressed sample reached 6.21 g/cm^3 , which was higher than 95 % of the theoretical density, and the T_c was 72 K. The grains were oriented with the c-axis along the pressing axis. After annealing at 835 °C in air for 40 h, the T_c was improved to be 108 K and the critical current density at 77 K was found to be 731 A/cm^2.

ACKNOWLEDGMENT

We are grateful to Dr. S. Kanzaki for the preparation of the hot-pressed sample.

REFERENCES

[1] H.Maeda, Y.Tanaka, M.Fukutomi and T.Asano: Jpn.J.Appl.Phys. 27, L290(1988).

[2] H.W.Zandbergen, Y.K.Huang, M.J.V.Menken, J.N.Li, K.Kadowaki, A.A.Menovsky, G.van Tendeloo and S.Amelinckx: Nature 332, 620(1988).

[3] M.A.Subramanian, C.C.Torardi, J.C.Calabrese, J.Gopalakrishnan, K.J.Morrissey, T.R.Askew, R.B.Flippen, U.Chowdhry and A.W.Sleight: Nature 239, 1015(1988).

[4] K.Togano, H.Kumakura, H.Maeda, K.Takahashi and M.Nakao: Jpn.J.Appl.Phys. 27, L323(1988).

[5] M.Uehara, Y.Asada, H.Maeda and K.Ogawa: Jpn.J.Appl.Phys. 27, L665(1988).

[6] M.Takano, J.Takada, K.Oda, H.Kitaguchi, Y.Miura, Y.Ikeda, Y.Tomii and H.Mazaki: Jpn.J.Appl.Phys. 27, L1041(1988).

Preparation of High-Tc Bi-Sr-Ca-Cu-O and Tl-Ba-Ca-Cu-O Superconductors

TOSHIO USUI, NOBUYUKI SADAKATA, YOSHIMITSU IKENO, OSAMU KOHNO, and HIROSHI OSANAI

Tokyo Laboratory, Fujikura Ltd., Koto-ku, Tokyo, 135 Japan

ABSTRACT

Bulk superconductors of the Bi-Sr-Ca-Cu-O and Tl-Ba-Ca-Cu-O systems have been investigated to clarify their superconducting properties. In the Bi-Sr-Ca-Cu-O system, its superconducting properties were improved by doping PbO after the calcination; the high-T_c phase ($\sim$105 K) was formed during the sintering process and stabilized down to room temperature. Zero-resistance was realized at 104 K for 5wt% PbO doped $BiSrCaCu_2O_x$ sintered at 848°C for 72 h in air. In the Tl-Ba-Ca-Cu-O system, its superconducting properties were found to be significantly sensitive to the sintering temperature and the composition of the starting materials. Zero-resistance was achieved at 118 K for $Tl_2Ba_2Ca_2Cu_3O_x$ sintered at 880°C for 1 h in O_2. In addition, an ac susceptibility result also supports its superconductivity with high critical temperature.

INTRODUCTION

Since Bednorz and Müller discovered a new ceramic superconductor in the La-Ba-Cu-O system [1], a series of high-T_c superconductors has been extensively investigated. As a result, four groups of the high-T_c superconductors, that is, 30 K superconductor in the La-Sr-Cu-O system by Uchida et al. [2], 90 K superconductor in the Y-Ba-Cu-O system by Wu et al. [3], and 100 K superconductors in the Bi-Sr-Ca-Cu-O system by Maeda et al. [4] and in the Tl-Ba-Ca-Cu-O system by Sheng and Hermann [5], have been reported. To date, when investigating the highest critical temperature, zero-resistance was achieved at 125 K in the Tl-Ba-Ca-Cu-O system. By contrast, it is very difficult to realize zero-resistance above 100 K in the Bi-Sr-Ca-Cu-O system. However, recently, it has been reported that partial substitution of Pb for Bi in the Bi-Sr-Ca-Cu-O system promotes the formation of high-T_c phase (T_c~105 K) [6]. We have investigated the preparation of the superconductors in the Bi-Sr-Ca-Cu-O and Tl-Ba-Ca-Cu-O systems. In this paper, we discuss the effect of PbO doping on the superconductivity in the Bi-Sr-Ca-Cu-O system, and the influence of the sintering temperature and the composition of the starting materials on the superconducting properties in the Tl-Ba-Ca-Cu-O system.

EXPERIMENTAL

The PbO doped Bi-Sr-Ca-Cu-O samples were prepared by the following process. The appropriate amounts of Bi_2O_3, $SrCO_3$, $CaCO_3$ and CuO were mixed thoroughly, ground and calcined at 800°C for 12h in air. After calcination, obtained powder of $BiSrCaCu_2O_x$ was mixed with PbO, and subsequently pressed into a rectangular bar of typical size 20x3x1 mm. The rectangular bars were sintered at 848°C in air and subsequently cooled (about 200°C/h).

The Tl-Ba-Ca-Cu-O samples were prepared by the following process, similar to that reported by Sheng and Hermann [5]. The appropriate amounts of Tl_2O_3, CaO, and $BaCu_3O_4$ (or $Ba_2Cu_3O_5$) were mixed thoroughly, ground and pressed into a pellet 13 mm in diameter and 1.5 mm in thickness. The barium cuprate starting materials were prepared using a typical solid-state reaction of $BaCO_3$ and CuO at 880°C for 12 h in air [7]. The pellets were sintered at 870 or 880°C for 1 h in the furnace in flowing oxygen and subsequently cooled (about 200°C/h). The initial composition and sintering conditions for the samples used for the experiment are listed in Table 1.

The resistive transition was measured using a four-probe technique on a sintered rectangular bar or a sintered pellet: the polarity of the applied current of 1 or 2 mA was switched from positive to negative and vice versa, and the voltage was measured simultaneously. The zero-resistance was defined as a voltage less than 10 nV. Voltage and current leads were attached to the samples using an ultrasonic-soldering technique (Sunbonder, Asashi Glass Co., Ltd.). The temperature was measured by a calibrated Pt-Co thermometer. The variation of susceptibility with temperature was also measured by an inductive method.

RESULTS AND DISCUSSION

Pb doped Bi-Sr-Ca-Cu-O system

Figure 1 shows the temperature dependence of the electrical resistance for the samples of PbO doped $BiSrCaCu_2Ox$, sintered at 848°C for 72 h in air. As can be seen, the formation of the high-T_c phase ($T_c \sim 105$ K) in the Bi-Sr-Ca-Cu-O samples significantly depends on the doping amount of PbO. In the case of PbO undoped sample, resistance curve with temperature slightly changed at about 110 K, but zero-resistance temperature was 47 K. This result indicates that most of PbO undoped sample was consisted of low-T_c phases (T_c<80 K). By contrast, in the case of PbO doped samples, the high-T_c phase was formed with ease during the sintering process and stabilized down to room temperature; the highest zero-resistance temperature was 104 K for 5 wt% PbO doped sample. Thus, it is clear the effect of PbO on the formation of the high-T_c phase in the Bi-Sr-Ca-Cu-O system. Figure 2 shows the relationship between the sintering time at 848°C and the zero-resistance temperature of the PbO doped samples. Zero-resistance temparature for PbO doped samples varied as a function of the sintering time and the doping amount of PbO. Zero-resistance temperature for 5 wt% PbO doped sample was significantly varied and realized above 100 K when the sintering time at 848°C for more than 50 h, whereas those for 3 and 10 wt% PbO doped samples slightly increased with the sintering time, but much lower than 100 K. On the other hand, that for PbO undoped sample was virtually independent of the sintering time at 848°C, and constant about 47 K. As can be seen from the above discussion, PbO doping after calcination for the Bi-Sr-Ca-Cu-O system was found to be effective on the formation and stabilization of the high-T_c phase.

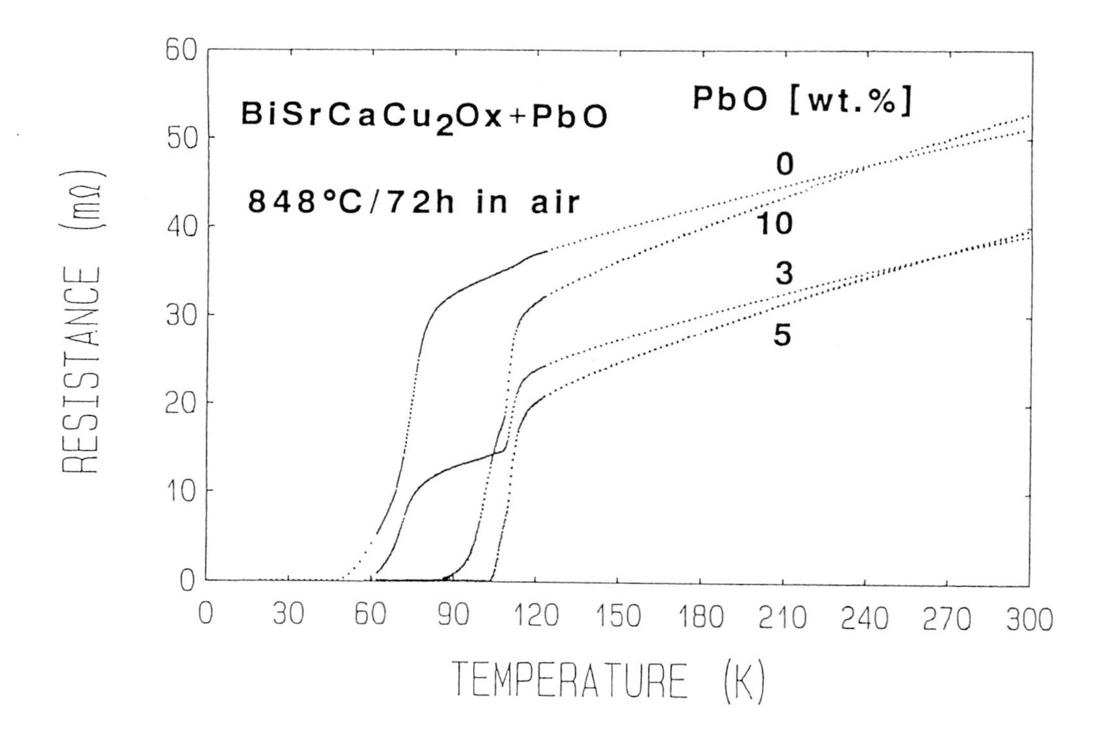

Fig. 1. Temperature dependence of the electrical resistance for the PbO doped Bi-Sr-Ca-Cu-O samples.

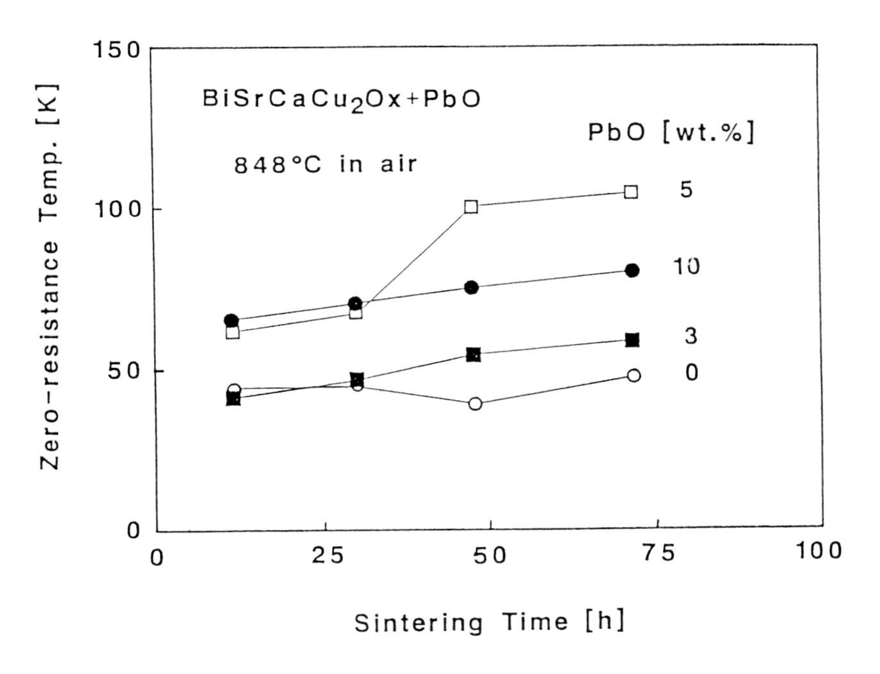

Fig. 2. Relationship between the zero-resistance temperature for the PbO doped Bi-Sr-Ca-Cu-O samples and the sintering time at 848°C.

Tl-Ba-Ca-Cu-O system [8]

Figure 3 shows the temperature dependence of the electrical resistance for the samples. As can been seen, the superconducting transition of the Tl-Ba-Ca-Cu-O samples significantly depends on the sintering temperature and the initial composition of the samples; thus the zero-resistance temperature, summarized in Table 1, varies for each sample. For sample (a), $Tl_2Ba_2Ca_2Cu_3O_x$ sintered at 880°C for 1 h in O_2, the highest zero-resistance temperature was achieved at 118 K, whereas the lowest zero-resistance temperature was 105 K for sample (b), being the same initial composition but sintered at 870°C. It should be noted that a mere 10°C-difference in the sintering temperature had a significant effect, e.g., 13 K, on the zero-resistance temperature. On the other hand, the zero-resistance temperatures for samples (c) and (d), the initial composition being $Tl_2BaCa_2Cu_3O_x$, were 112 K and 107 K, respectively; the influence of the sintering temperature on the zero-resistance temperature is relatively small.

Table I. Specifications and the superconducting properties of the Tl-Ba-Ca-Cu-O samples prepared in the present study.

Sample	Initial ratio of Tl:Ba:Ca:Cu	Sintering in flowing O_2		Zero-resistance temperature (K)	Jc (A/cm^2) at 77 K
		Temp. (°C)	Time (h)		
(a)	2:2:2:3	880	1	118	90
(b)	2:2:2:3	870	1	105	65
(c)	2:1:2:3	880	1	112	75
(d)	2:1:2:3	870	1	107	—

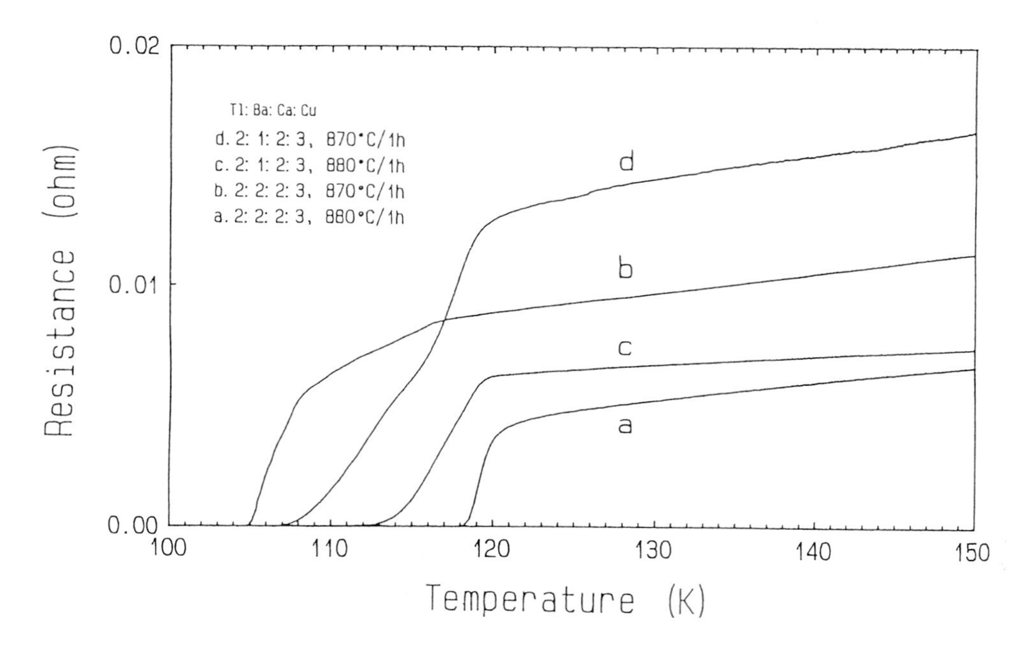

Fig. 3. Temperature dependence of the electrical resistance for the Tl-Ba-Ca-Cu-O samples.

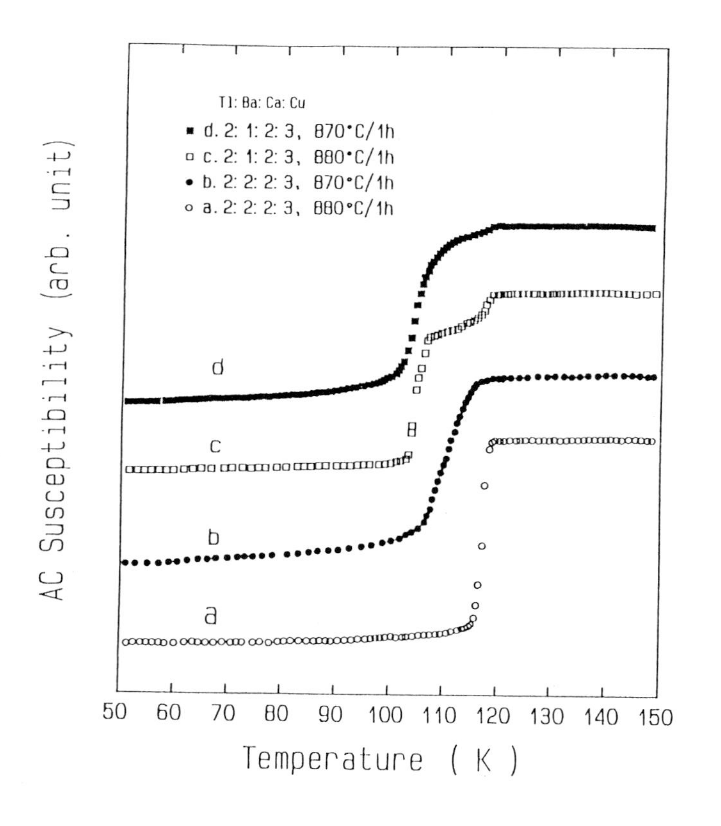

Fig. 4. Temperature dependence of the ac susceptibility for the Tl-Ba-Ca-Cu-O samples.

Figure 4 shows the variation of the ac susceptibility of the samples as a function of the temperature (50-150 K). Sample (a) has a sharp diamagnetic shift at 117-119 K, which is in close agreement with the zero-resistance temperature of 118 K. Samples (c) and (d) have a two-step diamagnetic shift: one started at 119 K and the other at about 107 K. These results suggest that these samples contain two superconducting phases, i.e., high-T_c phase ($T_c \sim 119$ K) and low-T_c phase ($T_c \sim 107$ K), similar to those in the Bi-Sr-Ca-Cu-O system [4]. By contrast, the susceptibility transition for sample (b) is rather broad, not showing a two-step diamagnetic shift. As predicted from the above discussion, the $Tl_2BaCa_{1.5}Cu_3O_x$ superconductor with the zero-resistance temperature of 103 K reported by Sheng and Hermann was probably associated with the low-T_c phase because the onset temperature, as determined by the derivation of the susceptibility of the sample, was about 105 K. In the case of Tl-Ba-Ca-Cu-O system, sample (a) consisted of the high-T_c phase as seen from Fig. 4, and thus the formation of the high-T_c phase is relatively easy, compared to that in the Bi-Sr-Ca-Cu-O system. As discussed above regarding preparation, it should be noted that the superconducting properties in the Tl-Ba-Ca-Cu-O system very much depend on the sintering temperature and the composition of the oxides.

The critical current density, J_c, is one of the most important characteristics of superconductivity from a practical viewpoint. Since the zero-resistance temperature in the Tl-Ba-Ca-Cu-O system is much higher than that in the Y-Ba-Cu-O system, the value of J_c in the Tl-Ba-Ca-Cu-O system is expected to be higher than that in the Y-Ba-Cu-O system. The values of J_c for sample pellets (a), (b) and (c) were measured by a four-probe method and were found to be 90, 65 and 75 A/cm^2, respectively, at 77 K in ambient magnetic field as listed in Table 1; not so large as expected. However, taking account of the fact that the samples are quite porous, the critical current density in the Tl-Ba-Ca-Cu-O system would be much improved by optimizing the preparation procedure.

References

[1] J.G. Bednorz and K.A. Müller: Z. Phys. **B64** (1986) 189.
[2] S. Uchida, H. Takagi, K. Kitazawa and S. Tanaka: Jpn. J. Appl. Phys. **26** (1987) L1.
[3] M.K. Wu, J.R. Ashburn, C.J. Torng, P.H. Hor, R.L. Meng, L. Gao, Z.J. Huang, Y.Q. Wang and C.W. Chu: Phys. Rev. Lett. **58** (1987) 908.
[4] H. Maeda, Y. Tanaka, M. Fukutomi and T. Asano: Jpn. J. Appl. Phys. **27** (1988) L209.
[5] Z.Z. Sheng and A.M. Hermann: Nature **332** (1988) 138.
[6] M. Takano, J. Takada, K. Oda, H. Kitaguchi, Y. Miura, Y. Ikeda, Y. Tomii and H. Mazaki: Jpn. J. Appl. Phys. **27** (1988) L1041.
[7] A.M. Hermann and Z.Z. Sheng: Appl. Phys. Lett. **51** (1987) 1854.
[8] T. Usui, N. Sadakata, O. Kohno and H. Osanai: Jpn. J. Appl. Phys. **27** (1988) L804.

Effects of Additional Elements (M=In, Al, Ga) on Superconductivity of Bi-Sr-Ca-Cu-O System

K. SHIBUTANI, T. MIYATAKE, and R. OGAWA

Superconducting and Cryogenic Technology Section, Kobe Steel, Ltd., Wakinohama-cho, Chuo-ku, Kobe, 651 Japan

ABSTRACT

The effects of additional elements on the superconductivity of a Bi-Sr-Ca-Cu-O system were studied by measuring the magnetic susceptibility and x-ray diffraction patterns of such a system. By observation of x-ray diffractions, a solid solution region in the low- and high-Tc (critical temperature) Bi-M-Sr-Ca-Cu-O phases were found to exist for $0 \leq x \leq 1$ (M=In, Ga) and for $0<x<0.5$ (M=Al) in a $Bi_{4-x}M_xSr_3Ca_3Cu_6O_y$ system (4-3-3-6 system). In a $Bi_{1-x}M_xSrCaCu_2O_y$ system (1-1-1-2 system), a solid solution region in the low- and high-Tc phases were found to exist for $0 \leq x<0.3$ (M=In, Ga, Al). Indium was found to be effective in raising the amount of both low- and high-Tc superconductivity phases and the critical current density. Aluminum and Gallium showed small effects on the formation of a high-Tc phase. Further addition of these elements broke the superconductivity of the Bi-Sr-Ca-Cu-O system. These results suggest that the crystal of high- or low-Tc Bi-Sr-Ca-Cu-O phase was stabilized by the controlled addition of these elements.

INTRODUCTION

Recently Maeda et al. [1] discovered a new high-Tc oxide superconductor Bi-Sr-Ca-Cu-O system which contains no rare earth elements. This oxide superconductor has a Tc of about 105K, higher than that of a Y-Ba-Cu-O system of more than 10K. However, some problems with this system have been identified. Firstly, it is difficult to separate the high-Tc phase from low-Tc phase [1,2,3]. Secondly, it needs a long time sintering to get a large amountof the high-Tc phase [4]. Thirdly, this system has very small critical current density [5].

The low-Tc superconducting materials in a Bi-Sr-Ca-Cu-O system have been investigated by powder X-ray diffraction [6,7,8,9], single-crystal X-ray diffraction [9], and electron diffraction measurements [6,8,10,11]. The results indicate that a superconducting material with a Tc of 80K has a modulated structure. The average structure has an orthorhombic unit cell with the dimensions of a=0.5396nm, b=0.5395nm and c=3.0643nm.

Nobumasa et al. [4] have reported that a superconducting phases with a Tc of above 105K in a Bi-Sr-Ca-Cu-O system was found to appear in proportion to the time of sintering just below the melting temperature. In order to stabilize the high Tc phase, several attempts have been reported [12,13,14].

The purpose of this study is to grasp the effect of additional elements (M=In, Ga, Al) on the superconductivity of a Bi-Sr-Ca-Cu-O system. For the effect of additional these elements, it is expected that the crystal structure of this system is stabilized by changing ionic radius at Bi-site. These elements are supposed to have 3+ valence and different ionic radius. In^{3+},Ga^{3+} and Al^{3+} have value of ionic radius, 0.12, 0.092, 0.062 and 0.057nm respectively. Especially, in this paper, we will report on the effects of adding these elements on the behavior of high-Tc phase formation and critical current density.

EXPERIMENTAL

All samples used in the present work were prepared by reacting the prescribed mixture of Bi_2O_3, M_2O_3, $SrCO_3$, $CaCO_3$ and CuO at 800 C in air for 12 hours, with various values for x : for $Bi_{4-x}M_xSr_3Ca_3Cu_6O_y$: x=0, 0.01, 0.05, 0.1, 0.5, 1.0 and for $Bi_{1-x}M_xSrCaCu_2O_y$: x=0, 0.0025,

Table 1 Result of chemical analysis

Sample	Bi : M
$Bi_{0.97}In_{0.03}SrCaCu_2O_y$	0.972:0.028
$Bi_{0.97}Ga_{0.03}SrCaCu_2O_y$	0.970:0.030
$Bi_{0.97}Al_{0.03}SrCaCu_2O_y$	0.972:0.028
$Bi_{3.9}In_{0.1}Sr_3Ca_3Cu_6O_y$	3.900:0.100
$Bi_{3.9}Ga_{0.1}Sr_3Ca_3Cu_6O_y$	3.916:0.084
$Bi_{3.9}Al_{0.1}Sr_3Ca_3Cu_6O_y$	3.908:0.092

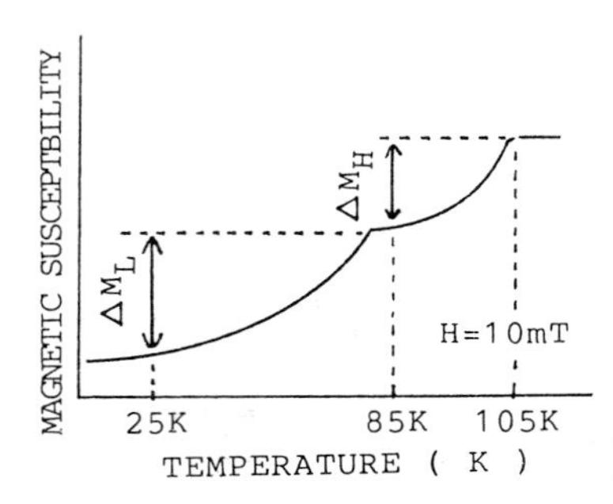

Fig.1 schematic explanation for ΔM_H and ΔM_L

0.005, 0.01, 0.03, 0.07, 0.1, 0.3. These powders were sintered in pellet form under the following four conditions in order to clarify the effect of the sintering time and temperature, 1) for 12 hours at 865°C, 2)for96 hours at 865°C, 3) for 96 hours at 870°C and, 4) for 0.5 hours at 880°C and then for 96 hours at 870°C. All samples were cooled to room temperature in a furnace after the sintering.

Magnetization measurements were taken for all samples using a vibrating sample magnetometer with a cryostat. The amounts of high- and low-Tc superconducting phases were estimated by the Meissner effect at 85K and 25K under 10mT. The superconducting transition temperature (on set) was also determined by the Meissner effect.

To evaluate the critical current density, M-H curve measurements were taken for 5 typical samples ($Bi_4Sr_3Ca_3Cu_6O_y$ and $Bi_{1-x}M_xSrCaCu_2O_y$: M=In, Ga, Al, x=0, 0.03) at 4.2K under less than H=2T.

All samples were analyzed by x-ray diffraction for identification of in crystallinephases. The X-ray powder diffraction patterns were obtained with Cu-Ka radiation, using a ground sample.

Table 1 shows the results of chemical analysis of the contents of the additional elements in comparison with Bi after sintering at 870°C for 96h. These additional elements do not evaporate after sintering.

RESULTS AND DISCUSSION

Figure 1 shows the schematic explanation for the determination of the amounts of high- and low-Tc phases. ΔM_H and ΔM_L in Fig.1 are considered to correspond to the amounts of high- and low-Tc phase respectively.

Figs.2-a and b show the results of the effect of adding In in $Bi_{4-x}In_xSr_3Ca_3Cu_6O_y$ system. All the samples which were sintered at 865°C for 96h showed larger amounts of superconducting phase than those sintered at 865°C for 12h. Furthermore, all the samples which were sintered at 870°C for 96h showed the largest amounts of high- and low-Tc superconducting phase of the three sintering conditions. There was small concentration dependence on the amounts of high and low Tc superconducting phases in the indium region of less than x=0.1. In cases where the sintering time is short (865°C,12h), a small peak was found at x=0.1 at both 25K and 85K. In the Indium region of more than x=0.05, the superconducting of all samples decreased rapidly. Similar behavior on amounts of high- and low-Tc phases were also observed in the $Bi_{4-x}Ga_{4-x}Sr_3Ca_3Cu_6O_y$ system and $Bi_{4-x}Al_xSr_3Ca_3Cu_6O_y$ systems. These resultsshows that a large amount of high-Tc superconducting phase can be obtained by starting with the nominal composition, 1112. In other words, a large amount of low-Tc superconducting phase can be obtained by starting with the nominal composition, 4336.

Figs.2-c and d show the results of effect of adding In in $Bi_{1-x}In_xSrCaCu_2O_y$ system. In Fig.2-c, the peak is found at x=0.03 which shows a larger amount of low-Tc superconducting phase than that obtained by starting with the nominal composition, 4336. In Fig.2-d, the sharp peak is found at x=0.03, which shows an amount of high-Tc superconducting phase about 2 times larger than that obtained by starting with the nominal composition, 1112. These results suggest that the In addition or substitution is effective in the formation of high- and low-Tc (mainly for high-Tc) superconducting phase at 870°C for 96h. In the case of sintering at 870°C for 96h after keeping at 880°C for 0.5h, the same tendency is observed in the formation of high-Tc superconducting phase.

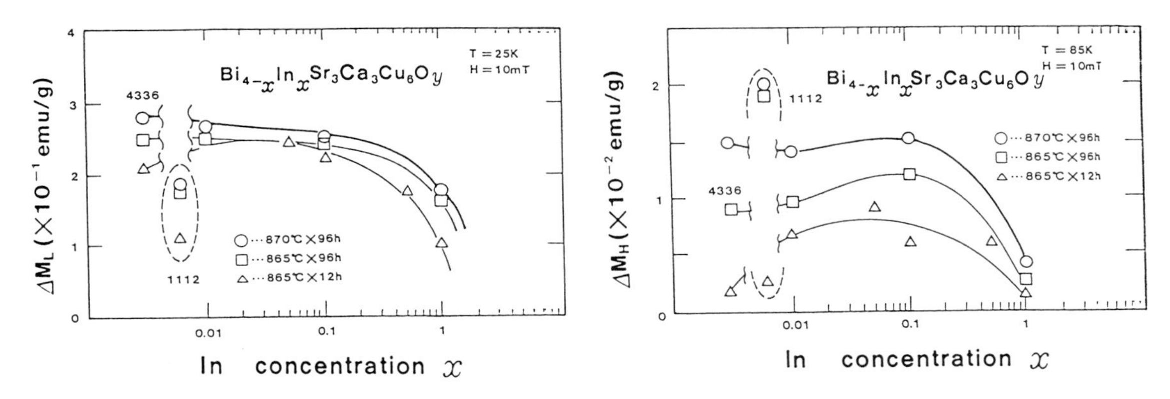

Fig.2-a,b the amount of diamagnetism
in the nominal composition $Bi_{4-x}In_xSr_3Ca_3Cu_6O_y$

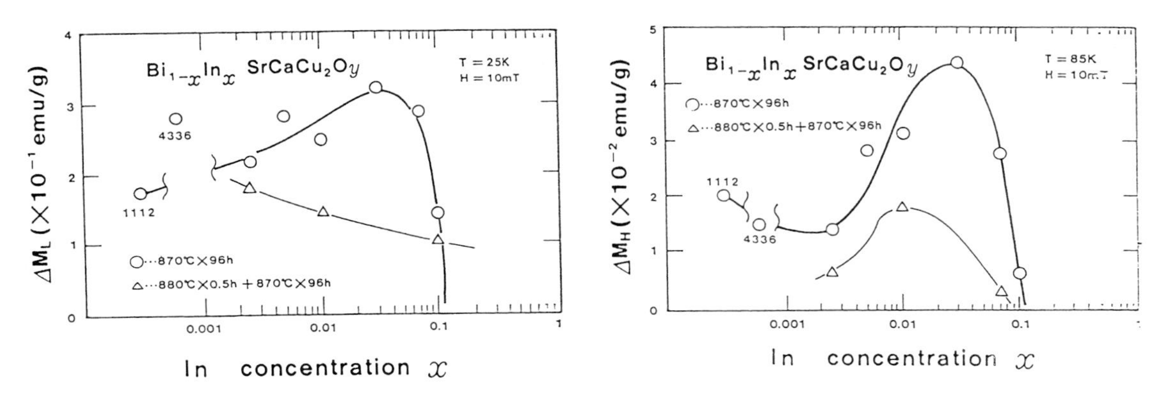

Fig.2-c,d the amount of diamagnetism
in the nominal composition $Bi_{1-x}In_xSrCaCu_2O_y$

Fig.3-a and b show the amounts of diamagnetism under 10mT in $Bi_{1-x}Ga_xSrCaCu_2O_y$ at 25K and 85K respectively, as a function of Ga concentration. The behavior of amounts of high- and low-Tc superconducting phase in Figs.3-a and b are similar to the behavior of amounts of high- and low-Tc superconducting phases in Figs.2-c and d, respectively. However, the height of peak at about x=0.05 in fig.3-b is smaller than that of fig.2-d.

In figs.4-a and b, ΔM_L and ΔM_H in $Bi_{1-x}Al_xSrCaCu_2O_y$ system were plotted against Al concentration. In fig.4-a, the amounts of diamagnetism of both samples which were sintered, one at 870 °C for 96h and the other at 870°C for 96h after being kept at 880°C at 0.5h, showed little or no dependence upon the Al concentration. A rapid decrease in the amount of diamagnetism seems to be visible only in the sample which was sintered at 870°C for 96h.

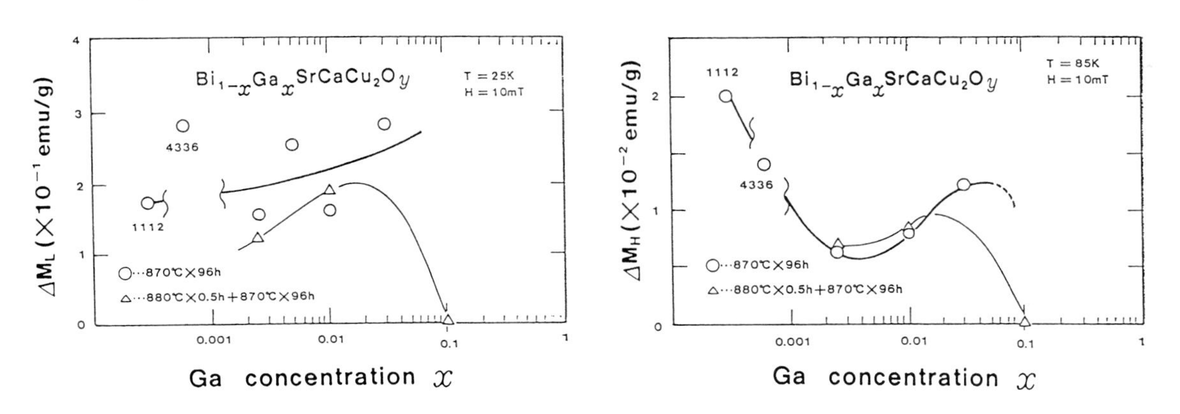

Fig.3-a,b the amount of diamagnetism
in the nominal composition $Bi_{1-x}Ga_xSrCaCu_2O_y$

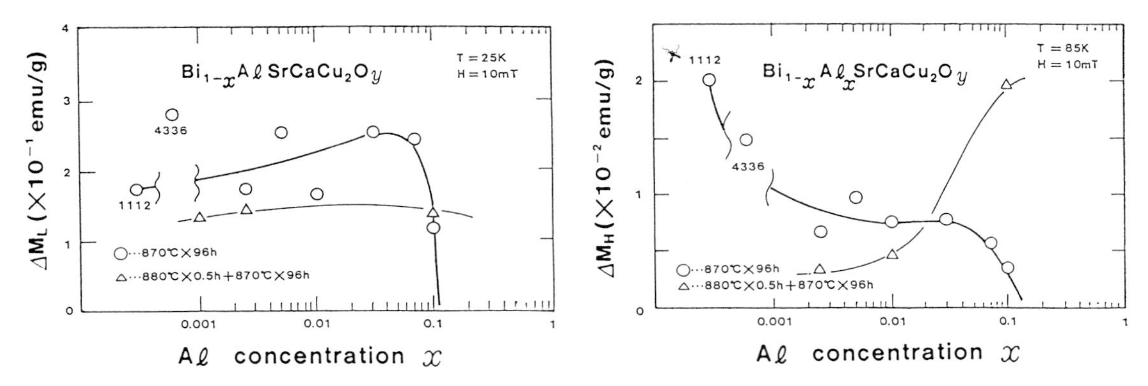

Fig.4-a,b the amount of diamagnetism in the nominal composition $Bi_{1-x}Al_xSrCaCu_2O_y$

In Fig 4-b, the samples sintered at 870°C for 96h show a negative dependence upon the Al concentration. In contrast, in the region of $x \lesssim 0.1$, the samples which were sintered at 870°C for 96h after being kept at 880°C for 0.5h show a positive dependence upon the Al concentration x as shown in fig.4-b. However both samples, 1112 and x=0.1, have the same amount of diamagnetism.

Tarascon et al. [9] have suggested that the Bi-O layers play an important role on the appearance of the high Tc superconductivity. The Bi-O bond was also found in the classical oxide superconductor $Ba(Pb_{1-x}Bi_x)O_3$. Although there is no doubt that the origin of the high Tc is mainly due to the Cu-O planes as proposed in the case of the lanthanum and yttrium systems, if additional elements are substituted for the Bi-site, the results of this work suggest that the Bi-O layers have the potential to help the superconductivity in Bi systems. All samples which were sintered at 865-880 C show almost the same critical transition temperature. This behavior is explained as below. The carrier density in these samples does not vary because additional elements M^{3+} are substituted for the Bi(3+) site. Bi^{3+}, In^{3+}, Ga^{3+} and Al^{3+} ions have the values of ionic radius of 0.12, 0.092, 0.062 and 0.057nm respectively. So appearance of the peak on M_H vs. in concentration of additional elements curves $Bi_{1-x}M_xSrCaCu_2O_y$ which were sintered at 870°C for 96h (Fig.2-d, Fig.3-b, Fig.4-b) might be due to the difference in ionic radius between Bi^{3+} and additional element. It is interesting that adding the element of with ionic radius is close to that of Bi, might be more effective in raising the amount of high-Tc superconducting phase.

An estimation of a critical current density was attempted for five typical samples at 0.5T, 1.0T and 1.5T from the M-H curve measurement under less than 2T at 4.2K, as shown in Table 2. The critical state model [15] was employed, in which the persistent microscopic current flows only on the surface of a whole sample. Here the ideal density of the sample is taken to be about 6.6g/cm^3. Thecritical current density was estimated on the assumption that these samples have the same grain size (=10μm), since these samples were treated under the same conditions. Three samples including In, Ga and Al show a larger value of Jc than that of a pure Bi system (1112, 4336) at 4.2K. Especially, indium was found to be effective to raise the value of critical current density.

CONCLUSION

The effects of additional elements (M=In, Ga, Al) on the superconductivity of Bi-M-Sr-Ca-Cu-O system were studied. Indium was found to be effective to raise the amount of superconductivity phase and the value of critical current density, showing the maximum at 0.03 substitution for Bi. In order to discover the origin of the superconductivity and role of Bi-O plane, it is important to clarify the relation between additional elements and superconducting property.

Table 2. critical current density from M-H curve measurement

sample	field (T)	M (emu/g)	Jc ($\times 10^5 A/cm^2$)
4336	0.5	3.23	4.26
	1.0	2.74	3.61
	1.5	2.27	2.99
1112	0.5	2.71	3.58
	1.0	2.02	2.66
	1.5	1.60	2.11
*1 In0.03	0.5	5.01	6.60
	1.0	3.73	4.91
	1.5	3.02	3.98
*2 Ga0.03	0.5	4.20	5.53
	1.0	2.82	3.72
	1.5	2.34	3.08
*3 Al0.03	0.5	4.03	5.31
	1.0	2.66	3.51
	1.5	2.16	2.85

*1 $Bi_{0.97}In_{0.03}SrCaCu_2O_y$
*2 $Bi_{0.97}Ga_{0.03}SrCaCu_2O_y$
*3 $Bi_{0.97}Al_{0.03}SrCaCu_2O_y$

REFERENCES

1) H.Maeda, Y.Tanaka, M.Fukutomi and T.Asano: Jpn. J. Appl. Phys. 27 (1988) L209.
2) Y.Tanaka, M.Fukutomi, T.Asano and H.Maeda: Jpn. J. Appl. Phys. 27 (1988) L548.
3) A.Maeda, T.Yabe, H.Ikuta, Y.Nakayama, T.Wada, S.Okuda, T.Itho, M.Izumi, K.Uchinokura, S.Tanaka: Jpn. J. Appl. Phys. 27 (1988) L661.
4) H.Nobumasa, K.Shimizu, Y.Kitano and T.Kawai: Jpn. J. Appl. Phys. 27 (1988) L846.
5) M.Uehara, Y.Asada, H.Maeda and K.Ogawa: Jpn. J. Appl. Phys. 27 (1988) L665.
6) E.Takayama-Muromachi, Y.Uchida, A.Ono, F.Izumi, M.Onoda, Y.Matsui, K.Kosuda, S.Takewaka and K.Kato: Jpn. J. Appl. Phys. 27 (1988) L365.
7) T.Kijima, J.Tanaka, Y.Bando, M.Onoda and F.Izumi: Jpn. J. Appl. Phys. 27 (1988) L369.
8) Y.Syono, K.Hiraga, N.Kobayashi, M.Kikuchi, K.Kusaba, T.Kajitani, D.Shindo, S.Hosoya, A.Tokiwa, S.Terada and Y.Muto: Jpn. J. Appl. Phys. 27 (1988) L569.
9) J.M.Tarascon, Y.Le Page, P.Barboux, B.G.Bagley, L.H.Greene, W.R.Mckinnon, G.W.Hull, M.Giroud and D.M.Hwang: Phys. Rev. B
10) M.Onoda, A.Yamamoto, E.Takayama-Muromachi and S.Takewaka: Jpn. J. Appl. Phys. 27 (1988) L833.
11) J.M.Tarascon, Y.Le Page, L.H.Greene, B.G.Bagley, P.Barboux, D.M.Hwang, G.W.Hull, W.R.Makinnon and M.Giroud: Phys. Rev. Lett.
12) M.Takano, J.Takada, K.Onoda, K.Oda, H.Kitaguchi, Y.Miura, Y.Ikeda, Y.Tomii and H.Mazaki: Jpn. J. Appl. Phys. 27 (1988) L1041.
13) Y.Yamada and S.Murase: Jpn. J. Appl. Phys. 27 (1988) L996.
14) N.Uno, N.Enomoto, Y.Tanakaand H.Takami: Jpn. J. Appl. Phys. 27 (1988) L1013.
15) C.B.Bean: Phys. Rev. Lett. 8 (1962) 250.

Optimization of Sintering and Metallurgical Studies on the Bi-Sr-Ca-Cu-O Superconductors

K. NUMATA[1], K. MORI[1], H. YAMAMOTO[1], H. SEKINE[2], K. INOUE[2], and H. MAEDA[2]

[1] Advanced Technology Research Center, Mitsubishi Heavy Industries, Ltd., 1-8-1, Sachiura, Kanazawa-ku, Yokohama, 236 Japan
[2] National Research Institute for Metals, 1-2-1, Sengen, Tsukuba, 305 Japan

ABSTRACT

Morphology, superconducting properties, possibility of orientation of the grains, workability and so on have been studied for the Bi-Sr-Ca-Cu-O system. Tapes and wires as well as bulk (pellet) specimens have been prepared. The results of the thermogravimetric analysis indicated that the phase transformation which accompanies absorption of oxygen does not occur in this material from ~860°C down to 400°C. Tape specimens prepared by a combination of cold work and sintering showed J_c of 1100A/cm^2 at 77K with a good reproductivity. In these tape specimens, c axis of the grains tends to align perpendicularly to the tape surface. A 1330-filament Bi-Pb-Sr-Ca-Cu-O wire with a Ag matrix has been successfully fabricated. This multifilamentary wire showed high-T_c transitions when sintered at relatively low temperature (~840°C). It has turned out that the concentration of Pb in the oxide must be controled (reduced) before the packing into a Ag sheath for the superconducting phase to be formed in a long wire. The results of this study indicate that the Bi-Sr-Ca-Cu-O system material has an advantage over other oxide superconductors with respect to practical use.

INTRODUCTION

A lot of works have been made on the Bi-Sr-Cu-Ca-O system since it's discovery in Dec. 1987[1], because this system includes a superconducting phase showing a high critical temperature, T_c, of ~ 110K. The formation of the single high-T_c phase, however, has been difficult due to appearence of other superconducting and non-superconducting phases. It has been reported more recently that partial substitution of Bi by Pb promotes the formation of the high T_c phase, and by this effect, single high-T_c phase has been formed[2-5]. Now, researches on the morphlogy of this material as well as those to improve critical current density, J_c, will be urgent as the next step to the formation of the single high-T_c phase.

In this study, we studied morphology, superconducting properties, possibility of orientation of the grains (when worked into wires or tapes), workability in a metal sheath, and so on.

EXPERIMENTAL PROCEDURE

Bi-Pb-Sr-Ca-Cu-O oxide powder prepared by a co-decomposition method was packed in a metal (mainly Ag) tube of 10 mm o.d. (outer diameter) and 6.7 mm i.d. (inner diameter), and was cold worked into a wire of 0.5 mm o.d. or into a square wire of 2 mm x 2 mm. The nominal cation ratio of this powder was Bi:Pb:Sr:Ca:Cu=0.8:0.2:0.8:1.0:1.4. (This ratio was determined according to the reference 5.) This powder was calcined and/or sintered before being packed. The conditions of calcination and sintering before packing were 800°C for 14h in air and 845°C for 100h in air, respectively. Disk pellets were also made both from Bi-Sr-Ca-Cu-O (1:1:1:2:x) oxide powder prepared by the conventional solid state reaction method and from the same Bi-Pb-Sr-Ca-Cu-O oxide powder (by the co-decomposition method) as was used for the wire fabrication. For fabrication of a multifilamentary wire, 70 pieces were cut from the above-mentioned monofilamentary wire of 0.5 mm o.d., packed into a silver sheath of 7.5 mm o.d. and 5 mm i.d., and cold worked into a wire of 0.7 mm o.d.. 19 pieces were cut from the 70 filament wire, packed again into a silver sheath of 5 mm o.d. and 3.5 mm i.d. and cold worked again into a wire of 1.5 mm o.d. which include 70x19=1330 oxide filaments. The monofilamentary wires, multifilamentary wires and disk pellets with Pb were sintered at 820-850°C for 12-300h while disk pellets without Pb were sintered at 860-880°C for 12-500h. Tape specimens were also prepared with the same (co-decomposition) Bi-Pb-Sr-Ca-Cu-O oxide powder as was used for the wire fabrication. The whole process for the tape specimens, however, includes a new technique besides the cold work process. All of the specimens here were slowly cooled down (about 100C/h) to the room temperature after sintering. Transition temperature, T_c, was measured by both the standard induction method and resistive method. The temperature was measured by Au-Fe/chromel thermocouples in the T_c measurements. Scannning electron microscopy (SEM) and back electron image (BEI) observations were made for these specimens. In order to investigate absorption or release of oxygen in the Bi-Sr-Ca-Cu-O system, thermal gravimetric and differential thermal analysis (TG-DTA) were made for the $Bi_1Sr_1Ca_1Cu_2O_x$ powder. TG-DTA were also made for the $Y_1Ba_2Cu_3O_x$ powder for comparison.

RESULTS AND DISCUSSION

Figure 1 shows results of the TG analysis for both (a) Bi-Sr-Ca-Cu-O and (b) Y-Ba-Cu-O systems in the temperature range from 800°C down to 400°C. In Fig. 1, the $Y_1Ba_2Cu_3O_x$ powder specimen shows a weight loss of about 1.5% during the decrease of temperature from 800°C to 400°C while the $Bi_1Sr_1Ca_1Cu_2O_x$ powder specimen didn't show a weight change of more than ~ 0.2%, which was not precisely determined due to change of the background, within the same temperature range. This indicates that a phase transformation that accompanies absorption of oxygen does not occur in the Bi-Sr-Ca-Cu-O system during the temperature decrease from 800°C to 400°C unlike the case of the Y-Ba-Cu-O system.

Figure 2 shows the results of the TG-DTA analysis in the temperature range of 800°C-900°C. It is seen that double peaks of endothermic reactions at ~ 860C and at 910°C appear during the increase of temperature, and a weight loss accompanies each endothermic reaction. As these reactions and weight losses are reversible, the weight losses are considered to result from release of oxygen. The weight change from 860°C to 950°C amounted to about 1.6%. The reason for the peak at ~ 860°C is not clear (maybe connected with formation of the high T_c phase[6]) although the peak at ~ 910°C can be connected with the melting point. Neither endothermic reaction nor weight loss was observed from 800°C to 860°C. After all, it seems unneccessary that the Bi-Sr-Ca-Cu-O system should absorb oxygen in order to undergo a phase transformation and form a superconducting phase at a low temperature around 600°C after sintering. This would be essential for the formation of the superconducting phase in a long wire of the Bi-Sr-Ca-Cu-O material within a metal sheath because external oxygen can scarcely pass through the metal sheath at such a low temperature around 600°C.

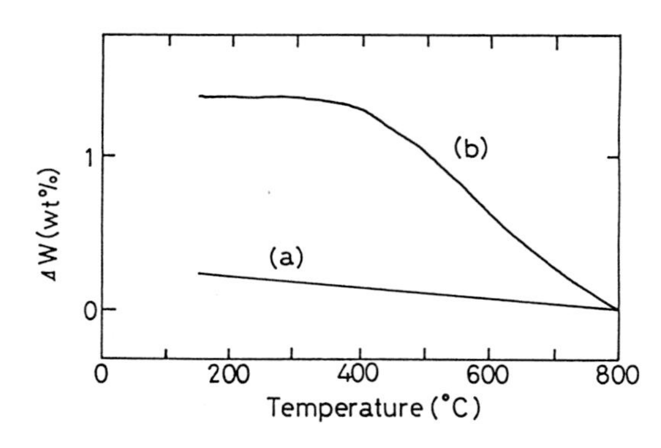

Fig. 1 TG curves for the (a) Bi-Sr-Ca-Cu-O and (b) Y-Ba-Cu-O specimens. (400-800°C)

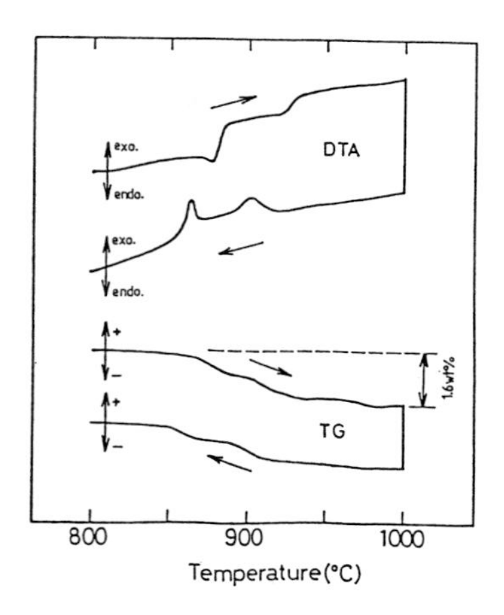

Fig. 2 TG-DTA curves for the Bi-Sr-Ca-Cu-O specimen.

Figure 3 shows SEM pictures taken on the fracture surfaces of (a) the $Bi_{0.8}Pb_{0.2}Sr_{0.8}Ca_{1.0}Cu_{1.4}O_x$ monofilamentary wire specimen (as-drawn) with reduction ratio R=50 (transverse cross section) and (b) the monofilamentary wire sintered at 845°C for 100h with R=50 (transverse cross section). The specimen (b) has the same nominal composition as (a). The pictures in Fig. 3 show that grains of the Bi-Sr-Ca-Cu-O wire specimens are very fine (less than ~ 2 um). The grain size of the bulk specimen sintered in the same condition as in Fig. 3 (b) was 5-10um. This indicates grain size of the Bi-Pb-Sr-Ca-Cu-O is remarkably reduced by the cold work process.

Figure 4 shows BEI pictures taken on the polished surface of the (a) bulk specimen without Pb (Bi:Sr:Ca: Cu=1:1:1:2) sintered at 875°C for 14h, (b) the same bulk specimen as (a) but sintered at 875°C for 100h, (c) the same bulk specimen as (a) but sintered at 875°C for 470h and (d) bulk specimen with Pb (Bi:Pb:Sr:Ca:Cu=0.8:0.2:0.8:1.0:1.4) sintered at 845°C for 100h. In these BEI pictures, gray and black regions correspond to the Bi-poor phase (non superconducting phase). Comparison of (a),(b) and (c) shows that the average size of the Bi-poor regions increase with increasing sintering time although the total volume fraction of the Bi-poor regions doesn't seem to change. On the other hand, comparison of Fig. 4 (d) with (a), (b) and (c) shows that the total volume fraction of the Bi-poor phase is remarkably reduced by the Pb addition.

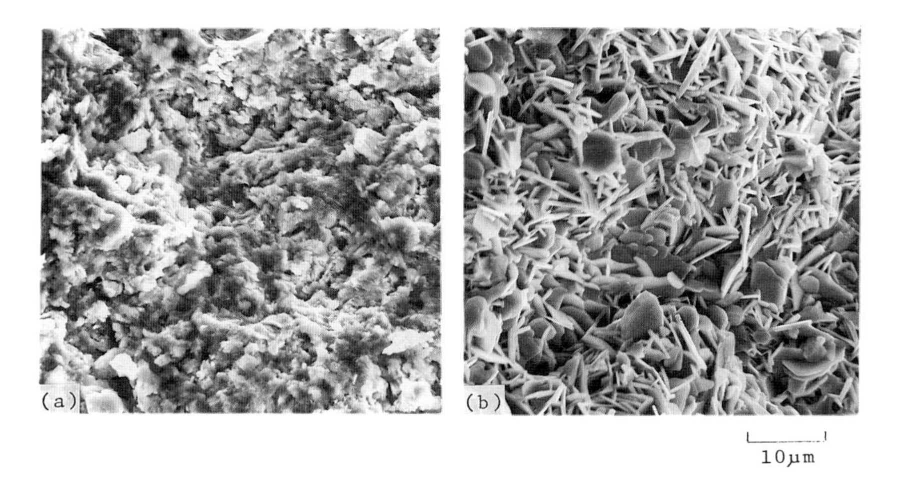

Fig. 3 SEM pictures taken on the fracture surfaces of monofilamentary Bi-Pb-Sr-Ca-Cu-O wire specimen with R=50; (a) as drawn (b) sintered at 845°C for 100h after drawing.

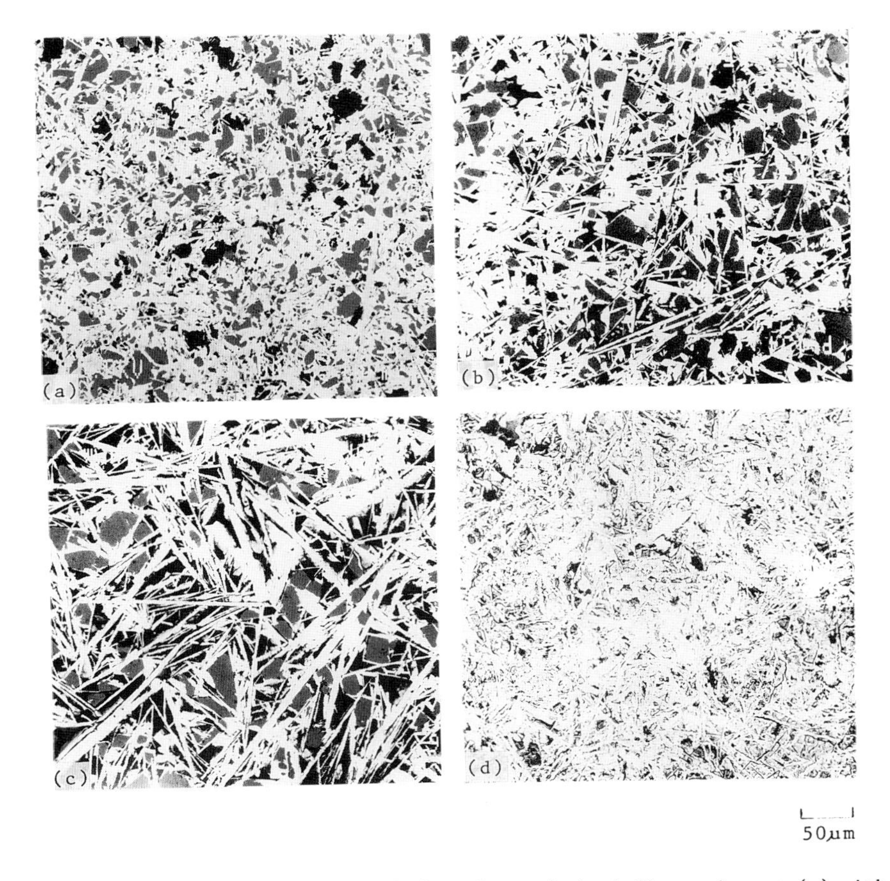

Fig. 4 BEI pictures taken on the polished surface of the bulk specimens; (a) without Pb sintered at 875°C for 14h, (b) without Pb sintered at 875°C for 100h, (c) without Pb sintered at 875°C for 475h and (d) with Pb sintered at 845°C for 100h.

Figure 5 shows a SEM picture taken on the longitudinal cross section of the wire specimen for which cold work has been done only by a groove roll. In Fig. 5, the horizontal direction of the picture corresponds to the direction of the wire elongation. The transverse cross section of this wire is a square of 2mm x 2mm. The sintering condition for this specimen and nominal composition are same as in Fig. 3 (b). Figure 5 indicates that c axis of the grains tend to align in two directions, i.e., perpendicularly to two sides of the square.

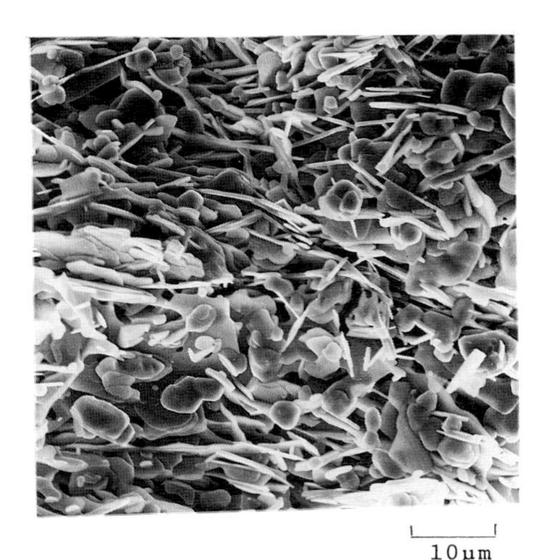

Fig. 5 A SEM picture taken on the longitudinal cross section of the Bi-Pb-Sr-Ca-Cu-O square wire specimen.

Figure 6 shows a SEM picture taken on the longitudinal cross section of the Bi-Pb-Sr-Ca-Cu-O tape specimen with the same nominal composition as in Fig. 3 (a) and (b) . This tape specimen has been prepared by a new process (i.e. a combination of cold work and sintering). In Fig. 6, the horizontal direction in the picture corresponds to the longitudinal direction of the wire (i.e. the direction of the tape elongation). The tape surface is perpendicular to the picture surface. Figure 6 shows that c axis of the planer grains tends to align perpendicularly to the tape surface.

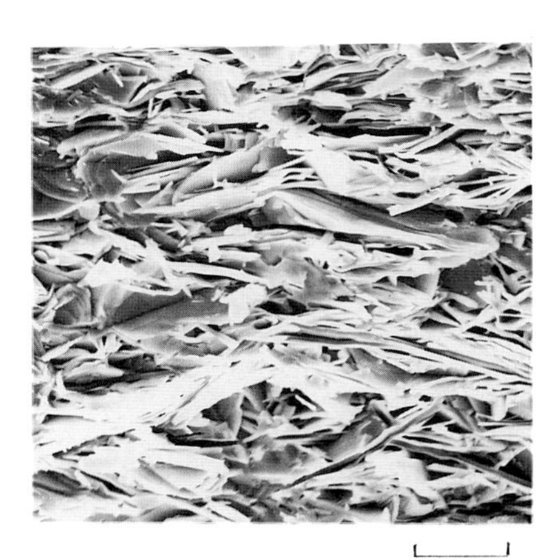

Fig. 6 A SEM picture taken on the longitudinal cross section of the tape specimen.

Figure 7 shows T_c transition curves measured by the resistive method for the (a) bulk specimen with Pb sintered at 845°C for 100h and (b) tape specimen (which is the same as shown in Fig. 6). J_c (critical current density) is also shown for the tape specimen (b). The bulk specimen shows slightly higher T_c than the wire specimen. J_c of the tape specimen, however, is much higher than J_c which has been obtained so far for the bulk specimen in our study. The tape specimens showed J_c higher than 1000A/cm^2 at 77K with a good reproductivity. The reason for the very high J_c of the tape specimen could be attributed to the alignment of c axis of the grains in the direction perpendicular to the tape surface as well as to the refinement of the grains which occurs during the cold work process.

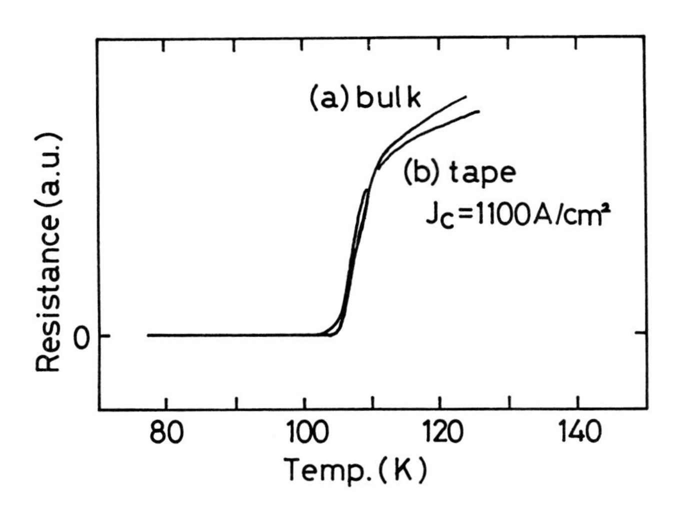

Fig. 7 T_c transition curves measured by the resistive method for the (a) bulk specimen sintered at 845°C for 100h and (b) the tape specimen (the same as shown in Fig. 6. J_c is is also shown for the tape specimen.

Figure 8 shows T_c transition curves measured by the inductive mehod for the (a) monofilamentary wire specimen with the oxide powder being calcined but not sintered before being packed into the Ag sheath and (b) monofilamentary wire specimen with the powder calcined and sintered at 845°C for 100h in air before packed. Both of the specimens have the nominal composition of $Bi_{0.8}Pb_{0.2}Sr_{0.8}Ca_{1.0}Cu_{1.4}O_x$ and they were sintered at 845°C for 50h in air after being drawn into the wire. The T_c transition curve of the specimen (b) shows a sharp transition near 100K while no transition can be seen in the curve of the specimen (a) above 50K. It is considered that too much amount of Pb is included in the specimen (a) even near the end of sintering probably because Pb cannot get away from the oxide through the Ag sheath during the sintering after the cold work. On the other hand, in the specimen (b), just appropriate amount of Pb remains in the oxide powder when it is packed into the Ag sheath because the sintering which is done before the packing reduces the Pb concentration. This could be the main reason for the results shown in Fig. 8. It can be concluded that the concentration of Pb in the oxide powder must be controled (reduced) before the powder is packed into the Ag sheath.

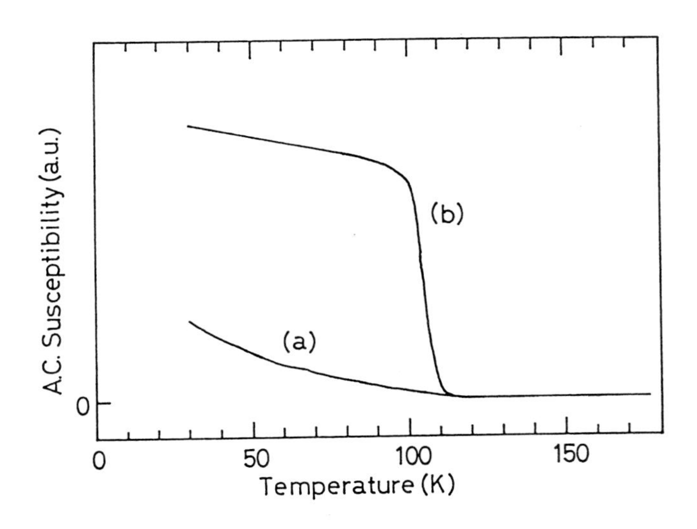

Fig. 8 T_c transition curves measured by the inductive method for the (a) monofilamentary wire specimen with the oxide powder being not sintered before the packing and (b) the wire specimen with the powder being sintered before the packing.

Figure 9 shows a picture of a cross section of the 1330-filament $Bi_{0.8}Pb_{0.2}Sr_{0.8}Ca_{1.0}Cu_{1.4}$ wire. The intermediate annealings at 150°C which was applied every time when the reduction ratio R= 10 gave appropriate hardness to the silver sheath and made cold work fabrication of this multifilamentary wire possible. Although grains of the Bi-Pb-Sr-Ca-Cu-O powder cannot be plastically deformed, the filaments of the oxide powder in Ag matrix can elongate probably because the Bi-Pb-Sr-Ca-Cu-O grains can be easily cracked by the stress produced during the cold work process. In other words, the refinement of the oxide grains makes the wire elongation possible. In this specimen, the oxide powder was calcined and sintered at 845°C for 100h in air before being packed into the Ag sheath.

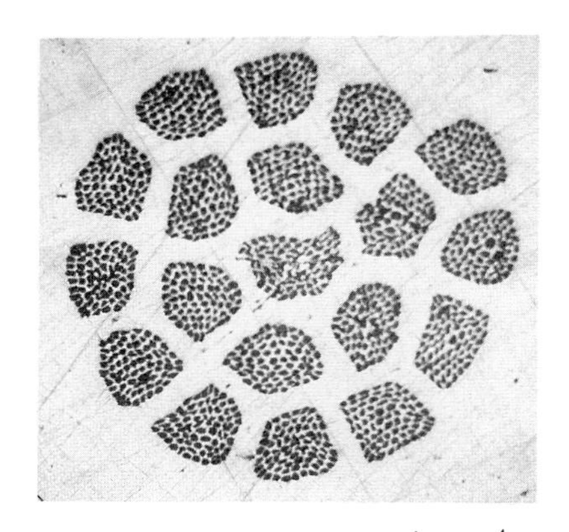

Fig. 9 A cross section of the 1330-filament Bi-Sr-Ca-Cu-O wire.

T_c measurements by the inductive method were made for the multifilamentary wires. The specimens sintered at 830°C-840°C for 50h after the cold work process showed T_c transition at relatively high temperature (95K-110K) while the specimen sintered at 850°C for 50h showed a broad transition from 70K-110K. The reaction of the oxide filaments with the Ag matrix would be the reason for the broad T_c transition. As a whole, the appropriate sintering temperature for the multifilamentary wires seems to be lower than that for bulk and monofilamentary wire specimens.

CONCLUSION

The superconducting properties, workability (with respect to the fabrication of a multifilamentary wire), possibility of forming the superconducting phase in a long wire with a metal sheath, morphology, possibility of alignment of the c axis of the grains and so on have been studied for the Bi-(Pb)-Sr-Ca-Cu-O system.

The results of the TG analysis indicated that the phase transformation which accompanies absorption of oxygen does not occur in the Bi-Sr-Ca-Cu-O material in the temperature range from 860°C to 400°C

The tape specimens prepared by a combination of cold work and sintering showed J_c of 1100 A/cm^2 at 77K with a good reproductivity. SEM revealed that c axis of the grains tends to align perpendicularly to the tape surface.

A 1330-filament Bi-Pb-Sr-Ca-Cu-O wire with a Ag matrix was successfully fabricated. This multifilamentary wire showed high T_c transitions when sintered at relatively low temperature (830°C-840°C).

The concentration of Pb in the oxide must be controled before the packing into Ag sheath in order to form the superconducting phase.

The results of this study indicates that the Bi-Sr-Ca-Cu-O system has an advantage over other oxide superconductors and has a potentiality for practical use.

REFERENCES

[1]. H. Maeda, Y. Tanaka, M. Fukutomi and T. Asano, Jpn. J. of Appl. Phys., 27, 209(1988).
[2]. R.J. Cava, 1988 Spring meeting of the American Physical Society.
[3]. C. Politis, MRS 1988 Spring Meeting, Reno, April 6, (1988).
[4]. M. Takano, J. Tanaka, K. Oda, H. Kitaguchi, Y. Miura, Y. Ikeda, Y. tomii and H. Mazaki, Jpn. J. of Appl. Phys., in press.
[5]. U. Endo, S. Koyama and T. Kawai, Jpn. J. Appl. Phys., to be published.

Structural Investigation of High Tc Bi-Sr-Ca-Cu-O

MASAAKI MATSUI, HAJIME ITOH, JUN LIU, TOSHIFUMI SHIMIZU,
HIROSHI MATSUOKA, KAZUHIKO OHMORI, and MASAO DOYAMA

Department of Iron and Steel Engineering, Faculty of Engineering, Nagoya University,
Furo-cho, Chikusa-ku, Nagoya, 464-01 Japan

ABSTRACT

High temperature X-ray diffraction, electrical resistivity and susceptibility measurements have been made for the Bi-Sr-Ca-Cu-O system of 1:1:1:2, 2:2:1:2 and 2:2:2:3 in the nominal composition ratio of Bi:Sr:Ca:Cu and for the system of 2:2:2:3 doped with Pb. No structural change has been observed in the temperature range from R.T. to 1123K for samples except for 1:1:1:2, where some extra peaks have appeared between 923K and 1073K. The crystal structure of a high Tc phase (Tc^{on}=115K, Tc^{zero}=110K) in the Pb-doped sample is tetragonal with a=b=3.82 Å and c=37.22 Å at R.T.. The temperature dependence of lattice parameters of the low Tc phase for undoped samples and of the high Tc phase for Pb-doped samples have been obtained. The mean thermal expansion coefficient between R.T. and 1100K of c-axis of the high Tc phase is 2.9×10^{-5} K^{-1} which is larger than that of low Tc phase.

INTRODUCTION

Since a discovery of a new high Tc superconductor, $BiSrCaCu_2O_y$ system without rare earth elements, by Maeda et al.[1], considerable efforts have been made to characterize the crystal structure and the superconductivity properties by many authors [2-10]. The superconductor contains a low Tc(=80K) and a high Tc(=105K) phases [1]. The structure of the low Tc phase reported previously [2,3] is modulated orthorhombic with a=5.4 Å, b=27.0 Å and c=30.8 Å in dimension of a unit cell, and the sub-lattice is tetragonal(or pseudo-tetragonal) structure with a=b=5.4 Å and c=30.8 Å. On the other hand, Tarascon et al. reported that they indexed the diffraction peaks observed to the space group of I4/mmm as a sub-structure of $Bi_4(Sr,Ca)_6Cu_4O_{16-x}$ and the lattice parameters were a=3.817 Å and c=30.5 Å [7]. These results were consistent with one another except for the difference of nomenclature of indexing. The structure of the high Tc phase is also orthorhombic with c=37.0 Å [9] , but the details of the structure has not been determined because of the difficulty of purifying the single phase by the heat treatment . The temperature range of the sintering to obtain the high Tc phase was very narrow (1143K to 1153K) [1,11]. Meanwhile, the high Tc phase of Tl-Ba-Ca-Cu-O system was easily purified and the structure was determined [12,13]. It had two more layers of CuO_2 and Ca than that of the low Tc phase of Tl-Ba-Ca-Cu-O. Furthermore, recently, Takano et al. reported that the nearly single phase with high Tc could be purified by means of doping of Pb to Bi-Sr-Ca-Cu-O [14] and suggested that the orthorhombic structure with a=5.37 Å, b=26.82 Å and c=37.26 Å. Thus the structure of high Tc phase of Bi-Sr-Ca-Cu-O system has not been determined so far because of non pure single phase. The high temperature experiment is necessary to obtain the single phase of the high Tc phase.

In this paper we report the high temperature X-ray diffraction experiment, electrical resistivity and susceptibility measurements for the Bi-Sr-Ca-Cu-O system of 1:1:1:2, 2:2:1:2 and 2:2:2:3 in the nominal composition ratio of Bi:Sr:Ca:Cu and for the system of 2:2:2:3 doped with Pb.

EXPERIMENTAL

High purity powders of 4N-Bi_2O_3, $SrCO_3$, $CaCO_3$ and CuO were mixed in appropriate ratios and ground. Then the powders were calcinated at 1073K for 16 hours. The sample were pressed into a disk of 20 mm in diameter and about 2 mm in thickness. Then the pellet was sintered at 1143K or 1153K for 16 hours. We named the samples prepared in the present work A, A', B, C for 1:1:1:2 sintered at 1153K, 1:1:1:2 sintered at 1143K, 2:2:1:2 sintered at 1143K and 2:2:2:3 sintered at 1143K , respectively. We also prepared the Pb-doped samples $Bi_{0.7}SrCaCu_{1.8}Pb_{0.3}O_y$ sintered at 1113K for 168 hours(sample D) and $BiSrCaCu_2Pb_{0.35}O_y$ sintered at 1113K for 45 hours (sample E) and at 1103K for 88 hours (sample E'). The oxide PbO_2 for doping were mixed with the other oxides at the initial stage before the first calcination. In this paper we use the nominal composition.

Electrical resistivity was measured by means of the four probe method, changing the current direction alternatively. The susceptibility was measured using SQUID magnetometer.

X-Ray diffraction(XRD) was made in air, using Cu target installed in a roterflex RU-200B (Rigaku Co.LTD) in the temperature range of R.T. to 1153K ,above which the sample was melted. We used a Cu-target and a high temperature furnace. The lattice parameters were estimated by means of the least mean square method using 27 peaks.

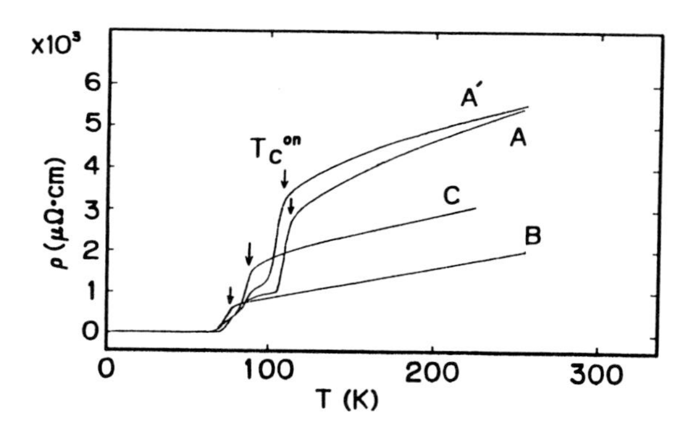

Fig. 1 Electrical resistivity as a function of temperature for undoped samples A, A', B and C.

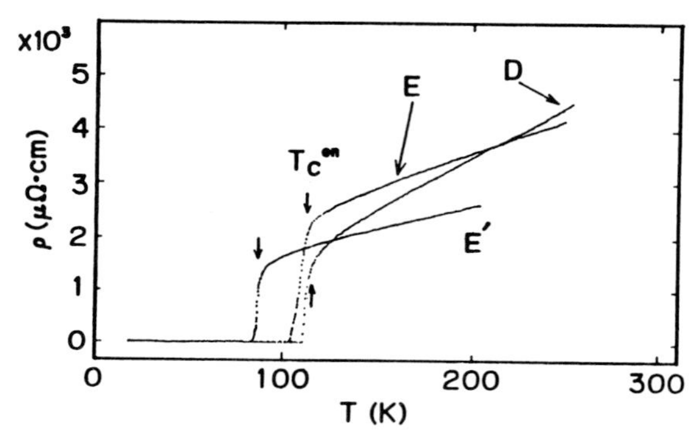

Fig.2 Electrical resistivity as a function of temperature for Pb-doped samples D, E and E'

RESULTS AND DISCUSSION

Electrical resistivity

The temperature dependence of resistivity of undoped samples is shown in Fig.1. The superconducting on-set transition temperature Tc^{on} and zero-resistivity temperature Tc^{zero} are listed in Table 1. The high Tc phase was contained in the samples A and A', whereas only low Tc phase was observed in samples C and D. It is noted that the larger resistivity above Tc, the higher Tc was observed. The tendency is different from the case of $YBa_2Cu_3O_y$ system [15]. The same tendency as Fig.1 is observed in Fig. 2, where the resistivity of doped samples D to E'is shown. In Fig. 2, the maximum transition temperatures Tc^{on}=115K and Tc^{zero}=110K was achieved for sample D. As the doping element Pb was almost depleted in the heat treatment process [14], the high Tc phase might come from the same structure as that of the undoped sample.

Magnetic susceptibility

Static magnetic susceptibility was measured with increasing temperature for samples A, A', B and C as shown in Fig.3. The transition temperature Tc^{m} where the susceptibility become zero is also listed in Table 1. The volume fraction of the superconducting phase to the total volume of the specimen was estimated from Meissner effect to be 41.5%, 35.8%, 40.3% and 30.9%, for sample A, A', B and C, respectively. The ratios of the volume of low Tc phase to that of high Tc phase were also estimated to be 1:0.85 and 1:0.38 for sample A and A', respectively. It suggests that the higher annealing temperature stabilizes the high Tc phase.

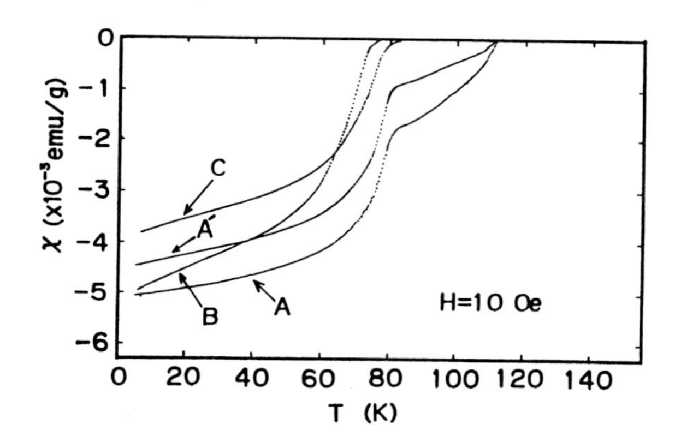

Fig.3 Static magnetic susceptibility as a function of temperature for undoped samples A, A', B and C.

Table 1 Transition temperatures Tc^{on} and Tc^{zero} obtained by the resistivity measurement and Tc^{m} obtained by the susceptibility measurement. The heat treatment for sintering of pellet is also shown.

Sample Name	Composition	Heat Treatment	Tc^{on} (K)	Tc^{zero} (K)	Tc^{m} (K)
A	$BiSrCaCu_2O_y$	1153K × 16hrs	111	61	112
A'	$BiSrCaCu_2O_y$	1143K × 16hrs	109	61	112
B	$Bi_2Sr_2CaCu_2O_y$	1143K × 16hrs	77	61	77
C	$Bi_2Sr_2Ca_2Cu_3O_y$	1143K × 16hrs	91	70	83
D	$Bi_{0.7}SrCaCu_{1.8}Pb_{0.3}O_y$	1113K × 168hrs	115	110	
E	$BiSrCaCu_2Pb_{0.35}O_y$	1113K × 45hrs	112	103	
E'	$BiSrCaCu_2Pb_{0.35}O_y$	1103K × 64hrs	88	83	

X-ray diffraction(XRD)

First we made the X-ray diffraction at room temperature. The XRD patterns for samples A, B, and C are shown in Figs.4 and 5. The XRD pattern of sample A' was same to that of sample A. Almost all the peaks observed at R.T. are well assigned to 2:2:1:2 sub-structure of the low Tc phase reported by Tarascon et al.[7]. In this paper we used the indices of sub-structure proposed by them. Despite the different nominal composition, there is no intrinsic difference among the patterns of sample A, B and C except for one peak denoted by asterisk in the pattern for sample A in Fig.5. The lattice spacing d=2.81 Å of the peak indicated by asterisk is in agreement with that of (109) plane of high Tc phase of the Pb-doped sample D (Fig.8) and is also consistent with the peak observed for as-melt-quenched amorphous sample reported by Komatsu et al [16]. The peak is most likely come from the high Tc phase because it only appeared in sample A which contained the high Tc phase.

High temperature X-ray diffraction was performed from R.T. to 1123K for these three samples. The structure of the low Tc phase did not changed, but the extra peaks were observed and the intensity of peaks of 2:2:1:2 sub-structure was decreased between 923K and 1073K only for samples A and A' as indicated by "E" in Fig.5. Above 1073K the peaks disappeared again and showed the same peaks to those observed at low temperatures below 923K. Interestingly, when we re-raised temperature very rapidly up to 1023K once cooling down to R.T. [17], the intensity of the extra peaks was grown up and the intensity of 2:2:1:2 peaks was decreased almost to zero. However, for another sample prepared to confirm the phenomenon, the intensity of extra peaks was decreased. These extra peaks are similar to peaks observed by Seebacher and Schindler [18], where they reported the XRD patterns at 1073K for nominal 1:1:1:2 sample. By means of many experiments, we found that the extra peaks were observed only for nominal 1:1:1:2 samples which contained the high temperature component in the resistivity. Accordingly the extra new phase might be formed by the diffusion process to reach the

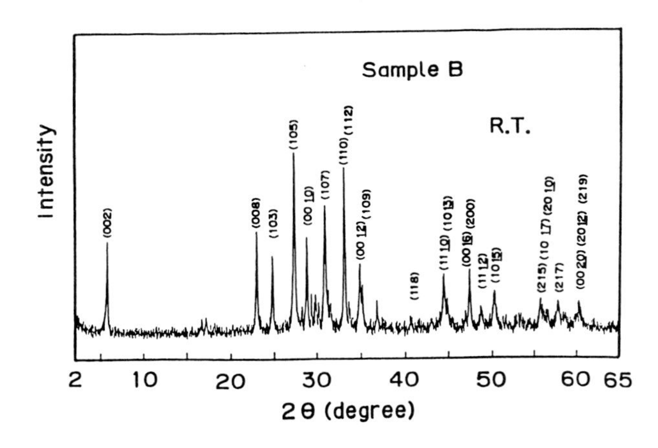

Fig.4 X-ray diffraction pattern at R.T. for sample B.

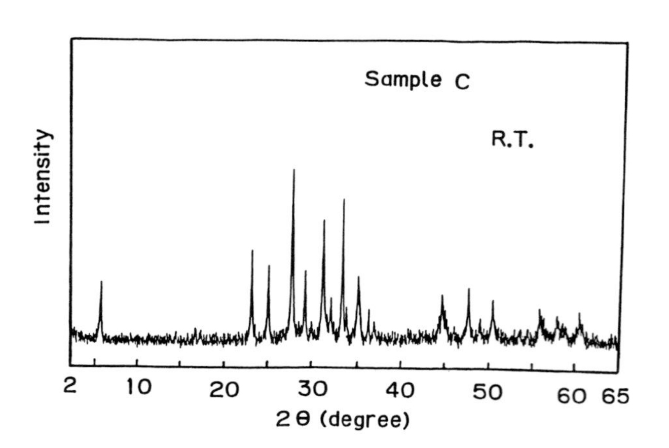

Fig.5 X-ray diffraction pattern at R.T. for sample C.

thermodynamical equilibrium state. Actually, long time sintering is necessary to obtain the high Tc phase [11].

In Fig. 7, lattice parameters for samples A, B and C are shown as a function of temperature. They increases monotonically with increasing temperature. The mean linear thermal expansion coefficient of the parameter c is 1.4×10^{-5} K^{-1} for the temperature range of R.T. to 850K and 2.3×10^{-5} K^{-1} for the range of 850K to 1100K and that of parameter a is 0.84×10^{-5} K^{-1} for the range of R.T. to 850K and 1.1×10^{-5} K^{-1} for the range of 850K to 1100K for stoichiometry 2:2:1:2 composition (sample B). It is interesting that the ratio c/a increases rapidly above about 850K which is very closed to the temperature at which the extra peaks were observed (Fig.5). The rapid expansion of c axis at high temperatures is profitable for the high Tc phase, whose c axis is longer than that of the low Tc phase, because the prolongation of the atomic distance along c axis makes layers such as CuO_2 and Ca easy to be inserted into the layer structure of the low Tc phase.

Next we made the XRD experiment for Pb-doped sample D. Fig.8 shows the diffraction pattern of sample D at R.T.. The result is consistent with that of Pb-doped sample reported by Takano et al. [14], where they suggested that a=5.37 Å, b=26.82 Å and c=37.26 Å in unit cell dimension referred to $Bi_4Ti_3O_{12}$ type structure. In Fig.8, it is noted that sample D contains both phases of the high Tc phase and the low Tc phase. We assigned many peaks to a low Tc phase and then we could assign almost of the remained peaks to the tetragonal structure with a=b=3.82 Å and c=37.22 Å Considering the result of resistivity (Fig.2), it seems that the tetragonal phase is responsible for high Tc. The intensity of peaks of the high Tc phase was also calculated, assuming 2:2:2:3 sub-structure which consist of the inserted two layers of CuO_2 and Ca to the structure of the low Tc phase submitted by

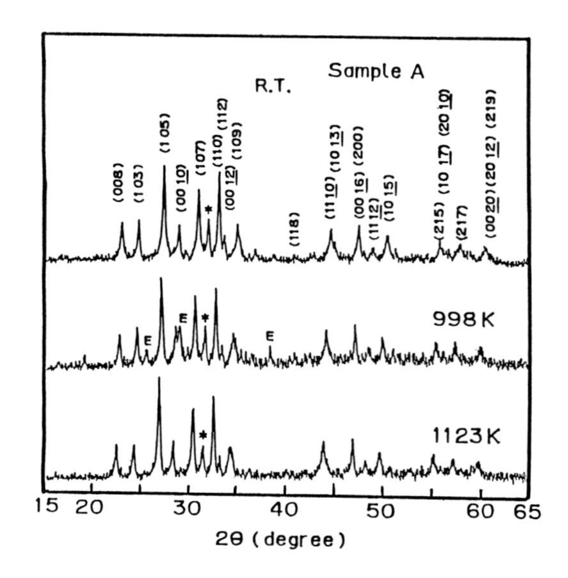

Fig.6 X-ray diffraction pattern for sample A at various temperatures. The peaks are indexed to 2:2:1:2 sub-structure, where the asterisk indicates the another phase. The extra peaks denoted as "E" are observed between 923K and 1073K.

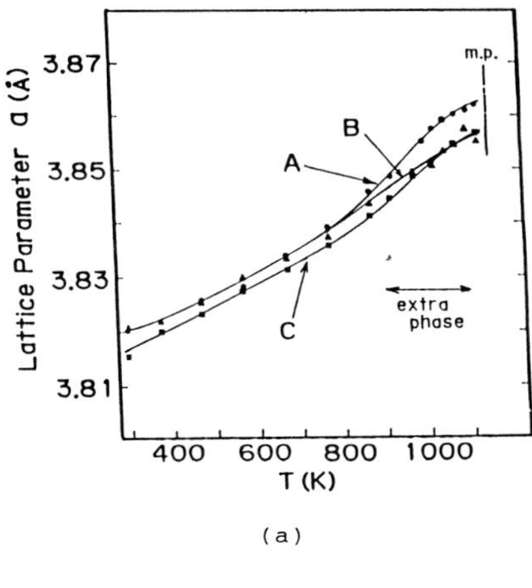

(a)

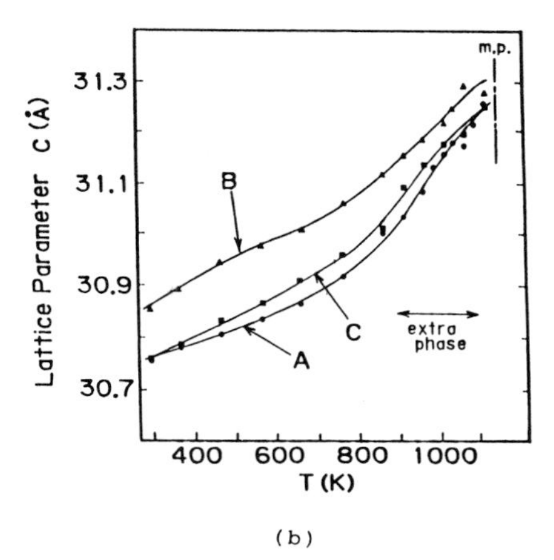

(b)

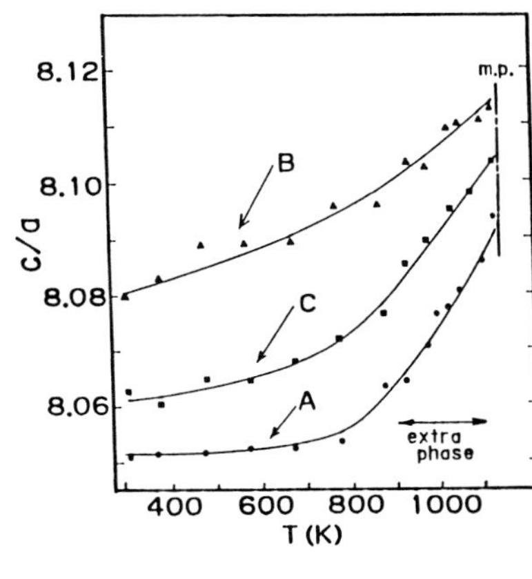

(c)

Fig.7 Lattice parameters, (a) a and (b) c, and (c) the ratio c/a as a function of temperature for sample A, B and C. The extra phase (peaks) was observed for sample A (see text)

Tarascon et al.. The calculation was consistent with the experimental results except for a peak denoted by asterisk in Fig.8. The lattice spacing of the peak can be fit to plane of Ca_2PbO_4. The observed intensities of (00L) planes were larger than the calculated ones. If the modulated structure in the a-b plane like the case of the low Tc phase exists in the high Tc phase, the calculated relative intensity of (00L) peaks to those of (HKL) peaks should be increased. Then the calculated intensity of (00L) peaks might be explained. The microscopic research of high Tc phase is expected.

Then the high temperature XRD experiment was also performed for sample D. We observed no structural change in the temperature range of R.T. to 1103K. Lattice spacings of (0010) and (110) planes are shown as a function of temperature in Fig.9. It is noted that the temperature dependence of the lattice spacing of (0010) or (110) is correspond to that of lattice parameter of c or a and b, respectively. Both of these d values increases monotonically with increasing temperature. The mean thermal expansion coefficients of the spacings were estimated to be 1.8×10^{-5} K^{-1} and 1.2×10^{-5} K^{-1} in the temperature range of R.T. to 800K and 2.9×10^{-5} K^{-1} and 1.7×10^{-5} K^{-1} in the range of 800K and 1100K for lattice spacing of (0010) and (110) plane, respectively. Since the cohesive force between atomic layers along c-axis of the high Tc phase is smaller than that of the low Tc phase due to the anomalous large thermal expansion, long time heat treatment is necessary to obtain the high Tc phase which is probably one of the thermodynamical equilibrium phases of the nominal composition of Bi-Sr-Ca-Cu-O system at high temperatures. It seems that one of the roles of dopant Pb is the stabilization of the high Tc phase, whose structure is prolonged along c-axis.

In conclusion, 2:2:1:2 structure responsible for the low Tc phase are stable up to near melting point and the lattice parameter c increases very rapidly at high temperatures above 850 K. The sample with nominal 1:1:1:2 composition sometimes decomposed in the temperature range of 923K to 1073K. However, the decomposition was not observed for the well heat treated sample. The high Tc phase can be obtained by the high temperature heat treatment. The long time sintering is necessary to insert extra layers of CuO_2 and Ca planes into the layer structure of the low Tc phase because of the weak cohesive force along c-axis of the high Tc phase. One of the roles of dopant Pb is the stabilization of the high Tc phase.

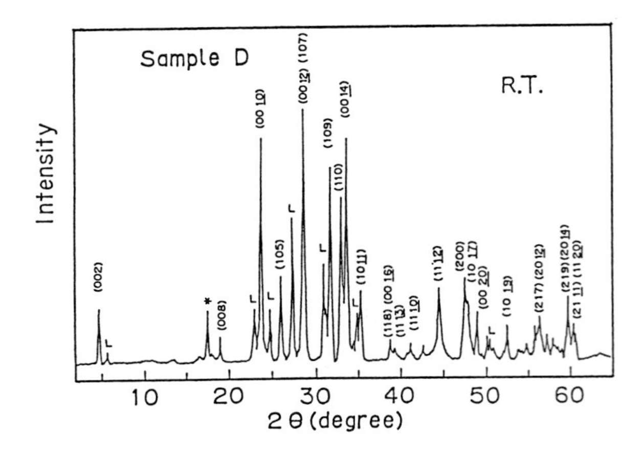

Fig.8 X-ray diffraction pattern for Pb-doped sample D. Peaks of the low Tc phase (2:2:1:2) are indicated as "L". Indices are diffraction planes of the high Tc phase.

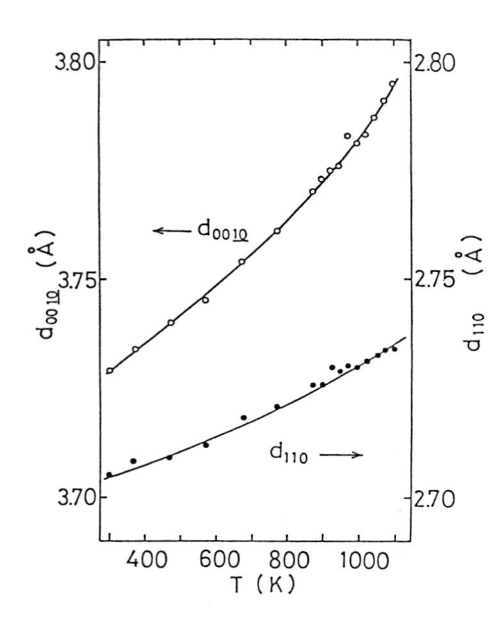

Fig.9 Lattice spacings d(0010) and d(110) of high Tc phase as a function of temperature.

ACKNOWLEDGEMENT

A part of this work was suported by New Functionality Materials-Design, Preparation and Control in Grand-in-Aid for Scientific Research on Priority Areas, Ministry of Education, Sience and Culture.

REFERENCES

[1] H.Maeda, Y.Tanaka, M.Fukutomi and T.Asano, Jpn. J. Appl. Phys. Lett. 4(1988)1209.
[2] Y.Matsui, H.Maeda, Y.Tanaka and S.Horiuchi, Jpn. J. Appl. Phys. 27(1988)L361.
[3] E.T.Muromachi, Y.Uchida, A.Ono, F.Izumi, M.Onoda, Y.Matsui, S.Takekawa and K.Kato, Jpn. J. Appl. Phys. 27(1988)L365.
[4] Y.Bando, T.Kijima, Y.Kitami, J.Tanaka F.Izumi and M.Yokoyama, Jpn. J. Appl. Phys. 27(1988)L358.
[5] K.Hiraga, M.Hirabayashi, M.Kikuchi and Y.Shono, Jpn. J. Appl. Phys. 27(1988)L573.
[6] D.R.Veblen, P.J.Heaney, R.J.Angel, L.W.Finger, R.M.Hazen, C.T.Prewitt, N.L.Ross, C.W.Chu, P.H.Hor and R.L.Menz , Nature 332(1988)334.
[7] J.M.Tarascon, Y.Le Page, G.W.Barboux, B.G.Bagley, L.H.Greene, W.R.Mckinnon, G.W.Hull, M.Groud and D.M. Hwang, Phys. Rev. B 37(1988)9382.
[8] M.Onoda, A.Yamamoto, E.T.Muromachi and S.Takekawa, Jpn. J. Appl. Phys., 27(1988)L833.
[9] E.T.Muromachi, Y.Uchida, Y.Matsui, M.Onoda and K.Kato, Jpn. J. Appl. Phys. 27(1988)L556.
[10] T.M.Shaw, S.A.Shivashankav, S.J. La Placa, J.J.Cuomo, T.R.Mcguire, R.A.Roy, K.H.Keller and D.S.Yee , Phys. Rev. B 37(1988)9856.
[11] H.Nobumasa, K.Shimizu, Y.Kitano and T.Kawai, Jpn. J. Appl. Phys., 27(1988)L846.
[12] Z.Z.Shen and A.M.Hermann, Nature 332(1988)55, 138.
[13] R.M.Hazen, C.T.Prewitt, R.A.Augel, N.L.Ross, L.W.Finger, C.G.Hadidiacos, D.R. Velben, P.J.Heaney, P.H. Hor, R.L.Meng, Y.Y. Sun, Y.Q.Wang, Y.Y.Xue, Z.J.Huang, L.Gao, J.Bechtold and C.W.Chu , Phys. Rev. Lett (1988) in press.
[14] M.Takano, J.Takada, K.Oda, H.Kitaguchi, Y.Miura Y.Ikeda, Y.Tomii and H.Mazaki, Jpn. J. Appl. Phys. 27(1988)L1041.
[15] M.Matsui, K.Ohmori, T.Shimizu and M.Doyama, Physica 148B(1987)432.
[16] T.Komatsu, R.Sato, K.Imai, K.Matsushita and T.Yamashita, Jpn. J. Appl. Phys., 27(1988)L550.
[17] M.Doyama, M.Matsui, J.Liu, H.Matsuoka, T.Shimizu and K.Ohmori, Proc. Special Symposium on Advanced Materials, p83, Tokyo, 1988.
[18] B.Seebacher and G.Schindler, Phisica C153(1988)615.

Microstructure and Critical Current Property in Pb-Doped BiSrCaCuO Bulk Superconductors

T. Asano[1], Y. Tanaka[1], M. Fukutomi[1], K. Jikihara[2], J. Machida[3], and H. Maeda[1]

[1] National Research Institute for Metals, Tsukuba Laboratories, 1-2-1, Sengen, Tsukuba, 305 Japan
[2] Sumitomo Heavy Industries, Ltd., Hiratsuka Research Laboratory, 63-30, Yuhigaoka Hiratsuka, 254 Japan
[3] Mitsui Mining & Smelting Co., Ltd., Central Research Laboratory, Ageo, Saitama, 362 Japan

ABSTRACT

Highly oriented microstructure in the (Bi,Pb)-Sr-Ca-Cu-O bulk samples has been obtained by modified powder method. Disk-shaped samples consisting of layered grains with the c-axis of the crystal normal to the disk-surface were prepared by applying an intermediate pressing between the sintering heat treatments. Enhancement in a formation rate of the high-T_c phase and a remarkable improvement in the critical current density have been achieved by this method. A high J_c value up to 700 A/cm^2 at 77 K in zero magnetic field was obtained although it decreased rapidly with application of a small magnetic field less than 100 Oe. Results of the magnetization measurements are also presented.

INTRODUCTION

Soon after the discovery of high temperature superconductivity in Bi-Sr-Ca-Cu-O system(1), the observation of superconductivity in Tl-Ba-Ca-Cu-O system(2) was reported. Those are multi-phase superconductors having more than two phases including the high-T_c phase of about 110 K and 125 K, respectively. The synthesis of the high-T_c single phase is, however, still difficult especially in the BiSrCaCuO system because the temperature region where the high-T_c phase forms is very narrow between 860-870 °C and requires a long sintering time(3). The effectiveness of partial substitution of Pb for Bi for increasing the volume fraction of the high-T_c phase has been reported.(4,5) Recently, it was found that the temperature region in which the high-T_c phase forms can be lowered and expanded under a low oxygen partial pressure of around 1/13 atmosphere(6).

On the other hand, bulk specimens such as sintered pellets, usually show rather small J_c values compared with those of films prepared by vapor deposition processes. Possible reasons to explain this difference are:one is misalignement of neighbouring plate-like grains in a porous microstructure of the bulk samples and the other is an inferior superconductivity around the grain boundaries. Therefore, it should be noticed that if the porous microstructure is changed to a dense one during sintering treatment more uniform and equilibrium superconducting phase will be formed due to an enhanced atom diffusion. We introduced an itermediate uniaxial pressing between the sintering treatment and found a remarkable change in the microstructure and superconducting properties. In the present report, the effect of this intermediate pressing on preparing highly oriented microstructure, enhanced formation rate of the high-T_c phase and an improvement in the superconducting properties of the bulk sample have been described.

EXPERIMENTAL

Samples with nominal composition of $Bi_1Pb_{0.3}Sr_1Ca_1Cu_{1.5}O_x$ and $Bi_{0.7}Pb_{0.3}Sr_1Ca_{1.5}Cu_2O_x$ were prepared from high purity powder reagents of PbO, Bi_2O_3, $SrCO_3$, $CaCO_3$ and CuO. The mixed powders were calcined at 800 °C for 10 h in air in alumina or silver boat, ground and cold-pressed into pellets 20 mm in diameter and 1 mm in thickness at a pressure of 5 ton/cm^2. Some samples were then sintered in air at 840-850 °C for 20-190h then cooled to room temperature in the furnace (non-pressed sample). Other samples were sintered at 840-850 °C for 20-96h, cooled to room temperature and re-pressed without grinding under the same pressure, then sintered again in air for more 20-96h followed by furnace-cooling to room temperature (pressed sample). Cooling rate in the furnace was about 100 C/h. The sintering atmosphere of Ar:O_2=12:1 was also applied. Fractured cross section was observed by scanning electron microscopy (SEM). X-ray diffraction analysis was performed on the sintered pellet and their powder samples. Chemical analyses by energy dispersive x-ray microanalysis (EDX) was carried out on the samples which were polished by Al_2O_3 and water. Critical temperature, T_c was measured on bar-shaped specimens cut from the sintered pellets. A four point probe technique was used to measure the T_c. The sample temperature was measured by a Au-Fe/chromel thermocouple.

Fig. 1 SEM micrograhs taken on the fractured cross sections of the smples with nominal composition of $Bi_1Pb_{0.3}Sr_1Ca_1Cu_{1.5}$, (a)-(c), and the sample $Bi_{0.7}Pb_{0.3}Sr_1Ca_{1.5}Cu_2$ (d). The sample(a) sintered at 845 °C for 192h in air, the sample (c): re-pressed between initial and final sintering treatments at 845 °C for 96h , the sample (b): as re-pressed days, cold pressed uniaxially then sintered again at 845 °C for 192h.

AC susceptibility change with temperature was also measured on bulk sample by applying alternating magnetic field of 700 H_z. Critical current, I_c, was measured at 77K on bridge-shaped specimens in a steady magnetic fields of 0-650 Oe generated using copper coil immersed in the liquid nitrogen. The I_c was defined as a current required to induce 5 uV acrosss a 10-mm length of the specimen. Critical current density, J_c, was calculated by dividing I_c with the cross-sectional area of the specimen. Magnetization measurement was carried out on the cubic-shaped specimens, 3x3x3 mm^3, in the magnetic field up to 12 KOe using a sample vibrating magnetometer.

RESULTS AND DISCUSSION

Microstructure

Figure 1 shows the SEM micrographs taken on the fractured cross sections for the sample with nominal composition of $Bi_1Pb_{0.3}Sr_1Ca_1Cu_{1.5}$, (a)-(c), and the sample $Bi_{0.7}Pb_{0.3}Sr_1Ca_{1.5}Cu_2$ (d). The non-pressed sample (a), which was sintered at 845 °C for 192h in air after pelletizing press, shows a dominant porous microstructure, while the pressed sample (c), which was re-pressed between initial and final sintering treatments at 845 °C for 96h, exhibits a highly oriented microstructure. Figure 1(b) shows the cross-section of an as-pressed state or before final sintering treatment. The re-pressing is clearly effective to crush the porous microstructure. It is also obvious that a smooth plate-like grains as shown in Fig.1(c) can be formed through recrystalizing process from the densified state as seen in Fig.1(b). In general, the re-pressing shorten the sintering time probably due to an enhanced atomic diffusion reaction. Furthermore, non-superconducting phase, whose composition analysed is $Sr_2Ca_2Cu_7$ and is usually coexisting with the layered superconducting phase, is considerably reduced in quantity by the re-pressing. The analysed chmeical compositions for the superconducting phase in the pressed and non-pressed sample were 19.7at%Bi, 2.3at%Pb, 24.1at%Sr, 20.6at%Ca and 33.4at%Cu and 19.6at%Bi, 1.0at%Pb, 21.1at%Sr, 22.4at%Ca and 36.1at%Cu, respectively. Figure 1(d) shows the fractured cross-section of the sample with a different nominal composition from that of (a)-(c) of this figure. Depending on the starting composition microstructure differs considerably. As is mentioned later, critical current density for the

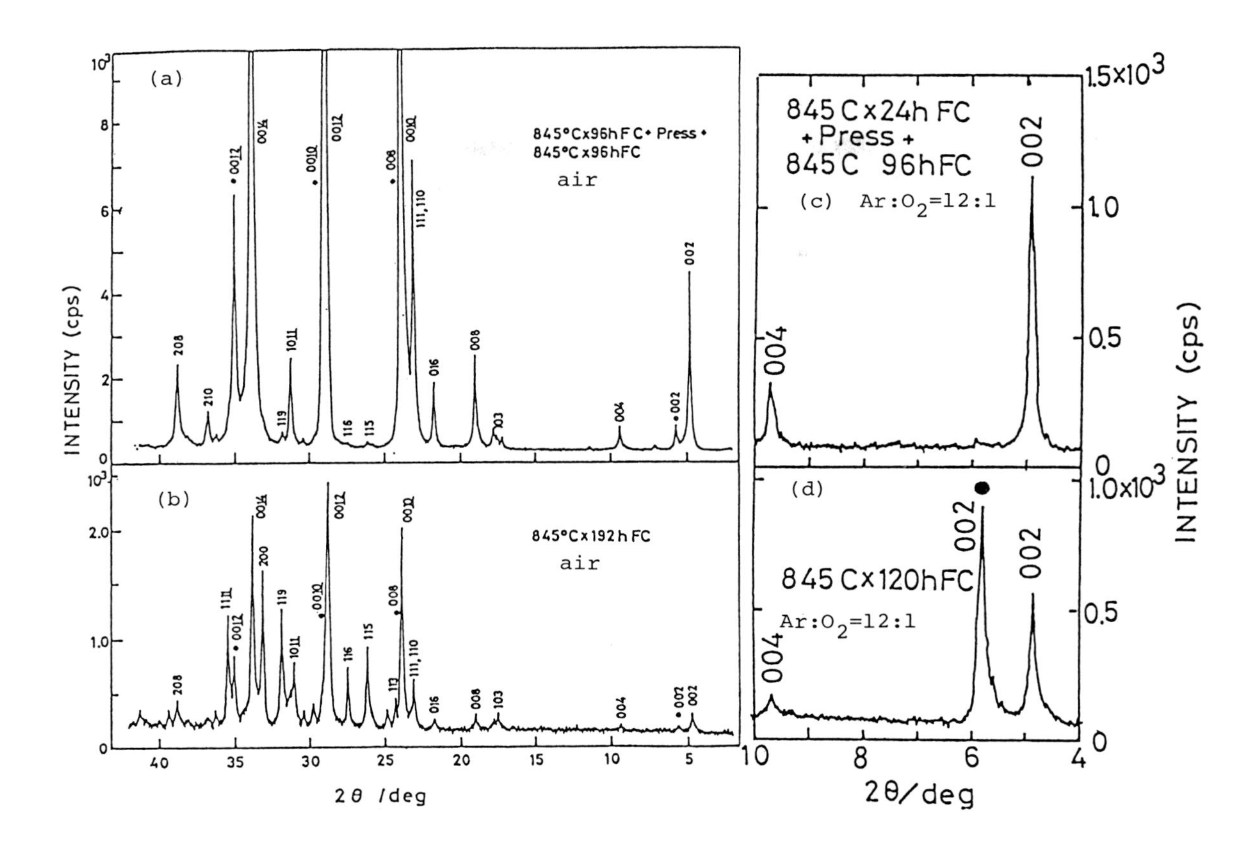

Fig. 2 X-ray diffraction patterns for the samples prepared by the quoted condition in the figure. The nominal composition is $Bi_1Pb_{0.3}Sr_1Ca_1Cu_{1.5}$.

sample (d) is larger than that for the sample (c).

Figure 2(a) and 2(b) show the X-ray diffraction patterns for the pressed (c) and the nonpressed sample (a) shown in Figure 1(a) and 1(b), respectively. Filled circles denote the refrections from the low-T_c phase. In the pressed sample(c), large increase in the intensities of the diffraction lines from 001 planes can be observed. The c-axes of the crystals in the sample(c) approximately parallel to the pressing direction. From this patterns, it is also shown even in the pressed sample (c), sintered in air, still contains the low-T_c phase although no clear transition corresponding to the low-T_c phase was not observed from the AC susceptibility measurement. On the other hand, for the re-pressed and sintered sample under a low oxygen partial pressure of 1/13 atom the low-Tc phase almost disapeared in the X-ray diffraction pattern. That is, an enhanced formation of the high-T_c phase seems to occur. The typical result is exhibited in Figure 2(c) and 2(d), which show the x-ray diffraction patterns of the samples with the same composition as Fig.1, but differing the sintering time and gaseous condition.

The critical temperature T_c^{on} (onset temperature) does not differ much with respect to not only re-pressing but also sintering atmosphere, although the volume fraction of the high-T_c phase are different as seen in the Fig.2(a)-2(d). The critical current density, however, varies considerably depending on these preparation condition. Figure 3 shows the magnetization vs. magnetic field curves for the re-pressed and non-pressed samples as in Fig.1(a) and 1(c), where the curves a and b correspond to the re-pressed and non-pressed sample, respectively. The magnetization measurement was carried out on cubic shaped samples as shown in Fig. 3. It is observed that the re-pressing increases the magnetization and the area enclosed by the hysteresis loop. If we assume that the similar pinning mechnism as to conventional superconductors is also valid to the oxide superconductors, a larger critical current density, J_c, for the re-pressed sample should be obtained. In fact, J_c value of about $50A/cm^2$ by more than one order larger than that for the non-pressed sample is achieved. However, this J_c value is rather low that may be attributed to the coexistence of the low-T_c phase or the non-superconducting second phase.

Superconducting properties

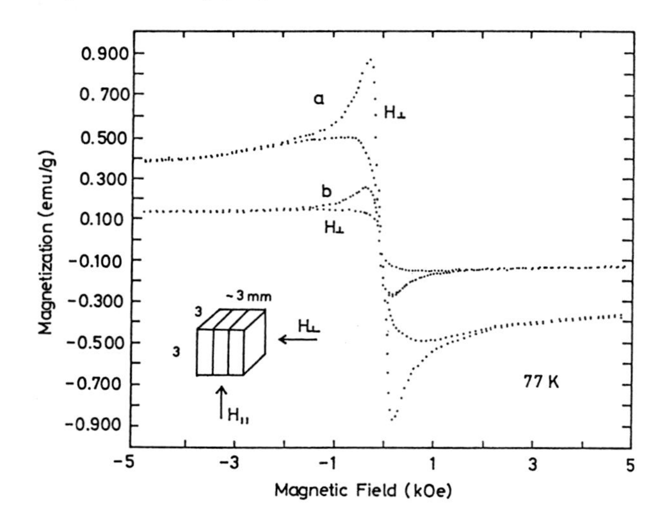

Fig. 3 Magnetization vs. magnetic field curves for the re-pressed and non-pressed samples as shown in Fig. 1(a) and 1(c).

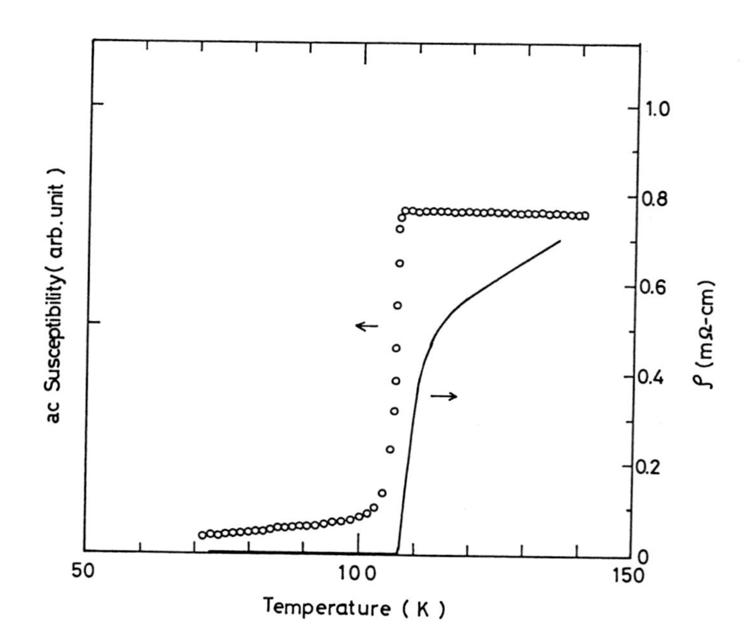

Fig. 4 AC susceptibility and resistivity vs. temperature curves for the re-pressed sample. The nominal composition is $Bi_{0.7}Pb_{0.3}Sr_1Ca_{1.5}Cu_2$ and sintered at 850 °C for 24h, re-pressed, then, sintered at 850 °C for more 96h in an atmosphere of $Ar:O_2$=12:1.

Figure 4 shows the ac susceptibility and electrical resistivity vs. temperature curves for the re-pressed sample as in Fig.1(d), which was sintered in a low oxygen partial pressure. This sample is almost consisted of single phase judging from these curves as well as the X-ray diffraction analysis. The J_c vs. magnetic field curves measured at 77K in a magnetic field of 0-650 Oe is shown in Fig.5. Sample shape is shown in the insert of Fig.5 and its dimensions are 20 mm in length, 3 mm in width and 0.5-0.6 mm in thickness with the central narrow bridge which cross-sectional area is 0.5-0.6 mm^2. The magnetic field was applied parallel and perpendicular to the sample surface, and normal to the current direction. A J_c as high as 700 A/cm^2 was obtained in the zero magnetic field. This value is about 8 times larger than that of non-pressed sample (c) shown in the Fig.5. It is noticed that the J_c decreases rapidly with increasing the magnetic field

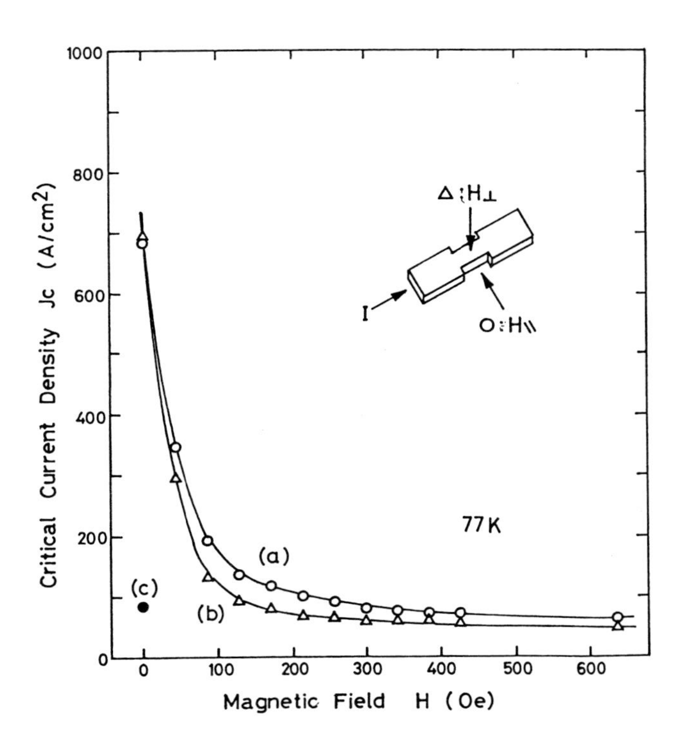

Fig. 5 J_c vs. magnetic field curves, (a) and (b), for the same sample as shown in Fig.4. Filled circle, (c), shows J_c for the non-pressed sample of the same composition as in Fig.4.

of less than 100 Oe. Furthermore, a slight anisotropy in J_c with respect to the direction of magnetic field can be observed.

CONCLUSIONS

Highly oriented microstructure in the high-T_c (Bi,Pb)-Sr-Ca-Cu-O bulk samples has been obtained by modified powder method. A porous microstructure can be changed to a dense one by this method, which re-press the pellet after an initial sintering treatment and sinter again, then an enhanced atomic diffusion and formation rate of the high-T_c phase have been obtained. A remarkable improvement in the critical current density was achieved. If the more highly oriented microstructre is realized, critical current density higher than 700 A/cm^2 at 77 K and zero magnetic field will be achieved.

ACKNOWLEDGEMENT

The authors would like to thank Mr. T. Yamamoto for EDX analysis and Dr. M. Uehara for the magnetization measurement.

REFERENCES

(1) H. Maeda, Y. Tanaka, M. Fukutomi and T. Asano, Jpn. J. Appl. Phys. 27,L209(1988)
(2) Z. Z. Sheng and A. M. Hermann, Nature 332, 138(1988)
(3) H. Nobumasa, K. Shimizu, Y. Kitano and T. Kawai, Jpn. J. Appl. Phys. 27,L846(1988)
(4) R. J. Cava, B. Batlogg, S. A. Sunshine, T. Siegrist, R. M. Fleming, K. Rabe, L. F. Snhneemeyer, D. W. Murphy, R. B. van Dover, P. K. Gallagher, S. H. Glarum, S. Nakahara, R. C. Farrow, J. J. Krajewski, S. M. Zahurak, J. V. Waszczak, J. H. Marshall, P. Marsh, L. W. Rupp, Jr., W. F. Peck, and E. A. Rietman, Physica C 153-155,560(1988)
(5) M. Takano, J. Takada, K. Oda, H. Kitaguchi, Y. Miura, Y. Ikeda, Y. Tomii and H. Mazaki, Submitted to Jpn. J. Appl. Phys.
(6) U. Endo, S. Koyama and T. Kawai, Submitted to Jpn. J. Appl. Phys.
Phys. 27,L1460

5
Research Policy and Technology Trends

Policy and Technology Trends: The Organisation of High Tc Superconductor Research in the UK

MARTIN WOOD

The Oxford Instruments Group PLC, Eynsham, Oxford OX8 1TL, UK

Introduction

Since Professor K.A. Muller and Dr J.G. Bednorz, working in the IBM laboratory in Zurich in 1986, discovered superconducting properties above 30K in various copper oxide ceramic materials, there has been a tremendous increase in interest in the phenomenon of superconductivity. This is partly because of the intrinsic scientific interest in the materials and the new processes which make superconductivity possible at these relatively high temperatures, and partly because of the exciting technical and commercial applications to which they may lead.

Universities in the UK have a long tradition of research at low temperatures and into superconductivity, and UK industry has played a substantial role in the development of the applications of niobium-based superconductors. It was therefore natural for interest to spring up spontaneously at many locations throughout the country when the dramatic increase in Tc was first reported.

For the UK to make a significant contribution to this field and for UK industry to participate effectively in future commercial exploitation, it is vital to bring academic institutions, industry and Government together into a single, cohesive, coordinated National Programme. The key element in this should be genuine multi-disciplinary collaborative work encompassing universities, Government laboratories (civil and defence), and industry. The aim is to achieve close working relationships across the whole spectrum of activities within the programme, minimising the traditional barriers between people working in different organisations and funded from different sources, and developing the rapid and effective communication of results and ideas.

Responsibility for preparing and promoting this National Programme has been vested in the National Committee for Superconductivity (NCS). This is a joint committee formed by the Department of Trade and Industry (DTI) and the Science and Engineering Research Council (SERC) with a balanced membership from industry and the universities.

The Programme Objectives

The four core R and D objectives of the National Programme, all of comparable importance in the long term, may be broadly stated as:

1. To investigate production and processing techniques which improve the performance of existing materials, and in particular their critical current densities and mechanical properties.

2. To discover and research new materials with higher critical temperatures, improved critical current densities, and better mechanical properties.

3. To obtain a theoretical and empirical understanding of the mechanisms of, and limitations to, superconductivity.

4. To study selected potential applications, and to research the appropriate materials and associated manufacturing techniques to enable their exploitation.

These broad objectives are, of course, common to the research programmes of all major countries. At one level the UK programme will be part of a European effort, on which discussions have already started with a view to coordinating the research programmes so as to avoid undesirable duplication of effort and to ensure complementarity. At the international level exchanges of the results of academic research through conferences and publications and individual visits will continue to be the norm.
Coordination of national effort on this scale has been stated as being desirable on many previous occasions and lip service has been paid to it but without much real progress. Superconductivity has been chosen as an area where we may actually make it happen and this initiative has the active support of the Prime Minister through the Advisory Committee for Science and Technology in which she takes an active interest.

The Main Participating Organisations

1. Institutes of Higher Education (HEI) - Universities, Polytechnics and Institutes of Technology

   Most of the basic research on superconductivity in the UK is conducted at the 43 institutions shown in Table 1. The scale of work varies from relatively minor involvement of some institutes which have, perhaps, some particular analytical instrument, to that of the major universities which have research groups ranging from theoretical departments through materials science and physics departments to applications laboratories.
   Allowing for some overlap, these HEI's can be divided into three groups:

   At the top lies the new Interdisciplinary Research Centry (IRC) which has been located within Cambridge University. Its programme will take into account the basic research necessary for both 'electronic' and 'power' applications so that it will, for example work towards bulk and filamentary conductors as well as towards thin film based devices. A significant fraction of its work will be collaborative, and it will be expected to seek funds from industry to support some of its projects. The management structure has been planned, from the outset, to give it an outward-looking character. As a national focus for research in this subject the IRC will play an important role in the coordination of the national programme by stimulating cooperation between all the parties to the programme. The IRC in Cambridge has, itself, been created from an earlier collaborative initiative involving five existing university departments - Earth Sciences, Engineering, Chemistry, Materials Science and Metallurgy and Physics.

   Below this come several informal groupings of laboratories involving about 25 universities which because of their geographical location relative to each other will be encouraged to coordinate their activities so as to become broadly self-sufficient in terms of equipment and expertise for their chosen programme.

   Finally, this leaves some 15 HEI's not formally associated with any other institute, but pursuing their own work in which they have some particular ability.

   Although most of the available funds from the SERC will be channelled into the major research groups, the SERC always retains the possibility of funding small projects elsewhere if applications of exceptional quality are submitted.

2. Industry

   Before 1986 there were only about five companies in the UK which had research groups working on superconductivity and superconducting materials, although in several of these the work did not have high priority. Since 1986 all these groups have been strengthened and another 20 companies have become involved to varying degrees. These new entries to the field include large automotive and general engineering firms, specialist materials companies, and young 'high tech' companies who see some potential for completely new product areas.

The total research effort involved is still relatively modest, perhaps 50 man years/year, but is increasing now as the DTI support programme gathers momentum. Under this scheme the DTI will provide up to 50% funding over three years for pre-competitive, collaborative research which embodies the free exchange of the results of research between the participants within each programme.

This high Tc superconductivity industrial programme is implemented with the technical advice of an Industrial Programme Advisory Group within the DTI, which has representation from both the electronics and electrical engineering industry, and with the advice of the National Committee for Superconductivity. It is currently expected that about 40% of the funds will go into electrical engineering and 60% into electronics.

There are currently some five collaborative projects within this industrial programme funded and under way.

Table 1 High Temperature Superconductivity at HEI

UNIVERSITY or HEI	SERC support	Basic Chemistry & Processing	Thermal and Calorimetric Studies	Micro & Macro Structural Studies	SEM, TEM, HREM	Characterisation & Bulk Material Studies	Single Crystals	Wires or tapes	Thick Films	Thin Films	SQUIDS & Junctions	Detectors, sensors, etc	RF, μWave	Power, Engineering	Theory
ABERDEEN		•	•			•									
BATH		•		•		•									
BIRKBECK COLLEGE		•		•		•						•			•
BIRMINGHAM	Y	•	•	•	•	•	•	•	•	•	•	•	•	•	•
BRADFORD		•		•		•			•				•		
BRISTOL POLY.				•											
BRISTOL UNIV.	Y	•	•	•	•	•									•
BRUNEL		•				•									
CAMBRIDGE (IRC)	Y	•	•	•	•	•	•	•	•	•	•	•	•	•	•
CARDIFF, UNIV. COLL.		•		•	•										
DURHAM		•		•		•						•		•	
EDINBURGH				•											
ESSEX				•	•										
EXETER		•		•											•
GLASGOW COLL. (POLY)		•	•	•											
GLASGOW UNIVERSITY		•	•	•	•	•					•				
HERIOT-WATT		•	•							•					•
IMPERIAL COLLEGE	Y	•	•	•	•	•	•		•	•				•	•
KEELE		•		•											•
KING'S COLLEGE				•											•
LANCASTER										•			•		
LEEDS	Y	•	•	•		•		•		•					
LIVERPOOL POLY						•									
LIVERPOOL UNIVERSITY	Y	•	•	•	•	•	•	*						•	
MANCHESTER		•	•	•	•	•	•								
NEWCASTLE-UPON-TYNE		•				•									
NORTH STAFFS POLY		•													
NOTTINGHAM				•	•	•				•					•
OXFORD	Y	•	•	•	•	•	•	•	•	•			•		•
READING		•		•		•				•					•
ROYAL HOLLOWAY & BNC				•											
QUEEN MARY COLLEGE	Y	•		•		•									•
St ANDREWS		•		•		•							•		
SHEFFIELD		•	•			•									
SOUTHAMPTON	Y	•	•	•	•	•		•	•	•				•	
STRATHCLYDE	Y	•				•				•	•		•		
SURREY				•	•					•					
SUSSEX	Y		•			•									
SWANSEA, U.COLL.		•		•		•									
ULSTER					•	•				•					
UMIST (MANCHESTER)		•			•	•				•					
UNIVERSITY COLLEGE															•
WARWICK	Y	•	•	•	•	•		•	•	•					•

3. Government supported laboratories

The following Government departments and ministries have laboratories which are active in high Tc research : the Department of Education and Science (DES) through its Rutherford and Appleton Laboratory and Daresbury Laboratory (under the SERC); the Department of Trade and Industry (DTI) through its National Physical Laboratory; the Atomic Energy Authority (AEA) through its Atomic Energy Research Establishment; and the Ministry of Defence (MOD) through its Royal Signals Research Laboratory and the Admiralty Research Establishments.

All of these laboratories are involved, to a greater or lesser extent, in collaborative projects with HEI's concerning high Tc superconductivity research. In addition, the two SERC laboratories have major facilities in the way of neutron scattering, in the ISIS facility, and synchrotron radiation in the SRS facility at Daresbury, which are widely used for structural research.

Resources Available

The resources currently available to the national programme are:

a. £5.3M over six years from the SERC for the IRC at Cambridge;
b. £2M/year (commitment limit) from SERC for research grants to HEI's;
c. £8M over 3 years from DTI for industrial collaborative research;
d. as yet unquantified sums from other government departments;
e. funds made available by industry, in excess of £8M over three years.

These figures do not include funds supplied by the University Grants Committee to finance the basic university staff and infrastructure. In many cases work is pursued on high Tc research using basic laboratory facilities without the benefit of special SERC grants.

International Cooperation

On the European level a standing committee now exists with representatives from : France (The Centre Nationale de la Researche Scientifique); Italy (The Consiglio Nazionale della Ricerche); West Germany (The Bundesministerium fur Forchung und Technology/Max Planck Gesellschaft); The Netherlands (The Zuiver Wetenschappelijk Onderzoek); Switzerland (The Fond Nationale Suisse); and the UK (The Science and Engineering Research Council). Sweden will send a representative to future meetings.

Discussions have taken place, among these countries, on a range of measures to improve cooperation and information transfer, and the activities of this committee will complement any coordination which may develop through the offices of the European Economic Commission in Brussels.

As yet there is no formalised cooperation on a national level outside Europe, but many individuals and laboratories have working relationships with colleagues overseas and it is hoped that these will be increased and extended to mutual benefit in the field of high Tc superconductivity research.

Closing Remarks

Summary

SADAO NAKAJIMA

Department of Physics, Tokai University, Hiratsuka, Kanagawa, 259-12 Japan

When Professor Kitazawa, our program chairman, honored me with an invitation to deliver this summery talk, I wondered how I could survey all the papers presented at a meeting of vast scope as ISS'88. I could not decline either, because Professor Kitazawa is one of my closest colleagues having collaborated for many years in our Japanese program of basic research into high-Tc superconductivity. I decided not to try to survey all the papers, but to focus on my own impressions of the present status of reseach and development which was throughly considered at ISS'88.

Since this Symposium started with two excelent lectures on the history of the science of superconductivity, I would also like to go back to the early 80's, when the above-mentioned Japanese research program was launched toward high-Tc materials. At that time, the BCS theory, applied to simple superconductors, centered around magnets, were fully developed, too, though within the limits of helium refrigeration. There seemed to remain only one fundamental question unsetteled: in which materials and through what mechanisms could one raise Tc substantially?

Infact, since the middle of the 70's, exotic as well as complex materials including oxides and organics had been developed and at the same time raised novel problems concerning the mechanisms of superconductivity in these materials, e.g. heavy Fermion superconductors and cubic perovskite BPBO. It appeared that this trend might lead us to high-Tc materials, but nobody could foresee that Tc would exceed 100K in few years.

The great discovery of high-Tc Cu oxides has changed the situation completely and we are now all busy dealing with high-Tc oxides lest we should be classified into nonexperts in the who's who of superconductivity. Its social impact is also such that tens of millions of Japanese people watched Professor Schrieffer speaking at this Symposium on the exciting news and, next morning, my taxi driver, like many politicians, was wondering which coutry would lead in this superconductivity competition.

For us the great discovery posed the two new problems of making high-Tc materials available to superconductivity technology at liquid nitrogen or even higher temperatures on one hand and of exploring the mechanism of high-Tc superconductivity on the other. Since these two inseparable at present, it was most appropriate and timely that ISTEC organized ISS'88 for researchers from both science and engineering fields to meet together to review progress made so far in solving the problems. Progress is quite remarkable if only for the greatly improved quality of data and samples reported at this Symposium. More quantitatively, the value or Tc in which the press and the public is most interested exceeds 120K in the Tl-Ca-Ba-Cu-O system. It also shows rather clean physical properties; we have heard of the Josephson effect in the junction with Pb. It is a pity that this beauty is poisonous.

From the studies of La-, Y-, Bi-, and Tl- based Cu oxides, we are beginning, though still partially, to understand the relation between Tc and atomic structures - stacking of Cu-O planes, roles of oxygen atoms and other metal ions, etc. The study of cubic perovskite BKBO will add important information.

The short coherence distance is an almost inevitable consequence of high-Tc and results in large critical fluctuations, depression of superconductivity near boudaries, week flux pinning, and giant flux creep. we must be careful not only in attempting to improve Jc, but also in interpreting the data of physical properties.

As for the problems of making high-Tc films and wires, we have witnessed significant progress, from which one might suggest the possibility of simple commercialization in the near future. This is important because it will encourage our sponsors and therefore stimulate us further.

Personally, however, I was much impressed by the degree of sophistication achieved with the use of low-Tc metal superconductors both in large scale and in microelectronic apprecations. We should be cautious to choose our strategy in applied high-Tc superconductivity.

There still remain many questions to be answered by basic research and development. For examples, no answer is available as yet to the possibility of any materials with higher Tc, or higher Jc, or more mechanical flexibility than the recently uncovered Cu oxides. The organic superconductor might be a candidate in view of the steep increase of Tc, but we are not quite sure. In the case of high-Tc Cu oxides, too, we don't know precisely why Tc is high and why the perovskite is layered.

Unfortunately, inspite of many theoretical proposals we have heard at this Symposium, theory of high-Tc superconductivity is not strong enough to provide basic assumptons on which the experimenter can plan his experiment and interprete the result. In the case of Cu oxides, high-Tc superconductivity is obtained by doping holes into a Mott type insulator which shows the antiferromagnetic spin odering. We also know that the superconducting current is carried in uits of $^{+}2e$.

But it is not obvious whether we can think of a simple-minded pair of holes similar to the usual Cooper pair in BCS theory. Probably we cannot; it is quite possible that the pair in our case might be a more complex bound state of more complex quasiparticles interacting through a more complex, non-phonon mechanism. We have a number of theoretical possibilities for the mechanism, depending upon the way we kill the spin ordering and whether we put emphasis on the spin fluctuation or on the charge fluctuation.

Similarly we have no definite physical pictures of excitations either in normal or superconducting phases. We might even ask whether the well known T-linear normal resistivity observed in all high-Tc oxides is really intrinsic.

Thus there are many things to do on the level of basic reseach and development before traveling on the superconducting train. Compare our situation with the case of semiconductors. There we had the simple concept of electron and hole porvided by the band theory from the outset. The major problem was how to control impurities in order to make the simple concept applicable to the actual semiconductor.

One important conclusion is that the scientific cooperation between researchers of different disciplines, in different institutions, and in different countries is extremely important and effective. Cooperation does not conflict with competitiveness at the present stage of development, where we are still far from full commercialization of high-Tc superconductivity.

I am quite sure all the participants will join me in concluding that ISS'88 was a great success and in expressing thanks to the organizers, Professor S. Tanaka, Professor K. Kitazawa and other members of the organizing committee, and the staff members of ISTEC in Tokyo and Nagoya.

Author Index

Subject Index

The page numbers refer to the page on which contribution begins.